Jesús Medina · Manuel Ojeda-Aciego
José Luis Verdegay · David A. Pelta
Inma P. Cabrera · Bernadette Bouchon-Meunier
Ronald R. Yager (Eds.)

Information Processing and Management of Uncertainty in Knowledge-Based Systems

Theory and Foundations

17th International Conference, IPMU 2018
Cádiz, Spain, June 11–15, 2018
Proceedings, Part I

Editors
Jesús Medina
Universidad de Cádiz
Cádiz, Cadiz
Spain

Manuel Ojeda-Aciego
Universidad de Málaga
Málaga, Málaga
Spain

José Luis Verdegay
Universidad de Granada
Granada
Spain

David A. Pelta
Universidad de Granada
Granada
Spain

Inma P. Cabrera
Universidad de Málaga
Málaga, Málaga
Spain

Bernadette Bouchon-Meunier
LIP6
Université Pierre et Marie Curie, CNRS
Paris
France

Ronald R. Yager
Iona College
New Rochelle, NY
USA

ISSN 1865-0929 ISSN 1865-0937 (electronic)
Communications in Computer and Information Science
ISBN 978-3-319-91472-5 ISBN 978-3-319-91473-2 (eBook)
https://doi.org/10.1007/978-3-319-91473-2

Library of Congress Control Number: 2018944294

Printed on acid-free paper

This Springer imprint is published by the registered company Springer International Publishing AG part of Springer Nature
The registered company address is: Gewerbestrasse 11, 6330 Cham, Switzerland

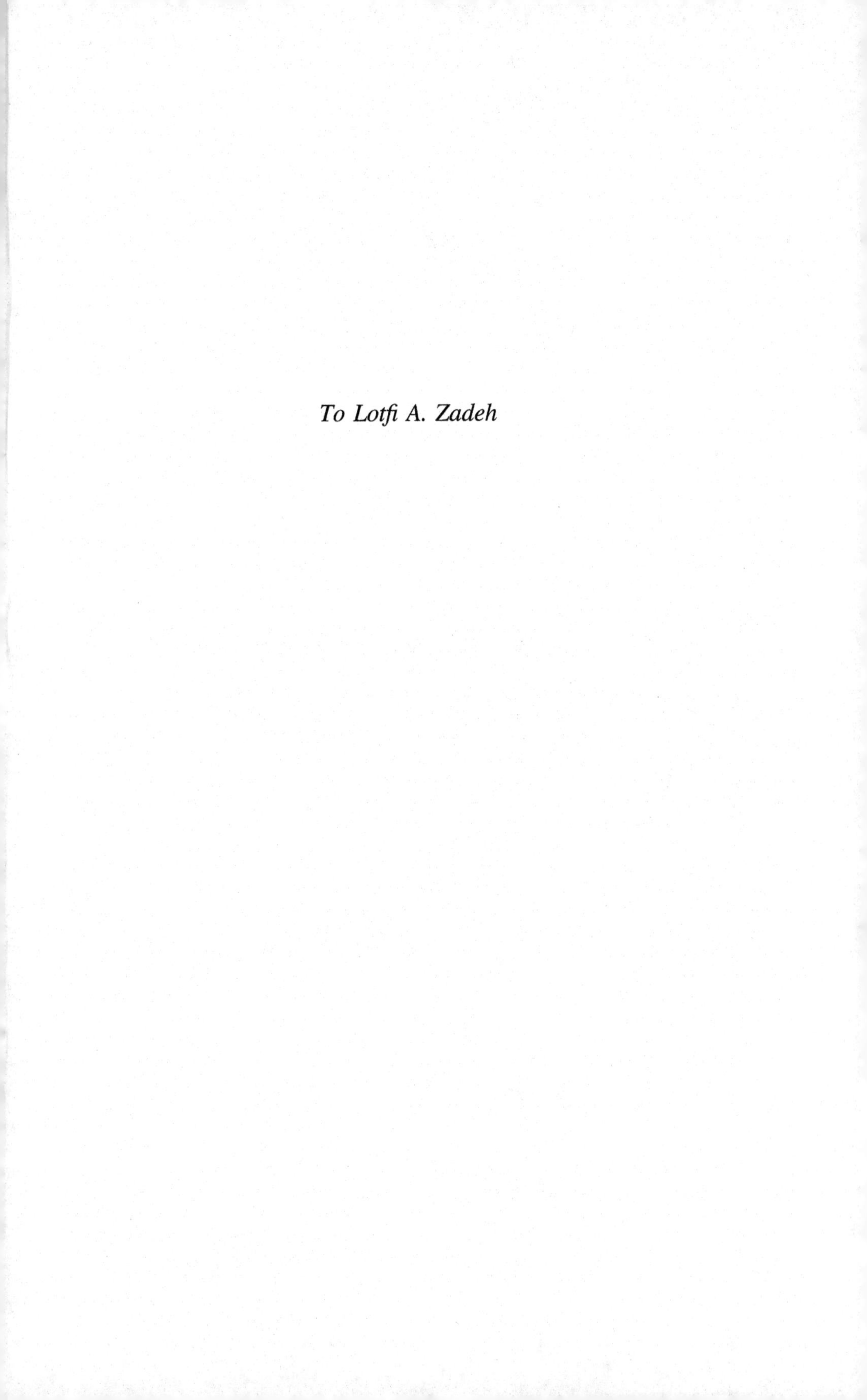

To Lotfi A. Zadeh

Preface

These are the proceedings of the 17th International Conference on Information Processing and Management of Uncertainty in Knowledge-Based Systems, IPMU 2018. The conference was held during June 11–15, in Cádiz, Spain.

The IPMU conference is organized every two years with the aim of bringing together scientists working on methods for the management of uncertainty and aggregation of information in intelligent systems. Since 1986, the IPMU conference has been providing a forum for the exchange of ideas between theoreticians and practitioners working in these areas and related fields.

This IPMU edition held special meaning since one of its co-founders, Lotfi A. Zadeh, passed away on September 6, 2017. To pay him a well-deserved tribute, and in memory of his long relationship with IPMU participants, a special plenary panel was organized to discuss the scientific legacy of his ideas. Renowned researchers and Lotfi's good friends made up the panel: it was chaired by Ronald Yager, while Bernadette Bouchon-Meunier, Didier Dubois, Janusz Kacprzyk, Rudolf Kruse, Rudolf Seising, and Enric Trillas acted as panelists. Besides this, a booklet of pictures with Lotfi Zadeh and friends was compiled and distributed at the conference.

Following the IPMU tradition, the Kampé de Fériet Award for outstanding contributions to the field of uncertainty and management of uncertainty was presented. Past winners of this prestigious award were Lotfi A. Zadeh (1992), Ilya Prigogine (1994), Toshiro Terano (1996), Kenneth Arrow (1998), Richard Jeffrey (2000), Arthur Dempster (2002), Janos Aczel (2004), Daniel Kahneman (2006), Enric Trillas (2008), James Bezdek (2010), Michio Sugeno (2012), Vladimir N. Vapnik (2014), and Joseph Y. Halpern (2016). In this 2018 edition, the award was given to Glenn Shafer (Rutgers University, Newark, USA) for his seminal contributions to the mathematical theory of evidence and belief functions as well as to the field of reasoning under uncertainty. The so-called Dempster–Shafer theory, an alternative to the theory of probability, has been widely applied in engineering and artificial intelligence.

The program consisted of the keynote talk of Glenn Shafer, as recipient of the Kampé de Feriet Award, five invited plenary talks, two round tables, and 30 special sessions plus a general track for the presentation of the 190 contributed papers that were authored by researchers from more than 40 different countries. The plenary presentations were given by the following distinguished researchers: Gloria Bordogna (IREA CNR – Institute for the Electromagnetic Sensing of the Environment of the Italian National Research Council), Lluis Godo (Artificial Intelligence Research Institute of the Spanish National Research Council, Barcelona, Spain), Enrique Herrera-Viedma (Department of Computer Science and Artificial Intelligence, University of Granada, Spain), Natalio Krasnogor (School of Computing Science at Newcastle University, UK), and Yiyu Yao (Department of Computer Science, University of Regina, Canada).

The conference followed a single-blind review process, respecting the usual conflict-of-interest standards. The contributions were reviewed by at least three reviewers. Moreover, the conference chairs further checked the contributions in those cases were conflicting reviews were obtained. Finally, the accepted papers are published in three volumes: Volumes I and II focus on "Theory and Foundations," while Volume III is devoted to "Applications."

The organization of the IPMU 2018 conference was possible thanks to the assistance, dedication, and support of many people and institutions. In particular, this renowned international conference owes its recognition to the great quality of the contributions. Thank you very much to all the participants for their contributions to the conference and all the authors for the high quality of their submitted papers. We are also indebted to our colleagues, members of the Program Committee, and the organizers of special sessions on hot topics, since the successful organization of this international conference would not have been possible without their work. They and the additional reviewers were fundamental in maintaining the excellent scientific quality of the conference. We gratefully acknowledge the local organization for the efforts in the successful development of the multiple tasks that a great event like IPMU involves.

We also acknowledge the support received from different areas of the University of Cádiz, including the Department of Mathematics, the PhD Program in Mathematics, the Vice-Rectorate of Infrastructures and Patrimony, and the Vice-Rectorate for Research; the International Global Campus of Excellence of the Sea (CEI·Mar) led by the University of Cádiz and composed of institutions of three different countries; the European Society for Fuzzy Logic and Technology (EUSFLAT); and the Springer team who managed the publication of these proceedings. Finally, J. Medina, M. Ojeda-Aciego, J. L. Verdegay, I. Cabrera, and D. Pelta acknowledge the support of the following research projects: TIN2016-76653-P, TIN2015-70266-C2-P-1, TIN2014-55024-P, TIN2017-86647-P, and TIN2017-89023-P (Spanish Ministery of Economy and Competitiveness, including FEDER funds).

June 2018

Jesús Medina
Manuel Ojeda-Aciego
Irina Perfilieva
José Luis Verdegay
Bernadette Bouchon-Meunier
Ronald R. Yager
Inma P. Cabrera
David A. Pelta

Organization

General Chair

Jesús Medina	Universidad de Cádiz, Spain

Program Chairs

Manuel Ojeda-Aciego	Universidad de Málaga, Spain
Irina Perfilieva	University of Ostrava, Czech Republic
José Luis Verdegay	Universidad de Granada, Spain

Executive Directors

Bernadette Bouchon-Meunier	LIP6 - Université Pierre et Marie Curie, CNRS, Paris, France
Ronald R. Yager	Iona College, USA

Sponsors and Publicity Chair

Martin Stepnicka	University of Ostrava, Czech Republic

Special Sessions Chair

Inma P. Cabrera	Universidad de Málaga, Spain

Publication Chair

David A. Pelta	Universidad de Granada, Spain

Local Organizing Committee

María José Benítez-Caballero	Universidad de Cádiz, Spain
María Eugenia Cornejo	Universidad de Cádiz, Spain
Juan Carlos Díaz-Moreno	Universidad de Cádiz, Spain
David Lobo	Universidad de Cádiz, Spain
Óscar Martín-Rodríguez	Universidad de Granada, Spain
Eloísa Ramírez-Poussa	Universidad de Cádiz, Spain

International Advisory Board

Anne Laurent
Benedetto Matarazzo
Christophe Marsala
Enric Trillas
Eyke Hüllermeier
Giulianella Coletti
João Paulo Carvalho
José Luis Verdegay
Julio Gutierrez-Rios
Laurent Foulloy
Llorenç Valverde
Lorenza Saitta
Luis Magdalena
Manuel Ojeda-Aciego
Maria Amparo Vila
Maria Rifqi
Marie-Jeanne Lesot
Mario Fedrizzi
Miguel Delgado
Olivier Strauss
Salvatore Greco
Uzay Kaymak

Program Committee

Michal Baczynski, Poland
Rafael Bello-Pérez, Cuba
Jim Bezdek, USA
Isabelle Bloch, France
Ulrich Bodenhofer, Austria
Bernadette Bouchon-Meunier, France
Humberto Bustince, Spain
Inma P. Cabrera, Spain
Joao Carvalho, Portugal
Giulianella Coletti, Italy
Oscar Cordon, Spain
Inés Couso, Spain
Keeley Crockett, UK
Fabio Cuzzolin, UK
Bernard De Baets, Belgium
Guy De Tré, Belgium
Sébastien Destercke, France
Antonio Di Nola, Italy
Didier Dubois, France
Fabrizio Durante, Italy
Francesc Esteva, Spain
Juan C. Figueroa-García, Colombia
Sylvie Galichet, France
Lluis Godo, Spain
Fernando Gomide, Brazil
Gil González-Rodríguez, Spain
Michel Grabisch, France
Przemysław Grzegorzewski, Poland
Lawrence Hall, USA
Francisco Herrera, Spain
Enrique Herrera-Viedma, Spain
Ludmilla Himmelspach, Germany
Janusz Kacprzyk, Poland
Uzay Kaymak, The Netherlands
James Keller, USA
Laszlo Koczy, Hungary
Vladik Kreinovich, USA
Tomas Kroupa, Italy
Rudolf Kruse, Germany
Christophe Labreuche, France
Weldon A. Lodwick, USA
Jean-Luc Marichal, Luxembourg
Trevor Martin, UK
Sebastian Massanet, Spain
Gaspar Mayor, Spain
Jesús Medina, Spain
Radko Mesiar, Slovakia
Ralf Mikut, Germany
Enrique Miranda, Spain
Javier Montero, Spain
Jacky Montmain, France
Serafín Moral, Spain
Zbigniew Nahorski, Poland
Pavel Novoa, Ecuador
Vilem Novak, Czech Republic
Hannu Nurmi, Finland
Manuel Ojeda-Aciego, Spain
Gabriella Pasi, Italy

Witold Pedrycz, Canada
David A. Pelta, Spain
Irina Perfilieva, Czech Republic
Fred Petry, USA
Vincenzo Piuri, Italy
Olivier Pivert, France
Henri Prade, France
Marek Reformat, Canada
Daniel Sánchez, Spain
Mika Sato-Ilic, Japan
Ricardo C. Silva, Brazil
Martin Stepnicka, Czech Republic
Umberto Straccia, Italy
Eulalia Szmidt, Poland
Settimo Termini, Italy
Vicenc Torra, Sweden/Spain
Linda van der Gaag, The Netherlands
Barbara Vantaggi, Italy
José L. Verdegay, Spain
Thomas Vetterlein, Austria
Susana Vieira, Portugal
Slawomir Zadrozny, Poland

Additional Reviewers

Jesús Alcalá-Fernández
José Carlos R. Alcantud
Svetlana Asmuss
Laszlo Aszalos
Mohammad Azad
Cristobal Barba Gonzalez
Gleb Beliakov
María J. Benítez-Caballero
Pedro Bibiloni
Alexander Bozhenyuk
Michal Burda
Ana Burusco
Camilo Alejandro Bustos Téllez
Francisco Javier Cabrerizo
Yuri Cano
Andrea Capotorti
J. Manuel Cascón
Dagoberto Castellanos
Francisco Chicano
María Eugenia Cornejo
Susana Cubillo
Martina Dankova
Luis M. de Campos
Robin De Mol
Yashar Deldjoo
Pedro Delgado-Pérez
Juan Carlos Díaz
Susana Díaz
Marta Disegna
Alexander Dockhorn
Paweł Drygaś
Talbi El Ghazali
Javier Fernández
Joao Gama
José Gámez
M. D. García Sanz
José Luis García-Lapresta
José García Rodríguez
Irina Georgescu
Manuel Gómez-Olmedo
Adrián González
Antonio González
Manuel González-Hidalgo
Jerzy Grzymala Busse
Piotr Helbin
Daryl Hepting
Michal Holcapek
Olgierd Hryniewicz
Petr Hurtik
Atamanyuk Igor
Esteban Induráin
Vladimir Janis
Andrzej Janusz
Sándor Jenei
Pascual Julian-Iranzo
Aránzazu Jurío
Katarzyna Kaczmarek
Martin Kalina
Gholamreza Khademi
Margarita Knyazeva
Martins Kokainis
Galyna Kondratenko

Oleksiy Korobko
Piotr Kowalski
Oleksiy Kozlov
Anna Król
Sankar Kumar Roy
Angelica Leite
Tianrui Li
Ferenc Lilik
Nguyen Linh
Hua Wen Liu
Bonifacio Llamazares
David Lobo
Marcelo Loor
Ezequiel López-Rubio
Gabriel Luque
Rafael M. Luque-Baena
M. Aurora Manrique
Nicolás Marín
Ricardo A. Marques-Pereira
Stefania Marrara
Davide Martinetti
Víctor Martínez
Raquel Martínez España
Miguel Martínez-Panero
Tamás Mihálydeák
Arnau Mir
Katarzyna Miś
Miguel A. Molina-Cabello
Michinori Nakata
Bac Nguyen
Joachim Nielandt
Wanda Niemyska
Juan Miguel Ortiz De Lazcano Lob
Sergio Orts Escolano
Iván Palomares
Esteban José Palomo
Manuel Pegalajar-Cuéllar
Barbara Pękala
Renato Pelessoni
Tomasz Penza
Davide Petturiti
José Ramón Portillo
Cristina Puente
Eloísa Ramírez-Poussa
Ana Belén Ramos Guajardo
Jordi Recasens
Juan Vicente-Riera
Rosa M. Rodríguez
Luis Rodríguez-Benítez
Estrella Rodríguez-Lorenzo
Maciej Romaniuk
Jesús Rosado
Clemente Rubio-Manzano
Pavel Rusnok
Hiroshi Sakai
Sancho Salcedo-Sanz
José María Serrano
Ievgen Sidenko
Gerardo Simari
Julian Skirzynski
Andrzej Skowron
Grégory Smits
Marina Solesvik
Anna Stachowiak
Sebastian Stawicki
Michiel Stock
Andrei Tchernykh
Jhoan S. Tenjo García
Luis Terán
Karl Thurnhofer Hemsi
S. P. Tiwari
Joan Torrens
Gracian Trivino
Matthias Troffaes
Shusaku Tsumoto
Diego Valota
Matthijs van Leeuwen
Sebastien Varrette
Marco Viviani
Pavel Vlašánek
Yuriy Volosyuk
Gang Wang
Anna Wilbik
Andrzej Wójtowicz

Special Session Organizers

Stefano Aguzzoli	University of Milan, Italy
José M. Alonso	University of Santiago de Compostela, Spain
Michal Baczynski	University of Silesia, Poland
Isabelle Bloch	University Paris-Saclay, France
Reda Boukezzoula	LISTIC- Université de Savoie Mont-Blanc, France
Humberto Bustince	Universidad Pública de Navarra, Spain
Inma P. Cabrera	Universidad de Málaga, Spain
Tomasa Calvo	Universidad de Alcalá, Spain
Ciro Castiello	University of Bari Aldo Moro, Italy
Juan Luis Castro	University of Granada, Spain
Yurilev Chalco-Cano	Universidad de Tarapacá, Chile
Davide Ciucci	University of Milano-Bicocca, Italy
Didier Coquin	LISTIC- Université de Savoie Mont-Blanc, France
Pablo Cordero	University of Málaga, Spain
Rocío de Andrés Calle	University of Salamanca, Spain
Bernard De Baets	Ghent University, Belgium
Juan Carlos de la Torre	University of Cádiz, Spain
Graçaliz Dimuro	Institute of Smart Cities and Universidade Federal do Rio Grande, Brazil
Enrique Domínguez	University of Málaga, Spain
Bernabe Dorronsoro	University of Cádiz, Spain
Krzysztof Dyczkowski	Adam Mickiewicz University in Poznań, Poland
Ali Ebrahimnejad	Islamic Azad University, Iran
Tommaso Flaminio	Artificial Intelligence Research Institute (CSIC), Barcelona, Spain
Pilar Fuster-Parra	Universitat de les Illes Balears, Spain
M. Socorro García Cascales	Polytechnic University of Cartagena, Spain
Brunella Gerla	University of Insubria, Italy
Juan Gómez Romero	University of Granada, Spain
Teresa González-Arteaga	University of Valladolid, Spain
Balasubramaniam Jayaram	Indian Institute of Technology Hyderabad, India
László Kóczy	Budapest University of Technology and Economics, Hungary
Yuriy Kondratenko	Petro Mohyla Black Sea National University, Ukraine
Weldon Lodwick	University of Colorado, USA
Vincenzo Loia	University of Salerno, Italy
Nicolás Madrid	University of Málaga, Spain
Luis Magdalena	Universidad Politécnica de Madrid, Spain
María J. Martín-Bautista	University of Granada, Spain
Sebastián Massanet	University of the Balearic Islands, Spain
Corrado Mencar	University of Bari Aldo Moro, Italy
Radko Mesiar	University of Technology, Slovakia
Enrique Miranda	University of Oviedo, Spain
Ignacio Montes	University of Oviedo, Spain

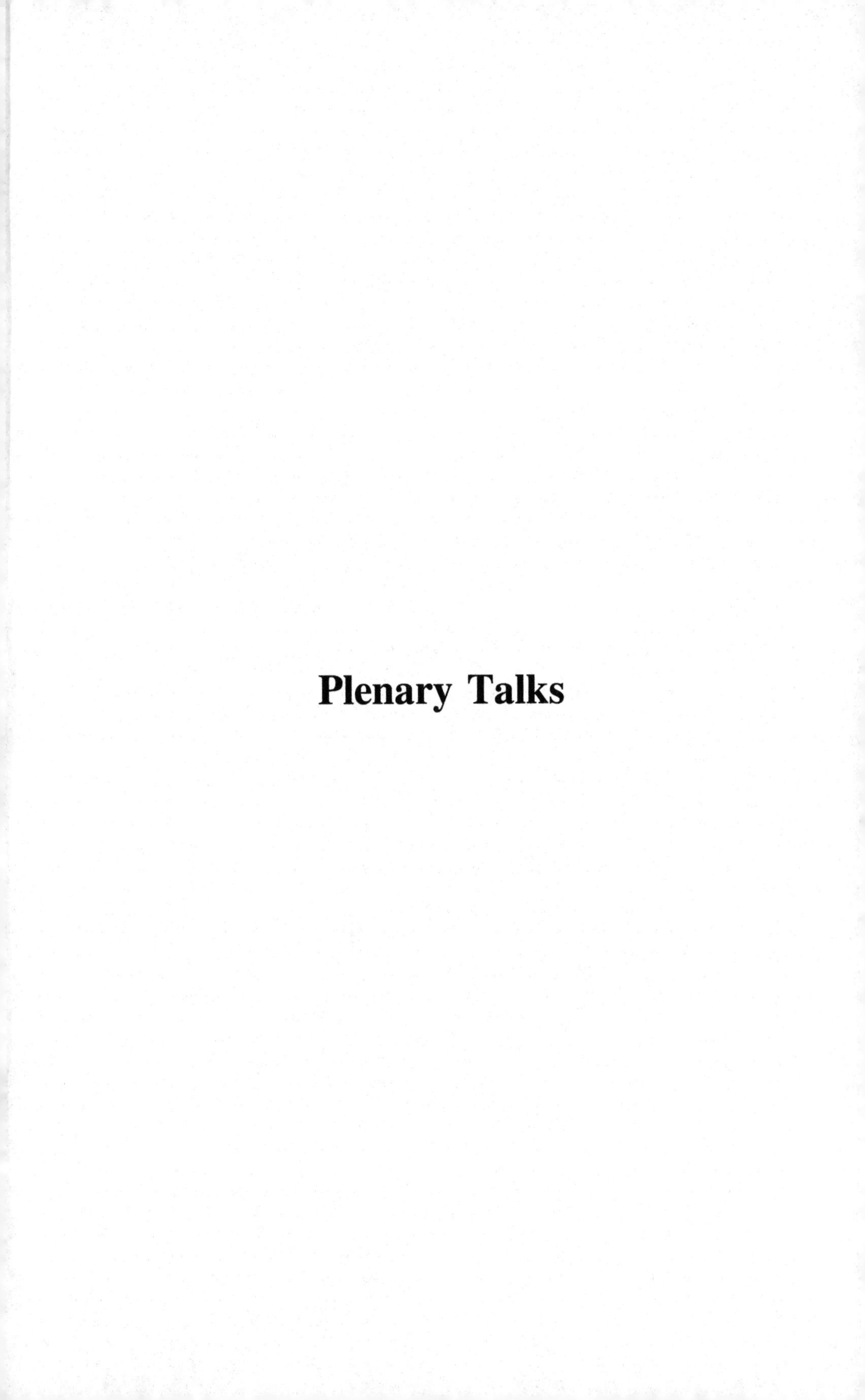

Plenary Talks

Hypothesis Testing as a Game

Glenn Shafer

Department of Accounting and Information Systems, Rutgers Business School–Newark and New Brunswick, 1 Washington Park, Newark, NJ 071022, USA
gshafer@business.rutgers.edu

Abstract. The correspondence between Blaise Pascal and Pierre Fermat is one of the most famous milestones in the history of probability. They differed on how to solve the problem of dividing stakes, but they arrived at the same answer. Their approaches were both rooted in a long history of reasoning about games, but Pascal's was more purely game-theoretic and has something to tell us about our persistent troubles with p-values and reproducibility.

References

1. Bru, M.F., Bru, B.: On dice games and contracts, by Glenn Shafer. In: Statistical Science (2018, to appear). http://www.glennshafer.com/assets/downloads/articles/article98_brus.pdf
2. Game-Theoretic Significance Testing, by Glenn Shafer. Working paper. http://www.probabilityandfinance.com/articles/49.pdf

Biography: Prof. Glenn Shafer is Board of Governors Professor at the Rutgers Business School – Newark and New Brunswick.

He earned two degrees from Princeton University: an A.B. in mathematics in 1968 and a Ph.D. in mathematical statistics in 1973.

In 1976 he published *A Mathematical Theory of Evidence*, launching a widely used methodology for handling uncertainty in expert systems, the Dempster-Shafer theory of belief functions.

Together with Volodya Vovk, he published *Probability and Finance: It's Only a Game!* where they showed how game theory can replace measure theory as a mathematical foundation for probability. Prof. Shafer believes that the game-theoretic foundation for probability will reshape the interpretation and use of probability in many fields. It leads to new methods of forecasting, clarifies how speculation can lead to the apparent random behavior of market prices, and deepens the analysis of causality posed in his previous 1996 book, *The Art of Causal Conjecture*.

Prof. Shafer has published in journals in statistics, philosophy, history, psychology, computer science, economics, engineering, accounting, and law. He has won teaching awards in both mathematics and business. He was a Guggenheim fellow in 1983–84, a fellow at the Center for Advanced Study in the Behavioral Sciences in 1988–89, and a

Fulbright fellow at the Free University of Berlin in Spring 2001. He is a fellow of both the Institute of Mathematical Statistics and the American Association for Artificial Intelligence.

For more information, please visit: http://www.glennshafer.com/

Uncertainty Management with Fuzzy Sets, Rough Sets, Interval Sets, Shadowed Sets, and Three-Way Decisions

Yiyu Yao

Department of Computer Science, University of Regina,
Regina, Saskatchewan S4S 0A2, Canada
yyao@cs.uregina.ca

Abstract. Uncertainty is a multifaceted notion that comes from many sources and calls for various interpretations. In this talk, we discuss uncertainty management in a set-theoretic setting and in a context of concept analysis. In the case of certainty, set theory provides a means to specify extensions of concepts (i.e., the instances of concepts) and the corresponding Boolean logic gives a language to describe intensions of concepts (i.e., the meaning of concepts). We can precisely describe a concept by a pair of a set of instances and a logic formula. In the case of uncertainty, several non-classical set theories have been proposed for modeling different types of uncertainty. Fuzzy sets and the corresponding many-valued logics deal with concepts with a non-sharp and gradually changing boundary. Rough sets and the corresponding modal logics study the approximations of indefinable concepts by definable concepts. Interval sets and the corresponding three-valued logics investigate partially known concepts. Shadowed sets focus on three-way approximations of fuzzy sets in order to gain cognitive advantages in concept analysis.

Each of these theories captures a particular type or a specific aspect of uncertainty. Their integrations offer additional tools for managing uncertainty. For instance, fuzzy rough sets and rough fuzzy sets enable us to model both fuzzy concepts and approximations of indefinable fuzzy concepts. A theory of three-way decision unifies rough sets, interval sets, and shadowed sets on a common ground. The main idea of three-way decision is thinking and processing in threes. When facing with uncertainty, we move from true/false, black/white, and yes/no to true/unsure/false, black/grey/white, and yes/maybe/no. The third option provides the flexibility and universality of using three parts, regions, or degrees in uncertainty management.

Biography: Prof. Yiyu Yao is a Professor with the Department of Computer Science, University of Regina, Canada. His research interests include Three-way Decisions, Granular Computing, Rough Sets, Uncertainty in Artificial Intelligence, Web Intelligence, Information Retrieval, Data Analysis, Machine Learning, and Data Mining. He proposed a theory of three-way decisions, a triarchic theory of granular computing, interval sets, and decision-theoretic rough set models. In 2015, 2016 and 2017, he was selected as a Highly Cited Researcher (Thomson Reuters and now Clarivate Analytics). In 2014, he received the University of Regina Alumni Association Faculty Award for

Research Excellence. In 2008, he received PAKDD Most Influential Paper Award (1999–2008). He is an Area Editor of International Journal of Approximate Reasoning, an Associate Editor of Information Sciences, an Advisory Board Member of Knowledge-based Systems, and a Track Editor of Web Intelligence. He is also an Editorial Board Member of Granular Computing, LNCS Transactions on Rough Sets, International Journal of Intelligent Information Systems, and several others. He is the elected President of International Rough Set Society (IRSS).

Consensus in Group Decision Making and Social Networks

Enrique Herrera Viedma

Department of Computer Science and Artificial Intelligence, E.T.S. de Ingenierías Informática y de Telecomunicación, Universidad de Granada, 18014 Granada, Spain
viedma@decsai.ugr.es

Abstract. The consensus reaching process is the most important step in a group decision making scenario. This step is most frequently identified as a process consisting of some discussion rounds in which several decision makers, which are involved in the problem, discuss their points of view with the purpose of obtaining the maximum agreement before making the decision.

Consensus reaching processes have been well studied and a large number of consensus approaches have been developed. In recent years, the researchers in the field of decision making have shown their interest in social networks since they may be successfully used for modelling communication among decision makers. However, a social network presents some features differentiating it to the classical scenarios in which the consensus reaching processes have been applied. The objective of this talk is to investigate the main consensus methods proposed in social networks and bring out the new challenges that should be faced in this research field.

Biography: Prof. Enrique Herrera-Viedma received the M.Sc. and Ph.D. degrees in Computer Science from the University of Granada, Granada, Spain, in 1993 and 1996, respectively. He is currently a Professor of Computer Science with the Department of Computer Science and Artificial Intelligence, University of Granada, and also the new Vice-President for Research and Knowledge Transfer. His current research interests include intelligent decision making, group decision making, consensus models, fuzzy linguistic modeling, aggregation of information, information retrieval, bibliometric, digital libraries, web quality evaluation, recommender systems, and social media.

Dr. Herrera-Viedma is an Associate Editor of several core international journals indexed in Journal Citation Reports such as the IEEE Transactions on Systems, Man and Cybernetics: Systems, IEEE Trans. On Fuzzy Systems, Knowledge Based Systems, etc. From 2014 he is member of the government of the IEEE SMC Society. He has published in Science [339:6126, p. 1382, 2013] on the new role of the public libraries and he is distinguished as Highly Cited Researcher by Clarivate Analytics in both scientific fields, Engineering and Computer Science (in 2014, 2015, 2016, and 2017), therefore, being considered one of the world's most influential scientific researchers.

Uncertainty Quantification of Many-Valued Events: Betting Methods, Geometry and Reasoning

Lluis Godo

AI Research Institute, IIIA-CSIC, Bellaterra, Spain
godo@iiia.csic.es

Abstract. Betting methods, of which de Finetti's Dutch Book is by far the most well-known, are uncertainty modelling/quantification devices which accomplish a twofold aim. Whilst providing an (operational) interpretation of the relevant measure of uncertainty, they also provide a formal definition of coherence. Indeed, based on the very simple idea that the probability of an unknown event is the fair price a rational gambler is willing to pay in a certain betting game, de Finetti showed that all theorems of probability theory may be derived as consequences of a coherence condition on probability assignments on logically connected events. This coherence condition has two other interpretations. On the one hand, it has a natural geometric interpretation: once the set of events E in the betting game is decided, the collection of coherent probability assignments on E forms a convex polytope P(E) in the real Euclidean space, and hence deciding whether a probability assignment is coherent is equivalent to decide whether it belongs to P(E). Finally, there is also a logical interpretation, where coherence can be understood as logical satisfiability of a set of probability formulas describing an assignment in a suitable probability logic.

This three-fold foundational framework has already been generalised in the classical setting to uncertainty measures other than probability like belief functions. In this talk, starting from Mundici's pioneering work on de Finetti's betting framework for probability on MV-algebras, we will discuss extensions of the basic framework when moving (i) from a classical to a many-valued or fuzzy setting and (ii) from probability to other classes of uncertainty measures on many-valued events. In these general scenarios, we will show how betting frameworks can be properly defined in such a way the notions of coherence arising from them also allow for equivalent (under suitable adjustments) logico-algebraic and geometrical interpretations.

Biography: Lluis Godo is a research professor at IIIA-CSIC, the Artificial Intelligence Research Institute (IIIA) of the Spanish National Research Council (CSIC), Barcelona, Spain. He obtained his MSc degree in Mathematics from the University of Barcelona (1979) and the PhD in Mathematics from the Technical University of Catalunya (1990). His main research interests are on mathematical fuzzy logic, logics to reason under uncertainty, similarity-based reasoning and computational argumentation systems, as well as their combinations. He is author of near 200 publications in international journals and conferences. He is co-editor-in-chief of the journal Fuzzy Sets and Systems, associate editor of Artificial Intelligence, Soft Computing and of

Autonomous Agents and Multi-Agent Systems (Springer). He is an ECCAI Fellow and an IFSA Fellow. He is past vice-president of the European Society for Fuzzy Logic Technologies (EUSFLAT) and of the Catalan AI association (ACIA).

Geo Big Data for Earth Observation: Challenges and Opportunities of Fuzzy Approaches

Gloria Bordogna

CNR – IREA Istituto per il Rilevamento Elettromagnetico dell'Ambiente, Milano, Italy
bordogna.g@irea.cnr.it

Abstract. Geo big data are heterogeneous by nature traditionally comprising both georeferenced images acquired by remote sensing and their derived products, and cartographic maps published as open data by public and private organizations. Furthermore, thanks to the Web 2.0 revolution and wide spread diffusion of IoT and smart devices equipped with GNSS sensors, the availability of new and real-time sources of geo big data is rapidly increasing. Let us think at Volunteered Geographic Information created by citizens eager to participate in citizen science initiatives, at crowdsourced geotagged posts, created by users of social networks, and at the great variety of low-cost sensor data. This heterogeneous multisource geo data constitutes a challenge for Earth Observation, i.e., for describing the planet Earth's physical, chemical, biological and anthropic systems to monitor and assess the status of and changes in the natural, built and social environment, although to convert data into value, we need to face some open issues related with the effective management of geo big data. Specifically, we need new methods for the *representation* and *discovery* of the relevant geo data among huge repositories, the *assessment of the questionable geo data quality*, and, finally, the *cross-analysis and synthesis of geo data* to provide decision makers with consistent and comprehensible information. All such tasks involve the management of the imprecision and the uncertainty of both geo data and user needs. The talk will analyze some applications of multisource geo big data management for Earth Observation focusing on the opportunities of fuzzy approaches to represent, discover, assess the quality, and synthesize geo big data.

Biography: Gloria Bordogna is a senior researcher of IREA CNR – Institute for the Electromagnetic Sensing of the Environment of the Italian National Research Council – is member of EUSFLAT, and IFSA Fellow since 2017. In 1984 she received the Laurea degree in Physics from "Università degli Studi di Milano", Italy, and since 1986 she was with different research institutes of CNR. From 2003 to 2010, she was adjunct professor of Information Systems within Bergamo University. Since 2008 she co-organizes the special track on Information Access and Retrieval at ACM SAC and is in the editorial board of ACM SIGAPP – "Applied Computing Review" and of the Int. Journal of "Intelligent Decision Technologies". Her research activity mainly concerns the representation and management of imprecision and uncertainty within information retrieval systems (IRSs) database management systems (DBMSs) and Geographic

Information Systems (GIS), by means of soft computing. Her current research concerns spatio-temporal analytics of social networks, quality assessment of Volunteered Geographic Information, and remote sensing image synthesis.

Biological Apps: Rapidly Converging Technologies for Living Information Processing

Natalio Krasnogor

Professor of Computing Science and Synthetic Biology, School of Computing Science, Claremont Tower, Newcastle University, UK
natalio.krasnogor@ncl.ac.uk

Abstract. There is a rapid convergence of biological, nanotechnological and computational technologies that will soon make biological "apps" programming a reality. In this talk I will overview some of the recent progress done by my group and others on information processing in vitro and in vivo and will mention some of the challenges that computer scientists face will need to overcome to fully exploit the possibilities afforded by living matter programming.

Biography: Prof. Natalio Krasnogor is a Professor of Computing Science and Synthetic Biology at the School of Computing Science at Newcastle University (he moved to Newcastle, from the University of Nottingham, on the 1st-September-2013). He has co-directed Newcastle's Interdisciplinary Computing and Complex BioSystems (ICOS) research group. He is also affiliated with the Centre for Synthetic Biology and Bio-exploitation as well as with the Centre for Bacterial Cell Biology.

His research activities lie at the interface of Computing Science and the Natural Sciences, e.g. Biology, Chemistry and Physics. He applies his expertise on Machine Intelligence (e.g. optimisation, data mining, big data, evolutionary learning), Complex Systems and Unconventional Computing (e.g. biocomputing) to Bioinformatics, Systems and Synthetic Biology. His research is generously funded from a number of external sources that supports multiple projects.

Contents – Part I

Aggregation Operators, Fuzzy Metrics and Applications

Belief Function Theory and Its Applications

Formal Concept Analysis and Uncertainty

Fuzzy Implication Functions

Fuzzy Logic and Artificial Intelligence Problems

Fuzzy Mathematical Analysis and Applications

Contents – Part II

Imprecise Probabilities: Foundations and Applications

Mathematical Fuzzy Logic and Mathematical Morphology

Rough and Fuzzy Similarity Modelling Tools

Soft Computing for Decision Making in Uncertainty

Soft Computing in Information Retrieval and Sentiment Analysis

Tri-partitions and Uncertainty

Contents – Part III

Decision Making Modeling and Applications

Optimization Models for Modern Analytics

Uncertainty in Medicine

Uncertainty in Video/Image Processing (UVIP)

General Track

Advances on Explainable Artificial Intelligence

A Bibliometric Analysis of the Explainable Artificial Intelligence Research Field

Jose M. Alonso[1(✉)], Ciro Castiello[2], and Corrado Mencar[2]

[1] Centro Singular de Investigación en Tecnoloxías da Información (CiTIUS), Universidade de Santiago de Compostela, Santiago de Compostela, Spain
josemaria.alonso.moral@usc.es

[2] Department of Informatics, University of Bari "Aldo Moro", Bari, Italy
{ciro.castiello,corrado.mencar}@uniba.it

Abstract. This paper presents the results of a bibliometric study of the recent research on eXplainable Artificial Intelligence (XAI) systems. We took a global look at the contributions of scholars in XAI as well as in the subfields of AI that are mostly involved in the development of XAI systems. It is worthy to remark that we found out that about one third of contributions in XAI come from the fuzzy logic community. Accordingly, we went in depth with the actual connections of fuzzy logic contributions with AI to promote and improve XAI systems in the broad sense. Finally, we outlined new research directions aimed at strengthening the integration of different fields of AI, including fuzzy logic, toward the common objective of making AI accessible to people.

Keywords: Interpretability · Understandability · Comprehensibility
Explainable AI · Interpretable Fuzzy Systems

1 Introduction

In the era of the Internet of Things and Big Data, data scientists are required to extract valuable knowledge from the given data. They first analyze, cure and pre-process data; then, they apply Artificial Intelligence (AI) techniques to automatically extract knowledge from data [1]. Getting AI into widespread real-world usage requires to think carefully of many important issues. Among them, we would like to highlight (1) Ethics, (2) Law and (3) Technology.

Recently, ACM issued a Statement on "Algorithmic Transparency and Accountability", which establishes a set of principles, consistent with the ACM Code of Ethics, to support the benefits of algorithmic decision-making while addressing ethical and legal concerns [2]. Among such principles, *Explanation* is of relevance for this study. According to ACM: "Systems and institutions that use algorithmic decision-making are encouraged to produce explanations regarding both the procedures followed by the algorithm and the specific decisions that are made. This is particularly important in public policy contexts."

J. Medina et al. (Eds.): IPMU 2018, CCIS 853, pp. 3–15, 2018.
https://doi.org/10.1007/978-3-319-91473-2_1

In addition, a new European General Data Protection Regulation (GDPR[1]) is expected to take effect in 2018 [3]. It takes care of the protection of natural people when personal data have to be processed and freely moved. Moreover, it emphasizes the "right to explanation" of European citizens: "[...] decision-making based on such processing, including profiling, should be allowed [...] In any case, such processing should be subject to suitable safeguards, which should include [...] the right to obtain human intervention, to express his or her point of view, to obtain an explanation of the decision reached after such assessment and to challenge the decision."

Regarding technological issues, the theme of explainability in AI is also remarked in the last challenge stated by the USA Defense Advanced Research Projects Agency (DARPA) [4]: "Even though current AI systems offer many benefits in many applications, their effectiveness is limited by a lack of explanation ability when interacting with humans."

Accordingly, non-expert users, i.e., users without a strong background on AI, require a new generation of explainable AI (XAI) systems. Such systems are expected to naturally interact with humans by providing comprehensible explanations of decisions that are automatically made. XAI systems can be also considered as an important step forward toward Collaborative Intelligence [5] which promises a fully accepted integration of AI in our society.

In this paper, we report the results of a bibliometric study of the recent research on XAI systems. We are interested in assessing the contributions of AI scholars in XAI, as well as in the subfields of AI that are mostly involved in the development of XAI systems. More specifically, we are interested in the role of the fuzzy logic community in the progress of XAI, exploring the connections of fuzzy logic contributions with AI to promote and improve systems explainability. While moving along this way, we hope to outline new research directions aimed at strengthening the integration of different fields of AI, including fuzzy logic, toward the common objective of making AI accessible to people.

The rest of the manuscript is organized as follows. Section 2 introduces material and methods. Section 3 presents our bibliometric analysis focused on XAI. Section 4 introduces additional details while focusing on Interpretable Fuzzy Systems (IFS) only. Finally, Sect. 5 remarks the main points of the study and pinpoints future work.

2 Material and Methods

2.1 Bibliometric Techniques

Scientometrics is informally defined as the discipline that studies the quantitative features and characteristics of science and scientific research, technology and innovation. Within Scientometrics, Bibliometrics copes with the statistical analysis of books, articles, or other kinds of publications [6].

[1] http://eur-lex.europa.eu/legal-content/en/TXT/?uri=CELEX%3A32016R0679.

Usually, bibliographical data are treated by statistical mathematical methods and results are visualized in form of tables and graphs. For example, Vargas-Quesada and Moya-Anegón [7] proposed a methodology for creating visual representations of scientific domains. They focused on illustrating interactions among authors and papers through citations and co-citations. Later, other authors generalized the idea and developed alternative methods and tools (e.g., [8,9]) to create maps of linked items (scientific publications, scientific journals, researchers, research organizations, countries, or keywords).

Different types of links between pairs of items can be considered. As an example, let us briefly introduce the concept of item co-citation. Given a set of items, all potential links among pairs of items can be characterized by the standardized co-citation measure [10] as follows:

$$MCN_{ij} = \frac{Cc_{ij}}{\sqrt{c_i \cdot c_j}} \quad (1)$$

where Cc means co-citation, c stands for citation, i and j are two different items.

The link values (MCN_{ij}) define the adjacency matrix of a graph which can be analyzed and visualized with social network analysis (SNA) techniques [11]. These techniques have been already applied to multiple fields of research, such as software development (e.g., debugging multi-agent systems [12]), scientometrics (e.g., analyzing large scientific domains [13]), or fuzzy modeling (e.g., analyzing fuzzy rule-bases with fingrams [14]).

There are many metrics designed to assess the importance of a node in a bibliographical graph (e.g., centrality degree, closeness, betweenness or page rank) [7]. In addition, there are many different methods for graph visualization [15]. Among them, force-directed algorithms are the most widely used in information science [16]. Their purpose is to locate the nodes of a graph in a 2D or 3D space, so that all the edges are approximately of equal length and there are as few crossing edges as possible, trying to obtain the most aesthetically pleasing view. There are also many clustering techniques aimed at discovering communities (or bunches of highly related nodes) in accordance with the importance of each single node and how it is connected to the others [17].

2.2 Bibliographic Repositories

Bibliographic data can be read from different sources such as Web of Science (WoS) or Scopus. WoS appears not to be adequate for assessing publications and citations in Informatics[2]. In addition, some other sources may be too large (e.g., Google Scholar) or too specialized (e.g., ACM DL, IEEEXplore, etc.). Therefore, in this work we focus on Scopus which also offers advanced search functionalities useful to select meaningful sets of items which can be considered as a ground to build our bibliometric analysis. Anyway, the selection of Scopus as a bibliographical source comes without any loss of generalization. We performed

[2] Informatics Research Evaluation (Draft), An Informatics Europe Report. http://www.informatics-europe.org/working-groups/research-evaluation.html.

a preliminary study on data collected from WoS: the main trends and the general conclusions remained unchanged (only slight minor variations were detected).

Finally, it should be highlighted that data collected from Scopus have been cured in order to remove spurious information that would have hampered the subsequent steps of our analysis.

2.3 Bibliographic Analysis Tools

We used a couple of tools to analyze the results of search queries from Scopus:

Bibliometrix [18] - An R package for performing comprehensive quantitative research in Scientometrics and Bibliometrics. It allows importing bibliographic data from several sources (including Scopus and WoS). In addition, it evaluates co-citation as well as other kinds of measures, such as coupling, scientific collaboration and co-word analyses.

VOS viewer [9] - A software tool for constructing and visualizing bibliometric networks which can be related to citation, co-citation, bibliographic coupling, co-authorship or co-occurrence of words relations. Some clustering methods to identify related groups or communities are also provided.

3 A Global Overview on XAI

On October 20th, 2017, we ran the following query through the "Advanced Search" tool provided by Scopus:

```
Q1 = TITLE (``*interpretab*'') OR TITLE (``*comprehensib*'')
OR TITLE (``*understandab*'') OR TITLE (``*explainab*'')
OR TITLE (``*self-explanat*'') OR KEY (``*interpretab*'')
OR KEY (``*comprehensib*'') OR KEY (``*understandab*'')
OR KEY (``*explainab*'') OR KEY (``*self-explanat*'')
```

As a result, we found out 5735 documents. It is worthy to note this query is intentionally very general in order to broaden the global picture of the research field under examination. We identified only 5 general terms and their variants represented by the * symbol. We required at least one of these terms to be present in the title of the retrieved document or in the associated keywords (provided by authors or automatically indexed).

Figure 1 depicts the number of XAI publications since 1960 (top picture) and the distribution of publications in the top-10 ranking of subject fields (bottom picture). The number of publications started to grow significantly since 2000. Accordingly, we decided to focus our analysis only on the years ranging from 2000 to 2017.

XAI represents a multidisciplinary research field, as witnessed by the variety of subject areas. Anyway, three of them (Computer Sciences, Mathematics, and Engineering) collect most of the publications. Therefore, we are going to pay attention only to publications in these research areas. In this way, the final

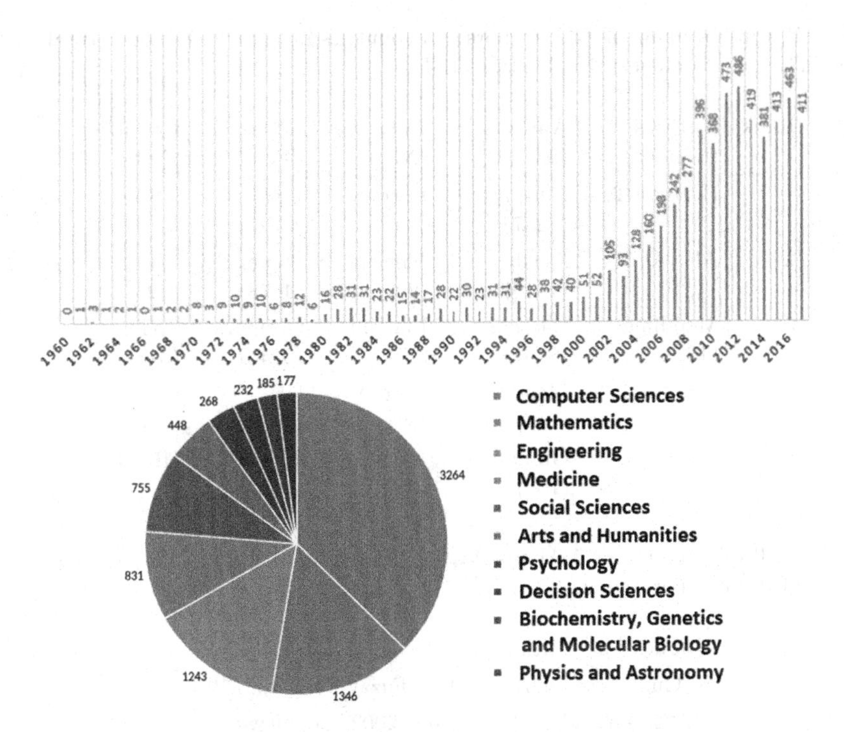

Fig. 1. Histogram of XAI publications and their distribution by subject areas.

number of publications to analyze is 3737. We downloaded from Scopus all the related bibliographical information in form of csv and bib files.

Table 1 presents the Top-5 rankings of authors (columns 2–4) and countries (columns 5 and 6) with respect to h-index, total number of citations (TC) and publications (NP). Herrera stands as the leading author in terms of h-index, TC and NP. USA is by far the leading country in terms of both TC and NP.

The leading publications are listed in Table 2 in terms of TC and average citations per year (ACY). Guillaume [19] reviewed methods for automatically designing IFS. This is the most cited publication being also the fifth as for ACY. Aleven and Koedinger [20] described how to improve students' learning with a computer-based approach endowed with self-explanation. This is the second most cited publication, ranked fourth in terms of ACY. Jin [21] authored the third most cited publication, presenting a fuzzy modeling approach designed to improve the interpretability of high-dimensional systems. This publication is out of the Top-5 in terms of ACY. García et al. [22] reviewed statistical techniques to get a good interpretability-accuracy trade-off in genetics-based machine learning. This is the fourth most cited paper and the second one in terms of ACY. The fifth publication in terms of TC (out of the ACY Top-5) comes from Ishibuchi and Nojima [23] who applied a multi-objective genetics-based machine learning approach to build fuzzy systems with a good interpretability-accuracy

Table 1. Top-5 ranking of XAI authors and countries in terms of h-index, Total Citations (TC) and Number of Publications (NP).

Rank.	Authors (h-index)	Authors (TC)	Authors (NP)	Countries (TC)	Countries (NP)
1	Herrera (22)	Herrera (2287)	Herrera (45)	USA (6737)	USA (604)
2	Alcalá (13)	Alcalá (1216)	Mencar (29)	Spain (4628)	China (323)
3	Mendling (13)	Baesens (846)	Piattini (29)	Germany (2242)	Spain (318)
4	Alonso (12)	Mendling (717)	Alonso (26)	UK (1703)	Germany (270)
5	Piattini (11)	Guillaume (707)	Alcalá (25)	China (1692)	UK (175)

Table 2. Top-5 ranking of XAI publications in terms of Total Citations (TC) and Average Citations per Year (ACY).

TC rank.	Publication (Authors, Year, Source, TC) [Ref]
1	S. Guillaume, 2001, IEEE T Fuzzy Syst, 435 [19]
2	V. Aleven and K. Koedinger, 2002, Cognitive Sci, 418 [20]
3	Y. Jin, 2000, IEEE T Fuzzy Syst, 333 [21]
4	S. García et al., 2009, Soft Comput, 308 [22]
5	H. Ishibuchi and Y. Nojima, 2007, Int J Approx Reason, 266 [23]
ACY rank.	**Publication (Authors, Year, Source, ACY) [Ref]**
1	L. Martínez and F. Herrera, 2012, Inform Sciences, 39.8 [24]
2	S. García et al., 2009, Soft Comput, 38.5 [22]
3	M.J. Gacto et al., 2011, Inform Sciences, 33.3 [25]
4	V. Aleven and K. Koedinger, 2002, Cognitive Sci, 27.9 [20]
5	S. Guillaume, 2001, IEEE T Fuzzy Syst, 27.2 [19]

trade-off. The scenario is completed by Martínez and Herrera [24] (first paper in terms of ACY), who proposed a linguistic model for solving decision-making problems, and Gacto et al. [25] (third paper in terms of ACY), who reviewed interpretability indexes for assessing IFS. Notice that Herrera co-authored 3 of the Top-5 publications in terms of ACY: this emphasizes his leading role in the XAI research field (see Table 1).

The leading sources in XAI are depicted in the pie chart on the left of Fig. 2. Most papers are published in conference proceedings. Nevertheless, the Top-5 papers (see Table 2) appear in well-recognized journals.

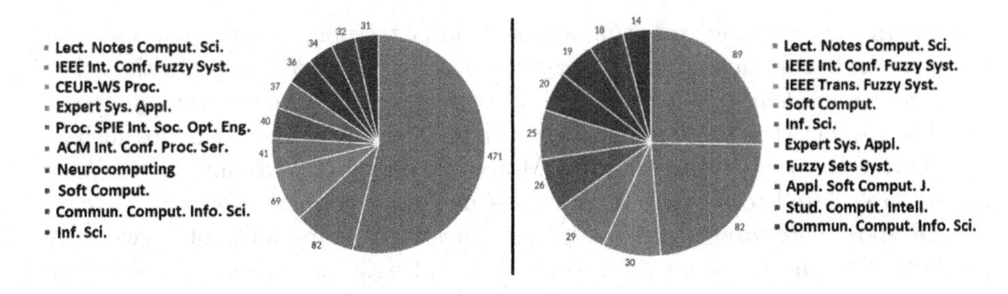

Fig. 2. Pie charts with the leading sources in XAI (left) and IFS (right).

Figure 3 shows a graph with the most popular author keywords in the publications under study. Each node is associated to a keyword and its size is proportional to the number of documents where the keyword appears. *Interpretability* is the main keyword since it is associated to the larger node. *Understandability* and *classification* are the second and the third main keywords. Links between nodes relate keywords which usually appear together in the same documents.

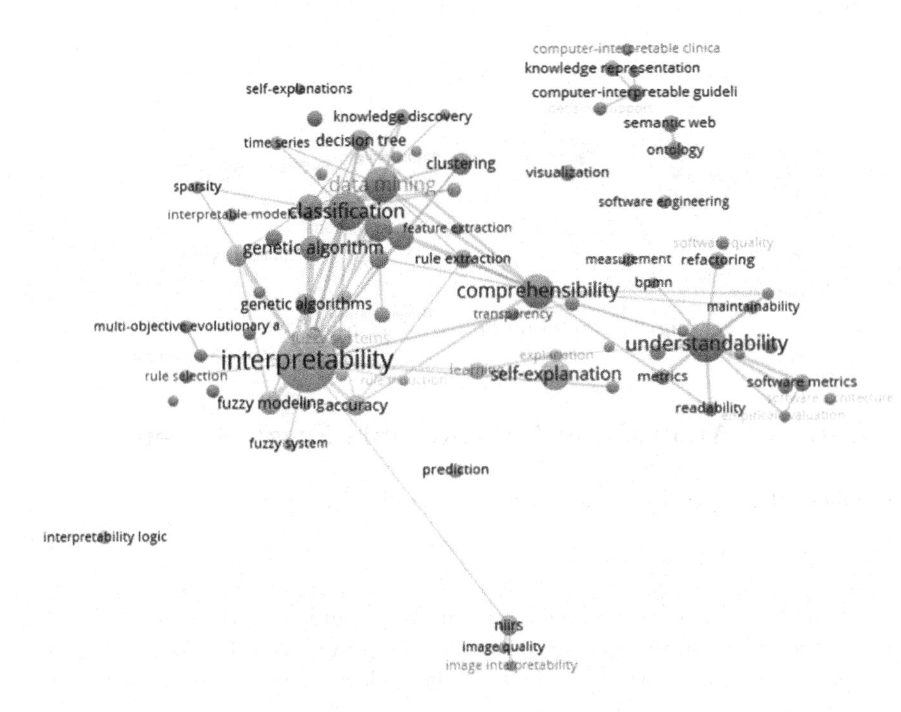

Fig. 3. Map of author keywords in XAI. (Color figure online)

This graph gives a global overview about the main topics of interest in the XAI research field, with groups of closely related nodes painted in the same color[3]. On the one hand, *interpretability* is closer to topics usually addressed in the fuzzy logic community (e.g., *fuzzy modeling* or *rule selection*). On the other hand, *understandability* is surrounded by keywords related to software engineering. The gap between the main keywords is filled by other relevant nodes such as *comprehensibility* or *self-explanation.* Moreover, *interpretability* and *comprehensibility* are related to a group of keywords including popular topics in AI (e.g., *classification, data mining* or *knowledge discovery*). A community of keywords is partially disconnected from the rest of the graph (e.g., *semantic web, ontology,* and so on), and some single nodes lie away from others (e.g., *interpretability logic* or *image interpretability*). That is due to their relatedness to some specific research lines. Notice that NIIRS stands for National Imagery Interpretability Rating Scale which is a subjective scale for rating the quality of images.

When we turn to consider author co-citation, we look for pairs of authors being cited by the same publications. Figure 4 shows the co-citation map obtained by the VOS viewer (the minimum number of total citations by author is set to 50). Size of nodes is proportional to the number of citations, while link weights come from the co-citation index defined by Eq. (1). Most nodes are concentrated in the left-hand side of the map. Again, Herrera stands out as the main node.

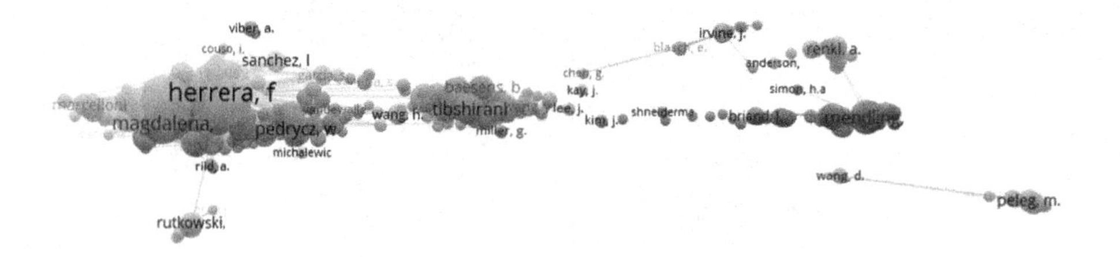

Fig. 4. Map of author co-citation in XAI.

4 Detailed Analysis on Interpretable Fuzzy Systems

We replicated the previous analysis with a modified query:

```
Q2 = Q1 AND ''fuzz*''
```

By adding "fuzz*" to Q1 we focus our search on publications in the XAI field that are related to fuzzy sets and systems. Hereafter, we refer to this field of research as IFS. In addition, we filtered the collected results by adopting the same constraints imposed in the previous section: (1) years range [2000–2017] and (2) subject areas [Computer Sciences, Mathematics and Engineering]. As

[3] The graph was generated by the VOS viewer employing the suggested default parameters for layout visualization and clustering of nodes. Other clustering approaches may be applied, but choosing the best approach is out of the scope of this paper.

a result, we got 1054 documents, consisting in about 28% of the whole set of documents previously analyzed.

The Top-5 rankings of authors and countries is detailed in Table 3 concerning IFS. Most authors in Table 3 are present also in Table 1, thus certifying the relevance of the fuzzy community in the context of XAI. However, USA (the leading country in Table 1) is now out of the Top-5. Moreover, European countries take up the Top-5 in terms of TC. China and India appear only when looking at NP. These data reflect the relevance of European scholars in the fuzzy community and their outstanding leadership in IFS.

Table 3. Top-5 ranking of IFS authors and countries.

Rank.	Authors (h-index)	Authors (TC)	Authors (NP)	Countries (TC)	Countries (NP)
1	Herrera (20)	Herrera (1900)	Herrera (42)	Spain (3255)	Spain (177)
2	Alcalá (13)	Alcalá (1216)	Mencar (29)	Italy (850)	China (116)
3	Alonso (12)	Guillaume (707)	Alonso (26)	France (766)	Italy (81)
4	Magdalena (11)	Ishibuchi (646)	Alcalá (25)	UK (749)	UK (68)
5	Mencar (10)	Nojima (568)	Fanelli (24)	Poland (552)	India (42)

Table 4. Top-5 ranking of IFS publications.

TC ranking	Publication (Authors, Year, Source, TC) [Ref]
1	S. Guillaume, 2001, IEEE T Fuzzy Syst, 435 [19]
2	Y. Jin, 2000, IEEE T Fuzzy Syst, 333 [21]
3	H. Ishibuchi and Y. Nojima, 2007, Int J Approx Reason, 266 [23]
4	M.J. Gacto et al., 2011, Inform Sciences, 200 [25]
5	L. Martínez and F. Herrera, 2012, Inform Sciences, 199 [24]
ACY ranking	**Publication (Authors, Year, Source, ACY) [Ref]**
1	L. Martínez and F. Herrera, 2012, Inform Sciences, 39.8) [24]
2	M. Fazzolari et al., 2013, IEEE T Fuzzy Syst, 34 [26]
3	M.J. Gacto et al., 2011, Inform Sciences, 33.3 [25]
4	S. Guillaume, 2001, IEEE T Fuzzy Syst, 27.2 [19]
5	H. Ishibuchi and Y. Nojima, 2007, Int J Approx Reason, 26.6 [23]

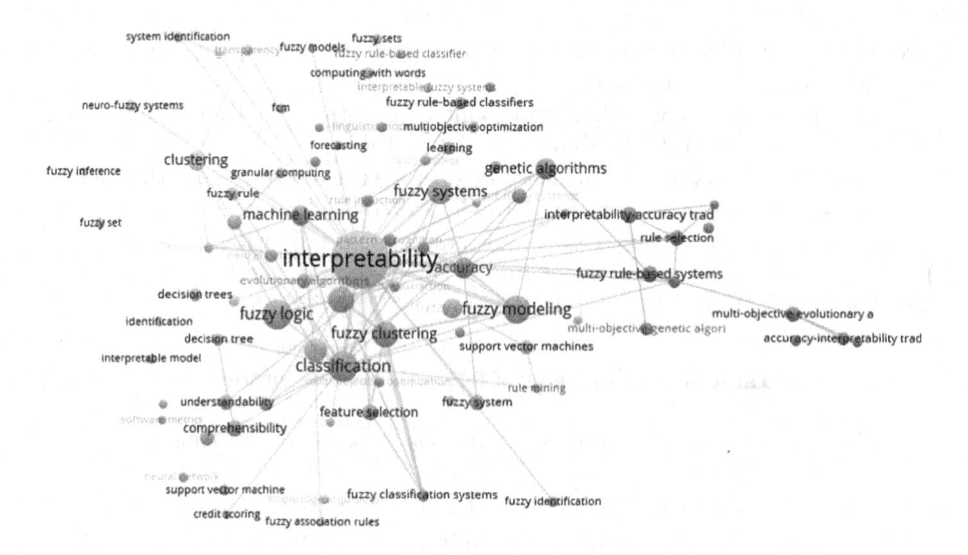

Fig. 5. Map of author keywords in IFS.

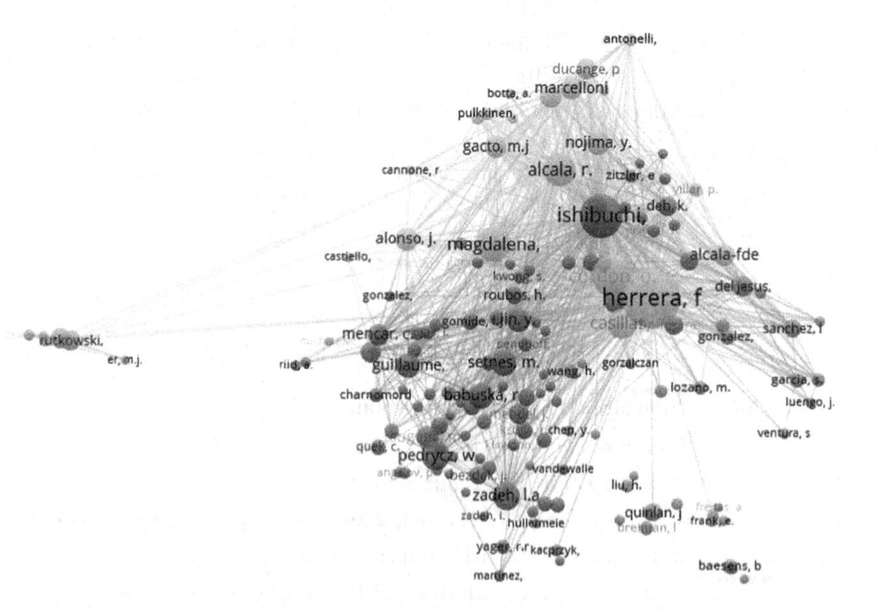

Fig. 6. Map of author co-citation in IFS.

Table 4 lists the leading publications in IFS and reflects once again the relevance of the fuzzy community in the context of XAI. All the papers in the TC ranking already appeared in Table 2: the current Top-3 is included in the TC ranking related to XAI, while [24,25] appear in the Top-3 of the ACY ranking of Table 2. Actually, only the work authored by Fazzolari et al. [26] (a review of multi-objective evolutionary fuzzy systems devoted, among other things, to find a good interpretability/accuracy trade-off) is a new entry with respect to Table 2.

Looking carefully at the map of author keywords in Fig. 5, we miss some of the important topics highlighted in Fig. 3. For example, *comprehensibility* and *understandability* seem to play a much more prominent role in XAI than in IFS. It could be argued that Fig. 5 may be read as a zoom produced in a specific area of Fig. 3, namely the one related to the *interpretability* node. This suggests that many important issues in XAI are still to be addressed by IFS scholars.

Finally, Fig. 6 shows the map of author co-citation in IFS. Once again, the current map looks like a zoom of the left-hand side of Fig. 4.

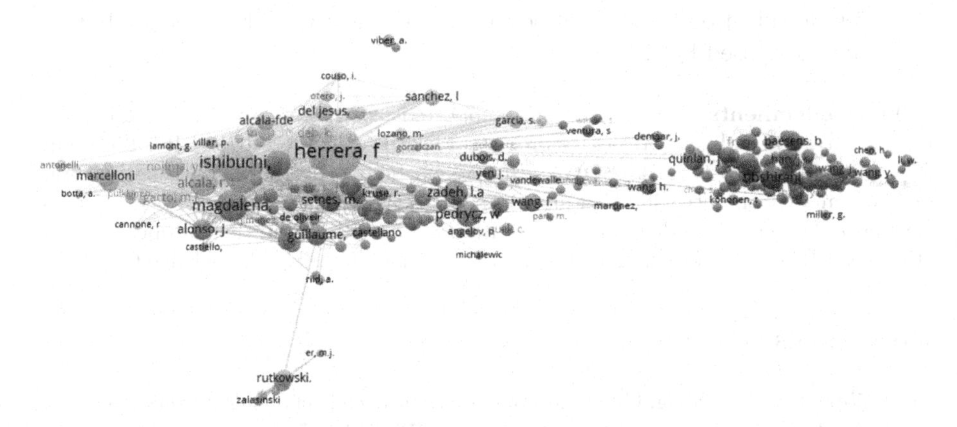

Fig. 7. Map of author co-citation in XAI (zoom in the left hand side of Fig. 4). (Color figure online)

5 Concluding Remarks

The results reported in the previous sections allow a number of considerations. First, there is a strong community of scholars in fuzzy logic addressing their study to the theme of XAI, with special emphasis on interpretability. In fact, interpretability studies in fuzzy logic started from pioneering works in 1999 and about one third of the selected papers belong to the fuzzy logic mainstream. As a result, many of the most influential authors and papers in XAI refer to the fuzzy community. However, if we compare Fig. 3 with Fig. 5 we observe that, within the fuzzy community, the main notions of *interpretability*, *comprehensibility*, *understandability* and *explainability* are not clearly distinct as in XAI. Rather, *interpretability* has a major role while the other keywords are either treated as synonyms or distinctly used in very specialized studies only.

Furthermore, a deeper analysis of the co-citation graph let us observe a neat separation between authors in the fuzzy community and authors in XAI not related to fuzzy logic. This can be appreciated in Fig. 7 where a zoom of the left-hand side of Fig. 4 is provided: authors related to fuzzy logic appear to be aggregated in two compact clusters (the yellow and the green ones in Fig. 7) demonstrating also a tight interconnection of the related research activities. On the other hand, authors in XAI not related to fuzzy logic appear to be loosely distributed, as a sign of a more scattered collaboration.

This analysis suggests at least two lines of development. Firstly, there is a need to clarify and distinguish the notions of *interpretability*, *comprehensibility*, *understandability* and *explainability* to provide a common terminological ground inside the varied XAI context. This could also shed light on refined conceptualizations where fuzzy logic could significantly contribute. Moreover, an opportunity emerges to tighten the connections of studies between fuzzy and non-fuzzy worlds of XAI, which now appear unnecessarily separated. We strongly believe that cross-fertilization between these communities is needed to successfully face the challenges posed by XAI.

Acknowledgements. This work was supported by RYC-2016-19802 (Ramón y Cajal contract), and two MINECO projects TIN2017-84796-C2-1-R (BIGBISC) and TIN2014-56633-C3-3-R (ABS4SOW). All of them funded by the Spanish "Ministerio de Economía y Competitividad". Financial support from the Xunta de Galicia (Centro singular de investigación de Galicia accreditation 2016–2019) and the European Union (European Regional Development Fund - ERDF), is gratefully acknowledged.

References

1. Philip Chen, C., Zhang, C.: Data-intensive applications, challenges, techniques and technologies: a survey on big data. Inf. Sci. **275**, 314–347 (2014)
2. ACM US Public Policy Council: Statement on Algorithmic Transparency and Accountability (2017)
3. Goodman, B., Flaxman, S.: European union regulations on algorithmic decision-making and a "right to explanation". In: ICML Workshop on Human Interpretability in Machine Learning (WHI), New York, NY, pp. 1–9 (2016)
4. Gunning, D.: Explainable Artificial Intelligence (XAI). Technical report, Defense Advanced Research Projects Agency, Arlington, USA, DARPA-BAA-16-53 (2016)
5. Epstein, S.L.: Wanted: collaborative intelligence. Artif. Intell. **221**, 36–45 (2015)
6. De Bellis, N.: Bibliometrics and Citation Analysis: From the Science Citation Index to Cybermetrics. Scarecrow Press, Lanham (2009)
7. Vargas-Quesada, B., Moya-Anegón, F.: Visualizing the Structure of Science. Springer, Heidelberg (2007). https://doi.org/10.1007/3-540-69728-4
8. Cobo, M., López-Herrera, A., Herrera-Viedma, E., Herrera, F.: Science mapping software tools: review, analysis, and cooperative study among tools. J. Assoc. Inf. Sci. Tech. **62**, 1382–1402 (2011)
9. Van Eck, N., Waltman, L.: Software survey: vosviewer, a computer program for bibliometric mapping. Scientometrics **84**, 523–538 (2010)
10. Salton, G., Bergmark, D.: A citation study of computer science literature. IEEE Trans. Prof. Commun. **22**, 146–158 (1979)
11. Wasserman, S., Faust, K.: Social Network Analysis: Methods And Applications (Structural Analysis in the Social Sciences). Cambridge University Press, Cambridge (1994)

12. Serrano, E., Quirin, A., Botia, J., Cordón, O.: Debugging complex software systems by means of pathfinder networks. Inf. Sci. **180**(5), 561–583 (2010)
13. Moya-Anegón, F., Vargas-Quesada, B., Herrero-Solana, V., Chinchilla-Rodríguez, Z., Corera-Álvarez, E., Muñoz-Fernández, F.J.: A new technique for building maps of large scientific domains based on the cocitation of classes and categories. Scientometrics **61**(1), 129–145 (2004)
14. Pancho, D., Alonso, J., Cordón, O., Quirin, A., Magdalena, L.: FINGRAMS: visual representations of fuzzy rule-based inference for expert analysis of comprehensibility. IEEE Trans. Fuzzy Syst. **21**(6), 1133–1149 (2013)
15. di Battista, G., Eades, P., Tamassia, R., Tollis, I.: Graph Drawing: Algorithms for the Visualization of Graphs. Prentice Hall, Upper Saddle River (1998)
16. Kobourov, S.G.: Force-directed drawing algorithms. In: Tamassia, R. (ed.) Handbook of Graph Drawing and Visualization. CRC Press, Boca Raton (2012)
17. Porter, M., Onnela, J., Mucha, P.: Communities in networks. Not. Am. Math. Soc. **56**(9), 1082–1166 (2009)
18. Aria, M., Cuccurullo, C.: *bibliometrix*: an R-tool for comprehensive science mapping analysis. J. Informetr. **11**(4), 959–975 (2017)
19. Guillaume, S.: Designing fuzzy inference systems from data: an interpretability-oriented review. IEEE Trans. Fuzzy Syst. **9**(3), 426–443 (2001)
20. Aleven, V., Koedinger, K.: An effective metacognitive strategy: learning by doing and explaining with a computer-based cognitive tutor. Cogn. Sci. **26**(2), 147–179 (2002)
21. Jin, Y.: Fuzzy modeling of high-dimensional systems: complexity reduction and interpretability improvement. IEEE Trans. Fuzzy Syst. **8**(2), 212–221 (2000)
22. García, S., Fernandez, A., Luengo, J., Herrera, F.: A study of statistical techniques and performance measures for genetics-based machine learning: accuracy and interpretability. Soft. Comput. **13**(10), 959–977 (2009)
23. Ishibuchi, H., Nojima, Y.: Analysis of interpretability-accuracy tradeoff of fuzzy systems by multiobjective fuzzy genetics-based machine learning. Int. J. Approx. Reason. **44**(1), 4–31 (2007)
24. Martínez, L., Herrera, F.: An overview on the 2-tuple linguistic model for computing with words in decision making: extensions, applications and challenges. Inf. Sci. **207**, 1–18 (2012)
25. Gacto, M., Alcalá, R., Herrera, F.: Interpretability of linguistic fuzzy rule-based systems: an overview of interpretability measures. Inf. Sci. **181**(20), 4340–4360 (2011)
26. Fazzolari, M., Alcalá, R., Nojima, Y., Ishibuchi, H., Herrera, F.: A review of the application of multiobjective evolutionary fuzzy systems: current status and further directions. IEEE Trans. Fuzzy Syst. **21**(1), 45–65 (2013)

Do Hierarchical Fuzzy Systems Really Improve Interpretability?

Luis Magdalena(✉)

Escuela Técnica Superior de Ingenieros Informáticos,
Universidad Politécnica de Madrid, Campus de Montegancedo,
28660 Boadilla del Monte, Madrid, Spain
luis.magdalena@upm.es

Abstract. Fuzzy systems have demonstrated a strong modeling capability. The quality of a fuzzy model is usually measured in terms of its accuracy and interpretability. While the way to measure accuracy is in most cases clear, measuring interpretability is still an open question.

The use of hierarchical structures in fuzzy modeling as a way to reduce complexity in systems with many input variables has also shown good results. This complexity reduction is usually considered as a way to improve interpretability, but the real effect of the hierarchy on interpretability has not really been analyzed.

The present paper analyzes that complexity reduction comparing it with that of other techniques such as feature extraction, to conclude that only the use of intermediate variables with meaning (from the point of view of model interpretation) will ensure a real interpretability improvement due to the hierarchical structure.

Keywords: Fuzzy · Hierarchical · Interpretability · Semantics
Complexity

1 Introduction

A model is the representation of a system (a part of the world). As a consequence, modeling is the process of creating a representation (model) of a certain system. The model can take quite different forms ranging from physical (a mockup) to formal models. Formal models use rules, concepts, mathematical equations, etc. to describe the system; and represent a powerful analysis tool.

As the model is a representation of the system, evaluating its quality usually encompasses different aspects that strongly relate to the purpose of the model. If the model was simply built as a demonstration tool, to show a client how the final system will look, it will only need to capture the essence, the idea of the real system. Other models are designed to know how the system will behave in the presence of a certain stimulus, as the model of an airplane wing to be tested in a wind tunnel. That kind of situation requires the model behaving as close as possible to the real system.

J. Medina et al. (Eds.): IPMU 2018, CCIS 853, pp. 16–26, 2018.
https://doi.org/10.1007/978-3-319-91473-2_2

When analyzing a computer model to evaluate its quality, it is also possible to consider different aspects. If the purpose of the model is similar to that of the airplane wing model tested on a wind tunnel, the idea will be to know how will the wing behave under certain wind conditions. Consequently, the closer the behavior of model and system was, the better the model will be. When that is the situation, it can be properly managed with a formal model that *simply* replicates the input-output relations of the system, with no particular interest on how it does. This task is well suited for many modeling tools including those known as black-box models. A completely different situation is that of a modeling process in which we are interested not only in *what* will be the output, but also in *why* will it be such. It is clear that the pure input-output relation is not enough in the latter case. The presentation of pieces of knowledge describing or explaining that input-output relation is needed, and consequently the internal structure of the model will be capital to cope with this kind of situations.

Summarizing the previous ideas, we can say that the quality of a model can be measured in terms of **how accurately reproduces** the stimulus/response relation of the modeled system, but also in terms of **how clearly it explains** or describes the underlying mechanism producing, or the knowledge justifying, those input-output relations.

Among the many tools that have been used for modeling, fuzzy systems have demonstrated great performance when applied to many real world problems. System modeling with fuzzy rule-based systems (FRBSs) is usually known as fuzzy modeling (FM) [3]. A fuzzy model, as any other model, can be evaluated in terms of those two previously described concepts: how accurately reproduces the behavior, and how clearly describes the underlying knowledge. Fuzzy models are well suited for both questions: the *accuracy*, capability to faithfully represent the real system, and the *interpretability*, capability to express the behavior of the real system in an understandable way. But when both of them are jointly considered, they mostly appear as two contradictory requirements. In fact, literature initially established subareas focusing on one or the other. While linguistic FM (mainly developed by linguistic FRBSs) was focused on the interpretability, precise FM (mainly developed by Takagi-Sugeno-Kang FRBSs) was focused on the accuracy. At present, both criteria are considered of vital importance, so that the balance between them has gained a significant attention in the field [5,6].

While accuracy can easily be measured (e.g., in terms of errors), interpretability evaluation still represents an open question where many different concepts and metrics offer a wide repertory of options. There is at least a certain level of agreement in considering the existence of two types of interpretability [8]: related to complexity and related to semantics. Semantic based metrics [2,9,17] have recently appeared to complement or complete the preexisting complexity based metrics [18]. Different overviews and comparisons of interpretability approaches can be considered [8,10], but only recently, the question of interpretability has been considered in the framework of type-2 FRBS [1,14,15] or hierarchical fuzzy systems (HFS) [19,23].

Taking into account that the primal idea for introducing hierarchical fuzzy systems was related to the reduction of structural complexity, namely, to avoid the course of dimensionality appearing in conventional FRBSs, it is clear that the interaction between interpretability and hierarchical fuzzy systems is a question to be considered. Nevertheless, only a few authors have studied it, as previously said. The present paper will focus on this question by analyzing the relation between hierarchy and interpretability, concentrating on the semantic component of interpretability.

2 Hierarchical Fuzzy Systems

When designing a FRBS to model a complex problem (particularly those with a large number of input variables) designers must cope with what is usually known as the *curse of dimensionality*, i.e., the exponential growth in the number of rules related to the number of input variables. Different options have been considered to manage this challenging questions: the use of compact rule structures, sparse rule bases, or a hierarchical fuzzy system (HFS) among others.

The way to create the hierarchical structure is not unique, and it is possible to define different kinds of hierarchy in the fuzzy system (different structures). The main difference relates to the components of the overall system being affected by the hierarchical decomposition. Three main options can be described: decompose at the level of fuzzy partitions, at the level of variables, or at the level of rules.

A *hierarchy of rules* produces a prioritization in the use of the rules in such a way that more specific rules receive a higher priority, while priority is lower for more generic rules [21,22]. In this approach a generic rule is applied only when no applicable specific rule is available, and the rules are grouped into prioritized levels to design an HFS. This structure has clear effects from the point of view of output explanation, where interpreting the output involves using the concept of *level of specificity* of the rule.

Other authors establish a *hierarchy of partitions* for each variable, with different levels of granularity. With this concept, the hierarchical structure is composed of a set of layers where each one contains linguistic partitions (concerning all the same set of variables) with different granularity levels, and linguistic rules whose linguistic variables take values in these partitions. The idea is clearly related to that of generic/specific rules, where the specificity of the rule relies on the specificity of the partition. The main difference concerns the design process that in this case is systematic, based either in reduction [11,13] or expansion [7] methods.

But the most common approach to HFSs, and the one we will consider in this paper, is that of the *hierarchy of variables*. The idea for these HFSs is to split a large system into a cascade of several smaller systems, by decomposing the input space into several input spaces with a reduced number of variables, where each input variable is only considered at a certain level of the hierarchy (Fig. 1). To involve all variables in the generation of the overall output, the output of each level is considered as one of the inputs to the following level [12].

It is clear that the main effect achieved with this hierarchical process is the reduction of the number of rules of the FRBS, i.e., the palliation of the curse

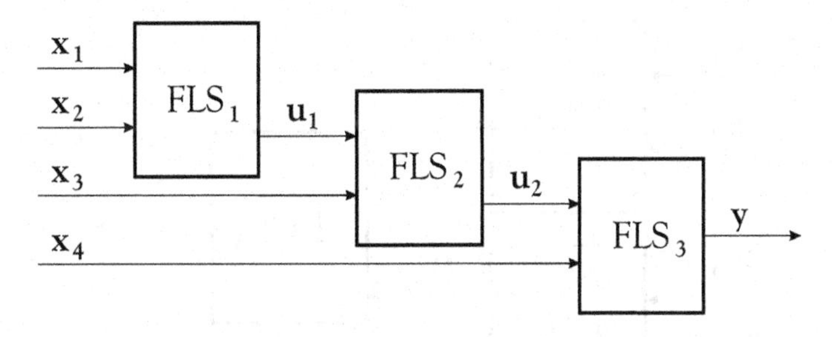

Fig. 1. Hierarchical fuzzy system (serial)

of dimensionality problem. As an example, a system with n input variables and m linguistic labels per variable will have m^n rules in a conventional FRBS. Transformed in a hierarchical fuzzy controller where the n variables are divided into L different levels, with n_k variables (including the output variable of the previous level) as inputs to the kth level of the hierarchy, the total number of rules is given by

$$T = \sum_{k=1}^{L} m^{n_k} \tag{1}$$

with

$$n_1 + \sum_{k=2}^{L} (n_k - 1) = n \tag{2}$$

And this number of rules will take on its minimum value when $n_k = 2$ (Figs. 1 and 2), being this minimum equal to

$$T = (n-1) * m^2$$

In summary, the number of rules in a complete hierarchical rule base could be reduced to a linear function of the number of variables, while in a conventional FRBS it was an exponential function of the number of variables.

In addition to the hierarchical structure shown in Fig. 1, usually known as incremental or serial HFS, where each level contains a single Fuzzy Systems, it is possible to define other hierarchical structures. The so called parallel or aggregated HFS receives all input variables in the fuzzy systems located at first level, having only output variables from the previous level as inputs of the subsequent level (Fig. 2).

Finally, cascade fuzzy systems [16,20] represent another option where all input variables are considered at every level of the hierarchy (in addition to previous levels outputs) somehow loosing the potential to cope with dimensionality problems.

Quite recently, the concept of cascade fuzzy systems have been revisited [23] to define the stacked hierarchical structure to improve its interpretability by additional complexity reduction. But this approach does not focus on what will be discussed below.

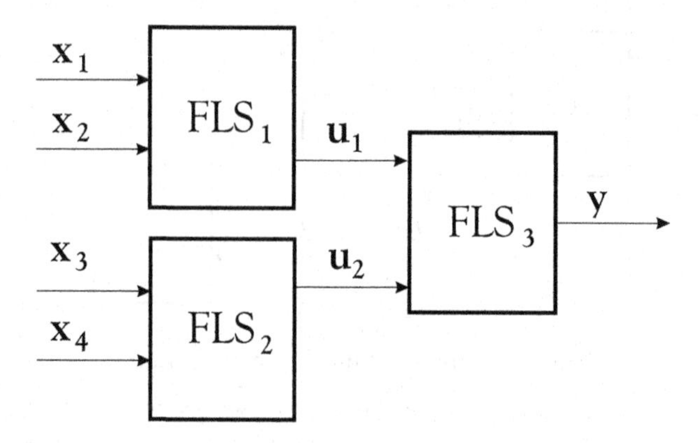

Fig. 2. Hierarchical fuzzy system (parallel)

3 Interpretability in HFSs

As interpretability has been widely linked to complexity (the higher the complexity, the lower the interpretability), the complexity reduction provided by HFSs has been usually viewed as the proof of interpretability improvement produced by these systems. However, the attempts to directly analyze HFSs in terms on interpretability (without putting it behind complexity) started only quite recently. What probably is the first approach to interpretability analysis for parallel and serial HFSs is presented in [19].

The idea of that paper is to use conventional interpretability measures to evaluate interpretability at the level of each of the FLSs composing the HFS (three in Figs. 1 and 2), and then aggregate the obtained values by means of a weighted sum that finally produces a value between 0 and 1. The aggregation works first with FLSs in the same layer to which the average is applied. Once obtained a single value per layer, different weights are applied to each layer, being higher for layers closer to the input and lower for those closer to the output (so descending when advancing through the hierarchy). The rationale behind this structure of weights is that usually the most influential variables are considered at first layer, and each new layer applies the most influential of the remaining ones, so that the output layer considers the least influential variables. In that way, the layers that apply more influential variables have a higher contribution to overall interpretability than those using less influential variables.

The main idea underlying the approach is that a hierarchical structure allows the independent analysis of the different blocks building up the hierarchy. The subsystems are considered as decoupled structures that can be independently interpreted. But the question is: Is it true? Is it really possible to interpret each subsystem as a single entity? We think that decoupling the analysis is only possible under certain circumstances, as will be considered below.

3.1 Structuring the Variables to Reduce Complexity

Hierarchical approaches are not the only way to cope with complexity in FRBSs. Many other options are possible and have been widely considered in literature. But if we focus on the idea of a hierarchy of variables, i.e., structuring the variables in different levels, it seems that the closest approaches (from a conceptual point of view) are those centered on feature extraction and selection. Complexity reduction through feature extraction and feature selection has a large presence not only in fuzzy systems but in almost any modeling technique. But there is a significant difference between feature extraction and feature selection. While feature extraction creates new *synthetic variables* encompassing the information proceeding from several variables, feature selection does not create any new variable, it simply picks up a few of the preexisting variables, those that apparently better represent the overall system.

Feature extraction generates a reduced set of new features from the original set, by means of a mapping function, trying to represent the original data more concisely. But this process is computationally expensive and, what is more interesting in this scope, produces a loss of interpretability since in most cases (probably always) no explicit and intuitive (semantic) relation exists between the original and the new features, being the original features the only ones having a *physical* explanation.

This point is somehow implicitly accepted by any designer of *interpretable fuzzy systems*, but, if we consider the many different interpretability metrics available in literature, to the author's knowledge no one will support this idea. If we generate a model with the same number of variables, rules, linguistic labels, etc, no index will consider how meaningful were the input variables, and consequently no one will distinguish between a model using selected variables and a second one using extracted (meaningless) variables.

It can be argued that the different measures and criteria are designed to compare several models designed in a similar context, or under similar boundary conditions, i.e., if variables are meaningless, that is something that can not be solved and will similarly affect any possible model. In that sense, we want to obtain the *best possible* model assuming the starting point. Consequently, feature extraction/selection is considered as a preliminary step were we can decide to avoid the meaningless variables.

Apparently, the approach better suited to perceive the differences between selected and extracted variables will be the *logical view index* based on cointension [4,17]. In this approach, cointension refers to a relation between concepts such that two concepts are cointensive if they refer to the same objects. Thus, a knowledge base will be interpretable if its semantics is cointensive with the knowledge a user builds in his/her mind after reading the knowledge representation (expressed in natural language). And we can consider that synthetic (extracted) variables will not be cointensive with any knowledge *in the mind of the reader*. In any case, there is not a clear way to measure how meaningful/meaningless are the variables. In addition, the implementation of this method focus on internal aspects of the designed model and not in the selection of input variables. Further exploration of cointension as a way to evaluate semantic quality of synthetic variables could be an option to cope with this question.

3.2 The Need for Using Meaningful Intermediate Variables

When measuring the interpretability of a linguistic variable, existing metrics only pay attention to questions as number of terms, distinguishability of fuzzy sets, coverage of the universe of discourse, etc. Properties that do not rely on the conceptual interpretation of the variable. Consequently there is no formal way to assert that a fuzzy system with selected features is more interpretable than the one with extracted variables, since no interpretability measure will make any difference. In fact, this conceptual interpretability seems to be a rather subjective question, and consequently, almost impossible to capture with the kind of *objective* metrics used to measure interpretability. But it is commonly accepted that selected variables are more interpretable than extracted features.

And what is the relation between the feature selection/extraction question, and hierarchical fuzzy systems design? The parallelism is quite simple, intermediate variables in hierarchical fuzzy systems are, at the end, equivalent to extracted features. The only difference is the kind of functional relations between original and extracted variables (mostly arithmetical) or input and intermediate variables (defined in terms of fuzzy rules). In that way, the fuzzy system with feature extraction will somehow be equivalent to a parallel two-levels hierarchical fuzzy system where the first level comprises the synthesis of the extracted variables, and the second level is made up of the fuzzy system itself (Fig. 3). So, why do we assume that the use of extracted variables reduces interpretability while the use of hierarchical systems increases it?

The most plausible answer is that we assume intermediate variables in a hierarchical fuzzy systems are *meaningful* as they are generated as the output of an *interpretable fuzzy system*. But most HFSs use intermediate variables without any conceptual/semantic support, i.e., synthetic variables.

Let us assume we have a four inputs system with three terms per partition. According to previous analysis, the minimal hierarchical structure will only need $3{\times}3^2 = 27$ rules, while a conventional system will contain $3^4 = 81$ rules. It should be much easier to interpret 27 rules with two variables per rule than 81 rules with four variables per rule. But in the first case we are assuming that each subsystem (FLS_n) can be interpreted by itself. And how can by independently interpreted a system where one (several) of the variables *has no meaning*? We can imagine a rule like

> When *First variable* is *Low* and *Second variable* is *Medium* then *Temperature* is *High*

It looks impossible to interpret this rule alone, without knowing what do *First* and *Second variable* mean. So, we need to analyze the system as a whole, including the definition of the intermediate variables (First and Second). The possibility of decoupling the analysis of a complex system into several simpler systems

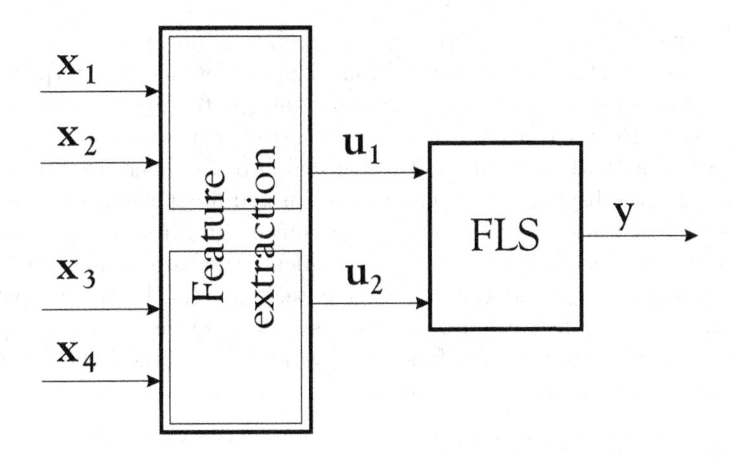

Fig. 3. Feature extraction as a hierarchical structure

seems to be lost. The different blocks are so tightly coupled that the only option is analyzing them as a whole. Let us then consider the complexity of the system as a whole.

For simplicity we will assume that the considered structure is that of Fig. 2, but the results would be identical with a serial structure. Intermediate variables will have three associated linguistic terms (the same for input variables) being $\{B_{11}, B_{12}, B_{13}\}$ the term set for u_1 and $\{B_{21}, B_{22}, B_{23}\}$ the term set for u_2. The fuzzy systems at first hierarchical level (FLS_1 and FLS_2) will contain nine rules per system, each of those rules having a consequent of the form u_i is B_{ij} with $i = 1, 2$ and $j = 1, 2, 3$. Let us define n_{ij} as the number of rules in FLS_i that refer as output to term B_{ij}. It is clear that $\sum_{j=1}^{3} n_{ij} = 9$ (the number of rules in the subsystem).

If we need to consider the hierarchical system as a single block, each rule in the second level of the hierarchy should be connected to the corresponding rules in first level, to be interpreted. And how should we connect them? We must expand each second level rule with all first level rules activating it. Consequently the rule

$$\text{IF } u_1 \text{ is } B_{11} \text{ and } u_2 \text{ is } B_{21} \text{ then } y \text{ is } C_k,$$

should be expanded with every rule from FLS_1 having B_{11} as output, and with every rule from FLS_2 having B_{21} as output. And the result is that the first rule in FLS_3 will be expanded to $n_{11} \times n_{21}$ rules considering four input variables each. If we repeat the process with the nine rules in FLS_3, the result is that the overall number of rules will be

$$\sum_{i=1}^{3} \sum_{j=1}^{3} n_i \times n_j = \sum_{i=1}^{3} n_i \times \sum_{j=1}^{3} n_j = 9 \times 9 = 3^4,$$

i.e., the same number of rules than the original (non hierarchical) fuzzy system. Then, the conclusion is that a hierarchical fuzzy system where the intermediate variables do not allow us to decouple the interpretation into subsystems, does not produce a real reduction of complexity from the point of view of interpretability, and consequently does not improve structural interpretability.

In summary, the use of a hierarchical structure only improves interpretability when provides us with the capability of decoupling the overall system into a family of simpler subsystems that can be interpreted independently. And this is only possible when intermediate variables are meaningful from the point of view of interpretation. Otherwise, using a hierarchical structure is not significantly different than the use of feature extraction techniques. The use of synthetic features as intermediate variables, regardless how those features were created (either with fuzzy rules or with a function), avoids decoupling and consequently makes impossible a proper interpretation of subsystems as independent entities.

4 Conclusions

The only way to ensure an actual improvement of interpretability in an HFS is by means of a semantic-guided design of the hierarchy, where the selected blocks of variables produce subsystems with independent meaning characterized by the appearance of intermediate variables linked to properties of the represented system, i.e., intermediate variables with meaning. Any *blind* approach synthesizing intermediate variables without any semantic relation to the modeled system, simply hides the real complexity of the system. From the interpretation point of view, an HFS using meaningless variables maintains the same complexity (number of variables, rules, terms, etc) than the non-hierarchical one.

When considering a hierarchical fuzzy system to analyze its interpretability, there is a key question:

> Is it possible to independently interpret each of the multiple fuzzy systems building up the hierarchy?

And there are only two options. If we can interpret each FLS in the hierarchy as a single entity with their inputs and outputs, and understand the role of that piece of knowledge in the overall system, the hierarchy is really improving interpretability by means of an actual complexity reduction. If the FLSs are not interpretable alone, mainly due to the fact that their input and output variables are not linked to the problem under consideration, the hierarchy does not really improve interpretability. And the way to connect those variables to the problem is by linking the intermediate variables added when building the hierarchical structure, to properties, features, characteristics, etc., of the modeled system.

In early times of fuzzy modeling, some designers considered that *any* fuzzy system was interpretable. Later on it was commonly accepted that interpretability was not an intrinsic property of fuzzy models, but something achieved through a suitable structure and design process. Now this same idea should be extended to HFSs. They are not intrinsically more interpretable than *conventional* fuzzy systems. Its interpretability relates, at least, to the appropriate selection of intermediate variables. Further analysis, as well as the definition of metrics adapted to measure interpretability of HFSs will be the matter for future works.

Acknowledgements. This paper was partially supported by Universidad Politécnica de Madrid (Spain).

References

1. Alhaddad, M., Mohammed, A., Kamel, M., Hagras, H.: A genetic interval type-2 fuzzy logic-based approach for generating interpretable linguistic models for the brain P300 phenomena recorded via brain–computer interfaces. Soft. Comput. **19**(4), 1019–1035 (2015)
2. Alonso, J.M., Magdalena, L., Guillaume, S.: HILK: a new methodology for designing highly interpretable linguistic knowledge bases using the fuzzy logic formalism. Int. J. Intell. Syst. **23**(7), 761–794 (2008)
3. Babuska, R.: Fuzzy Modeling and Control. Kluwer, Norwell (1998)
4. Cannone, R., Alonso, J.M., Magdalena, L.: Multi-objective design of highly interpretable fuzzy rule-based classifiers with semantic cointension. In: IEEE Symposium Series on Computational Intelligence (IEEE-SSCI), IV International Workshop on Genetic and Evolutionary Fuzzy Systems (GEFS), Paris, pp. 1–8 (2011)
5. Casillas, J., Cordon, O., Herrera, F., Magdalena, L.: Interpretability Issues in Fuzzy Modeling. Springer, Heidelberg (2003). https://doi.org/10.1007/978-3-540-37057-4
6. Casillas, J., Cordón, O., Herrera, F., Magdalena, L. (eds.): Accuracy Improvements in Linguistic Fuzzy Modeling. Springer, Heidelberg (2003). https://doi.org/10.1007/978-3-540-37058-1
7. Cordón, O., Herrera, F., Zwir, I.: Linguistic modeling by hierarchical systems of linguistic rules. IEEE Trans. Fuzzy Syst. **10**(1), 2–20 (2002)
8. Gacto, M.J., Alcala, R., Herrera, F.: Interpretability of linguistic fuzzy rule-based systems: an overview of interpretability measures. Inf. Sci. **181**(20), 4340–4360 (2011)
9. Galende, M., Gacto, M., Sainz, G., Alcalá, R.: Comparison and design of interpretable linguistic vs. scatter FRBSs: GM3M generalization and new rule meaning index for global assessment and local pseudo-linguistic representation. Inf. Sci. **282**, 190–213 (2014)
10. Guillaume, S.: Designing fuzzy inference systems from data: an interpretability-oriented review. IEEE Trans. Fuzzy Syst. **9**(3), 426–443 (2001)
11. Guillaume, S., Charnomordic, B.: Generating an interpretable family of fuzzy partitions from data. IEEE Trans. Fuzzy Syst. **12**(3), 324–335 (2004)
12. Raju, G.V.S., Zhou, J., Kisner, R.A.: Hierarchical fuzzy control. Int. J. Control **54**(5), 1201–1216 (1991)
13. Ishibuchi, H., Nozaki, K., Yamamoto, N., Tanaka, H.: Selecting fuzzy if-then rules for classification problems using genetic algorithms. IEEE Trans. Fuzzy Syst. **3**(3), 260–270 (1995)
14. Juang, C.F., Chen, C.Y.: Data-driven interval type-2 neural fuzzy system with high learning accuracy and improved model interpretability. IEEE Trans. Cybern. **43**(6), 1781–1795 (2013)
15. Lucas, L., Centeno, T., Delgado, M.: Towards interpretable general type-2 fuzzy classifiers. In: 9th International Conference on Intelligent Systems Design and Applications, ISDA 2009, pp. 584–589 (2009)
16. Mar, J., Lin, H.T.: A car-following collision prevention control device based on the cascaded fuzzy inference system. Fuzzy Sets Syst. **150**(3), 457–473 (2005)

17. Mencar, C., Castiello, C., Cannone, R., Fanelli, A.: Interpretability assessment of fuzzy knowledge bases: a cointension based approach. Int. J. Approx. Reason. **52**(4), 501–518 (2011)
18. Nauck, D.: Measuring interpretability in rule-based classification systems. In: Proceedings of 12th IEEE International Conference on Fuzzy Systems, vol. 1, pp. 196–201. IEEE (2003)
19. Razak, T., Garibaldi, J., Wagner, C., Pourabdollah, A., Soria, D.: Interpretability indices for hierarchical fuzzy systems. In: IEEE International Conference on Fuzzy Systems. Institute of Electrical and Electronics Engineers Inc. (2017)
20. Wang, S., Chung, F., HongBin, S., Dewen, H.: Cascaded centralized tsk fuzzy system: universal approximator and high interpretation. Appl. Soft Comput. J. **5**(2), 131–145 (2005)
21. Yager, R.R.: On a hierarchical structure for fuzzy modeling and control. IEEE Trans. Syst. Man Cybern. **23**(4), 1189–1197 (1993)
22. Yager, R.R.: On the construction of hierarchical fuzzy systems models. IEEE Trans. Syst. Man Cybern. **28**(1), 55–66 (1998)
23. Zhang, Y., Ishibuchi, H., Wang, S.: Deep takagi-sugeno-kang fuzzy classifier with shared linguistic fuzzy rules. IEEE Trans. Fuzzy Syst. (in Press)

Human Players Versus Computer Games Bots: A Turing Test Based on Linguistic Description of Complex Phenomena and Restricted Equivalence Functions

Clemente Rubio-Manzano[1,2](✉), Tomás Lermanda-Senoceaín[2], Christian Vidal-Castro[2], Alejandra Segura-Navarrete[2], and Claudia Martínez-Araneda[3]

[1] Department of Mathematics, University of Cádiz, Cádiz, Spain
[2] Department of Information Systems, University of the Bío-Bío, Concepción, Chile
clrubio@ubiobio.cl
[3] Department of Engineering Informatics, Catholic University of the Santísima Concepción, Concepción, Chile

Abstract. This paper aims to propose a new version of the well-known Turing Test for computer game bots based on Linguistic Description of Complex Phenomena and Restricted Equivalence Functions whose goal is to evaluate the "believability" of the computer games bots acting in a virtual world. A data-driven software architecture based on Linguistic Modelling of Complex Phenomena is also proposed which allows us to automatically generate bots behavior profiles. These profiles can be compared with human players behavior profiles in order to provide us with a similarity measure of believability between them. In order to show and explore the possibilities of this new turing test, a web platform has been designed and implemented by one of authors.

Keywords: Computer game bots · Turing test
Linguistic modelling of complex phenomena · Behavior analysis

1 Introduction

System evaluation allows an observer to obtain information about a system's behavior. In the case of Artificial Intelligence (AI), system evaluation is mostly performed by checking whether machines correctly do tasks for which they were designed and programmed. This kind of evaluation is due to that most AI research is better identified by Minsky's definition: *"AI is the science of making machines capable of performing task that would require intelligence if done by humans"*, however the challenge is to get a kind of AI evaluation based on the McCarthy's definition: *"AI is the science and engineering of making intelligent machines"*, hence AI evaluation should be on evaluating the "human intelligence" of the agents [7].

J. Medina et al. (Eds.): IPMU 2018, CCIS 853, pp. 27–39, 2018.
https://doi.org/10.1007/978-3-319-91473-2_3

On the other hand, it has proved that computer game bots (non-player characters (NPCs) controlled by AI algorithms in computer games) trained to play like a human are more fun [19], hence a similar challenge appears when AI computer games bots must be evaluated for impersonating human players [9]. This needed property of the intelligent agents is called "believability" and it is usually evaluated by using the well-known Turing test.

In computer games, a new version of the Turing test has been proposed in order to evaluate the ability of non-player characters [8] (*"Suppose we are playing an interactive video game with some entity. Could you tell, solely from the conduct of the game, whether the other entity was a human player or a bot? If not, then the bot is deemed to have passed the test"*). This test was designed to evaluate the abilities of the computer game bots to impersonate a human player. This test can be considered as a task-oriented evaluation based on human discrimination. However, human discrimination is usually informal and subjective due to the large amount of data generated in the execution of AI algorithms.

In this paper an ability-oriented Turing test based on Linguistic Description of Complex Phenomena (LDCP) and Restricted Equivalence Functions (REFs) is proposed. It is based on the concept of algorithm behavior profile. This profile is formed by the set of linguistic descriptions automatically generated by analyzing the data generated during the AI algorithm execution, hence "computer game bot behavior" can be defined as the behavior of a bot which has been implemented by using AI algorithms.

The structure of the paper is as follows. Section 2 provides a very brief review of the state of art on the different involved disciplines. Section 3 details the data-driven software architecture for providing designers with algorithm behavior profiles. Afterwards, Sect. 5 explains the experimentation and evaluation carried out on the projects of the student by employing an adaptation of the Turing test. Finally, Sect. 6 provides some concluding remarks.

2 Preliminary Concepts

2.1 Linguistic Description of Complex Phenomena

Computational systems based on Linguistic Description of Complex Phenomena collect and interpret data coming from complex phenomena, yielding messages in natural language which are easy to understand even by non-expert users [10]. The goal is finding out the set of expressions most commonly used in order to describe the most relevant aspects of the phenomenon under consideration. It uses a number of modelling techniques taken from the soft computing domain (fuzzy sets and relations, linguistic variables, etc.) that are able to adequately manage the inherent imprecision of the natural language in the generated texts [12]. LDD models and techniques have been used in a number of fields of application for textual reporting in domains such as: Deforestation Analysis [3], Big Data [4], Advices for saving energy at home [5], Self-Tracking Physical Activity [18], cosmology [1,17], driving simulation environments [6], air quality index textual forecasts [11], weather forecasts [13]. It is a subfield of Artificial Intelligence (AI) which allows us to produce language as output on the basis of data input.

2.2 Project-Based Learning in Artificial Intelligence

From the year 2011 a project-based learning methodology based on computer games is applied into the intelligence artificial course at the University of the Bío-Bío. This methodology aims to provide students with a better understanding of the heuristic algorithms which can be employed in real world applications. To this end, the project aims to develop a computer game in which the bot's ability should be like that human players.

In particular, the project proposed in 2017 was the development of a set of computer game bots which should be designed and implemented by using the Java programming language. A computer game bot aims to remain itself inside of a scenario based on cells during the most time possible. The student must take into account that the bots can lose energy in each movement performed (1 point of energy each five seconds). Three Bots opponent (also programmed by the students) will treat to stole its energy and a set of rewards will be distributed at the scenario which for providing bots with additional energy.

2.3 Restricted Equivalent Functions

A restricted equivalent function (REF) [2] is a similarity measure which allows to establish a similarity between the elements of a domain. A REF can be formally defined as follows:

Definition 1. *A REF, f, is a mapping $[0,1]^2 \longrightarrow [0,1]$ which satisfies the following conditions:*

1. $f(x,y) = f(y,x)$ *for all* $x, y \in [0,1]$
2. $f(x,y) = 1$ *if and only if* $x = y$
3. $f(x,y) = 0$ *if and only if* $x = 1$ *and* $y = 0$ *or* $x = 0$ *and* $y = 1$
4. $f(x,y) = f(c(x), c(y))$ *for all* $x, y \in [0,1]$, *c being a strong negation.*
5. *For all* x, y, $z \in [0,1]$, *if* $x \leq y \leq z$, *then* $f(x,y) \geq f(x,z)$ *and* $f(y,z) \geq f(x,z)$

For example, $g(x,y) = 1 - |x - y|$ satisfies conditions (1)-(5) with $c(x) = 1 - x$ for all $x \in [0,1]$. A similarity measure based on REFs between linguistic terms has been recently proposed in order to enhance the inference engine of Bousi Prolog [16].

3 A Data-Driven Software Architecture Based on Linguistic Modelling of Complex Phenomena

The software architecture proposed is formed for four modules: Tracing, Computational Perception Network, Behavior profile report generator and Evaluation module.

3.1 Tracing Module

In our case, traces of execution have been employed as computational procedure for capturing and storing data. Tracing recording, or tracing is a commonly used technique useful in debugging and performance analysis. Concretely, trace recording implies detection and storage of relevant events during run-time, for later off-line analysis. We use a trace recording which stores the following metrics:

- **Protection:** Number of obstacles between the agent and the $opponent_i$, a rectangular area is created from the position of the agent and the $opponent_i$, respectively
- **Distance:** Distance between two entities E_1 and E_2, $d(E_1, E_2) = \sqrt{x - x')^2 + (x - x')^2}$, being (x, y) the position of E_1 and (x', y') the position of E_2
- **Energy:** Energy of the player in an instant of time during the play session
- **Time:** Time registered from the start of the play session to the end of it
- **Reward:** True or false if a reward was captured at this instant of time
- **Iterations:** Number of iterations performed for the execution of the heuristic algorithm (It is executed in each move)
- **Memory:** Amount of memory required for the execution of the heuristic algorithm (It is executed in each move)

An example of trace can be seen in Fig. 1.

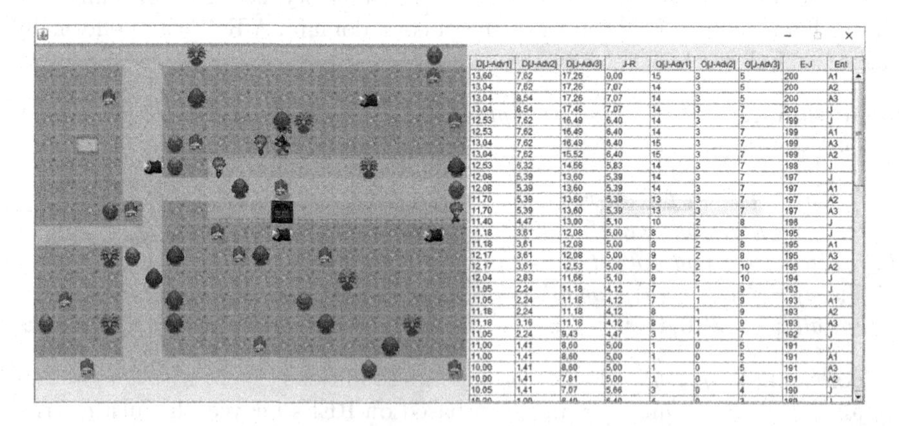

Fig. 1. Trace of execution created from the values captured during the execution of the algorithms

3.2 Computational Perception Network

We use here the concept of Declarative Computational Perception (DCP) network which is inspired by the definition of CP network proposed [20] and it allows to model the problem in a declarative way. A declarative CP can be recursively defined as follows:

- **Base case.** A CP = (A, $(u_1, \ldots, u_k)$), being A = $(a_1, \ldots, a_n)$ a vector of linguistic expressions that represents the whole linguistic domain of CP whose values are calculated by aggregating each u_i to either one or several elements of A.
- **Inductive case.** A CP = (A, $(CP_1, \ldots, CP_k)$), being A = $(a_1, \ldots, a_n)$ a vector of linguistic expressions whose values are calculated by aggregating each CP_i to either one or several elements of A.

Note that, the base case is produced when a CP is defined in terms of a real numbers set which belongs to a particular domain. The recursive case is produced when a CP is defined in terms of linguistic terms from a set of CPs. We say that a set of sub-CPs $\{CP_1, \ldots, CP_k\}$ completely define a CP or that a CP can be defined in terms of a sets of sub-CPs $\{CP_1, \ldots, CP_k\}$.

The computational perception network presented in [15] is enhanced for the problem previously presented. In this case, additional variables must be considered and hence the computational perceptions network must be enhanced, rules and templates must be also updated for these new requirements. The computational network can be declaratively defined as follows.

3.3 CP Situation

Currently, a computer game bot can stay in a safe, easy, dangerous or risky situation, it depends on three factors, the bot's protection with respect to the opponent, the distance to the opponent and the bot's energy in this moment. This is formalized as follows:

$$CP_{Situation} = ((\text{Safe, Easy, Dangerous, Risky}), (CP^{player,opponent}_{Protection}, CP^{player,opponent}_{Distance}, CP^{player}_{Energy}))$$

where:

- $CP^{player,opponent}_{Protection}$ = ((low, intermediate, high), (0, 1, ..., 380))), with low(0, 0, 0, 2), intermediate(1, 3, 3, 5) and high(4, 6, 380, 380);
- $CP^{player,opponent}_{Distance}$ = ((close, normal, far), (0, 1, ..., 38)), with close(0, 0.4, 7), normal(6, 9, 11, 14) and far(13, 16, 38, 38).
- CP^{player}_{Enery} = ((low, normal, high), (0, 1, ..., size(scenario)))

The corresponding values for this CP are computed by using the following rules[1]: {Risky ← Intermediate, Close, Normal; Dangerous ← Low, Close, Normal; Safe ← Intermediate, Normal, Normal; Easy ← Low, Normal, Normal; Dangerous ← Low, Normal, Low; Dangerous ← Normal, Close, Low; Dangerous ← Normal, Normal, Low;}

[1] Each rule has the form $A \leftarrow B_1, \ldots, B_n$ where A is the consequent and $B_1, \ldots, B_n$ the antecedent.

3.4 CP Attitude

Four attitudes can be detected for a computer game bot: wise, brave, cautious and passive. This depend on two factors, the distance between the bot and the closest reward and the distance between the opponent and the closest reward. This is formalized as follows:

$$CP_{Attitude} = ((\text{Wise, Brave, Cautious, Passive}), (CP_{Distance}, CP_{Distance}, CP_{Energy}))$$

where:

- $CP^{Oponnet,R*}_{Distance}$ = ((close, normal, far), (0, 1, ..., size(scenario))) with close(0, 0.4, 7), normal(6, 9, 11, 14) and far(13, 16, 38, 38), being R* the closest reward to the
- $CP^{player,R*}_{Distance}$ = ((close, normal, far), (0, 1, ..., size(scenario))) with close(0, 0.4, 7), normal(6, 9, 11, 14) and far(13, 16, 38, 38), being R* the closest reward to the agent.
- CP^{player}_{Enery} = ((low, normal, high), (0, 1, ..., size(scenario))) with low(0, 0, 3, 6), normal(4, 7, 9, 12) and high(10, 13, 100, 100).

The corresponding values for this CP are computed by using the following rules: {Wise ← Close, Normal; Brave ← Close, Close; Cautious ← Normal, Close; Passive ← Normal, Normal;}

3.5 CP Movement

A computer game bot can perform four types of movements: good, bad, scare, kamikaze. This depend on three factors, the distance between player and the closest reward, the distance between the bot and the opponent and the energy of the bot. This is formalized as follows:

$$CP_{Movement} = ((\text{Good, Bad, Scare, Kamikaze}), (CP^{player,R*}_{Distance}, CP^{player,opponent}_{Distance}, CP^{player}_{Enery}))$$

where:

- $CP^{player,R*}_{Distance}$ = ((close, normal, far), (0, 1, ..., size(scenario))) with close(0, 0.4, 7), normal(6, 9, 11, 14) and far(13, 16, 38, 38), being R* the closest reward to the agent;
- $CP^{player,opponent}_{Distance}$ = ((close, normal, far), (0, 1, ..., size(scenario))) with close(0, 0.4, 7), normal(6, 9, 11, 14) and far(13, 16, 38, 38);
- CP^{player}_{Enery} = ((low, normal, high), (0, 1, ..., size(scenario))) with low(0, 0, 3, 6), normal(4, 7, 9, 12) and high(10, 13, 100, 100).

The corresponding values for this CP are computed by using the following rules: {Good ← Close, Normal, Normal; Good ← Close, Close, Low; Scare ← Normal, Normal, Normal; Kamikaze ← Close, Close, Normal; Bad ← Normal, Close, Normal;}

3.6 CP Resources

The resources are the required time and space that a heuristic algorithm needs for its execution. This is formalized as follows:

$$CP_{Resources} = ((\text{Very\_Efficient}, \text{Efficient}, \text{Inefficient}, \text{Very\_Inefficient}), (CP_{Iterations}, CP_{Memory}))$$

where:

- $CP_{Iterations}$ = ((Little, Normal, Large), (0, 1, ..., max(iterations))) with little(0, 0, 18, 30), normal(18, 30, 42, 54) and large(42, 54, 104857600, 104857600);
- CP_{Memory} = ((Low, Normal, High), (0,1,..., max(memory))) with low(0, 0, 768, 1280), normal(768, 1280, 1792, 2304), high(1792, 2304, 104857600, 104857600);

The corresponding values for this CP are computed by using the following rules: {Very_Efficient ← Little, Low; Efficient ← Normal, Normal; Inefficient ← Normal, High; Very_Inefficient ← Large, High;}

3.7 CP Ability

The ability of a computer game bot depends on its attitude, kind of movement performed and the time.

$$CP_{Ability} = ((\text{Expert}, \text{Intermediate}, \text{Basic}, \text{Dummy}), (CP_{Attitude}, CP_{Movement}, CP_{Time}))$$

with CP_{Time} = ((little, normal, large), (0,1,, max_time)) with little(0, 0, 90000, 150000), normal(90000, 150000, 210000, 270000), and large(210000, 270000, 104857600, 104857600). The corresponding values for this CP are computed by using the following rules: {Expert ← Wise, Good, Small; Intermediate ← Brave, Good, Normal; Basic ← Passive, Bad, Much; Dummy ← Passive, Scare, Much;}

3.8 CP Skill

The skill of a computer game bot depends on its attitude, kind of movement performed and situations detected.

$$CP_{Skill} = ((\text{Very\_Skilled}, \text{Skilled}, \text{Improvable}, \text{Very\_Improvable}), (CP_{Attitude}, CP_{Movement}, CP_{Situation}))$$

The corresponding values for this CP are computed by using the following rules: {Very_Skilled ← Wise, Good, Easy; Skilled ← Cautious, Good, Safe; Improvable ← Brave, Bad, Dangerous; Improvable ← Passive, Bad, Risky;}

4 Behavior Profile Generation

In order to provide information about the whole play session, we calculate the summaries of each CP. The process consists in adding the values of a particular CP during the play session by using the concept of fuzzy cardinality [14]. We call these CPs, Behavior CPs (ΣCP). We use these ΣCP to generate the behavior profile.

A behavior profile is a set of $\Sigma CP_1, \ldots, \Sigma CP_n$ which are associated to the entities acting in a virtual world. For example, a bot behavior profile is a set of ΣCP_i which are determining the behavior of the bot: $\Sigma CP_{Attitude}$, $\Sigma CP_{Situation}$, $\Sigma CP_{Movements}$, ΣCP_{Skill}, $\Sigma CP_{Ability}$ and $\Sigma CP_{Resources}$.

The generation of the report is performed by using the set of ΣCP. For each CP a linguistic description is created in function of the pair $(a_i, w_i) \in \Sigma CP$. Percentages are calculated for each ΣCP. The percentages p_i is then transformed in a linguistic term of quantity as follows: few is when $p_i \in [0, 1/3]$; several when $p_i \in [1/3, 2/3]$ or many when $p_i \in [2/3, 1]$. Then, we are going to consider four cases:

1. There exists an only one pair $(a_i, p_i) \in \Sigma CP$ whose p_i is greater than 66%.
2. There exists an only one pair $(a_i, p_i) \in \Sigma CP$ whose p_i is greater than 33%.
3. There are two pairs $(a_1, p_1), (a_2, p_2) \in \Sigma CP$ whose p_i is greater than 33%.
4. There not exists any pair $(a_i, p_i) \in \Sigma CP$ whose is greater 33%.

Example 1. A complete execution approximately generates a trace of 10000 lines. For sake of simplicity, we suppose here that we only have one data row of the execution trace described in the Fig. 1.

1, 13, 4, 12, 2, 12, 3.60, 3.16, 1.41, 17, 5.0, 2.0, 1.0, 15.26, 17.08, 13.0, 13.89, 15995, false.42, 924, J

First, the input data generated by the entities are grouped in variables: (1, 13) is the player position (P); (3, 16) is the position of the opponent 1 (O1); (4, 12) is the position of the opponent 2 (O2); (2, 12) is the position of the opponent 3 (O3). Next, the values for each metric are computed from them (distance is calculated from positions for example) and the membership degree for each value is computed by using linguistic variables whose meaning is given by a set of fuzzy sets. For example, if the distance between the player and the closest reward is 15.26 then the corresponding membership degree by using the fuzzy set "high distance" is 0.75. This value is computed by using a trapezoidal function (13, 16, 38, 38), that is, $\mu_{high}(15.26) = 0.75$. This is performed for the rest of the values.

- Distance(P, O1) = 3.69 $\Rightarrow$ Close (3.69, 0, 0, 4, 7) = 1
- Distance(P, O2) = 3.16 $\Rightarrow$ Close (3.16, 0, 0, 4, 7) = 1
- Distance(P, O3) 1.41 $\Rightarrow$ Close (1.41, 0, 0, 4, 7) = 1
- Energy(P) = 17 $\Rightarrow$ High (17, 10, 13, 100, 100) = 1
- Protection(P, o1) = 5.0 $\Rightarrow$ High (5.0, 4, 6, 380, 380) = 0.5
- Protection(P, o2) = 2.0 $\Rightarrow$ Normal (2.0, 1, 3, 3, 5) = 0.5
- Protection(P, o3) = 1.0 $\Rightarrow$ Low (1.0, 0, 0, 0, 2) = 0.5

- Distance(P, closest(R)) = 15.26 ⇒ High (15.26, 13, 16, 38, 38) = 0.75
- Distance(o1, closest(R)) = 17.08 ⇒ High (17.08, 13, 16, 38, 38) = 1
- Distance(o2, closest(R)) = 13.0 ⇒ Normal (13.0, 6, 9, 11, 14) = 0.33
- Distance(o3, closest(R)) = 13.89 ⇒ High (13.89, 13, 16, 38, 38) = 0.29
- Time = 15995 ⇒ Small (15995, 0, 0, 90000, 150000) = 1
- Iterations at this movement = 42 ⇒ Normal (42, 18, 30, 42, 54) = 1
- Memory occupied (Bytes) = 924 ⇒ Low (924, 0, 0, 768, 1280) = 0.69

Parallel, a set of if-then rules are generated by aggregating these linguistic terms. Average is employed as an aggregation operator. For example, if the distance between the opponent 3 and the closest reward is high with 0.29, the distance between the player and the same reward is 0.33 and the energy of the player is high with 1.0, then the attitude of the player is Cautious with 0.54 $((0.29 + 0.33 + 1.0)/3)$.

$$\frac{\text{Distance(O3, R*)} = \text{(High, 0.29)} \qquad \text{Distance(P, R*)} = \text{(Normal, 0.33)} \qquad \text{Energy} = \text{(High, 1)}}{\text{Attitude} = \text{(Cautious, 0.54)}}$$

$$\frac{\text{Protection} = \text{(Low, 0.5)} \qquad \text{Distance(P, O)} = \text{(Close, 1)} \qquad \text{Energy} = \text{(High, 1)}}{\text{Situation} = \text{(Dangerous, 0.83)}}$$

$$\frac{\text{Distance(J, R*)} = \text{(Normal, 0.33)} \qquad \text{Distance(P, O)} = \text{(Close, 1)} \qquad \text{Energy} = \text{(High, 1)}}{\text{Movement} = \text{(Bad, 0.91)}}$$

$$\frac{\text{Attitude} = \text{(Cautious, 0.54)} \qquad \text{Situation} = \text{(Dangerous, 0.83)} \qquad \text{Movement} = \text{(Bad, 0.91)}}{\text{Ability} = \text{(Dummy, 0.82)}}$$

$$\frac{\text{Attitude} = \text{(Cautious, 0.54)} \qquad \text{Movement} = \text{(Bad, 0.91)} \qquad \text{Time} = \text{(Small, 1)}}{\text{Skill} = \text{(Improvable, 0.76)}}$$

$$\frac{\text{Memory} = \text{(Low, 0.69)} \qquad \text{Iteration} = \text{(Normal, 1)}}{\text{Resources} = \text{(Efficient, 0.76)}}$$

Then, a summarization of the results will be performed for finally automatically generating a linguistic report from all this process. An example of template and its corresponding instantiation is shown in the Fig. 1.

Template	Bot behavior profile
The bot showed $d_{Attitude}$ $a_{Attitude}$ attitudes. Definitely, $d_{Situation}$ $a_{Situation}$ were safe. The bot proved capable of performing *degree value* movements. The bot displayed an *value* skill level *degree* times. The bot proved to be *value degree* times. During most of the execution, the measured use of resources demonstrates an operation that is *degree* times *value*.	The bot showed **several brave** attitudes. Definitely, **many** situations were **safe**. The bot proved to be capable of performing **several good** movements. The bot displayed an **expert** skill level **several** times. The agent proved to be **skillful several** times. During most of the execution, the measured use of resources demonstrates an operation that is **many** times very **efficient**

5 Turing Test for Computer Games Bots Based on LLD and REF

A web platform has been implemented for performing turing test based on LLD and REF. A quick test of the application can be performed by downloading examples of traces at the following URL:

http://youractionsdefineyou.com/assess/web/examples_traces

First, the user must access to the URL:

http://www.youractionsdefineyou.com/assess

The main window shows two options: log in and register. The register of a user consists in introducing email, user name, full name, RUT and password. A confirmation via email will be sent to the user if the registration was correct. The log in of a user consists in introducing the user name and the password. Second, behavior profile report can be obtained by selecting and loading an execution trace file, then the behavior profile report is automatically generated.

In this kind of Turing test, a behavior profile is automatically generated for a human expert player ([15]), and for the evaluated computer game bot. The human player profile was created after that an expert human player played to the computer game: (i) Attitude is mainly brave during the most part of the time; (ii) Situation is mainly safe during the most part of the time; (iii) Movements were mainly good during the most part of the time; (iv) The player is expert; (v) The player is skilled; (vi) The use of computational resource is efficient in time and space.

Now, the bot's behavior profile is compared with the human player behavior profile (see Fig. 2) by using a similarity measure based on restricted equivalence functions. An AI algorithm is near-behavioral when its associated behavior profile report is similar to the behavior profile report generated by the human expert player. The final grade (from 1 to 7) is computed by using the similarity between the human behavior profile and the bot one. The equation for calculating the final grade is as follows:

$$FG = G_{Min} + S_{Attitude} + S_{Situation} + S_{Movement} + S_{Ability} + S_{Skill} + S_{Efficiency} \quad (1)$$

1. G_{Min}: 1 point (minimum score).
2. $S_{Attitude} = S_{REF}(\Sigma CP^{Human}_{Attitude}, \Sigma CP^{Bot}_{Attitude})$ is the similarity between human player and bot attitude.
3. $S_{Situation} = S_{REF}(\Sigma CP^{Human}_{Situation}, \Sigma CP^{Bot}_{Situation})$: is the similarity between human player and bot situation.
4. $S_{Movement} = S_{REF}(\Sigma CP^{Human}_{Movement}, \Sigma CP^{Bot}_{Movement})$: is the similarity between human player and bot movements.
5. $S_{Ability} = S_{REF}(\Sigma CP^{Human}_{Ability}, \Sigma CP^{Bot}_{Ability})$: is the similarity between human player and bot ability.

HUMAN BEHAVIOR PROFILE

The human player showed several brave attitudes. Definitely, many situations were safe. The human player proved to be capable of performing several good movements. The human player displayed an expert skill level several times. The human player proved to be skillful several times. During the most of the execution, the measured use of resources demonstrates an operation that is many times very efficient.

Grade: 7

BOT BEHAVIOR PROFILE

During the most of the game, the bot showed many brave attitudes. Definitely, many situations were dangerous. The bot proved to be capable of performing several kamikaze movements. The bot displayed an beginner skill level several times. With great certainty, the bot proved to be improvable many times. During the most of the execution, the measured use of resources demonstrates an operation that is many times efficient.

Grade: 5.15

Fig. 2. Similarity between behavior profiles reports: human player versus computer game bots

6. $S_{Skill} = S_{REF}(\Sigma CP^{Human}_{Skill}, \Sigma CP^{Bot}_{Skill})$: is the similarity between human player and bot skill.
7. $S_{Efficiency} = S_{REF}(\Sigma CP^{Human}_{Efficiency}, \Sigma CP^{Bot}_{Efficiency})$: is the similarity between human player and bot efficiency.

where S_{REF} is a similarity measure between computational perceptions. The following definition formalizes this measure.

Definition 2. *Given two ΣCP_i, ΣCP_j whose percentage linguistic vectors $\{(a_1, p_1) \dots, (a_n, p_n)\}$ and $\{(b_1, q_1) \dots, (b_n, q_n)\}$ respectively. A similarity measure between ΣCP_i and ΣCP_j is defined as:*

$$S_{REF}(\Sigma CP_i, \Sigma CP_i) = \sum\nolimits_{i=0}^{n} (REF(p_i, q_i))/n$$

being $REF(p_i, q_i) = 1 - |p_i - q_i|$

Example 2. Let $CP^{Human}_{Attitude}, CP^{Bot}_{Attitude}$ be two summation computational perceptions for the human player and the computer game bot, respectively:

- $\Sigma CP^{Human}_{Attitude}$ = {(wise, 122.35), (brave, 289), (cautious, 87.59), (passive, 8.75)}
- $\Sigma CP^{Bot}_{Attitude}$ = {(wise, 17.53), (brave, 101.55), (cautious, 14.05), (passive, 10.78)}

Then, the percentages linguistic vectors are calculated for each ΣCP by using their totals $Total_{\Sigma CP^{Human}_{Attitude}}$(507.69) and $Total_{\Sigma CP^{Bot}_{Attitude}}$(143.61), respectively:

- $\Sigma CP^{Human}_{Attitude}$ = {(wise, 0.240), (brave, 0.569), (cautious, 0.172), (passive, 0.017)}
- $\Sigma CP^{Bot}_{Attitude}$ = {(wise, 0.122), (brave, 0.709), (cautious, 0.097), (passive, 0.075)}

Now, the similarity $S_{REF}(\Sigma CP^{Human}_{Attitude}, \Sigma CP^{Bot}_{Attitude})$ can be calculated:

- $REF(0.240, 0.122) = 1 - |0.240 - 0.122| = 0.882$
- $REF(0.569, 0.172) = 1 - |0.569 - 0.172| = 0.882$
- $REF(0.172, 0.097) = 1 - |0.172 - 0.097| = 0.925$
- $REF(0.017, 0.075) = 1 - |0.017 - 0.075| = 0.942$

Hence, $S_{REF}(\Sigma CP^{Human}_{Attitude}, \Sigma CP^{Bot}_{Attitude}) = \frac{3.402}{4} = 0.838$. The rest of the similarities is computed in a similar way. The final grade together with the linguistic reports generated for the human player and the bot designed by an anonymous student are shown in the Fig. 2.

6 Conclusions and Future Work

In this paper a Turing test based on the similarity between computer bot and human player behavior profiles has been defined and implemented. The concept of the algorithm behavior profile has been proposed as a computational technique for evaluating computer game bots ability.

This work is a first approximation towards a broader approach and much work remains to be done in this direction. As future work we would like to incorporate our technology into more complex computer games in which several players can participate in the game.

Acknowledgments. Clemente Rubio-Manzano is partially supported by the State Research Agency (AEI) and the European Regional Development Fund (FEDER) project TIN2016-76653-P. This work has been done in collaboration with the research group SOMOS (SOftware-MOdelling-Science) funded by the Research Agency and the Graduate School of Management of the Bío-Bío University.

References

1. Arguelles, L., Trivino, G.: I-struve: automatic linguistic descriptions of visual double stars. Eng. Appl. Artif. Intell. **26**(9), 2083–2092 (2013)
2. Bustice, H., Barrenechea, E., Pagola, M.: Restricted equivalence functions. Fuzzy Sets Syst. **157**, 2333–2346 (2006)
3. Conde-Clemente, P., Alonso, J.M., Nunes, É.O., Sanchez, A., Trivino, G.: New types of computational perceptions: linguistic descriptions in deforestation analysis. Expert Syst. Appl. **85**, 46–60 (2017)
4. Conde-Clemente, P., Trivino, G., Alonso, J.M.: Generating automatic linguistic descriptions with big data. Inf. Sci. **380**, 12–30 (2017)
5. Conde-Clemente, P., Alonso, J.M., Trivino, G.: Toward automatic generation of linguistic advice for saving energy at home. Soft Comput. **22**(2), 345–359 (2018)
6. Eciolaza, L., Pereira-Fariña, M., Trivino, G.: Automatic linguistic reporting in driving simulation environments. Appl. Soft Comput. **13**(9), 3956–3967 (2013)
7. Hernández-Orallo, J.: Evaluation in artificial intelligence: from task-oriented to ability-oriented measurement. Artif. Intell. Rev. **48**(3), 397–447 (2017)
8. Hingston, P.: A turing test for computer game bots. IEEE Trans. Comput. Intell. AI in Games **1**(3), 169–186 (2009)

9. Livingstone, D.: Turing's test and believable AI in games. Comput. Entertainment (CIE) **4**(1), 6 (2006)
10. Phebes 2017 Phebes Lab: Linguistic Description of Complex Phenomena (2017). http://phedes.com/
11. Ramos-Soto, A., Bugarín, A., Barro, S., Gallego, N., Rodríguez, C., Fraga, I., Saunders, A.: Automatic generation of air quality index textual forecasts using a data-to-text approach. In: Puerta, J.M., Gámez, J.A., Dorronsoro, B., Barrenechea, E., Troncoso, A., Baruque, B., Galar, M. (eds.) CAEPIA 2015. LNCS (LNAI), vol. 9422, pp. 164–174. Springer, Cham (2015). https://doi.org/10.1007/978-3-319-24598-0_15
12. Ramos-Soto, A., Pereira-Fariña, M., Bugarín, A., Barro, S.: A model based on computational perceptions for the generation of linguistic descriptions of data. In: 2015 IEEE international conference on Fuzzy systems (FUZZ-IEEE), pp. 1–8. IEEE, August 2015
13. Ramos-Soto, A., Bugarin, A., Barro, S., Taboada, J.: Automatic generation of textual short-term weather forecasts on real prediction data. In: Larsen, H.L., Martin-Bautista, M.J., Vila, M.A., Andreasen, T., Christiansen, H. (eds.) Flexible Query Answering Systems, FQAS 2013. LNCS, vol. 8132, pp. 269–280. Springer, Heidelberg (2013). https://doi.org/10.1007/978-3-642-40769-7_24
14. Rubio-Manzano, C., Trivino, G.: Automatic linguistic feedback in computer games. In: IFSA-EUSFLAT (2015)
15. Rubio-Manzano, C., Trivino, G.: Improving player experience in computer games by using players' behavior analysis and linguistic descriptions. Int. J. Hum.-Comput. Stud. **95**, 27–38 (2016)
16. Rubio-Manzano, C.: Similarity measure between linguistic terms by using restricted equivalence functions and its application to expert systems. In: 9th European Symposium on Computational Intelligence and Mathematics, 4th–7th October. Faro (Portugal) (2017)
17. Sanchez-Valdes, D., Alvarez-Alvarez, A., Trivino, G.: Linguistic description about circular structures of the Mars' surface. Appl. Soft Comput. **13**(12), 4738–4749 (2013)
18. Sanchez-Valdes, D., Trivino, G.: Linguistic and emotional feedback for self-tracking physical activity. Expert Syst. Appl. **42**(24), 9574–9586 (2015)
19. Soni, P., Hingston, P.: Bots trained to play like a human are more fun. In: IEEE International Joint Conference on Neural Networks (IEEE World Congress on Computational Intelligence), IJCNN 2008, pp. 363–369. IEEE, June 2008
20. Trivino, G., Sugeno, M.: Towards linguistic descriptions of phenomena. Int. J. Approximate Reasoning **54**(1), 22–34 (2013)

Reinterpreting Interpretability for Fuzzy Linguistic Descriptions of Data

A. Ramos-Soto[1,2(✉)] and M. Pereira-Fariña[3,4]

[1] Centro Singular en Investigación en Tecnoloxías da Información (CiTIUS), Universidade de Santiago de Compostela, Rúa de Jenaro de la Fuente Domínguez, Santiago de Compostela, Spain
alejandro.ramos@usc.es

[2] Department of Computing Science, University of Aberdeen, Aberdeen, UK
alejandro.soto@abdn.ac.uk

[3] Departamento de Filosofía e Antropoloxía, Universidade de Santiago de Compostela, Santiago, Spain
martin.pereira@usc.es

[4] Centre for Argument Technology (ARG-tech), University of Dundee, Dundee, UK
m.z.pereirafarina@dundee.ac.uk

Abstract. We approach the problem of interpretability for fuzzy linguistic descriptions of data from a natural language generation perspective. For this, first we review the current state of linguistic descriptions of data and their use contexts as a standalone tool and as part of a natural language generation system. Then, we discuss the standard approach to interpretability for linguistic descriptions and introduce our complementary proposal, which describes the elements from linguistic descriptions of data that can influence and improve the interpretability of automatically generated texts (such as fuzzy properties, quantifiers, and truth degrees), when linguistic descriptions are used to determine relevant content within a text generation system.

Keywords: Fuzzy sets · Linguistic summarization
Fuzzy linguistic descriptions of data · Interpretability
Natural language generation · Data-to-text

1 Introduction

Among the different tools that fuzzy sets theory encompasses, the generation of linguistic summaries or descriptions on data (LDD) provides a way to synthesize numeric datasets into compact sentences (protoforms), such as *"Most days of the year the energy consumption is high"* from the corresponding raw data. These descriptive techniques, whose origins can be found in the works of Yager and Zadeh more than three decades ago [17,18], have been extensively researched from a formal perspective, but also many different use cases have been proposed in more recent years [2].

J. Medina et al. (Eds.): IPMU 2018, CCIS 853, pp. 40–51, 2018.
https://doi.org/10.1007/978-3-319-91473-2_4

Among the main advantages of LDD, we can highlight the management of imprecision in human language through the modeling of linguistic terms as fuzzy sets, and their compositionality (an inherent feature of fuzzy sets in general), which allows to aggregate and summarize data from different numeric variables into a single description. However, today there exists a general consensus about the limited expressiveness of these linguistic descriptions [7,10,14]. In short, the linguistic realization of the protoforms is poor for being presented to human users in a real context. Therefore, in order to improve the applicability of LDD in the real world, an important effort oriented to enhance its linguistic quality must be made.

In order to address this challenge, some authors have proposed to incorporate a natural language generation (NLG) layer that converts the instanced protoforms into texts [7]. Others take this interpretation one step further, and understand fuzzy linguistic descriptions of data not in isolation, but rather as a tool that can be used as part of the content determination stage of an NLG system [10,14]. This means that such linguistic descriptions need to be integrated with other tasks within the system, such as lexicalization or referring expression generation.

Be it as standalone descriptions or integrated as one of the many parts of an NLG system, the interpretability of LDD is one of the less explored characteristics of this tool beyond the initial discussion about their adequacy for human users in their original form. In fact, interpretability is one of the features that differentiates fuzzy systems from other types of computational intelligence approaches, and research on interpretable fuzzy systems (IFS) has been extensive in this regard, especially for fuzzy rule-based systems.

In this context, the aim of this paper is to approach the interpretability of LDD from an end-user's perspective, with a special focus on the crucial task that involves converting the original protoform structures into texts adapted to end-users. For this, in Sect. 2 we will describe the essential concepts about linguistic descriptions of data and their use in different contexts, and Sects. 3 and 4 will respectively discuss the problem of their interpretability from a classic point of view and from our own perspective.

2 Linguistic Descriptions of Data in a Real Application Context

Linguistic descriptions of data are collections of short linguistic propositions used to quantify certain properties on numeric datasets that follow the structure of a protoform, such as *"Q Xs are A"* in its simplest form (also known as type-1 fuzzy quantified sentence), where Q is a (fuzzy) quantifier, X is the reference set to be described, and A is a linguistic label that represents a fuzzy property used to describe the reference set. For instance, in "A few men are short", $Q =$ 'A few', $X =$ 'men', and $A =$ 'short'. Likewise, linguistic variables represent the partitioning of a numeric domain into several fuzzy properties, e.g., "height"= {*short, medium, tall*}.

Starting with the simplest type of fuzzy quantified sentence, more complex linguistic descriptions can be composed, for instance, by relating two different properties using type-2 quantified sentences *"Q DXs are A"* (e.g. "Some cold days are very humid") or quantifying over time (e.g., "most patients have a constant heart-rate most of the time"). In fact, time series data have been a recurring target of many use cases in LDD.

In addition to fuzzy quantifiers and linguistic labels, another important element to consider in linguistic descriptions is the truth degree associated to each sentence which is computed. This degree is calculated through the use of a fuzzy quantification model, such as Zadeh's [18] or the F^A [4] model, among many others [3]. In short, these models take into account the aggregation of truth degrees resulting from evaluating the fuzzy properties against the dataset, and then apply a fuzzy quantifier to provide a single truth degree that characterizes the whole sentence. For instance, T(*"Most days of the year the energy consumption is high"*) = 1 means that 'most' values of the reference set fulfill the property 'high' in the highest degree.

In their original conception, protoforms were proposed as a more human-friendly approach to perform queries on databases. However, their later application in the generation of LDD has raised some questions about their adequacy as a descriptive tool *per se*. As Fig. 1 (left side) shows, typical use cases of LDD involve the researcher alone. This means that the knowledge base (i.e., the definition of quantifiers and linguistic terms based on fuzzy sets, among others), the types of protoforms that are used, and even the datasets that are described are all selected by the researcher, whose purpose is to illustrate how the LDD technique works and provide some insights about its potential. Thus, in such cases, there does not exist a communicative act, and other actors such as domain experts or end-users are not considered at all.

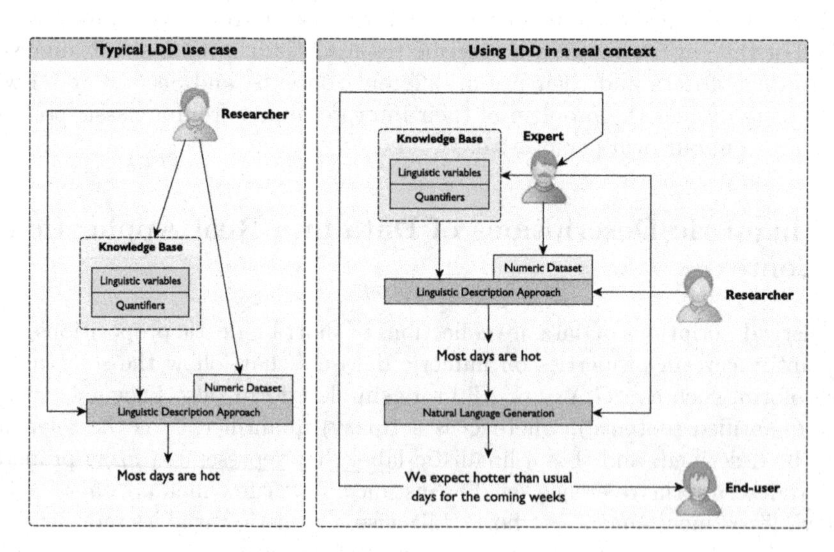

Fig. 1. Contexts of use for linguistic descriptions of data: use cases and use in a real setting.

When other actors such as end-users were finally considered, the direct utility of LDD was put into question. Despite their use of linguistic terms, it is a major consensus nowadays that the language that protoforms convey is too restricted and generic to be provided directly to other users [7,10,14]. In contraposition to the use case scenario, Fig. 1 (right side) shows the schema of a setting where LDD would be applied in a real problem, and there exists an actual need for providing textual descriptions of data.

The first consideration in such case is ensuring that the language we are using to convey the information that the linguistic descriptions hold is adapted to the domain requirements and fit for end-users [6,10,14]. This can be achieved through a natural language generation layer that transforms the linguistic description into an actual text. The second forethought consists in the more than likely presence of an expert and/or specific domain guidelines. These will determine our knowledge, and will also provide the kind of input data to be used and set the language requirements of the generated texts [15]. A third consideration is that the role of the researcher or developer is to determine whether using LDD is suitable for addressing the problem (or some of its parts) and, if so, to implement this technique and devise how to convert the protoform-like linguistic descriptions into texts that match the target language [13]. This scenario actually describes the setting of the GALiWeather system [15], which generates textual short-term weather forecasts.

Of course, other scenarios are also possible. For instance, the end-users of the generated texts might be the experts themselves (suppose that they need a tool to easily generate reports from data, to help them save time for other duties), the knowledge base could be built based on the preferences of the readers rather than the writers (because in some cases the language of experts is too technical for end-users). However, the presence of actors that need to interpret the textual information that is given by the system is common for all real context scenarios.

Although this more realistic scenario has not been considered until recent times, one of the main concerns of research on LDD has addressed how to devise criteria that allows to ensure the quality of the protoforms in their intrinsic form, leading to the concept of interpretability. Despite not having a clear definition, interpretability in fuzzy systems has been studied quite extensively, and it has been under this same light that interpretability on LDD has been discussed.

3 Interpretability of Linguistic Descriptions of Data

The traditional conception of interpretability in fuzzy sets theory has its roots in fuzzy rule-based systems, such as those built on Mamdani rules "if X_1 is P_1 and X_2 is P_2 ... then Y is R". In this case, there are several properties that are considered important to improve the interpretability of the rule-sets, such as compactness (a system with less rules will be easier to interpret), completeness (the system rules should cover all possible cases) and consistency (the definitions of the fuzzy sets should be the same for all the rules). However, this idea of interpretability is rather intrinsic and closer to the concept of legibility, and its influence is rather limited to the system designer (who, of course, should be able to unpack what is being designed and implemented).

In the case of LDD, interpretability has been explored mainly in [8]. This study provides an interesting explanation of several properties that should be taken into account in order to assess the quality of the linguistic descriptions. More interestingly, the authors distinguish between features that are important for determining the interpretability of individual protoforms and properties that are useful when considering a set composed of several protoforms.

For instance, for single protoforms, truth degrees are interpreted as "quality degrees". Thus, if T(*"Some cold days are very humid"*) = 1 and T(*"Some hot days are dry"*) = 0.8, then the quality of the first sentence is higher. However, in [8] it is also acknowledged that a truth degree is not the only property that determines the quality of a protoform, and other quality measures are also described. Among them, the degree of appropriateness, the relevance, the degree of informativeness or the differentiation score appear as additional features that can prove useful to refine the filtering of good linguistic descriptions.

Linguistic descriptions often involve over-generation, i.e., given several linguistic variables composed of several fuzzy properties each, a fuzzy quantifier partition and different types of target protoforms, one needs to generate all possible sentence combinations and then filter those with the highest quality. In this context, [8] also describes properties that a set of sentences should fulfill as a whole. For example, the properties of non-contradiction and double negation are described to ensure the consistency among the different sentences. Likewise, redundancy among the different sentences is also an issue that is related to the double negation property, the inclusion (when quantifiers, or properties are included in others) or the similarity.

4 Interpretability from a Text Generation Perspective

Recent works in fuzzy rule-based systems extend the classical idea of interpretability to consider not only researchers or designers, but experts or non-specialized users. In short, the objective is to provide systems whose decisions can be understood by those mainly affected by them. For this, research on interpretable fuzzy systems has already been made at an explanation level [1,11,12].

In the case of linguistic descriptions of data, users do not need to understand how the descriptions are computed. Instead, they need to properly interpret the result itself, i.e., the information represented by protoform-based sentences. In [8], this view is also briefly considered from a standalone perspective. Namely, users defining the vocabulary of the descriptions and the "linguistic rendering" of the sentences (their verbalization) are depicted to have a positive influence on the interpretability of the descriptions.

Our approach to the interpretability of LDD is made under a different, albeit complementary, view, which corresponds to the real setting context shown in Fig. 1. First of all, we assume that there exists a communicative purpose and that the act of communication is made in the form of an automatically generated text [6]. Secondly, we assume that the mechanisms of linguistic descriptions of data are tools meant to be used as part of an NLG system, and thus cannot be studied only in isolation [14]. Thirdly, under this light, users are meant to receive textual information fully adapted to their language, and not just descriptions composed of protoform-like sentences. This means that any information gathered by linguistic descriptions could be mixed within the text with other kinds of information (numerical, statistical, etc.) in order to fulfill the users' information needs, as was the case in [15].

Under these assumptions, the actual entities which are subject to an interpretation from users are the texts that NLG systems generate, instead of the raw linguistic descriptions. Thus, the problem shifts from focusing just on the intrinsic interpretability of the linguistic descriptions towards the problem of how to use LDD in the generation of textual explanations for an end-user. In other words, we need to devise how to translate the semantic depth that linguistic descriptions hold (the relations between their components, e.g., quantifiers, fuzzy labels, truth degrees...) into a textual element that can ultimately help users understand the whole text and the underlying data that is being described.

Some questions and ideas regarding the use of fuzzy sets and LDD within an NLG system have already been given in [13]. However, these were not considered strictly from an interpretability perspective. Thus, our purpose in what follows is to discuss and illustrate how some of the different elements that compose a linguistic description can be verbalized to improve the interpretability of an automatically generated text. Namely, we will consider the role that the following elements play:

- Fuzzy properties.
- Fuzzy quantifiers.
- Truth degrees.
- Fuzzy quantification mechanisms.

For reasons of clarity we will refer in what follows to the simplest type of linguistic descriptions, composed of type-I quantified sentences, as the issues described in this paper relate to elements that are common for all different kinds of protoforms.

4.1 Interpretability from Fuzzy Properties

As we said, in LDD, a linguistic variable categorizes a numerical domain in concepts, which are modeled by means of fuzzy sets. For instance, the notion of "temperature" could be divided into the fuzzy sets "cold", "mild" and "warm", whose membership functions cover the temperature numeric domain. For enhancing the interpretability of the output text containing these concepts (and in this we coincide with [9]), their definition is not just a decision of the designer of the system, but it must follow an expert's criteria, as NLG systems do, in order to suit in addressees' background.

In many occasions, it is not possible to use fuzzy properties if there exist strict guidelines to follow, but in others it might be necessary to capture the meaning of words and model their imprecision. For instance, when there are several experts with different perceptions and understandings it can be useful to aggregate them by creating fuzzy models. Likewise, if the aim of the NLG system is to provide texts whose meaning is more adapted to what readers understand, surveys can be made to ascertain the semantics of the words to be used.

Furthermore, in our context it is possible that the terms defined for a specific linguistic variable do not end up being reflected in the output text in their original form. For example, a fuzzy property "mild" related to the temperature could be expressed in the final text using a synonym as "soft". This kind of changes help improve the final text with richer style and variation, as long as the semantic similarity between the synonyms is high (and thus the conveyed concept remains the same).

4.2 Interpretability from Quantifiers

The case of fuzzy quantifiers is similar to the fuzzy properties in linguistic variables in the sense that both quantifiers and properties are assigned a specific linguistic term or expression, which can be included in the generated text. However, quantifiers in protoforms provide richer alternatives for text purpose generations, because they can lead to sentences that do not express the quantifier explicitly, but improve the interpretability for end-users thanks to the use of a more adapted language.

For instance, the GALiWeather system makes use of type-I fuzzy quantified sentences to create a description of the cloud coverage variable. However, the information represented by the protoforms using linguistic terms and quantifiers is further processed to be adapted to the target language requirements. In some cases, quantifiers are also included in the generated text (although different words are used to convey them), and sometimes they are omitted, as the text itself conveys the same meaning implicitly. Figure 2 shows four different linguistic

Linguistic description (set of protoforms)			Generated text
Most values are very cloudy	**A few** values are partly cloudy	**A few** values are clear	Very cloudy skies **in general** for the coming days [, **occasionally** partly cloud/clear].
Several values are very cloudy	**Several** values are partly cloudy	**A few** values are clear	Skies **alternating** very cloudy and partly cloudy moments [, **occasionally** clear].
A few values are very cloudy	**A few** values are partly cloudy	**Several** values are clear	There will be clear skies, with partly cloudy [and very cloudy] **moments.**
A few values are very cloudy	**A few** values are partly cloudy	**A few** values are clear	Very variable cloudiness throughout the coming short term.

Fig. 2. Examples of linguistic descriptions obtained by GALiWeather and associated verbalizations (translated from Spanish). Optional parts that depend on the coverage criterion of the protoforms are displayed in square brackets.

descriptions (composed of three protoforms) that GALiWeather can produce, and their corresponding realization in text.

The decision of when to make quantifiers implicit or explicit will largely depend on the requirements of the target language, as was the case in GALiWeather. For instance, if the NLG system generates technical reports to be reviewed by experts, providing explicit numerical quantifiers such as "around 40%" can provide enough information to the expert and improve the interpretability of the text compared to showing raw numbers. In more casual settings, quantifiers that indicate a wide coverage can be omitted, in order to improve the fluidity of the texts, e.g., a protoform like "most days last month were cold" could be realized as "last month was cold".

4.3 Interpretability from Truth Degrees

One of the main differences between LDD and other kinds of techniques that can be used for content determination in NLG resides in the truth degree which is associated to each computed protoform. As we have reviewed in Sect. 3, the truth degree of a protoform is considered its main quality indicator, so in general it is assumed that sentences with higher truth degrees are better than others with a lower degree. In this sense, the truth degree is also the first criterion which is usually taken into account to filter sentences. However, to our knowledge its influence on how the protoform can be linguistically realized has never been taken into account.

From our perspective, the truth degree should also be considered an important part of the underlying semantic of the protoform, even though it is not originally associated to a linguistic term. For this, we present a more general interpretation of this element, that understands a truth degree not as a degree of quality, but as a metric that measures the level of evidence that supports a sentence or, in other words, that provides a measure of how certain the system can be about the statement.

In order to provide a proper interpretation of the truth degree of a protoform, one must understand what a fuzzy quantification model actually does to calculate the truth degree. Particularly, we will focus on Zadeh's model for type-I fuzzy quantified sentences, which is shown in Eq. 1, where Q is a fuzzy quantifier, X is the data referential to be evaluated, A is a fuzzy property (summarizer), μ a membership function (different for Q and A), and v_i a data value in a data set in the referential, which is composed of n elements.

$$T(Q\ X's\ are\ A) = \mu_Q\left(\frac{1}{n}\sum_{i=1}^{n}\mu_A(v_i)\right) \tag{1}$$

We can determine this model involves obtaining the cardinality of the fuzzy property that is being evaluated. This cardinality is then evaluated against a fuzzy quantifier. This means that truth degrees corresponding to single data values aggregate into a single value. For instance, if we interpret *T(few cities have high population)* from a logical point of view, we could say that T it is "the degree in which the 'high population' cities fulfill that they are 'few' within all cities". However, this can also be interpreted from a language use perspective which can be useful for our purpose: to generate texts that are more interpretable.

Our interpretation of T, which is meant to have an influence on deciding how to realize protoforms linguistically, is that the fuzzy nature of the linguistic terms that compose them (both fuzzy properties and quantifiers) ends up causing a lack of certainty about the statement, which is reflected in the truth degree. For example, under this interpretation T*(few cities have high population)* = *0.4* would mean that we can not be quite certain (or, alternatively, that there does not exist a strong evidence) that "few cities have high population".

This fact led us to conclude that truth degrees can be useful in one of the current challenges in NLG systems: generating texts from non-linguistic input data (commonly known as data-to-text systems) that communicate *uncertainty about the reliability of the input or the system's analysis* [16]. Thus, an NLG system can use the truth degree obtained from a protoform as an indicator for applying a modal operator (i.e., *might*, *can*, *must*, etc.) or selecting a different quantifier (i.e., *most* - *in general*, *several* - *alternating*, etc.) for the generation of the final statement. For example, suppose that there are CO_2 sensors in every room of a house, and an intelligent assistant is able to retrieve their data in real time and inform us about their current state using text-to-speech. The assistant could communicate the information from a single protoform in different ways depending on its truth degree, as Fig. 3 shows. A simple syntactical-semantical structure like a protoform can actually lead to generating more sophisticated sentences with different variations.

Although in a real context the assistant would be likely to provide information associated to high truth degrees and high coverage quantifiers, there can appear situations where a statement can be considered relevant despite having a low degree (e.g., a protoform that includes a fuzzy property which is considered very important or exceptional). Likewise, situations of conflict or ambiguity could also

Protoform	T	Generated text
Most places in the house have a low CO2 concentration	1	The concentration of CO2 is low in most parts of the house.
	0.8	Based on the data gathered from the sensors, I am quite certain that the concentration of CO2 is low in most parts
	0.6	The sensors show there is some evidence that the concentration of CO2 is low in most parts of the house.
	0.4	I can not say for certain that the concentration of CO2 is low in most parts of the house.
	0.2	There is little evidence of a low concentration of CO2 in most parts of the house currently.
	0	I have no evidence about the concentration of CO2 being low in most parts of the house.

Fig. 3. Examples of verbalizations of the same protoform according to different truth degrees.

be communicated, when two competing sentences end up having similar truth degrees, in a similar fashion to what occurs when two rules of a fuzzy rule-based system with different consequents activate at the same time.

For our ultimate purpose, using the truth degree as a means to adding information about certainty or evidence in the generated texts can help provide a more familiar language that improves the interpretability of the texts for end-users [16], since these elements are usually present in our daily language use. Furthermore, this can also be useful for specialized users in some domains, where the NLG system can generate texts that indicate different possibilities to guide experts for the interpretation of the input data (e.g. in health domains, where there is a strong dependence on the expertise of doctors and strict guidelines are not always followed).

4.4 Influence of Fuzzy Quantification Models

The existence of different fuzzy quantification models means that the truth degree of a protoform can be calculated in several ways with different results depending on the model we choose. While we have not considered this in the truth degree discussion above, this issue will undoubtedly influence how one can interpret the truth degree of a protoform in order to achieve a proper linguistic realization afterwards.

It is not the purpose of this paper to enter into technical details about fuzzy quantification models or their use. For this, the reader can find very useful the extensive review of models in [3] and a behavioral guide with some practical guidelines in [5]. However, to illustrate the importance of the influence of a quantification model in the interpretation of truth degrees we will refer to the issue of aggregative behavior which is present in some models.

Referencing [5], *aggregative behavior makes reference to the tendency of a model to confuse one 'high degree' membership element with a large quantity of 'low degree' membership elements.* For instance, suppose a protoform "a few men are tall" and a dataset with 100 height values from different men. Under Zadeh's model, which is affected by this issue, the truth degree resulting from having 10 values with a truth degree of 1 is the same as the one that results from having all 100 values with a truth degree of 0.1. Thus, for both cases we could obtain that T*(a few men are tall)* = *1*, despite that all values in the second case fulfill the property 'tall' in an extremely low degree. Such cases would hurt the interpretability of the generated text and result in a misleading interpretation of the original data.

5 Conclusions

In this paper we have proposed a novel understanding of interpretability in the context of fuzzy linguistic descriptions of data. Instead of focusing on the classical notion of interpretability for fuzzy systems and LDD, we have approached this concept from an NLG perspective, where the end-user is key. We have discussed how the different elements in a protoform could be taken into account during the text generation process to improve the interpretability of the output texts.

The ideas here described are based on previous experiences of using LDD and fuzzy sets in the development of NLG systems, but still represent a starting point in this regard. As future work, we believe that the concepts here discussed should be further explored under an empirical setting. For instance, small controlled experiments with subjects could be done to verify whether different truth degrees for a same protoform should result into different verbalizations. Also, it would be interesting to study the influence of using different quantification models when calculating truth degrees for protoforms. Under this setting, evaluating the interpretability of the texts could be achieved through intrinsic evaluation methods to assert if the generated texts can be properly understood by users [14].

Acknowledgments. This work has been funded by TIN2014-56633-C3-1-R and TIN2014-56633-C3-3-R projects from the Spanish "Ministerio de Economía y Competitividad" and by the "Consellería de Cultura, Educación e Ordenación Universitaria" (accreditation 2016–2019, ED431G/08) and the European Regional Development Fund (ERDF). A. Ramos-Soto is funded by the "Consellería de Cultura, Educación e Ordenación Universitaria" (under the Postdoctoral Fellowship accreditation **ED481B 2017/030**). M. Pereira-Fariña is funded by the "Consellería de Cultura, Educación e Ordenación Universitaria" (under the Postdoctoral Fellowship accreditation **ED481B 2016/048-0**).

References

1. Márquez, A., Márquez, F.A., Peregrín, A.: A mechanism to improve the interpretability of linguistic fuzzy systems with adaptive defuzzification based on the use of a multi-objective evolutionary algorithm. Int. J. Comput. Intell. Syst. **5**(2), 297–321 (2012)
2. Boran, F.E., Akay, D., Yager, R.R.: An overview of methods for linguistic summarization with fuzzy sets. Expert Syst. Appl. **61**, 356–377 (2016)
3. Delgado, M., Ruiz, M.D., Sánchez, D., Vila, M.A.: Fuzzy quantification: a state of the art. Fuzzy Sets Syst. **242**, 1–30 (2014). Theme: Quantifiers and Logic
4. Díaz-Hermida, F., Losada, D.E., Bugarín, A., Barro, S.: A probabilistic quantifier fuzzification mechanism: the model and its evaluation for information retrieval. IEEE Trans. Fuzzy Syst. **13**(1), 688–700 (2005)
5. Diaz-Hermida, F., Pereira-Fariña, M., Vidal, J.C., Ramos-Soto, A.: Characterizing quantifier fuzzification mechanisms: a behavioral guide for applications. Fuzzy Sets Syst. (2017). https://doi.org/10.1016/j.fss.2017.07.017
6. Gatt, A., Krahmer, E.: Survey of the state of the art in natural language generation: core tasks, applications and evaluation. J. Artif. Intell. Res. **61**, 65–170 (2018)
7. Kacprzyk, J.: Computing with words is an implementable paradigm: fuzzy queries, linguistic data summaries, and natural-language generation. IEEE Trans. Fuzzy Syst. **18**(3), 451–472 (2010)
8. Lesot, M.J., Moyse, G., Bouchon-Meunier, B.: Interpretability of fuzzy linguistic summaries. Fuzzy Sets Syst. **292**, 307–317 (2016)
9. Liétard, L.: A functional interpretation of linguistic summaries of data. Inf. Sci. **188**, 1–16 (2012)

10. Marín, N., Sánchez, D.: On generating linguistic descriptions of time series. Fuzzy Sets Syst. **285**, 6–30 (2016). Special Issue on Linguistic Description of Time Series
11. Pancho, D.P., Alonso, J.M., Cordón, O., Quirin, A., Magdalena, L.: FINGRAMS: visual representations of fuzzy rule-based inference for expert analysis of comprehensibility. IEEE Trans. Fuzzy Syst. **21**(6), 1133–1149 (2013)
12. Pancho, D.P., Alonso, J.M., Magdalena, L.: Enhancing fingrams to deal with precise fuzzy systems. Fuzzy Sets Syst. **297**(Supplement C), 1–25 (2016)
13. Ramos-Soto, A., Bugarín, A., Barro, S.: Fuzzy sets across the natural language generation pipeline. Prog. Artif. Intell. **5**(4), 261–276 (2016). https://doi.org/10.1007/s13748-016-0097-x
14. Ramos-Soto, A., Bugarín, A., Barro, S.: On the role of linguistic descriptions of data in the building of natural language generation systems. Fuzzy Sets Syst. **285**, 31–51 (2016)
15. Ramos-Soto, A., Bugarín, A., Barro, S., Taboada, J.: Linguistic descriptions for automatic generation of textual short-term weather forecasts on real prediction data. IEEE Trans. Fuzzy Syst. **23**(1), 44–57 (2015)
16. Reiter, E.: An architecture for data-to-text systems. In: Busemann, S. (ed.) Proceedings of the 11th European Workshop on Natural Language Generation, pp. 97–104 (2007). http://www.csd.abdn.ac.uk/~ereiter/papers/enlg07.pdf
17. Yager, R.R.: A new approach to the summarization of data. Inf. Sci. **28**(1), 69–86 (1982)
18. Zadeh, L.A.: A computational theory of dispositions. Int. J. Intell. Syst. **2**(1), 39–63 (1987)

Personality Determination of an Individual Through Neural Networks

J. R. Sanchez[1(✉)], M. I. Capel[2], Celina Jiménez[3], Gonzalo Rodriguez-Fraile[4], and M. C. Pegalajar[1(✉)]

[1] Department of Computer Science and Artificial Intelligence, University of Granada, 18071 Granada, Spain
sanchezjr@ugr.es, mcarmen@decsai.ugr.es

[2] Software Engineering Department, University of Granada, ETSI Informatics and Telecommunication, 18071 Granada, Spain
manuelcapel@ugr.es

[3] Psychological Clinic Altea, Altea, Spain
celina@alteapsicologos.com

[4] Foundation for the Development of Consciousness and Development Chair, University of Granada, Granada, Spain

Abstract. The use of neural networks is proposed in this article as a means of determining the personality of an individual. This research work comes in view of the necessity of combining two psychological tests for carrying out personnel selection. From the assessment of the first test known as 16 Personality Factor we can directly obtain an appraisal of the individual's personality type as the one given by the Enneagram Test, which now does not need to be done. The two chosen tests are highly accepted by Human Resources Department in big companies as useful tools for selecting personnel when new recruitment comes up, for personnel promotion internal to the firm, for employees' personal development and growing as a person. The (mathematical/computer science) model chosen to attain the research objectives is based on Artificial Neuron Networks.

Keywords: Machine Learning · Eneagrama · 16PF · Psychology Regression · Neural networks

1 Introduccion

Nowadays, companies are aware of the importance of selecting the optimal candidate for any position in the firm; moreover, it becomes mandatory for responsibility positions such as team leaders or heads of area. Personnel selection is an issue that needs to be addressed by HR departments in companies, which have to face the problem of selecting the optimal candidate, as well as to tackle other problems such as to choose among a great number of applicants for the same job position, or to get to know well the firm's employees when it comes to perform

J. Medina et al. (Eds.): IPMU 2018, CCIS 853, pp. 52–61, 2018.
https://doi.org/10.1007/978-3-319-91473-2_5

a procedure for promotion and in this way to determine which one is the best candidate for a job position.

To solve the problems discussed, we can deploy a personnel selection test. These tests will help filter people and enable the HR professional to meet the candidate better. There are different kinds of tests used in personnel selection process; we can mention, among the most used ones, aptitude tests, projective and intellectual tests [8]. In addition to the latter test types, we can find the ones known as personality tests, whose objective is to obtain an accurate profile of the candidate, which can help the HR professional in decision making processes. In the study carried out in [4], we can see how the deployment of personality predictors improve candidate selection results and cut off expenses when it comes to conduct a personnel selection process.

One the best known and deployed personality tests is the Cattell's 16 Personality Factor (16PF) test. We can find in the related literature several references of studies documenting the results yielded by the 16PF test, as in the case of an airline company that used the test to correlate personality and work performance of the employees [5]. The 16PF test also helps to understand personality characteristics of different personality profiles, as the study conducted in [1] shows, in which the psychological-profile differences among executive staff in companies. Other references worthwhile reviewing on the subject are the following ones, [2,9].

On the other hand, there is also a test called the Enneagram of personality [10] that shows a suitable functionality to classify people according to their personality, and it is being widely used in that respect. Currently, many international companies are using tools based on the Enneagram for different purposes. For example, some of these companies are Toyota, DAimler/Mitsubishi, Genentech / Roche, Milling Hotels, Huron Hospital, Parker Hannifin Corporation, etc. Some of the objectives pursued by the companies that applied the Enneagram wanted to obtain focus on real needs, such as to address business and person needs, to get a strong leadership by promoting organization commitment, diagnosis of the personality type for attaining a balanced and accurate personal self-discovery of each employee. In the reference [7], we can see how the Enneagram is applied to manage and motivate employees by identifying each worker's internal motivations by means of applying the Enneagram principles. By studying the enneatypes of employees and the basic motivations that drive these, other Enneagram-related studies [6] help organizations and most specifically to HR departments to motivate, to attract and to keep the employees longer in the company. The work described in [12] shows the relationships among the Enneagram types, the employee behaviors and the key-cognitions at the workplace. The study was conducted using a sample of 416 candidates in total, and it finally demonstrates how the Enneagram provides a way to describe an individual person and it also demonstrates an interesting feature for predicting personal values of individuals.

The main objective of the work described here is to put together the knowledge extracted from both tests (16PF and Enneagram) by deploying Artificial

Neural Networks for combine in a novel way techniques that come from Psychology and Machine Learning for getting to predict the individual's personality and then to help HR professionals in their job of personnel selection and professional development of employees within a company. For this purpose, prediction techniques will be used. By the fusion of both tests, we are able to create a (new) model that allows us to anticipate the results that an Enneagram yields, and thus to save time and effort to the individuals who take the tests and to the evaluator who set the tests. In addition to this saving of time and effort, a more complete individual profile will be made available to the HR professional since these two tests are complementary. Whereas the 16PF provides a more specific vision of the emotional qualities of an individual, the Enneagram places more emphasis on the personality type, potential-development characteristics and self-improvement.

This paper is organized in the following way, Sect. 2 will show a description of the two tests used in the research; Sect. 3 describes the Artificial Neural Networks-based model that is used for solving the proposed problem; Sect. 4 discusses the experiments carried out together with the results obtained from these experiments. Finally, we present the conclusions and future work.

2 Description of the Deployed Tests

As mentioned in section above, there are several research works carried out to make possible personality determination of individuals of a set by busing specific processes of recruitment, training and promotion of HR. Most recent works, among all of them, make use of the 16PF test and the Enneagram.

On one hand, the test of 16 Personality Factor (PF) - started by Cattell and collaborators at the University of Illinois - is intended to determine the personality of a person. Therefore, the 16PF shows a classification of 16 factors, where each factor represents a quality (e.g., independent, perfectionist, insecure, etc.) and gives the fulfillment rate to each of the qualities of anyone who takes the test. It must be noticed that the test makes use of a set of 187 of compulsory choice questions.

On the other hand, Riso Hudson, classified the personality of an individual according to a number of types that would represent the role of the candidate, instead of resorting to more specific factors for personality determination. To this end, it was proposed the use of a test that classifies an individual according to 9 enneatypes, which represent a candidate's personality types, such as leader, winner, researcher, etc., and helps an expert to determine strengths and weaknesses of an applicant's personality.

Whereas the 16PF is responsible for determining the one candidate's personality 16 factors, in our study, the Enneagram test allows us to obtain a more dynamic profile of one individual's personality by taking into account other aspects that 16PF test does not address. The Enneagram test is based on the one of "Riso Hudson Enneagrama Type IndicatorRHETI", which is composed of 144 compulsory questions and each one of them has 2 possible answers. As output, the Enneagram test provides an assessment of 9 types of personality, of which the most valued one is the one assigned to a person.

3 Artificial Neural Networks

The mentioned networks (Artificial Neural Networks -ANN-) are inspired by the structure of biological neural networks, where a set of neurons are connected to each other and can interact among them through connections, transmitting information via electrical signals. ANN have been characterized as good tools for solving both regression and classification problems.

3.1 Feedforward ANN

We will use feedforward neural networks Fig. 1 (FNN) to solve the problems derived from the composition of the two types of tests that we address in this paper, and which were introduced in the prior section. FNN operation is based on the deployment of a number of simple processing units, which are known as neurons and are organized in layers. Each of the neurons, which is located on the same layer, is connected with all the neurons of the previous layer. The connections between neurons have different weights associated with them, which determines the influence degree that one neuron has on another. Neuron connections are a means to cast knowledge to the entire network.

Within intermediate layers of the network, to compute one neuron's output, a sigmoidal function is applied to the weighted sum of the inputs, whereas a linear function is applied in the last layer. The data input in the first layer correspond to the independent variables of the problem, while the neuron's output data on the last layer represent the network response, which can be understood as an approximation of the problem's dependent variables values. This procedure is described in functions 1 and 2, where $f_i^l(x_i^l)$ is the output value of the neuron i-th on the layer *l*, w_{ij}^1 is the weight associated to neuron j-th connection on the layer *l-1* to the neuron i-th on the layer *l*, N^l is the number of neurons on the layer *l*, l_j^p represents the value of the independent variable j-th within the sample *P*, *L* is the number of layers that make up the network and b_i^l is a bias value that matches the neuron i-th on the layer *l*.

$$f_i^1(x_i^1) = \begin{cases} \frac{1}{1+e^{-x_i^l}}; x_i^l; 1 \leq 1 \leq L \\ x_i^1; 1 = L \end{cases} \tag{1}$$

$$x_i^1 = \begin{cases} \sum_{j=1}^{N^{1-1}} (w_{ij}^1 I_j^P) + b_i^1; 1 = 1 \\ \sum_{j=1}^{N^{1-1}} (w_{ij}^1 f_j^{1-1}(x_j^{1-1})) + b_i^1; 1 < 1 \leq L \end{cases} \tag{2}$$

Training of an ANN consists of finding the optimized weights associated to the entries and the bias b_i^l of the neuron w_{ij}^l so that the network can provide

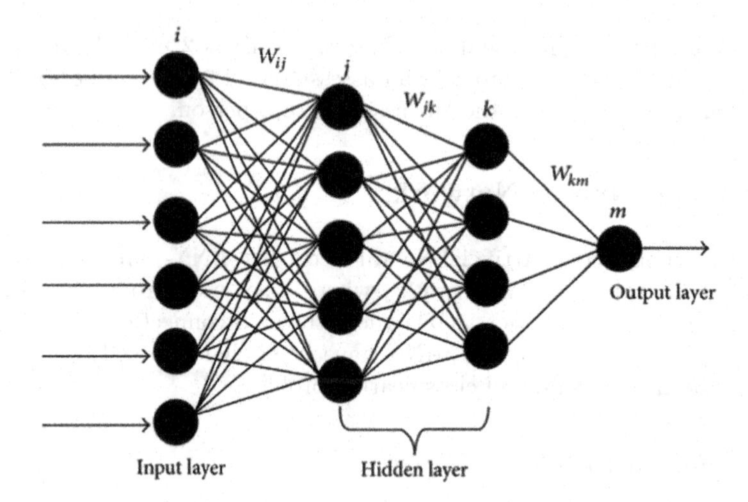

Fig. 1. Neural network feed-forward

the most accurate possible approximation results to the problem's dependent variables, according to a given error measure. The most commonly used error criterion for a network's output is the Minimum Mean Square Error (MSE), though to solve the problem discussed here the Root Mean Square Error (RMSE) was used, as described in Fig. 3. RMSE helps us in the measurement of the adjustment correlation between the real value $\hat{y}_t$ and the predicted value y_t. The best-known training algorithm is called backpropagation, though there are also other Quasi - Newton optimization techniques, such as the ones by Levenberg-Marquart [11] and Broyden - Fletcher - Goldfarb - Shanno (BFGS) [3], which are being used in recent years with high success rates. The training algorithm, deployed as our main optimization technique here, has been BFGS.

$$\mathrm{RMSE} = \sqrt{\frac{\sum_{t=1}^{n}(\hat{y}_t - y_t)^2}{n}} \tag{3}$$

3.2 Artificial Neural Network Model for Setting the Relationship Between both Tests

As we already mentioned, to provide a more accurate information to an individual expert in this subject, we established as the main objective of this work to prove a connection between both tests: 16pf and Enneagram. The model that we used to carry out such proof consists of 9 artificial neural networks, each one of them will calculate the assessment that corresponds to an individual's personality type. All the deployed networks are made up of 16 inputs for the assessment of each one of the characteristics of the 16pf test, 1 hidden layer that consists of 3 neurons and an output layer with a single neuron. Once evaluated the inputs by each network, the personality chosen for the individual is the enneatype represented by the network that provides a higher output value.

4 Conducted Experiments Discussion

4.1 Data Input and Pre-processing

As this is a novel study, in which there was no data collected, we decided to develop a web system that can collect and store the test results. The mentioned tests can be completed at website[1]. The development of this web system allowed us to show to the users who perform the test their results in real time and provide an assessment of their personality. Social networks, especially Facebook were used to give the maximum web site's diffusion[2]. In order to collect data in the social network, the method used was snowball sampling. This is a non-probabilistic (no random numbers are used). In order to generate new sources of primary data to be used in the research, the method gets references of initial subjects. More specifically, within the different snowball sampling methods, we can say that we used exponential non-discriminatory snowball sampling in this study, i.e., the first recruited subject provided multiple references to the search and each new reference was then explored until a sufficient number of samples were collected for carrying out the study. By this sampling method, we have found a number of elements that is large enough to consider the sample size as representative even for inference making.

Once stored the data was preprocessed, eliminating the noise generated by web export. The responses to both tests had been processed using standard expert criteria. Finally, both tests were unified to create the neural network training dataset.

In Table 1, we can see the results corresponding to the pre-processed Enneagram test data, which also shows the extreme values, i.e., the maximum and minimum assessment values that enneatypes can reach.

Table 1. Value range for eneatyping evaluations.

Range of outputs									
Enneatype	*1*	*2*	*3*	*4*	*5*	*6*	*7*	*8*	*9*
Highest	27	28	22	23	23	30	25	29	24
Lowest	9	8	6	4	6	6	4	5	4

4.2 Results and Experimental Configuration

After conducting the pre-processing described above, there are a set of 98 instances, which were used to train and assess the model. In order to verify the built model robustness, the data set was divided into two subsets: one "train" DataSet (90%) for training the neural network and the other called "test" (10%) for model validation. Finally, cross validation with 10 partitions has been used

[1] http://personalityandconsciousness.institute/personalityandconsciousness/.
[2] https://www.facebook.com/personalityconsciousness/.

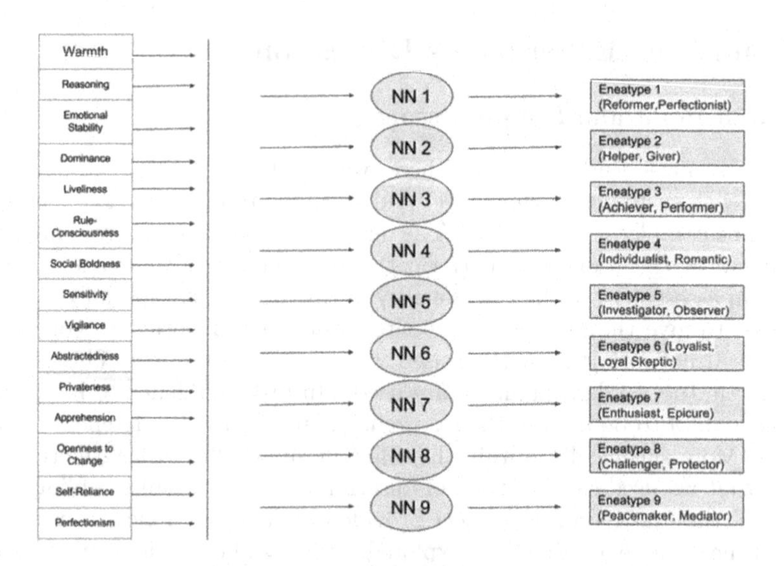

Fig. 2. Neural network feed-forward 16PF - Enneagram

Table 2. NNET algorithms parameters

Parameters ANN	
Input:	16pf test outputs (numerical)
Output:	1 Enneagram test output (numerical)
Size:	3 (No. of neuron units in the occult layer)
Decay:	0.01 (Decline in weight)
Maxit:	1000 (Maximum number of iterations)

Table 3. NNT results described as RMSE

	RMSE	Max	Min	Deviation
Reformer, perfectionist:	4.157	5.840	3.046	1.050
Helper, giver:	2.976	3.980	1.880	0.582
Achiever, performer:	3.537	4.453	2.324	0.735
Individualist, romantic:	3.579	5.671	2.930	0.836
Investigator, observer:	3.946	4.727	2.242	0.746
Loyalist, loyal skeptic:	4.797	7.084	3.102	1.233
Enthusiast, epicure:	4.073	5.613	3.163	0.847
Challenger, protector:	3.586	5.508	2.302	1.086
Peacemaker, mediator:	3.856	4.948	2.778	0.638

to ensure that there is no over-fitting in our model. Figure 2 shows the neural network built for unifying the 16pf and the enneagram tests. The Table 2 parameters have been used for each of the obtained networks.

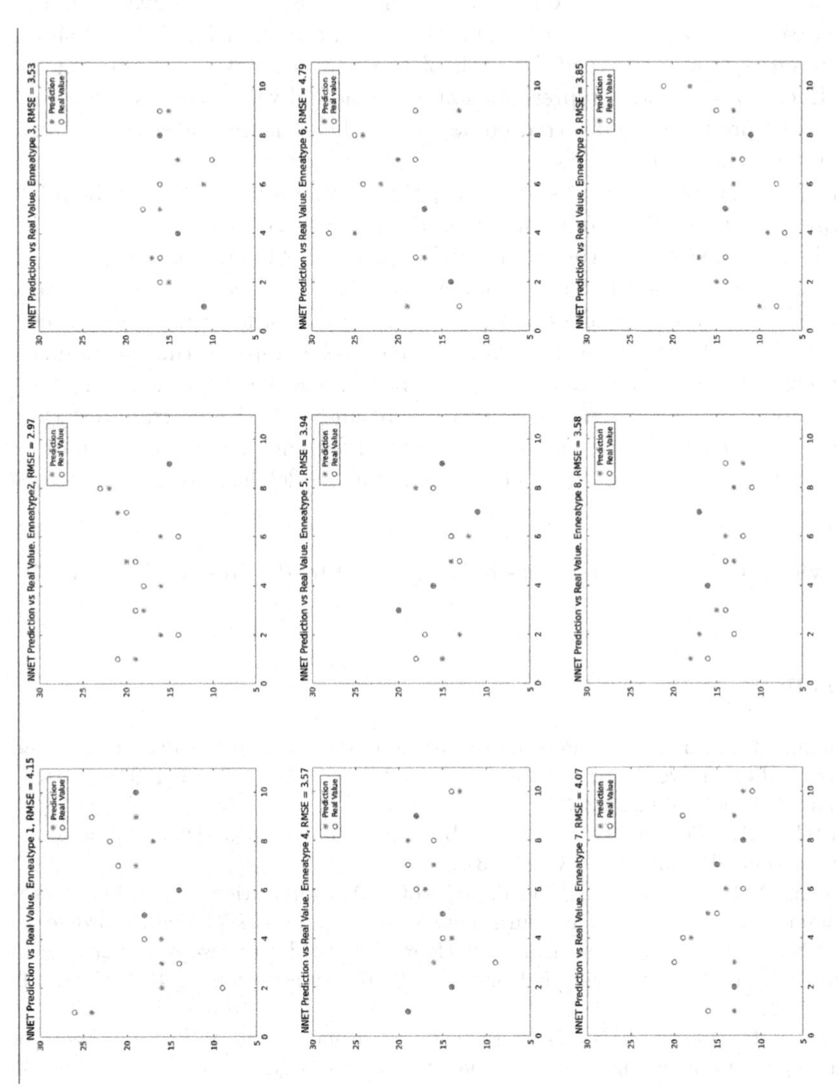

Fig. 3. Obtained results with NNET for the predicted values compared to the real values for each one of the 9 obtained models (Color figure online)

Finally, Table 3 shows the results obtained by the neural network. The first column shows the enneatypes of the Enneagram, followed by the next values: mean square error (RMSE), average, maximum, minimum and deviation, which are obtained for each one of these models. Taking into account the range of values with which we had worked in Table 1, the experts confirmed that the error shown in computations, on average, can be considered relatively low risk for the assessment of each of these enneatypes. On the other hand, Fig. 3 shows the different settings obtained for each of the neural networks created in the study. Each graph shows a representation of the actual values (blue circle) compared to the predicted values (red circle), it can be seen that adjustment fairly approximates the actual values.

This paper shows the results of a study conducted on how ANN can help HR departments in their efforts to improve recruitment procedures and the successful promotion of employees in the company. To this end, we have combined two of the tests currently used by most companies (16PF and Enneagrama), whose combined use allows a saving of time, since the results of the Enneagrama tests can be predicted from the 16PF tests if our method is applied, thus obtaining a more precise profile of the candidate's personality. The artificial neural network model proposed here achieves the predictable results of the Enneagram test from the 16PF test scores. Based on the results obtained and the expert evaluation, we can state that the use of an artificial neural model has given satisfactory results.

Acknowledgements. This work has been supported by the project TIN201564776-C3-1-R.

References

1. Austin, J.F., Murray, J.N.: Personality characteristics and profiles of employed and outplaced executives using the 16PF. J. Bus. Psychol. **8**(1), 57–65 (1993). https://doi.org/10.1007/BF02230393
2. Cattell, R.B.: Personality and Mood by Questionnaire. American Psychological Association, Washington, D.C. (1973)
3. Cuéllar, M.P., Delgado, M., Pegalajar, M.C.: An application of non-linear programming to train recurrent neural networks in time series prediction problems. In: Chen, C.S., Filipe, J., Seruca, I., Cordeiro, J. (eds.) Enterprise Information Systems VII, pp. 95–102. Springer, Dordrecht (2007). https://doi.org/10.1007/978-1-4020-5347-4_11
4. Ferris, G.R., Bergin, T.G., Gilmore, D.C.: Personality and ability predictors of training performance for flight attendants. Group Organ. Stud. **11**(4), 419–435 (1986). https://doi.org/10.1177/0364108286114008
5. Furnham, A.: Personality and occupational success: 16PF correlates of cabin crew performance. Pers. Individ. Differ. **12**(1), 87–90 (1991). http://www.sciencedirect.com/science/article/pii/019188699190135X
6. Hebenstreit, R.K.: Using the Enneagram to help organizations attract, motivate, and retain their employees. Ph.D. thesis (2007). oCLC: 632392974

7. Kale, S., Shrivastava, S.: Applying the enneagram to manage and motivate people. Bond Business School Publications, June 2001. http://epublications.bond.edu.au/business_pubs/52
8. Kline, P.: Handbook of Psychological Testing. Routledge, Abingdon (2013). google-Books-ID: JggVAgAAQBAJ
9. Krug, S.E., Johns, E.F.: A large scale cross-validation of second-order personality structure defined by the 16PF. Psychol. Rep. **59**(2), 683–693 (1986). https://doi.org/10.2466/pr0.1986.59.2.683
10. Riso, D.R., Hudson, R.: Understanding the Enneagram: The Practical Guide to Personality Types. Houghton Mifflin Harcourt, Boston (2000). google-Books-ID: vuIW_TIw8BIC
11. Saini, L.M., Soni, M.K.: Artificial neural network based peak load forecasting using Levenberg-Marquardt and quasi-Newton methods. IEE Proc. - Gener. Trans. Distrib. **149**(5), 578–584 (2002), http://digital-library.theiet.org/content/journals/10.1049/ip-gtd_20020462
12. Sutton, A., Allinson, C., Williams, H.: Personality type and work-related outcomes: an exploratory application of the Enneagram model. Eur. Manag. J. **31**(3), 234–249 (2013). http://www.sciencedirect.com/science/article/pii/S0263237312001442

Fuzzy Rule Learning for Material Classification from Imprecise Data

Arnaud Grivet Sébert(✉) and Jean-Philippe Poli

CEA, LIST, Data Analysis and System Intelligence Laboratory,
91191 Gif-sur-Yvette cedex, France
{arnaud.grivet.sebert,jean-philippe.poli}@cea.fr

Abstract. To address the problem of illicit substance detection at borders, we propose a complete method for explainable classification of materials. The classification is performed using imaprecise chemical data, which is quite rare in the literature. We follow a two-step workflow based on fuzzy logic induction. Firstly, a clustering approach is used to learn the suitable fuzzy terms of the various linguistic variables. Secondly, we induce rules for a justified classification using a fuzzy decision tree. Both methods are adaptations from classic ones to the case of imprecise data. At the end of the paper, results on simulated data are presented in the expectation of real data.

Keywords: Fuzzy partitioning · Clustering · Fuzzy decision tree
Fuzzy rules · Imprecise data · Explainable material classification

1 Introduction

Customs and ports security is a major issue in Europe. Indeed, many illegal or dangerous substances such as drugs, weapons, explosives pass through customs. Unfortunately, systematic container inspections are impossible in practice because of the cost and time that would be required. The volumes passing daily through the major European ports such as Rotterdam, Antwerp or Hamburg are indeed enormous: for example, 461.2 million tons of goods passed through the port of Rotterdam in 2016.

In this paper, we use tagged neutrons to obtain the chemical composition of a volume of the container. From this chemical composition, we want to determine the materials present in the container. In order to bring more credibility to the final software used by the customs officers, we will also provide a justification for this classification. Fuzzy rules allow to avoid the "black box" effect since customs officers have access to a real explanation of the classification made by the software, and close to natural language: "the container may contain drug (confidence degree: x) because the quantity of carbon is high, the quantity of nitrogen is low and the quantity of oxygen is medium" for instance.

The proportions are obtained by different treatments that are beyond our control. Thus, the input data are imprecise and are accompanied by a measure

J. Medina et al. (Eds.): IPMU 2018, CCIS 853, pp. 62–73, 2018.
https://doi.org/10.1007/978-3-319-91473-2_6

of this inaccuracy. Fuzzy logic thus seems appropriate for the exploitation of such data.

Given the time required and the authorizations needed to use a neutron generator, we will have a small learning dataset, even if all classes of relevant materials will obviously be represented. The idea is therefore to use fuzzy decision trees [4], which have been applied successfully on various classification problems [1,13]. The scarcity of training data and the intrinsic inaccuracy due to the preprocessing preclude conventional statistical learning approaches such as neural networks, SVM, etc. which also do not provide an explanation to users.

In this article, we adapt to imprecise data the classic two-step workflow consisting in using clustering methods to create relevant terms from data and then in building a decision tree to get fuzzy classification rules.

The paper is structured as follows: Sect. 2 describes the context of the application that motivates this work. Then, Sect. 3 describes the method to induce rules from imprecise data. Section 4 presents the results of the different experiments we conducted. Finally, Sect. 5 draws the conclusions of this paper.

2 Application Context

2.1 Neutron Inspection for Container Digging

The H2020 project C-BORD (effective Container inspection at BORDer control points) aims at facilitating the digging of containers at borders by exploiting different technologies: e-noses, X-rays, photo-fission and tagged neutrons. The goal is to detect dangerous (explosives, nuclear materials, ...) or illicit (drugs, contraband, ...) substances.

In this paper, we focus on the classification of materials with tagged neutron inspection. As shown in Fig. 1a, a device produces a neutron beam to focus on a certain volume (called voxel) of the container. The neutrons interact with the nuclei of the atoms contained in the voxel, producing new particles that can be detected by the matrix sensors which are positioned on the side of the container. These particles are thus characteristic of the atoms encountered in the examined voxel.

The processing of the raw data is not the topic of this article but we quickly describe the principle in Fig. 1b. After different pretreatments, a global spectrum is obtained. In comparison with the characteristic spectra of each of the studied atoms, this spectrum is decomposed into individual spectra by a Bayesian process which makes it possible to deduce the chemical composition of the voxel, expressed in percentages. This process is based on simulation and we can easily get the mean and the standard deviation for each proportion in order to characterize the inaccuracy of the reconstruction. Figure 2 shows the result of these treatments for an exposure to ceramics, and in which the inaccuracy is represented by “box plots”.

Our work consists in exploiting this information in order to recognize the materials contained in the voxel.

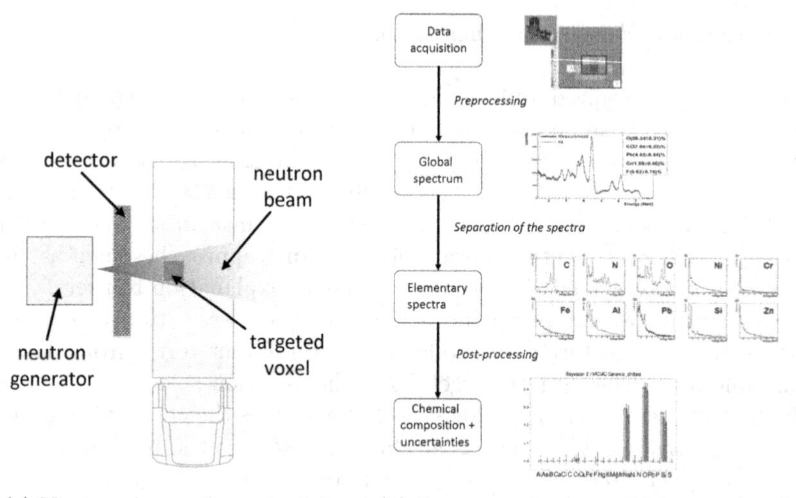

(a) Neutron inspection principle (b) From raw data acquisition to chemical composition assessment

Fig. 1. Neutron inspection workflow

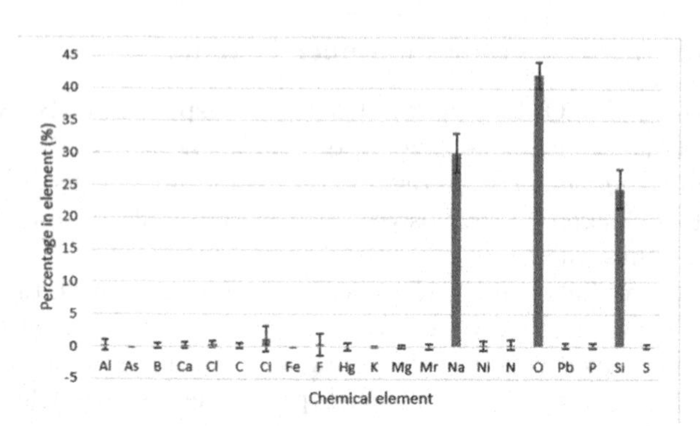

Fig. 2. Chemical composition obtained from ceramics simulation

2.2 Previous Work

The difficulty of recognizing materials from the chemical composition lies in several points:

- Data is scarce because few acquisition campaigns can be conducted. That also explains why few papers address the recognition of materials from their chemical composition.
- The inaccuracy of the chemical composition makes the task difficult for conventional statistical models.
- The device is insensitive to hydrogen atoms.

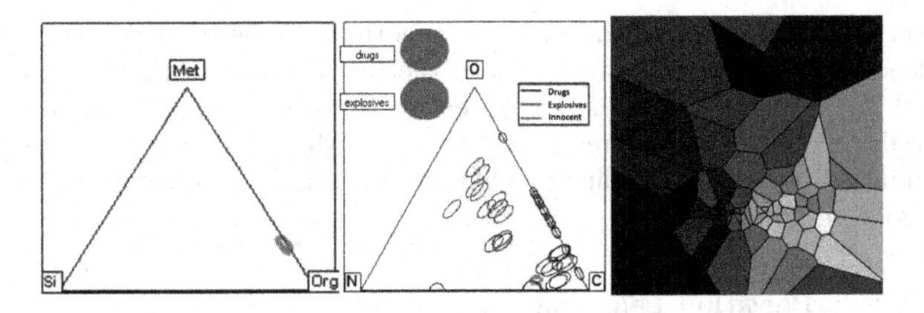

Fig. 3. Screenshots of the different visualizations introduced in [2] (from left to right: materials triangle, alert triangle, Voronoi diagram)

For these reasons, previous works proposed visual analytics methods to represent the content of the voxels. In [2], the authors proposed a Voronoi diagram to highlight the proximity, in terms of chemical composition, of the current voxel with voxels previously inspected and whose content is known (see Fig. 3). Thus, it is not a question of recognizing the materials present in the container but of displaying visually close and known containers in order to deduce the contents.

This approach has the advantage of not requiring learning or parameterization since it relies on the manual selection of a neighborhood. In Fig. 3, we can also see two classical representations that have been used in conjunction with this method. These are projections of the current voxel on two triangles: a triangle called "materials triangle" indicates the proximity of the voxel with metals, ceramics and organic materials, while a so-called "alert triangle" presents the ratios between carbon, nitrogen and oxygen. This last triangle makes it possible to distinguish between drugs and explosives. The main drawback of the visual analytics approach is that the operator must be able to interpret the different representations himself.

In practice, the mastering of these representations, particularly the Voronoi graph, can be difficult for operators who are not familiar with these visualization techniques. As part of the C-BORD project, we want to go further and propose a list of materials, and an explanation of the decision. It is to overcome these different difficulties that we want to use a fuzzy expert system.

3 Rule Induction for Classification

In order to effectively classify the materials while generating rules which are understandable by a human, we chose to use a fuzzy decision tree inspired by the one defined by Janikow in [11]. Each node is split, by a feature which was not used yet, into N child-nodes corresponding to the associated fuzzy terms. The maximum depth of the tree is thus the number of available features if no pruning is performed. In our case, the features are the different chemical elements the system can detect.

The rule induction follows a classic 2-step workflow but the different algorithms have been adapted to take into account data imprecision: firstly, the method extracts from data the various terms that will be involved in the decision tree with clustering methods. Then, the tree is induced. Finally, rules are created and used in a fuzzy expert system to perform the classification of new materials. To model the imprecision of the data, the inputs are represented as probability distributions, leading to a hybrid fuzzy/probabilistic approach which gives good results in practice.

3.1 Fuzzy Partition Learning

To create rules which will provide the end-user with an explanation of the material classification, we have to define the fuzzy terms which will be involved in the premises of the rules. Nevertheless, crisp terms would not represent the physical reality for which strict boundaries between different compounds are irrelevant. That is why fuzzy terms are used. Except for the extreme (first and last) terms which are trapezoidal, we opted for classical triangular overlapping terms that form a strong partition of the definition domains of the stoichiometric percentage of each element, namely [0, 100]. Triangular-shaped terms are used because more complex shapes, like trapezoidal or pentagonal shapes, reduce the performance and even the accuracy in some cases. An example of a partition with five terms is shown in Fig. 4.

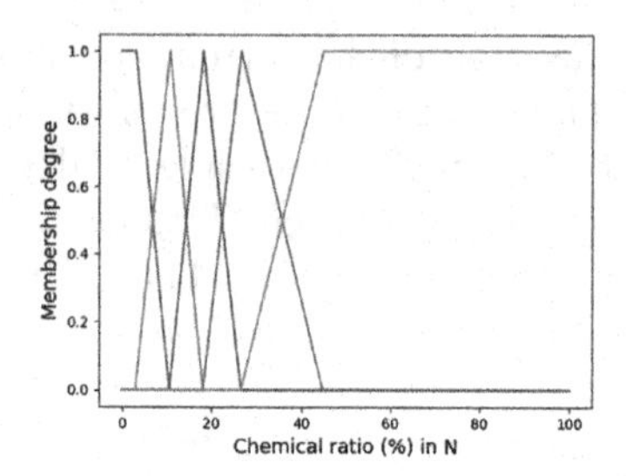

Fig. 4. Five fuzzy terms

The best methods to learn linguistic variables terms turned out to be clustering approaches. For example, using entropy-minimization-based methods gave poorer results, due to the difficulty to minimize the fuzzy entropy (see definition 4 in the following) which is not a convex function of the fuzzy sets parameters.

The basic idea is to cluster the data, feature by feature (chemical element), as if they were one-dimensional data, and build fuzzy terms over the resulting clusters. Once the clusters are built, the mode of a fuzzy triangle is simply set to the mean of its corresponding cluster; the spread is then induced by the other terms' modes through the constraint of the strong partition.

We tested existing algorithms and adapted them to the fuzzy case by the use of dissimilarities between distributions able to take into account the whole distribution of the data and not only an aggregated value. Distances or dissimilarities have already been defined and used with imprecise or uncertain data.

A distance consisting in the sum of the center Euclidean distance and the spread Euclidean distance of the imprecise data was used in [5, 7] but this implies a loss of the information provided by the whole distribution. Moreover, even though it might have been replaced by the deviation distance, the spread distance does not make a lot of sense in the case of normal distributions whose spreads are theoretically infinite, and in our case imposed by the domain boundaries. Gullo et al. [9] proposed an uncertain dissimilarity which corresponds, in the univariate case, to compute the double integral of the Euclidean distance of each pair $((x, f_1(x)), (x, f_2(x)))$ of points of the two distributions. We defined a simpler dissimilarity, in order to improve the performance.

The dissimilarity we use is inspired by the symmetric difference of two sets A and B: $d(A, B) = |A \cup B| - |A \cap B|$. The sets are here replaced by the probability distributions representing two imprecise data, the cardinality by the integral of the distribution, the intersection and union respectively by the minimum and maximum. Though, since in our case, the integral of a distribution on [0,100] is equal to 1, we have:

$$\int_0^{100} max(f_x(t), f_y(t))dt + \int_0^{100} min(f_x(t), f_y(t))dt = \int_0^{100} f_x(t)dt + \int_0^{100} f_x(t)dt = 2$$

We then obtain the following formula: $2 - 2\int_0^{100} min(f_x(t), f_y(t))dt$. For the sake of simplicity, we define the dissimilarity, without any influence on the results, as the half of this value:

$$d(x, y) = 1 - \int_0^{100} min(f_x(t), f_y(t))dt \quad (1)$$

This dissimilarity is actually a distance for continuous distributions but we will not show it in this paper for reasons of space. Figure 5 shows the associated similarity (integral of the minimum) between two distributions, drawn in red.

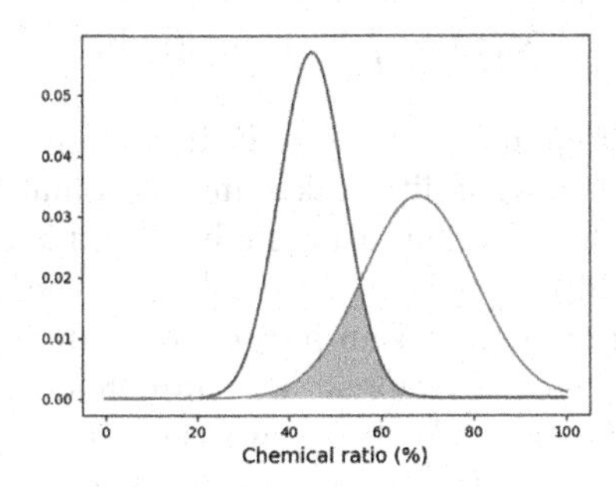

Fig. 5. Similarity between two imprecise data (Color figure online)

We also tried to use the T-norm product instead of the min in the definition of the dissimilarity and it gave very similar results.

We then ran the k-medoids algorithm [12] using this dissimilarity instead of the traditional Euclidean distance which would have only used the information given by the aggregated value of each distribution - namely the mean.

3.2 Fuzzy Decision Tree Induction

To build the tree considering the imprecision of the data, we represent these data by Gaussian probability distributions. Indeed, we will see in the following that the algorithm requires many integral computations and Gaussian distributions enable to do them simply. Since the definition domain of the data (proportion in each chemical element) is bounded ([0, 100]), the Gaussian distributions are transformed so that their integral on this bounded domain equals 1 and that the distribution thus keeps its probabilistic sense on the interval [0, 100] beyond which the values, which are percentages, do not have any sense. To do so, each distribution is divided by its integral on [0, 100]: $f(t) = norm(t) * \frac{\int_{-\infty}^{+\infty} norm}{\int_0^{100} norm} = \frac{norm(t)}{\int_0^{100} norm}$ where $norm$ is the probability density of the normal law.

We also tried a possibilistic approach, representing the imprecise data by triangular possibilistic distributions but the accuracy was lower, that is why we chose to focus on the probabilistic approach in this paper.

Once the fuzzy terms are learnt, we have to define how to compute the membership degrees of the imprecise examples, modeled by distributions, to each of these terms, which represent the nodes of the tree. We adapted integration techniques for uncertain data and crisp terms [6,15] to our imprecise data and fuzzy terms - the membership degree is defined as the integral of the product of the density of the imprecise distribution with the membership function of the fuzzy term. If f is the density of the distribution representing the imprecision associated with the proportion of an element e in an example x, and $\mu_v(t)$ the membership degree of the value t to the fuzzy term v, the membership degree of the imprecise example x to the fuzzy term v is:

$$\tilde{\mu}_v(x) = \int_0^{100} f(t)\mu_v(t)dt$$

This integral, illustrated in Fig. 6 where the imprecision distribution is dilated by a factor 10 for a better visibility, takes into account the whole continuous spectrum of probable values for the ratio of e in x, and is basically the weighted average of the membership degrees of these values to the fuzzy term n.

Given the membership degree of an imprecise example x to the fuzzy terms corresponding to each chemical element, we can define the membership degree of x to each node of the tree. To do so, we chose the product as a T-norm, because it gave better results than other T-norms, like the minimum. Hence, the membership degree of an example x to a node n is the product of the membership degrees of x to the fuzzy terms associated to n and to the ascendants of n. Thus, we have the recursive definition:

$$\begin{cases} d_{n_0}(x) = 1 \\ d_n(x) = d_N(x) \times \tilde{\mu}_v(x) \end{cases} \tag{2}$$

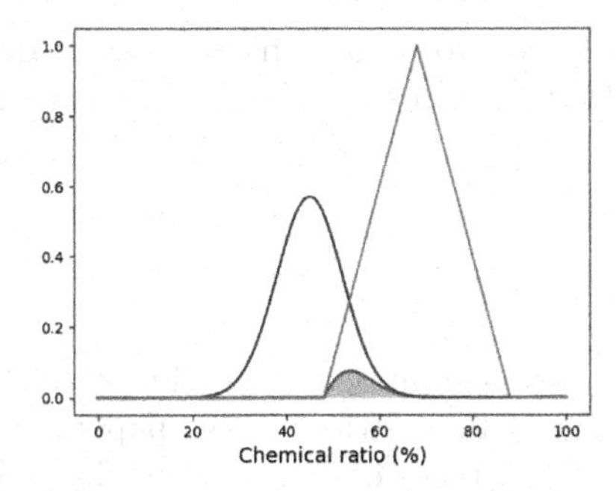

Fig. 6. Integral of the product imprecision-fuzzy term

where n_0 is the root of the tree, N is n's parent node and v is the fuzzy term associated to the node n.

To select which feature will divide the current node, we use the fuzzy entropy introduced by Peng and Flach in [14].

We consider the "fuzzy frequency" of a class c in a node n:

$$fr_{c/n} = \frac{\sum_{x_i \in c} d_n(x_i)}{\sum_{x_i} d_n(x_i)} \tag{3}$$

where $\{x_i,\ i \in [|1,T|]\}$ is the training set, and the "membership frequency" of the examples to the node n: $fr_n = \frac{\sum_{x_i} d_n(x_i)}{\sum_{n_j} \sum_{x_i} d_{n_j}(x_i)}$ where $\{n_j,\ j \in [|1,F|]\}$ is the set of sibling nodes of n, including n.

The fuzzy entropy, according to an element e, used at a node N having the children $\{n_j,\ j \in [|1,F|]\}$, these nodes corresponding to each fuzzy term relative to e, is then defined as:

$$E = -\sum_{n_j} fr_{n_j} \sum_{c_k} fr_{c_k/n_j} \log(fr_{c_k/n_j}) \tag{4}$$

where $\{c_k,\ k \in [|1,K|]\}$ is the set of the classes of the problem.

When the tree is built, we create a rule for each of its leaves: the premise is the conjunction of the fuzzy terms associated with the nodes leading from the root to the leaf, and the consequence is the most represented class in the leaf.

3.3 Recognition of New Samples

Once the tree has been constructed using the training data, the classification of the testing data is quite simple. Given an imprecise example x, we compute the membership degree of x to each leaf l of the tree, using the very same definition as for the training data (Eq. 2). For each class c, this membership degree is multiplied by the fuzzy frequency $fr_{c/l}$ of c conditionally to l (see definition 3), using the same formula as in [3], $fr_{c/l}$ being in fact the certainty factor defined in [3]. Using the weighted voting method [10], these results are summed up for each class, over all the leaves, to obtain the confidence degree $conf(x \in c)$ of $x \in c$. The algorithm classifies x in the class c maximizing $conf(x \in c)$.

4 Experiments

4.1 Data Simulation

Since the physical experiments supposed to provide us with real data could not be realized at the moment of the writing of this paper, we simulated data. We used the stoichiometric percentages of each element in each material (class) as reference values for each pair (class, element). To create an example x of a class c, a value m is randomly generated for each element e in an interval around the reference value of the pair (c, e), whose span is proportional to the reference value, the proportionality coefficient being called *degree of imprecision* in the following. The mean of the Gaussian distribution representing the imprecise proportion of e in x is set to m. A standard deviation is then randomly generated in an interval whose span is proportional to the one of the interval used to generate the mean.

To generate the data, we used the chemical formulæ of 17 explosives and 9 drugs, with a degree of imprecision of 15%. Since the real data will be few due to the financial and temporal costs of the physical experiments, we generated only ten examples per class for each data set. In the following, we present the results of five-fold cross-validations averaged on ten different data sets.

4.2 Comparison of Fuzzy Partitioning Algorithms

To study the influence of the size of the fuzzy partitions on the accuracy, we ran tests with k-medoids algorithm using the previously defined dissimilarity (see definition 1). We imposed five-term partitions for all the elements except carbon (C), nitrogen (N) and oxygen (O) which are the most discriminative elements in our problem. For C, N and O, we tested all the combinations of numbers of terms between 2 and 14. Figure 7 shows the correct classification rate as a function of the numbers of fuzzy terms partitioning C, N and O ratio domains. To display this dependence in a 3D graph, we merge N and O features on one axis. Each test with n_C, n_N and n_O fuzzy terms, for C, N and O respectively, and whose correct classification rate is r, is represented on the graph by the point of coordinates $(n_C, n_O + 14n_N, r)$. Thanks to this representation, two different tests will not be represented by the same point. We then link into a surface the points corresponding to a same value of n_N. In order to let a gap between two consecutive surfaces for better visibility, we multiply n_N by 14 and not 13 which would have been sufficient to have an injective representation.

The accuracy increases with the number of fuzzy sets, until a certain threshold of about 12 for N and O, and 6 for C, which is surprisingly low. The best results are obtained with values between 5 and 7 for C, and between 11 and 14 for N and O - several of these combinations give classification rates around 84%.

We then tested the clustering algorithm called affinity propagation [8], using the similarity associated with the previously defined dissimilarity: $s(x, y) =$

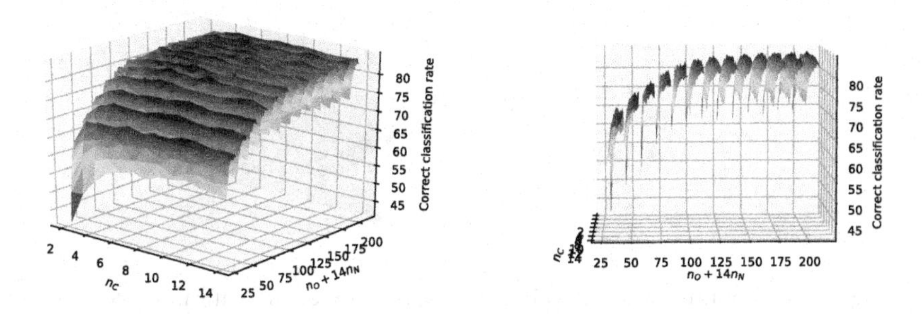

Fig. 7. Accuracy against number of fuzzy terms (two views of the same graph)

$\int_a^b min(f_x(t), f_y(t))dt$. Affinity propagation does not take the number of clusters (thus fuzzy terms) in parameter but a parameter called preference that increases the number of clusters when it is high. Giving the same preference to the clustering for the partitions of C, N and O ratios give worse results than the best obtained with k-medoids. But in the light of k-medoids tests, we tried to give smaller preference to C partition and greater for N and O partitions, and results similar to the best ones of k-medoids were obtained (around 84%). We can conclude that k-medoids algorithm is preferable for this problem since it gives equally good results but the parameter to be optimized - the list of the numbers of fuzzy terms - is discrete whereas affinity propagation preference is continuous.

In order to improve the robustness of our method, it would have been interesting to split the training set in two parts: one for the fuzzy sets learning and one to the tree induction. But since our data are very few, we decided to learn the fuzzy sets and the tree from the whole training set.

4.3 Results Analysis

From the grid-based test over the combinations of numbers of fuzzy sets for C, N and O, we determined the best combination in terms of accuracy: respectively 5, 14 and 12 terms for C, N and O with an accuracy of 84.92%. The confusion matrix in Fig. 9 summarizes the results of the classification using k-medoids with this optimal combination of numbers of fuzzy sets regarding C, N and O, and 5 fuzzy terms for the other elements. The algorithm classes quite correctly both the explosives (seventeen first classes from the top in ordinate) and the drugs (nine last classes). This method represents an improvement compared to the classic algorithms which do not take the imprecision of the data into account, this difference being higher for more imprecise simulated data (see Fig. 8).

We ran the method on a data set composed of the same classes as before and 12 classes of benigns (38 classes in total). With the same numbers of fuzzy sets, we obtained an accuracy of 75% (see the confusion matrix on Fig. 10).

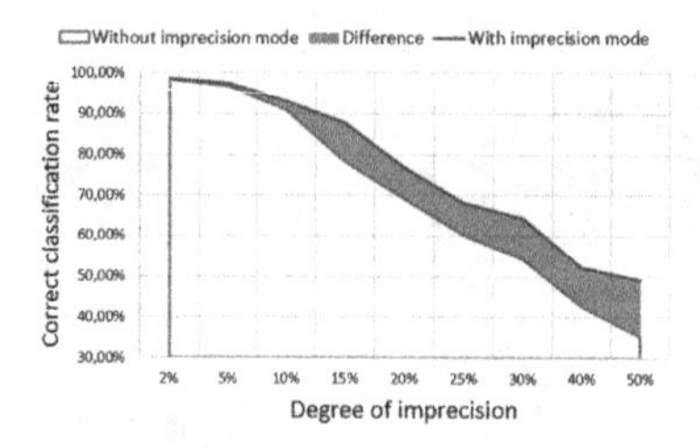

Fig. 8. Gain obtained when taking the imprecision of the data into account

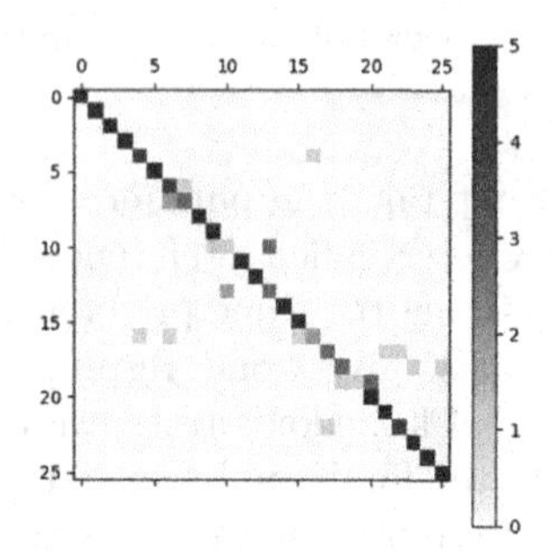

Fig. 9. Confusion matrix for drugs and explosives

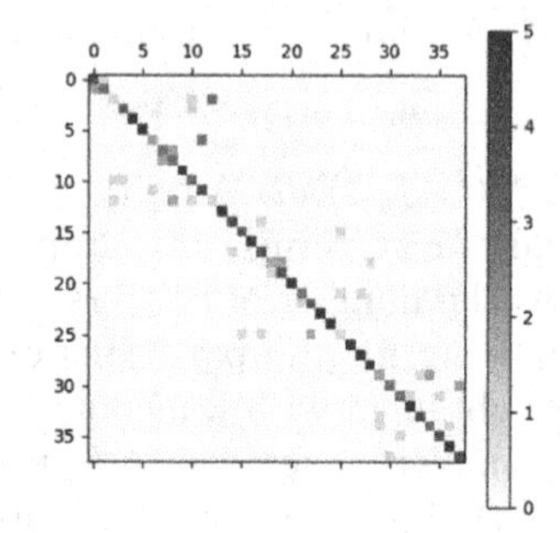

Fig. 10. Confusion matrix for drugs, explosives and benigns

5 Conclusion

In this paper, we introduced a workflow for explainable material recognition from their chemical composition and based on both decision tree induction and linguistic variables learning. The method takes into account the imprecision of data during the learning phase and the exploitation of the decision tree. The gain obtained by the imprecision treatment is higher for high imprecision. Using k-medoids for the fuzzy partition, the accuracy is up to 84% for a 26-class problem.

We have several opportunities to enhance our method's accuracy in further work. We may improve our clustering technique taking the samples' labels into account. Using pruning techniques but also random forests could make the method more accurate. We will also study more sophisticated techniques than the sum we are using for aggregating the results of each leaf of the tree. We might generate simpler rules corresponding to intern nodes of the tree. We shall also test the possibilistic approach with other measures more adapted to it.

At this stage of our work, the method has been evaluated on simulated data while we are performing real data acquisition on pure elements, chemical products, drug and explosive simulants. This will be a first step since in real life container voxels, several materials will be present. We will then have to deal with recognition of mixtures and not only mono-material samples.

Acknowledgment. This research has been funded by the project H2020 C-BORD. We warmly thank S. Moretto, C. Fontana, F. Pino, A. Sardet, C. Carasco, B. Pérot and V. Picaud for their contributions before our work, for their availability and their expertise.

References

1. Akiyama, T., Inokuchi, H.: Application of fuzzy decision tree to analyze the attitude of citizens for wellness city development. In: 2016 17th International Symposium on Advanced Intelligent Systems. IEEE, August 2016
2. Aupetit, M., Allano, L., Espagnon, I., Sannié, G.: Visual analytics to check marine containers in the eritr@c project. In: Proceedings of International Symposium on Visual Analytics Science and Technology, pp. 3–4 (2010)
3. Bounhas, M., Prade, H., Serrurier, M., Mellouli, K.: A possibilistic rule-based classifier. In: Greco, S., Bouchon-Meunier, B., Coletti, G., Fedrizzi, M., Matarazzo, B., Yager, R.R. (eds.) IPMU 2012. CCIS, vol. 297, pp. 21–31. Springer, Heidelberg (2012). https://doi.org/10.1007/978-3-642-31709-5_3
4. Chang, R.L.P., Pavlidis, T.: Fuzzy decision tree algorithms. IEEE Trans. Syst. Man Cybern. **7**(1), 28–35 (1977)
5. Coppi, R., D'Urso, P.: Fuzzy time arrays and dissimilarity measures for fuzzy time trajectories. In: Kiers, H.A.L., Rasson, J.P., Groenen, P.J.F., Schader, M. (eds.) Data Analysis, Classification, and Related Methods, pp. 273–278. Springer, Heidelberg (2000). https://doi.org/10.1007/978-3-642-59789-3_44
6. Duch, W.: Uncertainty of data, fuzzy membership functions, and multilayer perceptrons. IEEE Trans. Neural Netw. **16**(1), 10–23 (2005)
7. D'Urso, P., de Giovanni, L.: Robust clustering of imprecise data. Chemometr. Intell. Lab. Syst. **136**, 58–80 (2014)
8. Frey, B.J., Dueck, D.: Clustering by passing messages between data points. Science **315**(5814), 972–976 (2007)
9. Gullo, F., Ponti, G., Tagarelli, A.: Clustering uncertain data via K-medoids. In: Greco, S., Lukasiewicz, T. (eds.) SUM 2008. LNCS (LNAI), vol. 5291, pp. 229–242. Springer, Heidelberg (2008). https://doi.org/10.1007/978-3-540-87993-0_19
10. Ishibuchi, H., Nakasima, T., Morisawa, T.: Voting in fuzzy rule-based systems for pattern classification problems. Fuzzy Sets Syst. **103**, 223–238 (1999)
11. Janikow, C.Z.: Fuzzy decision trees: issues and methods. IEEE Trans. Syst. Man Cybern. - Part B: Cybern. **28**(1), 1–14 (1998)
12. Kaufman, L., Rousseeuw, P.J.: Clustering by means of medoids. In: Dodge, Y. (ed.) Statistical data analysis based on the L_1 norm and related methods, pp. 405–416 (1987)
13. Olaru, C., Wehenkel, L.: A complete fuzzy decision tree technique. Fuzzy Sets Syst. **138**, 221–254 (2003)
14. Peng, Y., Flach, P.A.: Soft discretization to enhance the continuous decision tree induction. Integrating Aspects of Data Mining, Decision Support and Meta-Learning (2001)
15. Tsang, S., Kao, B., Yip, K., Ho, W., Lee, S.: Decision trees for uncertain data. IEEE Trans. Knowl. Data Eng. (1), 64–78 (2011)

Tell Me Why: Computational Explanation of Conceptual Similarity Judgments

Davide Colla, Enrico Mensa, Daniele P. Radicioni(✉), and Antonio Lieto

Dipartimento di Informatica, Università di Torino, Turin, Italy
{colla,mensa,radicion,lieto}@di.unito.it

Abstract. In this paper we introduce a system for the computation of explanations that accompany scores in the conceptual similarity task. In this setting the problem is, given a pair of concepts, to provide a score that expresses in how far the two concepts are similar. In order to explain how explanations are automatically built, we illustrate some basic features of COVER, the lexical resource that underlies our approach, and the main traits of the MeRaLi system, that computes conceptual similarity and explanations, all in one. To assess the computed explanations, we have designed a human experimentation, that provided interesting and encouraging results, which we report and discuss in depth.

Keywords: Explanation · Lexical semantics
Natural language semantics · Conceptual similarity · Lexical resources

1 Introduction

In the Information Age an ever-increasing number of text documents are being produced over time [3]; herein, the growth of the Web and the tremendous spread of social networks exert a strong pressure on Computational Linguistics to refine methods and approaches in areas such as Text Mining and Information Retrieval.

One chief feature for systems being proposed in these areas would be that of providing some sort of explanation on the ways their output was attained, so to both unveil the intermediate steps of the computation process and to justify the obtained results. Different kinds of explanation can be drawn, ranging from argumentation based approaches [23] to inferential processes triggered in formal ontologies categorisation [15]. Almost since its inception, explanation has involved expert systems and dialogue systems. In particular, the pioneering research on knowledge-based systems and decision support systems revealed that in many tasks of problem-solving "when experts disagree, it is not easy to identify the 'right answer' [...]. In such domains, the process of argumentation between experts plays a crucial role in sharing knowledge, identifying inconsistencies and focusing attention on certain areas for further examination" [21]. Explanation is thus acknowledged to be an epistemologically relevant process, and a precious feature to build robust and informative systems.

J. Medina et al. (Eds.): IPMU 2018, CCIS 853, pp. 74–85, 2018.
https://doi.org/10.1007/978-3-319-91473-2_7

Within the broader area of Natural Language Semantics, we single out Lexical Semantics (that is, the study of word meanings and their relationships), and illustrate how COVER [18]—a resource developed in the frame of a long-standing attempt at combining ontological and commonsense reasoning [7,13]—can be used to the ends of building simple explanations that may be beneficial in the computation of conceptual similarity. COVER, so named after 'COmmonsense VEctorial Representation',[1] is a lexical resource resulting from the blend of BabelNet [22], NASARI [2] and ConceptNet [9]. As a result COVER combines, in a synthetic and cognitively grounded way, lexicographic precision and common sense aspects. We presently consider the task of automatically assessing the conceptual similarity (that is, given a pair of concepts the problem is to provide a score that expresses in how far the two concepts are similar), meantime providing an explanation to the score. This task has many relevant applications, in that identifying the proximity of concepts is helpful in various tasks, such as documents categorisation [25], conceptual categorisation [14], keywords extraction [4,16], question answering [8], text summarization [10], and many others. The knowledge representation adopted in COVER allows a uniform access to concepts via BabelNet synset IDs. The resource relies on a vector-based semantic representation which is, as shown in [12], also compliant with the Conceptual Spaces, a geometric framework for common-sense knowledge representation and reasoning [5].

In this paper we show that COVER, which is different in essence from all previously existing lexical resources, can be used to build explanations accompanying the similarity scores. To the best of our knowledge, COVER is the only lexical resource that 'natively' produces explanations: after a brief introduction of the resource (Sect. 2), we illustrate the main traits of the MeRaLi system, that has been designed to compute the conceptual similarity [17], and presently extended to build explanations (Sect. 3). We then illustrate the experimentation we conducted to assess the quality of the produced explanations (Sect. 4). Although the experimentation is a preliminary one, the automatic explanation has been found acceptable in many cases by the participants we interviewed. Additionally, formulating explanations seems to trigger some subtle though significant variation in the similarity judgement w.r.t. the condition in which no explanation is required, thus confirming the relevant role of the explanation in many complex tasks.

2 The COVER Lexical Resource

Let us start by quickly recalling the COVER resource [12,18]. COVER is a list of vectors, each representing and providing knowledge about a single concept. The representation of concepts rather then just terms requires the adoption of a set of concept identifiers (so to define a uniform naming space), and COVER relies on the sense inventory provided by BabelNet [22].

[1] COVER is available for download at http://ls.di.unito.it.

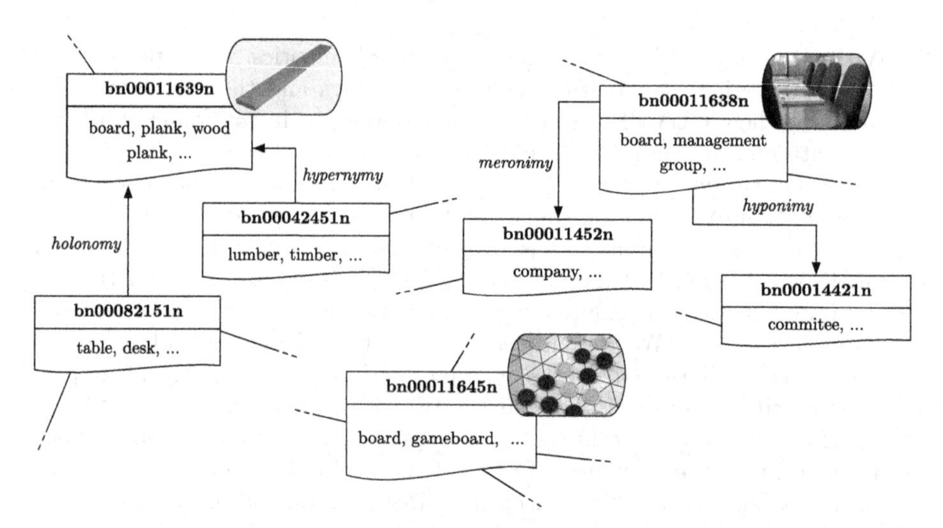

Fig. 1. A portion of BabelNet representing the possible meanings for the term *board.* Each meaning is represented as a synset, which is in turn identified uniquely by its own BabelNet synset ID. Synsets are connected via named semantic relationships.

BabelNet is a *semantic network* in which each node –called synset, that is 'set of synonyms', as originally conceived in WordNet [19]– represents a unique meaning, identified through a BabelNet synset ID (e.g., BN:00008010N). Furthermore, each node provides a list of multilingual terms that can be used in order to express that meaning. The synsets in BabelNet are also connected via a set of semantic relationships such as hyponymy, homonymy, meronymy, *etc.* As anticipated, COVER adopts BabelNet synset identifiers in order to uniquely refer to concepts and their attached vectors. Figure 1 illustrates an excerpt of the BabelNet graph, focusing on the different meanings underlying the term *board.*

The conceptual information borrowed from BabelNet has been coupled to common-sense knowledge, that has been extracted from ConceptNet [9]. ConceptNet is a network of terms and compound words that are connected via a rich set of relationships.[2] As an example, Fig. 2 shows the ConceptNet node for the term *board.*

The ConceptNet relationships have been set as the skeleton of the vectors in COVER, that is the set of D dimensions upon which a vector describes

[2] INSTANCEOF, RELATEDTO, ISA, ATLOCATION, DBPEDIA/GENRE, SYNONYM, DERIVEDFROM, CAUSES, USEDFOR, MOTIVATEDBYGOAL, HASSUBEVENT, ANTONYM, CAPABLEOF, DESIRES, CAUSESDESIRE, PARTOF, HASPROPERTY, HASPREREQUISITE, MADEOF, COMPOUNDDERIVEDFROM, HASFIRSTSUBEVENT, DBPEDIA/FIELD, DBPEDIA/KNOWNFOR, DBPEDIA/INFLUENCEDBY, DBPEDIA/INFLUENCED, DEFINEDAS, HASA, MEMBEROF, RECEIVESACTION, SIMILARTO, DBPEDIA/INFLUENCED, SYMBOLOF, HASCONTEXT, NOTDESIRES, OBSTRUCTEDBY, HASLASTSUBEVENT, NOTUSEDFOR, NOTCAPABLEOF, DESIREOF, NOTHASPROPERTY, CREATEDBY, ATTRIBUTE, ENTAILS, LOCATIONOFACTION, LOCATEDNEAR.

board

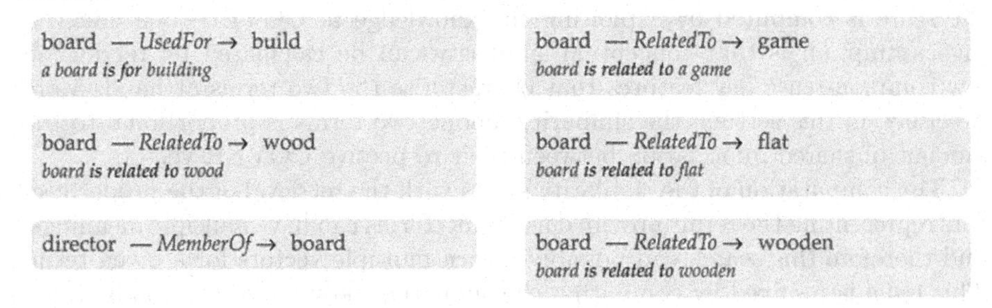

Fig. 2. Example of the ConceptNet node for the term *board*. The common-sense knowledge is encoded via a series of connections to other terms.

the represented concept. More precisely, each vector dimension contains a set of values that are concepts themselves, identified through their own BabelNet synset IDs. So a concept c_i has a vector representation $\vec{c}_i$ that is defined as

$$\vec{c}_i = [s_1^i, \ldots, s_N^i]. \tag{1}$$

Namely, each s_h^i is the set of concepts filling the dimension $d_h \in D$. Each s can either contain an arbitrary number of values, or be empty.

For instance, the concept *headmaster* (BN:00043259N) is represented in COVER by a vector that has nine filled dimensions (RelatedTo, IsA, HasContext, SimilarTo, Antonym, DerivedFrom, AtLocation, Synonym, FormOf), and therefore it has nine non-empty sets of values (Fig. 3).

Exemplar BN:00043259N (head, headmaster)

BN:00043259N$_{\text{RelatedTo}}$ = [prefect, college, rector, teacher, university, ...]
BN:00043259N$_{\text{IsA}}$ = [educator, head teacher]
BN:00043259N$_{\text{AtLocation}}$ = [school]
BN:00043259N$_{\text{Antonym}}$ = [student]
...

Fig. 3. A portion of the COVER vector for the *headmaster* concept. The values filling the dimensions are concepts identifiers (BabelNet synset IDs); for the sake of the readability they have been replaced with their corresponding terms.

3 Computing Conceptual Similarity

In order to compute the conceptual similarity, we designed the MeRaLi system [17]. In the conceptual similarity task, the system is provided with a pair of terms and it is required to provide a score of similarity between the two. Since the score is computed by exploiting the knowledge in COVER, one underlying assumption is that conceptual similarity can be calculated by relying on few common-sense key features that characterise the two terms at hand. More precisely, in this setting, the similarity among two terms is proportional to the amount of shared information between their respective COVER vectors.

The computation of the similarity starts with the retrieval of the proper vectors representing the terms provided as input. Terms can have multiple meanings, and therefore this search can possibly return multiple vectors for a given term. This issue is resolved by computing the similarity between all the combination of pairs of retrieved vectors, and then by choosing the highest similarity score: that is, given two terms w_1 and w_2, each with an associated list of senses $s(w_1)$ and $s(w_2)$, we compute

$$\mathrm{sim}(w_1, w_2) = \max_{\vec{c}_1 \in s(w_1), \vec{c}_2 \in s(w_2)} [\mathrm{sim}(\vec{c}_1, \vec{c}_2)] . \tag{2}$$

The similarity computation can be formally expressed as follows: given two input terms t_i and t_j, the corresponding COVER vectors $\vec{c}_i$ and $\vec{c}_j$ are retrieved. The similarity is then calculated by counting, dimension by dimension, the set of values (concepts) that $\vec{c}_i$ and $\vec{c}_j$ have in common. The scores obtained upon every dimension are then combined, thus obtaining an overall similarity score, that is our final output. So, given N dimensions in each vector, the similarity value

$$\mathrm{sim}(\vec{c}_i, \vec{c}_j)$$

is computed as

$$\mathrm{sim}(\vec{c}_i, \vec{c}_j) = \frac{1}{N} \sum_{k=1}^{N} |s_k^i \cap s_k^j|. \tag{3}$$

MeRaLi actually employs a more sophisticated formula in order to account for the possibility that the two COVER vectors may present very unequal amounts of information. Specifically, the similarity within each dimension is computed by means of the Symmetrical Tversky's Ratio Model [11], which is a symmetrical reformulation for the Tversky's ratio model [26],

$$\mathrm{sim}(\vec{c}_i, \vec{c}_j) = \frac{1}{N^*} \cdot \sum_{k=1}^{N^*} \frac{|s_k^i \cap s_k^j|}{\beta \left(\alpha a + (1-\alpha)\, b\right) + |s_k^i \cap s_k^j|} \tag{4}$$

where $|s_k^i \cap s_k^j|$ counts the number of shared concepts that are used as fillers for the dimension d_k in the concept $\vec{c}_i$ and $\vec{c}_j$, respectively; a and b are defined as $a = \min(|s_k^i - s_k^j|, |s_k^j - s_k^i|)$, $b = \max(|s_k^i - s_k^j|, |s_k^j - s_k^i|)$; finally N^* counts the dimensions actually filled with at least two concepts in both vectors. This formula allows tuning the balance between cardinality differences (through the parameter α), and between $|s_k^i \cap s_k^j|$ and $|s_k^i - s_k^j|, |s_k^j - s_k^i|$ (through the parameter β).[3]

[3] The parameters α and β were set to .8 and .2 for the experimentation, based on a parameter tuning performed on the RG, MC and WS-Sim datasets [17].

Example: computation of the similarity between atmosphere and ozone. As an example, we report the similarity computation between the concepts *atmosphere* and *ozone*. Firstly, the COVER resource is searched in order to find vectors suitable for the representation of the two terms. The best fit resulted to be the pair of concepts ⟨BN:00006803N, BN:00060040N⟩. The similarity was then computed on a scale [0, 1] by adopting Eq. 4, and lately mapped onto the range [0, 4]. The final similarity score was 00.63, (converted to 2.52). The gold standard for this pair of terms was instead 2.58 [1]. Figure 4 illustrates the comparison table between the two vectors selected for the computation. □

Similarity calculation for 'atmosphere' (bn:00006803n) and 'ozone' (bn:00060040n - ozone).

VDimension name	Sim	V1-V2 count	Shared	Direct	Values
InstanceOf	00.00	[000 | 000]	0	-	-
RelatedTo	00.57	[107 | 021]	8	✓	stratosphere, air, ozone, atmosphere layer, atmosphere, oxygen, gas
IsA	00.49	[004 | 005]	1	✓	gas
AtLocation	00.00	[001 | 001]	0	-	-
DBP_Genre	00.00	[000 | 000]	0	-	-
Synonym	00.00	[004 | 001]	0	-	-
DerivedFrom	00.00	[001 | 000]	0	-	-
Causes	00.00	[000 | 000]	0	-	-
UsedFor	00.00	[000 | 000]	0	-	-
MotivatedByGoal	00.00	[000 | 000]	0	-	-
HasSubevent	00.00	[000 | 000]	0	-	-
Antonym	00.00	[000 | 000]	0	-	-
CapableOf	00.00	[000 | 000]	0	-	-
Desires	00.00	[000 | 000]	0	-	-
CausesDesire	00.00	[000 | 000]	0	-	-
PartOf	00.00	[003 | 000]	0	-	-
HasProperty	00.00	[000 | 000]	0	-	-
HasPrerequisite	00.00	[000 | 000]	0	-	-
MadeOf	00.00	[000 | 000]	0	-	-
CompoundDerivedFrom	00.00	[000 | 000]	0	-	-
HasFirstSubevent	00.00	[000 | 000]	0	-	-
DBP_Field	00.00	[000 | 000]	0	-	-
DBP_KnownFor	00.00	[000 | 000]	0	-	-
influencedBy	00.00	[000 | 000]	0	-	-
DefinedAs	00.00	[000 | 000]	0	-	-
HasA	00.00	[007 | 000]	0	-	-
MemberOf	00.00	[000 | 000]	0	-	-
ReceivesAction	00.00	[000 | 000]	0	-	-
SimilarTo	00.00	[000 | 000]	0	-	-
SymbolOf	00.00	[000 | 000]	0	-	-
HasContext	00.83	[002 | 002]	1	✓	chemistry
NotDesires	00.00	[000 | 000]	0	-	-
ObstructedBy	00.00	[000 | 000]	0	-	-
HasLastSubevent	00.00	[000 | 000]	0	-	-
NotUsedFor	00.00	[000 | 000]	0	-	-
NotCapableOf	00.00	[000 | 000]	0	-	-
DesireOf	00.00	[000 | 000]	0	-	-
NotHasProperty	00.00	[000 | 000]	0	-	-
CreatedBy	00.00	[000 | 000]	0	-	-
Attribute	00.00	[000 | 000]	0	-	-
Entails	00.00	[000 | 000]	0	-	-
LocationOfAction	00.00	[000 | 000]	0	-	-
LocatedNear	00.00	[000 | 000]	0	-	-
FormOf	00.00	[001 | 000]	0	-	-

Fig. 4. Log of the comparison between the concepts *atmosphere* and *ozone* in MERALI. The 'V1-V2 count' column reports the number of concepts for a certain dimension in the first and second vector, respectively; the column 'Shared' indicates how many concepts are shared in the two conceptual descriptions along the same dimension; and the column 'Values' illustrates (the nominalization of) the concepts actually shared along that dimension.

Explaining Conceptual Similarity

The score of similarity provided by a system can often seem like an obscure number. It is difficult to demonstrate on which accounts two concepts are similar, especially if the score computation relies on complex networks or synthesised representations. However, thanks to the fact that COVER vectors contain explicit and human-readable knowledge, the explanation of the score is in this case allowed. Specifically, the COVER vectors adopted by the MERALI system provide human-readable features that are compared in order to obtain a similarity score. The explanation for this score can thus be obtained by simply reporting which values were a match in the two compared vectors. Ultimately, a simple Natural Language Generation approach has been devised: at this stage, a template is filled with the features in common between the two vectors, dimension by dimension (please refer to Fig. 4). For instance, the explanation for the previously introduced example, can be directly obtained by extracting the shared values among the two considered vectors, thus obtaining:

`The similarity between` *`atmosphere`* `[bn:00006803n] and` *`ozone`* `[bn:00060040n] is 2.52 because they are` *`gas`*`; they share the same context` *`chemistry`*`; they are related to` *`stratosphere, air, atmosphere, layer, ozone, atmosphere, oxygen, gas`*`.`

4 Experimentation

The experimentation is a preliminary pilot study, aimed at assessing the quality of the explanations. Since the language generation process itself is less relevant in the present approach, we focus on the content of the explanation rather than on the linguistic realisation.

4.1 Experimental Design

Overall 40 pairs of terms were randomly selected from the data-set designed for the shared task 'SemEval-2017 Task 2: Multilingual and Cross-lingual Semantic Word Similarity' [1] (Table 1).[4] Such pairs have been arranged into 4 questionnaires, that were administered to 33 volunteers, aged from 20 to 23. All recruited subjects were students from the Computer Science Department of the University of Turin (Italy); none of them was an English native speaker.

Questionnaires were split into 3 main sections:

- in the *task* 1 we asked the participants to assign a similarity score to 10 term pairs (in this setting, scores are continuous in the range [0, 4], as it is customary in the international shared tasks on conceptual similarity [1]);

[4] Actually the pair ⟨*mojito,mohito*⟩ was dropped in that 'mojito' was not recognised as a morphological variant of 'mohito' by most participants.

Table 1. The pairs of terms employed in each questionnaire, referred to as Q1–Q4.

Q1	desert, dune videogame, pc game	palace, skyscraper medal, trainers	mojito, mohito butterfly, rose	city center, bus Wall Street, financial market	beach, coast Apple, iPhone
Q2	lizard, crocodile demon, angel	sculpture, statue income, quality of life	window, roof underwear, body	agriculture, plant Boeing, plane	flute, music Caesar, Julius Caesar
Q3	basilica, mosaic myth, satire	snowboard, skiing sodium chloride, salt	pesticide, pest coach, player	level, score Zara, leggings	snow, ice Cold War, Soviet Union
Q4	car, bicycle digit, number	democracy, monarchy coin, payment	pointer, slide surfing, water sport	flag, pole Harry Potter, wizard	lamp, genie Mercury, Jupiter

- in the *task* 2 we asked them to explain in how far the two terms at stake were similar, and then to indicate a new similarity score (either the same or different) to the same 10 pairs as above;
- in the *task* 3 the subjects were given the automatically computed score along with the explanation built by our system. They were requested to evaluate the explanation by expressing a score in a [0, 10] Likert scale, and also to provide some comments on missing/wrong arguments, collected as open text comments.

Each volunteer compiled one questionnaire (containing 10 term pairs), which on average took 20 min.

4.2 Results and Discussion

The focus of the present experimentation was the assessment of the automatically computed explanations (addressed in the *task* 3): MeRaLi's explanations obtained, on average, the score of 6.62 (standard deviation: 1.92). Our explanations and the scores computed automatically have been overall judged to be reasonable.

By examining the 18 pairs that obtained an averaged poor score (≤ 6), we observe that either few information was available, or it was basically wrong. Regarding the first case, we counted 12 pairs with only one or two shared concepts (please refer to Eqs. 3 and 4): almost always these explanations were evaluated with low scores (on average, 4.48). We found only one notable exception about the pair ⟨*Boeing*, *plane*⟩ whose explanation was

The similarity between *Boeing* and *plane* is 2.53 because they are related to *airplane*, *aircraft*.

This explanation obtained an average score of 8.63. We hypothesise that this greater appreciation is due to the fact that even if only two justifications are provided, they match the most salient (based on common-sense[5] accounts) traits

[5] We refer to common-sense as to a portion of knowledge that is both widely accessible and elementary [20], and reflecting typicality traits encoded as prototypical knowledge [24].

Table 2. Correlation between the similarity scores provided by the subjects interviewed and the scores in the Gold standard. The bottom row shows the correlations between the scores gold standard and the scores computed by our system

	Spearman's ρ	Person's r
Gold - avg scores (task 1)	0.83	0.82
Gold - avg scores (task 2)	0.85	0.83
COVER - avg scores (task 1)	0.71	0.72
COVER - avg scores (task 2)	0.72	0.73
Gold - COVER	0.79	0.78

between the two considered concepts. It would seem thus that the quality of a brief explanation heavily depends on the presence of those particular and meaningful traits. In the remaining 6 pairs, vice versa, there is enough though wrong information, possibly due to the selection of the wrong meaning for input terms. In either cases, we observe that the resource still needs being improved for what pertains its coverage and the quality of the hosted information (since it is automatically built by starting from BabelNet, it contains all noise therein). This is the target of our present and future efforts.

The first and second task in the questionnaire can be thought of as providing evidence to support the result in the third one. In particular, the judgements provided by the volunteers closely approach the scores in the gold standard, as it is shown by the high (over 80%) Spearman's (ρ) and Person's (r) correlations (Table 2). The first two rows show the average agreement between the scores *before* producing an explanation for the score itself (Gold - avg scores (*task* 1)), and *after* providing an explanation (Gold - avg scores (*task* 2)). These figures show that even human judgement can benefit from producing explanations, as the scores in *task* 2 showcase a higher correlation with the gold standard scores. Additionally, the output of the system exhibits a limited though significantly higher correlation with the similarity scores provided after trying to explain the scores themselves (COVER - avg scores (*task* 1) condition *vs.* COVER - avg scores (*task* 2)).

In order to further assess our results we also performed a qualitative analysis on some spot cases. For the pair $\langle Mercury, Jupiter \rangle$ the MERALI system computed a semantic similarity score of 2.29 (the gold standard score was 3.17), while the average score indicated by the participants was 3.43 (*task* 1) and 3.29 (*task* 2). First of all, this datum corroborates our approach (Sect. 3) that computes the similarity between the closest possible senses (please refer to Eq. 2): it never happened that any participant raised doubts on the meaning of Mercury (always intended as the planet), whilst *Mercury* can be also a metallic chemical element, a Roman god, the Marvel character who can turn herself into mercury, and several further entities.

The open text comments report explanations such as that Mercury and Jupiter are *'both planets, though different'*. In this case, the participants acknowledge that the two entities at stake are planets but rather different (e.g., the first one is the smallest planet in the Solar System, whilst the second one is the largest). The explanation provided by our system is:

> **The similarity between Mercury and Jupiter is 2.29 because they are *planet*; they share the same context *deity*; they are semantically similar to *planet*; they are related to *planet*, *Roman_deity*, *Jupiter*, *deity*, *solar_System*.**

In this case, our explanation received an average score of 9.57 out of 10. Interestingly enough, even though the participants indicated different similarity scores, they assigned a high quality score to our explanation, thus showing that it is basically reasonable.

As a second example we look at the pair ⟨*myth*, *satire*⟩. The similarity score and the related explanation of such terms are:

> **The similarity between *myth* and *satire* is 0.46 because they are *aggregation*, *cosmos*, *cognitive_content*; they are semantically similar to *message*; they form *aggregation*, *division*, *message*, *cosmos*, *cognitive_content*.**

In this case, the gold standard similarity value was 1.92, the average scores provided by the participants 1.57 (*task* 1) and 1.71 (*task* 2). Clearly, the explanation was not satisfactory, and it was rated 4.49 out of 10. The participants gave no clear explanation about their judgement (in *task* 2) nor informative comments/-criticisms on the explanation above (in *task* 3). One possible reason behind the poor assessment might be found in the interpretation of the *satire* term. If we consider satire as the ancient literary genre where characters are ridiculed, the explanation becomes more coherent: they are forms of *aggregation* as it was for any sort of narrative in the ancient (mostly Latin) culture; they also both deliver some message, either explaining some natural or social phenomenon and typically involving supernatural beings (like myth), or criticising people's vices, particularly in the context of contemporary politics (like satire). This possible meaning has been considered only by 2 out of 8 participants, that mostly intended satire as a generic ironic sort of text. Even in this case, where the output of MERALI was rather unclear and questionable, the explanation shows some sort of coherence, although not immediately sensible for human judgement. In such cases, by resorting to an inverse engineering approach, the explanation can be used to figure out which senses (underlying the terms at hand) are actually intended.

5 Conclusions

In this paper we have illustrated how COVER can be used to build explanations for the conceptual similarity task. Furthermore, we have shown that in our approach two concepts are similar insofar as they share values on the same dimension, such as when they share the same function, parts, location, synonyms, prerequisites, and so forth; this approach is intrinsically ingrained with explanation, to such an extent that building an explanation in MeRaLi simply amounts to listing the elements actually used in the computation of the conceptual similarity score. We have then reported the experimental results obtained in a test involving human subjects over a data-set devised for an international challenge on semantic word similarity: the participants were requested to provide conceptual similarity scores, to produce explanations, and to assess the explanations computed through MeRaLi. The experimentation provided interesting and encouraging results, basically showing that when the COVER lexical resource has enough information on the concepts at hand it produces reasonable explanations. Moreover, the experimentation suggested that explanation can be beneficial also for human judgements, that tend to be more accurate (more precisely: statistically correlated to gold standard scores) after having produced explanation to justify some score in the conceptual similarity task. Such results confirm that systems for building explanations can be useful in many other semantics-related tasks, where it may be convenient (if necessary) to shepherd results and their justification.

Extending the present approach by adopting a realisation engine (such as, e.g., [6]) to improve the generation step, and devising a more extensive experimentation will be the object of our future work.

Acknowledgements. We desire to thank Simone Donetti and the Technical Staff of the Computer Science Department of the University of Turin, for their support.

References

1. Camacho-Collados, J., Pilehvar, M.T., Collier, N., Navigli, R.: SemEval-2017 task 2: multilingual and cross-lingual semantic word similarity. In: Proceedings of the 11th International Workshop on Semantic Evaluation (SemEval 2017), Vancouver, Canada, pp. 15–26 (2017)
2. Camacho-Collados, J., Pilehvar, M.T., Navigli, R.: NASARI: a novel approach to a semantically-aware representation of items. In: Proceedings of NAACL, pp. 567–577 (2015)
3. Cambria, E., Speer, R., Havasi, C., Hussain, A.: SenticNet: a publicly available semantic resource for opinion mining. AAAI fall Symp. Commonsense Knowl. **10**, 14–18 (2010)
4. Colla, D., Mensa, E., Radicioni, D.P.: Semantic measures for keywords extraction. In: Esposito, F., Basili, R., Ferilli, S., Lisi, F. (eds.) AI*IA 2017. LNCS, vol. 10640, pp. 128–140. Springer, Cham (2017). https://doi.org/10.1007/978-3-319-70169-1_10
5. Gärdenfors, P.: The Geometry of Meaning: Semantics Based on Conceptual Spaces. MIT Press, Cambridge (2014)
6. Gatt, A., Reiter, E.: SimpleNLG: a realisation engine for practical applications. In: Proceedings of the 12th European Workshop on Natural Language Generation, pp. 90–93. Association for Computational Linguistics (2009)
7. Ghignone, L., Lieto, A., Radicioni, D.P.: Typicality-based inference by plugging conceptual spaces into ontologies. In: AIC@ AI* IA, vol. 1100, pp. 68–79 (2013)
8. Harabagiu, S., Moldovan, D.: Question answering. In: The Oxford Handbook of Computational Linguistics (2003)

9. Havasi, C., Speer, R., Alonso, J.: ConceptNet: a lexical resource for common sense knowledge. Recent Adv. Nat. Lang. Process. V: Sel. Pap. RANLP **309**, 269–280 (2007)
10. Hovy, E.: Text summarization. In: The Oxford Handbook of Computational Linguistics 2nd edition (2003)
11. Jimenez, S., Becerra, C., Gelbukh, A., Bátiz, A.J.D., Mendizábal, A.: Softcardinality-core: improving text overlap with distributional measures for semantic textual similarity. In: Proceedings of *SEM 2013, vol. 1, pp. 194–201 (2013)
12. Lieto, A., Mensa, E., Radicioni, D.P.: A resource-driven approach for anchoring linguistic resources to conceptual spaces. In: Adorni, G., Cagnoni, S., Gori, M., Maratea, M. (eds.) AI*IA 2016. LNCS (LNAI), vol. 10037, pp. 435–449. Springer, Cham (2016). https://doi.org/10.1007/978-3-319-49130-1_32
13. Lieto, A., Minieri, A., Piana, A., Radicioni, D.P.: A knowledge-based system for prototypical reasoning. Connection Sci. **27**(2), 137–152 (2015)
14. Lieto, A., Radicioni, D.P., Rho, V.: Dual PECCS: a cognitive system for conceptual representation and categorization. J. Exp. Theor. Artif. Intell. **29**(2), 433–452 (2017)
15. Lombardo, V., Piana, F., Mimmo, D., Mensa, E., Radicioni, D.P.: Semantic models for the geological mapping process. In: Esposito, F., Basili, R., Ferilli, S., Lisi, F. (eds.) AI*IA 2017, vol. 10640, pp. 295–306. Springer, Cham (2017). https://doi.org/10.1007/978-3-319-70169-1_22
16. Marujo, L., Ribeiro, R., de Matos, D.M., Neto, J.P., Gershman, A., Carbonell, J.: Key phrase extraction of lightly filtered broadcast news. In: Sojka, P., Horák, A., Kopeček, I., Pala, K. (eds.) TSD 2012. LNCS (LNAI), vol. 7499, pp. 290–297. Springer, Heidelberg (2012). https://doi.org/10.1007/978-3-642-32790-2_35
17. Mensa, E., Radicioni, D.P., Lieto, A.: MERALI at SemEval-2017 task 2 subtask 1: a cognitively inspired approach. In: Proceedings of SemEval-2017, pp. 236–240. ACL (2017)
18. Mensa, E., Radicioni, D.P., Lieto, A.: TTCS$^{\mathcal{E}}$: a vectorial resource for computing conceptual similarity. In: EACL 2017 Workshop on Sense, Concept and Entity Representations and their Applications, pp. 96–101. ACL (2017)
19. Miller, G.A.: WordNet: a lexical database for English. Commun. ACM **38**(11), 39–41 (1995)
20. Minsky, M.: A framework for representing knowledge. In: Winston, P. (ed.) The Psychology of Computer Vision, pp. 211–277. McGraw-Hill, New York (1975)
21. Moulin, B., Irandoust, H., Bélanger, M., Desbordes, G.: Explanation and argumentation capabilities: towards the creation of more persuasive agents. Artif. Intell. Rev. **17**(3), 169–222 (2002)
22. Navigli, R., Ponzetto, S.P.: BabelNet: the automatic construction, evaluation and application of a wide-coverage multilingual semantic network. Artif. Intell. **193**, 217–250 (2012)
23. Resnick, L.B., Salmon, M., Zeitz, C.M., Wathen, S.H., Holowchak, M.: Reasoning in conversation. Cogn. Instr. **11**(3–4), 347–364 (1993)
24. Rosch, E.: Cognitive representations of semantic categories. J. Exp. Psychol. Gen. **104**(3), 192–233 (1975)
25. Sebastiani, F.: Machine learning in automated text categorization. ACM Comput. Surveys (CSUR) **34**(1), 1–47 (2002)
26. Tversky, A.: Features of similarity. Psychol. Rev. **84**(4), 327–352 (1977)

Multi-operator Decision Trees for Explainable Time-Series Classification

Vera Shalaeva(✉), Sami Alkhoury, Julien Marinescu, Cécile Amblard, and Gilles Bisson

Univ. Grenoble Alpes, CNRS, Grenoble INP, LIG, 38000 Grenoble, France
{vera.shalaeva,sami.alkhoury,julien.marinescu,cecile.amblard-girard, gilles.bisson}@univ-grenoble-alpes.fr

Abstract. Analyzing time-series is a task of rising interest in machine learning. At the same time developing interpretable machine learning tools is the recent challenge proposed by the industry to ease use of these tools by engineers and domain experts. In the paper we address the problem of generating interpretable classification of time-series data. We propose to extend the classical decision tree machine learning algorithm to Multi-operator Temporal Decision Trees (MTDT). The resulting algorithm provides interpretable decisions, thus improving the results readability, while preserving the classification accuracy. Aside MTDT we provide an interactive visualization tool allowing a user to analyse the data, their intrinsic regularities and the learned tree model.

Keywords: Temporal Decision Trees · Time-series classification Interpretability

1 Introduction

Over the last decade, not just data volume has been growing, but also data complexity and diversity. Today, each industry is keen to leverage their data analysis. Machine learning (ML) researchers develop novel powerful algorithms in order to satisfy a high demand in data mining. However, as ML methods are becoming highly sophisticated being capable of handling data of different complexity levels, industry experts face some difficulties when trying to use them and a ML specialist is often needed to understand their outputs. This raises the problem of ML models interpretability and the interest in developing algorithms able to generate 'domain level' knowledge that is closer to experts' activities.

Current research activity has been concentrated on time-series data classification [9]. It can solve the problems that appear in a wide range of real world domains, e.g., in biology [13], economic forecasting [10], energy, etc. Many machine learning algorithms have been developed to classify time-series [5,6], however, just a few of them can be applied to produce a comprehensible classification.

J. Medina et al. (Eds.): IPMU 2018, CCIS 853, pp. 86–99, 2018.
https://doi.org/10.1007/978-3-319-91473-2_8

A typical example of this problem can be illustrated with, e.g., COTE, a method proposed by Bagnall et al. [2] that is based on an ensemble of heterogeneous classifiers, each of them using different data transformations. It outperforms other approaches in terms of accuracy, however, it generates neither explainable nor readable output. Another algorithm was proposed by Kate [8] that first extracts features from time-series data by using Dynamic Time Warping (DTW) and then does the classification by applying the SVM algorithm. While yielding good accuracy, this algorithm also lacks the interpretability.

Some algorithms outputs have a certain notion of interpretability, but the ML expert has to be involved in the analysis process. For instance, in the BOSS algorithm proposed by Schäfer [15], substructures from time-series data are extracted and then used as high-level features in the ensemble classification method. At the final step of the algorithm, it is nevertheless difficult to make the connections between initial data and the results. Senin and Malinchik [16] propose the SAX-VSM algorithm that is based on symbolic representation of time-series (SAX proposed by Lin et al. [10]). It builds weight vectors that represents each class by applying the vector model and cosine similarity to associate time-series with a class. A domain expert will not be able to easily interpret the results.

There are algorithms that are easily interpretable and that work with time-series classification using shapelets [12,18]. A shapelet is a time-series subsequence that is a discriminating representative of a given class. To discover shapelets, a variety of methods were proposed from brute-force searching [18] to learning them by formulating and resolving an optimization problem [7]. Once top-K shapelets are discovered, they can be used as the input for different classifiers. However, in most studies, a decision tree classifier is used due to its high level of interpretability.

In order to assess the interpretability of an algorithm, the term itself must be well-defined: Lipton [11] explains that interpretability concept is ambiguous and classifies the properties of interpretable algorithms into two categories. The first category, *transparency*, aims to make more easily understandable how the model itself works while the other category, *post-hoc explanations*, gives practical information to the expert about the results.

Our research focuses on creating a transparent framework for analysis and classification of time-series data. There are two main steps:

1. Analyzing raw materials, i.e. non-processed time-series data and then on their discriminating patterns (i.e. small parts of the TS). The importance of this step is clear. By using the knowledge about the field a domain expert can make the initial hypothesis about the intrinsic regularities in the data. Then by looking at the dictionary of patterns they receive very first information about the discriminating ranges in time-series. And already at this stage they can correct the prior beliefs about the data.
2. Exploring and understanding a model generated by ML tools. The results reveal captured underlying insights from the data. A user must interpret these recommendations about the data with respect to their own knowledge

and validate the different parts of the model. This allows an expert to set-up some feedback to change the data description (step 1) or learning parameters.

In this work, we propose an ML algorithm that is able to handle temporal data and provides the user from any domain with interpretable results. Additionally, we put forward the interactive visualization tools to explore the data and the learned model.

One common algorithm that is favoured for a wide range of applications due to its interpretability is the Decision Tree algorithm [14]. This manipulates static data represented by observations and a set of features, which can be categorical, continuous or discrete values. At each non-terminal node the data is separated by the split operator that is the most discriminating feature and its split threshold according to an evaluation criterion, typically entropy or the Gini index. Handling in the decision tree data that comprises attributes coding time-series requires either the extraction of features or the establishment of a new split operator.

In the Temporal Decision Trees (TDT) algorithm proposed by Chouakria and Amblard [4], the classical decision tree algorithm was modified to be able to deal with time-series data. Here the split operator of a node is the most discriminating pair of time-series along with a distance measure, which defines a hyperplane to separate the data. This operator is interpretable since split decisions are based on a visual percept of analogy. From a geometrical point of view using such operator assumes that the data classes are linearly separable. Clearly not all data possess this property, or the data includes different sub-classes. Being unable to capture different geometrical properties leads to trees with large number of nodes and classically lower accuracy.

In this work we generalize the TDT algorithm to the Multi-operator Temporal Decision Trees (MTDT) algorithm that is able to capture different geometrical structures in the data. Furthermore, our approach improves the model readability by decreasing the size of the generated decision tree. The proposed range of split operators keeps the link with raw data that is valuable for a domain expert who either has some prior knowledge about the data or hypothesizes underlying patterns. We also provide a tool to interactively visualize and explore the data and model.

In Sect. 2 we describe the TDT algorithm and provide some background on time-series metrics. In Sect. 3 we propose the generalized TDT algorithm, MTDT, by introducing two new decision operators. We also present a visual tool that can be used to analyse time-series independently or supplementary with MTDT. Ultimately, in Sect. 4 we discuss the experimental results.

2 Temporal Decision Trees

The TDT algorithm uses a similarity-based approach to split data. In this section we provide the description of this algorithm and the definitions of metrics that are used in the TDT and MTDT algorithm.

Let a set $\mathcal{X} = \{ts_1, \dots, ts_N\}$ be a set of N time series. A set $\mathcal{Y} = \{1, \dots, J\}$ is a set of J classes that a time-series can be associated with. The distance between two time-series ts_i and ts_j is denoted as $d(ts_i, ts_j)$.

2.1 Mono-operator Temporal Decision Tree (TDT)

Chouakria and Amblard [4] proposed a modified algorithm of the classical Decision Trees that is able to deal with time-series data. The input data for the algorithm is a set of labelled time-series. Constructing a temporal decision tree involves a series of decisions on how to make a binary split of data at each node by minimizing impurity between classes. The process terminates when a node contains a set of time series belonging to one class. A binary split of each node is based on adaptive metrics $\mathrm{DTW}_k^{\mathrm{CORT}}$ that captures both the value and shape similarities between time-series data. At each non-terminal node of a tree the metric can change focusing attention to the most discriminating distance component.

The Algorithm 1 shows the steps required for making a split of the data belonging to a node. Searching for the best split in a node given a distance requires evaluation of all time-series pairs that belong to two different classes. The partition of time-series of the current node into two sub-nodes is based on their similarities with each series of the candidate pair x_{left} or x_{right} (line 5–11). Once a binary assignment is done, the Gini impurity index is measured to evaluate the quality of the partition generated by the split candidate.

Algorithm 1. Temporal Decision Tree

```
 1: Input: (X, Y), range of metrics M
 2: best splitter, S_left, S_right ← ∅
 3: for each pair of time-series (x_left, x_right) do
 4:     for each distance m in M and x ∈ X do
 5:         if d(x, x_left) <= d(x, x_right) then
 6:             S_left ← x
 7:             else S_right ← x
 8:         end if
 9:         evaluate split by Gini impurity index
10:         if Gini impurity index is improved then
11:             best splitter = (x_left, x_right)
12:         end if
13:     end for
14: end for
15: Return: best splitter
```

2.2 Metrics

To measure a similarity between time-series we need a metric. We will distinguish between value based, shape based, and value-shape based metrics being able to capture different attributes of time-series. A metric also can include or not an alignment of time stamps between two time-series. If the time-series measurements are not shifted in time, that is, an observation u_i of ts_i corresponds to v_i of ts_j, we can speak about static alignment between them, otherwise, the dynamic alignment is required to measure similarity.

Let $ts_i = \{u_1, \ldots, u_m\}$ and $ts_j = \{v_1, \ldots, v_n\}$ are times-series of length m and n. A warping path π is a dynamic alignment between two time-series ts_i and ts_j defined as a sequence of s pairs $p_1 = (u_{a_1}, v_{b_1}), \ldots, p_s = (u_{a_s}, v_{b_s})$ with $a_i \in \{1, \ldots, m\}$, $b_i \in \{1, \ldots, n\}$, $i \in \{1, \ldots, s\}$ and satisfying the constraints: boundary $a_1 = b_1 = 1$, $a_s = m$, $b_s = n$; monotonicity $a_{i+1} = a_i$ or $a_i + 1$, $b_{i+1} = b_i$ or $b_i + 1$; step size condition says that no time-stamp can be omitted. There exist many possible alignments between two times-series. The *optimal path* π minimizes the distance value between two time-series. The distance between two time-series ts_i and ts_j is defined by:

- l_p is a value-based distance $\mathrm{l}_p(ts_i, ts_j) = (\sum_{t=1}^{m} |u_t - v_t|^p)^{\frac{1}{p}}$
- Dynamic Time Warping (DTW) is a l_1 value-based distance

$$\mathrm{DTW}(ts_i, ts_j) = \min_{\pi}(\sum_{t=1}^{s} |u_{a_t} - v_{b_t}|).$$

- CORT is a shape-based distance

$$\mathrm{CORT}_\pi(ts_i, ts_j) = 1 - \min_{\pi}\left(\frac{\sum_{t=1}^{s}(u_{a_{t+1}} - u_{a_t})(v_{b_{t+1}} - v_{b_t})}{\sqrt{\sum_i (u_{a_{t+1}} - u_{a_t})^2}\sqrt{\sum_i (v_{b_{t+1}} - v_{b_t})^2}}\right).$$

- $\mathrm{DTW}_k^{\mathrm{CORT}}$ is a value-shape based distance

$$\mathrm{DTW}_k^{\mathrm{CORT}}(ts_i, ts_j) = \min_{\pi}\left(\frac{2}{(1 + \exp(k * \mathrm{CORT}))} \sum_{t=1}^{s} |u_{a_t} - v_{b_t}|\right).$$

The value of CORT belongs to $[-1, 1]$, where values 1 and -1 indicate similar and opposite shape between two time-series respectively. The value of 0 means two time-series are neither similar or opposite shape. The parameter $\mathrm{k} \in [0, 6]$ $\mathrm{DTW}_k^{\mathrm{CORT}}$ modulates the role of value ($\mathrm{k} = 0$) and shape ($\mathrm{k} = 6$) components. Table 1 summarizes metric arrangement by their characteristics.

Table 1. Metrics arranged by their categories and by the type of time stamps alignment.

		Value	Shape	Value-shape
Alignment	Static	l_p	CORT	(1-CORT)*l_p
	Dynamic	DTW	CORT_π	$\mathrm{DTW}_k^{\mathrm{CORT}}$, (1-CORT)*DTW

3 Multi-operator Temporal Decision Tree (MTDT)

We now introduce two new split operators for the tree. Adding them allows us to widen the range of geometrical properties of the data that the algortihm is able to capture. By using a split pair of time-series at each node we construct hyperplane decision trees assuming linear separability in the data. This leads to bias in the algorithm decision process and, moreover, to misleading interpretation about the fact that two time-series are really discriminative or relevant. It also induces some limitations in the learning process. Therefore, it is interesting to make other types of split operators available to the algorithm.

3.1 Spherical Split Operator

The first operator we propose is the spherical operator that captures ball-shaped geometrical structure of classes in the input data. The operator is represented by one time-series and the distance threshold. Each time-series is considered as a split candidate [17]. The assessment of the best split threshold for a candidate and the best split candidate of a node is done by the Gini index. As with the time-series pair split operator of TDT, we are still based on visual notion of similarity, even if the notion of threshold here is more difficult to assess for the user due to the lack of referential.

3.2 Heuristic Approach of Recognizing Patterns

The second split operator for MTDT is based on time-series discriminative patterns that the user can easily validate and understand since they are existing subparts of real TS. The patterns represent chunks of time-series with different length W and they are learned during the pre-processing step of the decision tree construction. These patterns are the outputs of the Heuristic Approach of Recognizing Patterns (HARP) algorithm, a supervised algorithm that uses the one-against-all method to learn distinctive patterns. The steps of HARP are:

1. **Time-series discretization:** HARP discretizes at first the values of each ts_i into d values (the number of bins) using the SAX method.
2. **Patterns generation:** In this step, for each class $y_i \in \mathcal{Y}$: (1) we store all patterns of size W in a hashtable; (2) for each pattern p we compute its purity rate, that is to say, for a given class C_i the ratio between the sum of pattern occurrences p in C_i and the sum of occurrences of p in all classes; (3) eventually, only the discriminating patterns that have a purity rate above a given threshold t (typically $\geq 90\%$) are kept for the next step.
3. **Pattern generalization:** The aim is to find more general patterns (i.e., a combination of patterns with similar shape and coverage), and for this, we use a heuristic approach: at each iteration we try (1) to generalize on one level each position of the selected patterns (for instance, if each letter corresponds to an interval of values, a pattern "B, E" is turned into "[BC], E", "[AB], E", "B, [EF]", "B, [DE]" and (2) to combine most similar patterns into a new more

general pattern (i.e. "[BC], E" and "B, [EF]" become "[BC], [EF]"). As these steps are costly (i.e., complete search involves a generalization lattice), we use a greedy approach that tests patterns by decreasing purity. This approach doesn't aim to be complete but to find rapidly simple patterns if they exist.
4. **Tiling:** In the final step, the algorithm aims to choose a subset of patterns that covers the same amount of time-series as the whole set does. We again use a greedy approach: first, we list all the time-series that the patterns cover and let call this list $L_t \subseteq \mathcal{X}$. We then loop over the patterns and select the one pattern that covers the most time-series L_{st}. We then subtract L_{st} from L_t and repeat the process again until L_t is empty. This procedure is repeated for each class $y_i \in \{1, \ldots, J\}$.

The final set of patterns is the representation of the HARP model, but to be used as a part of the MTDT process, we need to create a new set of variables that are encoding for the patterns. The variables are based on statistics collected for each pattern inside each time-series: the number of occurrences for a pattern; first/last/average position of the pattern; min/max/average gap length between two sequential occurrences.

The combination of the pattern and their features are used as split candidates. These features are used in the decision tree using the classical approach which is based on finding the best splitting value. Each variable (pattern, feature) is evaluated using Gini impurity index.

3.3 Data and Results Presentation

We developed a graphical interface that supports the proposed MTDT algorithm to make each step of time-series classification simple and interpretable [1]. The tool allows the user intuitively see the data that their learning about and results of each analysis step. It includes three modes of visualization.

"TSplorer" includes graphical representation of learned patterns by HARP. Two different views "Curve" and "Heatmap" are available and depicted on the Fig. 1. "Curve" view lets user drag-and-drop time-series into canvas called "container" and explore how they are segmented by its patterns (left side of the Fig. 1). With the view "Heatmap" users can get insights about the general distribution off all patterns in the learning set among all the classes and variables (right side of the Fig. 1). Horizontal axis represents time-series and vertical axis corresponds to the learned patterns. Intensity of a heat map pixel colour means how frequently a pattern appears in a time-series. Red mark in the top left corner of a pixel points to the pattern that is the representative of another class. A user can navigate between these two views.

The second tool "TDecisionTree" is designed to visualize the tree learned by the MTDT algorithm (Fig. 2). By visualizing MTDT tree a user is able to analyze the decision process of data separation and interpret class prediction for a new time-series. Graphical representation of each non-terminal node includes an interactive view on a selected split-operator allowing quick understanding what is the most discriminating time-series or pattern.

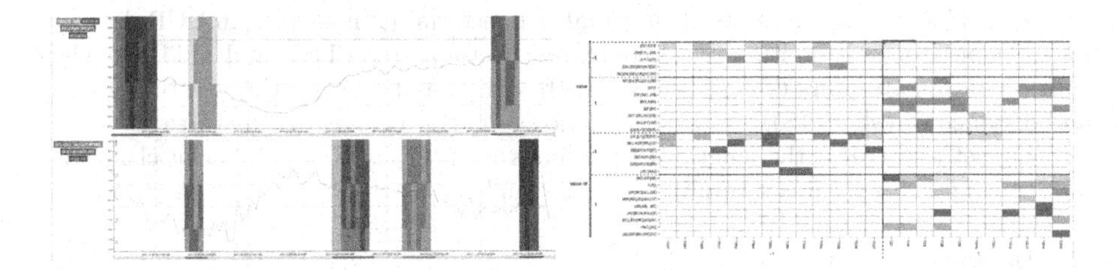

Fig. 1. Visualization of the patterns extracted by the HARP for ECG200 dataset. On the left side two containers with selected time-series and their patterns are shown. On the right side, a heatmap with the patterns distribution over all time-series is shown. (Color figure online)

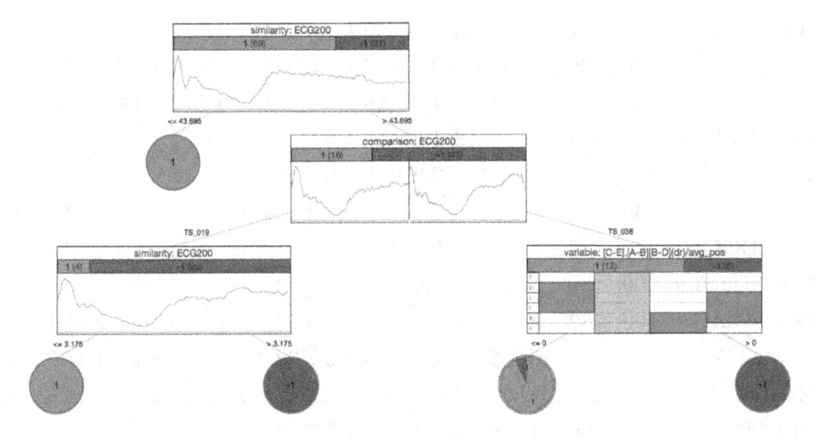

Fig. 2. Graphical representation of Multi-operator Temporal Decision Tree (MTDT) constructed for ECG200 dataset. Each non-terminal node depicts the selected split operator, number of classes, number of time-series related to each class. Circles represents the leaves, its color and the number in the middle indicates the label. (Color figure online)

4 Experimental Results

4.1 Experimental Settings

We did a set of experiments to evaluate whether Multi-operator Temporal Decision Trees (MTDT) improves output readability in the terms of size of the trees when compared with TDT while keeping result accuracy. For our experiments we used 46 univariate time-series datasets provided by the UCR benchmark [3]. We run 10 folds on each of dataset, the first fold corresponds to the same train and test partition as for UCR. The rest 9 folds keep the class distribution of original UCR split. For TDT we used the algorithm proposed by Chouakria and Amblard [4] (denoted H_b in the tables). We used the adaptive metric DTW_k^{Cort} as was proposed in the [4] with three possible values of the parameter k. The value $k = 0$ corresponds to value based distance DTW, $k = 6$ to shape based distance while $k = 3$ takes into account both notions.

Table 2. The tree size in terms of the number non-terminal nodes for 46 UCR dataset for one-operator and multi-operator temporal decision tree (TDT and MTDT). The split operators are pairs of time-series (H, H_b - the baseline algorithm), a time-series and distance threshold (S), and a combination of patterns and their features (P). The training set size is $|S|$, the length of time-series is $|TS|$, and the number of classes is $|C|$. For highlighted dataset results for H+S equals to H+S+P.

#	Datasets	$\|S\|$	$\|TS\|$	$\|C\|$	TDT			MTDT	
					H_b	H	S	H+S	H+S+P
Total					1233.9	926.5	699.6	656.4	529.2
Avg					26.8	20.1	15.2	14.3	11.6
1	**Adiac**	390	176	37	144.3 ± 6.9	92.8 ± 5.8	81.0 ± 4.5	76.7 ± 4.0	76.7 ± 4.0
2	ArrowHead	36	251	3	7.9 ± 2.1	5.5 ± 1.0	4.9 ± 0.8	4.4 ± 0.7	4.1 ± 0.7
3	Beef	30	470	5	12.8 ± 2.3	9.2 ± 1.6	9.0 ± 1.5	7.4 ± 1.1	7.6 ± 1.1
4	BeetleFly	20	512	2	4.3 ± 1.6	2.6 ± 1.0	2.0 ± 0.6	2.0 ± 0.6	1.7 ± 0.5
5	BirdChicken	20	512	2	3.2 ± 1.0	2.9 ± 0.8	2.7 ± 0.8	2.4 ± 0.7	1.5 ± 0.5
6	Car	60	577	4	14.5 ± 2.9	10.1 ± 2.1	9.7 ± 1.5	7.8 ± 1.5	6.6 ± 1.4
7	CBF	30	128	3	2.1 ± 0.3	2.1 ± 0.3	2.5 ± 0.7	2.0 ± 0.0	2.1 ± 0.3
8	**CincECGtor**	40	1639	4	13.7 ± 2.4	9.5 ± 1.9	7.5 ± 1.6	7.5 ± 1.3	7.5 ± 1.3
9	DisPhOutAge	400	80	3	26.3 ± 4.6	21.2 ± 3.3	13.8 ± 1.3	13.0 ± 1.5	8.3 ± 1.6
10	DisPhOutCor	600	80	2	53.5 ± 4.6	41.4 ± 2.5	23.1 ± 4.0	23.6 ± 3.2	14.1 ± 4.0
11	DisPhTW	400	80	6	37.2 ± 5.4	28.5 ± 2.7	22.4 ± 1.4	20.9 ± 1.4	17.0 ± 3.8
12	Earthquakes	322	512	2	28.4 ± 5.1	21.9 ± 2.5	14.0 ± 1.1	13.8 ± 1.9	10.0 ± 2.0
13	ECG200	100	96	2	14.0 ± 2.4	10.8 ± 1.7	6.6 ± 1.2	6.5 ± 1.5	4.5 ± 1.0
14	ECG5000	500	140	5	38.0 ± 6.1	27.8 ± 5.1	21.6 ± 3.3	19.9 ± 4.1	7.9 ± 4.0
15	ECGFiveDs	23	136	2	4.5 ± 1.6	2.7 ± 0.9	3.0 ± 0.8	2.5 ± 0.8	2.5 ± 0.5
16	FaceFour	24	350	4	3.3 ± 0.5	3.1 ± 0.3	3.2 ± 0.4	3.0 ± 0.0	3.0 ± 0.0
17	FacesUCR	200	131	14	27.5 ± 3.8	29.3 ± 3.7	27.8 ± 3.1	24.4 ± 2.1	22.0 ± 3.9
18	GunPoint	50	150	2	4.1 ± 1.4	3.4 ± 1.3	3.1 ± 0.7	2.6 ± 0.7	1.8 ± 0.9
19	**Ham**	109	431	2	21.0 ± 3.2	17.5 ± 2.8	12.6 ± 1.1	11.3 ± 0.8	11.3 ± 0.8
20	**Herring**	64	512	2	18.3 ± 2.2	12.5 ± 3.7	8.7 ± 0.6	8.8 ± 0.6	8.8 ± 0.6
21	**InsectWngS**	220	256	11	93.5 ± 4.2	59.8 ± 4.4	48.9 ± 4.3	47.6 ± 2.7	47.6 ± 2.7
22	ItalyPowDem	67	24	2	3.1 ± 1.7	1.9 ± 1.3	2.7 ± 0.9	1.4 ± 0.5	1.2 ± 0.4
23	**Lighting2**	60	637	2	8.5 ± 1.9	8.2 ± 2.7	5.4 ± 1.2	5.3 ± 1.3	5.3 ± 1.3
24	Lighting7	70	319	7	12.5 ± 1.3	12.1 ± 1.4	10.9 ± 0.8	10.3 ± 1.1	9.3 ± 1.4
25	MALLAT	55	1024	8	7.5 ± 0.9	7.2 ± 0.4	7.2 ± 0.4	7.2 ± 0.4	7.4 ± 0.5
26	Meat	60	448	3	3.5 ± 1.2	2.0 ± 0.0	4.0 ± 0.8	2.0 ± 0.0	2.0 ± 0.0
27	MedicalImag	381	99	10	96.8 ± 8.4	84.7 ± 7.4	59.5 ± 4.4	58.6 ± 2.6	47.3 ± 3.8
28	MidPhAge	400	80	3	41.4 ± 6.4	31.1 ± 5.9	22.0 ± 3.4	21.2 ± 2.6	14.1 ± 2.9
29	**MidPhCor**	600	80	2	66.4 ± 7.4	44.5 ± 6.1	29.4 ± 3.4	27.0 ± 3.0	27.0 ± 3.0
30	MiddPhTW	399	80	6	55.5 ± 4.4	45.5 ± 3.0	34.5 ± 1.6	33.2 ± 1.5	25.8 ± 4.5

Table 2. (*continued*)

#	Datasets	$\|S\|$	$\|TS\|$	$\|C\|$	TDT			MTDT	
					H_b	H	S	H+S	H+S+P
Total					1233.9	926.5	699.6	656.4	529.2
Avg					26.8	20.1	15.2	14.3	11.6
31	**MoteStrain**	20	84	2	1.6 ± 0.7	1.7 ± 1.0	1.7 ± 0.6	1.5 ± 0.5	1.5 ± 0.5
32	OliveOil	30	570	4	5.6 ± 1.4	4.8 ± 1.5	3.6 ± 0.7	3.6 ± 0.7	3.5 ± 0.5
33	ProxPhAge	400	80	3	75.0 ± 7.0	52.5 ± 5.5	31.4 ± 3.7	31.1 ± 2.9	10.5 ± 3.5
34	ProxPhCor	600	80	2	93.3 ± 7.1	68.5 ± 7.4	41.9 ± 3.9	40.5 ± 2.7	21.6 ± 3.7
35	ProxPhTW	400	80	6	46.6 ± 5.4	33.5 ± 3.5	22.1 ± 2.1	21.1 ± 1.4	13.3 ± 2.1
36	ShapeletSim	20	500	2	3.3 ± 1.5	3.4 ± 0.5	2.2 ± 0.4	2.4 ± 0.5	1.0 ± 0.0
37	SonyAIBO1	20	70	2	1.2 ± 0.4	1.3 ± 0.6	1.3 ± 0.5	1.2 ± 0.4	1.0 ± 0.0
38	**SwedishLeaf**	500	128	15	57.4 ± 4.2	42.5 ± 4.8	36.6 ± 3.1	32.0 ± 1.3	32.0 ± 1.3
39	SynControl	300	60	6	5.8 ± 0.9	5.6 ± 0.7	8.5 ± 0.9	5.6 ± 0.7	5.0 ± 0.0
40	ToeSegm1	40	277	2	6.1 ± 2.4	5.9 ± 1.8	3.4 ± 1.0	3.3 ± 1.0	3.2 ± 0.9
41	ToeSegm2	36	343	2	5.4 ± 1.9	4.6 ± 2.0	3.2 ± 1.0	3.2 ± 1.0	3.2 ± 1.2
42	Trace	100	275	4	3.5 ± 0.5	3.5 ± 0.5	4.2 ± 1.0	3.5 ± 0.5	3.1 ± 0.3
43	TwoLdECG	23	82	2	1.5 ± 1.0	1.0 ± 0.0	2.1 ± 0.7	1.0 ± 0.0	1.0 ± 0.0
44	**Wine**	57	234	2	14.6 ± 2.5	10.0 ± 2.6	7.5 ± 0.7	6.9 ± 1.4	6.9 ± 1.4
45	Worms	181	900	5	27.6 ± 2.9	23.1 ± 2.3	17.5 ± 1.2	16.7 ± 1.4	14.5 ± 1.3
46	Worms2Class	181	900	2	18.0 ± 3.1	12.7 ± 3.4	8.5 ± 1.1	8.5 ± 1.0	6.9 ± 1.7

For MTDT we run the experiments with variations of split operators: hyperplane operator (H) presented by a pair of time-series, spherical operator (S) defined by a time-series and its threshold, combination both of them (H+S), and combination with HARP (H+S+P). As metrics we used the l_p norm distance with p being 1 or 2, DTW, CORT, and value-shape based $(1 - CORT) * DTW$ (it is comparable to $\mathrm{DTW}_k^{\mathrm{Cort}}$ with $k = 3$ and avoids the computation of the exponential component).

4.2 Results

The experimental results reveal the average number of non-terminal nodes in a tree (Table 2) and accuracy (Table 3) with its variance over 10 resamples for each dataset and for each approach. By highlighting datasets in the resulting tables we point some of them for which the patterns weren't found by HARP, therefore results H+S equals to H+S+P.

It is interesting to note that the average total number of nodes is reduced by 25% (1233.9 versus 926.5) for H compared with H_b. By including l_p in the range of used metrics improves the readability of built trees for 41 out of 46 datasets. Higher decrease (by 43%, 699.6 over 1233.9) in the number of nodes

Table 3. The classification accuracy (%) of trees obtained on the test set of 46 UCR datasets for one-operator and multi-operator temporal decision tree (TDT and MTDT). The split operators are time-series pairs (H, H_b denotes the baseline algorithm), a time-series and distance threshold (S), a combination of patterns and their features (P). For highlighted dataset results for H+S equals to H+S+P.

#	Datasets	TDT			MTDT	
		H_b	H	S	H+S	H+S+P
Average		0.731	0.736	0.728	0.753	0.768
1	**Adiac**	0.47 ± 0.02	0.53 ± 0.04	0.57 ± 0.02	0.58 ± 0.03	0.58 ± 0.03
2	ArrowHead	0.69 ± 0.04	0.74 ± 0.05	0.65 ± 0.04	0.70 ± 0.06	0.73 ± 0.08
3	Beef	0.49 ± 0.10	0.53 ± 0.07	0.48 ± 0.10	0.60 ± 0.10	0.56 ± 0.06
4	BeetleFly	0.68 ± 0.14	0.73 ± 0.14	0.71 ± 0.12	0.71 ± 0.12	0.76 ± 0.06
5	BirdChicken	0.76 ± 0.10	0.75 ± 0.10	0.70 ± 0.07	0.74 ± 0.11	0.83 ± 0.08
6	Car	0.65 ± 0.07	0.64 ± 0.09	0.64 ± 0.04	0.68 ± 0.07	0.68 ± 0.07
7	CBF	0.94 ± 0.03	0.92 ± 0.05	0.85 ± 0.04	0.91 ± 0.04	0.92 ± 0.02
8	**CincECGtor**	0.53 ± 0.05	0.62 ± 0.04	0.62 ± 0.04	0.60 ± 0.07	0.60 ± 0.07
9	DisPhOutAge	0.74 ± 0.03	0.75 ± 0.03	0.77 ± 0.02	0.77 ± 0.02	0.78 ± 0.05
10	DisPhOutCor	0.72 ± 0.03	0.74 ± 0.02	0.77 ± 0.03	0.77 ± 0.02	0.81 ± 0.02
11	DisPhTW	0.70 ± 0.03	0.71 ± 0.03	0.71 ± 0.02	0.72 ± 0.02	0.74 ± 0.02
12	Earthquakes	0.68 ± 0.04	0.68 ± 0.04	0.68 ± 0.04	0.68 ± 0.03	0.71 ± 0.09
13	ECG200	0.76 ± 0.04	0.78 ± 0.04	0.80 ± 0.03	0.82 ± 0.04	0.80 ± 0.04
14	ECG5000	0.89 ± 0.01	0.90 ± 0.01	0.92 ± 0.01	0.92 ± 0.01	0.92 ± 0.01
15	ECGFiveDs	0.71 ± 0.04	0.82 ± 0.06	0.64 ± 0.07	0.81 ± 0.08	0.75 ± 0.10
16	FaceFour	0.79 ± 0.06	0.75 ± 0.08	0.69 ± 0.11	0.74 ± 0.10	0.81 ± 0.07
17	FacesUCR	0.76 ± 0.02	0.74 ± 0.02	0.71 ± 0.03	0.74 ± 0.03	0.70 ± 0.07
18	GunPoint	0.91 ± 0.03	0.87 ± 0.05	0.83 ± 0.06	0.85 ± 0.04	0.86 ± 0.06
19	**Ham**	0.68 ± 0.07	0.69 ± 0.04	0.64 ± 0.05	0.73 ± 0.06	0.73 ± 0.06
20	**Herring**	0.61 ± 0.07	0.54 ± 0.06	0.59 ± 0.04	0.61 ± 0.05	0.61 ± 0.05
21	**InsectWngS**	0.39 ± 0.01	0.51 ± 0.02	0.52 ± 0.02	0.55 ± 0.02	0.55 ± 0.02
22	ItalyPowDem	0.90 ± 0.03	0.95 ± 0.02	0.92 ± 0.02	0.95 ± 0.01	0.95 ± 0.02
23	**Lighting2**	0.76 ± 0.06	0.73 ± 0.06	0.76 ± 0.03	0.74 ± 0.04	0.74 ± 0.04
24	Lighting7	0.68 ± 0.06	0.62 ± 0.07	0.64 ± 0.09	0.62 ± 0.06	0.62 ± 0.06
25	MALLAT	0.87 ± 0.03	0.90 ± 0.04	0.88 ± 0.04	0.90 ± 0.03	0.85 ± 0.04
26	Meat	0.94 ± 0.05	0.96 ± 0.02	0.87 ± 0.07	0.96 ± 0.02	0.92 ± 0.05
27	MedicalImag	0.66 ± 0.02	0.61 ± 0.04	0.65 ± 0.02	0.67 ± 0.03	0.64 ± 0.03
28	MidPhAge	0.65 ± 0.04	0.65 ± 0.05	0.69 ± 0.03	0.69 ± 0.03	0.71 ± 0.06
29	**MidPhCor**	0.68 ± 0.03	0.71 ± 0.02	0.73 ± 0.02	0.76 ± 0.03	0.76 ± 0.03
30	MiddPhTW	0.52 ± 0.04	0.53 ± 0.02	0.56 ± 0.03	0.56 ± 0.04	0.59 ± 0.02
31	**MoteStrain**	0.85 ± 0.05	0.85 ± 0.03	0.78 ± 0.10	0.84 ± 0.03	0.84 ± 0.03
32	OliveOil	0.81 ± 0.05	0.77 ± 0.07	0.78 ± 0.10	0.81 ± 0.05	0.78 ± 0.07

Table 3. (*continued*)

#	Datasets	TDT			MTDT	
		H_b	H	S	H+S	H+S+P
Average		0.731	0.736	0.728	0.753	0.768
33	ProxPhAge	0.75 ± 0.03	0.76 ± 0.03	0.81 ± 0.03	0.80 ± 0.04	0.89 ± 0.06
34	ProxPhCor	0.78 ± 0.03	0.80 ± 0.03	0.82 ± 0.03	0.82 ± 0.03	0.84 ± 0.04
35	ProxPhTW	0.69 ± 0.02	0.70 ± 0.03	0.74 ± 0.03	0.75 ± 0.03	0.79 ± 0.04
36	ShapeletSim	0.63 ± 0.05	0.57 ± 0.06	0.64 ± 0.07	0.58 ± 0.09	0.99 ± 0.02
37	SonyAIBO1	0.81 ± 0.07	0.80 ± 0.04	0.85 ± 0.08	0.80 ± 0.06	0.80 ± 0.05
38	**SwedishLeaf**	0.81 ± 0.01	0.82 ± 0.03	0.82 ± 0.02	0.84 ± 0.01	0.84 ± 0.01
39	SynControl	0.96 ± 0.01	0.96 ± 0.01	0.95 ± 0.01	0.97 ± 0.01	0.94 ± 0.01
40	ToeSegm1	0.75 ± 0.03	0.74 ± 0.04	0.78 ± 0.06	0.80 ± 0.05	0.79 ± 0.07
41	ToeSegm2	0.78 ± 0.06	0.76 ± 0.07	0.73 ± 0.08	0.72 ± 0.10	0.75 ± 0.06
42	Trace	0.98 ± 0.02	0.95 ± 0.02	0.96 ± 0.02	0.95 ± 0.02	0.96 ± 0.03
43	TwoLdECG	0.91 ± 0.03	0.90 ± 0.06	0.77 ± 0.06	0.90 ± 0.05	0.93 ± 0.02
44	**Wine**	0.83 ± 0.13	0.79 ± 0.07	0.76 ± 0.06	0.79 ± 0.07	0.79 ± 0.07
45	Worms	0.46 ± 0.05	0.46 ± 0.05	0.49 ± 0.05	0.49 ± 0.05	0.48 ± 0.05
46	Worms2Class	0.62 ± 0.04	0.62 ± 0.03	0.65 ± 0.03	0.66 ± 0.03	0.65 ± 0.04

is observed by introducing a spherical split operator, showing the importance of capturing different geometry than a hyperplane split operator. Although, the average accuracy is slightly decreased (72.8% for S compare with 73.6% for H and 73.1% for H_b). This indicates that using one from two operators independently is not sufficient to obtain consistent results in the both terms of accuracy and trees size.

By using the combination of spherical and hyperplane split operators (H+S) the total number of non-terminal nodes reduced to 656.4, that is two times less than for baseline algorithm (H_b). And these results show growth in the accuracy (75.3% for H+S). Eventually by adding the patterns extracted by HARP as a MTDT operator, the total number of nodes drops to 529.2, that's is on 57% less than result for H_b. This shows strong positive influence on tree's readability and the algorithm performance (76.8% for H+S+P).

5 Conclusion

In this work we proposed a generalized algorithm to build the temporal decision tree with multiple split operators. We got a significant reduction of the number of nodes in constructed trees. That confers compactness of our model making its analysis easier and its visual representation better. MTDT is able to capture different geometrical structures in the data by making a split decision with operators that permit an intuitive explanation of a split. Similarity-based data separation allows an expert to understand the process by analogy. It preserves a better connection between initial data and learned model implying easy comprehension of new data classification and validation of obtained results. The provided visualization tool supports the data and model presentation. In combination with the interpretable learning tool it supports transparent framework from the user's perspective to understand how a classification conclusion was reached.

The first direction of future work includes the accuracy improvement of the MTDT algorithm with preserving the current level of interpretability by weighting discriminative time-series sub-intervals. The second is to reduce computational complexity by finding a way to avoid all split candidates evaluation.

Acknowledgments. The study is funded by IKATS project (an Innovative Toolkit for Analysing Time Series), which is a Research and Development project funded by BPIfrance in the frame of the french national PIA program.

References

1. Ikats visualization tool. http://ama.liglab.fr/~software/ikats/demo/
2. Bagnall, A., Lines, J., Hills, J., Bostrom, A.: Time-series classification with COTE: the collective of transformation-based ensembles. In: 32nd IEEE, ICDE 2016, Helsinki, Finland, 16–20 May 2016, pp. 1548–1549 (2016)
3. Chen, Y., Keogh, E., Hu, B., Begum, N., Bagnall, A., Mueen, A., Batista, G.: The UCR time series classification archive, July 2015. www.cs.ucr.edu/~eamonn/time_series_data/
4. Douzal-Chouakria, D., Amblard, C.: Classification trees for time series. Pattern Recogn. **45**, 1076–1091 (2012)
5. Esling, P., Agón, C.: Time-series data mining. ACM Comput. Surv. 12:1–12:34 (2012)
6. Fu, T.: A review on time series data mining. Eng. Appl. Artif. Intell. **24**, 164–181 (2011)
7. Grabocka, J., Schilling, N., Wistuba, M., Schmidt-Thieme, L.: Learning time-series shapelets. In: The 20th ACM SIGKDD, KDD 2014, New York, NY, USA, 24–27 August 2014, pp. 392–401 (2014)
8. Kate, R.J.: Using dynamic time warping distances as features for improved time series classification. Data Min. Knowl. Discov. **30**, 283–312 (2016)
9. Keogh, E.J., Kasetty, S.: On the need for time series data mining benchmarks: a survey and empirical demonstration. Data Min. Knowl. Discov. **7**, 349–371 (2003)
10. Lin, J., Keogh, E.J., Wei, L., Lonardi, S.: Experiencing SAX: a novel symbolic representation of time series. Data Min. Knowl. Discov. **15**, 107–144 (2007)
11. Lipton, Z.C.: The mythos of model interpretability. CoRR (2016)
12. Mueen, A., Keogh, E.J., Young, N.E.: Logical-shapelets: an expressive primitive for time series classification. In: Proceedings of the 17th ACM SIGKDD, San Diego, CA, USA, 21–24 August 2011, pp. 1154–1162 (2011)
13. Qian, L., Zheng, H., Zhou, H., Qin, R., Li, J.: Classification of time series gene expression in clinical studies via integration of biological network. PLOS ONE 1–12 (2013)
14. Quinlan, J.R.: Induction of decision trees. Mach. Learn. **1**(1), 81–106 (1986)

15. Schäfer, P.: The BOSS is concerned with time series classification in the presence of noise. Data Min. Knowl. Discov. **29**(6), 1505–1530 (2015)
16. Senin, P., Malinchik, S.: SAX-VSM: interpretable time series classification using SAX and vector space model. In: 2013 IEEE 13th International Conference on Data Mining, Dallas, TX, USA, 7–10 December 2013, pp. 1175–1180 (2013)
17. Yamada, Y., Suzuki, E., Yokoi, H., Takabayashi, K.: Decision-tree induction from time-series data based on a standard-example split test. In: ICML 2003, Washington, DC, USA, 21–24 August 2003, pp. 840–847 (2003)
18. Ye, L., Keogh, E.J.: Time series shapelets: a new primitive for data mining. In: Proceedings of the 15th ACM SIGKDD, pp. 947–956 (2009)

Comparison-Based Inverse Classification for Interpretability in Machine Learning

Thibault Laugel[1(✉)], Marie-Jeanne Lesot[1], Christophe Marsala[1], Xavier Renard[2], and Marcin Detyniecki[1,2,3]

[1] Laboratoire d'Informatique de Paris 6, LIP6, Sorbonne Université, CNRS, 75005 Paris, France
thibault.laugel@lip6.fr
[2] AXA – Data Innovation Lab, 48 rue Carnot, 92150 Suresnes, France
[3] Polish Academy of Science, IBS PAN, Warsaw, Poland

Abstract. In the context of post-hoc interpretability, this paper addresses the task of explaining the prediction of a classifier, considering the case where no information is available, neither on the classifier itself, nor on the processed data (neither the training nor the test data). It proposes an inverse classification approach whose principle consists in determining the minimal changes needed to alter a prediction: in an instance-based framework, given a data point whose classification must be explained, the proposed method consists in identifying a close neighbor classified differently, where the closeness definition integrates a sparsity constraint. This principle is implemented using observation generation in the *Growing Spheres* algorithm. Experimental results on two datasets illustrate the relevance of the proposed approach that can be used to gain knowledge about the classifier.

Keywords: Post-hoc interpretability · Comparison-based Inverse classification · Local explanation

1 Introduction

Making machine learning systems interpretable, i.e. explaining to the user the decision made by a classifier, can take multiple forms [6,7], depending on the intuition of what 'interpretable' means and the way the explanation is expressed. A basic characterisation distinguishes between *in-model* and *post-hoc* approaches: the former modifies the learning process so as to obtain, by design, understandable classifiers. Among these, many methods have been proposed in the framework of fuzzy systems, see e.g. [2,3]: the use of fuzzy logic favors a fluid interface to human beings, although raising many challenges.

Post-hoc approaches build a posteriori explainer systems, using the results of a classifier to interpret its predictions for particular observations.

They can be further distinguished depending on the inputs they require and on the forms of explanation they provide: some methods exploit the classifier

J. Medina et al. (Eds.): IPMU 2018, CCIS 853, pp. 100–111, 2018.
https://doi.org/10.1007/978-3-319-91473-2_9

type [5,12] or the training set [1,13]. Regarding the outputs, some methods offer visual [15] or linguistic [9,12] explanations, others use observations as explanations, in an instance-based framework [13,19,22]. Some other differences relate to the very definition of interpretability: for instance, feature importance analyzes [4,21] identify the attributes that play a major role on the classifier prediction, inverse classification [5,18] identify the minimal change that would change the prediction.

In this context, this paper proposes a method that can be characterised as (i) a post-hoc approach, i.e. explaining individual predictions of a classifier, (ii) in a model- and data-agnostic framework, i.e. considering that no information about the classifier to be explained nor about the training data is made available to the user, (iii) within the instance-based paradigm, i.e. explaining through comparison, (iv) applying an inverse classification principle.

More precisely, the principle of the proposed approach to explain the prediction for a given observation consists in exhibiting a close point classified differently: the reasons for the obtained prediction are characterised through the production of this neighbor counter-example. The closeness constraint integrates a sparsity constraint, to match the interpretability requirement that the explanation need to be simple and easy to understand for the user.

The paper is organised as follows: Sect. 2 presents related works in the framework of post-hoc interpretability, comparison-based approaches and inverse classification. Section 3 details the principle and formalisation of the proposed approach, as well as the *Growing Spheres* algorithm that implements this principle. Section 4 illustrates the results it obtains in two real-world applications.

2 Related Works

Post-hoc interpretability approaches aim at explaining the behavior of a classifier around particular observations to let the user understand their associated predictions, generally disregarding the actual learning process. They have received a lot of interest recently (see e.g. [14]), especially as black-box models such as deep neural networks and ensemble models are being more and more used for classification despite their complexity.

The variety of existing methods comes from the lack of consensus regarding the definition, and a fortiori the formalization, of the very notion of interpretability. Depending on the task performed by the classifier and the needs of the end-user, explaining a result can take multiple forms. Interpretability approaches rely on the meeting the following objectives to design explanations:

1. The explanations should be an accurate representation of what the classifier is doing.
2. The explanations should be understandably read by the user.

This section briefly discusses the hypotheses that are made about available inputs and details two categories especially related to the proposed method: instance-based approaches and inverse classification.

Available Inputs. To illustrate this discussion, let us consider the case of a physician using a diagnostic tool. It is natural to speculate that (s)he does not have any information about the machine learning model used to make disease predictions, neither may (s)he have any idea about what patients were used to train it. This raises the question of what knowledge (about the machine learning model and the training or other data) an end-user has, and hence what inputs a post-hoc explainer should use.

Several approaches rely specifically on the knowledge of the algorithm used to make predictions, taking advantage of the classifier structure to generate explanations [5,12]. However, in other cases, no information about the classifier is available (the model might be only accessible as an oracle for instance): model-agnostic interpretability methods that can explain predictions without making any hypotheses on the classifier are then required [1,4,21]. These approaches, sometimes called *sensitivity analyzes*, generally try to analyze how the classifier locally reacts to small perturbations: they for instance perform local approximation of the classifier decision boundary, e.g. using linear functions (LIME [21]) or Parzen window-based gradients [4].

Instance-Based Approaches. Instance-based approaches constitute a family of post-hoc methods that bring interpretability by comparing an observation to relevant neighbors [1,13,22]. They use other observations, from the train set, from the test set or generated ones as explanations to bring transparency to a prediction of a black-box classifier.

One of the motivations for instance-based approaches lies in the fact that in some cases the two aforementioned objectives 1 and 2 are contradictory and cannot be both reached in a satisfying way. In these complex situations, finding examples is an easier and more accurate way to describe the classifier behavior than trying to force a specific inappropriate explanation representation, which would result in incomplete, useless or misleading explanations for the user.

As an illustration, the Parzen window-based approach [4] is shown to not succeed well in providing explanations for individual predictions that are at the boundaries of the training data, giving explanation vectors (gradients) actually pointing in the (wrong) opposite direction from the decision boundary. In such a case, seeing this problem as an instance-based one, and more particularly using comparisons with observations from the other class, would probably make more sense and give more useful insights.

Inverse Classification. Inverse classification (see e.g. [16]) is a machine learning task that aims at identifying the minimal changes that can be applied to an observation to as to modify its associated prediction: it has been introduced as an approach to perform sensitivity analysis [18] and later formulated as an interpretability approach [5]. In this view, it belongs to the post-hoc framework and approaches can be categorised using the same characteristics, in particular regarding the assumptions about the available inputs: they can for instance use specific knowledge of the model [5,16] or the training data [18].

Existing approaches for inverse classification consider modifications on the data features, making them related to the feature importance family of methods. In the specific case of text classification, where texts are represented by possibly weighted bags of words, a related approach studies the terms whose removal would lead to modify the observation classification [19], whereas removing features cannot be considered in a general setting.

It can be underlined that inverse classification can also be related to the task of adversarial learning [23], which aims at 'fooling' a classifier by generating close variations of observations to change their predictions. However, adversarial learning focuses on the classifier robustness and exploits its sensitivity.

3 Proposed Approach

This section presents the principles and formalisation of the proposed approach, as well as its implementation in the *Growing Spheres* algorithm.

3.1 Motivations and Characteristics of the Proposed Approach

Motivations. In the light of the axes of discussion presented in the previous section, the justifications for the proposed approach are the following:

First regarding the available inputs, we consider a model- and data-agnostic approach, not requiring any knowledge from the user about the model nor processed data. The only hypotheses we make is that the numerical representation of the data as a feature vector is known, as well as the meaning of the attributes, and that the user can use the classifier to make predictions at will. These weak assumptions seem realistic: for instance, in the aforementioned example of a physician using a diagnostic tool, (s)he is supposed to know what features the system requires to be run, regardless whether the system used performs attribute rescaling or combinations (e.g. through PCA).

Secondly, we consider the instance-based framework and add to the motivations detailed in the previous section the strong justification provided by cognitive sciences of learning through examples [10,20,24]. For instance in [24], it is shown through experiments that generated examples help students 'see' abstract concepts that they had trouble understanding with more formal explanations.

Finally, we propose to apply this paradigm to the task of inverse classification in a hybrid approach to take advantage of their respective benefits.

It is important to note that our primary goal here is to give insights about the classifier, not the reality it is approximating. This approach thus aims at understanding a prediction regardless of whether the classifier is right or wrong, or of the distribution of the original data.

Principle. Given a black-box classifier and an observation, the explanation we propose to provide is based on a data point, in the light of instance-based interpretability; furthermore, in the light of inverse classification, this data point must belong to the other class. The final explanation is expressed in the form of the displacement vector between the observation and the identified data point.

Following the dual objective of interpretability mentioned in Sect. 2, the explaining data point must additionally be close to the considered observation. The closeness definition, discussed in the next section, is a key factor for the relevance of the proposed method.

3.2 Formalisation: Proposed Cost Function

We use the following classical notation: we consider a binary classifier f mapping the input space $\mathcal{X}$ of dimension d to an output space $\mathcal{Y} = \{0, 1\}$ (extension to multiclass classification is straightforward), and suppose that no information is available about this classifier. Let $x = (x_i)_i \in \mathcal{X}$ be the observation to be interpreted and $f(x) \in \mathcal{Y}$ its associated prediction. The goal of the proposed approach is to explain x through another observation $e \in \mathcal{X}$, belonging to another class, i.e. such that $f(e) \neq f(x)$. The final form of explanation is the difference vector $e - x$.

For simplification purposes, in the following we call *ally* an observation belonging to the same class as x by the classifier, and *enemy* if it is classified differently.

Recalling objective 1 mentioned earlier, the explanation $e - x$ should be an accurate representation of what the classifier is doing. This is why we decide to transform this problem into a minimization problem by defining the function $c : \mathcal{X} \times \mathcal{X} \to \mathbb{R}^+$ such that $c(x, e)$ is the cost of moving from observation x to enemy e.

Using this notation, we focus on solving the following minimization problem:

$$e^* = \arg\min_{e \in \mathcal{X}} \{c(x, e) \quad | \quad f(e) \neq f(x)\} \tag{1}$$

$$\text{with } c(x, e) = ||x - e||_2 + \gamma ||x - e||_0 \tag{2}$$

where $||.||_2$ the Euclidean norm, $||.||_0$ the l_0 norm defined as the number of non-zero coordinates, $||e - x||_0 = \sum_{i \leq d} 1_{x_i \neq e_i}$, and $\gamma \in \mathbb{R}^+$ a hyperparameter weighting the two terms.

Indeed, looking up to [22], we choose to use the l_2 norm of the vector $e - x$ as a component of the cost function to measure the proximity between e and x. However, recalling objective 2, we need to make sure that this cost function guarantees a final explanation that can be easily read by the user. In this regard, we consider that users intuitively find explanations of small dimension to be simpler. Hence, we decide to integrate vector sparsity, measured by the l_0 norm, as another component of the cost function c and combine it with the l_2 norm as a weighted average.

3.3 The Growing Spheres Algorithm

Due to the cost function c being discontinuous and the hypotheses made (black-box classifier and no existing data) solving problem defined in Eq. (1) is difficult.

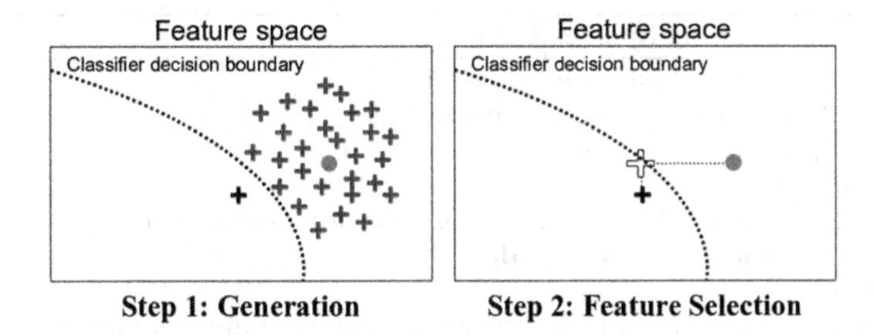

Fig. 1. Illustration of Growing Spheres: the red circle represents the observation to interprete, the plus signs the generated observations (blue for allies, black for ennemies). The white plus is the final enemy e^* used to generate explanations. (Color figure online)

Hence, we choose to solve sequentially the two components of the cost function and propose *Growing Spheres*, a two-step heuristic approach that approximates the solution of this problem. These two steps, namely Generation and Feature Selection, are described in turn below and illustrated in Fig. 1.

Generation. The instance generation, detailed in Algorithm 1, is performed without relying on existing data. Thus, for the considered observation x, we ignore in which direction the closest classifier boundary might be. A greedy approach to find the closest enemy is to explore the input space $\mathcal{X}$ by generating instances in all possible directions further and further until the decision boundary of the classifier is crossed, thus minimizing the l_2-component of function c. More precisely, the algorithm generates observations in the feature space in l_2-spherical layers around x until an enemy is found.

Formally, given two positive numbers a_0 and a_1, we define a (a_0, a_1)-spherical layer SL around x as: $SL(x, a_0, a_1) = \{z \in \mathcal{X} \ : \ a_0 \leq ||x - z||_2 \leq a_1\}$. To generate observations following a uniform distribution over these subspaces, we use the YPHL algorithm [11] which generates observations uniformly distributed over the surface of the unit sphere. We then draw $\mathcal{U}_{[a_0, a_1]}$-distributed values and use them to rescale the distances between the generated observations and x. As a result, we obtain observations that are uniformly distributed over $SL(x, a_0, a_1)$.

The first step of the algorithm consists in generating uniformly n observations in the l_2-ball of radius η and center x, which corresponds to $SL(x, 0, \eta)$ (line 1 of Algorithm 1), with n and η hyperparameters of the algorithm.

In case this initial generation step already contains ennemies, we need to make sure that the algorithm did not miss the closest decision boundary. This is done by updating the value of the initial radius: $\eta \leftarrow \eta/2$ and repeating the initial step until no enemy is found in the initial ball $SL(x, 0, \eta)$ (lines 2 to 5).

However, if no enemy is found in $SL(x, 0, \eta)$, we update a_0 and a_1 using η, generate over $SL(x, a_0, a_1)$ and repeat this process until the first enemy is found (as detailed in lines 6 to 11).

In the end, Algorithm 1 returns the l_2-closest generated enemy e from the observation to be interpreted x (as represented by the black plus in Fig. 1). Once this is done, we focus on making the associated explanation as easy to understand as possible through feature selection.

Algorithm 1. Growing Spheres Generation

Require: $f : \mathcal{X} \rightarrow \{-1; 1\}$ a binary classifier
Require: $x \in \mathcal{X}$ an observation to be interpreted
Require: Hyperparameters: η, n
Ensure: enemy e
1: Generate $(z_i)_{i \leq n}$ in $SL(x, 0, \eta)$ following a uniform distribution
2: **while** $\exists\, e \in (z_i)_{i \leq n} \mid f(e) \neq f(x)$ **do**
3: $\quad \eta = \eta/2$
4: $\quad$ Generate $(z_i)_{i \leq n}$ in $SL(x, 0, \eta)$ following a uniform distribution
5: **end while**
6: Set $a_0 = \eta$, $a_1 = 2\eta$
7: **while** $\nexists\, e \in (z_i)_{i \leq n} \mid f(e) \neq f(x)$ **do**
8: $\quad$ Generate $(z_i)_{i \leq n}$ uniformly in $SL(x, a_0, a_1)$
9: $\quad a_0 = a1$
10: $\quad a_1 = a1 + \eta$
11: **end while**
12: **Return** e, the l_2-closest generated enemy from x

Algorithm 2. Growing Spheres Feature Selection

Require: $f : \mathcal{X} \rightarrow \{-1; 1\}$ a binary classifier
Require: $x \in \mathcal{X}$ the observation to be interpreted
Require: $e \in \mathcal{X} \mid f(e) \neq f(x)$ the solution of Algorithm 1
Ensure: enemy e^*
Set $e' = e$
2: **while** $f(e') \neq f(x)$ **do**
$\quad e^* = e'$
4: $\quad i = \underset{j \in [1:d],\, e'_j \neq x_j}{\arg\min} |e'_j - x_j|$
$\quad$ Update $e'_i = x_i$
6: **end while**
Return e^*

Feature Selection. In the second step, in order to make the difference vector of the closest enemy sparse, we simplify it by reducing the number of features used when moving from x to e (thus minimizing the l_0 component of the cost function $c(x, e)$ and generating the final solution e^*), as explained in the Feature Selection part. To do so, we consider again a naive heuristic based on the idea that the smallest coordinates of $e - x$ might be less relevant locally regarding the classifier decision boundary and should thus be the first ones to be ignored. Thus, the algorithm tries to align as many coordinates of e with x as possible, as long as the predicted class does not change. The proposed feature selection algorithm we use is detailed in Algorithm 2.

The final explanation provided to interprete the observation x and its associated prediction is the vector $x - e^*$, with e^* the final enemy identified by the algorithms (represented by the white plus in Fig. 1).

4 Experimental Results

Although many numerical criteria for interpretability have been proposed (see e.g. [9]), there is no consensus about a global measure for the quality of an explanation. Evaluations based on user satisfaction [4,7,21], although ideal, also depend on the global task the explanations are integrated to and require difficult definitions of experimental protocol. Moreover, the variety of interpretability methods (see Sects. 1 and 2), both in terms of required inputs and of result forms, makes it difficult to compare them.

As a preliminary experiment, this section presents the results obtained when applying the proposed approach to news and image classification. It illustrates the explanations provided by *Growing Spheres* and shows, at a higher level, how a user can exploit them to derive knowledge about the characteristics of the considered classifier, including its possible weaknesses. It also examines whether the explanations can be easily read by a user by measuring the sparsity.

4.1 Prediction of News Popularity

Experimental Protocol. The news popularity dataset [8] is made of 39644 online articles from website Mashable, described by 58 numerical features. The latter encode information about the format and content of the articles, they for instance include the number of words in the title, a measure of the content subjectivity or the popularity of the used keywords. The binary classification task aims at predicting whether an article is popular or not, where popularity is defined as having been shared more than 1400 times.

We apply *Growing Spheres*, to explain the predictions of a classifier. Although it is of no importance for the proposed approach, the experimental protocol consisted in training a random forest (RF) on 70% of the data, after applying a grid search to select the best hyperparameters of RF (number of trees). Tested on the rest of the data, the RF achieved 0.7 AUC and 0.69 accuracy.

Regarding *Growing Spheres*, we use $\gamma = 1$ to define the cost function c (see Eq. 2) and set the hyperparameters of Algorithm 1 to $\eta = 0.001$ and $n = 10000$.

The hypothesis that no information is available about the classifier can be illustrated considering an online journalist writing for Mashable, who would like to predict whether the articles (s)he wrote are going to be popular or not and understand why. The journalist uses a black-box machine learning tool to make the prediction, and has hence no idea about what algorithm was used nor what data was used to train it. The user thus decides to use *Growing Spheres* to generate explanations for the prediction.

Illustrative Examples. We consider two observations from the test set: Article A1, entitled 'The White House is Looking for a Few Good Coders', that is predicted to be not popular by the considered classifier, and article A2, entitled '8 Vendors You Didn't Know Accepted Bitcoins', predicted to be popular. The explanation vector given by *Growing Spheres* are shown in Table 1.

Table 1. Output example of *Growing Spheres*

Article/class	Feature	Move
A1	Min. shares of referenced articles in Mashable	+2016
not popular	Avg. keyword (max. shares)	+913
A2	Avg. keyword (max. shares)	−911
popular	Min. shares of referenced articles in Mashable	−3557
	Rate of positive words (content)	−0.01

For article A1, among the articles it refers, the least popular of them would need to have 2016 more shares and the most popular article associated to its keywords would need to have 913 more shares in order to change the prediction. In other words, article A1 would be predicted to be popular by the considered classifier if the references and the keywords it uses were more popular themselves.

As for article A2, its associated prediction can be explained by three characteristics: to be predicted as unpopular, the same features relevant for A1 would need to be reduced; moreover, the feature 'rate of positive words in the content' would need to be reduced by 0.01. This means that a slightly less positive writing angle would contribute to have article A2 predicted as being not popular.

Sparsity Evaluation. In order to check whether the proposed approach fulfills its goal of finding explanations that can be easily understood by the user, we evaluate the global sparsity of the generated explanations. We measure sparsity as the number of non-zero coordinates of the explanation vector $||x - e^*||_0$.

Figure 2 shows the smoothed cumulative distribution of this value for all 11893 test data points. We observe that the maximum value over the whole test set is 20, meaning that each observation of the test dataset only needs to change 20 or less among the 58 available coordinates to cross the decision boundary. Moreover, 80% of them only need to move in 10 directions or less, that is 17% of

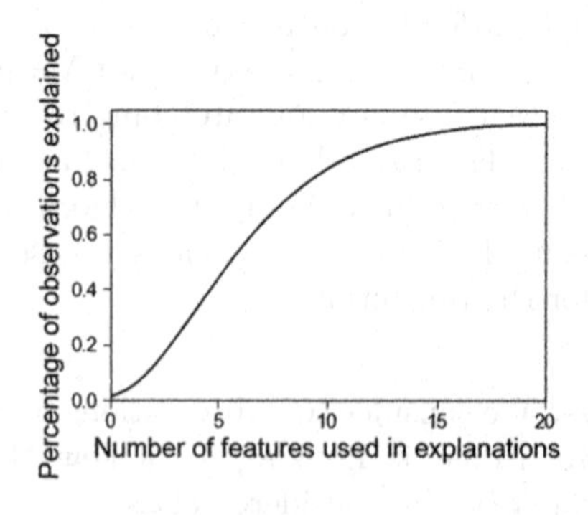

Fig. 2. Sparsity distribution over the news test dataset. Reading: '40% of the observations of the test dataset have explanations that use 5 features or less'.

the features only. This shows that the proposed method indeed achieves sparsity in order to make explanations more readable. It is important to note that this does not mean that only 20 features are needed to explain all the observations, since nothing guarantees different explanations use the same features.

4.2 Applications to Digit Classification

Experimental Protocol. We now use the MNIST handwritten digits database [17] and apply *Growing Spheres* to the binary classification problem of recognizing the digits 8 and 9 vs each other. The dataset contains 60000 instances of 784 features (28 by 28 pictures of digits). We use a support vector machine classifier with a RBF kernel and parameter $C = 15$. Once again, the choice of this model is arbitrary, since *Growing Spheres* is model-agnostic. We train the model on 50% of the data and test it on the rest (0.98 AUC score). As in the first experiment, we use $\gamma = 1$, $\eta = 0.001$ and $n = 10000$.

Illustrative Example. Given a picture of an 8, our goal is to understand, according to the classifier, why it is predicted to be an 8 (and reciprocally). Our intuition would be that closing the bottom loop of a 9 should be the most influential change needed to make it become an 8, and hence features provoking a class change should include pixels found in the bottom-left area of the digits. Output examples to interprete an 8 and a 9 predictions are shown in Fig. 3.

The first thing we observe is that the closest ennemies found by *Growing Spheres* in both cases are not proper 9 and 8 digits respectively. In fact, a human observer would probably still identify the generated enemies as noised versions of the original digits: (i) the pixels involved in the move vector are not all located around the digit but all accross the picture, and (ii) the pixels located in the bottom-left area of the digits do not form a line, and are not 'dark enough'.

This is consistent with the principle of our proposed approach: as mentioned in Sect. 3, *Growing Spheres* is trying to understand the classifier decision, not the reality it is approximating. In this case, the fact that the classifier apparently considers these pixels to be influential the classification of these digits could be an evidence of the learned boundary inaccuracy. Contrary to feature importances, these pixels are not an indication of the contribution of each pixel to the prediction, but rather of the shape of the local decision bourder.

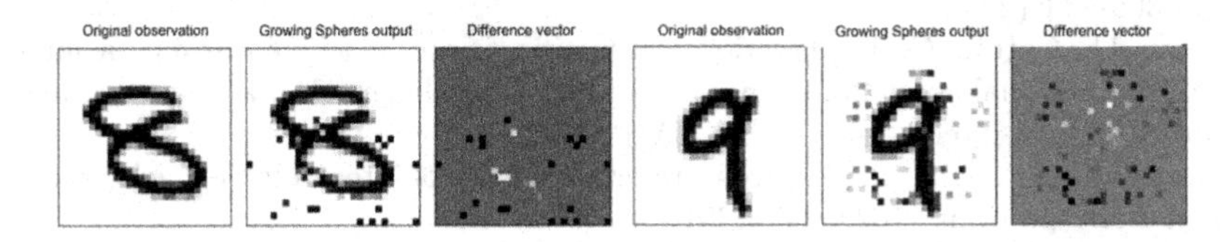

Fig. 3. *Growing Spheres* output examples. From left to right: example of the original instance x, closest enemy found e^*, explanation vector $x - e^*$. First for an 8, then for a 9. A white pixel indicates a 0 value, black a 1.

5 Conclusion and Future Works

The proposed post-hoc interpretability approach *Growing Spheres* provides explanations of a single prediction through the comparison of its considered observation with its closest enemy. In the case where no information is available, neither about the classifier nor about the data, it offers an instance-based inverse classification tool taking into account the objective of sparse explanations. Preliminary experiments illustrate its relevance for explaining predictions and providing insights about the classifier.

Ongoing works aim at experimentally studying the influence of the algorithm hyperarameters and validating the relevance of the explanations it provides in the framework of real-world applications. Another perspective consists in relaxing some of the strong constraints *Growing Spheres* relies on, in particular so as to cases where some information about the data is available: any knowledge about the data distribution might help to guide the generation process, thus for instance minimizing the risk of exploring irrelevant areas of the input space.

Acknowledgements. This work has been done as part of the Joint Research Initiative (JRI) project "Interpretability for human-friendly machine learning models" funded by the AXA Reseach Fund.

References

1. Adler, P., Falk, C., Friedler, S.A., Rybeck, G., Scheidegger, C., Smith, B., Venkatasubramanian, S.: Auditing black-box models for indirect influence. In: 2016 IEEE 16th International Conference on Data Mining (ICDM), pp. 1–10 (2016)
2. Alonso, J., Magdalena, L.: Special issue on interpretable fuzzy systems. Inf. sci. **181**(20) (2011)
3. Alonso, J.M., Castiello, C., Mencar, C.: Interpretability of fuzzy systems: current research trends and prospects. In: Kacprzyk, J., Pedrycz, W. (eds.) Springer Handbook of Computational Intelligence, pp. 219–237. Springer, Heidelberg (2015). https://doi.org/10.1007/978-3-662-43505-2_14
4. Baehrens, D., Schroeter, T., Harmeling, S., Kawanabe, M., Hansen, K., Mueller, K.R.: How to explain individual classification decisions. J. Mach. Learn. Res. **11**, 1803–1831 (2009)
5. Barbella, D., Benzaid, S., Christensen, J., Jackson, B., Qin, X.V., Musicant, D.: Understanding support vector machine classifications via a recommender system-like approach. In: Proceedings of the International Conference on Data Mining, pp. 305–311 (2009)
6. Biran, O., Cotton, C.: Explanation and justification in machine learning: a survey. In: International Joint Conference on Artificial Intelligence Workshop on Explainable Artificial Intelligence (IJCAI-XAI) (2017)
7. Doshi-Velez, F., Kim, B.: Towards a rigorous science of interpretable machine learning. arXiv preprint 1702.08608 (2017)
8. Fernandes, K., Vinagre, P., Cortez, P.: A proactive intelligent decision support system for predicting the popularity of online news. In: Pereira, F., Machado, P., Costa, E., Cardoso, A. (eds.) EPIA 2015. LNCS (LNAI), vol. 9273, pp. 535–546. Springer, Cham (2015). https://doi.org/10.1007/978-3-319-23485-4_53

9. Gacto, M., Alcal, R., Herrera, F.: Interpretability of linguistic fuzzy rule-based systems: an overview of interpretability measures. Inf. Sci. **181**(20), 4340–4360 (2011). Special issue on interpretable fuzzy systems
10. van Gog, T., Kester, L., Paas, F.: Effects of worked examples, example-problem, and problem-example pairs on novices' learning. Contemp. Educ. Psychol. **36**(3), 212–218 (2011)
11. Harman, R., Lacko, V.: On decompositional algorithms for uniform sampling from n-spheres and n-balls. J. Multivar. Anal. **101**(10), 2297–2304 (2010)
12. Hendricks, L.A., Akata, Z., Rohrbach, M., Donahue, J., Schiele, B., Darrell, T.: Generating visual explanations. In: Leibe, B., Matas, J., Sebe, N., Welling, M. (eds.) ECCV 2016. LNCS, vol. 9908, pp. 3–19. Springer, Cham (2016). https://doi.org/10.1007/978-3-319-46493-0_1
13. Kabra, M., Robie, A., Branson, K.: Understanding classifier errors by examining influential neighbors. In: Proceedings of the IEEE Computer Society Conference on Computer Vision and Pattern Recognition, pp. 3917–3925 (2015)
14. Kim, B., Doshi-Velez, F.: Interpretable machine learning: the fuss, the concrete and the questions. In: ICML Tutorial on Interpretable Machine Learning (2017)
15. Krause, J., Perer, A., Bertini, E.: Using visual analytics to interpret predictive machine learning models. In: ICML Workshop on Human Interpretability in Machine Learning, pp. 106–110 (2016)
16. Lash, M.T., Lin, Q., Street, W.N., Robinson, J.G.: A budget-constrained inverse classification framework for smooth classifiers. In: 2017 IEEE International Conference on Data Mining Workshops (ICDMW17) (2017)
17. LeCun, Y., Bottou, L., Bengio, Y., Haffner, P.: Gradient-based learning applied to document recognition. Proc. IEEE **86**, 2278–2324 (1998)
18. Mannino, M.V., Koushik, M.V.: The cost minimizing inverse classification problem: a genetic algorithm approach. Decis. Support Syst. **29**(3), 283–300 (2000)
19. Martens, D., Provost, F.: Explaining data-driven document classifications. Mis Q. **38**(1), 73–99 (2014)
20. Mvududu, N., Kanyongo, G.Y.: Using real life examples to teach abstract statistical concepts. Teach. Stat. **33**(1), 12–16 (2011)
21. Ribeiro, M.T., Singh, S., Guestrin, C.: Why should I trust you? In: Proceedings of the 22nd ACM SIGKDD International Conference on Knowledge Discovery and Data Mining - KDD 2016, pp. 1135–1144 (2016)
22. Štrumbelj, E., Kononenko, I., Robnik Šikonja, M.: Explaining instance classifications with interactions of subsets of feature values. Data and Knowl. Eng. **68**(10), 886–904 (2009)
23. Tygar, J.D.: Adversarial machine learning. IEEE Internet Comput. **15**(5), 4–6 (2011)
24. Watson, A., Shipman, S.: Using learner generated examples to introduce new concepts. Educ. Stud. Math. **69**(2), 97–109 (2008)

Aggregation Operators, Fuzzy Metrics and Applications

Efficient Binary Fuzzy Measure Representation and Choquet Integral Learning

Muhammad Aminul Islam[1(✉)], Derek T. Anderson[2], Xiaoxiao Du[2], Timothy C. Havens[3], and Christian Wagner[4]

[1] Mississippi State University, Starkville, MS, USA
mi160@msstate.edu
[2] University of Missouri, Columbia, MO, USA
andersondt@missouri.edu, xdy74@mail.missouri.edu
[3] Michigan Technological University, Houghton, MI, USA
thavens@mtu.edu
[4] University of Nottingham, Nottingham, UK
christian.wagner@nottingham.ac.uk

Abstract. The *Choquet integral* (ChI), a parametric function for information aggregation, is parameterized by the *fuzzy measure* (FM), which has 2^N real-valued variables for N inputs. However, the ChI incurs huge storage and computational burden due to its exponential complexity relative to N and, as a result, its calculation, storage, and learning becomes intractable for even modest sizes (e.g., $N = 15$). Inspired by empirical observations in multi-sensor fusion and the more general need to mitigate the storage, computational, and learning limitations, we previously explored the *binary ChI* (BChI) relative to the *binary fuzzy measure* (BFM). The BChI is a natural fit for many applications and can be used to approximate others. Previously, we investigated different properties of the BChI and we provided an initial representation. In this article, we propose a new efficient learning algorithm for the BChI, called EBChI, by utilizing the BFM properties that add at most one variable per training instance. Furthermore, we provide an *efficient representation of the BFM* (EBFM) scheme that further reduces the number of variables required for storage and computation, thus enabling the use of the BChI for "big N". Finally, we conduct experiments on synthetic data that demonstrate the efficiency of our proposed techniques.

Keywords: Binary Choquet integral · Binary fuzzy measure

1 Introduction

Data/information fusion can be described as the process of combining of multiple inputs to provide a more accurate, concise, and/or reliable result than what a single source can achieve on its own. Driven by the need for better results,

J. Medina et al. (Eds.): IPMU 2018, CCIS 853, pp. 115–126, 2018.
https://doi.org/10.1007/978-3-319-91473-2_10

countless applications in many fields, such as computer vision and remote sensing, have long been applying fusion at different "levels" (signal, feature, decision etc.). Furthermore, the daily advancement in engineering technologies like smart cars, which operate in complex and dynamic environments using multiple sensors, are raising both the demand for and complexity of fusion.

While there is a multitude of *fuzzy integral* (FI) variants for fusion, we focus in this paper on the *Choquet Integral* (ChI), a well-known, demonstrated, and flexible aggregation function. The ChI has been used in numerous applications (mostly focused on decision level fusion), e.g., humanitarian demining [1], computer vision [2], pattern recognition [3–7], multi-criteria decision making [8,9], control theory [10], and multiple kernel learning [1,11–14]. The ChI is a nonlinear aggregation function parameterized by a *fuzzy measure* (FM), a normal and monotone capacity. The FM is defined on the power set of the sources, i.e., on the sets of all possible combinations of sources, and therefore has 2^N variables for N sources. With the flexibility of choosing values for these $2^N - 2$ parameters (excluding the null set and X, which have fixed values by definition) in the FM, the ChI covers a wide range of aggregation operators. However, this advantage comes at a price: the requirement to specify (by human) or learn (from data) the FM. This means that the complexity of a learning problem, both in terms of storage and computation, is on an exponential order in respect to the number of sources. Therefore, a learning problem with the full set of variables becomes intractable at a relatively small N. Different approaches exist to learn the ChI from data, e.g., *quadratic programming* (QP) [15], gradient descent [16], penalty/reward [17], Gibbs sampler [18], linear programming [19], and efficient optimization with only data-supported variables [20].

In [21], we explored the *binary fuzzy measure* (BFM). The need to investigate the BFM was driven by experimental findings in binary decision making for machine learning [22]. Specifically, multiple instance learning was used to acquire the ChI for signal processing and the learned FM had values approximately in $\{0, 1\}$ versus $[0, 1]$. This suggests that the underlying FM in some applications can be binary, which motivates the use of a BFM directly rather than the real-valued FM due to its simplicity and efficient computation. Thus, many problems are a natural fit for the BFM and others are likely approximatable.

The BFM also has nice properties and computational advantages over the FM. In [21], we showed that the ChI relative to the BFM is equivalent to the Sugeno integral. We also showed that only one variable is effectively used for the ChI computation of an observation in comparison to N variables for the real valued FM. Moreover, only one-valued variables need be stored since zero-valued variables can be discarded. These features make the ChI computation and its storage (the BFM) less expensive compared to the real valued FM/ChI.

Herein, we first put forth an efficient data-driven learning method for the BFM and subsequently the BChI, which we refer to as *efficient BChI* (EBChI). Based on the fact that only one variable contributes to the BChI computation of an instance, we can explain this for variable selection during learning. Thus, each training instance adds at most one variable and the learning problem consequently becomes scalable to the problem size. That is, the number of variables to be optimized is no longer exponential of N, but rather linear to the number of instances (in the worst case). This not only lessens the computation burden, but it also provides a more robust and generalized solution since the number of unknowns (variables) are always fewer then the number of equations (training instances). In contrast, the learning problem with the entire set of FM variables is prone to overfitting as the number of training samples now becomes much smaller than the number of variables ($2^N - 2$) [23].

Next, we provide a representation scheme for the BFM with the minimum set of variables, which we call the *efficient BFM* (EBFM). The BFM variables can be partitioned into two groups, one-valued variables and zero-valued variables. Among the one-valued variables, some of them can deduce their values from others using the FM's monotonicity property (which we call dependent variables) while others cannot (which we call independent variables). The dependent variables can be eliminated without any loss of information and the BFM can be represented with only the independent variables, of which there can be at most $\binom{N}{N/2}$ variables. The full set of FM variables can be retrieved from these independent variables and vice-versa. Therefore, the independent variables constitute the minimal BFM or EBFM.

In Sect. 2 we give the preliminaries of the FM, BFM, ChI, and BChI. Section 3 describes the efficient data-driven BFM learning followed by the representation in Sect. 4. In Sect. 5, we conduct experiments on synthetic data to demonstrate the performance of our proposed learning method.

2 Fuzzy Measure and the Choquet Integral

Let $X = \{x_1, x_2, \ldots, x_N\}$ be a discrete set of N sources. A FM is a monotonic function defined on the power set of X, 2^X, as $\mu : 2^X \rightarrow \Re^+$ that satisfies: (i) (boundary condition:) $\mu(\emptyset) = 0$, and $\mu(X) > 0$ and (ii) (monotonicity:) if $A, B \subseteq X, A \subseteq B$, $\mu(A) \leq \mu(B)$. Often an additional constraint is imposed to limit the upper bound to 1, i.e., $\mu(X) = 1$. Let $h(x_i)$ be the data/information from the ith source. The discrete ChI (finite X) is

$$\int_C \mathbf{h} \circ \mu = C_\mu(\mathbf{h}) = \sum_{i=1}^{N} h(x_{\pi(i)}) \left[\mu(S_{\pi(i)}) - \mu(S_{\pi(i-1)})\right], \tag{1}$$

where π is a permutation of X, such that $h(x_{\pi(1)}) \geq h(x_{\pi(2)}) \geq \ldots \geq h(x_{\pi(N)})$, $S_{\pi(i)} = \{x_{\pi(1)}, \ldots, x_{\pi(i)}\}$, and $\mu(S_0) = 0$. Equation (1) is often referred to as the difference-in-measure form since the integral is represented as the sum of difference-in-measure weighted by the input-values. The ChI can equivalently be written in difference-in-inputs form as

$$\int_C \mathbf{h} \circ \mu = C_\mu(\mathbf{h}) = \sum_{i=1}^{N} [h(x_{\pi(i)}) - h(x_{\pi(i+1)})] \mu(S_{\pi(i)}), \tag{2}$$

where $h(x_{\pi(N+1)}) = 0$. The latter weighted-measure form at (2) is suitable for the FM learning problem herein, where μ is unknown. Equation 2 can also be written in matrix form to facilitate optimization as

$$C_\mu(\mathbf{h}) = \mathbf{c}^T \mathbf{u}_B, \tag{3}$$

where $\mathbf{u}_B$ is the vector of all variables except $\mu(\emptyset)$ and has a length of $2^N - 1$, and $\mathbf{c}$ holds the coefficients of $\mathbf{u}_B$ for observation $\mathbf{h}$.

The FM can be visualized with respect to its underlying Hasse diagram (shown in Fig. 1). Each instance yields a sort, π, which produces a walk up the lattice. The walk starts with $\mu(\emptyset)$ followed by N other variables, each of different size cardinality. For example, an observation $\mathbf{h}$ with $h(\{x_2\}) \geq h(\{x_1\}) \geq h(\{x_4\}) \geq h(\{x_3\})$ walks along the path shown in Fig. 1(b) and the corresponding ChI has variables $\mu(\{x_2\}), \mu(\{x_1, x_2\}), \mu(\{x_1, x_2, x_4\})$, and $\mu(X)$.

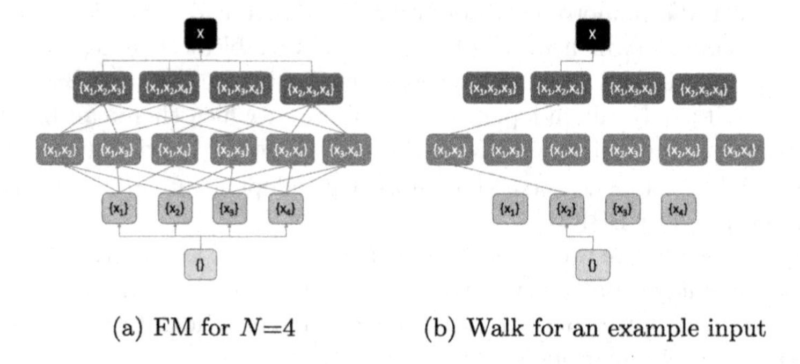

(a) FM for N=4 (b) Walk for an example input

Fig. 1. (a) FM for four inputs. Arrows indicate monotonicity conditions on immediate subsets. (b) Path taken by observation $\mathbf{h}$ with $h(\{x_2\}) \geq h(\{x_1\}) \geq h(\{x_4\}) \geq h(\{x_3\})$. Only four variables $\mu(\{x_2\})$, $\mu(\{x_1, x_2\})$, $\mu(\{x_1, x_2, x_4\})$, and $\mu(X)$ are used for the ChI.

2.1 The Binary Fuzzy Measure

As already stated, a BFM, μ_B, is a special case of the real-valued FM, μ, that restricts μ_B to $\{0, 1\}$ instead of $[0, 1]$. Obviously, this drastically reduces the search space for an optimization problem. In article [21], we proved that the BChI and Sugeno Integral are equivalent. The BChI is simply the standard Choquet integral with respect to the BFM. Suppose the BFM values along the walk for $\mathbf{h}$ are given by $\mu_B(S_{\pi(i)}) = 0$ if $i < k$, else 1, where $\mu_B(S_{\pi(k)})$ is the first variable encountered along this path with value 1. Replacing the FM with this BFM in Eq. (1) and then expanding it, the BChI can be written as [21].

$$\begin{aligned}
&\int_C \mathbf{h} \circ \mu_B = C_{\mu_B}(\mathbf{h}) = \sum_{i=1}^{N} h(x_{\pi(i)}) \left[\mu_B(S_{\pi(i)}) - \mu_B(S_{\pi(i-1)})\right] \\
&= \left(\sum_{i=1}^{k-1} h(x_{\pi(i)}) \left[\mu_B(S_{\pi(i)}) - \mu_B(S_{\pi(i-1)})\right] \right) + h(x_{\pi(k)}) \left[\mu_B(S_{\pi(k)}) - \mu_B(S_{\pi(k-1)})\right] \\
&+ \left(\sum_{i=k+1}^{N} h(x_{\pi(i)}) \left[\mu_B(S_{\pi(i)}) - \mu_B(S_{\pi(i-1)})\right] \right) = h(x_{\pi(k)}) \mu_B(S_{\pi(k)}),
\end{aligned} \tag{4}$$

since $\mu_B(S_{\pi(i)}) - \mu_B(S_{\pi(i-1)})$ is zero except for $i = k$ and $\mu_B(S_{\pi(k-1)}) = 0$. It is trivial to show with some mathematical manipulation that the BChI in difference-in-inputs form also can be written as

$$C_{\mu_B}(\mathbf{h}) = \sum_{i=1}^{N}[h(x_{\pi(i)}) - h(x_{\pi(i+1)})]\mu_B(S_{\pi(i)}) = h(x_{\pi(k)})\mu_B(S_{\pi(k)}). \quad (5)$$

According to (4) and (5), the BChI of an instance uses only one variable $\mu_B(S_{\pi(k)})$. This fact allows us to use significantly fewer variables than the standard ChI (which we will show in Sect. 3), thus enabling the learning of a BFM for larger number of inputs/sources problems, which otherwise would be intractable to solve on most personal computers.

3 BChI Learning

Let $O = \{\mathbf{h}_j, y_j\}$, $j = 1, 2, \ldots, M$, be a training data set with M instances. Here $\mathbf{h}_j$ represents the jth instance with data from N inputs and y_j is the associated label or ground-truth for $\mathbf{h}_j$. For example, $\mathbf{h}_j$ could be an image, $h_j(\{x_i\})$ could be the soft-max normalized decision of N different deep learners and $y_j = 0$ if $\mathbf{h}_j$ is not the category of interest, e.g., person, or $y_j = 1$ if it is.

The goal is to learn the BFM such that the aggregation results of the training instances optimize a criteria, which is usually specified by a function of error relative to a label, y_i, called an objective function. Common functions include the *sum of squared error* (SSE) [4,23,24] and the sum of absolute error. Without loss of generality, we focus on the SSE, which is widely used due to its continuity, differentiability, and non-linearity relative to errors. The sum of squared error for training data, O, is $E(O, \mathbf{u}_B) = \sum_{j=1}^{M}(C_{\mu_B}(\mathbf{h}_j) - y_j)^2$, where $C_{\mu_B}(\mathbf{h}_j)$ is the BChI for instance $\mathbf{h}_j$ and y_j is associated label.

3.1 Learning with the Full Set of FM Variables

Traditionally, FM learning is formulated with a full set of variables without extracting any knowledge from the training data to reduce the number of variables. In this case, the BChI for a training instance $\mathbf{h}_j$ is represented with a $2^N - 1$ dimensional vector $\mathbf{u}_B$, and, consequently, the SSE for training data, O, is

$$E(O, \mathbf{u}_B) = \sum_{j=1}^{M} e^j = \sum_{j=1}^{M}(\mathbf{c}_j^T \mathbf{u}_B - y_j)^2 = ||D\mathbf{u}_B - \mathbf{y}||_2^2,$$

where $D = [\mathbf{c}_1\ \mathbf{c}_2\ \ldots\ \mathbf{c}_M]^T$, $\mathbf{y} = [y_1\ y_2\ \ldots\ y_M]^T$, $||x||_2$ is norm-2 operation on x, and $\mathbf{u}_B$ is the vector of the $2^N - 1$ binary variables excluding the null set. Based on this, the SSE optimization problem is

$$\min_{\mathbf{u}_B} f(\mathbf{u}_B) = ||D\mathbf{u}_B - \mathbf{y}||^2,$$

$$u_B(k) \leq u_B(l), \text{ if } u_B(k) = \mu_B(A), u_B(l) = \mu_B(B),$$
$$\text{and } A \subset B, \forall k, l \text{ (monotonicity conditions)}$$
$$u_B(p) = 1, \text{ if } u_B(p) = \mu_B(X) \text{ (normality conditions)}$$
$$u_B(l) \in \{0, 1\}, \forall l \text{ (BFM value restriction)},$$

which can be solved by any mixed-integer, integer, or binary quadratic programming library [25,26]. It is obvious that this optimization problem does not scale well since the number of variables is exponential with respect to N regardless of the training sample size. Moreover, its complexity would be higher than the standard FM due to the use of integer-programming, which is in general costlier than the real-valued counterpart.

3.2 Efficient ChI Learning

Herein we propose an efficient learning algorithm that selects variables for optimization from the training data as opposed to using the full set of variables in a standard method. The proposed method can drastically reduce the number of variables; however, the amount of savings depends on different factors such as training data volume, underlying FM, the context of the problem (noisy or no-noise), and the acceptable error in the objective function. As shown in the previous section, given a set of inputs, the BChI needs only one variable to determine the output. The converse is also true, i.e., given a set of inputs, the actual output (which has no noise or fluctuation) can help retrieve the variable responsible for the BChI. In this case, the output equals the kth sorted input, and the variable is associated with this kth input.

Suppose a training instance, $\mathbf{h}_j$, is not affected by noise, i.e., $e_j = 0$, then the output is, $y_j = h_j(x_{\pi_j(k)})$, and the variable corresponding to $h_j(x_{\pi_j(k)})$ is $\mu_B(S_{\pi_j(k)})$. Thus, just by inspecting each training instance, we can retrieve for a noise-free problem the BChI variables with their exact values without requiring any optimization. However, real world problems often–if not always–are affected by noise. When noise variance is small, we can choose to select one variable per instance by finding the sorted input k closest to the output label (which also results in minimum error, e_j), where k can be determined as

$$k = \arg\min_i ||y_j - h_j(x_{\pi_j(i)})||^2. \tag{6}$$

Since the noise can be random, training instances with the same walk in the lattice can be affected by varying magnitude of noise. Consequently, they can pick different k's and hence more than one variable for the same walk as opposed to a single variable for an ideal case. Suppose, the instances $\mathbf{h}_l, l \in \{t_1, \ldots, t_p\} \subseteq \{1, \ldots, M\}$ have the same permutation, π, for their sorting order and the set of variables picked by them using Eq. (6) are $\mu_B(S_{\pi(k)}), k = \{k_1, k_2, \ldots, k_Q\}$ with $S_{\pi(k_1)} \subset S_{\pi(k_2)} \cdots \subset S_{\pi(k_Q)}$. Then the BChI for $\mathbf{h}_l$ w.r.t. these variables is

$$C_{\mu_B}(\mathbf{h}_l) = \sum_{q=1}^{Q} [h_l(x_{\pi(k_q)}) - h_l(x_{\pi(k_{q+1})})] \mu_B(S_{\pi(k_q)}), \tag{7}$$

where $h_l(x_{\pi(k_{Q+1})}) = 0$. The number of selected variables is bounded in $[1, N]$.

For a highly noisy system, the user can choose to select multiple variables vs. one variable per training instance and use different criteria instead of squared of Euclidean distance as in Eq. (6), e.g., selecting variables associated with those inputs that fall within a certain threshold (or standard deviation) of the training label y_j. While increasing the number of variables will have no impact for a system with little noise with sufficient training samples, it can lower the SSE for a highly noisy system with limited data. However, there is an optimum balance between the complexity and the error as increasing variables even for high noise case has diminishing results.

Let the set of variables selected by all the training instances (using Eq. (6)) be represented in vector form as $\mathbf{v}_B$. Then the EBChI of $\mathbf{h}_j$ can be written in matrix form as $C_{\mu_B}(\mathbf{h}_j) = \mathbf{a}_j^T \mathbf{v}_B$, where $\mathbf{a}_j$ be the coefficient of $\mathbf{v}_B$ calculated according to Eq. (7). Based on this, the SSE minimization problem becomes

$$\min_{\mathbf{v}_B} f(\mathbf{v}_B) = ||W\mathbf{v}_B - \mathbf{y}||_2^2,$$

$$\begin{aligned}
&v_B(i) \leq v_B(j), \text{ if } v_B(i) = \mu(A), v_B(j) = \mu(B) \\
&\text{and } A \subset B, \forall i, j \in \{1, 2, \ldots, Q\} \text{ (monotonicity conditions)} \\
&v_B(j) = 1, \text{ if } v_B(j) = \mu_B(X) \text{ (normality conditions)} \\
&v_B(i) \in \{0, 1\}, \ \forall i \text{ (BFM value restriction)}
\end{aligned} \tag{8}$$

$$W = [\mathbf{a}_1 \ \mathbf{a}_2 \ \ldots \ \mathbf{a}_M]^T.$$

4 Efficient BFM Data Structure

In the last section, we provided an algorithm to efficiently learn a BFM that uses fewer variables. Anderson et al. introduced a simple approach to represent the BFM that can be applied here to further reduce the learned variables for efficient storage and representation. In that method, the variable elimination process considers only the values of the variables and does not take into account the monotonicity property of the FM, which can greatly enhance the representation technique. By taking into consideration both values and properties, herein we propose a new way to efficiently represent the full-fledged BFM and then we provide an upper bound on the minimum number of variables required.

4.1 Representation

Since the variables of a BFM are binary valued, the variables can take either zero or one values. The zero-valued variables do not contribute to the BChI, so they can be discarded. Thus, only one-valued variables can be considered as candidates for representation. Due to the monotonicity property, if a variable $\mu_B(A)$ is one-valued, then all variables that are supersets of A are also one-valued. As such, the one-valued variables can be divided into two parts: (i) independent variables whose values cannot be derived from another one-valued variable using monotonicity condition and (ii) dependent variables whose values can be retrieved using the independent variables and monotonicity condition, and therefore, can be discarded. Consequently, only the one-valued independent variables are necessary to represent a BFM, which we refer to as EBFM. Figure 2 illustrates the EBFM representation technique for an example with $N = 4$.

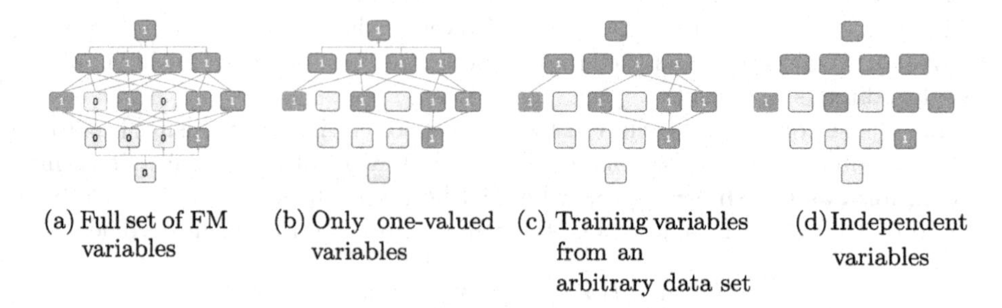

(a) Full set of FM variables (b) Only one-valued variables (c) Training variables from an arbitrary data set (d) Independent variables

Fig. 2. An example of the BFM representation for four inputs case. Light gray nodes with zeros represent zero-valued variables while dark-grey nodes with one's denote one-valued variables. Empty nodes are for placeholders only, and indicate that their variables are removed. (a) shows the full set of FM variables, (b) only one-valued variables, (c) EBChI variables selected from a noise free training data set, and (d) EBFM represented with independent variables. The full FM lattice (a) can be simply derived from (d) using the FM's monotonicity property.

From the EBFM with independent variables that correspond to sets $B = \{B_1, B_2, \ldots, B_l\}$, the BChI of an observation $\mathbf{h}_j$ can be computed as follows:

1. Sort inputs in descending order. Let the sorting order be $x_{\pi_j(i)}, i = 1, 2, \ldots, N$, and the associated variables for the BChI are $\mu_B(S_{\pi_j(i)}), i = 1, 2, \ldots, N$.
2. Find the minimum k for which $S_{\pi_j(k)} \supseteq B_l \in B, \forall B_l$.
3. Return $\mathbf{h}_j(x_{\pi_j(k)})$ as the output.

4.2 Upper Bound

To determine the upper bound on the number of variables in EBFM, we use a theorem by Sperner [27]. The theorem proves that if $B_1, B_2, \ldots, B_t$ are subsets of an N-element set B, such that no B_i is a subset of any other B_j, then

$$t \leq \binom{N}{[N/2]}, \tag{9}$$

where $[x]$ denotes the rounded integer value. As the independent variables in EBFM have the same definitions as the B_is above, Eq. (9) also gives the upper bound on the number of variables in the EBFM. For 20 inputs, there can be at most 184,756 independent variables (in median-like aggregation case), thus using only 17.62% of the total variables in the worst case. However, the actual usage (or saving) depends on the specific BFM; for example, min and max aggregation operators require 1 and N variables respectively.

5 Experiments

Experiments are conducted on synthetic data set for the following reasons. First and foremost, we know the true underlying BFM, which facilitates the comparison and investigation of the proposed method's behaviour, whereas in real world applications/data the true FM may never be known. Moreover, it is quite challenging to find real world examples of varying complexity while it is far easier to create a FM in synthetic data with different complexity. Synthetic experiments also give us insight into how the learning method will behave in different noisy contexts. The experiments are designed to compare the computational complexity as well as to measure the performance in terms of MSE from the predicted test labels using the learned BFM in a no noise as well as in a noisy environment.

5.1 No Noise Scenario

First, a training data set of $M = 500$ and $N = 8$ is generated pseudo-randomly from a uniform distribution, which is then partitioned to create five data-sets for five fold cross-validation. Each cross-validation data-set contains 400 training samples and 100 test samples. Then we created training data-sets of sample sizes 150, 75, 30, and 15 via random selection of instances from those of 400, 150, 75, and 30 respectively. Test data for all sample sizes remains the same. We specified three BFMs–BFM1, BFM2 and BFM3–in Table 1 using the EBFM representation with independent variables. In the table, each independent variable is denoted with the inputs' indices in the set, e.g., 12 stands for independent variable $\mu_B(\{x_1, x_2\})$. The independent variables in BFM1 lie in the lower part of the lattice (hence largest number of one valued variables) while those of BFM3 reside on the upper part. The BFM2 independent variables spread across the lattice from top to bottom. The labels for these BFMs are created without adding any noise, which also serve as the ground-truth for noisy system.

Table 1. The FMs used in the experiment

BFM	Independent variables														#var	# 1-var
1	8	56	67	57	345	346	347	1234	1235	1245	1236	1246	1237	1247	255	211
2	3568	3578	3678	4568	4578	4678	5678	12568	12578	12678	123458	123468	123478	1234567	255	59
3	67	68	78	123456	123457	123458									255	131

Figure 3 shows the results for different sample sizes. As can be seen, the number of variables increases linearly at small M, then remains constant for large sample sizes, which is still far fewer than the standard optimization method. The average of the MSE as well as the number of variables correctly learned are approximately the same for both standard and EBChI (Fig. 3(b) and (c)). An interesting observation from Fig. 3(d) is that the EBChI has far fewer independent variables for 15 training samples, meaning the learned FM from the EBChI is less complex than the standard one.

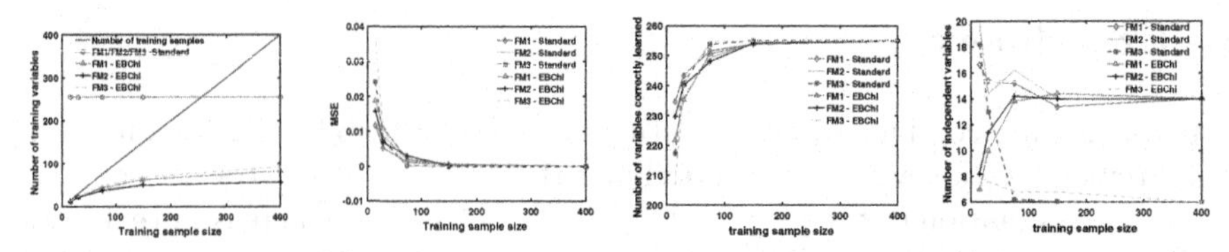

(a) Average number of training variables over five iterations (b) Average mean squared error on test data (c) Average number of variables correctly learned (d) Average number of independent variables

Fig. 3. Noise-free training data of different sizes, M = 15, 30, 75, 150, and 400.

5.2 Noisy Scenario

In most analysis of noisy systems, noise is modeled as Gaussian distribution, which provides a good approximation in many scenarios. Herein, we model the output as $y = C_{\mu_B}(\mathbf{h}) + \epsilon$, $\epsilon \sim \mathcal{N}(0, \sigma_n^2)$. The observations in the training data are the same as those for the noise-free system; however, the labels are created by adding randomly generated values from a normal distribution of variance σ_n^2. We conducted experiments with five different variances, $\sigma_n/\sigma_y = \{0, 0.01, 0.05, 0.1, 0.3, 0.5\}$, where σ_y^2 is the variance of the true labels (actual label without noise). The standard deviations for FM1, FM2, and FM3 are 0.184, 0.1804, and 0.2055. The MSE was measured with respect to the true test labels.

Figure 4 compares the results for noisy data with 400 training samples. More noise means more variations around the true value, which results in selection of multiple variables for instances with the same sorting order. Thus, the number of training variables in EBChI increases with the noise level. The presence of noise equally affects both the EBChI and the standard BChI. When the FMs are learned with sufficient number of samples, the error is minimal–on the order of 10^{-3} (Fig. 4(c))–and there is a mismatch of only 4 out of 255 variables in the worst scenario; see Fig. 4(c). As the result shows, both the standard method and EBChI are resilient to relatively moderate level of noise ($\sigma_n/\sigma_y = 0.3$).

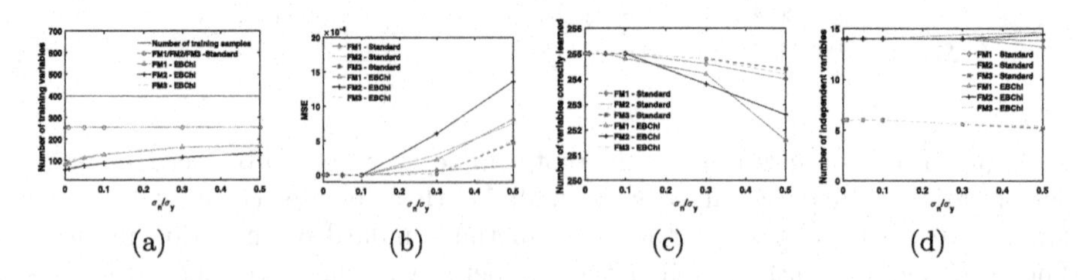

(a) (b) (c) (d)

Fig. 4. Results for data-driven learning with noise, with standard deviation, $\sigma_n = \{0.0\sigma_y\}$. (a) Average number of training variables over five iterations, (b) Average mean squared error on test data, (c) Average number of variables correctly learned, (d) Corresponding independent variables

6 Conclusion

In this paper, we proposed an efficient method to learn the BFM. Variable selection in the EBChI is driven by the observed instances whereas the standard learning method uses full set of variables. This makes the EBChI tractable for a relatively large problem (large N) in contrast to the standard approach. As demonstrated by the results, learning with the EBChI is approximately equivalent to the standard BChI learning method for noisy and noise-free scenarios; therefore, it provides an efficient alternative for data-driven learning of the BFM. Moreover, we introduced a representation technique called the EBFM to describe a BFM minimally via independent variables. In future work, we will apply our technique to real world problems. Additionally, we will study which problems can be natural fit to BFM and which problems can be approximated by a BFM.

References

1. Price, S.R., Murray, B., Hu, L., Anderson, D.T., Havens, T.C., Luke, R.H., Keller, J.M.: Multiple kernel based feature and decision level fusion of iECO individuals for explosive hazard detection in FLIR imagery. In: SPIE, vol. 9823, pp. 98231G–98231G–11 (2016)
2. Tahani, H., Keller, J.: Information fusion in computer vision using the fuzzy integral. IEEE Trans. Syst. Man Cybern. **20**, 733–741 (1990)
3. Grabisch, M., Nicolas, J.M.: Classification by fuzzy integral: performance and tests. Fuzzy Sets Syst. **65**(2–3), 255–271 (1994)
4. Grabisch, M., Sugeno, M.: Multi-attribute classification using fuzzy integral. In: IEEE International Conference on Fuzzy Systems, pp. 47–54. IEEE (1992)
5. Mendez-Vazquez, A., Gader, P., Keller, J.M., Chamberlin, K.: Minimum classification error training for Choquet integrals with applications to landmine detection. IEEE Trans. Fuzzy Syst. **16**(1), 225–238 (2008)
6. Keller, J.M., Gader, P., Tahani, H., Chiang, J., Mohamed, M.: Advances in fuzzy integration for pattern recognition. Fuzzy Sets Syst. **65**(2–3), 273–283 (1994)
7. Gader, P.D., Keller, J.M., Nelson, B.N.: Recognition technology for the detection of buried land mines. IEEE Trans. Fuzzy Syst. **9**(1), 31–43 (2001)
8. Grabisch, M.: The application of fuzzy integrals in multicriteria decision making. Eur. J. Oper. Res. **89**(3), 445–456 (1996)
9. Labreuche, C.: Construction of a Choquet integral and the value functions without any commensurateness assumption in multi-criteria decision making. In: EUSFLAT Conference, pp. 90–97 (2011)
10. Tomlin, L., Anderson, D.T., Wagner, C., Havens, T.C., Keller, J.M.: Fuzzy integral for rule aggregation in fuzzy inference systems. In: Carvalho, J.P., Lesot, M.-J., Kaymak, U., Vieira, S., Bouchon-Meunier, B., Yager, R.R. (eds.) IPMU 2016. CCIS, vol. 610, pp. 78–90. Springer, Cham (2016). https://doi.org/10.1007/978-3-319-40596-4_8
11. Pinar, A.J., Rice, J., Hu, L., Anderson, D.T., Havens, T.C.: Efficient multiple kernel classification using feature and decision level fusion. IEEE Trans. Fuzzy Syst. **PP**(99), 1 (2016)
12. Pinar, A., Havens, T.C., Anderson, D.T., Hu, L.: Feature and decision level fusion using multiple kernel learning and fuzzy integrals. In: 2015 IEEE International Conference on Fuzzy Systems, FUZZ-IEEE, pp. 1–7, August 2015

13. Hu, L., Anderson, D.T., Havens, T.C., Keller, J.M.: Efficient and scalable nonlinear multiple kernel aggregation using the Choquet integral. In: Laurent, A., Strauss, O., Bouchon-Meunier, B., Yager, R.R. (eds.) IPMU 2014. CCIS, vol. 442, pp. 206–215. Springer, Cham (2014). https://doi.org/10.1007/978-3-319-08795-5_22
14. Hu, L., Anderson, D.T., Havens, T.C.: Multiple kernel aggregation using fuzzy integrals. In: 2013 IEEE International Conference on Fuzzy Systems, FUZZ-IEEE, pp. 1–7, July 2013
15. Grabisch, M., Nguyen, H.T., Walker, E.A.: Fundamentals of Uncertainty Calculi with Applications to Fuzzy Inference, vol. 30. Springer Science & Business Media, Dordrecht (2013)
16. Keller, J.M., Osborn, J.: Training the fuzzy integral. Int. J. Approx. Reason. **15**(1), 1–24 (1996)
17. Keller, J.M., Osborn, J.: A reward/punishment scheme to learn fuzzy densities for the fuzzy integral. In: Proceedings of the International Fuzzy Systems Association World Congress, pp. 97–100 (1995)
18. Mendez-Vazquez, A., Gader, P.: Sparsity promotion models for the Choquet integral. In: IEEE Symposium on Foundations of Computational Intelligence, FOCI 2007, pp. 454–459. IEEE (2007)
19. Beliakov, G.: Construction of aggregation functions from data using linear programming. Fuzzy Sets Syst. **160**(1), 65–75 (2009)
20. Islam, M.A., Anderson, D.T., Pinar, A.J., Havens, T.C.: Data-driven compression and efficient learning of the Choquet integral. IEEE Trans. Fuzzy Syst. **PP**(99), 1 (2017)
21. Anderson, D.T., Islam, M., King, R., Younan, N.H., Fairley, J.R., Howington, S., Petry, F., Elmore, P., Zare, A.: Binary fuzzy measures and Choquet integration for multi-source fusion. In: 6th International Conference on Military Technologies, May 2017
22. Du, X., Zare, A., Keller, J.M., Anderson, D.T.: Multiple instance Choquet integral for classifier fusion. In: 2016 IEEE Congress on Evolutionary Computation, CEC, pp. 1054–1061, July 2016
23. Anderson, D.T., Price, S.R., Havens, T.C.: Regularization-based learning of the Choquet integral. In: 2014 IEEE International Conference on Fuzzy Systems, FUZZ-IEEE, pp. 2519–2526, July 2014
24. Cho, S.B., Kim, J.H.: Combining multiple neural networks by fuzzy integral for robust classification. IEEE Trans. Syst. Man Cybern. **25**(2), 380–384 (1995)
25. Olsson, C., Eriksson, A.P., Kahl, F.: Solving large scale binary quadratic problems: spectral methods vs. semidefinite programming. In: IEEE Conference on Computer Vision and Pattern Recognition, CVPR 2007, pp. 1–8. IEEE (2007)
26. Glover, F., Kochenberger, G.A., Alidaee, B.: Adaptive memory tabu search for binary quadratic programs. Manag. Sci. **44**(3), 336–345 (1998)
27. Sperner, E.: Ein satz über untermengen einer endlichen menge. Math. Z. **27**(1), 544–548 (1928)

Mapping Utilities to Transitive Preferences

Thomas A. Runkler(✉)

Siemens AG, Corporate Technology, Otto–Hahn–Ring 6, 81739 München, Germany
thomas.runkler@siemens.com

Abstract. This article deals with the construction of (pairwise) preference relations from degrees of utilities, e.g. ratings. Recently, the U2P method has been proposed for this purpose, but U2P is neither additively nor multiplicatively transitive. This paper proposes the U2PA and the U2PM methods. U2PA is additively transitive, and U2PM is multiplicatively transitive. Moreover, both U2PA and U2PM have linear preference over ambiguity.

Keywords: Preference relations · Utility theory · Rating
Decision making

1 Introduction

Utilities and preferences play an important role in rating and decision making processes [2]. Degrees of utilities are used to quantify how favorable each option is. Degrees of preferences are used to quantify to which degree each option is preferred against each other option. In some cases utility ratings are more convenient to specify and use, and in other cases preference ratings are more convenient [4]. To treat both cases in a unified decision making environment, we need to be able to convert utilities to preferences, and to convert preferences to utilities. For such conversions the U2P and P2U mappings have been recently proposed [9]. In experiments with movie rating data [1,7] U2P produced reasonable results. Preference matrices generated by U2P have interesting mathematical properties, but they are neither additively nor multiplicatively transitive. However, many authors have argued that additive or multiplicative transitivity are important properties of preference relations [6,8,10,11].

In this paper we propose two new functions to map utilities to preferences: U2PA generates preferences that are additively transitive, and U2PM generates preferences that are multiplicatively transitive. Both methods have the additional property that preference over ambiguity increases linearly with utility. We also introduce the corresponding inverse mappings P2UA and P2UM and present an extensive study of mathematical properties comparing the U2P, U2PA, and U2PM mappings.

This paper is structured as follows: Sect. 2 briefly reviews properties of preference relations and mappings of utilities to preferences. This includes properties

J. Medina et al. (Eds.): IPMU 2018, CCIS 853, pp. 127–139, 2018.
https://doi.org/10.1007/978-3-319-91473-2_11

from the literature as well as the discussion of a new property called linear preference over ambiguity. Section 3 introduces U2PA that generates preferences with additive transitivity, and Sect. 4 introduces U2PM that generates preferences with multiplicative transitivity. Section 5 presents a study of the mathematical properties of U2P, U2PA, and U2PM. Section 6 summarizes our conclusions and lists some open problems for future research.

2 Properties of Preference Relations

Mathematically we express (normalized) degrees of utility u and (normalized) degrees of preference p as numerical values in the unit interval $[0, 1]$, where 0 refers to no utility or no preference, 1 refers to full utility or full preference, 1/2 refers to ambiguity, and all other values between 0 and 1 are used to gradually interpolate between these two extremes [5]. Given a set of n options, the degrees of utility can be written as a n–dimensional vector $u = (u_1, \ldots, u_n) \in [0, 1]^n$, and the degrees of preference can be written as an $n \times n$ matrix

$$P = \begin{pmatrix} p_{11} & \cdots & p_{1n} \\ \vdots & \ddots & \vdots \\ p_{n1} & \cdots & p_{nn} \end{pmatrix} \tag{1}$$

where each matrix element $p_{ij} \in [0, 1]$ quantifies the degree of preference of option i over option j, $i, j \in \{1, \ldots, n\}$, and where no option is preferred over itself, so $p_{ii} = 1/2$ for all $i = 1, \ldots, n$ [4]. Mapping of utilities to preferences, or short *preference mapping*, can be mathematically expressed by a function

$$p_{ij} = f(u_i, u_j) \tag{2}$$

and the *reverse preference mapping* can be expressed by a function

$$u_i = g(u_j, p_{ij}) \tag{3}$$

The preference mapping called U2P proposed in [9] is defined as

$$p_{ij} = \frac{u_i(1 - u_j)(1 - u_i + u_j)}{u_i(1 - u_i) + u_j(1 - u_j)} \tag{4}$$

for $u_i, u_j \in (0, 1)$. For the examination of the properties of U2P in Sect. 5 it is sometimes convenient to write U2P as

$$p_{ij} = \frac{1 - u_i + u_j}{\dfrac{u_j}{u_i} + \dfrac{1 - u_i}{1 - u_j}} \tag{5}$$

In this section we present a number of desirable mathematical properties of preference mapping [6] which in the following sections will be used to construct transitive preference mapping functions.

2.1 Crisp Cases

We consider gradual degrees of utility $u_i, u_j \in [0, 1]$. For the special cases of crisp (non–gradual) utilities $u_i, u_j \in \{0, 1\}$ we require ambiguity if the utilities of both options are equal,

$$u_i = 0, \quad u_j = 0 \quad \Rightarrow \quad p_{ij} = 0.5 \tag{6}$$

$$u_i = 1, \quad u_j = 1 \quad \Rightarrow \quad p_{ij} = 0.5 \tag{7}$$

zero preference of an option with zero utility over an option with full utility

$$u_i = 0, \quad u_j = 1 \quad \Rightarrow \quad p_{ij} = 0 \tag{8}$$

and full preference of an option with full utility over an option with zero utility

$$u_i = 1, \quad u_j = 0 \quad \Rightarrow \quad p_{ij} = 1 \tag{9}$$

If these four conditions hold, then we call the preference mapping *compatible with the crisp cases.*

2.2 Zero and One Conditions

If the utility of one option is zero and the utility of the other option is nonzero, then we expect zero or full preference:

$$u_i = 0, \quad u_j > 0 \quad \Rightarrow \quad p_{ij} = 0 \tag{10}$$

$$u_i > 0, \quad u_j = 0 \quad \Rightarrow \quad p_{ij} = 1 \tag{11}$$

If the utility of one option is one and the utility of the other option is less than one, then we also expect zero or full preference:

$$u_i = 1, \quad u_j < 1 \quad \Rightarrow \quad p_{ij} = 1 \tag{12}$$

$$u_i < 1, \quad u_j = 1 \quad \Rightarrow \quad p_{ij} = 0 \tag{13}$$

We call these four conditions the *zero and one conditions.* Notice that each of (10) and (13) implies (8), and that each of (11) and (12) implies (9).

2.3 Reciprocity

A preference relation is called *reciprocal* if and only if

$$p_{ij} + p_{ji} = 1 \tag{14}$$

for all $i, j \in \{1, \ldots, n\}$.

2.4 Monotonicity

A preference mapping is called *monotonic* if and only if

$$u_j \leq u_k \quad \Rightarrow \quad p_{ij} \geq p_{ik} \quad \text{and} \quad p_{ji} \leq p_{ki} \tag{15}$$

for all $i \in \{1, \ldots, n\}$. It was shown in [9] that every monotonic mapping satisfies the triangle condition [8]

$$p_{ij} + p_{jk} \geq p_{ik} \tag{16}$$

2.5 Positivity

We call a preference mapping *positive* if and only if

$$p_{ij} \geq 0.5 \quad \Leftrightarrow \quad u_i \geq u_j \tag{17}$$

Obviously, every monotonic preference mapping is positive if and only if

$$p_{ij} = 0.5 \quad \Leftrightarrow \quad u_i = u_j \tag{18}$$

2.6 Transitivity

A preference relation is called *weakly transitive* [11] if and only if

$$p_{ij} \geq 0.5 \quad \wedge \quad p_{jk} \geq 0.5 \quad \Rightarrow \quad p_{ik} \geq 0.5 \tag{19}$$

Positivity implies weak transitivity because

$$p_{ij} \geq 0.5 \;\wedge\; p_{jk} \geq 0.5 \;\Rightarrow\; u_i \geq u_j \geq u_k \;\Rightarrow\; u_i \geq u_k \;\Rightarrow\; p_{ik} \geq 0.5 \quad \square \tag{20}$$

A preference relation is called *max–min transitive* [3,12] if and only if

$$p_{ik} \geq \min(p_{ij}, p_{jk}) \tag{21}$$

A preference relation is called *max–max transitive* [3,12] if and only if

$$p_{ik} \geq \max(p_{ij}, p_{jk}) \tag{22}$$

A preference relation is called *restricted max–min transitive* [11] if and only if

$$p_{ij} \geq 0.5 \quad \wedge \quad p_{jk} \geq 0.5 \quad \Rightarrow \quad p_{ik} \geq \min(p_{ij}, p_{jk}) \tag{23}$$

A preference relation is called *restricted max–max transitive* [11] if and only if

$$p_{ij} \geq 0.5 \quad \wedge \quad p_{jk} \geq 0.5 \quad \Rightarrow \quad p_{ik} \geq \max(p_{ij}, p_{jk}) \tag{24}$$

Reciprocity and positivity implies restricted max–min transitivity and restricted max–max transitivity because

$$p_{ij} \geq 0.5 \quad \Rightarrow \quad u_i \geq u_j \quad \Rightarrow \quad p_{ik} \geq p_{jk} \tag{25}$$

$$p_{jk} \geq 0.5 \quad \Rightarrow \quad u_j \geq u_k \quad \Rightarrow \quad p_{ik} \geq p_{ij} \tag{26}$$

$$p_{ik} \geq p_{jk} \quad \wedge \quad p_{ik} \geq p_{ij} \quad \Rightarrow \quad p_{ik} \geq \max(p_{ij}, p_{jk}) \geq \min(p_{ij}, p_{jk}) \quad \square \tag{27}$$

A preference relation is called *additively transitive* [10] if and only if

$$(p_{ij} - 0.5) + (p_{jk} - 0.5) = (p_{ik} - 0.5) \tag{28}$$

A preference relation is called *multiplicatively transitive* [11] if and only if

$$\frac{p_{ji}}{p_{ij}} \cdot \frac{p_{kj}}{p_{jk}} = \frac{p_{ki}}{p_{ik}} \tag{29}$$

2.7 Linear Preference over Ambiguity

We introduce a new requirement for preference mapping which we call *linear preference over ambiguity* where we require that the preference p_{ik} of an option with utility $u_i \in [0, 1]$ over an option with ambiguous preference $u_k = 0.5$ increases linearly with u_i, so

$$u_k = 0.5 \quad \Rightarrow \quad p_{ik} = \alpha u_i + \beta \tag{30}$$

with suitable parameters $\alpha, \beta \in \mathbb{R}$.

3 Mapping to Additively Transitive Preferences

To construct a mapping of utilities to preferences with additive transitivity we begin with the equation of additive transitivity (28).

$$(p_{ij} - 0.5) + (p_{jk} - 0.5) = (p_{ik} - 0.5) \tag{31}$$

We further require linear preference over ambiguity (30)

$$u_k = 0.5 \quad \Rightarrow \quad p_{ik} = \alpha u_i + \beta \tag{32}$$

Inserting this into (31) for $u_k = 0.5$ yields

$$(p_{ij} - 0.5) + (\alpha u_j + \beta - 0.5) = (\alpha u_i + \beta - 0.5) \quad \Rightarrow \quad p_{ij} = \alpha(u_i - u_j) + 0.5 \tag{33}$$

We further require compatibility with the crisp cases, so with (8) we obtain

$$u_i = 0 \quad \wedge \quad u_j = 1 \quad \Rightarrow \quad p_{ij} = -\alpha + 0.5 = 0 \quad \Rightarrow \quad \alpha = 0.5 \tag{34}$$

or with (9) we obtain the same result

$$u_i = 1 \quad \wedge \quad u_j = 0 \quad \Rightarrow \quad p_{ij} = \alpha + 0.5 = 1 \quad \Rightarrow \quad \alpha = 0.5 \tag{35}$$

Inserting $\alpha = 0.5$ into (33) finally yields

$$p_{ij} = \frac{1}{2}(u_i - u_j + 1) \tag{36}$$

which we call the *U2PA* preference mapping, where A is for *additive transitivity*. The corresponding reverse preference mapping *P2UA* can be easily obtained as

$$u_i = 2p_{ij} + u_j - 1 \tag{37}$$

4 Mapping to Multiplicatively Transitive Preferences

To construct a mapping of utilities to preferences with multiplicative transitivity we begin with the equation of multiplicative transitivity (29)

$$\frac{p_{ji}}{p_{ij}} \cdot \frac{p_{kj}}{p_{jk}} = \frac{p_{ki}}{p_{ik}} \tag{38}$$

We further require reciprocity (14)

$$p_{ij} + p_{ji} = 1 \tag{39}$$

and linear preference over ambiguity (30)

$$u_k = 0.5 \quad \Rightarrow \quad p_{ik} = \alpha u_i + \beta \tag{40}$$

Inserting these into (38) for $u_k = 0.5$ yields

$$\left(\frac{1}{p_{ij}} - 1\right) \frac{1 - \alpha u_j - \beta}{\alpha u_j + \beta} = \frac{1 - \alpha u_i - \beta}{\alpha u_i + \beta} \tag{41}$$

$$\Rightarrow \quad \frac{1}{p_{ij}} - 1 = \frac{(1 - \alpha u_i - \beta)(\alpha u_j + \beta)}{(1 - \alpha u_j - \beta)(\alpha u_i + \beta)} \tag{42}$$

$$\Rightarrow \quad \frac{1}{p_{ij}} = \frac{(1 - \alpha u_i - \beta)(\alpha u_j + \beta) + (1 - \alpha u_j - \beta)(\alpha u_i + \beta)}{(1 - \alpha u_j - \beta)(\alpha u_i + \beta)} \tag{43}$$

$$\Rightarrow \quad p_{ij} = \frac{(1 - \alpha u_j - \beta)(\alpha u_i + \beta)}{(1 - \alpha u_i - \beta)(\alpha u_j + \beta) + (1 - \alpha u_j - \beta)(\alpha u_i + \beta)} \tag{44}$$

Now we additionally require the zero and one conditions and obtain for (10)

$$u_i = 0, \quad u_j > 0 \quad \Rightarrow \quad p_{ij} = \frac{(1 - \alpha u_j - \beta)\beta}{(1 - \beta)(\alpha u_j + \beta) + (1 - \alpha u_j - \beta)\beta} \tag{45}$$

$$= \frac{1}{1 + \frac{(1 - \beta)(\alpha u_j + \beta)}{(1 - \alpha u_j - \beta)\beta}} \tag{46}$$

which approaches 0 for $\beta \to 0$, and obtain for (13)

$$u_i < 1, \quad u_j = 1 \quad \Rightarrow \quad p_{ij} = \frac{(1 - \alpha)\alpha u_i}{(1 - \alpha u_i)\alpha + (1 - \alpha)\alpha u_i} \tag{47}$$

$$= \frac{1}{1 + \frac{1 - \alpha u_i}{(1 - \alpha)u_i}} \tag{48}$$

which approaches 0 for $\alpha \to 1$. Inserting $\alpha = 1$ and $\beta = 0$ into (44) finally yields

$$p_{ij} = \frac{u_i(1 - u_j)}{u_i(1 - u_j) + u_j(1 - u_i)} \tag{49}$$

for $u_i, u_j \in (0,1)$, which we call the *U2PM* preference mapping, where *M* is for *multiplicative transitivity*. U2PM looks similar to U2P (4), but the factor $(1 - u_i + u_j)$ in the nominator is dropped, and the indices i and j in the denominator are changed. For the examination of the properties of U2PM in the following section it is sometimes convenient to write U2PM as

$$p_{ij} = \frac{1}{1 + \dfrac{u_j(1-u_i)}{u_i(1-u_j)}} = \frac{1}{1 + \left(\dfrac{1}{u_i} - 1\right)\left(\dfrac{1}{1-u_j} - 1\right)} \tag{50}$$

We obtain the reverse preference mapping *P2UM* by solving (49) for u_i.

$$u_i(1-u_j)p_{ij} + u_j(1-u_i)p_{ij} = u_i(1-u_j) \tag{51}$$

$$\Rightarrow \frac{1-u_i}{u_i} u_j p_{ij} = (1-u_j)(1-p_{ij}) \tag{52}$$

$$\Rightarrow \frac{1}{u_i} - 1 = \frac{(1-u_j)(1-p_{ij})}{u_j p_{ij}} \tag{53}$$

$$\Rightarrow \frac{1}{u_i} = \frac{(1-u_j)(1-p_{ij}) + u_j p_{ij}}{u_j p_{ij}} = \frac{1 - u_j - p_{ij} + 2u_j p_{ij}}{u_j p_{ij}} \tag{54}$$

$$\Rightarrow u_i = \frac{u_j p_{ij}}{1 - u_j - p_{ij} + 2u_j p_{ij}} \tag{55}$$

This is a much simpler expression than the reverse preference mapping P2U proposed in [9].

5 Properties of the U2P, U2PA, and U2PM Mappings

In this section we examine U2P (4), (5), U2PA (36), and U2PM (49), (50) with respect to the properties that were defined in Sect. 2.

U2P and U2PM are not compatible with the crisp cases $u_i = 0$, $u_j = 0$ and $u_i = 1$, $u_j = 1$, since they are undefined and unsteady in these points. U2PA is compatible with the crisp cases since

$$u_i = 0, \quad u_j = 0 \quad \Rightarrow \quad p_{ij} = \frac{1}{2}(0 - 0 + 1) = 0.5 \tag{56}$$

$$u_i = 1, \quad u_j = 1 \quad \Rightarrow \quad p_{ij} = \frac{1}{2}(1 - 1 + 1) = 0.5 \tag{57}$$

$$u_i = 0, \quad u_j = 1 \quad \Rightarrow \quad p_{ij} = \frac{1}{2}(0 - 1 + 1) = 0 \tag{58}$$

$$u_i = 1, \quad u_j = 0 \quad \Rightarrow \quad p_{ij} = \frac{1}{2}(1 - 0 + 1) = 1 \quad \square \tag{59}$$

Notice that U2PA is not only compatible with the crisp case, but satisfies an even stronger property:

$$u_i = u_j \quad \Rightarrow \quad p_{ij} = \frac{1}{2} \tag{60}$$

U2P satisfies the zero and one conditions (sometimes only in the limit) since

$$u_i \to 0, \quad u_j > 0 \quad \Rightarrow \quad p_{ij} \to \lim_{\varepsilon \to 0} \frac{1 - 0 + u_j}{\frac{u_j}{\varepsilon} + \frac{1-0}{1-u_j}} = 0 \tag{61}$$

$$u_i > 0, \quad u_j = 0 \quad \Rightarrow \quad p_{ij} = \frac{1 - u_i + 0}{\frac{0}{u_i} + \frac{1-u_i}{1-0}} = \frac{1-u_i}{1-u_i} = 1 \tag{62}$$

$$u_i = 1, \quad u_j < 1 \quad \Rightarrow \quad p_{ij} = \frac{1 - 1 + u_j}{\frac{u_j}{1} + \frac{1-1}{1-u_j}} = \frac{u_j}{u_j} = 1 \tag{63}$$

$$u_i < 1, \quad u_j \to 1 \quad \Rightarrow \quad p_{ij} \to \lim_{\varepsilon \to 0} \frac{1 - u_i + 1}{\frac{1}{u_i} + \frac{1-u_i}{1-1+\varepsilon}} = 0 \quad \square \tag{64}$$

U2PA violates all zero and one conditions, for example for

$$u_i = 0, \quad u_j = 0.5 \quad \Rightarrow \quad p_{ij} = \frac{1}{2}(0 - 0.5 + 1) = 0.25 \neq 0 \tag{65}$$

$$u_i = 0.5, \quad u_j = 0 \quad \Rightarrow \quad p_{ij} = \frac{1}{2}(0.5 - 0 + 1) = 0.75 \neq 1 \tag{66}$$

$$u_i = 1, \quad u_j = 0.5 \quad \Rightarrow \quad p_{ij} = \frac{1}{2}(1 - 0.5 + 1) = 0.75 \neq 1 \tag{67}$$

$$u_i = 0.5, \quad u_j = 1 \quad \Rightarrow \quad p_{ij} = \frac{1}{2}(0.5 - 1 + 1) = 0.25 \neq 0 \quad \square \tag{68}$$

U2PM satisfies the zero and one conditions (sometimes only in the limit) since

$$u_i \to 0, \quad u_j > 0 \quad \Rightarrow \quad p_{ij} \to \lim_{\varepsilon \to 0} \frac{1}{1 + \left(\frac{1}{\varepsilon} - 1\right)\left(\frac{1}{1-u_j} - 1\right)} = 0 \tag{69}$$

$$u_i > 0, \quad u_j = 0 \quad \Rightarrow \quad p_{ij} = \frac{(1-0)u_i}{(1-u_i)0 + (1-0)u_i} = \frac{u_i}{u_i} = 1 \tag{70}$$

$$u_i = 1, \quad u_j < 1 \quad \Rightarrow \quad p_{ij} = \frac{(1-u_j)1}{(1-1)u_j + (1-u_j)1} = \frac{1-u_j}{1-u_j} = 1 \tag{71}$$

$$u_i < 1, \quad u_j \to 1 \quad \Rightarrow \quad p_{ij} \to \lim_{\varepsilon \to 0} \frac{1}{1 + \left(\frac{1}{u_i} - 1\right)\left(\frac{1}{1-1+\varepsilon} - 1\right)} = 0 \quad \square \tag{72}$$

U2P satisfies reciprocity since

$$p_{ij} + p_{ji} = \frac{u_i(1-u_j)(1-u_i+u_j)}{u_i(1-u_i)+u_j(1-u_j)} + \frac{u_j(1-u_i)(1-u_j+u_i)}{u_j(1-u_j)+u_i(1-u_i)} \tag{73}$$

$$= \frac{u_i(1-u_j)(1-u_i+u_j)+u_j(1-u_i)(1-u_j+u_i)}{u_i(1-u_i)+u_j(1-u_j)} \tag{74}$$

$$= \frac{u_i(1-u_i)(1-u_j)+u_j(1-u_j)u_i+u_j(1-u_j)(1-u_i)+u_i(1-u_i)u_j}{u_i(1-u_i)+u_j(1-u_j)} \tag{75}$$

$$= \frac{u_i(1-u_i)+u_j(1-u_j)}{u_i(1-u_i)+u_j(1-u_j)} = 1 \quad \square \tag{76}$$

U2PA also satisfies reciprocity since

$$p_{ij} + p_{ji} = \frac{1}{2}(u_i - u_j + 1) + \frac{1}{2}(u_j - u_i + 1) = \frac{2}{2} = 1 \quad \square \tag{77}$$

And also U2PM satisfies reciprocity since

$$p_{ij} + p_{ji} = \frac{u_i(1-u_j)}{u_i(1-u_j)+u_j(1-u_i)} + \frac{u_j(1-u_i)}{u_j(1-u_i)+u_i(1-u_j)} \tag{78}$$

$$= \frac{u_i(1-u_j)+u_j(1-u_i)}{u_i(1-u_j)+u_j(1-u_i)} = 1 \quad \square \tag{79}$$

It has been shown in [9] that U2P is monotonic. U2PA is also monotonic since

$$p_{ij} = \frac{1}{2}(u_i - u_j + 1) \tag{80}$$

obviously increases with u_i and decreases with u_j. And also U2PM is monotonic since the denominator of

$$p_{ij} = \frac{1}{1 + \left(\frac{1}{u_i} - 1\right)\left(\frac{1}{1-u_j} - 1\right)} \tag{81}$$

decreases with u_i and increases with u_j. Since all three methods are monotonic, they also satisfy the triangle inequality according to (16).

It has been shown in [9] that U2P is weakly transitive, restricted max–min transitive, restricted max–max transitive, but not (strictly) max–min transitive, not (strictly) max–max transitive, not multiplicatively transitive, and not additively transitive. U2PA satisfies condition (18) since

$$p_{ij} = \frac{1}{2}(u_i - u_j + 1) = 0.5 \quad \Rightarrow \quad u_i - u_j + 1 = 1 \quad \Rightarrow \quad u_i = u_j, \tag{82}$$

$$u_i = u_j \quad \Rightarrow \quad p_{ij} = \frac{1}{2}(u_i - u_j + 1) = \frac{1}{2}(u_i - u_i + 1) = 0.5 \quad \square \tag{83}$$

Also U2PM satisfies condition (18) since

$$p_{ij} = \frac{u_i(1-u_j)}{u_i(1-u_j)+u_j(1-u_i)} = 0.5 \tag{84}$$

$$\Rightarrow \quad 2u_i(1-u_j) = u_i(1-u_j) + u_j(1-u_i) \tag{85}$$

$$\Rightarrow \quad u_i(1-u_j) = u_j(1-u_i) \quad \Rightarrow \quad u_i - u_iu_j = u_j - u_iu_j \quad \Rightarrow \quad u_i = u_j, \tag{86}$$

$$u_i = u_j \quad \Rightarrow \quad p_{ij} = \frac{u_i(1-u_j)}{u_i(1-u_j) + u_j(1-u_i)} \tag{87}$$

$$= \frac{u_i(1-u_i)}{u_i(1-u_i) + u_i(1-u_i)} = 0.5 \quad \square \tag{88}$$

So, U2PA and U2PM satisfy condition (18) and they are both monotonic, so they are both positive, and so the are both weakly transitive (20), restricted max–min transitive (25)–(27), and restricted max–max transitive (25)–(27). For the utility values $u_i = 0.3$, $u_j = 0.6$, $u_k = 0.8$, U2PA yields

$$p_{ik} = \frac{1}{2}(0.3 - 0.8 + 1) = 0.25 \tag{89}$$

$$p_{ij} = \frac{1}{2}(0.3 - 0.6 + 1) = 0.35 \tag{90}$$

$$p_{jk} = \frac{1}{2}(0.6 - 0.8 + 1) = 0.4 \tag{91}$$

so U2PA is not (strictly) max–min transitive because

$$p_{ik} = 0.25 \not\geq \min(p_{ij}, p_{jk}) = \min(0.35, 0.4) = 0.35 \quad \square \tag{92}$$

and U2PA is not (strictly) max–max transitive because

$$p_{ik} = 0.25 \not\geq \max(p_{ij}, p_{jk}) = \max(0.35, 0.4) = 0.4 \quad \square \tag{93}$$

For the same utility values U2PM yields

$$p_{ik} = \frac{0.3(1-0.8)}{0.3(1-0.8) + 0.8(1-0.3)} = 3/31 \approx 0.097 \tag{94}$$

$$p_{ij} = \frac{0.3(1-0.6)}{0.3(1-0.6) + 0.6(1-0.3)} = 2/9 \approx 0.222 \tag{95}$$

$$p_{jk} = \frac{0.6(1-0.8)}{0.6(1-0.8) + 0.8(1-0.6)} = 3/11 \approx 0.273 \tag{96}$$

so U2PM is not (strictly) max–min transitive because

$$p_{ik} \approx 0.097 \not\geq \min(p_{ij}, p_{jk}) \approx \min(0.222, 0.273) = 0.222 \quad \square \tag{97}$$

and U2PM is not (strictly) max–max transitive because

$$p_{ik} \approx 0.097 \not\geq \max(p_{ij}, p_{jk}) \approx \max(0.222, 0.273) = 0.273 \quad \square \tag{98}$$

When we constructed U2PA we required additive transitivity, and when we constructed U2PM we required multiplicative transitivity. The example from above shows that U2PA is not multiplicatively transitive because

$$\frac{p_{ji}}{p_{ij}} \cdot \frac{p_{kj}}{p_{jk}} = \frac{0.65}{0.35} \cdot \frac{0.6}{0.4} = 39/14 \approx 2.786 \neq \frac{p_{ki}}{p_{ik}} = \frac{0.75}{0.25} = 3 \quad \square \tag{99}$$

The same example also shows that U2PM is not additively transitive because

$$(p_{ij} - 0.5) + (p_{jk} - 0.5) = (2/9 - 0.5) + (3/11 - 0.5) = -50/99 \approx -0.505$$
$$\neq (p_{ik} - 0.5) = 3/31 - 0.5 = -25/62 \approx -0.403 \quad \square \tag{100}$$

When we constructed U2PA and U2PM we required linear preference over ambiguity. U2P does not have linear preference over ambiguity because

$$u_k = 0.5 \quad \Rightarrow \quad p_{ik} = \frac{0.5u_i(1 - u_i + 0.5)}{u_i(1 - u_i) + 0.5(1 - 0.5)} \tag{101}$$

$$= \frac{0.5u_i(1.5 - u_i)}{u_i(1 - u_i) + 0.25} = \frac{-0.5u_i^2 + 0.75u_i}{-u_i^2 + u_i + 0.25} \tag{102}$$

$$= \frac{u_i(u_i - 1.5)}{2(u_i - 0.5 + 0.5\sqrt{2})(u_i - 0.5 - 0.5\sqrt{2})} \tag{103}$$

which is nonlinear in u_i.

The properties of the U2P, U2PA, and U2PM methods are summarized in Table 1. None of the three methods satisfies restricted max–min or restricted max–max transitivity. Beyond these strict transitivities, U2P violates crisp case

Table 1. Properties of the U2P, U2PA, and U2PM methods.

	U2P	U2PA	U2PM
Crisp case compatibility		•	
Zero and one conditions	•		•
Reciprocity	•	•	•
Monotonicity	•	•	•
Positivity	•	•	•
Weak transitivity	•	•	•
Max–min transitivity			
Max–max transitivity			
Restricted max–min transitivity	•	•	•
Restricted max–max transitivity	•	•	•
Additive transitivity		•	
Multiplicative transitivity			•
Linear preference over ambiguity		•	•

compatibility, additive transitivity, multiplicative transitivity, and linear preference over ambiguity; U2PA violates the zero and one conditions and multiplicative transitivity; and U2PM violates crisp case compatibility and additive transitivity.

6 Conclusions

We have introduced the U2PA and U2PM methods for mapping utilities to preferences, and the corresponding reverse mappings P2UA and P2UM. In contrast to the previously suggested U2P method, U2PA is additively transitive and U2PM is multiplicatively transitive. Moreover, both U2PA and U2PM have linear preference over ambiguity, and U2PA is compatible with the crisp cases. Additive and multiplicative transitivity are considered important properties of preference relations. Therefore, the new U2PA and U2PM are useful tools for the generation of preferences from utilities. Many questions are left open for future research, for example: Is there a preference relation that satisfies both additive and multiplicative transitivity? What other mappings with additive or multiplicative transitivity can be constructed without demanding linear preference over ambiguity? Can we satisfy the zero and one conditions together with additive transitivity? Can we satisfy crisp case compatibility together with multiplicative transitivity? How can we achieve (strict) max–min or max–max transitivity?

References

1. Bell, R.M., Koren, Y.: Lessons from the Netflix prize challenge. ACM SIGKDD Explor. Newsl. **9**(2), 75–79 (2007)
2. Bellman, R., Zadeh, L.: Decision making in a fuzzy environment. Manag. Sci. **17**(4), 141–164 (1970)
3. Dubois, D., Prade, H.: Fuzzy Sets System. Academic Press, London (1980)
4. Fishburn, P.C.: Utility Theory. Wiley, Hoboken (1988)
5. Fodor, J.C., Roubens, M.R.: Fuzzy Preference Modelling and Multicriteria Decision Support, vol. 14. Springer Science & Business Media, Heidelberg (1994). https://doi.org/10.1007/978-94-017-1648-2
6. Herrera-Viedma, E., Herrera, F., Chiclana, F., Luque, M.: Some issues on consistency of fuzzy preference relations. Eur. J. Oper. Res. **154**(1), 98–109 (2004)
7. Jeon, T., Cho, J., Lee, S., Baek, G., Kim, S.: A movie rating prediction system of user propensity analysis based on collaborative filtering and fuzzy system. In: IEEE International Conference on Fuzzy Systems, Jeju, Korea, pp. 507–511 (2009)
8. Luce, R.D., Suppes, P.: Preferences utility and subject probability. In: Luce, R.D., Bush, R.R., Eugene, G.E. (eds.) Handbook of Mathematical Psychology, vol. III, pp. 249–410. Wiley, New York (1965)
9. Runkler, T.A.: Constructing preference relations from utilities and vice versa. In: Carvalho, J.P., Lesot, M.-J., Kaymak, U., Vieira, S., Bouchon-Meunier, B., Yager, R.R. (eds.) IPMU 2016. CCIS, vol. 611, pp. 547–558. Springer, Cham (2016). https://doi.org/10.1007/978-3-319-40581-0_44

10. Tanino, T.: Fuzzy preference orderings in group decision making. Fuzzy Sets Syst. **12**(2), 117–131 (1984)
11. Tanino, T.: Fuzzy preference relations in group decision making. In: Kacprzyk, J., Roubens, M. (eds.) Non-Conventional Preference Relations in Decision Making. LNE, vol. 301, pp. 54–71. Springer, Heidelberg (1988). https://doi.org/10.1007/978-3-642-51711-2_4
12. Zimmermann, H.J.: Fuzzy Set Theory and Its Applications. Kluwer Academic Publishers, Boston (1985)

Mortality Rates Smoothing Using Mixture Function

Samuel Hudec[1] and Jana Špirková[2(✉)]

[1] Faculty of Natural Sciences, Matej Bel University, Tajovského 40, 974 09 Banská Bystrica, Slovakia
samuel.hudec@umb.sk

[2] Faculty of Economics, Matej Bel University, Tajovského 10, 975 90 Banská Bystrica, Slovakia
jana.spirkova@umb.sk

Abstract. The paper offers a description of the new method of a smoothing of the mortality rates using the so-called moving mixture functions. Mixture functions represent a special class of weighted averaging functions where weights are determined by continuous, input values dependent, weighting functions. If they are increasing, they form an important class of aggregation functions. Such mixture functions are more flexible than the standard weighted arithmetic mean, and their weighting functions allow one to penalize or reinforce inputs based on their magnitude. The advantages of this method are that the weights of the input values depend on ourselves and coefficients of weighting functions can be changed each year so that the mean square error is minimized. Moreover, the paper offers the impact of this method on the amount of whole life pension annuities.

Keywords: Mixture function · Moving mixture function
Mortality rate · Smoothing · Annuity

1 Introduction

Mortality rate is a measure of the number of deaths in a particular population, scaled to the size of that population, per unit of time. There are several different mortality rates used to monitor the level of mortality in populations. So-called crude mortality rates are most commonly used. It counts all deaths, all causes, all ages and both sexes. These data are usually available on the websites of the statistical offices of individual countries [12,22]. More complete image of mortality is given by life tables, which show the mortality rates separately for each age, and therefore smoothing of discrete time series and subsequent predictions of future mortality rates is a problem of fundamental importance in a demography, but also in an insurance and pension schemes. Several methods are known which are used to smoothing the mortality curves [7,14,22].

J. Medina et al. (Eds.): IPMU 2018, CCIS 853, pp. 140–150, 2018.
https://doi.org/10.1007/978-3-319-91473-2_12

The main motivation of our investigation is to offer a method of curve smoothing that is as close as possible to the Lee-Carter model which is presented in [14]. The Lee–Carter model is a numerical algorithm used in mortality forecasting and life expectancy forecasting and solves so-called longevity as a risk loading of financial and insurance products. That is reason we have tried to apply so-called mixture functions to a smoothing of mortality rates.

Mixture functions represent a special class of weighted averaging functions where weights are determined by continuous, input values dependent, weighting functions. For more information see, for example [1–6,9,16,17]. If they are increasing, they form an important class of aggregation functions. Such mixture functions are more flexible than the standard weighted arithmetic mean, and their weighting functions allow one to penalise or reinforce inputs based on their magnitude. Because, mixture functions have a lot of very good properties, and they are also weighted averages with special weights, we want to propose new smoothing method of mortality rates just using mixture aggregation. This contribution recalls sufficient conditions of the standard monotonicity of mixture functions with selected weighting functions.

As [15,27] emphasize, a socially responsible approach to entrepreneurship and the maintenance of good morals must be explored at the level of all enterprises, not just the largest ones. Developing and innovative entrepreneurial activities are currently one of the most important elements of business and are pillars of their competitiveness and stability.

Therefore smoothing of mortality rate curves is one of the very important tools for risk management of insurance companies and represents the basic building block in all actuarial calculations [11,21,29].

The remainder of this paper is structured as follows. In Sect. 2, we provide the necessary properties of aggregation functions with special stress on the monotonicity of mixture functions. We also describe in this section the methodology of smoothing of mortality rates which is used on mortality rates smoothing by the Statistical Office of the Slovak Republic.

Our main results are concentrated in Sect. 3, where we introduce new discrete time series smoothing method using so-called moving mixture functions with selected weighting functions instead of standard moving weighted averaging. Moreover, in this section, we compare our results with a known methods. On the basis of obtained results, we point out the impact of mortality rates smoothing on whole life annuities. Our conclusions are presented in Sect. 4.

We have done the whole modelling using the libraries of the R software [18,20,28].

2 Preliminaries

In this section, we offer basic knowledge about aggregation functions and mixture functions as a subset of aggregation functions. Also, we recall mortality rates smoothing as a basic building block of demographic and insurance calculations.

We would like to emphasize that due to standard notations, we use in this section the same notation for input values $x_i, i = 1, 2, \ldots, n$, or independent variable x, respectively (Subsect. 2.1) and x as an age of an individual (Subsect. 2.2).

2.1 Mixture Function as an Aggregation Function

Throughout the paper, we give standard monotonicity of mixture functions on the interval $[0,1]$. The choice of the unit interval is not restrictive. In general, we could study these functions on any arbitrary closed non-empty interval $\mathbf{I} = [a,b] \subset \overline{\mathbf{R}} = [-\infty, \infty]$, $\mathbf{I}^n = \{\mathbf{x} = (x_1, \ldots, x_n) \mid x_i \in \mathbf{I}, i = 1, \ldots, n\}$.

Definition 1. ***Aggregation function.*** [2,9] *A function $F : [0,1]^n \to [0,1]$ is called an n-ary aggregation function if the following conditions hold:*

(A1) F satisfies the boundary conditions $F(0,0,\ldots,0)=0$ and $F(1,1,\ldots,1)=1$,
(A2) F is standard monotone increasing.

Definition 2. ***Standard monotonicity.*** *A function $F : [0,1]^n \to [0,1]$ is monotone increasing if for every $(x_1,\ldots,x_n),(y_1,\ldots,y_n) \in [0,1]^n$ such that $x_i \geq y_i$ for every $i = 1,\ldots,n$, the inequality $F(x_1,\ldots,x_n) \geq F(y_1,\ldots,y_n)$ holds.*[1]

Definition 3. ***Mixture function.*** [9] *A function $M_g : [0,1]^n \to [0,1]$ given by*

$$M_g(x_1, \ldots, x_n) = \frac{\sum_{i=1}^{n} g(x_i) \cdot x_i}{\sum_{i=1}^{n} g(x_i)}, \tag{1}$$

where $g : [0,1] \to]0,\infty[$ is a continuous weighting function, is called a mixture function.

Now, we give sufficient conditions of monotonicity of the mixture function.

Theorem 1. [24] *Mixture function $M_g : [0,1]^n \to [0,1]$, given by (1), is monotone increasing if for an increasing, piecewise differentiable weighting function $g : [0,1] \to]0,\infty[$ at least one from the following conditions is satisfied:*

$$g(x) \geq g'(x), \tag{2}$$

$$g(x) \geq g'(x) \cdot (1-x). \tag{3}$$

Other sufficient conditions of the standard monotonicity of mixture functions and their generalizations can be found, for instance, in [24].

On the basis of previous conditions, we give the set of coefficients for weighting functions $g(x) = cx + 1 - c$, $g(x) = 1 + \gamma x^2$ and $g(x) = a\left(\frac{1}{a}\right)^x$ to be mixture functions monotone increasing.

[1] The term "increasing" is understood in a non-strict sense.

Proposition 1. *Let $M_g : [0,1]^n \to [0,1]$ be a mixture function with the weighting function $g_c(x) = cx + 1 - c$, $c \in [0,1[$.*

Then M_g is monotone increasing with respect to (2) and (3) for $c \in [0, 0.5]$.

Proposition 2. *Let $M_g : [0,1]^n \to [0,1]$ be a mixture function with the weighting function $g_\gamma(x) = 1 + \gamma x^2$, $\gamma > 0$.*

Then M_g is monotone increasing with respect to (2) and (3) for $\gamma \in [0,1]$, or $\gamma \in [0,3]$.

Proposition 3. *Let $M_g : [0,1]^n \to [0,1]$ be a mixture function with the weighting function $g_a(x) = a\left(\frac{1}{a}\right)^x$, $0 < a < 1$.*

Then M_g is monotone increasing with respect to (2) and (3) for $a \in [\frac{1}{e}, 1[$.

2.2 Mortality Rates Smoothing

A number of ways exist for smoothing of mortality rates. They are very often described by statistical offices of individual countries. One of these methods is described on the web page of the Statistical Office of the Slovak Republic, too [22].

It consists from the following basic steps.

1. There we have real numbers L_x of the living at age x and D_x of deaths at age x, middle condition.
2. Point estimate of the force of mortality at each age x is given by

$$\mu_x = \frac{D_x}{L_x} \tag{4}$$

The force of mortality reflects the probability that an individual at age x will die within a small time interval converging to zero.

3. Therefore, according to [8], the probability q_x that an individual at age x dies before age $(x+1)$ is as follows:

$$q_x = 1 - \exp\{-\mu_x\}. \tag{5}$$

4. These data do not form a smooth curve, therefore one can apply moving averages for probability of death q_x of an individual at age x for $4 \leq x \leq 84$ which are given by formula [22]

$$\hat{q}_x^* = \frac{105q_x + 90(q_{x-1} + q_{x+1}) + 45(q_{x-2} + q_{x+2}) - 30(q_{x-3} + q_{x+3})}{315}. \tag{6}$$

5. Subsequently one needs to apply so-called interpolation of the second degree and for higher ages smoothing by Gompertz-Makeham's Formula and special King-Hardy method.

3 Smoothing Using Moving Mixture Functions

As we have already mentioned, mixture functions are more flexible than the standard weighted arithmetic mean, and their weighting functions allow one to penalise or reinforce inputs based on their magnitude. So we decided to investigate the effect of weighting functions of the mixture functions on smoothing of mortality rates.

3.1 Methodology Using Mixture Functions

In this section, we develop new method on smoothing of mortality rates using so-called moving mixture function.

Our approach is as follows:

1. *Calculation of basic probabilities.* Based on available data, for entry ages x from the interval $[0, 100]$, we calculate point estimate of the force of mortality and probability of death with respect to formulas (4) and (5), respectively.
2. *Calculation of "moving mixture function".* We smooth the values q_x for ages $3 \leq x \leq 97$ by moving mixture function which is given as follows

$$\hat{q}_x = \frac{\sum_{j=0}^{3} g(q_{x\pm j}) \cdot q_{x\pm j}}{\sum_{j=0}^{3} g(q_{x\pm j})}. \tag{7}$$

   As weighting functions, we use already mentioned functions $g(x) = cx + 1 - c$, $g(x) = 1 + \gamma x^2$ and $g(x) = a\left(\frac{1}{a}\right)^x$.
3. *Moving mixture fit.* For probabilities obtained using moving mixture functions, we extend since age 80 the right tail by Gompertz-Makeham formula in the form $\hat{q}_x = A + B \times c^x$. We look for the best intersection with moving mixture functions.
4. *Model selection.* For all coefficients of linear weighting function of moving mixture functions, we calculate Weighted Mean Square Error MSE by formula as follows, [13]

$$MSE = \frac{\sum_{x=0}^{100} \left(q_x - \frac{\sum_{j=0}^{3} (cq_{x\pm j} + 1 - c) q_{x\pm j}}{\sum_{j=0}^{3} (cq_{x\pm j} + 1 - c)} \right)^2 \times (L_x + D_x)}{\sum_{x=0}^{100} (L_x + D_x)}. \tag{8}$$

   We think that in this case it was natural to use a Weighted Mean Square Error, because at any age it lives, but also die a different number of people. By the similar way, we calculate MSE using quadratic and exponential weighting functions.
   In the next part, we explain our procedure in details.

3.2 Analysis of Smoothing Curves

In this part, we give an explanation of Fig. 1 which contains three separate images which follow each other. We start with the top image. There are original data of mortality rate obtained using formula (5) (Legend -qx) and smoothing data obtained using moving mixture functions (7), (Legend - Mixture).

In the middle image are illustrated data obtained using moving mixture functions (Legend - Mixture) and Gompertz-Makeham smoothing which is implemented in the R software (Legend - Gompertz-Makeham). In this case the intersection of both curves at age 84 is very important. In the third image is illustrated our result - smoothing mortality rates using moving mixture functions (7) up to age 84 and the remaining part is obtained using Gompertz-Makeham function of the R software (Legend - Gompertz-Makeham). Moreover, in this image also original data q_x can be seen. We would like to emphasize that distances between individual values of moving mixture functions calculated using linear, quadratic and exponential weighting functions are so small that the curves merge in the images. Therefore, we added selected specific data in Table 1. In all our calculations, we work with the so-called unisex life tables on the basis of the European Union rules on gender-neutral pricing in insurance industry[2].

Table 1. Probability of death $\hat{q}_x$ fitted by moving mixture functions and Gompertz-Makeham formula using the R software

x	$g(x) = cx + 1 - c$	$g(x) = 1 + \gamma x^2$	$g(x) = a\left(\frac{1}{a}\right)^x$
62	0.0131247396059544	0.0131227418161407	0.013125107464233
63	0.0140982331743885	0.0140967990550384	0.014098499506204
64	0.0151751934123956	0.0151739111052197	0.015175434779201
⋮	⋮	⋮	⋮
105	0.568989096849097	0.559262199520758	0.558049007950153
106	0.605655119081360	0.595058953280515	0.593854212851836
107	0.642573640373201	0.631189123587682	0.630003782437894
108	0.679381366041013	0.667322346697244	0.666167885317843
109	0.715680484716906	0.703094275968703	0.701982388674047
110	0.751048956102780	0.738114852634371	0.737057058650676

In Fig. 2 is illustrated final smoothing curve of mortality rates which is calculated by our proposed methodology using moving mixture function with

[2] Council Directive 2004/113/EC, Under new rules which are entering into force, insurers in Europe will have to charge the same prices to women and men for the same insurance products without distinction on the grounds of sex.

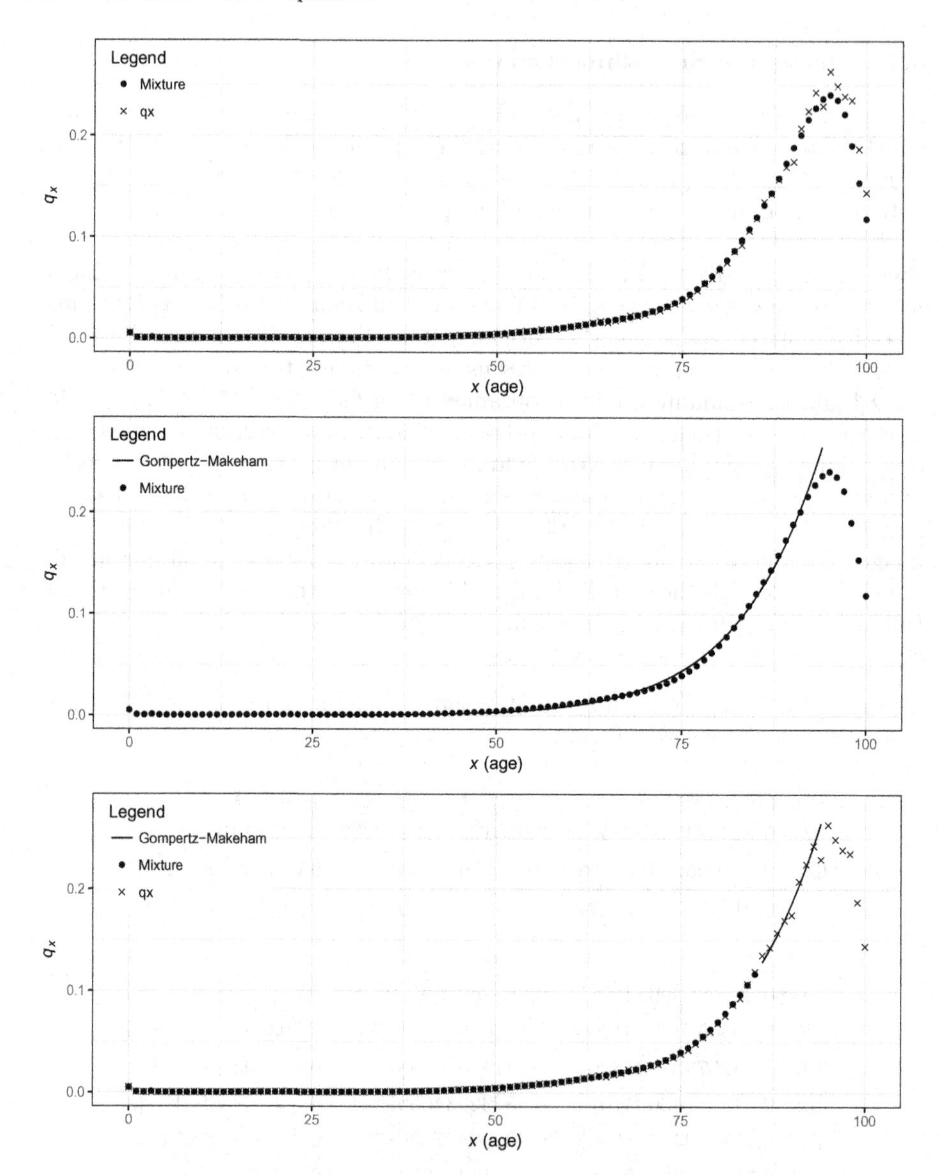

Fig. 1. Process of mortality rates smoothing

exponential weighting function and the R software Gompertz-Makeham function (Legend - MixtureFit) in comparison with curves calculated with respect to (5) (Legend - qx) and (6) (Legend - Statistics).

In Fig. 3 are shown mean square errors of estimated values using moving mixture function (Legend - MixtureError) and moving mixture function supplemented by Gompertz-Makeham function (Legend - MixtureFitError) from the original value q_x. We calculated mean square errors for all ranges of coefficients

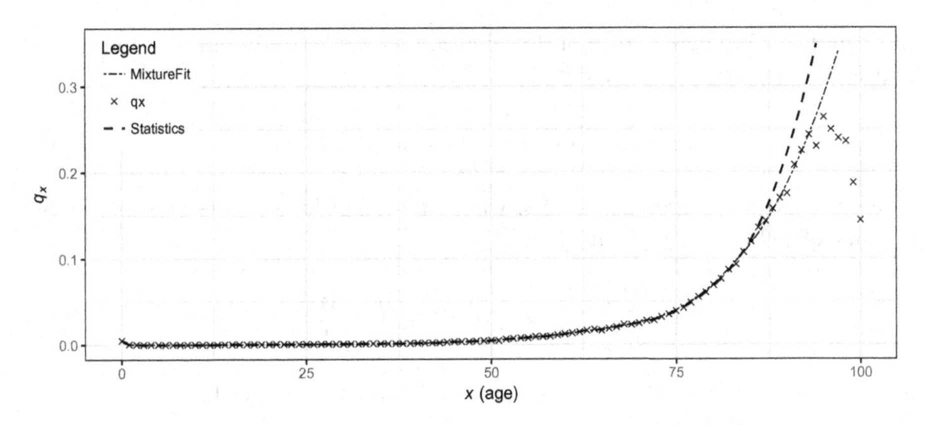

Fig. 2. Comparison of mortality rates

of weighting functions c, γ and a. On the basis of "the shape" of these errors, we decided to use middle values of the corresponding intervals, hence, we have chosen fitting parameters $c = 0.25$, $\gamma = 1.5$ and $a = 0.683959$. This problem is well known from statistics as the Bias-Variance trade-off [13]. On the edges on the intervals of corresponding coefficients, values are either biased or overfitted. Of course, the choice of "the best" coefficient depends on the expert.

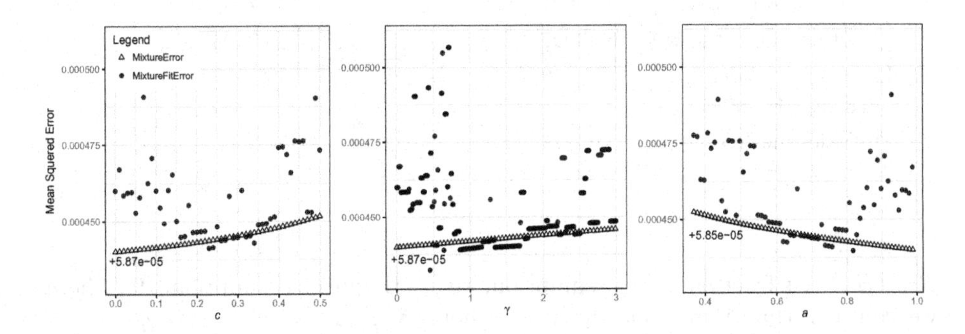

Fig. 3. Mean square errors for individual weighting functions

On the basis of MSE, in our personal opinion, the best choice is to apply into moving mixture function exponential weighting function with $a = 0.683959$.

3.3 The Impact of Mortality Rates Smoothing on Whole Life Annuity

In this part, we want to point out on the impact of mortality rate smoothing on an amount of life insurance premium, especially on the amount of the whole life annuities. We calculate monthly pension as a whole life annuity for x-aged pensioner with the accumulated sum S, by formula

$$MA_x = \frac{S}{12 \times a_x^{(12)}}, \tag{9}$$

where the expected present value of whole life annuities payable monthly in the amount of 1/12 m.u. is given by

$$a_x^{(12)} = \sum_{t=1}^{12(\omega-x)} \frac{1}{12} \times {}_{\frac{t}{12}}p_x \times \exp\left(-\frac{R(\frac{t}{12})}{100} \times \frac{t}{12}\right) \tag{10}$$

and $R(z)$ is annual yield (in %) of the bond. The euro area yield curve shows all euro area central government bonds, including AAA-rated (December 7, 2017). These yield curves are required by the European Parliament in all life insurance calculations [10][3].

Table 2. The impact of smoothing methods on whole life pension annuity, the accumulated sum $S = 10,000$ euros

Smoothing by	Expected present value of whole life annuity (euros) (10)	Monthly pension annuity (euros) (9)
$\hat{q}_x$ with linear w.f	16.43076	50.7179
$\hat{q}_x$ with quadratic w.f	16.42821	50.7258
$\hat{q}_x$ with exponential w.f	16.43027	50.7194
Lee-Carter model of longevity [14,23]	16.74022	49.7803
Statistics methodology [22]	15.94728	52.2555

Remark 1. As fractional age assumption, we use Balducci assumption [19], where we consider that yearly mortality rate holds ${}_{1-s}q_{x+s} = (1-s) \times q_x$, $p_x = {}_sp_x \times {}_{1-s}p_{x+s}$, $x \in N$, $0 < s < 1$, from where

$${}_sp_x = \frac{1}{1 + s \times \frac{q_x}{p_x}}. \tag{11}$$

In Table 2 can be seen comparison of monthly pension annuities on the basis of different approaches. We would like to emphasize that monthly pension annuity on the basis of Statistics methodology was calculated only for available entry ages from 0 to 100 years. Other results are modeled for interval of ages from 0 to 110 years.

[3] Directive 2009/138/EC of the European Parliament and of the Council of 25 November 2009 on the taking-up and pursuit of the business of Insurance and Reinsurance (Solvency II).

4 Conclusion

In this paper, we proposed a new method of mortality rates smoothing using so-called moving mixture function and Gompertz-Makeham function in the R software. In our personal opinion, this method is very suitable and it is easy to use for different development years. We can observe how weighting factors behave and, therefore, which are most suitable for that year of determination of mortality rates.

In future, we plan to investigate the impact of more methods on premium for all life insurance products and pension annuities. Moreover, we also plan to study the impact of another weighting functions on mortality rates mixture aggregation and so-called generalized mixture functions and to determine the most fitting smooth mortality rates curve [25, 26].

All calculations and figures were developed using the statistical R software.

Acknowledgement. The paper was supported by the Slovak Scientific Grant Agency VEGA no. 1/0093/17 Identification of risk factors and their impact on products of the insurance and savings schemes, and VEGA no. 1/0618/17 Modern tools for modelling and managing risks in life insurance.

References

1. Beliakov, G., Bustince Sola, H., Calvo Sánchez, T.: A Practical Guide to Averaging Functions. Studies in Fuzziness and Soft Computing, vol. 329. Springer, Cham (2016). https://doi.org/10.1007/978-3-319-24753-3
2. Beliakov, G., Pradera, A., Calvo, T.: Aggregation Functions: A Guide for Practitioners. Springer, Berlin (2007). https://doi.org/10.1007/978-3-540-73721-6
3. Beliakov, G., Calvo, T., Wilkin, T.: Three types of monotonicity of averaging functions. Knowl.-Based Syst. **72**, 114–122 (2014)
4. Beliakov, G., Calvo, T., Wilkin, T.: On the weak monotonicity of Gini means and other mixture functions. Inf. Sci. **300**, 70–84 (2015)
5. Beliakov, G., Špirková, J.: Weak monotonicity of Lehmer and Gini means. Fuzzy Sets Syst. **299**, 26–40 (2016)
6. Bustince, H., Fernandez, J., Kolesárová, A., Mesiar, R.: Fusion functions and directional monotonicity. In: Laurent, A., Strauss, O., Bouchon-Meunier, B., Yager, R.R. (eds.) IPMU 2014. CCIS, vol. 444, pp. 262–268. Springer, Cham (2014). https://doi.org/10.1007/978-3-319-08852-5_27
7. Currie, I.D., Durban, M., Eilers, P.H.C.: Smoothing and forecasting mortality rates. Stat. Model. **4**, 279 (2004). https://doi.org/10.1191/1471082X04st080oa. http://www.maths.ed.ac.uk/~mthdat25/mortality/Smoothing-and-forecasting-mortality-rates.pdf
8. Dickson, D.C.M., et al.: Actuarial Mathematics for Life Contingent Risks. Cambridge University Press, New York (2009)
9. Grabisch, M., Marichal, J.L., Mesiar, R., Pap, E.: Aggregation Functions. Cambridge University Press, Cambridge (2009)
10. Euro area yield curves. https://www.ecb.europa.eu/stats/financial_markets_and_interest_rates/euro_area_yield_curves/html/index.en.html

11. Hunt, A., Blake, D.: Modelling mortality for pension schemes. ASTIN Bull. **47**(2), 601–629 (2017)
12. The Human Mortality Database. https://mortality.org/
13. James, G., Witten, D., Hastie, T., Tibshirani, R.: An Introduction to Statistical Learning with Applications in R. Springer, New York (2013). https://doi.org/10.1007/978-1-4614-7138-7
14. Lee, R.D., Carter, L.R.: Modeling and forecasting U. S. mortality. J. Am. Stat. Assoc. **87**(419), 659–671 (1992). http://pagesperso.univ-brest.fr/~ailliot/doc_cours/M1EURIA/regression/leecarter.pdf
15. Marková, V., Lesníková, P., Kaščáková, A., Vinczeová, M.: The present status of sustainability concept implementation by businesses in selected industries in the Slovak Republic. E&M Econ. Manag. **20**(3), 101–117 (2017). https://doi.org/10.15240/tul/001/2017-3-007
16. Mesiar, R., Špirková, J.: Weighted means and weighting functions. Kybernetika **42**(2), 151–160 (2006)
17. Mesiar, M., Špirková, J., Vavríková, L.: Weighted aggregation operators based on minimization. Inf. Sci. **17**(4), 1133–1140 (2008)
18. Nakazawa, M.: Package 'fmsb'. https://cran.r-project.org/web/packages/fmsb/fmsb.pdf
19. Potocký, R.: Models in life and non-life insurance. In: Slovak: Modely v životnom a neživotnom poistení, Bratislava, STATIS (2012)
20. R Core Team. R: A language and environment for statistical computing. R Foundation for Statistical Computing, Vienna, Austria. http://www.R-project.org/
21. Richards, S.J.: Detecting year-of-birth mortality patterns with limited data. J. R. Stat. Soc. Ser. Stat. Soc. **171**(Part: 1), 279–298 (2008)
22. Statistical Office of the Slovak Republic. https://slovak.statistics.sk/
23. Szűcs, G., Špirková, J., Kollár, I.: Detailed View of a Payout Product of the Old-Age Pension Saving Scheme in Slovakia. J. Econ. (2018, submitted). Institute of Economic Research SAS, Slovakia
24. Špirková, J.: Dissertation thesis. In: Weighted Aggregation Operators and Their Applications, Bratislava (2008)
25. Špirková, J.: Weighted operators based on dissimilarity function. Inf. Sci. **281**, 172–181 (2014)
26. Špirková, J.: Induced weighted operators based on dissimilarity functions. Inf. Sci. **294**, 530–539 (2015)
27. Vinczeová, M.: The relationship between corporate social responsibility and business performance (in Slovak: Vzt'ah medzi spoločenskou zodpovednost'ou podniku a jeho výkonnost'ou). In: Proceedings of Scientific Studies from the Project VEGA No. 1/0934/16 Cultural Intelligence as an Important Presumption of Slovakia's Competitiveness in a Global Environment (in Slovak: Zborník vedeckých štúdií z projektu VEGA 1/0934/16 Kultúrna inteligencia ako dôležitý predpoklad konkurencieschopnosti Slovenska v globálnom prostredí). Faculty of Economics, Matej Bel Univerzity, CD-ROM, Banská Bystrica (2016)
28. Wickham, H.: ggplot2 Elegant Graphics for Data Analysis, 2nd edn. Springer, New York (2016). https://doi.org/10.1007/978-3-319-24277-4
29. Zimmermann, P.: Modeling mortality at old age with time-varying parameters. Math. Popul. Stud. **24**(3), 172–180 (2017). https://doi.org/10.1080/08898480.2017.1330013

Aggregation Functions Based on Deviations

Marián Decký, Radko Mesiar, and Andrea Stupňanová(✉)

Faculty of Civil Engineering, Slovak University of Technology in Bratislava,
Radlinského 11, 810 05 Bratislava, Slovak Republic
{marian.decky,radko.mesiar,andrea.stupnanova}@stuba.sk

Abstract. After recalling penalty and deviation based constructions of idempotent aggregation functions, we introduce the concept of a general deviation function and related construction of aggregation functions. Our approach is exemplified in some examples, illustrating the ability of our method to model possibly different aggregation attitudes in different coordinates of the aggregated score vectors.

Keywords: Aggregation function · Deviation function · Mean
Moderate deviation function · Penalty function

1 Introduction

Two basic mathematical tools in numerous applications when the information is expressed in a form of some function are related to finding the roots, i.e., to solve the equation $f(x_1, \ldots, x_n, y) = 0$, and to finding of extremes (either extremal values, or, rather often, to find maximizers/minimizers of the function f).

In the domain of aggregation functions, the optimization problems, in particular, the minimizers of appropriate functions, were used to construct aggregation functions. Here, the considered function is called a penalty, thus we will use notation P, and it measures the penalty we have to pay if a score vector $(x_1, \ldots, x_n)$ is replaced by unanimous score vector $(y, \ldots, y)$. Clearly, penalty functions are closely related to distance functions (metrics), and the aggregation based on minimizers of penalty functions can be traced to the ancient Greece.

The other approach, related to finding the roots of a special function D indicating the deviation of a score vector $(x_1, \ldots, x_n)$ from a unanimous vector $(y, \ldots, y)$, was proposed to construct particular means by Daróczy [6] more than fifty years ago. Note that Daróczy means need not be monotone and thus they need not be aggregation functions. Recently, we have proposed a modification of Daróczy approach, considering moderate deviation functions [7], and then the monotonicity of the constructed means is guaranteed. Note that in particular cases, when penalty functions and deviation functions are related by a derivative, both approaches coincide (compare the finding of minimizers of a smooth function f and of roots of the related derivative f'). In earlier works, penalty

J. Medina et al. (Eds.): IPMU 2018, CCIS 853, pp. 151–159, 2018.
https://doi.org/10.1007/978-3-319-91473-2_13

functions P (for any arity $n \geq 2$) were obtained from appropriate binary penalty functions LP, considering

$$P(x_1, \ldots, x_n, y) = \sum_{i=1}^{n} LP(x_i, y),$$

see [5]. This approach was further generalized, either considering weights, or considering different binary penalty functions LP_i for i-th input x_i, as well as considering P as a particular function of $n+1$ variables, $P : I^{n+1} \to \mathbb{R}^+$, where I is the real interval from which the considered inputs $x_1, \ldots, x_n$, as well as the value y we aim for replacing the given inputs are taken [4]. This approach allows to obtain an arbitrary idempotent aggregation function A acting on I to be constructed. We recall some basics of penalty-based approach to construct aggregation functions in the next section. Observe that a deep discussion concerning pros and cons of the general approach proposed in [4] was recently presented in [3]. The aim of this paper is a generalization of the deviation functions based approach to construct aggregation functions, in a similar manner as it was done with penalty functions in [4]. Note that, though being related by the idea of finding minimizers by means of roots of derivatives, both approaches differ, in general.

The paper is organized as follows. In the next section, we recall some basic notions and results concerning the penalty based approach. In Sect. 3, moderate deviation functions are recalled and related construction of aggregation functions is discussed. The Sect. 4 brings our proposal of general deviation functions and shows how the aggregation functions can be seen as roots of these functions. In Sect. 5, some particular examples are considered, showing how we can aggregate inputs related to criteria with different attitudes (e.g., some of criteria are linked to the arithmetic mean, while the others are linked to the quadratic mean). Finally, some concluding remarks are added.

2 Aggregation Functions Based on Penalties

The original idea of penalty functions-based aggregation due to Yager [14] has suffered from some problems (the existence of minimizers, non-convex set of minimizers, etc.). These problems were solved in the later concept proposed by Calvo et al. in [5].

Definition 1. *Let $K : \mathbb{R} \to \mathbb{R}^+$ be a convex function such that $K(x) = 0$ if and only if $x = 0$, and $s : I \to \mathbb{R}$ be a continuous strictly monotone function. Then dissimilarity L is given in the form*

$$L(x, y) = K(s(x) - s(y)).$$

and penalty function $P : I^{n+1} \to \mathbb{R}^+$ is defined, for any $\mathbf{x} \in I^n$ and $y \in I$ as:

$$P(\mathbf{x}, y) = \sum_{i=1}^{n} L(x_i, y).$$

A function $f_P : I^n \to I$*, defined for all* $\mathbf{x} \in I^n$ *and* $n \in \mathbb{N}$ *by*

$$f_P(\mathbf{x}) = \frac{l_{\mathbf{x}} + r_{\mathbf{x}}}{2},$$

where

$$l_{\mathbf{x}} = \inf\{u \in I \mid \forall v \in I : P(\mathbf{x}, u) \leq P(\mathbf{x}, v)\}$$
$$r_{\mathbf{x}} = \sup\{u \in I \mid \forall v \in I : P(\mathbf{x}, u) \leq P(\mathbf{x}, v)\}$$

is called a penalty-based function.

Note that the set $[l_{\mathbf{x}}, r_{\mathbf{x}}]$ of all minimizers of $P(\mathbf{x}, y)$ always exists and it is a subinterval of $[\min \mathbf{x}, \max \mathbf{x}]$.

Observe that convex real functions having a unique minimum $K(0) = 0$ can be constructed easily. Moreover, the penalty function P is, in fact, related to a one-variable function, which can be seen as a distance function. In this case, we always obtain an idempotent aggregation function, which in the case of linearity of function s, is also shift invariant.

Later, Calvo and Beliakov [4] have proposed a rather general concept of penalty functions allowing to introduce any idempotent aggregation function.

Definition 2. *The function* $P : I^{n+1} \to [0, \infty]$ *is a penalty function if and only if it satisfies for all* $\mathbf{x} \in I^n$ *and* $y \in I$*:*

(i) $P(\mathbf{x}, y) \geq 0$,
(ii) $P(\mathbf{x}, y) = 0$ *if* $x_i = y$ *for all* i*;*
(iii) For every fixed $\mathbf{x}$*, the set of minimizers of* $P(\mathbf{x}, y)$ *is either a singleton or an interval.*

Moreover, a penalty-based function $f_P : I^n \to I$ *is defined, for all* $\mathbf{x} \in I^n$*, by*

$$f_P(\mathbf{x}) = \arg\min_y P(\mathbf{x}, y), \tag{1}$$

if y *is the unique minimizer, and* $y = \frac{a+b}{2}$ *if the set of minimizers is the interval with boundaries* a *and* b*.*

Obviously, for any idempotent aggregation function $A : I^n \to I$, the function $P : I^{n+1} \to [0, \infty[$ given by $P(\mathbf{x}, y) = (y - A(\mathbf{x}))^2$ is a penalty function in the sense of the above definition and the related aggregation functions derived by means of formula (1) is just A. For a deeper discussion concerning the construction of aggregation functions based on penalty functions we recommend a recent paper [3]. There also several problems concerning the above definition are discussed and mentioned shortcuts are corrected.

In particular, observe that, to ensure the validity of item *(iii)* in the above definition, it is enough to consider the quasi-convexity of the penalty function P in the last coordinate, i.e.,

$$P(\mathbf{x}, \lambda y_1 + (1 - \lambda) y_2) \leq \max\{P(\mathbf{x}, y_1), P(\mathbf{x}, y_2)\}$$

for all $x \in I^n$ and $y_1, y_2 \in I$ and $\lambda \in [0,1]$ and its lower semicontinuity in the last coordinate, i.e., for any $\mathbf{x} \in I^n$ and $y_0 \in I$ it holds

$$P(\mathbf{x}, y_0) \leq \lim \inf_{y \to y_0} P(\mathbf{x}, y).$$

Considering the above mentioned quasi-convexity and lower semicontinuity of P in the last coordinate instead of axiom *(iii)* in the above definition leads to a sound approach to construct aggregation functions, see [3].

Example 1. Define a function $P : [0,1]^3 \to [0, \infty[$ by

$$P(x_1, x_2, y) = (x_1 - y)^2 + |x_2 - y|^3.$$

Then P is a penalty function and the corresponding idempotent aggregation function $f_P : [0,1]^2 \to [0,1]$ is given by

$$f_P(x_1, x_2) = \begin{cases} \frac{3x_2+1-\sqrt{1+6(x_2-x_1)}}{3} & \text{if } x_1 \leq x_2 \\ \frac{3x_2-1+\sqrt{1+6(x_1-x_2)}}{3} & \text{otherwise.} \end{cases}$$

3 Deviation Functions

The history of a deeper investigation of means goes back to Kolmogoroff [8] and Nagumo [13], and their overview can be found, e.g., in Bullen's monograph [2]. Evidently, each idempotent aggregation function is a mean, however, the opposite claim is not valid. Therefore, several means introduced and studied so far do not belong to the class of aggregation functions (the monotonicity can be violated, see, e.g., Lehmer mean $L : [0,1]^2 \to [0,1]$ given by $L(x_1, x_2) = \frac{x_1^2+x_2^2}{x_1+x_2}$, with convention $\frac{0}{0} = 0$.)

About 50 years ago, Daróczy has introduced means based on finding roots of deviation functions, see [6]. In the case of Daróczy means, also the monotonicity is not guaranteed.

Recall that, considering an arbitrary real interval I, a function $D : I^2 \to \mathbb{R}$ is called a deviation function whenever

- all sections $D(x, .) : I \to \mathbb{R}$, $x \in I$, are continuous and strictly increasing,
- and $D(x, x) = 0$ for all $x \in I$.

Then, for any n-tuple $\mathbf{x} = (x_1, \ldots, x_n) \in I^n$, there is a unique root of the function $H(\mathbf{x}, .) : I \to \mathbb{R}$,

$$H(\mathbf{x}, y) = \sum_{i=1}^{n} D(x_i, y)$$

denoted as $y_{\mathbf{x}} \in I$. Letting $\mathbf{x}$ free and denoting $A_D(\mathbf{x}) = y_{\mathbf{x}}$, a function $A_D : I^n \to I$ is well defined and it is not difficult to check that

$$\min(\mathbf{x}) \leq A_D(\mathbf{x}) \leq \max(\mathbf{x})$$

for all $\mathbf{x} \in I^n$, i.e., A_D is a symmetric mean on I.

Example 2. Define $D : [0,1]^2 \to \mathbb{R}$ by

$$D_\varepsilon(x,y) = (x+\varepsilon)(y-x), \qquad \text{where}\, \varepsilon \in]0,\infty[.$$

Then D_ε is a deviation function, and the related Daróczy mean $M_{D_\varepsilon} : [0,1]^2 \to [0,1]$ is given by

$$M_{D_\varepsilon}(x_1,x_2) = \frac{(x_1+\varepsilon)x_1 + (x_2+\varepsilon)x_2}{x_1+x_2+2\varepsilon}.$$

Note that M_{D_ε} is a mixture operator [12], and it is an aggregation function only if $\varepsilon \geq 1$. For example, $M_{D_{0.5}}(0,1) = 0.75$ but $M_{D_{0.5}}(0.1,1) = 0.743$.

Among several modifications of Daróczy means (mostly dealing with several deviation functions, or with deviation functions and weights, and thus violating the symmetry), we recall Losonczi means [9,10], Bajraktarevič means [1] and quasi-deviation means due to Páles [11]. To avoid the lack of monotonicity, we have modified the approach of Daróczy and we have introduced so called moderate deviation functions, see [7].

Definition 3. *A moderate deviation function is a mapping $D : [0,1]^2 \to \mathbb{R}$ satisfying*

(i) for all $x \in [0,1]$, $D(x,\cdot) : [0,1] \to \mathbb{R}$ is increasing (not necessarily strictly);
(ii) for all $y \in [0,1]$, $D(\cdot,y) : [0,1] \to \mathbb{R}$ is decreasing (not necessarily strictly);
(iii) $D(x,y) = 0$ if and only if $x = y$.

The set of all moderate deviation functions is denoted as $\mathcal{D}$.

Observe that due to the possible non-continuity and non-strict monotonicity of moderate deviation functions, we hardly can expect a unique root of a function $f(y) = \sum_{i=1}^{n} D(x_i,y)$. Therefore, a modified approach to derive an aggregation function from a moderate deviation function D was proposed in [7].

Definition 4. *For a given $D \in \mathcal{D}$, and any $n \in \mathbb{N}$, the mapping $M_D : [0,1]^n \to [0,1]$ given by*

$$M_D(\mathbf{x}) = \frac{1}{2}\left(\sup\left\{ y \in [0,1] \,\middle|\, \sum_{i=1}^{n} D(x_i,y) < 0 \right\} + \inf\left\{ y \in [0,1] \,\middle|\, \sum_{i=1}^{n} D(x_i,y) > 0 \right\} \right) \tag{2}$$

is called a D-mean.

The next result was shown in [7].

Theorem 1. *Let $D \in \mathcal{D}$. Then the D-mean $M_D : [0,1]^n \to [0,1]$ is an idempotent symmetric aggregation function.*

Note that also in this case one can modify the approach described above, considering either different moderate deviation functions for different coordinates, or considering some weights, and thus violating the symmetry of the introduced idempotent aggregation functions. For more details see [7].

4 General Deviation Functions

Similarly as in the case of penalty-based construction of idempotent aggregation functions, also in the case when moderate deviation functions are considered, not all idempotent aggregation functions can be obtained by this method. This is caused by the fact, that either in the case of penalty functions or in the case of deviation functions, their coordinate-wise decomposition into a sum cannot cover all idempotent aggregation functions (for more arguments see [4]). To avoid this problem, we propose an extension of till now discussed approaches to deviation functions, considering the concept of a general deviation function.

Definition 5. *For a fixed $n \geq 2$ and real interval I, a function $D : I^{n+1} \to \mathbb{R}$ is called a general deviation function (GDF, in short), if and only if it satisfies the next conditions:*

(i) for all $\mathbf{x} \in I^n$, $D(\mathbf{x}, .) : I \to \mathbb{R}$ is increasing (not necessarily strictly);
(ii) for all $y \in I, D(., y) : I^n \to \mathbb{R}$ is decreasing (not necessarily strictly);
(iii) for all $c, y \in I$, $D(\mathbf{c}, y) = 0$ if and only if $c = y$, where $\mathbf{c} \in I^n$ is a constant n-tuple.

Definition 6. *For given GDF $D : I^{n+1} \to \mathbb{R}$, the mapping $M_D : I^n \to I$ given by*

$$M_D(\mathbf{x}) = \frac{1}{2}\left(\sup\left\{ y \in [0,1] \middle| D(\mathbf{x}, y) < 0 \right\} + \inf\left\{ y \in [0,1] \middle| D(\mathbf{x}, y) > 0 \right\} \right) \quad (3)$$

is called a D-mean.

Evidently, for any idempotent aggregation function $A : I^n \to I$, denoting $D^A(\mathbf{x}, y) = y - A(\mathbf{x})$, $D^A : I^{n+1} \to \mathbb{R}$ is a well defined function which is a general deviation function and $A = M_{D^A}$.

Theorem 2. *Let $D : I^{n+1} \to \mathbb{R}$ be a GDF. Then the function $M_D : I^n \to I$ given by (3) is an idempotent aggregation function.*

The proof of the above theorem can be done in similar steps as the proof of Theorem 3.1 in [7] and thus we omit its details. Obviously, M_D is symmetric only if the GDF D is invariant under any permutation of inputs such that the last coordinate remains unchanged. Observe also that, for any moderate deviation function $H : I^2 \to \mathbb{R}$, and for any $n \geq 2$, the function $D : I^{n+1} \to \mathbb{R}$ given by $D(\mathbf{x}, y) = \sum_{i=1}^{n} H(x_i, y)$ is a GDF, and thus the concept of general deviation functions extends our previous concept of moderate deviation functions.

5 Particular Examples

Standard symmetric means can be obtained by means of the next moderate deviation functions:

- arithmetic mean: $H(x, y) = y - x$;
- geometric mean: $H(x, y) = \log y - \log x$;
- quadratic mean: $H(x, y) = y^2 - x^2$;
- median: $H(x, y) = \text{sign}(y - x)$,

and their weighted forms are simply related to a GDF $D : I^{n+1} \to \mathbb{R}$ given by

$$D(\mathbf{x}, y) = \sum_{i=1}^{n} w_i \cdot H(x_i, y).$$

Note that the same idempotent aggregation function A can be obtained from different GDFs. Obviously, $M_D = M_{c \cdot D}$ for any GDF D and positive constant c (then also $c \cdot D$ is a GDF).

Coming back to the geometric mean, say on $I = [0, 1]$, we can consider either

$$D(\mathbf{x}, y) = \sum_{i=1}^{n} (\log y - \log x_i), \text{or } D(\mathbf{x}, y) = y^n - x_1 \cdot x_2 \cdot \ \ldots \ \cdot x_n,$$

and in both cases the related M_D is just the standard geometric mean. Our approach allows to mix different aggregation approaches. Consider, for example, that two experts insist that the observed data should be aggregated by two different methods. The first one is in favour of the arithmetic mean, while the other one prefers the geometric mean. Obviously, one can applied a convex combination of these approaches. However, based on deviation functions, new possibilities are open. For the case of simplicity, suppose $n = 2$ and $I = [0, 1]$. Not having any preference concerning the experts, one can consider the arithmetic mean of both considered methods, i.e. then

$$A_1(x_1, x_2) = \frac{\frac{x_1 + x_2}{2} + \sqrt{x_1 \cdot x_2}}{2}.$$

Based on deviation approach, consider a GDF $D_2 : [0, 1]^3 \to \mathbb{R}$ given by

$$D_2(x_1, x_2, y) = y - x_1 + y - x_2 + y^2 - x_1 \cdot x_2,$$

mixing together the GDF $y - x_1 + y - x_2$ related to the arithmetic mean, and the GDF $y^2 - x_1 \cdot x_2$ related to the geometric mean. Then the resulting idempotent aggregation function A_2 is given by

$$A_2(x_1, x_2) = \sqrt{(1 + x_1) \cdot (1 + x_2)} - 1.$$

However, one can mix the GDFs $y - x_1 + y - x_2$ and $\log y - \log x_1 + \log y - \log x_2$ (the second one being again related to the geometric mean). Then the value of $A_3(x_1, x_2)$ is implicitly given by

$$y + \log y = \frac{x_1 + x_2}{2} + \log \sqrt{x_1 \cdot x_2}.$$

Put $x_1 = 0.1$ and $x_2 = 0.9$. Then

$$A_1(0.1, 0.9) = \frac{0.5 + 0.3}{2} = 0.4;$$
$$A_2(0.1, 0.9) = \sqrt{2.09} - 1 \sim 0.446;$$
$A_3(0.1, 0.9)$ is the unique solution (in $[0, 1]$) of the equation
$$y + \log y = 0.5 + \log 0.3, \text{ and thus}$$
$$A_3(0.1, 0.9) \sim 0.349.$$

Another related example aims to force arithmetic mean of some coordinates, and the quadratic mean on some other coordinates. For example, considering $n = 4$ and $I = [0, 1]$, for any positive constant c let $D_c : [0, 1]^5 \to \mathbb{R}$ be given by

$$D_c(\mathbf{x}, y) = c \cdot (y - x_1 + y - x_2) + (y^2 - x_3^2 + y^2 - x_4^2).$$

Then D_c is a GDF, first two coordinate have the attitude of the arithmetic mean (i.e., for $n = 2$, considering GDF given by $c \cdot (y - x_1 + y - x_2)$ yields the arithmetic mean $\frac{x_1+x_2}{2}$ (independently of c)), while the last two coordinates have the attitude of the quadratic mean (i.e., GDF given by $y^2 - {x_3}^2 + y^2 - {x_4}^2$ yields the quadratic mean $\sqrt{\frac{{x_3}^2+{x_4}^2}{2}}$.) Applying Theorem 2, we obtain an idempotent aggregation function $M_{D_c} : [0, 1]^4 \to [0, 1]$ given by

$$M_{D_c}(\mathbf{x}) = \frac{\sqrt{c^2 + 2c(x_1 + x_2) + 2{x_3}^2 + 2{x_4}^2} - c}{2}.$$

Then is not difficult to check that

$$\lim_{c \to 0^+} M_{D_c}(\mathbf{x}) = \sqrt{\frac{{x_3}^2 + {x_4}^2}{2}},$$

i.e., the quadratic mean of inputs x_3 and x_4 is obtained. On the other hand,

$$\lim_{c \to \infty} M_{D_c}(\mathbf{x}) = \frac{x_1 + x_2}{2},$$

i.e., the arithmetic mean of inputs x_1 and x_2 is obtained.

6 Concluding Remarks

We discussed penalty and deviation based approaches to construct idempotent aggregation functions and we have developed the notion of a general deviation function. Our approach is rather general, covering all possible idempotent aggregation functions. GDF approach offers interesting combination of experts different feelings related to aggregation one should use, or forcing different aggregation attitudes in different coordinates. This idea was illustrated on simple examples. We believe that, similarly as penalty based approaches, also deviation based approach will attract the attention of experts working in several applied fields, including multicriteria decision support, image processing, etc.

Acknowledgments. The support of the grants APVV-14-0013 and VEGA 1/0682/16 is kindly announced.

References

1. Bajraktarevič, M.: Über die Vergleichbarkeit der mit Gewichtsfunktionen gebildeten Mittelwerte. Stud. Math. Hungar. **4**, 3–8 (1969)
2. Bullen, P.S.: Handbook of Means and Their Inequalities: Mathematics and Ist Applications, vol. 560. Kluwer Academic Publishers Group, Dordrecht (2003)
3. Bustince, H., Beliakov, G., Dimuro, G.P., Bedregal, B., Mesiar, R.: On the definition of penalty functions in aggregations. Fuzzy Sets Syst. **323**, 1–18 (2017). https://doi.org/10.1016/j.fss.2016.09.011
4. Calvo, T., Beliakov, G.: Aggregation functions based on penalties. Fuzzy Sets Syst. **161**(10), 1420–1436 (2010). https://doi.org/10.1016/j.fss.2009.05.012
5. Calvo, T., Mesiar, R., Yager, R.R.: Quantitative weights and aggregation. IEEE Trans. Fuzzy Syst. **12**(1), 62–69 (2004). https://doi.org/10.1109/TFUZZ.2003.822679
6. Daróczy, Z.: Über eine Klasse von Mittelwerten. Publ. Math. Debrecen **19**, 211–217 (1972)
7. Decký, M., Mesiar, R., Stupňanová, A.: Deviation-based aggregation functions. Fuzzy Sets Syst. **332**, 29–36 (2018). https://doi.org/10.1016/j.fss.2017.03.016
8. Kolmogoroff, A.N.: Sur la notion de la moyenne. Atti Accad. Naz. Lincei. **12**(6), 388–391 (1930)
9. Losonczi, L.: General inequalities of non-symmetric means. Aequationes Math. **9**, 221–235 (1973)
10. Losonczi, L.: Hölder-type inequalities. GI3, pp. 91–105 (1981)
11. Páles, Z.: On homogeneous quasideviation means. Aequationes Math. **36**(2–3), 132–152 (1988)
12. Ribeiro, R.A., Marques Pereira, R.A.: Aggregation with generalized mixture operators using weighting functions. Fuzzy Sets Syst. **137**, 43–58 (2003). https://doi.org/10.1016/S0165-0114(02)00431-1
13. Nagumo, M.: Über eine Klasse der Mittelwerte. Jpn. J. Math. **7**, 71–79 (1930)
14. Yager, R.R.: Toward a general theory of information aggregation. Inf. Sci. **68**(3), 191–206 (1993)

Nullnorms and T-Operators on Bounded Lattices: Coincidence and Differences

Slavka Bodjanova[1] and Martin Kalina[2](✉)

[1] Department of Mathematics, Texas A&M University-Kingsville, MSC 172, Kingsville 78363, TX, USA
kfsb000@tamuk.edu

[2] Department of Mathematics, Faculty of Civil Engineering, Slovak University of Technology in Bratislava, Radlinského 11, 810 05 Bratislava, Slovakia
kalina@math.sk

Abstract. T-operators were defined on $[0,1]$ by Mas et al. in 1999. In 2001, Calvo et al. introduced the notion of nullnorms, also on $[0,1]$. Both of these operations were defined as generalizations of t-norms and t-conorms. As Mas et al. in 2002 pointed out, t-operators and nullnorms coincide on $[0,1]$. Afterwards, only nullnorms were studied and later generalized as operations on bounded lattices. Our intention is to introduce also t-operators as operations on bounded lattices. We will show that, on bounded lattices, nullnorms and t-operators need not coincide. We will explore conditions under which one of these operations is necessarily the other one, and conditions under which they differ.

Keywords: Bounded lattice · Sequential convergence
Maximal chain · Nullnorm · T-operator

1 Introduction

T-operators [10] and nullnorms [2] are special types of associative, commutative and increasing operations on $[0,1]$. They are generalizations of both, t-norms and t-conorms. Nullnorms were introduced in such a way that the absorbing element $a \in [0,1]$ is arbitrary. On the other hand, t-operators were introduced to be continuous on the border, i.e., to have continuous partial functions $\mathrm{Op}(0,\cdot)$ and $\mathrm{Op}(1,\cdot)$. As Mas et al. in [12] pointed out, these two types of operations, defined on $[0,1]$, coincide. Particularly, under constrained that $\mathrm{Op} : [0,1]^2 \to [0,1]$ is a commutative, associative and monotone operation then properties

(a) 0 and 1 are idempotent elements of Op and functions $\mathrm{Op}(0,\cdot)$ and $\mathrm{Op}(1,\cdot)$ are continuous,
(b) there exists $a \in [0,1]$ such that 0 is a partial neutral element of Op on $[0,a]$, and 1 is a partial neutral element of Op on $[a,1]$,

are equivalent to each other.

J. Medina et al. (Eds.): IPMU 2018, CCIS 853, pp. 160–170, 2018.
https://doi.org/10.1007/978-3-319-91473-2_14

In this contribution we generalize the notion of a t-operator to be an operation on an arbitrary bounded lattice. Nullnorms were already generalized by Karaçal et al. in [7]. The authors showed that they can be constructed on arbitrary bounded lattice and having an arbitrarily chosen absorbing element. Our intention is to show that conditions (a) and (b) are not necessarily equivalent on bounded lattices. We will point out the difference between these two conditions.

Mas et al. in [11] introduced t-operators on finite totally ordered sets using the 1-smoothness condition.

Definition 1. *Let $\tilde{L} = \{x_0, x_1, \ldots, x_n\}$ be a finite chain such that $x_i \leq x_j$ for $0 \leq i \leq j \leq n$, and $F : \tilde{L}^2 \to \tilde{L}$. Function F is said to be 1-*smooth *if the following is fulfilled*

$$F(x_i, x_j) = x_k \quad \Rightarrow \quad F(x_{i-1}, x_j) = x_\ell \;\&\; F(x_i, x_{j-1}) = x_k,$$

where $k - 1 \leq \ell \leq k$ and $k - 1 \leq m \leq k$.

1-smoothness condition is a quite natural definition of continuity for finite chains (finite totally ordered sets). Moreover, as Mas et al. pointed out in [11], 1-smoothness is connected with directed algebras in such a way that if $(L, \leq, T, S, N)$ is a directed algebra then T and S are 1-smooth. The importance of directed algebras lies in the fact that, on $[0, 1]$, the only directed algebra structures are continuous de Morgan triplets. However, the approach in [11] is different from the one presented in this contribution.

In our contribution, we will use the sequential continuity for partial functions $V(0, \cdot)$ and $V(1, \cdot)$ which is one possibility of defining continuity on bonded lattices (sequential convergence is a natural way of defining a topology on bounded lattices).

We will show that the relationship between these two types of operations depends on the structure of a particular bounded lattice L on which they are defined. In [6], some examples were published that illustrate relationships between nullnorms and t-operators on bounded lattices. Our paper is organized as follows: Sect. 2 provides some preliminary information. It is split into two parts – Subsect. 2.1 presents basic notions on topology induced by lattice-theoretical operations. Subsection 2.2 covers some known facts on nullnorms and t-operators as operations on $[0, 1]$, as well as operations on bounded lattices. A generalization of the definition of a t-operator in the case of bounded lattices is proposed. Section 3 is devoted to exploration of a relationship between nullnorms and t-operators.

2 Preliminaries

We provide some basic information on lattices and topological spaces that will be needed further in our paper. For more information we refer to the monographs by Birkhoff [1] and by Kelley [8]. Further, we recall some known facts on nullnorms and t-operators.

2.1 Sequential Convergence and Lattice-Topology

Bounded lattices will be considered. L will denote the set of all elements of the lattice. If it will cause no confusion, by L will be denoted also the lattice itself. Every bounded lattice $(L, \wedge, \vee, \mathbf{0}_L, \mathbf{1}_L)$ is equipped with a partial order $\leq_L$ given by

$$(\forall a, b \in L)(a \leq_L b \Leftrightarrow a \wedge b = a).$$

Then the notion of a closed interval can be introduced as follows

$$\text{for all } a \leq_L b \quad [a, b] = \{x \in L; a \leq_L x \leq_L b\}.$$

The lattice-theoretical operations $\wedge, \vee$ generate a convergence.

Definition 2 (See, e.g., [8]). *A sequence of elements of a lattice L, $\{c_i\}_{i=1}^{\infty}$,* converges to $c \in L$ *if*

$$\bigvee_{i=1}^{\infty} \bigwedge_{j=i}^{\infty} c_j = c = \bigwedge_{i=1}^{\infty} \bigvee_{j=i}^{\infty} c_j \tag{1}$$

In fact, the left-hand-side of formula (1) is $\liminf_{i \to \infty} c_i$ and the right-hand-side is $\limsup_{i \to \infty} c_i$. Hence, Definition 2 is just the standard definition of a convergent sequence.

In what follows we will use the notation

$$\liminf_{i \to \infty} c_i = \bigvee_{i=1}^{\infty} \bigwedge_{j=i}^{\infty} c_j, \quad \limsup_{i \to \infty} c_i = \bigwedge_{i=1}^{\infty} \bigvee_{j=i}^{\infty} c_j.$$

Continuity of a function can be defined in the following way.

Definition 3 (See, e.g., [8]). *A function $f : L \to L$ is said to be* continuous *at $c \in L$ if for every sequence $\{c_i\}_{i=1}^{\infty}$ that converges to c we have*

$$f\left(\liminf_{i \to \infty} c_i\right) = \liminf_{i \to \infty} f(c_i) = f(c) = f\left(\limsup_{i \to \infty} c_i\right) = \limsup_{i \to \infty} f(c_i).$$

A function $f : L \to L$ is said to be continuous, *if it is continuous at every point $c \in L$.*

The convergence introduced in Definition 2, induces a topology.

Definition 4 (See, e.g., [8]). *A set $A \subset L$ is said to be open if for every $c \in A$ and for every sequence $\{c_i\}_{i=1}^{\infty}$ of elements of L that converges to c, the following holds*

$$(\exists i \in \mathbb{N})(\forall j > i)(c_j \in A).$$

Directly by Definition 4 we get the following:

Lemma 1. *Let $\mathcal{T}$ be the system of all open sets in a lattice L. Then $(L, \mathcal{T})$ is a topological space, i.e., the following properties are satisfied:*

(i) $\emptyset \in \mathcal{T}$ *and* $L \in \mathcal{T}$,
(ii) *for all* $A, B \in \mathcal{T}$, $A \cap B \in \mathcal{T}$,
(iii) *if* $\{T_i\}_{i \in \mathcal{I}}$ *is an arbitrary system of open sets, then* $\bigcup_{i \in \mathcal{I}} T_I \in \mathcal{T}$.

Definition 5. *Let $\mathcal{T}$ be the system of all open subsets of L induced by the convergence given by formula (1). Then $\mathcal{T}$ will be called the* lattice-topology.

Definition 6 (See, e.g., [1]). *Let $C \subset L$, then C is said to be a* chain in L *if elements of C are linearly ordered by $\leq_L$.*

Definition 7. *A chain M is said to be a* maximal chain in L *if $M \cup \{x\}$ is not a chain regardless of which $x \notin M$ is chosen.*

The system of all maximal chains containing a fixed element $a \in L$ *will be denoted by $\mathcal{M}(a)$.*

Definition 8. *Let $M_1, M_2 \subset L$ be maximal chains in L. A transformation $\varphi : M_1 \to M_2$ is said to be a σ-homomorphism with respect to $\wedge$ and $\vee$ if for any sequence $\{c_i\}_{i=1}^{\infty}$ of elements of M_1*

$$\varphi\left(\liminf_{i\to\infty} c_i\right) = \liminf_{i\to\infty} \varphi(c_i), \quad \varphi\left(\limsup_{i\to\infty} c_i\right) = \limsup_{i\to\infty} \varphi(c_i),$$

meaning that the right-hand-side of the equations exists if the left-hand-side does, and then they are equal to each other.

Definition 9 (See, e.g., [8]). *Let $C \subset L$, $C \neq \emptyset$, and let $\mathcal{T}$ be the lattice-topology. Then*

(i) *the set M is said to be* connected in $\mathcal{T}$, *if for arbitrary pair of open sets $A, B \in \mathcal{T}$ such that $A \cup B \supset M$, we have if $A \cap M \neq \emptyset$ and $B \cap M \neq \emptyset$, then $A \cap B \neq \emptyset$.*
(ii) *The set C is said to be* discrete in $\mathcal{T}$ *if for arbitrary sequence $\mathbf{c} = \{c_i\}_{i=1}^{\infty}$ of elements of C, $\mathbf{c}$ is convergent only if it is constant up to a finite number of elements.*

The following assertion is important in our further study.

Lemma 2. *Let $D \subset L$ be a discrete set. Then arbitrary function $f : D \to L$ is continuous in the lattice-topology $\mathcal{T}$.*

Further Notations

(1) Let $a \in L$ be a fixed element. Then $\|_a$ denotes the set of all elements of L that are incomparable with a.
(2) For all $x \in \|_a$ let $x_a = x \wedge a$ and $x^a = x \vee a$. Then $\tilde{\|}_a$ denotes the set $\|_a \cup \{x_a; x \in \|_a\} \cup \{x^a; x \in \|_a\}$.

2.2 Nullnorms and T-Operators on Bounded Lattices

First, we recall some basic properties of binary operations.

Definition 10. *Let L be a bounded lattice and $*$ be a binary commutative operation on L. Then*

(i) *element $c \in L$ is said to be* idempotent *if $c * c = c$,*
(ii) *element $e \in L$ is said to be* neutral *if $e * x = x$ for all $x \in L$,*
(iii) *element $a \in L$ is said to be* absorbing *if $a * x = a$ for all $x \in L$.*

Lemma 3. *Let $*$ be a commutative and associative operation on L. Further, let c be an idempotent element. Assume that there exist elements $x, y \in L$ such that $x * c = y$. Then also $y * c = y$.*

Proof. By the assumptions, $y = x * c = x * (c * c) = (x * c) * c = y * c$. □

For more information on associative (and monotone) operations on $[0, 1]$ refer to the monographs [3,5].

As previously mentioned, t-operators were defined in [10], and independently, nullnorms were defined in [2]. When Calvo et al. were solving Frank functional equation (see [4]) with unknown operations U and V, hence

$$U(x, y) + V(x, y) = x + y,$$

it turned out that V has to be a nullnorm whenever U is a uninorm.

Definition 11 ([10])**.** *An operation* $\mathrm{Op} : [0, 1]^2 \to [0, 1]$ *is said to be a t-operator if* Op *is associative, commutative, monotone, and moreover*

(1a) $\mathrm{Op}(0, 0) = 0$, $\mathrm{Op}(1, 1) = 1$,
(2a) *functions* $f_0(x) = \mathrm{Op}(0, x)$ *and* $f_1(x) = \mathrm{Op}(1, x)$ *are continuous.*

Definition 12 ([2])**.** *An operation $V : [0, 1]^2 \to [0, 1]$ is said to be a nullnorm if V is associative, commutative, monotone, and moreover if there exists an element $a \in [0, 1]$ such that*

(1b) *$V(0, x) = x$ for all $x \in [0, a]$,*
(2b) *$V(1, x) = x$ for all $x \in [a, 1]$.*

Remark 1

(a) It is well-known that the element a in Definition 12 is the absorbing element of nullnorm V. Further, 0 is a partial neutral element of V on the interval $[0, a]$ and 1 is a partial neutral element of V on the interval $[a, 1]$. Particularly, we have
$$V(0, 0) = 0, \quad V(1, 1) = 1. \tag{2}$$
(b) Setting $a = 0$ ($a = 1$) the nullnorm V becomes a t-norm (t-conorm). For properties of t-norms and t-conorms see, e.g., the monograph [9].

Remark 2. Comparing Definitions 11 and 12, we see that they differ only in the fact that, for a t-operator, continuity of function f_0 and f_1 is required (property (2a)), while for a nullnorm the elements 0 and 1 are partial neutral elements on $[0, a]$ and $[a, 1]$, respectively (properties (1b), (2b)).

In [7] nullnorms were defined on bounded lattices and it was shown that on every bounded lattice it is possible to choose arbitrarily an element a, and to construct a nullnorm where a is the absorbing element.

In order to define t-operators on bounded lattices, we modify Definition 11 as follows.

Definition 13. *Let L be a bounded lattice. An operation* $\mathrm{Op} : L^2 \to L$ *is a t-operator if* Op *is associative, commutative, monotone, and moreover*

(i) $\mathrm{Op}(\mathbf{0}_L, \mathbf{0}_L) = \mathbf{0}_L$, $\mathrm{Op}(\mathbf{1}_L, \mathbf{1}_L) = \mathbf{1}_L$,
(ii) *functions* $f_{\mathbf{0}_L}(x) = \mathrm{Op}(\mathbf{0}_L, x)$ *and* $f_{\mathbf{1}_L}(x) = \mathrm{Op}(\mathbf{1}_L x)$ *are continuous in the lattice-topology $\mathcal{T}$.*

3 Relationship Between T-Operators and Nullnorms on Bounded Lattices

The fact that on the unit interval nullnorms and t-operators coincide can be generalized in the following way.

Proposition 1. *Let L be a connected chain. Then every nullnorm is a t-operator and vice-versa, every t-operator is a nullnorm.*

Lemma 4. *Let $\mathcal{T}$ be the lattice topology of a bounded lattice L. Assume that $a \in L$ is comparable with all elements of L. Then functions $f_{\mathbf{0}_L} : L \to L$ and $f_{\mathbf{1}_L} : L \to L$ defined by*

$$f_{\mathbf{0}_L}(x) = \begin{cases} x & \text{for } x \le a, \\ a & \text{otherwise,} \end{cases} \qquad f_{\mathbf{1}_L}(x) = \begin{cases} x & \text{for } x \ge a, \\ a & \text{otherwise,} \end{cases} \tag{3}$$

are continuous in the topology $\mathcal{T}$.

The proof is straightforward.

Proposition 2 is a direct consequence of Lemma 4.

Proposition 2. *Let L be a bounded lattice and $a \in L$ be comparable with all elements of L. Then every nullnorm $V : L^2 \to L$ is a t-operator.*

Corollary 1. *Let L be a chain. Then every nullnorm is a t-operator.*

Since it is known that every t-norm T and every t-conorm S are a nullnorm (which we get just letting $a = \mathbf{0}_L$, or $a = \mathbf{1}_L$, respectively), Proposition 2 implies the next corollary.

Corollary 2. *Every t-norm $T : L^2 \to L$ and every t-conorm $S : L^2 \to L$ are t-operators.*

As illustrated in examples below, Proposition 2, as well as Corollary 2, cannot be written as equivalences.

Example 1. Let $L_1 = \{(x,1); x \in]0,1[\}$, $L_2 = \{(x,2); x \in]0,1[\}$, and $L = L_1 \cup L_2 \cup \{0,1\}$. The set L is ordered in the following way: $c_1 \leq_L c_2$ if one of the following properties holds

1. $c_1 = 0$ and c_2 is arbitrary,
2. $c_2 = 1$ and c_1 is arbitrary,
3. $c_1 = (x_1,1) \in L_1$, $c_2 = (x_2,1) \in L_1$ and $x_1 \leq x_2$,
4. $c_1 = (x_1,2) \in L_2$, $c_2 = (x_2,2) \in L_2$ and $x_1 \leq x_2$.

Then obviously $(L, \leq_L)$ is a bounded lattice. Define a transformation $\varphi : L_2 \cup \{0,1\} \to L_1 \cup \{0,1\}$ by

$$\varphi(c) = \begin{cases} c & \text{for } c \in \{0,1\}, \\ (x,1) & \text{for } c = (x,2). \end{cases}$$

Then φ is a σ-homomorphism of the maximal chain $M_2 = L_2 \cup \{0,1\}$ onto $M_1 = L_1 \cup \{0,1\}$. Further, we set

$$\bar{\varphi}(c) = \begin{cases} \varphi(c) & \text{for } c \in M_2, \\ c & \text{otherwise.} \end{cases}$$

Choose a t-norm T on M_1 and define the following operation $\mathrm{Op} : L^2 \to L$ by

$$\mathrm{Op}(c_1, c_2) = T\big(\bar{\varphi}(c_1), \bar{\varphi}(c_2)\big).$$

Operation Op is a t-operator with absorbing element $a = 1$, but the range of $*$ is the maximal chain M_1, and therefore Op has no neutral element. This means that $*$ is not a t-norm and thus nor a nullnorm.

Example 2. Assume $L = \left[0, \frac{1}{2}\right[\cup \left]\frac{1}{2}, 1\right]$. Further, let $T : \left[\frac{3}{4}, 1\right]^2 \to \left[\frac{3}{4}, 1\right]$ be a t-norm and denote $a = \frac{3}{4}$. Then the operation Op given by

$$\mathrm{Op}(x,y) = \begin{cases} T(x,y) & \text{if } (x,y) \in \left[\frac{3}{4}, 1\right]^2, \\ 0 & \text{if } (x,y) \in \left[0, \frac{1}{2}\right]^2, \\ a & \text{otherwise} \end{cases}$$

is a t-operator which is not a nullnorm.

Remark 3. If the lattice L has incomparable elements, we may have an operation $V : L^2 \to L$ which is a nullnorm but not a t-operator. We illustrate this fact in the next example.

Example 3. Let L be the lattice from Example 1 (which is the horizontal sum of two copies of $[0,1]$). Let $a = (0.5, 2)$, $S : [0, (0.5, 2)]^2 \to [0, (0.5, 2)]$ be a t-conorm and $T : [(0.5, 2), 1]^2 \to [(0.5, 2), 1]$ be a t-norm. Then the operation

$$V(x,y) = \begin{cases} S(x,y) & \text{if } (x,y) \in [0,(0.5,2)]^2, \\ T(x,y) & \text{if } (x,y) \in [(0.5,2),1]^2, \\ a & \text{otherwise,} \end{cases}$$

is a nullnorm which is not a t-operator, since the functions f_0 and f_1 are not continuous.

Lemma 2 states that addition of a discrete set D to any lattice L, will not violate the relationship between nullnorms and t-operators provided the absorbing element a is incomparable with any element of D.

Proposition 3. *Let L be a bounded lattice and $a \in L$ be such that the set $\widetilde{\|}_a$ is discrete in the lattice-topology $\mathcal{T}$. Then every nullnorm V on the lattice L is a t-operator.*

Proof. Choose $a \in L$ such that the set $\widetilde{\|}_a$ is discrete. Denote $\tilde{L} = L \setminus \|_a$. Then $\tilde{L}$ is a sublattice of L. Let V be an arbitrary nullnorm on L. Realize that for $(x,y) \in \tilde{L}^2$ also $V(x,y) \in \tilde{L}$ which follows from monotonicity of V. This means that $V|_{\tilde{L}^2}$ is a nullnorm on $\tilde{L}$ and, by Proposition 2, also a t-operator. Since every function defined on a discrete set is continuous, we conclude that V is a t-operator. □

Proposition 4. *Let $L = M \cup P$ where M is a connected maximal chain, $a \in M$ and $P = \|_a$. Then every t-operator* Op *on L is a nullnorm.*

Proof. Let Op be a t-operator on L. Using the same reasoning as in the proof of Proposition 3, $\text{Op}|_{M^2}$ is a t-operator by Proposition 1. Since, neither $\mathbf{0}_L$ nor $\mathbf{1}_L$ are partial neutral elements on the set $\|_a$, we get immediately that every t-operator Op on $L = M \cup P$ is a nullnorm. □

Propositions 3 and 4 describe the relationship between nullnorms and t-operators. However, they cannot be generalized as we will show in Examples 4 and 5. Before proceeding to those examples, we generalize Proposition 1.

Proposition 5. *Let $L = M \cup D$, where M is a maximal chain that is connected, $a \in M$, and the set $D = \widetilde{\|}_a$ is discrete. Then an operation $*$ on L is a nullnorm if and only if it is a t-operator.*

Remark that the lattice L in Proposition 5 fulfills the assumptions of Proposition 3 as well as of Proposition 4. That is why the proof is omitted.

Example 4. Let $L = [0,1] \cup \{c\}$ and $c \parallel x$ for all $x \in]0,1[$. Consider operation Op given by

$$\mathrm{Op}(x,y) = \begin{cases} 1 & \text{if } x = 1 \text{ or } y = 1, \\ \max\{x,y\} & \text{if } x \in [0,1], \\ x & \text{if } y = c, x \in [0,1], \\ y & \text{if } x = c, y \in [0,1], \\ c & \text{if } x = y = c. \end{cases}$$

Then Op is a t-operator whose absorbing element is $a = 1$ and is not t-conorm and hence neither a nullnorm.

Example 5. Assume the lattice L from Example 4 and the operation V given by

$$V(x,y) = \begin{cases} 0 & \text{if } x = y = 0, \\ 1 & \text{if } x = y = 1, \\ c & \text{otherwise.} \end{cases}$$

Then V is a nullnorm with absorbing element c and it is not a t-operator. Moreover, there exists no t-operator Op on the presented lattice L with absorbing element c.

Example 5 illustrates that on some lattices L when constructing a t-operator with the chosen absorbing element $a \in L$, one cannot select a arbitrarily. In Proposition 6 we formulate a necessary condition under which it is possible to construct such t-operator. Recall that, for arbitrary $a \in L, \mathcal{M}(a)$ denotes the family of all maximal chains containing a.

Proposition 6. *Let $a \in L$ be chosen such that $\|_a \neq \emptyset$. Assume that there exists a t-operator on L whose absorbing element is a. Then for every maximal chain $C \notin \mathcal{M}(a)$ there exists a maximal chain $M \in \mathcal{M}(a)$ and a σ-homomorphism $\varphi : C \to M$ with respect to $\wedge$ and $\vee$.*

Proof. Any t-operator Op on L is monotone. Functions $f_{\mathbf{0}_L}(x) = \mathrm{Op}(\mathbf{0}_L, x)$ and $f_{\mathbf{1}_L}(x) = \mathrm{Op}(\mathbf{1}_L, x)$ are continuous and

$$f_{\mathbf{0}_L}(\mathbf{0}_L) = \mathbf{0}_L, f_{\mathbf{0}_L}(\mathbf{1}_L) = a, \ f_{\mathbf{1}_L}(\mathbf{1}_L) = \mathbf{1}_L, f_{\mathbf{1}_L}(\mathbf{0}_L) = a.$$

Therefore for arbitrary $C \notin \mathcal{M}(a)$ there must exist a maximal chain $M \in \mathcal{M}(a)$ and a σ-homomorphism $\varphi : C \to M$ with respect to $\wedge$ and $\vee$. □

A sufficient condition for the existence of a t-operator on L is formulated in Proposition 7.

Proposition 7. *Let $a \in L$ be chosen such that $\|_a \neq \emptyset$. Assume there exists a maximal chain $M \in \mathcal{M}(a)$ and an increasing mapping $\psi : L \to M$ which is continuous in the lattice-topology $\mathcal{T}$ and for all $x \in M$, $\psi(x) = x$. Then there exists a t-operator $\overline{\mathrm{Op}}$ on L whose absorbing element is a.*

Proof. First, by Proposition 2 we know that there exists a t-operator $\mathrm{Op} : M^2 \to M$ (since arbitrary nullnorm can serve as a t-operator). Realize that for arbitrary $x, y \in L$

$$\psi(\mathrm{Op}(\psi(x), \psi(s))) = \mathrm{Op}(\psi(x), \psi(s)).$$

Then, associativity of Op implies that for all $x, y, z \in L$

$$\begin{aligned} &\mathrm{Op}(\psi(\mathrm{Op}(\psi(x), \psi(y))), \psi(z)) = \mathrm{Op}(\mathrm{Op}(\psi(x), \psi(y)), \psi(z)) \\ &= \mathrm{Op}(\psi(x), \mathrm{Op}(\psi(x), \psi(y))) = \mathrm{Op}(\psi(x), \psi(\mathrm{Op}(\psi(x), \psi(y)))). \end{aligned}$$

Hence, the operation $\overline{\mathrm{Op}}(x, y) = \mathrm{Op}(\psi(x), \psi(y))$ is associative. Commutativity of $\overline{\mathrm{Op}}$ follows from commutativity of Op and increasingness of $\overline{\mathrm{Op}}$ follows from increasingness of Op and of ψ. Continuity of functions

$$f_{\mathbf{0}_L}(x) = \overline{\mathrm{Op}}(\mathbf{0}_L, x) = \mathrm{Op}(\mathbf{0}_L, \psi(x)), \quad f_{\mathbf{1}_L}(x) = \overline{\mathrm{Op}}(\mathbf{1}_L, x) = \mathrm{Op}(\mathbf{1}_L, \psi(x))$$

follows from continuity of ψ and of $\mathrm{Op}(\mathbf{1}_L, \cdot)$ and $\mathrm{Op}(\mathbf{0}_L, \cdot)$. In summary, $\overline{\mathrm{Op}}$ is a t-operator on L. □

4 Conclusions

The notion of a t-operator on bounded lattices was proposed and a necessary and a sufficient condition for the existence of a t-operator on a bounded lattice L with a chosen absorbing element $a \in L$ was formulated. A relationship between nullnorms and t-operators on bounded lattices was characterized. It was shown that, under constrained that $\mathrm{Op} : L^2 \to L$ is a commutative, associative and monotone operation then properties

(a) $\mathbf{0}_L$ and $\mathbf{1}_L$ are idempotent elements of Op and functions $\mathrm{Op}(\mathbf{0}_L, \cdot)$ and $\mathrm{Op}(\mathbf{1}_L, \cdot)$ are continuous,
(b) there exists $a \in L$ such that $\mathbf{0}_L$ is a partial neutral element of Op on $[\mathbf{0}_{L,a}]$, and $\mathbf{1}_L$ is a partial neutral element of Op on $[a, \mathbf{1}_L]$,

are not equivalent.

Acknowledgements. The work of Martin Kalina has been supported from the Science and Technology Assistance Agency under contract No. APVV-14-0013, and from the VEGA grant agency, grant No. 2/0069/16.

References

1. Birkhoff, G.: Lattice Theory. American Mathematical Society Colloquium Publishers, Providence (1967)
2. Calvo, T., De Baets, B., Fodor, J.: The functional equations of Frank and Alsina for uninorms and nullnorms. Fuzzy Sets Syst. **120**, 385–394 (2001)

3. Calvo, T., Mayor, G., Mesiar, R. (eds.): Aggregation Operators. Physica-Verlag, Heidelberg (2002)
4. Frank, M.: On the simultaneous associativity of $F(x; y)$ and $x + y - F(x; y)$. Aeq. Math. **19**, 194–226 (1979)
5. Grabisch, M., Pap, V., Marichal, J.L., Mesiar, R.: Aggregation Functions. University Press, Cambridge (2009)
6. Kalina, M.: Nullnorms and t-operators on bounded lattices: a comparison. In: Proceeding of ISFS 2017, The 3rd International Symposium on Fuzzy Sets - Uncertainty Modelling, Rzeszów, 19–20 May Poland, pp. 23–26 (2017)
7. Karaçal, F., Ince, M.A., Mesiar, R.: Nullnorms on bounded lattices. Inf. Sci. **325**, 227–236 (2015)
8. Kelley, J.L.: General Topology. D. van Nostrand Company, Inc., Princeton (1964)
9. Klement, E.P., Mesiar, R., Pap, E.: Triangular Norms. Kluwer Academic Publisher, Dordrecht (2000)
10. Mas, M., Mayor, G., Torrens, J.: T-operators. J. Uncertain. Fuzziness Knowl.-Based Syst. **7**, 31–50 (1999)
11. Mas, M., Mayor, G., Torrens, J.: T-operators and uninorms in a finite totally ordered set. Int. J. Intell. Syst. **14**(9), 909–922 (1999)
12. Mas, M., Mayor, G., Torrens, J.: The distributivity condition for uninorms and t-operators. Fuzzy Sets Syst. **128**, 209–225 (2002)

Steinhaus Transforms of Fuzzy String Distances in Computational Linguistics

Anca Dinu[1,3], Liviu P. Dinu[1,3], Laura Franzoi[1,3(✉)], and Andrea Sgarro[2,3]

[1] University of Bucharest, Bucharest, Romania
ancaddinu@gmail.com, liviu.p.dinu@gmail.com, laura.franzoi@gmail.com

[2] DMG, University of Trieste, Trieste, Italy
sgarro@units.it

[3] Human Language Technologies Research Center,
University of Bucharest, Bucharest, Romania

Abstract. In this paper we deal with distances for fuzzy strings in $[0,1]^n$, to be used in distance-based linguistic classification. We start from the fuzzy Hamming distance, anticipated by the linguist Muljačić back in 1967, and the taxicab distance, which both generalize the usual crisp Hamming distance, using in the first case the standard logical operations of minimum for conjunctions and maximum for disjunctions, while in the second case one uses Łukasiewicz' T-norms and T-conorms. We resort to the Steinhaus transform, a powerful tool which allows one to deal with linguistic data which are not only fuzzy, but possibly also irrelevant or logically inconsistent. Experimental results on actual data are shown and preliminarily commented upon.

Keywords: Steinhaus transform · Language classification trees

1 Introduction: Fuzziness in Linguistic

This paper continues work on fuzzy string distances and linguistic classification started in [4–6], and inspired by the path-breaking ideas put forward back in 1967 [10] by the Croat linguist Muljačić. The technical tool which will be used in this paper is *Steinhaus transform*, applied both to fuzzy and crisp string distances.

In his 1967 paper Muljačić, even if only rather implicitly, had introduced what appears to us as a natural *fuzzy* generalization of crisp Hamming distances between binary strings of fixed length n, and this only two years after Zadeh's seminal work [12]: the aim was showing that Dalmatic, now an extinct language, is a bridge between the Western group of Romance languages and the Eastern group, mainly Romanian. The situation is the following: Romance languages $L, \Lambda, \ldots$ are each described by means of n features, which can be present or absent, and so are encoded by strings as $s(L) = \underline{x} = x_i \ldots x_n$, where x_i is the truth value of the proposition *feature i is present in language L*; however, presence/absence is sometimes only vaguely defined and so each $x = x_i$ is rather a truth value $x \in [0,1]$ in a multi-valued logic as is fuzzy logic; $x = x_i$ is *crisp*

J. Medina et al. (Eds.): IPMU 2018, CCIS 853, pp. 171–182, 2018.
https://doi.org/10.1007/978-3-319-91473-2_15

only when either $x = 0 = false = absent$ or $x = 1 = true = present$, else x is *strictly fuzzy.* So, the mathematical objects we shall deal with are *strings* $\underline{x}, \underline{y}, \ldots$ of length n, each of the n components being a real number in the interval [0,1], and moreover *distances* between such objects, since the classifications we tackle are all distance-based. Below, beside Muljačić distance, we define more string distances, obtained by use of the *Steinhaus transform,* cf. below, and comment on them; they are all *metric* distances, in particular they verify the triangle inequality; cf. [4–6]. Unlike the case of Muljačić distances, which span the interval $[0, n]$, these distances are *normalized* to the interval [0,1].

The reason to use Steinhaus transforms is that it allows one to deal also with *irrelevance* and *inconsistency* in linguistics, as we now argue, and not only with vagueness, or fuzziness, as did Muljačić.

Based on arguments defended by the linguist Longobardi and his school, cf. [1,7–9], if a linguistic feature i has a low truth value in two languages L and Λ, then that feature is scarcely relevant: in fact, in the practice of linguistics the values 0 and 1 have a very *asymmetric* use, and the fact that languages L and Λ both have a zero in a position i means that such an irrelevant feature i should *not* really contribute to the distance between the two languages. Technically, one should move from Hamming distances to (normalized) Jaccard distances, but all of this in a fuzzy rather in a crisp context as is usual. To achieve the goal, the convenient tool to be used is the *Steinhaus transform,* cf. Sect. 3, which is known to preserve metricity and which is general enough so as to amply cover also the fuzzy situation: one starts from a distance like Muljačić distance $d_M(\underline{x}, \underline{y})$, and obtains its Steinhaus transform, in this case a *fuzzy Jaccard distance* $d_J(\underline{x}, \underline{y})$ for fuzzy strings $\underline{x}$ and $\underline{y}$ (starting from the usual *crisp* Hamming distance the transform gives the usual *crisp* Jaccard distance).

Actually, such an approach to irrelevance is maybe too radical (a zero in both languages "kills"that position), and so one might prefer to resort to a convex combination after normalizing also Muljačić distance $d_M(\underline{x}, \underline{y})$:

$$(1 - \lambda)\,\frac{d_M(\underline{x}, \underline{y})}{n} + \lambda\, d_J(\underline{x}, \underline{y}), \;\; \lambda \in [0, 1] \tag{1}$$

e.g. with the weight $\lambda = \frac{2}{3}$ as suggested by Longobardi and based on linguistic arguments (personal communication; alternatively, the weight might be learned from available linguistic data).

In general, to apply a Steinhaus transformation one needs a *pivot string,* which in the Jaccard case is the all-0 string $\underline{z} = \underline{0} = (0, \ldots, 0)$. Actually, any other string $\underline{z}$ might be used: this we shall do in this paper, so as to try to cover not only the case of fuzziness and irrelevance, but also the case of *logical inconsistency,* which we now comment upon.

Longobardi's strings are *ternary,* cf. e.g. [1,7–9]; the two crisp values 0 and 1 are there, but there is also a third symbol $*$ which signals *logical inconsistency*: one has situations as e.g. *if feature 17 is present and feature 21 is absent, then feature 35 is logically undefined,* it does not make any sense. All this establishes an extremely complex network of logical dependences in his data, and makes it necessary, if one wants to cover also this new intriguing facet, to suitably generalize crisp Hamming distances, or crisp Jaccard distances, respectively: in Longobardi's approach, cf. e.g. [1,7–9], the two distances for ternary strings one defines and uses (or a convex combination of the two as above) are quite useful, but unfortunately they violate the triangle property, and so are not metric (fortunately, with Longobardi's actual data, metricity is violated only mildly, and so clustering methods were all the same used even when they would assume metricity). In this paper we propose two *metric* alternatives based on Steinhaus transforms: the star $*$ will be replaced by the totally ambiguous truth value $\frac{1}{2}$, and the pivot string in the transform will be the all-$\frac{1}{2}$ string, i.e. the totally ambiguous string $\underline{z} = (\frac{1}{2}, \ldots, \frac{1}{2})$, rather than the all-0 string, i.e. the totally false string. The idea is to play down not the contribution of 0's, as in the case of irrelevance, but rather the contribution of the $\frac{1}{2}$-positions. It will turn out that in this case, which is not genuinely fuzzy, rather than to Muljačić distances, the Steinhaus transform had been better applied to the usual *taxicab distance* (Manhattan distance, Minkowski distance), re-found when the standard fuzzy logical operators of *min* and *max* for conjunctions and disjunctions are replaced by Łukasiewicz T-norms and T-conorms, cf. Sect. 2 below. In the experimental Sect. 4 we shall try to give a first-hand evaluation of this metricity-preserving choice.

In Sect. 2 we re-take both fuzzy Hamming distances, or Muljačić distances, and taxicab distances stressing how the latter relate to Łukasiewicz T-norms. Section 3 introduces Steinhaus transforms, while Sect. 4 is devoted to experiments on actual linguistic data. This paper is meant to discuss and offer technical tools to be used in computational linguistics, more specifically in distance-based linguistic classification; the strictly linguistic purport of our results is the object of the current activities of the Human Language Technologies Research Center, University of Bucharest.

2 Fuzzy Hamming Distances, or Muljačić Distances

We need some notations and definitions: we set $x \wedge y \doteq \min[x, y]$, $x \vee y \doteq \max[x, y]$ and $\overline{x} \doteq 1 - x$; these are the truth values of conjunction AND, disjunction OR and negation NOT w.r. to propositions with truth values x and y in *standard fuzzy logic*, a relevant form of multi-valued logic; $x \in [0, 1]$. Define the *fuzzines* of the truth value x to be $f(x) \doteq x \wedge (1 - x)$. For two truth values x and y in [0,1] we say that x and y are *consonant* if either $x \vee y \le \frac{1}{2}$ or $x \wedge y \ge \frac{1}{2}$, else they are *dissonant*; let $\mathcal{D}$ and $\mathcal{C}$ denote the set of dissonant and consonant positions i, respectively. We define the following distance for strings $\underline{x}, \underline{y} \in [0, 1]^n$:

$$d_M(\underline{x}, \underline{y}) \doteq \sum_{i \in \mathcal{D}} [1 - [f(x_i) \vee f(y_i)]] + \sum_{i \in \mathcal{C}} [f(x_i) \vee f(y_i)] \tag{2}$$

Expression (2) stresses the link with *crisp* Hamming distances for binary strings $\in \{0, 1\}^n$, but its meaning is better understood due to the following fact: each of the n *additive* terms summed in (2) is the truth value of the statement $\big[$(*feature*

f_i is present in L and absent in Λ) or (feature f_i is absent in L and present in Λ)], since, as soon proved, cf. e.g. [4], for two truth values x and y one has $(x \wedge \overline{y}) \vee (\overline{x} \wedge y)$ equal to $f(x_i) \vee f(y_i)$ or to $1 - [f(x_i) \vee f(y_i)]$ according whether there is consonance or dissonance. This distance, called henceforth *Muljačić distance* (and rather inappropriately called *Sgarro distance* in [2]; cf. also [11]) is simply a natural generalization of crisp Hamming distances to a fuzzy setting. As for alternative logical operators for conjunctions and disjunctions (different T-norms and T-conorms, for which cf. e.g. [3]), they have been discussed in [5]. From a metric point of view, the only attractive choice, beside fuzzy Hamming distances, turned out to be Łukasiewicz T-norms for conjunctions and the corresponding T-conorms for disjunctions:

$$x \top y \doteq (x + y - 1) \vee 0, \quad x \perp y \doteq (x + y) \wedge 1$$

One soon checks that in this case, rather curiously, $(x \top \overline{y}) \perp (\overline{x} \top y)$ turns out to be simply $|x - y|$, and so the string distance one obtains is nothing else but the very well-known taxicab distance $d_T(\underline{x}, \underline{y}) = \sum_i |x_i - y_i|$, which in our context, when it is applied to fuzzy strings of length n, might be also legitimately called *Łukasiewicz distance.*

The distance in (2) is a *fuzzy metric distance*, cf. the Appendix, from which a standard metric distance is soon obtained by imposing that self-distances $d_M(\underline{x}, \underline{x})$ should be 0, while, unless $\underline{x}$ is crisp (i.e. belongs to $\{0,1\}^n$, the set of the 2^n binary strings of length n), the value given by (2) would be strictly positive, cf. next section.

As for taxicab or Łukasiewicz distances, the self-distance $d_T(\underline{x}, \underline{y})$ is always zero even when the argument $\underline{x}$ is not crisp, a possibly unpleasant fact in a fuzzy context, as argued in [5].

3 Steinhaus Transforms and Steinhaus Fuzziness of Truth Values

The Steinhaus transform $\mathcal{S}(d)$ of a metric distance $d(x, y)$, possibly fuzzy, cf. also the Appendix, is defined as:

$$\mathcal{S}\big(d(x,y)\big) = \mathcal{S}_z\big(d(x,y)\big) \doteq \frac{2d(x,y)}{d(x,y) + d(x,z) + d(y,z)} \tag{3}$$

Above one needs a *pivot element* z, in our case a *constant* string $\underline{z} = (z, \ldots, z)$, $z_i = z \ \ \forall i, \ z \in [0,1]$; with Muljačić distances $d_M(\underline{x}, \underline{y})$, we shall simply write $d_z(\underline{x}, \underline{y})$ for their Steinhaus transforms:

$$d_z(\underline{x}, \underline{y}) \doteq \mathcal{S}_{\underline{z}}\big(d_M(\underline{x}, \underline{y})\big) = \frac{2d_M(\underline{x}, \underline{y})}{d_M(\underline{x}, \underline{y}) + d_M(\underline{x}, \underline{z}) + d_M(\underline{y}, \underline{z})} \tag{4}$$

For generality's sake, in this section we have found it convenient to cover the general case of any constant pivot string $\underline{z}$ with $z \in [0,1]$, but actually, to no loss of generality, we shall always assume $z \in [0, \frac{1}{2}]$, because, setting $\neg \underline{x} \doteq (1 - x_1, \ldots, 1 - x_n) = (\overline{x_1}, \ldots, \overline{x_n})$, one has $d_M(\underline{x}, \underline{y}) = d_M(\neg \underline{x}, \neg \underline{y})$, cf. the Appendix, and so

$$\mathcal{S}_z\big(d_M(\underline{x}, \underline{y})\big) = \mathcal{S}_{1-z}\big(d_M(\neg \underline{x}, \neg \underline{y})\big)$$

We recall that one has always to impose $d_z(\underline{x}, \underline{x}) \doteq 0$ for all strings $\underline{x}$ if one wants to obtain a standard metric distance rather than a *fuzzy* metric distance, else formulas (3) and (4) would give a strictly positive value for "self-distances" $d_z(\underline{x}, \underline{x})$ as soon as one of the components x_i is strictly fuzzy; once chosen a distance as in (2), (3) or (4), the original value of the self-distance, possibly positive, will be called the *fuzziness* of the string $\underline{x}$ and will be denoted by $f_M(\underline{x})$ or $f_z(\underline{x})$ for Muljačić or Steinhaus distances, respectively. For $n = 1$ one has $f_M(x) = f(x)$ as anticipated in Sect. 2. One has to set explicitly $d_0(\underline{0}, \underline{0}) = f_0(\underline{0}) \doteq 0$ when the pivot string is the all-0 string used with irrelevant features.

Even if Steinhaus transforms yield normalized distances which are *not* additive, unlike normalized and unnormalized Muljačić distances, we find it meaningful to consider explicitly the case $n = 1$, i.e. the case of the Muljačić-Steinhaus fuzziness $f_z(x)$ of truth values $x \in [0, 1]$, $z \in [0, \frac{1}{2}]$. Elementary arguments of standard calculus show that $f_z(x)$ is a *continuous* function of x such that:

* $f_z(x)$ is convex-cap on the interval $[0, z]$ where it increases from $f_z(0) = 0$ to $f_z(z) = \frac{2}{3}$
* it is constant on the interval $[z, \frac{1}{2}]$ where $f_z(x) = \frac{2}{3}$
* it is convex-cup on the interval $[\frac{1}{2}, 1 - z]$ where it decreases from $f_z(\frac{1}{2}) = \frac{2}{3}$ to $f_z(1 - z) = \frac{2z}{2-z}$
* it is convex-cap on the interval $[1 - z, 1]$ where it decreases to 0.

Relevant are the limit cases when the pivot truth value z is 0, and then the first and the fourth cases vanish save for $f_0(0) = 0$, or when z is $\frac{1}{2}$ where it is the second and the third cases which vanish. In the last situation, $z = \frac{1}{2}$, one has a convex-cap function on [0,1] which is symmetric around $\frac{1}{2}$; cf. Figs. 1, 2 and 3, which show the behavior of a typical fuzziness $f_z(x)$ (we have chosen $z = \frac{1}{4}$), and the two limit-case behaviors for $z = 0$ and $z = \frac{1}{2}$.

We take the chance to stress that a term like $d_M(\underline{x}, \underline{z})$ which appears in the denominator of (4) is equal to the *fuzzy Hamming weight* $w(\underline{x}) \doteq \sum_i x_i$ for $z = 0$ (irrelevance), while it is equal to $\frac{n}{2}$ independent of $\underline{x}$ when $z = \frac{1}{2}$. The fact that $d_M(\underline{x}, \underline{z})$ with $z = \frac{1}{2}$ is independent of $\underline{x}$ is a serious drawback, indeed. This is why in the case of Longobardi's data, while still keeping the pivot string $\underline{z}$ with totally ambiguous components all equal to $\frac{1}{2}$, we will apply the Steinhaus transform, rather than to the fuzzy Hamming distance or Muljačić distance, directly to the taxicab distance or Łukasiewicz distance $d_T(\underline{x}, \underline{y})$ to obtain a new normalized distance $d_{TS}(\underline{x}, \underline{y})$. In the case of $d_{TS}(\underline{x}, \underline{y})$, in the denominator of the corresponding Steinhaus transform the fuzzy Hamming weight $w(\underline{x})$ is replaced by $\sum_i |x_i - \frac{1}{2}|$.

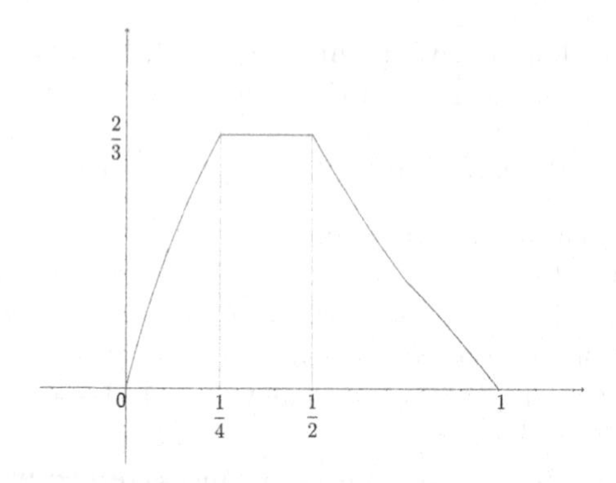

Fig. 1. Fuzziness $f_z(x) = f_{\frac{3}{4}}(x)$

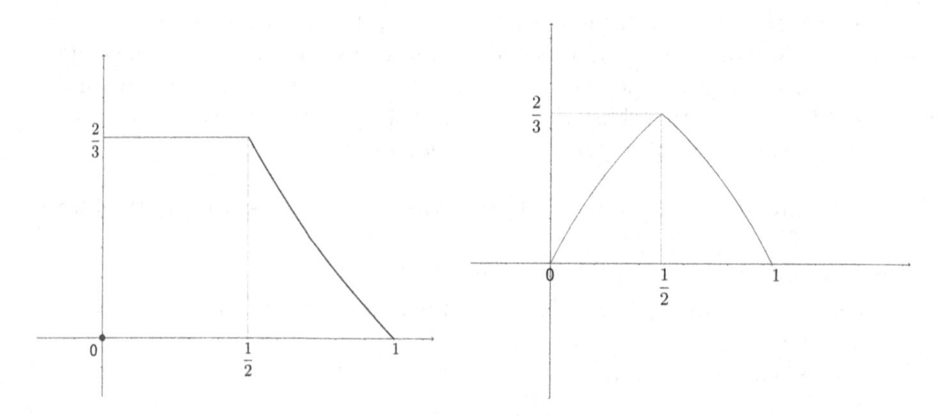

Fig. 2. Fuzziness $f_z(x) = f_0(x)$

Fig. 3. Fuzziness $f_z(x) = f_{\frac{1}{2}}(x)$

4 Experiments and a Final Comment

The preceding sections describe fuzzy tools to be used in computational linguistics (but not only there, cf. e.g. [4]). We shall tackle two situations.

To begin with, we go back to Muljačić original data [10], which are available also in [5,13]. He uses 12 ternary strings of length 40, and makes a very sparse use of fuzziness: the only strictly fuzzy value is .5, in [5] and below replaced by a star $*$ to help readability. The 12 languages are, respectively: R = Romanian, D = Dalmatic, I = Italian, Sa = Sardinian, Frl = Frioulan, spoken in North-East Italy, Rsh = the 4th language of Switzerland, Romansh, Pr = Provençal, FP = Francoprovençal spoken across Italian and French Alps, F = French, C = Catalan, S = Spanish (Castilian) and P = Portuguese.

A simple Python code has been used to compute our distances, cf. again [13]. After showing a few sample distances, we reproduce the UPGMA tree for the original Muljačić distances $d_M(\underline{x}, \underline{y})$ (even if trees were not used explicitly by Muljačić) and for its Jaccard-like variant $d_J(\underline{x}, \underline{y}) = d_0(\underline{x}, \underline{y})$ meant to cope with irrelevance; cf. also [6]. In this case use of the totally ambiguous pivot $(\frac{1}{2}, \ldots, \frac{1}{2})$ would be out of place, since there is no problem of logical inconsistency in Muljačić data. Together with the 12 strings, the two complete matrices exhibiting distances are to be found in [13]; the most fuzzy string corresponds to Provençal with four star. For example, the three strings corresponding to Italian I, Romanian R and Spanish S are

$$I = ([1,1,1,0,0,0,1,0,1,0,0,0,1,*,0,0,1,1,1,1,1,0,1,0,1,0,1,1,1,1,1,1,1,1,0,1,0,0,1,0)$$
$$R = (1,0,1,1,0,0,0,0,0,0,1,1,1,0,1,0,1,1,1,0,0,1,1,0,1,0,0,1,0,1,1,0,0,0,1,1,0,1,1,1)$$
$$S = (1,1,1,1,0,0,0,1,1,0,0,1,1,1,1,0,0,1,1,1,0,1,1,0,1,0,1,*,0,0,0,0,1,0,0,1,0,0,0,0)$$

and with these languages Muljačić distances (fuzzy Hamming distances) and Jaccard fuzzy distances are, respectively (Table 1):

Table 1. Muljačić and Jaccard fuzzy distances

d_M	I	R	S
I		18.5	15
R	18.5		15.5
S	15	15.5	

d_J	I	R	S
I		.587	.517
R	.587		.553
S	.517	.553	

(We have left blank the uninformative secondary diagonal.) Note that we are implicitly assuming, as did Muljačić himself, that features are non-interactive (independent) and are equally important: when performing clustering on larger real-world data, we would have to resort to methods which bioinformatics has nowadays made popular, bootstrapping, say.

Figures 4 and 5 exhibit the UPGMA trees of distances; linguists might object to the use of outdated material, and so the (actually weak) differences between the two trees might be felt to have only dubious linguistic significance. Be as it may, the trees agree all rather well with current classifications and clearly support Muljačić objective to prove the tight kinship of Dalmatic and Romanian.

We move to Longobardi's data (or rather to a sample of his languages, since the data he and his school are providing are steadily improving and extending), data which are not really fuzzy, even if we have decided to "simulate" logical inconsistency by *total fuzziness*, cf. [13]. In this case the number of features is 53, and the language we have selected are: Wo = Wolof, spoken mainly in Senegal, Hu = Hungarian, Finn = Finnish, Ar = Arabic, Heb = Hebrew or 'ivrit, Hi = Hindi, Po = Polish, Rus = Russian, Sc = Serbo (Croatian), Slo = Slovenian, Lat = Latin, Grk = Greek, NTG = New Testament Greek, Gri = Grico, a variant of Greek spoken in South Italy, Ir = Gaelic Irish, Wel = Welsh, E = English,

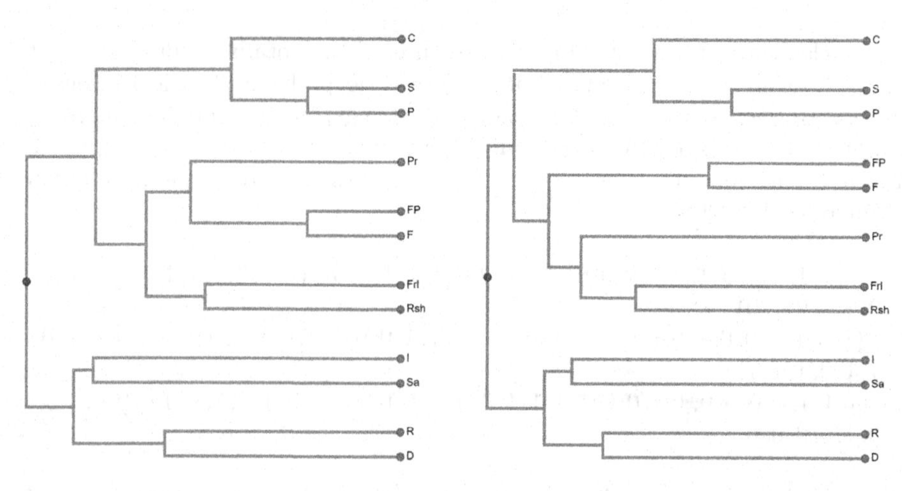

Fig. 4. Muljačić distance-tree for Muljačić data

Fig. 5. Jaccard distance-tree for Muljačić data

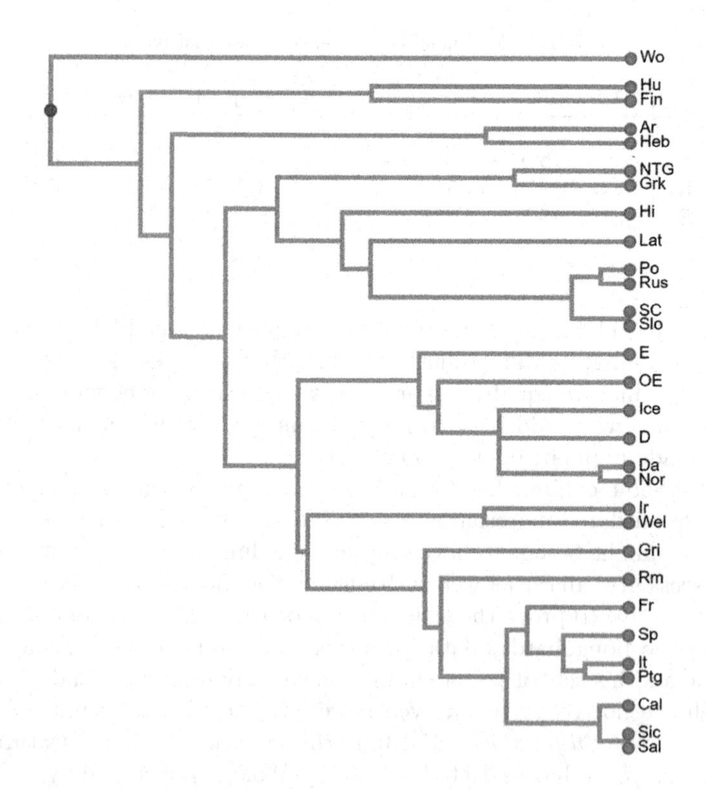

Fig. 6. Taxicab tree with Longobardi data

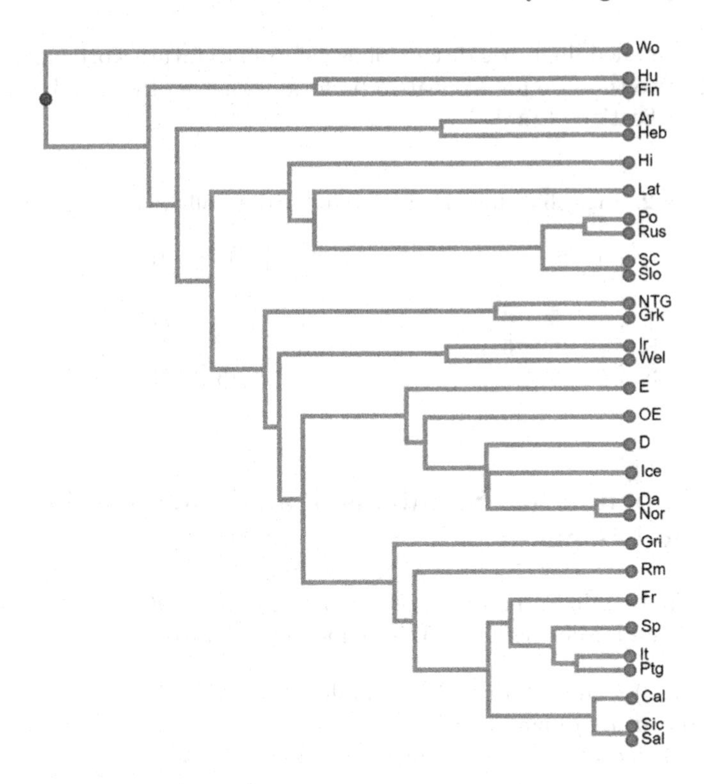

Fig. 7. Steinhaus-taxicab tree with Longobardi data

OE = Old English, D = German, Ice = Icelandic, Da = Danish, Nor = Norwegian, Rm = Romanian, Fr = French, Sp = Spanish, IT = Italian, PTG = Portuguese, Cal = Calabrese as spoken in South Italy, Sic = Sicilian, Sal = Salentine as spoken in Salento, South Italy. As an example, we write the three strings corresponding again to Italian It = I, Romanian Rm = R and Spanish Sp = S, and the two sub-matrices of distances (complete matrices are to be found in [13]).

$$I = (1,1,1,0,1,1,1,1,1,0,0,0,*,0,0,1,0,1,*,0,1,1,1,*,1,*,1,0,1,0,1,0,0,\\ 0,1,0,0,0,*,0,1,0,0,0,1,*,*,*,*,*,0,0,1)$$
$$R = (1,1,1,0,1,1,1,1,1,0,0,1,*,0,0,1,0,1,*,*,1,1,1,*,1,*,1,0,0,*,*,0,0,\\ 0,1,*,0,0,*,1,1,0,0,0,1,*,*,*,*,*,0,0,1)$$
$$S = (1,1,1,0,1,1,1,1,1,0,0,0,*,0,0,1,0,1,*,1,1,1,1,*,1,*,1,0,1,0,1,0,0,\\ 1,1,0,0,0,*,1,1,0,0,0,1,*,*,*,*,*,0,0,1)$$

Figures 6 and 7 show the UPGMA trees for the taxicab distance or Łukasiewicz distance $d_T(\underline{x},\underline{y})$, and for its Steinhaus transform $d_{TS}(\underline{x},\underline{y})$, with the all-$\frac{1}{2}$ pivot-string as meant to cope with logical inconsistency. Limited to Romance languages, one may compare the trees with Muljačić data versus those with Longobardi data.

Again a detailed linguistic discussion is deferred to future work; again we just observe the two trees do not appear to perform rather well w.r. to Longobardi's original classifications (Table 2).

Table 2. Taxicab distances without and with totally ambiguous pivot

d_T	I	R	S
I		5	3
R	5		5
S	3	5	

d_{TS}	I	R	S
I		.217	.13
R	.217		.217
S	.13	.217	

5 An Appendix on Steinhaus Transforms and Fuzzy Metric Distances

Given a finite or infinite non-void set $\mathcal{X}$, a *fuzzy metric distance* $d(\ ,\ )$ on the elements x, y and z belonging to $\mathcal{X}$ is defined by the axioms:

(i) $0 \leq d(x,x) \leq d(x,y)$, $d(x,y) = 0$ implies $x = y$
(ii) $d(x,y) = d(y,x)$ (symmetry)
(iii) $d(x,z) + d(z,y) \geq d(x,y)$ (triangle inequality).

If $d(x,x) > 0$ the element x is called fuzzy, else x is crisp. A standard metric distance, where axiom *(i)* is replaced by $d(x,y) \geq 0$, $d(x,y) = 0$ if and only if $x = y$, is soon obtained by forcing self-distances to be 0; in this case the original value of the self-distance is kept to gauge the fuzziness $f(x)$ of the element x. Note that in the literature much more ambitious and interesting definitions of fuzzy metric spaces are given, but our simple, or even simple-minded, choice will do in our context. Muljačić distances as in (2) are fuzzy metric distances, as soon checked, cf. e.g. [4]: as a matter of fact, given the additive nature of definition (2), it is enough to make checks for $n = 1$. This way one checks also criteria for equality, cf. again [4]. E.g. $d(x,x) = 0$ if and only $\underline{x}$ is crisp in the usual sense, i.e. all of its components are crisp, $d(x,y) \leq d(x,x)$ whenever y is a crisper copy of x, i.e. the two strings are consonant in all positions i, and moreover $f(y_i) \leq f(x_i)$, $1 \leq i \leq n$. The maximum value for the Muljačić distance is n, which is found for $\underline{x}$ and $\underline{y}$ crisp and distinct in all positions. A further remarkable property is $d_M(\underline{x}, \neg\underline{y}) = n - d_M(\underline{x}, \underline{y})$ which soon implies $d_M(\neg\underline{x}, \neg\underline{y}) = d_M(\underline{x}, \underline{y})$. Muljačić distance and the related *Muljačić distinguishability* [4] might add to the comprehension of the geometry of the unit hypercube (*Kosko's fuzzy hypercube* as often called in a fuzzy context), in particular as concerns clustering.

Steinhaus transforms as in (3) can be defined by taking a *pivot element* $z \in \mathcal{X}$ to obtain a metric distance $d_z(x,y)$ starting from any metric distance $d(x,y)$; the transform has positive self-distances when so has $d(x,y)$, is standard when $d(x,y)$ is standard. Steinhaus transforms are a general mold which is inspired by the way crisp Hamming distances relate to crisp Jaccard distances. Steinhaus distances are normalized; actually the inequality $d_z(x, y) \leq 1$ re-writes as the triangle inequality for $d(x, y)$, and so the equality criterion is collinearity of x, z, y in the original metric space, where z is the pivot element. In the case of Muljačić distances, collinearity, as soon checked, holds if and only if in all positions i one has $x_i \wedge y_i \leq z_i \leq x_i \vee y_i$ and moreover at least two of the three (not necessarily distinct) numbers x_i, z_i, y_i are crisp. Note also that whenever distances to be transformed are quick to compute, so are their transforms: in our case, with Muljačić and taxicab distances, where computations are of course linear in the string length n, the time complexity to compute Steinhaus transforms is itself $\Theta(n)$. That Steinhaus distances are *metric* distances is soon shown, save for the triangle inequality $d_z(x, u) + d_z(u, y) \geq d_z(x, y)$, $x, y, z, u \in \mathcal{X}$, whose proof is here reproduced to enhance self-readability of the paper.

Proof of the Triangle Inequality for Steinhaus Transforms
First we observe that, for $a, b > 0$ and $c \geq 0$, if $a \leq b$ then $\frac{a}{b} \leq \frac{a+c}{b+c}$. In our case a is $d(x, y)$, the initial distance to be transformed, while b is the numerator in (3), $b \geq a$, and $c \doteq d(x, u) + d(y, u) - d(x, y) \geq 0$ $\big(d(x, y)$ is triangular$\big)$. One obtains:

$$\frac{d_z(x, y)}{2} \leq \frac{d(x, u) + d(y, u)}{d(x, z) + d(y, z) + d(x, u) + d(y, u) \pm d(u, z)}$$

This is the sum of two terms: in the first, with numerator $d(x, u)$, use the triangle inequality for $d(u, z)$ with triangulating element y, while in the second, with numerator $d(y, u)$, use the triangle inequality for $d(u, z)$ with triangulating element x to obtain precisely $d_z(x, y) \leq d_z(x, u) + d_z(y, u)$. □

6 Conclusions

Unlike Muljačić data, which are quite traditional, Longobardi's data are definitely unorthodox, being syntactic as e.g.: *The ergative case is present in the declension of substantives*, as happens in Basque. This requires use of new and "unorthodox" distances, preferably metric as here (but not in Longobardi's case). As for Jaccard distances vs. possibly normalized Hamming distances, the situation is well understood, and can be re-phrased using Steinhaus transforms with a pivot which represents total irrelevance. To deal with Longobardi's inconsistent features we rather used here a totally ambiguous pivot meant to represent inconsistency. Linguistic results compare quite well with those obtained by Longobardi's school on the basis of their non-metric distance.

Acknowledgment. Authors A. Dinu and L. P. Dinu are supported by HerCoRe project (no. 91970), funded by Volkswagen Foundation; L. Franzoi and A. Sgarro are with the INdAM research group GNCS.

References

1. Bortolussi, L., Sgarro, A., Longobardi, G., Guardiano, C.: How many possible languages are there? In: Biology, Computation and Linguistics, pp. 168–179. IOS Press, Amsterdam (2011). https://doi.org/10.3233/978-1-60750-762-8-168
2. Deza, M.M., Deza, E.: Dictionary of Distances. Elsevier B.V., New York City (2006)
3. Dubois, D., Prade, H.: Fundamentals of Fuzzy Sets. Kluwer Academic Publishers, Dordrecht (2000)
4. Franzoi, L., Sgarro, A.: Fuzzy Hamming distinguishability. In: IEEE International Conference on Fuzzy Systems, FUZZ-IEEE, pp. 1–6 (2017). https://doi.org/10.1109/FUZZ-IEEE.2017.8015434
5. Franzoi, L., Sgarro, A.: Linguistic classification: T-norms, fuzzy distances and fuzzy distinguishabilities. Proc. Comput. Sci. **112**, 1168–1177 (2017). https://doi.org/10.1016/j.procs.2017.08.163. KES
6. Franzoi, L.: Jaccard-like fuzzy distances for computational linguistics. In: Presented at SYNASC 2017 (2017, in press)
7. Longobardi, G., Ceolin, A., Bortolussi, L., Guardiano, C., Irimia, M.A., Michelioudakis, D., Radkevich, N., Sgarro, A.: Mathematical modeling of grammatical diversity supports the historical reality of formal syntax. In: Proceedings of the Leiden Workshop on Capturing Phylogenetic Algorithms for Linguistics, pp. 1–4. University of Tübingen, Tübingen DEU (2016). https://doi.org/10.15496/publikation-10122
8. Longobardi, G., Guardiano, C., Silvestri, G., Boattini, A., Ceolin, A.: Toward a syntactic phylogeny of modern Indo-European languages. J. Hist. Linguist. **3**(11), 122–152 (2013). https://doi.org/10.1075/bct.75.07lon
9. Longobardi, G., Ghirotto, S., Guardiano, C., Tassi, F., Benazzo, A., Ceolin, A., Barbujani, G.: Across language families: genome diversity mirrors language variation within Europe. Am. J. Phys. Anthropol. **157**, 630–640 (2015). https://doi.org/10.1002/ajpa.22758
10. Muljačić, Z.: Die Klassifikation der romanischen Sprachen. Rom. J. Buch. **XVIII**, 23–37 (1967)
11. Sgarro, A.: A fuzzy Hamming distance. Bull. Math. de la Soc. Sci. Math. de la R. S. de Romanie **69**(1–2), 137–144 (1977)
12. Zadeh, L.: Fuzzy sets. Inf. Control **8**(3), 338–353 (1965)
13. Steinhaus transforms of fuzzy string distances in computational linguistics (Support material). goo.gl/3p1sY7

Merging Information Using Uncertain Gates: An Application to Educational Indicators

Guillaume Petiot(✉)

CERES, Catholic Institute of Toulouse, 31 rue de la Fonderie, 31062 Toulouse, France
guillaume.petiot@ict-toulouse.fr
http://www.ict-toulouse.fr

Abstract. Knowledge provided by human experts is often imprecise and uncertain. The possibility theory provides a solution to handle these problems. The modeling of knowledge can be performed by a possibilistic network but demands to define all the parameters of Conditional Possibility Tables. Uncertain gates allow us, as noisy gates in probability theory, the automatic calculation of Conditional Possibility Tables. The uncertain gates connectors can be used for merging information. We can use the T-norm, T-conorm, mean, and hybrid operators to define new uncertain gates connectors. In this paper, we will present an experimentation on the calculation of educational indicators. Indeed, the LMS Moodle provides a large scale of data about learners that can be merged to provide indicators to teachers. Therefore, teachers can better understand their students' needs and how they learn. The knowledge about the behavior of learners can be provided by teachers but also by the process of datamining. The knowledge is modeled by using uncertain gates and evaluated from the data. The indicators can be presented to teachers in a decision support system.

Keywords: Uncertain gates · Theory of possibility · Education
Decision making · Possibilistic network

1 Introduction

The modeling of knowledge is often a complex task because human experts describe problems with imprecision and uncertainty. The possibility theory introduced in [23] is a solution to the problem of knowledge modeling. Since the knowledge can be evaluated by a Directional Acyclic Graph, it can be evaluated by possibilistic networks. The possibilistic networks [3,7] are adaptations of the Bayesian Network [20,21] to the possibility theory. The building of the Conditional Possibility Tables requires too many parameters. Indeed, the number of conditional possibilities to define in the CPTs is growing exponentially depending proportionally on the number of variables. So it can be more easy to use

J. Medina et al. (Eds.): IPMU 2018, CCIS 853, pp. 183–194, 2018.
https://doi.org/10.1007/978-3-319-91473-2_16

logical gates between the variables as the noisy gates in the probability theory in order to build automatically the CPT. The noisy gates provide another advantage which is the modeling of noise. Moreover, it is difficult to have a perfect model of knowledge. There are often unknown variables which contribute to the spurious behavior of models. This problem is often encountered in complex systems. A leakage variable can be added to the models to represent the unknown knowledge.

Uncertain gates can be used to model the knowledge but we need to have a large set of connectors because the expected behavior of variables combination is rarely only conjunctive or disjunctive. In [24] there is an example of one mean noisy gates connector but there is no related work for uncertain gates. The variables are often qualitative as in [11] but to use uncertain gates we have to encode the modalities in numerical values. The use of variable intensity in $[0, 1]$ allows us to inherit the operators from the T-norm, T-conorm, and hybrid operators. So it is very interesting for merging information [5].

In this paper, we would like to perform an experimentation of indicator calculation using uncertain gates. Several studies have already focused on the objective to better understand students and how learners learn in order to highlight the pedagogy which contributes to learning [6,8,16]. The approach uses Bayesian Networks, Neural Networks, Support Vector Machines. They often try to detect the student with difficulties who risks to drop out or fail at the examination.

For the experimentation, we have chosen a course of Spreadsheet in a bachelor degree with resources in Moodle. Moodle is a Learning Management System which allows us to retrieve the data of the learners. The knowledge about the indicators can be provided by teachers or extracted from the data by datamining [14].

The available information concerns attendance to courses, groups, and the student's results at the examination. Moodle provides information about the results of the quiz and the resources consulted. We would like to build an indicator which merges information about all resources. This indicator would take into account the importance of the resources. Another important indicator that we would like to obtain concerns the indicator of the acquired skills which can be calculated by using the results of the quiz if the questions are categorized by skill.

The goal of this paper is to evaluate knowledge in order to build educational indicators in a decision support system. In the first part, we will present uncertain gates, then we will study the modeling of knowledge and finally, we will present the results.

2 Uncertain Gates

Uncertain gates are an analogy of noisy gates in the possibility theory. The possibility theory was developed in 1978 by Zadeh in [23]. In this theory, the imprecise and uncertain knowledge can be modeled by a possibility distribution. For example, if X is a variable and π a possibility distribution defined from its domain R in $[0,1]$, then $\pi(x) = 0$ means that $X = x$ is not possible and $\pi(x) = 1$ means that the value $X = x$ is possible. If we have a set A of the set of parts $P(X)$, then we can define the possibility measure Π and the necessity measure N as in [12]. The possibility measure is a function defined from $P(X)$ in $[0,1]$:

$$\forall A \in P(X), \Pi(A) = \sup_{x \in A} \pi(x). \tag{1}$$

The dual necessity measure is a function from $P(X)$ in $[0,1]$:

$$\forall A \in P(X), N(A) = 1 - \Pi(\neg A). \tag{2}$$

The possibility theory is not additive but maxitive:

$$\forall A \in P(X), \forall B \in P(X), \Pi(A \bigcup B) = \max(\Pi(A), \Pi(B)). \tag{3}$$

The possibilistic networks [3,4,9] are based on d-separation, conditional independence [1], and factoring property. The factoring property can be defined from the joint possibility distribution $\Pi(V)$ for a DAG $G = (V, A)$. $\Pi(V)$ can be factorized toward the graph G:

$$\Pi(V) = \bigotimes_{X \in V} \Pi(X/Pa(X)). \tag{4}$$

With Pa the parents of the node X. The combination rule must be associative, we have chosen the minimum for $\bigotimes$. The Independence of Causal Influence is presented in [10,15]. In the noisy ICI model [10] there is a set of causal variables $X_1, \ldots, X_n$ which influence the result of another variable Y called effect variable. We can introduce an intermediate variable Z_i between each $X_i s$ and Y which represent uncertainty in the causal influence of $X_i s$ on Y. For example, even if a cause is met, it is possible that an inhibitor will not produce Y. The relation between X_i and Z_i is probabilistic. The leaky ICI model (Fig. 1) is derived from the noisy model by adding a leakage variable Z_l which represents the unknown knowledge in the model. The noisy ICI model is based on a deterministic model. So we obtain the equation $Y = f(Z_1, \ldots, Z_n, Z_l)$ where f is a deterministic function. To resume, in the probability theory we have the following graph:

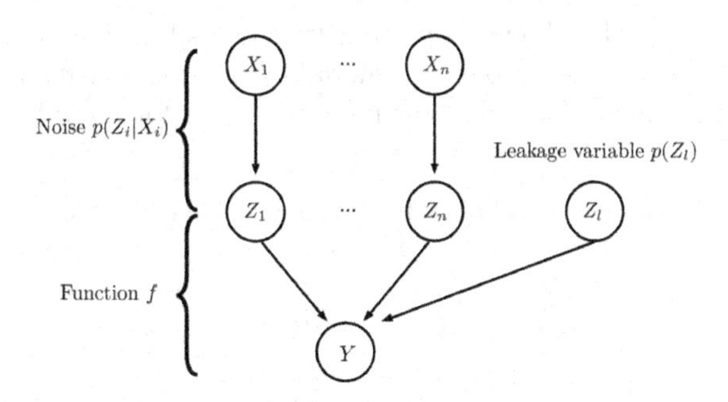

Fig. 1. Leaky ICI model.

ICI means that there is no causal interaction in the effects of the variables X_i on the variable Y. We can calculate $P(Y|X_1, \ldots, X_n)$ by marginalizing the variables Z_i as below:

$$P(y|x_1, \ldots, x_n) = \sum_{z_1,\ldots,z_n} P(y|z_1, \ldots, z_n) \times P(z_1, \ldots, z_n|x_1, \ldots, x_n) \quad (5)$$

$$P(y|x_1, \ldots, x_n) = \sum_{z_1,\ldots,z_n} P(y|z_1, \ldots, z_n) \times \prod_{i=1}^{n} P(z_i|x_i) \quad (6)$$

$$\text{where } P(y|z_1, \ldots, z_n) = \begin{cases} 1 \text{ if } y = f(z_1, \ldots, z_n) \\ 0 \text{ else} \end{cases} \quad (7)$$

As a result we have:

$$P(y|x_1, \ldots, x_n) = \sum_{z_1,\ldots z_n : y = f(z_1,\ldots,z_n)} \prod_{i=1}^{n} P(z_i|x_i) \quad (8)$$

Authors in [10] provide several examples of functions f which can be AND, OR, NOT, INV, XOR, MAX, MIN, MEAN, and linear combination. We can by analogy propose the same formula for the possibilistic model with ICI:

$$\pi(y|x_1, \ldots, x_n) = \max_{z_1,\ldots,z_n : y = f(z_1,\ldots,z_n)} \otimes_{i=1}^{n} \pi(z_i|x_i) \quad (9)$$

The $\otimes$ is the minimum. The CPT is obtained by the calculation of the above formula. For Boolean variables, the possibility table (Table 1) between the variables X_i and Z_i is:

Table 1. Possibility table for Boolean variables.

$\pi(Z_i\|X_i)$	x_i	$\neg x_i$
z_i	1	0
$\neg z_i$	κ_i	1

If we have three ordered levels of intensity such as low, medium and high, as in our application, we can encode the modality by an intensity level as in [11] (Table 2). We can have 0 for low, 1 for medium and 2 for high. So we obtain:

Table 2. Possibility table for multivalued variables.

$\pi(Z_i\|X_i)$	$x_i = 2$	$x_i = 1$	$x_i = 0$
$z_i = 2$	1	$\kappa_i^{2,1}$	0
$z_i = 1$	$\kappa_i^{1,2}$	1	0
$z_i = 0$	$\kappa_i^{0,2}$	$\kappa_i^{0,1}$	1

If we consider as in [11] that a cause of weak intensity cannot produce a strong effect, then $\kappa_i^{2,1} = 0$. In our application this parameter is greater than 0. Another constraint is that $\kappa_i^{1,2} \geq \kappa_i^{0,2}$. So we have 4 parameters per variable. If we add a leakage variable Z_l in the previous model, we obtain the following equation:

$$\pi(y|x_1, \ldots, x_n) = \max_{z_1,\ldots,z_n,z_l:y=f(z_1,\ldots,z_n,z_l)} \otimes_{i=1}^{n} \pi(z_i|x_i) \otimes \pi(z_l) \tag{10}$$

Several uncertain logical gate connectors AND, OR, MIN and MAX were described in [11]. A mathematical simplification has been performed leading to optimized connectors. These connectors are useful in several applications but sometimes you need a different behavior. We can use T-norm, T-conorm, and compromise operators commonly used to merge fuzzy sets to combine the variables Z_i. The T-norm $\top$ is a function of $[0,1] \times [0,1] \rightarrow [0,1]$ with several properties such as commutativity, associativity, monotony and neutral element is 1. There is the same property for the T-conorm $\bot$. The function min is a T-norm and the function max is a T-conorm. These functions will be used in our experimentation. The compromise operators have a behavior between the $\bot$ and $\top$. It includes the mean family operators. It is also possible to find operators which combine reinforcement and compromise. This hybrid behavior consists to adapt the behavior to the data in order to perform a reinforcement if the data are consonant and a compromise if the data are dissonant. The Mycin like operator is commonly used in the range $[-1,1]$ but can be easily adapted to the range $[0,1]$:

$$\hbar(x,y) = \begin{cases} x + y + x * y & \text{if } x \leq 0 \text{ and } y \leq 0 \\ x + y - x * y & \text{if } x > 0 \text{ and } y > 0 \\ \quad x + y & \text{if } x > 0 \text{ and } y < 0 \text{ or } x < 0 \text{ and } y > 0 \end{cases} \tag{11}$$

Many of the operators are defined in the range $[0,1]$. This implies that the intensity scale of the variables to merge must be ordered and in the range $[0,1]$ if we want to use them. For example, we can use the following table (Table 3) for $\pi(Z_i|X_i)$ with three intensity levels in $[0,1]$:

Table 3. Possibility table for multivalued variables with intensities in $[0,1]$.

$\pi(Z_i\|X_i)$	$x_i = 1$	$x_i = 0.5$	$x_i = 0$
$z_i = 1$	1	$\kappa_i^{1,0.5}$	0
$z_i = 0.5$	$\kappa_i^{0.5,1}$	1	0
$z_i = 0$	$\kappa_i^{0,1}$	$\kappa_i^{0,0.5}$	1

The operators must be combined to a threshold function because f must return a value compatible with the variable y except for *min* and *max*. More generally, if the modality of the variable defines an ordered scale $E = \{\vartheta_0 < \vartheta_1 < \ldots < \vartheta_m\}$, the function f of $y = f(z_1, \ldots, z_n)$ must return compatible values with the coding of the qualitative variable Y. To perform this constraint we have to use a scale function named f_e as in [13] which realizes a threshold as follows:

$$f_e(x) = \begin{cases} \vartheta_0 & \text{if } x \leq \theta_0 \\ \vartheta_1 & \text{if } \theta_0 < x \leq \theta_1 \\ \vdots & \vdots \\ \vartheta_m & \text{if } \theta_{m-1} < x \end{cases} \tag{12}$$

The coefficient θ_i allows us to define the expected behavior of the threshold for rounding. The function f is $f = f_e \circ g$ where g is a combination operator of the variables Z. If we want to introduce the leakage parameter Z_l, we have $f = \max(f_e \circ g, z_l)$. These connectors are useful in several applications but sometimes you need a different behavior which takes into account the importance of the variables as in weighted average. This cannot be performed by T-norm, T-conorm or compromise operators such as mean. To do this, we propose to use a linear combination for g. The function g is $g(z_1, \ldots, z_n) = \omega_1 z_1 + \ldots + \omega_n z_n$. The model of the new connector is the following (Fig. 2):

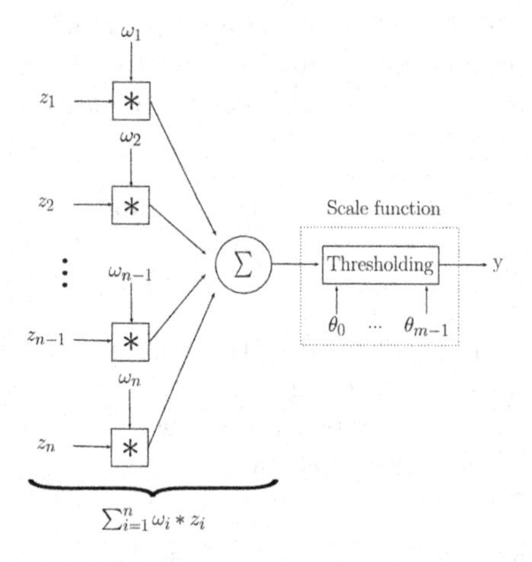

Fig. 2. The connector of linear combination.

The function g has n parameters which are the weights of the linear combination. If all weights are equal to $\frac{1}{n}$, then we calculate the mean of the intensity of the variables and we return the value of Y closest to the mean. If $\forall_{i \in [1,n]} \omega_i = 1$, then we realize the sum of the causal variable intensities. We can associate for each modality of Y a threshold for the sum of intensities. If the weights are different for all variables, then we can take into account the importance of each variable. We can perform in this case a weighted average. It is also possible to use the OWA operator [22] for the function g. In the OWA operator, the data are sorted before applying a weighted combination. Depending on the weights, we can also perform the mean of the Z_i if all weights are equal to $\frac{1}{n}$. But it is not possible to define a weighted average because the rank of the variables in the combination is unknown. The general algorithm for the computation of the uncertain gate connector is below:

Algorithm 1. Algorithm of uncertain gate connectors.

Input :
- Y: CPT to calculate.
- $X_1, \ldots, X_n$: the n parents of Y.
- ω: the vector of weights $(\omega_1, \ldots, \omega_n)$.
- $\kappa[i][Z][X]$: the coefficients $\pi(Z_i|X_i)$.
- f_e: a threshold function.
- g: a merging function.

Output:
- The result is $\pi(Y|X_1, \ldots, X_n)$.

```
1  begin
2     forall the (y, x_1, ..., x_n) ∈ Y × X_1 × ... × X_n do
3        π(y|x_1, ..., x_n) ⟵ 0
4        forall the (z_1, ..., z_n) ∈ Z_1 × ... × Z_n do
5           Result ⟵ g(z_1, ..., z_n)
6           V[z_1, ..., z_n] ⟵ f_e(Result)
7        K ⟵ {(z_1, ..., z_n) ∈ Z_1 × ... × Z_n | V[z_1, ..., z_n] = y}
8        γ ⟵ 0
9        forall the (z_1, ..., z_n) ∈ K do
10          γ ⟵ max(γ, min_{i∈[1,n]} κ[i][z_i][x_i])
11       π(y|x_1, ..., x_n) ⟵ γ
```

3 Experimentation

We propose to apply uncertain gates in education. In our experimentation, we focused our interest on a Spreadsheet course at bachelor level proposed in face-to-face enriched. This means that the course is face-to-face but with resources on Moodle. The knowledge about the indicators is provided by teachers or can be extracted from the data by datamining. The pedagogical indicators are evaluated through quantitative data such as the use of Moodle resources, results in a quiz, presence,... The questions of the quiz are categorized by skills. When the data are missing, we perform an imputation of these data by an iterative PCA [2]. To represent the knowledge we have chosen to use a Directional Acyclic Graph (DAG). This graph is as follows (Fig. 3):

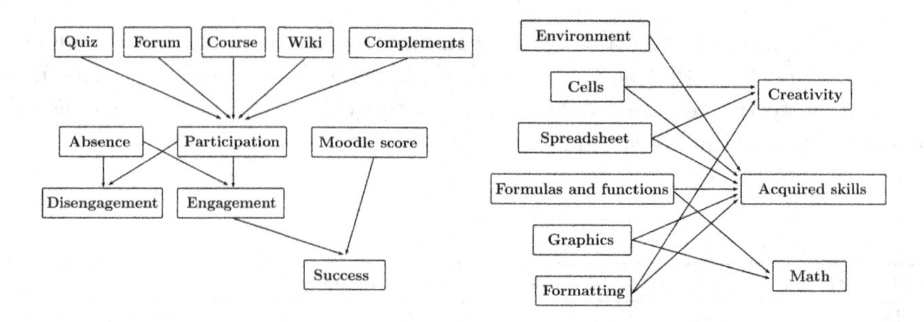

Fig. 3. Modeling of knowledge by a DAG.

The qualitative variables have 3 ordered modalities (low, medium, high) which can be encoded with numerical values. As often happens in human descriptions, knowledge is uncertain and imprecise. The use of a possibility distribution for each modality is a solution to these problems.

The possibilistic network can be used to evaluate the indicators but it requires the definition of all CPTs. This is time-consuming. For example, for the acquired skills indicator which has 5 parent variables, we have $3^{5+1} = 729$ parameters. The elicitation of all these parameters cannot be performed. The use of uncertain gates needs fewer parameters, so it can contribute to complex knowledge modeling. Therefore, we can merge information on the resources consultation in Moodle to build an indicator which takes into account the importance of the resources. The Uncertain Weighted Average connector can be used. The weights are provided by teachers and shown in Fig. 4. If we want to merge information about skills and to model an effect of reinforcement, it may be interesting to use the $\sum$ connector as in Fig. 5.

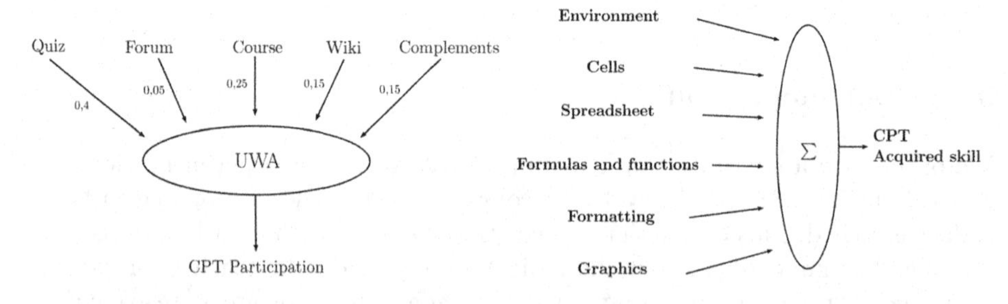

Fig. 4. Weights of the UWA.

Fig. 5. Sum of the variable intensities.

We used the uncertain MIN connector for conjunctive behavior and the uncertain hybrid connector for indicators which need a compromise in case of conflict and a reinforcement if the values are concordant. As a result we obtain the following model (Fig. 6):

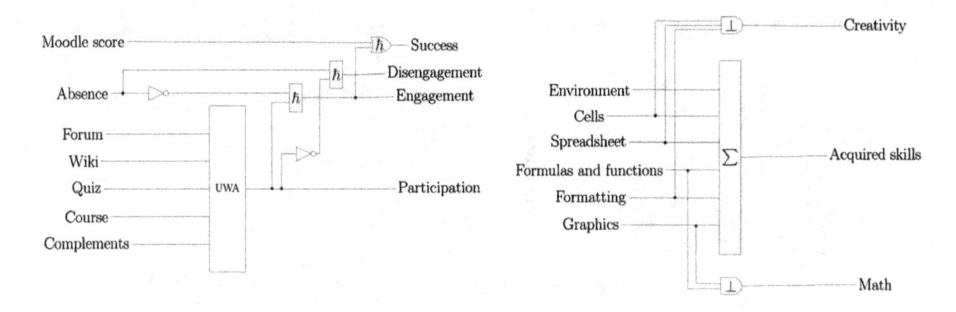

Fig. 6. Knowledge modeling with uncertain connectors.

Before the propagation of new information, we have to build the CPTs of all the uncertain gates. Then, we can apply the junction tree message passing algorithm [19] of Bayesian Networks adapted to Possibilistic Networks. The junction tree is composed of cliques and separators. The cliques are extracted by using the Kruskal algorithm [17] after the generation of the moral graph and the triangulated graph [18]. Therefore, we can propagate new information. The propagation algorithm can be resumed in three steps. The initialization phase with the injection of evidence (new information), then, the collect phase with the propagation of evidence from leaf to root and the distribution phase with the propagation of evidence from root to leaf.

4 Results

We can present the result of the most interesting indicator (the success indicator) which represents the synthesis of several variables and the prediction of student success. We can compare the levels of the indicator and the real success result at the exam. We present below the results of this indicator:

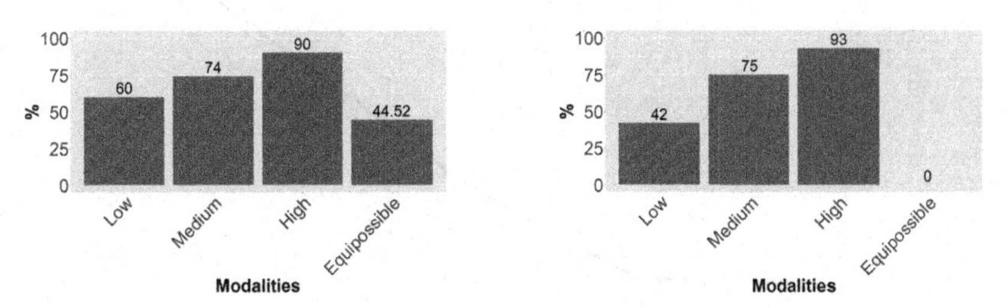

(a) Without the estimation of missing data. (b) With the estimation of missing data.

Fig. 7. Result of the success indicator.

In Fig. 7(a) we can see a lot of equipossible results due to no estimation. This means that all modalities are possible. To reduce the equipossible variables, we have performed an imputation of missing data using an iterative PCA algorithm

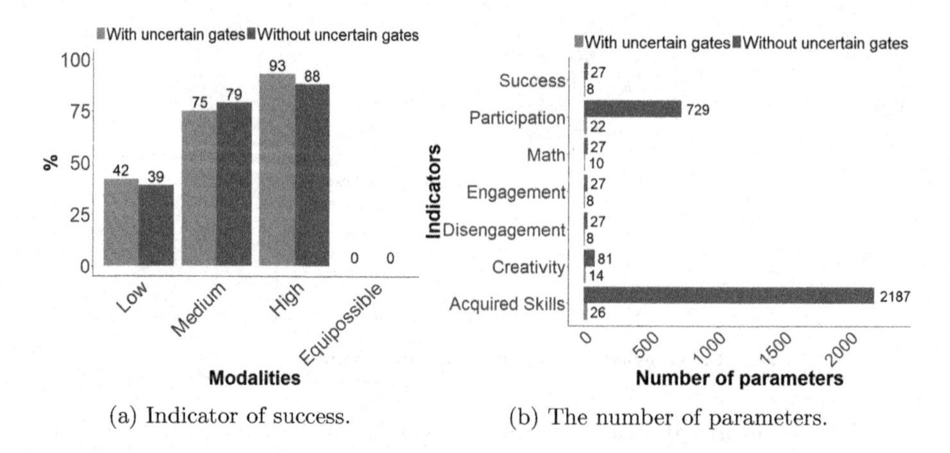

(a) Indicator of success.

(b) The number of parameters.

Fig. 8. Comparison of the results with and without uncertain gates.

[2]. We present the results in Fig. 7(b). We can now compare the results with and without uncertain gates:

In Fig. 8(a) the results are very close nevertheless the uncertain gates require fewer parameters than CPTs elicited by human expert. In Fig. 8(b) the number of parameters is highly decreased by using uncertain gates. We have also measured the performance of the uncertain connectors. We have developed small networks with the number of parents from 2 to 10 and we have generated the CPTs. The results are as follows (Fig. 9):

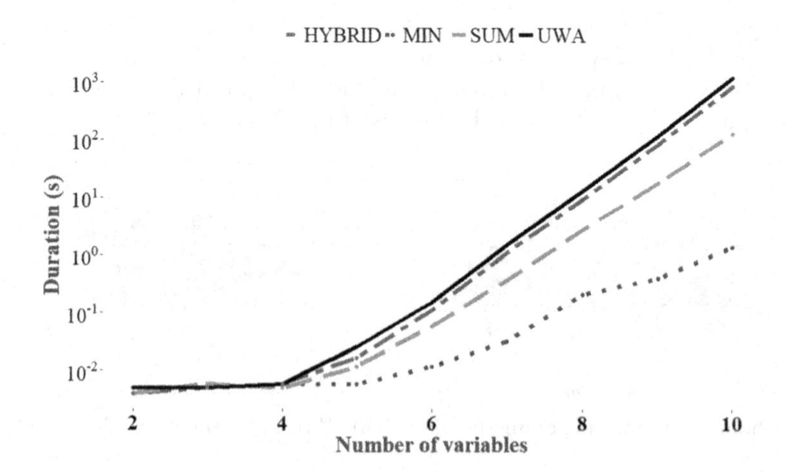

Fig. 9. The calculation time of the uncertain gates.

The computation time is better for the uncertain MIN because of the mathematical simplification demonstrated in [11]. In the benchmark, the calculation time is growing exponentially when the number of variables is increased. Our solution to improve the performance of our system is to calculate only once the CPTs and save the results.

5 Conclusion

The eliciting of the CPT by a human expert is very complex and time consuming. Uncertain gates allow us to generate automatically the CPT. Uncertain gates can be used for merging information. Nevertheless, the behavior of the uncertain MIN and MAX gates is not sufficient. So to complete the existing connectors, we have proposed a bridge between uncertain gates and the operators used to combine the fuzzy sets such as T-norm, T-conorm, mean, hybrid,... These operators have often special behaviors of reinforcement or compromise which are very useful. The family of mean operators can be used to perform the compromise. We have also proposed a new connector which allows a combination of the variables taking into account the importance of the variable as with Weighted Average. We used this connector to merge the information about the consultation of Moodle resources in order to elaborate the indicator of Moodle participation. We have also proposed a connector which takes into account the reinforcement of the intensity of the variables. We have computed the indicators of acquired skills with this connector. The results of the experimentation highlight as expected the students with difficulties. We have to continue the improvement of the performance of computation. An experimentation is also needed to evaluate the pedagogical impact of this study on students and teachers but also on skills.

References

1. Amor, N.B., Benferhat, S.: Graphoid properties of qualitative possiblistic independance relations. Int. J. Uncertain. Fuzzyness Knowl. Based Syst. **5**, 59–96 (2005)
2. Audigier, V., Husson, F., Josse, J.: A principal components method to impute missing values for mixed data. Adv. Data Anal. Classif. **10**(1), 5–26 (2015)
3. Benferhat, S., Dubois, D., Garcia, L., Prade, H.: Possibilistic logic bases and possibilistic graphs. In: Proceedings of the Conference on Uncertainty in Artificial Intelligence, pp. 57–64 (1999)
4. Borgelt, C., Gebhardt, J., Kruse, R.: Possibilistic graphical models. In: Della Riccia, G., Kruse, R., Lenz, H.-J. (eds.) Computational Intelligence in Data Mining. ICMS, vol. 408, pp. 51–67. Springer, Vienna (2000). https://doi.org/10.1007/978-3-7091-2588-5_3
5. Dubois, D., Prade, H., Yager, R.R.: Merging fuzzy information. In: Bezdek, J.C., Dubois, D., Prade, H. (eds.) Fuzzy Sets in Approximate Reasoning and Information Systems. The Handbooks of Fuzzy Sets Series, vol. 5, pp. 335–401. Springer, Boston (1999). https://doi.org/10.1007/978-1-4615-5243-7_7
6. Bousbia, N., Labat, J.M., Balla, A., Reba, I.: Analyzing learning styles using behavioral indicators in web based learning environments. In: EDM 2010 International Conference on Educational Data Mining, Pittsburgh, USA, pp. 279–280 (2010)

7. Benferhat, S., Smaoui, S.: Représentation hybride des réseaux causaux possibilistes. Rencontres francophones sur la Logique Floue et ses Applications, pp. 43–50. Cépaduès Editions, Nantes (2004)
8. Baker, R.S.J.D., Yacef, K.: The state of educational data mining in 2009: a review and future visions. J. Educ. Data Min. **1**(1), 3–17 (2009)
9. Caglioni, M., Dubois, D., Fusco, G., Moreno, D., Prade, H., Scarella, F., Tettamanzi, A.G.B.: Mise en oeuvre pratique de réseaux possibilistes pour modéliser la spécialisation sociale dans les espaces métropolisées. In: Rencontres Francophones sur la Logique Floue et ses Applications (LFA 2014), pp. 267–274. Cépaduès Editions, Cargèse (2014)
10. Dìez, F., Drudzel, M.: Canonical probabilistic models for knowledge engineering. Technical report CISIAD-06-01 (2007)
11. Dubois, D., Fusco, G., Prade, H., Tettamanzi, A.: Uncertain logical gates in possibilistic networks. An application to human geography. In: Beierle, C., Dekhtyar, A. (eds.) SUM 2015. LNCS (LNAI), vol. 9310, pp. 249–263. Springer, Cham (2015). https://doi.org/10.1007/978-3-319-23540-0_17
12. Dubois, D., Prade, H.: Possibility Theory: An Approach to Computerized Processing of Uncertainty. Plenum Press, New York (1988)
13. van Gerven, M.A.J., Jurgelenaite, R., Taal, B.G., Heskes, T., Lucas, P.J.F.: Predicting carcinoid heart disease with the noisy-threshold classifier. Artif. Intell. Med. **40**, 45–55 (2006)
14. Petiot, G.: Calcul d'indicateurs pédagogiques par des réseaux possibilites. Rencontres Francophones sur la Logique Floue et ses Applications (LFA 2016), pp. 195–202. Cépaduès Editions, La Rochelle (2016)
15. Heckerman, D., Breese, J.: A new look at causal independence. In: Proceedings of the Tenth Annual Conference on Uncertainty in Artificial Intelligence (UAI94) CA, pp. 286–292. Morgan Kaufmann Publishers, San Francisco (1994)
16. Huebner, R.A.: A survey of educational data mining research. Res. Higher Educ. J. **19**, 4–15 (2013)
17. Kruskal, J.B.: On the shortest spanning subtree of a graph and the travelling salesman problem. Proc. Am. Math. Soc. **7**, 48–50 (1956)
18. Kjaerulff, U.: Reduction of computational complexity in Bayesian Networks through removal of week dependences. In: Proceeding of the 10th Conference on Uncertainty in Artificial Intelligence. Morgan Kaufmann publishers (1994)
19. Lauritzen, S., Spiegelhalter, D.: Local computation with probabilities on graphical structures and their application to expert systems. J. Royal Stat. Soc. **50**, 157–224 (1988)
20. Neapolitan, R.E.: Probabilistic Reasoning in Expert Systems: Theory and Algorithms. Wiley, New York (1990)
21. Pearl, J.: Probabilistic Reasoning in Intelligent Systems: Networks of Plausible Inference, 2nd edn. Morgan Kaufman Publishers Inc., San Mateo (1988)
22. Yager, R.R.: On ordered weighted averaging aggregation operators in multi-criteria decision making. IEEE Trans. Syst. Man Cybern. **18**, 183–190 (1988)
23. Zadeh, L.A.: Fuzzy sets as a basis for a theory of possibility. Fuzzy Sets Syst. **1**, 3–28 (1978)
24. Zagorecki, A., Druzdzel, M.J.: Probabilistic Independence of Causal Influences. In: Probabilistic Graphical Models, pp. 325–332 (2006)

On the Use of Fuzzy Preorders in Multi-robot Task Allocation Problem

José Guerrero(✉), Juan-José Miñana, and Óscar Valero

Mathematics and Computer Science Department, Universitat de les Illes Balears,
Carr. Valldemossa km. 7.5, Palma de Mallorca, Spain
{jose.guerrero,jj.minana,o.valero}@uib.es

Abstract. This paper addresses the multi-robot task allocation problem. In particular, given a collection of tasks and robots, we focus on how to select the best robot to execute each task by means of the so-called response threshold method. In the aforesaid method, each robot decides to leave a task and to perform another one (decides to transit) following a probability (response functions) that depends mainly of a stimulus and the current task. The probabilistic approaches used to model the transitions present several handicaps. To solve these problems, in a previous work, we introduced the use of indistinguishability operators to model response functions and possibility theory instead of probability. In this paper we extend the previous work in order to be able to model response functions when the stimulus under consideration depends on the distance between tasks and the utility of them. Thus, the resulting response functions that model transitions in the Markov chains must be asymmetric. In the light of this asymmetry, it seems natural to use fuzzy preorders in order to model the system's behaviour. The results of the simulations executed in Matlab validate our approach and they show again how the possibilistic Markov chains outperform their probabilistic counterpart.

Keywords: Fuzzy preorders · Markov chain · Multi-robot
Possibility · Swarm Intelligence · Task allocation · Asymmetric distance

1 Introduction

Systems with two ore more robots, or agents, which cooperatively carry out a mission provide several advantages compared to systems with only one robot. For example, they can perform the mission faster or in a more flexible way than systems with a single robot. In order to make profit of these systems, from now on referred to as Multi-Robot Systems, several issues must be faced up, such as task decomposition, task assignment or movement coordination. This paper focuses on the "Multi-robot Task Allocation" (MRTA for short) problem which consists in selecting the best robot or set of robots to execute each one of the tasks that must be performed.

In general, MRTA can be a NP-hard problem and, therefore, it can be intractable by a multi-robot system [1]. Due to this fact, a lot of research has been

J. Medina et al. (Eds.): IPMU 2018, CCIS 853, pp. 195–206, 2018.
https://doi.org/10.1007/978-3-319-91473-2_17

done in order to get a good enough solution in a reasonable computation time (see [2]). Some of the proposed solutions are based on the so-called Swarm Intelligence approaches, where a cooperative performance emerges from the interaction of very simple behaviours running in each robot. The main advantages of swarm systems are their simplicity, scalability and robustness. Probably, nowadays one of the most widely used swarm-like methods are those based on the so-called Response Threshold Methods (RTM for short). In these methods each robot has a stimulus associated with each task to execute which represents how suitable is the mentioned task for the robot. According to [3], the classical response threshold method assigns to each robot r_i and to each task t_j, a stimulus $s_{r_i,t_j} \in \mathbb{R}$ which represents how appropriate t_j is for r_i. For example, the stimulus can be the inverse of the distance between the task and the robot. When s_{r_i,t_j} exceeds a threshold value θ_{r_i} ($\theta_{r_i} \in \mathbb{R}$), the robot r_i starts the execution of t_j following a given probability $p(r_k, ij)$. There are different kinds of probabilities response functions that can define a transition. In this paper we will focus on one of the most widely used (see [4]) which is given by the following expression:

$$p(r_k, ij) = \frac{s^n_{r_i,t_j}}{s^n_{r_i,t_j} + \theta^n_{r_i}}, \tag{1}$$

where $n \in \mathbb{R}^+$ (here $\mathbb{R}^+$ stands for the set of positive real numbers).

Recently, in [5] indistinguishability operators were shown to be an appropriate mathematical tool to model such response functions whenever the stimulus only depends on the (Euclidean) distance among tasks (in fact whenever the stimulus only depends of the inverse of the Euclidean distance). Of course, these situation is an admisible hypothesis in those cases in which each robot senses the need only to handle the closest task (see [6]).

Let us recall that, according to [7], given a non-empty set X and a t-norm T (for the basics of t-norms we refer the reader to [8]), a T-indistinguishability operator is a fuzzy set $E : X \times X \to [0, 1]$ satisfying for each $x, y, z \in X$ the following:

(i) $E(x, x) = 1$; (Reflexivity)
(ii) $E(x, y) = E(y, x)$; (Symmetry)
(iii) $E(x, z) \geq T(E(x, y), E(y, z))$. (Transitivity)

If, in addition, E satisfies condition **(i')** then E is said to be a T-indistinguishability operator that separates points, where the aforesaid condition is given as follows:

(i') given $x, y \in X$, then $E(x, y) = 1 \Rightarrow x = y$.

Concretely, in [5], it was proved, from the theoretical point of view and by means of different implementations, that indistinguishability operators provides a mathematical tool to model such kind of response functions, whenever the stimulus only depends on the inverse of the Euclidean distance among tasks.

In fact, under the aforesaid hypothesis, the probability response function (1) can be transformed into the following response function

$$p(r_k, ij) = \frac{\theta^n}{\theta^n + d^n(r_i, t_j)}. \tag{2}$$

Notice that the response function (1) can be modeled by the next indistinguishability operator:

$$E(x, y) = \frac{\theta^n}{\theta^n + d^n(x, y)} \text{ for each } x, y \in \mathbb{R}^2 \tag{3}$$

where d denotes the Euclidean metric on $\mathbb{R}^2$ and x and y denotes the coordinates of the allocation t_i and t_j, respectively. Of course, in this case the numerical value provided by the indistinguishability operator E given by (3) must be interpreted as a possibility instead of a probability. This is the reason for which we call possibilistic response functions to those response functions that matches up with indistinguishability operators.

Nevertheless, in general, the stimulus of a robot depends on more factors than the inverse of the distance (see [1]). Motivated for this fact, we are interested on extending the use of indistinguishability operators for modeling in a suitable way these type of situations. Concretely, we focus our efforts on those cases in which for each task there is an associated utility value which represents the importance of performing it. However, in this situation indistinguishability operators do not seem the best mathematical tool to describe the response functions. This is due to the fact that the transition possibility must reflect that there is a task with more associated utility and, thus, the robot must perceive as more attractive to leave the current task in order to start the one with more associated utility. Taking into account this fact, it seems natural to consider fuzzy preorders (asymmetric indistinguishability operators in [7]) for modeling response functions.

Let us recall, according to [9], that a T-preorder (fuzzy preorder in [9]) is a fuzzy set $E : X \times X \to [0, 1]$ satisfying for each $x, y, z \in X$ the conditions **(i)** and **(iii)** of the indistinguishability operator axiomatic. Moreover a T-preorder E is said to separate points provided that it satisfies the following condition **(i')**:

(i') given $x, y \in X$, then $E(x, y) = E(y, x) = 1 \Rightarrow x = y$.

On account of the fact that fuzzy preorders seem to be a suitable tool to model possibilistic response functions, the main goal of this paper is to propose a Response Threshold Method in which the transitions are modeled via fuzzy preorders. Specifically, we will consider a response function of type (2) in which the Euclidean distance is replaced by a new information. In particular, the aforesaid information is obtained by means of an aggregation of the information relative to the space distance and improvement made in utility when the robot leaves the task t_i and start to perform the task t_j. The aggregated information will be yielded by an asymmetric distance which is induced by means of a function, an asymmetric distance aggregation function in the sense of [10], that merges the Euclidean distance and an asymmetric distance that measures the aforementioned utility improvement. The obtained new response function will be shown formally to be a fuzzy preorder. Finally, several numerical experiments will be made using Matlab to validate the utility of the developed theory to model the evolution of the decision process given as a possibilistic Markov chain. A comparison between our results and those provided by the probabilistic framework will be made.

2 Possibilistic Task Allocation Problem

As pointed out before, in the modeling of the Task Allocation Problem, the probability of executing the next task (referenced as response function) depends strongly on the current task (state). Therefore, from the classical viewpoint, the decision about the next task to be executed is a memoryless process that holds the conditions of a probabilistic Markov chain. Such classical probabilistic approach presents a huge number of inconveniences, such as problems with the selection of the probability response function when more than two tasks are under consideration, asymptotic converge, and so on (see [11]). In order to overcome the aforementioned problems, we proposed a new possibilistic theoretical formalism for implementing the RTM algorithms based on the use of indistinguishability operators in [5,11]. The possibilistic approach improves the probabilistic one. In particular, in [11] we demonstrated that possibilistic Markov chains converge to stationary possibilistic distribution 10 times faster and they can predict in a better way the system's behaviour under imprecise information (for instance, when the data is affected by errors). Next we recall the basics about possibilistic Markov chains with the aim of incorporating asymmetric response functions in the RTAM algorithm.

2.1 Possibilistic Markov Chains

Following [12,13], a possibilistic Markov (memoryless) process can be defined as follows: let $S = \{s_1, \ldots, s_m\}$ ($m \in \mathbb{N}$) denote a finite set of states. If the system is in the state s_i at time τ ($\tau \in \mathbb{N}$), then the system will move to the state s_j with possibility p_{ij} at time $\tau + 1$. Let $x(\tau) = (x_1(\tau), \ldots, x_m(\tau))$ be a fuzzy state set, where $x_i(\tau)$ is defined as the possibility that the state s_i will occur at time τ for all $i = 1, \ldots, m$. It must be noted that $\bigvee_{i=1}^{m} x_i(\tau) \leq 1$, where $\vee$ stands for the maximum operator on $[0, 1]$. Following the preceding facts, the evolution of the possibilistic Markov chain in time is given, for all $\tau \in \mathbb{N}$, by

$$x_i(\tau) = \bigvee_{j=1}^{m} p_{ji} \wedge x_j(\tau - 1),$$

where $\wedge$ stands for the minimum operator on $[0, 1]$. The preceding expression admits a matrix formulated as follows:

$$x(\tau) = x(\tau - 1) \circ P = x(0) \circ P^{\tau}, \tag{4}$$

where $P = \{p_{ij}\}_{i,j=1}^{m}$ is the fuzzy transition matrix, $\circ$ is the matrix product in the max-min algebra $([0,1], \vee, \wedge)$ and $x(\tau) = (x_1(\tau), \ldots, x_m(\tau))$ for all $\tau \in \mathbb{N}$ is the possibility distribution at time τ.

According to the aforesaid matrix notation, and following [12], a possibility distribution $x(\tau)$ of the system states at time τ is said to be stationary, or stable, whenever $x(\tau) = x(\tau) \circ P$.

One of the main advantages of the possibilistic Markov chains with respect to their probabilistic counterpart is given by the fact that under certain conditions, provided in [14], the system converges to a stationary distribution in a finite number of steps. Contrarily, the convergence of probabilistic Markov chains is, in general, only guaranteed asymptotically.

2.2 The Asymmetric Response Function and Fuzzy Preorders

Next we construct a new response function in order to model those situations in which the stimulus of a robot depends on two factors, the distance between tasks and the utility associated to them. Thus we will show that this new possibilistic response function will be able to reflect that, in general, the robot will perceive as more attractive those tasks with better associated utility.

To achieve the target, let us fix, for the shake of simplicity, a few aspect of the mission under consideration. From now on, we will assume that the tasks are randomly placed in an environment and the robots are initially randomly placed too. Besides, each robot is always assigned to a task and only one robot per task can be assigned at the same time. Moreover, each task t_j has associated an utility, $U_j \in \mathbb{R}^+$ (here $\mathbb{R}^+$ denotes the set of non-negative real numbers) which indicates how useful is the task for that robot. Hence each task t_j can be identified with a triple (U_j, x_j, y_j), where the first coordinate represents the utility task and, in addition, the remainder two coordinates denotes the allocation of the task. Furthermore, each robot stimulus depends on both, the distance between the robot (current task allocation) and the task to perform and the improvement made in utility.

In order to construct the response function the tasks distance will be measured via the Euclidean distance $d((x_i, y_i), (x_j, y_j))$, where (x_i, y_i) and (x_j, y_j) denotes the allocation coordinates of tasks t_i and t_j respectivaly. In addition, the utility improvement will be measure via the so-called upper asymmetric distance $q(U_i, U_j)$, where $q(U_i, U_j) = \max\{U_i - U_j, 0\}$. Note that an asymmetric distance satisfies the metric axiomatic excepts the symmetry.

Clearly $q(U_i, U_j) = 0$ provides that the task t_i is more attractive than the task t_j. However, positive values of $q(U_i, U_j)$ (which means $U_i \leq U_j$) measures the improvement in utility made when the task t_i is leaved by the robot and it starts to perform the task t_j.

Since the stimulus must depend on the Euclidean distance and the upper asymmetric distance we merge both information in oder to obtain a global measure between tasks that incorporates the information coming from both different sources. Such an information fusion is provided by the function $\Phi : (\mathbb{R}^+)^2 \rightarrow \mathbb{R}^+$ given by $\Phi(x, y) = \alpha_u \cdot x + y$, where α_u will be a system's parameter that, on the one hand, makes that the utility value has the same dimension and scale as the distance and, on the other hand, indicates how important is the utility with respect to the distance (see Sect. 3.3 for a detailed discussion). Since Φ is an asymmetric aggregation function we have, by Theorem 6 in [10], that the non-negative real valued function Q_Φ, given by $Q_\Phi((U_i, x_i, y_i), (U_j, x_j, y_j)) = \alpha_u \cdot q(U_i, U_j) + d(x_i, x_j)$, is an asymmetric distance function.

Under this considerations, the decision process of a robot r_k will follow a possibilistic Markov chain in such a way that it will leave the task t_i (where is allocated) in order to perform the task t_j according to the transition $p(r_k, ij)$ given by

$$p(r_k, ij) = \frac{\theta^n}{\theta^n + Q_\Phi^n((U_i, x_i, y_i), (U_j, x_j, y_j))} \tag{5}$$

Notice that the expression (5) is obtained replacing in the expression of (2) the Euclidean distance by the asymmetric distance Q_Φ.

Next we prove that the following fuzzy set E_{Dom}^n is a fuzzy preorder on $\mathbb{R}^3$, where

$$E_{Q_\Phi, Dom}^n(x, y) = \frac{\theta^n}{\theta^n + Q_\Phi^n(x, y)} \text{ for each } x, y \in \mathbb{R}^3. \tag{6}$$

Clearly the possibility $p(r_k, ij)$, given by (5), matches up with the value $E_{Dom}^n(x, y)$, where x and y denotes the coordinates (utility and allocation) of tasks t_i and t_j respectively.

Let us recall that the family of Dombi t-norms $\{T_{Dom}^\lambda\}_\lambda$ is given as follows:

$$T_{Dom}^\lambda(a, b) = \begin{cases} 0, & \text{if } a = 0 \text{ or } b = 0 \\ \frac{1}{1 + \left(\left(\frac{1-a}{a}\right)^\lambda + \left(\left(\frac{1-b}{b}\right)^\lambda\right)\right)^{\frac{1}{\lambda}}}, & \text{elsewhere} \end{cases} . \tag{7}$$

Taking into account the exposed information we have the next result.

Proposition 1. *Let $n \in \mathbb{N}$ and let q be an asymmetric distance on a non-empty set X. The fuzzy set $E_{q,Dom}^n$ on $X \times X$, defined by $E_{q,Dom}^n(x, y) = \frac{1}{1 + (q(x,y))^n}$, is a $T_D^{\frac{1}{n}}om$-preorder that separates points.*

Proof. Next we show that E_q^n satisfies **(i)**, **(i')** and **(iii)** for each $x, y, z \in X$ when the t-norm $T_D^{\frac{1}{n}}om$ is considered.

(i) $E_{q,Dom}^n(x, x) = \frac{1}{1 + (q(x,x))^n} = 1$, since q is an asymmetric distance on X and so $q(x, x) = 0$. Moreover, $E_{q,Dom}^n$ also satisfies **(i')**, since $q(x, y) = q(y, x) = 0 \Leftrightarrow x = y$ and, thus, $E_{q,Dom}^n(x, y) = E_{q,Dom}^n(y, x) = 1 \Leftrightarrow x = y$.

(iii) Let $x, y, z \in X$. We will show that

$$E_{q,Dom}^n(x, z) \geq T_{Dom}^{\frac{1}{n}}\left(E_{q,Dom}^n(x, y), E_{q,Dom}^n(y, z)\right).$$

First, observe that

$$\frac{1 - E_{q,Dom}^n(x, y)}{E_{q,Dom}^n(x, y)} = \frac{1 - \frac{1}{1 + (q(x,y))^n}}{\frac{1}{1 + (q(x,y))^n}} = \frac{\frac{(q(x,y))^n}{1 + (q(x,y))^n}}{\frac{1}{1 + (q(x,y))^n}} = (q(x, y))^n, \text{ for each } x, y \in X.$$

It follows that

$$T_{Dom}^{\frac{1}{n}}\left(E_{q,Dom}^{n}(x,y), E_{q,Dom}^{n}(y,z)\right) = \frac{1}{1+(q(x,y)+q(y,z))^n} \leq \frac{1}{1+q(x,z)^n} = E_{q,Dom}^{n}(x,z),$$

since q is an asymmetric distance on X and, so, it satisfies $q(x,z) \leq q(x,y) + q(y,z)$.

In the light of the preceding result we obtain immediately that $E_{Q_\Phi,Dom}^{n}$ is a $T_{Dom}^{\frac{1}{n}}$-preorder and, thus, that the transition value $p(r_k, ij)$, given by (5), can be understood as a fuzzy preorder. Clearly, the response function provided by (5) is asymmetric. Moreover it must be stressed that if we take $\alpha_u = 0$ in (5), then indistinguishability operator given by (3) is retrieves as a particular case of the fuzzy preorder given by (6). So this new framework allows to model the new situations and those explored in [5,11]. Furthermore, observe that the obtained transition possibilities does not fulfill equality $\sum_{j=1}^{m} p(r_k, ij) = 1$ that is assumed for probability distributions. So the new transitions does not meet the axioms of the probability theory.

Finally it must be pointed out that if the asymmetric response function given by (5) is incorporated in the RTAM algorithm which implement a possibilistic Markov chain in order to describe the evolution of the decission process, then the conditions that ensure the finite convergence of the chain are hold. Concretely, the column diagonally dominant and power dominant conditions (for more information about the conditions we refer the reader to [14]). Therefore, the possibilistic Markov chain whose transition possibilities are given by (5) converges to a stationary distribution in at most $m-1$ steps, where m is the number of tasks.

3 Experimental Results

In this section we will analyze the results of experiments executed to study the number of steps needed to converge to stationary possibilistic distribution with the fuzzy Markov chains induced from the transitions possibilities (5), or equivalently by the fuzzy preorder (6).

3.1 Experimental Framework

The experiments have been carried out with different positions of the objects in the environment (placement of tasks). We assume that the power value n will always be equal to 2. All the experiments have been carried out using Matlab with different synthetic environments following a uniform distribution to generate the position of the tasks. Figure 1 represents one of these environments, where each blue dot represents a task. Furthermore, all the environments have the same dimension (width = 600 units and high = 600 units), the threshold value θ_{rk} will always be equal to $\frac{d_{max}}{4}$, where d_{max} is the maximum distance between two tasks. In our case d_{max} is equal to 800.5 units of distance. Moreover, all the experiments have been performed with 500 different environments, all of them with 100 tasks ($m = 100$).

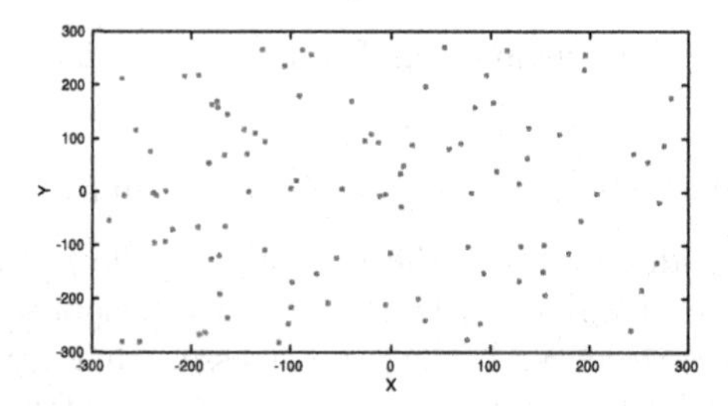

Fig. 1. An example of environment. (Color figure online)

Finally, each task has a randomly utility generated following an uniform distribution between 100 and 200. In order to aggregate the distance and the utility in the possibilistic response function (5) (that is to obtain the asymmetric distance Q_Φ), the parameter α_u has been split into two components as follows:

$$\alpha_u = \alpha_c \cdot \alpha_w.$$

α_c is a weighting factor that makes the utility component of Q_Φ has the same dimension and scale as the Euclidean distance component. Thus, α_c is the same for all the experiments and is equal to

$$\alpha_c = \frac{d_{max}}{U_{max}},$$

where U_{max} is the maximum value of the utility, i.e. 200, and d_{max} is the aforesaid maximum distance (800.5 units of distance). The second parameter, α_w, is a weighting factor that indicates how important is the improvement made in utility with respect to the distance during the allocation process. The impact of this value on the system's performance will be evaluated in the following sections.

3.2 Results: Probabilistic/Possibilistic Markov Chains

This section analyses the number of steps required, during the allocation process, to converge to a stationary distribution when probabilistic Markov chains are under consideration. In order to transform the possibilitic transition matrix (P_{r_k}) (the matrix P in Subsect. 2.1, here r_k points out that the Markov chain models the allocation process for that robot), obtained from the transitions possibilities given by (5), we make use of the transformation proposed in [15], where each element of the matrix (P_{r_k}) is normalized (divided by the sum of all the elements in its row). Obviously, the resulting matrix meets all the conditions of a probability distribution. Notice that (see Subsect. 2.1) that the convergence to stationary distribution in a finite number of steps is not guaranteed for probabilistic Markov chains. Along the experiments we assume that if the convergence is not reached after 500 steps, the system does not converge.

Table 1 shows some results related to the number of steps that a probabilistic Markov chain required to converge to stationary probabilistic distribution for several values of the parameter α_w. The first table's column shows the percentage of experiments that does not converge. The number of steps required to converge, when the system converges, are shown in the second column. As can be seen, the percentage of experiments that do converge stays stable (it is similar in all considered cases) when the value of α_w is low ($\alpha_w \leq 10$) and dramatically decreases when α_w is high. Actually, if α_w is greater than 400 there are not any experiment that converge for the probabilistic approach. Likewise, when the system converges, the number of steps is clearly increases. Thus, we can conclude that, for the experiments carried in this paper, the parameter α_w (the importance of the improvement made in utility with respect to the distance between tasks) has a great impact on the probabilistic Markov chain behaviour.

Table 1. Steps required to converge for probabilistic/possibilistic Markov chains.

α_w	% Convergence prob.	Steps prob.	% Convergence fuzzy	Steps fuzzy
0	32.6%	274.30	100%	23.28
2	41.6%	319.03	100%	22.17
10	30.04%	313.52	100%	25.34
40	13.8%	402.72	100%	27.56
100	0.2%	499	100%	28
400	0%	-	100%	11.17

The third column of Table 1 shows the percentage of simulations that do converge with possibilistic Markov chain. As was pointed out in Subsect. 2.1, the convergence of these chains is always guaranteed and, therefore, in all cases this percentage is 100%. The last column of Table 1 shows the number of steps that possibilistic Markov chains need to converge to stationary distribution. As can be seen, contrary to its probabilistic counterpart, possibilistic Markov chains require a lower number of steps to converge when the parameter α_w is higher. Moreover, in all cases, possibilistic Markov chains need much lower number of steps to converge than the probabilistic ones.

3.3 More on Possibilistic Markov Chains Results

In this section we analyze in a more detailed way the impact of the parameter α_w on the system's behaviour for the possibilistic case.

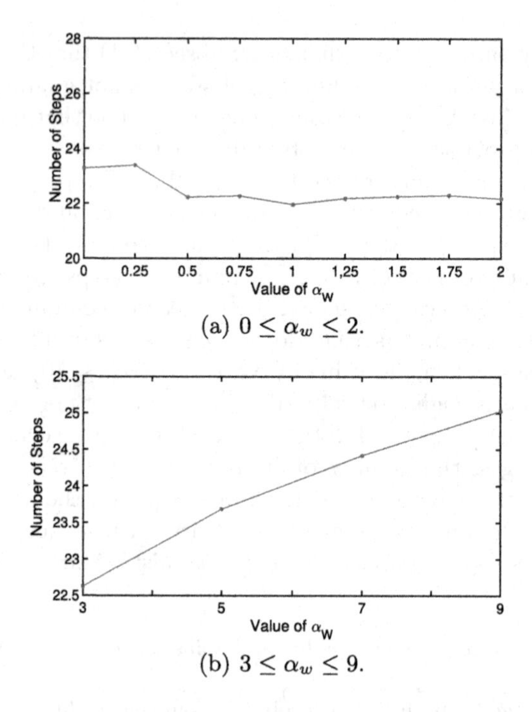

(a) $0 \leq \alpha_w \leq 2$.

(b) $3 \leq \alpha_w \leq 9$.

Fig. 2. Number of steps to converge to stationary possibilistic distribution for $0 \leq \alpha_w \leq 9$.

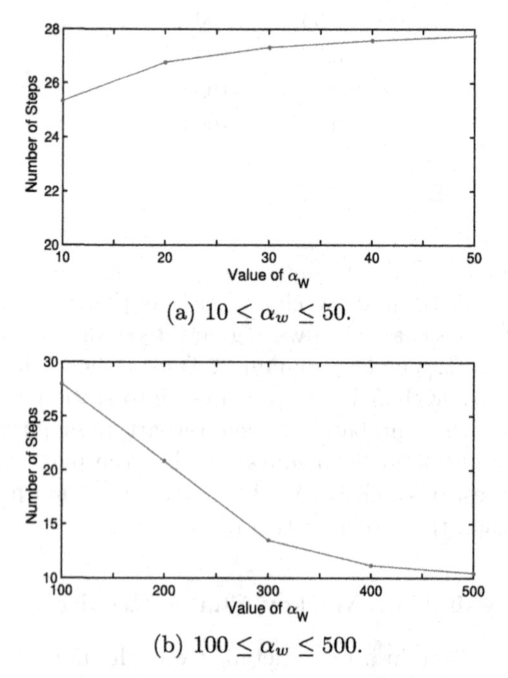

(a) $10 \leq \alpha_w \leq 50$.

(b) $100 \leq \alpha_w \leq 500$.

Fig. 3. Steps to converge to stationary possibilistic distribution for $10 \leq \alpha_w \leq 500$.

Figures 2 and 3 show the mean number of steps required to converge to stationary distribution for possibilistic Markov chains with respect to the parameter α_w. As this number of steps clearly depends on the value of the parameter α_w, the results have been split into 4 figures: Fig. 2(a) shows the results when $0 \leq \alpha_w \leq 2$; in Fig. 2(b) can be seen the number of steps if $3 \leq \alpha_w \leq 9$; Fig. 3(a) shows the values when $10 \leq \alpha_w \leq 50$; and, finally, Fig. 3(b) shows the number of steps required to converge when $100 \leq \alpha_w \leq 500$. In the shake of improving the visualization of the results, the minimum value of the Figs. 2(a), (b) and 3(a) is 20 and the minimum value for the Fig. 3(b) is 10 steps. As can be seen, if the value of the parameter α_w is lower than 50, the number of steps required to converge increases slightly with respect to the increments of the value α_w. When α_w is greater than 100, the number of steps start to dramatically decrease. It must be stressed that, as was said in Subsect. 3.2, the probabilistic Markov chain show the opposite behaviour.

4 Conclusions and Future Work

For the first time this paper apply the concept of fuzzy preorders to implement a multi-robot task allocation method. The task allocation algorithm implemented in this work is based on the Response Threshold methods (RTM), where a robot selects the next task to execute according to a probabilistic Markov chain. Authors' previous works show that the possibilistic Markov chains outperform their probabilistic counterparts when the possibility of transition only depends on the inverse of the Euclidean distance between tasks. This paper also considers the utility of a task as a criteria to make the task allocation. Therefore, the utility value of the tasks has been included into the possibility transition function. The new resulting function is asymmetric and has been shown to be a fuzzy preorder with respect to a t-norm belonging to the Dombi family. The results of the simulations show that, in all cases, the possibilistic Markov chains converge in a lower number of steps than its probabilistic counterpart. Moreover, the weight of the utility (parameter α_w) has critical impact on the system's behavior. In the light of these first results new challenges and questions arises. For example, a deeper study of the impact of the utility on the system, different ways of aggregating the utility and distance component in the construction of the response function and the implantation of these methods on a physical multi-robot simulator are under consideration for a future work.

Acknowledgments. This research was funded by the Spanish Ministry of Economy and Competitiveness under Grants DPI2014-57746-C03-2-R, TIN2014-53772-R, TIN2016-81731-REDT (LODISCO II) and AEI/FEDER, UE funds, by the Programa Operatiu FEDER 2014-2020 de les Illes Balears, by project ref. PROCOE/4/2017 (Direccio General d'Innovacio i Recerca, Govern de les Illes Balears), and by project ROBINS. The latter has received research funding from the EU H2020 framework under GA 779776. This publication reflects only the authors views and the European Union is not liable for any use that may be made of the information contained therein.

References

1. Gerkey, B.P.: On multi-robot task allocation. Ph.D. thesis, Center of Robotics and Embedded Systems, University of Southern California, Los Angeles, USA (2003)
2. Guerrero, J., Oliver, G., Valero, O.: Multi-robot coalitions formation with deadlines: complexity analysis and solutions. PLoS ONE **12**(1), 1–26 (2017). https://doi.org/10.1371/journal.pone.0170659
3. Agassounon, W., Martinoli, A.: Efficiency and robustness of threshold-based distributed allocation algorithms in multi-agent systems. In: AAMAS 2012, Bolonia, Italy, pp. 1090–1097, July 2002. https://doi.org/10.1145/545056.545077
4. Castello, E., Yamamoto, T., Libera, F.D., Liu, W., Winfield, A.F.T., Nakamura, Y., Ishiguro, H.: Adaptive foraging for simulated and real robotic swarms: the dynamical response threshold approach. Swarm Intell. **10**(1), 1–31 (2016). https://doi.org/10.1007/s11721-015-0117-7
5. Guerrero, J., Miñana, J.J., Valero, O., Oliver, G.: Indistinguishability operators applied to task allocation problems in multi-agent systems. Appl. Sci. **7**(10), 963 (2017). https://doi.org/10.3390/app7100963
6. Kalra, N., Martinoli, A.: A comparative study of market-based and threshold-based task allocation. In: Gini, M., Voyles, R. (eds.) DARS 2006, vol. 2, pp. 91–102. Springer, Tokyo (2006). https://doi.org/10.1007/4-431-35881-1_10
7. Recasens, J.: Indistinguishability Operators: Modelling Fuzzy Equalities and Fuzzy Equivalence Relations. Springer, Heidelberg (2010). https://doi.org/10.1007/978-3-642-16222-0
8. Klement, R., Mesiar, R., Pap, E.: Triangular Norms. Kluwer Academic Publishers, Dordrecht (2000). https://doi.org/10.1142/S0218488503002053
9. Zadeh, L.: Similarity relations and fuzzy orderings. Inf. Sci. **1**, 177–200 (1971). https://doi.org/10.1016/S0020-0255(71)80005-1
10. Mayor, G., Valero, O.: Aggregation of asymmetric distances in computer science. Inf. Sci. **180**, 803–812 (2010). https://doi.org/10.1016/j.ins.2009.06.020
11. Guerrero, J., Valero, O., Oliver, G.: Toward a possibilistic swarm multi-robot task allocation: theoretical and experimental results. Neural Process. Lett. **46**(3), 881–897 (2017). https://doi.org/10.1007/s11063-017-9647-x
12. Avrachenkov, K., Sanchez, E.: Fuzzy Markov chains and decision making. Fuzzy Optim. Decis. Mak. **1**, 143–159 (2002). https://doi.org/10.1023/A:1015729400380
13. Zadeh, L.: Fuzzy sets as a basis for a theory of possibility. Fuzzy Sets Syst. **3**, 177–200 (1971). https://doi.org/10.1016/S0165-0114(99)80004-9
14. Duan, J.: The transitive clousure, convegence of powers and adjoint of generalized fuzzy matrices. Fuzzy Sets Syst. **145**, 301–311 (2004). https://doi.org/10.1016/S0165-0114(03)00165-9
15. Vajargah, B.F., Gharehdaghi, M.: Ergodicity of fuzzy Markov chains based on simulation using sequences. Int. J. Math. Comput. Sci. **11**(2), 159–165 (2014)

On the Problem of Aggregation of Partial T-Indistinguishability Operators

Tomasa Calvo Sánchez[1], Pilar Fuster-Parra[2(✉)], and Óscar Valero[2]

[1] University of Alcalá de Henares, Alcalá de Henares, Madrid, Spain
tomasa.calvo@uah.es
[2] Universitat Illes Balears, Palma de Mallorca, Spain
{pilar.fuster,o.valero}@uib.es

Abstract. In this paper we focus our attention on exploring the aggregation of partial T-indistinguishability operators (relations). Concretely we characterize, by means of (T-$T_{\min}$)-tuples, those functions that allow to merge a collection of partial T-indistinguishability operators into a single one. Moreover, we show that monotony is a necessary condition to ensure that a function aggregates partial T-indistinguishability operators into a new one. We also provide that an inter-exchange composition function condition is a sufficient condition to guarantee that a function aggregates partial T-indistinguishability operators. Finally, examples of this type of functions are also given.

Keywords: Aggregation operator
Partial indistinguishability operator · t-norm
Inter-exchange composition function condition · (T-$T_{\min}$)-tuple · Mean

1 Introduction

Aggregation functions constitute an important tool in the field of information fusion (see [5,6]). The information can be given by means of fuzzy relations that depend on different applications (see [1,11,17,23,24,31]). According to the definition provided by Trillas in [29], given a t-norm $T : [0,1]^2 \to [0,1]$, a T-indistinguishability operator on a (non-empty) set X is a fuzzy relation $E : X \times X \to [0,1]$ verifying for all $x, y, z \in X$ the following properties: (i) $E(x,x) = 1$, (ii) $E(x,y) = E(y,x)$, (iii) $T(E(x,y), E(y,z)) \leq E(x,z)$. A T-indistinguishability operator is a mathematical tool for classifying objects when a measure presents some kind of uncertainty. They are also known as measures of similarity, in fact, the greater $E(x,y)$ the most similar are x and y. Throughout this paper we will assume that the reader is familiar with the basics of triangular norms (see [20] for a deeper treatment of the topic).

In the literature, one can find another measures used to classify objects, namely the so-called pseudo-metrics. Let us recall, according to [10], that a

In Memoriam of Lofti Zadeh.

J. Medina et al. (Eds.): IPMU 2018, CCIS 853, pp. 207–218, 2018.
https://doi.org/10.1007/978-3-319-91473-2_18

pseudo-metric on a (non-empty) set X is a function $d : X \times X \rightarrow [0, \infty]$ which satisfies the following axioms for all $x, y, z \in X$: (i) $d(x, x) = 0$, (ii) $d(x, y) = d(y, x)$, (iii) $d(x, z) \leq d(x, y) + d(y, z)$. In contrast to indistinguishability operators, pseudo-metrics are measures of dissimilarity, the smaller $d(x, y)$ the most similar are x and y.

T-indistinguishability operators and pseudo-metrics are closely related. In fact, a technique to generate pseudo-metrics from indistinguishability operators, and vice-versa, has been explored in [2,3,15,18,20,22,27,30].

The functions that allow us to merge a collection of indistinguishability operators into a single one have been addressed in [19,25–27]. In particular, in [25] a characterization of those functions that aggregate indistinguishability operators have been given in terms of triangle triplets. Recently, the aforementioned description of the functions that aggregates indistinguishability operators have been extended by Mayor and Recasens in [21], now, in terms of T-triangular triplets. Let us recall that, given a t-norm T and $n \in \mathbb{N}, n > 1$ ($\mathbb{N}$ stands for the positive integer numbers), a triplet $(\mathbf{a}, \mathbf{b}, \mathbf{c})$ such that $\mathbf{a}, \mathbf{b}, \mathbf{c} \in [0, 1]^n$ is said to be a n-dimensional T-triangular triplet whenever $T(a_i, b_i) \leq c_i$ $T(a_i, c_i) \leq b_i$ and $T(b_i, c_i) \leq a_i$ for all $i = 1, \ldots, n$. A characterization in terms of T-triangular triplets of those functions that aggregate the so-called relaxed indistinguishability operators in the sense of [13] (i.e., fuzzy relations verifying only properties (ii) and (iii) of a T-indistinguishability operator) has also been given in [8].

In 1994, a generalization of the metric notion was introduced by Matthews in order to develop a suitable mathematical tool for quantitative models in denotational semantics (see [7]). This new metric notion was called partial pseudo-metric and it is defined as follows: a partial pseudo-metric on a non-empty set X is a function $p : X \times X \rightarrow [0, \infty[$ which satisfies for all x, y, z the following: (i) $p(x, x) \leq p(x, y)$, (ii) $p(x, y) = p(y, x)$ (iii) $p(x, y) + p(z, z) \leq p(x, z) + p(z, y)$.

Similarly to the pseudo-metric case, in the partial framework a closely related notion of indistinguishability operator, the so-called multivalued equivalence, is considered in the works by Demirci [9] and by Bukatin et al. [7]. Inspired by the correspondence between the axiomatic of partial pseudo-metrics and indistinguishability operators stated in the aforesaid references, in [14] the notion of partial T-indistinguishability operator was introduced.

Let us recall that a partial T-indistinguishability operator E on a nonempty set X is a fuzzy relation $E : X \times X \longrightarrow [0, 1]$ satisfying the following properties for any $x, y, z \in X$: (i) $E(x, y) \leq E(x, x)$, (ii) $E(x, y) = E(y, x)$, and (iii) $T(E(x, z), E(z, y)) \leq T(E(x, y), E(z, z))$. A partial T-indistinguishability operator E on a nonempty set X will be called a partial T-equality provided that it satisfies the following additional property for any $x, y, z \in X$: (i') $E(x, y) = E(x, x) = E(y, y)$ if and only if $x = y$, i.e., it is a partial T-indistinguishability operator on a set X that separates points.

The following is an instance of partial indistinguishability operator that is not an indistinguishability one.

Example 1. Let Σ be a non-empty alphabet. Denote by Σ^{∞} the set of all finite and infinite sequences over Σ. Given $v \in \Sigma^{\infty}$ denote by $l(v)$ the length of v. Thus $l(v) \in \mathbb{N} \cup \{\infty\}$ for all $v \in \Sigma^{\infty}$. Moreover, if $\Sigma_F = \{v \in \Sigma^{\infty} : l(v) \in \mathbb{N}\}$ and $\Sigma_{\infty} = \{v \in \Sigma^{\infty} : l(v) = \infty\}$, then $\Sigma^{\infty} = \Sigma_F \cup \Sigma_{\infty}$ and we will write $v = v_1 v_2 \ldots v_{l(v)}$ and $w = w_1 w_2 \ldots$ whenever $v \in \Sigma_F$ and $w \in \Sigma_{\infty}$, respectively.

Next, given $x, y \in \Sigma_{\infty}$, denote by $l(x, y)$ the longest common prefix of x and y (of course if x and y have not a common prefix then $l(x, y) = 0$).

Define the fuzzy relation E_{Σ} on $\Sigma^{\infty} \times \Sigma^{\infty}$ by

$$E_{\Sigma}(u, v) = 1 - 2^{-l(v,w)}$$

for all $u, v \in \Sigma^{\infty}$. Then it is a simple matter to check that E_{Σ} is a partial $T_{\min}$-indistinguishability operator on Σ^{∞} that separates points, i.e., a partial $T_{\min}$-equality on Σ^{∞}. Note that $T_{\min}$ denotes the minimum t-norm.

Note that E_{Σ} is not a indistinguishability operator, since $E_{\Sigma}(u, u) = 1 \Leftrightarrow u \in \Sigma_{\infty}$.

Due to the fact that the problem of aggregating fuzzy relations has received considerable attention from the community researching in fuzzy mathematics (see, for instance [12,19,20,25–27]), in this work, we focus on the preservation of the properties of partial T-indistinguishability operators by means of aggregation. Thus, we characterize those functions that define a new partial T-indistinguishability operator as output whenever they receive a collection of partial T-indistinguishability operators as input. Concretely, a characterization of such functions is provided through the notion of (T-$T_{\min}$)-tuples (see Definitions 2 and 3).

The remainder of this paper is organized as follows. In Sect. 2, we provide a characterization of aggregation functions based on the transformation of n-dimensional (T-$T_{\min}$)-tuple into 1-dimensional (T-$T_{\min}$)-tuple. Also, we show that monotony is a necessary condition to ensure that a function aggregates partial T-indistinguishability operators into a new one. Besides, we provide that an inter-exchange composition function condition is a sufficient condition to guarantee that a function aggregates partial T-indistinguishability operators. Some examples of this type of functions are also provided. Finally, in Sect. 3, some conclusions are given and future work is proposed.

2 Aggregation of Partial *T*-indistinguishability Operators

In this section we address the problem of how to combine by means of a function a collection of partial T-indistinguishability operators into a single one as a result. To this aim, let us introduce the notion of partial T-indistinguishability operator aggregation function.

Definition 1. *Let $n \in \mathbb{N}$. A function $F : [0,1]^n \to [0,1]$ aggregates partial T-indistinguishability operators if $F(E_1, \ldots, E_n)$ is partial T-indistinguishability operators on X for any set X and any collection $(E_1, \ldots, E_n)$ of partial T-indistinguishability operators on X, where $F(E_1, \ldots, E_n)$ is the fuzzy binary relation given by $F(E_1, \ldots, E_n)(x, y) = F(E_1(x, y), \ldots, E_n(x, y))$.*

The following example gives an instance of partial T-indistinguishability operator aggregation function when the t-norm under consideration is the product t-norm T_P.

Example 2. Let $n \in \mathbb{N}$. Consider a collection $(E_1, \ldots, E_n)$ of partial T_P-indistinguishability operators. A straightforward computation yields that the Product function $F : [0,1]^n \to [0,1]$, given by

$$F(E_1, \ldots, E_n)(x,y) = F(E_1(x,y), \ldots, E_n(x,y)) = \prod_{i=1}^{n} E_i(x,y),$$

provides that $F(E_1, \ldots, E_n)$ is also a partial T_P-indistinguishability operator.

Inspired by the preceding example we introduce in the next result an easy technique to generate functions that aggregate partial T-indistinguishability operators.

Proposition 1. *Let $n \in \mathbb{N}$. If $(E_1, \ldots, E_n)$ is a collection of partial T-indistinguishability operators on a set X, then the relation E_T defined for all $x, y \in X$ by*

$$E_T(x,y) = T(E_1(x,y), \ldots, E_n(x,y))$$

is a partial T-indistinguishability operator.

Proof. The fact that T is non-decreasing and that each E_i is a partial T-indistinguishability operator on X we have that $E_T(x,y) \leq E_T(x,x)$ for all $x, y \in X$. The commutativity of T and the symmetry of each E_i gives that $E_T(x,y) = E_T(y,x)$ for all $x, y \in X$. Next we show that $T(E_T(x,z), E_T(z,y)) \leq T(E_T(x,y), E_T(z,z))$ for all $x, y, z \in X$. To this aim we know that

$$T(E_i(x,z), E_i(z,y)) \leq T(E_i(x,y), E_i(z,z))$$

for all $x, y, z \in X$ and for all $i = 1, \ldots, n$. Since T is a non-decreasing function we have that

$$\begin{aligned} &T\left(T(E_1(x,z), E_1(z,y)), T(E_2(x,z), E_2(z,y)), \ldots, T(E_n(x,z), E_n(z,y))\right) \leq \\ &T\left(T(E_1(x,y), E_1(z,z)), T(E_2(x,y), E_2(z,z)), \ldots, T(E_n(x,y), E_n(z,z))\right) \end{aligned}$$

Now due to the associativity and commutativity of T we conclude that

$$\begin{gathered} T(E_T(x,z), E_T(z,y)) = T\left(T(E_1(x,z), \ldots, E_n(x,z)), T(E_1(z,y), \ldots, E_n(z,y))\right) \\ \leq \\ T\left(T(E_1(x,y), \ldots, E_n(x,y)), T(E_1(z,z), \ldots, E_n(z,z))\right) = T(E_T(x,y), E_T(z,z)). \end{gathered}$$

Note that Proposition 1 ensures that any t-norm T aggregates partial T-indistinguishability operators.

The next result yields a necessary condition to ensure that a function aggregates partial T-indistinguishability operators.

Proposition 2. *Let $n \in \mathbb{N}$. If a function $F : [0,1]^n \to [0,1]$ aggregates partial T-indistinguishability operators, then it is a non-decreasing function.*

Proof. Let $x, y \in [0,1]^n$ such that $x_i \leq y_i$ for all $i = 1, 2, \ldots, n$. Consider $X = \{a, b\}$ with $a \neq b$. Define the fuzzy relations E_i on $X \times X$ by $E_i(b,b) = y_i$ and $E_i(a,a) = E_i(a,b) = E_i(b,a) = x_i$ for all $i = 1, 2, \ldots, n$. Clearly, each E_i is a partial T-indistinguishability operator. Since $F : [0,1]^n \to [0,1]$ aggregates partial T-indistinguishability operators we have that

$$F(x_1, \ldots, x_n) = F(E_1(a,b), \ldots, E_n(a,b)) = F(E_1, \ldots, E_n)(a,b)$$
$$\leq$$
$$F(E_1, \ldots, E_n)(b,b) = F(E_1(b,b), \ldots, E_n(b,b)) = F(y_1, \ldots, y_n).$$

The reciprocal of Proposition 2 is not necessarily true as it is shown in the following example.

Example 3. Consider a set $X = \{x, y, z\}$ of different elements. Define the fuzzy sets $E_i : X \times X \to [0,1]$, for $i = 1, 2, 3$, by $E_1(x,y) = E_1(x,z) = 0.2$, $E_1(z,y) = E_1(z,z) = E_1(y,y) = E_1(x,x) = 0.4$, $E_2(x,y) = E_2(z,y) = 0.2$, $E_2(x,z) = E_2(x,x) = E_2(y,y) = E_2(z,z) = 0.4$, and $E_3(x,y) = E_3(x,z) = 0.2$, $E_3(z,y) = E_3(x,x) = E_3(y,y) = E_3(z,z) = 0.4$. Besides, $E_i(x,y) = E_i(y,x)$, $E_i(x,z) = E_i(z,x)$, and $E_i(z,y) = E_i(y,z)$ for all $i = 1, 2, 3$. It is not hard to check that (E_1, E_2, E_3) is a collection of partial $T_{\min}$-indistinguishability operators.

Let us take the non-decreasing function $F : [0,1]^3 \to [0,1]$, $F(a,b,c) = ab+c$, then we will see that $F(E_1, E_2, E_3)$ does not preserve $T_{\min}$-transitivity. Indeed, note that

$$T_{\min}(F(E_1(y,z), E_2(y,z), E_3(y,z)), F(E_1(z,x), E_2(z,x), E_3(z,x))) = T_{\min}(0.48, 0.28)$$
$$>$$
$$T_{\min}(0.24, 0.56) = T_{\min}(F(E_1(y,x), E_2(y,x), E_3(y,x)), F(E_1(z,z), E_2(z,z), E_3(z,z)))$$

and, therefore, F does not aggregate partial $T_{\min}$-indistinguishability operators.

In order to give a characterization, in terms of tuples in the sense of Mayor and Recasens [21], of those functions that allow to merge a collection of partial T-indistinguishability operators into a single one, the following two concepts will play a crucial role.

Definition 2. *Let T be a t-norm. We say that $(a, b, c, d, d', d'') \in [0,1]^6$ is a (T-$T_{\min}$)-tuple if and only if $T(a,b) \leq T(c,d)$, $T(a,c) \leq T(b,d')$, $T(c,b) \leq T(a,d'')$, $a \leq \min\{d, d'\}$, $b \leq \min\{d, d''\}$, and $c \leq \min\{d'', d'\}$.*

Definition 3. *If $n \in \mathbb{N}$. Let T be a t-norm and $\mathbf{a}, \mathbf{b}, \mathbf{c}, \mathbf{d}, \mathbf{d}', \mathbf{d}'' \in [0,1]^n$, $n > 1$, we say that $(\mathbf{a}, \mathbf{b}, \mathbf{c}, \mathbf{d}, \mathbf{d}', \mathbf{d}'')$ is a (n-dimensional) (T-$T_{\min}$)-tuple provided that $(a_i, b_i, c_i, d_i, d_i', d_i'')$ is a (T-$T_{\min}$)-tuple for all $i = 1, \ldots, n$, where $\mathbf{a} = (a_1, \ldots, a_n)$, $\mathbf{b} = (b_1, \ldots, b_n)$, $\mathbf{c} = (c_1, \ldots, c_n)$, $\mathbf{d} = (d_1, \ldots, d_n)$, $\mathbf{d}' = (d_1', \ldots, d_n')$, and $\mathbf{d}'' = (d_1'', \ldots, d_n'')$.*

In the light of the preceding notions we are able to describe those functions that aggregate partial T-indistinguishability operators.

Theorem 1. *Let $n \in \mathbb{N}$ and let $F : [0,1]^n \to [0,1]$. The following assertions are equivalent:*

(1) F aggregates partial T-indistinguishability operators,
(2) F transforms a n-dimensional (T-$T_{\min}$)-tuple into 1-dimensional (T-$T_{\min}$)-tuple.

Proof (1) $\Rightarrow$ (2). Let us suppose that F aggregates partial T-indistinguishability operators. We have to prove that F transforms a n-dimensional (T-$T_{\min}$)-tuple into 1-dimensional (T-$T_{\min}$)-tuple.

To this aim, let us prove that $(F(\mathbf{a}), F(\mathbf{b}), F(\mathbf{c}), F(\mathbf{d}), F(\mathbf{d}^{'}), F(\mathbf{d}^{''}))$ is a (T-$T_{\min}$)-tuple whenever $(\mathbf{a}, \mathbf{b}, \mathbf{c}, \mathbf{d}, \mathbf{d}^{'}, \mathbf{d}^{''})$ also is. Take $X = \{x, y, z\}$ where x, y, z are different elements. Define a collection of fuzzy sets $(E_1, E_2, \ldots, E_n)$ on $X \times X$ as follows: $E_i(x,y) = E_i(y,x) = a_i$, $E_i(x,z) = E_i(z,x) = b_i$, $E_i(y,z)) = E_i(z,y) = c_i$, $E_i(x,x) = d_i$, $E_i(y,y) = d_i^{'}$ and $E_i(z,z) = d_i^{''}$ for all $i = 1, \ldots, n$. The fact that $(\mathbf{a}, \mathbf{b}, \mathbf{c}, \mathbf{d}, \mathbf{d}^{'}, \mathbf{d}^{''})$ is a (T-$T_{\min}$)-tuple yields that the collection $(E_1, E_2, \ldots, E_n)$ is of partial T-indistinguishability operators on X. Hence, taking into account that F aggregates partial T-indistinguishability operators, we obtain that

$$\begin{aligned} T(F(\mathbf{a}), F(\mathbf{b})) &= T\left(F(E_1, \ldots, E_n)(y,x), F(E_1, \ldots, E_n)(x,z)\right) \\ &\leq T\left(F(E_1, \ldots, E_n)(y,z), F(E_1, \ldots, E_n)(x,x)\right) \\ &= T(F(\mathbf{c}), F(\mathbf{d})) \end{aligned}$$

and

$$\begin{aligned} F(\mathbf{a}) = F(E_1, \ldots, E_n)(x,y) &\leq \min\{F(E_1, \ldots, E_n)(x,x), F(E_1, \ldots, E_n)(y,y)\} \\ &= \min\{F(\mathbf{d}), F(\mathbf{d}^{'})\} \end{aligned}$$

Similarly, we can see the rest of cases to ensure that

$$(F(\mathbf{a}), F(\mathbf{b}), F(\mathbf{c}), F(\mathbf{d}), F(\mathbf{d}^{'}), F(\mathbf{d}^{''}))$$

is a (T-$T_{\min}$)-tuple.

(2) $\Rightarrow$ (1). Assuming F transforms n-dimensional (T-$T_{\min}$)-tuple into 1-dimensional (T-$T_{\min}$)-tuple, we must prove that $F(E_1, \ldots, E_n)$ is a partial T-indistinguishability operator for every collection $(E_1, \ldots, E_n)$ of partial T-indistinguishability operators.

Let us consider a set $X = \{x, y, z\}$ where x, y, z are different elements and a collection $(E_1, \ldots, E_n)$ of partial T-indistinguishability operators on X. Then the tuple $(\mathbf{a}, \mathbf{b}, \mathbf{c}, \mathbf{d}, \mathbf{d}^{'}, \mathbf{d}^{''})$ is a (T-$T_{\min}$)-tuple, where $a_i = E_i(x,y), b_i = E_i(x,z), c_i = E_i(z,y), d_i = E_i(x,x), d_i^{'} = E_i(y,y)$ and $d_i^{''} = E_i(z,z)$ for all $i = 1, \ldots, n$. So we have that $(F(\mathbf{a}), F(\mathbf{b}), F(\mathbf{c}), F(\mathbf{d}), F(\mathbf{d}^{'}), F(\mathbf{d}^{''}))$ is a (T-$T_{\min}$)-tuple and, thus,

$$\begin{aligned} &T\left(F(E_1(y,x), \ldots, E_n(y,x)), F(E_1(x,z), \ldots, E_n(x,z))\right) = T(F(\mathbf{a}), F(\mathbf{b}) \\ &\leq \\ &T(F(\mathbf{c}), F(\mathbf{d}) = T\left(F(E_1(y,z), \ldots, E_n(y,z)), F(E_1(x,x), \ldots, E_n(x,x))\right). \end{aligned}$$

Hence, $F(E_1,\ldots,E_n)$ satisfies condition (iii) of a partial T-indistinguishability operator. Symmetry follows immediately by symmetry of E_i for all $i = 1,\ldots,n$,

$$\begin{aligned} F(E_1,\ldots,E_n)(x,y) &= F(E_1(x,y),\ldots,E_n(x,y)) \\ &= F(E_1(y,x),\ldots,E_n(y,x)) \\ &= F(E_1,\ldots,E_n)(y,x). \end{aligned}$$

To check that $F(E_1,\ldots,E_n)$ satisfies condition (i) of a partial T-indistinguishability operator, we only need to prove that

$$F(E_1,\ldots,E_n)(x,y) \leq F(E_1,\ldots,E_n)(x,x)$$

which follows from the fact that $(F(\mathbf{a}),F(\mathbf{b}),F(\mathbf{c}),F(\mathbf{d}),F(\mathbf{d}'),F(\mathbf{d}''))$ is a (T-$T_{\min}$)-tuple as $F(E_1,\ldots,E_n)(x,y) = F(\mathbf{a}) \leq \min\{F(\mathbf{d}),F(\mathbf{d}')\} \leq F(\mathbf{d}) = F(E_1,\ldots,E_n)(x,x)$. Therefore, $F(E_1,\ldots,E_n)$ is a partial T-indistinguishability operator on X.

Observe that we retrieve, as a particular case of Theorem 1, the main result of [21] (Theorem 2 below) whenever partial T-indistinguishability operators are exactly T-indistinguishability operators.

Theorem 2. *Let $n \in \mathbb{N}$ and let $F : [0,1]^n \to [0,1]$. The following assertions are equivalent:*

(1) F aggregates T-indistinguishability operators,
(2) F transforms a n-dimensional T-tuple into 1-dimensional T-tuple.

Next we present sufficient conditions to guarantee that a function aggregates partial T-indistinguishability operators. With this aim we need to introduce the following notion.

Definition 4. *Let $n \in \mathbb{N}$. We will say that a t-norm T and a function $F : [0,1]^n \to [0,1]$ satisfy the inter-exchange composition function equality (ICFE, for short) provided that $T(F(\mathbf{a}),F(\mathbf{b})) = F(T(a_1,b_1),\ldots,T(a_n,b_n))$ for all $\mathbf{a},\mathbf{b} \in [0,1]^n$.*

Notice that a t-norm T and a function $F : [0,1]^n \to [0,1]$ satisfying the inter-change composition equality are said to be commutative in [28].

In the light of the preceding notion, the next result gives the announced sufficient conditions.

Proposition 3. *Let $n \in \mathbb{N}$ and let $(E_1,\ldots,E_n)$ be a collection of partial T-indistinguishability operators on a non-empty set X. If $F : [0,1]^n \to [0,1]$ is a non-decreasing function and, in addition, T and F satisfy the ICFE, then F aggregates partial T-indistinguishability operators.*

Proof. Set $a_i = E_i(y,x)$, $b_i = E_i(x,z)$, $c_i = E_i(y,z)$, $d_i = E_i(x,x)$, $d_i^{'} = E_i(y,y)$, $d^{''} = E_i(z,z)$ for all $i = 1,\ldots,n$. As each E_i is a partial T-indistinguishability operator we have that $T(a_i,b_i) \leq T(c_i,d_i)$, $T(a_i,c_i) \leq T(b_i,d_i^{'})$, and $T(b_i,d_i) \leq T(a_i,d_i^{''})$, $a_i \leq \min\{d_i,d_i^{'}\}$, $b_i \leq \min\{d_i,d_i^{''}\}$ and $c_i \leq \min\{d_i^{'},d_i^{''}\}$ for all $i = 1,\ldots,n$. It follows that $(\mathbf{a},\mathbf{b},\mathbf{c},\mathbf{d},\mathbf{d}',\mathbf{d}'')$ is a n-dimensional $(T\text{-}T_{\min})$-tuple.

Next we show that $(F(\mathbf{a}),F(\mathbf{b}),F(\mathbf{c}),F(\mathbf{d}),F(\mathbf{d}^{'}),F(\mathbf{d}^{''}))$ is a $(T\text{-}T_{\min})$-tuple. To this end we only show that

$$T(F(\mathbf{a}),F(\mathbf{b})) \leq T(F(\mathbf{c}),F(\mathbf{d}))$$

and that $F(\mathbf{a}) \leq \min\{F(\mathbf{d}),F(\mathbf{d}^{'})\}$, since the remainder inequalities will follow applying similar reasonings.

Since F is non-decreasing we have that

$$\begin{aligned} F(E_1,\ldots,E_n)(x,y) = F(\mathbf{a}) &\leq F(\min\{F(\mathbf{d}),F(\mathbf{d}^{'})\}) \\ &\leq \min\{F(E_1,\ldots,E_n)(x,x),F(E_1,\ldots,E_n)(y,y)\} \end{aligned}$$

and

$$F(T(a_1,b_1),\ldots,T(a_n,b_n)) \leq F(T(c_1,d_1),\ldots,T(c_n,d_n)).$$

Now taking into account that, in addition, F and T satisfy the IFCE we have that

$$\begin{aligned} T(F(\mathbf{a}),F(\mathbf{b})) &= F(T(a_1,b_1),\ldots,T(a_n,b_n)) \\ &\leq F(T(c_1,d_1),\ldots,T(c_n,d_n)) = T(F(\mathbf{c}),F(\mathbf{d})) \end{aligned}$$

Therefore F transforms a n-dimensional $(T\text{-}T_{\min})$-tuple into a 1-dimensional $(T\text{-}T_{\min})$-tuple. By Theorem 1 we can ensure that F aggregates partial T-indistinguishability operators.

Observe that every t-norm T and every i-th projection are examples of functions that aggregate partial T-indistinguishability operators, since they are non-decreasing and satisfy the ICFE condition.

The next example shows that there are functions that aggregate partial T-indistinguishability operators but they do not satisfy the ICFE.

Example 4. Fix $k \in]0,1[$. Consider the function $F_k : [0,1]^n \to [0,1]$ defined by

$$F_k(\mathbf{a}) = k$$

for all $\mathbf{a} \in [0,1]^n$. It is not hard to verify that F_k aggregates partial T_P-indistinguishability operators. However, F_k and T_P does not satisfy the ICFE. Indeed,

$$k^2 = T_P(k,k) = T_P(F_k(\mathbf{a}),F_k(\mathbf{b})) < F_k(T_P(a_1,b_1),\ldots,T_P(a_n,b_n)) = k$$

for all $\mathbf{a},\mathbf{b} \in [0,1]^n$.

We end the paper providing instances of functions that aggregate partial indistinguishability operators which have been obtained by means of Propositions 3 and 4, whose easy proof we omit, below.

Proposition 4. *Let $n \in \mathbb{N}$. Assume that $F : [0,1]^n \to [0,1]$ is a function and T is a continuous strict Archimedean t-norm with additive generator t and let $\mathbf{t} : [0,1]^n \to [0,1]^n$ be the function given by $\mathbf{t}(\mathbf{a}) = (t(a_1), \ldots, t(a_n))$. If there exists a non-decreasing additive function $s : [0,\infty[^n \to [0,\infty[$ such that $F = t^{-1} \circ s \circ \mathbf{t}$ and Range$(s \circ \mathbf{t}) \subset$ Range(t), then F and T satisfy the ICFE.*

In the next example, all functions aggregate partial T-indistinguishability operators. For a detailed study about the below listed functions we refer the reader to [4,5,16].

Example 5. With the aim of being able to apply Proposition 4 we consider in the following a weighting list $(w_1, \ldots, w_n)$ of non-negative real numbers satisfying $\sum_{i=1}^n w_i = 1$ and the function $s(x_1, \ldots, x_n) = \sum_{i=1}^{n} w_i x_i$.

1. Weighted quasi-arithmetic means F such that

$$F(a_1, \ldots, a_n) = t^{-1}(\sum_{i=1}^{n} w_i t(a_i))$$

   where t is an additive generator of a given continuous Archimedean t-norm T.
   (a) Weighted geometric means F such that

$$F(x_1, \ldots, x_n) = \prod_{i=1}^{n} x_i^{w_i}.$$

   Note that, in order to apply Proposition 4, we must consider the additive generator t of product t-norm T_P, i.e., $t(x) = -\ln(x)$.
   (b) Weighted harmonic means F such that

$$F(x_1, \ldots, x_n) = \left(\sum_{i=1}^{n} \frac{w_i}{x_i} \right)^{-1}.$$

   Observe that, in order to apply Proposition 4, we must consider the additive generator t of the Hamacher t-norm T_0^H, i.e., $t(x) = \frac{1-x}{x}$.
   (c) Weighted power means F such that

$$F(x_1, \ldots, x_n) = \left(\sum_{i=1}^{n} w_i \cdot x_i^{\lambda} \right)^{1/\lambda}.$$

   Note that, in order to apply Proposition 4, we must consider the additive generator t of the Schweizer-Sklar t-norms T_λ^{SS} with $\lambda \in]-\infty, 0[$, i.e., $t(x) = \frac{1-x^\lambda}{\lambda}$.
   (d) $F(x_1, \ldots, x_n) = e^{-\left(\ln \prod_{i=1}^{n} (x_i^{w_i})^{1/\lambda}\right)^{1/\lambda}}$. Observe that, in order to apply Proposition 4, we must consider the additive generator t of the Aczél-Alsina t-norms T_λ^{AA} with $\lambda \in]0, +\infty[$, i.e., $t(x) = (-\ln x)^\lambda$.

2. Ordered weighted quasi-arithmetic means F such that

$$F(a_1, \ldots, a_n) = t^{-1}(\sum_{i=1}^{n} w_i t(a_{(n-i)})),$$

where $a_{(k)}$ denotes the k-largest input in the list $(a_1, \ldots, a_n)$ and t is an additive generator of a given continuous Archimedean t-norm T.

3 Conclusions

We have carried out the problem of the aggregation of partial T-is indistinguishability operators. Namely we have characterized, by means of the new concept of $(T\text{-}T_{\min})$-tuple, those functions that allow to merge a collection of partial T-indistinguishability operators into a new one. Also we have shown that monotony is a necessary condition to ensure that a function aggregates partial T-indistinguishability operators into a new one. In addition we have seen that an interexchange composition function condition is a sufficient condition to guarantee that a function aggregates partial T-indistinguishability operators. Besides a few examples of the functions under consideration were also provided, among which we can highlight several sort of means. As a further work, we will focus our efforts on the aggregation problem when a collection of partial indistinguishability operators is considered in such a way that each member of the collection to be merged is a partial indistinguishability operator with respect to a different t-norm. In this direction we will try to extend to this new framework, on the one hand, Theorem 1 and, on the other hand, a few techniques that involve the additive generator of a continuous Archimedean t-norm, metric transforms and the pseudo-inverse of the aforementioned additive generator in the spirit of Proposition 4 and those results given in [19].

Acknowledgements. This work was partially supported by the Spanish Ministry of Economy and Competitiveness under Grants DPI2017-86372-C3-3-R, TIN2016-81731-REDT (LODISCO II) and AEI/FEDER, UE funds, by Programa Operatiu FEDER 2014-2020 de les Illes Balears, by project ref. PROCOE/4/2017 (Direcció General d'Innovació i Recerca, Govern de les Illes Balears), and by project ROBINS. The latter has received research funding from the EU H2020 framework under GA 779776. This publication reflects only the authors views and the European Union is not liable for any use that may be made of the information contained therein.

References

1. De Baets, B., Mesiar, R.: T-partitions. Fuzzy Sets Syst. **97**, 211–223 (1998). https://doi.org/10.1016/S0165-0114(96)00331-4
2. De Baets, B., Mesiar, R.: Pseudo-metrics and T-equivalences. J. Fuzzy Math. **5**, 471–481 (1997). http://hdl.handle.net/1854/LU-268970
3. De Baets, B., Mesiar, R.: Metrics and T-equalities. J. Math. Anal. Appl. **267**, 531–547 (2002). http://hdl.handle.net/1854/LU-157982

4. Beliakov, G., Bustince, H., Calvo-Sánchez, T.: A Practical Guide to Averaging Functions. Springer, Heidelberg (2016). https://doi.org/10.1007/978-3-319-24753-3
5. Beliakov, G., Pradera, A., Calvo, T.: Aggregation Functions: A Guide for Practitioners. Springer, Heidelberg (2007). https://doi.org/10.1007/978-3-540-73721-6
6. Bouchon-Meunier, B. (ed.): Aggregation and Fusion of Imperfect Information. Sudies in Fuzziness and Soft Computing, vol. 12. Physica-Verlag, Heidelberg (1998)
7. Bukatin, M., Kopperman, R., Matthews, S.G.: Some corollaries of the correspondence between partial metric and multivalued equalities. Fuzzy Set. Syst. **256**, 57–72 (2014). https://doi.org/10.1016/j.fss.2013.08.016
8. Calvo, T., Fuster-Parra, P., Valero, O.: On the problem of relaxed indistinguishability aggregation operators. In: Proceedings of the Workshop on Applied Topological Structures, WATS 2017, pp. 19–26 (2017)
9. Demirci, M.: The order-theoretic duality and relations between partial metrics and local equalities. Fuzzy Set. Syst. **192**, 45–57 (2012). https://doi.org/10.1016/j.fss.2011.04.014
10. Deza, M.M., Deza, E.: Encyclopedia of Distances. Springer, Heidelberg (2016). https://doi.org/10.1007/978-3-662-52844-0
11. Drewniak, J., Dudziak, U.: Aggregation preserving classes of fuzzy relations. Kybernetica **41**, 265–284 (2005)
12. Dudziak, U.: Preservation of t-norm and t-conorm based properties of fuzzy relations during aggregation process. In: Proceedings of the 8th Conference of European Society for Fuzzy Logic and Technology, EUSFLAT 2013, pp. 376–383 (2013)
13. Fuster-Parra, P., Martín, J., Miñana, J.J., Valero, O.: A study on the relationship between relaxed metrics and indistinguishability operators (2017, submitted)
14. Fuster-Parra, P., Martín, J., Recasens, J., Valero, O.: On the metric behavior of partial indistinguishability operators. preprint (2018)
15. Gottwald, S.: On t-norms which are related to distances of fuzzy sets. BUSEFAL **50**, 25–30 (1992)
16. Grabisch, M., Marichal, J.L., Mesiar, R., Pap, E.: Aggregation Functions. Cambridge University Press, New York (2009)
17. Grigorenko, O., Lebendinska, J.: On another view of aggregation of fuzzy relations. In: Galichet, S., Montero, J., Mauris, G. (eds) Proceedings of the 7th Conference on EUSFLAT 2011 and LFA 2011, pp. 21–27. Atlantis Press (2011)
18. Höhle, U.: Fuzzy equalities and indistinguishability. In: Proceedings of EUFIT 1993, Aachen, vol. 1, pp. 358–363 (1993)
19. Jacas, J., Recasens, J.: Aggregation of T-transitive relations. Int. J. Intell. Syst. **18**, 1193–1214 (2003). https://doi.org/10.1002/int.10141
20. Klement, P., Mesiar, R., Pap, E.: Triangular Norms. Kluwer, Dordrecht (2000)
21. Mayor, G., Recasens, J.: Preserving T-transitivity. In: Nebot, A., et al. (eds.) Artificial Intelligence Research and Development, vol. 288, pp. 79–87. IOS Press (2016)
22. Ovchinnikov, S.V.: Representations of transitive fuzzy relations. In: Skala, H.J., Termini, S., Trillas, E. (eds.) Aspects of Vageness. Theory and Decision Library (An International Series in the Philosophy and Methodology of the Social and Behavioral Sciences), vol. 39, pp. 105–118. Springer, Dordrecht (1984). https://doi.org/10.1007/978-94-009-6309-2_7
23. Ovchinnikov, S.: Similarity relations, fuzzy partitions, and fuzzy orderings. Fuzzy Sets Syst. **40**, 107–126 (1991). https://doi.org/10.1016/0165-0114(91)90048-U

24. Peneva, V., Popchev, I.: Properties of the aggregation operators related with fuzy relations. Fuzzy Sets Syst. **139**, 615–633 (2003). https://doi.org/10.1016/S0165-0114(03)00141-6
25. Pradera, A., Trillas, E., Castiñeira, E.: On distances aggregation. In: Bouchon-Meunier, B. et al. (eds.) Proceedings of Information Processing and Management of Uncertainty in Knowledge-Based Systems International Conference, vol. II, pp. 693–700. Universidad Politécnica de Madrid Press (2000)
26. Pradera, A., Trillas, E., Castiñeira, E.: On the aggregation of some classes of fuzzy relations. In: Bouchon-Meunier, B., Gutierrez, J., Magdalena, L., Yager, R. (eds.) Technologies for Constructing Intelligent Systems. STUDFUZZ, vol. 90, pp. 125–136. Physica, Heidelberg (2002). https://doi.org/10.1007/978-3-7908-1796-6_10
27. Recasens, J.: Indistinguishability Operators: Modelling Fuzzy Equalities and Fuzzy Equivalence Relations. Springer, Heidelberg (2010). https://doi.org/10.1007/978-3-642-16222-0
28. Saminger-Platz, S., Mesiar, R., Dubois, D.: Aggregation operators and commuting. IEEE Trans. Fuzzy Syst. **12**, 1032–1045 (2007). https://doi.org/10.1109/TFUZZ.2006.890687
29. Trillas, E.: Assaig sobre les relacions d'indistingibilitat. In: Proceedings of the Primer Congrés Catalá de Lógica Matemática, Barcelona, pp. 51–59 (1982)
30. Valverde, L.: On the structure of F-indistinguishability operators. Fuzzy Set. Syst. **17**, 313–328 (1985). https://doi.org/10.1016/0165-0114(85)90096-X
31. Zadeh, L.A.: Similarity relations and fuzzy orderings. Inf. Sci. **3**, 177–200 (1971). https://doi.org/10.1016/S0020-0255(71)80005-1

Size-Based Super Level Measures on Discrete Space

Jana Borzová, Lenka Halčinová(✉), and Jaroslav Šupina

Institute of Mathematics, P. J. Šafárik University in Košice,
Jesenná 5, 040 01 Košice, Slovakia
jana.borzova@student.upjs.sk, {lenka.halcinova,jaroslav.supina}@upjs.sk

Abstract. We continue in the investigation of a concept of size introduced by Do and Thiele [3]. Our focus is a computation of corresponding super level measure, a key component of size application, on discrete space, i.e., a finite set with discrete topology. We found critical numbers which determine the change of a value of super level measure and we present an algorithm for super level measure computation based on these numbers.

Keywords: Size · Super level measure · Non-additive measure Possibility measure

1 Introduction

Do and Thiele [3] introduced a concept of size to build an L_p-theory as a unifying language for both Carlson measures and time-frequency analysis. After its definition in [3] they defined an auxiliary notion of outer essential supremum and a key notion of their L_p-theory, the super level measure. They briefly presented its basic properties and the rest of the paper is focused on applications of their notion. The paper [5] is an attempt to study the concept of size and corresponding notions of outer essential supremum as well as super level measure in its own. In [7], the author used the concept of size to generalize sublinear means into the infinite spaces. In [4] the author introduced super level measure-based non-additive integrals together with examples. Especially, the well-known Choquet [2] integral as well as the Shilkret [8] and the Sugeno integral [9] are discussed. The present paper is a continuation of all these efforts.

In contrast to [3,5], we are considering only discrete space, i.e., a finite set with discrete topology. Such simplified model led us to several reductions. For instance, we consider only specific basic sets $\{1, \dots, n\}$, every supremum is the greatest element, results as formulas for computing super level measure and corresponding integrals are simplified etc. The second difference with [5], the one more essential, is the focus only on super level measure computation. We study neither outer essential supremum nor its computation.

Supported by the grants APVV-16-0337, VVGS-2016-255, VVGS-PF-2017-255.

J. Medina et al. (Eds.): IPMU 2018, CCIS 853, pp. 219–230, 2018.
https://doi.org/10.1007/978-3-319-91473-2_19

The main result of the paper is the discovery of critical numbers which completely describe the super level measure. The super level measure is a non-increasing function defined on positive half-line taking finitely many values. Thus sets with one value of super level measure assigned are in fact intervals. We show that the endpoints of such interval are exactly certain values of outer essential supremum. Moreover, we characterize both these endpoints and the corresponding value of super level measure. Finally, based on these findings, we describe an algorithm for super level measure computation.

Similar idea was also used in [1], in the context of integral equivalence of couples $(\mu, \mathbf{x})$ and $(\mu, \mathbf{y})$ with μ being a possibility (necessity) measure and $\mathbf{x}, \mathbf{y} \in [0,1]^n$. The authors describe the survival function (a special case of super level measure) for a fixed score vector $\mathbf{x} \in [0,1]^n$, a fixed possibility measure Π and they characterize the set of all Π-integral equivalent vectors to $\mathbf{x}$. Subsequently, this procedure ensures the equality of their universal integrals on $[0,1]$ which the authors were focused on. Universal integrals, introduced in [6], together with other appropriate mappings on discrete space form one class of utilitity functions used in multiple-criteria decision making. Therefore the restriction to discrete space makes a sense.

The paper is divided into three parts. The first one, Sect. 2, is only preliminary part recalling necessary terminology mainly from [3,5], containing basic settings for the paper and repeating simplified formulas for outer essential supremum computation from [5]. This part can be skipped for the first reading. Section 3 is devoted to the presentation of algorithm for super level measure computation. There is an illustrative example of how the algorithm is applied and the modified version of algorithm for possibility measure. In the last section, Sect. 4, we show the correctness of the algorithm. It contains all the necessary theoretical background. Moreover, we derive several auxiliary observations for outer measure computation and we study the summation size in more depth. The simplified version of the main result for possibility measure is stated there as well.

2 Preliminaries

In order to be self-contained as far as possible, we recall in this section necessary definitions and all basic notations adapted for discrete spaces. In comparison with more general settings in [3–5] where the basic set X is considered to be any topological or metric space, we restrict our attention to the set

$$X = [n] := \{1, \dots, n\}$$

with discrete topology. Then we immediately get that each non-negative real-valued function $f : [n] \to [0, \infty)$ is a non-negative real-valued vector. Therefore we shall use $\mathbf{x} = (x_1, \dots, x_n)$, $x_i \in [0, \infty)$, $i = 1, 2, \dots, n$ instead of the function symbol f. The set $[0, +\infty)^M$ is the family of all non-negative real-valued functions on a non-empty set M. Further, each subset of $[n]$ is Borel and each function $\mathbf{x}$ is Borel measurable. Hence, we shall omit adjective "Borel" and in accordance with denotation in [3–5], the family of all Borel sets $\mathbf{E}_{\mathrm{B}}$ is $2^{[n]}$.

Under a *measure* we shall understand any set function vanishing on empty set, i.e., $\mu : 2^{[n]} \to [0, +\infty)$ with the only (natural) condition $\mu(\emptyset) = 0$. Measure μ is called monotone if $\mu(A) \leq \mu(B)$ whenever $A, B \in 2^{[n]}$, $A \subseteq B$. In the whole paper we shall consider only monotone measures $\mu : 2^{[n]} \to [0, \infty)$ with p different values

$$0 = u_1 < u_2 < \cdots < u_p, \quad p \in \mathbb{N}, p \leq 2^n.$$

The set of all values will be denoted by $\mathscr{H}(\mu) = \{u_1, \ldots, u_p\}$. Normed monotone measures, i.e., $\mu : 2^{[n]} \to [0, 1]$, with $\mu([n]) = 1$ will be called *capacities.* A capacity Π is called *a possibility measure* whenever it is maxitive, i.e., $\Pi(A \cup B) = \max\{\Pi(A), \Pi(B)\}$ for any $A, B \subseteq 2^{[n]}$. One can easily compute $\Pi(A)$ by formula $\Pi(A) = \max_{i \in A}\{\pi(i)\}$, where $\pi : [n] \to [0, 1]$, $\pi(i) = \Pi(\{i\})$ is usually called a *possibility distribution* (of Π), see [1,10].

In the following we present the definition of sizes originally introduced in [3] and slightly modified in [5].

Definition 2.1. Let $n \in \mathbb{N}$. A *size* on $[n]$ is a map $\mathsf{s} : [0, +\infty)^{[n]} \to [0, +\infty)^{(2^{[n]})}$ such that for any $\mathbf{x}, \mathbf{y} \in [0, +\infty)^{[n]}$ and $E \in 2^{[n]}$ it holds that[1,2]

(i) if $\mathbf{x} \leq \mathbf{y}$, then $\mathsf{s}(\mathbf{x})(E) \leq \mathsf{s}(\mathbf{y})(E)$: (monotonicity)
(ii) $\mathsf{s}(\lambda \mathbf{x})(E) = \lambda\, \mathsf{s}(\mathbf{x})(E)$ for each $\lambda \in [0, +\infty)$: (scaling)
(iii) $\mathsf{s}(\mathbf{x} + \mathbf{y})(E) \leq C_{\mathsf{s}} \cdot (\mathsf{s}(\mathbf{x})(E) + \mathsf{s}(\mathbf{y})(E))$, $C_{\mathsf{s}} \geq 1$. (quasi-sublinearity)

The concept of size can be viewed as a form of averaging the non-negative functions from the class $[0, +\infty)^{[n]}$ over the collection $\mathbf{E} \subseteq 2^{[n]}$. This approach involves averaging such as the classical arithmetic mean, generalized arithmetic mean, as well as the supremum of a function over a set, see [3,5]. The last mentioned example does not reflect averaging in its original sense, but it is a size and it plays an important role in the theory of super level measures.

Example 1. The most used sizes in this paper are the following mappings. Let $\mathbf{x} = (x_1, x_2, \ldots, x_n)$ be a function on $[n]$, $\mathbf{E} \subseteq 2^{[n]}$.

(a) *Sum.* $\mathsf{s}_{\mathrm{sum}}(\mathbf{x})(E) := \sum_{i \in E} x_i, \quad x_i \in [0, \infty)$,
(b) *The Choquet integral.* $\mathsf{s}^{(\mathrm{Ch})}_{\mathrm{int},\mu}(\mathbf{x})(E) := \sum_{i \in E} (x_{(i)} - x_{(i-1)})\mu(A_{(i)})$,
where $(\cdot) : [n] \to [n]$ is a permutation such that $x_{(1)} \leq x_{(2)} \leq \cdots \leq x_{(n)}$ and $x_{(0)} = 0$ by convention. Sets $A_{(i)}$ are given by $A_{(i)} = \{(i), \ldots, (n)\}$, μ is a monotone measure.

In the definition of super level measure the so-called outer essential supremum plays a crucial role.[3]

[1] Size was originally introduced for complex valued functions. However, non-negative values are sufficient, compare [3,5].
[2] The constant C_{s} depends only on s.
[3] By $\mathbf{1}_F : [n] \to \{0, 1\}$ we denote the characteristic function of the set F.

Definition 2.2. The *outer essential supremum* of a function $\mathbf{x} \in [0, +\infty)^{[n]}$ over a set $F \in 2^{[n]}$ with respect to a size s and a collection $\mathbf{E} \subseteq 2^{[n]}$ is defined by

$$\operatorname*{outsup}_{F} \mathsf{s}(\mathbf{x})\langle\mathbf{E}\rangle := \max_{E\in\mathbf{E}} \mathsf{s}(\mathbf{x}\mathbf{1}_F)(E).$$

It is easy to see that different collections $\mathbf{E}$ can lead to different results, see [4, Example 2]. Note that the computation of outer essential supremum by definition forces us to consider the size values of all sets in the collection $\mathbf{E}$. However, by Theorem 4.6 in [5], for many interesting examples of sizes and collections the computation can be done in much easier way. For instance, if the collection $\mathbf{E}$ contains the whole $[n]$[4] then $\operatorname{outsup}_F \mathsf{s}_{\text{sum}}(\mathbf{x})\langle\mathbf{E}\rangle = \mathsf{s}_{\text{sum}}(\mathbf{x})(F)$, $\operatorname{outsup}_F \mathsf{s}^{(\mathrm{N})}_{\text{int},\mu}(\mathbf{x})\langle\mathbf{E}\rangle = \mathsf{s}^{(\mathrm{N})}_{\text{int},\mu}(\mathbf{x})(F)$ for $\mathrm{N} \in \{\mathrm{Ch}, \mathrm{Sh}\}$ with Sh being the Shilkret integral.

Finally, we are prepared to recall the definition of super level measure as a size analogy of the measure of super level sets $\{i \in [n] : \; x_i > \alpha\}$.

Definition 2.3. The quantity

$$\mu(s(\mathbf{x})\langle\mathbf{E}\rangle > \alpha) := \min\left\{\mu(F) : \operatorname*{outsup}_{[n]\setminus F} \mathsf{s}(\mathbf{x})\langle\mathbf{E}\rangle \le \alpha\right\}, \quad \alpha \ge 0. \tag{1}$$

is called a *super level measure* of $\mathbf{x} \in [0, +\infty)^{[n]}$ with respect to monotone measure μ, size s and collection $\mathbf{E}$.

The super level measure may or may not coincide with the measure of super level sets, see [3,5]. Basic example of the first type is a supremum size, for more details see [5, Proposition 5.3]. Finally, the quadraple $([n], \mathsf{s}, \mathbf{E}, \mu)$ will be called discrete size space throughout the paper.

3 Algorithm for Super Level Measure Computation

We present a simple algorithm for super level measure computation. Given a discrete size space $([n], \mathsf{s}, \mathbf{E}, \mu)$ and given a function $\mathbf{x}$, the algorithm computes all the values taken by super level measure and it finds the corresponding intervals where the values are taken. Namely, the output is a number q expressing the amount of taken values, decreasing sequence $\gamma_1, \ldots, \gamma_q$ and increasing sequence $v_1, \ldots, v_q$ such that $\gamma_q = v_1 = 0$ and

$$\mu(s(\mathbf{x})\langle\mathbf{E}\rangle > \alpha) = \sum_{i=1}^{q} v_i \mathbf{1}_{[\gamma_i, \gamma_{i-1})}(\alpha).$$

We assume that the computation of outer essential supremum is implemented in its own procedure. We are not interested in this procedure in the frame of the present paper. We believe that although a general procedure, the one

[4] Which is equivalent to covering property (COV) in [5].

dependent on the input size, could be implemented, the knowledge of specific size and collection could lead to more efficient algorithm, see Sect. 2 or [5]. Furthermore, we do not specify details of searching the minimal element of the family of outer essential supremums.

Data: n, s, $\mathbf{E}$, μ, $\mathbf{x}$
Result: q, $\gamma_1, \ldots, \gamma_q$, $v_1, \ldots, v_q$
$u := 0$, $\beta := \min\left\{\text{outsup}_{[n]\setminus F}\, \mathsf{s}(\mathbf{x})\langle \mathbf{E}\rangle : \mu(F) = 0\right\}$;
$i := 1$, $v_1 := u$, $\gamma_1 := \beta$;
while $\beta > 0$ **do**
  $u := \min\{v \in \mathscr{H}(\mu) : v > u\}$;
  $\beta := \min\left\{\text{outsup}_{[n]\setminus F}\, \mathsf{s}(\mathbf{x})\langle \mathbf{E}\rangle : \mu(F) = u\right\}$;
  if $\beta < \gamma_i$ **then**
    $i := i + 1$;
    $v_i := u$;
    $\gamma_i := \beta$;
  end
end
q := i;

The application of the algorithm is presented in the following example.

Example 2. Consider $X = [3]$ and the monotone measure $\mu : 2^{[3]} \to [0, 1]$ given by

$$\mu(A) = \begin{cases} 0, & \text{if } A = \emptyset; \\ \frac{1}{2i}, & \text{if } A = \{i\}; \\ \frac{1}{\min A}, & \text{otherwise.} \end{cases}$$

Consider function $\mathbf{x} = (0.3; 0.4; 0.5)$, and collections $\mathbf{E}_1 = \{\{1\}, \{2\}, \{3\}\}$ and $\mathbf{E}_2 = \{\{1, 2\}, \{2, 3\}, \{1, 2, 3\}\}$. Let Π be a possibility measure related to the possibility distribution $\pi : [3] \to [0, 1]$ given by

$$\pi(1) = 0.6, \;\; \pi(2) = 1 \;\text{ and }\; \pi(3) = 0.5.$$

Consider the Choquet integral-based size $\mathsf{s}^{(\mathrm{Ch})}_{\mathrm{int},\Pi}(\mathbf{x})(E)$, see Example 1. Our aim is to compute values of the super level measure on $\mathbf{E}_1$ and $\mathbf{E}_2$ and show that values of super level measure μ are dependent on the chosen collection.

The capacity μ achieves values $u_1 = 0, u_2 = \frac{1}{6}, u_3 = \frac{1}{4}, u_4 = \frac{1}{2}, u_5 = 1$.

(1) $\beta_1 = \text{outsup}_{[n]}\, s(\mathbf{x})\langle \mathbf{E}_1\rangle = \max\{0.18, 0.4, 0.25\} = 0.4 = \gamma_1$, since the capacity μ achieves zero value only for the empty set. Then

$$\mu(s(\mathbf{x})\langle \mathbf{E}_1\rangle > \alpha) = 0 = v_1 \quad \text{for any } \alpha \in [0.4, \infty).$$

(2) $\beta_2 = \min\left\{\text{outsup}_{[n]\setminus F}\, \mathsf{s}(\mathbf{x})\langle\mathbf{E}\rangle : m(F) = \frac{1}{6}\right\} = 0.4$, therefore the super level measure μ never achieves value $\frac{1}{6}$.

(3) $\beta_3 = \min\left\{\text{outsup}_{[n]\setminus F}\, \mathsf{s}(\mathbf{x})\langle\mathbf{E}\rangle : \mu(F) = \frac{1}{4}\right\} = 0.25 = \gamma_2$. Then

$$\mu(s(\mathbf{x})\langle\mathbf{E}_1\rangle > \alpha) = u_3 = \frac{1}{4} = v_2 \quad \text{for any } \alpha \in [0.25, 0.4).$$

(4) $\beta_4 = \min\left\{\text{outsup}_{[n]\setminus F}\, \mathsf{s}(\mathbf{x})\langle\mathbf{E}\rangle : \mu(F) = \frac{1}{2}\right\} = \min\{0.4, 0.18\} = 0.18 = \gamma_3$. Then

$$\mu(s(\mathbf{x})\langle\mathbf{E}_1\rangle > \alpha) = u_4 = \frac{1}{2} = v_3 \quad \text{for any } \alpha \in [0.18, 0.25).$$

(5) $\beta_5 = 0$, therefore

$$\mu(s(\mathbf{x})\langle\mathbf{E}_1\rangle > \alpha) = u_5 = 1 = v_4 \quad \text{for any } \alpha \in [0, 0.18).$$

All together $q = 4$,

$$\mu(s(\mathbf{x})\langle\mathbf{E}_1\rangle > \alpha) = \frac{1}{4} \cdot \mathbf{1}_{[0.25,0.4)}(\alpha) + \frac{1}{2} \cdot \mathbf{1}_{[0.18,0.25)}(\alpha) + 1 \cdot \mathbf{1}_{[0,0.18)}(\alpha).$$

If we consider the collection $\mathbf{E}_2$, then by computing we get $q = 5$, $v_i = u_i$, $(\gamma_1, \ldots, \gamma_5) = (0.45, 0.4, 0.28, 0.18, 0)$ and consequently

$$\mu(s(\mathbf{x})\langle\mathbf{E}_2\rangle > \alpha) = \frac{1}{6} \cdot \mathbf{1}_{[0.4,0.45)}(\alpha) + \frac{1}{4} \cdot \mathbf{1}_{[0.28,0.4)}(\alpha) + \frac{1}{2} \cdot \mathbf{1}_{[0.18,0.28)}(\alpha) + 1 \cdot \mathbf{1}_{[0,0.18)}(\alpha).$$

Therefore in contrast to the collection $\mathbf{E}_1$, the super level measure achieves all values of the given measure μ and the values are achieved on different intervals.

We add a simplified algorithm for possibility measures. The task to compute the minimum of all outer essential supremum values on complements of sets with certain value of measure is substituted by selecting the set of all elements with possibility distribution value greater than certain number.

Data: n, s, $\mathbf{E}$, π, $\mathbf{x}$
Result: q, $\gamma_1, \ldots, \gamma_q$, $v_1, \ldots, v_q$
$u := 0$, $F := \{j : \pi(j) > 0\}$, $\beta := \text{outsup}_F\, \mathsf{s}(\mathbf{x})\langle\mathbf{E}\rangle$;
$i := 1$, $v_1 := u$, $\gamma_1 := \beta$;
while $\beta > 0$ **do**
  $u := \min\{v \in \mathscr{H}(\mu) : v > u\}$;
  $F := \{j : \pi(j) > u\}$;
  $\beta := \text{outsup}_F\, \mathsf{s}(\mathbf{x})\langle\mathbf{E}\rangle$;
  if $\beta < \gamma_i$ **then**
    $i := i + 1$;
    $v_i := u$;
    $\gamma_i := \beta$;
  end
end
q := i;

The correctness of both algorithms is confirmed by theoretical results in the next section.

4 Theoretical Background

It is obvious that the super level measure (1) may have at most p values $u_1, \ldots, u_p$ and the value $u_1 = 0$ is always taken. Indeed, from the monotonicity of size mapping we have that

$$\operatorname*{outsup}_{[n]\setminus\emptyset} \mathsf{s}(\mathbf{x})\langle\mathbf{E}\rangle \geq \operatorname*{outsup}_{[n]\setminus F} \mathsf{s}(\mathbf{x})\langle\mathbf{E}\rangle,$$

for each $F \in 2^{[n]}$. Then we get

$$\mu(s(\mathbf{x})\langle\mathbf{E}\rangle > \alpha) = \min\left\{\mu(F) : \operatorname*{outsup}_{[n]\setminus F} \mathsf{s}(\mathbf{x})\langle\mathbf{E}\rangle \leq \alpha\right\} = 0$$

for any $\alpha \in [\mathrm{outsup}_{[n]} \mathsf{s}(\mathbf{x})\langle\mathbf{E}\rangle, \infty)$. So, the zero value is taken at least on interval $[\mathrm{outsup}_{[n]} \mathsf{s}(\mathbf{x})\langle\mathbf{E}\rangle, \infty)$. If we consider monotone measures μ such that $\mu(F) = 0$ if and only if $F = \emptyset$, then this interval will be the greatest possible (we prove it later). If there exists a set $F \in 2^{[n]}$, $F \neq \emptyset$ with $\mu(F) = 0$, then the situation may differ, i.e. the interval $[\mathrm{outsup}_{[n]} \mathsf{s}(\mathbf{x})\langle\mathbf{E}\rangle, \infty)$ need not be the greatest possible. Let us denote

$$\alpha_1 := \min\left\{\operatorname*{outsup}_{[n]\setminus F} \mathsf{s}(\mathbf{x})\langle\mathbf{E}\rangle : \mu(F) = 0\right\}.$$

Proposition 4.1. *Let $([n], \mathsf{s}, \mathbf{E}, \mu)$ be a discrete size space and $\mathbf{x}$ be a function on $[n]$. Then*

$$\mu(s(\mathbf{x})\langle\mathbf{E}\rangle > \alpha) = 0 = u_1 \text{ for any } \alpha \in [\alpha_1, \infty)$$

and this interval is the greatest possible.

Proof. First, taking $G \in 2^{[n]}$ such that $\mu(G) = 0$, we show that super level measure takes zero value for each $\alpha \in [\mathrm{outsup}_{[n]\setminus G} \mathsf{s}(\mathbf{x})\langle\mathbf{E}\rangle, \infty)$. Indeed, it is enough to realize that

$$\mu(G) \in \left\{\mu(F) : \operatorname*{outsup}_{[n]\setminus F} \mathsf{s}(\mathbf{x})\langle\mathbf{E}\rangle \leq \alpha\right\}$$

and $\mu(G)$ is the minimum. Therefore,

$$\mu(s(\mathbf{x})\langle\mathbf{E}\rangle > \alpha) = 0 \text{ for each } \alpha \in [\operatorname*{outsup}_{[n]\setminus G} \mathsf{s}(\mathbf{x})\langle\mathbf{E}\rangle, \infty).$$

Taking $G_1 \in 2^{[n]}$ such that $\mu(G_1) = 0$ and $\mathrm{outsup}_{[n]\setminus G_1} \mathsf{s}(\mathbf{x})\langle\mathbf{E}\rangle = \alpha_1$ then

$$\mu(s(\mathbf{x})\langle\mathbf{E}\rangle > \alpha) = 0 \text{ for each } \alpha \in [\alpha_1, \infty).$$

Moreover, this interval is the best possible. By contradiction, let there exist $\beta < \alpha_1$ such that the super level measure vanishes, i.e., $\mu(s(\mathbf{x})\langle\mathbf{E}\rangle > \beta) = 0$. From the definition of super level measure there exists a set $G \in 2^{[n]}$ with $\mu(G) = 0$ such that $\operatorname{outsup}_{[n]\setminus G} \mathsf{s}(\mathbf{x})\langle\mathbf{E}\rangle \leq \beta$. This is a contradiction. □

Arising from previous ideas, we shall derive in the following the whole algorithm how to compute the super level measure. Although the super level measure always takes the zero value, the positive values $u_2, \dots, u_p$ need not be achieved. Let us denote

$$\alpha_k := \min\left\{\operatorname*{outsup}_{[n]\setminus F} \mathsf{s}(\mathbf{x})\langle\mathbf{E}\rangle : \mu(F) \leq u_k\right\}, \quad k = 2, \dots, p. \tag{2}$$

Since the values $\{u_k : k = 1, \dots, p\}$ are enumerated increasingly, by Proposition 4.1, the definition of $\{\alpha_k : k = 1, \dots, p\}$ and the fact that $\mathsf{s}(\mathbf{1}_\emptyset)(E) = 0$ we have

$$0 = \alpha_p \leq \alpha_{p-1} \leq \cdots \leq \alpha_2 \leq \alpha_1.$$

By Proposition 4.1 one can see that the super level measure $\mu(s(\mathbf{x})\langle\mathbf{E}\rangle > \alpha)$ has non-zero values exactly on interval $(\alpha_p, \alpha_1) = (0, \alpha_1)$. The following proposition describes these values and sets which they are assigned on. Let us repeat that the only possible values for $\mu(s(\mathbf{x})\langle\mathbf{E}\rangle > \alpha)$ are the values of measure μ, i.e., the values $\{u_k : k = 1, \dots, p\}$.

Proposition 4.2. *Let $([n], \mathsf{s}, \mathbf{E}, \mu)$ be a discrete size space, $\mathbf{x}$ be a function on $[n]$ and $k = 2, \dots, p$. Then either*

$$\mu(s(\mathbf{x})\langle\mathbf{E}\rangle > \alpha) = u_k \quad \textit{for any } \alpha \in [\alpha_k, \alpha_{k-1})$$

and this interval is the greatest possible with value u_k or $\alpha_k = \alpha_{k-1}$ and then the value u_k will not be taken.

Proof. In accordance with (2) we can write

$$\begin{aligned}\alpha_k &= \min\left\{\operatorname*{outsup}_{[n]\setminus F} \mathsf{s}(\mathbf{x})\langle\mathbf{E}\rangle : \mu(F) \leq u_{k-1} \text{ or } \mu(F) = u_k\right\} \\ &= \min\left\{\alpha_{k-1} : \min\left\{\operatorname*{outsup}_{[n]\setminus F} \mathsf{s}(\mathbf{x})\langle\mathbf{E}\rangle : \mu(F) = u_k\right\}\right\} \leq \alpha_{k-1}.\end{aligned} \tag{3}$$

Let $\alpha_k < \alpha_{k-1}$. Thus by (3) there exists set $G \in 2^{[n]}$ with $\mu(G) = u_k$ such that $\alpha_k = \operatorname{outsup}_{[n]\setminus G} \mathsf{s}(\mathbf{x})\langle\mathbf{E}\rangle < \alpha_{k-1}$. We shall show that $\mu(s(\mathbf{x})\langle\mathbf{E}\rangle > \alpha) = u_k$ for any $\alpha \in [\alpha_k, \alpha_{k-1})$. For each $\alpha \geq \alpha_k$

$$u_k = \mu(G) \in \left\{\mu(F) : \operatorname*{outsup}_{[n]\setminus F} \mathsf{s}(\mathbf{x})\langle\mathbf{E}\rangle \leq \alpha\right\},$$

therefore $\mu(\mathsf{s}(\mathbf{x})\langle\mathbf{E}\rangle > \alpha) \leq u_k$. If $\alpha < \alpha_{k-1}$ then $\mu(s(\mathbf{x})\langle\mathbf{E}\rangle > \alpha) > u_{k-1}$, therefore $\mu(s(\mathbf{x})\langle\mathbf{E}\rangle > \alpha) \geq u_k$. Indeed, let

$$\alpha < \alpha_{k-1} = \min\left\{\operatorname*{outsup}_{[n]\setminus F} \mathsf{s}(\mathbf{x})\langle\mathbf{E}\rangle : \mu(F) \leq u_{k-1}\right\}.$$

Thus if $H \in 2^{[n]}$ with $\mu(H) = u_j \leq u_{k-1}$ then $\operatorname{outsup}_{[n]\setminus H} \mathsf{s}(\mathbf{x})\langle\mathbf{E}\rangle > \alpha$. Consequently,

$$u_j \notin \left\{\mu(F) : \operatorname*{outsup}_{[n]\setminus F} \mathsf{s}(\mathbf{x})\langle\mathbf{E}\rangle \leq \alpha\right\}$$

and we obtain that $\mu(\mathsf{s}(\mathbf{x})\langle\mathbf{E}\rangle > \alpha) \neq u_j$.

By Proposition 4.1 we have $\mu(\mathsf{s}(\mathbf{x})\langle\mathbf{E}\rangle > \alpha) = 0$ for $\alpha \geq \alpha_1$. Moreover, $\alpha_p = 0$. Using already proven part of this proposition for $\alpha_p, \ldots, \alpha_1$, if $\alpha_k = \alpha_{k-1}$ then the value u_k cannot be taken. □

Summarizing Propositions 4.1 and 4.2 we have the following result.

Corollary 4.3. *Let $([n], \mathsf{s}, \mathbf{E}, \mu)$ be a discrete size space and $\mathbf{x}$ be a function on $[n]$. Then*

$$\mu(s(\mathbf{x})\langle\mathbf{E}\rangle > \alpha) = \sum_{i=1}^{q} v_i \mathbf{1}_{[\gamma_i, \gamma_{i-1})}(\alpha),$$

where

(a) $q = \left|\left\{\min\left\{\operatorname*{outsup}_{[n]\setminus F} \mathsf{s}(\mathbf{x})\langle\mathbf{E}\rangle : \mu(F) \leq v\right\} : v \in \mathscr{H}(\mu)\right\}\right| \leq p,$

(b) $\{\gamma_i : i = 1, \ldots, k\} = \left\{\min\left\{\operatorname*{outsup}_{[n]\setminus F} \mathsf{s}(\mathbf{x})\langle\mathbf{E}\rangle : \mu(F) \leq v\right\} : v \in \mathscr{H}(\mu)\right\},$
where $+\infty > \gamma_1 > \gamma_2 > \cdots > \gamma_q = 0,$

(c) $v_i = \min\left\{v \in \mathscr{H}(\mu) : \gamma_i = \min\left\{\operatorname*{outsup}_{[n]\setminus F} \mathsf{s}(\mathbf{x})\langle\mathbf{E}\rangle : \mu(F) \leq v\right\}\right\},$
for any $i = 1, \ldots, q$.

Let us denote

$$\beta_k := \min\left\{\operatorname*{outsup}_{[n]\setminus F} \mathsf{s}(\mathbf{x})\langle\mathbf{E}\rangle : \mu(F) = u_k\right\}, \quad k = 1, \ldots, p.$$

Note that $\alpha_{k+1} = \min\{\alpha_k, \beta_{k+1}\} = \min\{\alpha_1, \beta_2, \ldots, \beta_{k+1}\}$. Moreover, if $\alpha_k > \alpha_{k+1}$ then $\alpha_{k+1} = \beta_{k+1}$ and $\alpha_{k+1} = \alpha_k$ if and only if $\alpha_k \leq \beta_{k+1}$.

Corollary 4.4. *Let $([n], \mathsf{s}, \mathbf{E}, \mu)$ be a discrete size space and $\mathbf{x}$ be a function on $[n]$.*

(a) $\max\{\mu(\mathsf{s}(\mathbf{x})\langle\mathbf{E}\rangle > \alpha) : \alpha > 0\} = \min\{\mu(F) : \operatorname{outsup}_{[n]\setminus F} \mathsf{s}(\mathbf{x})\langle\mathbf{E}\rangle = 0\}$
$= \min\{\mu(F) : (\forall E \in \mathbf{E})\, \mathsf{s}(\mathbf{x}\mathbf{1}_{[n]\setminus F})(E) = 0\}$

(b) *Let $k > i$. If $\beta_k \geq \alpha_i$ then $u_k \notin \{\mu(\mathsf{s}(\mathbf{x})\langle\mathbf{E}\rangle > \alpha) : \alpha > 0\}$, i.e., the value u_k is not taken.*

(c) *Let $k > i$. If $\operatorname{outsup}_{[n]\setminus G} \mathsf{s}(\mathbf{x})\langle\mathbf{E}\rangle \geq \operatorname{outsup}_{[n]\setminus F} \mathsf{s}(\mathbf{x})\langle\mathbf{E}\rangle$ for all G and F such that $\mu(F) = u_i$, $\mu(G) = u_k$ then $u_k \notin \{\mu(\mathsf{s}(\mathbf{x})\langle\mathbf{E}\rangle > \alpha) : \alpha > 0\}$, i.e., the value u_k is not taken.*

(d) *Let $k > i$. If the inequality $\operatorname{outsup}_{[n]\setminus F} \mathsf{s}(\mathbf{x})\langle\mathbf{E}\rangle < \alpha_i$ holds and $\mu(F) = u_k$ then we have $[u_{i+1}, u_k] \cap \{\mu(\mathsf{s}(\mathbf{x})\langle\mathbf{E}\rangle > \alpha) : \alpha > 0\} \neq \emptyset$.*

(e) *If $p = 2^n$ and $\operatorname{outsup}_{[n]\setminus F} \mathsf{s}(\mathbf{x})\langle\mathbf{E}\rangle > \operatorname{outsup}_{[n]\setminus G} \mathsf{s}(\mathbf{x})\langle\mathbf{E}\rangle$ for every $\mu(F) < \mu(G)$ then $\{\mu(\mathsf{s}(\mathbf{x})\langle\mathbf{E}\rangle > \alpha) : \alpha > 0\} = \{u_i : i = 1, \ldots, p\}$.*

An interesting consequence of Corollary 4.4 a) is that if $\mathbf{x}$ has positive values, then

$$\max\{\mu(\mathsf{s}_{\text{sum}}(\mathbf{x})\langle 2^{[n]}\rangle > \alpha) : \alpha > 0\} = \max \mathscr{H}(\mu) = u_p.$$

In [5], Example 5.7, there was constructed an example of a function on three element set such that any measure with 8 distinct values produces a super level measure of a summation size s_{sum} which again takes 8 distinct values. This is possible due to concept of super level measure since the measure of super level sets can take at most 4 values. An attempt to describe this behavior of super level measure is part (e) of our Corollary 4.4, which is a sufficient condition for a function such that super level measure takes all 2^n values. In fact, we can state more.

Corollary 4.5. *Let $([n], \mathsf{s}, \mathbf{E}, \mu)$ be a discrete size space with 2^n values and $\mathbf{x}$ be a function on $[n]$. Then $|\{\mu(\mathsf{s}(\mathbf{x})\langle\mathbf{E}\rangle > \alpha) : \alpha > 0\}| = 2^n$ if and only if it holds that $\operatorname{outsup}_{[n]\setminus F} \mathsf{s}(\mathbf{x})\langle\mathbf{E}\rangle > \operatorname{outsup}_{[n]\setminus G} \mathsf{s}(\mathbf{x})\langle\mathbf{E}\rangle$ for every $\mu(F) < \mu(G)$.*

Proof. If μ has 2^n values, then $2^n = p$. Thus if $\operatorname{outsup}_{[n]\setminus F} \mathsf{s}(\mathbf{x})\langle\mathbf{E}\rangle > \operatorname{outsup}_{[n]\setminus G} \mathsf{s}(\mathbf{x})\langle\mathbf{E}\rangle$ for every $\mu(F) < \mu(G)$ then by Corollary 4.4 (e) we have $|\{\mu(\mathsf{s}_{\text{sum}}(\mathbf{x})\langle 2^{[n]}\rangle > \alpha) : \alpha > 0\}| = 2^n$. To prove the reversed implication, let us assume that $|\{\mu(\mathsf{s}_{\text{sum}}(\mathbf{x})\langle 2^{[n]}\rangle > \alpha) : \alpha > 0\}| = 2^n$. If there were sets F, G such that $\mu(F) < \mu(G)$ and $\operatorname{outsup}_{[n]\setminus F} \mathsf{s}(\mathbf{x})\langle\mathbf{E}\rangle \leq \operatorname{outsup}_{[n]\setminus G} \mathsf{s}(\mathbf{x})\langle\mathbf{E}\rangle$ then by Corollary 4.4 (c) there should have been a value of measure μ which is not taken by $\mu(\mathsf{s}_{\text{sum}}(\mathbf{x})\langle\mathbf{E}\rangle)$. $\square$

We can restate Corollary 4.5 in terms of outer essential supremum of a summation size s_{sum} and a collection $\mathbf{E}$ containing $[n]$.[5] Let μ be a measure on $[n]$ with 2^n values. $|\{\mu(\mathsf{s}_{\text{sum}}(\mathbf{x})\langle\mathbf{E}\rangle > \alpha) : \alpha > 0\}| = 2^n$ if and only if $\sum_{i\in[n]\setminus F} \mathbf{x}_i > \sum_{i\in[n]\setminus G} \mathbf{x}_i$ for every $\mu(F) < \mu(G)$. Hence, one can see that the measure in Example 5.7 in [5] has to possess additional properties.

[5] I.e., satisfying condition (COV) in [5].

Example 3. Let μ be a measure on $[n]$ with 2^n values such that $\mu(\{k\}) < \mu(\{l\})$ and $\mu([n]\setminus\{k\}) < \mu([n]\setminus\{l\})$. Then for every function $\mathbf{x}$, the super level measure of a summation size s_{sum} takes less than 2^n values. Indeed, to get a contradiction let us assume that there is a function $\mathbf{x}$ such that $|\{\mu(\mathsf{s}_{\text{sum}}(\mathbf{x})\langle\mathbf{E}\rangle > \alpha) : \alpha > 0\}| = 2^n$. Denote $S = \sum_{i\in[n]} \mathbf{x}_i$. By Corollary 4.5 and $\mu([n]\setminus\{k\}) < \mu([n]\setminus\{l\})$ we have $x_k > x_l$. However, Corollary 4.5 and $\mu(\{k\}) < \mu(\{l\})$ implies that $S - x_k = \sum_{i\in[n]\setminus\{k\}} \mathbf{x}_i > \sum_{i\in[n]\setminus\{l\}} \mathbf{x}_i = S - x_l$. Thus $x_l > x_k$, a contradiction.

If we consider a possibility measure Π, then numbers α_k, $k = 1,\ldots,p$ may be defined in a simpler way.

Corollary 4.6. *Let $([n], \mathsf{s}, \mathbf{E}, \Pi)$ be a discrete size space with possibility measure Π related to the possibility distribution π and $\mathbf{x}$ be a function on $[n]$. Then*

$$\alpha_k := \operatorname*{outsup}_{F_k} \mathsf{s}(\mathbf{x})\langle\mathbf{E}\rangle, \ k = 1,\ldots,p$$

where $F_k = \{i \in [n] : \pi(i) > u_k\}$.

Proof. For any $k = 1,\ldots,p$ and $F_k^* = \{i \in [n] : \pi(i) \leq u_k\}$ we can write

$$\begin{aligned}\alpha_k &= \min\left\{\operatorname*{outsup}_{[n]\setminus F} \mathsf{s}(\mathbf{x})\langle\mathbf{E}\rangle : \max_{i\in F} \pi(i) \leq u_k\right\}\\ &= \min\left\{\max_{E\in\mathbf{E}} s(\mathbf{x}\cdot\mathbf{1}_{[n]\setminus F})(E) : \max_{i\in F} \pi(i) \leq u_k\right\}\\ &= \max_{E\in\mathbf{E}} s(\mathbf{x}\cdot\mathbf{1}_{[n]\setminus F_k^*})(E),\end{aligned}$$

the last equality holds since for any $F \in 2^{[n]}$ such that $\max_{i\in F}\pi(i) \leq u_k$ we have $F \subseteq F_k^*$ and consequently $s(\mathbf{x}\cdot\mathbf{1}_{[n]\setminus F_k^*})(E) \leq s(\mathbf{x}\cdot\mathbf{1}_{[n]\setminus F})(E)$. Denote $F_k = [n]\setminus F_k^* = \{i \in [n] : \pi(i) > u_k\}$. Then for any $k = 1,\ldots,p$ we have

$$\alpha_k = \max_{E\in\mathbf{E}} s(\mathbf{x}\mathbf{1}_{F_k})(E) = \operatorname*{outsup}_{F_k} \mathsf{s}(\mathbf{x})\langle\mathbf{E}\rangle.$$

□

Corollary 4.6 confirms the correctness of our second algorithm and it concludes the last section of the paper.

5 Conclusion

In this paper we have provided a general algorithm for size-based super level measure computation on discrete spaces. The universality of this algorithm lies not only in the fact that size-based super level measures generalize standard level measure but also is useful in the very concept of standard level measures. Indeed, we believe that special case of algorithm computing standard level measure may be helpful as well, probably when one works with many inputs.

We consider size-based super level measures to be beneficial because they take into account more interactions between elements on basic set $[n]$. In contrast to standard level measures, super level measures can achieve up to 2^n values. This number corresponds to the number of interactions that have been taken into account. Therefore, our further research efforts will be directed to study how sizes and collections act on number of interactions.

References

1. Chen, T., Mesiar, R., Li, J., Stupňanová, A.: Possibility and necessity measures and integral equivalence. Int. J. Approx. Reason. **86**, 62–72 (2017)
2. Choquet, G.: Theory of capacities. Ann. Inst. Fourier **5**, 131–295 (1953)
3. Do, Y., Thiele, C.: L^p theory for outer measures and two themes of Lennart Carleson united. Bull. Amer. Math. Sci. **52**(2), 249–296 (2015)
4. Halčinová, L.: Sizes, super level measures and integrals. In: Torra, V., Mesiar, R., De Baets, B. (eds.) AGOP 2017. AISC, vol. 581, pp. 181–188. Springer, Cham (2018). https://doi.org/10.1007/978-3-319-59306-7_19
5. Halčinová, L., Hutník, O., Kisel'ák, J., Šupina, J.: Beyond the scope of super level measures. Fuzzy Sets Syst. (2018). https://doi.org/10.1016/j.fss.2018.03.007
6. Klement, E.P., Mesiar, R., Spizzichino, F., Stupňanová, A.: Universal integrals based on copulas. Fuzzy Optim. Decis. Making **13**(3), 273–286 (2014)
7. Pap, E.: Sublinear means. In: 36th Linz Seminar on Fuzzy Sets Theory: Functional Equations and Inequalities, Linz, 2–6 February 2016, pp. 75–88 (2016)
8. Shilkret, N.: Maxitive measure and integration. Indag. Math. **33**, 109–116 (1971)
9. Sugeno, M.: Theory of fuzzy integrals and its applications. Ph.D. thesis, Tokyo Institute of Technology (1974)
10. Zadeh, L.A.: Fuzzy sets as a basis for a theory of possibility. Fuzzy Sets Syst. **1**(1), 3–28 (1978)

What Is the Aggregation of a Partial Metric and a Quasi-metric?

Juan-José Miñana and Óscar Valero(✉)

Mathematics and Computer Science Department, Universitat de les Illes Balears, Palma de Mallorca, Spain
{jj.minana,o.valero}@uib.es

Abstract. Generalized metrics have been shown to be useful in many fields of Computer Science. In particular, partial metrics and quasi-metrics are used to develop quantitative mathematical models in denotational semantics and in asymptotic complexity analysis of algorithms, respectively. The aforesaid models are implemented independently and they are not related. However, it seems natural to consider a unique framework which remains valid for the applications to the both aforesaid fields. A first natural attempt to achieve that target suggests that the quantitative information should be obtained by means of the aggregation of a partial metric and a quasi-metric. Inspired by the preceding fact, we explore the way of merging, by means of a function, the aforementioned generalized metrics into a new one. We show that the induced generalized metric matches up with a partial quasi-metric. Thus, we characterize those functions that allow to generate partial quasi-metrics from the combination of a partial metric and a quasi-metric. Moreover, the relationship between the problem under consideration and the problems of merging partial metrics and quasi-metrics is discussed. Examples that illustrate the obtained results are also given.

Keywords: Aggregation · Quasi-metric · Partial metric
Partial quasi-metric · Denotational semantics
Asymptotic complexity of algorithms

1 Introduction

In Computer Science there are two fields in which generalized metrics have been shown to be useful. Concretely, partial metrics have been applied successfully to denotational semantics and quasi-metrics have been used in asymptotic complexity analysis of algorithms. Let us recall briefly the role of such dissimilarities in the aforementioned fields.

In denotational semantics, one of the aims consists in analysing the correctness of recursive algorithms by means of mathematical models of the programming languages in which the algorithm has been written. Moreover, in many programming languages one can construct recursive algorithms through procedures in such a way that the meaning of such a procedure is expressed in terms

J. Medina et al. (Eds.): IPMU 2018, CCIS 853, pp. 231–243, 2018.
https://doi.org/10.1007/978-3-319-91473-2_20

of its own meaning. An easy, but illustrative, example is the procedure which computes the factorial function. In fact a procedure which computes the factorial of a positive integer number typically uses the following recursive denotational specification, where $\mathbb{N}$ denotes the set of positive integer numbers:

$$fact(n) = \begin{cases} 1 & \text{if } n = 1 \\ nfact(n-1) & \text{if } n > 1 \end{cases}. \tag{1}$$

In order to analyze whether a recursive denotational specification of a procedure is meaningful it is usual to make use of fixed point mathematical techniques in which the meaning of such recursive denotational specification is obtained as the fixed point of a nonrecursive mapping associated to the denotational specification. In the particular case of the factorial function the aforesaid nonrecursive mapping ϕ_{fact} will be given as follows:

$$\phi_{fact}(f)(n) = \begin{cases} 1 & \text{if } n = 1 \\ nf(n-1) & \text{if } n > 1 \text{ and } n-1 \in \text{ dom } f \end{cases}. \tag{2}$$

Of course, the entire factorial function is given by the unique fixed point of the nonrecursive mapping ϕ_{fact}.

Notice that ϕ_{fact} is defined on the set of partial functions (see for a detailed discussion [6]). With the aim of developing quantitative fixed point techniques which will be able to analyze the meaning of recursive denotational specification, Matthews introduced the notion of partial metric in [3]. Le us recall that a partial metric space is a pair (X, p) such that X is a non-empty set and $p : X \times X \to \mathbb{R}_+$ is a function satisfying for all $x, y, z \in X$ the following (here $\mathbb{R}_+$ denotes the set of positive real numbers): (P1) $x = y \Leftrightarrow p(x, x) = p(x, y) = p(y, y)$, (P2) $p(x, x) \leq p(x, y)$, (P3) $p(x, y) = p(y, x)$ and (P4) $p(x, z) \leq p(x, y) + p(y, z) - p(y, y)$. Among different partial metrics introduced in the literature, the so-called Baire partial metric plays a central role in the application of the fixed point framework of Matthews to denotational semantics.

According to [3], let us recall that the Baire partial metric space consists of the pair (Σ_∞, p_B), where Σ_∞ is the set of finite and infinite sequences over a non-empty alphabet Σ and the partial metric p_B is given by $p_B(x, y) = 2^{-l(x,y)}$ for all $x, y \in \Sigma_\infty$ with $l(x, y)$ denoting the longest common prefix of the words x and y. The success of the Baire partial metric and the Matthews fixed point method in denotational semantics is given by the fact that the natural order between words, the prefix order, is encoded by p_B in the sense that x is a prefix of y if and only if $p_B(x, y) = p_B(x, x)$.

Often the running time of computing of the recursive algorithm that performs the computation of the meaning of a recursive denotational specification is discussed in conjunction with the correctness of such a recursive denotational specification. In this direction, Schellekens introduced the so-called complexity space which allows to develop quantitative fixed point techniques in order to determine the complexity of recursive algorithms whose running time of computing fulfills a recurrence equation (see [5]).

Going back to the example of the factorial function, it is clear that the running time of computing of an algorithm that computes the factorial of a nonnegative integer number, through the recursive denotational specification (1), is solution to the following recurrence equation

$$T_{fact}(n) = \begin{cases} c & \text{if } n = 0, 1 \\ T_{fact}(n-1) + c & \text{if } n > 1 \end{cases}, \tag{3}$$

where $c \in \mathbb{R}_+$ ($c > 0$) is the time taken by the algorithm to obtain the solution to the problem on the base case.

In contrast to the Matthews approach, Schellekens framework is based on the use of quasi-metrics. Let us recall that a quasi-metric space (X, q) is a pair such that X is a non-empty set and $q : X \times X \to \mathbb{R}_+$ is a function satisfying for all $x, y, z \in X$ the following: (Q1) $x = y \Leftrightarrow q(x, y) = q(y, x) = 0$ and (Q2) $q(x, z) \leq q(x, y) + q(y, z)$.

Concretely, the complexity space is the pair $(\mathcal{C}, q_{\mathcal{C}})$, where $\mathcal{C} = \{f : \mathbb{N} \to (0, \infty] : \sum_{n=0}^{\infty} 2^{-n} \frac{1}{f(n)} < \infty\}$ and $q_{\mathcal{C}}$ is the quasi-metric on $\mathcal{C}$ given as follows

$$q_{\mathcal{C}}(f, g) = \sum_{n=0}^{\infty} 2^{-n} \max\left(\frac{1}{g(n)} - \frac{1}{f(n)}, 0\right).$$

Obviously we adopt the convention that $\frac{1}{\infty} = 0$.

The success of the Schellekens fixed point method in complexity analysis of algorithms is provided by the fact that the running time of computing of an algorithm can be associated to a function belonging to $\mathcal{C}$. Moreover, given two functions $f, g \in \mathcal{C}$, the numerical value $q_{\mathcal{C}}(f, g)$ (the complexity distance from f to g) can be interpreted as the relative progress made in lowering the complexity by replacing any program P with complexity function f by any program Q with complexity function g. In fact, if $f \neq g$, the condition $q_{\mathcal{C}}(f, g) = 0$ can be read as f is "at least as efficient" as g on all inputs. Observe that $q_{\mathcal{C}}(f, g) = 0$ implies that $f(n) \leq g(n)$ for all $n \in \mathbb{N}$, and this is key to state an asymptotic bound of the complexity of an algorithm. Furthermore, notice that the asymmetry of the complexity distance $q_{\mathcal{C}}$ is crucial in order to provide information about the increase of complexity whenever a program is replaced by another one. A metric (symmetric) will be able to yield information on the increase but it, however, will not reveal which program is more efficient.

In the light of the exposed facts we have that quantitative fixed point techniques are used in Computer Science in order to discuss the complexity analysis of algorithms and the meaning of recursive denotational specifications for programming languages. Both techniques are independent and they are used separately without any relationship between them. Inspired by the preceding fact it seems natural to consider $(\Sigma_{\infty} \times \mathcal{C}, p_B + q_{\mathcal{C}})$ as a first attempt to develop a framework to analyze simultaneously, by means of fixed point methods, the running time of computing of an algorithm that performs a computation using a recursive denotational specification and the meaning of such a specification.

However, the previous proposal presents a handicap. Indeed, it is not clear what kind of generalized metric is the function $p_B + q_{\mathcal{C}}$. It is obvious that it is neither a partial metric nor a quasi-metric on $\Sigma_\infty \times \mathcal{C}$. Actually one can prove that $p_B + q_\mathcal{C}$ is a partial quasi-metric (see Example 1 in Sect. 2) on $\Sigma_\infty \times \mathcal{C}$. Motivated by the preceding fact, in this paper we will focus our efforts on stating formally the problem of how to induce partial quasi-metrics merging a partial metric and a quasi-metric, i.e., from an aggregation perspective. To this end, we introduce the notion of partial quasi-metric generating function, those functions that allow to generate a partial quasi-metric from the combination of a partial metric and a quasi-metric. Moreover, we provide a characterization of such functions. Furthermore, we discuss the relationship between the problem under consideration and the partial metric aggregation problem, studied in [2], and the quasi-metric aggregation problem explored in [4]. Finally, examples that illustrate the obtained results are also given.

2 Partial Quasi-metric Generating Functions and Their Characterization

In [1], the notions of partial metric and quasi-metric were generalized. Specifically, Künzi et al. introduced the concept of partial quasi-metric. Let us recall that a partial quasi-metric space is a pair (X, pq), where X is a non-empty set and $pq : X \times X \to \mathbb{R}_+$ is a function satisfying for all $x, y, z \in X$ the following: (PQ1) $pq(x,x) \leq pq(x,y)$ and $pq(x,x) \leq pq(y,x)$, (PQ2) $pq(x,z) + pq(y,y) \leq pq(x,y) + pq(y,z,)$ and (PQ3) $x = y \Leftrightarrow pq(x,x) = pq(x,y)$ and $pq(y,y) = pq(y,x)$.

With the aim of providing a solution to the problem stated in the preceding section we introduce the notion of partial quasi-metric generating function. Thus, we will say that a function $\Phi : \mathbb{R}_+^2 \to \mathbb{R}_+$ is a partial quasi-metric generating function (*pqmg*-function for short) provided that for each partial metric space (X, p) and each quasi-metric space (Y, q), the function $PQ_\Phi : (X \times Y) \times (X \times Y) \to \mathbb{R}_+$ is a partial quasi-metric on $X \times Y$, where $PQ_\Phi((x,y),(u,v)) = \Phi(p(x,u), q(y,v))$ for each $(x,y),(u,v) \in X \times Y$.

Later, Proposition 1 and Example 2 will provide non-trivial instances of *pqmg*-functions. Furthermore, an instance of a function that is not a *pqmg*-function, and that arises in a natural way in aggregation operator theory, will be given by Example 3.

Before, we yield a characterization of pqmg-functions. To this end, let us recall that $\mathbb{R}_+^2$ becomes a partial ordered set when we endow it with the pointwise partial order $\preceq$, i.e., $(a, \alpha) \preceq (b, \beta) \Leftrightarrow a \leq b$ and $\alpha \leq \beta$. Moreover, a function $\Phi : \mathbb{R}_+^2 \to \mathbb{R}_+$ is increasing provided that $\Phi(a, \alpha) \leq \Phi(b, \beta)$ whenever $(a, \alpha) \preceq (b, \beta)$.

The following result will be useful in our subsequent work.

Lemma 1. *If $\Phi : \mathbb{R}_+^2 \to \mathbb{R}_+$ is a pqmg-function, then it is increasing.*

Proof. Let $a, b, \alpha, \beta \in \mathbb{R}_+$ such that $(a, \alpha) \preceq (b, \beta)$. Consider the partial metric space $(\mathbb{R}_+, p_m)$ and the quasi-metric space $(\mathbb{R}, q_u)$, where $p_m(x,y) = \max\{a, b\}$

and $q_u(x,y) = \max\{y-x,0\}$ for all $x,y \in \mathbb{R}$. By our assumption, the function $PQ_\Phi : (\mathbb{R}_+ \times \mathbb{R}) \times (\mathbb{R}_+ \times \mathbb{R}) \to \mathbb{R}_+$ defined by $PQ_\Phi((x,y),(u,v)) = \Phi(p_m(x,u), q_u(y,v))$ is a partial quasi-metric on $\mathbb{R}_+ \times \mathbb{R}$. Then, for each $(x_1,y_1),(x_2,y_2),(x_3,y_3) \in \mathbb{R}_+ \times \mathbb{R}$, we have that

$$\begin{aligned}
&\Phi(p_m(x_1,x_3), q_s(y_1,y_3)) + \Phi(p_m(x_2,x_2), q_s(y_2,y_2)) = \\
&PQ_\Phi((x_1,y_1),(x_3,y_3)) + PQ_\Phi((x_2,y_2),(x_2,y_2)) \leq \\
&PQ_\Phi((x_1,y_1),(x_2,y_2)) + PQ_\Phi((x_2,y_2),(x_3,y_3)) = \\
&\Phi(p_m(x_1,x_2), q_s(y_1,y_2)) + \Phi(p_m(x_2,x_3), q_s(y_2,y_3)).
\end{aligned}$$

Taking in the preceding inequalities $x_1 = a$, $x_2 = b$, $x_3 = 0$, $y_1 = 0$, $y_2 = \beta$ and $y_3 = \alpha$ we obtain that $\Phi(a,\alpha) + \Phi(b,0) \leq \Phi(b,\beta) + \Phi(b,0)$, since $p_m(x_1,x_3) = a$, $p_m(x_2,x_2) = p_m(x_1,x_2) = p_m(x_2,x_3) = b$, $q_s(y_1,y_3) = \alpha$, $q_s(y_1,y_2) = \beta$, $q_s(y_2,y_2) = 0$ and $q_s(y_2,y_3) = 0$. So $\Phi(a,\alpha) \leq \Phi(b,\beta)$.

The next theorem provides a characterization of those functions $\Phi : \mathbb{R}_+ \times \mathbb{R}_+ \to \mathbb{R}_+$ which are a *pqmg*-functions.

Theorem 1. *A function $\Phi : \mathbb{R}_+^2 \to \mathbb{R}_+$ is a pqmg-function if and only if for each $a,b,c,d,\alpha,\beta,\gamma \in \mathbb{R}_+$ the following assertions hold:*

(i) If $b \geq a$, then $\Phi(a,0) \leq \Phi(b,\alpha)$;
(ii) If $c \geq \max\{a,b\}$, $\Phi(a,0) = \Phi(c,\alpha)$ and $\Phi(b,0) = \Phi(c,\beta)$, then $a = b = c$ and $\alpha = \beta = 0$;
(iii) If $b \leq \min\{c,d\}$, $a+b \leq c+d$ and $\alpha \leq \beta+\gamma$, then $\Phi(a,\alpha) + \Phi(b,0) \leq \Phi(c,\beta) + \Phi(d,\gamma)$.

Proof. Assume that $\Phi : \mathbb{R}_+^2 \to \mathbb{R}_+$ is a *pqmg*-function. Next we show that Φ satisfies (i), (ii) and (iii).

(i) It is fulfilled by Lemma 1.
(ii) Let $a,b,c,\alpha,\beta \in \mathbb{R}_+$ such that $c \geq \max\{a,b\}$, $\Phi(a,0) = \Phi(c,\alpha)$ and $\Phi(b,0) = \Phi(c,\beta)$. We will show that $a = b = c$ and $\alpha = \beta = 0$. On the one hand, consider the set $X = \{a,b,c\}$ and we define the function p_X on $X \times X$ as follows: $p_X(a,b) = p_X(b,a) = p_X(a,c) = p_X(c,a) = p_X(b,c) = p_X(c,b) = c$ and, $p_X(a,a) = a$, $p_X(b,b) = b$ and $p_X(c,c) = c$. By our assumption on $a,b,c \in \mathbb{R}_+$, one can verify that p_X is a partial metric on X.
On the other hand, we will distinguish two cases on $\alpha,\beta \in \mathbb{R}_+$:
Case 1. Suppose that $\alpha = \beta = 0$. Then the function $PQ_\Phi : (X \times \mathbb{R}) \times (X \times \mathbb{R}) \to \mathbb{R}_+$ given, for each $(x,y),(u,v) \in X \times \mathbb{R}$, by $PQ_\Phi((x,y),(u,v)) = \Phi(p_X(x,u), d_e(y,v))$ is a partial quasi-metric, where d_e denotes the euclidean metric (note that every metric is a quasi-metric). Moreover,

$$\begin{aligned}
PQ_\Phi((a,0),(a,0)) &= \Phi(p_X(a,a), d_e(0,0)) = \Phi(a,0), \\
PQ_\Phi((a,0),(b,0)) &= \Phi(p_X(a,b), d_e(0,0)) = \Phi(c,0), \\
PQ_\Phi((b,0),(b,0)) &= \Phi(p_X(b,b), d_e(0,0)) = \Phi(b,0), \\
PQ_\Phi((b,0),(a,0)) &= \Phi(p_X(b,a), d_e(0,0)) = \Phi(c,0).
\end{aligned}$$

Since $\Phi(a,0) = \Phi(c,0)$ and $\Phi(b,0) = \Phi(c,0)$ we deduce that

$$PQ_\Phi((a,0),(a,0)) = PQ_\Phi((a,0),(b,0)) \text{ and}$$
$$PQ_\Phi((b,0),(b,0)) = PQ_\Phi((b,0),(a,0)).$$

Therefore, by axiom (PQ3), we deduce that $(a,0) = (b,0)$ and so $a = b$. Furthermore, if we repeat the above process using now a and c, we deduce that $a = c$ and so, $a = b = c$ as we claimed.

Case 2. Suppose that either $\alpha \neq 0$ or $\beta \neq 0$. We will show that this case cannot be given. Consider the partial metric space (X, p_X) introduced in Case 1 and the quasi-metric space (Y, q_Y), where $Y = \{1,2,3\}$ and the quasi-metric q_Y on Y is given by $q_Y(1,2) = q_Y(2,3) = q_Y(1,3) = \alpha$, $q_Y(2,1) = q_Y(3,2) = q_Y(3,1) = \beta$ and $q_Y(i,i) = 0$ for all $i \in \{1,2,3\}$. Then the function $PQ_\Phi : (X\times Y)\times(X\times Y) \to \mathbb{R}_+$ given by $PQ_\Phi((x,y),(u,v)) = \Phi(p_X(x,u), q_Y(y,v))$, for each $(x,u),(y,v) \in X \times Y$, is a partial quasi-metric. Moreover

$$PQ_\Phi((a,1),(a,1)) = \Phi(p_X(a,a), q_Y(1,1)) = \Phi(a,0),$$
$$PQ_\Phi((a,1),(b,2)) = \Phi(p_X(a,b), q_Y(1,2)) = \Phi(c,\alpha),$$
$$PQ_\Phi((b,2),(b,2)) = \Phi(p_X(b,b), q_Y(2,2)) = \Phi(b,0),$$
$$PQ_\Phi((b,2),(a,1)) = \Phi(p_X(b,a), q_Y(2,1)) = \Phi(c,\beta).$$

Since $\Phi(a,0) = \Phi(c,\alpha)$ and $\Phi(b,0) = \Phi(c,\beta)$ we deduce that

$$PQ_\Phi((a,1),(a,1)) = PQ_\Phi((a,1),(b,2)) \text{ and}$$
$$PQ_\Phi((b,2),(b,2)) = PQ_\Phi((b,2),(a,1)).$$

So, by axiom (PQ3), $(a,1) = (b,2)$ which gives a contradiction. Thus, such a case cannot be given.

Hence, we have shown that $a = b = c$ and $\alpha = \beta = 0$.

(iii) Let $a,b,c,d,\alpha,\beta,\gamma \in \mathbb{R}_+$, with $b \leq \min\{c,d\}$, $a+b \leq c+d$ and $\alpha \leq \beta+\gamma$. First, we will show that $\Phi(c+d-b,\alpha) + \Phi(b,0) \leq \Phi(c,\beta) + \Phi(d,\gamma,)$. To this end we distinguish two cases:

Case 1. $\alpha = \beta = \gamma = 0$. Set $X = \{x_1,x_2,x_3\}$ with $x_1 = a$, $x_2 = b$ and $x_3 = c$ and, besides, define the partial metric p'_X on X as follows: $p'_X(x_1,x_3) = p'_X(x_3,x_1) = c+d-b$, $p'_X(x_1,x_2) = p'_X(x_2,x_1) = c$, $p'_X(x_2,x_3) = p'_X(x_3,x_2) = d$, $p'_X(x_1,x_1) = p'_X(x_3,x_3) = 0$ and $p'_X(x_2,x_2) = b$. Then the function $PQ_\Phi : (X\times\mathbb{R})\times(X\times\mathbb{R}) \to \mathbb{R}_+$ defined, for each $(x,y),(u,v) \in X \times \mathbb{R}$, by $PQ_\Phi((x,y),(u,v)) = \Phi(p'_X(x,u), d_e(y,v))$, is a partial quasi-metric. By axiom (PQ2), we have that

$$\Phi(p'_X(x_1,x_3), d_e(0,0)) + \Phi(p'_X(x_2,x_2), d_e(0,0)) =$$
$$PQ_\Phi((x_1,0),(x_3,0)) + PQ_\Phi((x_2,0),(x_2,0)) \leq$$
$$PQ_\Phi((x_1,0),(x_2,0)) + PQ_\Phi((x_2,0),(x_3,0)) =$$
$$\Phi(p'_X(x_1,x_2), d_e(0,0)) + \Phi(p'_X(x_2,x_3), d_e(0,0)).$$

So $\Phi(c+d-b,0) + \Phi(b,0) \leq \Phi(c,0) + \Phi(d,0)$. Thus, in this case, $\Phi(c+d-b,\alpha) + \Phi(b,0) \leq \Phi(c,\beta) + \Phi(d,\gamma)$.

Case 2. The condition $\alpha = \beta = \gamma = 0$ does not hold, i.e., at least one element of $\{\alpha, \beta, \gamma\}$ is different of 0. Notice that $\alpha \leq \beta + \gamma$ and, thus, that either $\beta \neq 0$ or $\gamma \neq 0$. Suppose, without loss of generality, that $\beta \neq 0$.
Consider the quasi-metric space (Y, q'_Y) such that $Y = \{1, 2, 3\}$ and the quasi-metric q'_Y is defined as follows:
$q'_Y(1,3) = \alpha$, $q'_Y(1,2) = q'_Y(3,1) = q'_Y(3,2) = \beta$, $q'_Y(2,3) = q'_Y(2,1) = \gamma$ and $q'_Y(i,i) = 0$ for each $i \in \{1,2,3\}$. Then the function $PQ_\Phi : (X \times Y) \times (X \times Y) \to \mathbb{R}_+$ defined by $PQ_\Phi((x,y),(u,v)) = \Phi(p'_X(x,u), q'_Y(y,v))$, is a partial quasi-metric. Here (X, p'_X) is the partial metric introduced in the preceding Case 1. By axiom (PQ2) we have that

$$\Phi(p'_X(x_1,x_3), q'_Y(1,3)) + \Phi(p'_X(x_2,x_2), q'_Y(2,2)) =$$
$$PQ_\Phi((x_1,1),(x_3,3)) + PQ_\Phi((x_2,y_2),(2,2)) \leq$$
$$PQ_\Phi((x_1,1),(x_2,2)) + PQ_\Phi((x_2,2),(x_3,3)) =$$
$$\Phi(p'_X(x_1,x_2), q'_Y(1,2)) + \Phi(p'_X(x_2,x_3), q'_Y(2,3)).$$

Thus, $\Phi(c+d-b, \alpha) + \Phi(b,0) \leq \Phi(c,\beta) + \Phi(d,\gamma)$ in this case too.

Now, taking into account that $a \leq c+d-b$, we have, by Lemma 1, that $\Phi(a,\alpha) \leq \Phi(c+d-b,\alpha)$ and so $\Phi(a,\alpha) + \Phi(b,0) \leq \Phi(c+d-b,\alpha) + \Phi(b,0) \leq \Phi(c,\beta) + \Phi(d,\gamma)$, as we claimed.

Hence, we have proved that if Φ is a *pqmg*-function then it satisfies (i), (ii) and (iii).

Next we assume that $\Phi : \mathbb{R}^2_+ \to \mathbb{R}_+$ satisfies (i), (ii) and (iii). We will prove that Φ is a *pqmg*-function. To this end, (X, p) be a partial metric space and let (Y, q) be a quasi-metric space. Define the function $PQ_\Phi : (X \times Y) \times (X \times Y) \to \mathbb{R}_+$ by $PQ_\Phi((x,y),(u,v)) = \Phi(p(x,u), q(y,v))$ for each $(x,y),(u,v) \in X \times Y$. We will prove that PQ_Φ is a partial quasi-metric on $X \times Y$.

(PQ1). Let $(x_1,y_1),(x_2,y_2) \in X \times Y$. Since p is a partial metric on X, we have that $p(x_1,x_2) = p(x_2,x_1) \geq p(x_1,x_1)$. The fact that Φ fulfills (i) gives $\Phi(p(x_1,x_1),0) \leq \Phi(p(x_1,x_2), q(y_1,y_2))$. Thus

$$PQ_\Phi((x_1,y_1),(x_1,y_1)) = \Phi(p(x_1,x_1), q(y_1,y_1)) = \Phi(p(x_1,x_1),0) \leq$$
$$\Phi(p(x_1,x_2), q(y_1,y_2)) = PQ_\Phi((x_1,y_1),(x_2,y_2)).$$

Similarly we can show that $PQ_\Phi((x_1,y_1),(x_1,y_1)) \leq PQ_\Phi((x_2,y_2),(x_1,y_1))$, since $\Phi(p(x_1,x_1),0) \leq \Phi(p(x_1,x_2), q(y_2,y_1))$.

(PQ2). Let $(x_1,y_1),(x_2,y_2),(x_3,y_3) \in X \times Y$. Then $p(x_2,x_2) \leq \min\{p(x_1,x_2), p(x_2,x_3)\}$, $p(x_1,x_3) + p(x_2,x_2) \leq p(x_1,x_2) + p(x_2,x_3)$ and $q(y_1,y_3) \leq q(y_1,y_2) + q(y_2,y_3)$.
Since Φ satisfies (iii) we obtain that

$$\Phi(p(x_1,x_3), q(y_1,y_3)) + \Phi(p(x_2,x_2),0) \leq$$
$$\Phi(p(x_1,x_2), q(y_1,y_2)) + \Phi(p(x_2,x_3), q(y_2,y_3)).$$

Therefore,

$$PQ_\Phi((x_1,y_1),(x_3,y_3)) + PQ_\Phi((x_2,y_2),(x_2,y_2)) \leq$$
$$PQ_\Phi((x_1,y_1),(x_2,y_2)) + PQ_\Phi((x_2,y_2),(x_3,y_3)).$$

(PQ3). Obviously if $(x_1, y_1) = (x_2, y_2)$, then we have that $PQ_\Phi((x_1, y_1), (x_1, y_1)) = PQ_\Phi((x_1, y_1), (x_2, y_2))$ and $PQ_\Phi((x_2, y_2), (x_2, y_2)) = PQ_\Phi((x_2, y_2), (x_1, y_1))$. Conversely, suppose that $PQ_\Phi((x_1, y_1), (x_1, y_1)) = PQ_\Phi((x_1, y_1), (x_2, y_2))$ and, in addition, $PQ_\Phi((x_2, y_2), (x_2, y_2)) = PQ_\Phi((x_2, y_2), (x_1, y_1))$ for some $(x_1, y_1), (x_2, y_2) \in X \times Y$. It follows that $\Phi(p(x_1, x_1), 0) = \Phi(p(x_1, x_1), q(y_1, y_1)) = \Phi(p(x_1, x_2), q(y_1, y_2))$ and $\Phi(p(x_2, x_2), 0) = \Phi(p(x_2, x_2), q(y_2, y_2)) = \Phi(p(x_2, x_1), q(y_2, y_1))$. Since p is a partial metric on X we have that $\max\{p(x_1, x_1), p(x_2, x_2)\} \leq p(x_1, x_2) = p(x_2, x_1)$. The fact that Φ satisfies (ii) yields that

$$p(x_1, x_1) = p(x_1, x_2) = p(x_2, x_2) \text{ and } q(y_1, y_2) = q(y_2, y_1) = 0.$$

Whence we conclude that $x_1 = x_2$ and $y_1 = y_2$ and, hence, that $(x_1, y_1) = (x_2, y_2)$.

Therefore PQ_Φ is a partial quasi-metric on $X \times Y$.

The following result will be crucial in order to provide examples of pqmg-functions.

Lemma 2. *Let $a, b, c, d \in \mathbb{R}_+$ with $a + b \leq c + d$ and $b \leq \min\{c, d\}$. Then, $a \cdot b \leq c \cdot d$.*

Proof. First, we will show that $(c + d - b) \cdot b \leq c \cdot d$. To this end, we can suppose, with out loss of generality, that $d \geq c$. Then,

$$0 \leq (c - b)^2 \leq (c - b) \cdot (d - b) = c \cdot d - c \cdot b - d \cdot b + b^2 = c \cdot d - (c + d - b) \cdot b.$$

Therefore, $(c + d - b) \cdot b \leq c \cdot d$. It follows that $a \cdot b \leq (c + d - b) \cdot b \leq c \cdot d$, since $0 \leq a \leq c + d - b$.

The next result provides instances of pqmg-functions.

Proposition 1. *Let $M, N, L \in \mathbb{R}_+$. Then the following functions $\Phi : \mathbb{R}_+^2 \to \mathbb{R}_+$ are pqmg-functions:*

(1) $\Phi(x, y) = Mx + Ny + L$ *for each* $(x, y) \in \mathbb{R}_+^2$;
(2) $\Phi(x, y) = \sqrt{Mx + Ny + L}$ *for each* $(x, y) \in \mathbb{R}_+^2$;
(3) $\Phi(x, y) = \log(Mx + Ny + L)$ *for each* $(x, y) \in \mathbb{R}_+^2$ *and* $L \geq 1$.

Proof. (1) Consider the function given by $\Phi(x, y) = \sqrt{Mx + Ny + L}$. Next we show that it fulfills (i), (ii) and (iii) in Theorem 1.

(i) Suppose that $b \geq a$. Then, $\Phi(a, 0) = Ma + L \leq Mb + L \leq Mb + N\alpha + L = \Phi(b, \alpha)$.

(ii) Now, suppose that $c \geq \max\{a, b\}$. Then, on the one hand,

$$\Phi(a, 0) = Ma + L = Mc + N\alpha + L = \Phi(c, \alpha) \Leftrightarrow a = c \text{ and } \alpha = 0,$$

and, on the other hand,

$$\Phi(b, 0) = Mb + L = Mc + N\beta + L = \Phi(c, \beta) \Leftrightarrow b = c \text{ and } \beta = 0.$$

Therefore $\Phi(a, 0) = \Phi(c, \alpha)$ and $\Phi(b, 0) = \Phi(c, \beta)$ implies $a = b = c$ and $\alpha = \beta = 0$.

(iii) Finally, suppose that $b \leq \min\{c,d\}$, $a+b \leq c+d$ and $\alpha \leq \beta+\gamma$. Then

$$\Phi(a,\alpha)+\Phi(b,0) = Ma+N\alpha+L+Mb+L = M(a+b)+N\alpha+2L \\ \leq M(c+d)+N(\beta+\gamma)+2L = Mc+N\beta+L+Md+N\gamma+L = \\ \Phi(c,\beta)+\Phi(d,\gamma).$$

(2) Consider the function given by $\Phi(x,y)=\sqrt{Mx+Ny+L}$. We will only prove that Φ satisfies condition (iii) in Theorem 1, since the fact that Φ satisfies conditions (i) and (ii) can be proved following similar reasoning to those given in statement (1) in this result.
Suppose that $b \leq \min\{c,d\}$, $a+b \leq c+d$ and $\alpha \leq \beta+\gamma$. Then,

$$\Phi(a,\alpha)+\Phi(b,0) \leq \Phi(c,\beta)+\Phi(d,\gamma),$$

if and only if $\sqrt{Ma+N\alpha+L}+\sqrt{Mb+L} \leq \sqrt{Mc+N\beta+L}+\sqrt{Md+N\gamma+L}$, which is equivalent to $\left(\sqrt{Ma+N\alpha+L}+\sqrt{Mb+L}\right)^2 \leq \left(\sqrt{Mc+N\beta+L}+\sqrt{Md+N\gamma+L}\right)^2$, since $\sqrt{Ma+N\alpha+L}, \sqrt{Mb+L}, \sqrt{Mc+N\beta+L}, \sqrt{Md+N\gamma+L} \in \mathbb{R}_+$. Therefore, we need to prove that

$$Ma+N\alpha+L+Mb+L+2\cdot\left(\sqrt{Ma+N\alpha+L}\right)\cdot\left(\sqrt{Mb+L}\right) \leq \\ Mc+N\beta+L+Md+N\gamma+L+2\cdot\left(\sqrt{Mc+N\beta+L}\right)\cdot\left(\sqrt{Md+N\gamma+L}\right).$$

We have shown that $Ma+N\alpha+L+Mb+L \leq Mc+N\beta+L+Md+N\gamma+L$ in the proof of (1). So it remains to prove that

$$(Ma+N\alpha+L)\cdot(Mb+L) \leq (Mc+N\beta+L)\cdot(Md+N\gamma+L), \text{ or equivalently,} \\ M^2ab+MaL+N\alpha Mb+N\alpha L+LMb+L^2 \leq \\ \leq M^2cd+McN\gamma+McL+N\beta Md+N^2\beta\gamma+N\beta L+LMd+LN\gamma+L^2.$$

It is clear that

$$MaL+LMb = ML(a+b) \leq ML(c+d) = McL+LMd, \\ N\alpha Mb = NMb\alpha \leq NM\min\{c,d\}(\beta+\gamma) \leq McN\gamma+N\beta Md, \\ N\alpha L = NL\alpha \leq NL(\beta+\gamma) = N\beta L+LN\gamma.$$

By Lemma 2, we have that $M^2ab \leq M^2cd$, and so the proof is concluded.

(3) Consider the function given by $\Phi(x,y)=\sqrt{Mx+Ny+L}$ with $L \geq 1$. We will only prove that Φ satisfies condition (iii) in Theorem 1, since the fact that Φ satisfies conditions (i) and (ii) can be proved following similar reasoning to those given in statement (1) in this result.
It is clear that $\Phi(a,\alpha)+\Phi(b,0)=\log\left((Ma+N\alpha+L)\cdot(Mb+L)\right)$ and $\Phi(c,\beta)+\Phi(d,\gamma)=\log\left((Mc+N\beta+L)\cdot(Md+N\gamma+L)\right)$. Suppose that $b \leq \min\{c,d\}$, $a+b \leq c+d$ and $\alpha \leq \beta+\gamma$. Then, in statement (2) in this result we have shown that

$$(Ma+N\alpha+L)\cdot(Mb+L) \leq (Mc+N\beta+L)\cdot(Md+N\gamma+L).$$

Since log is an increasing function we have that $\Phi(a,\alpha)+\Phi(b,0) \leq \Phi(c,\beta)+\Phi(d,\gamma)$ and the proof is finished.

In the light of Proposition 1 we immediately obtain the following instances of partial quasi-metrics. The example below is relevant because of the lack of instances of partial quasi-metrics in the literature which limits its applicability.

Example 1. Let (X, p) be a partial metric space and let (X, q) be a quasi-metric space. If $M, N, L \in \mathbb{R}_+$, then the following functions are partial quasi-metrics on $X \times Y$.

1. $PQ_\Phi((x,y),(u,v)) = M \cdot p(x,u) + N \cdot q(y,v) + L$.
2. $PQ_\Phi((x,y),(u,v)) = \sqrt{M \cdot p(x,u) + N \cdot q(y,v) + L}$.
3. $PQ_\Phi((x,y),(u,v)) = \log(M \cdot p(x,u) + N \cdot q(y,v) + L)$ provided that $L \geq 1$.

Notice that the preceding example shows that the partial quasi-metric induced by a *pqmg*-function is not, in general, neither a partial metric nor a quasi-metric. Indeed, consider the partial metric space $(\mathbb{R}_+, p_m)$ and the quasi-metric space $(\mathbb{R}, q_u)$. Then we have that $PQ_\Phi((x,y),(u,v)) = \max\{x,u\} + \max\{v-y, 0\}$. Now take the *pqmg*-function $\Phi(a, \alpha) = a + \alpha$. Clearly $PQ_\Phi((1,1),(1,1)) = 1$ and, thus, PQ_Φ is not a quasi-metric. Moreover, we have that $PQ_\Phi((1,0),(0,1)) = 2$ and $PQ_\Phi((0,1),(1,0)) = 1$. So PQ_Φ is not a partial metric.

Observe that this is the reason for which the addition of the Baire partial and the complexity quasi-metric, $p_B + q_C$, is neither a partial metric nor a quasi-metric such as it has been pointed out in Sect. 1.

In [2], the functions that aggregate a collection of partial metrics into a new one were characterized. Such functions were called partial metric aggregation functions (*pma*-functions for short) and their characterization is given by the next theorem.

Theorem 2 [2]. *A function $\Phi : \mathbb{R}_+^2 \to \mathbb{R}_+$ is a pma-function if and only if it satisfies the following two properties for all $a, b, c, \alpha, \beta, \gamma, \delta \in \mathbb{R}_+$:*

(1) $\Phi(a,\alpha) + \Phi(b,\beta) \leq \Phi(c,\gamma) + \Phi(d,\delta)$ whenever $(a+b, \alpha+\beta) \preceq (c+d, \gamma+\delta)$, $(b,\beta) \preceq (c,\gamma)$ and $(b,\beta) \preceq (d,\delta)$.
(2) If $(b,\beta) \preceq (a,\alpha)$, $(c,\gamma) \preceq (a,\alpha)$ and $\Phi(a,\alpha) = \Phi(b,\beta) = \Phi(c,\gamma)$, then $(a,\alpha) = (b,\beta) = (c,\gamma)$.

Note that, the preceding characterization was presented for a collection of partial metrics in [2]. However, we have stated such a result only for two partial metrics with the aim of studying the relationship between *pma*-functions and *pqmg*-functions. In the next proposition we show that each *pma*-function is a *pqmg*-function. To this end, let us recall that, according to Proposition 9 in [2], every *pma*-funtion is increasing with respect to $\preceq$.

Proposition 2. *Let $\Phi : \mathbb{R}_+^2 \to \mathbb{R}_+$ be a pma-function. Then Φ is a pqmg-function.*

Proof. We will show that Φ satisfies conditions (i), (ii) and (iii) in Theorem 1.

(i) Suppose that $b \geq a$. Since $(a,0) \preceq (b,\alpha)$ and Φ is increasing we obtain that $\Phi(a,0) \leq \Phi(b,\alpha)$.
(ii) Now, suppose that $c \geq \max\{a,b\}$ and, $\Phi(a,0) = \Phi(c,\alpha)$ and $\Phi(b,0) = \Phi(c,\beta)$. The monotony of Φ gives that $\Phi(c,0) \leq \Phi(c,\alpha) = \Phi(a,0) \leq \Phi(c,0)$. It follows that $\Phi(a,0) = \Phi(c,\alpha) = \Phi(c,0)$ with $(a,0) \preceq (c,\alpha)$ and $(c,0) \preceq (c,\alpha)$. By assertion (2) in Theorem 2 we deduce that $(a,0) = (c,\alpha) = (c,0)$. Thus $a = c$ and $\alpha = 0$. Analogously, we can prove that $b = c$ and $\beta = 0$. Thus, $a = b = c$ and $\alpha = \beta = 0$.
(iii) Suppose that $b \leq \min\{c,d\}$, $a+b \leq c+d$ and $\alpha \leq \beta+\gamma$. Then $(a,\alpha)+(b,0) \preceq (c,\beta)+(d,\gamma)$, $(b,0) \preceq (c,\beta)$ and $(b,0) \preceq (d,\gamma)$. Since Φ satisfies assertion (1) in Theorem 2 we deduce that $\Phi(a,\alpha) + \Phi(b,0) \leq \Phi(c,\beta) + \Phi(d,\gamma)$.

The following example shows that there are *pqmg*-functions which are not *pma*-function.

Example 2. Consider the function $\Phi : \mathbb{R}_+^2 \to \mathbb{R}_+$ given by

$$\Phi(a,\alpha) = \begin{cases} a, & \text{if } \alpha = 0 \\ a+1, & \text{if } \alpha \neq 0 \end{cases}.$$

It is not hard to check that Φ is a *pqmg*-function. Nevertheless, Φ is not a *pma*-function. Indeed, $(1,1) \preceq (1,3)$, $(1,2) \preceq (1,3)$ and $\Phi(1,3) = \Phi(1,1) = \Phi(1,2) = 2$ but $(1,3) \neq (1,2)$. Therefore, Φ does not satisfy assertion (2) in Theorem 2 and, hence, it is not a *pma*-function.

In [4], the functions that aggregate a collection of quasi-metrics into a new one were characterized. Such functions were called quasi-metric aggregation functions (*qma*-functions for short) and their characterization is given by the next theorem.

Theorem 3 [4]. *Let $a,b,c,\alpha,\beta,\gamma \in \mathbb{R}_+$. A function $\Phi : \mathbb{R}_+^2 \to \mathbb{R}_+$ is a qma-function if and only it has the following properties:*

(1) $\Phi^{-1}(0) = \{(0,0)\}$.
(2) If $(a,\alpha) \preceq (b,\beta) + (c,\gamma)$, *then* $\Phi(a,\alpha) \leq \Phi(b,\beta) + \Phi(c,\gamma)$.

As in the partial metric case, the preceding characterization was presented for a collection of quasi-metrics in [4]. However, we have stated such a result only for two partial quasi-metrics with the aim of studying the relationship between *qma*-functions and *pqmg*-functions.

Concerning the relationship between *qma*-functions and *pqmg*-functions, the situation is different. Indeed, we will show that both classes of functions are not comparable. The next examples show that there are *qma*-functions that are not *pqmg*-functions and vice-versa.

Example 3. Consider the function $\Phi : \mathbb{R}_+^2 \to \mathbb{R}_+$ given by $\Phi(a,\alpha) = \max\{a,\alpha\}$. By Corollary 13 in [4], Φ is a *qma*-function. However, it is not a *pqmg*-function. Indeed, take $a = b = c = 2$, $\alpha = 1$ and $\beta = 2$. Clearly $\Phi(a,0) = \Phi(c,\alpha) = 2$ and $\Phi(b,0) = \Phi(c,\beta) = 2$ but $\alpha = 1 \neq 2 = \beta$. Thus, Φ does not fulfill assertion *(ii)* in Theorem 1. Hence, it is not a *pqmg*-function.

Example 4. According to Proposition 1, the function $\Phi : \mathbb{R}_+^2 \to \mathbb{R}_+$ given by $\Phi(a, \alpha) = a + \alpha + 1$ is a *pqmg*-function. Nevertheless it is not a *qma*-function, since $\Phi(0, 0) = 1$ and so Φ does not satisfy assertion (1) in Theorem 3.

The next result provides a sufficient condition that ensures that a *pqmg*-function is a *qma*-function.

Proposition 3. *Let $\Phi : \mathbb{R}_+^2 \to \mathbb{R}_+$ be a pqmg-function such that $\Phi(0, 0) = 0$. Then, Φ is a qma-function.*

Proof. We will see that Φ fulfills assertions (1) and (2) in Theorem 3. Let $a, \alpha \in \mathbb{R}_+$ such that $\Phi(a, \alpha) = 0$. By Lemma 1 we deduce that $\Phi(a, \frac{\alpha}{2}) = 0$, since $0 = \Phi(0, 0) \leq \Phi(a, \frac{\alpha}{2}) \leq \Phi(a, \alpha) = 0$. Whence we have that $\Phi(a, \alpha) = \Phi(0, 0)$ and $\Phi(a, \frac{\alpha}{2}) = \Phi(0, 0)$. Then, by assertion (*ii*) in Theorem 1, we obtain that $a = 0$ and that $\alpha = 0$. Thus $\Phi^{-1}(0) = \{(0, 0)\}$.

Let $a, b, c, \alpha, \beta, \gamma \in \mathbb{R}_+$ such that $(a, \alpha) \preceq (b, \beta) + (c, \gamma)$. It follows that $\alpha \leq \beta + \gamma$, $a + 0 \leq b + c$ and that $0 \leq \min\{b, c\}$. By assertion (*iii*) in Theorem 1 and the fact that $\Phi(0, 0) = 0$, we have that $\Phi(a, \alpha) + \Phi(0, 0) \leq \Phi(c, \beta) + \Phi(d, \gamma)$.

Hence, by Theorem 3 we have that Φ is a *qma*-function.

Example 3 shows that the converse of Proposition 3 is not true, in general.

Notice that, by Proposition 3, every *pqmg*-function such that $\Phi(0, 0) = 0$ is a metric aggregation function (see [4]).

3 Conclusions

Partial metrics and quasi-metrics have been shown to be useful to develop quantitative mathematical models in denotational semantics and in asymptotic complexity analysis of algorithms, respectively. The aforesaid models are implemented independently and they are not related. A first natural attempt to develop a framework which remains valid for modeling in denotational semantics and in complexity analysis of algorithms suggests to construct a generalized metric by means of the aggregation of a partial metric and a quasi-metric. Inspired by the preceding fact, we have studied the way of merging, by means of a function, the aforementioned generalized metrics into a new one. We have showed that the induced generalized metric matches up with a partial quasi-metric. Thus, we have characterized those functions that allow to generate partial quasi-metrics from the combination of a partial metric and a quasi-metric. Moreover, the have explored the relationship between the problem under consideration and the problems of merging partial metrics and quasi-metrics. The exposed theory has been illustrated by appropriate examples.

Acknowledgements. This research was partially supported by the Spanish Ministry of Economy and Competitiveness under Grant TIN2016-81731-REDT (LODISCO II) and AEI/FEDER, UE funds, by the Programa Operatiu FEDER 2014-2020 de les Illes Balears, by project ref. PROCOE/4/2017 (Direccio General d'Innovacio i Recerca, Govern de les Illes Balears), and by project ROBINS. The latter has received research funding from the EU H2020 framework under GA 779776. This publication reflects only the authors views and the European Union is not liable for any use that may be made of the information contained therein.

References

1. Künzi, H.-P.A., Pajoohesh, H., Schellekens, M.P.: Partial quasi-metrics. Theor. Comput. Sci. **365**(3), 237–246 (2006). https://doi.org/10.1016/j.tcs.2006.07.050
2. Massanet, S., Valero, O.: New results on metric aggregation. In: Proceedings of the 17th Spanish Conference on Fuzzy Technology and Fuzzy Logic (Estylf), Valladolid, pp. 558–563 (2012)
3. Matthews, S.G.: Partial metric topology. Ann. N. Y. Acad. Sci. **728**, 183–197 (1994). https://doi.org/10.1111/j.1749-6632.1994.tb44144.x
4. Mayor, G., Valero, O.: Aggregation of asymmetric distances in computer science. Inf. Sci. **180**, 803–812 (2010). https://doi.org/10.1016/j.ins.2009.06.020
5. Schellekens, M.: The Smyth completion: a common foundation for denotational semantics and complexity analysis. Electron. Notes Theor. Comput. Sci. **1**, 211–232 (1995). https://doi.org/10.1016/S1571-0661(04)00029-5
6. Scott, D.S.: Outline of a mathematical theory of computation. In: Proceedings of the 4th Annual Princeton Conference on Information Sciences and Systems, Princeton, pp. 169–176 (1970)

Generalized Farlie-Gumbel-Morgenstern Copulas

Anna Kolesárová[1], Radko Mesiar[2(✉)], and Susanne Saminger-Platz[3]

[1] Faculty of Chemical and Food Technology, Slovak University of Technology, Radlinského 9, 812 37 Bratislava, Slovakia
anna.kolesarova@stuba.sk

[2] Faculty of Civil Engineering, Slovak University of Technology, Radlinského 11, 810 05 Bratislava, Slovakia
radko.mesiar@stuba.sk

[3] Department of Knowledge-Based Mathematical Systems, Johannes Kepler University, 4040 Linz, Austria
susanne.saminger-platz@jku.at

Abstract. The Farlie-Gumbel-Morgenstern copulas are related to the independence copula Π and can be seen as perturbations of Π. Based on quadratic constructions of copulas, we provide a new look at them. Starting from any 2-dimensional copula and an appropriate real function, we introduce new parametric families of copulas which in the case of the independence copula Π coincide with the Farlie-Gumbel-Morgenstern family. Using the proposed approach, we also obtain as a particular case a subclass of the Fréchet family of copulas containing all three basic copulas W, Π and M, i.e. a comprehensive family of copulas. Finally, based on an iterative approach, we introduce copula families $(C_r)_{r\in[-\infty,\infty]}$ complete w.r.t. dependence parameters, resulting in the case of the independence copula and parameters $r \in [-1,1]$ in the Farlie-Gumbel-Morgenstern family.

Keywords: Copula · FGM copula · Comprehensive family of copulas
Quadratic construction of copulas
Copula family complete w.r.t. dependence parameters
Extended FGM family

1 Introduction

One of the most frequently used copula families—both in theory and applications—is the Farlie-Gumbel-Morgenstern family (FGM family for short), see, e.g., [10,19]. If the FGM family is denoted by $\mathfrak{C}^{FGM}$ and its members by C_θ^{FGM}, then we can write $\mathfrak{C}^{FGM} = \left(C_\theta^{FGM}\right)_{\theta\in[-1,1]}$, where

$$C_\theta^{FGM}(x,y) = xy + \theta xy(1-x)(1-y). \tag{1}$$

J. Medina et al. (Eds.): IPMU 2018, CCIS 853, pp. 244–252, 2018.
https://doi.org/10.1007/978-3-319-91473-2_21

Several generalizations of this family, including the related dependence parameters, were introduced and studied in [2,3,12,13,15]. Except for the proposals concerning the perturbation of copulas in [12,13] all these generalizations are related to some modifications of the independence copula Π. Inspired by [12,13] and using an approach based on quadratic constructions of copulas recently introduced in [11], we propose a rather general method for constructing parametric families of copulas based on an arbitrary copula C and a function $f\colon [-1,1] \to [0,1]$ satisfying some mild restrictions. When the considered copula C is equal to Π, then all newly proposed families coincide with the FGM family, which justifies to call these copulas generalized FGM copulas. However, neither the FGM copulas nor the proposed generalizations can cover the full possible ranges of dependence parameters. For example, the ranges of Kendall's tau and Spearman's rho in the case of the FGM copulas are equal to the interval $[-1/3, 1/3]$ while the full generally possible ranges of these parameters are $[-1, 1]$. To remove this drawback, we extend our proposal and define families of copulas $(C_r)_{r\in[-\infty,\infty]}$ which are so-called complete w.r.t. dependence parameters and generalize the FGM family. These families are increasing in their parameter and fulfil $C_{-\infty} = W$ and $C_\infty = M$.

The paper is organized as follows. In the next section, some preliminary notions concerning copulas as well as quadratic constructions of copulas are given. In Sect. 3, parametric families $\left(C_{\theta,f(\theta)}\right)_{\theta\in[-1,1]}$ of copulas generalizing the FGM family are introduced and exemplified. Note that as a particular case, a subset of the Fréchet copulas [10,19] is obtained. In Sect. 4 we introduce parametric families $(C_r)_{r\in[-\infty,\infty]}$ of copulas which are complete w.r.t. dependence parameters of copulas and contain members $C_{-\infty} = W$, $C_\infty = M$, and $C_0 = C$, where C is an a priori given copula. Finally, some concluding remarks are provided.

2 Quadratic Constructions of Copulas

In this contribution we will deal with 2-dimensional copulas only. We will call them simply copulas. The class of all copulas will be denoted by $\mathfrak{C}$. Recall that a copula is a bivariate cumulative distribution function (restricted to $[0,1]^2$) with uniform $[0,1]$ marginals capturing the dependence properties of two random variables. Formally, a function $C\colon [0,1]^2 \to [0,1]$ is called a copula if

(i) $C(x,0) = C(0,x) = 0$ and $C(x,1) = C(1,x) = x$ for each $x \in [0,1]$ (boundary conditions),
(ii) $C(x_2,y_2) - C(x_1,y_2) + C(x_1,y_1) - C(x_2,y_1) \geq 0$ for all $0 \leq x_1 \leq x_2 \leq 1$ and $0 \leq y_1 \leq y_2 \leq 1$ (2-increasing property).

Due to the Sklar theorem [23], for any random vector $Z = (X, Y)$ there is a copula $C\colon [0,1]^2 \to [0,1]$ such that for all $u, v \in \mathbb{R}$ we have

$$F_Z(u,v) = C\left(F_X(u), F_Y(v)\right). \tag{2}$$

The symbols F_Z, F_X, F_Y denote the respective distribution functions of random variables Z, X and Y. Formula (2) is not only a representation but a construction as well. In fact, for any 1-dimensional distribution functions F, G and any copula C, the function $H\colon \mathbb{R}^2 \to [0,1]$ given by $H(u,v) = C(F(u), G(v))$ is a 2-dimensional distribution function.

Recall that for each copula C we have $W \le C \le M$, where W denotes the lower Fréchet-Hoeffding bound (the copula of countermonotone dependence) given by $W(x,y) = \max\{0, x+y-1\}$, and M the upper Fréchet-Hoeffding bound (the copula of comonotone dependence) given by $M(x,y) = \min\{x,y\}$. The third basic copula is the copula Π given by $\Pi(x,y) = xy$ (the copula of independence). For more details on copulas we recommend, e.g., the monographs [10,19].

In the literature one can observe lots of effort devoted to the proposals of new construction methods of copulas [4,6–8,14,16–18,20,22]. Here we only focus on quadratic constructions which have recently been introduced in [11].

Theorem 2.1 ([11]). *Let $\Omega \subset \mathbb{R}^2$ be a convex set generated by the vertices $(-1,0)$, $(0,0)$, $(1,1)$ and $(0,1)$ and let $C\colon [0,1]^2 \to [0,1]$ be a copula. Then, for any $(c,d) \in \Omega$, the function $C_{c,d}\colon [0,1]^2 \to [0,1]$ given by*

$$C_{c,d}(x,y) = cC^2(x,y) - cxC(x,y) - cyC(x,y) + dxy + (1+c-d)C(x,y) \quad (3)$$

is a copula.

The set Ω can be described as $\Omega = \{(c,d) \in \mathbb{R}^2 \mid 0 \le d \le 1,\ 0 \le d-c \le 1\}$ and is illustrated in Fig. 1 (left).

Formula (3) allows to introduce a 2-parametric family $(C_{c,d})_{(c,d)\in\Omega}$ of copulas and several 1-parametric subfamilies. From many distinguished copula families let us recall at least the following two ones:

- the Fréchet family $\mathfrak{C}^F = \left(C^F_{a,b}\right)_{(a,b)\in\Gamma}$, where

  $$C^F_{a,b}(x,y) = aW + bM + (1-a-b)\Pi \text{ and } \Gamma = \{(a,b) \in [0,1]^2 \mid a+b \le 1\};$$

- the Farlie-Gumbel-Morgenstern family $\mathfrak{C}^{FGM} = \left(C^{FGM}_\theta\right)_{\theta\in[-1,1]}$ given in (1).

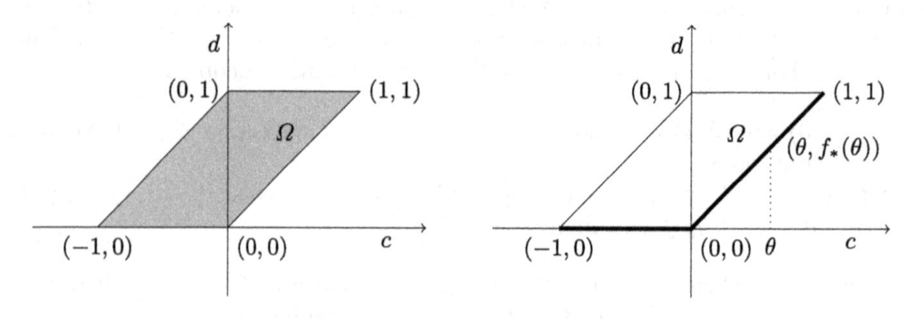

Fig. 1. The set Ω of all possible parameters (c,d) in quadratic constructions of copulas (left) and the illustration of parameters $(\theta, f_*(\theta))$, $\theta \in [-1,1]$, corresponding to the copula given in (5) (right)

3 Generalized FGM Copulas

Let $\mathcal{F}$ be the set of all functions $f\colon [-1,1] \to [0,1]$ such that, for each $c \in [-1,1]$, we have $(c, f(c)) \in \Omega$. Observe that $\mathcal{F}$ is a convex class which is also a lattice with the top element f^* and bottom element f_*, given by

$$f^*(c) = \min\{c+1, 1\}, \quad f_*(c) = \max\{0, c\}.$$

Based on Theorem 2.1, we have the following corollary.

Corollary 3.1. *For any fixed $C \in \mathfrak{C}$ and $f \in \mathcal{F}$, the system*

$$\mathfrak{C}_{C,f} = \big(C_{\theta,f(\theta)}\big)_{\theta\in[-1,1]} \tag{4}$$

is a family of copulas.

Let us now consider the independence copula Π, i.e., $C = \Pi$, and any $f \in \mathcal{F}$. Then

$$\begin{aligned}\Pi_{\theta,f(\theta)}(x,y) &= \theta x^2y^2 - \theta x^2 y - \theta x y^2 + f(\theta)xy + (1 - f(\theta) + \theta)xy \\ &= xy + \theta xy(1-x)(1-y) = C_\theta^{FGM}(x,y),\end{aligned}$$

i.e., $\mathfrak{C}_{\Pi,f} = \mathfrak{C}^{FGM}$, independently of $f \in \mathcal{F}$. This fact justifies the following definition.

Definition 3.1. *Let $C \in \mathfrak{C}$ and $f \in \mathcal{F}$. Then the family $\mathfrak{C}_{C,f}$ given in (4) is called a (C, f)-generalization of the FGM family.*

Remark 3.1. Observe that the family $\mathfrak{C}_{C,f_*} = \big(C_{\theta,f_*(\theta)}\big)_{\theta\in[-1,1]}$ is given by

$$C_{\theta,f_*(\theta)} = \begin{cases} C_{\theta,0} & \text{if } \theta \le 0, \\ C_{\theta,\theta} & \text{if } \theta \ge 0, \end{cases}$$

i.e.,

$$C_{\theta,f_*(\theta)}(x,y) = \begin{cases} C(x,y) + \theta C(x,y)(C(x,y) + 1 - x - y) & \text{if } \theta \le 0, \\ C(x,y) + \theta(x - C(x,y))(y - C(x,y)) & \text{if } \theta \ge 0. \end{cases} \tag{5}$$

This family was discussed in [12,13], and its related parameters $(\theta, f_*(\theta))$ are visualized in Fig. 1 (right).

Note that no family $\mathfrak{C}_{C,f}$ can cover the full range $[-1,1]$ of the dependence parameters such as Kendall's tau or Spearman's rho. For example, in the case of the FGM family both these parameters attain their values in the interval $[-1/3, 1/3]$, while for the Blomquist beta the interval $[-1/4, 1/4]$ is obtained. The above claim is based on the following facts:

- Each copula $C_{c,d}$ is a convex combination of four copulas $C_{-1,0}, C_{0,0} = C, C_{1,1}$ and $C_{0,1} = \Pi$.

- If $C = M$ then $M_{-1,0} = \Pi$ and thus any copula $M_{c,d} \geq \Pi$.
- If $C = W$ then $W_{1,1} = \Pi$ and thus any copula $W_{c,d} \leq \Pi$.
- If $C \notin \{M, W\}$ then both $C_{1,1}$ and $C_{-1,0} \notin \{M, W\}$ and thus any $C_{c,d} \notin \{M, W\}$.

As an additional fact, it is not difficult to check that for each copula $C \neq W$ we have $C_{-1,0} < C$ and for each copula $C \neq M$, $C < C_{1,1}$. Thus, for any copula C we have $C_{-1,0} < C_{1,1}$. Based on this property, for any $C \in \mathfrak{C}$ and the function $f \in \mathcal{F}$ given by $f(c) = \frac{c+1}{2}$, we obtain that the generalized FGM family $\mathfrak{C}_{C,f} = \left(C_{\theta,f(\theta)}\right)_{\theta\in[-1,1]}$ is continuous and strictly increasing in parameter θ.

On the other hand, for any continuous functions $f, g \in \mathcal{F}$, the families $\left(M_{a,f(a)}\right)_{a\in[-1,1]}$ and $\left(W_{b,g(b)}\right)_{b\in[-1,1]}$ are continuous in parameters, and it holds that $M_{-1,f(-1)} = \Pi = W_{1,g(1)}$. Then the family $\mathcal{M}_{f,g} = (K_\theta)_{\theta\in[-1,1]}$ of copulas defined by

$$K_\theta = \begin{cases} W_{2\theta+1,g(2\theta+1)} & \text{if } \theta \leq 0, \\ M_{2\theta-1,f(2\theta-1)} & \text{if } \theta \geq 0, \end{cases}$$

is a family continuous in parameter θ, such that $K_{-1} = W$, $K_0 = \Pi$, and $K_1 = M$. Note that copula families containing all three basic copulas W, Π and M are called comprehensive [19]. Evidently, for this family, the ranges of all dependence parameters are full, i.e., when considering all members of the family, each dependence parameter attains all its theoretically possible values.

Definition 3.2. *A copula family with full ranges of dependence parameters will be called a complete family w.r.t. dependence parameters (a complete family for short).*

The introduced family $\mathcal{M}_{f,g}$ is a subfamily of the Fréchet family $\mathfrak{C}^F$. Indeed,

$$K_\theta = \begin{cases} g(2\theta+1)\Pi + (1 - g(2\theta+1))W & \text{if } \theta \leq 0, \\ (f(2\theta-1) - 2\theta + 1)\Pi + (2\theta - f(2\theta-1))M & \text{if } \theta \geq 0. \end{cases}$$

Example 3.1. Consider the functions $f, g \in \mathcal{F}$, given by $f(c) = g(c) = \frac{c+1}{2}$, $c \in [-1, 1]$. Then

$$K_\theta = \max\{0, -\theta\}W + \max\{0, \theta\}M + (1 - |\theta|)\Pi = C^F_{\max\{0,-\theta\},\max\{0,\theta\}}$$

4 Complete Families of Copulas

In this section, based on the generalized FGM families of copulas proposed in Sect. 3, we introduce some complete families of copulas.

For an arbitrary copula C, we set:

$$C_0 = C, \quad C_1 = C_{1,1}, \quad C_{-1} = C_{-1,0}.$$

For any $n \in \mathbb{N}$ we define recursively:

$$C_{n+1} = (C_n)_1 \quad \text{and} \quad \mathrm{C}_{-(n+1)} = (\mathrm{C}_{-n})_{-1}. \tag{6}$$

Then $(C_n)_{n\in Z}$, where Z is the set of all integers, is an increasing family of copulas, and

$$C_\infty = \sup\{C_n \mid n \in \mathbb{N}\} = M, \quad C_{-\infty} = \inf\{C_{-n} \mid n \in \mathbb{N}\} = W.$$

For any $r \geq 0$, we put:

$$C_r = \left(C_{\lfloor r \rfloor}\right)_{r-\lfloor r \rfloor, f_*(r-\lfloor r \rfloor)} = \left(C_{\lfloor r \rfloor}\right)_{r-\lfloor r \rfloor, r-\lfloor r \rfloor},$$

where $\lfloor r \rfloor$ is the floor of r. Obviously, if $r \in \mathbb{N}$ then $C_r = C_{\lfloor r \rfloor}$ as defined in (6).
Similarly, for $r \leq 0$, we put

$$C_r = \left(C_{\lceil r \rceil}\right)_{r-\lceil r \rceil, f_*(r-\lceil r \rceil)} = \left(C_{\lceil r \rceil}\right)_{r-\lceil r \rceil, 0},$$

where $\lceil r \rceil$ is the ceiling of r. Again, if $-r \in \mathbb{N}$ then $C_r = C_{\lceil r \rceil}$ as defined in (6).

Theorem 4.1. *Let $C\colon [0,1]^2 \to [0,1]$ be a fixed copula. Then the family of copulas $(C_r)_{r\in[-\infty,\infty]}$ is complete, continuous and increasing in parameter.*

Consider the independence copula $C = \Pi$. Then $(\Pi_r)_{r\in[-\infty,\infty]}$ is a comprehensive family which is complete, continuous and strictly increasing in parameter, $\Pi_{-\infty} = W$, $\Pi_\infty = M$, and for each $r \in [-1,1]$, $\Pi_r = C_r^{FGM}$. Therefore this family deserves to be called a complete extended Farlie-Gumbel-Morgenstern family of copulas.

Further, considering $C = M$, we obtain the family $(M_r)_{r\in[-\infty,\infty]}$ satisfying the properties $M_r = M$ for any $r \in [0,\infty]$, $M_{-1} = \Pi$, and $M_{-\infty} = W$. Thus the family $(M_r)_{r\in[-\infty,0]}$ is also comprehensive, complete, continuous and strictly increasing in parameter.

Similarly, $W_r = W$ whenever $r \leq 0$, $W_1 = \Pi$, and $W_\infty = M$. Hence the family $(W_r)_{r\in[0,\infty]}$ is comprehensive, complete, continuous and strictly increasing in parameter.

Remark 4.1

(i) For any $r \leq -1$, we have $M_r = \Pi_{r+1}$. Similarly, for any $r \geq 1$, $W_r = \Pi_{r-1}$. Thus the family $(M_r)_{r\in[-\infty,0]}$ contains the negative quadrant dependent part of the FGM family, while $(W_r)_{r\in[0,\infty]}$ contains its positive quadrant dependent part.

(ii) All proper members Π_r of the extended FGM family $(\Pi_r)_{r\in[-\infty,\infty]}$ (i.e., for $r \in]-\infty,\infty[$) are polynomial copulas. For example,

$$\begin{aligned}
\Pi_0(x,y) &= xy, \\
\Pi_1(x,y) &= xy(1 + (1-x)(1-y)), \\
\Pi_2(x,y) &= xy\left(1 + (1-x)(1-y)(1 + (1-x+xy)(1-y+xy))\right), \\
\Pi_{-1}(x,y) &= xy(1 - (1-x)(1-y)), \\
\Pi_{-2}(x,y) &= xy(1 - (1-x)(1-y))(x^2y^2 + (x+y)(1-xy)), etc.
\end{aligned}$$

(iii) In a similar way we can define complete extended FGM families $\left(\Pi^f_{(r)}\right)_{r\in[-\infty,\infty]}$, where $f \in \mathcal{F}$ is a fixed continuous function satisfying $f(0) = 0$. For example, consider $f \in \mathcal{F}$ given by

$$f(c) = \begin{cases} -c - c^2 & \text{if } c \le 0, \\ 2c - c^2 & \text{if } c \ge 0. \end{cases}$$

For an integer r or $r \in \{-\infty, \infty\}$ we put $\Pi^f_{(r)} = \Pi_r$. For non-integer real r we define:

- if $r > 0$ then $\Pi^f_{(r)} = \left(\Pi_{\lfloor r\rfloor}\right)_{r-\lfloor r\rfloor, f(r-\lfloor r\rfloor)}$,
- if $r < 0$ then $\Pi^f_{(r)} = \left(\Pi_{\lceil r\rceil}\right)_{r-\lceil r\rceil, f(r-\lceil r\rceil)}$.

Note that independently of f, $\left(\Pi^f_{(r)}\right)_{r\in[-1,1]}$ is just the FGM family of copulas. Evidently, the family $\left(\Pi^f_{(r)}\right)_{r\in[-\infty,\infty]}$ is complete and continuous in parameter. However, it need not be monotone in parameter.

5 Concluding Remarks

We have introduced and discussed several new families of copulas which generalize the famous Farlie-Gumbel-Morgenstern family. Our approach was based on quadratic constructions of copulas as introduced in [11]. Based on an iterative approach, we have introduced several copula families complete w.r.t. dependence parameters, i.e., families whose members can attain any possible value of the standard dependence parameters—such as Kendall's tau, Spearman's rho, Gini's gamma, etc. Moreover, we have introduced complete extended FGM families $(\Pi_r)_{r\in[-\infty,\infty]}$ and $\left(\Pi^f_{(r)}\right)_{r\in[-\infty,\infty]}$ which for $r \in [-1, 1]$ coincide with the FGM family, and if r is an integer, $\Pi^f_{(r)} = \Pi_r$. Our approach can easily be implemented into several fitting softwares. For example, all copulas Π_r and $\Pi^f_{(r)}$, $r \in]-\infty, \infty[$, are polynomials and thus absolutely continuous copulas which is an aspect rather promising for applications.

Acknowledgments. The first two authors kindly acknowledge the support of the project of Science and Technology Assistance Agency under the contract No. APVV–14–0013. The work of A. Kolesárová was also supported by the grant VEGA 1/0891/17. The second and third author also acknowledge the support of the "Technologie-Transfer-Förderung" of the Upper Austrian Government (Wi-2014-200710/13-Kx/Kai).

References

1. Amblard, C., Girard, S.: Symmetry and dependence properties within a semiparametric family of bivariate copulas. J. Nonparametric Stat. **14**, 715–727 (2002). arXiv:1103.5953
2. Bekrizadeh, H., Parham, A.G., Zadkarmi, R.M.: The new generalization of Farlie-Gumbel-Morgenstern copulas. Appl. Math. Sci. **6**, 3527–3533 (2012)
3. Cuadras, C.M., Díaz, W.: Another generalizations of the bivariate FGM distribution with two-dimensional extensions. Acta et Commentationes Univ. Tartuensis de Math. **16**(1), 3–12 (2012)
4. De Baets, B., De Meyer, H.: Orthogonal grid constructions of copulas. IEEE Trans. Fuzzy Syst. **15**, 1053–1062 (2007). https://doi.org/10.1109/TFUZZ.2006.890681
5. Dolati, A., Úbeda-Flores, M.: Constructing copulas by means of pairs of order statistics. Kybernetika **45**, 992–1002 (2009)
6. Durante, F., Saminger-Platz, S., Sarkoci, P.: On patchwork techniques for 2-increasing aggregation functions and copulas. In: Dubois, D., Lubiano, M.A., Prade, H., Gil, M.Á., Grzegorzewski, P., Hryniewicz, O. (eds.) Soft Methods for Handling Variability and Imprecision. Advances in Soft Computing, vol. 48, pp. 349–356. Springer, Heidelberg (2008). https://doi.org/10.1007/978-3-540-85027-4_42
7. Durante, F., Saminger-Platz, S., Sarkoci, P.: Rectangular patchwork for bivariate copulas and tail dependence. Commun. Stat. Theory Methods **38**, 2515–2527 (2009). https://doi.org/10.1080/03610920802571203
8. Durante, F., Rodríguez-Lallena, J.A., Úbeda-Flores, M.: New constructions of diagonal patchwork copulas. Inf. Sci. **179**, 3383–3391 (2009). https://doi.org/10.1016/j.ins.2009.06.007
9. Huang, J.S., Kotz, S.: Modifications of the Farlie-Gumbel-Morgenstern distributions. A tough hill to climb. Metrika **49**, 135–145 (1999). https://doi.org/10.1007/s001840050030
10. Joe, H.: Multivariate Model and Dependence Concept. Monographs on Statistics and Applied Probability, vol. 73. Chapman and Hall, London (1997)
11. Kolesárová, A., Mayor, G., Mesiar, R.: Quadratic constructions of copulas. Inf. Sci. **310**, 69–76 (2015). https://doi.org/10.1016/j.ins.2015.03.016
12. Komorník, J., Komorníková, M., Kalická, J.: Families of perturbation copulas generalizing the FGM family and their relations to dependence measures. In: Torra, V., Mesiar, R., De Baets, B. (eds.) AGOP 2017. AISC, vol. 581, pp. 53–63. Springer, Cham (2018). https://doi.org/10.1007/978-3-319-59306-7_6
13. Komorník, J., Komorníková, M., Kalická, J.: Dependence measures for perturbations of copulas. Fuzzy Sets Syst. **324**, 100–116 (2017). https://doi.org/10.1016/j.fss.2017.01.014
14. Mesiar, R., Jágr, V., Juráňová, M., Komorníková, M.: Univariate conditioning of copulas. Kybernetika **44**, 807–816 (2008)
15. Mesiar, R., Najjari, V.: New families of symmetric/asymmetric copulas. Fuzzy Sets Syst. **252**, 99–110 (2014). https://doi.org/10.1016/j.fss.2013.12.015
16. Mesiar, R., Sempi, C.: Ordinal sums and idempotents of copulas. Aequationes Math. **79**, 39–52 (2010). https://doi.org/10.1007/s00010-010-0013-6
17. Mesiar, R., Szolgay, J.: W-ordinal sums of copulas and quasi-copulas. In: Magia 2004, Conference, Kočovce, pp. 78–83 (2004)
18. Morillas, P.M.: A method to obtain new copulas from a given one. Metrika **61**, 169–184 (2005). https://doi.org/10.1007/s001840400330

19. Nelsen, R.B.: An Introduction to Copulas, 2nd edn. Springer, New York (2006). https://doi.org/10.1007/0-387-28678-0
20. Nelsen, R.B., Quesada-Molina, J.J., Rodríguez-Lallena, J.A., Úbeda-Flores, M.: On the construction of copulas and quasi-copulas with given diagonal sections. Insur. Math. Econ. **42**, 473–483 (2008). https://doi.org/10.1016/j.insmatheco.2006.11.011
21. Rodríguez-Lallena, J.A., Úbeda-Flores, M.: A new class of bivariate copulas. Stat. Probab. Lett. **66**, 315–325 (2004). https://doi.org/10.1016/j.spl.2003.09.010
22. Siburg, K.F., Stoimenov, P.A.: Gluing copulas. Commun. Stat. Theory Methods **37**, 3124–3134 (2008). https://doi.org/10.1080/03610920802074844
23. Sklar, A.: Fonctions de répartition à n dimensions et leurs marges. Publ. Inst. Stat. Univ. Paris **8**, 229–231 (1959)

Extracting Decision Rules from Qualitative Data via Sugeno Utility Functionals

Quentin Brabant[1(✉)], Miguel Couceiro[1], Didier Dubois[2], Henri Prade[2], and Agnès Rico[3]

[1] Université de Lorraine, CNRS, Inria, LORIA, 54000 Nancy, France
{quentin.brabant,miguel.couceiro}@loria.fr
[2] IRIT, CNRS & Université Paul Sabatier, 118 route de Narbonne, 31062 Toulouse, France
{dubois,prade}@irit.fr
[3] ERIC & Université Claude Bernard Lyon 1, 43 bld du 11-11, 69100 Villeurbanne, France
agnes.rico@univ-lyon1.fr

Abstract. Sugeno integrals are qualitative aggregation functions. They are used in multiple criteria decision making and decision under uncertainty, for computing global evaluations of items, based on local evaluations. The combination of a Sugeno integral with unary order preserving functions on each criterion is called a Sugeno utility functionals (SUF). A noteworthy property of SUFs is that they represent multi-threshold decision rules, while Sugeno integrals represent single-threshold ones. However, not all sets of multi-threshold rules can be represented by a single SUF. In this paper, we consider functions defined as the minimum or the maximum of several SUFs. These max-SUFs and min-SUFs can represent all functions that can be described by a set of multi-threshold rules, i.e., all order-preserving functions on finite scales. We study their potential advantages as a compact representation of a big set of rules, as well as an intermediary step for extracting rules from empirical datasets.

Keywords: Sugeno integral · Sugeno utility functional
Piecewise unary function · Decision rules · Qualitative representation

1 Introduction

Sugeno integrals [12] are aggregation functions that are used in multiple criteria decision making and in decision under uncertainty [7,9]. They are qualitative aggregation functions because they can be defined on non-numerical scales (more precisely on distributive lattices [5]). In this paper we only consider Sugeno integrals defined on completely ordered scales, and denote such a scale by L. A noteworthy property of Sugeno integrals is that their output is always comprised between the minimum and the maximum of their parameters. Moreover, Sugeno

J. Medina et al. (Eds.): IPMU 2018, CCIS 853, pp. 253–265, 2018.
https://doi.org/10.1007/978-3-319-91473-2_22

integrals on L are a subclass of lattice polynomials on L. More precisely, they correspond to all idempotent functions from L^n to L that can be formulated using min and max operations, variables and constants. From a decision making point of view, they also can be regarded as functions whose result depends on an importance value assigned to each subset of criteria. Sugeno integrals are known to represent any set of single-threshold if-then rules [8,10] of the form

$$x_1 \geq \alpha \text{ and } x_2 \geq \alpha \ldots \text{ and } x_n \geq \alpha \Rightarrow y \geq \alpha \text{ (selection rules), or}$$
$$x_1 \leq \alpha \text{ and } x_2 \leq \alpha \ldots \text{ and } x_n \leq \alpha \Rightarrow y \leq \alpha \text{ (deletion rules).}$$

Sugeno utility functionals (SUFs) are a generalization of Sugeno integrals [6] where each criterion value is mapped to an element of L by an order preserving function. They allow to represent multi-threshold rules of the form:

$$x_1 \geq \alpha_1 \text{ and } x_2 \geq \alpha_2 \ldots \text{ and } x_n \geq \alpha_n \Rightarrow y \geq \delta \text{ (selection rules), or}$$
$$x_1 \leq \alpha_1 \text{ and } x_2 \leq \alpha_2 \ldots \text{ and } x_n \leq \alpha_1 \Rightarrow y \leq \delta \text{ (deletion rules).}$$

However, although any *single* multi-threshold rule can be represented by a SUF, not all *sets* of multi-threshold rules can be represented by a single SUF [3].

In this paper, we consider functions defined as disjunctions or conjunctions of SUFs, recently introduced in [3]. They capture all order-preserving piecewise unary functions on finite scales, this is to say, all functions that can be represented by means of a set of multi-threshold rules. We study the potential advantages of this representation based on combinations of SUFs. In particular, we investigate whether it is possible to represent a large set of rules by means of only a few SUFs, and whether combinations of SUFs can help learning models which offer a good trade-off between simplicity and predictive accuracy.

In the next section we present combinations of SUFs as a framework with equivalent expressiveness to that of decision rules. In Sect. 3, we deal with the problem of finding a minimal combination of SUFs that represents a set of rules. In Sect. 4, we propose a method for approximately representing real datasets by means of a disjunction of SUFs. We show that it achieves predictive accuracy scores similar to the rule sets learned by the rough set-based method VC-DomLEM [1]. We also look at the compactness of the obtained model, and at the relation between the compactness of the model and its predictive accuracy. We omit the proof of most results because of space limitation.

2 Preliminaries

We use the terminology of multiple criteria decision-making where some objects are evaluated according to several criteria. We denote by $C = \{1, \ldots, n\}$ a set of criteria, by 2^C its power set, and by $X_1, \ldots, X_n$ and L totally ordered scales with top $1_{X_1}, \ldots, 1_{X_n}$ and 1_L and bottom $0_{X_1}, \ldots, 0_{X_n}$ and 0_L, respectively. We denote by ν the order reversing operation on L (ν is involutive and such that $\nu(0) = 1$ and $\nu(1) = 0$. We denote the Cartesian product $X_1 \times \cdots \times X_n$ by $\mathbf{X}$. An object is represented by a vector $\mathbf{x} = (x_1, \ldots, x_n) \in \mathbf{X}$ where x_i is the evaluation of x w.r.t. criterion i. Let $f : \mathbf{X} \to L$ be an evaluation function.

2.1 Decision Rules

Order-preserving functions from $\mathbf{X}$ to L can always be described by a set of selection rules or of deletion rules [3]. For the sake of simplicity, in this paper we mainly focus on selection rules. When there is no risk of ambiguity, we simply refer to selection rules as rules. Since any selection rule $r \in R$ has the form

$$x_1 \geq \alpha_1^r, \ldots, x_n \geq \alpha_n^r \Rightarrow f(\mathbf{x}) \geq \delta^r,$$

we will use the following abbreviation for defining a rule $r : \alpha_1^r, \ldots, \alpha_n^r \Rightarrow \delta^r$. Moreover, the left hand side of a rule r will be denoted by $\boldsymbol{\alpha}^r = (\alpha_1^r, \ldots, \alpha_n^r)$. We say that a function $f : \mathbf{X} \to L$ is *compatible* with a selection rule r if $f(\mathbf{x}) \geq \delta^r$ for all $\mathbf{x}$ such that $x_i \geq \alpha_i^r$ for each $i \in C$. For any selection rule r, we will denote by f_r the least function from $\mathbf{X}$ to L that is compatible with r, i.e., the function defined by

$$f_r(\mathbf{x}) = \delta^r \text{ if } [x_i \geq \alpha_i^r \ \forall i \in C], \ 0 \text{ otherwise},$$

for all $x \in \mathbf{X}$. We say that a criterion $i \in C$ is *active* in r if $\alpha_i^r > 0_{X_i}$. Moreover, we denote by A^r the set of criteria active in a rule r. For any set of rules R, we denote by f_R the least function compatible with all rules in R, defined by $f_R = \max_{r \in R} f_r(\mathbf{x})$ for all $\mathbf{x} \in \mathbf{X}$. We say that a function $f : \mathbf{X} \to L$ *represents* a set of selection rules R (or equivalently, that R represents f) if $f = f_R$. We say that a set of selection rules is *redundant* if there exists $r, s \in R$ such that $r \neq s$,

$$\alpha_i^r \geq \alpha_i^s \text{ for all } i \in C \quad \text{and} \quad \delta^s \geq \delta^r, \tag{1}$$

or, equivalently, if there exists $r \in R$ such that $f_R = f_{R \setminus \{r\}}$. A set of rules that is not redundant is said to be *irredundant*. We define the *equivalence class* of a set of rules R as $[R] = \{R' \mid f_R = f_{R'}\}$.

Proposition 1. *The equivalence class of R has one minimal element, which is*

$$R^{\min} = \{r \in R \mid \nexists s \in R : f_r \leq f_s\} = \bigcap_{R' \in [R]} R', \tag{2}$$

and is the only irredundant element of $[R]$.

2.2 Sugeno Integrals and Their Generalizations

A Sugeno integral is defined with respect to a capacity which is a set function $\mu : 2^C \to L$ that satisfies $\mu(\emptyset) = 0$, $\mu(C) = 1$ and $\mu(I) \leq \mu(J)$ for all $I \subseteq J \subseteq C$. The conjugate capacity of μ is defined by $\mu^c(I) = \nu(\mu(I^c))$, where I^c is the complement of I. The capacity can be seen as a function assigning an importance level to each subset of criteria. The Sugeno integral S_μ associated to μ can be defined in two ways:

$$S_\mu(\mathbf{x}) = \max_{I \subseteq C} \min(\mu(I), \min_{i \in I}(x_i)) = \min_{I \subseteq C} \max(\mu(I^c), \max_{i \in I}(x_i)),$$

for all $\mathbf{x} \in L^n$ [11]. The inner qualitative Möbius transform of a capacity μ is a mapping $\mu_\# : 2^C \to L$ defined by $\mu_\#(I) = \mu(I)$ if $\mu(I) > \max_{J \subset I} \mu(J)$ and 0 otherwise. A set $I \subseteq C$ such that $\mu_\#(I) > 0$ is called a focal set. The set of focal sets of μ is denoted by $\mathcal{F}(\mu)$. Since $\mu(A) = \max_{I \subseteq A} \mu_\#(I)$ for all $A \subseteq C$, the set function $\mu_\#$ contains the minimal amount of information needed to reconstruct μ. The qualitative Möbius transform provides a concise representation of the Sugeno integral as:

$$S_\mu(\mathbf{x}) = \max_{I \in \mathcal{F}(\mu)} \min(\mu_\#(I), \min_{i \in I}(x_i)) = \min_{I \in \mathcal{F}(\mu^c)} \max(\nu(\mu^c_\#(I)), \max_{i \in I}(x_i)).$$

A Sugeno utility functional (SUF) is a combination of a Sugeno integral and unary order preserving maps on each criterion [3,10]. Formally, a SUF is a function $S_{\mu,\varphi}$ defined by

$$S_{\mu,\varphi}(\mathbf{x}) = \max_{I \in \mathcal{F}(\mu)} \min(\mu(I), \min_{i \in I} \varphi_i(x_i)), \quad \text{for all } \mathbf{x} \in \mathbf{X},$$

where μ is a capacity, $\varphi = (\varphi_1, \ldots, \varphi_n)$ and, for all $i \in C$, $\varphi_i : X_i \to L$ is order preserving, with $\varphi(0_{X_i}) = 0_L$ and $\varphi(1_{X_i}) = 1_L$.

It is shown in [10] that any SUF can be represented in terms of single-thresholded rules of the form "$\varphi_1(x_1) \geq \alpha, \ldots, \varphi_n(x_n) \geq \alpha \Rightarrow f(\mathbf{x}) \geq \alpha$" (in other words, for all SUFs $S_{\mu,\varphi}$ there is a set of rules R such that $f_R = S_{\mu,\varphi}$). On the contrary, some sets of rules cannot be represented by a SUF. This justifies the use of combinations of SUFs. A max-SUF (resp. min-SUF) is defined by

$$f(\mathbf{x}) = \max_{i \in \{1,\ldots,k\}} \underline{S}^i(\mathbf{x}) \quad \left(\text{resp. } f(\mathbf{x}) = \min_{j \in \{1,\ldots,\ell\}} \overline{S}^j(\mathbf{x})\right),$$

for all $\mathbf{x} \in \mathbf{X}$, where $\underline{S}^1, \ldots, \underline{S}^k, \overline{S^1}, \ldots \overline{S}^\ell$ are SUFs.

Remark 1. It is shown in [3] that min-SUFs and max-SUFs can represent any set of deletion or selection rules, respectively. In other words, any order-preserving function from $\mathbf{X}$ to L can be expressed by a max-SUF and by a min-SUF. Also, since any SUF is such that $S_{\mu,\varphi}(0_{X_1}, \ldots, 0_{X_n}) = 0_L$ and $S_{\mu,\varphi}(1_{X_1}, \ldots, 1_{X_n}) = 1_L$, note that no combination of SUFs from $\mathbf{X}$ to L can represent a set of rules R such that $f_R(0_{X_1}, \ldots, 0_{X_n}) > 0_L$ or $f_R(1_{X_1}, \ldots, 1_{X_n}) < 1_L$. However, one can take $L' = \{y \in L \mid \exists \mathbf{x} \in X : f_R(\mathbf{x}) = y\}$ and find a combination of SUFs from $\mathbf{X}$ to L' that represents R.

Based on these results, we can try to model any order-preserving function defined on finite scales by means of max-SUFs or min-SUFs. Moreover, datasets that can be conveniently approximated by means of rules could also be approximated by means of a max-SUF or a min-SUF. In the next section, we focus on the exact representation of an order-preserving function.

3 Representing a Set of Rules by a (max-)SUF

In this section we will address the following problems. Given a set of rules R:

1. Determine whether R is *SUF-representable*, i.e., whether there is a SUF such that $f_R = S_{\mu,\varphi}$.
2. Find such a SUF if it exists.
3. In the case it does not exist, find a max-SUF that represents R while involving the least possible number of SUFs.

3.1 From a SUF to a Rule Set and Back

We already know that any SUF can be represented by a set of rules. Let us denote by $R_{\mu,\varphi}$ the set of rules defined from focal sets in $\mathcal{F}(\mu)$ as

$$\bigcup_{F\in\mathcal{F}(\mu)} \bigcup_{\delta^r\leq\mu(F)} \{r \mid A^r = F \text{ and } \forall i \in A^r : \alpha_i^r = \min\{x_i \in X_i \mid \varphi_i(x_i) \geq \delta^r\}\}.$$

As it will be shown in Lemma 1, this set of rules is equivalent to $S_{\mu,\varphi}$.

Although it is not always possible to find a SUF that represents a given set of rules, we can give a method for constructing such a SUF, when it exists. For any set of rules R, let $S_R^{\max}$ be the SUF defined by $S_R^{\max} = S_{\mu,\varphi}$,

$$\forall I \subset C \text{ such that } I \neq \emptyset : \qquad \mu(I) = \max_{\substack{r\in R,\\ A^r\subseteq I}} \delta^r, \tag{3}$$

$$\forall i \in [n],\ \forall \mathbf{x} \in \mathbf{X} \text{ such that } x_i \neq 1_{X_i} : \qquad \varphi_i(x_i) = \max_{\substack{r\in R,\\ 0<\alpha_i^r\leq x_i}} \delta^r, \tag{4}$$

and $\varphi_i(1_{X_i}) = 1_L$ for all $i \in C$. From this definition, we easily see that we always have $S_R^{\max} \geq f_R$. However it can be the case that $S_R^{\max} > f_R$ as shown in the following example.

Example 1. Consider a set of three criteria $C = \{1, 2, 3\}$, the scales $X_1 = X_2 = X_3 = L = \{0, a, b, 1\}$, with $0 < a < b < 1$, and the rule set $R = \{r^1, r^2, r^3\}$, with

$$r^1 : 0, b, 1 \Rightarrow 1, \quad r^2 : a, a, 0 \Rightarrow a, \quad r^3 : a, 0, b \Rightarrow b.$$

Let $S_{\mu,\varphi} = S_R^{\max}$. The function μ is such that $\mu_\#(\{1, 2\}) = a \quad \mu_\#(\{1, 3\}) = b, \quad \mu_\#(\{2, 3\}) = 1$ and, for all other $I \subset C$, $\mu_\#(I) = 0$. Moreover, we have $\varphi_1(a) = \varphi_1(b) = b, \quad \varphi_2(a) = a, \ \varphi_2(b) = 1, \quad$ and $\quad \varphi_3(a) = 0, \ \varphi_3(b) = b$. One can check that $S_{\mu,\varphi}(0, b, b) = b$, while $f_R(0, b, b) = 0$.

We will show that, when $S_R^{\max} > f_R$, there exists no SUF that represents R (see Proposition 2).

Lemma 1. *For any SUF $S_{\mu,\varphi}$, we have $S_{\mu,\varphi} = f_{R_{\mu,\varphi}} = S^{\max}_{R_{\mu,\varphi}}$.*

Proof. Let $S_{\mu,\varphi}$ be a SUF, and let μ^* and φ^* be such that $S^{\max}_{R_{\mu,\varphi}} = S_{\mu^*,\varphi^*}$. First, take any $\mathbf{x} \in \mathbf{X}$, $y \in L$ such that $S_{\mu,\varphi}(\mathbf{x}) \geq y$. Necessarily, there is $F \in \mathcal{F}(\mu)$ such that $\min(\mu(F), \min_{i \in F} \varphi_i(x_i))) \geq y$. Therefore $\mu(F) \geq y$ and $\forall i \in F$: $\varphi_i(x_i) \geq y$. From the definition of $R_{\mu,\varphi}$ it follows that there is $r \in R_{\mu,\varphi}$ such that $\delta^r = \mu(F) \geq y$, $A^r = F$ and $\forall i \in A^r : \alpha^r_i \leq x_i$. So we have $f_r(\mathbf{x}) \geq y$ and thus $f_{R_{\mu,\varphi}}(\mathbf{x}) \geq y$. Now, from the definition of $S^{\max}_{R_{\mu,\varphi}}$, it follows that $\mu^*(A^r) \geq y$ and $\forall i \in A^r : \varphi^*_i(x_i) \geq y$. Therefore $S^{\max}_{R_{\mu,\varphi}} \geq y$. Summing up, we have proven $S_{\mu,\varphi} \leq S^{\max}_{R_{\mu,\varphi}}$ and $S_{\mu,\varphi} \leq f_{R_{\mu,\varphi}}$.

Now we will prove $S_{\mu,\varphi} \geq f_{R_{\mu,\varphi}}$. Let $\mathbf{x} \in \mathbf{X}$, $y \in L$ such that $f_{R_{\mu,\varphi}}(\mathbf{x}) \geq y$. Necessarily there is $r \in R_{\mu,\varphi}$ such that $f_r(\mathbf{x}) \geq y$. Therefore $\delta^r \geq y$, and from the definition of $R_{\mu,\varphi}$ it follows that $\mu(A^r) \geq \delta^r$. Moreover for all $i \in A^r$ we have $x_i \geq \alpha^r_i = \min\{z_i \in X_i \mid \varphi_i(z_i) \geq \delta^r\}$, and thus $\varphi(x_i) \geq \delta^r$. Finally we get that for all y s.t. $f_{R_{\mu,\varphi}}(\mathbf{x}) \geq y$:

$$S_{\mu,\varphi}(\mathbf{x}) \geq \min(\mu(A^r), \min_{i \in A^r} \varphi_i(x_i)) \geq \delta^r \geq y.$$

Therefore $S_{\mu,\varphi} \geq f_{R_{\mu,\varphi}}$.

We still have to prove $S^{\max}_{R_{\mu,\varphi}} \leq S_{\mu,\varphi}$. Let $\mathbf{x} \in \mathbf{X}$ and $y \in L$ be such that $S^{\max}_{R_{\mu,\varphi}}(\mathbf{x}) \geq y$. Necessarily there is $F \in \mathcal{F}(\mu^*)$ such that

$$\min(\mu^*(F), \min_{i \in F} \varphi^*_i(x_i)) \geq y \tag{5}$$

Thus $\mu^*(F) \geq y$ and from the definition of $S^{\max}_{R_{\mu,\varphi}}$ it follows that there is $r \in R_{\mu,\varphi}$ such that $A^r \subseteq F$ and $\delta^r \geq y$. From the definition of $R_{\mu,\varphi}$ we obtain that $\mu(A^r) \geq y$. From (5) we also get that $\forall i \in F : \varphi^*_i(x_i) \geq y$. Again, from the definition of $S^{\max}_{R_{\mu,\varphi}}$, it follows that, for each $i \in F$, there is $r^i \in R_{\mu,\varphi}$ such that $\delta^{r^i} \geq y$ and $0 < \alpha^{r^i}_i \leq x_i$. So, for each $i \in F$: $x_i \geq \alpha^{r^i}_i = \min\{z_i \in X_i \mid \varphi_i(z_i) \geq \delta^{r^i}\}$, and thus $\varphi_i(x_i) \geq \delta^{r^i}$. Therefore we get $\min(\mu(A^r), \min_{i \in A^r} \varphi_i(x_i)) \geq y$. We have shown that $S^{\max}_{R_{\mu,\varphi}} \leq S_{\mu,\varphi}$, and the proof is complete. □

Lemma 2. *Let R and R' be two sets of selection rules belonging to the same equivalence class. Necessarily $S^{\max}_R = S^{\max}_{R'}$.*

Proposition 2. *For any set of selection rules R, if there exists a SUF S such that $S = f_R$, then $S = S^{\max}_R$.*

Relying on Proposition 2, we are able to identify sets of rules that can be represented by a single SUF, and to define such a SUF if it exists. A simple procedure for doing so, starting from R, is to compute μ and φ such that $S^{\max}_R = S_{\mu,\varphi}$ and then to compute $R_{\mu,\varphi}$. If $R_{\mu,\varphi}$ and R are in the same equivalence class, then $S^{\max}_R$ is the SUF equivalent to R, otherwise there is no such SUF.

3.2 From a Rule Set to a max-SUF

Now consider the case where R cannot be represented by a SUF, but we want to find a max-SUF that involves the least possible number of SUFs for representing R. In order to find such an "optimally parsimonious" max-SUF, one has to build the smallest partition (in terms of number of subsets) where each subset P is SUF-representable. All such partitions cannot be enumerated in reasonable time. Moreover note that, although in Example 1, r^1 and r^3 are responsible for the fact that $S_R^{\max} > f_R$, whether R is SUF-representable or not depends on larger combinations of rules. There are cases where each $A \subset R$ of size at most n is SUF-representable, while R is not. One example of such a case is the following.

Example 2. Let $n = 4$. Consider $L = \{0, a, b, 1\}$ and the set of selection rules $R = \{r^1, \dots, r^5\}$ with domain L^4 and codomain L, where $r^1 : a, a, 0, 0 \Rightarrow b$, $r^2 : a, 0, a, 0 \Rightarrow b$, $r^3 : a, 0, 0, a \Rightarrow b$, $r^4 : 0, a, a, a \Rightarrow a$, $r^5 : 0, b, b, b \Rightarrow 1$. One can check that $S_R^{\max} > f_R$, while each subset of R of size 4 is SUF-representable. □

Algorithm 1 presents a greedy method for extracting the max-SUF representing a set of rules. Although the resulting max-SUF is not necessarily minimal, it constitutes a first approximation.

Algorithm 1. Builds a set of SUFs $\boldsymbol{S}$ that represents a given set of rules R

```
1  P ← {}
2  for each r ∈ R do
3      covered ← false
4      for each P ∈ P do
5          if f_{P∪{r}} = S^max_{P∪{r}} then
6              add r to P
7              covered ← true
8              break foreach
9      if covered = false then
10         add {r} to P
11 S ← {S^max_P | P ∈ P}
```

Remark 2. In Algorithm 1, one does not have to compute $S_{P\cup\{r\}}^{\max}$ from scratch each time a rule r is added to P; the function $S_P^{\max}$ can be updated iteratively.

In order to get an idea of the number of SUFs required to represent a set of rules, we randomly generated sets of rules, and used Algorithm 1 to find their representations in terms of max-SUFs. Algorithm 2 describes the random generation of a rule set. The number of rules depends on the number of criteria and on the size of L. For the sake of simplicity, for our test we set $\mathbf{X} = L^n$.

Algorithm 2. Random generation of a set of rules R, for a given domain $\mathbf{X}$, codomain L, and real number $p << 1$

1 $R \leftarrow \{\}$
2 **for** i from 1 to $\lceil |\mathbf{X}| * p \rceil$ **do**
3 pick a random $(\alpha_1, \ldots, \alpha_n) \in \mathbf{X}$
4 $\delta^{\min} = \max\{\delta \mid ((\beta_1, \ldots, \beta_n) \Rightarrow \delta) \in R \text{ and } \forall i \in C,\ \beta_i \leq \alpha_i\}$
5 $\delta^{\max} = \min\{\delta \mid ((\beta_1, \ldots, \beta_n) \Rightarrow \delta) \in R \text{ and } \forall i \in C,\ \alpha_i \leq \beta_i\}$
6 pick a random δ in $\{y \in V \mid \delta^{\min} \leq y \leq \delta^{\max}\}$
7 $R \leftarrow R \cup \{(\alpha_1, \ldots, \alpha_n) \Rightarrow \delta\}$

Our aim was to approximate the number of utility functionals necessary to represent a set of rules, depending on the size of the set. Table 1 displays the variation of the number of rules and Sugeno-utility functionals obtained, depending on n and the size of L. These results suggest that in general, the representation of a set of selection rules in terms of a max-SUF is not very compact, since many SUFs have to be involved. However, these results only concern randomly generated rule sets. In the next section, we use combination of SUFs for empirical study on real datasets.

Table 1. Number of rules/Sugeno-utility functionals

		Size of L						
		2	3	4	5	6	7	8
n	2	2/1	2/1	3/1	2/1	3/1	4/1	4/2
	3	2/1	3/2	4/1	5/3	7/4	9/5	13/7
	4	2/1	4/2	7/4	14/8	26/15	43/23	70/38
	5	2/1	5/3	20/11	50/28	112/60	248/122	445/209
	6	3/1	13/7	60/31	202/103	600/295	1397/642	3204/1384
	7	3/1	31/13	197/93	873/410	3125/1386	8942/3859	21762/8780

4 Modeling Empirical Data

Formally, we represent a dataset by a multiset $\mathcal{D}$, whose elements belong to $\mathbf{X} \times L$. Elements of $\mathcal{D}$ are called *instances*, and values of L are referred to as *classes*. In this section we present a method for representing an empirical dataset by means of a max-SUF. This method is then applied on 12 datasets, the characteristics of which are given by Table 2. The process by which we extract a max-SUF from a dataset can be divided into four steps.

1. Select an order-preserving subset of data.
2. Associate a rule to each instance.
3. Group rules into a max-SUF.
4. Simplify the model by pruning some of the SUFs that compose it.

Table 2. Description of the datasets. Further information can be found in [1].

Id	Name	Instances	Criteria	Classes	Id	Name	Instances	Criteria	Classes
1	breast-c	286	8	2	7	denbosch	119	8	2
2	breast-w	699	9	2	8	ERA	1000	4	9
3	car	1296	6	4	9	ESL	488	4	9
4	CPU	209	6	4	10	LEV	1000	4	5
5	bank-g	1411	16	2	11	SWD	1000	10	4
6	fame	1328	10	5	12	windsor	546	10	4

Step 1. Selection of an order-preserving subset of data. Real life datasets often contain pairs of instances that are *order-reversing*, i.e., instances $(\mathbf{x}, y)$ and $(\mathbf{x'}, y')$ such that $\mathbf{x} \leq \mathbf{x'}$ and $y > y'$. In this step, we build a graph where the set of vertices is $\mathcal{D}$, and the set of edges contains each order-reversing pair of instances. Then, we iteratively remove from $\mathcal{D}$ the node with the largest number of neighbors, until no edge remains in the graph. In the next steps, we refer as $\mathcal{D}^-$ to the data remaining after Step 1. Table 3 presents the ratio of instances which are removed in each of the 12 datasets.

Table 3. Average ratio of data that is removed during Step 1, for each dataset.

1	2	3	4	5	6	7	8	9	10	11	12
.176	.008	.001	.0	.003	.025	.046	.657	.198	.333	.356	.237

Step 2. Associate a rule to each instance. A naive way to proceed would be to return the following set of rules: $R = \{r \mid \exists(\mathbf{x}, y) \in \mathcal{D}^- : [\delta^r = y \text{ and } \forall i \in C : \alpha_i^r = x_i]\}$. However, with most datasets, this would lead to all criteria being active in most rules. This is problematic because SUFs that will be learned by grouping such rules will only have focal sets of great size. Algorithm 3 provides an alternative solution, in which some criteria are set to 0 before rules are extracted. Its principle is that the ith criterion value of an instance can be set to 0 if by doing so the data are still order-preserving. Since the order in which criteria are considered can influence the result of the algorithm, we define this order w.r.t.

the discriminative power of each criterion in each instance. A rough estimation of this feature is given by the function $u : \bigcup_{i \in C} X_i \times L \to \mathbb{N}$ defined by

$$u(x_i, y) = \left|\{(\mathbf{x}', y') \in \mathcal{D}^- \mid [y > y' \text{ and } x_i > x'_i] \text{ or } [y < y' \text{ and } x_i < x'_i]\}\right|.$$

Intuitively, for any $(\mathbf{x}, y) \in \mathcal{D}^-$, $u(x_i, y)$ represents the number of other instances $(\mathbf{x}', y') \in \mathcal{D}^-$ such that the relation between y and y' could be explained (at least partially) by the ith criterion.

Algorithm 3. Builds a set of rules R covering all instances of a given $\mathcal{D}^-$

```
1 R ← {}
2 for each (x, y) ∈ D⁻ do
3     for x_i ∈ x in ascending order of u(x_i, y) do
4         if D⁻ ∪ {((x_1, ..., x_{i-1}, 0_{X_i}, x_{i+1}, ..., x_n), y)} is order-preserving
          then
5             x_i ← 0_{X_i}
6     add x_1, ..., x_n ⇒ y to R
```

Step 3. Group rules into a max-SUF. In this step we simply exploit a variant of Algorithm 1, where the condition in line 6 is replaced by "$S^{\max}_{P \cup \{r\}}(\mathbf{x}) \leq y$ for all $(\mathbf{x}, y) \in \mathcal{D}^-$". This guarantees that the max-SUF obtained is greater than or equal to f_R and does not misclassify any instance of $\mathcal{D}^-$.

Step 4. Simplify the model. In this step, we remove some of the SUFs from the model (see Algorithm 4). The algorithm depends on a parameter $\rho \in [0, 1]$, usually set close to 1, which represents the ratio of accuracy that has to be preserved when removing a SUF from the model. A lower value of ρ therefore allows to sacrifice more of the accuracy on the training data while removing a SUF from the model. We denote by accuracy($\boldsymbol{S}$,$\mathcal{D}$) the accuracy obtained by the model $\boldsymbol{S}$ on the data $\mathcal{D}$.

Algorithm 4. Removes useless SUFs from $\boldsymbol{S}$, for given $\mathcal{D}^-$ and $\rho \in [0, 1]$

```
1 end ← false
2 while end = false do
3     end ← true
4     for S ∈ S do
5         if accuracy(S\S, D⁻) ≥ ρ * accuracy(S, D⁻) then
6             remove S from S
7             end ← false
```

For each dataset and each value of $\rho \in \{0.95, 0.96, \ldots, 1.\}$, we tested this four-step process using ten-fold cross validation repeated several time. For each

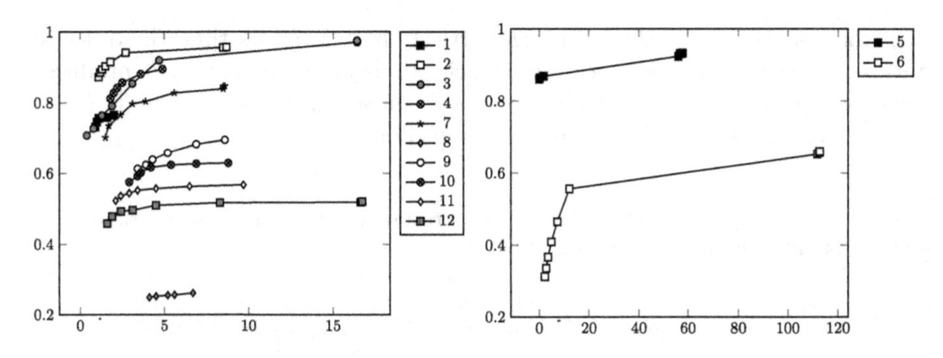

Fig. 1. Average accuracy (in ordinate) and number of SUF (in abscissa) of selected results. Each curve corresponds to one data set.

test, we computed the average accuracy (on validation data) and the average number of SUFs of the model. We then selected the best trade-offs between a high accuracy and a low number of SUFs involved in the model; those are represented in Fig. 1. We can see that, although involving more SUFs in the model can increase accuracy, a few SUFs are often enough to reach an accuracy close to that obtained by the best max-SUFs.

In order to get an idea of the accuracy that can be reached by our method, we selected the best accuracy obtained on each dataset. The 12 datasets considered in this paper have been used in [1] for evaluating the predictive accuracy of monotonic VC-DomLEM, which is a method for extracting decision rules based on the Dominance-based Rough Set Approach. Its overall accuracy on the 12 datasets is higher than those of several methods, such as SVM, OSDL [2] (instance based) or C4.5 (decision trees). Table 4 displays accuracy scores obtained by our method (with standard deviation) and those obtained by monotonic VC-DomLEM, reported from [1]. Both methods yield similar accuracy scores.

Table 4. Accuracy of VC-DomLEM and our method on each dataset.

	1	2	3	4	5	6	7	8	9	10	11	12	avg.
max-SUF	76	95.3	**97.2**	89.3	92.4	65.2	84.5	26.4	**69.4**	**63**	**56.7**	53.2	72.4
±	.57	.27	.19	1.33	0.52	.34	1.48	.75	.66	.53	.71	.75	
VC-DomLEM	**76.7**	**96.3**	97.1	**91.7**	**95.4**	**67.5**	**87.7**	**26.9**	66.7	55.6	56.4	**54.6**	72.7

Finally, Table 5 shows the distribution of rule lengths of the models with the best accuracy. Rule length distribution is an interesting feature for measuring interpretability. Indeed, each decision of a classifier based on selection (or deletion) is caused by one or a few rules only. The shorter these rules, the easier it is for a user to understand the classifier decision. Again, the rule lengths distributions we obtained can be compared to those obtained by VC-DomLEM in

[1]. None of the two methods provides strictly better results than the other in terms of rule length. However, a drawback of our approach is that it produces very long rules (albeit in a small percentage) on several datasets (e.g., 5, 6, 11, 12). Such long rules are less cognitively appealing.

Table 5. Rule length distributions, given as percentage. The number of criteria in each dataset is indicated by double vertical bars.

dataset	Rule length 1	2	3	4	5	6	7	8	9	10	11	12	13	14	15	16
1		1	64	23	11	1		‖								
2	13	72	13	1	1	1	1	1	‖							
3			3	24	41	32‖										
4	38	47	10	3	1	1‖										
5	3	25	31	13	5	2	3	2	1	1	3	5	3	1		1‖
6	2	20	37	22	6	2	1	1	1	9‖						
7		20	21	21	12	9	15	3‖								
8	6	62	16	16‖												
9	6	26	36	32‖												
10	5	26	42	27‖												
11	6	2	25	45	9	5	4	1		3‖						
12	3	24	44	21	5	1				2‖						

5 Conclusion

Keeping in mind that the results of the previous section are dependent of our algorithmic choices, we can draw the following conclusions: it seems that a single Sugeno integral, even with utility functions, is not enough to give an accurate representation of monotonic datasets. In our experiments, the number of SUFs required to achieve an accuracy similar to decision rule-based models can greatly vary from one dataset to another. Moreover, Table 5 shows that rule length distributions are quite uneven, even if Step 2 tries to privilege short rules. Progress may be achieved in two directions. On the one hand we could put a restriction on the capacities, limiting ourselves to k-maxitive capacities. On the other hand, one may start by constructing a decision tree of limited depth from a dataset, and use SUFs at its leaves.

Acknowledgements. This work has been partially supported by the Labex ANR-11-LABX-0040-CIMI (Centre International de Mathématiques et d'Infor-matique) in the setting of the program ANR-11-IDEX-0002-02, subproject ISIPA (Interpolation, Sugeno Integral, Proportional Analogy).

References

1. Blaszczynski, J., Slowinski, R., Szelag, M.: Sequential covering rule induction algorithm for variable consistency rough set approaches. Inf. Sci. **181**, 987–1002 (2011)
2. Cao-Van, K.: Supervised ranking - from semantics to algorithms. Ph.D. thesis, Ghent University, CS Department (2003)
3. Couceiro, M., Dubois, D., Prade, H., Rico, A.: Enhancing the expressive power of Sugeno integrals for qualitative data analysis. In: Kacprzyk, J., Szmidt, E., Zadrożny, S., Atanassov, K.T., Krawczak, M. (eds.) IWIFSGN/EUSFLAT -2017. AISC, vol. 641, pp. 534–547. Springer, Cham (2018). https://doi.org/10.1007/978-3-319-66830-7_48
4. Couceiro, M., Dubois, D., Prade, H., Waldhauser, T.: Decision making with Sugeno integrals. Bridging the gap between multicriteria evaluation and decision under uncertainty. Order **33**(3), 517–535 (2016)
5. Couceiro, M., Marichal, J.-L.: Characterizations of discrete Sugeno integrals as polynomial functions over distributive lattices. Fuzzy Sets Syst. **161**(5), 694–707 (2010)
6. Couceiro, M., Waldhauser, T.: Pseudo-polynomial functions over finite distributive lattices. Fuzzy Sets Syst. **239**, 21–34 (2014)
7. Dubois, D., Marichal, J.-L., Prade, H., Roubens, M., Sabbadin, R.: The use of the discrete Sugeno integral in decision making: a survey. Int. J. Uncertain. Fuzz. **9**, 539–561 (2001)
8. Dubois, D., Prade, H., Rico, A.: The logical encoding of Sugeno integrals. Fuzzy Sets Syst. **241**, 61–75 (2014)
9. Grabisch, M., Murofushi, T., Sugeno, M. (eds.): Fuzzy Measures and Integrals. Theory and Applications. Physica-verlag, Berlin (2000)
10. Greco, S., Matarazzo, B., Slowinski, R.: Axiomatic characterization of a general utility function and its particular cases in terms of conjoint measurement and rough-set decision rules. Eur. J. Oper. Res. **158**, 271–292 (2004)
11. Marichal, J.-L.: On Sugeno integrals as an aggregation function. Fuzzy Sets Syst. **114**(3), 347–365 (2000)
12. Sugeno, M.: Fuzzy measures and fuzzy integrals: a survey. In: Gupta, M.M., et al. (eds.) Fuzzy Automata and Decision Processes, North-Holland, pp. 89–102 (1977)

Image Feature Extraction Using OD-Monotone Functions

Cedric Marco-Detchart[1,2(✉)], Carlos Lopez-Molina[1], Javier Fernández[1,2], Miguel Pagola[1,2], and Humberto Bustince[1,2]

[1] Dpto. Automatica y Computacion, Universidad Publica de Navarra, Campus Arrosadia, 31006 Pamplona, Spain
cedric.marco@unavarra.com

[2] Institute of Smart Cities, Universidad Publica de Navarra, Campus Arrosadia, 31006 Pamplona, Spain

Abstract. Edge detection is a basic technique used as a preliminary step for, *e.g.*, object extraction and recognition in image processing. Many of the methods for edge detection can be fit in the breakdown structure by Bezdek, in which one of the key parts is feature extraction. This work presents a method to extract edge features from a grayscale image using the so-called ordered directionally monotone functions. For this purpose we introduce some concepts about directional monotonicity and present two construction methods for feature extraction operators. The proposed technique is competitive with the existing methods in the literature. Furthermore, if we combine the features obtained by different methods using penalty functions, the results are equal or better results than state-of-the-art methods.

Keywords: Edge detection · Feature extraction
Ordered directionally monotone functions · Penalty functions

1 Introduction

Most applications of image processing require, at some stage, automatic identification of objects on an image. In order to recognize objects, some sort of hints has to be extracted from the original image. One of the most useful hints is the boundaries of the own objects, which lead to a vast amount of studies on edge detection. The main difficulty in edge detection arises from the inherent process of capturing real word information; even the definition of edge itself is a fuzzy concept. An edge can be regarded as a location in which a big enough jump between neighbour pixel intensities happen. However, even this basic, loose definition can be criticized, since it would not consider situations as textures (in which no edges should be appointed) or *hallucinated boundaries* [21] (in which boundaries appear with little or no intensity contrast).

A great variety of edge detection methods are based on gradients, often computed by the convolution of a filter and an image. Examples are Sobel [25], Prewitt [23] and Canny [9]. Usually considered as a unique operation, edge detection is in fact reached through a sequence of operations.

J. Medina et al. (Eds.): IPMU 2018, CCIS 853, pp. 266–277, 2018.
https://doi.org/10.1007/978-3-319-91473-2_23

These sequence of operations has been studied by several authors, mainly attempting to develop a standard, multi-phase structure for edge detection [26]. For example, Law *et al.* [15] proposed a process consisting of three steps: *filtering*, *detection* and *tracing*. A few years later, Bezdek *et al.* [4] presented a framework to encompass a variety of methods in the literature (mainly those based on gradient extraction), proposing a process made up of four phases: *conditioning*, *feature extraction*, *blending* and *scaling*.

Although we stick to the Bezdek Breakdown Structure (BSS) for our experimentation [16], our work focuses on the *feature extraction* phase, *i.e.*, the way in which visual information at each pixel is converted into problem-specific information. This problem-specific information usually reduces to gradient magnitudes, but we also consider gradient direction. Moreover, we build feature maps considering information of pixel neighbourhood and fuse all the information using Ordered Directionally Monotone (ODM) functions. These functions are monotonic along different directions over the decreasingly ordered input vector. Furthermore, we test our method in combination with other techniques using penalty functions.

The structure of the document is as follows. Section 2 is devoted to explain some mathematical concepts applied to image processing, as well as some notions about aggregation theory to introduce ODM functions. Section 3 describes our proposal. In Sect. 4 an application to edge detection is presented, along with some preliminary results. Finally, in Sect. 5 we expose conclusions and future work.

2 Preliminaries

Given an image $\mathbb{I}_L$, we consider it as a grid of elements, the set of positions being $D = R \times C = \{1, \ldots, r\} \times \{1, \ldots, c\}$, where r represents the number of rows of the image, *i.e.*, the height, and c the columns of the image, *i.e.*, the width. Each of the elements (pixels) in the grid takes values in a scale L. The use of L permits us to represent different types of images. In the case of binary images pixels will take values in $L = \{0, 1\}$, whereas grey-scale images take $L = \{0, \ldots, 255\}$, and colour images in the RGB colour space are in the range $L = \{0, \ldots, 255\}^3$. This work has been carried out using grey-scale images.

As this work is related to the theory of aggregation and monotonicity we expose some of the necessary theory.

Definition 1 [2,8]. *A mapping $M : [0,1]^n \to [0,1]$ is an aggregation function if it is monotone non-decreasing in each of its components and satisfies $M(\mathbf{0}) = 0$ and $M(\mathbf{1}) = 1$, with $\mathbf{0} = (0, \ldots, 0)$ and $\mathbf{1} = (1, \ldots, 1)$*

An aggregation function M is an averaging or mean if

$$\min(x_1, \ldots, x_n) \leq M(x_1, \ldots, x_n) \leq \max(x_1, \ldots, x_n).$$

If an aggregation function is averaging then it is idempotent, and the converse is also true.

There are many types and classes of aggregation functions in the literature [3,13], where some of them have appeared to address specific applications. In this work we make use of Ordered Weighted Averaging (OWA) operators, defined by Yager [29].

Definition 2. *An OWA operator of dimension n is a mapping $\Phi : [0,1]^n \to [0,1]$ such that it exists a weighting vector $\boldsymbol{w} = (w_1, \ldots, w_n) \in [0,1]^n$ with $\sum_{i=1}^{n} w_i = 1$, and such that*

$$\Phi(x_1, \ldots, x_n) = \sum_{i=1}^{n} w_i \cdot x_{\sigma(i)},$$

where $\mathbf{x}_\sigma = (x_{\sigma(1)}, \ldots, x_{\sigma(n)})$ is a decreasing permutation on the input $\boldsymbol{x}$.

Another relevant family of aggregation functions used in this work are Choquet integrals. Particularly, we consider the discrete Choquet integral, related to fuzzy measures which are defined on finite dimensional spaces.

Definition 3. *Let $N = \{1, 2, \ldots, n\}$. A function $\mathfrak{m} : 2^N \to [0,1]$ is a fuzzy measure if, for all $X, Y \subseteq N$, it satisfies the following properties:*

($\mathfrak{m}$1) *Increasingness: if $X \subseteq Y$, then $\mathfrak{m}(X) \leq \mathfrak{m}(Y)$;*
($\mathfrak{m}$2) *Boundary conditions: $\mathfrak{m}(\emptyset) = 0$ and $\mathfrak{m}(N) = 1$.*

Definition 4 [3,13]**.** *Let $\mathfrak{m} : 2^N \to [0,1]$ be a fuzzy measure. The discrete Choquet integral is the function $C_\mathfrak{m} : [0,1]^n \to [0,1]$, defined, for each $\mathbf{x} \in [0,1]^n$, by*

$$C_\mathfrak{m}(\mathbf{x}) = \sum_{i=1}^{n} \left(x_{(i)} - x_{(i-1)}\right) \cdot \mathfrak{m}\left(A_{(i)}\right),$$

where $\left(x_{(1)}, \ldots, x_{(n)}\right)$ is an increasing permutation on the input $\mathbf{x}$, that is, $x_{(1)} \leq \ldots \leq x_{(n)}$, with the convention that $x_{(0)} = 0$, and $A_{(i)} = \{(i), \ldots, (n)\}$ is the subset of indices of the $n - i + 1$ largest components of $\mathbf{x}$.

Monotonicity can sometimes be too restrictive for some applications, excluding a large family of non-monotonic averaging functions, *e.g.*, the mode (most frequent element) is not a monotonic function thought it is widely used in certain applications like image filtering [27]. From this observation Wilkin and Beliakov [28] introduced the notion of weak monotonicity.

This definition was later extended into the notion of directional monotonicity [6].

Definition 5 [6]**.** *Let $\vec{r} = (r_1, \ldots, r_n)$ be a real n-dimensional vector, $\vec{r} \neq \mathbf{0}$. A function $F : [0,1]^n \to [0,1]$ is $\vec{r}$-increasing if for all points $(x_1, \ldots, x_n) \in [0,1]^n$ and for all $c > 0$ such that $(x_1 + cr_1, \ldots, x_n + cr_n) \in [0,1]^n$ it holds*

$$F(x_1 + cr_1, \ldots, x_n + cr_n) \geq F(x_1, \ldots, x_n).$$

That is, a $\vec{r}$-increasing function is a function which is increasing along the ray (direction) determined by the vector $\vec{r}$.

From this concept of directional monotonicity, we come to ordered directionally monotone functions, where the direction along which monotonicity is required varies depending on the relative size of the coordinates of the considered input.

Definition 6 [5]. *Let $F : [0,1]^n \to [0,1]$ be a function and let $\vec{r} \neq \mathbf{0}$. F is said to be ordered directionally (OD) $\vec{r}$-increasing if for any $\mathbf{x} \in [0,1]^n$, for any $c > 0$ and for any permutation $\sigma \in S_n$ with $x_{\sigma(1)} \geq \cdots \geq x_{\sigma(n)}$ and such that*

$$1 \geq x_{\sigma(1)} + cr_1 \geq \cdots \geq x_{\sigma(n)} + cr_n \geq 0,$$

it holds that

$$F(\mathbf{x} + c\vec{r}_{\sigma^{-1}}) \geq F(\mathbf{x}),$$

where $\vec{r}_{\sigma^{-1}} = (r_{\sigma^{-1}(1)}, \ldots, r_{\sigma^{-1}(n)})$.

As we are also building consensus feature images we need to introduce the concept of penalty functions.

Given a set of n numerical values $x_1, \ldots, x_n$ and q averaging aggregation functions $M_1, \ldots, M_q$, these functions allow us to select, between the q functions, the one that provides the least dissimilar output to all the inputs.

Taking into account the previous considerations, in this work we consider the following definition of a penalty function in a Cartesian product of lattices [7]:

Definition 7. *A function $P_\nabla : ([0,1]^n)^m \times [0,1]^m \to [0, \infty[$ is a penalty function if, for every $\mathbf{X} = (\mathbf{x^1}, \ldots, \mathbf{x^m}) \in ([0,1]^n)^m$ (with $\mathbf{x^i} = (x_1^i, \ldots, x_n^i)$ for every $i \in \{1, \ldots, m\}$) and for every $\mathbf{y} = (y_1, \ldots, y_m) \in [0,1]^m$, it satisfies that:*

1. *$P_\nabla(\mathbf{X}, \mathbf{y}) \geq 0$;*
2. *$P_\nabla(\mathbf{X}, \mathbf{y}) = 0$ if and only if $x_1^i = \cdots = x_n^i = y^i$ for every $i \in \{1, \ldots, m\}$;*
3. *P_∇ is convex in y_i or every $i \in \{1, \ldots, m\}$.*

3 Ordered Directionally Monotone Functions

In this section we present two alternative methods for obtaining ODM functions. Firstly we propose an affine construction and secondly we use the Choquet integral.

Theorem 1. *Let $G : [0,1]^n \to [0,1]$ be defined, for $\mathbf{x} \in [0,1]^n$ and $\sigma \in S_n$ such that $x_{\sigma(1)} \geq \ldots \geq x_{\sigma(n)}$, by*

$$G(\mathbf{x}) = a + \sum_{i=1}^{n} b_i x_{\sigma(i)},$$

for some $a \in [0,1]$ and $\vec{b} = (b_1, \ldots, b_n) \in \mathbb{R}^n$ such that $0 \leq a + b_1 + \cdots + b_j \leq 1$ for all $j \in \{1, \ldots, n\}$. Then G is OD $\vec{r}$-increasing for every non-null vector $\vec{r}$ such that $\vec{b} \cdot \vec{r} \geq 0$. In particular, for every non-null vector $\vec{r}$ which is orthogonal to $\vec{b}$.

Theorem 1 can be generalized taking into account the following lemma.

Lemma 1 [5]. *Let $\varphi : [0,1] \to [0,1]$ be an automorphism (i.e., an increasing bijection). Then, if $G : [0,1]^n \to [0,1]$ is an ordered directionally increasing function, the function $\varphi \circ G$ is also an ordered directionally increasing function.*

Corollary 1. *Let $p > 0$. Let $G : [0,1]^n \to [0,1]$ be defined, for $\mathbf{x} \in [0,1]^n$ and $\sigma \in S_n$ such that $x_{\sigma(1)} \geq \ldots \geq x_{\sigma(n)}$, by*

$$G(\mathbf{x}) = \left(a + \sum_{i=1}^{n} b_i x_{\sigma(i)} \right)^{\frac{1}{p}}, \tag{1}$$

for some $a \in [0,1]$ and $\vec{b} = (b_1, \ldots, b_n) \in \mathbb{R}^n$ such that $0 \leq a + b_1 + \cdots + b_j \leq 1$ for all $j \in \{1, \ldots, n\}$. Then G is OD $\vec{r}$-increasing for every non-null vector $\vec{r}$ such that $\vec{b} \cdot \vec{r} \geq 0$.

Proof. It follows from Lemma 1, since the function $\varphi(x) = x^{\frac{1}{p}}$ is an automorphism. ∎

From Theorem 1, the following two corollaries are straight.

Corollary 2. *Let $\mathfrak{m}$ be a fuzzy measure. Then the Choquet integral*

$$C_{\mathfrak{m}}(\mathbf{x}) = \sum_{i=1}^{n} \left(x_{(i)} - x_{(i-1)} \right) \cdot \mathfrak{m}\left(A_{(i)} \right),$$

is OD $\vec{r}$-monotone for every non-null n-dimensional vector $\vec{r}$ such that, for all maximal chains $A_1 \supset A_2 \supset \ldots \supset A_n \supset A_{n+1} = \emptyset$, it holds that

$$\sum_{i=1}^{n} r_{n-i+1}(\mathfrak{m}(A_{(i)}) - \mathfrak{m}(A_{(i+1)})) \geq 0,$$

where $\mathfrak{m}(A_{(n+1)}) = 0$.

Proof. The result follows from Theorem 1 after noting that the Choquet integral can be rewritten as

$$C_{\mathfrak{m}}(\mathbf{x}) = \sum_{i=1}^{n} \left(\mathfrak{m}(A_{(i)}) - \mathfrak{m}(A_{(i-1)})) \right) \cdot x_{(i)}.$$

∎

Corollary 3. *Let $A : [0,1]^n \to [0,1]$ be an OWA operator associated to the weighting vector $\boldsymbol{w} = (w_1, \ldots, w_n)$. Then A is OD $\vec{r}$-increasing for every non-null vector $\vec{r}$ such that $\boldsymbol{w} \cdot \vec{r} \geq 0$.*

4 Experimental Study

The aim of this experiment is to test the effectiveness of ODM functions to detect edges in images. To do so, we propose two possible constructions for ODM functions to extract feature images and then compare them to well-known methods in the literature. We also use the feature images obtained to build consensus solutions by means of penalty functions.

4.1 Proposed Method and Parameters

Given a grey-scale image $\mathbb{I}_g$ the primary step is to normalize the intensity values to the range $[0, 1]$. Then we use Algorithm 1 to obtain a feature image by means of ODM functions.

Algorithm 1. Constructing an edge feature image using ODM functions

Input: A normalized grey-scale image $\mathbb{I}_g$ and an ODM function G as in Corollary 1.
Output: A feature image $\mathbb{I}_f$.
1: **for** each pixel (x, y) of $\mathbb{I}_g$ **do**
2: Compute the corresponding values by means of the absolute value of the difference between $\mathbb{I}_g(x, y)$ and its 8-neighbourhood;
3: Order the eight values of step 2 in a decreasing way;
4: Apply the ODM function G, with its corresponding a, p values and $\vec{r}$, $\vec{b}$ vectors (see Eq. (1)), to the values obtained in step 3;
5: Assign as intensity of the pixel (x, y) of $\mathbb{I}_f$ the value obtained in step 4.
6: **end for**

As we consider an 8-neighbourhood around each position (x, y) (*i.e.* neighbours from $(x-1, y-1)$ to $(x+1, y+1)$). Each value needed for step 2 is computed as:

$$x_1 = |a_{(x,y)} - a_{(x-1,y-1)}|, \ldots, x_8 = |a_{(x,y)} - a_{(x+1,y+1)}|$$

for each of the values.

In step 3, these differences are ordered in a decreasing way; that is,

$$x_{\sigma(1)} \geq x_{\sigma(2)} \geq \ldots \geq x_{\sigma(7)} \geq x_{\sigma(8)}.$$

Finally, in step 4 an ODM function is applied to each position in the image, with different a, p, $\vec{r}$ and $\vec{b}$ parameters.

To test our method we propose two expressions to build ODM functions for step 4 of Algorithm 1. We construct these expressions by means of Corollary 1 using Eq. (1) and choosing specific values for the specified parameters.

Elaborating on the expression in Eq. (1), the parameter p adjusts the brightness level of the feature image. In this way, the image is lighter when $p > 1$ or darker if $0 < p < 1$ [11]. Then, we consider $\vec{r} = (x_{\sigma(1)}, x_{\sigma(2)}, \dots, x_{\sigma(8)})$; and for the first ODM construction (*Case 1*) we use the following $\vec{b}$:

$$\vec{b} = \left(\frac{x_{\sigma(1)}}{2\sum_{i=1}^{8} x_{\sigma(i)}}, \frac{x_{\sigma(2)}}{2\sum_{i=1}^{8} x_{\sigma(i)}}, \dots, \frac{x_{\sigma(8)}}{2\sum_{i=1}^{8} x_{\sigma(i)}} \right)$$

Considering the second proposed construction (*Case 2*) we maintain the same previous parameters and change $\vec{b}$ as follows:

$$\vec{b} = \left(\frac{\left|x_{\sigma(1)} - x_{\sigma(8)}\right|}{2\sum_{i=1}^{8} \left|x_{\sigma(i)} - x_{\sigma(8)}\right|}, \dots, \frac{\left|x_{\sigma(7)} - x_{\sigma(8)}\right|}{2\sum_{i=1}^{8} \left|x_{\sigma(i)} - x_{\sigma(8)}\right|}, 0 \right)$$

Regarding the value of $\frac{1}{p}$ we take for both cases $\frac{1}{p} = 0.35$ and $a = 0$. Choosing $a = 0$ is because we need $G(\mathbf{0}) = 0$ if and only if it takes this value. And then having $a = 0$, $G(\mathbf{1}) = 1$ if and only if $b_1 + \dots + b_n = 1$. Note that in the case of a flat region in the image, *i.e.*, when all the pixels have the same value we would obtain a zero denominator, so we mark directly the corresponding position in the feature image as not containing an edge.

Figure 1 displays the results obtained by applying our proposed algorithm with the two alternative construction methods for ODM.

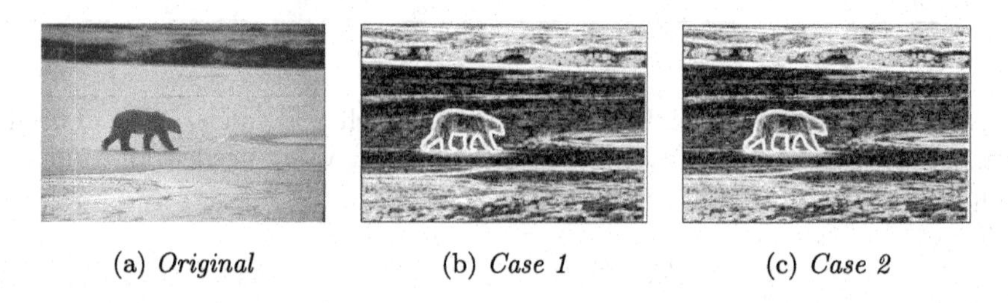

(a) *Original* (b) *Case 1* (c) *Case 2*

Fig. 1. Original image from BSDS [1] (100007) along with feature images obtained after applying Algorithm 1 with different ODM functions to original image.

4.2 Experimental Framework

In order to analyse the behaviour of our proposal we follow the scheme given by Bezdek *et al.* [4], adding the final step to quantify the results of the method:

($S1$) Smooth the image applying a Gaussian filter (with $\sigma = 1$) to $\mathbb{I}_g$;
($S2$) Obtain the feature image with Algorithm 1;
($S3$) Thin the feature image using non-maxima suppression [9];
($S4$) Binarize the thinned image using the hysteresis method [22].
($S5$) Compare the binary image with ground truth images [10].

As we want to evaluate the performance of ODM functions for edge detection, we carry out a fair comparison between different edge detection methods focusing on step ($S2$). Hence, the remaining steps are homogeneous for all contending methods.

For comparison purposes, we test our algorithm with the following well-known edge detection approaches: the Canny method [9] with $\sigma_C = 2.25$ for the derivative operator, as indicated in [19]; the Gravitational Edge Detection (GED) method [17], using the probabilistic sum (G_{S_P}) and the maximum (G_{S_M}); Fuzzy morphological edge detector [12] with two variants, using the Schweizer-Sklar [24] t-norm and t-conorm (FM_{SS}) and the minimum and maximum (FM_{MM}).

4.3 Feature Image Fusion Using Penalty Functions

As a complementary experiment, we consider different feature images obtained in step ($S2$), using different edge detection methods, and we build a consensus feature image based on the concept of penalty function. In this way, the corresponding value of a pixel is the least dissimilar to all the proposed feature images.

For the experimentation we use the following specific expression for the penalty function:

$$P_\nabla(\mathbf{X}, \mathbf{y}) = \sum_{q=1}^{m} \sum_{p=1}^{n} |x_p^q - y_q|^2. \tag{2}$$

To obtain the consensus feature image we follow the next steps for each pixel of the images at position (x, y):

(1) Take 3 pixels from each feature image in the same position;
(2) Apply a series of q aggregation functions, $M_1, \ldots, M_q$;
(3) Compute the penalty function given by Eq. (2) over the previous result;
(4) Get the values associated to the aggregation providing the smallest value of Eq. (2).

As an example of this procedure we propose a construction method, *Case 3*, where we aggregate a series of five feature images with five aggregation functions. Concretely we use the *minimum*, the *maximum*, the *arithmetic mean* and two *OWA operators*; the first one associated with the quantifier *the largest possible amount* and the second one with the quantifier *the largest part of*. With the aggregation functions we take the following feature images: *Case 1*, *Case 2*, Canny, FM_{SS}, G_{S_P}.

4.4 Dataset and Quantification of the Results

For our experiments we have used the images of Berkeley Segmentation Dataset (BSDS500) [1], specifically 100 natural images from the test set. Associated to each original image there exist several hand-labelled segmentations denoted as *ground truth images*.

As the ground truth are binary images, we are dealing with a classification problem where each pixel is considered in a confusion matrix as in the Martin *et al.* approach [20], where *True Positive, False Positive, etc.* are considered. The matching process between the solution obtained and all the human ground-truths is done comparing pixel-to-pixel in both images. Because of possible spatial differences we need some tolerance in the comparison, as an edge pixel can be slightly displaced w.r.t. the ground-truth. Concretely we apply a tolerance of 2.5% of the image diagonal. This comparison is done using the approach presented by Estrada and Jepson in [10] and available at [14].

In order to interpret the previous values, we use the following Precision/Recall measures:

$$Prec = \frac{TP}{TP + FP}, \quad Rec = \frac{TP}{TP + FN}, \quad F_\alpha = \frac{Prec \cdot Rec}{\alpha \cdot Prec + (1 - \alpha) \cdot Rec}.$$

In this work we stick to the usual value of $\alpha = 0.5$ for the F-measure according to [18,20], representing the harmonic mean between $Prec$ and Rec.

4.5 Experimental Results

In Table 1 the results of each edge detection method are indicated displaying the average of $Prec$, Rec and $F_{0.5}$. On the one hand, in terms of Rec we can infer that we have obtained as good results as the Canny method with *Case 1* and *Case 2*, i.e., not including a lot of false positives. On the other hand, we may observe that FM_{SS} combines a medium precision with a very high recall, therefore the majority of edges are detected at the cost of including a high number of false positives. Considering the overall measure, *i.e.*, $F_{0.5}$, the results achieved with *Case 1* and *Case 2* are competitive with the ones obtained with the Canny method and gravitational forces. In addition, if we observe *Case 3* we have substantially improved the results and even surpassed the best performer, namely, the Canny method.

As complementary comparative measure we consider the number of images being the best and worst performer in terms of $F_{0.5}$. In Table 2 we show the results of such a measure and we observe that with our proposed approaches we obtain a score of 0 in terms of worst result. Moreover, there exist a good number of best images obtained with *Case 1* and *Case 2* methods. The result is comparable to the one obtained with gravitational forces method taking into account that in some cases this method is the worst performer. Considering *Case 3* we can see that we slightly improve the result of best images obtained, taking them mainly from the Canny method and G_{S_P} methods.

As a visual demonstration, we show in Fig. 2 the results obtained with all the approaches considered in our experiments. As we can easily observe, the binary edges given by each approach are different, some of them depict more edges than others, giving more information but taking less care of false positives; therefore one approach can provide a better solution for a specific problem than other edge detection method.

Table 1. Comparison of ODM functions *Case 1, Case 2* approach, along with penalty functions *Case 3* respect to gravitational, Fuzzy Morphology and the Canny method in terms of *Prec*, *Rec* and $F_{0.5}$.

Edge detection methods	*Prec*	*Rec*	$F_{0.5}$
Case 1	0.61	0.67	0.61
Case 2	0.60	0.69	0.62
Case 3	0.61	0.73	**0.64**
FM_{SS}	0.44	**0.87**	0.57
FM_{MM}	**0.70**	0.40	0.45
Canny	0.66	0.65	0.63
G_{S_P}	0.59	0.71	0.62
G_{S_M}	0.60	0.68	0.60

Table 2. Comparison of best and worst approaches in terms of $F_{0.5}$ considering ODM functions *Case 1, Case 2* and penalty approaches with five feature images *Case 3.*

	Edge Detection Method											
	*		FM_{SS}		FM_{MM}		*Canny*		G_{S_P}		G_{S_M}	
	✓	✗	✓	✗	✓	✗	✓	✗	✓	✗	✓	✗
Case 1	18	**0**	15	18	2	76	**39**	1	18	2	8	3
Case 2	18	**0**	15	18	2	76	**39**	1	18	2	8	3
Case 3	29	**0**	14	18	2	76	**35**	1	12	2	8	3

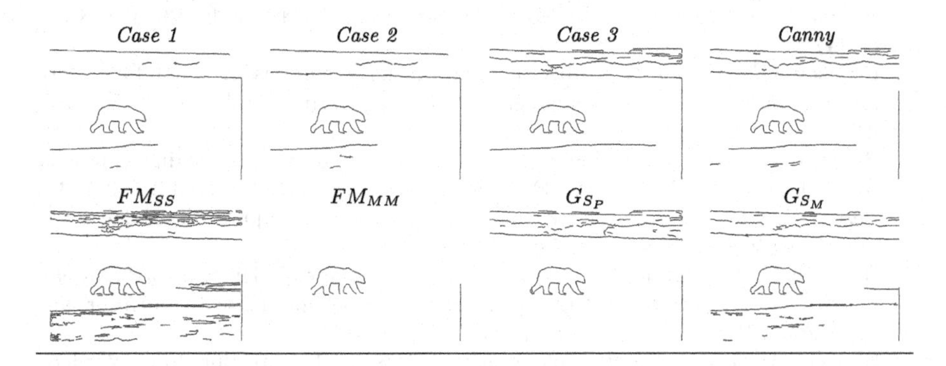

Fig. 2. Binary image obtained with ODM functions and penalty functions (*Case 1, Case 2, Case 3*), Canny, Fuzzy Morphology (FM_{SS}, FM_{MM}) and gravitational forces (G_{S_P}, G_{S_M}).

5 Conclusion

We have take a novel approach to edge detection using ODM functions. Given a neighbourhood we have measured the changes of intensity around each pixel taking into account the direction defined by the intensity variation vector, ordered in a decreasing way. Such functions are proven effective to determine the existence of an edge. Furthermore, using different ODM functions to obtain feature images and considering $F_{0.5}$ measure, we can conclude that the proposed methods are very competitive in comparison with the Canny method and gravitational forces, outperforming the Fuzzy Morphology approach. Moreover, if we use penalty functions to build a consensus feature image, we get the best value of the $F_{0.5}$ measure, overtaking the Canny method. However, regarding the number of images for which our methods get the best or the worst result, we see that, using penalty functions, the Canny method is the best method for more images, but our method provides the best result regarding worst count. Also, we would like to stress the simplicity of our method, in particular when the feature image is built using the expression in Eq. (1).

Regarding future research lines, we consider studying more possible values for the proposed parameters and optimize them, in order to confirm this preliminary results, and to analyse the spatio-temporal complexity of the algorithm in relation with other edge detection algorithms.

Acknowledgments. This work is supported by the Spanish Ministry of Science (Project TIN2016-77356-P) and the Research Services of Universidad Publica de Navarra.

References

1. Arbelaez, P., Maire, M., Fowlkes, C., Malik, J.: Contour detection and hierarchical image segmentation. IEEE Trans. Pattern Anal. Mach. Intell. **33**(5), 898–916 (2011)
2. Beliakov, G., Pradera, A., Calvo, T.: Aggregation Functions: A Guide for Practitioners, vol. 18. Springer, Heidelberg (2007). https://doi.org/10.1007/978-3-540-73721-6
3. Beliakov, G., Sola, H.B., Sánchez, T.C.: A practical guide to averaging functions, vol. 329. Springer, Heidelberg (2016). https://doi.org/10.1007/978-3-319-24753-3
4. Bezdek, J., Chandrasekhar, R., Attikouzel, Y.: A geometric approach to edge detection. IEEE Trans. Fuzzy Syst. **6**(1), 52–75 (1998)
5. Bustince, H., Barrenechea, E., Sesma-Sara, M., Lafuente, J., Dimuro, G.P., Mesiar, R., Kolesarova, A.: Ordered directionally monotone functions. Justification and application. IEEE Trans. Fuzzy Syst. **PP**(99), 1 (2017)
6. Bustince, H., Fernandez, J., Kolesárová, A., Mesiar, R.: Directional monotonicity of fusion functions. Eur. J. Oper. Res. **244**, 300–308 (2015)
7. Bustince, H., Beliakov, G., Pereira Dimuro, G., Bedregal, B., Mesiar, R.: On the definition of penalty functions in data aggregation. Fuzzy Sets Syst. **323**, 1–18 (2017)
8. Calvo, T., Kolesárová, A., Komorníková, M., Mesiar, R.: Aggregation Operators: Properties, Classes and Construction Methods. Aggreg. Oper. New Trends Appl. **97**(1), 3–104 (2002)
9. Canny, J.F.: A computational approach to edge detection. IEEE Trans. Pattern Anal. Mach. Intell. **8**(6), 679–698 (1986)
10. Estrada, F.J., Jepson, A.D.: Benchmarking image segmentation algorithms. Int. J. Comput. Vis. **85**(2), 167–181 (2009)

11. Forero-Vargas, M.G.: Fuzzy thresholding and histogram analysis. In: Nachtegael, M., Van der Weken, D., Kerre, E.E., Van De Ville, D. (eds.) Fuzzy Filters for Image Processing, pp. 129–152. Springer, Heidelberg (2003). https://doi.org/10.1007/978-3-540-36420-7_6
12. Gonzalez-Hidalgo, M., Massanet, S., Mir, A., Ruiz-Aguilera, D.: On the choice of the pair conjunction-implication into the fuzzy morphological edge detector. IEEE Trans. Fuzzy Syst. **23**(4), 872–884 (2015)
13. Grabisch, M., Marichal, J.L., Mesiar, R., Pap, E.: Aggregation Functions (Encyclopedia of Mathematics and Its Applications), 1st edn. Cambridge University Press, New York (2009)
14. Kermit Research Unit (Ghent University): The kermit image toolkit (kitt). www.kermitimagetoolkit.com
15. Law, T., Itoh, H., Seki, H.: Image filtering, edge detection, and edge tracing using fuzzy reasoning. IEEE Trans. Pattern Anal. Mach. Intell. **18**(5), 481–491 (1996)
16. Lopez-Molina, C.: The breakdown structure of edge detection - analysis of individual components and revisit of the overall structure. Ph.D. thesis, Universidad Publica de Navarra (2012)
17. Lopez-Molina, C., Bustince, H., Fernandez, J., Couto, P., De Baets, B.: A gravitational approach to edge detection based on triangular norms. Pattern Recogn. **43**(11), 3730–3741 (2010)
18. Lopez-Molina, C., De Baets, B., Bustince, H.: Quantitative error measures for edge detection. Pattern Recogn. **46**(4), 1125–1139 (2013)
19. Lopez-Molina, C., De Baets, B., Bustince, H.: A framework for edge detection based on relief functions. Inf. Sci. (Ny) **278**, 127–140 (2014)
20. Martin, D.R., Fowlkes, C.C., Malik, J.: Learning to detect natural image boundaries using local brightness, color, and texture cues. IEEE Trans. Pattern Anal. Mach. Intell. **26**(5), 530–549 (2004)
21. Martin, D.R.: An empirical approach to grouping and segmentation. EECS Department, University of California, Berkeley, August 2003. UCB/CSD-03-1268. http://www2.eecs.berkeley.edu/Pubs/TechRpts/2003/5252.html
22. Medina-Carnicer, R., Muñoz-Salinas, R., Yeguas-Bolivar, E., Diaz-Mas, L.: A novel method to look for the hysteresis thresholds for the Canny edge detector. Pattern Recogn. **44**(6), 1201–1211 (2011)
23. Prewitt, J.M.S.: Object enhancement and extraction. Pict. Process. Psychopictorics **10**(1), 75–149 (1970)
24. Schweiser, B., Sklar, A.: Associative functions and statistical triangle inequalities. Publ. Math. Debr. **8**, 169–186 (1961)
25. Sobel, I., Feldman, G.: A 3x3 isotropic gradient operator for image processing. In: Hart, P.E., Duda, R.O. (eds.) Pattern Classification Scene Analysis, pp. 271–272 (1973)
26. Torre, V., Poggio, T.: On edge detection. IEEE Trans. Pattern Anal. Mach. Intell. **8**(2), 147–163 (1986)
27. van de Weijer, J., van den Boomgaard, R.: Local mode filtering. In: 2001 IEEE Computer Society Conference on Computer Vision and Pattern Recognition (CVPR 2001), Kauai, HI, USA, 8–14 December 2001, pp. 428–433. IEEE Computer Society (2001)
28. Wilkin, T., Beliakov, G.: Weakly monotonic averaging functions. Int. J. Intell. Syst. **30**(2), 144–169 (2015)
29. Yager, R.R.: On Ordered weighted averaging aggregation operators in multicriteria decisionmaking. IEEE Trans. Syst. Man Cybern. **18**(1), 183–190 (1988)

Applying Suitability Distributions in a Geological Context

Robin De Mol(✉) and Guy De Tré

Faculty of Engineering and Architecture, Department of Telecommunication and Information Processing, Databases, Documents and Content Management Research Group, Ghent University, Sint-Pietersnieuwstraat 41, 9000 Ghent, Belgium
{robin.demol,guy.detre}@ugent.be

Abstract. Some industrial purposes require specific marine resources. Companies rely on information from resource models to decide where to go and what the cost will be to perform the required extractions. Such models, however, are typical examples of imprecise data sets wherein most data is estimated rather than measured. This is especially true for marine resource models, for which acquiring real data samples is a long and costly endeavor. Consequently, such models are largely computed by interpolating data from a small set of measurements. In this paper, we discuss how we have applied fuzzy set theory on a real data set to deal with these issues. It is further explained how the resulting fuzzy model can be queried so it may be used in a decision making context. To evaluate queries, we use a novel preference modeling and evaluation technique specifically suited for dealing with uncertain data, based on *suitability distributions*. The technique is illustrated by evaluating an example query and discussing the results.

Keywords: Imperfect information · Decision support
Preference modeling · Suitability distributions

1 Introduction

Sand is one of the most important resources available to the industry. Towards long-term sustainable exploitation, the Transnational and Integrated Long-term Marine Exploitation Strategies (TILES)[1] project partners aim to create a geological knowledge base that can be consulted for a broad spectrum of applications. Creating such a knowledge base is difficult for multiple reasons, but it is particularly challenging because there is only little data available. This in turn is due to the fact that it is expensive and time-consuming to perform offshore measurements. As a result, the available data are sparse, (geographically) non-uniformly distributed across the region of interest and span multiple decades. Making decisions based off this data must be done with these things in mind to avoid arriving at false conclusions.

[1] https://odnature.naturalsciences.be/tiles/.

J. Medina et al. (Eds.): IPMU 2018, CCIS 853, pp. 278–288, 2018.
https://doi.org/10.1007/978-3-319-91473-2_24

In this paper, we discuss how these issues are treated in the TILES knowledge base. Fuzzy set theory is used to add data quality indicators, which are taken into account during querying. Query evaluation relies on a novel evaluation technique that is particularly well suited for such data sets. This is illustrated on a case study towards the extraction of sand.

The remainder of this paper is structured as follows. In Sect. 2, the data set from the TILES project is introduced. It is shown how it deals with uncertainty regarding attribute values. Section 3 explains what suitability distributions are and how they can be used to evaluate fuzzy queries on data sets containing imperfect information. These are then applied to the TILES data set in Sect. 4. A fictive, fuzzy query is evaluated on the model to show how different ways of dealing with the uncertainty in the model have an impact on the results, which are also visually represented. This shows the viability of the approach and the richness of the TILES data set. Section 5 concludes the paper by summarizing its content.

2 The TILES Project and Voxel Modeling

TILES is a project in which the partners collaborate to create a state-of-the-art knowledge base of geological information of the subsurface of the Southern part of the North Sea, off the coast of Belgium and the Southern half of the Netherlands. The intention is to use the knowledge base for a multitude of purposes regarding resource availability, long-term ecological impact of exploitation, industrial decision making, and so on. Data is available in the form of borehole samples and multibeam echosounder information which both contain information of several geological properties such as lithology and lithostratigraphy. Throughout the paper we will focus only on lithology. Lithology is essentially the study of sediment and its characteristics. Sediment is typically classified based on its average grain size. There exist predefined classifications which map specific grain size ranges to named lithological classes (see Table 1). In TILES, a simplified Wentworth classification is used, combining some of the original Wentworth classes into fewer, larger classes, until only 6 remain: *clay*, *silt*, *fine sand*, *medium sand*, *coarse sand* and *gravel*.

To best meet the needs of the project partners and stakeholders, the TILES knowledge base is designed as a *voxel model*. A voxel model is essentially a spatial partitioning of a three dimensional region into a regular grid. Each cell in the grid is called a *voxel* (short for *volume element*, similar to the pixel in 2D) and represents a unique volume of space of the original region. From a data storage point of view, a voxel is a vector of attribute values indicating the (geological) properties in a specific area. It is assumed the property values are homogeneous per voxel. In TILES, each voxel represents a cuboid space of 200 by 200 by 1 m. This cell size is deliberately chosen to strike a balance between computational requirements and model accuracy. Alternatively, a model with irregular cell sizes could be used but this discussion is outside the scope of the paper.

To construct such a voxel model from the available data, a statistical approach called kriging [9] is applied. Kriging relies on *variograms*. Briefly put, the

Table 1. Wentworth classification table

Size range	Aggregate name
>256 mm	Boulder
64–256 mm	Cobble
32–64 mm	Very coarse gravel
16–32 mm	Coarse gravel
8–16 mm	Medium gravel
4–8 mm	Fine gravel
2–4 mm	Very fine gravel
1–2 mm	Very coarse sand
0.5–1 mm	Coarse sand
0.25–0.5 mm	Medium sand
125–250 μm	Fine sand
62.5–125 μm	Very fine sand
3.9–62.5 μm	Silt
0.98–3.9 μm	Clay
0.95–977 nm	Colloid

variogram of an attribute captures the directional trends between attribute values throughout a spatial data set. The kriging process then predicts attribute values voxel per voxel based on these variograms and on measurements that lie in the voxel's vicinity. In TILES, this technique is used to predict, among others, the lithological class of each voxel. A rendering of the voxel model showing the predicted lithological class for each voxel is given in Fig. 1.

Due to the scarceness and non-uniformous distribution of the available data, the reliability of the predicted lithological class for many of the voxels is questionable. In an attempt to mitigate this, each voxel is enriched with data quality indicators during the prediction process. Example data quality indicators are *variability/entropy* (in case multiple prediction runs are made) and *borehole density* (amount of true measurements in a given vicinity of the voxel), but also imprecise indicators that are derived from metadata related to the measurements themselves are stored, such as *reliability of sample analysis method*, *reliability of the measurement vintage* and so on. These metadata are often available only in the form of free-text (*core descriptions*), if they are available at all. The interpretation of these descriptions are carried out by experienced geologists and recorded in the data set. Even then, their estimated level of quality is subjective, which calls for fuzzy logic. Some examples of such data quality indicators are: "the sample was analyzed with outdated techniques", and "the positional information of the measurement is very reliable". During the prediction process, these data quality indicators are taken into account and derived quality indicator values are added to the voxels. When predicting the lithological class based

Fig. 1. An example visualization of the TILES voxel model. The voxels are colored based on their predicted lithological class. (Color figure online)

on measurements whose sample analysis quality are annotated as "outdated", "reliable" and "unknown", the resulting voxel might receive a value of "possibly inaccurate" for that same data quality indicator. A voxel might look like this:

1. position: (51.1215768, 2.9187675)
2. coordinate reference system: WGS 84
3. depth from mean sea level: 37 m
4. predicted lithological class: fine sand
5. prediction entropy: 0.57
6. amount of prediction runs: 100
7. borehole density: *very low*
8. vintage quality: *high*
9. sample analysis quality: *mediocre*
10. reliability description of original analyst: *unknown*
11. ...

Linguistic terms such as *high*, *mediocre* and *very low* are stored as possibility distributions over the unit interval. The inclusion of these data quality indicators essentially implies the TILES voxel model is a fuzzy database [1,11].

3 Fuzzy Querying and Suitability Distributions

In this section we discuss how the TILES fuzzy voxel model can be queried. To that end, a special querying tool was implemented. The querying tool relies on concepts from fuzzy logic [5,14] in order to allow decision makers to model their preferences using fuzzy sets and advanced aggregation operators [2,6–8,10,12,13]. The preference model expects that a criterion regarding a specific attribute is expressed by means of a fuzzy set that maps each value from the domain of this attribute to a number between 0 (unwanted) and 1 (preferred). Evaluating a datum is done by using this mapping to arrive at the degree associated to the datum's value. This is sometimes called the degree of suitability (of that datum for the query purpose). It is common to omit unwanted values from the mapping, and we will do so in this paper. Essentially, a fuzzy query specified by a preference model imposes a complete order on all values from the attribute's domain. The order may be partial, i.e. different values may be mapped to the same suitability degree to indicate they are considered equally suitable. Evaluating such a criterion on a set of data effectively corresponds to sorting the data according to the imposed order. It is easy to see that a regular query is a special case of a fuzzy query where all values are mapped to either 0 or 1.

Consider the following example fuzzy query preference model regarding lithological class:

- Coarse sand: 0.2
- Medium sand: 1.0
- Fine sand: 0.5
- Silt: 0.2.

This can be interpreted as follows: for the purpose of the decision maker, medium sand is ideal, fine sand is good, and coarse sand and silt are equally poor, yet still acceptable. All other sediment types are considered unacceptable. The numbers used in the mapping might be based on external factors, such as a cost related to processing the resource before it can be used to serve its purpose, or arbitrarily chosen with the sole purpose of implying the desired order. In case of the latter, a different mapping that implies the same order would be equally expressive. Clearly, a preference model is subjective and is therefore best interpreted by the decision maker that created it.

In order to evaluate the fuzzy data quality indicators present in the TILES fuzzy voxel model, a novel evaluation technique specifically for dealing with imperfect information [4] was used. The premise of the technique is that evaluating a fuzzy value leads to a fuzzy degree of suitability, except in some specific cases. More precisely, evaluating a fuzzy preference model on a fuzzy datum results in a possibility distribution over degrees of suitability, called a *suitability distribution*. Computing the suitability distribution essentially comes down to evaluating all possible worlds and representing the results in a concise but complete way, without losing information. Essentially, when considering the evaluation of a datum as an operator, the computing of a suitability distribution comes down to applying Zadeh's extension principle.

Consider for example the data quality indicator *sample analysis quality*. Assume a decision maker wants to express a preference for voxels for which the sample analysis quality should be at least "reliable" such that higher quality levels are considered more preferable. This preference might be modelled by a fuzzy set with the following membership function (visualized in Fig. 2):

$$p(x) = \begin{cases} 0 & x \leq 0.2 \\ (5x-1)/4 & x > 0.2 \end{cases}$$

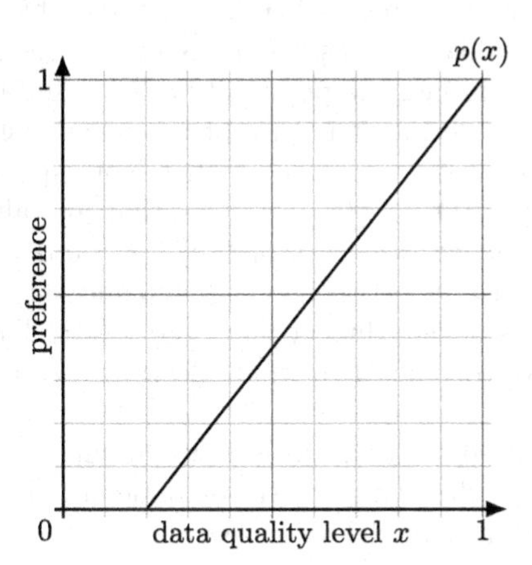

Fig. 2. An example fuzzy criterion regarding "sample analysis quality". The indicated preference model reflects that a higher value is considered more preferable.

To illustrate the evaluation process which leads to the construction of a suitability distribution, consider a voxel whose value for sample analysis quality is a fuzzy set labelled "inaccurate or possibly reliable". A possible membership function for this fuzzy set is given:

$$\pi(x) = \begin{cases} 1 & x \leq 1/2 \\ (8-10x)/3 & x > 1/2 \wedge x \leq 4/5 \\ 0 & x > 4/5 \end{cases}$$

This membership function and the suitability distribution that results from the evaluation of the previously defined fuzzy criterion are shown in Fig. 3. The construction of the suitability distribution, $s(q)$, is now elaborated. First, the membership function of the fuzzy criterion is analyzed piecewise.

The first piece, over the range $[0, 1/5]$, denotes values that are considered equi-suitable for the purpose of the decision maker. In this case these values, denoting low quality, are all considered unacceptable. This piece, combined with the information regarding the voxel's value, can be used to derive information regarding the possibility that the voxel is not suitable for the decision maker. Therefore, we look at the fuzzy set "inaccurate or possibly reliable" in the range $[0, 1/5]$. The possibility that the voxel's quality level is such that it is not suitable regarding the criterion, is then equal to the maximal possibility that it takes a value from this range. For the voxel under evaluation, all values in this range are fully possible, hence it is fully possible that this voxel would be deemed unsuitable by the decision maker.

The second piece of the preference model implies a total order on the remaining quality values, reflecting that the decision maker wants to sort voxels by their quality from highest to lowest. This piece maps to the entire suitability range, so it will allow us to derive a possibility for each possible degree of suitability. By construction, suitability degrees not mapped by the preference model are considered impossible. Finding the possibility of each suitability degree is fairly straight forward due to the linear nature of the membership functions used in this example. A quality value of 4/5 (or higher) corresponds to a preference of $p(4/5) = 3/4$ (or higher). From $\pi(x)$ we know that all values of this quality or higher are impossible. Hence, $s(q) = 0, \forall q > 3/4$. For quality values between 1/5 and 1/2, the preference varies between $p(1/5) = 0$ and $p(1/2) = 3/8$ respectively. These values are fully possible, thus $s(q) = 1, \forall q \leq 3/8$. The remaining suitability degrees are linearly correlated to $\pi(x)$. Between 1/2 and 4/5, $s(q)$ declines linearly from 1 to 0.

The results of the piecewise analysis is now combined by taking the pointwise maximum over the suitability range, with the understanding that the possibility is 0 where it is not defined. In this example, the overall suitability distribution is given by:

$$s(q) = \begin{cases} 1 & q \leq 3/8 \\ (6-8q)/3 & q > 3/8 \wedge q \leq 3/4 \\ 0 & q > 3/4 \end{cases}$$

The suitability distribution immediately conveys the following information. It is clear that this voxel can not be more suitable than to a degree of 3/4. However, it is mostly plausible that its suitability is less than 3/8. Moreover, it is fully possible that it is absolutely not suitable at all. Overall, the voxel is clearly not very suitable for the decision maker.

Note that the similarity between the suitability distribution and the voxel's fuzzy value for the data quality indicator can be explained by the shape of the preference function, which closely resembles the identity function. A fully detailed description of the suitability distribution technique can be found in [4].

Finally, one might wonder how these suitability distributions can be used in a decision support setting. Traditionally, in decision support, a preference model is used to evaluate a (possibly very large) set of systems by computing a suitability degree for each system so they can be sorted from "best" to "worst". Using a suitability based approach, aggregating and sorting, two key concepts from decision support, is not as straight forward. There exist techniques to aggregate and compare possibility distributions, but a detailed analysis in the context of the semantics of suitability distributions has not yet been performed. Another approach, which has been studied in this context [3], is to use defuzzification. Defuzzifying a suitability distribution results in a suitability degree. In [3], it is established that different defuzzification strategies can lead to different represen-

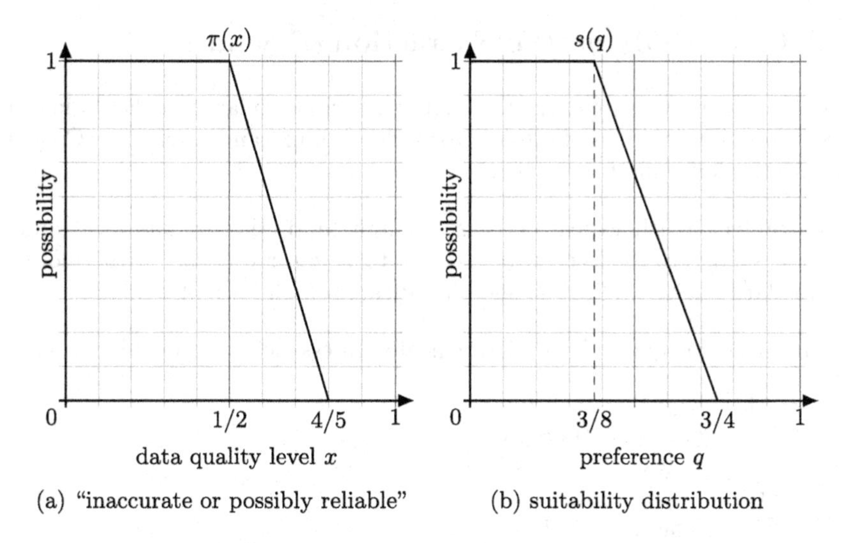

(a) "inaccurate or possibly reliable"

(b) suitability distribution

Fig. 3. A fictive voxel which has the value "inaccurate or possibly reliable" for the data quality indicator "sample analysis quality" on the left (3a) and the suitability distribution resulting from the evaluation of the fuzzy criterion from Fig. 2 on this voxel on the right (3b).

tative suitability degrees. On the one hand, these strategies can be used to reflect a certain tolerance for uncertainty (or rather, mistakes in the suitability prediction), while on the other hand certain strategies have properties that render them particularly interesting for a specific purpose. Consider for example defuzzification through computing the area under the possibility distribution. Whereas, in general, the area under a possibility distribution may be arbitrarily large, the surface under a suitability distribution can always be scaled without modification of the semantics such that it is bound between 0 and 1. Note that this is naturally the case if the unit interval is used for representing possibility degrees and also for preference degrees, and that it is common to do so for both. This defuzzification strategy may be especially interesting to sort the distributions by uncertainty (a larger surface means a larger uncertainty) and consequently, in decision making, to identify the systems that require additional measurements. This strategy, however, would not be useful to find the "best" system. Using defuzzification, suitability distributions can be mapped to suitability degrees, thus making it possible to include them in existing decision support systems. Additionally, the decision maker can choose a different defuzzification strategies per criterion, providing an additional layer of control specifically for dealing with uncertainty.

4 A Case Study on the Extraction of Sand

In this section, we will apply the techniques described so far on real data in order to identify potential areas of interest towards medium sand extraction. Consider therefore the following fuzzy query:

> The preferred lithological class is medium sand, though fine sand and coarse sand are also acceptable, albeit to a lesser degree. Furthermore, most data quality indicators should indicate a high level of reliability.

We can break this query down hierarchically. In essence, it is a conjunction of...

- a fuzzy lithological class criterion
- the OWA aggregation (most) of...
  - sample analysis quality is high
  - vintage quality is high
  - positional accuracy is high
  - ...

The fuzzy lithological class criterion is implemented using a discrete mapping that associates medium sand to 1 and light and coarse sand to 0.5. The criteria on the data quality indicators are evaluated using the fuzzy set *high*, which is implemented by the membership function:

$$p(x) = \begin{cases} 0 & x < 1/2 \\ 2x - 1 & x \geq 1/2 \end{cases}$$

In order to be able to aggregate the suitability distributions resulting from the data quality criteria evaluations, they are defuzzified by using a "cautious optimism" strategy. This strategy takes the ordinate of the center of mass under the suitability distribution as representative suitability degree and has the property that its distinctive power is higher towards the bounds of the suitability degree domain. In other words, the distributions that are defuzzified onto suitability degrees near 0 (or 1) are reliably bad (respectively good). The downside of this strategy is that, for distributions that are defuzzified onto suitability degrees near 0.5, it is unclear whether the datum is uncertain or if it is known to be of mediocre suitability. We choose this defuzzification strategy because we are only interested in the top-k best voxels and thus do not care about the voxels with mediocre suitability degrees. A more detailed analysis of this approach can be found in [3].

The aggregation of the suitability degrees regarding data quality is translated into an OWA operator with the semantics of "most", which is implemented using the shape function $y = \sqrt{x}$. For each voxel, the final, global suitability degree is computed using a pure conjunction, corresponding to the mathematical *minimum* of the overall data quality suitability and lithological class suitability. This global suitability degree is representative of the overall degree to which the voxel is suitable for our purpose of finding and extracting medium sand with low risk of the model giving false information, as per the preference for high data quality indicators. For the TILES data, it makes most sense to explore the results visually, using a three dimensional *suitability map*. The suitability map is simply a rendering of the original voxel model wherein each voxel is colored based on its global suitability degree. In this example, the suitability map (shown in Fig. 4) uses a gray scale where white corresponds to unsuitable and black corresponds to maximally suitable.

Fig. 4. The suitability map for the case study.

Keeping in mind the chosen defuzzification strategies, we can interpret the suitability map as follows. The dark areas indicate regions where it is reliably certain that medium sand can be found. The white areas indicate regions where it is reliably certain that nor medium sand, fine sand nor coarse sand can be found. The remaining gray areas are either too uncertain for our purpose or they are reliably of lesser suitability, containing mostly fine or coarse sand rather than medium sand. These regions are ideal candidates to be examined first if the need for medium sand exceeds the amount that is available.

5 Conclusions

This paper illustrates a suitability distribution based data evaluation technique by applying it to a real data set of sediment information of the subsurface of the North sea off the coast of Belgium and the Southern half of the Netherlands. The data set contains fuzzy information, including data quality indicators, in a rasterized voxel model. It is shown how using suitability distributions makes it possible to evaluate uncertain data while still obtaining interpretive results that convey reliable information that can be important for decision makers.

References

1. Bosc, P., Pivert, O.: Fuzzy queries against regular and fuzzy databases. In: Andreasen, T., Christiansen, H., Larsen, H.L. (eds.) Flexible Query Answering Systems, pp. 187–208. Springer, Boston (1997). https://doi.org/10.1007/978-1-4615-6075-3_10
2. Choquet, G.: Theory of capacities. In: Annales de l'institut Fourier, vol. 5, pp. 131–295. Institut Fourier (1954)
3. De Mol, R., De Tré, G.: Representing uncertainty regarding satisfaction degrees using possibility distributions. In: Kacprzyk, J., Szmidt, E., Zadrożny, S., Atanassov, K.T., Krawczak, M. (eds.) IWIFSGN/EUSFLAT -2017. AISC, vol. 641, pp. 597–604. Springer, Cham (2018). https://doi.org/10.1007/978-3-319-66830-7_53
4. De Mol, R., Bronselaer, A., De Tré, G.: Evaluating flexible criteria on uncertain data. Fuzzy Sets Syst. **328**, 122–140 (2017)
5. Dubois, D.J.: Fuzzy Sets and Systems: Theory and Applications, vol. 144. Academic Press, Cambridge (1980)
6. Dujmović, J.J.: A comparison of andness/orness indicators. In: Proceedings of the 11th Information Processing and Management of Uncertainty International Conference, IPMU 2006, pp. 691–698 (2006)
7. Dujmovic, J.J.: Continuous preference logic for system evaluation. IEEE Trans. Fuzzy Syst. **15**(6), 1082–1099 (2007)
8. Grabisch, M.: The application of fuzzy integrals in multicriteria decision making. Eur. J. Oper. Res. **89**(3), 445–456 (1996). https://doi.org/10.1016/0377-2217(95)00176-X. ISSN 03772217
9. Krige, D.G.: A statistical approach to some basic mine valuation problems on the Witwatersrand. J. South Afr. Inst. Min. Metall. **52**(6), 119–139 (1951)
10. Sugeno, M.: Theory of fuzzy integrals and its applications. In: Theory of Fuzzy Integrals and Its Applications (1975)
11. Umano, M.: FREEDOM-0: a fuzzy database system. In: Fuzzy Information and Decision Processes, pp. 339–347 (1982)
12. Yager, R.R.: On ordered weighted averaging aggregation operators in multicriteria decisionmaking. IEEE Trans. Syst. Man Cybern. **18**(1), 183–190 (1988)
13. Yager, R.R.: Families of OWA operators. Fuzzy Sets Syst. **59**, 125–148 (1993). https://doi.org/10.1016/0165-0114(93)90194-M. ISSN 01650114
14. Zadeh, L.A.: Fuzzy sets. Inf. Control **8**(3), 338–353 (1965)

Metrics for Tag Cloud Evaluation

Úrsula Torres-Parejo[1(✉)], Jesús R. Campaña[2], Maria-Amparo Vila[2], and Miguel Delgado[2]

[1] Department of Statistics and Operational Research, University of Cádiz, Cádiz, Spain
ursula.torres@uca.es

[2] Department of Computer Science and Artificial Intelligence, University of Granada, Granada, Spain

Abstract. Since their appearance Tag Clouds are widely used tools in Internet. The main purposes of these textual visualizations are information retrieval, content representation and browsing of text. Despite their widespread use and the large number of research that has been carried out on them, the main metrics available in the literature evaluate the quality of the tag cloud based only on the query results. There are no adequate metrics when the tag cloud is extracted from text and used to represent information content. In this work, three new metrics are proposed for the evaluation of tag clouds when their main function is to represent information content: coverage, overlap and disparity, as well as a fourth metric: the balance, in which we propose a way to calculate it by using OWA operators.

Keywords: Tag cloud · Content representation · Metrics · Coverage Overlap · Balance · OWA operators

1 Introduction

In [6–8] we widely studied tag clouds and proposed a standard method to generate them. These works were focused on the extraction of valuable information stored in textual databases. The main aim was to represent this information to users with no previous knowledge of a textual database content.

With this objective in mind, we established a complete methodology for text processing. This methodology includes the tasks of syntactic and semantic preprocessing, generation of an intermediate form, postprocessing and visualization through a tag cloud. The novelty of the proposal was the preservation of text semantics, due to the fact that related terms could remain together in the visualization, that is, the tag cloud generated was a multi-term tag cloud.

To evaluate the tag cloud obtained through this methodology when used for text retrieval, we used the precision, recall and $F1$ Score metrics [2]. But, to evaluate it as a tool of content representation, we did not find adequate metrics in the literature, thus modifying some of those proposed in [9] to assess the tag cloud obtained from the query results.

J. Medina et al. (Eds.): IPMU 2018, CCIS 853, pp. 289–296, 2018.
https://doi.org/10.1007/978-3-319-91473-2_25

In this work, we establish a formal definition for the modified metrics, "coverage" and "overlap", and propose a new one, the "disparity". The last one fixes the inconveniences found in other existing metrics such as balance or entropy. We expose these inconveniences in the next section. In addition, we give a formal definition for the balance, establishing a new way of calculating it through OWA operators.

This paper is organized as follows: Sect. 2 presents a brief summary of the existing metrics to evaluate the tag cloud, exposing their inconveniences for evaluation of content representation in tag clouds. Section 3 proposes new metrics and illustrates the proposal with an example. Finally, Sect. 4 gives some conclusions and future work.

2 Existing Metrics

There are not a lot of metrics in literature to evaluate the goodness of a tag cloud and much less when it works as a tool for content representation.

We found the first in [1]. The authors define the **entropy** of a tag cloud as follows:

Let t ∈ T be a tag in a tag cloud T:

$$Entropy(T) = -\sum_{t \in T} p(t)log\{p(t)\}$$

where

$$p(t) = \frac{weight(t)}{\sum_{t \in T} weight(t)}$$

Entropy quantifies the weight disparity between tags. If it is low, the tag cloud is significant or effective. If, on the contrary, it is high, the weights of the tags are uniform, which visually is not very informative. A tag cloud will be effective if it consists of significant tags.

The inconvenience of this metric is that it is unbounded, so it is difficult to know when its value can be considered high or low.

A set of metrics to capture the structural properties of the tag cloud generated from the query results is defined in [3,9]. These metrics are:

1. **Coverage.** Gives the fraction of the query set C_q covered by the tag cloud S. This metric takes values between 0 and 1. If it is close to 0, the tag cloud covers few objects in the query set, but if it is close to 1, it covers many objects.
2. **Overlap.** Different tags in S may be associated with the same objects in C_q. With this metric the extension of such redundancies is evaluated. This metric also takes values in the interval [0, 1]. If it is close to 0, there is little overlap. If, on the contrary, the value of this metric is close to 1, the overlap is high and the tags are not very different from each other.
3. **Cohesiveness.** Measures the closeness of the objects in each query set associated to each tag in the tag cloud, according to the relationships between these objects.

4. **Relevance.** It is defined as the overlap between the set of the results obtained with the query (C_q) and the set of the objects retrieved with each tag (C_t). It is calculated as the fraction of results in C_t that are also in C_q.
5. **Popularity.** A tag in S is popular in C_q if it is associated with many objects in C_q.
6. **Independence.** Two tags in S are independent if the objects they recover are not similar to each other. The metric in this case is similar to the cohesiveness, but the latter is calculated for each pair of tag sets.
7. **Balance.** A tag cloud S is balanced if its tags represent a similar number of objects in C_q. The balance takes values in the range $[0, 1]$. A tag cloud is considered to be balanced if the value of this metric is close to 1. This metric is calculated as the fraction between the minimum and the maximum number of objects retrieved through a tag, so only two values are considered for its calculation.

From these metrics, only the coverage, overlap, and balance are suitable for its adaptation to evaluate the tag cloud generated with the purpose of representing text content. The other metrics are only useful for tag clouds coming from query sets or if they are calculated for one or two isolated tags.

Furthermore, the balance metric has the inconvenience that for its calculation, only two tags are considered, the tag which more objects represents and the one representing less. A tag cloud is said to be unbalanced if this metric is close to 0, but if we think for example in a tag cloud with all tags representing the same number of objects except for one tag which differs so much from the others (see tag cloud in Fig. 1 where *tag8* is much smaller than the others), we could say that this tag cloud is unbalanced when it is not. For this reason, this metric does not seem appropriate to us.

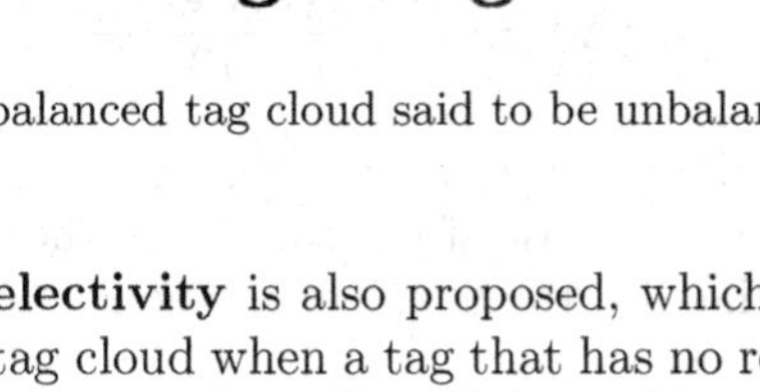

tag1
tag2 tag3 tag4 tag5
tag6 tag7 tag8 tag9
tag10 tag11

Fig. 1. Example of balanced tag cloud said to be unbalanced according to [9]

In [5] the metric **Selectivity** is also proposed, which measures the number of objects filtered in a tag cloud when a tag that has no relation with the former tag is selected. Other metrics such as **Simplicity** or **Detailedness** can be found in [4] to evaluate the cluster grouping into the tag clouds.

3 Metrics Proposed for the Evaluation of the Tag Cloud as a Tool of Content Representation in Textual Databases

In [7,8] we defined a methodology of text processing in databases that had as final step the visualization of the text through a tag cloud. This representation helps in content identification and in querying and browsing tasks, for textual data.

To evaluate the goodness of the information retrieved through the tags, we used the precision, recall and F-Score metrics [2], which are very standardized measures for these purposes. But, for evaluating the tag cloud as a content representation tool, we find few specific metrics that do not meet our requirements.

We take the metrics "coverage" and "overlap" from [9] and adapt them to evaluate the tag cloud as a tool of content representation. The metric "balance" has the inconvenience of taking only two values for its calculation, the minimum and maximum weights, being a metric very influenced by the extreme values, which could lead us to erroneous conclusions.

A way to calculate the balance avoiding these abnormal values in the weight is excluding first the outliers. Values that exceed 1.5 times the interquartile range are considered to be outliers. In this way the influence of values due to errors in the data would be avoided, but still only two values would be considered for the calculation of the balance.

Next, we define three metrics to evaluate the tag cloud when it is used as a tool for content representation in textual databases: the coverage, the overlap and the disparity. The third one is related with the balance and the entropy, but is bounded and considers all the tags for its calculation. In addition, a new way of calculating the balance through OWA operators is proposed, avoiding first the outliers as we exposed above.

3.1 Coverage, Overlap and Disparity

Let x_i be a tag of a cloud X and t_i be a tuple in a set of tuples T, then we call $T(x_i)$ to the set of tuples associated to a tag t_i.

We calculate the coverage of X as:

$$cov(X) = \frac{card(\cup_{x_i \in X} T(x_i))}{card(T)} \tag{1}$$

This metric takes values in the interval [0, 1]. A value close to 1 indicates that the tag cloud represents most of the content of the database.

The overlap is calculated through the expression:

$$over(X) = avg_{i \neq j} \left(\frac{card(T(x_i) \cap T(x_j))}{min\{card(T(x_i)), card(T(x_j))\}} \right) \tag{2}$$

This metric also takes values between 0 and 1. A value close to 0 means that tags in the tag cloud represent different objects in the database.

And finally, we calculate the disparity with the next expression:

$$dis(X) = avg_{i \neq j} \left(\frac{|card(T(x_i)) - card(T(x_j))|}{max\{card(T(x_i)), card(T(x_j))\}} \right) \tag{3}$$

As the two other metrics, the disparity takes values between 0 and 1. A value close to 1 indicates that the disparity between the weights of the tags is high, which is necessary in order to highlight the importance of the tags through the weight.

3.2 Balance Calculated Through OWA Operators

Ordered Weighted Averaging (OWA) Operators. An OWA operator of dimension n is a mapping [11]:

$$f : R^n \rightarrow R$$

that has an associated n vector W:

$$W = [w_1 \;\; w_2 \;\; w_3]^T$$

such that

$$1.\, w_i \in [0, 1]$$
$$2.\, \sum_i w_i = 1$$

Furthermore $f(a_1, \ldots, a_n) = \sum_j w_j b_j$ where b_j is the jth largest of the a_i.

An important aspect to consider in this operation is the re-ordering step, since a weight is associated with a particular ordered position of aggregate, that is, the $a_1, \ldots, a_n$ vector has to be previously ordered in a descending way.

In [10] Yager pointed out three important special cases of OWA aggregations:

1. F^*: In this case $W = W^* = [1 \;\; 0 \;\; \ldots \;\; 0]^T$. Then $F^*(a_1, \ldots, a_n) = Max_i(a_i)$,
2. F_*: In this case $W = W_* = [0 \;\; 0 \;\; \ldots \;\; 1]^T$. Then $F_*(a_1, \ldots, a_n) = Min_i(a_i)$,
3. F_{Ave}: In this case $W = W_{Ave} = [1/n \;\; \ldots \;\; 1/n]^T$. Then $F_{Ave}(a_1, \ldots, a_n) = \frac{1}{n} \sum_i a_i$.

Balance Calculated Through OWA Operators. If we consider the two first special cases of OWA operators pointed out in [10], we can express the balance as:

$$bal(S) = \frac{F_*(C)}{F^*(C)} \tag{4}$$

where F_* and F^* are OWA operators of size n with associating weighting vectors $W_* = [0 \; 0 \; \ldots 1]^T$ and $W^* = [1 \; 0 \; \ldots 0]^T$, respectively and $C = \{card(T(x_1)), card(T(x_2)), \ldots, card(T(x_n)\}$ is the set of cardinals from the associated sets $T(x_i)$ to each tuple t_i.

Before applying this formula we have to exclude the outliers from C, identifying these as the values that exceed 1.5 times the interquartile range.

The balance takes values in the interval [0, 1]. A value close to 0 indicates that the tag cloud is considered to be unbalanced.

3.3 Example of Calculation of Coverage, Overlap, Disparity and Balance Through OWA Operators

Table 1 presents several movie titles selected from FilmAffinity[1] in the category of romance.

Table 1. Titles of movies selected from FilmAffinity

	Movie titles		Movie titles
1	A true love story never ends	11	True love
2	This is not your story	12	Love and air sex
3	True romance	13	Holiday affair
4	Shakespeare in love	14	A holiday to remember
5	Love and other cults	15	Love on the air
6	A holiday for love	16	My true love story
7	Love actually	17	Our story needs no filter
8	West side story	18	Love story
9	A taste of romance	19	Love is in the air
10	The history of love	20	Roman holiday

After cleaning the text in Table 1 and following the methodology explained in [8], two tag clouds have been generated considering different supports in terms of absolute frequency. We can see them in Fig. 2.

For both tag clouds, metrics proposed in Subsects. 3.1 and 3.2 have been calculated. We can see the values obtained in Table 2.

The details of the calculation of balance through OWA operators is in Table 3, where B is the ordered vector of items in C.

As we can see, Tag Cloud 1 performs better in coverage and disparity, but worst in overlap. The Tag Cloud 1 has smaller support than Tag Cloud 2, so a

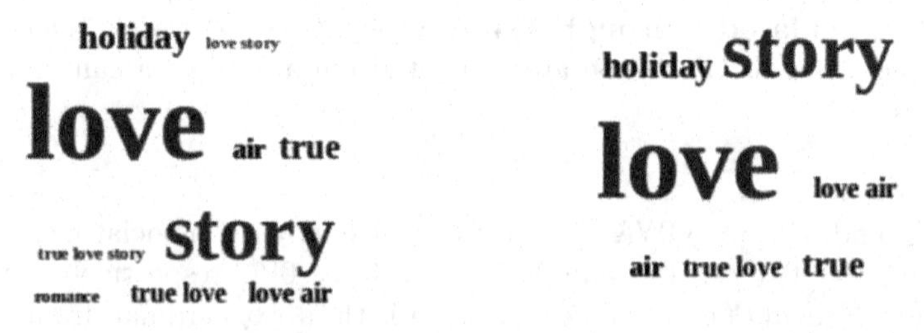

Fig. 2. Tag clouds for text in Table 1

[1] https://www.filmaffinity.com/, last accessed March 2018.

Table 2. Coverage, overlap and disparity for tag clouds in Fig. 2

Metrics	Tag Cloud 1	Tag Cloud 2
Coverage	1	0.95
Overlap	0.393	0.365
Disparity	0.406	0.369
Balance	0.5	0.75

Table 3. Details of the balance calculation

	Tag Cloud 1	Tag Cloud 2
C	{2, 3, 4}	{3, 4}
Outliers	None	None
B	[4 3 2]	[4 3]
W_*	$[0\ 0\ 1]^T$	$[0\ 1]^T$
W^*	$[1\ 0\ 0]^T$	$[1\ 0]^T$
$F_*(C)$	$[4\ 3\ 2] * [0\ 0\ 1]^T = 2$	$[4\ 3] * [0\ 1]^T = 3$
$F^*(C)$	$[4\ 3\ 2] * [1\ 0\ 0]^T = 4$	$[4\ 3] * [1\ 0]^T = 4$
Balance	$2/4 = 0.5$	$3/4 = 0.75$

greater number of tags appears in its visualization, increasing the coverage. The bigger number of tags causes redundant information to be represented, so the overlap is increased as well. The disparity is also increased with the appearance of tags with smaller weights and the tag cloud is more unbalanced. Apparently, disparity is a softer measure than balance, due to the fact that it considers all the tags for its calculation and the balance only two.

The choice of one or another tag cloud will depend on the preferences in each case.

4 Conclusions

In order to evaluate the tag cloud as a content representation tool, three metrics have been proposed: coverage, overlap and disparity. The three take values in the interval $[0, 1]$ and use all the tags in the visualization for its calculation. In addition, a new way to calculate balance through OWA operators has been proposed.

Coverage and overlap are usually in confrontation with each other. When the coverage increases, the overlap also increases as more tags appear in the cloud, representing a large amount of information which brings more redundancies. Disparity and balance can also be affected with the variations in coverage and overlap. Disparity seems to be a softer measure than balance. The proportion between these values will be established according to the requirements in each case.

As future work we plan to continue researching with tag clouds generated over textual databases, the inclusion of fuzzy logic in tags is considered, their grouping in clusters, the adoption of OWA operators for tag cloud comparison, as well as the study of the properties of said operators. We intend also to use the tag cloud for searching entities and the introduction of other languages to create a multilingual tool based on ontologies.

Acknowledgements. This work has been partially supported by the "Plan Andaluz de Investigación, Junta de Andalucía" (Spain) under research project P10- TIC6019.

References

1. Aouiche, K., Lemire, D., Godin, R.: Web 2.0 OLAP: from data cubes to tag clouds. In: Cordeiro, J., Hammoudi, S., Filipe, J. (eds.) WEBIST 2008. LNBIP, vol. 18, pp. 51–64. Springer, Heidelberg (2009). https://doi.org/10.1007/978-3-642-01344-7_5
2. Goutte, C., Gaussier, E.: A probabilistic interpretation of precision, recall and *F*-score, with implication for evaluation. In: Losada, D.E., Fernández-Luna, J.M. (eds.) ECIR 2005. LNCS, vol. 3408, pp. 345–359. Springer, Heidelberg (2005). https://doi.org/10.1007/978-3-540-31865-1_25
3. Leone, S., Geel, M., Müller, C., Norrie, M.C.: Exploiting tag clouds for database browsing and querying. In: Soffer, P., Proper, E. (eds.) CAiSE Forum 2010. LNBIP, vol. 72, pp. 15–28. Springer, Heidelberg (2011). https://doi.org/10.1007/978-3-642-17722-4_2
4. Morik, K., Kaspari, A., Wurst, M., Skirzynski, M.: Multi-objective frequent termset clustering. Knowl. Inf. Syst. **30**(3), 715–738 (2012)
5. Skoutas, D., Alrifai, M.: Tag clouds revisited. In: Proceedings of the 20th ACM International Conference on Information and Knowledge Management, CIKM, pp. 221–230 (2011)
6. Torres-Parejo, U., Campaña, J.R., Vila, M.-A., Delgado, M.: Text retrieval and visualization in databases using tag clouds. In: Greco, S., Bouchon-Meunier, B., Coletti, G., Fedrizzi, M., Matarazzo, B., Yager, R.R. (eds.) IPMU 2012. CCIS, vol. 297, pp. 390–399. Springer, Heidelberg (2012). https://doi.org/10.1007/978-3-642-31709-5_40
7. Torres-Parejo, U., Campaña, J., Delgado, M., Vila, M.: MTCIR: a multi-term tag cloud information retrieval system. Expert Syst. Appl. **40**, 5448–5455 (2013)
8. Torres-Parejo, U., Campaña, J., Vila, M., Delgado, M.: A theoretical model for the automatic generation of tag clouds. Knowl. Inf. Syst. **40**(2), 315–347 (2014)
9. Venetis, P., Koutrika, G., Garcia-Molina, H.: On the selection of tags for tag clouds. In: Proceedings of the 4th ACM International Conference on Web Search and Data Mining, WSDM, pp. 835–844 (2011)
10. Yager, R.: On ordered weighted averaging aggregation operators in multicriteria decisionmaking. IEEE Trans. Syst. Man Cybern. **18**(1), 183–190 (1988)
11. Yager, R.: Families of OWA operators. Fuzzy Sets Syst. **59**(2), 125–148 (1993)

Evidential Bagging: Combining Heterogeneous Classifiers in the Belief Functions Framework

Nicolas Sutton-Charani(✉), Abdelhak Imoussaten, Sébastien Harispe, and Jacky Montmain

LGI2P, IMT Mines Alès, University of Montpellier, Alès, France
{nicolas.sutton-charani,abdelhak.imoussaten, sebastien.harispe,jacky.montmain}@mines-ales.fr, http://lgi2p.mines-ales.fr

Abstract. In machine learning, *Ensemble Learning* methodologies are known to improve predictive accuracy and robustness. They consist in the learning of many classifiers that produce outputs which are finally combined according to different techniques. *Bagging*, or Bootstrap Aggregating, is one of the most famous Ensemble methodologies and is usually applied to the same classification base algorithm, i.e. the same type of classifier is learnt multiple times on bootstrapped versions of the initial learning dataset. In this paper, we propose a *bagging* methodology that involves different types of classifier. Classifiers' probabilist outputs are used to build mass functions which are further combined within the belief functions framework. Three different ways of building mass functions are proposed; preliminary experiments on benchmark datasets showing the relevancy of the approach are presented.

Keywords: Belief functions · Information fusion · Bagging
Supervised learning

1 Introduction

As the amount of learning algorithms and methodologies in the literature has reached a point where it is almost impossible to stay up to date on all of them, many users or even researchers tend to use them as black boxes, without focusing much on their understanding or interpretation, often setting model parameters to default values. Beside some obvious computational time differences between them, it has been proven that there is no *optimal* learning algorithm in the sense that most models are optimal for certain types of learning data and have advantages and drawbacks [1]. Indeed, dataset dimensions, attribute types, variance and noise make each learning dataset more suited to some learning algorithms than others. To illustrate this fact, six standard learning algorithms have been applied to six benchmark datasets and the resulting mean accuracies (i.e. the

J. Medina et al. (Eds.): IPMU 2018, CCIS 853, pp. 297–309, 2018.
https://doi.org/10.1007/978-3-319-91473-2_26

correct predictions rate) from 1000 10-fold cross validation simulations are presented in Fig. 1. We can easily observe some disparities between the different types of classifier (decision trees, SVM, etc) used on those six datasets. For example, *neural network* seems to be one of the less accurate classifier on almost all datasets except for "Balance scale" where it is the most accurate one. We can also observe that decision trees and neural network have the highest accuracy variances.

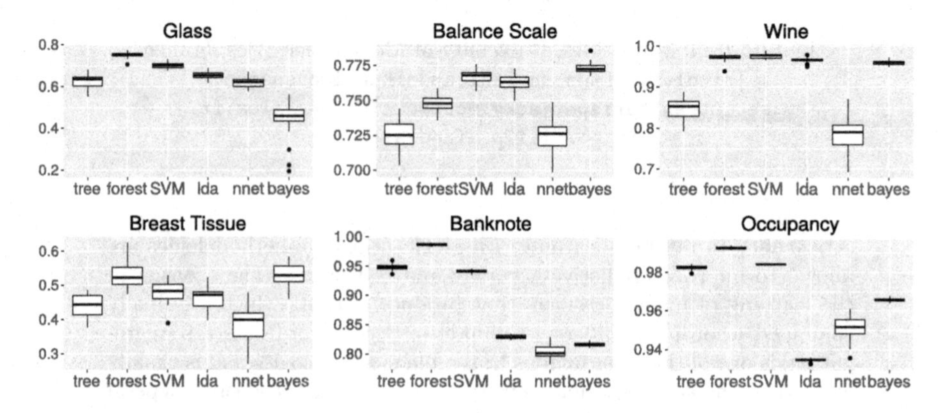

Fig. 1. Learning algorithms mean accuracies on 1000 simulations with default parameters tuning in R.

Ensemble learning methodologies consist in the learning of many classifiers from the same initial dataset. In that context, classifiers are then aggregated, or the predictions they provide (their outputs) are combined in order to get final predictions. The resulting classifier is usually more accurate and robust [2–4]. One particular ensemble method, called *bagging* or bootstrap aggregating, has an additional advantage: it tends to avoid over-fitting [3]. Bagging uses some re-sampling methods in order to decrease the dependency of classifiers on the learning data, i.e. to decrease the learning data's bias and variance. Unlike other ensemble methods as *Bucket-of-models* [2,5], bagging usually involves the simultaneous and multiple use of a single algorithm, to further combine their predictions with a simple *vote* procedure.

This problem can be seen as an information fusion problem if we consider the trained classifiers as information sources and their predictions as the information, or evidence, to fuse. Traditionally, in bagging methods classifiers' outputs are combined through a vote procedure as they usually involve the same type of classifiers. With that approach, fusing the same type of classifiers should indeed lead to a uniform weighting.

In the evidential framework, researchers have proposed some evidential bagging methods, handling uncertain data and the combination procedure is also performed between the same type of classifiers [6–8]. To the best of our knowledge, no work has been proposed that combine heterogeneous types of classifier with belief functions.

In this paper we present a bagging method that involves the fusion of different types of classifiers' outputs, or predictions. We propose to take into account the classifiers' outputs reliability during the aggregation step within the formalism of the belief functions theory [9,10]. This choice is motivated by its generalization power (including probabilistic and possibilistic cases) and the flexible tools it provides. In addition many approaches have been developed in different contexts, especially for information fusion problems when we have evidence about the sources reliability [11]. In this work, we chose to use predictive performance of the classifiers as reliability evidences. Then the classifiers' probabilistic outputs and their reliabilities are used to build one predictive belief function per classifier. Finally, a suitable combination procedure is used to merge those belief functions into a global predictive mass function.

After defining our classification formalism, recalling bagging basis, the theory of belief functions with some basic fusion tools are presented in Sect. 2; our evidential bagging model is described in Sect. 3, experiments are provided on benchmark datasets in Sect. 4, and finally results and perspectives are discussed in Sect. 5.

2 Related Works and Positioning

In this section, first the general classification formalism is given, then a succinct overview of *Ensemble learning* methods is presented; finally *Bagging* is more precisely described.

2.1 Classification Formalism

Starting from a dataset D containing N learning examples $(x, y)_{i=1,\ldots,N}$, classification tasks aim at learning a model f able to predict the *class label* y^* of any new unlabeled example from its *attribute* (or feature) values x such that $y^* = f(x)$. The attributes $X = (X^1, \ldots, X^J)$ take their values in $\mathcal{X} = \mathcal{X}^1 \times \cdots \times \mathcal{X}^J$, the class Y in a finite set Ω. Spaces $\mathcal{X}^j$ can be categorical or numerical.

$$D = \begin{pmatrix} x_1, y_1 \\ \vdots \\ x_N, y_N \end{pmatrix} = \begin{pmatrix} x_1^1 & \ldots & x_1^J & y_1 \\ \vdots & & \vdots & \vdots \\ x_N^1 & \ldots & x_N^J & y_N \end{pmatrix}.$$

Samples are assumed to be i.i.d. but in practice data are often noisy, casual correlations can randomly occur and some outliers or very rare examples can occur even in very small datasets. In order to discount those outliers' influence and to reduce bias, bootstrapping techniques [12] can be applied. By doing so, overfitting can be decreased without the need of large datasets but the different bootstrapped subsamples' results (i.e. predictions) have to be aggregated in a conservative way so that the most uncertain predictions can be weakened.

In this context, a classifier is a function $c : \mathcal{X} \to \Omega$, the notation $c(x)$ corresponds to the prediction obtained by c from attribute values $x \in \mathcal{X}$.

2.2 Ensemble Learning Methods

As almost each learning algorithm is optimal for specific types of learning data and problems [1], many researchers do not confine in single classifiers and tend to use many of them when dealing with real data. In this context, *Ensemble Learning* methods have become very common tools to improve classifiers performance (in term of accuracy and robustness) by using predictive algorithms' heterogeneity and resampling techniques to avoid overfitting. In fact, Ensemble Learning has become a whole research field [3,4] where we can distinguish three main types of methodology: *boosting*, *stacking* and *bagging*. Boosting [13] is an iterative procedure where the classifier is *re*-learnt in order to better classify misclassified examples. *AdaBoost* is the most famous boosting algorithm [14]. Stacking [15] focuses on the classifiers outputs aggregation step with the learning of an aggregation algorithm. *Bagging*, or Bootstrap-Aggregation, is based on resampling and was first presented by Breiman [16] with the well known *Random forest* algorithm which consists in the learning of several decision trees from different bootstrapped subsamples and in their aggregation through a *vote* procedure. In this paper, a *Bagging* methodology is proposed based on a bucket of diverse learning algorithms. The aggregation step is formalized in the evidential framework and involves belief functions generated from the single classifiers' predictions and evaluations.

2.3 Bagging

In most applications, getting a dataset truly representative of the actual population is a complex task as sampling often implies noise intrusion and thus bias and variance creation [17]. As a matter of fact, in most learning datasets, some rare examples of the reality can be over-represented which often leads to overfitted models. During the learning process, those examples should not be given too much weight as they can be considered outliers. It is for that matter that *Bagging* involves resampling, in order to discount that kind of bias [12].

Nevertheless, in most bagging approaches, the same learning algorithm is used to fill the *bag* because of the heterogeneity of the classifiers' natural outputs (scores, class probability, etc). During the aggregation step, trainable combiners have shown some asymptotically optimal properties [18] but involves a common training set for all single classifiers. In this work we use a combiner learnt on such a validation set which will be also used for the single classifiers evaluations.

2.4 Theory of Belief Functions

The theory of *belief functions*, also known as *Evidence* or *Dempster-Shafer* theory was first presented by Dempster in 1967 in a statistical context [9] as an attempt to reconcile frequentists and Bayesians. Dempster laid the foundation of a mathematical theory that deals with uncertainty in a much more general framework than standard probability theory and which handles aleatory (or *objective*) uncertainties and epistemic (or *subjective*) ones.

Even if some evidential works are related to statistical and classification contexts, many researchers use belief functions as a convenient framework for information fusion problems. As a matter of fact, fusion problems occur in many contexts, even in the classification ones. In the evidential framework two levels are considered: the credal and the decision ones. The credal level is dedicated to the uncertainty representation whereas the decision step uses the credal one to make decision.

Credal Level. Let Y be an uncertain quantity whose value y lies in a finite set Ω called the *frame of discernment.*

Definition 1. *A mass function m regarding Y's value $y \in \Omega$ is a function defined on the set of subsets of Ω, which is usually written 2^Ω or $\mathcal{P}(\Omega)$ and is called the powerset, with its values in $[0,1]$ and verifying $\sum_{B \in 2^\Omega} m(B) = 1$.*

The quantity $m(\emptyset)$ is usually fixed to 0 in many cases. From m, two uncertainty measures can be defined about y, the *belief* function Bel and the *plausibility* function Pl, which express the information contained in m in different ways, more conservatively for Bel than for Pl. Since there is *1 to 1* correspondances between m, Bel and Pl, belief functions can indifferently refer to a mass function m or its corresponding inferior uncertainty measure Bel.

Decision Level. In a practical matter, the decision step of many problems is often handled by transforming belief functions into *pignistic* probabilities [19] (cf. Definition 2) to further consider the most *probable* event in the *pignistic sense.*

Definition 2. *The pignistic probability distribution attached to a mass function m is defined by:*

$$\forall \omega \in \Omega,\ BetP(\omega) = \sum_{A \subseteq \Omega | \omega \in A} \frac{m(A)}{|A|}. \tag{1}$$

Sources Reliability and Evidential Fusion. Information fusion problems aim at combining different informative contents coming from different sources. In the evidential framework, sources are represented by mass functions.

Three main concepts must be taken into account when combining evidential sources: dependence, reliability and conflict. The first combination method proposed in the belief function theory was Dempster's conjunctive combination rule. Nevertheless, this rule handles conflict in a way that has been criticized [20,21]. Moreover, this rule requires independence between sources.

To take more conveniently into account the conflict and the dependence between sources, several combination rules have been proposed [20–22]. To avoid dependence hypothesis, some works [11,23] use the *average* operator to combine beliefs functions, which is in line with voting procedures. In this paper we chose this operator as a basis.

3 Evidential Bagging

This section presents the three evidential generative models defined in this paper, as well as the fusion approach proposed to aggregate the predictions provided by the different single classifiers. Finally the general scheme of our approach is given.

3.1 Generative Models

Even if the different learning algorithms may provide outputs of different structures, most data science softwares enable a calibration step in order to get a probability distribution on class labels from those outputs. Those probabilities express an uncertainty which should be integrated in any *bagging* approach that presents an uncertainty focus. Nevertheless, those uncertainty measures are computed on the learning data and are therefore subject to over-fitting. Better estimators of classifiers' performance can be computed on other dataset. In this paper we compute single classifiers' accuracy on such a separate dataset which we recall as the *validation set.*

If we consider the single classifiers as information sources, and their probabilist outputs as the uncertain information to merge, one way to evaluate the classifiers reliability is to evaluate their accuracy on the *validation set.* Our approach consists in discounting the classifiers' outputs according to their reliabilities. To this aim, belief functions are generated in order to enrich the probabilist outputs of the single classifiers. Those belief functions are based both on class probabilist predictions, and accuracies computed on validation sets. They are finally merged into a final belief function and its pignistic probability gives class predictions. Proposed models are introduced hereafter:

1. **Simple discounting ($EBag_{SD}$):** using this model we consider that the less accurate classifiers should be the most discounted during the aggregation step. Otherwise stated, we consider that the *reliability* of a classifier c is a function of its global accuracy denoted acc_c (computed on the validation set). To this aim, the mass function associated to c is defined as $\forall i \in \{1, \dots, N\}$, $\forall \omega \in \Omega$:

$$\begin{cases} m_i^c(\{\omega\}) = P_i^c(\omega) \times acc_c \\ m_i^c(\Omega) \quad = 1 - acc_c \end{cases} . \qquad (2)$$

   where $P_i^c(\omega)$ stands for the probability of the class label ω provided by the classifier c on the example x_i.
2. **Class-dependent model ($EBag_{CD}$):** some classifiers are more accurate in predicting some specific class labels. Such accuracy variations can be observed by analysing the confusion matrix associated to each classifier. We therefore propose to take advantage of this information to spread uncertainty about classification involving specific class labels from Ω. Using this model the following definitions of mass functions are considered; $\forall c \in \{1, \dots, C\}$, $\forall i \in \{1, \dots, N\}$, $\forall \omega \in \Omega$:

$$\begin{cases} m_i^c(\{\omega\}) = P_i^c(\omega) \times P^c(Y = y_i | c(x_i) = y_i) \\ m_i^c(\Omega) \quad = \quad P^c(Y \neq y_i | c(x_i) = y_i) \end{cases}. \tag{3}$$

Note that for a given classifier c, $P^c(Y|c(X))$ is empirically estimated on the confusion matrix of the classifier c computed on the validation set. We have $P^c(Y = y_i|c(x_i) = y_i) \approx \frac{|(x,y) \colon \{c(x_i) = y_i\} \cap \{Y = y_i\}|}{|(x,y) \colon \{c(x) = y_i\}|}$.

3. **Contextual model** (**$EBag_{con}$**): we consider two types of regions of $\mathcal{X}$: one containing instances often misclassified and its complementary. In addition, we consider that not all classifiers will misclassify instances of the same regions of $\mathcal{X}$. We are therefore interested by the definition of a contextual model that will consider an estimation of the single classifiers' misclassification risk or *probability* of each examples. Formally, to estimate this risk, that is both dependent on the classifier and the processed instance, we consider that $\forall c \in \{1, \dots, C\}$, we have $mis_c : \mathcal{X} \to [0, 1]$ a function used to assess the misclassification risk of a classifier c regarding a given $x_i \in \mathcal{X}$ whose predicted label is y_i, i.e. $y_i = c(x_i)$; Intuitively, higher is the estimated risk, lower the confidence on the provided class will be. Based on this function, we consider the following definition of the mass function, $\forall c \in \{1, \dots, C\}$, $\forall i \in \{1, \dots, N\}, \forall \omega \in \Omega$:

$$\begin{cases} m_i^c(\{\omega\}) \quad = P_i^c(\omega) \times (1 - mis_c(x_i)) \\ m_i^c(\Omega \setminus \{y_i\}) = \quad mis_c(x_i) \end{cases}. \tag{4}$$

We consider that mis_c is obtained using a binary regressor (SVM in our case) that will be trained using examples defined by the misclassifications of c during the validation phase. Otherwise stated the regressor predicts a numerical output in $[0, 1]$ standing for the misclassifications observed in the validation set (1 for misclassified examples and 0 for well classified ones).

Whereas generative models $EBag_{SD}$ and $EBag_{CD}$ are based on the single classifiers global accuracy, model $EBag_{con}$ uses their local misclassification risk to discount their predictions. Therefore, the focal elements of models $EBag_{SD}$ and $EBag_{CD}$ are the class labels and the frame of discernment Ω. Model $EBag_{con}$ considers the class labels and their complementary sets. Intuitively, if a single classifier misclassifies an instance, our belief should be focused on its prediction's complementary.

3.2 Combination and Prediction

Classifiers' Mass Combination: As evidential independence is needed to apply Dempster's combination rule and disjunctive semantic has no sense in case of different classifiers predicting different class labels, the aggregation step of our model was done by averaging the mass functions generated by the single classifiers. Actually, in the evidential bagging context, most authors have combined the classifiers' outputs through a vote procedure or with the average operator applied to their class probabilities because of the dependencies between classifiers [6,8,24,25].

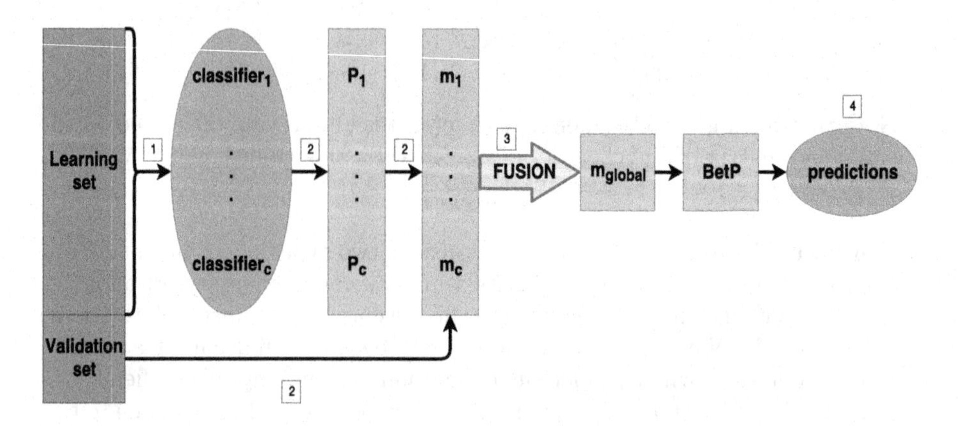

Fig. 2. Global process

Final Prediction: At the decision or prediction step, we chose the most likely class labels in the *pignistic* sense, i.e. the average belief function (computed from all single classifiers outputs) is transformed into its *pignistic* probability and then the most likely class label is predicted.

$$prediction(m_i^c) = \underset{\omega \in \Omega}{\operatorname{argmax}}\, BetP_i^c(\{\omega\}).$$

3.3 Global Procedure

Considering a set of classifiers, and a set of labelled data, the global procedure of our evidential bagging model is composed of four main steps illustrated in Fig. 2:

1 ***Training*** of each classifier. This step is used to estimate the best parameters of each classifiers based on training data. A set of trained classifiers, i.e. tuned models, is obtained.

2 ***Belief function generation*** through a post-analysis of training performance evaluations: the step used to compute data that will be used to define the mass functions. The treatments applied for each model vary; all rely on the analysis of the single classifiers' performance on the validation set. For the simple discounting model (Eq. 2), the accuracy of each classifier is computed on the validation set. For the class-dependent model (Eq. 3) the confusion matrix is computed on the validation set. The conditional probabilities that will be used to define the mass functions are estimated based on the confusion matrix. Finally, using the contextual model (Eq. 4) the misclassification risk is estimated by training a SVM classifier in the aim of distinguishing cases for which the classifier fails to provide a good classification.

3 ***Mass combination***: given an evidential generative model, the various evidential predictions provided by the classifiers are combined to output a single global mass function with the *average* operator.

[4] ***Prediction*** (or decision) step: the global mass function is transformed into its corresponding pignistic probability in order to predict the most *probable* class labels.

4 Experiments

In this section, we evaluate our evidential bagging approach on several UCI[1], Kaggle[2] and KEEL[3] benchmark datasets. Number of examples, attributes and class labels of those datasets are summarized in Table 1.

Table 1. Number of examples (N), attributes (J) and class labels (K) of several benchmark datasets from *UCI*, *Kaggle* and *KEEL*

Dataset	N	J	K
Balance scale	625	3	3
Banana	5300	2	2
Banknote	1372	4	2
Breast tissue	106	9	6
Contraceptive method	1473	9	3
E.coli	336	5	8
Glass	214	9	6
Iris	150	4	3
Mammographic	830	5	2
Nursery	12958	8	4
Occupancy	8143	6	2
Pima	768	8	2
Satimage	6435	36	6
Tic tac toe	958	9	2
Titanic	2201	3	2
Wine	178	13	3

The considered learning algorithms were: decision tree (*'tree'*), random forest (*'forest'*), support vector machine (*'SVM'*), linear discriminant analysis (*'lda'*) and naive Bayes classifier (*'bayes'*). The implementation was handled in R with the following functions (and packages) *rpart* (rpart), *randomForest* (randomForest), *svm* (e1071), *nnet* (nnet), *naiveBayes* (e1071) and *lda* (MASS). All the learning algorithms were implemented in R with their default parameters. For each dataset, 100 10-fold cross validation procedures were implemented for different bagging methods involving different aggregation approaches:

[1] https://archive.ics.uci.edu/ml/datasets.html.
[2] https://www.kaggle.com/datasets.
[3] http://sci2s.ugr.es/keel/category.php?cat=clas.

- *voteBag*: simple vote procedure from the precise single classifiers' predictions
- P_{mean}: averaging of the single classifiers' probabilistic predictions and prediction of most probable class label
- $EBag_{SD}$: *simple discounting* model (see Eq. (2))
- $EBag_{CD}$: *class-dependent model* (see Eq. (3))
- $EBag_{con}$: *contextual model* (see Eq. (4)).

For each fold, a learning set and a validation set were built as follows: a quarter of the examples contained in the nine other folds were randomly selected as the validation set; the remaining examples were then used to build the learning set. The accuracy means and standard deviations results are summarised in Table 2. T-tests were performed between the two most accurate models, bold accuracies stand for the significantly highest ones. R implementations of tested methods, links to datasets, as well as complete technical details about the evaluation are provided at https://github.com/lgi2p/evidentialBagging.

Table 2. Mean accuracies on 100 10-fold cross validation procedures

Dataset	*voteBag*	P_{mean}	$EBag_{SD}$	$EBag_{CD}$	$EBag_{con}$
Balance scale	**0.767**	0.766	0.766	0.761	0.765
Banana	0.765	0.831	0.843	0.853	0.855
Banknote	0.911	0.918	0.925	**0.936**	0.932
Breast Tissue	0.519	0.519	0.523	0.508	**0.538**
Contraceptive method	0.560	0.559	0.560	0.559	**0.562**
E. Coli	0.863	0.862	0.863	0.658	**0.865**
Glass	0.689	0.688	0.689	0.675	**0.701**
Iris	0.959	0.959	0.959	0.960	0.959
Mammographic	0.833	0.834	0.834	0.834	**0.836**
Nursery	0.957	0.969	**0.970**	0.968	0.970
Occupancy	0.983	0.983	0.983	0.984	**0.984**
Pima	0.765	0.765	0.765	0.712	0.765
Satimage	**0.881**	0.873	0.875	0.880	0.880
Tic tac toe	0.809	0.839	0.846	**0.887**	0.852
Titanic	0.782	0.782	0.782	0.782	0.782
Wine	0.979	0.977	0.978	0.979	0.979

Globally, there is no systematically and significantly outperformance between models even if $EBag_{con}$ seems to be the most accurate model. This is not too surprising as the general spirit of classifiers is to link attributes (i.e. context) and class labels, whereas inferring global reliability or treating class labels non-symmetrically is not at the basis of standard classification tasks. For most of the datasets containing the more class labels (i.e. for $K \geq 6$: 'Breast Tissue',

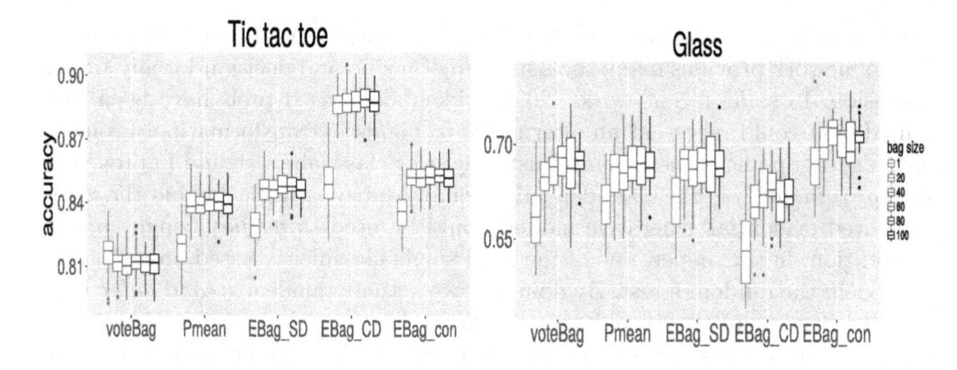

Fig. 3. Bag size effect

'E.Coli' and 'Glass'), $EBag_{CD}$ model is the less accurate and robust one. This suggests that expressing single classifiers' reliability on specific class labels has a sense for limited number of labels.

As in any bagging methods, the number of bagged classifiers has an impact on the resulting classifier's accuracy, bigger bags implying higher accuracies. In Fig. 3, the accuracies over 100 10-fold cross validations are represented for different bag sizes (1 to 100 per learning algorithm) on the 'Tic tac toe' and 'Glass' datasets. For the dataset 'Tic tac toe', the most accurate model is $EBag_{CD}$ whereas it is $EBag_{con}$ for the dataset 'Glass'. This corroborates the fact that our *class-dependent* model ($EBag_{CD}$) is more accurate on dataset containing few class labels (2 for 'Tic tac toe', 6 for 'Glass'). As enhanced by Table 2, for datasets containing many class labels, our *contextual model* ($EBag_{con}$) is preferable.

5 Conclusions and Perspectives

A general *bagging* approach has been proposed that involves belief mass generation from each classifiers and evidential fusion between classifiers' evidential predictions. Different aspects of the classifiers (global accuracy, confusion matrix or local misclassification risk) can be separately evaluated on a validation set and used to generate those belief functions. Experiments on benchmark datasets show encouraging results especially for the model $EBag_{con}$, which is based on a misclassification risk learnt on a separated validation set and depending on attributes values. That approach can refer to the concepts of *taking into account the learning context* or *learning the local context* (before the actual learning process) and should be further studied.

A more complete study of this evidential bagging approach should include a global sensitivity analysis over the type of classifiers to bag, the combination rule (between single classifiers' evidential predictions) and the prediction, or *decision* step. It is noticeable that considering the pignistic transform after averaging the mass functions is equivalent to some straightforward probabilistic modelling.

Nevertheless, since probabilities are some particular belief functions, the evidential framework provides many tools in term of fusion and decision for any future extension. In some recent works [26,27], likelihood-based tools have been presented that could represent an alternative to pignistic transformation. From an optimisation point of view, some clustering over class labels should improve the $EBag_{CD}$ model (i.e. the class-dependent one). Moreover, ideas behind the three presented evidential generative models could be used to define a single model. In addition, in this paper, one of the used single classifier is based on a bagging approach: the random forest. By doing so, we actually made a *second order* bagging (or bagging of classifiers bags). To go further on this aspect, mathematical properties of bagging approaches should be taken into account in order to solve one pragmatic dilemma: should we make larger bags or should we nest single classifiers bags?

References

1. Wolpert, D.H.: The supervised learning no-free-lunch theorems. In: Roy, R., Köppen, M., Ovaska, S., Furuhashi, T., Hoffmann, F. (eds.) Soft Computing and Industry: Recent Applications, pp. 25–42. Springer, London (2002). https://doi.org/10.1007/978-1-4471-0123-9_3
2. Qu, G., Wu, H.: Bucket learning: improving model quality through enhancing local patterns. Knowl.-Based Syst. **27**, 51–59 (2012)
3. Zhou, Z.H.: Ensemble Methods: Foundations and Algorithms, 1st edn. Chapman & Hall/CRC, Boca Raton (2012)
4. Polikar, R.: Ensemble based systems in decision making. IEEE Circ. Syst. Mag. **6**(3), 21–45 (2006)
5. Džeroski, S., Ženko, B.: Is combining classifiers with stacking better than selecting the best one? Mach. Learn. **54**(3), 255–273 (2004)
6. Vannoorenberghe, P.: On aggregating belief decision trees. Inf. Fus. **5**(3), 179–188 (2004)
7. Xu, P., Davoine, F., Zha, H., Denœux, T.: Evidential calibration of binary SVM classifiers. Int. J. Approx. Reason. **72**, 55–70 (2016)
8. Ma, L., Sun, B., Li, Z.: Bagging likelihood-based belief decision trees. In: 2017 20th International Conference on Information Fusion (Fusion), pp. 1–6 (2017)
9. Dempster, A.P.: Upper and lower probabilities induced by a multivalued mapping. Ann. Math. Stat. **38**, 325–339 (1967)
10. Shafer, G.: A Mathematical Theory of Evidence. Princeton University Press, Princeton (1976)
11. Denoeux, T., El Zoghby, N., Cherfaoui, V., Jouglet, A.: Optimal object association in the Dempster-Shafer framework. IEEE Trans. Cybern. **44**(22), 2521–2531 (2014)
12. Efron, B.: Bootstrap methods: another look at the jackknife. Ann. Stat. **7**, 1–26 (1979)
13. Freund, Y., Schapire, R.E.: A decision-theoretic generalization of on-line learning and an application to boosting. J. Comput. Syst. Sci. **55**, 119–139 (1997)
14. Schapire, R.E.: Explaining adaboost. In: Schölkopf, B., Luo, Z., Vovk, V. (eds.) Empirical Inference, pp. 37–52. Springer, Heidelberg (2013). https://doi.org/10.1007/978-3-642-41136-6_5
15. Wolpert, D.H.: Stacked generalization. Neural Netw. **5**(2), 241–259 (1992)

16. Breiman, L.: Random forests. Mach. Learn. **45**, 5–32 (2001)
17. Cortes, C., Mohri, M., Riley, M., Rostamizadeh, A.: Sample selection bias correction theory. In: Freund, Y., Györfi, L., Turán, G., Zeugmann, T. (eds.) ALT 2008. LNCS (LNAI), vol. 5254, pp. 38–53. Springer, Heidelberg (2008). https://doi.org/10.1007/978-3-540-87987-9_8
18. Duin, R.P.W.: The combining classifier: to train or not to train? In: Object Recognition Supported by User Interaction for Service Robots vol. 2, pp. 765–770 (2002)
19. Smets, P., Kennes, R.: The transferable belief model. Artif. Intell. **66**(2), 191–234 (1994)
20. Yager, R.R.: On the dempster-shafer framework and new combination rules. Inf. Sci. **41**(2), 93–137 (1987)
21. Smets, P.: Belief functions: the disjunctive rule of combination and the generalized Bayesian. Int. J. Approx. Reason. **9**(1), 1–32 (2005)
22. Dubois, D., Prade, H.: Representation and combination of uncertainty with belief functions and possibility measures. Comput. Intell. **4**(3), 244–264 (1988)
23. Florea, M.C., Dezert, J., Valin, P., Smarandache, F., Jousselme, A.: Adaptative combination rule and proportional conflict redistribution rule for information fusion. CoRR abs/cs/0604042 (2006)
24. François, J., Grandvalet, Y., Denceux, T., Roger, J.M. In: Bagging improves uncertainty representation in evidential pattern classification. Physica-Verlag HD, pp. 295–308 (2002)
25. Xu, P., Davoine, F., Denoeux, T.: Evidential combination of pedestrian detectors. In: British Machine Vision Conference, pp. 1–14. Nottingham (2014)
26. Denœux, T.: Maximum likelihood estimation from uncertain data in the belief function framework. IEEE Trans. Knowl. Data Eng. **25**, 119–130 (2011)
27. Sutton-Charani, N., Destercke, S., Denœux, T.: Learning decision trees from uncertain data with an evidential EM approach. In: International Conference on Machine Learning and Applications (ICMLA) (2013)

Uninorms That Are Neither Conjunctive Nor Disjunctive on Bounded Lattices

Gül Deniz Çaylı(✉)

Department of Mathematics, Faculty of Sciences,
Karadeniz Technical University, 61080 Trabzon, Turkey
guldeniz.cayli@ktu.edu.tr

Abstract. In this paper, we demonstrate that on some bounded lattices L, there exist elements $e \in L\backslash\{0,1\}$ such that all uninorms having e as the neutral element are only conjunctive or disjunctive. And we introduce two new construction methods to obtain uninorms that are neither conjunctive nor disjunctive on a bounded lattice with a neutral element under some additional constraints. Furthermore, an illustrative example showing that our methods differ slightly from each other is added.

Keywords: Bounded lattice · Uninorm · Conjunctive uninorm
Disjunctive uninorm · Neutral element

1 Introduction

Yager and Rybalov [18] introduced uninorms on the unit interval as an important generalization of triangular norms (t-norms for short) and triangular conorms (t-conorms for short) and Fodor et al. [14] studied the general structure of these operators. In this generalization, the neutral element (sometimes called identity) of uninorm can be any number from the unit interval $[0,1]$. If the neutral element is 1, we obtain t-norms and if the neutral element is 0, we obtain t-conorms. Uninorms were proved to be useful in many fields like fuzzy logic, expert systems, neural networks, aggregation and fuzzy system modeling [10,17–20]. It is well known that a uninorm U can be conjunctive or disjunctive whenever $U(0,1) = 0$ or 1, respectively.

Karaçal and Mesiar [15] showed that considering a bounded lattice L, it is possible to choose arbitrary elements $e \in L\backslash\{0,1\}$ as the neutral one and to construct the greatest uninorm based on a t-norm T on L and the least uninorm based on a t-conorm S on L. In addition, they constructed the weakest and the strongest uninorm on L with the neutral element e. Çaylı et al. [6] presented two new methods to generate uninorms on bounded lattices different from proposed in [15]. As a by-product of these methods, it was shown the existence of idempotent uninorms on a bounded lattice L for any element $e \in L\backslash\{0,1\}$ playing the role of a neutral element, i.e., it is obtained the smallest idempotent uninorm and the greatest idempotent uninorm on L with a neutral element. Some other related constructions of uninorms on bounded lattices can be found also in [4,7].

J. Medina et al. (Eds.): IPMU 2018, CCIS 853, pp. 310–318, 2018.
https://doi.org/10.1007/978-3-319-91473-2_27

Deschrijver [11] demonstrated that there exist uninorms which are neither conjunctive nor disjunctive in interval-valued fuzzy set theory. In the same paper, Deschrijver comprehensively studied such uninorms and researched the structure of these uninorms in interval-valued fuzzy set theory. In addition, Bodjanova and Kalina [5] studied uninorms which are neither conjunctive nor disjunctive on bounded lattices and investigated some necessary and some sufficient conditions, under which it is possible to construct such uninorms on a bounded lattice L.

Note that the approaches for constructing uninorms on general bounded lattices described by [4,6,7,15] generate either conjunctive uninorm or disjunctive uninorm. The aim of this paper is to give new constructions of uninorms on bounded lattices that are neither conjunctive nor disjunctive. The paper is organized as follows. First, some basic constructs and their resulting elements are briefly discussed in Sect. 2. In Sect. 3, we show that on some bounded lattices L, given a properly chosen neutral element e different from the top and bottom elements of L, a uninorm with that neutral element e, which is neither conjunctive nor disjunctive, does not exist. Furthermore, we propose two different methods for building uninorms that are neither conjunctive nor disjunctive on a bounded lattice L for an element $e \in L\backslash\{0,1\}$ playing the role of a neutral element under some additional assumptions. And we give an illustrative example to obtain such uninorms by means of our methods. Finally, some concluding remarks are added.

2 Preliminaries

In this section, some preliminaries concerning bounded lattices and uninorms (t-norms, t-conorms) on them are recalled.

A bounded lattice $(L, \leq)$ is a lattice which has the top and bottom elements, which are written as 1 and 0, respectively, that is, there exist two elements $1, 0 \in L$ such that $0 \leq x \leq 1$, for all $x \in L$.

Definition 1 ([3]). *Given a bounded lattice $(L, \leq, 0, 1)$ and $a, b \in L$, if a and b are incomparable, we use the notation $a \parallel b$.*

Definition 2 ([3]). *Given a bounded lattice $(L, \leq, 0, 1)$ and $a, b \in L$, $a \leq b$, a subinterval $[a, b]$ of L is defined as*

$$[a, b] = \{x \in L \mid a \leq x \leq b\}.$$

Similarly, we define $]a, b] = \{x \in L \,|\, a < x \leq b\}$, $[a, b[= \{x \in L \,|\, a \leq x < b\}$ and $]a, b[= \{x \in L \,|\, a < x < b\}$.

Let $(L, \leq, 0, 1)$ be a bounded lattice and $e \in L$. Let $A(e) = [0, e] \times [e, 1] \cup [e, 1] \times [0, e]$.

Definition 3 ([6,8,15]). *Let $(L, \leq, 0, 1)$ be a bounded lattice. An operation $U : L^2 \to L$ is called a uninorm on L (shortly a uninorm, if L is fixed) if it is commutative, associative, increasing with respect to both variables and there exist some elements $e \in L$ called the neutral element such that $U(e, x) = x$ for all $x \in L$.*

Definition 4 ([1,2,9]). *Let $(L, \leq, 0, 1)$ be a bounded lattice. An operation $T : L^2 \to L$ $(S : L^2 \to L)$ is called a t-norm (t-conorm) if it is commutative, associative, increasing with respect to both variables and has a neutral element $e = 1$ $(e = 0)$.*

Proposition 1 ([15]). *Let $(L, \leq, 0, 1)$ be a bounded lattice, $e \in L \backslash \{0, 1\}$ and U be a uninorm on L with the neutral element e. Then*

(i) $T^* = U|_{[0,e]^2} : [0, e]^2 \to [0, e]$ *is a t-norm on* $[0, e]$.
(ii) $S^* = U|_{[e,1]^2} : [e, 1]^2 \to [e, 1]$ *is a t-conorm on* $[e, 1]$.

Proposition 2 ([15]). *Let $(L, \leq, 0, 1)$ be a bounded lattice, $e \in L \backslash \{0, 1\}$ and U be a uninorm on L with the neutral element e. The following properties hold:*

(i) $x \wedge y \leq U(x, y) \leq x \vee y$ *for all* $(x, y) \in A(e)$.
(ii) $U(x, y) \leq x$ *for* $(x, y) \in L \times [0, e]$.
(iii) $U(x, y) \leq y$ *for* $(x, y) \in [0, e] \times L$.
(iv) $x \leq U(x, y)$ *for* $(x, y) \in L \times [e, 1]$.
(v) $y \leq U(x, y)$ *for* $(x, y) \in [e, 1] \times L$.

Definition 5 ([6]). *Let $(L, \leq, 0, 1)$ be a bounded lattice, $e \in L \backslash \{0, 1\}$ and U be a uninorm on L with the neutral element e. U is called an idempotent uninorm if $U(x, x) = x$ for all $x \in L$.*

Proposition 3 ([6]). *Let $(L, \leq, 0, 1)$ be a bounded lattice, $e \in L \backslash \{0, 1\}$ and U be an idempotent uninorm on L with the neutral element e.*

(i) If $(x, y) \in [e, 1]^2$, *then* $U(x, y) = x \vee y$.
(ii) If $(x, y) \in [0, e]^2$, *then* $U(x, y) = x \wedge y$.

Definition 6 ([6]). *Let $(L, \leq, 0, 1)$ be a bounded lattice, $e \in L \backslash \{0, 1\}$ and U be a uninorm on L with the neutral element e.*

(i) U is called conjunctive uninorm if $U(0, 1) = 0$.
(ii) U is called disjunctive uninorm if $U(0, 1) = 1$.

Proposition 4 ([6]). *If $(L, \leq, 0, 1)$ be a bounded lattice, $e \in L \backslash \{0, 1\}$ and U be an idempotent uninorm on L with the neutral element e, then we have that*

(i) $U(x, y) \leq x \wedge y$ *for all* $(x, y) \in [0, e]^2$.
(ii) $x \vee y \leq U(x, y)$ *for all* $(x, y) \in [e, 1]^2$.
(iii) $x \wedge y \leq U(x, y) \leq x \vee y$ *for all* $(x, y) \in L^2 \backslash ([0, e]^2 \cup [e, 1]^2)$.

Theorem 1 ([6]). *Let $(L, \leq, 0, 1)$ be a bounded lattice and $e \in L \backslash \{0, 1\}$. Then the uninorm $U_t : L^2 \to L$ is the greatest idempotent uninorm and the uninorm $U_s : L^2 \to L$ is the smallest idempotent uninorm on L with the neutral element e.*

$$U_t(x, y) = \begin{cases} x \wedge y \,, & if \ (x, y) \in [0, e]^2 \\ y \,, & if \ x \in [0, e], \ y \parallel e \\ x \,, & if \ y \in [0, e], \ x \parallel e \\ x \vee y \,, & otherwise \end{cases} \tag{1}$$

$$U_s(x,y) = \begin{cases} x \vee y\ , if\ (x,y) \in [e,1]^2 \\ y \quad\ \ , if\ x \in [e,1],\ y \parallel e \\ x \quad\ \ , if\ y \in [e,1],\ x \parallel e \\ x \wedge y\ , otherwise \end{cases} \tag{2}$$

3 Uninorms That Are Neither Conjunctive Nor Disjunctive

In this section, we investigate the existence of uninorms on a bounded lattice L with the neutral element $e \in L \backslash \{0,1\}$ which are neither conjunctive nor disjunctive and propose two construction methods to obtain such uninorms with some constraints. Recall the uninorms U_t ($U_t(0,1) = 1$) and U_s ($U_s(0,1) = 0$) on a bounded lattice L with the neutral element $e \in L \setminus \{0,1\}$ which are introduced in Theorem 1 are disjunctive uninorm and conjunctive uninorm, respectively.

Proposition 5. *Let $(L, \leq, 0, 1)$ be a bounded lattice, $e \in L \backslash \{0,1\}$ and U be a uninorm on L with the neutral element e. Then $U(0,1) = 0$ or $U(0,1) = 1$ or $U(0,1) \parallel e$.*

Proof. Take an arbitrary uninorm U on a bounded lattice L with the neutral element e. The proof is split into all possible cases.

1. Let $U(0,1) = e$. Then we have that $U(U(0,1),1) = U(e,1) = 1$ and $U(0, U(1,1)) = U(0,1) = e$. This is a contradiction with the associativity of U. So, it can not be $U(0,1) = e$.

2. Let $U(0,1) = p$ such that $p \in]e,1[$. Then we have that $U(U(0,1),1) = U(p,1) = 1$ and $U(0, U(1,1)) = U(0,1) = p$. This is a contradiction with the associativity of U. So, it can not be $U(0,1) = p$.

3. Let $U(0,1) = q$ such that $q \in]0,e[$. Then we have that $U(0, U(0,1)) = U(0,q) = 0$ and $U(U(0,0),1) = U(0,1) = q$. This is a contradiction with the associativity of U. So, it can not be $U(0,1) = q$.

Therefore, we obtain that $U(0,1) = 0$ or $U(0,1) = 1$ or $U(0,1) \parallel e$.

Corollary 1. *Consider a bounded lattice L, $e \in L \setminus \{0,1\}$ such that all elements in L are comparable with e. Then every uninorm on L with the neutral element e is either conjunctive or disjunctive.*

The proof is immediate from Proposition 5.

Proposition 6. *Let $(L, \leq, 0, 1)$ be a bounded lattice, $e \in L \backslash \{0,1\}$ and there is only one element incomparable with e in L. If we denote by k the element incomparable with e and there are the elements $m, n \in]0,e[$ such that $m \wedge n < k, m \parallel k$ and $n \parallel k$, then every uninorm U on L with the neutral element e is either conjunctive or disjunctive.*

Proof. We can see the fact that $U(0,1) \notin]0,e]$ and $U(0,1) \notin]e,1[$ from Proposition 5.

Suppose that $U(0,1) = k$. By the monotonicity of U, we have that $U(n,1) \geq U(n,e) = n$ and $U(n,1) \geq U(0,1) = k$, i.e. $U(n,1) \geq n \vee k$. Since $n \vee k > e, U(n,1) > e$. By the associativity of U, we have that $k = U(0,1) = U(0, U(1,1)) = U(U(0,1),1) = U(k,1)$, i.e. $U(k,1) = k$. Then we have that $U(U(m,n),1) \leq U(k,1) = k$ and $U(m, U(n,1)) \geq U(m,e) = m$. Then it holds $m \leq k$ from the associativity of U. This is a contradiction. So, it can not be $U(0,1) = k$.

Therefore, we obtain that $U(0,1) = 0$ or $U(0,1) = 1$, i.e. a uninorm U on the bounded lattice L contains such elements with the neutral element e is either conjunctive or disjunctive.

Proposition 7. *Let $(L, \leq, 0, 1)$ be a bounded lattice, $e \in L \backslash \{0,1\}$ and there is only one element incomparable with e in L. If we denote by k the element incomparable with e and there are the elements $m, n \in]e,1[$ such that $m \vee n > k, m \parallel k$ and $n \parallel k$, then every uninorm U on L with the neutral element e is either conjunctive or disjunctive.*

It can be proved similarly to Proposition 6.

Proposition 8. *Let $(L, \leq, 0, 1)$ be a bounded lattice, $e \in L \backslash \{0,1\}$ and there is only one element incomparable with e in L. If we denote by k the element incomparable with e and there are the elements $m, t \in]0,e[$ such that $m \wedge t = 0$, $m \parallel k$ and $y, z \in]e,1[$ such that $y \vee z = 1$, $y \parallel k$, then every uninorm U on L with the neutral element e is either conjunctive or disjunctive.*

Proof. We can see the fact that $U(0,1) \notin]0,e]$ and $U(0,1) \notin]e,1[$ from Proposition 5.

Suppose that $U(0,1) = k$. In this case, we obtain that $m = U(m,e) \leq U(m,y) \leq U(e,y) = y$. Let $U(m,y) = p$ such that $p \in [m,e]$. Then $p \parallel k$. Due to the associativity of U, we have that $k = U(0,1) = U(U(m,0),1) = U(m, U(0,1)) = U(m,k)$, i.e. $U(m,k) = k$. By the associativity of U, we have that $U(p,z) = U(U(m,y),z) = U(m, U(y,z)) \geq U(m,k) = k$, i.e. $U(p,z) \geq k$. In addition, it holds $z \geq U(p,z) \geq k \vee p$ since $z = U(e,z) \geq U(p,z) \geq U(p,e) = p$. Since $k \vee p > e$, it holds $z \geq U(p,z) > e$. By the monotonicity of $U, U(U(p,z),1) \geq U(e,1) = 1$, i.e. $U(U(p,z),1) = 1$. By the associativity of $U, 1 = U(U(p,z),1) = U(p, U(z,1)) = U(p,1)$, i.e. $1 = U(p,1)$. From the associativity and commutativity of $U, 1 = U(p,1) = U(U(y,m),1) = U(m, U(y,1)) = U(m,1)$, i.e. $1 = U(m,1)$. By the associativity of $U, 1 = U(m,1) = U(m, U(y,z)) = U(U(m,y),z) = U(p,z)$, i.e. $1 = U(p,z)$. Since $z \geq U(p,z)$, we have that $z = 1$. This is a contradiction. So, it can not be $U(m,y) = p$ such that $p \in [m,e]$. Similarly, it is shown that it can not be $U(m,y) = q$ such that $q \in]e,y]$. It is clearly seen that it can not be $U(m,y) = k$. Therefore, we have that it can not be $U(0,1) = k$.

Hence, we obtain that $U(0,1) = 0$ or $U(0,1) = 1$, i.e. a uninorm U on the bounded lattice L contains such elements with the neutral element e is either conjunctive or disjunctive.

Due to the above mentioned propositions, we can see that, choosing properly the neutral element e, on some bounded lattices L only conjunctive and disjunctive uninorms with the neutral element e exist. Note that in the bounded lattices which are given in Propositions 6–8, there is only one element incomparable with e and at least one of the subintervals $[0, e]$ and $[e, 1]$ is not a chain. Taking into account all of these, we research in the following theorem that there is always a uninorm that is neither conjunctive nor disjunctive on which bounded lattices and the structure of such uninorms. For this purpose, we present two methods to construct uninorms on a bounded lattice L with the neutral element $e \in L\backslash\{0, 1\}$ which are neither conjunctive nor disjunctive under some additional assumptions on L. It is easy to see that these proposed methods differ from each other and besides from the existing methods fundamentally by reason of their structures that are neither conjunctive nor disjunctive.

Theorem 2. *Let $(L, \leq, 0, 1)$ be a bounded lattice, $e \in L\backslash\{0, 1\}$, the subintervals $[0, e]$ and $[e, 1]$ be chains and there be only one element incomparable with e in L. If we denote by k the element incomparable with e, then the function $U_e^k : L^2 \to L$ defined as*

$$U_e^k(x, y) = \begin{cases} x \vee y \,, & \begin{array}{l} if\ x \in [e, 1]\ and\ y \in [e, 1] \\ or\ x \in]0, e]\,,\ x \parallel k\ and\ y \in]e, 1]\,,\ y \in]k, 1] \\ or\ x \in]e, 1]\,,\ x \in]k, 1]\ and\ y \in]0, e]\,,\ y \parallel k \end{array} \\ k \,, & \begin{array}{l} if\ x \in [0, e[\,,\ x \in [0, k[\ and\ y \in]e, 1]\,,\ y \in]k, 1] \\ or\ x \in]e, 1]\,,\ x \in]k, 1]\ and\ y \in [0, e[\,,\ y \in [0, k[\\ or\ x \in [e, 1]\ and\ y = k\ or\ x = k\ and\ y \in [e, 1] \\ or\ x \in [0, e]\ and\ y = k\ or\ x = k\ and\ y \in [0, e] \end{array} \\ x \wedge y \,, & otherwise \end{cases} \tag{3}$$

is a uninorm on L with the neutral element e that is neither conjunctive nor disjunctive.

Theorem 3. *Let $(L, \leq, 0, 1)$ be a bounded lattice, $e \in L\backslash\{0, 1\}$, the subintervals $[0, e]$ and $[e, 1]$ be chains and there be only one element incomparable with e in L. If we denote by k the element incomparable with e, then the function $U_k^e : L^2 \to L$ defined as*

$$U_k^e(x, y) = \begin{cases} x \wedge y \,, & \begin{array}{l} if\ x \in [0, e]\ and\ y \in [0, e] \\ or\ x \in [e, 1[\,,\ x \parallel k\ and\ y \in [0, e[\,,\ y \in [0, k[\\ or\ x \in [0, e[\,,\ x \in [0, k[\ and\ y \in [e, 1[\,,\ y \parallel k \end{array} \\ k \,, & \begin{array}{l} if\ x \in]e, 1]\,,\ x \in]k, 1]\ and\ y \in [0, e[\,,\ y \in [0, k[\\ or\ x \in [0, e[\,,\ x \in [0, k[\ and\ y \in]e, 1]\,,\ y \in]k, 1] \\ or\ x \in [e, 1]\ and\ y = k\ or\ x = k\ and\ y \in [e, 1] \\ or\ x \in [0, e]\ and\ y = k\ or\ x = k\ and\ y \in [0, e] \end{array} \\ x \vee y \,, & otherwise \end{cases} \tag{4}$$

is a uninorm on L with the neutral element e that is neither conjunctive nor disjunctive.

Consider a bounded lattice L, $e \in L \backslash \{0, 1\}$ such that the subintervals $[0, e]$ and $[e, 1]$ are chains and there is only one element incomparable with e in L. Note that the uninorms U_e^k and U_k^e which are given by the formula (3) in Theorem 2 and the formula (4) in Theorem 3, respectively, are idempotent uninorms on L with the neutral element e. In addition, it can be clearly seen that since the idempotent uninorms U_e^k and U_k^e are not conjunctive or disjunctive, these idempotent uninorms are different from the uninorms U_t and U_s which are given by the formulas (1) and (2), respectively, in Theorem 1. Furthermore, the idempotent uninorms U_e^k and U_k^e differ slightly from each other. Let us demonstrate these with the following example:

Example 1. Given a bounded lattice $L = \{0, p, q, e, r, s, k, 1\}$ with the given order in Fig. 1. It is seen that the subintervals $[0, e]$ and $[e, 1]$ are chains and there is only one element incomparable with e in L. Define the functions $U_e^k, U_k^e : L^2 \to L$ as Tables 1 and 2, respectively, by using Theorems 2 and 3. Then the functions U_e^k, U_k^e are uninorms on L with the neutral element e which are neither conjunctive nor disjunctive. In addition, the uninorms U_e^k and U_k^e differ from each other, since $U_e^k(q, r) = q \neq r = U_k^e(q, r)$ for $q, r \in L$.

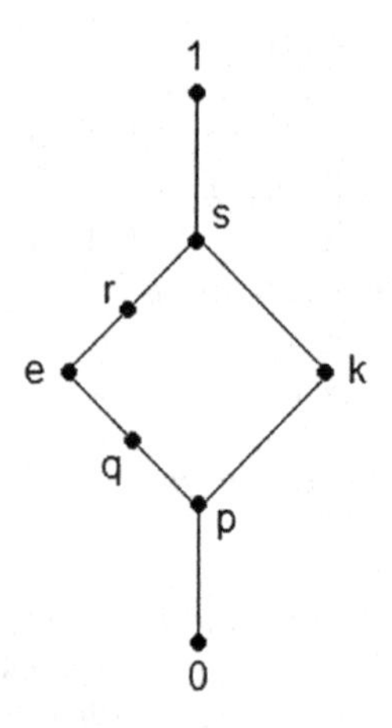

Fig. 1. The lattice L

Table 1. The uninorm U_e^k on L

U_e^k	0	p	q	e	r	s	k	1
0	0	0	0	0	0	k	k	k
p	0	p	p	p	p	k	k	k
q	0	p	q	q	q	s	k	1
e	0	p	q	e	r	s	k	1
r	0	p	q	r	r	s	k	1
s	k	k	s	s	s	s	k	1
k	k	k	k	k	k	k	k	k
1	k	k	1	1	1	1	k	1

Table 2. The uninorm U_k^e on L

U_k^e	0	p	q	e	r	s	k	1
0	0	0	0	0	0	k	k	k
p	0	p	p	p	p	k	k	k
q	0	p	q	q	r	s	k	1
e	0	p	q	e	r	s	k	1
r	0	p	r	r	r	s	k	1
s	k	k	s	s	s	s	k	1
k	k	k	k	k	k	k	k	k
1	k	k	1	1	1	1	k	1

4 Concluding Remarks

Following the notation of uninorm introduced on the unit interval as generalization of t-norms and t-conorms by Yager and Rybalov [18], the notation of uninorm on bounded lattices have recently attracted much attention. However, the structure of uninorms on bounded lattices were discussed by Karaçal and Mesiar [15], Çaylı et al. [6], the characterization of uninorms that are neither conjunctive nor disjunctive on general bounded lattices is still active research. The uninorms were also studied by many authors in other papers [12,13,16]. In this paper, we show that a uninorm on some bounded lattices with the given neutral element that is neither conjunctive nor disjunctive does not exist. We introduce two new construction methods to obtain uninorms on a bounded lattice L with the neutral element $e \in L\backslash\{0,1\}$ which are neither conjunctive nor disjunctive once there is only one element in L incomparable with e and the subintervals $[0,e]$ and $[e,1]$ are chains. In addition, we add an illustrative example to clearly understand the structure of the uninorms obtained by means of these methods.

References

1. Aşıcı, E.: Some remarks on F-partial order and properties. Rom. J. Math. Comput. Sci. **7**, 72–79 (2017)
2. Aşıcı, E.: On the properties of the F-partial order and the equivalence of nullnorms. Fuzzy Sets Syst. (in press). https://doi.org/10.1016/j.fss.2017.11.008
3. Birkhoff, G.: Lattice Theory. American Mathematical Society Colloquium Publishers, Providence (1967)
4. Bodjanova, S., Kalina, M.: Construction of uninorms on bounded lattices. In: IEEE 12th International Symposium on Intelligent Systems and Informatics, SISY 2014, Subotica, Serbia (2014). https://doi.org/10.1109/SISY.2014.6923558
5. Bodjanova, S., Kalina, M.: Uninorms on Bounded Lattices – Recent Development. In: Kacprzyk, J., Szmidt, E., Zadrożny, S., Atanassov, K.T., Krawczak, M. (eds.) IWIFSGN/EUSFLAT -2017. AISC, vol. 641, pp. 224–234. Springer, Cham (2018). https://doi.org/10.1007/978-3-319-66830-7_21

6. Çaylı, G.D., Karaçal, F., Mesiar, R.: On a new class of uninorms on bounded lattices. Inf. Sci. **367–368**, 221–231 (2016). https://doi.org/10.1016/j.ins.2016.05.036
7. Çaylı, G.D., Karaçal, F.: Construction of uninorms on bounded lattices. Kybernetika **53**(3), 394–417 (2017). https://doi.org/10.14736/kyb-2017-3-0394
8. Çaylı, G.D., Drygaś, P.: Some properties of idempotent uninorms on a special class of bounded lattices. Inf. Sci. **422**, 352–363 (2018). https://doi.org/10.1016/j.ins.2017.09.018
9. Çaylı, G.D.: On a new class of t-norms and t-conorms on bounded lattices. Fuzzy Sets Syst. **332**, 129–143 (2018). https://doi.org/10.1016/j.fss.2017.07.015
10. De Baets, B., Fodor, J.: Van Melle's combining function in MYCIN is a representable uninorm: an alternative proof. Fuzzy Sets Syst. **104**, 133–136 (1999). https://doi.org/10.1016/S0165-0114(98)00265-6
11. Deschrijver, G.: Uninorms which are neither conjunctive nor disjunctive in interval-valued fuzzy set theory. Inf. Sci. **244**, 48–59 (2013). https://doi.org/10.1016/j.ins.2013.04.033
12. Drewniak, J., Drygaś, P.: On a class of uninorms. Int. J. Uncertain. Fuzziness. Knowl.-Based Syst. **10**, 5–10 (2002). https://doi.org/10.1142/S021848850200179X
13. Drygaś, P.: On properties of uninorms with underlying t-norm and t-conorm given as ordinal sums. Fuzzy Sets Syst. **161**, 149–157 (2010). https://doi.org/10.1016/j.fss.2009.09.017
14. Fodor, J., Yager, R.R., Rybalov, A.: Structure of uninorms. Int. J. Uncertain. Fuzziness Knowl.-Based Syst. **5**, 411–427 (1997). https://doi.org/10.1142/S0218488597000312
15. Karaçal, F., Mesiar, R.: Uninorms on bounded lattices. Fuzzy Sets Syst. **261**, 33–43 (2015). https://doi.org/10.1016/j.fss.2014.05.001
16. Mesiarová-Zemanková, A.: Multi-polar t-conorms and uninorms. Inf. Sci. **301**, 227–240 (2015). https://doi.org/10.1016/j.ins.2014.12.060
17. Pedrycz, W., Hirota, K.: Uninorm-based logic neurons as adaptive and interpretable processing constructs. Soft. Comput. **11**(1), 41–52 (2016). https://doi.org/10.1007/s00500-006-0051-0
18. Yager, R.R., Rybalov, A.: Uninorms aggregation operators. Fuzzy Sets Syst. **80**, 111–120 (1996). https://doi.org/10.1016/0165-0114(95)00133-6
19. Yager, R.R.: Uninorms in fuzzy systems modeling. Fuzzy Sets Syst. **122**(1), 167–175 (2001). https://doi.org/10.1016/S0165-0114(00)00027-0
20. Yager, R.R., Kreinovich, V.: Universal approximation theorem for uninorm-based fuzzy systems modeling. Fuzzy Sets Syst. **140**, 331–339 (2003). https://doi.org/10.1016/S0165-0114(02)00521-3

On the Migrativity Property for Uninorms and Nullnorms

Emel Aşıcı(✉)

Department of Software Engineering, Faculty of Technology,
Karadeniz Technical University, 61830 Trabzon, Turkey
emelkalin@hotmail.com

Abstract. In this paper the notions of α-migrative uninorms over a fixed nullnorm and α-migrative nullnorms over a fixed uninorm are introduced and studied. All solutions of the migrativity equation for all possible combinations of uninorms and nullnorms are investigated. So, (α, T)-migrative nullnorm and (α, T)-migrative uninorm ((α, S)-migrative nullnorm and (α, S)-migrative uninorm) for a given t-norm (t-conorm) are extended to a more general form.

Keywords: Uninorm · Nullnorm · Migrativity

1 Introduction

The notion of α-migrativity of triangular norms was introduced by Durante and Sarkoci [10]. The definition is given as follows:

Definition 1. *Let* $\alpha \in]0,1[$ *be given. A binary operation* $T : [0,1]^2 \to [0,1]$ *is said to be* α*-migrative if we have*

$$T(\alpha x, y) = T(x, \alpha y)$$

for all $x, y \in [0,1]$.

Note that the product αx can be replaced by any t-norm T_0 obtaining the property for t-norms called (α, T_0) migrativity, that can be written as

$$T(T_0(\alpha, x), y) = T(x, T_0(\alpha, y))$$

for all $x, y \in [0,1]$.

Many authors investigated α-migrative property. The migrativity property was studied for t-norm in [14–16,26], for t-subnorms in [29], for semicopulas, quasi-copulas and copulas in [11–13,25].

Nullnorms and t-operators were introduced in [4,21], respectively, which are also generalizations of the notions of t-norms and t-conorms. On the other hand, uninorms were first introduced by Yager and Rybalov [30] and studied by Fodor et al. [17].

J. Medina et al. (Eds.): IPMU 2018, CCIS 853, pp. 319–328, 2018.
https://doi.org/10.1007/978-3-319-91473-2_28

In [23] it was introduced the definition of (α, U_0)-migrative uninorm analyzing some properties. (α, U_0)-migrative uninorms were characterized when U_0 lies in one of the following classes of uninorms: $\mathcal{U}_{min}$ or $\mathcal{U}_{max}$, idempotent uninorms, representable uninorms.

In [31] it was discussed and characterized the migrative property for the following four cases:

- Nullnorms with the same absorbing element.
- A nullnorm with the absorbing element $k \in]0, 1[$ is α-migrative over a t-norm.
- A nullnorm with the absorbing element $k \in]0, 1[$ is α-migrative over a t-conorm.
- Nullnorms F_1, F_2 with the different absorbing elements $k_1, k_2 \in]0, 1[$.

In [22] it was introduced the definition of (α, T)-migrative uninorm for a given t-norm T, analyzing some of its initial properties. The authors continue with the characterization of those (α, T)-migrative uninorms, that lay in each one of the most usual classes of uninorms, i.e., uninorms in $\mathcal{U}_{min}$ and $\mathcal{U}_{max}$, idempotent uninorms, representable uninorms and uninorms continuous in the open square $]0, 1[^2$.

In [28] it was studied α-migrative uninorms over a fixed uninorm, where those two uninorms have different neutral elements. All cases when both uninorm lay in any one of the most usual classes of uninorms are analyzed, characterizing all solutions of the migrativity equation for some possible combinations. The nullnorms, uninorms and t-norms were also studied by many authors in other papers [1–3, 5–9, 18–20, 24].

In the present paper, we introduce the migrativity of uninorms over nullnorms and the migrativity of nullnorms over uninorms. The paper is organized as follows. We shortly recall some basic notions in Sect. 2. In Sect. 3, we introduce the definition of (α, F)-migrative uninorm for a given nullnorm F, analyzing some of its properties. So, (α, T)-migrative uninorm ((α, S)-migrative uninorm) for a given t-norm (t-conorm) is extended to a more general form. In Sect. 4, an analogous study is done for nullnorms that are (α, U)-migrative for a given uninorm U. So, (α, T)-migrative nullnorm ((α, S)-migrative nullnorm) for a given t-norm (t-conorm) is extended to a more general form.

2 Preliminaries

Definition 2. [17] *A binary function $U : [0, 1]^2 \to [0, 1]$ is called a uninorm if it is associative, commutative, increasing with respect to the both variables and there is a neutral element $e \in [0, 1]$ such that $U(e, x) = x$ for all $x \in [0, 1]$.*

Evidently, a uninorm with neutral element $e = 1$ is a t-norm and a uninorm with neutral element $e = 0$ is a t-conorm. $A(e) =]0, e]x[e, 1[U[e, 1[x]0, e]$ for $e \in]0, 1[$.

For any uninorm it is satisfied that $U(0, 1) \in \{0, 1\}$ and a uninorm U is called conjunctive if $U(1, 0) = 0$ and disjunctive when $U(1, 0) = 1$.

Theorem 1. [17] *Let* $U : [0,1]^2 \to [0,1]$ *be a uninorm with neutral element* $e \in]0,1[$. *Then the sections* $x \mapsto U(x,1)$ *and* $x \mapsto U(x,0)$ *are continuous in each point except perhaps for* e *if and only if* U *is given by one of the following formulas.*

(a) If $U(0,1) = 0$, *then*

$$U(x,y) = \begin{cases} eT(\frac{x}{e}, \frac{y}{e}), & (x,y) \in [0,e]^2 \\ e + (1-e)S(\frac{x-e}{1-e}, \frac{y-e}{1-e}), & (x,y) \in [e,1]^2 \\ min(x,y), & (x,y) \in A(e), \end{cases} \quad (1)$$

where T *is a t-norm and* S *is a t-conorm.*

(b) If $U(0,1) = 1$, *then the same structure holds, changing minimum by maximum in* $A(e)$.

The set of uninorms as in case (a) will be denoted by $\mathcal{U}_{min}$ and the set of uninorms as in case (b) by $\mathcal{U}_{max}$. We will denote a uninorm U in $\mathcal{U}_{min}$ with underlying t-norm T, underlying t-conorm S and neutral element e by $U \equiv \langle T, e, S\rangle_{min}$ and in a similar way, a uninorm in $\mathcal{U}_{max}$ by $U \equiv \langle T, e, S\rangle_{max}$.

Definition 3. [17] *Consider* $e \in]0,1[$. *A binary operation* $U : [0,1]^2 \to [0,1]$ *is a representable uninorm if and only if there exists a continuous strictly increasing function* $h : [0,1] \to [-\infty, +\infty]$ *with* $h(0) = -\infty$, $h(e) = 0$ *and* $h(1) = +\infty$ *such that*

$$U(x,y) = h^{-1}(h(x) + h(y))$$

for all $(x,y) \in [0,1]^2 \setminus \{(0,1),(1,0)\}$ *and* $U(0,1) = U(1,0) = \{0,1\}$. *The function* h *is usually called an additive generator of* U.

Remark 1. [17] We will denote by $\mathcal{U}_{rep}$ the class of representable uninorms. Any representable uninorm U with neutral element e and additive generator h, will be denoted by $U \equiv \langle h, e\rangle_{rep}$.

Theorem 2. [27] *Suppose* U *is a uninorm continuous in* $]0,1[^2$ *with neutral element* $e \in]0,1[$. *Then either one of the following cases is satisfied:*

(a) There exist $u \in [0,e]$, $\lambda \in [0,u]$, *two continuous t-norms* T_1 *and* T_2 *and a representable uninorm* R *such that* U *can be represented as*

$$U(x,y) = \begin{cases} \lambda T_1(\frac{x}{\lambda}, \frac{y}{\lambda}), & if\ x,y \in [0,\lambda] \\ \lambda + (u-\lambda)T_2(\frac{x-\lambda}{u-\lambda}, \frac{y-\lambda}{u-\lambda}), & if\ x,y \in [\lambda,u] \\ u + (1-u)R(\frac{x-u}{1-u}, \frac{y-u}{1-u}), & if\ x,y \in]u,1[\\ 1, & if\ min(x,y) \in]\lambda,1]\ and\ max(x,y) = 1 \\ \lambda\ or\ 1, & if\ (x,y) \in \{(\lambda,1),(1,\lambda)\} \\ min(x,y), & otherwise \end{cases} \quad (2)$$

(b) There exist $v \in]e,1]$, $w \in [v,1]$, two continuous t-conorms S_1 and S_2 and a representable uninorm R such that U can be represented as

$$U(x,y) = \begin{cases} vT_1(\frac{x}{v}, \frac{y}{v}), & if\ x,y \in]0,v[\\ v + (w-v)S_1(\frac{x-v}{w-v}, \frac{y-v}{w-v}), & if\ x,y \in [v,w] \\ w + (1-w)S_2(\frac{x-w}{1-w}, \frac{y-w}{1-w}), & if\ x,y \in [w,1] \\ 0, & if\ max(x,y) \in [0,w[\ and\ min(x,y) = 0 \\ w\ or\ 0, & if\ (x,y) \in \{(0,w),(w,0)\} \\ max(x,y), & otherwise. \end{cases} \tag{3}$$

The class of all uninorms continuous in $]0,1[^2$ will be denoted by $\mathcal{U}_{cos}$. A uninorm as in (2) will be denoted by $U \equiv \langle T_1, \lambda, T_2, u, (R,e)\rangle_{cos,min}$ and the class of all uninorms continuous in the open unit square of this form will be denoted by $\mathcal{U}_{cos,min}$. Analogously, a uninorm as in (3) will be denoted by $U \equiv \langle (R,e), v, S_1, w, S_2\rangle_{cos,max}$ and the class of all uninorms continuous in the open unit square of this form will be denoted by $\mathcal{U}_{cos,max}$.

Definition 4. [4] *A function $F : [0,1]^2 \to [0,1]$ is called nullnorm if it is commutative, associative, increasing with respect to the both variables and there exists $k \in [0,1]$ called absorbing (zero) element that verifies $F(k,x) = k$ for all $x \in [0,1]$ and*

$$F(0,x) = x\ forall\ x \le k\ and\ F(1,x) = x\ forall\ x \ge k.$$

In that case, when $k = 0$ we obtain a t-norm and when $k = 1$ we obtain a t-conorm. In general, the absorbing element is always given by $k = F(1,0)$. The structure of nullnorms is given as follows.

Theorem 3. [21] *Let $F : [0,1]^2 \to [0,1]$ be a nullnorm with absorbing element $F(1,0) = k \notin \{0,1\}$. Then*

$$F(x,y) = \begin{cases} kS(\frac{x}{k}, \frac{y}{k}), & if\ (x,y) \in [0,k]^2 \\ k + (1-k)T(\frac{x-k}{1-k}, \frac{y-k}{1-k}), & if\ (x,y) \in [k,1]^2 \\ k, & otherwise. \end{cases} \tag{4}$$

where S is a t-conorm and T is a t-norm.

A nullnorm F with absorbing element k, underlying t-norm T will be denoted by $F \equiv \langle S, k, T\rangle$.

3 Migrativity of Uninorms over Nullnorms

In this section, we will introduce the definition of migrativity of a uninorm U over a nullnorm.

Definition 5. *Given $\alpha \in]0,1[$, let F be a nullnorm with absorbing element a and U be a uninorm with neutral element e. U is called α-migrative over F or simply (α, F)-migrative if*

$$U(F(\alpha,x),y) = U(x, F(\alpha,y))\ for\ all\ x,y \in [0,1]. \tag{5}$$

Remark 2. If we choose $e = 1$ ($e = 0$) and $a = 1$ ($a = 0$), then we have a t-norm (t-conorm) and a t-conorm (t-norm), respectively. In this case $T(S)$ is not α-migrative over $S(T)$.

Remark 3. Let $e = 1$ ($e = 0$) and $a \in]0, 1[$, then we have a t-norm (t-conorm). In this case we have the following lemma:

Lemma 1. *Consider $\alpha \in]0, 1[$. Let $T(S)$ be a t-norm (t-conorm), F be a nullnorm with absorbing element a. Then T is not α-migrative over F.*

Remark 4. If $e = 1$ and $a = 0$ or $e = 0$ and $a = 1$, then t-norms and t-conorms are already known we will consider only uninorms with neutral element $e \in]0, 1[$ and nullnorms with absorbing element $a \in]0, 1[$.

Lemma 2. *Consider $0 < a = e < 1$. Let U be a uninorm with neutral element e and F be a nullnorm with absorbing element a. Then U is not α-migrative over F.*

Proof. (i) Let $\alpha = a = e$. If U is α-migrative over F, then

$$U(0, F(a, e)) = U(0, e) = 0 < a = U(a, e) = U(F(a, 0), e).$$

It leads a contradiction that U is not α-migrative over F.

(ii) Let $0 < \alpha < a = e < 1$. If U is α-migrative over F, then

$$U(0, F(\alpha, e)) = U(0, e) = 0 < \alpha = U(\alpha, e) = U(F(\alpha, 0), e).$$

It leads a contradiction that U is not α-migrative over F.

(iii) Let $0 < a = e < \alpha < 1$. If U is α-migrative over F, then

$$U(F(\alpha, 1), e) = U(\alpha, e) = \alpha < 1 = U(1, e) = U(1, F(\alpha, e)).$$

It leads a contradiction that U is not α-migrative over F.

So, if $0 < a = e < 1$, then it is obtained that U is not α-migrative over F.

Lemma 3. *Consider $0 < \alpha < a < e < 1$. Let U be a uninorm with neutral element e and F be a nullnorm with absorbing element a. Then U is not α-migrative over F.*

Proof. By the monotonicity of F, we have that $F(\alpha, e) = a$. Also, since $a < e$, by the monotonicity of U, we get $U(a, 0) = 0$.
Suppose that U is α-migrative over F. Then,

$$U(0, F(\alpha, e)) = U(0, a) = 0 < \alpha = U(\alpha, e) = U(F(\alpha, 0), e).$$

It leads a contradiction that U is not α-migrative over F.

Lemma 4. *Consider $0 < e < a < \alpha < 1$. Let U be a uninorm with neutral element e and F be a nullnorm with absorbing element a. Then U is not α-migrative over F.*

Proposition 1. *Let U be a conjunctive uninorm with neutral element e and F be a nullnorm with absorbing element a and $\alpha \in]0, a[$. Then U is not α-migrative over F.*

Proof. Let U is (α, F)-migrative. Since U is a conjunctive, we have that $U(F(\alpha, e), 0) = 0$. On the other hand, we have that

$$U(e, F(\alpha, 0)) = U(e, \alpha) = \alpha.$$

So, it is obtained that $\alpha = 0$, is a contradiction.

Proposition 2. *Let U be a disjunctive uninorm with neutral element e and F be a nullnorm with absorbing element a and $\alpha \in]a, 1[$. Then U is not α-migrative over F.*

Theorem 4. *Let U be a representable uninorm and $\alpha \in]a, 1[$. Then U is not α-migrative over F.*

Proof. Let U be representable uninorm. Then we have that $U(\alpha, 1) = 1$ for all $\alpha \in]0, 1[$. In this case,

$$U(e, F(\alpha, 1)) = U(e, \alpha) = \alpha < 1 = U(F(\alpha, e), 1).$$

It is obtained that a contradiction.

Theorem 5. *Let U be a uninorm with neutral element e, F be a nullnorm with absorbing element a, $a \neq e$ and $\alpha \in]0, 1[$. Then the following statements are equivalent:*

(i) $F(\alpha, e) = \alpha$ and U is (α, F)-migrative
(ii) $U(\alpha, y) = F(\alpha, y)$ for all $y \in [0, 1]$.

Proof. For the first part, let $F(\alpha, e) = \alpha$ and U is α-migrative over F. In this case, we have that

$$U(\alpha, y) = U(F(\alpha, e), y) = U(e, F(\alpha, y)) = F(\alpha, y)$$

for all $y \in [0, 1]$.
For the second part, let $U(\alpha, y) = F(\alpha, y)$ for all $y \in [0, 1]$.
$F(\alpha, e) = U(\alpha, e) = \alpha$ and by commutativity and associativity of U

$$U(F(\alpha, x), y) = U(U(\alpha, x), y) = U(U(x, \alpha), y) = U(x, U(\alpha, y)) = U(x, F(\alpha, y)).$$

Proposition 3. *Let U be a conjunctive uninorm with neutral element e, F be a nullnorm with absorbing element a, $\alpha \in]a, 1[$ and $F(\alpha, e) = \alpha$. If $e \leq \alpha$, then U is not α-migrative over F.*

Proof. We suppose U is α-migrative over F. It is obtained that $U(\alpha, y) = F(\alpha, y)$ for all $y \in [0, 1]$ by Theorem 5. Since $e \leq \alpha$, then we have that $U(\alpha, 1) = 1$, $F(\alpha, 1) = \alpha$, a contradiction.

Proposition 4. *Let U be a disjunctive uninorm with neutral element e, F be a nullnorm with absorbing element a, $\alpha \in]0, a[$ and $F(\alpha, e) = \alpha$. If $\alpha \leq e$, then U is not (α, F)-migrative.*

Theorem 6. *Let $U \equiv \langle T_1, \lambda, T_2, u, (R, e)\rangle_{cos,min}$ be a uninorm in $\mathcal{U}_{cos,min}$, F be a nullnorm, $\alpha \in]a, 1[$ and $F(\alpha, e) = \alpha$. Then U is not α-migrative over F.*

Proof. Let U be (α, F)-migrative. Since $F(\alpha, e) = \alpha$, we obtained that $U(\alpha, y) = F(\alpha, y)$ for all $y \in [0, 1]$ by Theorem 5. Let $\alpha > \lambda$. Since $min(\alpha, 1) = \alpha > \lambda$ and $max(\alpha, 1) = 1$, it is obtained that $U(\alpha, 1) = 1$. On the other hand, we have that $F(\alpha, 1) = \alpha$, a contradiction.

Theorem 7. *Let $U \equiv \langle (R, e), v, S_1, w, S_2\rangle_{cos,max}$ be a uninorm in $\mathcal{U}_{cos,max}$, F be a nullnorm, $\alpha \in]0, a[$ and $F(\alpha, e) = \alpha$. Then U is not α-migrative over F.*

4 Migrativity of Nullnorms over Uninorms

Now, we will introduce the definition of migrativity of a nullnorm F over a uninorm.

Definition 6. *Given $\alpha \in]0, 1[$, let U be a uninorm with neutral element e and F be a nullnorm with absorbing element a. F is called α-migrative over U or simply (α, U)-migrative if*

$$F(U(\alpha, x), y) = F(x, U(\alpha, y)) \ for \ all \ x, y \in [0, 1]. \quad (6)$$

Similar arguments can be suggested for migrativity of uninorms over nullnorms.

Remark 5. Let $a = 1$ ($a = 0$) and $e \in]0, 1[$, then we have a t-conorm (t-norm). In this case we have the following lemma:

Lemma 5. *Consider $S(T)$ be a t-conorm (t-norm) and U be a uninorm with neutral element e. Then, S is e-migrative over U.*

As migrativity of uninorms over nullnorms, we will consider only nullnorms with absorbing element $a \in]0, 1[$ and uninorms with neutral element $e \in]0, 1[$.

Lemma 6. *Consider F be a nullnorm with absorbing element a and U be a uninorm with neutral element e. Then, F is e-migrative over U.*

Proof. $F(U(e, x), y) = F(x, y) = F(x, U(e, y))$ for all $x, y \in [0, 1]$. So, F is e-migrative over U.

Lemma 7. *Consider $0 < a = e < 1$. Let F be a nullnorm with absorbing element a and U be a uninorm with neutral element e. Then F is not α-migrative over U.*

Proof. (i) Let $\alpha < a = e$ and F is α-migrative over U, then

$$F(0, U(\alpha, a)) = F(0, \alpha) = \alpha < a = F(U(\alpha, 0), a).$$

It leads a contradiction that U is not α-migrative over F.

(ii) Let $a = e < \alpha$ and F is α-migrative over U, then

$$F(U(\alpha, 1), a) = a < \alpha = F(1, \alpha) = F(1, U(\alpha, a)).$$

It leads a contradiction that U is not α-migrative over F.

Lemma 8. *Consider $0 < \alpha < e < a < 1$. Let F be a nullnorm with absorbing element a and U be a uninorm with neutral element e. Then F is not α-migrative over U.*

Proof. If F is α-migrative over U, then

$$F(0, U(\alpha, e)) = F(0, \alpha) = \alpha < e = F(0, e) = F(U(\alpha, 0), e).$$

It leads a contradiction that U is not α-migrative over F.

Lemma 9. *Consider $0 < e < a < \alpha < 1$. Let F be a nullnorm with absorbing element a and U be a uninorm with neutral element e. Then F is not α-migrative over U.*

Lemma 10. *Consider $0 < \alpha < a < e < 1$. Let F be a nullnorm with absorbing element a and U be a uninorm with neutral element e. Then F is not α-migrative over U.*

Proof. If F is α-migrative over U, then

$$F(0, U(\alpha, e)) = F(0, \alpha) = \alpha < a < F(0, e) = F(U(\alpha, 0), e).$$

It leads a contradiction that U is not α-migrative over F.

Lemma 11. *Consider $0 < a < e < \alpha < 1$. Let F be a nullnorm with absorbing element a and U be a uninorm with neutral element e. Then F is not α-migrative over U.*

Lemma 12. *Consider $0 < e < \alpha < a < 1$. Let F be a nullnorm with absorbing element a and U be a conjunctive uninorm with neutral element e. Then F is not α-migrative over U.*

Proof. Assume that F is α-migrative over U. Since U is a conjunctive, we have that

$$F(U(\alpha, 0), e) = F(0, e) = e < \alpha = F(0, \alpha) = F(0, U(\alpha, e)).$$

It leads a contradiction that F is not α-migrative over U.

Lemma 13. *Consider $0 < a < \alpha < e < 1$. Let F be a nullnorm with absorbing element a and U be a disjunctive uninorm with neutral element e. Then F is not α-migrative over U.*

5 Conclusions

We have introduced and studied the migrativity of uninorms over nullnorms and the migrativity of nullnorms over uninorms. Specially the migrative is studied for the most usual classes of uninorms including uninorms in $\mathcal{U}_{min}$ or $\mathcal{U}_{max}$, representable uninorms, conjunctive and disjunctive uninorms. We have characterized all uninorms in these classes that are (α, F)-migrative where $\alpha \in]0, 1[$ and F is a nullnorm. Similarly we have characterized all nullnorms in these classes that are (α, U)-migrative where $\alpha \in]0, 1[$ and U is a uninorm.

Acknowledgement. We are grateful to the anonymous reviewers and editors for their valuable comments which have enabled us to improve the original version of our paper.

References

1. Aşıcı, E.: On the properties of the F-partial order and the equivalence of nullnorms. Fuzzy Sets Syst. (in press). https://doi.org/10.1016/j.fss.2017.11.008
2. Aşıcı, E.: Some remarks on F-partial order and properties. Rom. J. Math. Comput. Sci. **7**, 72–79 (2017)
3. Aşıcı, E.: Some remarks on an order induced by uninorms. In: Kacprzyk, J., Szmidt, E., Zadrożny, S., Atanassov, K.T., Krawczak, M. (eds.) IWIFSGN/EUSFLAT - 2017. AISC, vol. 641, pp. 69–77. Springer, Cham (2018). https://doi.org/10.1007/978-3-319-66830-7_7
4. Calvo, T., De Baets, B., Fodor, J.: The functional equations of Frank and Alsina for uninorms and nullnorms. Fuzzy Sets Syst. **120**, 385–394 (2001). https://doi.org/10.1016/S0165-0114(99)00125-6
5. Casasnovas, J., Mayor, G.: Discrete t-norms and operations on extended multisets. Fuzzy Sets Syst. **159**, 1165–1177 (2008). https://doi.org/10.1016/j.fss.2007.12.005
6. De Baets, B., Mesiar, R.: Triangular norms on the real unit square. In: Proceedings of the 1999 EUSFLAT-ESTYLF Joint Conference, Palma de Mallorca, Spain, pp. 351–354 (1999)
7. Çaylı, G.D.: On a new class of t-norms and t-conorms on bounded lattices. Fuzzy Sets Syst. **332**, 129–143 (2018). https://doi.org/10.1016/j.fss.2017.07.015
8. Çaylı, G.D., Drygaś, P.: Some properties of idempotent uninorms on a special class of bounded lattices. Inf. Sci. **422**, 352–363 (2018). https://doi.org/10.1016/j.ins.2017.09.018
9. Drewniak, J., Drygaś, P., Rak, E.: Distributivity between uninorms and nullnorms. Fuzzy Sets Syst. **159**, 1646–1657 (2008). https://doi.org/10.1016/j.fss.2007.09.015
10. Durante, F., Sarkoci, P.: A note on the convex combinations of triangular norms. Fuzzy Sets Syst. **159**, 77–80 (2008). https://doi.org/10.1016/j.fss.2007.07.005
11. Durante, F., Fernández-Sánchez, J., Quesada-Molina, J.J.: On the α-migrativity of multivariate semi-copulas. Inf. Sci. **187**, 216–223 (2012). https://doi.org/10.1016/j.ins.2011.10.026
12. Durante, F., Ricci, R.G.: Supermigrative semi-copulas and triangular norms. Inf. Sci. **179**, 2689–2694 (2009). https://doi.org/10.1016/j.ins.2009.04.001
13. Fernández-Sánchez, F., Quesada-Molina, J.J., Úbeda-Flores, M.: On (α, β)-homogeneous copulas. Inf. Sci. **221**, 181–191 (2013). https://doi.org/10.1016/j.ins.2012.09.048

14. Fodor, J., Rudas, I.J.: An extension of the migrative property for triangular norms. Fuzzy Sets Syst. **168**, 70–80 (2011). https://doi.org/10.1016/j.fss.2010.09.020
15. Fodor, J., Rudas, I.J.: On continuous triangular norms that are migrative. Fuzzy Sets Syst. **158**, 1692–1697 (2007). https://doi.org/10.1016/j.fss.2007.02.020
16. Fodor, J., Rudas, I.J.: Migrative t-norms with respect to continuous ordinal sums. Inf. Sci. **181**, 4860–4866 (2011). https://doi.org/10.1016/j.ins.2011.05.014
17. Fodor, J., Rudas, I.J., Rybalov, A.: Structure of uninorms. Int. J. Uncertain. Fuzziness Knowl.-Based Syst. **5**, 411–427 (1997). https://doi.org/10.1142/S0218488597000312
18. Karaçal, F., Aşıcı, E.: Some notes on T-partial order. J. Inequalities Appl. **2013**, 219 (2013). https://doi.org/10.1186/1029-242X-2013-219
19. Klement, E.P., Mesiar, R., Pap, E.: Triangular Norms. Kluwer Academic Publishers, Dordrecht (2000)
20. Liang, X., Pedrycz, W.: Logic-based fuzzy networks: a study in system modeling with triangular norms and uninorms. Fuzzy Sets Syst. **160**, 3475–3502 (2009). https://doi.org/10.1016/j.fss.2009.04.014
21. Mas, M., Mayor, G., Torrens, J.: t-operators. Int. J. Uncertain. Fuzziness Knowl.-Based Syst. **7**, 31–50 (1999). https://doi.org/10.1142/S0218488599000039
22. Mas, M., Monserrat, M., Ruiz-Aquilera, D., Torrens, J.: Migrative uninorms and nullnorms over t-norms and t-conorms. Fuzzy Sets Syst. **261**, 20–32 (2015). https://doi.org/10.1016/j.fss.2014.05.012
23. Mas, M., Monserrat, M., Ruiz-Aquilera, D., Torrens, J.: An extension of the migrative property for uninorms. Fuzzy Sets Syst. **246**, 191–198 (2013). https://doi.org/10.1016/j.ins.2013.05.024
24. Martin, J., Mayor, G., Torrens, J.: On locally internal monotonic operations. Fuzzy Sets Syst. **137**, 27–42 (2003). https://doi.org/10.1016/S0165-0114(02)00430-X
25. Mesiar, R., Bustince, H., Fernandez, J.: On the α-migrativity of semicopulas, quasi-copulas and copulas. Inf. Sci. **180**, 1967–1976 (2010). https://doi.org/10.1016/j.ins.2010.01.024
26. Ouyang, Y.: Generalizing the migrativity of continuous t-norms. Fuzzy Sets Syst. **211**, 73–83 (2013). https://doi.org/10.1016/j.fss.2012.03.008
27. Ruiz-Aquilera, D., Torrens, J.: Distributivity and conditional distributivity of a uninorm and a continuous t-conorm. IEEE Trans. Fuzzy Syst. **14**, 180–190 (2006). https://doi.org/10.1109/TFUZZ.2005.864087
28. Su, Y., Zong, W., Liu, H.W.: Migrative property for uninorms. Fuzzy Sets Syst. **287**, 213–226 (2016). https://doi.org/10.1016/j.fss.2015.05.018
29. Wu, L., Ouyang, Y.: On the migrativity of triangular subnorms. Fuzzy Sets Syst. **226**, 89–98 (2013). https://doi.org/10.1016/j.fss.2012.12.013
30. Yager, R.R., Rybalov, A.: Uninorm aggregation operators. Fuzzy Sets Syst. **80**, 111–120 (1996). https://doi.org/10.1016/0165-0114(95)00133-6
31. Zong, W., Su, Y., Liu, H.W.: Migrative property for nullnorms. Int. J. Uncertain. Fuzziness Knowl.-Based Syst. **5**, 749–759 (2014). https://doi.org/10.1142/S021848851450038X

Comparison of Fuzzy Integral-Fuzzy Measure Based Ensemble Algorithms with the State-of-the-Art Ensemble Algorithms

Utkarsh Agrawal[1(✉)], Anthony J. Pinar[2], Christian Wagner[1], Timothy C. Havens[2,3], Daniele Soria[4], and Jonathan M. Garibaldi[1]

[1] School of Computer Science, The University of Nottingham, Nottingham NG8 1BB, UK
utkarsh.agrawal@nottingham.ac.uk
[2] Department of Electrical and Computer Engineering, Michigan Technological University, Houghton, MI, USA
[3] Department of Computer Science, Michigan Technological University, Houghton, MI, USA
[4] Department of Computer Science, University of Westminster, London WIW 6UW, UK

Abstract. The Fuzzy Integral (FI) is a non-linear aggregation operator which enables the fusion of information from multiple sources in respect to a Fuzzy Measure (FM) which captures the worth of both the individual sources and all their possible combinations. Based on the expected potential of non-linear aggregation offered by the FI, its application to decision-level fusion in ensemble classifiers, i.e. to fuse multiple classifiers outputs towards one superior decision level output, has recently been explored. A key example of such a FI-FM ensemble classification method is the Decision-level Fuzzy Integral Multiple Kernel Learning (DeFIMKL) algorithm, which aggregates the outputs of kernel based classifiers through the use of the Choquet FI with respect to a FM learned through a regularised quadratic programming approach. While the approach has been validated against a number of classifiers based on multiple kernel learning, it has thus far not been compared to the state-of-the-art in ensemble classification. Thus, this paper puts forward a detailed comparison of FI-FM based ensemble methods, specifically the DeFIMKL algorithm, with state-of-the art ensemble methods including Adaboost, Bagging, Random Forest and Majority Voting over 20 public datasets from the UCI machine learning repository. The results on the selected datasets suggest that the FI based ensemble classifier performs both well and efficiently, indicating that it is a viable alternative when selecting ensemble classifiers and indicating that the non-linear fusion of decision level outputs offered by the FI provides expected potential and warrants further study.

J. Medina et al. (Eds.): IPMU 2018, CCIS 853, pp. 329–341, 2018.
https://doi.org/10.1007/978-3-319-91473-2_29

Keywords: Ensemble classification comparison · Fuzzy measures
Fuzzy Integrals · Adaboost · Bagging · Majority Voting
Random Forest

1 Introduction

Ensemble classifiers are a set of classification algorithms with the objective of classifying data objects by combining the outcome of each individual classifier, generally using weights. Many combination techniques exist in the literature including Boosting, Bagging, Random Forest, Majority Voting, etc. [1]. These ensemble methods have been very popular in the machine learning community due to their ability of producing more accurate results than individual classifiers [2] in a very wide range of application areas [3].

Another approach to obtain ensemble classification is the use of the Fuzzy Integral (FI) aggregation defined with respect to a Fuzzy Measure (FM) [4–8]. The FI is a non-linear aggregation operator to fuse weighted information from multiple sources, where the weights are captured by a FM. The FM not only captures the worth of the individual sources, but also the weights of all subset of sources. Recently, a FM-FI based ensemble classification algorithm Decision-level Fuzzy Integral Multiple Kernel Learning (DeFIMKL) [4] was introduced, which aggregates the results of kernel-SVMs through the use of Choquet Fuzzy Integral (CFI) with respect to a FM learned by a regularised quadratic programming approach. Upon initial investigation in [4,9], the accuracy of FI-FM based ensemble classification method were found to be better than classifiers based on multiple kernel learning. These works further concluded that DeFIMKL was the best among decision level fusion based FI-FM based ensemble classifiers and thus it has been selected in the work as representative of the FI-FM based ensemble classifier family. However, no in-depth comparison of the FI-FM based ensemble classifier with other ensemble methods have been found in the literature [4–9] (discussed in detail in Sect. 2.3). Thus, the motivation of this study is to determine the performance of FM-FI based ensemble methods for the purpose of ensemble classification. We therefore compare DeFIMKL (FI-FM based ensemble classifier) with the state-of-the art ensemble methods including Adaboost, Bagging, Random Forest and Majority Voting on 20 datasets from UCI machine learning repository.

We focus on the FI-FM ensembles as FIs are powerful non-linear aggregation functions (unlike most of the ensemble methods which perform linear combinations) which are capable of exploiting interactions between the models in the ensemble (through FM). FI-FM based ensembles have found applications in numerous domains including software defect prediction [10], Multi-Criteria Decision Making (MCDM) [11], Brain Computer Interface (BCI) [12], Face Recognition in Computer Vision [13,14], Forensic science [15], Explosive Hazard Detection [16].

In the following section (Sect. 2) we discuss the literature on FI-FM based ensemble classification methods. In this section we also discuss the background of

Table 1. Acronyms and notation

FM	Fuzzy Measure
FI	Fuzzy Integral
CFI	Choquet Fuzzy Integral
RAV	Recursive Weighted Power Mean Aggregation Operator
SSE	Sum of squared error
SVM	Support Vector Machines
nf	number of features in each dataset
DeFIMKL	Decision-level Fuzzy Integral Multiple Kernel Learning
MJSVM	Majority Voting with Support Vector Machines ensemble classifier
X	Set of information sources i.e. $X = \{x_1, \ldots, x_n\} \subset \mathbb{R}^d$
$h(x_i)$	Support of the question for the source x_i
g	Fuzzy Measure
$g(A)$	Fuzzy Measure for subset A
$f_k(x)$	Output by the kth classifier in the ensemble
$f^g(x)$	Final decision by the ensemble using CFI with respect to the FM g

Adaboost, Bagging, Majority Voting and Random Forest ensemble classification methods. In Sect. 3 the UCI datasets are described along with the experimental settings of the selected algorithms. Section 4 presents the results and discussion followed by conclusions and future works in Sect. 5. Table 1 lists the most commonly used acronyms in the paper.

2 Background

2.1 Fuzzy Measure

Fuzzy Measure (FM) captures the worth of each information source and all their possible combinations i.e. every subset in a power set [4,17].

Let $X = \{x_1, \ldots, x_n\}$ be a discrete and finite set of information sources and $g : 2^X \to [0, 1]$ be a FM having the following properties:

P1: Boundary condition, i.e., $g(\emptyset) = 0, g(X) = 1$, and
P2: Monotonic and non-decreasing, i.e., $g(A) \leq g(B) \leq 1$, if $A \subseteq B \subseteq X$.

For an infinite domain X there is an additional property to ensure continuity; however, it is not applicable in this paper as X is finite and discrete. In the context of multi-source data fusion, $g(A)$ represents the weight or importance of subset A. The FM values of the singletons i.e. $g(x_i)$ are commonly called the densities. Three major approaches have been used to determine FMs: (a) Experts: FMs could be specified by the experts, although it would be virtually impossible to specify FM for large collection of sources. (b) Algorithms: Several algorithmic methods including Sugeno λ -measure and S-decomposable measure have been proposed in the literature [18,19]. This method needs the weights of the individual sources to be defined in advance i.e. this method builds FMs from given source densities. (c) Optimisation: Various methods including evolutionary algorithms and quadratic programming have been used to generate FMs [17]. FMs derived using optimisation methods have been used in this work (described in Sect. 2.4) as they extract weights from the training data, where the worth of the sources are not known in advance. The information quantified by these FMs are combined using the aggregation operators, defined in the next subsection.

2.2 Fuzzy Integral

Fuzzy Integrals (FIs) are often used as non-linear aggregation functions which combine information from multiple sources using the worth of each subset of sources (provided by a FM 'g') and the support of the question (the evidence) [4,17]. In the context of ensemble classifiers, FIs together with FMs extend the concept of weighted average ensembles and are able to capture the interactions among the classifiers in the ensemble, resulting in a non-linear ensemble classifier. The two most commonly used FIs in the literature include Choquet Fuzzy Integral (CFI) and Sugeno Fuzzy Integral (SFI) [20], although in this work CFI is in focus which is defined as follows:

Choquet Fuzzy Integral: Let $h : X \rightarrow [0, \infty)$ be a real valued function that represents the evidence or support of a hypothesis. The discrete Choquet Fuzzy Integral (CFI) [4,5,17,20] can be defined as:

$$\int_{CFI} h \circ g = CFI_g(h) = \sum_{i=1}^{n} h(x_{\pi(i)})[g(A_i) - g(A_{i-1})] \quad (1)$$

where π is a permutation of X such that $h(x_{\pi(1)}) \geq h(x_{\pi(2)}) \geq \ldots \geq h(x_{\pi(n)})$, $A_i = [x_{\pi(1)}, \ldots, x_{\pi(n)}]$ and $g(A_0) = 0$. More detail on the property of FIs and the CFI can be found in [21]. The next subsection discusses the literature on FI-FM based ensemble. It also presents the gap and motivation of the current study.

2.3 Related Work

In the past decade, researchers have turned their attention towards FI-FM based ensemble classifiers, and proposed a number of FI-FM ensembles generating FM from fuzzy densities i.e. algorithmic FM [14,18,19]. For example, Wang et al. [19] proposed the use of posterior probabilities to obtain the fuzzy densities from the ensemble of heterogeneous classifiers. Subsequently, the $\lambda-$measure was used to obtain the FM from the densities followed by aggregation using the CFI. The ensemble model was compared with five individual classifiers on the Satimage dataset. In another work, Fakhar et al. [18] proposed the use of the training accuracy and the fuzzy entropy (the reliability of information provided by each information source) to generate fuzzy densities followed by aggregation using the CFI. The proposed FI-FM based ensemble model was compared with seven fuzzy set theory based fusions, all of which are multibiometric identification systems [22]. The accuracy of the proposed FI-FM based ensemble outperformed the previously used classification models on the NIST database. Similarly, Wang and Xiao [14] proposed a FI-FM ensemble where the fuzzy densities are generated using the accuracy rate, error distance and the failure extent of the Neural Network models. The model was tested on the JAFFE facial expression database and compared with five Neural Network models.

In another set of studies, Anderson et al. [23] proposed the use of Genetic Algorithm (GA) (optimisation method) to learn FMs. This study indicated that FI-FM based ensemble classifiers could learn the FMs from the training dataset, leading to efficient data-driven FMs. Hu et al. [9] extended the work and proposed a Fuzzy Integral-Genetic Algorithm (FIGA) ensemble classifier which aggregates results of SVM classifiers using the CFI w.r.t. FM learned through the hybrid of Sugeno λ-measure and GA. FIGA generated the initial measure through the use of Sugeno λ-measure followed by GA to search for an optimal FM through the error optimisation. The resultant ensemble was compared with the Multiple Kernel Learning Group Lasso (MKLGL) ensemble classifier on three datasets. FIGA performed better than MKLGL on all the three datasets.

Pinar and colleagues [4–8] built upon the previous works on data-driven FMs and proposed Decision-level Fuzzy Integral Multiple Kernel Learning (DeFIMKL) algorithm as an alternative to algorithmic and algorithm-optimisation hybrid FMs, which aggregates the outputs of SVM classifiers through the use of CFI with respect to a FM learned through a regularised quadratic programming approach. DeFIMKL was compared to FIGA, MKGL and other FI-FM based ensemble classifiers for six datasets. FIGA and DeFIMKL were best among the feature level fusion classifiers and decision level fusion classifiers respectively.

The researchers in all the above works concluded that FI-FM ensemble performed better than individual classifiers, but they left two important questions unanswered. First the comparison of FI-FM based ensemble with other state-of-the-art ensemble methods and secondly the performance on multiple datasets. In this paper we aim to answer these two questions and thus, compare FI-FM ensemble classifiers with other state-of-the-art ensemble classifiers. Since DeFIMKL was best among the FM-FI based ensemble classifier family for decision level fusion, it was selected as the representative of the FI-FM based ensemble classifiers, described in the next subsection.

2.4 Non-linear FM-FI Ensemble Classifier: DeFIMKL

Let $f_k(\mathbf{x_i})$ be the normalised output generated by the kth classifier in an ensemble on a feature vector $\mathbf{x_i}$. The overall decision of the ensemble is computed by the Choquet Integral, where g encodes the relative worth of each classifier in the ensemble. Thus, the output of the ensemble with respect to the FM g on feature-vector $\mathbf{x_i}$ is produced by $f^g(\mathbf{x_i})$, mathematically described as follows,

$$f^g(\mathbf{x_i}) = \sum_{k=1}^{m} f_{\pi(k)}(\mathbf{x_i})[g(A_k) - g(A_{k-1})], \tag{2}$$

where $A_k = f_{\pi(1)}(\mathbf{x_i}), \ldots, f_{\pi(k)}(\mathbf{x_i})$, such that $f_{\pi(1)}(\mathbf{x_i}) \geq f_{\pi(2)}(\mathbf{x_i}) \geq \ldots \geq f_{\pi(m)}(\mathbf{x_i})$. It can be shown that (2) can be reformulated as

$$f^g(\mathbf{x_i}) = \sum_{k=1}^{m} [f_{\pi(k)}(\mathbf{x_i}) - f_{\pi(k+1)}(\mathbf{x_i})]g(A_k). \tag{3}$$

where $f_{\pi(m+1)} = 0$. Pinar et al. [4] proposed to learn FM g using a regularised sum of squared error (SSE) optimisation, described as follows,

$$E^2 = \sum_{i=1}^{n} (f^g(\mathbf{x_i}) - y_i)^2 + v(u), \tag{4}$$

where y_i is the class label for $\mathbf{x_i}$ and $v(u)$ is a regularisation function. Equation (4) can be further expanded as

$$E^2 = \sum_{i=1}^{n} (H_{x_i}^T * u - y_i)^2 + v(u), \tag{5}$$

where y_i is the actual class label for $\mathbf{x_i}$, u is lexicographically ordered FM g i.e. $u = (g\{x_1\}, g\{x_2\}, \ldots, g\{x_1 \cup x_2\}, g\{x_1 \cup x_3\}, \ldots, g\{x_1 \cup x_2 \cup \ldots \cup x_m\})$, and

$$H_{x_i} = \begin{pmatrix} \cdot \\ f_{\pi(1)}(\mathbf{x_i}) - f_{\pi(2)}(\mathbf{x_i}) \\ \cdot \\ \cdot \\ 0 \\ \cdot \\ \cdot \\ f_{\pi(m)}(\mathbf{x_i}) - 0 \end{pmatrix}, \tag{6}$$

where H_{x_i} is of size $(2^m - 1)$ and contains all the difference terms $f_{\pi(k)}(\mathbf{x_i}) - f_{\pi(k+1)}(\mathbf{x_i})$ at the corresponding locations of A_k in u. We can fold out the square terms from (5), producing

$$\begin{aligned} E^2 &= \sum_{i=1}^{n} (u^T H_{x_i} H_{x_i}^T u - 2y_i H_{x_i}^T u + y_i^2) + v(u) \\ &= (u^T D u + f^T u + \sum_{i=1}^{n} y_i^2) + v(u), \end{aligned} \tag{7}$$

where D and f are

$$D = \sum_{i=1}^{n} H_{x_i} H_{x_i}^T, \quad f = -\sum_{i=1}^{n} 2y_i H_{x_i}$$

Equation (7) is a quadratic function and thus we can add the constraints on u such that it represents a FM, producing a constraint QP. We can add the monotonicity constraint on u according to the properties P1 and P2 as $Cu \leq 0$, such that

$$C = \begin{pmatrix} M_1^T \\ M_2^T \\ . \\ . \\ M_{n+1}^T \\ . \\ . \\ M_{m(2^{m-1}-1)}^T \end{pmatrix}, \tag{8}$$

where $M_1^T..M_{m(2^{m-1}-1)}^T$ are vectors representing monotonicity constraint such as the one used in this work i.e. $g\{x_1\} - g\{x_1 \cup x_2\} \leq 0$ (see [5] for more details on C). Thus, the full QP to learn FM u is

$$\min_u \quad 0.5u^T\hat{D}u + f^Tu + v(u), \quad Cu \leq 0, \quad (0,1)^T \leq u \leq 1, \tag{9}$$

where $\hat{D} = 2D$. We test the performance using ℓ_1 regularisation, i.e.

$$\min_u \quad 0.5u^T\hat{D}u + f^Tu + \lambda||u||_1, \tag{10}$$

where λ is the regularisation weight. The QPs at (9) and (10) provide a method to learn the FM u (i.e. g) from the training data. A new feature vector x', from a test set, can thus be classified using the following steps:

1. Compute the normalised SVM decision value $f_k(x')$,
2. Apply the CFI at Eq. 1 with respect to the learned FM g,
3. Compute the class label using $sign(f_k(x'))$.

2.5 State-of-the-Art Ensemble Methods

Adaboost. Adaboost was introduced by Freund and Schapire [24] in 1997 which uses training sets to serially train each classifier and accords higher weight to the instances which are difficult to classify, with the objective of correctly classifying these in the next iteration [25]. Hence, after each iteration the weights of the misclassified instances are increased (which was initially equal for all instances) and the weights of the correctly classified instances are decreased. Moreover depending upon the overall accuracy, an additional weight (higher weight is assigned to more accurate classifiers) is assigned to each individual classifier, which is further used in the test phase. The sum of the weighted predictions is the final output of the ensemble model. The experimental settings and the base algorithm for the Adaboost are further discussed in the Sect. 3.2.

Bagging. Bagging (Bootstrap Aggregation) is an ensemble method introduced by Breiman in 1996 [26], which aims to increase accuracy by combining the outputs of the classifiers in the ensemble. Sampling with replacement is used to train all the classifiers in the ensemble and thus some of the instances may appear more than once in the training set. Each classifier returns the class predictions for the test instances, and combines them using majority voting over all the class labels. Bagging is effective on unstable learning algorithms such as neural networks and decision trees [27] and thus we have chosen decision trees as the base classifier, discussed later in Sect. 3.2.

Majority Voting with SVM (MJSVM). Let x be an instance and S_i (where $i = 1, 2, \ldots k$) be a set of base classifiers (Support Vector Machines) that output class labels $m_i(x, c_j)$ for each class label c_j (where $j = 1 \ldots n$). The output of the final classifier $y(x)$ for instance x is given by

$$y(x) = \max_{c_j} \sum_{i=1}^{k} m_i(x, c_j). \tag{11}$$

More details on the MJSVM are described in the Experimental settings Sect. 3.2.

Random Forest. A random forest is a collection of randomised decision trees where each decision tree is learned from different subsets of samples. The random forest classifier, in particular, needs two parameters: the number of classification trees (k), and the number of prediction variables to grow the trees (m) [28]. To classify a test sample each tree is traversed and a vote is assigned to the class based on the probability score. The output is selected by choosing the mode i.e. the output with most votes, of all the 'k' classification outputs. Reducing the number of predictive variables 'm' reduces the correlation between trees, which stops ensemble model from converging to similar generalisation error and in turn helps in increasing the accuracy. Thus, 'm' needs to optimised to minimise the generalisation error.

3 Materials and Methods

3.1 Datasets and Pre-processing

20 benchmark datasets from the UCI machine learning repository [29] were selected to compare the performance of the selected algorithms, as shown in Table 2. These selected datasets contain different range of number of instances with different type of datasets. Not all the UCI datasets are binary and thus in some cases multiple classes are joined together for the purpose of binary classification [30].

The next step is data pre-processing to standardise the datasets for an unbiased comparison. All the data which had missing values were deleted and made

Table 2. Comparison of ensemble classification methods

Dataset name	Binary classes	No. of features	No. of instances
Dermatology	{1,2,3} vs {5,6,7}	33	366
Wine	{1} vs {2,3}	13	178
Ecoli	{1,2,5,8} vs {3,4,6,7}	7	336
Glass	{1,2,3} vs {5,6,7}	9	214
Sonar	{1} vs {2}	60	208
Ionosphere	{0} vs {1}	34	351
SPECTF Heart	{0} vs {1}	44	267
Bupa	{1} vs {2}	6	345
WDBC	{M} vs {B}	30	569
Haberman	{+} vs {−}	3	306
Pima	{+} vs {−}	8	768
Australian	{0} vs {1}	14	690
SA Heart	{0} vs {1}	9	462
Satimage	{1,2,3} vs {4,5,6,7}	36	6,435
Segmentation	{1,2,3,4} vs {5,6,7}	19	2,310
Mammographic	{0} vs {1}	5	830
Credit-approval	{+} vs {−}	15	653
Ozone	{0} vs {1}	72	1,848
Tic-tac-toe	{+} vs {−}	9	958
Ilpd	{1} vs {2}	7	583

homogeneous, i.e. all numeric. This might also help to locate inconsistencies among the data. Each dataset was processed using z-score [31] normalisation i.e. zero mean and unit standard-deviation. No further processing techniques were used as: (1) the aim of this work is not to report the best possible result for each dataset, but to compare the performance of the classifiers. (2) to improve the classification results, further processing specific to each dataset wold be required, leading to more challenging comparison [30].

3.2 Experiments

The results were produced by running each dataset for 100 trials. In each trial, 80% of the data were randomly used for training the classifiers and the remaining 20% for testing. Subsequently, the accuracies were statistically compared using a two-sample t-test.

The Adaboost ensemble was run with 200 decision trees, Bagging also with 200 but Random Forest ensemble method with 100 decision trees. The DeFIMKL and the MJSVM ensemble methods used the Support Vector Machine (SVM) algorithms with Radial Basis Function (RBF) kernels as their base classifiers. Five RBF kernels with their width (σ) evenly spaced between $0.5 - 1.5/(number\,of\,features)$ were used for both the ensemble methods. Additionally, $L1$ regularisation with $\lambda = 0.5$ was used for all the datasets. The focus of this work is to show the effect on the final output with change in the aggregation models (DeFIMKL and MJSVM), and thus the settings for the RBF kernels and other ensemble methods were kept the same for all the datasets.

An underlying issue with ensemble classification algorithms is determining the size of ensemble. Not much discussion is given to this parameter selection as they are not in the scope of the paper.

4 Results and Discussions

Table 3 reports the average accuracies of the DeFIMKL, MJSVM, Adaboost, Bagging and Random Forest ensemble classification algorithms with standard deviations over 100 runs. A series of two-sample t-tests were conducted, which compared the accuracy of each algorithm against that of the highest performing algorithm for each dataset. To illustrate the results of these tests, both the absolute highest performing algorithm, along with any further algorithms that were found not to have a significantly lower (at $p < .05$) accuracy, are highlighted in bold. Thus, the values highlighted in bold should represent the algorithm(s) with the highest accuracy for each dataset, and will include all values that were not found to significantly differ from the best-performing model.

Table 3. Accuracy comparison of ensemble classification methods for the benchmark datasets*

Datasets	DeFIMKL	MJSVM	Adaboost With trees	Bagging	Random Forest
Dermatology	**97.35 (1.86)**	**97.47 (1.75)**	96.69 (2.05)	**97.31 (1.77)**	95.47 (2.65)
Wine	**99.44 (1.18)**	**99.42 (1.44)**	96.81 (2.91)	97.78 (2.68)	96.44 (3.14)
Ecoli	**96.77 (1.84)**	**96.84 (1.85)**	95.57 (2.01)	**96.34 (2.36)**	95.82 (2.29)
Glass	**94 (3.8)**	**94.16 (3.73)**	**93.91 (3.92)**	**94.28 (3.13)**	92.72 (3.78)
Sonar	**84.57 (4.7)**	**83.76 (5.4)**	**83.17 (5.89)**	**83.21 (6.2)**	79.31 (6.2)
Ionosphere	**94.61 (2.71)**	**94.34 (2.59)**	90.41 (3.72)	92.76 (2.74)	91.15 (3.57)
SPECTF Heart	79.19 (4.4)	79.48 (4.18)	79.78 (4.98)	**81.85 (5.01)**	**80.81 (4.23)**
Bupa	**69.88 (4.95)**	**69.77 (4.71)**	**70.93 (5.32)**	**70.43 (5.02)**	**69.77 (5.65)**
WDBC	**97.22 (1.64)**	**97.28 (1.62)**	96.59 (2.03)	95.51 (1.94)	95.11 (2.15)
Haberman	**73.77 (4.84)**	**73.92 (4.88)**	**73.4 (4.25)**	69.34 (4.99)	69.24 (4.52)
Pima	**76.12 (3.17)**	**76.49 (3.14)**	**75.97 (3.22)**	**76.61 (3.16)**	**75.97 (3.21)**
Australian	85.75 (2.42)	85.85 (2.35)	85.94 (2.79)	**86.96 (2.68)**	**86.37 (2.81)**
SA Heart	**71.16 (4.04)**	**71.66 (3.89)**	69.31 (4.23)	69.39 (4.23)	68.63 (4.07)
Satimage	**95.69 (0.58)**	95.53 (0.56)	94.09 (0.6)	95.6 (0.47)	95.17 (0.69)
Segmentation	91.81 (1.15)	91.46 (1.08)	92.86 (1.53)	**96.03 (1.1)**	**96.29 (1.26)**
Mammographic	**82.16 (3.65)**	**82.88 (4.02)**	**81.8 (5.71)**	78.6 (5.39)	77.21 (5.02)
Credit-approval	**86.89 (2.38)**	**86.66 (2.73)**	86.47 (2.84)	**87.4 (3.15)**	**87.02 (2.83)**
Ozone	**97.12 (0.79)**	**97.1 (0.8)**	96.76 (0.77)	96.85 (0.85)	96.64 (0.83)
Tic-tac-toe	89.64 (2.25)	88.54 (2.39)	84.34 (2.5)	94.38 (1.85)	**95.35 (1.95)**
Ilpd	**72.18 (3.58)**	**71.79 (3.58)**	68.18 (4.49)	68.01 (4.38)	67.77 (4.15)

*The numbers report the average accuracies with standard deviations over 100 runs

The results for the comparison of FM-FI based ensemble classification algorithm DeFIMKL show promising results on the selected UCI datasets. It can be observed that DeFIMKL has good performance in 16 out of 20, whereas MJSVM, Adaboost, Bagging and Random Forest were best in 15, 6, 10 and 7 datasets respectively. DeFIMKL can more closely be compared with MJSVM ensemble classification algorithm as the difference between the two algorithms is in the decision level fusion. MJSVM was one of the best algorithm in 15 out of the 20 selected datasets, one less than DeFIMKL. However, MJSVM has an advantage since it simply takes a majority vote rather than learning a (typically) very large FM whose length scales exponentially with the number of classifiers. It can thus be inferred that the non-linear CFI aggregation can squeeze out a better performance at the cost of additional memory complexity, although a detailed comparison is need to fully answer this question.

5 Conclusions and Future Works

FIs are aggregation operators which are capable of exploiting all the possible interactions among the inputs through the use of FMs. Thus, in the context of ensemble classification FI-FM based ensemble classifiers not only consider individual classifier in the ensemble but all the possible combinations of classifiers in the ensemble. FI-FM ensemble classifiers have been used in various applications over the past decade, yet in most papers, a dataset oriented comparison with state-of-the-art ensemble classification has been made. Thus, in this paper we addressed this gap and compared a FI-FM based ensemble classification algorithm DeFIMKL with the state-of-the-art ensemble classification algorithms Adaboost, Bagging, MJSVM and Random Forest over 20 datasets.

DeFIMKL was one of the best algorithms among 16 out of 20 selected datasets and fell very close to the best in the remaining datasets, with the exception of the Segmentation and tic-tac-toe dataset. On comparing the accuracies, it can be observed that DeFIMKL is either equivalent or better in performance than MJSVM. However, MJSVM has an advantage as it reduces the memory complexity without having to learn the huge 2^n FM vector i.e. taking the majority voting of the SVMs results in similar performance to the non-linear CFI aggregation of the SVMs.

It is important to note that the DeFIMKL and the MJSVM are only compared for one base classifier i.e. SVM. Thus, in future, we aim to include different base classifiers such as Neural Networks, Decision Trees, etc, to fully answer the question raised in the paper. The results suggest that DeFIMKL achieves good accuracy over a number of datasets, although a time complexity comparison is also needed to answer the effect of base classifiers on the FI-FM ensemble methods. Thus, in future we also aim to compare the time complexity among the selected ensemble classifiers.

Acknowledgements. The research was supported by The University of Nottingham Vice-Chancellor's Scholarship for Research Excellence (International).

References

1. Krawczyk, B., Minku, L.L., Gama, J., Stefanowski, J., Woźniak, M.: Ensemble learning for data stream analysis: a survey. Inf. Fusion **37**, 132–156 (2017)
2. Domingos, P.: A few useful things to know about machine learning. Commun. ACM **55**(10), 78 (2012)
3. Islam, M.A., Anderson, D.T., Pinar, A.J., Havens, T.C.: Data-driven compression and efficient learning of the Choquet integral. IEEE Trans. Fuzzy Syst. (2017). https://doi.org/10.1109/TFUZZ.2017.2755002
4. Pinar, A.J., Rice, J., Hu, L., Anderson, D.T., Havens, T.C.: Efficient multiple kernel classification using feature and decision level fusion. IEEE Trans. Fuzzy Syst. **25**(6), 1403–1416 (2016)
5. Anderson, D.T., Price, S.R., Havens, T.C.: Regularization-based learning of the Choquet integral. In: IEEE International Conference on Fuzzy Systems, pp. 2519–2526. IEEE, July 2014
6. Pinar, A.J., Havens, T.C., Islam, M.A., Anderson, D.T.: Visualization and learning of the Choquet integral with limited training data. In: 2017 IEEE International Conference on Fuzzy Systems (FUZZ-IEEE), pp. 1–6, July 2017
7. Pinar, A.J., Anderson, D.T., Havens, T.C., Zare, A., Adeyeba, T.: Measures of the Shapley index for learning lower complexity fuzzy integrals. Granular Comput. **2**(4), 303–319 (2017)
8. Pinar, A., Havens, T.C., Anderson, D.T., Hu, L.: Feature and decision level fusion using multiple kernel learning and fuzzy integrals. In: 2015 IEEE International Conference on Fuzzy Systems (FUZZ-IEEE), pp. 1–7, August 2015
9. Hu, L., Anderson, D.T., Havens, T.C.: Multiple kernel aggregation using fuzzy integrals. In: IEEE International Conference on Fuzzy Systems (FUZZ-IEEE) (2013)
10. Li, K., Chen, C., Liu, W., Fang, X., Lu, Q.: Software defect prediction using fuzzy integral fusion based on GA-FM. Wuhan Univ. J. Nat. Sci. **19**(5), 405–408 (2014)
11. Zhang, L., Zhou, D.Q., Zhou, P., Chen, Q.T.: Modelling policy decision of sustainable energy strategies for Nanjing city: a fuzzy integral approach. Renew. Energy **62**, 197–203 (2014)
12. Cavrini, F., Bianchi, L., Quitadamo, L.R., Saggio, G.: A fuzzy integral ensemble method in visual P300 brain-computer interface. Comput. Intell. Neurosci. **2016**, 1–9 (2016)
13. Karczmarek, P., Pedrycz, W., Reformat, M., Akhoundi, E.: A study in facial regions saliency: a fuzzy measure approach. Soft. Comput. **18**, 379–391 (2014)
14. Wang, Z., Xiao, N.: Fuzzy integral-based neural network ensemble for facial expression recognition. In: Proceedings of the International Conference on Computer Information Systems and Industrial Applications (2015)
15. Anderson, D.T., Havens, T.C., Wagner, C., Keller, J.M., Anderson, M.F., Wescott, D.J.: Extension of the fuzzy integral for general fuzzy set-valued information. IEEE Trans. Fuzzy Syst. **22**(6), 1625–1639 (2014)
16. Pinar, A.J., Rice, J., Havens, T.C., Masarik, M., Burns, J., Anderson, D.T.: Explosive hazard detection with feature and decision level fusion, multiple kernel learning, and fuzzy integrals. In: 2016 IEEE Symposium Series on Computational Intelligence, SSCI 2016, pp. 1–8, December 2017
17. Wagner, C., Havens, T.C., Anderson, D.T.: The arithmetic recursive average as an instance of the recursive weighted power mean. In: IEEE International Conference on Fuzzy Systems (2017)

18. Fakhar, K., El Aroussi, M., Saidi, M.N., Aboutajdine, D.: Applying the upper integral to the biometric score fusion problem in the identification model. In: International Conference on Electrical and Information Technologies (ICEIT) (2015)
19. Wang, Q., Zheng, C., Yu, H., Deng, D.: Integration of heterogeneous classifiers based on choquet fuzzy integral. In: 7th International Conference on Intelligent Human-Machine Systems and Cybernetics, IHMSC 2015, vol. 1, pp. 543–547 (2015)
20. Murofushi, T., Sugeno, M.: A learning model using fuzzy measures and the Choquet integral. In: Proceedings of the 5th Fuzzy System Symposium, vol. 29, pp. 213–218 (1989)
21. Grabisch, M.: The application of fuzzy integrals in multicriteria decision making. Eur. J. Oper. Res. **89**(3), 445–456 (1996)
22. Ko, Y.C., Fujita, H., Tzeng, G.H.: An extended fuzzy measure on competitiveness correlation based on WCY 2011. Knowl.-Based Syst. **37**, 86–93 (2013)
23. Anderson, D.T., Keller, J.M., Havens, T.C.: Learning fuzzy-valued fuzzy measures for the fuzzy-valued Sugeno fuzzy integral. In: Hüllermeier, E., Kruse, R., Hoffmann, F. (eds.) IPMU 2010. LNCS (LNAI), vol. 6178, pp. 502–511. Springer, Heidelberg (2010). https://doi.org/10.1007/978-3-642-14049-5_52
24. Freund, Y., Schapire, R.E.: A desicion-theoretic generalization of on-line learning and an application to boosting. J. Comput. Syst. Sci. **55**, 119–139 (1997)
25. Galar, M., Fernandez, A., Barrenechea, E., Bustince, H., Herrera, F.: A review on ensembles for the class imbalance problem: bagging-, boosting-, and hybrid-based approaches. IEEE Trans. Syst. Man Cybern. Part C: Appl. Rev. **42**(4), 463–484 (2012)
26. Breiman, L.: Bagging predictors. Mach. Learn. **24**(2), 123–140 (1996)
27. Sun, Q., Pfahringer, B.: Bagging ensemble selection. In: Wang, D., Reynolds, M. (eds.) AI 2011. LNCS (LNAI), vol. 7106, pp. 251–260. Springer, Heidelberg (2011). https://doi.org/10.1007/978-3-642-25832-9_26
28. Rodriguez-Galiano, V.F., Ghimire, B., Rogan, J., Chica-Olmo, M., Rigol-Sanchez, J.P.: An assessment of the effectiveness of a random forest classifier for land-cover classification. ISPRS J. Photogramm. Remote Sens. **67**(1), 93–104 (2012)
29. Lichman, M.: UCI Machine Learning Repository. University of California, Irvine, School of Information and Computer Sciences (2013)
30. Fernández-Delgado, M., Cernadas, E., Barro, S., Amorim, D., Amorim Fernández-Delgado, D.: Do we need hundreds of classifiers to solve real world classification problems? J. Mach. Learn. Res. **15**, 3133–3181 (2014)
31. Patro, S.G.K., Sahu, K.K.: Normalization: a preprocessing stage. arXiv preprint arXiv:1503.06462 (2015)

Application of Aggregation Operators to Assess the Credibility of User-Generated Content in Social Media

Gabriella Pasi and Marco Viviani(✉)

Università degli Studi di Milano-Bicocca,
Department of Informatics, Systems and Communication (DISCo),
Edificio U14, Viale Sarca, 336, 20126 Milano, Italy
{pasi,marco.viviani}@disco.unimib.it
http://www.ir.disco.unimib.it

Abstract. Nowadays, User-Generated Content (UGC) spreads across social media through Web 2.0 technologies, in the absence of traditional trusted third parties that can verify its credibility. The issue of assessing the credibility of UGC is a recent research topic, which has been tackled by many approaches as a classification problem: information is automatically categorized into genuine and fake, usually by employing data-driven solutions, based on Machine Learning (ML) techniques. In this paper, to address some open issues concerning the use of ML, and to give to the decision maker a major control on the process of UGC credibility assessment, the importance that the Multi-Criteria Decision Making (MCDM) paradigm can have in association with the use of aggregation operators is discussed. Some potential aggregation schemes and their properties are illustrated, as well as some interesting research directions.

Keywords: Credibility · User-Generated Content
Multi-Criteria Decision Making · Aggregation operators

1 Introduction

The risk of running into misinformation is nowadays higher than in the past, due to Web 2.0 technologies that promote interaction between users and the diffusion of information in the form of User-Generated Content (UGC) without traditional forms of trusted external control. For this reason, mining the credibility of UGC constitutes nowadays a fundamental issue for users.

Credibility, also referred as believability, is a concept that has been studied for a long time, in many disciplines, such as communication, psychology, and social sciences. Credibility is perceived by individuals, who generally do not have sufficient cognitive capacities to discern between genuine information from fake one. In the on-line context, characterized by a multiplicity of possibly anonymous sources spreading heterogeneous contents of various quality, the need

J. Medina et al. (Eds.): IPMU 2018, CCIS 853, pp. 342–353, 2018.
https://doi.org/10.1007/978-3-319-91473-2_30

of developing systems for helping users in automatically assessing credibility of information and information sources is a particularly urgent research topic.

In the last years, numerous approaches have been proposed to tackle this issue. Most of the state-of-the-art solutions consider multiple features connected to the source of information, the content, and the media across which information diffuses, and employ Machine Learning (ML) techniques that classify UGC as credible and not credible based on these features [32]. Some issues connected to the use of these data-driven approaches concern dataset dependency, necessity of big volume of data to ameliorate classification performances, and inscrutability to observers. In fact, in some ML approaches, it is not always clear the contribution that single or interacting features have (in this case in terms of credibility) in the final classification process.

For these reasons, in this paper we describe the potentialities that model-driven approaches based on Multi-Criteria Decision Making (MCDM) and aggregation operators can have, and their adequacy to the considered research issue [31,33]. In fact, each considered piece of UGC can be considered as an alternative, characterized by different credibility criteria, i.e., its features. An overall credibility estimate can be obtained via aggregation for each alternative, and these aggregated values can allow both to classify credible UGC with respect to not credible one, and to rank alternatives with respect to credibility. Some aggregation schemes and their properties are illustrated in the paper, describing how they can consider a different importance associated with criteria and represent interactions among them.

2 The Notion of Credibility

The notion of *credibility* has been studied since the 4th century BC, by Plato and Aristotle. In the modern era, depending on the context, credibility has been in turn associated with believability, trustworthiness, perceived reliability, expertise, accuracy, and with numerous other concepts or combinations of them [26]. In particular, the attention raised by both credibility and credibility assessment has gradually moved from traditional communication environments, characterized by interpersonal and persuasive communication, to mass communication and interactive mediated communication, with particular reference to on-line communication [21,26]. The research undertaken by Hovland and colleagues in the 1950s [15] constitutes the first systematic work about credibility and mass media.

As illustrated by Fogg and Tseng [9], credibility is a *perceived* quality of the *information receiver*, and it is composed of multiple dimensions. In this sense, the process of assessing *perceived credibility* involves different *characteristics*, which can be connected to: (*i*) the *source* of the message, (*ii*) the *message* itself, i.e., its structure and its content, and (*iii*) the *media* used to diffuse the message [21]. While in the "real world" the focus was more on the study of the characteristics connected to the source of information, in the digital realm the emphasis has shifted from speakers to messages. Furthermore, attention must be paid to the fact that new media can introduce new factors into credibility assessment [34], and to the suitable conceptualization of credibility in on-line environments [22].

2.1 From Off-Line to On-Line Credibility

Users have always been confronted with the problem of trusting information obtained via different kinds of media, even before the diffusion of the Web [4]. In real life, people usually deal with distinct sources of information: (*i*) *organization-oriented*, i.e., provided by well known organizations such as reputed newspapers or popular enterprises; (*ii*) *independent*, i.e., provided by third parties such as no profit organizations or individuals considered as experts in a given field; (*iii*) *interpersonal*, i.e., based on direct communication and knowledge among individuals. Considering these sources of information, people have traditionally reduced uncertainty about credibility by relying on both the *reputation* connected to them, and the personal *trust* based on first-hand experiences with the information providers. Nowadays, the Web represents for many people the primary means to access information; in this context, several of the traditional intermediaries have been removed through a process of 'disintermediation' [8]. Thus, the assessment of credibility in the on-line environment is often more complex with respect to previous media contexts, for a number of different reasons: "the multiplicity of sources embedded in the numerous layers of on-line dissemination of content" [29], the absence of standards for information quality, the ease in manipulating and altering the information, the lack of clarity of the context, and the presence of many potential targets of credibility evaluation interacting together in users' perceptions, i.e., the content, the source, the medium [22].

2.2 Credibility and Social Media

In the Social Web, were users are encouraged to interact and collaborate with each other through social media, the previously described issues to assess credibility are even exacerbated. In this context, personal knowledge is replaced by virtual relationships, through which it is easier and inexpensive for users to directly exchange information in the form of User-Generated Content [16]. In such a scenario, the process of evaluating the most credible information differs both from traditional media and Web 1.0 technologies.

Evaluating UGC credibility in social media deals with the analysis of both the content and the author's characteristics [23], and the multiple kinds of relationships established in social media platforms [3,25]. This means to consider characteristics of credibility both connected to UGC and to information sources, as well as to the social network connecting these entities. In this context, the users' credibility perceptions can be exploit the notion of *crowd consensus*, which is typical in interacting communities [13]. It is important to underline that, even if credibility is a characteristic perceived by individuals, assessing credibility should not be up to users, which have to be assisted in this process, especially in the on-line environment [21]. In fact, humans have only limited cognitive capacities to effectively evaluate the information they receive, especially in situations where the complexity of the features to be taken into account increases [17]. For this reason, there is nowadays the need of developing interfaces, tools or systems that are designed to help users in automatically or semi-automatically assess information credibility, in particular UGC credibility in social media.

3 Multi-Criteria Decision Making and UGC Credibility Assessment

In the context of assessing the credibility of UGC, a user is confronted with multiple contents (alternatives) of which s/he does not know a priori the level of believability. As illustrated in the previous section, in the on-line context users must be supported in assessing information credibility by automatic tools, which can help them in identifying genuine alternatives with respect to fake ones. The cognitive process that determines the choice to take between different alternatives is known as *decision making*, and in the considered context it is connected to the evaluation of multiple characteristics connected to UGC, users and social media in terms of credibility.

These characteristics, namely *features*, can be simple *linguistic features* associated with the text of the UGC, they can be additional *meta-data features* associated for example with the content of a review or a tweet, they can also be extracted from the behavior of the users in social media, i.e., *behavioral features*, or they can be connected to the user profile, if available. Furthermore, different *product-based features* can be considered, in the case of products or services reviewed, as well as *social features*, which exploit the network structure and the relationships connecting entities in social media platforms [14]. These features have been employed so far in different ways and under different configurations by several data-driven approaches [14,32].

3.1 Appropriateness of the MCDM Paradigm to Represent the Considered Problem

In this paper, the adequacy and the potentialities of *Multi-Criteria Decision Making* (MCDM) are illustrated, since it allows to explicitly evaluate multiple conflicting criteria in decision making problems [30], which is exactly the case of the considered research issue. In fact, in assessing the credibility of UGC by using MCDM, the components of the decision making problem can be illustrated as follows:

- *alternatives* are the pieces of UGC that the decision maker have to evaluate in terms of credibility;
- *multiple criteria*, which are also referred in the literature as *attributes* or *goals*, are the credibility features previously illustrated, and which can vary according to the considered task, i.e., the features used for fake review detection could be different from the ones used for fake news detection. These features can be *independent*, *conflicting*, or *synergistic*, and a way to represent these concepts is necessary.
- *importance weights* are indications about the different importance that distinct or groups of features can have, and the way of defining them can allow to represent interaction among criteria.

In the literature, many solutions employing *numeric techniques* to assist decision makers in choosing among a finite set of alternatives have been proposed

[30]. To solve the credibility assessment issue, a possible solution consists in employing the *cardinal* approach from *Multi-Attribute Utility Theory* (MAUT) [7], where distinct absolute scores are given to each alternative with respect to each criterion. These scores represents a *degree of satisfaction* that expresses to what extent a given alternative is satisfactory with respect to the considered criterion. In the considered context, they represents a sort of "degree of credibility" associated with each feature with respect to the considered piece of UGC. These multiple scores, i.e, one score for each feature, are then *aggregated* to obtain an *overall credibility score*, which in the literature it is known as the *overall utility* of the considered alternative for the decision maker(s).

3.2 Formal Representation of the Problem

The solution of an MCDM problem as an aggregation problem can be formally defined for the considered context by following the definition provided in [35]. Let us assume that:

- $C = \{c_1, c_2, \dots, c_n\}$ is the set of *criteria*, i.e., features;
- $A = \{a_1, a_2, \dots, a_m\}$ is the set of *alternatives*, i.e., pieces of UGC;
- s_i is the *satisfaction function* that, for each criterion c_i $(1 \leq i \leq n)$, returns the degree, namely the *performance score* $s_i(a_j) \in I$, $I = [0, 1]$, to which the alternative a_j $(1 \leq j \leq m)$ satisfies the criterion c_i, i.e., the "degree of credibility" that a particular piece of UGC has with respect to a considered feature.

The solution of an MCDM problem is formulating an *overall decision function* d, such that for any alternative a_j, $d(a_j) \in I$ indicates the degree, i.e., the *global performance score*, to which a_j meet the decision maker's preferences with respect to all criteria. Formally:

$$d(a_j) = \mathcal{A}(s_1(a_j), s_2(a_j), \dots, s_n(a_j)) \tag{1}$$

where $\mathcal{A}$ is a function called *aggregation operator* (or *aggregation function*).

Definition 1. *An (n-ary)* aggregation operator $\mathcal{A}^{(n)}$ *is a mapping*

$$\mathcal{A}^{(n)} : [0, 1]^n \rightarrow [0, 1] \tag{2}$$

where $n \in \mathbb{N}_0$.

When the number of the input values to be aggregated is not known, extended aggregation operators have been defined [2,12]. Since in MAUT aggregation is performed on a *finite* number n of performance scores, only the definition of n-ary aggregation operators has been provided, and in the rest of the paper the notation $\mathcal{A}^{(n)}$ will be simply replaced by $\mathcal{A}$.

It is worth to be underlined that, according to [1,2], basic mathematical properties must be satisfied by $\mathcal{A}$ to be defined an aggregation operator, i.e., *boundary conditions* and *monotonicity.* These properties and other potential useful properties in MCDM will be better illustrated in Sect. 4.2.

3.3 Main Phases of the Problem

By using the cardinal approach, it is possible to divide the MCDM problem in three main phases, which has been summarized as follows in [19].

1. *Modeling phase:* in this phase, criteria and the associated satisfaction functions, i.e., each s_i, are defined, as well as appropriate models for determining the importance of each criterion (i.e., the importance weights). With respect to the considered research issue, this is the phase where features are selected and modeled in a way to make them usable to be interpreted from a credibility point of view.
2. *Aggregation phase:* in this phase, one or more suitable aggregation operators are selected or defined, to obtain a global performance score for each alternative, on the basis of the partial performance scores and the importance weights obtained in the previous phase. This phase allows to generate an overall credibility score for each piece of UGC; this way, the overall score will be used in the exploitation phase to help the user in her/his decision making process about UGC credibility.
3. *Exploitation phase:* in this last phase, global performance scores can be used to provide a complete ranking of the alternatives in A, or to provide a choice of the best alternatives in A. With respect tot the considered problem, it will be possible to provide both a binary classification of UGC into genuine or fake, or to provide a ranking of pieces of UGC from the most credible to the less credible, based on the value of the overall credibility scores. This way the user could only refer to the top-ranked pieces of UGC.

4 Possible Solutions

In this section, we present some possible ways of defining the three main phases of the MCDM problem illustrated in Sect. 3.3, by providing some examples and considerations taken from prior research in the considered context.

4.1 Modeling Satisfaction Functions

When considering data-driven approaches, multiple features of different nature, using different scales, and referring to different concepts, are coupled together to solve the problem under consideration. In the context of UGC credibility assessment, they are usually employed to provide a binary classification into genuine and fake UGC. By using the MCDM paradigm and the cardinal approach, it is necessary to design suitable satisfaction functions that can produce, for each considered feature associated with a piece of UGC, a normalized value in the interval [0, 1], which can be easily interpreted in terms of credibility.

To clarify this concept, in this section an example from [31] is provided, where a subset of features in the context of review sites were considered. A review site allows users to write reviews concerning business activities, such as restaurants, hotels, etc.; due to their nature, it is possible to extract from these kinds of site features connected to reviews, reviewers, and relationships between them.

The features that were considered in [31], considering some prior literature in the field of fake reviews detection [18,24], were the number of friends of a reviewer, denoted as n_f, the number of the reviews written by a reviewer, denoted as n_r, the length of the review, in terms of words employed, denoted as l_r, the rating (in terms of number of 'stars') assigned to a restaurant in a review, denoted as m_r, the distance between the rating (in terms of 'stars') given by a reviewer to a restaurant, with respect to the global evaluation of the restaurant itself, denoted as d_e, and the presence of the standard picture in the reviewer profile rather then a customized one, denoted as s_p.

In the approach described in [31], different satisfaction functions s_{f_i} for each feature f_i (each criterion) with respect to a review r (the alternative) were defined as follows:

- $s_{n_f}(r) = f\left(\frac{n_f}{avg_f}\right)$, if $n_f \leq avg_f$, $s_{n_f}(r) = 1$, otherwise,
- $s_{n_r}(r) = f\left(\frac{n_r}{avg_r}\right)$, if $n_r \leq avg_r$, $s_{n_r}(r) = 1$, otherwise,
- $s_{l_r}(r) = f\left(\frac{l_r}{avg_l}\right)$, if $l_r \leq avg_l$, $s_{l_r}(r) = 1$, otherwise,
- $s_{d_e}(r) = 1 - \frac{d_e}{4}$,
- $s_{m_r}(r) = 0$, if $m_r \in \{1, 5\}$, $s_{m_r}(r) = 1$, otherwise,
- $s_{s_p}(r) = 0$, if $s_i = \text{standard}$, $s_{s_p}(r) = 1$, otherwise.

where avg_f is the average number of friends in the dataset, avg_r is the average number of reviews in the dataset, and avg_l is the average length of reviews in the dataset in terms of number of words employed. In the three cases, $f(x) = \sqrt[3]{x}$. The satisfaction functions designed in this way allow, for each review, to produce performance scores which are more close to 1 (which indicates full credibility) by interpreting as more credible those reviews written by reviewers having a high number of friends, writing a high number of long reviews, and having a customized picture in their profile. In addition, performance scores closer to 1 are produced for reviews not characterized by extreme ratings (in terms of 'stars') associated, and whose ratings are not so different from the overall ratings provided by other reviewers. Please refer to [31] for further details.

4.2 Suitable Aggregation Operators

In previous works [31,33], it has been analyzed which are the most suitable aggregation operators, and associated properties, that could be useful to address the issue of the assessment of the credibility of UGC in social media.

In general, according to [1,2], some *mathematical* and *behavioral* properties are necessary to define and select suitable aggregation operators in the context of MCDM. Basic mathematical properties are the preservation of the bounds of the domain and of the range, i.e., the *boundary conditions*, *monotonicity* (non-decreasingness) with respect to each argument, *continuity*, *idempotence*, and in some cases *symmetry*. Concerning behavioral properties, the most suitable aggregation operators should: (*i*) allow to express the *behavior* or the decision maker, i.e., *optimistic*: *at least 1* criterion is satisfied by the alternative, or *pessimistic*: *all* criteria are satisfied by the alternative; (*ii*) provide the possibility to express a *compensatory* effect among criteria, i.e., a low score of a given alternative with respect to a criterion may be compensated by a high score on another criterion [6,20]; (*iii*) give the possibility to express *interaction* among criteria; (*iv*) allow to express *importance weights* associated with criteria if necessary; (*v*) be easily interpretable from a semantic point of view.

In the literature, several *averaging operators* have been used for solving problems of Multi-Criteria Decision Making. In the considered context of UGC credibility assessment, families of aggregation operators which could be of potential interest are those of *Ordered Weighted Averaging* (OWA) operators, defined by Yager in [35], and of *fuzzy integrals*.

Ordered Weighted Averaging (OWA) Operators. The principal interest in employing OWA operators is the possibility to guide the aggregation by *linguistic quantifiers*, which encompass the *all* quantifier (represented by the *min* operator) and the *at least one* quantifier (represented by the *max* operator). Linguistic quantifiers can be modeled by distributing the weights different from zero close to the first, central, or last positions of the weighting vector W, which characterizes OWA operators [35]. Yager, in [36], has defined an elegant solution to the problem of computing the weights in W, by exploiting an analytical definition of a membership function associated with a linguistic quantifier. In particular, *Regular Increasing Monotone* (RIM) quantifiers can express constraints like *most* or *at least k criteria must be satisfied*; *Regular Decreasing Monotone* (RDM) quantifiers can express a constraints like *few*, *at most k*; finally, *Regular UniModal* (RUM) quantifiers can express the constraint *about k*.

Considering six features, as in the example related to UGC credibility assessment provided in Sect. 4.1, an example of the membership functions representing the RIM quantifiers *at least 3* and *most* is illustrated in Fig. 1.

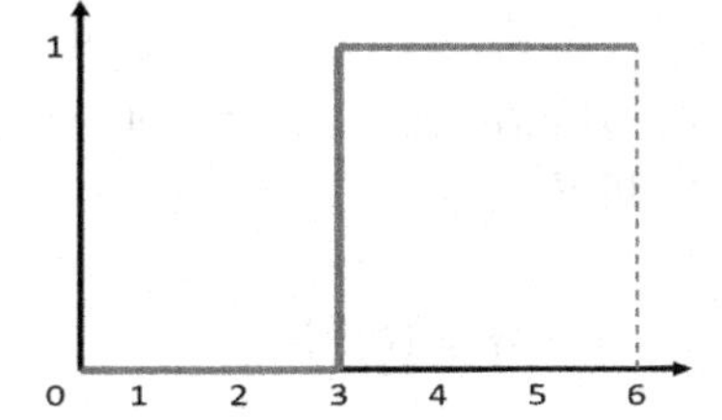

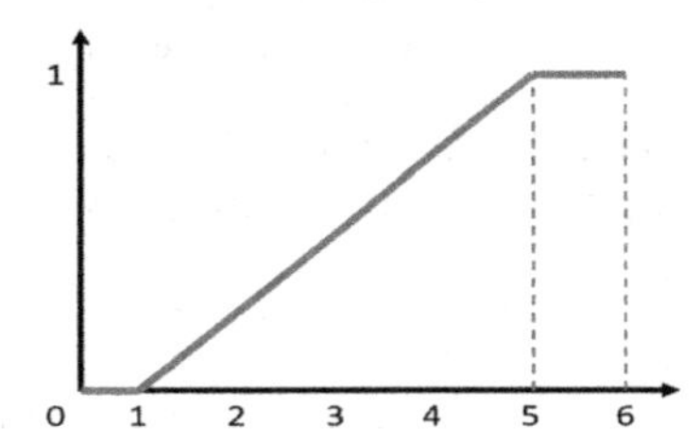

Fig. 1. Examples of membership functions representing the linguistic quantifiers *at least* 3 and *most*.

Another interesting aspect of OWA operators is the possibility to associate a different *importance* with distinct criteria, as illustrated in [35].

OWA operators have been successfully employed to address the considered research issue in [31], in the case of the six features previously detailed.

In this paper, it was demonstrated that by considering the linguistic quantifier *around 50%*, and different importance associated with distinct criteria, let to good results in identifying genuine UGC with respect to fake one. Despite this, OWA operators do not allow to model in an intuitive and understandable way the concept of *interaction* among criteria [10].

Fuzzy Integrals. To be able to represent in an easy way the interaction among criteria, the use of *fuzzy measures* was proposed by Sugeno in 1974 [27], together with a new family of aggregation operators, i.e., *fuzzy integrals* [27,28]. Since for the discussed research issue only discrete spaces are considered, it is sufficient to provide basic definitions referred to the discrete case.

Definition 2. *A (discrete)* fuzzy measure *on the set C of criteria is a set function $\mu : \mathcal{P}(C) \rightarrow [0,1]$ satisfying the following conditions:*

i. $\mu(\emptyset) = 0$, $\mu(C) = 1$,
ii. $S \subseteq T \Rightarrow \mu(S) \leq \mu(T)$ *(monotonicity),*

for any $S, T \subseteq C$.

In this context, $\mu(S)$ can be interpreted as the importance weight of the set of criteria S for the considered decision problem. Whenever $S \cap T = \emptyset$, a fuzzy measure is said to be [10]:

- *additive*, if $\mu(S \cup T) = \mu(S) + \mu(T)$,
- *superadditive*, if $\mu(S \cup T) \geq \mu(S) + \mu(T)$,
- *subadditive*, if $\mu(S \cup T) \leq \mu(S) + \mu(T)$.

Fuzzy integrals are built with respect to a fuzzy measure, and they are able to represent different kinds of interaction among criteria, from *redundancy*, i.e., negative interaction, to *synergy*, i.e., positive interaction. The most representative classes of fuzzy integrals are those of Sugeno and Choquet [5,28]. When scores are given on a cardinal scale, the Choquet integral is the best choice with respect to the Sugeno integral [19]. For this reason, only the definition of the Choquet integral will be provided, as reported in [10].

Definition 3. *Let μ be a fuzzy measure on C. The (discrete)* Choquet integral *of $x_1, x_2, \ldots, x_n \in [0,1]$ with respect to μ is defined by*

$$\mathcal{C}_\mu(x_1, x_2, \ldots, x_n) = \sum_{i=1}^{n} \left(x_{(i)} - x_{(i-1)}\right) \mu\left(S_{(i)}\right) \tag{3}$$

where $\cdot_{(i)}$ indicates that the indexes have been permuted so that $0 \leq x_{(1)} \leq x_{(2)} \leq \cdots \leq x_{(n)} \leq 1$, and $S_{(i)} = \{c_{(i)}, c_{(i+1)}, \ldots, c_{(n)}\}$. In addition, $x_{(0)} = 0$, and $\mu\left(S_{(1)}\right) = 1$.

As illustrated in Fig. 2, the Choquet integral encompasses several aggregation operators, such as the arithmetic mean, the weighted sum, and Ordered Weighted Aggregation operators. Considering this aspect, in the context of UGC

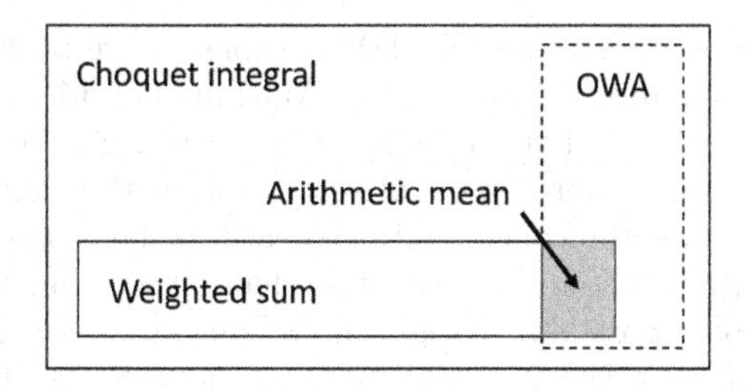

Fig. 2. Relationships between some aggregation operators and the Choquet integral.

credibility assessment, the Choquet integral could be of potential interest due in particular to its flexibility in defining interactions among credibility features. The main drawback when using fuzzy integrals is that, given a certain number of features (criteria), say n, to define the fuzzy measure means to define the $2^n - 2$ coefficients corresponding to the 2^n subsets of C (except $\emptyset$ and C, whose coefficients are provided according to Definition 2) taking into account the monotonicity relations between the coefficients [10,19]. Since it is practically impossible to manually define all these coefficients in a meaningful way in case of a high number of features, as in the case of the considered research issue, over the years some optimization techniques based on learning from available data have been proposed [11]. Due to their complexity, further research is still needed to be able to use in an effective way fuzzy integrals related to complex MCDM problems.

4.3 Binary Classification and Ranking

Once obtained, by using suitable aggregation operators, the overall scores associated with each alternative, these values can be used to evaluate which alternative or alternatives better satisfy the decision maker. In the context of UGC credibility assessment, this means to provide the user either with a *binary classification* of pieces of UGC into genuine or fake, or with a complete ranking of UGC based on the values of the overall credibility estimates.

To follow the first solution, a simple way could consist in selecting a *threshold*, with respect to the overall credibility estimates, above which the corresponding alternatives are considered as credible. To optimize the choice of the threshold, it should be necessary to have some labeled data with respect to which evaluate the effectiveness of the classification using different thresholds. Since in the UCG credibility assessment context few labeled dataset are provided, or they are biased [32], the ranking solution could be more interesting and useful for the user. In fact, by having a complete ranking of UGC where more credible UGC is on top of the list, the user could just concentrate on the top-ranked results, without evaluating each piece of UGC in the set of alternatives.

5 Conclusions

In this paper, the adequacy and the potentialities to describe the issue of the assessment of the credibility of User-Generated Content (UGC) in social media as a Multi-Criteria Decision Making (MCDM) problem have been discussed. With respect to data-driven approaches based on Machine Learning (ML) techniques, the use of the MCDM paradigm in association with aggregation operators can make the decision maker more aware of some choices that led to the proposed solutions, and can make the considered problem less data-dependent.

Some considerations about the way of modeling the problem have been provided, as well as some possible aggregation schemes which could be suitable in the considered context of UGC credibility assessment. Some of these considerations are based on prior research, others are useful for developing future works.

References

1. Beliakov, G., Pradera, A., Calvo, T.: Aggregation Functions: A Guide for Practitioners, vol. 221. Springer, Heidelberg (2007). https://doi.org/10.1007/978-3-540-73721-6
2. Calvo, T., Kolesárová, A., Komorníková, M., Mesiar, R.: Aggregation operators: properties, classes and construction methods. In: Calvo, T., Mayor, G., Mesiar, R. (eds.) Aggregation Operators. Studies in Fuzziness and Soft Computing, vol. 97, pp. 3–104. Springer, Heidelberg (2002). https://doi.org/10.1007/978-3-7908-1787-4_1
3. Carminati, B., Ferrari, E., Viviani, M.: A multi-dimensional and event-based model for trust computation in the social web. In: Aberer, K., Flache, A., Jager, W., Liu, L., Tang, J., Guéret, C. (eds.) SocInfo 2012. LNCS, vol. 7710, pp. 323–336. Springer, Heidelberg (2012). https://doi.org/10.1007/978-3-642-35386-4_24
4. Carminati, B., Ferrari, E., Viviani, M.: Security and trust in online social networks. Synth. Lect. Inf. Secur. Priv. Trust **4**(3), 1–120 (2013)
5. Choquet, G.: Theory of capacities. In: Annales de l'institut Fourier, vol. 5, pp. 131–295 (1954)
6. Dubois, D., Prade, H.: On the use of aggregation operations in information fusion processes. Fuzzy Sets Syst. **142**(1), 143–161 (2004). Aggregation Techniques
7. Dyer, J.S.: MAUT - multi-attribute utility theory. In: Greco, S. (ed.) Multiple Criteria Decision Analysis: State of the Art Surveys, pp. 265–292. Springer, New York (2005). https://doi.org/10.1007/0-387-23081-5_7
8. Eysenbach, G.: Credibility of health information and digital media: new perspectives and implications for youth. In: Metzger, M.M., Flanagin, A.J. (eds.) Digital Media, Youth, and Credibility, pp. 123–154. The MIT Press, Cambridge (2008)
9. Fogg, B.J., Tseng, H.: The elements of computer credibility. In: Proceedings of the SIGCHI Conference on Human Factors in Computing Systems, pp. 80–87. ACM (1999)
10. Grabisch, M.: The application of fuzzy integrals in multicriteria decision making. Eur. J. Oper. Res. **89**(3), 445–456 (1996)
11. Grabisch, M., Kojadinovic, I., Meyer, P.: A review of methods for capacity identification in Choquet integral based multi-attribute utility theory: applications of the Kappalab R package. Eur. J. Oper. Res. **186**(2), 766–785 (2008)
12. Grabisch, M., Marichal, J.-L., Mesiar, R., Pap, E.: Aggregation functions: means. Inf. Sci. **181**(1), 1–22 (2011)
13. Hajli, M.N., Sims, J., Featherman, M., Love, P.E.: Credibility of information in online communities. J. Strateg. Mark. **23**(3), 238–253 (2015)
14. Heydari, A., Ali Tavakoli, M., Salim, N., Heydari, Z.: Detection of review spam: a survey. Expert Syst. Appl. **42**(7), 3634–3642 (2015)

15. Hovland, C.I., Janis, I.L., Kelley, H.H.: Communication and Persuasion. Yale University Press, New Haven (1953)
16. Kaplan, A.M., Haenlein, M.: Users of the world, unite! The challenges and opportunities of social media. Bus. Horiz. **53**(1), 59–68 (2010)
17. Lang, A.: The limited capacity model of mediated message processing. J. Commun. **50**(1), 46–70 (2000)
18. Luca, M., Zervas, G.: Fake it till you make it: reputation, competition, and yelp review fraud. Manag. Sci. **62**(12), 3412–3427 (2016)
19. Marichal, J.-L.: Aggregation operators for multicriteria decision aid. Ph.D. thesis, Universtié de Liège - Faculté des Sciences, Department of Management, FEGSS, Boulevard du Rectorat 7–B31 B-4000 Liège, Belgium (1999)
20. Mesiar, R.: Basic classification of aggregation operators and some construction methods. In: Reusch, B. (ed.) Computational Intelligence, Theory and Applications, pp. 545–553. Springer, Heidelberg (2005). https://doi.org/10.1007/3-540-31182-3_50
21. Metzger, M.J.: Making sense of credibility on the web: models for evaluating online information and recommendations for future research. J. Am. Soc. Inf. Sci. Technol. **58**(13), 2078–2091 (2007)
22. Metzger, M.J., Flanagin, A.J.: Credibility and trust of information in online environments: the use of cognitive heuristics. J. Pragmat. Part B **59**, 210–220 (2013)
23. Moens, M.-F., Li, J., Chua, T.-S. (eds.): Mining User Generated Content. Social Media and Social Computing. Chapman and Hall/CRC, Boca Raton (2014)
24. Mukherjee, A., Venkataraman, V., Liu, B., Glance, N.S.: What yelp fake review filter might be doing? In: Proceedings of ICWSM (2013)
25. Safko, L.: The Social Media Bible: Tactics, Tools, and Strategies for Business Success, 2nd edn. Wiley, Hoboken (2010)
26. Self, C.S.: Credibility. In: Salwen, M.B., Stacks, D.W. (eds.) An Integrated Approach to Communication Theory and Research, 2nd edn, pp. 435–456. Routledge, Taylor and Francis Group, Abingdon (2008)
27. Sugeno, M.: Theory of fuzzy integrals and its applications. Doctorial thesis (1974)
28. Sugeno, M.: Fuzzy measures and fuzzy integrals - a survey. In: Readings in Fuzzy Sets for Intelligent Systems, pp. 251–257. Elsevier (1993)
29. Sundar, S.S.: The main model: a heuristic approach to understanding technology effects on credibility. In: Digital Media, Youth, and Credibility, pp. 73–100 (2008)
30. Triantaphyllou, E.: Multi-criteria decision making methods. In: Triantaphyllou, E. (ed.) Multi-criteria Decision Making Methods: A Comparative Study, pp. 5–21. Springer, New York (2000). https://doi.org/10.1007/978-1-4757-3157-6
31. Viviani, M., Pasi, G.: Quantifier guided aggregation for the veracity assessment of online reviews. Int. J. Intell. Syst. **32**(5), 481–501 (2016)
32. Viviani, M., Pasi, G.: Credibility in social media: opinions, news, and health information - a survey. WIREs Data Min. Knowl. Discov. **7**(5), 1–25 (2017)
33. Viviani, M., Pasi, G.: A multi-criteria decision making approach for the assessment of information credibility in social media. In: Petrosino, A., Loia, V., Pedrycz, W. (eds.) WILF 2016. LNCS (LNAI), vol. 10147, pp. 197–207. Springer, Cham (2017). https://doi.org/10.1007/978-3-319-52962-2_17
34. Wathen, C.N., Burkell, J.: Believe it or not: factors influencing credibility on the web. J. Am. Soc. Inf. Sci. Technol. **53**(2), 134–144 (2002)
35. Yager, R.R.: On ordered weighted averaging aggregation operators in multicriteria decisionmaking. IEEE Trans. Syst. Man Cybern. **18**(1), 183–190 (1988)
36. Yager, R.R.: Quantifier guided aggregation using OWA operators. Int. J. Intell. Syst. **11**(1), 49–73 (1996)

Belief Function Theory and Its Applications

Measuring Features Strength in Probabilistic Classification

Rosario Delgado[1(✉)] and Xavier-Andoni Tibau[2]

[1] Department of Mathematics, Universitat Autònoma de Barcelona, Edifici C- Campus de la UAB. Av. de l'Eix Central s/n., 08193 Bellaterra (Cerdanyola del Vallès), Barcelona, Spain
delgado@mat.uab.cat

[2] Institute for Data Science, German Aerospace Center, 07745 Jena, Germany
xavier.tibau@dlr.de

Abstract. Probabilistic classifiers output a probability of an input being a member of each of the possible classes, given some of its feature values, selecting most probable class as predicted class. We introduce and compare different measures of the feature strength in probabilistic confidence-weigthed classification models. For that, we follow two approaches: one based on conditional probability tables of the classification variable with respect to each feature, using different statistical distances and a correction parameter, and the second one based on accuracy in predicting classification from evidences on each isolated feature. On a case study, we compute these feature strength measures and rank features attending to them, comparing results.

Keywords: Probabilistic classifier · Feature strength
Statistical distance · Prediction accuracy

1 Introduction

In machine learning and statistics, classification is the problem of identifying to which of a set of classes or categories (sub-populations) a new observation belongs, on the basis of a training set of data containing observations whose classification membership is known. Any individual observation is analyzed into a set of quantifiable properties or characteristics, termed features, from which its category or class is to be predicted. In this work we allow features to be binary, categorical or discrete. An algorithm that implements classification, mapping input data to a category or class (output) is known as a classifier. A good classifier is one that predicts that output accurately.

A common subclass of classification is *probabilistic classification*. Algorithms of this nature use statistical inference to find the best class for a given instance. Unlike other algorithms, which simply output a "best" class, probabilistic algorithms output a probability of the instance being a member of each of the possible classes. The best class is then selected as the one with the highest probability,

J. Medina et al. (Eds.): IPMU 2018, CCIS 853, pp. 357–369, 2018.
https://doi.org/10.1007/978-3-319-91473-2_31

and is the "predicted class" for the input. Such algorithms have advantages over non-probabilistic classifiers. Among them, it can output a confidence value associated with its choice and, correspondingly, it can abstain when its confidence of choosing any particular output is too low; in this way, it allows to adapt both sensitivity and specificity of the model depending on priorities.

Bayesian classifiers are probabilistic classification procedures that provide a natural way of taking into account any available information about the composition of the sub-populations associated with the different groups within the overall population. If the elements of the population are grouped into sub-populations or classes because they have common values of they features, then it could be natural to try to predict the values of the features or attributes for the members of any fixed class. On the opposite, if the class is unknown, Bayes' rule can be used to predict the class given (some of) the feature values. The Bayesian classifier is a probabilistic model including the class variable and the features, and perhaps other (latent) variables as well. This model, after construction and validation, can be used to predict (infer) the classification of any new element. The simplest case is the naive Bayesian classifier, which makes the assumption that the features are conditionally independent of each other given the classification, but other models can be considered. We will assume that the considered features, as well as the class variable, are all discrete. Although many classification methods have been developed specially for binary classification, we do not restrict ourselves to this scenario but extend our study to the multiclass setting. More specifically, we consider a probabilistic classifier for class variable C with feature variables $F_1, \ldots, F_r$.

To build effective models, data as accurate as possible is needed, but in real life, with limited resources, obtaining accurate data could have a huge associated cost. In this context, the need to decide on what features we focus on is clear. At the very first steps of the process, this is solved by passing the sieve of feature selection techniques (see Friedman et al. [2]), which attempt to shrink the dimensionality of the dataset to improve both accuracy (e.g. by avoiding overfitting or reducing variance), and interpretability. But later on, once the model is build, analyzing the importance of each feature is still a matter of importance not only because of the comments above, but because in a certain way, a model is a simplistic approximation of reality, so knowing what features are revealed as fundamental in a given model may be a clue about what features we should zoom in when looking for a deeper knowledge of the phenomenon. That is, feature strength focuses on the interpretability of the model and not on its simplification or reduction. After feature selection and the construction of the model of the desired dimension, it is a deeper step in the interpretation of the model, in which it is intended to analyze the influence of the features in the classifier.

Our goal is to introduce and compare different measures of the features strength for classification the classifier. As far as we know, there are no precedents on this type of study. There are, however, some works related in a certain sense. Indeed, different software packages deal with the question of measure the strength of influence between neighboring nodes in a Bayesian network through some kind of strength measure of the arcs among them in the Directed Acyclic Graph (DAG). Only to mention two of them:

1. SMILE (Structural Modeling, Inference, and Learning Engine) is a fully platform independent portable library of C++ classes implementing graphical decision-theoretic methods, such as Bayesian networks, and influence diagrams and structural equation models. Among its tools, we find the *strength of influence* tool of a directed arc, which is always calculated from the CPT of the child node and essentially expresses some form of distance between the probability distributions of the child node conditional on the state of the parent node. With respect to this work, we introduce two families of measures of influence features in classification, not only these which are children nodes of the class. The first one is of the type considered in GeNIe, while the second one is inherently different in nature.
2. Similar to the *strength of influence* of SMILE is the *magnitude of influence* that the software Elvira computes. Paper [5] introduces the *magnitude of influence* of a link (MI) of Elvira for ordinal variables. This definition has no sense if variables are discrete but not ordered. We introduce measures of influence named *strength measures*, which apply for discrete variables (ordinal or not), and not only refer to the directed arc from a parent to a child, but they apply to any pair of variables.

We have followed two different approaches to the problem of defining measures of the strength of a feature in a probabilistic classifier: one based on conditional probability tables (CPT) of the classification variable C to the feature, using different statistical distances, and the other based on accuracy in prediction. With respect to the first one, which is the subject of Sect. 2, statistical distances, divergences, and similar quantities have a large history and play a fundamental role in statistics, machine learning and associated scientific disciplines. Statistical distances are defined in a variety of ways, by comparing probability mass distributions in the discrete probability models context, as is the case at hand. We will choose four of them as an example in our case study (Sect. 4). We will assign a measure of strength to each feature, say F_i, as the *"maximum discrepancy"* observed on the CPT of C conditioned to feature F_i, which is defined as the maximum distance in the pairwise comparisons corresponding to the conditional probability distribution of C to different fixed values of F_i.

In the second case, considered in Sect. 3, the strength measure of each feature F_i for the classification variable is defined as the corresponding accuracy when predicting classification from an evidence expressed exclusively in terms of F_i. In both cases we obtain a ranking of the features in classification (in the first case, a possible different ranking is obtained for each statistical distance that has been considered). In Sect. 4 we apply the different measures we have introduced in previous sections to a case study, and compare features rankings by using both Hamming distance and the degree of consistency.

2 Strength Measures Based on CPT

In this section we introduce some strength measures to deeply analyze to what extent class variable C is affected by the different features in the classifier. Fix a feature variable F. We would like to compute a measure of the effect of induced changes in F on the conditional probability distribution of variable C. For that, we consider evidences of the form $F = a$ and estimate from sample probabilities for the query variable C given the evidence, $P_a^F(x) = P(C = x \,/\, F = a)$ with $x \in \mathcal{C}$ ($\mathcal{C}$ being the set of possible outcomes of variable C), if $P(F = a) > 0$. These probabilities conform the Conditional Probability Table (CPT) of C conditioned to F. We propose an approach based on a statistical distance.

Different statistical distances or divergence measures have been introduced in the literature between two discrete probability distributions. For example, and only to mention four of them, the Kullback-Leibler divergence, the Pearson chi-square distance, the Hellinger measure or the Kolmogorov distance (see [6]). Some of them are asymmetric, provoking that changing order of the arguments can yield substantially different values. For that, we consider symmetrized versions of them. We denote by KL the Kullback-Leibler distance or divergence, also known as *relative entropy*, that is, given two discrete probability distributions taking values $x \in \mathcal{X}$, Q_1 and Q_2, and with the understanding that there is not $x \in \mathcal{X}$ such that $Q_1(x) = Q_2(x) = 0$, $KL(Q_1, Q_2) = \sum_{x \in \mathcal{X}} Q_1(x) \left(\log Q_1(x) - \log Q_2(x) \right)$, with the convention that $0 \log(0) = 0$. Note that $KL(Q_1, Q_2) \geq 0$ although it could be $+\infty$ (if $Q_2(x) = 0$ for some x). Following [4], we symmetrize this distance by means of the harmonic sum, that is, the half the harmonic mean, of the component Kullback-Leibler divergences. Pearson chi-squared and Hellinger distances have also been symmetrized.

Distance name	Formula
Kullback-Leibler	$d_1(Q_1, Q_2) = 1/\left(\frac{1}{KL(Q_1, Q_2)} + \frac{1}{KL(Q_2, Q_1)}\right)$
Pearson chi-squared	$d_2(Q_1, Q_2) = \sum_{x \in \mathcal{X}} 2 \frac{(Q_1(x) - Q_2(x))^2}{Q_1(x) + Q_2(x)}$
Squared blended Hellinger	$d_3(Q_1, Q_2) = \sqrt{\sum_{x \in \mathcal{X}} 2 \left(\sqrt{Q_1(x)} - \sqrt{Q_2(x)}\right)^2}$
Kolmogorov-Smirnov	$d_4(Q_1, Q_2) = \max_{x \in \mathcal{X}} \lvert Q_1(x) - Q_2(x) \rvert$

Note that $d_i \geq 0$, but that if there exist $x_1, x_2 \in \mathcal{X}$ such that $Q_1(x_1) = 0$ and $Q_2(x_2) = 0$, then $d_1(Q_1, Q_2) = +\infty$.

We introduce a strength measure for feature F based on a statistical distance or symmetrised divergence measure d, which could be any of the previous distances $d_1, \ldots, d_4$, or even another, and name it *Strength Distance (SD)*, in this way:

$$SD(F) = \max_{a,\, b \in \mathcal{F}} d_{a,\, b}^F$$

where $\mathcal{F}$ is the set of the possible outcomes of variable F, and $d^F_{a,\,b}$ denotes the statistical distance between P^F_a and P^F_b, that is, $d^F_{a,\,b} = d(P^F_a,\, P^F_b)$. Therefore, we compute strength distance from pairwise comparatives through the statistical distance or symmetrised divergence d.

Proposition 1. *$SD(F) \geq 0$, and $SD(F) = 0$ if and only if F and C are independent.*

Proof. $SD(F) \geq 0$ by definition. On the other hand, $SD(F) = 0$ if and only if $d^F_{a,b} = 0$ for any $a,\, b \in \mathcal{F}$, but this fact is equivalent to say that $P^F_a = P^F_b$ for any $a,\, b \in \mathcal{F}$, which is equivalent to the independency between F and C. □

Note that although we have defined SD through the maximum, we could have chosen any other aggregation function of the distances $d^F_{a,b}$ that verified Proposition 1. The maximum is the least robust (jointly with the minimum) option, since it is maximally sensitive to extreme values, which represents an advantage if extreme values are real (not measurement errors), as in our case, where they are of great importance to assess the strength of a feature for classification.

Because previous measure does not consider if different instantiations of a feature variable produce different predictions for class variable C, it seems appropriate to introduce a correction that does take account of this fact.

Let $\alpha = \#\mathcal{C}$ and $\beta(F) = \#\mathcal{F}$, where # denotes the cardinal of a finite set, and let $\gamma(F)$ denote the number of different predictions obtained from the classifier for C given the evidences $E = \{F = a\}$, with a varying in $\mathcal{F}$, that is,

$$\gamma(F) = \#\{\arg\max_{x \in \mathcal{C}} P(C = x \,/\, F = a),\; a \in \mathcal{F}\}.$$

Then, define

$$\delta(F) = \frac{\gamma(F)}{\min(\alpha,\, \beta(F))} \in (0,\, 1]\,,$$

which is the proportion of different predictions actually obtained by the classifier for class C among the possible we could obtain from an evidence on F. Therefore, $\delta(F)$ is a measure of the influence of feature F on C, and we can use it to correct strength measure SD by introducing the *Corrected Strength Distance (CSD)* in this way: $CSD(F) = SD(F) \times \delta(F)$, which is $\leq SD(F)$. Note that as SD, $CSD(F) \geq 0$, and $CSD(F) = 0$ if and only if F and C are independent.

3 Strength Measures Based on Accuracy in Prediction

In general, after constructing the classifier from the dataset, we perform validation of the model, and once validated, we can use it for future predictions. Validation consists of a procedure for assessing how the classifier performs in the sense of correctly predict the query variable C from any evidence given in terms of the features. The most elementary validation procedure is based on splitting the dataset into two parts, *training* and *test sets*, what is known as *split-validation*. Cross-validation procedure is one of the most widely used methods for estimating prediction error, most common forms for implementation are *k-fold cross-validation* and its particular case ($k = r$) *leave-one-out cross-validation*, We will apply *leave-one-out cross-validation*, from which we obtain *accuracy*, defined as the success rate in prediction, that is, $\frac{\#\{Matches\}}{\#\{Validation\ set\}}$. In addition, for each fixed feature F, we can apply the procedure by predicting the class outcome from single evidences on F, and estimate the accuracy in predicting class C. We denote it by $Acc(F)$.

Nevertheless, this strength measure for each feature does not take into account the following fact: if feature F were *independent* of class variable C, $\gamma(F) = 1$ since for any $a \in \mathcal{F}, \arg\max_{x \in \mathcal{C}} P(C = x \,/\, F = a) = Mode(C)$, with $Mode(C)$ the most frequent class in dataset, that is, if F and C were independent, prediction for C will always be its more likely outcome, independently of the instantiation of F. Denote by p_{mode} the relative frequency of this value in the dataset, $p_{mode} \geq 1/\alpha$. In general, $Acc(F) \geq 1/\alpha$ but if F and C were independent, we would have $Acc(F) = p_{mode}$. Therefore, it seems natural to scale *accuracy* and introduce the *Relative Increment in Accuracy (RIC)* (with respect to p_{mode}) of any feature F by

$$RIA(F) = \frac{Acc(F) - p_{mode}}{p_{mode}}.$$

Proposition 2. *RIA verifies the following properties:*

(a) $RIA(F) = 0$ if F and C are independent, but the reciprocal is not true.
(b) $-1 < c_1 - c_2 \leq RIA(F) \leq c_1 \leq \alpha - 1$, with $c_1 = \frac{1-p_{mode}}{p_{mode}}$, $c_2 = (c_1 + 1)\,\frac{\alpha-1}{\alpha}$.

Proof

(a) If F and C are independent, the predicted class given any evidence on F will be always the same, $Mode(C)$. Therefore, the proportion of correct prediction, which is $Acc(F)$, has to be equal to p_{mode} by definition.
(b) First, since $Acc(F) \leq 1, RIA(F) \leq \frac{1-p_{mode}}{p_{mode}} = c_1$, and $c_1 \leq \alpha - 1$ due to the fact that $p_{mode} \geq 1/\alpha$. Secondly, since $Acc(F) \geq 1/\alpha$, we have that

$$RIA(F) \geq \frac{1/\alpha - p_{mode}}{p_{mode}} = \frac{(1/\alpha - 1) + (1 - p_{mode})}{p_{mode}} = c_1 - \frac{1 - 1/\alpha}{p_{mode}}$$

and $c_2 = \frac{1-1/\alpha}{p_{mode}}$ can be written as $c_2 = (c_1 + 1)\,\frac{\alpha-1}{\alpha}$ if we use that by definition of c_1 we can isolate and obtain $p_{mode} = \frac{1}{c_1+1}$. □

Interpretation of $RIA(F) < 0$ is that F is a feature that as predictor is worse that choosing the most common class. That is, to make classification, it is worst to use evidence on F than nothing, just the opposite that if $RIA(F) > 0$, case in which the higher the value of $RIA(F)$, the stronger the influence of feature F in classification. Therefore, this measure allows to make a ranking of the features, taking into account the strength of their influence in the classification process.

Particular Cases:

(i) Uniform distribution of class C in the dataset. Then, $p_{mode} = 1/\alpha$ and $c_1 = c_2 = \alpha - 1$, obtaining $0 \leq RIA(F) \leq \alpha - 1$.
(ii) Binary classification ($\alpha = 2$). Then, $c_2 = (c_1 + 1)/2$ and $-1 < \frac{c_1 - 1}{2} \leq RIA(F) \leq c_1 \leq 1$.
(iii) If our situation is a combination of both, that is, binary classification and uniform distribution of C into the database, then $0 \leq RIA(F) \leq 1$.

4 Case Study

We consider a dataset of 1,597 policing clarified arson-caused wildfires (for which the alleged offenders have been identified), that has been feeding since 2008 by

Table 1. Variables in the dataset of the arson-caused wildfires.

Forest fire features	Outcomes
C_1 = season	Spring/winter/summer/autumn
C_2 = risk level	High/medium/low
C_3 = start time	Morning/afternoon/evening
C_4 = starting point	Pathway/road/houses/crops/interior/forest track/others
C_5 = use burned surface	Agricultural/forestry/ livestock/interface/recreational
C_6 = number of seats	One/more
C_7 = related offense	Yes/no
C_8 = pattern	Yes/no
C_9 = traces	Yes/no
C_{10} = who denounces	Guard/particular/vigilance
Arsonist characteristics	Outcomes
A_1 = age	≤34/35–45/46–60/>60
A_2 = way of living	Parents/in couple/single/others
A_3 = kind of job	Handwork/qualified
A_4 = employment status	Employee/unemployed/sporadic/retired
A_5 = educational level	Illiterate/elementary/middle/upper
A_6 = income level	High/medium/low/without incomes
A_7 = sociability	Yes/no
A_8 = prior criminal record	Yes/no
A_9 = history subst. abuse	Yes/no
A_{10} = history psychol. probl.	Yes/no
A_{11} = stays in the scene	No/remains there/remains and gives aid
A_{12} = distance home-scene	Short/medium/long/very long
A_{13} = displacement means	On foot/by car/all terrain/others
A_{14} = residence type	Village/house/city/town
$\mathbf{A_{15}}$ = motivation (Class)	Slight negligence/gross negligence/impulsive/profit/revenge

the Secretary of State for Security throughout the entire Spanish territory, under the leadership of the Prosecution Office of Environment and Urbanism of the Spanish state, and contains information obtained from a specific questionnaire concerning authors that have been arrested or imputed. This dataset is an update of that considered in [1]. A total number of $n = 25$ categorical variables are consigned, from which 10 refer to forest fire features, $C_1, \ldots, C_{10}$, and the rest to arsonist characteristics, $A_1, \ldots, A_{15}$, and are described in Table 1.

Bayesian network classifier has been constructed from the dataset with the restriction that directed arcs from forest fire features to arsonist characteristics are forbidden, and is used for classification variable A_{15}, which is author motivation and has proved to be the most significant author variable (see [1]) and forest fire features $C_1, \ldots, C_{10}$.

All calculations, as well as the process of model construction, validation and inference, have been carried out with **R** (https://cran.r-project.org). Two packages of R has been adopted: bnlearn, for network and parameter learning, and gRain, for making inference by probability propagation.

4.1 Ranking Features by Strength Using Measures Based on CPT

The CPT of class variable A_{15} with respect to any of the features $C_1, \ldots, C_{10}$, are learned from the dataset and given in Tables 6, 7, 8, 9, 10 and 11 in the Appendix. Fixed the evidence in terms of a feature variable and one of its values (that is, fixing a column in a CPT), the corresponding predicted class is the most likely, that is, that with the highest probability, which is highlighted in boldface.

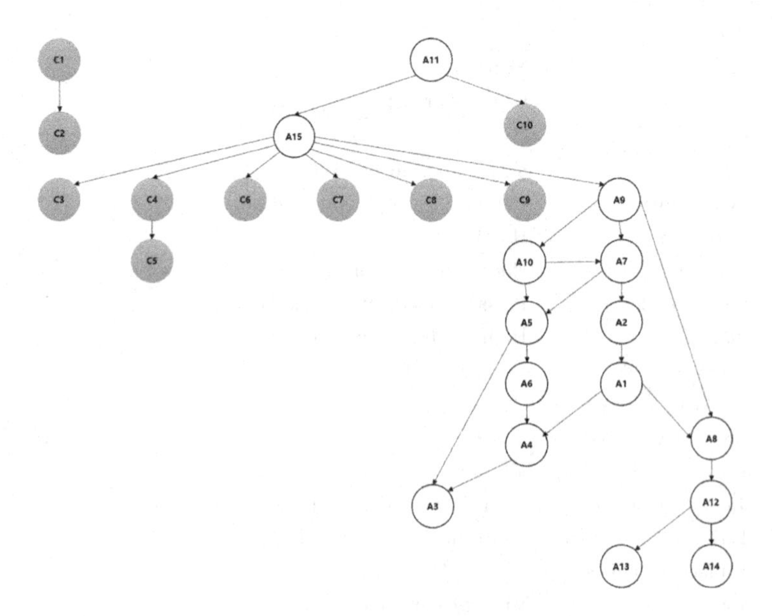

Fig. 1. Learned structure (DAG) of the BN from the dataset of arson-caused wildfires.

Table 2. SD and CSD for the feature variables, using different distances.

Feature	$SD(C_i)$				$\delta(C_i)$	$CSD(C_i)$			
	d_1	d_2	d_3	d_4		d_1	d_2	d_3	d_4
C_3	0.2147	0.7506	0.6430	0.3892	2/3	0.1431	0.5004	0.4287	0.2595
C_4	0.4748	1.3499	0.9584	0.3603	2/5	0.1899	0.5399	0.3833	0.1441
C_5	0.0205	0.0801	0.2028	0.0821	1/5	0.0041	0.0160	0.0406	0.0164
C_6	0.2235	0.8069	0.6611	0.3181	2/2	0.2235	0.8069	0.6611	0.3181
C_7	0.5368	1.6341	0.9956	0.4470	2/2	0.5368	1.6341	0.9956	0.4470
C_8	0.2508	0.8646	0.7022	0.2597	1/2	0.1254	0.4323	0.3511	0.1299
C_9	0.0561	0.2073	0.3348	0.1238	1/2	0.0281	0.1037	0.1674	0.0619
C_{10}	0.0520	0.2041	0.3217	0.2233	2/3	0.0347	0.1361	0.2145	0.1488

Table 3. Ranking of the feature variables by SD and by CSD, using different distances, from the strongest (top) to the weakest (bottom).

Ranking by SD				Ranking by CSD			
d_1	d_2	d_3	d_4	d_1	d_2	d_3	d_4
C_7	C_7	C_7	C_7	C_7	C_7	C_7	C_7
C_4	C_4	C_4	C_3	C_6	C_6	C_6	C_6
C_8	C_8	C_8	C_4	C_4	C_4	C_3	C_3
C_6	C_6	C_6	C_6	C_3	C_3	C_4	C_{10}
C_3	C_3	C_3	C_8	C_8	C_8	C_8	C_4
C_9	C_9	C_9	C_{10}	C_{10}	C_{10}	C_{10}	C_8
C_{10}	C_{10}	C_{10}	C_9	C_9	C_9	C_9	C_9
C_5	C_5	C_5	C_5	C_5	C_5	C_5	C_5

In this way we see in Table 6, for example, that regardless of the value of feature C_1, the prediction for A_{15} is always "slight negligence", which is consistent with the fact that both are independent variables, as is deduced from the fact that in the DAG they appear as disconnected (see Fig. 1). Instead, evidences on feature C_3 can lead to predict "slight negligence" or "gross negligence", depending on if C_3 ="morning" or "afternoon", or C_3 ="evening" (see Table 7). In Table 2 we have the values of SD and CSD for the features $C_3 - C_{10}$ and the four statistical distances introduced in Sect. 2. Both are zero for C_1 and C_2, since they are disconnected from A_{15}.

Glancing at Table 3, we realize that rankings of the features do not match for all the distances, although C_7 is the number one, and C_5 is always at the end of the classification. But if we restrict ourselves to CSD, C_6 is the top second for the four considered distances, while C_9 is the bottom second one.

4.2 Ranking Features by Strength Using Measures Based on Accuracy in Prediction

We perform *leave-one-out cross-validation* and the accuracy value $Acc(C_i)$ is obtained by dividing the number of correct predictions using as evidence the value of C_i, by the total number of predictions (excluding blanks). Both Acc values and RIA are recorded in Table 4. Take into account that p_{mode} is the probability of the most likely non-missing class in the dataset, normalizing probability after eliminating missing values. In this case, $p_{mode} = 697/1463 \simeq 0.47642$.

Finally, we compare rankings obtained from SD and CSD and that obtained by applying RIA criterion, by using both the Hamming distance and the degree of consistency indicator c (see [3]), as consigned in Table 5. In information theory, the Hamming distance between two strings of equal length is the number of positions at which the corresponding symbols are different, that is, it measures the minimum number of substitutions required to change one string into the other. For two measures f and g on a domain Ψ, let $R = \{(a, b) \in \Psi \times \Psi : f(a) > f(b), g(a) > g(b)\}$ and $V = \{(a, b) \in \Psi \times \Psi : f(a) > f(b), g(a) < g(b)\}$. Then, the *degree of consistency* c *of* f *and* g is $c(f, g) = \frac{|R|}{|R|+|V|}$, where $|A|$ denotes the number of elements of the (finite) set A. We apply this indicator with f and g ranking functions. We observe that Hamming distance is minimized with CSD and distances d_3 = Squared blended symmetric Hellinger distance, and

Table 4. Acc and RIA for the feature variables, which have been ranked from top to bottom in descending order.

Feature	$Acc(C_i)$	$RIA(C_i)$
C_7	0.4990	0.0474
C_{10}	0.4949	0.0387
C_3	0.4919	0.0316
C_4	0.4846	0.0172
C_6	0.4826	0.0129
C_8	0.4764	0.0000
C_9	0.4764	0.0000
C_5	0.4764	0.0000

Table 5. Hamming distance and degree of consistency indicator c between SD and CDS, with distance d_i, and RIA.

	Hamming SD-RIA	Hamming CSD-RIA	c(SD, RIA)	c(CSD, RIA)
d_1	5	5	19/28	21/28
d_2	5	5	19/28	21/28
d_3	5	3	19/28	22/28
d_4	5	3	24/28	24/28

d_4 = Kolmogorv-Smirnov, while d_4 maximizes the consistency indicator, being CSD more consistent with RIA than SD. This reinforces the hypothesis that the correction in SD obtained multiplying by factor δ, improves it.

5 Conclusion

We introduce different measures of features strength in a probability classifier. From them, Corrected Strength Distance (CSD), which is based on CPT of class conditioned to each feature, seems to outperform Strength Distance (SD) since it is more consistent with Relative Increment in Accuracy (RIA), which is a measure based on accuracy in prediction. From the chosen distances, the best options have been Hellinger and Kolmogorov-Smirnov, both after correction.

Acknowledgments. The authors are supported by Ministerio de Economía y Competitividad, Gobierno de España, project ref. MTM2015 67802-P, and belong to the "Quantitative Methods in Criminology" research group of the Universitat Autònoma de Barcelona. They wish to express their acknowledgment to the Secretary of State for Security and the Prosecution Office of Environment and Urbanism of the Spanish state, for providing dataset used in the case study.

Appendix: Conditional Probability Tables of Features with Respect to Class Variable A_{15}

Table 6. CPT of A_{15} conditioned to C_1 (in %).

$C_1 \rightarrow$	Spring	Summer	Autumn	Winter
Pulsional	10.05	10.05	10.05	10.05
Gross negligence	31.31	31.31	31.31	31.31
Slight negligence	**47.64**	**47.64**	**47.64**	**47.64**
Profit	7.59	7.59	7.59	7.59
Revenge	3.42	3.42	3.42	3.42

Table 7. CPT of A_{15} conditioned to C_2, and conditioned to C_3 (in %).

	$C_2 \downarrow$			$C_3 \downarrow$		
	High	Medium	Low	Morning	Afternoon	Evening
Pulsional	10.05	10.05	10.05	11.04	7.33	25.82
Gross negligence	31.31	31.31	31.31	19.63	33.18	**30.22**
Slight negligence	**47.64**	**47.64**	**47.64**	**57.06**	**51.07**	18.13
Profit	7.59	7.59	7.59	10.43	6.44	12.09
Revenge	3.42	3.42	3.42	1.84	1.97	13.74

Table 8. CPT of A_{15} conditioned to C_4 (in %).

$C_4 \rightarrow$	Pathway	Road	Houses	Crops	Interior	F. Track	Others
Pulsional	20.00	**36.94**	3.90	0.90	6.22	16.38	6.12
Gross negligence	24.19	16.22	32.47	35.75	33.97	26.72	35.61
Slight negligence	**33.49**	26.13	**59.74**	**59.50**	**50.24**	**30.17**	**51.80**
Profit	14.88	10.81	1.30	3.62	8.61	15.52	5.04
Revenge	7.44	9.91	2.60	0.23	0.96	11.21	1.44

Table 9. CPT of A_{15} conditioned to C_5 (in %).

$C_5 \rightarrow$	Agricultural	Forestry	Livestock	Interface	Recreational
Pulsional	6.28	12.16	11.88	13.40	10.58
Gross negligence	33.00	30.26	30.60	29.44	31.55
Slight negligence	**52.51**	**44.30**	**44.61**	**45.96**	**45.88**
Profit	6.13	8.96	8.81	6.96	7.92
Revenge	2.09	4.32	4.10	4.24	4.08

Table 10. CPT of A_{15} conditioned to C_6, to C_7 and to C_8 (in %).

	$C_6 \downarrow$		$C_7 \downarrow$		$C_8 \downarrow$	
	One	More	Yes	No	Yes	No
Pulsional	7.77	**25.13**	**51.39**	7.72	27.57	3.65
Gross negligence	32.31	21.99	19.44	31.92	18.48	34.95
Slight negligence	**52.23**	20.42	5.56	**50.26**	**28.74**	**54.71**
Profit	5.18	23.04	6.94	7.65	19.65	3.65
Revenge	2.51	9.42	16.67	2.45	5.57	3.04

Table 11. CPT of A_{15} conditioned to C_9 and to C_{10} (in %).

	$C_9 \downarrow$		$C_{10} \downarrow$		
	Yes	No	Guard	Vigilance	Particular
Pulsional	6.55	12.11	11.69	12.08	8.85
Gross negligence	34.06	29.58	37.39	**41.17**	26.18
Slight negligence	**55.68**	**43.30**	**38.26**	33.09	**55.41**
Profit	2.40	10.61	8.69	9.22	6.65
Revenge	1.31	4.39	3.96	4.44	2.90

References

1. Delgado, R., González, J.L., Sotoca, A., Tibau, X.-A.: A Bayesian network profiler for wildfire arsonists. In: Pardalos, P.M., Conca, P., Giuffrida, G., Nicosia, G. (eds.) MOD 2016. LNCS, vol. 10122, pp. 379–390. Springer, Cham (2016). https://doi.org/10.1007/978-3-319-51469-7_31
2. Friedman, J., Hastie, T., Tibshirani, R.: The Elements of Statistical Learning. Springer Series in Statistics, vol. 1, pp. 337–387. Springer, New York (2001). https://doi.org/10.1007/978-0-387-21606-5
3. Huang, J., Ling, C.: Using AUC and accuracy in evaluating learning algoritms. IEEE Trans. Knowl. Data Eng. **17**, 299–310 (2005). https://doi.org/10.1109/TKDE.2005.50
4. Johnson, D., Sinanovic, S.: Symmetrizing the Kullback-Leibler distance. Technical report, ECE Publications, Rice University (2001). https://www.ece.rice.edu/~dhj/resistor.pdf
5. Lacave, C., Luque, M., Díez, F.J.: Explanation of Bayesian networks and influence diagrams in Elvira. IEEE Trans. Syst. Man Cybern.-Part B: Cybern. **37**(4), 952–965 (2007). https://doi.org/10.1109/TSMCB.2007.896018
6. Markatou, M., Chen, Y., Afendras, G., Lindsay, B.G.: Statistical distances and their role in robustness. In: Chen, D.-G., Jin, Z., Li, G., Li, Y., Liu, A., Zhao, Y. (eds.) New Advances in Statistics and Data Science. IBSS, pp. 3–26. Springer, Cham (2017). https://doi.org/10.1007/978-3-319-69416-0_1

DETD: Dynamic Policy for Case Base Maintenance Based on EK-NNclus Algorithm and Case Types Detection

Safa Ben Ayed[1(✉)], Zied Elouedi[1(✉)], and Eric Lefevre[2(✉)]

[1] LARODEC, Institut Supérieur de Gestion de Tunis, Université de Tunis, Tunis, Tunisia
safa.ben.ayed@hotmail.fr, zied.elouedi@gmx.fr
[2] Univ. Artois, EA 3926, LGI2A, 62400 Béthune, France
eric.lefevre@univ-artois.fr

Abstract. Case Based Reasoning (CBR) systems know a success in various domains. Consequently, we find several works focusing on Case Base Maintenance (CBM) that aim to preserve CBR systems performance. Thus, CBM tools are generally offering techniques to select only the most potential cases for problem-solving. However, cases are full of imperfection since they represent real world situations, which makes this task harder. In addition, new problems having substantially new solutions will be found in case bases over the time. Hence, we aim, in this paper, to propose a new CBM approach having the ability to manage uncertainty and the dynamic aspect of maintenance using the evidential clustering technique called EK-NNclus based on belief function theory, where clusters' number is fixed automatically and changes from one maintenance application to another. Finally, the maintenance task is performed through selecting only two types of cases.

Keywords: Case Based Reasoning · Case base maintenance
Belief function theory · Uncertainty · Clustering · EK-NNclus

1 Introduction

Case Based Reasoning is an analogy-based problem-solving paradigm which learns from old experiences using a memory called case base [1]. The strength of CBR systems can be summed up through their ability to offer a high-quality solution even with a weak-understanding domains. Besides, CBR systems are characterized by an incremental learning since each solved problem will be stored in order to serve for future problems resolution. There is a wide range of successful CBR applications in several domains such that diagnosis [2], design [3], help desk [4], decision support [5], etc. Like all other applications that are designed to work over long period of time, CBR systems need a maintenance task, especially regarding their case bases, since their quality presents the key success of all the system. Actually, the case base of a CBR system should contain only relevant

J. Medina et al. (Eds.): IPMU 2018, CCIS 853, pp. 370–382, 2018.
https://doi.org/10.1007/978-3-319-91473-2_32

cases in order to improve its competence on problems resolution on the one hand, and its performance by reducing the research time on the other hand. For this reason, case base maintenance operations are generally opting to select from the case base the most competent ones. To obtain that, a number of aspects should be taken into account while maintenance.

First, the maintenance task should take into account the uncertainty aspect. In fact, within real world situations, information are often imprecise, uncertain and/or incomplete. Hence, the most powerful cases in problem resolution cannot be well defined without considering the uncertainty aspect that can be caused by unquantifiable data, user ignorance and/or overlapping of data regions.

Second, CBR systems are exposed with time and with users requirement evolution to new types of solutions for new problems. Hence, the dynamicity in solutions should be managed while the maintenance of case bases which reflects modern and contemporary environment. To manage these solutions along with their problems in the case base, some CBM approaches [10–12] opt to partition cases using a clustering technique as a preprocessing task so as to learn from the case base and devise it into a number of small ones. Then, they select the most representative cases from each cluster.

However, the dynamic aspect consists, in our context, on the capacity of the CBM approach to fix dynamically and automatically the number of clusters regardless which time we perform the maintenance. Besides, existing CBM approaches are not offering a dynamic maintenance because they are suitable only for static collections of cases, and their offered maintenance should be accompanied every time by prior information to be well re-applicable. Actually, and to the best of our knowledge, the dynamicity in case bases maintenance for CBR systems is the most neglected aspect in CBM field.

To manage these two aspects while maintenance, we propose, in this paper, a new approach for case base maintenance that uses the belief function theory [8,9] as one of the most powerful tools for handling uncertainty, more accurately a dynamic evidential machine learning technique called EK-NNclus [7].

The rest of this paper is organized as follows. In Sect. 2, we review some of CBM methods based on clustering technique. The necessary background regarding the belief function theory and the used evidential clustering technique called EK-NNclus [7] are offered in Sect. 3. Section 4 describes the different steps of our proposed approach called **DETD** for "**D**ynamic policy for case base maintenance based on **E**K-NNclus algorithm and case **T**ypes **D**etection". Finally, the experimentation is shown during Sect. 5.

2 Case Base Maintenance: Partitioning Based Policies

Case Base Maintenance represents a fundamental task aiming to give CBR systems the ability to solve effectively new problems within a reasonable time since they are faced to a large number of cases with a continuous evolution [6]. Conspicuously, the most intuitively way to deal with large case bases while maintenance is to divide them into a number of small ones. Consequently, it will be easier to handle them. Nevertheless, we find in the literature several policies belonging to other strategies as shown in [10]. In the remaining of this Section, we review some CBM policies that use the partition strategy, more accurately, the clustering as a machine learning technique.

On the one hand, with considering partition strategy, we find in the literature hard CBM policies that are not able to deal with uncertainty in data. For instance, we cite COID method [11] encoding *Clustering, Outliers and Internal case Deletion* which is based on a density-based clustering technique for cases gathering and noisy cases detection. Then, it computes cases-clusters distances to flag outliers and internal cases so as to perform the maintenance. As an extension of COID, we find among others, WCOID method [12] which appends a feature weighting technique to give more importance to the most "informative" features in term of problem solving while the maintenance. However, this type of policies is generally reducing the case bases competence since they suffer from their disability to manage uncertainty in cases involving real world situations.

On the other hand, we find a number of CBM policies based on soft clustering techniques which are able to deal with imperfection. Using fuzzy set theory [13] for uncertainty management, SCBM method [14] denoting *Soft CBM Competence Based Model* was able to handle vagueness in real data by applying foremost the soft clustering technique called Soft DBSCAN-GM (SDG) [15]. Thus, SCBM detects three types of cases in order to maintain case bases by removing noisy and redundant cases. Furthermore, we cite one more CBM policy that tries to deal with all levels of uncertainty in cases, from the complete ignorance to the total certainty, using belief function theory [8,9]. This approach called ECTD encoding *"Evidential Clustering and case Types Detection for case base maintenance"*. In a nutshell, ECTD approach goes through three main steps: First, it uses the Evidential C-Means (ECM) [16] for the uncertainty management regarding the membership of cases to the different clusters. The partitions centers offered by ECM as well as the different degrees of belief will serve during the second step at the detection of four types of cases: *Noisy cases* have a high degree of belief to not belonging to any one of clusters, *Similar cases* are considered as redundant experiences and situated on the core of the different clusters, *Isolated cases* are situated on clusters borders, so they can only be solved through themselves, and finally *Internal cases* as the representatives of the different clusters. Ultimately, ECTD accomplishes the maintenance by selecting only cases flagged as *Internal* or *Isolated.*

Obviously, ECTD approach [10] follows a good strategy of maintenance with the ability to manage uncertainty. Besides, it showed practically good results. However, this approach is not able to deal with the dynamic aspect of maintenance where cases are grouped according to their solutions with a predefined and static number of clusters. Hence, it does not take into account the dynamicity of the encountered solutions in the case bases knowing that they contain real experiences. In addition, if the case base contains a high number of distinct solutions categories, ECTD approach suffers from a high complexity when dealing with uncertainty towards all possible subsets of solutions. In what follows, we present, therefore, our proposed approach for this paper dealing with these matters, but we offer before that some background related to the belief function theory as well as the used clustering technique.

3 Background: Belief Function Theory

In order to manage uncertainty in cases as well as the dynamic aspect of maintenance, our contribution is based on Belief function theory and the evidential clustering technique called EK-NNclus. Hence, we show during this section the basic concepts of this theory as well as the corresponding clustering technique.

3.1 Basic Concepts

Belief function theory [8,9] is one of the most used theoretical frameworks for reasoning under uncertainty. It is based on the explicit representation and combination of pieces of evidence. Thus, the problem domain is represented through the frame of discernment (Θ) and containing a finite set of elementary events. Hence, each variable ω takes values from Θ. In this theory, a *mass function* m represents the uncertain evidence about ω on Θ. Actually, m is the mapping function from the powerset of Θ containing all possible subsets, denoted 2^{Θ}, to $[0, 1]$ such that:

$$\sum_{A \subseteq \Theta} m(A) = 1 \tag{1}$$

where each mass $m(A)$ is the evidence that supports exactly the ascertain $\omega \in A$. In particular, if $A = \Theta, m(\Theta)$ is interpreted as the probability that the evidence does not give us any information about the variable ω from the frame of discernment. If $m(A) > 0$, the event A is called a focal element.

Given a mass function m, the corresponding belief (bel) and plausibility (pl) functions from 2^{Θ} to $[0, 1]$ are defined such that:

$$bel(A) = \sum_{\emptyset \neq B \subseteq A} m(B) \qquad \forall A \subseteq \Theta \tag{2}$$

and

$$pl(A) = \sum_{A \cap B \neq \emptyset} m(B) \qquad \forall A \subseteq \Theta \tag{3}$$

Actually, $bel(A)$ represents the entire belief allocated to support only the event A. However, $pl(A)$ measures the maximum amount of belief that can be assigned to A.

Within belief function theory, several combination rules of evidences can be used. Dempster's rule of combination [9] is one of the most used ones to combine two pieces of evidence (m_1 and m_2) induced from two independent and reliable sources of information. This rule is defined as follows:

$$(m_1 \oplus m_2)(A) = \frac{1}{1 - \kappa} \sum_{B \cap C = A} m_1(B)\, m_2(C) \tag{4}$$

where κ is called the conflict of the global combination and defined such that:

$$\kappa = \sum_{A \cap B = \emptyset} m_1(A)\, m_2(B) \tag{5}$$

One of the techniques that allow us to make decision within the belief function framework is the pignistic probabilities transformation and defined as follows:

$$BetP(A) = \sum_{B \subseteq \Theta} \frac{|A \cap B|}{|B|} \frac{m(B)}{1 - m(\emptyset)} \qquad \forall A \in \Theta \tag{6}$$

The decision making is therefore done through the variable having the highest pignistic probability.

3.2 Evidential Clustering: EK-NNclus

Evidential clustering techniques are aiming to generate credal partition for managing uncertainty in cases' membership to clusters. Among the most known ones, we cite Evidential c-means [16], EVCLUS [17] and EK-NNclus [7] which is based on EKNN rule [18].

The Evidential K-Nearest Neighbor (EKNN) Rule: In EKNN rule, the knowledge that an object o is distant from an object o_j with a value d_j produces the following piece of evidence m_j on $\Theta = \{\omega_1, \ldots, \omega_c\}$:

$$m_j(\{\omega_k\}) = u_{jk}\varphi(d_j), \qquad k = 1, \ldots, c \tag{7a}$$

$$m_j(\Theta) = 1 - \varphi(d_j) \tag{7b}$$

where $\lim_{d \to \infty} \varphi(d) = 0$, and $u_{jk} = 1$ if o_j is classified in ω_k and 0 otherwise.
The K mass functions obtained through the K nearest neighbors are combined then using Dempster's rule as defined in Eqs. 4 and 5. To make decision about the membership of cases, the combined contour function for $l = 1, \ldots, c$ is defined such that:

$$pl(\omega_l) \propto \prod_{j \in N_K} (1 - \varphi(d_j))^{1 - u_{jl}} \tag{8}$$

and its logarithm can be written as follows:

$$ln\ pl(\omega_l) = \sum_{j=1}^{n} v_j\ u_{jl} + C \tag{9}$$

where N_K denotes the indices set of K nearest neighbors, C is a constant, $v_j = -ln(1 - \varphi(d_j))$ if $j \in N_k$, and $v_j = 0$ otherwise.

EK-NNclus: The EK-NNclus algorithm is a decision-directed clustering procedure based on the above EKNN rule. For the initialization step, EK-NNclus starts with a randomly labeled objects (each object exists lonely in one cluster if the number of objects n is not too large, otherwise the number of clusters c

will be taken large but lower than n). For objects membership to clusters, we initialize u_{ik} to 1 if the object o_i belongs to cluster k and to 0 otherwise.

Using EKNN rule, the algorithm's iteration updates the object labels in some randomly order. Then, using Eq. 9, the logarithms of the membership plausibilities of each object to each cluster are computed as:

$$u_{ik} = \sum_{j \in N_K(i)} v_{ij} s_{jk}, \qquad k = 1, \ldots, c \tag{10}$$

where $N_K(i)$ is the set of indices of the K-NNs of the object o_i. The membership of the object o_i is then updated according to the highest plausibility such that:

$$s_{ik} = \begin{cases} 1 & if\ u_{ik} = \max_{k'} u_{ik'} \\ 0 & Otherwise. \end{cases} \tag{11}$$

A new iteration is started with a randomly reordered objects if exists at least one object that its label changes. In addition, we note a disappearance of clusters from an iteration to another. Finally, and after the convergence of the number of clusters (demonstrated in [7]), the resulting mass function is computed as:

$$m_i = \bigoplus_{j \in N_K(i)} m_{ij} \tag{12}$$

where the different bbas m_{ij} are calculated using Eq. 7.

To conclude, EK-NNclus algorithm provides a simpler credal partition compared to other techniques such as ECM [16] and EVCLUS [17] which yield mass functions with 2^{Θ} focal sets. Hence, EK-NNclus is able to avoid the exponential complexity when treating a large number of clusters and it has lower storage requirement. Actually, Ek-NNclus and ECM are equally effective. However, EK-NNclus has an additional major strength which is its automatically determination of clusters number which converges automatically after a number of iterations.

4 DETD: The New Proposed CBM Method

The purpose of our proposed approach is to use a dynamic clustering technique, which fixes automatically the number of clusters for each case base, in order to select carefully cases that should be maintained. So, what types of cases should be retained in a case base? Actually, we define specially the two following types of cases as the same spirit as those defined in [19]:

- **Isolated cases:** They are far to clusters centers but not noises. Hence, they can only solve themselves and no other cases can solve them. For this reason, their deletion from the case base leads to the decrease of their competence in problem resolution.

- **Internal cases:** Each internal case is a representative of a set of similar cases founded in the same cluster. Hence, deleting all similar cases do not affect the case base competence because they are close to each other. However, internal cases must be maintained to cover all of them.

On the opposite side, there are two other types of cases that should be removed, which are noisy and similar cases. In fact, noisy cases are irrelevant and distort the problem resolution process. Otherwise, similar cases are redundant and useless.

Our newly DETD approach aims to well distinguish between these types of cases while managing uncertainty and dynamicity of new occurred solutions over the time. To do that, we detail in what follows its different steps.

4.1 Step 1: DYNAMIC Evidential Clustering of Cases

To partition the case base, this first step of our proposed CBM policy aims at using the clustering algorithm EK-NNclus (see Subsect. 3.2) that response to a number of requirements and is characterized with the following properties:

- *Property 1:* Managing the uncertainty in cases' descriptions towards their membership to the different clusters as well as the total ignorance about them.
- *Property 2:* Managing the dynamicity in case bases by fixing automatically the number of clusters through stored cases learning. This property is important to well detecting the different groups of similar cases each time we apply the CBM policy.
- *Property 3:* Managing the scalability while handling uncertainty with a large number of clusters or distinct solutions in case bases.

In our context, EK-NNclus is performed on case bases as our first step in order to generate automatically and dynamically from them the different clusters reflecting cases solutions. Besides, it generates the degrees of belief towards the membership of cases to the different clusters as well as to the partition reflecting the total ignorance (the frame of discernment Θ). Indeed, the output offered by EK-NNclus will be exploited thereafter within case types detection steps.

4.2 Step 2: Isolated Cases Detection

To distinguish isolated cases from the whole case base, we have foremost to detect noisy cases in order to be eliminated since they seriously affect computations.

Noisy Cases Detection. Actually, uncertainty management and the credal partition offered by EK-NNclus allow us to detect noisy cases, especially through the degrees of belief assigned to the frame of discernment reflecting the complete ignorance. Accordingly, cases having high degree of belief to be assigned to the total ignorance are flagged as noises [7]. Thus, our idea is summed up by detecting noisy cases using the following way:

$$\boldsymbol{o_i} \in NoC \quad iff \quad m_i(\Theta) > \sum_{A_j \subset \Theta} m_i(A_j) \tag{13}$$

where $\boldsymbol{o_i}$ is a case instance and NoC represents the set of all the noisy cases.

Isolated Cases Detection. As it was defined earlier, Isolated cases are situated on the borders of the generated clusters. Hence, we detect them as cases having a distance to the different clusters centers higher than a predefined threshold (do not considering cases that already flagged as noisy -NoC-), otherwise they are considered as similar since they are so close to each others. By this way, we follow these different points:

- The center of each cluster is calculated as the mean of cases attributes values in which they belong. The decision about the membership of cases to the different clusters is achieved using the pignistic probability (Eq. 6).
- The threshold for each cluster is fixed as the mean of distances toward its center with excluding cases flagged as noisy.
- To manage uncertainty, also in distances calculation, we compute cases-clusters distances using the Belief Mahalanobis Distance (BMD) as has been used in [10].

Consequently, isolated cases are defined such that:

$$\boldsymbol{o_i} \in IsC \quad if \quad \forall\, k,\ BMD(\boldsymbol{o_i}, \boldsymbol{v_k}) > Threshold_k \tag{14}$$

where IsC represents the set of isolated cases, $\boldsymbol{o_i}$ is a case instance with $\boldsymbol{o_i} \notin NoC$, and $\boldsymbol{v_k}$ presents the center of cluster k.

4.3 Step 3: Detecting Internal Case for Each Generated Cluster

Since an internal case presents a prototype of one cluster, we fix it as the nearest case to the center of each generated cluster. Hence, a case is flagged as internal if it has the shortest Belief Mahalanobis Distance (BMD) [10] to one cluster's center. Accordingly, we define formally internal cases as follows:

$$\boldsymbol{o_i} \in InC \quad iff \quad \exists\, k;\ \nexists\, \boldsymbol{o_j} / BMD(\boldsymbol{o_j}, \boldsymbol{v_k}) < BMD(\boldsymbol{o_i}, \boldsymbol{v_k}) \tag{15}$$

where $\boldsymbol{o_i}$ and $\boldsymbol{o_j}$ are two cases instances, $\boldsymbol{v_k}$ is the center of cluster k, BMD presents the Belief Mahalanobis Distance [10] between cases and clusters, and InC presents the set of all internal cases.

4.4 Step 4: Updating the Case Base

By arriving to this last step, we have already detected the types of cases that should be selected and maintained for preserving case bases competence in future problem resolution. Therefore, our proposed approach updates ultimately the case base by holding back isolated and internal cases, and removing all the others. By this way, DETD method can be efficient in case bases alleviation while preserving or rather improving their competence in problem resolution.

5 Experimental Analysis

In this section, our aim is to evaluate the maintenance quality provided by our approach. Hence, we test it using a number of case bases from U.C.I repository of Machine Learning datasets. While developing, default parameters of EK-NNclus [7] technique are taken. Thus, we propose to measure our maintaining method's effectiveness through three evaluation criteria as done in [10–12,14]. Then, we compare results with those provided by the Initial non-maintained case bases (ICBR) as well as the non-dynamic CBM approach called ECTD [10].

5.1 Evaluation Criteria

- **Storage size [S (%)]:** The percentage of the remaining case base's size after maintenance. Hence, it is the rate of case base size reduction, and defined as follows:
$$S = \frac{Number\ of\ cases\ after\ maintenance}{Number\ of\ cases\ before\ maintenance} \times 100$$
- **Time [t (s)]:** The time of problem resolution exerted on 1-Nearest-Neighbor algorithm. This criterion allows to measure the performance of CBR systems in term of retrieval time reduction.
- **Accuracy [PCC (%)]:** The average percentage of correct classification criterion. It is applied using ten fold cross validation runned in front of 1-Nearest-Neighbor as a classification algorithm. Thus, it is defined such that:
$$PCC = \frac{Well\ solved\ problems}{Total\ solved\ problems} \times 100$$

5.2 Dynamic Aspect

To evaluate the ability of our approach in handling the dynamic aspect of maintenance, we measure the accuracies dynamically. Drawing to the actual logic provided by CBR systems towards their case bases, we present dynamicity, in our work, through evolving case bases within three consecutive times such that:

- t_1: We select randomly from the original case base a subset of cases (CB_1) with respecting the constraint of containing only two solutions (the minimal number that a case base can contain).
- t_2: We increment CB_1' size randomly with a probability to meet new solutions (without reaching the entire case base).
- t_3: We test on the totality of case bases with the totality of their solutions.

5.3 Results and Discussion

According to the evaluation criteria and the dynamicity aspect defined above, we expose the different results in Tables 1, 2 and 3. Actually, in term of reduction size rate (Table 1), our DETD approach has been able to shrink more than half of all the initial tested case bases that contain the totality of cases (100%). For instance, it provides a reduction rate about 35% for "Iris" data set and 37% for "Ionosphere" data set. Obviously, this is the result of redundant and non relevant cases removing. On the other hand, DETD and ECTD approaches are offering very close reduction rates which vary from about 35% to 50% for both of them.

Table 1. Storage size (S%)

Case bases		Storage size (S%)		
		ICBR	ECTD	DETD
1	Glass	100%	50%	46.1%
2	Indian	100%	51.21%	43.22%
3	Ionosphere	100%	41.03%	37.42%
4	Iris	100%	38.67%	35.33%
5	Vehicle	100%	46.12%	48.93%
6	Heberman	100%	39.87%	47.07%

For the retrieval time criterion, it is basically in relation with case bases density involving the storage size criterion. Actually, as shown in Table 2, the retrieval time is remarkably decreasing with the different case bases. For example, it moves on from 0.2825 s with ICBR to 0.0062 with the DETD for the "Heberman" data set. Even comparing with ECTD approach, we are noting a slight decreasing of time provided for almost all the different tested case bases.

Table 2. Retrieval time (s)

Case bases		Retrieval time (s)		
		ICBR	ECTD	DETD
1	Glass	0.0091	0.0050	0.0045
2	Indian	0.0125	0.0101	0.0083
3	Ionosphere	0.0156	0.0077	0.0057
4	Iris	0.0841	0.0068	0.0041
5	Vehicle	0.0716	0.0063	0.0061
6	Heberman	0.2825	0.0133	0.0062

Once our newly approach is able to reduce the case base along with the retrieval time and improves accordingly the CBR systems performance, we should now ascertain to their competence stability in problem resolution through the most important criterion called "Accuracy (PCC)". Actually, its results are shown in Table 3 within a dynamic way as defined in Subsect. 5.2. For ECTD approach, we fix the number of clusters equal to the number of solutions appearing during t_1 ($K = 2$). Indeed, we note that our DETD approach is more able to preserve the competence of CBs, each time we maintain them. For instance, the accuracy provided by DETD while maintaining "Vehicle" data set is almost stable as it moves on from 74.43% (t_1) to 74.05% (t_2) until 73.55% (t_3). However, it is more and more decreasing after applying ECTD (from about 75% (t_1) until 65% (t_3)). In fact, this is logically explained by the capacity of DETD to handle automatically the number of clusters whereas it is fixed for ECTD and not able to take into account this evolution. Furthermore, we note that DETD and ECTD are providing almost the same results for "Indian", "Ionosphere" and "Heberman" datasets. In fact, it is quite reasonable since they are binary, so the dynamicity in solutions is not really introduced for them. On the other hand, with considering the totality of case bases, we note that the maintained case bases with DETD offer precision even better than non-maintained ones (ICBR) such as for "Glass" data set where it reaches more than 25% of difference in t_2.

Table 3. Dynamic aspect influence in maintenance efficiency

	Case bases (CB)	Dynamic accuracy evaluation (PCC %)								
		t_1			t_2			t_3		
		ICBR	ECTD	DETD	ICBR	ECTD	DETD	ICBR	ECTD	DETD
1	Glass	76.11	77.15	80.1	61.14	70.4	88.24	86.92	63.64	**94.39**
2	Indian	75.98	76.33	76.64	73.57	73.22	74.04	**75.91**	73.78	73.8
3	Ionosphere	78.45	91.03	88.04	82.76	69.71	86.38	86.89	**87.5**	**87.5**
4	Iris	100	99.18	98.54	98.75	96.3	99.02	98	92.21	**98.28**
5	Vehicle	75.12	74.56	74.43	71.89	68.15	74.05	72.34	65.21	**73.55**
6	Heberman	68.76	77.16	75.36	69.2	59.68	73.45	74.18	**76.23**	**76.23**

6 Conclusion

In this paper, we proposed a new case base maintenance method called DETD approach where we focused on exceeding a limitation of some existing CBM policies related to the dynamicity in maintenance. While achieving its main purpose of case base maintenance, our DETD method was able to manage both of uncertainty in cases descriptions using belief function theory tools, and the aspect of dynamicity in CBR systems' case bases using a clustering method that offers a dynamic number of clusters each time we perform the maintenance.

Hence, the main idea of our work is summed up by selecting as well as possible only the most relevant types of cases for preserving CBR systems competence and performance. This is actually done, also, through handling uncertainty regarding cases positions towards the different generated clusters. As future work, we aim to propose a dynamic CBM approach in term of real-time mode of maintenance.

References

1. Aamodt, A., Plaza, E.: Case-based reasoning: foundational issues, methodological variations, and system approaches. Artif. Intell. Commun. **7**, 39–52 (1994)
2. Malek, M., Rialle, V.: Design of a case-based reasoning system applied to neuropathy diagnosis. In: Haton, J.-P., Keane, M., Manago, M. (eds.) EWCBR 1994. LNCS, vol. 984, pp. 255–265. Springer, Heidelberg (1995). https://doi.org/10.1007/3-540-60364-6_41
3. Maher, M., Garza, A.: Case-based reasoning in design. Int. J. IEEE Int. Syst. **12**, 34–41 (1997)
4. Kang, B., Yoshida, K., Motoda, H., Compton, P.: Help desk system with intelligent interface. Int. J. Appl. Artif. Intell. **11**, 611–631 (1997)
5. Varshavskii, P., Eremeev, A.: Analogy-based search for solutions in intelligent systems of decision support integrated models and flexible calculations in artificial intelligence. Int. J. Comput. Syst. Sci. **13**, 90–101 (2005)
6. Wilson, D.C., Leake, D.B.: Maintaining case-based reasoners: dimensions and directions. Comput. Intell. **17**, 196–213 (2001)
7. Denoeux, T., Kanjanatarakul, O., Sriboonchitta, S.: EK-NNclus: a clustering procedure based on the evidential K-nearest neighbor rule. Knowl.-Based Syst. **88**, 57–69 (2015)
8. Dempster, A.P.: Upper and lower probabilities induced by a multivalued mapping. Ann. Math. Stat. **38**, 325–339 (1967)
9. Shafer, G.: A Mathematical Theory of Evidence, vol. 1. Princeton University Press, Princeton (1976)
10. Ben Ayed, S., Elouedi, Z., Lefevre, E.: ECTD: evidential clustering and case types detection for case base maintenance. In: The 14th ACS/IEEE International Conference on Computer Systems and Applications (AICCSA), pp. 1462–1469 (2017)
11. Smiti, A., Elouedi, Z.: COID: maintaining case method based on clustering, outliers and internal detection. In: Lee, R., Ma, J., Bacon, L., Du, W., Petridis, M. (eds.) Software Engineering, Artificial Intelligence, Networking and Parallel/Distributed Computing, pp. 39–52. Springer, Heidelberg (2010). https://doi.org/10.1007/978-3-642-13265-0_4
12. Smiti, A., Elouedi, Z.: WCOID: maintaining case-based reasoning systems using weighting, clustering, outliers and internal cases detection. In: The 11th International Conference on Intelligent Systems Design and Applications (ISDA), pp. 356–361 (2011)
13. Zadeh, L.A.: Fuzzy sets. Inf. Control **8**, 338–353 (1965)
14. Smiti, A., Elouedi, Z.: SCBM: soft case base maintenance method based on competence model. J. Comput. Sci. **25**, 221–227 (2017)
15. Smiti, A., Elouedi, Z.: Fuzzy density based clustering method: soft DBSCAN-GM. In: The 8th International Conference on Intelligent Systems (IS), pp. 443–448 (2016)

16. Masson, M.H., Denœux, T.: ECM: an evidential version of the fuzzy c-means algorithm. Pattern Recogn. **41**, 1384–1397 (2008)
17. Denœux, T., Masson, M.H.: EVCLUS: evidential clustering of proximity data. IEEE Trans. Syst. Man Cybern. B **34**, 95–109 (2004)
18. Denœux, T.: A k-nearest neighbor classification rule based on Dempster-Shafer theory. IEEE Trans. Syst. Man Cybern. **25**, 804–813 (1995)
19. Smiti, A., Elouedi, Z.: WCOID-DG: an approach for case base maintenance based on Weighting, Clustering, Outliers, Internal Detection and DBSCAN-Gmeans. J. Comput. Syst. Sci. **80**(1), 27–38 (2014)

Ensemble Enhanced Evidential k-NN Classifier Through Rough Set Reducts

Asma Trabelsi[1,2(✉)], Zied Elouedi[1], and Eric Lefevre[2]

[1] LARODEC, Institut Supérieur de Gestion de Tunis, Université de Tunis, Tunis, Tunisia
trabelsyasma@gmail.com, zied.elouedi@gmx.fr

[2] Laboratoire de Génie Informatique et d'Automatique de l'Artois (LGI2A), Univ. Artois, EA 3926, 62400 Béthune, France
eric.lefevre@univ-artois.fr

Abstract. Data uncertainty is seen as one of the main issues of several real world applications that can affect the decision of experts. Several studies have been carried out, within the data mining and the pattern recognition fields, for processing the uncertainty that is associated to the classifier outputs. One solution consists of transforming classifier outputs into evidences within the framework of belief functions. To gain the best performance, ensemble systems with belief functions have been well studied for several years now. In this paper, we aim to construct an ensemble of the Evidential Editing k-Nearest Neighbors classifier (EEk-NN), which is an extension of the standard k-NN classifier for handling data with uncertain attribute values expressed within the belief function framework, through rough set reducts.

Keywords: Evidential Editing k-Nearest Neighbors classifier · Rough set reducts · Belief function theory · Uncertain attributes · Ensemble classifier

1 Introduction

A multiple classifier system, also referred to as a classifier ensemble, has been proven to be an effective and efficient way for solving complex classification problems and achieving high performance [17]. The construction of an ensemble of classifiers consists mainly on two distinct levels: the generation of a set of base classifiers and the combination of their output predictions. It should be emphasized that the process of improving ensemble accuracy requires the best choice of the base classifiers and also the combination operator. In this paper, we focus only on the generation of good base classifiers for enhancing accuracies.

Ensuring diversity between the base classifiers has been defended as a successful means for the production of a good ensemble of base classifiers. Although diversity can be achieved in several ways, the manipulation of the input feature space has been theoretically and experimentally proven to be one of the best

J. Medina et al. (Eds.): IPMU 2018, CCIS 853, pp. 383–394, 2018.
https://doi.org/10.1007/978-3-319-91473-2_33

methods for establishing high diversity between base classifiers [2,29,31]. In fact, it does not only allow the correlation reduction between the combined classifiers, but it also performs faster thanks to the reduced size of the input feature space [2,5,9,30]. The process of generating feature subsets with good predicting power is still under study. One commonly used solution is the random subspace method (RSM) oftenly called random subspacing. The major shortcoming of this latter technique is the random partition of the original input features. As a matter of fact, the random selection may potentially increase the risk of irrelevant and redundant features as part of the selected subsets.

The rough set theory, introduced by Pawlak [15], has been successfully applied in pattern recognition, data mining and machine learning domains, more particularly for attribute reduction problems. The reduced attribute set, representing the minimal subset of attributes that enables the discernation of objects with different decision values, is referred to as reduct. Since there have been usually multiple reducts for a given data set, the concept of ensemble classifiers through rough set reducts have been introduced and applied in a range of practical problems such as text classification [20], biomedical classification [21], tumor classification [32], web services classification [19], etc. It is important to emphasize that several real world application data suffer from some kinds of uncertainty, imprecision and also incompleteness that mainly pervade the attribute values. However, to the best of our knowledge, there are no rough set techniques allowing to obtain the possible reducts from data with uncertain attribute values.

In this paper, we aim to develop a classifier ensemble through rough set reducts (RSR) for dealing with uncertain data. More precisely where the uncertainty exists in the attribute values and is represented within the belief function theory, a flexible way for managing and representing all kinds of uncertainty. We therefore propose a new method for generating approximate reducts from such a kind of data. Since tens or hundreds of reducts may be generated, a selected subset of these reducts have to be used for constructing the base classifiers, notably the most diverse ones.

Herein, we have used the Evidential k-Nearest neighbors [28], an extension of the well known k-NN to handle the uncertainty that occurs in the attribute values within the belief function framework, as base classifiers. Given a query instance, the output beliefs of the base evidential classifiers will then be merged through a combination operator that is offered by the belief function framework [23].

The remaining of this paper is organized as follows: Sect. 2 is dedicated to recall some basic concepts of the belief function theory. Section 3 is committed to highlighting the fundamental concepts of the rough set theory. We describe, in Sect. 4, our proposed idea for constructing an ensemble of classifiers via rough set reducts for handling uncertain data. Our conducted experimentation on several synthetic databases is presented in Sect. 5. Finally, the conclusion and our main future work directions are reported in Sect. 6.

2 Belief Function Theory: Fundamental Concepts

The belief function theory has been shown to be a convenient way for representing, managing and reasoning under uncertainty. In this Section, we briefly recall some fundamental concepts underlying this theory.

2.1 Information Representation

Let $\Theta = \{\theta_1, \theta_2, \ldots, \theta_N\}$ be a frame of discernment with a finite non empty set of N elementary hypotheses that are assumed to be exhaustive and mutually exclusive. An expert's belief over a given subset of Θ has to be represented by the so-called basic belief assignment m (bba) as follows:

$$\sum_{A \subseteq \Theta} m(A) = 1 \tag{1}$$

Each subset A of 2^Θ having fulfilled $m(A) > 0$ is called a focal element.

2.2 Combination Operators

In several real-world problems, information has to be gathered from distinct sources. These latter have to be merged with the aim of obtaining the most accurate information possible. The belief function framework provides a set of combination rules for fusing such kinds of information. The conjunctive rule, proposed by Smets within the Transferable Belief Model (TBM) [25], is one of the best known rules. Given two information sources S_1 and S_2 with respectively m_1 and m_2 as bbas, the conjunctive rule, denoted by $\bigcirc\!\!\!\!\cap$, will be set as:

$$m_1 \bigcirc\!\!\!\!\cap m_2(A) = \sum_{B \cap C = A} m_1(B) m_2(C), \quad \forall A \subseteq \Theta. \tag{2}$$

The belief fully involved to the empty set reflects the conflictual mass. With the aim of retaining the basic properties of the belief function theory, Dempster have proposed in [4], a normalized version of the conjunctive rule. This latter allows to manage the conflict by redistributing the conflictual mass over all focal elements. It is obtained as follows:

$$m_1 \oplus m_2(A) = \frac{1}{1-K} \sum_{B \cap C = A} m_1(B) m_2(C), \quad \forall A \subseteq \Theta \tag{3}$$

where the conflictual mass K caused by the combination of the two bbas m_1 and m_2 through the conjunctive rule, is given as follows:

$$K = \sum_{B \cap C = \emptyset} m_1(B) m_2(C) \tag{4}$$

2.3 Decision Making

The pignistic probability $BetP$, proposed in [24], is an efficient and binding way for decision-making. It transforms beliefs into probability measures as follows:

$$BetP(A) = \sum_{B \cap A = \emptyset} \frac{|A \cap B|}{|B|} m(B), \quad \forall A \in \Theta \tag{5}$$

Making decision consists of selecting the most likely hypothesis, meaning the hypothesis H_s with the highest pignistic probability:

$$H_s = argmax_A BetP(A), \quad \forall A \in \Theta \tag{6}$$

2.4 Dissimilarity Between bbas

Numerous measures have been introduced for computing the dissimilarity degree between two given bbas [7,18,26]. The Jousselme distance [7] is regarded as one of the well-known ones. Formally, the Jousselme distance, for two given bbas m_1 and m_2, is defined by:

$$dist(m_1, m_2) = \sqrt{\frac{1}{2}(m_1 - m_2)^T D (m_1 - m_2)} \tag{7}$$

where the Jaccard similarity measure D is set to:

$$D(X,Y) = \begin{cases} 1 & \text{if } X = Y = \emptyset \\ \dfrac{|X \cap Y|}{|X \cup Y|} & \forall\, X, Y \in 2^{\Theta} \end{cases} \tag{8}$$

3 Rough Set Theory

The rough set theory, which is proposed by Pawlak [15], constitutes a valid mathematical solution for handling imperfect data for several machine learning applications. Examples include clustering [14], classification [6,8] and attribute reduction [1,11], etc. Attribute reduction within the rough set theory consists of discovering the minimal subsets of relevant features, also named reduct, from the original set. Authors in [22], have introduced the notation of discernibility matrix and function as a way for finding reducts for a given data T. Suppose that $T = \{x_1, \ldots, x_D\}$ is a data composed with D objects x_i ($i \in \{1, \ldots, D\}$) characterized by N attributes $A = \{a_1, \ldots, a_N\}$ having values $V = \{v_1^i, \ldots, v_N^i\}$ and a class label $Y_i \in C = \{c_A, \ldots, c_Q\}$ (i.e. Q is the number of classes). The discernibility matrix of T, denoted by DM, is a $|D| \times |D|$ matrix in which the element $DM(x_i, x_j)$ for an object pair (x_i, x_j) is defined as follows $\forall\ i, j = \{1, \ldots, D\}$ and $\forall\ n = \{1, \ldots, N\}$:

$$DM(x_i, x_j) = \{a_n \in A | v_n^i(x_i) \neq v_n^j(x_j) \text{ and } Y_i \neq Y_j\}$$

Each element $DM(x_i, x_j)$ represents the set of all condition attributes discerning objects x_i and x_j that have not the same class label. The notion of discernibility function can be defined from the discernibility matrix as follows:

$$f(DM) = \wedge\{\vee(DM(x_i, x_j)) | \forall x_i, x_j \in T, DM(x_i, x_j) \neq \emptyset\} \quad (9)$$

Reducts may be yielded by transforming the discernibility function from conjunctive normal form into disjunctive normal form. The major shortcoming of this solution is its costly computation which makes it impractical for large or even medium sized data sets. Therefore, several heuristics have been discussed to overcome this drawback. Johnson's heuristic algorithm and the hitting set approach are ones of the most known algorithms [3].

4 Classifier Ensemble Through Rough Set Reducts

In this paper, we aim to construct an ensemble of classifiers from data characterized by uncertain attribute values expressed within the belief function framework. Particularly, we propose to construct an ensemble of the Enhanced Evidential k Nearest Neighbors (EEk-NN), an extended version of the classical k-NN for handling evidential data, through rough set reducts. The general structure of our proposed idea is depicted in Fig. 1.

Given a training data with uncertain attribute values, we have to generate firstly all possible rough set reducts $R = \{r_1, \ldots, r_M\}$. Subsequently, we have to choose the ones enabling the construction of a good ensemble of EEk-NNs. Mainly, we have to pick out the most diverse ones. The decision yielded by each individual classifier, for a given query instance, will be merged using the Dempster operator, one well used belief function combination rules for merging distinct classifiers.

Numerous reduct generation methods have been proposed in the literature and the commonly used ones are mainly based on the information entropy and the discernibility matrix. Examples include the Johnson algorithm and the hitting set approach. The former one consists of a greedy search technique for piking out a single reduct which is generally close to the optimal, while the latter one allows the computation of multiple reducts. Within the hitting set approach, a multiset ζ will contain the non empty sets of a given discernability matrix and the minimal hitting sets of ζ are exactly the reducts.

Since the computation of reducts using the hitting set approach is an NP-hard problem, genetic algorithms have been used for generating approximate hitting sets, meaning approximate reducts. One example is the SAVGenetic algorithm reducer [16], a Rosetta toolkit algorithm providing multiple reducts on the basis of the hitting set paradigm [10]. Since this algorithm is widely used, it has not the ability to handle uncertain data. Herein, we propose to extend the SAVGenetic algorithm reducer for handling data with uncertain attribute values represented by belief functions.

In analogy with the SAVGenetic algorithm, our proposed algorithm consists firstly of computing the discernability matrix for data with uncertain attributes.

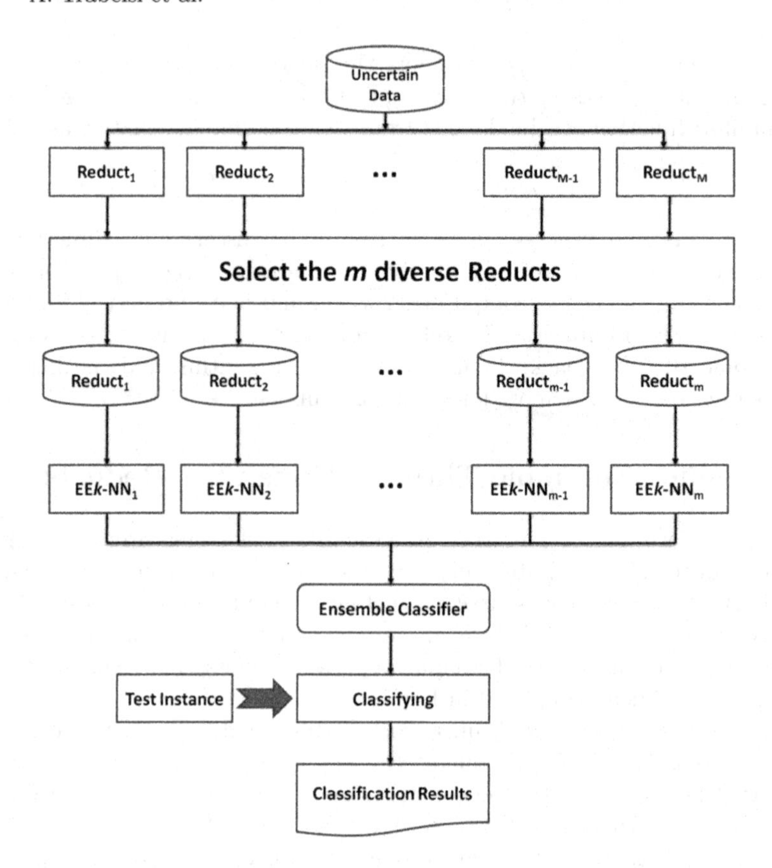

Fig. 1. The general structure of the proposed method.

We have already proposed, in [27], a belief discernability matrix for discerning pairs of objects with uncertain attribute values expressed in terms of belief functions.

Given a data set $T = \{x_1, \ldots, x_D\}$ with a finite set of D objects x_i ($i \in \{1, \ldots, D\}$). Every object x_i is defined by a set of N uncertain attributes $A = \{a_1, \ldots, a_N\}$ with values $uV^i = \{uv_1^i, \ldots, uv_N^i\}$ and a certain class label $Y_i \in C = \{c_1, \ldots, c_Q\}$. Each uncertain attribute value uv_n^i relative to an instance x_i (with n in $\{1, \ldots, N\}$) will be expressed by a basic belief assignment $m_i^{\Theta_n}$ where Θ_n reflects the frame of discernment relative to the attribute n. Let S denotes a tolerance threshold (i.e. in this paper S is set to 0.1 for maximizing the search space) and *dist* reflects the Jousselme distance. The entries of our proposed belief discernibility matrix, denoted by Λ', have been set as follows $\forall\, i, j \in \{1, \ldots, D\}$ and $n \in \{1, \ldots, N\}$:

$$\Lambda'(x_i, x_j) = \{a_n \in A | dist(m_i^{\Theta_n}, m_j^{\Theta_n}) > S \text{ and } Y_i \neq Y_j\} \quad (10)$$

A multiset ζ' will then contain the non empty set of Λ' and the approximate hitting sets of ζ' correspond exactly to the approximate reducts. Our algorithm's fitness function corresponds exactly to that of the standard SAVGenetic algorithm. It consists of two main parts. The former one rewards subsets with shortest size, while the latter one rewards subsets that are hitting sets (i.e. meaning subsets having a non empty intersection with all elements of the discernability matrix). It is set as follows for each subset $B \in 2^N$:

$$f(B) = (1-\alpha) \times \frac{|A| - |B|}{|B|} + \alpha \times min\{\varepsilon, \frac{[F \in \zeta' | F \cap B = \emptyset|]}{|\zeta'|}\} \quad (11)$$

where $\alpha \in [0,1]$ reflects the adaptive weighting between the two terms and ε expresses the minimal hitting set fraction.

It is important to note that rough set approaches may generate tens or even hundreds of reducts and the most diverse ones have to be chosen for ensemble learning. One simplest algorithm for picking out diverse reducts is introduced in [3]. It consists of choosing randomly a reduct from the initial list and then adding progressively reducts that are diverse as much as possible from the chosen ones. The diversity degree is set as:

$$Div_s = 1 - \frac{\sum_{k \in L} \frac{Red_s \cap Red_k}{Red_s \cup Red_k}}{L} \quad (12)$$

where L is the number of the chosen reducts and Red_s is the candidate reduct. The candidate reduct with the highest diversity will be added to the list of the chosen reducts.

5 Experimentations

More recently, we have introduced an ensemble of EEk-NNs through the random subspace method (RSM) and we have proven its performance compared with the individual EEk-NN that is trained in full feature space [28]. In this paper, we aim to evaluate the performance of the ensemble of EEk-NNs through rough set reducts (RSR). Thus, we propose to carry out a comparative study with the ensemble proposed in [28]. We relied mainly on the percentage of correct classification criterion (PCC). In what follows, we present our experimentation settings and results.

5.1 Experimentation Settings

As we are handling data with uncertain attribute values expressed in terms of belief functions, we have proposed to construct synthetic databases by injecting uncertainty on real databases obtained from the UCI machine learning repository [12]. We provide, in Table 1, a brief description of some categorical databases where #Instances, #Attributes and #Classes denote, respectively, the number of instances, the number of attributes and the number of classes. We have tackled different uncertainty levels P:

- Certain case: $P = 0$
- Low uncertainty case: $(0 < P < 0.4)$
- Middle uncertainty case: $(0.4 \leq P < 0.7)$
- High uncertainty case: $(0.7 \leq P \leq 1)$

Table 1. Description of databases.

Databases	#Instances	#Attributes	#Classes
Voting Records	435	16	2
Monks	432	7	2
Breast Cancer	286	9	2
Lymphography	148	18	4
Tic-Tac-Toa	958	9	2

Suppose that T is a database composed with D instances x_i ($i \in \{1, \dots, D\}$). Each instance x_i is characterized by N uncertain attribute values uv_i^n ($n \in \{1, \dots, N\}$). Suppose that Θ_n is the frame of discernment relative to the attribute n. Each attribute value uv_i^n relative to an instance x_i such that $uv_i^n \subseteq \Theta_n$ will be expressed in terms of belief functions as follows:

$$\begin{aligned} m^{\Theta_n}\{x_i\}(uv_i^n) &= 1 - P \\ m^{\Theta_n}\{x_i\}(\Theta_n) &= P \end{aligned} \tag{13}$$

It is important to note that some databases suffer from incompleteness. The belief function theory allows to represent and manage missing attribute values. In this paper, the missing attribute values will be modeled as follows:

$$\begin{aligned} m^{\Theta_n}\{x_i\}(uv_i^n) &= 0 \qquad \forall\, uv_i^n \subseteq \Theta_n \\ m^{\Theta_n}\{x_i\}(\Theta_n) &= 1 \end{aligned} \tag{14}$$

5.2 Experimentation Results

In this experimentations, we have relied on the 10-fold cross validation strategy for learning the individual Enhanced k- Nearest Neighbors classifiers. One key issue which has to be addressed is the number of neighbors that may give satisfactory results. In our experimentation tests, we evaluate five values of k which respectively correspond to 1, 3, 5, 7 and 9. Another substantial key element when designing an ensemble of classifier is the number of individual classifiers used to get the final decision. The conclusion conducted following to the study of [13] proves that ensembles of 25 classifiers are sufficient for reducing the error rate and consequently for improving performance. Thus, in our paper, the number of the merged classifiers will be equal to 25. The final PCCs, which are obtained

Table 2. Results for Voting Records database (%).

	$k = 1$		$k = 3$		$k = 5$		$k = 7$		$k = 9$	
	RSM	RSR	RSM	RSR	RSM	RSR	RSM	RSR	RSM	RSR
No	90.23	93.72	91.62	93.72	90.93	93.72	91.93	93.72	91.16	93.72
Low	90.46	93.95	91.62	93.95	91.16	93.95	90.06	93.95	91.39	93.95
Middle	91.39	93.95	91.39	93.95	91.86	93.95	91.62	93.95	90.93	93.95
High	88.37	95.53	89.53	89.76	89.76	89.76	89.76	90	89.30	90
Average	90.11	**95.03**	91.04	**92.84**	90.92	**92.84**	90.70	**93.65**	90.69	**93.65**

Table 3. Results for Lymphography database (%).

	$k = 1$		$k = 3$		$k = 5$		$k = 7$		$k = 9$	
	RSM	RSR	RSM	RSR	RSM	RSR	RSM	RSR	RSM	RSR
No	83.57	86.42	82.85	86.42	85.71	85	84.28	85	80.71	85
Low	81.42	77.85	90	77.85	81.42	73.85	78.85	78.57	79.28	77.85
Middle	83.57	81.42	82.85	81.42	83.57	80.71	81.42	80	82.85	80
High	62.24	72.85	62.85	73.75	61.42	73.57	62.85	73.57	61.42	73.57
Average	77.70	**79.62**	77.38	**79.81**	78.03	**79.28**	76.82	**79.10**	76.06	**79.10**

Table 4. Results for Tic-Tac-Toa database (%).

	$k = 1$		$k = 3$		$k = 5$		$k = 7$		$k = 9$	
	RSM	RSR	RSM	RSR	RSM	RSR	RSM	RSR	RSM	RSR
No	62.10	62	61.15	63.89	61.05	62.42	60.84	62	60	60.63
Low	55.36	56.73	55.78	57.05	55.89	54.52	55.89	55.26	55.57	55.68
Middle	55.57	57.57	56	57.47	56	55.47	56.21	56.10	56.21	56
High	57.78	57.89	57.68	58.10	58	57.89	58.31	58.21	50.31	59.15
Average	57.70	**58.54**	57.25	**59.2**	**57.73**	57.57	57.81	**57.89**	57.52	**57.86**

through the combination of the classifier outputs using the Dempster rule, will be given from Tables 2, 3, 4, 5 and 6, where RSM reflects the results yielded through the random subspace method and RSR reflects the results obtained with the rough set reducts method.

The PCC results given from Tables 2, 3, 4, 5 and 6 have proven the efficiency of the ensemble classifiers that are obtained through the rough set reduct approach over that yielded using the random subspace method. In fact, the average accuracies achieved by the RSR method for the different values of k are almost always greater than those achieved by the RSM method. Taking the Voting Records database as example, the average accuracy done by the RSR approach are equal to 95.03, 92.84, 92.84, 93.65 and 93.65, while there are equal to 90.11,

Table 5. Results for Monks database (%).

	$k = 1$		$k = 3$		$k = 5$		$k = 7$		$k = 9$	
	RSM	RSR	RSM	RSR	RSM	RSR	RSM	RSR	RSM	RSR
No	73.13	85.45	60.26	85.45	61.68	85.45	69.03	85.45	79.81	85.45
Low	71.01	84.36	59.49	84.36	94.16	84.36	70.65	84.36	76.54	84.36
Middle	69.85	85.27	60.26	85.27	68.9	85.27	72.84	85.27	70.72	85.27
High	56.14	64.72	53.68	64.72	52.03	64.72	53.72	64.72	54.18	64.72
Average	67.35	**79.95**	58.42	**79.95**	69.19	**79.95**	66.56	**79.95**	70.31	**79.95**

Table 6. Results for Breast Cancer database (%).

	$k = 1$		$k = 3$		$k = 5$		$k = 7$		$k = 9$	
	RSM	RSR	RSM	RSR	RSM	RSR	RSM	RSR	RSM	RSR
No	73.13	75.08	76.18	75.07	75.10	73.71	74.04	76.03	76.03	76.9
Low	73.10	73.57	76.10	76.18	74.89	75.10	73.91	74.04	75.8	76.9
Middle	73.92	73.75	75.59	76.18	74.32	75.10	73.80	74.04	76	76.9
High	73.01	73.75	75.45	76.18	74.62	75.10	72.12	74.04	76.13	76.9
Average	73.29	**74.03**	75.83	**75.90**	74.73	**74.75**	73.46	**74.53**	75.99	**76.9**

91.04, 90.92, 90.70 and 90.69 for respectively $k = 1, k = 3, k = 5, k = 7$ and $k = 9$. The conclusion derived from the carried out experimentation tests may be justified by the fact that random subspace methods may negatively affect the classification process as irrelevant and redundant features can part of the selected subsets.

6 Conclusion

The idea underlying this paper is to increase accuracy for a given classification system through ensemble systems. Herein, we have constructed an ensemble of the so-called Enhanced Evidential k-Nearest Neighbors for dealing with uncertain data, more precisely where the uncertainty pervades the attribute values and is represented with belief functions. With the aim of assessing the performance of our proposed technique, we have conducted a comparative study with ensemble constructed through random subspaces. The yielded results have shown the efficiency of the rough set reducts over random subspaces. As we combine distinct classifiers, in this paper, we have relied on the Dempster rule of combination. As there are other combination rules, in our future work, we intend to pick out the combination operator that yields the best classification results. With the aim of increasing accuracy, we look forward to take into consideration not only the diversity between reducts, but also the diversity between the merged classifiers and the accuracy of the individual classifiers to yield more performance.

References

1. Bhatt, R.B., Gopal, M.: On fuzzy-rough sets approach to feature selection. Pattern Recogn. Lett. **26**(7), 965–975 (2005)
2. Bryll, R., Gutierrez-Osuna, R., Quek, F.: Attribute bagging: improving accuracy of classifier ensembles by using random feature subsets. Pattern Recogn. **36**(6), 1291–1302 (2003)
3. Debie, E., Shafi, K., Lokan, C., Merrick, K.: Reduct based ensemble of learning classifier system for real-valued classification problems. In: IEEE Symposium on Computational Intelligence and Ensemble Learning (CIEL), pp. 66–73. IEEE (2013)
4. Dempster, A.P.: Upper and lower probabilities induced by a multivalued mapping. Ann. Math. Stat. **38**, 325–339 (1967)
5. Günter, S., Bunke, H.: Feature selection algorithms for the generation of multiple classifier systems and their application to handwritten word recognition. Pattern Recogn. Lett. **25**(11), 1323–1336 (2004)
6. Jensen, R., Cornelis, C.: Fuzzy-rough nearest neighbour classification and prediction. Theor. Comput. Sci. **412**(42), 5871–5884 (2011)
7. Jousselme, A., Grenier, D., Bossé, E.: A new distance between two bodies of evidence. Inf. Fusion **2**(2), 91–101 (2001)
8. Khoo, L., Tor, S., Zhai, L.: A rough-set-based approach for classification and rule induction. Int. J. Adv. Manuf. Technol. **15**(6), 438–444 (1999)
9. Kim, Y.: Toward a successful CRM: variable selection, sampling, and ensemble. Decis. Support Syst. **41**(2), 542–553 (2006)
10. Komorowski, J., Øhrn, A., Skowron, A.: The ROSETTA rough set software system. In: Handbook of Data Mining and Knowledge Discovery, pp. 2–3 (2002)
11. Kumar, P., Vadakkepat, P., Poh, L.A.: Fuzzy-rough discriminative feature selection and classification algorithm, with application to microarray and image datasets. Appl. Soft Comput. **11**(4), 3429–3440 (2011)
12. Murphy, P., Aha, D.: UCI repository databases. http://www.ics.uci.edu/mlear (1996)
13. Opitz, D., Maclin, R.: Popular ensemble methods: an empirical study. J. Artif. Intell. Res. **11**, 169–198 (1999)
14. Parmar, D., Wu, T., Blackhurst, J.: MMR: an algorithm for clustering categorical data using rough set theory. Data Knowl. Eng. **63**(3), 879–893 (2007)
15. Pawlak, Z.: Rough sets. Int. J. Comput. Inf. Sci. **11**(5), 341–356 (1982)
16. Phon-Amnuaisuk, S., Ang, S.-P., Lee, S.-Y. (eds.): MIWAI 2017. LNCS (LNAI), vol. 10607. Springer, Cham (2017). https://doi.org/10.1007/978-3-319-69456-6
17. Ponti Jr., M.P.: Combining classifiers: from the creation of ensembles to the decision fusion. In: 24th SIBGRAPI Conference on Graphics, Patterns and Images Tutorials (SIBGRAPI-T), pp. 1–10. IEEE (2011)
18. Ristic, B., Smets, P.: The TBM global distance measure for the association of uncertain combat id declarations. Inf. Fusion **7**(3), 276–284 (2006)
19. Saha, S., Murthy, C.A., Pal, S.K.: Classification of web services using tensor space model and rough ensemble classifier. In: An, A., Matwin, S., Raś, Z.W., Ślęzak, D. (eds.) ISMIS 2008. LNCS (LNAI), vol. 4994, pp. 508–513. Springer, Heidelberg (2008). https://doi.org/10.1007/978-3-540-68123-6_55
20. Shi, L., Ma, X., Xi, L., Duan, Q., Zhao, J.: Rough set and ensemble learning based semi-supervised algorithm for text classification. Expert Syst. Appl. **38**(5), 6300–6306 (2011)

21. Shi, L., Xi, L., Ma, X., Weng, M., Hu, X.: A novel ensemble algorithm for biomedical classification based on ant colony optimization. Appl. Soft Comput. **11**(8), 5674–5683 (2011)
22. Skowron, A., Rauszer, C.: The discernibility matrices and functions in information systems. In: Słowiński, R. (ed.) Intelligent Decision Support, pp. 331–362. Springer, Dordrecht (1992). https://doi.org/10.1007/978-94-015-7975-9_21
23. Smets, P.: The combination of evidence in the transferable belief model. IEEE Trans. Pattern Anal. Mach. Intell. **12**(5), 447–458 (1990)
24. Smets, P.: Decision making in the TBM: the necessity of the pignistic transformation. Int. J. Approx. Reas. **38**(2), 133–147 (2005)
25. Smets, P., Kennes, R.: The transferable belief model. Artif. Intell. **66**(2), 191–234 (1994)
26. Tessem, B.: Approximations for efficient computation in the theory of evidence. Artif. Intell. **61**(2), 315–329 (1993)
27. Trabelsi, A., Elouedi, Z., Lefevre, E.: Feature selection from partially uncertain data within the belief function framework. In: Carvalho, J.P., Lesot, M.-J., Kaymak, U., Vieira, S., Bouchon-Meunier, B., Yager, R.R. (eds.) IPMU 2016. CCIS, vol. 611, pp. 643–655. Springer, Cham (2016). https://doi.org/10.1007/978-3-319-40581-0_52
28. Trabelsi, A., Elouedi, Z., Lefevre, E.: Ensemble enhanced evidential k-NN classifier through random subspaces. In: Antonucci, A., Cholvy, L., Papini, O. (eds.) ECSQARU 2017. LNCS (LNAI), vol. 10369, pp. 212–221. Springer, Cham (2017). https://doi.org/10.1007/978-3-319-61581-3_20
29. Tumer, K., Ghosh, J.: Classifier combining: analytical results and implications. In: Proceedings of the National Conference on Artificial Intelligence, pp. 126–132. Citeseer (1996)
30. Tumer, K., Oza, N.C.: Input decimated ensembles. Pattern Anal. Appl. **6**(1), 65–77 (2003)
31. Turner, K., Oza, N.C.: Decimated input ensembles for improved generalization. In: Proceedings of International Joint Conference on Neural Network (IJCNN 1999), vol. 5, pp. 3069–3074. IEEE (1999)
32. Wang, S.-L., Li, X., Zhang, S., Gui, J., Huang, D.-S.: Tumor classification by combining PNN classifier ensemble with neighborhood rough set based gene reduction. Comput. Biol. Med. **40**(2), 179–189 (2010)

Towards a Hybrid User and Item-Based Collaborative Filtering Under the Belief Function Theory

Raoua Abdelkhalek(✉), Imen Boukhris(✉), and Zied Elouedi(✉)

LARODEC, Institut Supérieur de Gestion de Tunis,
Université de Tunis, Tunis, Tunisia
abdelkhalek_raoua@live.fr, imen.boukhris@hotmail.com, zied.elouedi@gmx.fr

Abstract. Collaborative Filtering (CF) approaches enjoy considerable popularity in the field of Recommender Systems (RSs). They exploit the users' past ratings and provide personalized recommendations on this basis. Commonly, neighborhood-based CF approaches focus on relationships between items (item-based) or, alternatively, between users (user-based). User-based CF predicts new preferences based on the users sharing similar interests. Item-based computes the similarity between items rather than users to perform the final predictions. However, in both approaches, only partial information from the rating matrix is exploited since they rely either on the ratings of similar users or similar items. Besides, the reliability of the information provided by these pieces of evidence as well as the final predictions cannot be fully trusted. To tackle these issues, we propose a new hybrid neighborhood-based CF under the belief function framework. Our approach tends to take advantage of the two kinds of information sources while handling uncertainty pervaded in the predictions. Pieces of evidence from both items and users are combined using Dempster's rule of combination. The performance of the new recommendation approach is validated on a real-world data set and compared to state of the art CF neighborhood approaches under the belief function theory.

Keywords: Recommender Systems · Collaborative filtering
User-based · Item-based · Belief function theory · Uncertainty

1 Introduction

Due to the plethora of available alternatives, offering suggestions to users and helping them in their decision making have become increasingly important nowadays. In fact, Recommender Systems (RSs) [1] have emerged for this purpose. Such systems have the effect of guiding users in a personalized way to interesting or useful items among a large space of possible options. More specifically, they filter data, predict users' preferences and provide them with the appropriate recommendations. RSs have become widely used in recent years and a panoply

J. Medina et al. (Eds.): IPMU 2018, CCIS 853, pp. 395–406, 2018.
https://doi.org/10.1007/978-3-319-91473-2_34

of recommendation approaches has been proposed. Collaborative Filtering (CF) [2] is among the most promising strategies commonly used in RSs. This latter exploits the users' past activities and preferences to perform recommendations. An important class for collaborative recommender is the neighborhood-based CF approaches [3] typically divided into user-based [4] and item-based [5] according to different assumptions. The core idea of a user-based approach consists on predicting the user's preferences based on the users having similar rating patterns. That is to say, it assumes that if two users rated some items similarly then, they would rate other items similarly. On the other hand, the idea behind item-based is to rely on the ratings of similar items rather than similar users to predict the final ratings. It assumes that two items are similar if many users in the system have rated these items in a similar way.

Several assumptions from different points of views result in different recommendation strategies. Despite their simplicity and popularity, these two approaches exhibit some limitations since they do not make use of the entire user-item matrix. They rely only on the ratings corresponding to the similar users or, alternatively, on those corresponding to the similar items. Hence, it is obvious that in both cases, an important part of the available information is often discarded.

Yet, some CF approaches have been proposed in this context aiming to unify the predictions provided by the two information sources [6–11]. Most of the proposed approaches have been exclusively conceived to deal only with binary ratings. This particular case of Collaborative Filtering is commonly referred to as One-Class Collaborative Filtering (OCCF) [12] where every value in the rating matrix would be either 1 (Like) or 0 (Dislike). However, in many real-world CF problems, discrete numeric scales are commonly adopted to judge items [1]. Furthermore, the uncertainty that reigns in the provided recommendations as well as the reliability of each information source should be taken into account to reflect more credible results. An effective recommendation approach must increase the intelligibility and the transparency of the predictions. This is a crucial challenge to improve the users' confidence as well as their satisfaction [1].

Aiming to overcome the limitations mentioned above, we propose in this paper a new hybrid neighborhood-based CF using the belief function theory [13–15]. Such theory represents a general framework to deal with uncertainty and offers a rich representation about all situations ranging from complete knowledge to complete ignorance. Our approach takes advantage of both similar users and similar items considered as different pieces of evidence. It tends to take into account both information imperfection and source reliability to form more valuable evidence. The final prediction is obtained by fusing pieces of evidence from both items and users using Dempster's rule of combination. The prediction process of our hybrid approach is represented through a basic belief assignment. This would be a convenient way for the users to help them taking a better decision.

The remainder of this paper is organized as follows: Sect. 2 recalls the belief function theory basic concepts and operations. Section 3 is dedicated to the related works. Our proposed recommendation approach is described in Sect. 4. Section 5 illustrates its experimental results conducted on a real-world data set. Finally, the contribution is summarized and the paper is concluded in Sect. 6.

2 Background on the Belief Function Theory

Under the belief function framework [13–15], a problem domain is represented by a finite set of events called the frame of discernment and denoted by Θ, where $\Theta = \{\theta_1, \theta_2, \cdots, \theta_n\}$. It contains hypotheses concerning the given problem.

All the possible values that can be associated to each subset of Θ is called the power set of Θ and denoted by 2^Θ, where $2^\Theta = \{A : A \subseteq \Theta\}$. The belief committed to each element of Θ is expressed by a basic belief assignment (*bba*) which is a mapping function $m : 2^\Theta \rightarrow [0, 1]$ such that:

$$\sum_{A \subseteq \Theta} m(A) = 1 \tag{1}$$

Each mass $m(A)$, called a basic belief mass (*bbm*), represents the degree of belief exactly attached to the event A of Θ. The *bba* that models the state of the total ignorance is called vacuous *bba* and defined such that: $m(\Theta) = 1$. The fusion of two *bba*'s m_1 and m_2 can be ensured using Dempster's rule of combination dealing with the closed world assumption. This rule, which is characterized by the commutativity and the associativity properties, assumes pieces of evidence to be reliable and distinct and it is defined as follows:

$$(m_1 \oplus m_2)(A) = k. \sum_{B,C \subseteq \Theta : B \cap C = A} m_1(B) \cdot m_2(C) \tag{2}$$

$$\textit{where} \quad (m_1 \oplus m_2)(\varnothing) = 0 \quad \textit{and} \quad k^{-1} = 1 - \sum_{B,C \subseteq \Theta : B \cap C = \varnothing} m_1(B) \cdot m_2(C)$$

The reliability of the piece of evidence can be evaluated by a discounting mechanism [15] which could therefore be performed on m. The discounted *bba* m^α is obtained as follows:

$$\begin{aligned} m^\alpha(A) &= (1 - \alpha) \cdot m(A), \text{ for } A \subset \Theta \\ m^\alpha(\Theta) &= \alpha + (1 - \alpha) \cdot m(\Theta) \end{aligned} \tag{3}$$

where the coefficient $\alpha \in [0, 1]$ is called the discounting factor.

3 Related Work

Neighborhood-based CF methods can be divided into user-based and item-based. These approaches compute the similarities between users or items and perform recommendations accordingly. A recent direction in Collaborative Filtering research combines user-based and item-based approaches in order to take advantage of both worlds. Miyahara and Pazzani [6] have defined a formulation of the Simple Bayesian Classifier for both user-based an item-based. Numerical ratings have been transformed into two labels namely, Like and Dislike. Then, the rating matrix has been filled out with pseudo-scores generated by item-based, and user-based has been applied. In the same context, a Hybrid Predictive Algorithm with Smoothing (HSPA) has been conceived by Hu and Lu [7] where the missing ratings have been estimated as a weighted average of similar items. On this basis, user-based has been implemented and only the users belonging to the similar clusters have been considered. Wang et al. [8] have proposed to unify user-based and item-based by fusing predictions. Actually, the final rating of the unifying approach is computed through a weighted sum of predictions from user-based and item-based. The choice of the weight parameter is still arguable since it appears to change under different circumstances. A deep research performed by Yamashita et al. [9] has proved this assumption where their several studies have showed that such parameter is affected by different system factors such as the frequency of the input ratings and the distributions of the user's preferences. Another hybrid CF approach has also been described by Tso-Sutter et al. [10] where they defined an adaptive fusion mechanism to capture the 3-dimensional correlations between users, items and tags. More recently, Verstrepen and Goethals [11] have proposed a reformulation that unifies user and item-based nearest neighbors for One-Class Collaborative Filtering. However, such approach is able to deal only with binary ratings which is not always the case in real-world CF situations. Working on a movie Recommender System, the authors have proposed to convert the ratings 4 and 5 to 1 (Like) and the ratings 1 and 2 to 0 (Dislike) while they ignored the rating 3. Even in such case, the performed prediction does not take into account the different degrees of the users' preferences.

In this work, we are rather interested in Multi-Class Collaborative Filtering (MCCF) where the multi grade ratings is taken into account in this model. We embrace the belief function theory in order to model conveniently the uncertainty related to the information sources and the final predictions. In fact, recent studies have emphasized the relevance of the adoption of the belief function theory in RSs area. Nguyen and Huynh [16] have proposed to model the user's preferences through the belief function tools and incorporate context information for generating suitable recommendations. An item-based collaborative filtering under the belief function theory has been conceived in [17]. The final prediction was an aggregation of the evidence provided by the similar items. The reliability of each similar item has been quantified in [18] using the belief function tools. In [19], similar users have been considered as different pieces of evidence contributing to the prediction process. However, the reliability of each piece of evidence has not been considered. Several methods have been proposed under the belief function theory. Nonetheless, they only take into consideration a very small portion of ratings since they focus only on similar users or, alternatively, on similar items.

In contrast to these evidential methods, the framework proposed in our paper extends these ideas to include both user-based CF and item-based CF into the final prediction. Hence, the new approach does not require to rely only on a particular information source to provide recommendations. The evidence collected from the similar items as well as the similar users is evaluated according to its reliability and aggregated using Dempster's rule of combination leading to the final predictions.

4 A Hybrid Neighborhood-Based CF Under the Belief Function Theory

Our proposed hybrid CF approach, that we denote by (E-HNBCF), is based on the intuition of the neighborhood-based CF commonly used in the Recommender Systems area. We want to be able to infer more reliable and trustworthy predictions based on the whole information available in the user-item matrix. To do this, we draw on a new recommendation method based on the ratings of both similar items and similar users. We adopt the belief function theory in order to model conveniently the uncertainty in the information provided by these two kinds of sources. We tend also to quantify the uncertainty pervaded throughout the prediction process which reflects more reality and transparency. Hence, the new hybrid approach allows the user to get involved in the final decision. The whole process of our new recommendation approach is illustrated in Fig. 1.

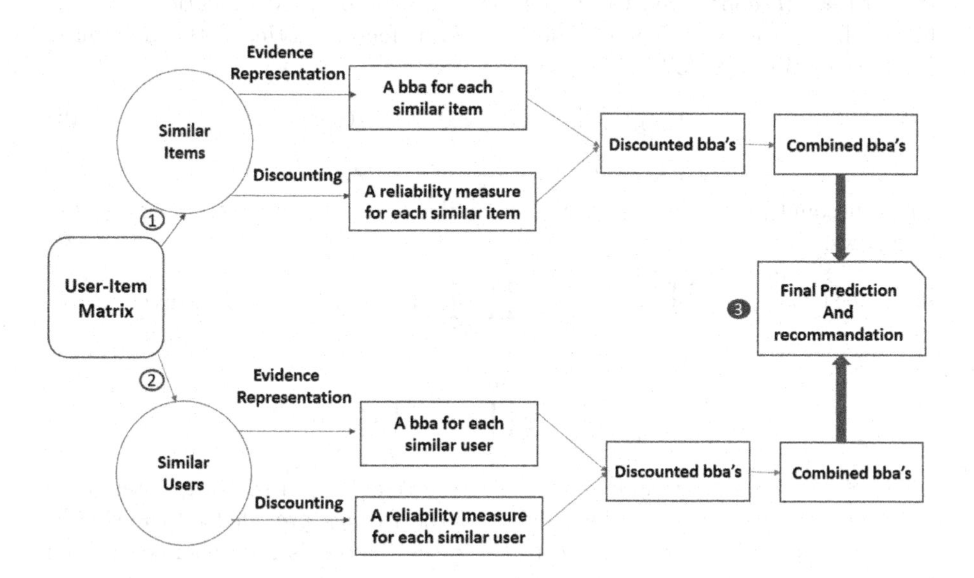

Fig. 1. A new evidential hybrid neighborhood-based collaborative filtering

4.1 Evidential Prediction Based on Items' Neighborhood

The evidential prediction process consists first in transforming the rating of each neighbor into a mass function as in the formalism proposed in [20]. The frame of discernment is defined as $\Omega_{pref} = \{p_1, p_2, \cdots, p_N\}$ where N is the number of the possible ratings p in the system. The *bba* is defined as:

$$m_{A,B}(\{p_i\}) = \alpha_0 \exp^{-(\gamma_{p_i}^2 \times (D(A,B))^2} \tag{4}$$

$$m_{A,B}(\Omega_{pref}) = 1 - \alpha_0 \exp^{-(\gamma_{p_i}^2 \times (D(A,B))^2}$$

Note that α_0 is initialized to the value 0.95 as invoked in [20]. To obtain the second parameter γ_{p_i}, the inverse of the mean distance between each couple of items sharing the same rating p_i should be computed. $D(A, B)$ is the normalized euclidean distance between the item A and the item B defined as:

$$D(A,B) = \frac{\sqrt{\sum_{u \in (u_A \cap u_B)} (p_{u,A} - p_{u,B})^2}}{|u_A \cap u_B|} \tag{5}$$

where $p_{u,A}$ and $p_{u,B}$ correspond to the ratings of the user u for the target item A and its neighbor B. Moreover, u_A and u_B are the users who have rated both items A and B. Once the *bba*'s are generated, the discounting technique [15] is applied to quantify the reliability of each similar item. We define the discounting factor ω_I as following: $\omega_I = D(A, B)/max(D)$ where $max(D)$ corresponds to the maximum value of the computed distances. The items having the highest similarities are considered to be the more reliable ones. We mention that the higher the reliability, the smaller the discounting factor ω_I. Hence, the discounted *bba*'s are obtained as follows:

$$m_{A,B}^{\omega_I}(\{p_i\}) = (1 - \omega_I) \cdot m_{A,B}(\{p_i\}) \tag{6}$$

$$m_{A,B}^{\omega_I}(\Omega_{pref}) = \omega_I + (1 - \omega_I) \cdot m_{A,B}(\Omega_{pref})$$

After discounting the evidence of each similar item, the final *bba* is aggregated as follows:

$$m^{\omega_I}(\{p_i\}) = \frac{1}{Z}(1 - \prod_{i \in S_K} (1 - \alpha_{p_i})) \cdot \prod_{p_j \neq p_i} \prod_{i \in S_K} (1 - \alpha_{p_j}) \qquad \forall p_i \in \{p_1, \cdots, p_N\} \tag{7}$$

$$m^{\omega_I}(\Omega_{pref}) = \frac{1}{Z} \prod_{i=1}^{N} (1 - \prod_{i \in S_K} (1 - \alpha_{p_i}))$$

where S_K is the set containing the K-nearest neighbors of the target item over the user-item matrix. N is the number of the ratings provided by the similar items, α_{p_i} is the belief committed to the rating p_i, α_{p_j} is the belief committed to the rating $p_j \neq p_i$, Z is a normalized factor defined by [20]:

$$Z = \sum_{i=1}^{N} (1 - \prod_{i \in S_K} (1 - \alpha_{p_i}) \prod_{p_j \neq p_i} \prod_{i \in S_K} (1 - \alpha_{p_j}) + \prod_{i=1}^{N} (\prod_{i \in S_K} (1 - \alpha_{p_j}))) \tag{8}$$

4.2 Evidential Prediction Based on Users' Neighborhood

Let us consider the same frame of discernment Ω_{pref}. In this phase, the K-similar users are considered as the pieces of evidence involved in the prediction process. The distance D(U, V) between two users U and V is defined as follows:

$$D(U,V) = \frac{\sqrt{\sum_{i \in (i_U \cap i_V)} (p_{U,i} - p_{V,i})^2}}{|i_U \cap i_V|} \quad (9)$$

where $p_{U,i}$ and $p_{V,i}$ are respectively the ratings of the user U and V for the item i while i_U and i_V are the items rated by both the user U and V.

Each selected similar user involves a particular hypothesis about the predicted rating. Similarly, we generate *bba*'s over the rating of each piece of evidence as well as the whole frame of discernment as following:

$$m_{U,V}(\{p_i\}) = \alpha_0 \exp^{-(\gamma_{p_i}^2 \times (D(U,V))^2} \quad (10)$$

$$m_{U,V}(\Omega_{pref}) = 1 - \alpha_0 \exp^{-(\gamma_{p_i}^2 \times (D(U,V))^2}$$

Where U denotes the active user and V corresponds to its neighbors, α_0 is fixed to the same value initialized in the previous phase. γ_{p_i} is the inverse of the average distance between each pair of users sharing the same interest towards a given item having a rating p_i.

When it comes to assessing the reliability of each similar user, a discounting factor ω_U is proposed as following: $\omega_U = D(U,V)/max(D)$. Based on the same assumption of the first phase, the discounted *bba*'s are represented as follows:

$$m_{U,V}^{\omega_U}(\{p_i\}) = (1 - \omega_U) \cdot m_{U,V}(\{p_i\}) \quad (11)$$

$$m_{U,V}^{\omega_U}(\Omega_{pref}) = \omega_U + (1 - \omega_U) \cdot m_{U,V}(\Omega_{pref})$$

Once the *bba*'s of each similar user are discounted, we aggregate these *bba*'s as follows:

$$m(\{p_i\})^{\omega_U} = \frac{1}{Y}(1 - \prod_{u \in S_K} (1 - \alpha_{p_i})) \cdot \prod_{p_j \neq p_i} \prod_{u \in S_K} (1 - \alpha_{p_j}) \qquad \forall p_i \in \{p_1, \cdots, p_N\} \quad (12)$$

$$m(\Omega_{pref}) = \frac{1}{Y} \prod_{u=1}^{N} (1 - \prod_{u \in S_K} (1 - \alpha_{p_i}))$$

$$Where\, Y = \sum_{u=1}^{N} (1 - \prod_{u \in S_K} (1 - \alpha_{p_i}) \prod_{p_j \neq p_i} \prod_{u \in S_K} (1 - \alpha_{p_j}) + \prod_{u=1}^{N} (\prod_{u \in S_K} (1 - \alpha_{p_j}))) \quad (13)$$

The set S_K represents the K-Nearest Neighbors of the active user. N is the number of the ratings given by the similar users, α_{p_i} is the belief committed to the rating p_i and α_{p_j} is the belief committed to the rating $p_j \neq p_i$.

4.3 Combining Neighbors' Predictions

The final step in the E-HNBCF approach is to combine the predictions issued from the evidential user-based and item-based process. In fact, the *bba*'s generated during the two previous steps encode, on the one hand, the belief of the K-similar users regarding the final prediction, and on the other hand, the belief of the K-similar items. In this phase, the *bba*'s of these two neighboring sources are fused using Dempster's rule of combination such that:

$$m_{Neighors} = (m^{\omega_I} \oplus m^{\omega_U}) \tag{14}$$

We remind that m^{ω_I} corresponds to the *bba* obtained from the evidential item-based CF while m^{ω_U} is the prediction performed using the evidential user-based approach. Note that in some cases where the evidential predictions are generated only by one side information, for instance, by item-based CF, the vacuous basic belief assignment representing the total ignorance is assigned to the other neighborhood-based approach. The combination then results in the individual predictions which would lead to richer recommendations. The new evidential prediction process is illustrated in Fig. 2.

5 Experimental Study and Analysis

To evaluate our proposal, we opt for the MovieLens[1] data set which contains in total 100.000 ratings collected from 943 users on 1682 movies. In our experiments, we compare our proposed evidential hybrid neighborhood-based CF approach (E-HNBCF), to three other individual neighborhood approaches under the belief function theory namely, the evidential user-based CF (E-UBCF) [19], the evidential item-based CF (E-IBCF) [17] and the discounting-based item-based CF (E-DIBCF) [18]. The protocol in [21] has been adopted for conducting our experiments. The movies in the data set are ranked according to the number of the provided ratings. Then, we extract 10 subsets each of which contains the ratings provided by the users for 20 movies. The selection of the subsets is performed by progressively increasing the number of the missing rates. In this way, each subset will contain a specific number of ratings leading to different degrees of sparsity. For each subset, we randomly extract 10% of the users as a testing data and the remaining users were considered as a training data.

5.1 Evaluation Metrics

Three evaluation metrics are used in our experiments namely the *Mean Absolute Error* (MAE) [22], the *Root Mean Squared Error* (RMSE) [23] and the *Distance criteron* (Dist_crit) [24] defined respectively by:

$$MAE = \frac{\sum_{u,i} |\widehat{p_{u,i}} - p_{u,i}|}{N}, \tag{15}$$

[1] http://movielens.org.

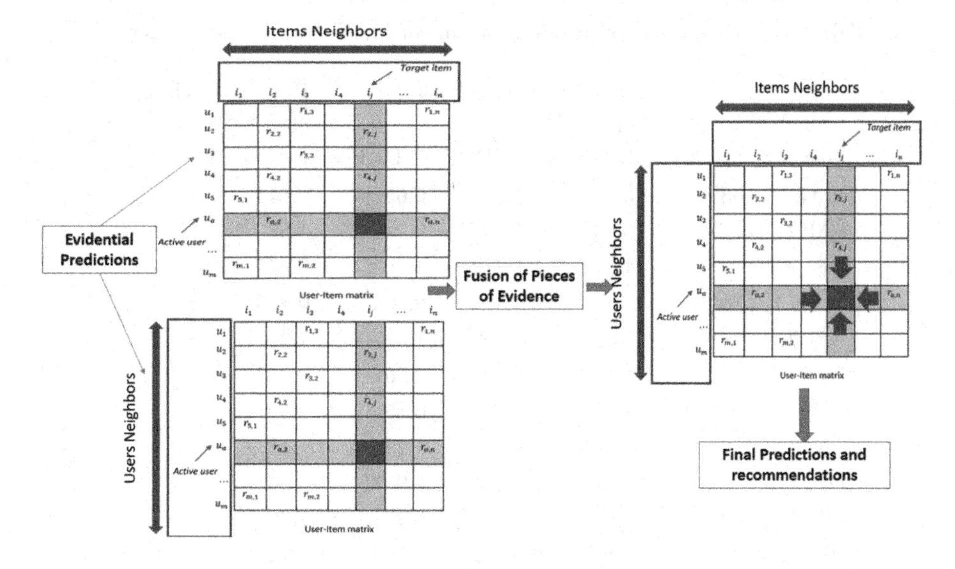

Fig. 2. Evidential prediction process of hybrid neighborhood-based CF

$$RMSE = \sqrt{\frac{\sum_{u,i}(\widehat{p_{u,i}} - p_{u,i})^2}{N}} \tag{16}$$

$$Dist\_crit = \frac{\sum_{u,i}(\sum_{i=1}^{n}(BetP(\{p_{u,i}\}) - \delta_i)^2)}{\text{N}} \tag{17}$$

$p_{u,i}$ is the real rating for the user u on the item i, $\widehat{p_{u,i}}$ is the predicted value of the rating, N is the total number of the predicted ratings over all the users, n is the number of the possible ratings that can be provided in the system and δ_i is equal to 1 if $p_{u,i}$ is equal to $\widehat{p_{u,i}}$ and 0 otherwise.

The smaller values of MAE, RMSE and Dist_crit include a better prediction accuracy and a higher performance.

5.2 Results

We carry on experiments over the different subsets corresponding to different sparsity levels. For each subset, the MAE, the RMSE and the Dist_crit results are the average of 10 repetitions corresponding to 10 different values of neighborhood size. K varies from 1 to 10 for item-based and for user-based, the values of K are: 15, 20, 25, 30, 35, 40, 45, 50, 55 and 60. Note that, in all our experiments, n is equal to 5 while the number of the predicted ratings N has a maximum value of 618 ratings. The experimental results are displayed in Table 1. The results show that the combination of user-based and item-based CF under the belief function theory leads to a better performance than the other individual neighborhood-based approaches. In fact, the hybrid method acquires the lowest average values in terms of MAE, RMSE and Dist_crit. For instance, the

Table 1. The comparison results in terms of MAE, RMSE and Dist_crit

		Individual neighborhood approaches			Proposed hybrid approach
Measure	Sparsity	E-UBCF	E-IBCF	E-DIBCF	E-HNBCF
MAE	53%	0.945	0.663	0.652	0.641
RMSE		0.949	1.09	0.996	1.53
Dist_crit		1.672	1.672	1.054	0.946
MAE	56.83%	0.826	0.875	0.864	0.815
RMSE		1.28	1.199	1.317	1.24
Dist_crit		1.452	1.598	0.962	0.933
MAE	59.8%	1.12	0.89	0.85	0.691
RMSE		1.369	1.279	1.292	1.11
Dist_crit		1.16	1.42	0.953	0.94
MAE	62.7%	0.889	0.741	0.758	797
RMSE		0.886	1.022	1.063	1.12
Dist_crit		1.043	1.327	0.944	0.932
MAE	68.72%	0.798	0.854	0.877	0.847
RMSE		1.147	1.087	1.303	1.13
Dist_crit		1.131	0.758	0.961	0.961
MAE	72.5%	0.86	0.886	0.864	0.822
RMSE		1.154	1.224	1.266	1.147
Dist_crit		2	1.385	0.952	0.936
MAE	75%	1	0.837	0.853	0.881
RMSE		1.402	1.16	1.339	1
Dist_crit		1.77	1.477	0.907	0.872
MAE	80.8%	1	0.932	0.922	0.539
RMSE		1	1.2	1.23	1.02
Dist_crit		2	1.98	1.206	1.08
MAE	87.4%	1	0.62	0.544	1
RMSE		1	0.946	0.916	0.85
Dist_crit		2	1.68	0.818	0.665
MAE	95.9%	1	1	1	1
RMSE		1	3	3	3
Dist_crit		2	2	1.9	1.93
Overall MAE		0.943	0.828	0.818	**0.773**
Overall RMSE		1.318	1.322	1.375	**1.314**
Overall Dist_crit		1.594	1.529	1.065	**1.01**

E-HNBCF approach achieves better results in term of MAE with a value of 0.773 compared to 0.818 for the E-DIBCF, 0.828 for the E-IBCF and 0.943 for the E-UBCF. If we consider the Dist_crit, the average value of the proposed approach (equal to 1.01) widely outperforms E-UBCF (equal to 1.594), E-IBCF (equal to 1.529) and E-DIBCF (equal to 1.065). Similarly, the RMSE values indicate a low value of 1.314 compared to 1.375 (for E-DIBC), 1.322 (for E-IBCF) and 1.318 (for E-UBCF).

6 Conclusion

In this paper, we have proposed a new evidential CF approach combining both user-based and item-based methods under the belief function framework. The basic idea is to transform the preferences extracted from the K-similar users and the K-similar items into basic belief assignments. From this representation, the predictions for both user-based and item-based approaches are performed and then fused into a unique one using Dempster's rule of combination. The fusion framework is effective in improving the prediction accuracy of single approaches (user-based and item-based) under certain and uncertain frameworks. As a future work, we intend to use other combination rules in the fusion process and to perform more complete comparisons with previous methods.

References

1. Ricci, F., Rokach, L., Shapira, B.: Recommender systems: introduction and challenges. In: Ricci, F., Rokach, L., Shapira, B. (eds.) Recommender Systems Handbook, pp. 1–34. Springer, Boston (2015). https://doi.org/10.1007/978-1-4899-7637-6_1
2. Su, X., Khoshgoftaar, T. M.: A survey of collaborative filtering techniques. Adv. Artif. Intell. 1–19 (2009)
3. Aggarwal, C.C.: Neighborhood-based collaborative filtering. Recommender Systems, pp. 29–70. Springer, Cham (2016). https://doi.org/10.1007/978-3-319-29659-3_2
4. Herlocker, J.L., Konstan, J.A., Borchers, A., Riedl, J.: An algorithmic framework for performing collaborative filtering. In: International ACM SIGIR Conference on Research and Development in Information Retrieval, pp. 230–237. ACM (1999)
5. Sarwar, B., Karypis, G., Konstan, J., Riedl, J.: Item-based collaborative filtering recommendation algorithms. In: International Conference on World Wide Web, pp. 285–295. ACM (2001)
6. Miyahara, K., Pazzani, M. J.: Improvement of collaborative filtering with the simple Bayesian classifier. Inf. Process. Soc. Jpn. **43** (2002)
7. Hu, R., Lu, Y.: A hybrid user and item-based collaborative filtering with smoothing on sparse data. In: International Conference on Artificial Reality and Telexistence, pp. 184–189. IEEE (2006)
8. Wang, J., De Vries, A. P., Reinders, M. J.: Unifying user-based and item-based collaborative filtering approaches by similarity fusion. In: International ACM SIGIR Conference on Research and Development in Information Retrieval, pp. 501–508. ACM (2006)

9. Yamashita, A., Kawamura, H., Suzuki, K.: Adaptive fusion method for user-based and item-based collaborative filtering. Adv. Complex Syst. **14**, 133–149 (2011)
10. Tso-Sutter, K.H., Marinho, L.B., Schmidt-Thieme, L.: Tag-aware recommender systems by fusion of collaborative filtering algorithms. In: ACM Symposium on Applied Computing, pp. 1995–1999. ACM (2008)
11. Verstrepen, K., Goethals, B.: Unifying nearest neighbors collaborative filtering. In: ACM Conference on Recommender systems, pp. 177–184. ACM (2014)
12. Pan, R., Zhou, Y., Cao, B., Liu, N.N., Lukose, R., Scholz, M., Yang, Q.: One-class collaborative filtering. In: IEEE International Conference on Data Mining, pp. 502–511. IEEE (2008)
13. Smets, P.: The transferable belief model for quantified belief representation. In: Smets, P. (ed.) Quantified Representation of Uncertainty and Imprecision. Handbook of Defeasible Reasoning and Uncertainty Management Systems, vol. 1, pp. 267–301. Springer, Dordrecht (1998). https://doi.org/10.1007/978-94-017-1735-9_9
14. Dempster, A.P.: A generalization of Bayesian inference. J. R. Stat. Soc. Ser. B (Methodol.) **30**, 205–247 (1968)
15. Shafer, G.: A Mathematical Theory of Evidence, vol. 1. Princeton University Press, Princeton (1976)
16. Nguyen, V.-D., Huynh, V.-N.: A reliably weighted collaborative filtering system. In: Destercke, S., Denoeux, T. (eds.) ECSQARU 2015. LNCS (LNAI), vol. 9161, pp. 429–439. Springer, Cham (2015). https://doi.org/10.1007/978-3-319-20807-7_39
17. Abdelkhalek, R., Boukhris, I., Elouedi, Z.: Evidential item-based collaborative filtering. In: Lehner, F., Fteimi, N. (eds.) KSEM 2016. LNCS (LNAI), vol. 9983, pp. 628–639. Springer, Cham (2016). https://doi.org/10.1007/978-3-319-47650-6_49
18. Abdelkhalek, R., Boukhris, I., Elouedi, Z.: Assessing items reliability for collaborative filtering within the belief function framework. In: Jallouli, R., Zaïane, O.R., Bach Tobji, M.A., Srarfi Tabbane, R., Nijholt, A. (eds.) ICDEc 2017. LNBIP, vol. 290, pp. 208–217. Springer, Cham (2017). https://doi.org/10.1007/978-3-319-62737-3_18
19. Abdelkhalek, R., Boukhris, I., Elouedi, Z.: A new user-based collaborative filtering under the belief function theory. In: Benferhat, S., Tabia, K., Ali, M. (eds.) IEA/AIE 2017. LNCS (LNAI), vol. 10350, pp. 315–324. Springer, Cham (2017). https://doi.org/10.1007/978-3-319-60042-0_37
20. Denoeux, T.: A k-nearest neighbor classification rule based on Dempster-Shafer theory. IEEE Trans. Syst. Man Cybern. **25**, 804–813 (1995)
21. Su, X., Khoshgoftaar, T.M.: Collaborative filtering for multi-class data using Bayesian networks. Int. J. Artif. Intell. Tools **17**, 71–85 (2008)
22. Pennock, D.M., Horvitz, E., Lawrence, S., Giles, C.L.: Collaborative filtering by personality diagnosis: a hybrid memory-and model-based approach. In: International Conference on Uncertainty in Artificial Intelligence, pp. 473–480. Morgan Kaufmann Publishers Inc. (2000)
23. Bennett, J., Lanning, S.: The netflix prize. In: KDD Cup and Workshop, p. 35 (2007)
24. Elouedi, Z., Mellouli, K., Smets, P.: Assessing sensor reliability for multisensor data fusion within the transferable belief model. IEEE Trans. Syst. Man Cybern. Part B Cybern. **34**, 782–787 (2004)

Evidential Top-k Queries Evaluation: Algorithms and Experiments

Fatma Ezzahra Bousnina[1,3](✉), Mouna Chebbah[2], Mohamed Anis Bach Tobji[2], Allel Hadjali[3], and Boutheina Ben Yaghlane[4]

[1] LARODEC, Institut Supérieur de Gestion, Université de Tunis, Tunis, Tunisie
fatmaezzahra.bousnina@gmail.com
[2] ESEN, Univ. Manouba, Manouba, Tunisie
mouna.chebbah@esen.tn, anis.bach@isg.rnu.tn
[3] LIAS, ENSMA - Université de Poitiers, Poitiers, France
allel.hadjali@ensma.fr
[4] Institut des Hautes Etudes Commerciales, Université de Carthage, Tunis, Tunisie
boutheina.yaghlane@ihec.rnu.tn

Abstract. Top-k queries represent a vigorous tool to rank-order answers and return only the most interesting ones. ETop-k queries were introduced to discriminate answers in the context of evidential databases. Due to their interval degrees, such answers seem to be difficult to rank-order and to interpret. Two methods of ranking intervals were proposed in the evidential context. This paper presents an efficient implementation of these methods and discusses the experimental results obtained.

Keywords: Evidence theory · Evidential databases
Evidential Top-k queries

1 Introduction

Querying imperfect databases received a lot of importance recently with the emergence of domains like sensor networks, data cleaning, recommendation and recommender systems, etc. Indeed, information generated from this type of applications is obviously pervaded with imperfection (uncertainty, imprecision, ignorance...). That is why, database models that handle imperfect data were introduced (Probabilistic, possibilistic and evidential databases [1,3,5,6,16]). The latter, models several types of imperfect data but also perfect information using theory of belief functions. In database management, querying is a fundamental step. As consequence, multiple types of 'imperfect' queries were introduced. We name the evidential skyline [9], the extended relational queries [1] and the evidential Top-k queries [4].

In general, Top-k queries are needed in real world applications. For an example, movies, music and books are ordered by the preferred ones, researchers by their H-index, etc. Imperfect top-k queries can be very challenging when it comes to their semantics but also when it comes to their practical implementation.

J. Medina et al. (Eds.): IPMU 2018, CCIS 853, pp. 407–417, 2018.
https://doi.org/10.1007/978-3-319-91473-2_35

In this paper, we present two algorithms of evidential Top-k queries: The first named *NaiETop-k* is based on the computation of the preference degrees (called evidential scores) as introduced in [18] and adapted in [4]. The second named *OptETop-k* is based on an optimized version of the preference degree calculation justified by the complementarity property as detailed in [4]. The proposed implementation allows the ranking of all evidential scores and finally, it provides the k most interesting results among all rank-ordered answers.

Table 1 is an example of an evidential table that stores some users' preferences about books: b_1, b_2, b_3, b_4. This relation includes three attributes: The ID which is a unique reader identifier. The $BookRate$ that includes the reader's appreciations about one book and/or several books modeled through the belief functions theory (in this context only few researches addressed the issue of preference elicitation using this theory [2,11]). The CL which is a specific attribute to evidential databases that stores intervals of confidence about user's responses.

Table 1. Books appreciations' table: EDB

ID	BookRate	CL
1	b_1 0.3	[0.5; 1]
	$\{b_2, b_3\}$ 0.7	
2	b_2 0.5	[0.3; 0.8]
	b_4 0.5	
3	$\{b_1, b_2, b_3\}$ 1	[1; 1]
4	b_3 1	[0.5; 0.9]

This paper is organized as follows: we recall, in Sect. 2 some basic concepts about the belief functions theory and evidential databases. In Sect. 3, we remind needs and challenges of evidential Top-k queries and we present the mathematical materials to compute and compare *Evidential Scores* and *Preference Degrees.* Section 4 is dedicated to the presentation of proposed algorithms. Experiments and results are shown in Sect. 5. Section 6 is devoted to the conclusion and the future works.

2 Evidence Theory and Evidential Databases

Evidence theory named *the belief functions theory or the Dempster-Shafer theory* [7,8,17], is a powerful tool to model ignorance and to represent uncertain, imprecise and inconsistent information.

In the theory of belief functions, a set $\Theta = \{\theta_1, \theta_2, \ldots, \theta_n\}$ is a finite, non empty and exhaustive set of n elementary and mutually exclusive hypotheses related to a given problem. The set Θ is called the *frame of discernment* or *universe of discourse.*

The *power set* $2^{\theta} = \{\varnothing, \theta_1, \theta_2, \ldots, \theta_n, \{\theta_1, \theta_2\}, \ldots, \{\theta_1, \theta_2, \ldots, \theta_n\}\}$ is the set of all subsets of Θ.

A *mass function*, noted m, is a mapping from 2^{Θ} to the interval $[0, 1]$. The *basic belief mass* of an hypothesis x is noted $m(x)$, it represents the belief on the truth of that hypothesis x. A mass function is also called *basic belief assignment* (*bba*). It is formalized such that:

$$\sum_{x \subseteq \Theta} m^{\Theta}(x) = 1 \tag{1}$$

If $m^{\Theta}(x) > 0$, x is called *focal element.* The set of all focal elements is denoted F and the couple $\{F, m\}$ is called *body of evidence.*

The belief function, denoted *bel*, is the minimal degree of support committed exactly to x such that:

$$bel(x) = \sum_{y \subseteq x; y \neq \varnothing} m^{\Theta}(y) \tag{2}$$

The plausibility function, denoted *pl*, is the maximal degree of support committed exactly to x such that:

$$pl(x) = \sum_{y \subseteq \Theta; x \cap y \neq \varnothing} m^{\Theta}(y) \tag{3}$$

An evidential database, denoted EDB, stores different types of data using the belief functions theory as shown in Table 2.

Table 2. The different types of information modeled in the evidential database

Information	Properties	Example
Certain	When the focal element is a singleton with a mass equal to 1 bba is Certain	b_3 1
Probabilistic	When focal elements are singletons bba is Bayesian	b_2 0.5 b_4 0.5
Possibilistic	When focal elements are nested bba is Consonant	b_1 0.2 $\{b_1, b_2\}$ 0.8
Evidential	When none of previous types is present bba is Evidential	$\{b_1, b_2, b_3\}$ 1

Definition 1. *[Compact Evidential Database]*
An EDB *has N objects and A attributes. An* evidential value, *noted V_{la}, is the value of an attribute a $(1 \leq a \leq A)$ for an object l $(1 \leq l \leq N)$ that represents a basic belief assignment.*

$$V_{la} : 2^{\Theta_a} \rightarrow [0, 1] \quad (4)$$

$$with\ m_{la}^{\Theta_a}(\varnothing) = 0 \quad and \quad \sum_{x \subseteq \Theta_a} m_{la}^{\Theta_a}(x) = 1 \quad (5)$$

The set of focal elements relative to the bba V_{la} is noted F_{la} such that:

$$F_{la} = \{x \subseteq \Theta_a / m_{la}(x) > 0\} \quad (6)$$

A confidence level, *CL, is a specific attribute that includes intervals. Each one represents the confidence about its object l in the evidential database. The confidence level is a pair of belief and plausibility [bel; pl] reflecting the pessimistic and the optimistic degrees of support about each object' existence in the database [1, 14, 15].*

Multiple types of queries can be applied over an EDB like the extended relational operators (select, project, join...) [1,14,15], skyline queries [9,10] and ranking queries [4].

3 Evidential Top-k Querying

Top-k queries represent a mighty tool to order queries' results and give only the most interesting answers. Top-k queries were firstly introduced in the multimedia systems [12,13]. They use a score function to rank answers where only results with the highest scores are returned.

Evidential Top-k queries, denoted ETop-k, rank answers using an evidential score function and return the most interesting ones (with the highest scores). Contrary to usual top-k queries that give a ranking based on a score function with precise values, the ETop-k queries give answers based on a score function with intervals. The latter reflect the minimal and the maximal amounts of confidence about each answer.

Definition 2. *[Evidential Score]*
Let R_i be a response generated from processing a query Q over an evidential database EDB of a size N and let $S(R_i)$ be the score function of that answer R_i and $bel(R_i)$ and $pl(R_i)$ are respectively its belief and plausibility in the table, such that:

$$S(R_i) = [bel(R_i); pl(R_i)] \quad (7)$$

$$where \quad bel(R_i) = \frac{\sum_{l=1}^{N} bel_l(R_i) * bel_l}{N}$$

$$pl(R_i) = \frac{\sum_{l=1}^{N} pl_l(R_i) * pl_l}{N}$$

The belief of an answer, $bel(R_i)$, is a disjunction of the response's beliefs in each object of the database. The belief of a response in one object l, denoted bel_l, *is the product of its belief in the attribute and the belief of that object. Same for the plausibility of an answer, $pl(R_i)$. It is the disjunction of the response's plausibilities in each object of the database where the plausibility of a response in one object l, denoted pl_l is the product of its plausibility in the attribute and the plausibility of that object [1,14].*

Example 1. *The Top-k query processed over the evidential database of Table 1 is the following [4]:*

Q: ***SELECT*** *BookRate* ***FROM*** *EDB* ***ORDER BY*** *S(BookRate)* ***LIMIT*** *k;*

Four possible responses are computed using the evidential score as detailed in Definition 2:

- $S(b_1) = [bel(b_1); pl(b_1)] = [0{,}0375; 0.325]$
- $S(b_2) = [bel(b_2); pl(b_2)] = [0{,}0375; 0.525]$
- $S(b_3) = [bel(b_3); pl(b_3)] = [0{,}125; 0.65]$
- $S(b_4) = [bel(b_4); pl(b_4)] = [0{,}0375; 0.1]$

Often a top-k query processed over an evidential database gives a large number of results. These latter need to be ranked in order to respond to the objective of the given query. In the evidential case, the result is a set of intervals that must be compared. Two methods were introduced to compare interval results in EDBs' context [4,18].

(i) The first method was introduced in [18] and adapted in [4]. It is about computing degrees of preference of two intervals and then compare their results to deduce the rank based on three cases:

Definition 3. *[Preference Degree]*
Let $S(R_i) = [bel_i; pl_i]$ and $S(R_j) = [bel_j; pl_j]$ be two evidential scores. Each one is an interval composed of a belief degree and a plausibility degree. The degree of one interval to be greater than the other one is called a degree of preference *and denoted P.*

The degree of preference that $S(R_i) > S(R_j)$ is defined such that:

$$P(S(R_i) > S(R_j)) = \frac{max(0, pl_i - bel_j) - max(0, bel_i - pl_j)}{(pl_i - bel_i) + (pl_j - bel_j)} \tag{8}$$

The degree of preference that $S(R_i) < S(R_j)$ is defined such that:

$$P(S(R_i) < S(R_j)) = \frac{max(0, pl_j - bel_i) - max(0, bel_j - pl_i)}{(pl_i - bel_i) + (pl_j - bel_j)} \tag{9}$$

The different cases of comparing intervals $S(R_i)$ and $S(R_j)$ are as follows:

- *If $P(S(R_i) > S(R_j)) > P(S(R_j) > S(R_i))$, then $S(R_i)$ is said to be superior to $S(R_j)$, denoted by $S(R_i) \succ S(R_j)$.*

- *If* $P(S(R_i) > S(R_j)) = P(S(R_j) > S(R_i)) = 0.5$, *then* $S(R_i)$ *is said to be indifferent to* $S(R_j)$, *denoted by* $S(R_i) \sim S(R_j)$.
- *If* $P(S(R_j) > S(R_i)) > P(S(R_i) > S(R_j))$, *then* $S(R_i)$ *is said to be inferior to* $S(R_j)$, *denoted by* $S(R_i) \prec S(R_j)$.

(ii) The second method optimizes the first one using the *complementarity proof**, results are compared in order to deduce their rank [4]:

Definition 4. *[Optimized Preference Degree]*
Let $S(R_i) = [bel_i; pl_i]$ *and* $S(R_j) = [bel_j; pl_j]$ *be two evidential scores. Every interval is composed of degrees of belief (bel) and plausibility (pl) and P is the calculated preference degree.*

$$P(S(R_i) > S(R_j)) = \frac{max(0, pl_i - bel_j) - max(0, bel_i - pl_j)}{(pl_i - bel_i) + (pl_j - bel_j)} = \lambda \tag{10}$$

The different cases of comparing intervals $S(R_i)$ *and* $S(R_j)$ *are as follows:*

- *If* $\lambda > 0.5$ *then* $S(R_i) \succ S(R_j)$.
- *If* $\lambda = 0.5$, *then* $S(R_i) \sim S(R_j)$.
- *If* $\lambda < 0.5$ *then* $S(R_i) \prec S(R_j)$.

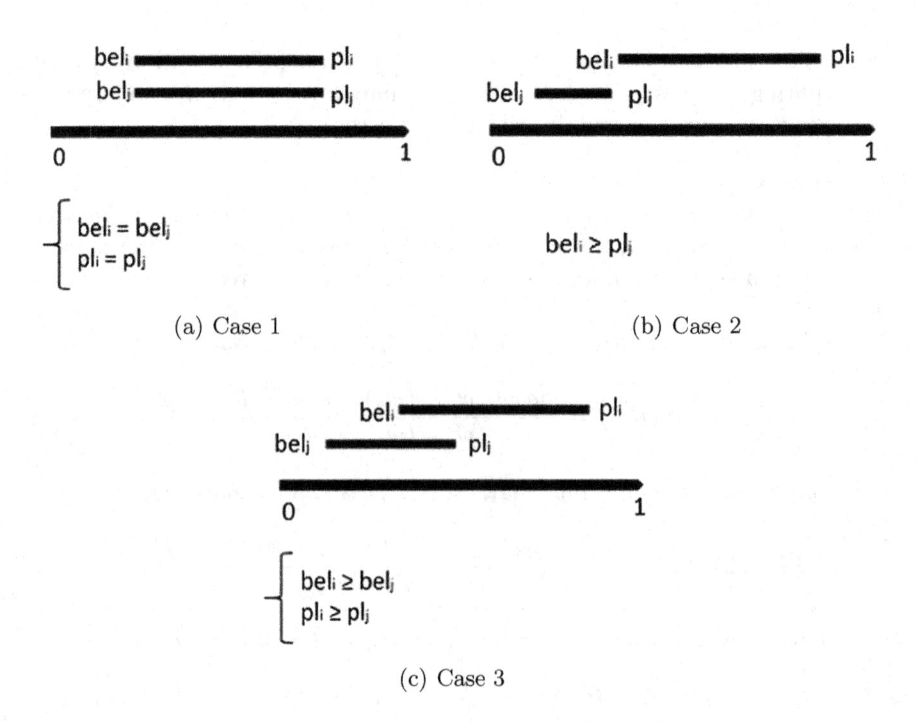

(a) Case 1 (b) Case 2

(c) Case 3

Fig. 1. Specific cases to deduce evidential scores [4]

Cases that permit to minimize computations before using Definitions 3 or 4 are illustrated in Fig. 1.

Note here the importance of *transitivity*** property detailed in [18] to give the final ranking.

4 Implementation of Evidential Top-*k* Query

As the best of our knowledge, there is no implementation of evidential top-k queries. Indeed, we present in this paper an object-oriented implementation of two methods to rank evidential scores (the evidential intervals). The first method is naive, it consists on computing the preference degree through three steps each time: (a) it computes the preference degree that the first interval is superior to the second one and then (b) it computes the preference degree that the first interval is inferior to the second one. Finally, (c) it compares results and give the partial rank. This algorithm is presented in Table 3.

The second method is an optimization of the first one. Indeed, it consists on computing only in one step the preference degree and then deduce the partial order between two intervals. This algorithm is detailed in Table 4.

Finally, the last order is treated using a sorting algorithm, that ranks all evidential intervals and provide the k most interesting ones. The presented implementations offer two methods of evidential intervals' ranking. Both algorithms use the object-oriented paradigm for its programming benefits.

Table 3. ETop-k naive algorithm

Naive Method	**Naive Evidential Top-*k* Algorithm**
Initialization	***Initialization***
Tuple a, b ;	Integer m;
begin	ArrayList Table;
if (a.Bel=b.Bel ***and*** a.Pl=b.Pl)	***begin***
return 0;	***for*** (int i ⟵ 0; i<Table.size()-1; i++)
if (a.Pl<b.Bel)	{ m⟵i;
return -1;	***for*** (int j⟵ i+1; j<Table.size(); j++)
if (b.Pl<a.Bel)	{
return 1;	***if*** (*NaiveMethod*(Table.get(j), Table.get(m))=1)
if (a.Bel>b.Bel ***and*** a.Pl>b.Pl)	{ m⟵j; }
return 1;	}
if (b.Bel>a.Bel ***and*** b.Pl>a.Pl)	***if*** (*NaiveMethod*(Table.get(m) ,Table.get(i))=1)
return -1;	{ Tuple c ⟵ Table.get(i);
if (score(a,b)>score(b,a))	Table.set(i,Table.get(m));
return 1;	Table.set(m,c); }
else return -1;	}
end	***end***

Table 4. ETop-k optimized algorithm

ETop-k Method

```
Initialization
Tuple a, b ;
begin
if (a.Bel=b.Bel and a.Pl=b.Pl)
return 0;
if (a.Pl<b.Bel)
return -1;
if (b.Pl<a.Bel)
return 1;
if (a.Bel>b.Bel and a.Pl>b.Pl)
return 1;
if (b.Bel>a.Bel and b.Pl>a.Pl)
return -1;
if (score(a,b)>0.5)
return 1;
else return -1;
end
```

Optimized ETop-k Algorithm

```
Initialization
Integer m;
ArrayList Table;
begin
for(int i←—0; i<Table.size()-1; i++)
{
for (int j←—i+1; j<Table.size(); j++)
{
if(EtopKMethod(Table.get(j), Table.get(m))=1)
{m←—j;}
}
if(EtopKMethod(Table.get(m) ,Table.get(i))=1)
{ Tuple c ←— Table.get(i);
Table.set(i,Table.get(m));
Table.set(m,c); }
}
end
```

5 Experimental Study

In this section, we evaluate both algorithms from a performance point of view. We used a windows 10 operating system with 2.10 GHz CPU and 4 GB RAM. We also used Java programming language and NetBeans platform.

5.1 Data Sets

We used synthetic data sets with the following parameters (a) N the size of the database, (b) S the evidential score which is an interval of belief and plausibility $[Bel; PL]$ with BEl, PL $\in$ [0;1] and BEL $\leq$ PL[1].

To generate a synthetic evidential database, the used algorithm uses a procedure that generates a synthetic S. Indeed, the procedure computes randomly a fixed number of evidential scores in the interval $[0, 1]$. Then one of the algorithms (naive or optimized) are processed in order to compare intervals. Finally, a sorting function is used to provide the final complete ranking of all intervals. Note that each interval is associated to a specific and unique item in the evidential database. In our example, the item is a specific book.

Experiments showed interesting results from a performance point of view. In fact, we varied the database size parameter (N) from 10 to 3000. The execution time did not exceed 4 min and 50 s for both algorithms. Results are presented in Table 5. Both algorithms showed interesting results. Moreover, $OptETopK$ gave better ones as shown in Fig. 2. For example, $OptETopK$ ranked 1500 tuples in

[1] Bel and Pl are two functions defined in the object-relational implementation of evidential databases in [5].

Table 5. Impact of the database size for methods: NaiTopK and OptTopK

Tuples number (N)	Execution time (s)	
	NaiETopK method	OptETopK method
10	1	0
50	2	0
100	2	0
200	3	0
300	4	1
500	8	5
800	19	13
1000	33	28
1500	69	60
2000	125	121
3000	279	277

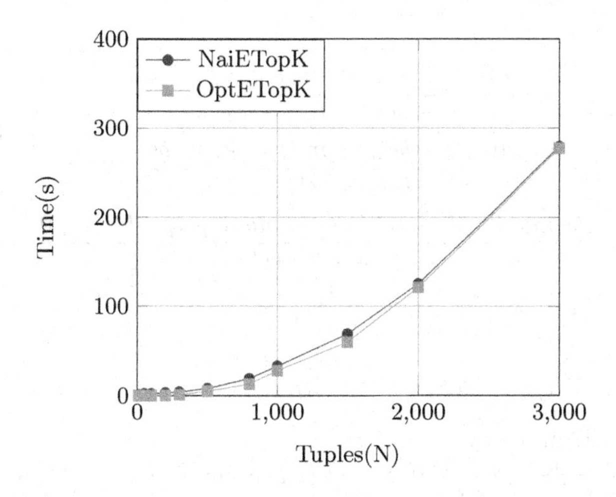

Fig. 2. Comparison of performance of NaiETopK and OptETopK

69 s against 60 s for $NaiETopKNote$. Note that complexity depends also on the intervals' nature generated randomly as detailed theoretically in Sect. 3.

6 Conclusion

Throughout this paper, we presented an implementation of the Evidential Top-k query, *ETop-k*. In fact, we proposed two algorithms *NaiETopK* and *OptETopK*. Both methods showed interesting results when we varied the database size but *OptETopK* showed best performance in practice as shown theoretically in [4].

The proposed implementation is an important achievement of the evidential Top-k querying fitting the semantics of returning the k most credible answers.

Other types of queries, in the evidential context, like aggregation, range, threshold remain as a promising future works.

A Appendix

Proof. *Complementarity:

$$P(S(R_i) < S(R_j)) = \frac{max(0, pl_j - bel_i) - max(0, bel_j - pl_i)}{(pl_i - bel_i) + (pl_j - bel_j)}$$

$$P(S(R_j) < S(R_i)) = \frac{max(0, pl_i - bel_j) - max(0, bel_i - pl_j)}{(pl_i - bel_i) + (pl_j - bel_j)}$$

$$P(S(R_i) < S(R_j)) + P(S(R_j) < S(R_i))$$

$$= \frac{max(0, pl_j - bel_i) - max(0, bel_j - pl_i)}{(pl_i - bel_i) + (pl_j - bel_j)} + \frac{max(0, pl_i - bel_j) - max(0, bel_i - pl_j)}{(pl_i - bel_i) + (pl_j - bel_j)}$$

$$= \frac{max(0, pl_j - bel_i) - 0 + max(0, pl_i - bel_j) - 0}{(pl_i - bel_i) + (pl_j - bel_j)}$$

$$= \frac{pl_j - bel_i + pl_i - bel_j}{pl_i - bel_i + pl_j - bel_j} = 1$$

$$P(S(R_i) < S(R_j)) + P(S(R_j) < S(R_i)) = 1$$

Property 1. **Transitivity

Let $S(R_i) = [bel_i; pl_i]$, $S(R_j) = [bel_j; pl_j]$ and $S(R_k) = [bel_k; pl_k]$ be three intervals. If $S(R_i) \succ S(R_j)$ and $S(R_j) \succ S(R_k)$ then $S(R_i) \succ S(R_k)$.

References

1. Bell, D.A., Guan, J.W., Lee, S.K.: Generalized union and project operations for pooling uncertain and imprecise information. Data Knowl. Eng. (DKE) **18**, 89–117 (1996)
2. Yaghlane, A.B., Denœux, T., Mellouli, K.: Elicitation of expert opinions for constructing belief functions. In: Uncertainty and Intelligent, Information Systems, pp. 75–88 (2008)
3. Bousnina, F.E., Bach Tobji, M.A., Chebbah, M., Liétard, L., Ben Yaghlane, B.: A new formalism for evidential databases. In: Esposito, F., Pivert, O., Hacid, M.-S., Raś, Z.W., Ferilli, S. (eds.) ISMIS 2015. LNCS (LNAI), vol. 9384, pp. 31–40. Springer, Cham (2015). https://doi.org/10.1007/978-3-319-25252-0_4

4. Bousnina, F.E., Chebbah, M., Bach Tobji, M.A., Hadjali, A., Ben Yaghlane, B.: On top-k queries over evidential data. In: 19th International Conference on Enterprise Information Systems (ICEIS), Porto, Portugal, vol. 1, pp. 106–113 (2017)
5. Bousnina F., Chebbah M., Bach Tobji M., Hadjali A. and Ben Yaghlane B.: Object-relational implementation of evidential databases. In: 1st International Conference on Digital Economy (ICDEc), La Marsa, Tunisia, pp. 80–87 (2016)
6. Cavallo, R., Pittarelli, M.: The theory of probabilistic databases. In: Proceedings of the 13th VLDB Conference, Brighton, UK, pp. 71–81 (1987)
7. Dempster, A.P.: Upper and lower probabilities induced by a multiple valued mapping. Ann. Math. Stat. **38**(2), 325–339 (1967)
8. Dempster, A.P.: A generalization of Bayesian inference. J. R. Stat. Soc. Ser. B **30**, 205–247 (1968)
9. Elmi, S., Benouaret, K., Hadjali, A., Bach Tobji, M.A., Ben Yaghlane, B.: Computing skyline from evidential data. In: Straccia, U., Calì, A. (eds.) SUM 2014. LNCS (LNAI), vol. 8720, pp. 148–161. Springer, Cham (2014). https://doi.org/10.1007/978-3-319-11508-5_13
10. Elmi, S., Benouaret, K., HadjAli, A., Bach Tobji, M.A., Ben Yaghlane, B.: Requêtes skyline en présence des données évidentielles. In: Extraction et Gestion des Connaissances (EGC), pp. 215–220 (2015)
11. Ennaceur, A., Elouedi, Z., Lefevre, E.: Multi-criteria decision making method with belief preference relations. Int. J. Uncertain. Fuzziness Knowl.-Based Syst. **22**(04), 573–590 (2014)
12. Fagin, R.: Combining fuzzy information from multiple systems. In: 15th ACM SIGACT-SIGMOD-SIGART Symposium on Principles of Database Systems, Montreal, Canada, pp. 216–226. ACM (1996)
13. Fagin, R.: Fuzzy queries in multimedia database systems. In: 17th ACM SIGACT-SIGMOD-SIGART Symposium on Principles of Database Systems, Seattle, WA, USA, pp. 1–10. ACM (1998)
14. Lee, S.K.: An extended relational database model for uncertain and imprecise information. In: 18th Conference on Very Large Data Bases (VLDB), Canada, pp. 211–220 (1992)
15. Lee, S.K.: Imprecise and uncertain information in databases: an evidential approach. In: 8th International Conference on Data Engineering (ICDE), Arizona, USA, pp. 614–621 (1992)
16. Prade, H., Testemale, C.: Generalizing database relational algebra for the treatment of incomplete or uncertain information and vague queries. Inf. Sci. **34**(2), 115–143 (1984)
17. Shafer, G.: A Mathematical Theory of Evidence. Princeton University Press, Princeton (1976)
18. Wang, Y.-M., Yang, J.-B., Dong-Ling, X.: A preference aggregation method through the estimation of utility intervals. Comput. Oper. Res. **32**(8), 2027–2049 (2005)

Independence of Sources in Social Networks

Manel Chehibi[1], Mouna Chebbah[2(✉)], and Arnaud Martin[3]

[1] ESEN, Univ. Manouba, La Mannouba, Tunisie
chehibimanel@gmail.com
[2] LARODEC, ESEN, Univ. Manouba, La Mannouba, Tunisie
mouna.chebbah@esen.tn
[3] IRISA, Université de Rennes1, Lannion, France
Arnaud.Martin@univ-rennes1.fr

Abstract. Online social networks are more and more studied. The links between users of a social network are important and have to be well qualified in order to detect communities and find influencers for example. In this paper, we present an approach based on the theory of belief functions to estimate the degrees of cognitive independence between users in a social network. We experiment the proposed method on a large amount of data gathered from the Twitter social network.

Keywords: Cognitive dependence · Theory of belief functions
Twitter social network · Independence measure

1 Introduction

Online social networks are online platforms that connect users. They have gained a lot of interest and popularity over the last decade. Many people rely on social networks particularly on information, news and opinions shared by users on diverse subjects.

An online social network, such as Twitter, helps users to share subjective information reflecting their personal opinions. In fact, in a social network, users become sources of information who produce different kinds of information (opinions, facts, news, rumors, etc.). However, some users are cognitively dependent on others. In addition, an online social network enable its users to interact with each other by several activities such as sharing, quoting, or commenting other users' posts. These users' interactions provide insights for the cognitive dependence/independence relationships among users in a social network. A user is supposed to be cognitively dependent on another user if he relies on and adopts information that he provides.

The aim of this paper is to study dependencies of sources in social networks. Information about sources' dependencies in a social network can be used to detect related groups, communities [1], ...

J. Medina et al. (Eds.): IPMU 2018, CCIS 853, pp. 418–428, 2018.
https://doi.org/10.1007/978-3-319-91473-2_36

The identification of communities can help for targeted marketing. It can also be used for influence propagation [2] to promote new products and define new marketing strategies. Indeed, a company wishing to launch a marketing campaign or a new product can use relations of dependencies to speed up the propagation.

In this paper, we propose an approach to estimate the degrees of independence/dependence between users of a social network. Twitter is chosen as an example of a directed social network; thus, we detail the proposed measure using Twitter vocabulary. The dependence relationship between users is an oriented relation; therefore, Twitter is very appropriate to illustrate our approach.

The proposed approach is based on the theory of belief functions to estimate uncertain degrees of independence between users. The theory of belief functions is used to asses uncertain degrees of belief on the independence of users. This theory is also chosen thanks to the great number of combination rules that merge subjective information.

The remainder of this paper is organized as follows: Sect. 2 recalls some basic concepts of the theory of belief functions; Sect. 3 details the proposed approach to estimate degrees of independence/dependence. Finally, Sect. 4 presents an experimental study of our approach before concluding in Sect. 5.

2 Theory of Belief Functions

The theory of belief functions, also called Dempster-Shafer theory, was first introduced by Dempster [3] and mathematically formalized by Shafer [4]. This theory models imprecise, uncertain and missing data.

In the theory of belief functions, a *frame of discernment*, noted $\Theta = \{H_1, \ldots, H_N\}$, is a set of N exhaustive and mutually exclusive hypotheses $H_i, 1 \leq i \leq N$. only one of them is likely to be true.

The *power set*, $2^\Theta = \{A/A \subseteq \Theta\} = \{\emptyset, H_1, \ldots, H_N, H_1 \cup H_2, \ldots, \Theta\}$, enumerates 2^N sub-assemblies of Θ. It includes not only hypotheses of Θ, but also, disjunctions of these hypotheses.

The true hypothesis in Θ is unknown; thus, a degree of belief is assessed to subsets of 2^Θ reflecting our degree of faith on the truth of each subset of 2^Θ.

A *basic belief assignment (bba)*, also called *mass function*, is noted m^Θ and defined such that:

$$\begin{aligned} &m^\Theta : 2^\Theta \rightarrow [0,1] \\ &m^\Theta(\emptyset) = 0 \\ &\sum_{A \subseteq \Theta} m(A) = 1 \end{aligned} \tag{1}$$

The mass $m^\Theta(A)$ represents the degree of belief on the truth of $A \in 2^\Theta$. When $m^\Theta(A) > 0$, A is called *focal element.*

In the theory of belief functions, decision is generally made using *pignistic probabilities* [5]. The pignistic probability, noted $BetP^\Theta$, is deduced from m^Θ as follows:

$$BetP(H_i) = \sum_{\substack{A \in 2^\Theta \\ H_i \subset A}} \frac{1}{|A|} m^\Theta(A) \qquad \forall H_i \in \Theta \tag{2}$$

where $|A|$ is the number of hypotheses which train it.

In the theory of belief functions, combination rules are proposed to merge distinct mass functions in order to produce a more reliable information. It consists on building an unique mass function by combining several elementary mass functions arising from multiple distinct sources of information.

Dempster's rule of combination [3] is the first rule that merges several mass functions provided by distinct and independent sources. The combination of two mass functions $m^\Theta_{S_1}$ and $m^\Theta_{S_2}$ provided by S_1 and S_2 is given as follows:

$$m^\Theta_{1\oplus 2}(A) = (m^\Theta_1 \oplus m^\Theta_2)(A) = \begin{cases} \dfrac{\sum_{B\cap C=A} m^\Theta_1(B) \times m^\Theta_2(C)}{1-\sum_{B\cap C=\emptyset} m^\Theta_1(B) \times m^\Theta_2(C)} & \forall A \subseteq \Theta,\ A \neq \emptyset \\ 0 & \text{if } A = \emptyset \end{cases} \tag{3}$$

The reliability of an evidential information is not always insured. In fact, an evidential data can be supplied by a partially reliable or an unreliable source. In order to take the source's reliability into account, its beliefs are discounted proportionally to its reliability. Let $\alpha \in [0, 1]$ be the reliability of a source S_1 and m^Θ a mass function provided by S_1. The *discounting* of m^Θ produces ${}^\alpha m^\Theta$ defined by:

$$\begin{cases} {}^\alpha m^\Theta(A) = \alpha \times m^\Theta(A) & \text{if, } \forall A \subset \Theta \\ {}^\alpha m^\Theta(\Theta) = 1 - \alpha \times (1 - m^\Theta(\Theta)) \end{cases} \tag{4}$$

3 Uncertain Measure of Independence in Twitter

Many researches are focused on measuring the independence in several social networks. Leenders [6] proposed and approach focused on the opinions and attitudes of users in a social system. These opinions and attitudes are shaped by social influence. The proposed approach depend partially on individual characteristics.

Kudělka et al. [1] makes use of the measurement of dependence between the network vertices for the detection of communities in social networks.

To predict a user actions (behaviors) in a social network, Tan et al. [7] consider diverse factors: the influence from his friends, the *correlation* between users' actions and his historic behaviors. They conducted an experiment on Twitter and they found that more friends perform the action, a user also tends to perform the action and the likelihood that two friends perform an action at the same time is always larger than the likelihood that randomly two users perform the same action at the same time.

Jendoubi et al. [2] propose to detect influencer in Twitter using the theory of belief functions. They consider three Twitter metrics to quantify the influence between users: followers, mention, retweet.

Twitter is a social network that enables its users to establish many types of relation between them. A relation between users of Twitter may be a *follow*, a *retweet*, a *mention* or a *citation*.

These ties are considered as dependence indexes for the several reasons: First, the retweet actions represent the amount of information tweeted by a user from the tweets of another user. This amount reflects the degree of adoption of the opinions of other users. Then, the mention represents the quantity of messages directly sent to other specific users in order to establish direct communications with them. These actions reflect the importance of a part of the Twitter users and their ideas for other users in the network. Finally, the citation represents the degree of reliance of some users on other users by citing them in their tweets.

Therefore, we consider that degrees of dependence between users of Twitter can be deduced from numbers of follows, retweets, mentions and citations. In this paper, we propose to estimate degrees of cognitive dependence between users of Twitter. Two users are cognitively dependent when information provided by a user are affected by the information produced by the other one. We note that the cognitive independence is matter of researches in the theory of belief functions [8]. Two variables [4] are assumed to be cognitively independent with respect to a belief function if any new evidence that appears on only one of them does not change the evidence of the other variable. In addition, two sources [8] are cognitively independent if they do not communicate and if their evidential corpora are different. Two sources are either positively are negatively dependent; in the case of negative dependence, sources are dependent but their ideas are different. Otherwise, influencers [2] are sources that have a maximum of impact in the ideas of others. Dependence and influence measures are different but quite similar. Thus, the dependence measure may be used for influence maximization.

A user u in Twitter is cognitively dependent on another user v if u is following v and u frequently retweets tweets of v or/and, u frequently mentions v in his tweets.

Figure 1 shows the proposed approach to estimate independence of users in Twitter. The proposed approach is in 2 steps:

1. In the first step, weights are estimated. Thus, we define a weight for each aspect of dependence: retweet, mention and citation.
2. In the second step, the independence estimation. In this step, we use the theory of belief functions to (i) model each independence aspect, (ii) to combine them and (iii) to make a decision regarding the independence of users.

3.1 Step 1: Estimation of Weights

In Twitter, a user u following a user v can retweet, mention or/and cite v. Each information about the retweet, mention or/and citation may reflect the

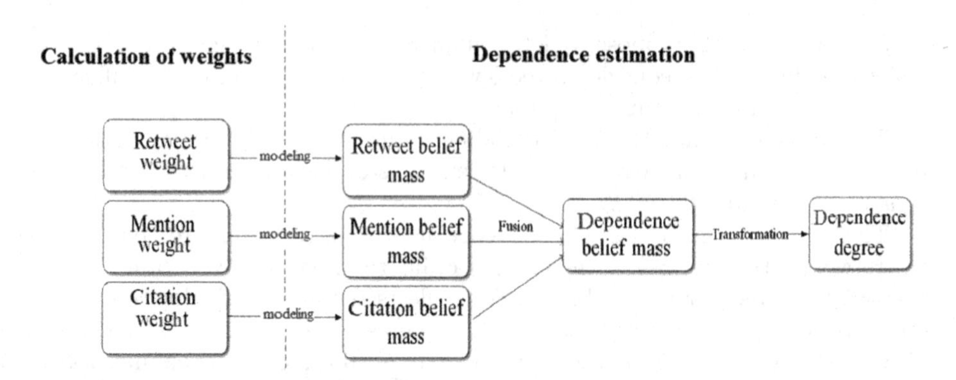

Fig. 1. The general framework of the proposed approach

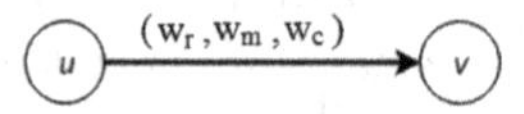

Fig. 2. Weight vector between u and v

dependence or the independence of u on v. Thus, a vector of weights (w_r, w_m, w_c) is assigned to each link (u, v) as shown in Fig. 2. Note that u is following v and the vector of weights will be used to learn the independence/dependence of u on v. Let $G = (V, E)$ be the social network where V is the set of nodes, E is the set of links, $u \in V$ is a follower of $v \in V$ in Twitter. The weights w_r, w_m and w_c of the link $(u, v) \in E$ are estimated using the following measures:

1. The retweet weight, $w_r(u, v) = \dfrac{Rt_u(v)}{Rt_u}$, is the weight defining the number of times that u has retweeted the tweets of v; $Rt_u(v)$ is the number of tweets of v that were retweeted by u and Rt_u is the total number of retweets of u.
2. The mention weight, $w_m(u, v) = \dfrac{Mt_u(v)}{Mt_u}$, is the weight defining the number of times that u mentioned v in his tweets; $Mt_u(v)$ is the number of tweets of u in which v was mentioned and Mt_u is the total number of mentions of u.
3. The citation weight, $w_c(u, v) = \dfrac{Ct_u(v)}{Ct_u}$, is the weight defining the number of times that u quoted the tweets of v; $Ct_u(v)$ is the number of tweets of v who have been quoted by u and Ct_u is the total number of citations of u.

3.2 Step 2: Independence Estimation

The dependence estimation is based on the defined weights. Let $G = (V, E, W)$ be a directed graph where W is the set of weights' vectors, such that $(w_r(u, v), w_m(u, v), w_c(u, v)) \in W$ is the weight vector associated to the link (u, v). The independence estimation process is in three basic steps:

1. In the first step, a mass function is built from each weight on the link. Let $\mathcal{I} = \{D, I\}$ be the frame of discernment of the independence where D is the hypothesis that users are dependent and I is the hypothesis that users are independent. Mass functions are estimated as follows:
   (a) First, the retweet weight justifies our belief on the independence of users. Therefore, $m^{\mathcal{I}}_{r_{(u,v)}}$ is defined as follows:

$$\begin{cases} m^{\mathcal{I}}_{r_{(u,v)}}(D) = \alpha_{r_u} \times w_r(u,v) \\ m^{\mathcal{I}}_{r_{(u,v)}}(I) = \alpha_{r_u} \times (1 - w_r(u,v)) \\ m^{\mathcal{I}}_{r_{(u,v)}}(I,D) = 1 - \alpha_{r_u} \end{cases} \tag{5}$$

   Note that $\alpha_{r_u} = \dfrac{Rt_u}{T_u}$ is a discounting coefficient that takes into account the total number of tweets T_u. The mass function $m^{\mathcal{I}}_{r_{(u,v)}}$ is more reliable when the number of retweets is enough big in comparison with the total number of tweets. For example, assume that a user u has posted twenty eight tweets in two weeks and that among these tweets there are ten retweets, seven of them are from v. Without discounting using α_{r_u}, the value of $m^{\mathcal{I}}_{r_{(u,v)}}(D)$ will be equal to 0.7 which does not reflect the reality. In fact, the number of tweets that u has retweeted v represents only the quarter of the total number of tweets of u.
   (b) Then, a mass function $m^{\mathcal{I}}_{m_{(u,v)}}$ is deduced from the mention weight as follows:

$$\begin{cases} m^{\mathcal{I}}_{m_{(u,v)}}(D) = \alpha_{m_u} \times w_m(u,v) \\ m^{\mathcal{I}}_{m_{(u,v)}}(I) = \alpha_{m_u} \times (1 - w_m(u,v)) \\ m^{\mathcal{I}}_{m_{(u,v)}}(I,D) = 1 - \alpha_{m_u} \end{cases} \tag{6}$$

   where $\alpha_{m_u} = \dfrac{Mt_u}{T_u}$ is a discounting coefficient. The discounting coefficient α_{m_u} is used to take into account the total number of tweets quoted by u with respect to the total number of tweets of u.
   (c) Finally, the mass function $m^{\mathcal{I}}_{c_{(u,v)}}$ is deduced from the citation weight as follows:

$$\begin{cases} m^{\mathcal{I}}_{c_{(u,v)}}(D) = \alpha_{x_u} \times w_c(u,v) \\ m^{\mathcal{I}}_{c_{(u,v)}}(I) = \alpha_{c_u} \times (1 - w_c(u,v)) \\ m^{\mathcal{I}}_{c_{(u,v)}}(I,D) = 1 - \alpha_{c_u} \end{cases} \tag{7}$$

   where $\alpha_{c_u} = \dfrac{Ct_u}{T_u}$ is a discounting coefficient that takes into account the total number of tweets of u mentioning v with respect to the total number of tweets of u.
2. Then, mass functions $m^{\mathcal{I}}_{r_{(u,v)}}, m^{\mathcal{I}}_{m_{(u,v)}}(D)$ and $m^{\mathcal{I}}_{c_{(u,v)}}$ are combined with Dempster's rule of combination as follows:

$$m^{\mathcal{I}}_{(u,v)} = m^{\mathcal{I}}_{r_{(u,v)}} \oplus m^{\mathcal{I}}_{m_{(u,v)}} \oplus m^{\mathcal{I}}_{c_{(u,v)}} \tag{8}$$

3. Finally, degrees of independence $Ind(u, v)$ and dependence $Dep(u, v)$ corresponds to pignistic probabilities computed from the combined mass function $m^{\mathcal{I}}_{(u,v)}$ as follows:

$$\begin{cases} Dep(u,v) = BetP(D) \\ Ind(u,v) = BetP(I) \end{cases} \tag{9}$$

We have:

$$Dep(u,v) + Ind(u,v) = 1 \tag{10}$$

The dependence degree $Dep(u, v)$ is non-negative, it is either positive or null. It is also normalized. In fact, the degree of dependence $Dep(u, v)$ is a degree that lies in the interval $[0, 1]$. When $Dep(u, v) = 1, u$ is totally dependent on v; $Dep(u, v) = 0$ implies that u is totally independent of v. Decision is made according to the maximum of pignistic probabilities. If $Dep(u, v) \geq Ind(u, v)$ then u is dependent on v, in the opposite case, if $Ind(u, v) > Dep(u, v), u$ is independent from v.

4 Experiments

The proposed approach is tested on data collected from Twitter; because it is a directed social network that provides a large number of messages published per day. Unlike other social media platforms like Facebook, the content of Twitter is public and accessible via programming interfaces. In our experimental study, we used the Twitter streaming API through a Python library called Tweepy. This library provides access to Twitter data *via* its programming interface, Twitter API. The Twitter Streaming API allows retrieving data in real-time. It allows also filtering tweets by several keywords or according to their geographical position. In our case, we are interested in collecting tweets written by specific users. For this purpose, we filtered tweets by a list of users IDs. We crawled Twitter data for the period between 05/06/2017 and 13/8/2017. We get an important number of tweets (205271 tweets) corresponding to 10350 users on this period. Experiments of the proposed approach detailed in this section are made on a large number on users, tweets, retweets, mention and citation as detailed in Table 1. Note that retweets, mentions and citations are considered as tweets.

Table 1. Data collected from 05/06/2017 to 13/8/2017

Users	Tweets	Retweets	Mentions	Citations
10350	205271	32842	71901	14613

- Table 2, shows that there are independent relationship between a part of users despite there are a follow relationship between them. For example, the user S_1 is independent from the user S_2 and the same for the user S_3 with S_{29} with a lower degree of dependence. All experiments are made on real data described on Table 1 which are collected from Tweeter. Users are numbered to respect the anonymity and privacy. Therefore, the follow relationship in Twitter does not necessarily imply the cognitive dependence between users. In an explicit way, a user u who follows another user v in Twitter can be either cognitively independent or dependent on v.

Table 2. Examples of independence relationship

Link	The degree of dependence
(S_1, S_2)	0.1
(S_3, S_{29})	0.3
(S_4, S_{37})	0.2

- Table 3 shows that in the case where a user u is dependent on a user v, v is not necessarily dependent on u. In the case where a user u is independent on a user v, v is not necessarily independent on u.

Table 3. Examples of asymmetrical relationships

Link	The degree of dependence
(S_8, S_{35})	0.6
(S_{35}, S_8)	0.2
(S_{10}, S_{13})	0.7
(S_{13}, S_{10})	0.3

- Table 4 shows that if users u and v are mutually independent or dependent, degrees of independence or dependence are not necessarily equal.

Table 4. Examples of mutual independence/dependence with different degrees of independence/dependence

Link	The degree of dependence
(S_{11}, S_5)	0.7
(S_5, S_{11})	0.6
(S_{12}, S_{23})	0.3
(S_{23}, S_{12})	0.1

Tests are made on data collected from 05/06/2017 to 13/08/2018 as detailed in Table 1. Degrees of independence and dependence are computed of each pair of users from the 10350. Thus, degrees of independence and dependence are computed for each couple of users (u, v) for all the 10350 users. Note that for each couple of users we compute $Ind(u, v)$ and $Ind(v, u)$. Therefor $10350! * 2$ values of independence are computed. In the complete graph there are 10350 nodes, each node represents a user and 2 values of independence for each couple of users. For tests, we have also estimated the degree of independence/dependence for users without any relationship of follow.

The dependence graph of Fig. 3 is a part of the complete graph. In Fig. 3, only 10 users from the 10350 users are represented. These 10 users are randomly chosen for simplicity seek and also to have a readable graph. Black links represent a follow link, the bold part links reflects the direction of follows. In other words, S_1 is following S_2 and S_4; S_2 is following S_9; S_3 is following S_8; S_4 is following S_3, S_{10} and S_9; S_5 is following S_6, S_{10} and S_1; S_6 is following S_1, S_2 and S_7; S_7 is following S_{10} and S_6; S_8 is following S_{10}; S_9 is following S_{10} and finally S_{10} is following S_4, S_5, S_7 and S_8. Note that (S_4, S_{10}), (S_5, S_{10}), (S_6, S_7), (S_7, S_{10}), (S_8, S_{10}) are mutually following each other.

Figure 3 shows that some users are cognitively dependent, for example S_1 is dependent on S_4 with a degree 0.64; S_4 is dependent on S_9 and $Dep(S_4, S_9) =$

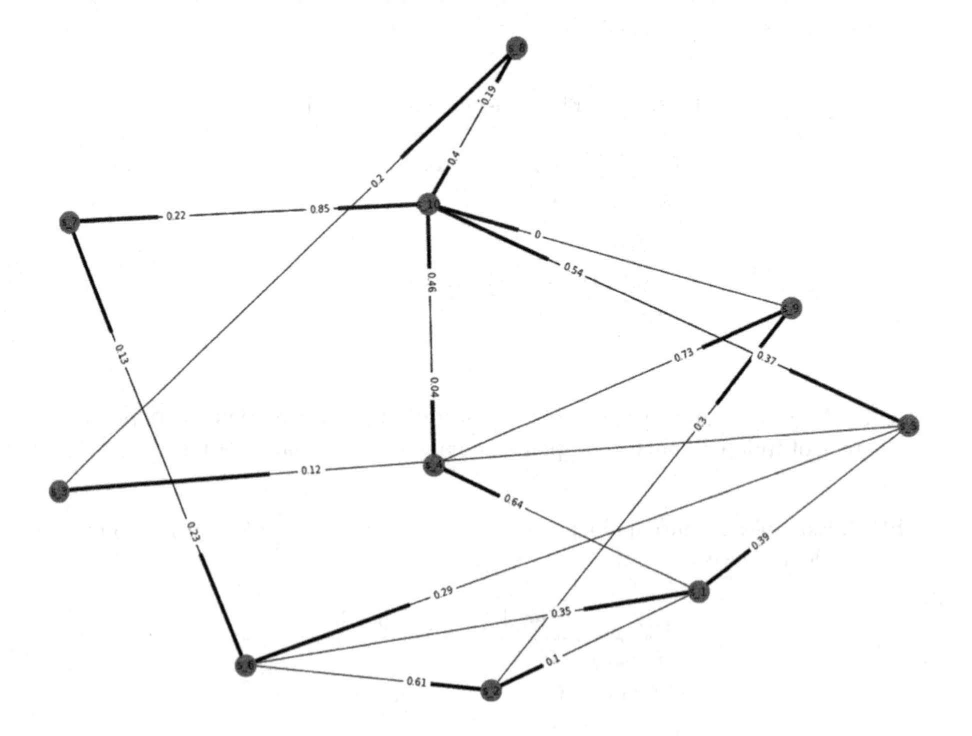

Fig. 3. Example dependence analysis between users.

0.73; S_5 is dependent on S_{10} and $Dep(S_5, S_{10}) = 0.54$; S_6 is dependent on S_2 and $Dep(S_6, S_2) = 0.61$; finally S_7 is dependent on S10 and $Dep(S_7, S_{10}) = 0.85$.

Finally, (S_1, S_2), (S_2, S_9), (S_3, S_8), (S_4, S_3), (S_4, S_{10}), (S_5, S_6), (S_5, S_1), (S_6, S_1), (S_6, S_7), (S_7, S_6), (S_8, S_{10}), (S_9, S_{10}), (S_{10}, S_4), (S_{10}, S_5), (S_{10}, S_7) and (S_{10}, S_8) are independent. Note that S_{10} and S_4, S_{10} and S_8, S_7 and S_6 are mutually independent.

Table 5. Dependence between users without any link of follow

Users	The degree of dependence
S_1, S_4	$Dep(S_4, S_1) = 0.21$
S_1, S_3	$Dep(S_1, S_3) = 0.08$
	$Dep(S_3, S_1) = 0.16$
S_2, S_4	$Dep(S_2, S_4) = 0.19$
	$Dep(S_4, S_2) = 0.25$
S_5, S_6	$Dep(S_6, S_5) = 0.15$
S_3, S_8	$Dep(S_8, S_3) = 0.13$
S_3, S_7	$Dep(S_3, S_7) = 0.1$
	$Dep(S_7, S_3) = 0.21$

Table 5 shows that users without any follow are independent. For example there is no follow between S_1 and S_3 because S_1 is not following S_3 and S_3 is not following S_1. Users S_1 and S_3 are mutually independent. Users that are not following others are independent. When a user u is not following another user v, u is necessarily independent from v.

5 Conclusion

Studying cognitive independence relationship among the Twitter social network users is a very important research topic since this online social network is widely used to post and share information. In fact, quantify the degrees of dependence between users can be very useful to disseminate information to the largest number of users which is a very important thing in many fields such as marketing.

Most of existing works that try to study the dependence between users in a social network, use only the network structure to measure the dependence of a user on another user and ignore many interesting dependence aspects. Nevertheless, the dependence measures that is based only on the network structure is not adequate to quantify the dependence between sources. In fact, in the twitter social network, a user can follow another user in the network without being necessarily cognitively dependent on him.

In this work, we propose an approach based on the theory of belief functions for measuring the dependence degrees between users in Twitter. We consider three dependence aspects witch are the retweets, the mentions and the citations and we use the Dempster-Shafer theory to model each dependence aspect, to combine them with taking into consideration the conflict that can arise between them and to make a decision with regard to the dependence a user on another user in the network.

The results of the experimental study of our proposed approach show that the follow relationship in twitter does not necessarily imply the cognitive dependence between users and that the more the number of retweets, citations or/and mentions increase, the more the degree of dependence of a user on an other user increases and vice versa. It shows also that the dependence relationship between two users is not necessarily mutual and the dependence degrees between them are not necessarily equal.

As a future work, we will use our approach to detect communities in social networks.

References

1. Kudělka, M., Dráždilová, P., Ochodková, E., Slaninová, K., Horák, Z.: Local community detection and visualization: experiment based on student data. In: Kudělka, M., Pokorný, J., Snášel, V., Abraham, A. (eds.) IHCI 2011. Advances in Intelligent Systems and Computing, vol. 179, pp. 291–303. Springer, Heidelberg (2013). https://doi.org/10.1007/978-3-642-31603-6_25
2. Jendoubi, S., Martin, A., Liétard, L., Hadji, H.B., Yaghlane, B.B.: Two evidential data based models for influence maximization in Twitter. Knowl.-Based Syst. **121**, 58–70 (2017)
3. Dempster, A.P.: Upper and lower probabilities induced by a multiple valued mapping. Ann. Math. Stat. **38**, 325–339 (1967)
4. Shafer, G.: A Mathematical Theory of Evidence. Princeton University Press, Princeton (1976)
5. Smets, P.: Decision making in the TBM: the necessity of the pignistic transformation. Int. J. Approx. Reason. **38**, 133–147 (2005)
6. Leenders, R.: Modeling social influence through network autocorrelation: constructing the weight matrix. Soc. Netw. **24**, 21–47 (2002)
7. Tan, C., Tang, J., Sun, J., Lin, Q., Wang, F.: Social action tracking via noise tolerant time-varying factor graphs. In: Proceedings of the 16th ACM SIGKDD International Conference on Knowledge Discovery and Data Mining, KDD 2010, pp. 1049–1058. ACM, New York (2010)
8. Chebbah, M., Martin, A., Yaghlane, B.B.: Combining partially independent belief functions. Decis. Support Syst. **73**, 37–46 (2015)

Current Techniques to Model, Process and Describe Time Series

Forecasting Energy Demand by Clustering Smart Metering Time Series

Christian Bock[1,2](✉)

[1] Institute of Computer Science, Heinrich-Heine-University, 40225 Düsseldorf, Germany
bock@cs.uni-duesseldorf.de
[2] BTU EVU Beratung GmbH, 40545 Düsseldorf, Germany

Abstract. Current demands on the energy market, such as legal policies towards green energy usage and economic pressure due to growing competition, require energy companies to increase their understanding of consumer behavior and streamline business processes. One way to help achieve these goals is by making use of the increasing availability of *smart metering time series*. In this paper we extend an approach based on *fuzzy clustering* using smart meter data to yield *load profiles* which can be used to forecast the energy demand of customers. In addition, our approach is built with existing business processes in mind. This helps not only to accurately satisfy real world requirements, but also to ease adoption by the industry. We also assess the quality of our approach using real world smart metering datasets.

Keywords: Big data · Data mining · Knowledge discovery
Clustering · Time series · Smart metering · Load profiles

1 Introduction

The energy market has been undergoing significant reorganizations for the last decade: legal obligations such as policies towards green energy usage have heavily impacted the way producers plan for and distribute energy, while growing competition due to the liberalization of the energy market and political commitments to expand digitization continue to increase the need of electricity companies to convert the information accessible thanks to new technologies into valuable knowledge [15]. One such technology is the increasing availability of *smart meters*. Pilot projects in the UK, France and the Netherlands, among others, demonstrated the benefits for both the electricity companies as well as the end consumer [11]. These projects also open up possibilities to better understand consumer behavior and streamline business processes.

To guarantee the security of the energy supply, it is important for electricity companies to estimate the energy demand of their customers in advance. This is because energy producers need lead time to prepare their capacities accordingly. This results in the need of electricity companies for capable forecast models.

J. Medina et al. (Eds.): IPMU 2018, CCIS 853, pp. 431–442, 2018.
https://doi.org/10.1007/978-3-319-91473-2_37

However, current models have not been adjusted to the technological advances in recent years. Because of this, they are considered an increasingly bad model to estimate the energy demand of customers [23]. Expected future changes in consumer behavior, e.g. caused by the growing usage of electric vehicles, further require the energy market to adapt. In this work we extend an existing approach [7] which uses *smart metering data* and *fuzzy clustering* to help electricity companies to improve their forecast models. In doing so our approach builds so called *load profiles* which strictly conform to current business processes. We also assess the quality our load profiles the same way electricity companies in Germany would grade them in a production environment.

2 Related Work

Research towards understanding and processing data has gained significant attention in the energy economy [10,19]. Areas of interest include outlier detection [24], marketing and tariff optimizations [21] as well as predicting long-term changes of the aggregated energy load [2]. Publications that make use of clustering often choose *K-Means* to analyze the data [5,16] or approaches based on it [20]. Newer research also increasingly use *fuzzy clustering* [19,22,24] and include newer technologies like *smart metering* [4,12,13].

In this work, we continue our previous work [7] and present an approach based on *fuzzy clustering* [6] for electricity companies to derive load profiles from smart metering time series, which are used to predict the energy demand in advance. To model long- and short-term periodic changes in consumer behavior, we focus not only on identifying *customer groups* to aggregate customers with similar consumption behavior, but also to dynamically find *day types* as a way to model different daily routines. Furthermore, we evaluate our approach using three real-world smart metering datasets and assess the quality of the energy forecast closely related to the requirements in a production environment.

3 Problem Statement

One of the main challenges of electricity companies is to accurately forecast the total energy demand. The importance of this task is due to the fact that energy must already be injected into the electrical grid when it is expected to be needed by the consumers. This means that energy producers need to know the energy demand ahead of time in order to ramp production up and down accordingly. Electricity companies currently achieve this feat by using so called *load profiles* to model the energy demand. In a nutshell, current business processes have evolved around load profiles and work by segmenting all customers into customer groups and all calendar days into so called day types. Day types are used to identify calendar days on which the total energy demand is as similar as possible compared to other calendar days belonging to the same day type. Similarly, customer groups describe customers with genuinely different consumer behavior. This way, day types can be used to model the different daily routines of each customer, as long as the sum of those daily routines over all customers has a sufficiently significant impact on the expected total energy demand. As a result, load profiles can be described as both a segmentation of calendar days into day types and customers into customer groups as well as a time series describing the expected consumer behavior for each combination of said customer groups and day types. By knowing which day type a given calendar day belongs to, electricity companies can concatenate the corresponding consumption patterns of a load profile in order to build a time series which represents the expected average consumption behavior of every customer who have said load profile assigned. What is important to emphasize here is that the time series contained by the load profiles are normalized, meaning they describe, similar to a weight function, how a given amount of energy a given customer is expected to consume over the course of a given time span is distributed over said time span, rather than directly representing the amount of energy consumed by the customer. In order to use a rolled out load profile to forecast the energy demand of a customer, the industry uses a scalar called the *year consumption forecast (YCF)*. The YCF is individually assigned for each customer and describes the amount of energy the corresponding customer is expected to consume over the course of one whole year. Typically, the YCF of a single customer is derived by either using the total amount of energy said customer has consumed in the previous year or by computing an average of the total energy consumed of the customer over several years. This way, by scaling the rolled out load profile associated with a given customer using the corresponding YCF, the consumer behavior can be predicted one year in advance. By aggregating all scaled and rolled out load profiles, electricity companies can successfully forecast the total energy demand of their customers and plan their buy-in of energy accordingly. This way of forecasting the total energy demand is also able to cope with the use case of customers switching their energy provider and therefore causing minor abrupt changes in the total amount of energy an electricity company has to provide, which can not be properly accounted for by forecasting methods which only focus on the total energy demand time series.

In our previous work [7] we presented an approach based on *fuzzy clustering* [6] using smart meter time series to build load profiles which obey the business processes described earlier but aim to optimize forecasting the total energy demand. Using this approach, we achieved a significant improvement in forecasting accuracy compared to the standard load profiles currently in use by the industry. At the same time, the load profiles derived that way share lot of the same weaknesses as the standard load profiles, namely the difference in amplitude of the consumption patterns causing block-like transitions when changing day types as illustrated in Fig. 1. This creates the problem of the energy forecast being bad near day type transitions. While this block-like structure of rolled out load profiles can be mitigated by increasing the number of day types, it does not solve the underlying problem. Optimally, it is desirable to have load profiles compare the energy consumption of customers based on their shape, rather than have them being influenced by energy offsets caused by predictable and periodic seasonal patterns. Some load profiles currently in use by the industry tackle this problem by multiplying a well-known polynomial, which is often referred to as a *dynamization function*, onto the forecast yielded by the load profiles. The goal of this function is to roughly mimic the sine-shaped patterns seen in total energy consumption time series in production environments. Means to derive such a dynamization function include statistical methods such as the method of least squares. As real world datasets, such as the ones described in Sect. 5.1, have different yearly periodic patterns depending on the overall customer base, it is highly unlikely for a single dynamization function to properly represent the seasonal patterns of any given customer behavior. In the next section, we extend our original approach for generating load profiles to account for yearly periodic changes while still maintaining backwards compatibility with existing business processes of electricity companies.

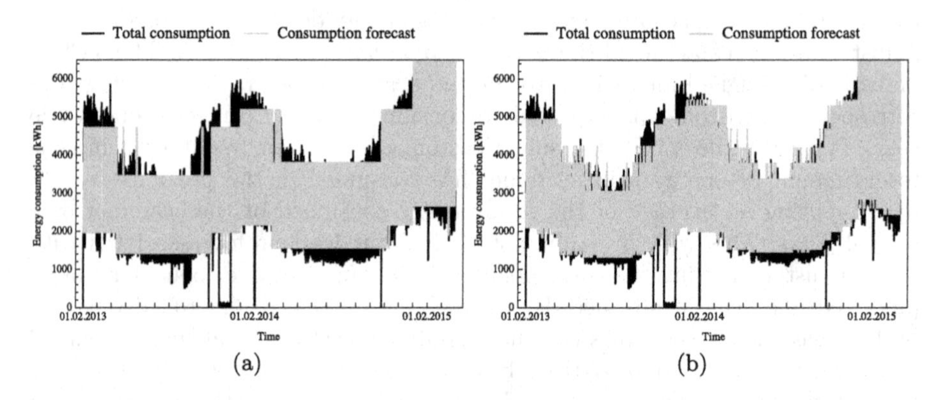

Fig. 1. Comparison of the actual total energy consumption of the *BTU-Dataset* (black graph) and the energy consumption predicted by the load profiles generated using the approach presented in [7] (light gray graph) with (a) 2 and (b) 8 day types.

4 Optimization Approach

One major reason for the block-like structures observed in load profiles such as those illustrated in Fig. 1 is due to the sine-shaped seasonal patterns ingrained in the consumer behavior. Optimally, a load profile would compare and predict energy usages of customers mostly based on their shape since the amount of energy consumed is accounted for by the YCF. While doing so, only the total energy demand time series is interesting to electricity companies and taken into account when assessing the quality of the forecast. Because of this, it is reasonable to assume the exact same energy usage pattern for each customer in a given customer group on multiple calendar days if the total energy demand is sufficiently similar to one another on those calendar days. This holds true even if the resulting forecast is extraordinarily bad on an individual customer basis. Without special consideration of interferences such as seasonal patterns however, this has caused the offsets in energy consumption created by these patterns to play the dominant role in the segmentation of calendar days into day types in our original approach [7] as seen in Fig. 1. This behavior of our approach is not bad fundamentally disadvantageous for building high-quality load profiles. However, we also observed the presence of yearly periodic patterns in other real world time series describing the total energy demand of electricity companies that don't use smart meters. At the same time, we would like to emphasize that their concrete features differs between datasets, including those introduced in Sect. 5.1. Thus, it can be assumed that these patterns are a characteristic of a given customer base, which makes them predictable after they have been identified. To identify these characteristics, we propose to apply a *high-pass filter* on the total consumption time series of a representative year. A candidate to achieve this is the *Fourier transformation*, as it enables the notation of the yearly periodic patterns using the sine and cosine coefficients of only the lowest-frequency terms. Dividing each measurement of the smart meter data using those low-frequency sine and cosine coefficients effectively yields the desired behavior of a high-pass filter.

In summary, our approach to build load profiles using smart metering time series is as follows:

1. Apply the *Fourier transformation* on the total consumption time series to extract the lowest-frequency sine and cosine coefficients. These coefficients form the *dynamization function* used to eliminate yearly periodic patterns.
2. Determine the optimal number of day types and their segmentation onto the individual calendar days. In addition, compose rules to classify future calendar days.
3. For each day type, determine the optimal number of consumption patterns and their characteristics.
4. Compile load profiles by combining former results and assign a profile to each customer.
5. After using a load profile to forecast the energy demand of a customer, reapply the yearly periodic patterns onto the forecast time series using the sine and cosine coefficients of the dynamization function from the first step.

Because using the Fourier transformation as a high-pass filter yields a *dynamization function* that is specific to a given dataset, its seasonal patterns can be more accurately described than by a well-known polynomial that is used for every dataset across the industry. This helps to focus on the shape of the consumption time series when identifying the optimal segmentations for day types and consumption patterns.

4.1 Day Type Segmentation

The main idea behind the concept of day types is to group calendar days on which the total energy demand is expected to be as similar as possible to other calendar days belonging to the same day type. At the same time, the expected energy demand of calendar days belonging to different day types should be genuinely dissimilar. A suitable approach to solve this problem is to use clustering algorithms such as *fuzzy clustering* [6].

To construct the dataset from whose segmentation the day type segmentation can be derived, we define a new time series $X = \{x_1, x_2, ..., x_T\}$ for each point in time $t_j, 1 \leq j \leq T$ using the individual customer's smart metering time series $S_i, 1 \leq i \leq N$ as follows:

$$x_j := \frac{1}{N_j \cdot DF(t_j)} \sum_{i=1}^{N} \frac{s_{i,j}}{YCF_{i,j}} \tag{1}$$

In Eq. 1, $s_{i,j}$ represents the measurement and $YCF_{i,j}$ describes the *year consumption forecast* assigned to S_i for the year that t_j belongs to and $DF(t_j)$ stands for the *dynamization function* evaluated at t_j. Furthermore, N represents the number of distinct consumers while N_j represents the number of distinct consumers measured at t_j. The distinction between N and N_j is necessary to account for customers leaving and joining to and from other energy providers, as well as to factor in *missing values* caused by temporary technical failures, e.g. by any of the customers smart metering devices becoming faulty or by network transmission errors. In short, X can be described as an average time series of all normalized smart metering time series. Afterwards, we use X to construct the dataset D as follows:

$$D := \left\{ d_l := (x_j, ..., x_{j+m}) \,\middle|\, \begin{array}{l} \forall a \text{ with } 1 \leq j \leq a \leq j+m \leq T: \\ t_a \text{ belongs to the } l\text{-th calendar day} \end{array} \right\} \tag{2}$$

Each d_l contains a slice of the time series X as a tuple. Smart metering time series are usually measured in intervals of 15, 30 or 60 min, which corresponds to d_l being a 96-, 48- or 24-tuple, respectively. A good day type segmentation can now be derived by applying clustering on D. In case a partitioning clustering algorithm like *Fuzzy-C-Means* [6] is chosen for this task, a suitable approach is to process the dataset repeatedly using different values for the number of clusters c and to determine the optimal segmentation using a selection of *Cluster Validity Indices* [8]. Determining which calendar days belong to the same day type can then be done by knowing which d_l got assigned to the same cluster and which calendar days they represent. For categorizing future calendar days we rely on the expertise of an analyst to review the day type segmentation and derive rulesets based upon observed regularities.

4.2 Identifying Typical Consumption Patterns

The concept behind determining consumption patterns is to identify representative daily routines of customers. In a production environment, daily routines of customers are only interesting if they remain relatively constant in terms of their impact on the total energy consumption. Because of this, it is advisable to look at the data for each day type separately since the day type segmentation as described in Sect. 4.1 will have already revealed which calendar days have a sufficiently similar total energy demand. With this in mind, let $K_n, 1 \leq n \leq L$ be the sets of day types built in Sect. 4.1 where each K_n contains the corresponding t_j. We then construct the disjoint sets $P_n, 1 \leq n \leq L$ as follows:

$$P_n := \left\{ p_{e,n} := (y_{i,j}, ..., y_{i,j+m}) \middle| \begin{array}{c} \forall a, b \text{ with } 1 \leq j \leq a, b \leq j+m \leq T: \\ t_a, t_b \in K_n \text{ and } y_{i,a}, y_{i,b} \text{ belong to} \\ \text{the same calendar day} \end{array} \right\} \quad (3)$$
$$\text{with } y_{i,j} := \frac{s_{i,j}}{YCF_{i,j} \cdot DF(t_j)}$$

Similar to dataset D in Eq. 2, each $p_{e,n}$ contains a slice of the normalized smart metering time series representing one calendar day for a given customer. The difference is that each P_n contains individual normalized measurements $y_{i,j}$ instead of the average time series X and that P_n only contains data for the day type K_n. We then proceed to apply clustering on each P_n separately using a centroid-based clustering algorithm. This is because the optimal cluster prototypes $C_{q,n}, 1 \leq q \leq c_{n,optimal}$ directly correspond to the desired typical consumption patterns for the day type K_n.

4.3 Compiling Load Profiles

To generate an energy forecast for a given customer, electricity companies need a single consumption pattern which represents the customer as accurately as possible on all calendar days of a given day type. Thus, a load profile can be described as an array of L-tuples where the n-th entry corresponds to the consumption pattern for calendar days assigned to K_n. While the *year consumption forecast (YCF)* and *dynamization function* introduced in Sect. 3 also play a vital role in computing the energy forecast, they are maintained separately from the load profiles. To build and assign a load profile to a given customer, represented by the corresponding smart metering time series S_i, we propose to iterate over all P_n and use a majority vote to choose the $C_{q,n}$ as the fitting consumption pattern for K_n which has the most $p_{e,n}$-tuples based on S_i with their highest membership degrees pointing to $C_{q,n}$.

Algorithm 1 outlines this approach in pseudocode. The $u_{q,e,n} \in U_n$ describe the fuzzy membership of $p_{e,n}$ to $C_{q,n}$ after the clustering process has terminated. In addition, H is an L-dimensional array which stores the load profile for the currently examined customer. By using this algorithm, it is possible to derive both the set of load profiles, indicated by G, as well as the set of profile assignments, notated by Z.

In combination with the customer-specific YCF and the dataset-specific dynamization function, this gives electricity companies all the tools they need to efficiently forecast the energy demand. In the next section, we evaluate this approach using real world data the same way they would be assessed in a production environment.

Algorithm 1. Compiling load profiles

Input: $S_i, P_n, K_n, C_{q,n}, U_n$
Output: set of all load profiles G, set of profile assignments Z
1: $G \leftarrow \emptyset$
2: $Z \leftarrow \emptyset$
3: **for** $i = 1$ **to** N **do**
4: **for** $n = 1$ **to** L **do**
5: $H[n] \leftarrow C_{q,n}$ with $q := \arg\max_{q'} \left| \left\{ p_{e,n} \middle| \begin{array}{l} \exists j : (y_{i,j}, ..., y_{i,j+m}) = p_{e,n} \\ \wedge \nexists q'' : u_{q'',e,n} > u_{q',e,n} \end{array} \right\} \right|$
6: **end for**
7: $G \leftarrow G \cup H$
8: $Z \leftarrow Z \cup (S_i, H)$
9: **end for**
10: **return** G, Z

5 Evaluation

5.1 Description of Datasets

To assess the accuracy of the forecast yielded by our approach we have used three real world smart metering datasets which are visualized in Fig. 2. Since all datasets consist of real world data they also contain *missing values* caused by temporary technical failures, e.g. by any of the customers smart metering devices becoming faulty or by network transmission errors. These *missing values* are illustrated by a non-constant value for the number of measurements.

The first dataset contains a total of 7668 distinct customers with a resolution of 1 measurement per hour over the course of 26 months. In this work, we refer to it as the *BTU-Dataset*. Because this dataset is provided in cooperation with a German electricity company who already had a complete rollout of smart meters, we are able to test our approach under realistic conditions. The dataset is maintained and made available by the *BTU EVU Beratung GmbH* [1].

The second dataset, which we will call the *CER-Dataset*, contains a total of 6445 distinct Irish customers with 1 measurement every 30 min over the course of 16 months. It is provided by the *Irish CER (Commission for Energy Regulation)* and accessed via the *Irish Social Science Data Archive (ISSDA)* [9]. Though the dataset is also accompanied by surveys containing additional information about the customers like salary and civil status, this extra data was not taken account for as in most real world scenarios electricity companies are unlikely to have accurate survey data available.

The third dataset, named the *IZES-Dataset*, covers the energy demand of 416 distinct consumers with 1 measurement every 15 min over the course of 18 months. The data was gathered as part of the field test *"Moderne Energiesparsysteme im Haushalt" (modern energy saving systems in household environments)* and is made available by the *IZES institute* [17].

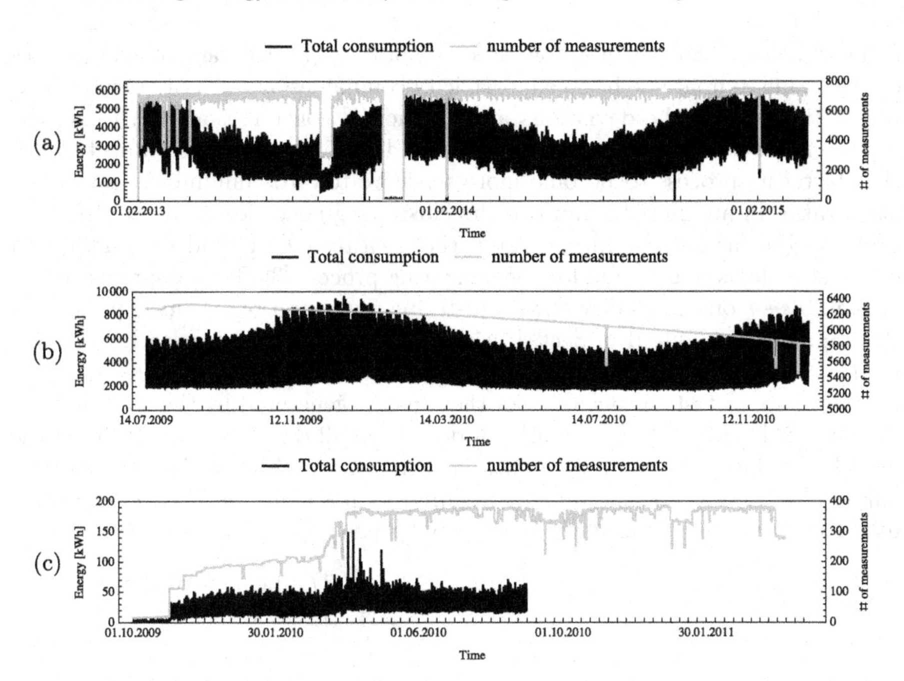

Fig. 2. Overview of the (a) *BTU-Dataset*, (b) *CER-Dataset* and (c) *IZES-Dataset*. The black colored graphs show the total energy consumption (primary axis); the grey colored graphs show the number of non-missing values per time slot (secondary axis).

5.2 Experimental Setup

For the experimental evaluation, we have opted to use *Fuzzy-C-Means* as the clustering algorithm for our approach. The reason for this is its tendency to build spherical clusters [8]; as customers belonging to the same customer group are modeled by only one consumption pattern per day type, using spherical clusters helps to better conform to the way load profiles are expected to even out deviations between the individual consumption behaviors when aggregating them to the total energy demand time series. To measure the dissimilarity of the data, we modified the *Partial Distance Strategy* [14] often used for datasets containing *missing values* to get what we call the *Partial Manhattan Distance*:

$$dist_{PMD}(a,b) = \frac{m+1}{I} \cdot \sum_{n=1}^{m+1} |a_n - b_n| \cdot I_n$$
$$\text{with } I_n = \begin{cases} 1 & \text{if } a_n \text{ is not a missing value} \\ 0 & \text{else} \end{cases} \quad \text{and } I = \sum_{n=1}^{m+1} I_n \tag{4}$$

In essence, where *Partial Distance Strategy* corresponds to an adaptation of the euclidean distance that is usually used for Fuzzy-C-Means, the *Partial Manhattan Distance* applies the same concept to the manhattan distance. In a production environment, load profiles aim to match the total energy demand as closely as possible, with deviations by both over- and underestimating the actual energy demand being equally undesirable. Because of this, the accuracy of load profiles can be made comparable between electricity companies of different sizes by looking at the ratio of the deviations and the actual consumption [18]. Using the *Partial Manhattan Distance* to measure the dissimilarity between objects helps the clustering process to become more sensitive towards minimizing said deviation ratio. In an effort to improve the cluster segmentation to derive the load profiles from further, we incorporated the *k-means++* method to generate an initial starting configuration for the clustering process [3]. To measure the effect of *k-means++* on the quality of the clustering segmentation, we have performed the approach introduced in Sect. 4 using both the standard approach of Fuzzy-C-Means of generating random coordinates as well as the *k-means++* method. Furthermore, to test the accuracy of the forecast generated by the load profiles, we have excluded the last 3 months of data from all datasets while training the model. The formerly excluded month is then used as the test data to compare our forecast to. As for the $YCF_{i,j}$ required to normalize each time series, we used the total energy consumed per customer per year:

$$YCF_{i,j} := \sum_{j' \in Z_j} \frac{s_{i,j'}}{DF\left(t'_j\right)} \quad \text{with} \quad Z_j := \left\{ j' \,\middle|\, \begin{array}{c} t_{j'} \text{ belongs to the} \\ \text{same year as } t_j \end{array} \right\} \tag{5}$$

5.3 Results

The results of our experimental evaluation are shown in Fig. 3. If used in a production environment, electricity companies would plan their buy-in of energy

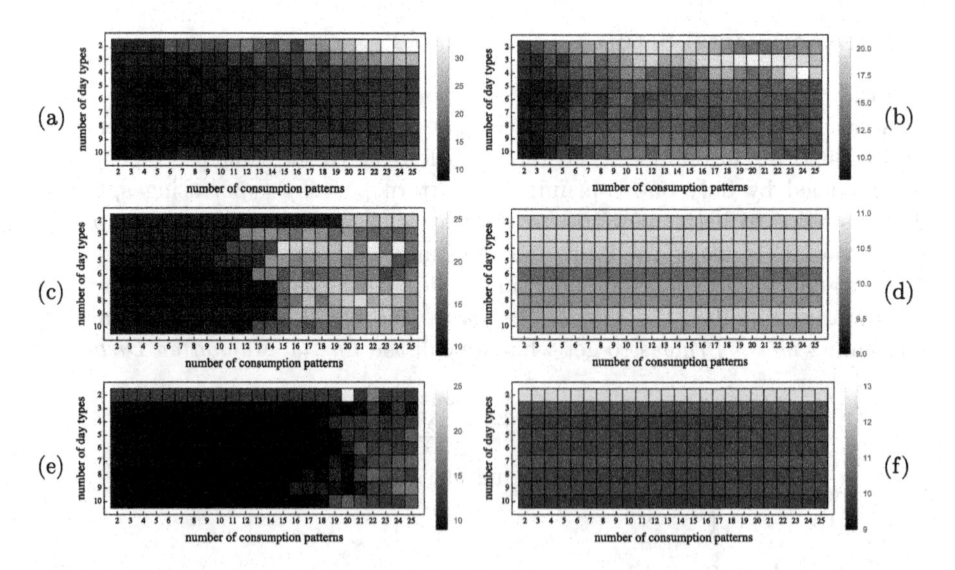

Fig. 3. Ratio of the deviations and the actual consumption in percent yielded by the load profiles generated using different values for the number of day types and consumption patterns. The graphs visualize the results for (a, b) the *BTU-Dataset*, (c, d) the *CER-Dataset* and (e, f) the *IZES-Dataset* using (a, c, e) *k-means++* and (b, d, f) *random coordinates* to generate the starting configuration of the clustering process.

according to the forecast yielded by their load profiles. Since the forecast is known ahead of time, the corresponding capacities from energy producers are cheap from a business standpoint. Deviations from the actual total energy consumption however require adjustments in real time. Because of their limited availability, these short-term adjustments are usually much more cost-intensive. This is why electricity companies aim to keep these deviations to a minimum by improving the long-term forecast as much as possible. Thus, load profiles with a lower ratio between the deviations and the actual consumption are more favorable. In our experiments, we achieved the best results when using not more than 10 consumption patterns per day type, in which case we achieved deviation ratios roughly in the range of 9% to 11%. The results pose a significant improvement compared to the standard load profile currently in use by most of the electricity companies, with which the typically achieved deviation ratio is roughly 14% [18].

6 Conclusion and Future Work

In this paper we have introduced and assessed extensions to an approach for deriving load profiles from smart metering time series using fuzzy clustering. Since the way of generating the load profiles is built with existing business processes in mind, it helps electricity companies to adopt the presented approach with only minor adjustments as well as to ensure that the load profiles are tailored to their specific needs. Our experimental evaluation shows that load profiles built using the presented approach can significantly improve the buy-in of energy, thus lowering costs and improving the security of the energy supply.

As of now, our approach requires the load profiles to be generated from scratch each time new smart metering data is available. While this is not a major concern for typical use cases since electricity companies usually change their load profiles at most once per year, there is an interest to reduce the required computation time, e.g. by incrementally adding new smart metering data as it becomes available. This might be one potential area for future research.

References

1. BTU EVU Beratung GmbH. http://www.btu-evu.de
2. Andersen, F., Larsen, H., Boomsma, T.: Long-term forecasting of hourly electricity load: identification of consumption profiles and segmentation of customers. Energy Convers. Manag. **68**, 244–252 (2013)
3. Arthur, D., Vassilvitskii, S.: K-means++: the advantages of careful seeding. In: Proceedings of the Eighteenth Annual ACM-SIAM Symposium on Discrete Algorithms, SODA 2007, Philadelphia, PA, USA, pp. 1027–1035. Society for Industrial and Applied Mathematics (2007)
4. Beckel, C., Sadamori, L., et al.: Revealing household characteristics from smart meter data. Energy **78**, 397–410 (2014)
5. Benítez, I., Quijano, A., et al.: Dynamic clustering segmentation applied to load profiles of energy consumption from Spanish customers. Int. J. Electr. Power Energy Syst. **55**, 437–448 (2014)

6. Bezdek, J.C.: Pattern Recognition with Fuzzy Objective Function Algorithms. Springer US, New York (1981). https://doi.org/10.1007/978-1-4757-0450-1
7. Bock, C.: Generating load profiles using smart metering time series. In: Kacprzyk, J., Szmidt, E., Zadrożny, S., Atanassov, K.T., Krawczak, M. (eds.) IWIFSGN/EUSFLAT 2017. AISC, vol. 641, pp. 211–223. Springer, Cham (2018). https://doi.org/10.1007/978-3-319-66830-7_20
8. Bouguessa, M., Wang, S., Sun, H.: An objective approach to cluster validation. Pattern Recogn. Lett. **27**(13), 1419–1430 (2006)
9. CER - The Commission for Energy Regulation: Accessed via the Irish social science data archive. http://www.ucd.ie/issda
10. Diamantoulakis, P.D., Kapinas, V.M., Karagiannidis, G.K.: Big data analytics for dynamic energy management in smart grids. CoRR, abs/1504.02424 (2015)
11. Edelmann, D.H., Kästner, T.: Kosten-Nutzen-Analyse für einen flächendeckenden Einsatz intelligenter Zähler, July 2013. https://www.bmwi.de/Redaktion/DE/Publikationen/Studien/kosten-nutzen-analyse-fuer-flaechendeckenden-einsatz-intelligenterzaehler.html
12. Fernandes, M.P., Viegas, J.L., Vieira, S.M., Sousa, J.M.C.: Seasonal clustering of residential natural gas consumers. In: Carvalho, J.P., Lesot, M.-J., Kaymak, U., Vieira, S., Bouchon-Meunier, B., Yager, R.R. (eds.) IPMU 2016. CCIS, vol. 610, pp. 723–734. Springer, Cham (2016). https://doi.org/10.1007/978-3-319-40596-4_60
13. Fusco, F., Wurst, M., Yoon, J.: Mining residential household information from low-resolution smart meter data. In: 2012 21st International Conference on Pattern Recognition (ICPR), pp. 3545–3548, November 2012
14. Hathaway, R.J., Bezdek, J.C.: Fuzzy c-means clustering of incomplete data. IEEE Trans. Syst. Man Cybern. Part B: Cybern. **31**(5), 735–744 (2001)
15. Hayn, M., et al.: Electricity load profiles in Europe: the importance of household segmentation. Energy Res. Soc. Sci. **3**, 30–45 (2014)
16. Hernández, L., Baladrón, C., et al.: Classification and clustering of electricity demand patterns in industrial parks. Energies **5**(12), 5215 (2012)
17. Hoffman, P., Frey, G., Friedrich, D.M., Kerber-Clasen, S., Marschall, J., Geiger, D.M.: Praxistest "Moderne Energiesparsysteme im Haushalt" (2012)
18. Kolo, A., Kretschmann, C.: Hebung finanzieller Potentiale in der Strombilanzierung. et - Energiewirtschaftliche Tagesfragen **7**, 43–45 (2015)
19. Zhou, K., Yang, S., et al.: A review of electric load classification in smart grid environment. Renew. Sustain. Energy Rev. **24**, 103–110 (2013)
20. López, J.J., Aguado, J.A., et al.: Hopfield-k-means clustering algorithm: a proposal for the segmentation of electricity customers. Electr. Power Syst. Res. **81**(2), 716–724 (2011)
21. Mahmoudi-Kohan, N., Moghaddam, M.P., Sheikh-El-Eslami, M.: An annual framework for clustering-based pricing for an electricity retailer. Electr. Power Syst. Res. **80**(9), 1042–1048 (2010)
22. Viegas, J.P.L., Vieira, S.M., Sousa, J.M.C.: Fuzzy clustering and prediction of electricity demand based on household characteristics. In: 2015 World Congress of the International Fuzzy Systems Association and Conference of the European Society for Fuzzy Logic and Technology (IFSA-EUSFLAT 2015) (2015)
23. von Roon, D.-I.S., et al.: Statusbericht zum Standardlastprofilverfahren Gas, November 2014. https://www.ffegmbh.de/kompetenzen/system-markt-analysen/508-statusbericht-standardlastprofile-gas
24. Zhang, X., Sun, C.: Dynamic intelligent cleaning model of dirty electric load data. Energy Convers. Manag. **49**(4), 564–569 (2008)

Linguistic Description of the Evolution of Stress Level Using Fuzzy Deformable Prototypes

Francisco P. Romero(✉), José A. Olivas, and Jesus Serrano-Guerrero

Department of Information Technologies and Systems,
University of Castilla La Mancha, Ciudad Real, Spain
{FranciscoP.Romero,JoseAngel.Olivas,Jesus.Serrano}@uclm.es

Abstract. The purpose of this paper is to show that it is possible to describe stress levels through a complete time-log analysis. For this purpose it has been developed a fuzzy deformable prototypes based model that uses a fuzzy representation of the prototypical situations. The proposed model has been applied to a database composed of time logs from students with and without stress. Preliminary results from the proposed model application have been validated by experts. Moreover, the model has been applied as a classifier obtaining good results for both sensitivity and specificity. Finally, the proposal has been validated and should be considered useful for the expert systems design to support the stress level description.

Keywords: Fuzzy prototypes · Linguistic descriptions · Stress level

1 Introduction

Nowadays, there is a proliferation of tools, devices and applications, such as smartphones and wearable devices, that allow capturing data of people activity in every moment of our life. With this data captured over a long period of time, vast archives of personal data where the totality of an individual's experiences, captured multi-modally through digital sensors are stored permanently as a personal archive.

This unified digital records, commonly referred to as *heterogeneous digital lifelogs* has been gathering increasing attention in recent years within the research community due to the need for systems that can automatically analyse this huge amounts of data in order to categorize, summarize and also query them to retrieve the information that the user may need [1].

Therefore, *Lifelogging* and the management of personal digital data is rapidly becoming a mainstream research topic which requires both advanced methods that can provide an insight of the activities of an individual, and systems capable of managing this huge amount of data. There are several contributions about

J. Medina et al. (Eds.): IPMU 2018, CCIS 853, pp. 443–452, 2018.
https://doi.org/10.1007/978-3-319-91473-2_38

solving methodological on learning and data analysis focused on active and adaptive learning [2], or focused on solving real-world challenges, like human behavior analysis [3].

For example, in [2] data reduction techniques are used to automatically generate a profile of the user's everyday behavior and activities. The used method to capture everyday activities is a wearable camera called SenseCam. On the other hand, some works like [4], try to solve the interoperability problems generated by diverse health self-tracking devices. The unified representation of lifelog terms facilitated by [4] help describe an individual's lifestyle and environmental factors, which can be included with user-generated data for clinical research and thereby enhance data integration and sharing.

Despite the increasing number of successful works in the existing literature, little empirical research, however, is available, from the perspective of the analysis of time tracking data, mainly numerical, from tools and wearable devices. Although less attention has been paid to this concept in lifelog literature, there is a reason to believe this data could provide important information and knowledge to the user about his/her health status, stress level, etc.

The main problem to be addressed when trying to analyse these data sources lies in the inherent inaccuracy in the recording process. With the aim of handling data of these characteristics, Fuzzy Logic [7] offers mechanisms that allow to introduce expert knowledge and formalize the uncertainty that exists on the subject. For this purpose, our proposal is based on the use of fuzzy logic techniques such as Fuzzy Deformable Prototypes [8] that allow not only the description of the phenomena but inferences to be made about the presented situations.

This paper presents an approach for the linguistic description of stress levels extracted from time logs based on Fuzzy Deformable Prototypes. The main purpose is to create a complex behaviour model that allows to offer recommendations regarding the current situation and predictions. The experimental results presented here illustrate the feasibility of the proposal, then our conclusions are that the technique we have presented for unobtrusively and ambiently characterizing stress levels across individuals is of sufficient accuracy to be usable in a range of applications.

The remainder of the paper is structured as follows: in Sect. 2, the background concepts are presented. In Sect. 3 we describe the different steps followed until getting the linguistic description model, which consist of the FPKD process, the proper prediction process and the validation of the prediction model. In Sect. 4 the model is tested on a classifier based trained with a database composed of with/without stress time logs. Finally, in Sect. 5 we present the main conclusions and the future research lines emerged from this work.

2 Background

The classical Pattern Recognition theory [5] provides a description of pattern or prototype like the notion of reconciling an element of reality with its abstract ideal truth (pattern, or prototype), and draws the comparison with the classical view; where everything present in reality is an imperfect representation of a higher form [6].

However, complete description of human behaviour from these primitives is infeasible, therefore the problem of linguistic description of human behaviour can be seen, in some sense, as finding the ideal descriptions of nebulous concepts for which mathematical descriptions do not exist and then measuring the similarity between the observed (i.e. unknown) object.

Common description sentences show uncertainty or vagueness associated to the fulfilment of this ideal. Then, in the context of descriptive techniques, a fact or a set of facts is associated with a paradigm so that the paradigm interprets the behaviour. Thus, it is possible to describe the behaviour according to this interpretation. To generalize, many of the descriptions depend on the way the most similar paradigm or prototype for the circumstances of the problem is found.

For finding a compact and synthesizing description of a phenomenon, previous approaches propose the construction of a fuzzy prototype [7]. Zadeh's idea suggests a concept that encompasses a set of prototypes, which represents different compatibilities of the samples with the concept. Then, a fuzzy prototype is a set of good, bad and borderlines elements of a category. The principle to obtain a fuzzy prototype of a population (ς) is to stratify it by grouping objects sharing the same membership degree. For finding a compact and synthesizing description of a phenomena, this paper proposes the construction of a fuzzy prototype. For this purpose, a population of objects not necessarily distinct is represented as a fuzzy multi-set (Eq. 1):

$$\varsigma = \mu_1/(m_1 * a_1) + \cdots + \mu_n/(m_n * a_n) \tag{1}$$

Then, a process of stratification of ς is carried out with the aim of grouping objects sharing the same membership degree (Eq. 2).

$$\varsigma = H/\varsigma_{good} + M/\varsigma_{border} + L/\varsigma_{poor} \tag{2}$$

where ς_{good}, ς_{border} and ς_{poor} are multisets of good, borderline and poor elements respectively and H, M and L are fuzzy numbers which represent the corresponding *high*, *medium* and *low* membership degrees respectively. For each level of stratification of ς this fuzzy prototype is obtained using an iterative process of clustering. During the iterative process, an object maximally summarized from each level of clustering is obtained which can be viewed as a fuzzy prototype. For example, $P_t(\varsigma_{good})$ is the fuzzy prototype of ς_{good}, etc. Therefore, the prototype of the population $P_t(\varsigma)$ is represented by the Eq. 3.

$$P_t(\varsigma) = H/P_t(\varsigma_{good}) + M/P_t(\varsigma_{border}) + L/P_t(\varsigma_{poor}) \tag{3}$$

Hence, there are many prototypes of behaviour according to their activities. Where the number of prototypes for a behaviour is given, it may be meaningful to compute the collective properties of the prototypes and consider them as the reference for the corresponding descriptions and recommendations. Such use of fuzzy prototypes has been suggested for this purpose.

In this case, the aim must be to generate conceptual prototypes (Zadeh's approach: fuzzy schemes) that allow us to evaluate new phenomena from these patterns, and to provide descriptions and make recommendations.

With the same purpose, but from a different point of view, Fuzzy Deformable Prototypes (from now on FDPs) [8], come from the confluence of Zadeh's approach and the "deformable prototypes" of Bremermann, introduced in the late seventies from the field of pattern recognition [9]. In the framework of "deformable prototypes" a real element is classified according to the minimum energies required for physically deforming the closest prototype.

The definition of FDPs inherits some features of Zadeh's fuzzy prototype approach but includes some extensions to manage the complexity of real-world problems. This work uses the concept of fuzzy deformable prototypes in order to model stress levels.

The definition of FDPs [8] includes the following extensions of Zadeh's fuzzy prototype approach: the number of fuzzy prototypes depends on the problem, categories are structured using typicality degrees and the shapes of the categories have not been defined. In addition, Fuzzy Deformable Prototypes can also be represented as fuzzy sets. It means that it is possible to calculate a membership degree between an element and the fuzzy set. As a result of this process, we can obtain some sentences to describe the describe exactly a new situation with different degrees of typicality instead of being assimilated to standard patterns.

The use of FDP's allows to evaluate new situations from these patterns, to deform [8] the most similar prototypes to this new behaviour $(w_1, w_2 \dots w_n)$ and describe it using a combination of prototypes $(v_1, v_2 \dots v_n)$ with the membership degrees (μp_i) as coefficients (Eq. 4).

$$C_{real}\left(w_1 \dots w_n\right) = \left|\sum \mu_{p_i}\left(v_1 \dots v_n\right)\right| \tag{4}$$

For example, we can use the prototypical way to describe the situation, or show some uncertainty or vagueness or show higher and lower relationships with the prototypical behaviour. In other words, a fact or a set of facts is associated with a paradigm so that the paradigm interprets the situation and the actions we carry out depend on it. To generalize, many of the actions we carry out depend on the way that we find the most similar paradigm or prototype for the circumstances of the problem.

3 Linguistic Description of Human Behaviours Using Fuzzy deformable Prototypes

The basic concept of the presented comprehensive approach is to describe behaviors from lifelog data automatically, by statistically identifying interdependencies between behavioral features, discovered prototypes and recorded feedback data.

The results obtainable from the proposed method employing this knowledge are improved during the first discovery process (first order perceptions), by updating the knowledge (fuzzy prototypes) based on feedback data (see Fig. 1).

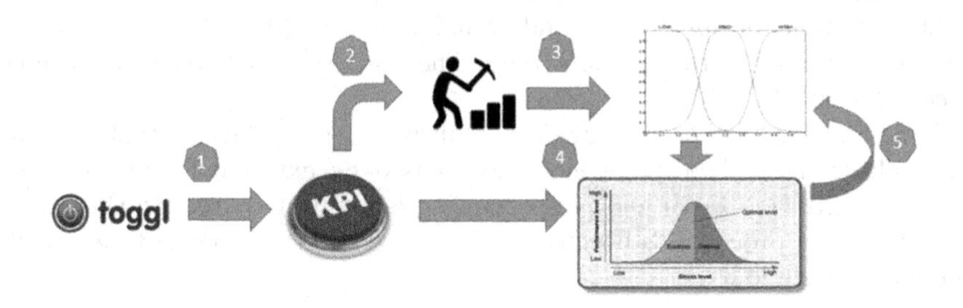

Fig. 1. Overview of the approach

The behavioral indicators to be included in the analysis are selected in step 1. The selected indicators and the associated feedback data from previous interactions are transformed and pre-processed in step 2. Fuzzy Deformable Prototypes are identified through data mining techniques in step 3. Hereby the knowledge base is created, which can be used by inference systems to provide descriptions, recommendation and predictions. The process to obtain a linguistic description based on the identified fuzzy deformable prototypes constitutes step 4. The updating of knowledge based on a comparison of described and real situation is carried out in step 5.

3.1 Analyzing the Past

The purpose of analyzing the past (steps 1 to 3) is to discover useful knowledge for further applications. In this section, the typical characteristics of the data using in our approach and the knowledge discovered in them is explained.

Data Sources and Applications. All the modeling algorithms used in this work required the input of data sources from lifelog data. Our system processes personal information from wearable devices, time tracking apps or weather apps. In a typical lifelog system, these data are delivered in real, streaming time from widely distributed data sources.

The first step of our process is to calculate the behavioral features from the data provided by these data sources. The information necessary to quantify these features must be discovered in the raw data and is often delivered as logs. These quantitative values can be specified by psicologists and generally correspond to indicators an expert would use to investigate personal behaviors. Also known as variables or features in the field of machine learning, they are usually extracted on a per-entity, per-time-segment basis. The result is a vector of indicators that describe the behavior of an entity over a predefined period of time. For this purpose, we analyze all relevant historical records and perform the aggregations that feature definition demands for example aggregating the sleep time in a day.

Modeling Procedure. The second step of the proposed approach includes the further selection, pre-processing and transformation of the input data for the statistical analysis of the previous step. These activities are part of the KDD process described in [10].

Time logs analysis process involves multiples context factors that influence the resulting data. The distortion noise caused by external factors obstructs the characterization tasks and pre-processing is highly required. In this case, a homogenization process has been considered based on the working time calculation of each time log.

Next, it is necessary to detect the relationships between the different time segments, i.e., the stratification process in order to discover fuzzy deformable prototypes. With this aim, this system learns a descriptive model of those features extracted from the data via unsupervised learning (hierarchical clustering, fuzzy clustering, soft clustering). To achieve confidence and robustness when detecting prototypical behaviors, we fuse multiple clustering results into a final one that represents the stratification of the population in order to define fuzzy deformable prototypes.

Finally, each of the prototypes is represented using fuzzy numbers. We obtain a score of each element, we aggregate the elements using a divisive clustering algorithm and solve the problem of overlapping of each prototype similar to the described in [11]. Moreover, each prototype is related to a template with the instantiation of the *dependent* component, for example, the template "You have to ≪*dependent*≫ ≪*hedge*≫ stress level", in the prototype related to "low stress" will be "You have to increase ≪*hedge*≫ your stress level".

3.2 Interpreting the Present

The next target (steps 4 and 5) is to provide descriptions and make recommendations about the present. Then, the aim is to simulate the capacity of interpretation of the situation, that is, to find the stress description more adapted to the real circumstances. For this purpose, the previously discovered knowledge based on fuzzy deformable prototypes is used.

Basic Descriptions and Recommendations. In order to obtain a linguistic description of the stress level we used the granular linguistic model of a phenomenon introduced in [12]. According to this methodology, it is necessary to define the "top-order perception", i.e., a definition of the general state of the phenomenon using natural language sentences. For this purpose, a template, like the proposed in [13], is bound to a fuzzy deformable prototype, for example.

$$Q \text{ the level of stress is } R$$

where

- Q is a fuzzy quantifier [14] applied on the cardinality of the perception the stress level is R.
- R is a summarizer. It is a constraint applied to the set of elements in the stress situation.

Therefore, our purpose is to describe the stress level of the person using non-expert friendly natural language sentences. Following the mentioned methodology, the process to obtain these sentences consists of the following steps or phases:

1. **Data Capture:** The system is fed with the real values of the factors that a priori determine the behaviour evolution: *fitbit* data, time logs, weather data, etc.
2. **Definition of the first order perceptions:** The combination of indicators will give a degree of compatibility with prototypes, those prototypes with a degree of non-zero compatibility will be modified to give rise to the characterization of the time segment by a combination whose coefficients are the degrees of compatibility previously obtained. For example, a situation could have a positive degree of compatibility with the low and medium stress level. The perception protoform of *stress level* is a tuple (U, O_x, g, T) where:
   - U is the empty set,
   - O_x is a linguistic label expressing the degree of stress level
   - g is a function based on membership functions
   - T is a set of the linguistic templates using to build the result. Two examples of templates of this set are: "You have to reduce—maintain—increase <hedge> level of stress" and "you stress level is <hedge> a high—medium—low". The component <hedge> represent a linguistic hedge (very, a little, etc.).
3. **Definition of the top-order perception:** The system describes the situation based on the detected stress level. To do this, it calculates the current stress state by dividing for each prototype the stress curve [15] into five stages: start, up, top, down, end. Identifying within the prototype the phase in which the user is, through a finite state machine, can be generated the linguistic description of the situation and therefore recommendations from it. Formally, the perception protoform of the *stress level* is a tuple (U, S_x, g, T) where:
   - U are the linguistic variables related to the stress stage.
   - S_x is a linguistic variable of the stress level.
   - g is the aggregation function 4.
   - T is the set of templates to provide recommendations.

   For example, if the degree of compatibility with prototypes are 0.65 to *low* and 0.3 to *medium*, the recommendations provided could be "your *stress_level* is *a bit* low" and "you have to *increase a_little* your stress level". The description provided to the current time segment would be valid if there are no influence of local factors. In this case, it is mandatory to modify any description or recommendation.

4 Experimental Phase: Real-Life Applications

To validate our method, we experimented with a real-world data set which corresponds to three months worth of logs, generated by the use of a Time Tracking tool by students of the University of Castilla la Mancha. This dataset contains records a bunch of time log lines per day, each corresponding to student specific tasks, and has about a hundred of daily active users. Time log analysis is aimed at the description of time management behaviour, while using *fitbit* information allows us to identify special event related to sleep time or physical exercise (workout).

The database comprised both known with-stress and without-stress students. A group of expert teachers established the level of stress with four possible results, namely, sure, probably, discarding, or negative. Those time logs showing unequivocal signs of stress are marked as sure. If a marker is not clear or does not appear, the time log is marked as probable. When there is only a sign but cannot guarantee the stress level it is marked as discarding. Finally, time logs which do not show any sign associated with stress problems are marked as negative.

The proposed model was applied to a set of 27 time logs with positive stress and 62 randomly selected with no stress. The first step of the analytics process is to calculate the behavioral features of the data provided by Time Tracking Tool. For this purpose, it is considered in a separate manner several indicators such as the mean, median and mode of the working-time, the number of interruptions, etc. These features values extracted from the time log in each case have been used as an input set to train an automatic classifier based on the FDP's model.

The purpose of this approach is to use the FDP-based model to analyse time logs for the stress level calculation. Moreover, since there are several time series analysis algorithms widely used for stress characterization oriented to the training of automatic classification systems, a comparison was made in the validation process between the proposed model versus the following methods: decision tress, kNN, naive bayes and random forests.

The training sets for the classifier system were divided as follows:

- Training set, used to optimize the model parameters to minimize the error. It employs 60% of the total set.
- Validation set, to use for overfitting detection. It employs 20% of the total set.
- Test Set, which is used to check model performance. These data are not used in the model training, so it could be possible to determine the generalization level from the model used. It employs 20% from the complete set.

In Table 1 the different values achieved of sensitivity and specificity are shown. The FDP's base models presents a sensitivity value of 95%, a specificity value of 89%.

A group of teachers/educators examined the linguistic descriptions resulting from the model application and used the parameters associated with the training that showed the highest level of stress. A set of linguistic descriptions was showed

Table 1. Results. Specificity(S), Sensitity(E)

Model	S	E
FDP's	95%	89%
kNN	86%	83%
Decision tree	87%	84%
Naive Bayes	92%	84%
Random forests	93%	87%

to the experts, containing the behaviour evolution of the students with positive stress indicators and with non-positive stress indicators and an overlap of the first two.

The experts' assessment showed that the linguistic representation of the student behavior is similar to the performed in teaching practice. Moreover, comparing the linguistic representation of the behavior from the students without performance problems, the indicators identified as characteristic of the proposed model are those that mark the main differences. This subjective assessment has been issued by all the experts that took part of the survey and is considered highly important due to their commitment to the teaching practice routine.

5 Conclusions and Future Work

Continuous analysis of time series captured from human activity allows us to describe changes in the people behaviour and offer recommendations. Fuzzy Deformable Prototypes of time segments show the different behavioural possibilities. Subsequently, each time segment type goes through a "deformation process" until a prototype match is achieved. The better we can describe the "time segment" during any window time, the easier it becomes to offer specific recommendations and predictions. It would be interesting in future work to devise a version of the analysis in which reinforcement learning could play an important role. For example, the user's opinion about the descriptions and recommendations provided by the system can feedback the application with this knowledge to improve its performance.

References

1. Kumar, G., Jerbi, H., Gurrin, C., O'Mahony, M.P.: Towards activity recommendation from lifelogs. In: Proceedings of the 16th International Conference on Information Integration and Web-based Applications & Services, pp. 87–96. ACM (2014). https://doi.org/10.1145/2684200.2684298
2. Wang, P., Smeaton, A.F.: Using visual lifelogs to automatically characterize everyday activities. Inf. Sci. **230**, 147–161 (2013). https://doi.org/10.1016/j.ins.2012.12.028

3. Dobbins, C., Rawassizadeh, R., Momeni, E.: Detecting physical activity within lifelogs towards preventing obesity and aiding ambient assisted living. Neurocomputing **230**, 110–132 (2017). https://doi.org/10.1016/j.neucom.2016.02.088
4. Kim, H.H., Lee, S.Y., Baik, S.Y., Kim, J.H.: MELLO: Medical lifelog ontology for data terms from self-tracking and lifelog devices. Int. J. Med. Inf. **84**(12), 1099–1110 (2015). https://doi.org/10.1016/j.ijmedinf.2015.08.005
5. Pavlidis, T.: Structural Pattern Recognition. Springer, Heidelberg (1977). https://doi.org/10.1007/978-3-642-88304-0
6. Grundy, E.: Extraction and Classification of Features in Accelerometry Data Doctoral dissertation, Swansea University (2008)
7. Zadeh, L.A.: A note on prototype theory and fuzzy sets. In: Fuzzy Sets, Fuzzy Logic, and Fuzzy Systems, pp. 587–593 (1996). https://doi.org/10.1142/9789814261302-0027. Selected Papers by Lotfi A Zadeh
8. Olivas, J.A., Sobrino, A.: An application of Zadeh's prototype theory to the prediction of forest fire in a knowledge-based system. In: Proceedings of the 5th International IPMU, vol. 2, pp. 747–752 (1994)
9. Bremermann, H.: Pattern recognition by deformable prototypes. In: Hilton, P. (ed.) Structural Stability, the Theory of Catastrophes, and Applications in the Sciences. LNM, vol. 525, pp. 15–57. Springer, Heidelberg (1976). https://doi.org/10.1007/BFb0077842
10. Fayyad, U., Piatetsky-Shapiro, G., Smyth, P.: From data mining to knowledge discovery in databases. AI Mag. **17**(3), 37 (1996)
11. Vazquez, M.R., Romero, F.P., Olivas, J.A., Orbe, E., Serrano-Guerrero, J.: An approach to academic performance prediction in tutoring systems based on fuzzy deformable prototypes. Prog. Artif. Intell. **5**(1), 55–64 (2016). https://doi.org/10.1007/s13748-015-0074-9
12. Trivino, G., Sugeno, M.: Towards linguistic descriptions of phenomena. Int. J. Approx. Reason. **54**(1), 22–34 (2013). https://doi.org/10.1016/j.ijar.2012.07.004
13. Yager, R.R.: Fuzzy summaries in database mining. In: Proceedings of the 11th Conference on Artificial Intelligence for Applications, CAIA 1995, p. 265 (1995). https://doi.org/10.1109/CAIA.1995.378813
14. Zadeh, L.A.: A computational approach to fuzzy quantifiers in natural languages. Comput. Math. Appl. **9**(1), 149–184 (1983). https://doi.org/10.1016/0898-1221(83)90013-5
15. Teigen, K.H.: Yerkes-Dodson: A law for all seasons. Theory Psychol. **4**(4), 525–547 (1994). https://doi.org/10.1177/0959354394044004

Model Averaging Approach to Forecasting the General Level of Mortality

Marcin Bartkowiak[1], Katarzyna Kaczmarek-Majer[2], Aleksandra Rutkowska[1](✉), and Olgierd Hryniewicz[2]

[1] University of Economics and Business, Niepodleglosci 10, Poznan, Poland
{marcin.bartkowiak,aleksandra.rutkowska}@ue.poznan.pl
[2] Systems Research Institute, Polish Academy of Sciences, Newelska 6, Warsaw, Poland
{k.kaczmarek,olgierd.hryniewicz}@ibspan.waw.pl

Abstract. Already a 1% improvement to the overall forecast accuracy of mortality rates, may lead to the significant decrease of insurers costs. In practice, Lee-Carter model is widely used for forecasting the mortality rates. Within this study, we combine the traditional Lee-Carter model with the recent advances in the weighted model averaging. For this purpose, first, the training database of template predictive models is constructed for the mortality data and processed with similarity measures, and secondly, competitive predictive models are averaged to produce forecasts. The main innovation of the proposed approach is reflecting the uncertainty related to the shortness (e.g., 14 observations) of available data by the incorporation of multiple predictive models. The performance of the proposed approach is illustrated with experiments for the Human Mortality Database. We analyzed time series datasets for women and men aged 0–100 years from 10 countries in the Central and Eastern Europe. The presented numerical results seem very promising and show that the proposed approach is highly competitive with the state-of-the-art models. It outperforms benchmarks especially when forecasting long periods (6–10 years ahead).

Keywords: Lee-Carter model · Mortality forecast · Model averaging Small samples · Time series

1 Introduction

One of the signs of the social progress is an increase in life expectancy, which at the same time causes increase of the longevity cost, especially the pension and health costs. This poses a challenge to governments, private pension plans and life insurers and, in consequence, new statistical techniques for the modeling and projection of mortality rates are needed.

The basic modeled variable in the mortality analysis is the central death rate. For a given population or cohort, the central death rate at age x during

J. Medina et al. (Eds.): IPMU 2018, CCIS 853, pp. 453–464, 2018.
https://doi.org/10.1007/978-3-319-91473-2_39

a given period of one year, is found by dividing the number of people, who died after they had reached the exact age x but before they reached the exact age $x + 1$, by the average number who were living in that age group during the period. The available mortality data are usually uncertain due to the fact, that the raw data are usually manipulated by different organizations to create the complete databases [33]. This manipulation consists of: splitting data into finer; aggregating them into wider age categories; smoothing the observed values in order to obtain an improved representation of the given demographic ratio. Furthermore, the available data may be very short (e.g., even 14 observations). All of the above make data uncertain, and therefore, the forecasting approach should appropriately reflect the uncertainty associated with this data.

One of the most influential publications about the stochastic approach to modeling mortality was written by Lee and Carter [17]. They modeled the logarithm of central mortality, taking into account the changes in mortality associated with age and calendar year. One of its parameters is the time trend parameter κ_t that shows the general level of mortality. The stochastic model proposed by Lee and Carter has gained popularity among actuaries and demographers because of its good performance. Lee and Carter developed their approach specifically for the U.S. mortality data, but method is now being applied to mortality data from many countries and time periods i.a.: Austria [1], Belgium [5], England and Wales [25], Spain [26], Sweden [23]. In the literature, there are multi-country comparisons of mortality models as well, e.g. the Nordic countries [15,27], the industrialized countries [24]. The overview of applications and extensions can be found in [3,18,28]. The above mentioned results, are handled mostly for the economically developed nations. The authors of this paper have identified the need for investigating datasets of countries with high[1] mortality rates, because, to the best of authors knowledge, the state-of-the-art research seems to miss the in-depth analysis of these countries.

Within this paper, we introduce a model averaging approach to forecast the general mortality levels κ_t, instead of modeling it as a single process. For this purpose, we adapt the 'Forecasting with Data-Mining weights' (F-DM) approach introduced by Kaczmarek and Hryniewicz [13]. First, the training database of template predictive models is constructed for the mortality data and processed with similarity measures, and secondly, competitive predictive models are averaged to produce forecasts.

Many researchers show that combining various forecasts leads on average to better results than applying individual ones [19]. Especially, models used for the prediction in short time series are based on the concept of weighted model averaging. However, finding proper weights is a serious practical problem. To overcome this, Hryniewicz and Kaczmarek [9] proposed methods inspired by the

[1] We can distinguish high mortality countries from low mortality countries with life expectancy at birth [United Nations, Department of Economic and Social Affairs, Population Division (2013). World Mortality Report 2013 (United Nations publication)]. This indicator for Euro Area overtakes 82, while for Central Europe and the Baltics equals 77 [The World Bank 2015].

computational intelligence [16] for the construction of the prior distribution on the prechosen set of models. Their algorithm appears to be highly competitive when compared to the best available algorithms used for the prediction in short time series [14]. In their recent papers devoted to the statistical process control field [11], they have adopted a similar approach for the construction of a new Shewhart-type control chart. This paper presents further development and new application of the forecasting approach introduced in [11,13].

2 Lee-Carter Model

Let $m_{x,t}$ denote the mortality rate in an age group $x = x_1, \ldots, x_N$ and time $t = 1, 2, \ldots, T$. The Lee-Carter (LC) model can be presented as follows:

$$\ln(m_{x,t}) = \alpha_x + \beta_x \kappa_t + \epsilon_{x,t}, \tag{1}$$

where α_x and β_x are age-specific parameters and κ_t is a time-variant parameter. α_x can be interpreted as the mean mortality at age x. The time trend parameter κ_t shows the general level of mortality and β_x describes the mortality change at a given age for a unit of yearly total mortality change. The error term $\epsilon_{x,t}$ is the corresponding residual term. Thus the LC model has $2N + T$ parameters. To obtain a unique solution to (1) the following constraints are imposed:

$$\alpha_x = \frac{1}{T}\sum_{t=1}^{T} \ln(m_{x,t}), \sum_{t=1}^{T} \kappa_t = 0, \sum_{x=x_1}^{x_N} \beta_x = 1 \tag{2}$$

Lee and Carter used the singular value decomposition (SVD) to find parameters. The disadvantage of their model is the assumption of the random component's homoscedasticity. This assumption is not confirmed in empirical studies - the intensity of deaths in older age groups is subject to greater variation than it is in younger groups. Therefore, SVD is often replaced by maximum likelihood estimation or weighted least squares method.

The corresponding future $m_{x,t+n}$ is calculated according to (1). The state-of-the-art approaches assume modeling κ_t as a random walk with trend [5,17,29] or as integrated autoregressive and moving average (ARIMA) process [30,31]. The common approach to forecast mortality models κ_t using the autoregressive integrated moving average (ARIMA) process. Let us denote this process as M_0. We follow the procedure of Hyndman and Khandakar [12] implemented in the 'forecast' package of R to estimate the M_0 model. It uses the unit root test and Akaike information criterion (AIC). We describe its parameters by a vector $(a_1^0, \ldots .a_{p_0}^0)$. However, the main disadvantage of the widely used ARIMA model is the requirement of at least 50 but preferably more than 100 observations [4], which is often hard to meet in case of mortality data, and this is our case of the East European countries.

3 Forecasting Mortality Level with Model Averaging

To diminish the risk of selecting one inadequate model for κ_t, multiple predictive models are combined through the model averaging inference. For this purpose, the 'Forecasting with Data-Mining weights' (F-DM) algorithm [13] is adapted to select the k-top performing models and calculate their weights. The main goal of the application of the model averaging is to compensate the lack of accuracy of the estimated model that is caused by mostly the shortness and uncertainty of the available mortality data.

We improve the F-DM approach by the addition of the M_0 estimated model basing on the monitored κ_t data. We assign to this estimated model M_0 a certain weight $w_0 \in [0, 1]$. Apart from model M_0, the k alternative models $M_j, j = 1, \ldots, k$, each described by a vector of parameters $(a_1^j, \ldots, a_{p_j}^j)$ are considered. The approach with this improvement is referred as F-DM* in the remaining of this paper.

The input for the F-DM* algorithm is the stationary time series for prediction κ_t and the set of template predictive models M, and its output are forecasts for κ_{t+h}. Thus, the preprocessing phase that ensures the transition of a time series to stationarity is needed. Hopefully, for the considered mortality data, already series of first or second order (depending on data) differences are stationary. The F-DM* algorithm consists of the following three steps preceded with the preprocessing phase: (i) construction of the training database of predictive models; (ii) calculation of similarities; (iii) weighted model averaging.

3.1 Construction of Database with Predictive Models

We adapt stationary autoregressive processes (AR) as predictive models $M = \{M_1, M_2, \ldots, M_J\}$

$$y_i = \sum_{i=1}^{p} \phi_i y_{t-i} + a_t \tag{3}$$

where $a_t \sim N(0, \sigma^2)$ are normally distributed independent random variables, $\sigma^2 \in (0, 1)$ and $\phi_i \in (-1, 1)$. The training database is constructed using all autoregressive models of order AR(2) and smaller. It consists of s realizations (training time series) $\{y_{j,i}\}_{i=1}^{s}$ for $j \in \{1, 2, \ldots, J\}$ models (processes).

Let $w_1^{'}, \ldots, w_k^{'}$ denote the weights assigned to models $M_1, \ldots, M_k$ by the F-DM [13] algorithm. In F-DM*, we assign the weight w_0 to the estimated model. Because the total weight of the chosen alternative models is $1 - w_0$, and thus, to each chosen alternative model we will assign a weight $w_j = (1 - w_0)w_j^{'}, j = 1, \ldots, k$.

3.2 Calculation of Similarities

Dynamic Time Warping (DTW) algorithm [2] for measuring the distance between two time series is adapted. The distances between each of the training time series and the time series for prediction $\mathbf{y}$ are calculated for all $j \in \{1, \ldots, J\}$ models and all their $i \in \{1, \ldots, s\}$ realizations. Then, the distances for various time series of the same predictive model are averaged, and the k most similar models are selected as alternative predictive models for further forecasting. Wang et al. [32] show that especially on small data sets elastic measures like DTW can be significantly more accurate than Euclidean distance and other lock-step measures because the elastic (non-linear) measures take into account the dilatation in time.

3.3 Weighted Model Averaging

Individual forecasts of the selected k most similar models are calculated $f_i(t_0)$ according to their prior definitions M. The final forecast is calculated as the weighted average of the individual forecasts $f_i(t_0)$ and the corresponding weights $\{w_1, \ldots, w_k\}$:

$$f(t_0) = \sum_{i=1}^{k} w_i f_i(t_0). \tag{4}$$

The concept of model averaging was promoted by Geweke, see i.e. Bayesian model averaging in [7]. For further details about the construction and performance of the weighted model averaging, we refer the reader to [13].

4 Empirical Results

The performance of the proposed method is illustrated with the empirical study.

4.1 About Datasets

The mortality data from the Human Mortality Database for the EU Member States joined after 2000[2] are of main interest. It is observed that there were differences in mortality patterns in East and West Europe. During the 1970s and 1980s mortality of the young and middle aged man increased significantly only in this region. The most frequently quoted reason for this phenomenon are civilizations diseases and cardiovascular mortality [20, 21].

Table 1. Mortality data considered in the experiments.

Country	Bulgaria	Croatia	Czechia	Estonia	Hungary	Latvia	Lithuania	Poland	Slovakia	Slovenia
Period	1947–2010	2002–2015	1950–2014	1959–2014	1950–2014	1959–2014	1959–2014	1958–2014	1950–2014	1983–2014
n	63	14	64	55	64	55	55	56	64	31

The study is conducted separately for the male and female population for ages 0 to 100 from ten countries from Central and East Europe. The detailed time periods for each country with number of observations (n) are shown in Table 1. We divide each data set into a fitting period and a forecasting period.

[2] Datasets are available for download from https://www.mortality.org.

The commencing years of fitting periods differ by country. The forecasting period is last 10 years, for most countries starting in 2005 and ending in 2014, with the exception of Bulgaria and Croatia. For Bulgaria we forecast over the period 2001–2010 due to the lack of the newer data. Croatia is a separate case, with the shortest time series, with only 14 observations, therefore we decide that the forecasting period is set to be last 3 years (2013–2015).

4.2 Forecast Evaluation

Using the data in the fitting period, we compute 10-step-ahead forecasts (three-step-ahead for Croatia) and calculate the forecast errors by comparing the forecast with actual out-of-the-sample data. As a benchmark, we use the LC model with the ARIMA forecasts denoted as M_0. Estimation is done in R via function fit in 'StMoMo' packages and the order of differencing and the numbers of AR or MA terms in an ARIMA model, that is identified automatically by the use of function auto.arima from R package 'forecast'. We estimate the results of the algorithm based on the errors of the basic variable - the mortality rate[3] ($m_{x,t}$). The accuracy of forecasts is evaluated with the following measures: the mean absolute error (MAE), the mean squared error (MSE), the median absolute error (MDAE), the median absolute percentage error (MDAPE), the symmetric mean absolute percentage error (SMAPE) and the symmetric median absolute percentage error (SMDAPE). We compute commonly used scale-dependent measures (MAE, MSE, MDAE) because they are useful when comparing different methods applied to the same set of data. Furthermore we use measures based on percentage errors (MDAPE, SMAPE, SMDAPE), which have the advantage of being scale-independent.

4.3 Results

Let us start from an illustrative example. Figure 1 shows observed and forecasted log mortality rates for men in Poland for each age in year 2005 and 2014. Visual analysis confirms that the trends are correctly defined. As observed, the quality of forecast is getting worse with lengthening time horizon (MAE is raising from 0.0023 to 0.0490 for female and from 0.0040 to 0.0640 for male). However, in comparison with M_0, the longer forecasts are clearly better in the proposed F-DM* method (cf. Table 3). One should be aware that the accuracy of forecasts is different for different age groups. In the Eastern European countries forecasting mortality for middle age is particularly troublesome. For example, the Fig. 2 shows mortality rate for female (left panel) and male (right panel) aged 40 in Poland. Though the forecast for female is quite good ($\text{MAE} = 5.18 \times 10^{-5}$), there is still a potential for improvement for male data ($\text{MAE} = 1.25 \times 10^{-3}$).

[3] Some authors calculate prediction errors based on the model's variable $\ln(m_{x,t})$. It is worth pointing out that this approach does not always generate the same conclusions (the results are not equivalent). From a practical point of view we are interested in the central death rate. The LC model is just a estimation tool. Therefore, we compute errors comparing the estimated values to the central death rate.

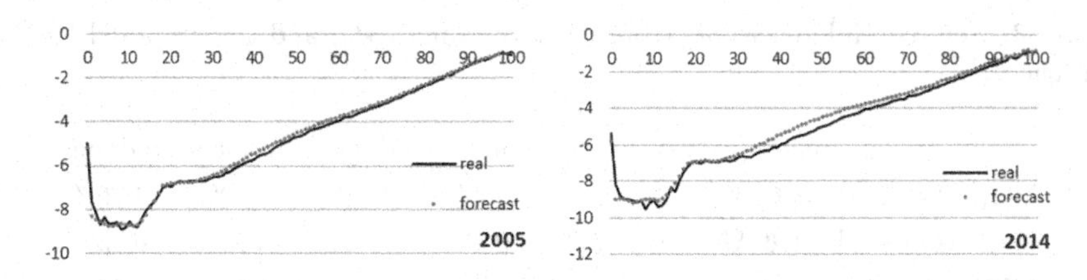

Fig. 1. Observed and forecast of log mortality rates for male of Poland by F-DM* model averaging method in 2005 (left) and 2014 (right)

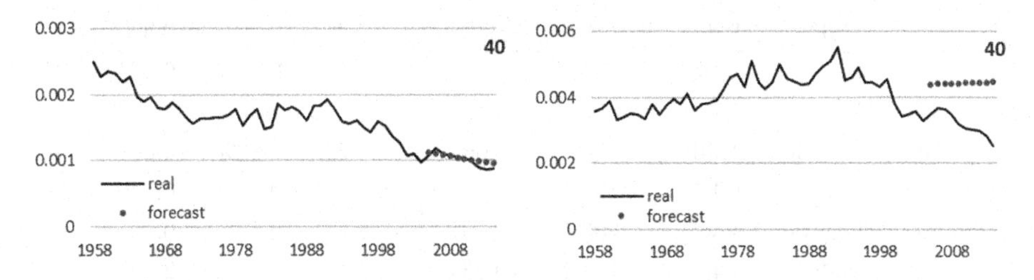

Fig. 2. Observed and forecast of mortality rates for female (left) and male (right) aged 40 in Poland

Table 2. Forecast errors for model auto.arima (M_0) and the F-DM* approach ($w_0 = 0.5$).

Country	Gender	M_0						F-DM* average					
		mae[a]	mse[a]	mdae[a]	mdape	smape	smdape	mae[a]	mse[a]	mdae[a]	mdape	smape	smdape
POL	Female	3.71	0.11	0.20	10.03	16.28	9.84	3.60	0.11	0.20	9.63	16.65	9.47
	Male	7.05	0.21	2.60	18.10	18.93	16.84	6.93	0.21	2.60	17.85	18.92	16.69
CZE	Female	3.94	0.13	0.20	12.02	20.13	11.65	3.72	0.13	0.20	11.24	20.38	10.83
	Male	6.32	0.38	1.00	13.55	22.19	12.85	6.39	0.38	1.00	13.75	22.10	13.17
SVK	Female	5.42	0.48	0.30	11.60	24.09	11.54	5.45	0.46	0.30	12.61	23.08	12.33
	Male	14.41	1.07	2.40	31.85	34.44	27.57	14.14	1.02	2.60	31.35	33.50	27.47
HUN	Female	2.90	0.07	0.40	10.97	23.83	10.59	2.87	0.07	0.40	10.77	23.84	10.52
	Male	13.36	0.78	3.70	34.67	38.86	29.70	13.70	0.83	3.40	35.18	39.34	30.06
BGR	Female	7.86	0.46	0.40	15.37	18.96	15.49	7.58	0.43	0.40	14.97	17.85	14.56
	Male	9.93	0.80	0.50	9.61	13.46	9.68	10.07	0.82	0.50	9.68	13.54	9.67
LTU	Female	9.20	0.79	0.50	14.49	22.68	14.24	9.23	0.80	0.50	14.60	22.72	14.32
	Male	11.72	1.27	1.90	17.07	25.74	16.25	11.67	1.31	1.70	16.39	25.30	15.85
LVA	Female	6.92	0.28	1.30	26.16	30.39	23.34	6.92	0.28	1.30	26.21	30.32	23.25
	Male	11.41	0.72	2.40	27.50	35.07	24.53	11.21	0.72	2.30	26.48	34.40	23.49
EST	Female	6.31	0.26	1.20	28.90	42.96	26.36	6.16	0.26	1.20	28.06	42.80	25.79
	Male	17.30	1.92	4.10	42.85	48.55	35.89	17.32	1.92	4.10	42.43	48.28	35.40
SVN	Female	4.73	0.28	0.30	16.82	37.84	16.74	4.69	0.27	0.30	17.04	37.84	16.91
	Male	15.71	3.63	0.60	17.65	32.38	16.93	15.47	3.46	0.60	18.00	32.49	17.51
HRV	Female	4.20	0.21	0.20	11.67	24.08	11.82	4.23	0.21	0.20	11.41	23.76	11.60
	Male	12.57	2.23	0.40	10.57	22.22	10.42	11.67	1.90	0.40	10.25	21.86	10.05
Avg		8.75	0.80	1.23	19.07	27.65	17.61	8.65	0.78	1.21	18.90	27.45	17.45
Median		7.45	0.47	0.55	16.10	24.09	15.87	7.25	0.01	0.00	15.68	23.80	15.21

[a] Error values multiplied by 10^3

Table 3. Relative differences of errors for auto.arima (M_0) and the proposed F-DM* approach.

		mae	mse	mdae	mdape	smape	smdape
POL	Female	**3.02%**	2.75%	0.00%	3.91%	−2.29%	3.79%
	Male	**1.65%**	1.43%	0.00%	1.34%	0.01%	0.93%
CZE	Female	**5.59%**	3.82%	0.00%	6.46%	−1.23%	7.04%
	Male	−1.08%	−0.53%	0.00%	−1.44%	0.42%	−2.52%
SVK	Female	−0.61%	4.60%	0.00%	−8.72%	4.20%	−6.81%
	Male	**1.84%**	4.76%	−8.33%	1.56%	2.76%	0.39%
HUN	Female	**1.07%**	1.35%	0.00%	1.84%	−0.06%	0.66%
	Male	−2.50%	−6.58%	8.11%	−1.47%	−1.25%	−1.22%
BGR	Female	**3.56%**	7.59%	0.00%	2.63%	5.88%	6.01%
	Male	−1.45%	−2.37%	0.00%	−0.74%	−0.55%	0.14%
LTU	Female	−0.36%	−0.89%	0.00%	−0.75%	−0.18%	−0.49%
	Male	0.45%	−3.31%	10.53%	4.00%	1.72%	2.48%
LVA	Female	0.04%	−0.36%	0.00%	−0.19%	0.24%	0.40%
	Male	**1.68%**	−0.14%	4.17%	3.70%	1.91%	4.22%
EST	Female	**2.30%**	1.91%	0.00%	2.90%	0.37%	2.16%
	Male	−0.14%	−0.21%	0.00%	0.99%	0.54%	1.36%
SVN	Female	0.68%	1.09%	0.00%	−1.26%	0.00%	−1.01%
	Male	**1.51%**	4.66%	0.00%	−1.98%	−0.34%	−3.43%
HRV	Female	−0.71%	1.44%	0.00%	2.26%	1.30%	1.84%
	Male	**7.12%**	14.75%	0.00%	3.00%	1.62%	3.54%
Full sample (1–10 years ahead)							
Avg		1.18%	1.79%	0.72%	0.90%	0.75%	0.97%
Median		0.87%	1.39%	0.00%	1.45%	0.31%	0.80%
1st period (1–5 years ahead)							
Avg		0.43%	0.73%	**1.11%**	0.19%	0.48%	−0.04%
Median		0.56%	0.66%	0.00%	0.40%	0.18%	0.11%
2nd period (6–10 years ahead)							
Avg		**1.31%**	**1.41%**	**1.28%**	0.89%	0.76%	1.19%
Median		**1.22%**	**1.46%**	0.00%	0.15%	0.28%	0.06%

We now compare the forecasting results of F-DM* and M_0 benchmark for the whole sample of countries. In Table 2, the forecast errors for M_0 and for the proposed F-DM* approach ($w_0 = 0.5$) is presented.

Table 3 and Fig. 3 present relative differences of obtained errors. In all samples and actually for almost all measures results show that the smaller errors are generated by the proposed F-DM* method. The biggest differences appear in the long-term forecast for 6–10 years. Significantly better forecasts were obtained for:

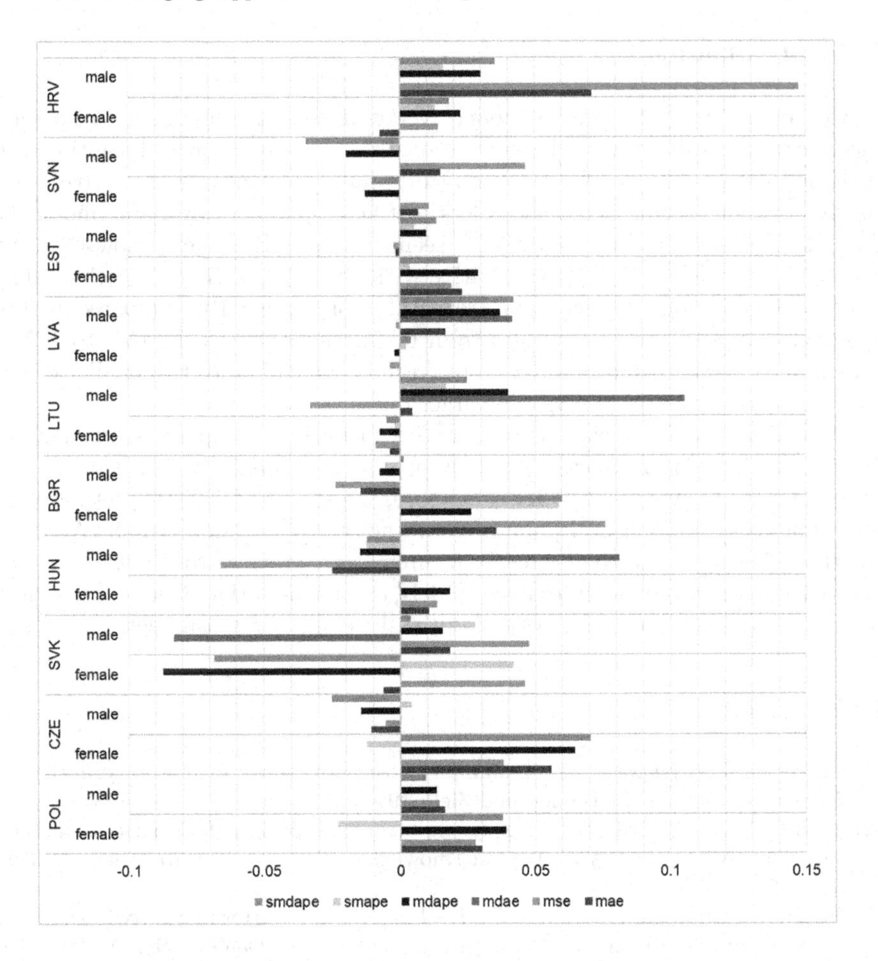

Fig. 3. Relative differences of errors for M_0 vs F-DM* approach.

women in Slovenia (forecast improved by 6.68%), Czechia (by 5.59%), Poland (3.02%), Bulgaria (3.56%), Estonia (2.30%) and men in Croatia (by 7.12%)[4]. The improvement of relative differences of MEA error above 1% is also observed for women in Hungary and men in Poland, Slovakia, Latvia, Slovenia.

To test the significance of differences in the forecast errors in the proposed approach and the standard approach, the Diebold-Mariano [6,8] test is carried out. The null hypothesis is the two forecasts have the same accuracy. The alternative hypothesis is that proposed method is more accurate than M_0. The Diebold-Mariano test rejected the null hypothesis in favor of the alternative hypothesis that proposed method is more accurate in 11 cases, i.e. for man in countries: Croatia, Poland, Slovenia, Slovakia and women in Estonia, Poland, Bulgaria, Czechia, Slovenia, Slovakia and Hungary.

[4] According to MAE relative differences.

5 Conclusions

In this paper, we introduce a model averaging F-DM* approach to forecast the general mortality levels. The main advantages of the proposed solution are: (i) handling uncertainty of κ_t by the incorporation of multiple predictive models; (ii) the forecasting accuracy unless to the shortness of available mortality data. The presented numerical results seem very promising and show that the proposed approach is highly competitive with the state-of-the-art method. The improvement in forecast accuracy (relative error for $h = 10$) compared to the commonly applied method, ranges from 0.72% according to MDAE to 1.79% according to MSE. Additionally we obtain in 55.83% of cases the smaller errors than auto.arima and in 30.83% the larger ones.

According the financial report of one insurance company - Prudential, improving the overall forecast accuracy of mortality rates by 1%, leads to the decrease of costs even by 3% [22]. Therefore, the proposed F-DM* method very appealing from the practical point of view and worth further research. We identify the following major challenges for future research related to this work: extension of the database with alternative models, extension of the LC model to reflect cohort effects and further analysis of optimal weights establishment.

References

1. Carter, L.R., Prskawetz, A.: Examining structural shifts in mortality using the Lee-Carter method. Methoden und Ziele **39** (2001)
2. Berndt, D.J., Clifford, J.: Using dynamic time warping to find patterns in time series. In: AAAI 1994 Workshop on Knowledge Discovery in Databases, pp. 359–370 (1994)
3. Booth, H., Hyndman, R., Tickle, L., De Jong, P.: Lee-Carter mortality forecasting: a multi-country comparison of variants and extensions. Demogr. Res. **15**, 289–310 (2006). https://doi.org/10.4054/DemRes.2006.15.9
4. Box, G.E., Tiao, G.C.: Intervention analysis with applications to economic and environmental problems. J. Am. Stat. Assoc. **70**(349), 70–79 (1975). https://doi.org/10.2307/2285379
5. Brouhns, N., Denuit, M., Vermunt, J.K.: A Poisson log-bilinear regression approach to the construction of projected lifetables. Insur. Math. Econ. **3**(31), 373–393 (2002). https://doi.org/10.1016/S0167-6687(02)00185-3
6. Diebold, F.X., Mariano, R.S.: Comparing predictive accuracy. J. Bus. Econ. stat. **20**(1), 134–144 (2002). https://doi.org/10.2307/1392185
7. Geweke, J.: Contemporary Bayesian Econometrics and Statistics. Wiley, Hoboken (2005)
8. Harvey, D., Leybourne, S., Newbold, P.: Testing the equality of prediction mean squared errors. Int. J. Forecast. **13**(2), 281–291 (1997). https://doi.org/10.1016/S0169-2070(96)00719-4
9. Hryniewicz, O., Kaczmarek, K.: Bayesian analysis of time series using granular computing approach. Appl. Soft Comput. J. **47**, 644–652 (2016). https://doi.org/10.1016/j.asoc.2014.11.024

10. Hryniewicz, O., Kaczmarek, K.: Monitoring of short series of dependent observations using a control chart approach and data mining techniques. In: Proceedings of the International Workshop ISQC 2016, Helmut Schmidt Universität, Hamburg, pp. 143–161 (2016)
11. Hryniewicz, O., Kaczmarek-Majer, K.: Monitoring of short series of dependent observations using a XWAM control chart. In: Knoth, S., Schmid, W. (eds.) Frontiers in Statistical Quality Control 12 (2018)
12. Hyndman, R.J., Khandakar, Y.: Automatic time series forecasting: the forecast package for R. J. Stat. Softw. **26**(3) (2008)
13. Kaczmarek-Majer, K., Hryniewicz, O.: Data-mining approach to finding weights in the model averaging for forecasting of short time series. In: Kacprzyk, J., Szmidt, E., Zadrożny, S., Atanassov, K.T., Krawczak, M. (eds.) IWIFSGN/EUSFLAT - 2017. AISC, vol. 642, pp. 314–327. Springer, Cham (2018). https://doi.org/10.1007/978-3-319-66824-6_28
14. Kaczmarek, K., Hryniewicz, O., Kruse, R.: Human input about linguistic summaries in time series forecasting. In: Proceedings of The Eighth International Conference on Advances in Computer-Human Interactions ACHI (2015)
15. Koissi, M.-C., Shapiro, A.F., Hgns, G.: Evaluating and extending the Lee-Carter model for mortality forecasting: bootstrap confidence interval. Insur. Math. Econ. **38**(1), 1–20 (2006)
16. Kruse, R., Borgelt, C., Braune, C., Mostaghim, S., Steinbrecher, M.: Computational Intelligence: A Methodological Introduction. TCS. Springer, London (2016). https://doi.org/10.1007/978-1-4471-7296-3
17. Lee, R.D., Carter, L.R.: Modeling and forecasting US mortality. J. Am. Stat. Assoc. **87**(419), 659–671 (1992). https://doi.org/10.2307/2290201
18. Lee, R.: The Lee-Carter method for forecasting mortality, with various extensions and applications. North Am. Actuar. J. **4**(1), 80–91 (2000). https://doi.org/10.1080/10920277.2000.10595882
19. Makridakis, S., Hibon, M.: The M3-competition: results, conclusions and implications. Int. J. Forecast. **16**, 451–476 (2000). https://doi.org/10.1016/S0169-070(00)00057-1
20. McKee, M., Zatoski, W.: How the cardiovascular burden of illness is changing in eastern Europe. Evid.-Based Cardiovasc. Med. **2**(2), 39–41 (1998)
21. Watson, P.: Explaining rising mortality among men in Eastern Europe. Soc. Sci. Med. **41**(7), 923–934 (1995). https://doi.org/10.1016/0277-9536(94)00405-I
22. Fund Pension Protection: The Pensions Regulator (2006), The Purple Book: DB Pensions Universe Risk Profile (2008)
23. Lundström, H., Qvist, J.: Mortality forecasting and trend shifts: an application of the Lee-Carter model to Swedish mortality data. Int. Statist. Rev. **72**, 37–50 (2004). https://doi.org/10.1111/j.1751-5823.2004.tb00222.x
24. Tuljapurkar, S., Li, N., Boe, C.: A universal pattern of mortality decline in the G7 countries. Nature **405**(6788), 789–792 (2000). https://doi.org/10.1038/35015561
25. Wang, D., Lu, P.: Modelling and forecasting mortality distributions in England and Wales using the Lee-Carter Model. J. Appl. Statist. **32**, 873–885 (2005). https://doi.org/10.1080/02664760500163441
26. Felipe, A., Guillén, M., Perez-Marin, A.M.: Recent mortality trends in the Spanish population. Br. Actuar. J. **4**(8), 757–786 (2002). https://doi.org/10.1017/S1357321700003901
27. Lovász, E.: Analysis of Finnish and Swedish mortality data with stochastic mortality models. Eur. Actuar. J. **2**(1), 259–289 (2011). https://doi.org/10.1007/s13385-011-0039-8

28. Cairns, A.J.G., Blake, D., Dowd, K., Coughlan, G.D., Epstein, D., Khalaf-Allah, M.: Mortality density forecasts: an analysis of six stochastic mortality models. Insur.: Math. Econ. **3**(48), 355–367 (2011). https://doi.org/10.1016/j.insmatheco.2010.12.005
29. Cairns, A.J.G., Blake, D., Dowd, K.: A two-factor model for stochastic mortality with parameter uncertainty: theory and calibration. J. Risk Insur. **73**(4), 687–718 (2006). https://doi.org/10.1111/j.1539-6975.2006.00195.x
30. Macdonald, A., Gallop, A., Miller, K., Richards, S., Shah, R., Willets, R.: Stochastic projection methodologies: Lee-Carter model features, example results and implications. Continuous Mortality Investigation Bureau, Working Paper No. 25 (2007)
31. Renshaw, A.E., Haberman, S.: A cohort-based extension to the Lee-Carter model for mortality reduction factors. Insur.: Math. Econ. **38**(3), 556–570 (2006). https://doi.org/10.1016/j.insmatheco.2005.12.001
32. Wang, X., Mueen, A., Ding, H., Trajcevski, G., Scheuermann, P., Keogh, E.: Experimental comparison of representation methods and distance measures for time series data. Data Min. Knowl. Disc. **26**, 275–309 (2013). https://doi.org/10.1007/s10618-012-0250-5
33. Wilmoth, J.R., Andreev, K., Jdanov, D., Glei, D.A., Boe, C., Bubenheim, M., Philipov, D., Shkolnikov, V., Vachon, P.: Methods protocol for the human mortality database. University of California, Berkeley, and Max Planck Institute for Demographic Research, Rostock (2017). http://www.mortality.org/Public/Docs/MethodsProtocol.pdf. Version 6: Accessed 27 Nov 2017

Discrete Models and Computational Intelligence

Robust On-Line Streaming Clustering

Omar A. Ibrahim, Yizhuo Du, and James Keller(✉)

Electrical Engineering and Computer Science Department,
University of Missouri, Columbia, MO 65211, USA
{oai9bc, ydypb}@mail.missouri.edu,
kellerj@missouri.edu

Abstract. With the explosion of ubiquitous continuous sensing, on-line streaming clustering continues to attract attention. The requirements are that the streaming clustering algorithm recognize and adapt clusters as the data evolves, that anomalies are detected, and that new clusters are automatically formed as incoming data dictate. In this paper, we extend an earlier approach, called Extended Robust On-Line Streaming Clustering (EROLSC), which utilizes both the Possibilistic C-Means and Gaussian Mixture Decomposition to perform this task. We show the superiority of EROLSC over traditional streaming clustering algorithms on synthetic and real data sets.

Keywords: Streaming clustering · Outlier detection · Change detection

1 Introduction

Monitoring systems, the internet of things (IoT), and mining content from social media are new emerging applications that rely on processing large amounts of streaming data. For any data analytics technique to be applied on these applications, it has to be unsupervised, online and temporal, i.e., adaptive over time. Clustering is a data mining technique that searches for specific structures on streaming data and detects abnormal patterns in the data [1]. To find clusters in a dataset, clustering algorithms usually require multiple runs over the data, which necessitates all of the dataset to be available before running the algorithm. In a streaming data problem, there is a desire to learn structure as the data arrives instead of waiting for processing by standard techniques. Therefore, developing efficient streaming (online) clustering algorithms is not an easy task.

Online clustering algorithms can be divided into two groups [2]. The first group is general clustering algorithms which do not assume any ordering on the data stream and work on any streaming data, such as sequential k-means (sk-means) [3]. Algorithms of this group requires the number of clusters to be known in advance. On the other hand, the second group relies on the assumption that close observations in time are highly related. Consequently, online clustering algorithms from this group assume natural ordering on the data stream as in time series. These algorithms have a change detection technique to detect new emerging clusters in the streaming information. This paper is an extension of our previous work [4]. In that paper, we only tested the algorithm with synthetic datasets that mimic the behavior of elder adults. Here, we extend the original algorithm by first investigating the ability of our algorithm to detect multiple emerging

J. Medina et al. (Eds.): IPMU 2018, CCIS 853, pp. 467–478, 2018.
https://doi.org/10.1007/978-3-319-91473-2_40

structures at the same time, whereas in [4], we looked for one new cluster only. Also, we use only online incremental update for all the parameters on the algorithm. Consequently, new data points are used to update the parameters and then removed from memory, which makes our algorithm applicable for big data applications. This new algorithm is called Extended Robust On-Line Streaming Clustering (EROLSC). Finally, we test EROLSC with synthetic and real-life datasets, and compare the clustering results with sk-means, Basic Sequential clustering algorithm (BSAS) [5], and Modified Basic Sequential clustering algorithm (MBSAS) [6]. The next section describes background information and related work. In Sect. 3, we present the datasets used on the evaluation process. In Sect. 4, we describe our online clustering algorithm and the other algorithms used in this paper. Section 5 shows numerical evaluation of our method and comparisons to previous approaches. A summary and conclusions are given in Sect. 6.

2 Background

There are two main categories to cluster streaming data [7]. The first category uses a window (S) of the data stream and clusters data points in S using batch-clustering techniques. Clustering results of adjacent windows are combined to get the final clustering results [8–10]. Computational complexity is one drawback of algorithms in this category due to the multiple passes over the data in each window. The second category uses incremental learning techniques and are known as online clustering or streaming clustering algorithms [11, 12]. After initialization, data points are processed one at a time, which makes them good candidates for big data. Online clustering algorithms that assume natural ordering of the data streams have two main parts. The first part is a change detection mechanism that helps the algorithm in identifying new structures in the data. The second part is an adaptive model for the clusters [2].

Change detection is a major part of the online clustering algorithm because it detects new emerging structures and finds outliers on the data streams. In [13], the change detection is based on the violation of the exchangeability condition using a randomized power Martingal that makes it unsuitable for time series.

The detection mechanism used in this paper is based on possibilistic c-means clustering and cluster dispersion. To the best of our knowledge, all existing online clustering algorithm look for one new structure at a time. However, EROLSC can detect multiple emerging structures at the same time because outliers are clustered with more than one cluster and only dense regions in the outlier set are flagged as new clusters.

3 Incremental Clustering Algorithms (Sequential Algorithms)

Sequential clustering algorithms form one approach to produce a single clustering by iterating through subsets of the data once or a few times. In our evaluation, we examine sk-means, BSAS, and MBSAS algorithms to compare with our algorithm. In the next section, we provide a brief overview of these incremental clustering algorithms.

3.1 Extended Robust On-Line Streaming Clustering Algorithm

In [4], we proposed a streaming clustering algorithm based on Gaussian Mixture Models (GMM) combined with possibilistic fuzzy clustering. The idea is to combine the Possibilistic C-Means (PCM) [14] and the Automatic Merging Possibilistic Clustering Method (AMPCM) [15] to initialize the cluster structure in a window S, initializing the GMM as can be seen in the initialization of Fig. 5. PCM is used to detect anomalies in that first window, S. When a new data point x_{n+1} arrives at time n + 1, its Mahalanobis distance is computed to all Gaussians as in Eq. 1. If the minimum distance falls within pre-specified threshold, x_{n+1} is incorporated into the winning Gaussian. The mean and covariance of the winning Gaussian are incrementally updated using Eqs. 2 and 3. After updating the Gaussian parameters, x_{n+1} is removed from the records.

$$d = \sqrt{(x - \mu)^T \Sigma^{-1} (x - \mu)} \quad (1)$$

$$\mu_{new} = \mu_{old} + \frac{x_{n+1} - \mu_{old}}{|\mu_{new}|} \quad (2)$$

```
Input: X - set of data points, choose a window size S for
initialization, select the minimal number of feature vectors
in a Gaussian cloud M, set the distance threshold Td, set
the membership threshold Tn, and set the outlier threshold
To = Td;
Initialization:
Compute the possibilistic partition UCXN of the initialization
data of size S using the PCM;
Find feature vectors whose memberships to all the clusters
are lower than Tn and log them into anomaly history;
Cluster the rest of the data using the AMPCM;

for each cluster do:
    if the number of feature vectors > M
        Calculate its mean and covariance for their
        corresponding Gaussian component;
        Set the counter of that cluster n_k = number
        of data points;
        Delete the data record in the cluster;
    else
        Log all data into the anomaly history ;
    endif
endfor

Update:
for each x_i (i = S + 1: N) do:
    Calculate x_i's Mahalanobis distances to each of
    the Gaussian clouds and find the minimum
    if the minimal distance < Td
        Incrementally update the mean and
        covariance of the winning Gaussian
        cloud (o) using equations 1 and 2;
        n_o = n_o + 1;
        Delete the data record of x_i;
    else
        Log x_t into anomaly history;
    endif
    Examine the anomaly history to see if any data falls into the
    updated Gaussian cloud:
    Calculate Mahalanobis distances of all data in anomaly
    history to each of the Gaussian clouds and find the
    minimum;

    if the minimal distance of any data in anomaly history < Td
        Incrementally update the mean and covariance of
        the corresponding Gaussian cloud using equations 1
        and 2;
        Update the counters of each Gaussian cloud;
        Delete the data record of the data in anomaly history
    endif

    Examine the anomaly history for an emergent new behavior
    pattern:
    Compute the possibilistic partition U of the anomaly record
    using the PCM with C ≥ 10;
    Find feature vectors whose memberships to all the clusters
    are lower than Tn and log them into anomaly history;
    Cluster the rest of the data using the AMPCM;
    Compute the dispersion of each of the detected clusters;

    for each detected cluster do:
        if the cluster is legitimate && the number of
        feature vectors on the cluster > M
            Spawn a new normal Gaussian component;
            Set the counter of the new cluster n_k =
            number of data points;
            Delete the data record of the cluster from
            the anomaly history;
        endif
    endfor

endfor
```

Fig. 1. Pseudo code of EROLSC

$$\Sigma_{new} = \frac{(|\mu_{new}| - 1) * \Sigma_{\text{old}} + (x_{n+1} - \mu_{\text{old}})^{\text{T}}(x_{n+1} - \mu_{\text{new}})}{|\mu_{new}|} \quad (3)$$

where $|\mu_{new}|$ is the cardinality of the winning cluster, μ and Σ are mean and covariance of the cluster.

On the other hand, the new input vector is flagged as an outlier and saved on the anomaly list if it does not meet the threshold. The points in the anomaly history may or may not indicate the emergence of a new cluster. We check the anomaly list in two different ways. First, we compute the Mahalanobis distance between the outliers and cluster centers. Points are assigned to their closest Gaussians if they are within a pre-specified threshold as can be noticed on Fig. 5 (the cluster has "grown" to encompass what was initially an anomaly). Second, we look for single or multiple emerging structures by clustering the outliers as shown in the pseudo code in Fig. 1. See [4] for more detailed description of the basic algorithm, along with details on initialization and new cluster formation. One significant feature of EROLSC is that the PCM can be used with C = 1, or with a larger value of C causing co-incident clusters if there are fewer actual groups. It worth mentioning that our approach is incremental, wherein new data are used to update the clustering parameters and then removed from memory. We keep cluster representatives such as means, covariance matrices, cluster cardinalities, as well as the outlier data points.

3.2 Sequential k-Means (sk-Means)

Sk-means is a well-known clustering algorithm, which was introduced by Macqueen [3] and the pseudo code can be seen in Fig. 2. Different forms of sk-means have been introduced in the literature [9]. One drawback of sk-means is that number of clusters, k, needs to be stated in advance in Macqueen's algorithm. It can be initialized in different ways: selecting the first k data points, randomly selecting k data points from a window of size S (S < N), or randomly selecting k data points from the whole dataset (size N). We use the second way to initialize the prototypes because we use a window size S to initialize EROLSC, which leads to a fair comparison. After that, those k data points

Input: X - set of data points, k - number of clusters, window size S;
Note: ||.|| is the Euclidean norm;
Initialization: Select cluster centers V_k as k-random data points from the window size S;
$V_k = \{s_1, s_2, \ldots, s_k\}$, where s_k is a random sample from S;
Set the counter for each cluster $\{n_1, n_2, \ldots, n_k\} \in 1^k$
for each x_i *in the stream* **do**

$$o = argmin_{o \in \{1,\ldots,k\}} \|v_o - x_i\|;$$
$$n_o = n_o + 1;$$
$$v_o = v_o + \frac{(x_i - v_o)}{n_o};$$

endfor

Fig. 2. Pseudo code of sequential k-means algorithm [3]

represent the cluster centers, $V_k = \{v_1, v_2, \ldots, v_k\}$, each with cluster cardinality of 1. When a new data point arrives at time n + 1, its distance is computed to all k prototypes, and it is assigned to the closest cluster center. Then the parameters of winning cluster are updated as shown in the pseudo code in Fig. 2.

3.3 Basic Sequential Algorithm (BSAS)

BSAS is a basic clustering technique where data points are presented to the algorithm only once and number of clusters is not known a priori [5]. Clustering results rely on the dissimilarity measure d (x, V), dissimilarity threshold Θ, and the number of maximum clusters allowed, q. The first data point is used to initialize the first cluster and it represents its cluster center v_1. When a new data point (X_{n+1}) comes in at time n + 1, the algorithm computes the distance between the new data point and existing clusters prototypes. If the distance to the closest cluster is within the Θ and maximum number of clusters (q) is not met, the new data point is assigned to the closest cluster. Then, the parameters of winning cluster are updated as can be seen in the pseudo code, shown in Fig. 3(a). Otherwise, the new data point spawns a new cluster.

```
Input: X - set of data points, select Θ and q
thresholds;
Initialization: Select the first data point as
the first cluster center:
v_1 = {x_1};
Set number of clusters m = 1;
Set the counter for each cluster n_1 = 1;

for each x_i in the stream from 2 to N do
        o = argmin_{o∈{1,...,m}} ||v_o − x_i|| ;
        if d (x_i, v_o ) > Θ and m < q
                Create a new cluster
                m = m + 1;
                v_m = {x_i};
                n_m = 1;
        else

                Update  cluster o
                parameters
                n_o = n_o + 1;
                v_o = v_o + (x_i − v_o) / n_o;

        endif

endfor
```

(a)

```
Input: X - set of data points, select Θ and q
thresholds;
Initialization: Select the first data point as
the first cluster center:
v_1 = {x_1};
Set number of clusters m = 1;
Set the counter for each cluster n_1 = 1;
for each x_i in the stream from 2 to N do
        o = argmin_{o∈{1,...,m}} ||v_o − x_i|| ;
        if d (x_i, v_o ) > Θ and m < q
                Create a new cluster
                m = m + 1;
                v_m = {x_i};
                n_m = 1;
        endif
endfor

for each x_i in the stream from 1 to N do
        if x_i has not assigned to a cluster
        O = argmin_{o∈{1,...,m}} ||v_o − x_i|| ;
        Update the cluster o parameters
        n_o = n_o + 1;
        v_o = v_o + (x_i − v_o) / n_o;
        endif
endfor
```

(b)

Fig. 3. Pseudo codes of: (a) BSAS algorithm [5], (b) MBSAS algorithm [6]

3.4 Modified Basic Sequential Algorithm (MBSAS)

MBSAS is a modified version of BSAS where it runs through the data samples twice [6]. It overcomes the drawback of BSAS where a sample is assigned to a cluster before all the clusters have been created. In the first phase, the clusters prototypes are determined by assigning only one data point to each cluster. The second phase of the algorithm assigns the remaining data samples to the nearest cluster center. The pseudo code of this algorithm is shown in Fig. 3(b).

4 Datasets

Synthetic and real-life datasets are used in this paper, all of which are presented in a streaming on-line fashion. The first synthetic dataset, S1, has 490 instances in 2 dimensional space, generated from five Gaussian distributions as in Fig. 4(a). S1 has five clusters (two with 100 samples each and 3 with 80 instances) and random noise (50 samples). The first real-life dataset used in our evaluation is the LG dataset, which is a collection of weather station nodes in the Le Genepi (LG) region in Switzerland [16]. We use two weeks of data at node 18 starting from October 10, 2007. Average surface

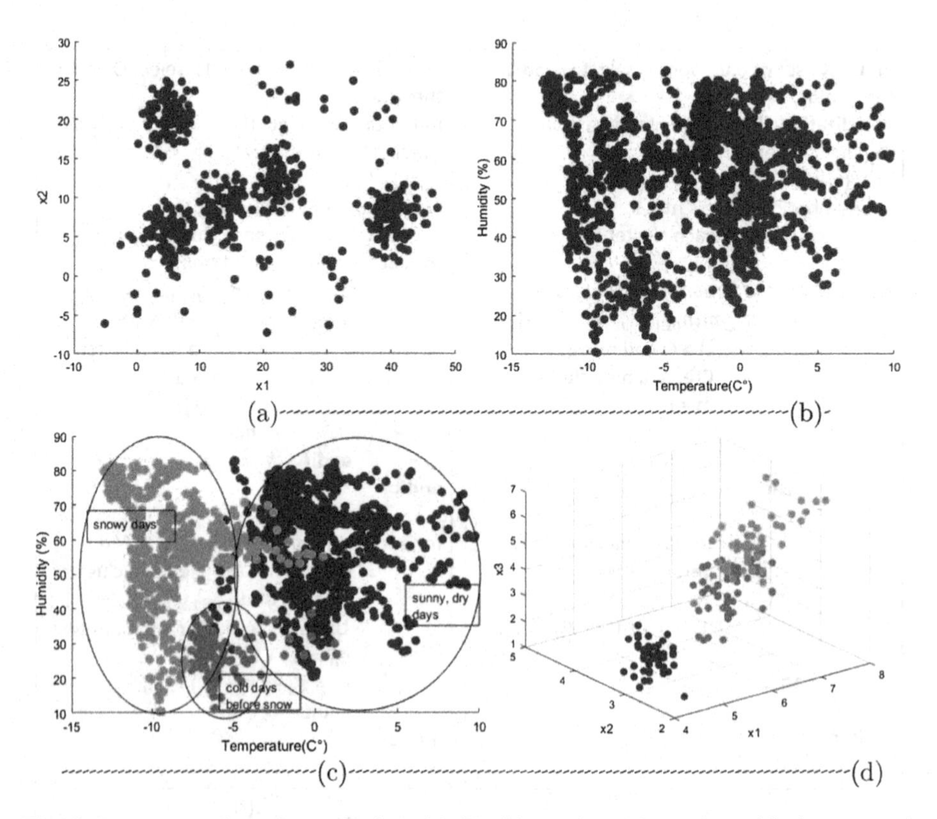

Fig. 4. Datasets used on the evaluation: (a) S1, (b) weather, (c) weather with the expected number of clusters, (d) Iris data

temperature (T) and humidity (H) readings over 10-min intervals were used to create a two dimensional feature vector $\{x_i\} = \{(T_i, H_i)\}$. The scatter plot of the data can be seen in Fig. 4(b). By looking at the scatter plot, it does not provide clear visual evidence about number of clusters in the LG data. Therefore, the imagery information from the site is used to show that there is a snowy day during the two week period. A windy and cold day precedes the snow. For that reason, we consider the LG data to have three different events: sunny days before and after the snow, cold front moving in, and the snowy day as in the "ground truth" seen in Fig. 4(c). We use k = 3 as the number of clusters for sk-means whereas EROLSC finds the expected number of clusters. BSAS and MBSAS on the other hand rely on the distance threshold to find clusters. We try multiple distance threshold values and select the best results (number of clusters) for comparative purposes.

The second real-life dataset is the wine dataset from UCI [17]. These data are the results of a chemical analysis of wines grown in the same region in Italy but derived from three different cultivars. The analysis determined the quantities of 13 constituents found in each of the three types of wines. In the Wine data, the number of instances are 59, 71 and 48 in classes 1, 2 and 3, respectively. The third real-life dataset is the Iris dataset [17]. It has three classes with 50 instances each in 4-dimensional space. Table 1 shows a summary.

Table 1. Summary characteristics of the datasets used in the evaluation.

Dataset	# instances	# dimensions	# clusters	Noise	Labeling
S1	490	2	5	50	Exact
Weather	1817	2	3	Unknown	Our estimate
Wine	178	13	3	None	Exact
IRIS	150	4	3	None	Exact

5 Experimental Results

5.1 Evaluation

During the evaluation process, we compare EROLSC with sk-means, BSAS, and MBSAS. The setting of each algorithm will be described first. Then, synthetic and real-life datasets are used to discuss the performance of the algorithms. Accuracy is used to compare the performance if class labels are available. Visual inspection in terms of finding the desired number of clusters and/or detecting outliers is used when there are no labels.

5.2 Parameter Settings

Sk-means is fed with number of clusters (K) expected in the dataset because it needs a priori knowledge of the number of clusters. We randomly select K data samples as the initial clusters centers from a window size S. BSAS and MBSAS are fed with the maximum number of desired clusters (K) and distance threshold, which we get by 3 experimenting and selecting the best results. For EROLSC, we use a window of size S to initialize the algorithm and we rely on the recommended table in [4] to select the distance threshold. Similar window size is used for sk-means.

5.3 Clustering Results

To demonstrate the ability of EROLSC in detecting multiple clusters at the same time, we use S1 dataset. In S1, data points from cluster 1 come first then data points from cluster 2, 3 and 4 arrive randomly. This enables us to evaluate the robustness of the algorithm to detect multiple structures at the same time. Figure 5(a) shows the initialization of the algorithm where PCM is used to detect outliers (red points). When the outliers become dense enough as in Fig. 5(b), the algorithm in [4] detects only one cluster because it looks for one new emerging structure as in Fig. 5(c). Clusters 2, 3 and 4 are merged into one cluster which could limit the ability of using the algorithm in real applications where multiple structures could emerge at the same time. EROLSC on the

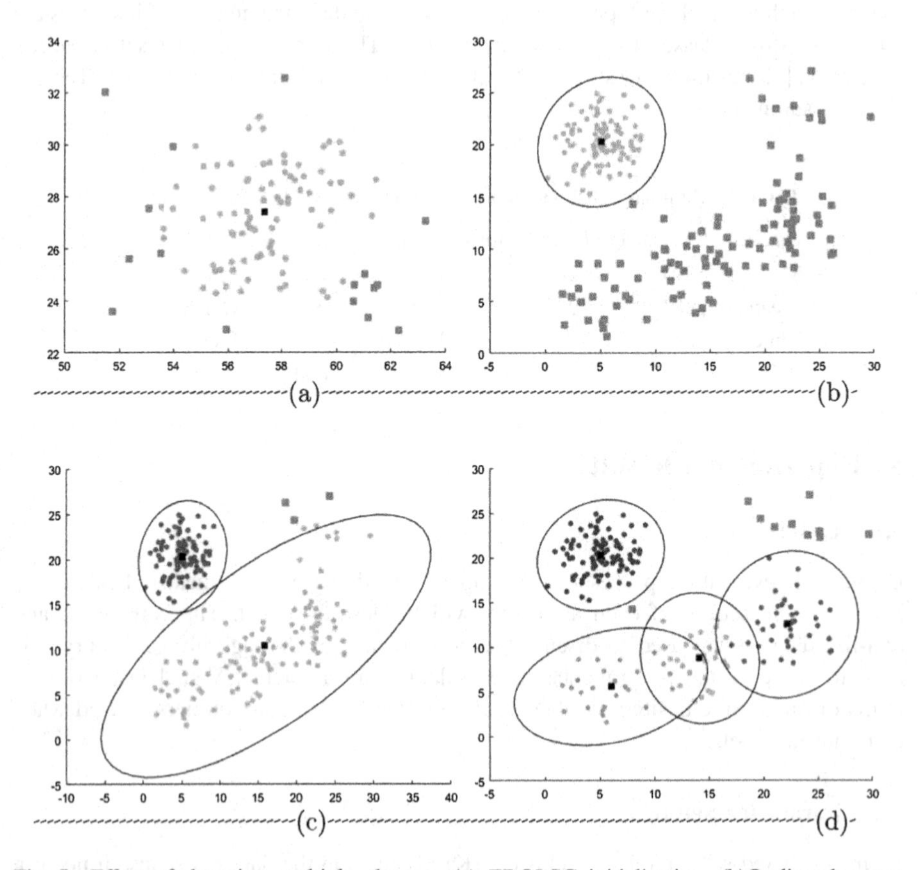

Fig. 5. Effect of detecting multiple clusters: (a) EROLSC initialization, (b)Outliers become dense enough to detect new cluster or clusters, (c) Using our algorithm in [4] to look for new structures in the outliers, (d) Using EROLSC to detect multiple structures in the outliers. (Color figure online)

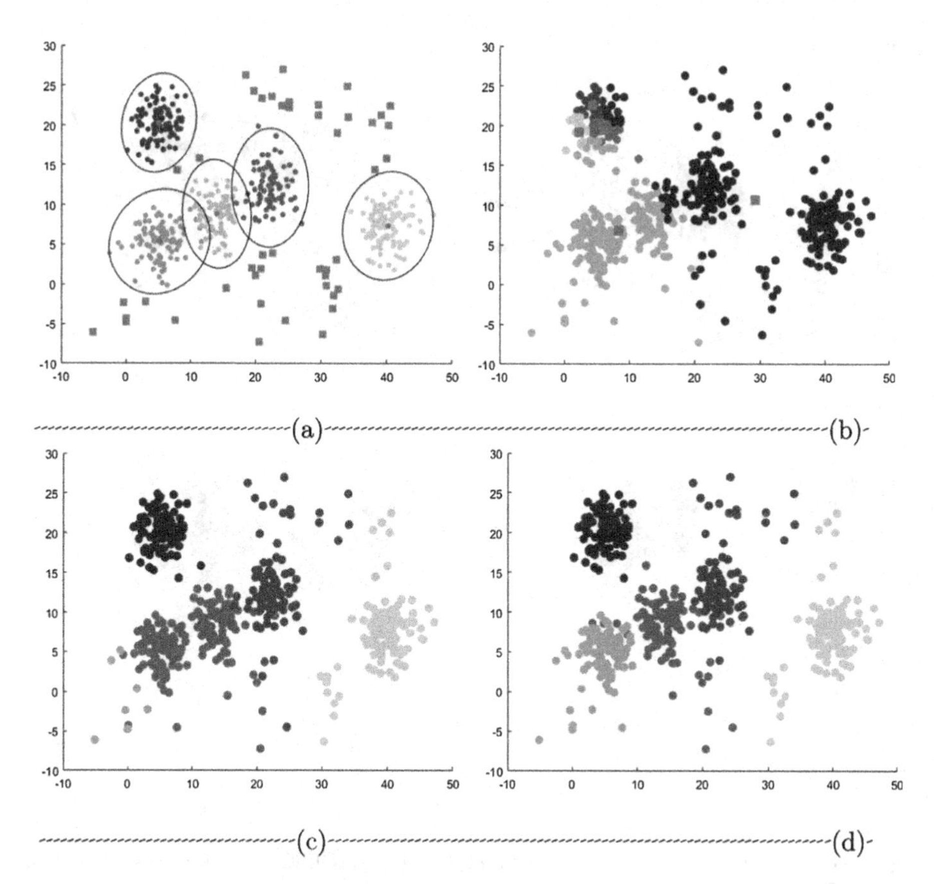

Fig. 6. Final hardened clustering results on S1 dataset for: (a) EROLSC, (b) sk-means algorithm, (c) BSAS algorithm, (d) MBSAS algorithm.

other hand detects the three structures as can be seen in Fig. 5(d) because it takes into account that multiple clusters could happen a similar time.

The clustering results on the first dataset S1 are shown in Fig. 6. This dataset has five different clusters where it starts: with one cluster, three clusters form at the same time, the fifth cluster forms after that and outlier's data points arrive in between. Result of our algorithm is shown in Fig. 6(a) where it detects all the clusters and finds 45 out of 50 outliers. Sk-means result can be seen in Fig. 6(b) where 3 centers get trapped in the first cluster due to the initialization. BSAS and MBSAS have better clustering results compared to sk-means as can be noticed in Fig. 6(c) and Fig. 6(d). MBSAS detected all the 5 clusters correctly. However, it fails to detect outliers because each sample must be assigned to the nearest prototype.

The weather dataset is more complicated and less obvious to cluster. EROLSC finds three clusters and matches the ground truth based on the explanation earlier. Figure 7(a) shows the cluster structures of our algorithm, which follows the data evolution as time progresses. We notice that EROLSC flagged some of the sunny cluster data points as outliers due to the way the data is presented to the algorithm. The

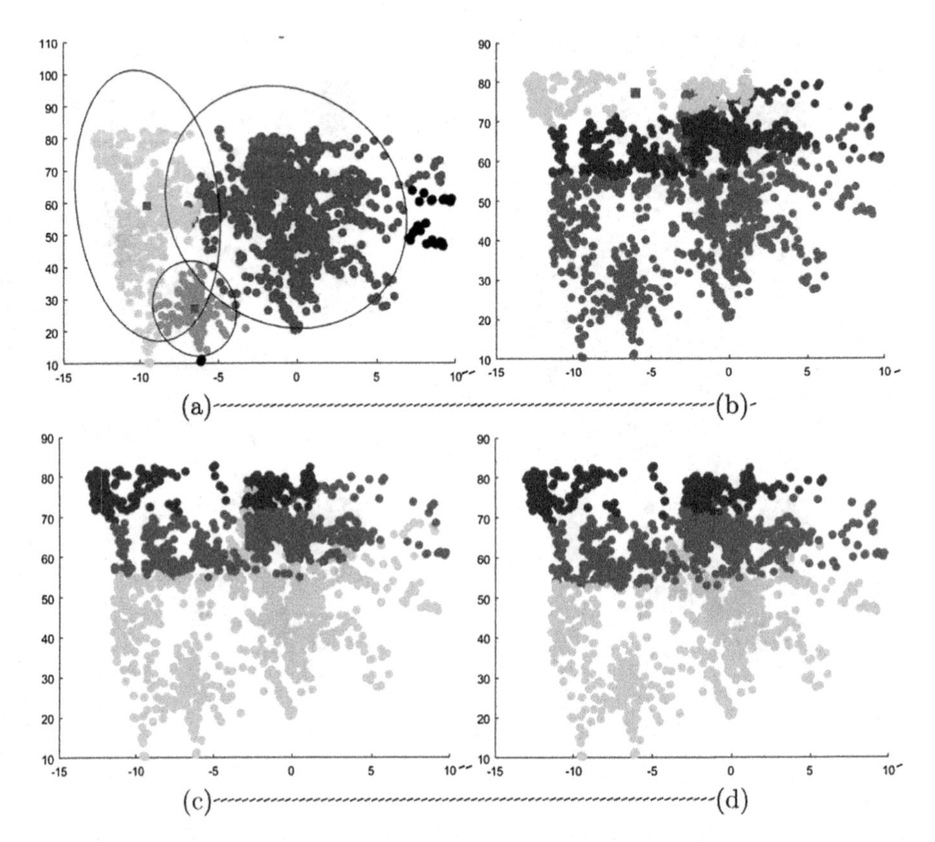

Fig. 7. Clustering results on weather dataset for: (a) EROLSC, (b) sk-means algorithm, (c) BSAS algorithm, (d) MBSAS algorithm.

algorithm initialized itself at the "sunny days" cluster and new samples dragged the cluster center to the left. After that, these data points arrived and are flagged as outliers because they are far from the cluster center. Sk-means detects three clusters as well, but it tries to achieve good separation between clusters as can be seen in Fig. 7(b). The results of BSAS and MBSAS are not much different from sk-means, where they also look for separated clusters as in Fig. 7(c) and (d). We conclude that only EROLSC finds the three expected clusters (sunny, windy and cold before snow, and snowy days).

Both the wine and Iris datasets are labeled, so accuracy can be used to evaluate the performance. EROLSC achieves the highest performance on the Iris dataset with 94% accuracy as in Table 2. The miss-assigned samples (around 9) are those which belong to one class, but they are closer to another, as depicted in Fig. 4(d). BSAS and MBSAS achieved better accuracy compared to sk-means due to the initialization, we think. Similarly, EROLSC has the highest accuracy in the Wine dataset where it finds the exact number of classes and misclassifies a few data points. This is normal for a clustering algorithm because our main goal is to detect the desired number of structures in the data. Sk-means, BSAS, and MBSAS do poorly on the wine dataset as expected, based on their performance on other datasets.

Table 2. Accuracy of the final clusters to match classification labels.

Dataset	EROLSC	Sk-means	BSAS	MBSAS
Iris	94%	49%	67%	67%
Wine	92%	51%	33%	33%

6 Conclusion

Streaming clustering, directed by change detection, can identify structures in online data whereas traditional batch approaches fail. General online clustering algorithms such as sk-means, BSAS and MBSAS suffer from the same drawback of batch clustering due to their lack of having an effective change detection mechanism. Our approach in this paper, the Extended Robust On-Line Streaming Clustering Algorithm has a change detection mechanism by flagging data points that are far from the existing state or states, and monitors them over time. If the outliers become dense enough, they are added to the current states as new clusters. Otherwise, they are treated as anomalies unless they fall into one of the data structures over time. Having this property of dealing with outliers enables EROLSC to identify anomalies during the process of clustering. For instance, the S1 dataset has multiple outliers, which was only detected by our algorithm. EROLSC detects the expected structures in the real-life datasets. In addition, it outperforms sk-means, BSAS and MBSAS in all datasets tested. More research needs to be done to automatically determine correct parameter settings and to adapt EROLSC to Big Data applications.

References

1. Silva, J.A., Faria, E.R., Barros, R.C., Hruschka, E.R., de Carvalho, A.C.P.L.F., Gama, J.: Data stream clustering: a survey. ACM Comput. Surv. **46**(1), 13–31 (2013)
2. Moshtaghi, M., Leckie, C., Bezdek, J.C.: Online clustering of multivariate time-series. In: Proceedings of SIAM International Conference on Data Mining, pp. 360–368 (2016)
3. MacQueen, J.: Some methods for classification and analysis of multivariate observations. In: Proceedings of the Fifth Berkeley Symposium on Mathematical Statistics and Probability, vol. 1, no. 14, pp. 281–297 (1967)
4. Ibrahim, O.A., Shao, J., Keller, J.M., Popescu, M.: A temporal analysis system for early detection of health changes. In: IEEE International Conference on Fuzzy Systems, FUZZ-IEEE, pp. 186–193 (2016)
5. Theodoridis, S., Koutroumbas, K.: Basic Sequential Algorithmic Scheme (BSAS). Academic Press, London (1999)
6. Theodoridis, S., Koutroumbas, K.: Pattern Recognition. Academic Press, London (2006)
7. Guha, S., Meyerson, A., Mishra, N., Motwani, R., O'Callaghan, L.: Clustering data streams: theory and practice. IEEE Trans. Knowl. Data Eng. **15**(3), 515–528 (2003)
8. Aggarwal, C.C., Han, J., Wang, J., Yu, P.S.: On clustering massive data streams: a summarization paradigm. In: Aggarwal, C.C. (ed.) data streams. Advances in Database Systems, vol. 31, pp. 9–38. Springer, Boston (2007). https://doi.org/10.1007/978-0-387-47534-9_2

9. Ailon, N., Jaiswal, R., Monteleoni, C.: Streaming k-means approximation. In: Neural Information Processing Systems, vol. 22, pp. 10–18 (2009)
10. Salehi, M., Leckie, Christopher A., Moshtaghi, M., Vaithianathan, T.: A relevance weighted ensemble model for anomaly detection in switching data streams. In: Tseng, V.S., Ho, T.B., Zhou, Z.-H., Chen, A.L.P., Kao, H.-Y. (eds.) PAKDD 2014. LNCS (LNAI), vol. 8444, pp. 461–473. Springer, Cham (2014). https://doi.org/10.1007/978-3-319-06605-9_38
11. Ackerman, M., Dasgupta, S.: Incremental clustering: the case for extra clusters. In: Advances in Neural Information Processing Systems, pp. 307–315 (2014)
12. Angelov, P., Zhou, X.: Evolving fuzzy-rule-based classifiers from data streams. IEEE Trans. Fuzzy Syst. **16**(6), 1462–1475 (2008)
13. Ho, S.-S.: A martingale framework for concept change detection in time-varying data streams. In: Proceedings of the International Conference on Machine Learning, pp. 321–327 (2005)
14. Krishnapuram, R., Keller, J.M.: A possibilistic approach to clustering. IEEE Trans. Fuzzy Syst. **I**(2), 98–110 (1993)
15. Yang, M.-S., Lai, C.-Y.: A robust automatic merging possibilistic clustering method. IEEE Trans. Fuzzy Syst. **19**(1), 26–41 (2011)
16. SensorScope (2007). http://lcav.epfl.ch/page-86035-en.html
17. Lichman, M.: UCI Machine Learning Repository. School of Information and Computer Science, University of California, Irvine, CA (2013). http://archive.ics.uci.edu/ml

T-Overlap Functions: A Generalization of Bivariate Overlap Functions by t-Norms

Hugo Zapata[1,2(✉)], Graçaliz Pereira Dimuro[3,4], Javier Fernández[4,5], and Humberto Bustince[4,5]

[1] Facultad de Ciencias, Universidad Central de Venezuela, Avenida Los Ilustres, Caracas 1020, Venezuela
hugo.zapata@unisucre.edu.co
[2] Departamento de Matemáticas, Universidad de Sucre, Sincelejo, Colombia
[3] Centro Ciências Computacionais, Universidade Federal do Rio Grande, Campus Carreiros, Rio Grande 96201-900, Brazil
[4] Institute of Smart Cities, Universidad Publica de Navarra, Campus Arrosadía, 31006 Pamploma, Spain
[5] Depto. of Automática y Computación, Universidad Publica de Navarra, Campus Arrosadía, 31006 Pamplona, Spain

Abstract. This paper introduces a generalization of overlap functions by extending one of the boundary conditions of its definition. More specifically, instead of requiring that "the considered function is equal to zero if and only if some of the inputs is equal to zero", we allow the range in which some t-norm is zero. We call such generalization by a t-overlap function with respect to such t-norm. Then we analyze the main properties of t-overlap function and introduce some construction methods.

Keywords: Aggregation function · Overlap function · t-norm

1 Introduction

The notion of overlap function [1,4,7–10] has shown itself very useful to deal with situations in which it is necessary to determine up to what extent a given element belongs to one or several classes whose boundaries are not crisp. It has been used, e.g., in image processing [12], classification problems [13,14] and decision making [11].

Our goal here is to generalize the notion of overlap function by relaxing one of the boundary condition. In particular, instead of demanding that "the considered function is equal to zero if and only if some of the inputs is equal to zero", we allow for some kind of threshold, defined in terms of a t-norm T. We call such generalization by a t-overlap function with respect to T.

We notice that, this simple generalization allows us to state several interesting properties, which may allow for application in fuzzy rule-based system in order to discard bad rules when computing the compatibility degree. Section 2 presents some preliminary concepts. In Sect. 3, besides studying the main properties, we also propose some construction methods. Section 4 is the Conclusion.

J. Medina et al. (Eds.): IPMU 2018, CCIS 853, pp. 479–489, 2018.
https://doi.org/10.1007/978-3-319-91473-2_41

2 Preliminaries

This section aims at introducing the background necessary to understand the paper.

Definition 1. *A fuzzy negation is a function $N\colon [0,1] \to [0,1]$ satisfying:* ***(N1)*** *the boundary conditions: $N(0) = 1$ and $N(1) = 0$;* ***(N2)*** *N is decreasing: if $x \leq y$ then $N(y) \leq N(x)$.*

A fuzzy negation N is said to be strong if: $\forall x \in [0,1] : N(N(x)) = x$ (the involutive property). The standard negation or the Zadeh's negation is given by $N_Z(x) = 1 - x$.

Definition 2. *[3,15] A function $A : [0,1]^n \to [0,1]$ is said to be an n-ary aggregation operator if the following conditions hold:*

(A1) *A is increasing*[1] *in each argument: for each $i \in \{1,\ldots,n\}$, if $x_i \leq y$, then $A(x_1,\ldots,x_n) \leq A(x_1,\ldots,x_{i-1},y,x_{i+1},\ldots,x_n)$;*
(A2) *A satisfies the Boundary conditions: $A(0,\ldots,0) = 0$ and $A(1,\ldots,1) = 1$.*

Definition 3. *A t-norm is a bivariate aggregation function $T : [0,1]^2 \to [0,1]$ satisfying the following properties, for all $x, y, z \in [0,1]$:*

(T1) *Commutativity: $T(x,y) = T(y,x)$;*
(T2) *Associativity: $T(x,T(y,z)) = T(T(x,y),z)$;*
(T3) *Boundary condition: $T(x,1) = x$.*

Example of t-norms are the Łukasiewicz and Yager t-norms, defined, respectively, by $T_Ł(x,y) = \max\{0, x + y - 1\}$ and $T_Y(x,y) = \max\{0, 1 - \sqrt{(1-x)^2 + (1-y)^2}$.

An element $x \in]0,1]$ is a non-trivial zero divisor of T if there exists $y \in]0,1]$ such that $T(x,y) = 0$. A t-norm is positive if and only if it has no non-trivial zero divisors, i.e., if $T(x,y) = 0$ then either $x = 0$ or $y = 0$. Examples of continuous and positive t-norms are the minimum and the product t-norms, defined, respectively, by $T_M(x,y) = \min\{x,y\}$ and $T_P(x,y) = xy$.

The main concern of this paper is the concept of overlap function [1,4,7–9,12].

Definition 4. *[4] An overlap function is a bivariate function $O\colon [0,1]^2 \to [0,1]$ satisfying the following properties, for all $x, y \in [0,1]$:*

(O1) *O is commutative: $O(x,y) = O(y,x)$;*
(O2) *$O(x,y) = 0$ if and only if $x = 0$ or $y = 0$;*
(O3) *$O(x,y) = 1$ if and only if $x = y = 1$;*
(O4) *O is increasing;*
(O5) *O is continuous.*

---

[1] In this paper, a increasing (decreasing) function does not need to be strictly increasing (decreasing).

3 Introducing T-Overlap Functions

This section generalizes the concept of overlap functions by changing the condition **(O2)** of Definition 4, namely, the property that requires that, for all $x, y \in [0,1]$ and overlap function $O : [0,1]^2 \to [0,1]$ it holds that $O(x,y) = 0 \Leftrightarrow xy = 0$. In our generalization, we replace the product operation by a t-norm $T : [0,1]^2 \to [0,1]$.

Definition 5. *Let $T : [0,1]^2 \to [0,1]$ be a t-norm. A function $O_T : [0,1]^2 \to [0,1]$ is said to be a t-overlap function with respect to T if the following conditions hold:*

(O_T1) $O_T(x,y) = O_T(y,x)$,
(O_T2) $O_T(x,y) = 0 \Leftrightarrow T(x,y) = 0$,
(O_T3) $O_T(x,y) = 1 \Leftrightarrow x = y = 1$,
(O_T4) O_T *is increasing,*
(O_T5) O_T *is continuous.*

Remark 1. Observe that, considering a fuzzy rule-based system, this generalization allows to discard bad rules when computing the compatibility degree. This is due to the fact that the membership degrees of the input with the antecedents would be low for bad rules and, consequently, t-overlap functions may return 0 instead of a low value, which can mislead the final prediction. Accordingly, we have maintained the third condition, since, intuitively, it is not interesting to give the same value to all the rules whose membership degrees are high, since it may imply a decrease in the predictive power.

Remark 2. Notice that the proposed generalization enlarge the use of overlap function. For example, consider the overlap function $O = \frac{\sqrt{xy}}{\sqrt{xy}+(1-xy)}$, which only becomes zero in the case where $x = 0$ or $y = 0$, by condition **(O2)**, which means that overlap functions are t-overlap functions with respect to t-norms without zero divisors. Our definition overcomes this limitation by changing the condition **(O2)** by the condition (O_T2), where T is a t-norm that can have zero divisors. See, for example, the t-overlap function with respect to the Łukasiewicz t-norm $T_Ł$ given by:

$$O_{T_Ł}(x,y) = \frac{\max\{0,(1+\lambda)(x+y-1)-\lambda xy\}}{\max\{0,(1+\lambda)(x+y-1)-\lambda xy\} + \min\{1, 1-(1+\lambda)(x+y-1)+\lambda xy\}}.$$

Example 1. Let $G : [0,1]^2 \to [0,1]$ be defined by $G(x,y) = (\min\{x,y\})^p$, with $p > 0$, and consider the Łukasiewicz and Yager t-norms, $T_Ł(x,y) = \max\{0, x + y - 1\}$ and $T_Y(x,y) = \max\{0, 1 - \sqrt{(1-x)^2 + (1-y)^2}$. Then, the functions $O^G_{T_Ł}, O^2_{T_Ł}, O_{T_Y} : [0,1]^2 \to [0,1]$, defined by $O^G_{T_Ł}(x,y) = G(x,y)T_Ł(x,y)$, $O^2_{T_Ł}(x,y) = 2^{T_Ł} - 1$ and $O_{T_Y}(x,y) = 2^{T_Y} - 1$ are t-overlap functions whit respect to $T_Ł$ and T_Y.

The previous example may be generalized as the following results:

Remark 3. Let $T : [0,1]^2 \to [0,1]$ be a continuous t-norm. Then T is a t-overlap function with respect to itself.

Proposition 1. *Let $O : [0,1]^2 \to [0,1]$ be an overlap function and $T : [0,1]^2 \to [0,1]$ be a continuous t-norm. Then the function $O_T : [0,1]^2 \to [0,1]$, defined, for all $x, y \in [0,1]$, by $O_T(x,y) = O(x,y)T(x,y)$ is a t-overlap function with respect to T.*

Proof. (O_T1) It is immediate.
(O_T2) For all $x, y \in [0,1]$, it follows that:

$$\begin{aligned} O_T(x,y) = 0 &\Leftrightarrow O(x,y)T(x,y) = 0 \\ &\Leftrightarrow O(x,y) = 0 \ \vee \ T(x,y) = 0 \\ &\Leftrightarrow x = 0 \ \vee \ y = 0 \ \vee \ T(x,y) = 0 \ \text{ by } (\mathbf{O2}) \\ &\Leftrightarrow T(x,y) = 0. \end{aligned}$$

(O_T3) For all $x, y \in [0,1]$, it follows that:

$$\begin{aligned} O_T(x,y) = 1 &\Leftrightarrow O(x,y)T(x,y) = 1 \\ &\Leftrightarrow O(x,y) = 1 \ \wedge \ T(x,y) = 1 \\ &\Leftrightarrow x = 1 \ \wedge \ y = 1 \ \wedge \ T(x,y) = 1 \ \text{ by } (\mathbf{O3}) \\ &\Leftrightarrow T(x,y) = 1. \end{aligned}$$

(O_T4-5) Since both O and T are continuous and increasing, then the results are immediate. □

The previous theorem may be generalized using a special t-norm T' instead of the product between the overlap function O and the t-norm T with which the function O_T is a t-overlap with respect to T.

Proposition 2. *Let $O : [0,1]^2 \to [0,1]$ be an overlap function and $T : [0,1]^2 \to [0,1]$ be a continuous t-norm. For any continuous and positive t-norm $T' : [0,1]^2 \to [0,1]$, one has that the function $O_T : [0,1]^2 \to [0,1]$, defined, for all $x, y \in [0,1]$, by $O_T(x,y) = T'(O(x,y), T(x,y))$, is a t-overlap function with respect to T.*

Proof. (O_T1) It is immediate.
(O_T2) For all $x, y \in [0,1]$, it follows that:

$$\begin{aligned} O_T(x,y) = 0 &\Leftrightarrow T'(O(x,y), T(x,y)) = 0 \\ &\Leftrightarrow O(x,y) = 0 \ \vee \ T(x,y) = 0 \ \text{ Since } T' \text{ is positive} \\ &\Leftrightarrow x = 0 \ \vee \ y = 0 \ \vee \ T(x,y) = 0 \ \text{ by } (\mathbf{O2}) \\ &\Leftrightarrow T(x,y) = 0. \end{aligned}$$

(O_T3) For all $x, y \in [0,1]$, it follows that:

$$\begin{aligned} O_T(x,y)(x,y) = 1 &\Leftrightarrow T'(O(x,y), T(x,y)) = 1 \\ &\Leftrightarrow O(x,y) = 1 \ \vee \ T(x,y) = 1 \\ &\Leftrightarrow x = y = 1 \ \vee \ T(x,y) = 1 \\ &\Leftrightarrow T(x,y) = 1. \end{aligned}$$

($O_T4 - 5$) It is immediate. □

Note that if a t-norm T is positive, then O_T is an overlap function.

Theorem 1. *Let $O_T^1, \ldots, O_T^n : [0,1]^2 \to [0,1]$ be t-overlap functions with respect to a t-norm $T : [0,1]^2 \to [0,1]$ and $\omega_1, \ldots, \omega_n \in [0,1]$ be weights with $\sum_{i=1}^n \omega_i = 1$. Then the function $O_T : [0,1]^2 \to [0,1]$, defined, for all $x, y \in [0,1]$, by $O_T(x,y) = \sum_{i=1}^n \omega_i O_T^i(x,y)$ is also a t-overlap function with respect to T.*

Proof. (O_T1) It is immediate.
(O_T2) For all $x, y \in [0,1]$, it follows that:

$$\begin{aligned} \mathrm{O}_T(x,y) = 0 &\Leftrightarrow \sum_{i=1}^n \omega_i O_T^i(x,y) = 0 \\ &\Leftrightarrow \omega_i O_T^i(x,y) = 0, \forall i = 1, \ldots, n. \end{aligned}$$

Since $\sum_{i=1}^n \omega_i = 1$, then there exists $k \in \{0, \ldots, n\}$ such that $\omega_k \neq 0$, and, thus $O_T^k(x,y) = 0$. By (O_T2), it holds that $T(x,y) = 0$. The reciprocal is analogous.
(O_T3) For all $x, y \in [0,1]$, it follows that:

$$\mathrm{O}_T(x,y) = 1 \Leftrightarrow \sum_{i=1}^n \omega_i O_T^i(x,y) = 1 = \sum_{i=1}^n \omega_i.$$

One has that $\sum_{i=1}^n \omega_i O_T^i(x,y) - \sum_{i=1}^n \omega_i = 0$, i.e., $\sum_{i=1}^n \omega_i (O_T^i(x,y) - 1) = 0$. This means that, for all $i = 1, \ldots, n$, it holds that $\omega_i O_T^i(x,y) - \sum_{i=1}^n \omega_i = 0$. However, since $\sum_{i=1}^n \omega_i \neq 0$, there exist $k \in \{1, \ldots, n\}$ such that $\omega_k \neq 0$. Thus, one has that $O_T^k(x,y) = 1$, and, by (O_T3), it follows that $x = y = 1$. The reciprocal is analogous.
($O_T4 - 5$) It is immediate. □

Let $T : [0,1]^2 \to [0,1]$ be a t-norm and denote $\mathrm{K}_T = \{(x,y) \in [0,1]^2 \mid T(x,y) = 0\}$. Obviously, any t-overlap function with respect to a t-norm T coincides with an overlap function if and only if $\mathrm{K}_T = \{(x,y) \in [0,1]^2 \mid x = 0 \vee y = 0\}$.

Denote by Θ the set of all t-overlap functions with respect of any t-norm T. The following result is immediate.

Theorem 2. *The ordered set $\mathfrak{S} = (\Theta, \leq_\Theta)$ is a lattice, where $\leq_\Theta$ is defined, for all $O_{T_1}, O_{T_2} \in \Theta$, by $O_{T_1} \leq_\Theta O_{T_2}$ if and only if $O_{T_1}(x,y) \leq O_{T_2}(x,y)$, for all $(x,y) \in [0,1]^2$.*

Theorem 3. *Let O_{T_i} be a t-overlap function with respect to the t-norms $T_1, \ldots, T_n : [0,1]^2 \to [0,1]$ and let $\omega_1, \ldots, \omega_n \in [0,1]$ be weights such that $\sum_{i=1}^{n} \omega_i = 1$. If $T = \sum_{i=1}^{n} \omega_i T_i : [0,1]^2 \to [0,1]$ is a t-norm, then O_{T_i} is a t-overlap function with respect to T.*

Proof. $(O_T 1)$ It is immediate.
$(O_T 2)$ $(\Rightarrow)$ Since O_{T_i} is a t-overlap function with respect to the t-norms $T_1, \ldots, T_n$, then, by $(O_T 2)$, for all $i = 1, \ldots, n$, it holds that whenever $O_{T_i}(x,y) = 0$ then $T_i(x,y) = 0$, for all $x, y \in [0,1]$. Then, it follows that $\sum_{i=1}^{n} \omega_i T_i(x,y) = 0$. $(\Leftarrow)$ If $\sum_{i=1}^{n} \omega_i T_i(x,y) = 0$, then, since $\sum_{i=1}^{n} \omega_i \neq 0$, there exists $k = 1, \ldots, n$ such that $\omega_k \neq 0$. It follows that $T_k(x,y) = 0$. Since O_{T_i} is a t-overlap function with respect to the t-norm T_k, one has that $O_{T_i}(x,y) = 0$. It follows that O_{T_i} is a t-overlap function with respect to the t-norm T.
$(O_T 3-5)$ It is immediate. □

Theorem 4. *Let $O_1, O_2 : [0,1]^2 \to [0,1]$ be t-overlap functions with respect to the t-norms $T_1, T_2 : [0,1]^2 \to [0,1]$, respectively. Consider $\omega_1, \omega_2 \in [0,1]$ such that $\omega_1 + \omega_2 = 1$. If $T'[0,1]^2 \to [0,1]$ is a positive t-norm then*

$$O_T(x,y) = \omega_1 O_1(x,y) + \omega_2 O_2(x,y)$$

is a t-overlap function with respect to the t-norm $T : [0,1]^2 \to [0,1]$, defined, for all $x, y \in [0,1]$, by $T(x,y) = T'(T_1(x,y), T_2(x,y))$.

Proof. $(O_T 1)$ It is immediate.
$(O_T 2)$ For all $x, y \in [0,1]$, it follows that:

$$\begin{aligned} O_T(x,y) = 0 &\Leftrightarrow \omega_1 O_1(x,y) + \omega_2 O_2(x,y) = 0 \Leftrightarrow \omega_1 O_1(x,y) = \omega_2 O_2(x,y) = 0 \\ &\Leftrightarrow \omega_1 = 0 \vee O_1(x,y) = 0 \text{ and } \omega_2 = 0 \vee O_2(x,y) = 0. \end{aligned}$$

Now suppose that $\omega_1 \neq 0$. Then one has that $O_1(x,y) = 0$ and, by $(O_T 2)$, it holds that $T_1(x,y) = 0$. It follows that $T(x,y) = T'(T_1(x,y), T_2(x,y)) = 0$. The reciprocal is analogous, taking into account that T' is a positive t-norm.
$(O_T 3)$ For all $x, y \in [0,1]$, it follows that:

$$\begin{aligned} O_T(x,y) = 1 &\Leftrightarrow \omega_1 O_1(x,y) + \omega_2 O_2(x,y) = 1 \\ &\Leftrightarrow \omega_1 O_1(x,y) + \omega_2 O_2(x,y) = \omega_1 + \omega_2 \\ &\Leftrightarrow \omega_1(1 - O_1(x,y)) + \omega_2(1 - O_2(x,y)) = 0 \\ &\Leftrightarrow \omega_1 = 0 \vee 1 - O_1(x,y) = 0 \text{ and } O_2 = 0 \vee 1 - O_2(x,y) = 0 \\ &\Leftrightarrow \omega_1 = 0 \vee O_1(x,y) = 1 \text{ and } O_2 = 0 \vee O_2(x,y) = 1 \\ &\Leftrightarrow \omega_1 = 0 \vee x = y = 1 \text{ and } \omega_2 = 0 \vee x = y = 1. \end{aligned}$$

Now, since $\omega_1 + \omega_2 = 1$ it holds that $x = y = 1$. The reciprocal is immediate.
$(O_T 4-5)$ It is immediate. □

Theorem 5. *The function* $O_T : [0,1]^2 \to [0,1]$ *is a t-overlap function with respect to a t-norm* $T : [0,1]^2 \to [0,1]$ *if and only if*

$$O_T(x,y) = \frac{f(x,y)}{f(x,y) + h(x,y)},$$

for all $x, y \in [0,1]$ *and some functions* $f, h : [0,1]^2 \to [0,1]$ *such that*

(i) f *and* h *are commutative.*
(ii) f *is increasing and* h *is decreasing.*
(iii) $f(x,y) = 0$ *if and only if* $T(x,y) = 0$.
(iv) $h(x,y) = 0$ *if and only if* $x = y = 1$.
(v) f *and* h *are continuous.*

Proof. ($\Rightarrow$) Suppose that O_T is a T-overlap function with respect to a t-norm T. Consider that $O_T(x,y) = f(x,y)$ and $h(x,y) = 1 - f(x,y)$. It is immediate that (i) f and h are symmetric, (ii) f is increasing and h is decreasing and (v) f and h are continuous. (iii) Now, by (O_T2), $f(x,y) = 0$ if and only if $T(x,y) = 0$. (iv) Similarly, by (O_T3), $h(x,y) = 0$ if and only if $f(x,y) = 1$ if and only if $x = y = 1$. Since $f(x,y) + h(x,y) = 1$ then

$$O_T(x,y) = f(x,y) = \frac{f(x,y)}{1} = \frac{f(x,y)}{f(x,y) + h(x,y)}.$$

($\Leftarrow$) Consider two functions $f, g : [0,1]^2 \to [0,1]$ satisfying the conditions (i)-(v), and the function $O_T : [0,1]^2 \to [0,1]$, defined, for all $x, y \in [0,1]$, by

$$O_T(x,y) = \frac{f(x,y)}{f(x,y) + h(x,y)}.$$

(O_T1) It is immediate.
(O_T2) For all $x, y \in [0,1]$, it follows that:

$$O_T(x,y) = 0 \Leftrightarrow \frac{f(x,y)}{f(x,y) + h(x,y)} = 0 \Leftrightarrow f(x,y) = 0 \Leftrightarrow T(x,y) = 0.$$

(O_T3) For all $x, y \in [0,1]$, it follows that:

$$\begin{aligned} O(x,y) = 1 \Leftrightarrow \frac{f(x,y)}{f(x,y) + h(x,y)} &= 1 \Leftrightarrow f(x,y) = f(x,y) + h(x,y) \\ &\Leftrightarrow h(x,y) = 0 \Leftrightarrow x = y = 1. \end{aligned}$$

(O_T4) Let $x, y, z \in [0,1]$ be such that $x \le y$, then $f(x,z) \le f(y,z)$ and $h(y,z) \le h(x,z)$. It follows that

$$\begin{gathered} f(x,z)h(y,z) \le f(y,z)h(x,z) \Rightarrow \\ f(x,z)h(y,z) + f(x,z)f(y,z) \le f(y,z)h(x,z) + f(x,z)f(y,z) \Rightarrow \\ f(x,z)(h(y,z) + f(y,z)) \le f(y,z)(h(x,z) + f(x,z)) \Rightarrow \\ \frac{f(x,z)}{h(x,z) + f(x,z)} \le \frac{f(y,z)}{h(y,z) + f(y,z)} \Rightarrow \\ O(x,z) \le O(y,z). \end{gathered}$$

(O_T5) It is immediate. □

From the previous theorem, one may consider the particular case where the function f is the t-norm T (with respect to the function O_T is a t-overlap function), and the function h is $N(T)$, where $N : [0,1] \to [0,1]$ is a strong negation. It is immediate that:

Corollary 1. *Let T be a continuous t-norm and N a strong negation. Then the function $O_T : [0,1]^2 \to [0,1]$, defined, for all $x, y \in [0,1]$, by*

$$O_T(x,y) = \frac{T(x,y)}{T(x,y) + N(T(x,y))}$$

is a t-overlap function.

Supported by the previous corollary, we give some examples of t-overlap functions that are not overlap functions, considering continuous and positive t-norms.

Example 2. The following functions are some examples of associative t-overlap functions that are not overlap functions, since the property **(O2)** does not hold:

(i) Consider the standard negation $N_Z(x) = 1-x$ and the family of Lukasiewicz t-norms $\mathrm{T}_Ł(x,y) = \max\{0, (1+\lambda)(x+y-1) - \lambda xy\}$, where $\lambda \geq -1$. The function $O_1 : [0,1]^2 \to [0,1]$, defined, for $x, y \in [0,1]$, by

$$O_1(x,y) = \frac{\max\{0,(1+\lambda)(x+y-1)-\lambda xy\}}{\max\{0,(1+\lambda)(x+y-1)-\lambda xy\} + \min\{1, 1-(1+\lambda)(x+y-1)+\lambda xy\}}$$

is a t-overlap function with respect to $\mathrm{T}_Ł$.

(ii) Consider the Lukasiewicz t-norm $T_Ł(x,y) = \max\{0, x+y-1\}$ and the strong negation $N(x) = \sqrt{1-x^2}$. The function $O_2 : [0,1]^2 \to [0,1]$, defined, for all $x, y \in [0,1]$, by

$$O_2(x,y) = \frac{\max\{0,(x+y-1)\}}{\max\{0,x+y-1\} + \min\{1, \sqrt{1-(x+y-1)^2}\}},$$

is a t-overlap function with respect to $T_Ł$. Now, if one takes the strong negation $N(x) = \frac{2}{\pi}\arcsin[1-\sin(\frac{\pi}{2}x)]$, then the function $O_3 : [0,1]^2 \to [0,1]$, defined, for all $x, y \in [0,1]$, by

$$O_3(x,y) = \frac{\max\{0,x+y-1\}\}}{\max\{0,x+y-1\} + \frac{2}{\pi}\arcsin[\frac{\pi}{2}\max\{o,x+y-1\}]}$$

is a t-overlap function with respect to $\mathrm{T}_Ł$.

(iii) Consider the Yager t-norm $T_Y(x,y) = \max\{0, 1-\sqrt{(1-x)^2+(1-y)^2}$ and the strong negation $N(x) = \sqrt{1-x^2}$. The function $O_4 : [0,1]^2 \to [0,1]$, defined, for all $x, y \in [0,1]$, by

$$O_4(x,y) = \frac{\max\{0, 1-\sqrt{(1-x)^2+(1-y)^2}\}}{\max\{0,1-\sqrt{(1-x)^2+(1-y)^2}\} + \sqrt{1-\max^2\{0,1-\sqrt{(1-x)^2+(1-y)^2}\}}}$$

is a t-overlap function with respect to T_Y. Now, if one takes the strong negation $N(x) = \frac{2}{\pi}\arcsin(1-\sin(x\frac{\pi}{2}))$, then the function $O_5 : [0,1]^2 \to [0,1]$, defined, for all $x, y \in [0,1]$, by

$$O_5(x,y) = \frac{\max\{0, 1-\sqrt{(1-x)^2+(1-y)^2}\}}{\max\{0, 1-\sqrt{(1-x)^2+(1-y)^2}\} + \frac{2}{\pi}\arcsin(1-\sin(\frac{\pi}{2}\max\{0, 1-\sqrt{(1-x)^2+(1-y)^2}\}))}$$

is a t-overlap function with respect to T_Y.

Corollary 2. *Let $O_T : [0,1]^2 \to [0,1]$ be a t-overlap function with respect to a t-norm $T : [0,1]^2 \to [0,1]$ and $h : [0,1]^2 \to [0,1]$ satisfying the conditions (1), (ii), (iv) and (v) of Theorem 5. Then it holds that $O_T(x,x) = x$, for some $x \in [0,1[$ if and only if*

$$f(x,x) = \frac{x}{1-x}h(x,x).$$

Proof. For $x \in [0,1[$, it follows that:

$$O_T(x,x) = x \Leftrightarrow x = \frac{f(x,x)}{f(x,x)+h(x,x)} \quad \text{by Theorem 5}$$
$$\Leftrightarrow xf(x,x) + xh(x,x) = f(x,x) \Leftrightarrow f(x,x) = \frac{x}{1-x}h(x,x).$$

□

Given two t-norms $T_1, T_2 : [0,1]^2 \to [0,1]$, define $T_1T_2 : [0,1]^2 \to [0,1]$ by $T_1T_2(x,y) = T_1(x,y)T_2(x,y)$, for all $[x,y] \in [0,1]$.

Theorem 6. *Let $O : [0,1]^2 \to [0,1]$ be a overlap function, $T_1, T_2 : [0,1]^2 \to [0,1]$ be continuous t-norms such that $T_1, T_2 : [0,1]^2 \to [0,1]$ is a t-norm. Then the function $O_T : [0,1]^2 \to [0,1]$ defined, for all $[x,y] \in [0,1]$, by $O_T(x,y) = O(T_1(x,y), T_2(x,y))$, is a t-overlap function with respect to T_1T_2.*

Proof. (O_T1) It is immediate.
(O_T2) For all $[x,y] \in [0,1]$, it follows that:

$$O_T(x,y) = 0 \Leftrightarrow O(T_1(x,y), T_2(x,y)) = 0 \Leftrightarrow T_1(x,y)T_2(x,y) = 0$$
$$\Leftrightarrow (T_1T_2)(x,y) = 0.$$

(O_T3 For all $[x,y] \in [0,1]$, it follows that:

$$O_T(x,y) = 1 \Leftrightarrow O(T_1(x,y), T_2(x,y)) = 1 \Leftrightarrow T_1(x,y) = T_2(x,y) = 1$$
$$\Leftrightarrow x = y = 1.$$

(O_T4-5) It is immediate. □

Theorem 7. *Let $O_1, O_2 : [0,1]^2 \to [0,1]$ be t-overlap functions with respect to the t-norms $T_1, T_2 : [0,1]^2 \to [0,1]$, respectively, and $M : [0,1]^2 \to [0,1]$ be a continuous and positive function such that $M(x,y) = 1 \Leftrightarrow x = y = 1$. Then the function $O_T : [0,1]^2 \to [0,1]$, defined, for all $x, y \in [0,1]$, by $O_T(x,y) = M(O_1(x,y), O_2(x,y))$, is a t-overlap function with respect to T_1 or T_2.*

Proof. (O_T1) It is immediate.
(O_T2) For all $[x, y] \in [0, 1]$, it follows that:

$$O_T(x,y) = 0 \Leftrightarrow M(O_1(x,y), O_2(x,y)) = 0 \Leftrightarrow O_1(x,y) = 0 \vee O_2(x,y) = 0$$
$$\Leftrightarrow T_1(x,y) = 0 \vee T_2(x,y) = 0.$$

(O_T3) For all $[x, y] \in [0, 1]$, it follows that:

$$O_T(x,y) = 1 \Leftrightarrow M(O_1(x,y), O_2(x,y)) = 1 \Leftrightarrow O_1(x,y) = O_2(x,y) = 1$$
$$\Leftrightarrow x = y = 1.$$

($O_T4 - 5$) It is immediate. □

4 Conclusion

In this work, we generalized the concept of overlap functions, by relaxing the requirement that "one of its inputs must be zero so that the overlap function is zero". For that, we considered overlap functions associated to positive t-norms, as the Luckasiewicz t-norm. Likewise, a method for constructing t-overlap functions based on certain simple conditions has been presented. Future work is concerned this generalization under an interval-valued approach, as in [2,5,6].

Acknowledgment. Supported by Caixa and Fundación Caja Navarra of Spain, the Brazilian National Counsel of Technological and Scientific Development CNPq (Proc. 307781/2016-0), the Spanish Ministry of Science and Technology (TIN2016-77356-P).

References

1. Bedregal, B.C., Dimuro, G.P., Bustince, H., Barrenechea, E.: New results on overlap and grouping functions. Inf. Sci. **249**, 148–170 (2013)
2. Bedregal, B.C., Dimuro, G.P., Santiago, R.H.N., Reiser, R.H.S.: On interval fuzzy S-implications. Inf. Sci. **180**(8), 1373–1389 (2010)
3. Beliakov, G., Pradera, A., Calvo, T.: Aggregation Functions: A Guide for Practitioners. Springer, Berlin (2007). https://doi.org/10.1007/978-3-540-73721-6
4. Bustince, H., Fernandez, J., Mesiar, R., Montero, J., Orduna, R.: Overlap functions. Nonlinear Anal. Theory Methods Appl. **72**(3–4), 1488–1499 (2010)
5. Dimuro, G.P.: On interval fuzzy numbers. In: 2011 Workshop-School on Theoretical Computer Science, WEIT 2011, pp. 3–8. IEEE, Los Alamitos (2011)
6. Dimuro, G.P., Bedregal, B.C., Reiser, R.H.S., Santiago, R.H.N.: Interval additive generators of interval t-norms. In: Hodges, W., de Queiroz, R. (eds.) WoLLIC 2008. LNCS (LNAI), vol. 5110, pp. 123–135. Springer, Heidelberg (2008). https://doi.org/10.1007/978-3-540-69937-8_12
7. Dimuro, G.P., Bedregal, B.: Archimedean overlap functions: the ordinal sum and the cancellation, idempotency and limiting properties. Fuzzy Sets Syst. **252**, 39–54 (2014)
8. Dimuro, G.P., Bedregal, B.: On residual implications derived from overlap functions. Inf. Sci. **312**, 78–88 (2015)

9. Dimuro, G.P., Bedregal, B., Bustince, H., Asiáin, M.J., Mesiar, R.: On additive generators of overlap functions. Fuzzy Sets Syst. **287**, 76–96 (2016). Theme: Aggregation Operations
10. Dimuro, G.P., Bedregal, B., Bustince, H., Jurio, A., Baczyński, M., Miś, K.: QL-operations and QL-implication functions constructed from tuples (O, G, N) and the generation of fuzzy subsethood and entropy measures. Int. J. Approx. Reason. **82**, 170–192 (2017)
11. Garcia-Jimenez, S., Bustince, H., Hüllermeier, E., Mesiar, R., Pal, N.R., Pradera, A.: Overlap indices: construction of and application to interpolative fuzzy systems. IEEE Trans. Fuzzy Syst. **23**(4), 1259–1273 (2015)
12. Jurio, A., Bustince, H., Pagola, M., Pradera, A., Yager, R.: Some properties of overlap and grouping functions and their application to image thresholding. Fuzzy Sets Syst. **229**, 69–90 (2013)
13. Lucca, G., Dimuro, G.P., Mattos, V., Bedregal, B., Bustince, H., Sanz, J.A.: A family of Choquet-based non-associative aggregation functions for application in fuzzy rule-based classification systems. In: 2015 IEEE International Conference on Fuzzy Systems (FUZZ-IEEE), pp. 1–8. IEEE, Los Alamitos (2015)
14. Lucca, G., Sanz, J.A., Dimuro, G.P., Bedregal, B., Asiain, M.J., Elkano, M., Bustince, H.: CC-integrals: Choquet-like copula-based aggregation functions and its application in fuzzy rule-based classification systems. Knowl.-Based Syst. **119**, 32–43 (2017)
15. Mayor, G., Trillas, E.: On the representation of some aggregation functions. In: Proceedings of IEEE International Symposium on Multiple-Valued Logic, pp. 111–114. IEEE, Los Alamitos (1986)

On the Existence and Uniqueness of Fixed Points of Fuzzy Cognitive Maps

István Á. Harmati[1](✉), Miklós F. Hatwágner[2], and László T. Kóczy[2,3]

[1] Department of Mathematics and Computational Sciences, Széchenyi István University, Győr, Hungary
harmati@sze.hu

[2] Department of Information Technology, Széchenyi István University, Győr, Hungary
{hatwagner,koczy}@sze.hu

[3] Department of Telecommunications and Mediainformatics, Budapest University of Technology and Economics, Budapest, Hungary

Abstract. Fuzzy Cognitive Maps (FCMs) are decision support tools, which were introduced to model complex behavioral systems. The final conclusion (output of the system) relies on the assumption that the system reaches an equilibrium point (fixed point) after a certain number of iteration. It is not straightforward that the iteration leads to a fixed point, since limit cycles and chaotic behaviour may also occur.

In this article, we give sufficient conditions for the existence and uniqueness of the fixed point for log-sigmoid and hyperbolic tangent FCMs, based on the weighted connections between the concepts and the parameter of the threshold function. Moreover, in a special case, when all of the weights are non-negative, we prove that fixed point always exists, regardless of the parameter of the threshold function.

Keywords: Fuzzy modelling · Fuzzy cognitive maps · Fixed point

1 Introduction

Decision making problems arise very often in various fields. For example, the manager of a company has to choose among possible strategic investments, a surgeon has to select an appropriate way of operation, a corporal has to instruct his platoon in order to decrease the possibility of own casualties but increase the chance of a successful mission, etc. All of these example cases have similar properties: a wrong decision may cause serious damage or financial losses, the high number of interrelated factors should be considered by the decision maker and these factors usually form a complex system [5].

Several quantitative models and knowledge-based methods are known to support the decision making process, e.g. Bayesian Networks or Petri nets. The application of these techniques is often troublesome for non-experts, because they require the knowledge of the mathematical background [12]. The situation

J. Medina et al. (Eds.): IPMU 2018, CCIS 853, pp. 490–500, 2018.
https://doi.org/10.1007/978-3-319-91473-2_42

can be even worse if more or less numerical data is missing. Numerous decision making techniques are based on cognitive or fuzzy models [13]. Fuzzy Cognitive Maps (FCM) can effectively represent causal expert knowledge and uncertain information of complex systems, while the application of them is simple, clear and user-friendly. They use direct causal representation, and make possible to perform quick simulation of complex models [15]. These properties make it suitable for use at the front-end knowledge engineering.

2 Basic Notion of Fuzzy Cognitive Maps

The first cognitive maps suggested by Axelrod were used to describe the relationships among political elites in order to analyse the consequences of possible decisions [1]. Cognitive maps use directed graphs. Kosko extended this technique later in [8,9] and assigned constant weights in the $[-1,1]$ interval to the edges of the graph to express the strength and direction of causal connections. The nodes represent specific factors of the modelled system, and are usually called 'concepts' in FCM theory. The current state of them are also characterized by numbers in the $[0,1]$ interval (in some rare cases $[-1,1]$ is also applicable [16]). These are the so-called 'activation values'. Concepts have their initial activation values but these values change usually very quickly during the consecutive discrete time steps of the simulation.

The system can be formally defined by a 4-tuple (C, W, A, f) where $C = C_1, C_2, \ldots, C_n$ is the set of n concepts, $W : (C_i, C_j) \rightarrow w_{ij} \in [-1; +1]$ is a function which associates a causal value (weight) w_{ij} to each edge connecting the nodes (C_i, C_j), describing how strongly influenced is concept C_i by concept C_j. The sign of w_{ij} is indicates whether the relationship between C_j and C_i is direct or inverse. So the connection or weight matrix $W_{n\times n}$ gathers the system causality which could be estimated by experts or automatically computed from historical data [8,12,14]. The function $A : (C_i) \rightarrow A_i$ assigns an activation value $A_i \in \mathbb{R}$ to each node C_i at each time step t ($t = 1, 2, \ldots, T$) during the simulation. A transformation or threshold function $f : \mathbb{R} \rightarrow [0,1]$ calculates the activation value of concepts and keeps them in the allowed range (sometimes a function $f : \mathbb{R} \rightarrow [-1,1]$ is applied). The iteration which calculates the values of the concept may or may not include self-feedback. In general form it can be written as

$$A_i(k) = f\left(\sum_{j=1, j\neq i}^{n} w_{ij}A_j(k-1) + d_iA_i(k-1)\right) \tag{1}$$

where $A_i(k)$ is the value of concept C_i at discrete time k, w_{ij} is the weight of the connection from concept C_j to concept C_i and d_i expresses the possible self-feedback. If we include the self-feedback into the weight matrix W, the equation can be rewritten in a simpler form:

$$A(k) = f\left(WA(k-1)\right) \tag{2}$$

Unfortunately it makes impossible to perform a quantitative analysis, but qualitative comparisons are still possible. There are several well-known and widely applied threshold functions [16], such as bivalent, trivalent and various kind of sigmoidal ones. In our paper, the sigmoid-like functions, the log-sigmoid and the hyperbolic tangent function are examined.

3 Problem Statement

According to [16], continuous FCM may behave chaotically, can produce limit cycles or reach a fixed-point attractor. Chaotic behaviour means that the activation vector never stabilizes. (Technically, the state vectors do not stabilize after a predefined high number of iterations.) If a limit cycle occurs, a specific number of consecutive state vectors turn up repeatedly. In case of a fixed-point attractor the state vector stabilizes after a certain number of iterations.

The nature of the iteration depends on

- the threshold function applied and its parameter(s);
- the elements (weights) of the extended weight matrix;
- the topology of the map.

Let's see the following example. The weight matrix is W and we do not apply self-feedback.

$$W = \begin{pmatrix} 0 & -0.5 & 0.5 & 0.5 & -0.5 & 0.5 & -0.5 \\ -0.5 & 0 & 0.5 & 0.5 & -0.5 & 0.5 & -0.5 \\ 0.5 & 0.5 & 0 & -0.5 & 0.5 & -0.5 & 0.5 \\ 0.5 & 0.5 & -0.5 & 0 & 0.5 & -0.5 & 0.5 \\ -0.5 & -0.5 & 0.5 & 0.5 & 0 & 0.5 & -0.5 \\ 0.5 & 0.5 & -0.5 & -0.5 & 0.5 & 0 & 0.5 \\ -0.5 & -0.5 & 0.5 & 0.5 & -0.5 & 0.5 & 0 \end{pmatrix} \tag{3}$$

The threshold function is the log-sigmoid function:

$$f(x) = \frac{1}{1 + e^{-\lambda x}} \tag{4}$$

We start the iteration at $A(0)$:

$$A(0) = \begin{pmatrix} 1 \, 0 \, 0 \, 0 \, 0 \, 0 \, 0 \end{pmatrix}^T \tag{5}$$

As we can see in Fig. 1, above a certain value of λ the system does not converge the an equilibrium point, but stays in a limit cycle. In the following we state conditions under a FCM has a fixed point.

In Sect. 4 we give sufficient conditions for which the fuzzy cognitive map equipped with sigmoid-like (log-sigmoid and hyperbolic tangent) threshold function has one and only one fixed point. These conditions include the parameter of the threshold function (λ) and the Frobenius norm of the extended weight matrix, too.

In Sect. 5 we prove the existence of fixed points if the extended weight matrix satisfies a nonnegativity condition (it has only non-negative entries).

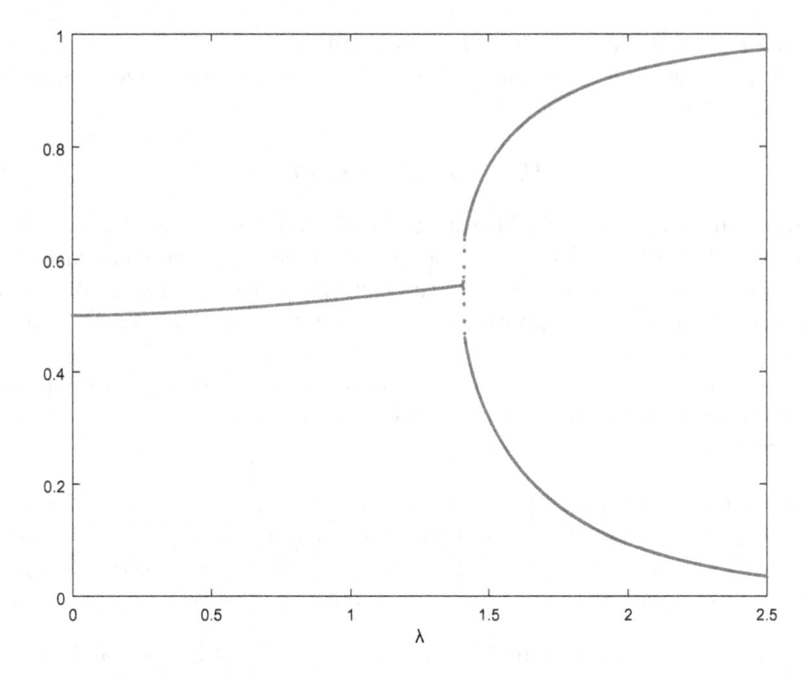

Fig. 1. The first coordinate of the concept vector after a huge number of iterations in function of the inclination parameter (λ) of the applied log-sigmoid function. We can observe that after a certain value of λ a bifurcation occurs, so the iteration does not converge to an equilibrium point, but arrives to a limit cycle.

4 Existence and Uniqueness of Fixed Points of Log-Sigmoid and Hyperbolic Tangent FCMs

The existence and uniqueness of fixed point of sigmoid fuzzy cognitive maps was firstly discussed by Boutalis et al. in [3] for the case when the parameter of the log-sigmoid threshold function is $\lambda = 1$, (so the function was $f(x) = 1/(1 + e^{-x})$. In [4] they tried to generalize this result to arbitrary log-sigmoid functions. Although their statement was cited several times, for example in [6,10,11], it was disproved by numerical counterexamples in [7]. Moreover, in [7] we state and prove theorems about the existence and uniqueness of fixed points of FCMs. In the present paper, two important special cases are concerned, which are not discussed in such a detailed form in [7].

In this section, we prove two theorems regarding of the existence and uniqueness of fixed points of an FCM (we note that the results of [3] are special cases of them when the parameter equals one). From mathematical point of view the updating process of the concept values is iterative application of an n dimensional mapping, $G\colon [0,1]^n \to [0,1]^n$ (or in some cases $G\colon [-1,1]^n \to [-1,1]^n$), so examining of an FCM with respect to its fixed points means analysis of mapping G. The approach applied here is based on the notion of the so-called *contraction*, so first we recall its definition.

Definition 1. *Let (X, d) be a metric space. A mapping $f\colon X \to X$ is a* contraction mapping *or* contraction *if there exists a constant c (independent from x and y), with $0 \leq c < 1$, such that*

$$d\left(f(x), f(y)\right) \leq cd(x, y) \tag{6}$$

The iteration process of an FCM stops when the concept vector does not change anymore (technically it means that the difference between two consecutive iterations is less than a given value), so the concept vector is a fixed point in the n dimensional space. More precisely, a fixed point of mapping $f\colon X \to X$ is a point $x^* \in X$ such that $f(x^*) = x^*$.

The following theorem ensures the existence and uniqueness of a fixed point of a contraction mapping. Moreover, it provides an iterative algorithm to determine the fixed point.

Theorem 1 (Banach's fixpoint theorem). *If $f\colon X \to X$ is a contraction mapping on a nonempty complete metric space (X, d), then f has only one fixpoint x^*. Moreover, x^* can be found as follows: start with an arbitrary $x_0 \in X$ and define the sequence $x_n = f(x_{n-1})$, then $\lim_{n\to\infty} x_n = x^*$.*

In the following we state conditions under which an FCM has one and only one fixed point. The philosophy of the proofs is that under the conditions stated, mapping G is a contraction, i.e. the distance of $G(A)$ and $G(A')$ measured by an appropriate metric (in our case the Euclidean distance) is less than the distance of A and A', where $G(A) = f\left(WA\right) = \left[f(w_1A), f(w_2A), \ldots, f(w_nA)\right]^T$, so according to Banach's fixpoint theorem, it has one and only one fixed point.

4.1 Log-Sigmoid Case

The most widely used threshold function in applications of FCMs is the log-sigmoid (sometimes mentioned simply as sigmoid) function. In the proof of the theorem regarding of existence and uniqueness of fixed points of FCMs with log-sigmoid threshold function, the following (maybe well-known) fact plays an important role.

Lemma 2. *The derivative of the log-sigmoid function $f\colon \mathbb{R} \to \mathbb{R}$, $f(x) = 1/(1+e^{-\lambda x})$ $(\lambda > 0)$ is bounded by $\lambda/4$.*

Proof: The derivative of the log-sigmoid function is:

$$f'(x) = \left(\frac{1}{1+e^{-\lambda x}}\right)' = \frac{\lambda e^{-\lambda x}}{(1+e^{-\lambda x})^2} = \lambda \cdot f(x) \cdot (1 - f(x)) \tag{7}$$

Since $0 < f(x) < 1$ for every $x \in \mathbb{R}$, the inequality $0 < f(x)(1 - f(x)) \leq 1/4$ holds for every $x \in \mathbb{R}$, which implies that $0 < f'(x) \leq \lambda/4$.

Theorem 3. *Let W be the extended (including possible feedback) weight matrix of a FCM, and let $\lambda > 0$ be the parameter of the log-sigmoid function. If the inequality*

$$\|W\|_F < \frac{4}{\lambda} \tag{8}$$

holds, then the FCM has one and only one fixed point. Here $\|\cdot\|_F$ stands for the Frobenius norm of the matrix, $\|A\|_F = \left(\sum_i \sum_j a_{ij}^2\right)^{1/2}$.

Proof: Let

$$G(A) = [f(w_1 A), f(w_2 A), \ldots, f(w_n A)]^T \tag{9}$$

where w_i is the ith row of matrix W and

$$f(x) = \frac{1}{1 + e^{-\lambda x}} \tag{10}$$

We give an upper estimation of the value of $\|G(A) - G(A')\|_2$:

$$\|G(A) - G(A')\|_2 = \left(\sum_{i=1}^{n} (f(w_i A) - f(w_i A'))^2\right)^{1/2} \tag{11}$$

From Lagrange mean value theorem and Lemma 2 we have

$$|f(w_i A) - f(w_i A')| \le \frac{\lambda}{4} |w_i A - w_i A'| \tag{12}$$

An upper estimation of the right hand side expression using Cauchy-Schwarz inequality:

$$\frac{\lambda}{4} |w_i A - w_i A'| = \frac{\lambda}{4} |w_i (A - A')| \le \frac{\lambda}{4} \|w_i\|_2 \cdot \|A - A'\|_2 \tag{13}$$

So we get the following upper estimation of $\|G(A) - G(A')\|_2$:

$$\|G(A) - G(A')\|_2 = \left(\sum_{i=1}^{n} (f(w_i A) - f(w_i A'))^2\right)^{1/2} \tag{14}$$

$$\le \left(\sum_{i=1}^{n} \left(\frac{\lambda}{4} \|w_i\|_2 \cdot \|A - A'\|_2\right)^2\right)^{1/2} \tag{15}$$

$$= \|A - A'\|_2 \cdot \left(\sum_{i=1}^{n} \left(\frac{\lambda}{4} \|w_i\|_2\right)^2\right)^{1/2} \tag{16}$$

$$= \|A - A'\|_2 \cdot \frac{\lambda}{4} \cdot \left(\sum_{i=1}^{n} \sum_{j=1}^{n} w_{ij}^2\right)^{1/2} \tag{17}$$

$$= \|A - A'\|_2 \cdot \frac{\lambda}{4} \cdot \|W\|_F \tag{18}$$

So we get that

$$\|G(A) - G(A')\|_2 \leq \|A - A'\|_2 \cdot \frac{\lambda}{4} \cdot \|W\|_F \tag{19}$$

If $\frac{\lambda}{4} \cdot \|W\|_F < 1$ (i.e. if $\|W\|_F < 4/\lambda$) then the coefficient of $\|A - A'\|_2$ is less than 1, so mapping G is a contraction (i.e. it has one and only one fixpoint), which completes the proof. □

We note here that this is a sufficient, but not necessary condition. The fact that $\|W\|_F < 4/\lambda$ implies that there is one and only one fixed point, while if $\|W\|_F \geq 4/\lambda$, one cannot ensure that there are more than one fixed points or limit cycles.

In the example of Sect. 3, the Frobenius norm of matrix W is 3.2404. By rearranging the condition in the theorem we get the if $\lambda < 4/\|W\|_F$ then the FCM has one and only one fixed point. In our case, this value is 1.2344 and we can observe in Fig. 1 that under this value there is no limit cycle, but only one value for each λ. It can be also observed that there is a small gap between this value (1.2344) and the point where the limit cycle starts. Since in the theorem we only used the norm of the matrix and not the whole structure of the fuzzy cognitive map, our conjecture is that this gap might be in connection with the current topology of the FCM.

4.2 Hyperbolic Tangent Case

In some cases, the hyperbolic tangent function is applied as a threshold function, which squashes the concept values into the interval $[-1, 1]$. Although the conditions, under which an FCM with hyperbolic tangent threshold function has one and only one fixed point, and the proof are similar to the case of the log-sigmoid function, they are not exactly the same. First, similarly to the previous case, we state the bound of the derivative of the threshold function.

Lemma 4. *The derivative of the hyperbolic tangent function* $f\colon \mathbb{R} \to \mathbb{R}$, $f(x) = \frac{e^{2\lambda x} - 1}{e^{2\lambda x} + 1}$ *(*$\lambda > 0$*) is bounded by* λ.

Proof: The derivative of $f(x)$ is

$$f'(x) = \frac{4\lambda e^{2\lambda x}}{(e^{2\lambda x} + 1)^2} = \lambda(1 - f^2(x))$$

Since $-1 < f(x) < 1$, it follows that $0 < f'(x) < \lambda$.

Theorem 5. *Let* W *be the extended (including possible feedback) weight matrix of a FCM, let* $\lambda > 0$ *be the parameter of the hyperbolic tangent threshold function. If the inequality*

$$\|W\|_F < \frac{1}{\lambda} \tag{20}$$

holds, then the FCM has one and only one fixed point.

Proof: The proof goes completely similarly to the previous one. Let

$$G(A) = [f(w_1 A), f(w_2 A), \ldots, f(w_n A)]^T \tag{21}$$

where w_i is the ith row of matrix W and

$$f(x) = \tanh(\lambda x) = \frac{e^{\lambda x} - e^{-\lambda x}}{e^{\lambda x} + e^{-\lambda x}} \tag{22}$$

We give an upper estimation of the value of $\|G(A) - G(A')\|_2$:

$$\|G(A) - G(A')\|_2 = \left(\sum_{i=1}^{n} (f(w_i A) - f(w_i A'))^2 \right)^{1/2} \tag{23}$$

From Lagrange mean value theorem and Lemma 4 we have that

$$|f(w_i A) - f(w_i A')| \leq \lambda \, |w_i A - w_i A'| \tag{24}$$

An upper estimation of the right hand side expression using Cauchy-Schwarz inequality:

$$\lambda \, |w_i A - w_i A'| = \lambda \, |w_i (A - A')| \leq \lambda \, \|w_i\|_2 \cdot \|A - A'\|_2 \tag{25}$$

So we get the following upper estimation of $\|G(A) - G(A')\|_2$:

$$\|G(A) - G(A')\|_2 = \left(\sum_{i=1}^{n} (f(w_i A) - f(w_i A'))^2 \right)^{1/2} \tag{26}$$

$$\leq \left(\sum_{i=1}^{n} (\lambda \, \|w_i\|_2 \cdot \|A - A'\|_2)^2 \right)^{1/2} \tag{27}$$

$$= \|A - A'\|_2 \cdot \left(\sum_{i=1}^{n} (\lambda \, \|w_i\|_2)^2 \right)^{1/2} \tag{28}$$

$$= \|A - A'\|_2 \cdot \lambda \cdot \left(\sum_{i=1}^{n} \sum_{j=1}^{n} w_{ij}^2 \right)^{1/2} \tag{29}$$

$$= \|A - A'\|_2 \cdot \lambda \cdot \|W\|_F \tag{30}$$

So we get that

$$\|G(A) - G(A')\|_2 \leq \|A - A'\|_2 \cdot \lambda \cdot \|W\|_F \tag{31}$$

If $\lambda \cdot \|W\|_F < 1$ then the coefficient of $\|A - A'\|_2$ is less than 1, so mapping G is a contraction (so it has one and only one fixed point), which completes the proof. □

Just like in the previous case, we note that this is a sufficient, but necessary condition for an FCM to have one and only one fixed point.

5 Existence of Fixed Points for Non-negative Weight Matrix

Definition 2. *A partially ordered set (also known as poset) is a tuple* $\langle S, \preccurlyeq \rangle$ *consisting of a set* S *and a relation* $\preccurlyeq$ *which is defined for all* $a, b, c \in S$ *such that*

- $a \preccurlyeq a$;
- *if* $a \preccurlyeq b$ *and* $b \preccurlyeq a$ *then* $a = b$;
- *if* $a \preccurlyeq b$ *and* $b \preccurlyeq c$ *then* $a \preccurlyeq c$.

Definition 3. *A partially ordered set* $\langle S, \preccurlyeq \rangle$ *is a complete lattice if every subset of* S *has a least upper bound (supremum) and a greatest lower bound (infimum).*

For $\boldsymbol{x} = (x_1, \ldots, x_n), \boldsymbol{y} = (y_1, \ldots, y_n) \in \mathbb{R}^n$ define the ordering $\boldsymbol{x} \preccurlyeq \boldsymbol{y}$ to mean that $x_1 \leq y_1, \ldots, x_n \leq y_n$ (coordinate-wise ordering). Let $a, b \in \mathbb{R}^n$ satisfy $a \preccurlyeq b$. Then the set

$$[a, b] = x \in \mathbb{R}^n : a \leq x \leq b = [a_1, b_1] \times \ldots \times [a_n, b_n]$$

is a complete lattice with respect to the coordinate-wise order. So as special cases the sets $[0, 1]^n = [0, 1] \times \ldots \times [0, 1]$ and $[-1, 1]^n = [-1, 1] \times \ldots \times [-1, 1]$ are complete lattices.

Theorem 6 (Tarski's fixpoint theorem). *Let* L *be a complete lattice and let* $f \colon L \to L$ *be an order-preserving function (i.e. if* $x \preccurlyeq y$ *then* $f(x) \preccurlyeq f(y)$ *for every* $x, y \in L$*). Then the set of fixed points of* f *in* L *is also a complete lattice.*

Since a complete lattice is a non-empty set, this theorem ensures the existence of fixpoints if the conditions are fulfilled.

Theorem 7. *Let* W *be the extended (including possible feedback) weight matrix of a FCM, and let the threshold function be*

1. *a log-sigmoid function,* $f(x) = \dfrac{1}{1 + e^{-\lambda x}}$, $\lambda > 0$, *or*
2. *a hyperbolic tangent function,* $f(x) = \dfrac{e^{2\lambda x} - 1}{e^{2\lambda x} + 1}$, $\lambda > 0$.

If W *has only non-negative elements then the FCM has fixpoint, regardless to the value of* λ.

Proof: Tarski's fixpoint theorem tells us that if f is an order preserving mapping, then it has a fixpoint. We have to show that the mapping $G \colon [0, 1]^n \to [0, 1]^n$ (in case of the hyperbolic tangent function it is $G \colon [-1, 1]^n \to [-1, 1]^n$) is an order preserving mapping, where $G(A) = [f(w_1 A), f(w_2 A), \ldots, f(w_n A)]$, w_i is the ith row of W and $A = (A_1, \ldots, A_n)$.

Let A and B be concept vectors such that $A \preccurlyeq B$, which means that $A_i \leq B_i$ for every $1 \leq i \leq n$. Since all of the elements of W are greater than or equal to zero ($w_{ij} \geq 0$ for every i, j) then the following inequality holds for every i:

$$w_i A = w_{i1} A_1 + \ldots + w_{in} A_n \leq w_{i1} B_1 + \ldots + w_{in} B_n = w_i B \tag{32}$$

Moreover, the threshold function f is a monotone increasing function, which implies that if $w_i A \leq w_i B$ then $f(w_i A) \leq f(w_i B)$. From this fact it follows that if $A \preccurlyeq B$ then $G(A) \preccurlyeq G(B)$, which means that G is an order preserving mapping and according to Tarski's theorem, it has a fix point. □

The theorem above is about the existence of a fixpoint. In the application of FCMs the fixpoint is determined by series of iterations. Kleene's theorem [2] provides a procedure by which we can determine the least fixed point.

Let $\perp$ be the least of the possible concept vectors (i.e. it is $(0, \ldots, 0)$ if the concept values are between 0 and 1, and $(-1, \ldots, -1)$ if the if the concept values are between -1 and 1), then the least fixed point of f is given by the limit

$$\lim_{n \to \infty} f^n(\perp) \tag{33}$$

where f^n stands for the n times iterated f function.

6 Summary

In modeling of complex systems by fuzzy cognitive maps, an iterative application of special functions (in most of the cases log-sigmoid or hyperbolic tangent) plays a crucial role. The iteration may lead to a fixed point, limit cycle or produces chaotic behaviour. In this article theorems regarding of the existence and uniqueness of fixed points were proved, using the weights of the interconnections in the FCM and the parameter of the threshold function. These results may be useful in the proper choice of the parameter of the threshold function applied. Naturally arises a question that how difficult is to fulfill the conditions of the theorems and is it feasible simultaneously determine the weights and the parameter of the threshold function, such that the FCM would have one and only fixed point. Of course, the answer depends on the nature of the concrete problem, but our numerical studies show that it is usually possible to obtain proper weights and parameters.

Moreover, in a special case, namely when all of the weights are non-negative, the existence of fixed points was proved, regardless of the parameter of the threshold function.

Acknowledgments. This work was supported by EFOP-3.6.2-16-2017-00015, HU-MATHS-IN – Intensification of the activity of the Hungarian Industrial Innovation Service Network and by National Research, Development and Innovation Office (NKFIH) K124055.

References

1. Axelrod, R.: Structure of Decision: The Cognitive Maps of Political Elites. Princeton University Press, Princeton (1976)
2. Baranga, A.: The contraction principle as a particular case of Kleene's fixed point theorem. Discret. Math. **98**(1), 75–79 (1991). https://doi.org/10.1016/0012-365X(91)90413-V
3. Boutalis, Y., Kottas, T.L., Christodoulou, M.: Adaptive estimation of fuzzy cognitive maps with proven stability and parameter convergence. IEEE Trans. Fuzzy Syst. **17**(4), 874–889 (2009). http://ieeexplore.ieee.org/document/4801671/
4. Boutalis, Y., Kottas, T.L., Christodoulou, M.: Bi-linear adaptive estimation of fuzzy cognitive networks. Appl. Soft Comput. **12**(12), 3736–3756 (2012). https://doi.org/10.1016/j.asoc.2012.01.025
5. Busemeyer, J.R.: Dynamic decision making. Int. Encycl. Soc. Behav. Sci., 3903–3908 (2001). https://doi.org/10.1016/B0-08-043076-7/00641-0
6. Felix, G., Nápoles, G., Falcon, R., Froelich, W., Vanhoof, K., Bello, R.: A review on methods and software for fuzzy cognitive maps. Artif. Intell. Rev., 1–31 (2017). https://doi.org/10.1007/s10462-017-9575-1
7. Harmati, I.A., Hatwágner, F.M., Kóczy, L.T.: On the existence and uniqueness of fixed point attractors of fuzzy cognitive maps. Appl. Soft Comput. (submitted)
8. Kosko, B.: Fuzzy cognitive maps. Int. J. Man-Mach. Stud. **24**, 65–75 (1986). https://doi.org/10.1016/S0020-7373(86)80040-2
9. Kosko, B.: Neural Networks and Fuzzy Systems. Prentice-Hall, Upper Saddle River (1992)
10. Nápoles, G., Papageorgiou, E., Bello, R., Vanhoof, K.: On the convergence of sigmoid Fuzzy Cognitive Maps. Inf. Sci. **349–350**, 154–171 (2016). https://doi.org/10.1016/j.ins.2016.02.040
11. Nápoles, G., Papageorgiou, E., Bello, R., Vanhoof, K.: Learning and convergence of fuzzy cognitive maps used in pattern recognition. Neural Process. Lett. **45**(2), 431–444 (2017). https://doi.org/10.1007/s11063-016-9534-x
12. Papageorgiou, E.I. (ed.): Fuzzy Cognitive Maps for Applied Sciences and Engineering. Intelligent Systems Reference Library, vol. 54. Springer, Heidelberg (2014). https://doi.org/10.1007/978-3-642-39739-4
13. Papageorgiou, E.I., Salmeron, J.L.: Methods and algorithms for fuzzy cognitive map-based modeling. In: Papageorgiou, E.I. (ed.) Fuzzy Cognitive Maps for Applied Sciences and Engineering. ISRL, vol. 54, pp. 1–28. Springer, Heidelberg (2014). https://doi.org/10.1007/978-3-642-39739-4_1
14. Sharif, A.M., Irani, Z.: Exploring fuzzy cognitive mapping for IS evaluation. Eur. J. Oper. Res. **173**, 1175–1187 (2006). https://doi.org/10.1016/j.ejor.2005.07.011
15. Stylios, C.D., Groumpos, P.P.: Modeling complex systems using fuzzy cognitive maps. IEEE Trans. Syst. Man Cybern.-Part A: Syst. Hum. **34**(1), 155–162 (2004). http://ieeexplore.ieee.org/document/1259444/
16. Tsadiras, A.K.: Comparing the inference capabilities of binary, trivalent and sigmoid fuzzy cognitive maps. Inf. Sci. **178**(20), 3880–3894 (2008). https://doi.org/10.1016/j.ins.2008.05.015

Searching Method of Fuzzy Internally Stable Set as Fuzzy Temporal Graph Invariant

Alexander Bozhenyuk[1(✉)], Stanislav Belyakov[1], Margarita Knyazeva[1], and Igor Rozenberg[2]

[1] Southern Federal University, Nekrasovskiy Str., 44, 347928 Taganrog, Russia
{avb002, beliacov}@yandex.ru,
margarita.knyazeva@gmail.com

[2] Public Corporation "Research and Development Institute of Railway Engineers", Nizhegorodskaya Str., 27/1, 109029 Moscow, Russia
i.yarosh@vniias.ru

Abstract. In this paper we consider the problem of finding the invariant of a fuzzy temporal graph, namely, a fuzzy internally stable set. Fuzzy temporal graph is a generalization of a fuzzy graph on the one hand, and a temporal graph on the other hand. In this paper, a temporal fuzzy graph is considered, in which the connectivity degree of vertices varies in discrete time. The notion of maximum internally stable subset of fuzzy temporal graph is considered. A method and an algorithm for finding all maximal internally stable sets are proposed which makes it possible to find a fuzzy internally stable set. The example of definition of internal stable fuzzy set is considered as well.

Keywords: Fuzzy temporal graph · Subgraph of fuzzy temporal graph
Fuzzy internally stable set · Internal stability degree

1 Introduction

Graph models attract the great attention of specialists in the various fields of knowledge. They are used as models of various complex objects and phenomena with some predefined structure. In addition, along with the applications of graph models in such sciences as chemistry, electrical engineering, physics, they are also used in economics, linguistics, and sociology. Graph models can be used to define the relations between the structures of the different nature of their elements [1–3]. In the case when the relations between the elements, or the elements of the system themselves are partially undefined or fuzzy, fuzzy graphs are used as a models of the system [4–6].

In this case the relations between the elements (vertices of the graph) are considered as permanent and cannot be changed during the simulation. Such graphs were called static [7]. However, if the relationship between elements of this structure may change over time the traditional graph models are not suitable to describe them. They cannot be used for modeling processes in time. Here we take into consideration other relevant graph models, i.e. graphs in which connections between vertices may change over discrete (or continuous) time period. These graphs are called temporal [8]. It should be

J. Medina et al. (Eds.): IPMU 2018, CCIS 853, pp. 501–510, 2018.
https://doi.org/10.1007/978-3-319-91473-2_43

noted that the concept of the temporal graph is well-known in literature and can be usually interpreted as time graphs, oriented acyclic graphs or Petri nets [9–14].

In this paper we consider a fuzzy temporal graph in which fuzzy connections between the vertices of a graph change over discrete time. The vertices themselves remain unchanged [15, 16].

When these graphs are used as models of complex systems it is actual to take into consideration their invariants, i.e. such characteristics of the graph which are not changed when graph isomorphic transformation is done. Such invariants for crisp graphs are presented by numbers of internal and external stability, chromatic number, chromatic class et al. In this paper we consider the notion of an internally stable set, as an invariant of a fuzzy temporal graph and the following approach to find it.

2 Basic Concepts and Definitions

Definition 1 [17]. Let a fuzzy temporal graph $\tilde{G} = (X, \{\tilde{\Gamma}_t\}, T)$ be given, where set X is a set of vertices ($|X| = n$), set of natural numbers T defines discrete time and set $\{\tilde{\Gamma}_t\}$ defines a family of sets, which display the vertices of X into itself at time $t = \overline{1, T}$. In other words:

$$(\forall x \in X)(\forall t \in T)\,[\tilde{\Gamma}_t(x) = \{<\mu_t(y)/y>\}],\; y \in X, \mu_t \in [0, 1].$$

Example 1. Let's consider the example of temporal fuzzy graph $\tilde{G} = (X, \{\tilde{\Gamma}_t\}, T)$ whose set of vertices $X = \{x_1, x_2, x_3, x_4\}$, discrete time $T = \{1, 2, 3\}$, and multi-valued mapping $\{\tilde{\Gamma}_t\}$ has the form:

$\tilde{\Gamma}_1(x_1) = \{<0.2/x_2>\}$, $\tilde{\Gamma}_2(x_1) = \{<0.5/x_2>\}$, $\tilde{\Gamma}_3(x_1) = \{<0.3/x_2>\}$
$\tilde{\Gamma}_2(x_2) = \{<0.4/x_3>\}$, $\tilde{\Gamma}_3(x_2) = \{<0.6/x_3>\}$, $\tilde{\Gamma}_1(x_3) = \{<0.2/x_4>\}$,
$\tilde{\Gamma}_2(x_3) = \{<0.3/x_4>\}$, $\tilde{\Gamma}_1(x_4) = \{<0.2/x_1>, <0.5/x_2>\}$, $\tilde{\Gamma}_2(x_4) = \{<0.9/x_2>\}$,
$\tilde{\Gamma}_3(x_4) = \{<0.1/x_1>, <1/x_2>\}$.

A graphical representation of this graph is shown in Fig. 1. Here the membership functions for the time $T = \{1, 2, 3\}$ are specified on the graph edges.
So, a temporal fuzzy graph can be represented as the union of T fuzzy subgraphs defined on the same set of vertices X.

Consider the fuzzy subgraph $\tilde{G}'_t = (X', \tilde{U}'_t)$ of the temporal fuzzy graph $\tilde{G}$, where $X' \subseteq X$ is the subset of vertices, and $\tilde{U}'_t = \{\mu_t(x_i, x_j) | (x_i, x_j) \in X'^2\}$ is the fuzzy set of edges over time t with the membership function $\mu_t : X'^2 \to [0, 1]$. Denote by $\tau = \max_{\forall x_i, x_j \in X'} \{\mu_t(x_i, x_j)\}$.

Definition 2. A subset of vertices X' is called an internally stable set for time t with internal stability degree $\alpha(X') = 1 - \tau$.

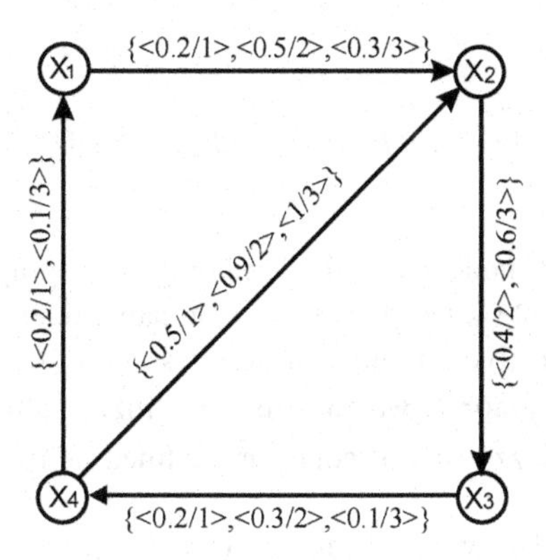

Fig. 1. Fuzzy temporal graph $\tilde{G}$

Example 2. For the fuzzy graph shown in Fig. 1, for time $t = 2$ and the subset of vertices $X^{'} = \{x_1, x_2, x_3\}$, the fuzzy subgraph is illustrated in Fig. 2:

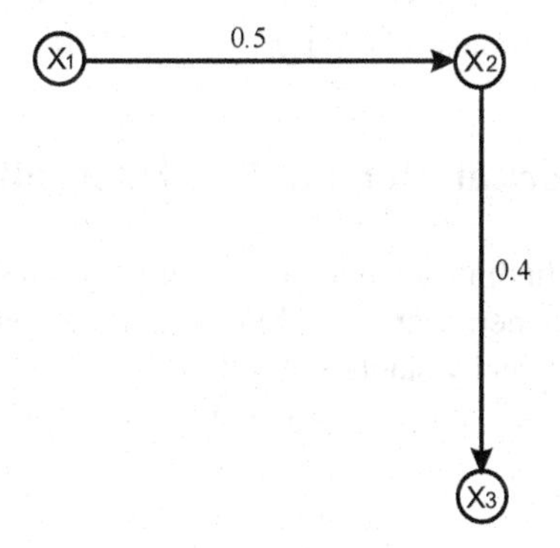

Fig. 2. Fuzzy subgraph for subset of vertices X', and $t = 2$

The quantity τ is $\max\{0.5, 0.4\} = 0.5$. Consequently, the degree of internal stability of the subset X' is equal to $\alpha(X') = 1 - 0.5 = 0.5$.

If $X' = X$ then the value $\alpha(X)$ determines the degree of internal stability of the fuzzy subgraph $\tilde{G}_t$.

Definition 3 [18]. A subset of vertices $X' \subseteq X$ is called a maximal internally stable set at time t with degree of internal stability $\alpha(X')$ if for any $X'' \supset X'$ the following inequality is fulfilled: $\alpha(X'') < \alpha(X')$.

Denote by $Y_k = \{X_{k_1}, X_{k_2}, \ldots, X_{k_l}\}$ the family of all subsets with k vertices for time t and with the degrees of internal stability $\alpha_{k1}, \alpha_{k2}, \ldots, \alpha_{kl}$, respectively. Let's $\alpha_k^{\max} = \max\{\alpha_{k1}, \alpha_{k2}, \ldots, \alpha_{kl}\}$. The value $\alpha_k^{\max}$ means that there is a subgraph with k vertices with a degree of internal stability in the graph $\tilde{G}_t$ and there is no other subgraph with k vertices whose degree of internal stability would be greater than $\alpha_k^{\max}$.

The set $\tilde{\Psi}_t = \{ < \alpha_1^{\max}/1 > , < \alpha_2^{\max}/2 > , \ldots, < \alpha_n^{\max}/n > \}$ is a fuzzy set of internal stability of subgraph $\tilde{G}_t$.

Definition 4 [18]. A set

$$\tilde{\Psi} = \bigcap_{t=\overline{1,T}} \tilde{\Psi}_t = \{<\beta_1/1>, <\beta_2/2>, .., <\beta_n/n>\} \quad (1)$$

is called fuzzy internally stable set of the temporal fuzzy graph $\tilde{G}$.

A fuzzy internally stable set determines the greatest degree of internal stability of temporal fuzzy graph for given number of vertices at any time.

For a fuzzy temporal graph $\tilde{G}$ we construct two fuzzy subgraphs $\tilde{G}_1 = (X, \tilde{U}_1)$ and $\tilde{G}_2 = (X, \tilde{U}_2)$ in which fuzzy sets of edges are defined as following:

$$\tilde{U}_1 = \{\mu_1(x_i, x_j) = \min_t\{\mu_t(x_i, x_j)\} | (x_i, x_j) \in X^2\};$$
$$\tilde{U}_2 = \{\mu_2(x_i, x_j) = \max_t\{\mu_t(x_i, x_j)\} | (x_i, x_j) \in X^2\}.$$

Let the fuzzy sets of internal stability of these graphs be $\tilde{\Psi}_1 = \{<\beta_1^1/1>, <\beta_2^1/2>, \ldots, <\beta_n^1/n>\}$ and $\tilde{\Psi}_2 = \{<\beta_1^2/1>, <\beta_2^2/2>, \ldots, <\beta_n^2/n>\}$ respectively.

The following property holds: $(\forall i \in \overline{1,n})[\beta_i^2 \leq \beta_i \leq \beta_i^1]$.

3 Method and Algorithm for Finding Internally Stable Set

Consider the method of finding all maximal internally stable sets with the greatest degree. The method is a generalization of Maghout's method for fuzzy graphs [19].

Let $X' \subseteq X$ be some internally stable set with a degree of internal stability $\alpha(X')$ of subgraph $\tilde{G}_t$. Then for arbitrary vertices $x_i, x_j \in X$, one of the following conditions must be true:

$$\text{(a)}\, x_i \notin X'; \ \text{(b)}\, x_j \notin X'; \ \text{(c)}\, x_i \in X' \text{ and } x_j \in X'.$$

In the latter case, the following inequality holds: $\alpha(X') \leq (1 - \mu_t(x_i, x_j))$

In other words, the following statement is true:

$$(\forall x_i, x_j \in X)[x_i \notin X' \vee x_j \notin X' \vee (\alpha(X') \leq (1 - \mu_t(x_i, x_j)))]. \quad (2)$$

We assign Boolean variable p_i to each vertex $x_i \in X$. Boolean variable p_i takes the value 1, if $x_i \in X'$ and 0 otherwise. To the statement $\alpha(X') \leq (1 - \mu_t(x_i, x_j))$ we consider the fuzzy variable $\xi_{ij} = (1 - \mu_t(x_i, x_j))$. Using analogy between generality and existence quantifiers on the one hand, both operations conjunction and disjunction on the other hand, we obtain a true logical proposition:

$$\Phi_t = \underset{\forall i, j \neq i}{\&} (\overline{p}_i \vee \overline{p}_j \vee \xi_{ij}). \quad (3)$$

We open the parentheses in the expression (3) and reduce the similar terms according to the following rules:

$$\begin{aligned} & a \vee a \,\&\, b = a, \\ & a \,\&\, b \vee a \,\&\, \overline{b} = a, \\ & \xi' \,\&\, a \vee \xi'' \,\&\, a \,\&\, b = \xi' \,\&\, a, \text{ if } \xi' \geq \xi''. \end{aligned} \tag{4}$$

Here $a, b \in \{0, 1\}$, and $\xi', \xi'' \in [0, 1]$.

As a result expression (3) will become:

$$\Phi_t = \underset{i=\overline{1,l}}{\vee} (\overline{p}_{1_i} \,\&\, \overline{p}_{2_i} \,\&\ldots\&\, \overline{p}_{k_i} \,\&\, \alpha_i). \tag{5}$$

The following property holds: If in expression (5) the further simplification on the basis of rules (4) is impossible, then for every i-th disjunctive term the set of all vertices corresponding to the variables that do not exist in it gives the maximum internally stable set with internal stability degree α_i.

This property allows us to propose the following method for finding the fuzzy internally stable set of the temporal fuzzy graph:

- write proposition (3) for all fuzzy subgraphs $\tilde{G}_t$, $t = \{1, 2, \ldots, T\}$;
- simplify proposition (3) by proposition (4) and present it as proposition (5);
- determine the maximal internally stable sets from the obtained disjunctive terms of the expansion (5), with the computed degrees.
- determine $\{\tilde{\Psi}_t\}$ and determine fuzzy internal stable set $\tilde{\Psi}$ by (1).

To construct the expression (5) we will convert a variable $\overline{p}_j$ from the expression (3) to a weighted binary vector $1\overline{P}_j$, and constant ξ to a weighted binary vector $\xi\overline{P}_0$. Here $\overline{P}_j = ||p_i^{(j)}||$ and $\overline{P}_0 = ||p_i^{(0)}||$ are binary vectors that have dimension of n. The elements of $\overline{P}_j$ and $\overline{P}_0$ are defined as:

$$p_i^{(j)} = \begin{cases} 1, & \text{if } i = j \\ 0, & \text{if } i \neq j \end{cases}, \text{ and } p_i^{(0)} = 0.$$

Example 3. Let n = 4, so, taking into account our rules of conversion, the expression $\overline{p}_2$ is equal to the vector *1.0(0, 1, 0, 0)* and constant $\xi = 0.4$ is equal to the vector *0.4 (0,0,0,0)*.

We define the operation $\leq$ "less or equal" between binary vectors. The binary vector $\overline{P}_1$ is less or equal than $\overline{P}_2$ if and only if each element of $\overline{P}_1$ is less or equal than corresponding element of vector $\overline{P}_2$. Or

$$(\overline{P}_1 \leq \overline{P}_2) \leftrightarrow (\forall i = \overline{1,n})[p_i^{(1)} \leq p_i^{(2)}].$$

Example 4. Let's consider the example to illustrate the paragraph above:

$$(0,1,0,1) \le (0,1,1,1).$$

Considering new algebra in the space of weighted binary vectors, we can construct an absorption rule:

$$a_1\overline{P}_1 \vee a_2\overline{P}_2 = a_1\overline{P}_1, \;\text{ if } a_1 \ge a_2 \text{ and } \overline{P}_1 \le \overline{P}_2. \tag{6}$$

Example 5. Let's consider the following example of absorption rule (6):

$0.6(0,1,0,1) \vee 0.4(0,1,1,1) = 0.6(0,1,0,1)$ and $0.6(0,1,0,0) \vee 0.7(0,0,0,0) = 0.7(0,0,0,0)$ correspond the rules of absorption in the space of binary elements $0.6 \,\&\, \overline{p}_2 \,\&\, \overline{p}_4 \vee 0.4 \,\&\, \overline{p}_2 \,\&\, \overline{p}_3 \,\&\, \overline{p}_4 = 0.6 \,\&\, \overline{p}_2 \,\&\, \overline{p}_4$ and $0.6 \,\&\, \overline{p}_2 \vee 0.7 = 0.7$.

Now we can construct the expression (5) using the conjunction operation and the rule of absorption of the weighted binary vectors using the following algorithm:

1°. Each element of the first bracketed expression ($j = 1$) of the expression (3) is converted to the weighted binary vectors. The result is to be written in the first n elements of the vector $\overline{V}_1 = ||v_i^{(1)}||, i = \overline{1, n^2}$.
2°. Increment j ($j: = 2$).
3°. Each element of the bracketed j of expression (3) is also converted to the weighted binary vectors. The result is to be written in the first n elements of the vector $\overline{V}_2 = ||v_i^{(2)}||, i = \overline{1, n}$.
4°. The next stage consists of conjunction of two vectors $\overline{V}_1$ and $\overline{V}_2$. The result is placed into the vector $\overline{V}_3 = ||v_i^{(3)}||, i = \overline{1, n^2}$. While placing elements into vector $\overline{V}_3$, each new element, is compared with previous ones with use of the rule (6).
5°. All the elements of the buffer $\overline{V}_3$ should be copied to the buffer $\overline{V}_1$ ($v_i^{(1)} := v_i^{(3)}, i = \overline{1, n^2}$).
6°. Increment j ($j: = j+1$).
7°. If $j \le n$, the next step is 3°, otherwise 8°.
8°. Expression (5) is to be built using element in the buffer $\overline{V}_1$.

4 Numerical Example

We find fuzzy internal stable set for the fuzzy temporal graph $\tilde{G}$ shown in Fig. 1. To do this, we define fuzzy internal stable sets $\{\tilde{\Psi}_t\}$ for fuzzy subgraphs $\tilde{G}_1$, $\tilde{G}_2$ and $\tilde{G}_3$ shown in Fig. 3:

The corresponding expression (3) for fuzzy subgraph $\tilde{G}_1$ has the following form:

$$\Phi_1 = (\overline{p}_1 \vee \overline{p}_2 \vee 0.8) \,\&\, (\overline{p}_1 \vee \overline{p}_4 \vee 0.8) \,\&\, (\overline{p}_2 \vee \overline{p}_4 \vee 0.5) \,\&\, (\overline{p}_3 \vee \overline{p}_4 \vee 0.8).$$

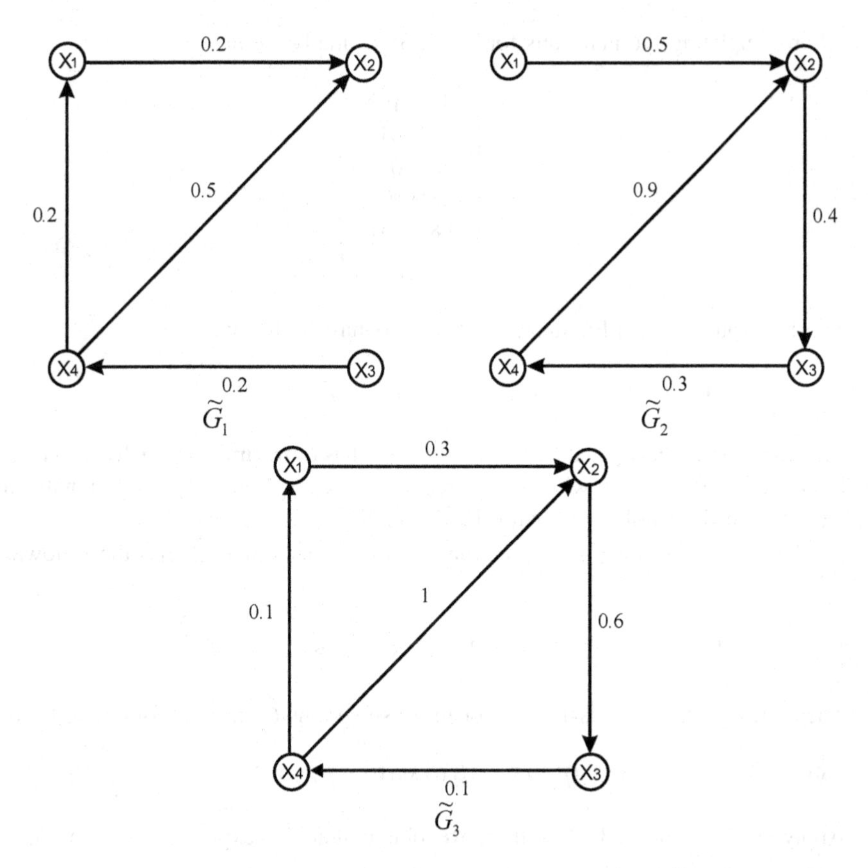

Fig. 3. Fuzzy subgraphs for fuzzy temporal graph $\tilde{G}$

Before the first iteration of the algorithm vectors $\overline{V}_1$ and $\overline{V}_2$ are the following:

$$\overline{V}_1 = \begin{pmatrix} 1(1000) \\ 1(0100) \\ 0.8(0000) \end{pmatrix} \overline{V}_2 = \begin{pmatrix} 1(1000) \\ 1(0001) \\ 0.8(0000) \end{pmatrix}$$

After the first iteration of the algorithm we have:

$$\overline{V}_1 = \begin{pmatrix} 1(1000) \\ 1(0101) \\ 0.8(0000) \end{pmatrix}.$$

After completing the iterations for $j = 2, 3, 4$, finally we have:

$$\overline{V}_1 = \begin{pmatrix} 1(1110) \\ 1(1001) \\ 1(0101) \\ 0.8(0100) \\ 0.8(0001) \\ 0.5(0000) \end{pmatrix}.$$

So, the expression (5) for fuzzy subgraph $\tilde{G}_1$ has the form:

$$\Phi_1 = 1\overline{p}_1\overline{p}_2\overline{p}_3 \vee 1\overline{p}_1\overline{p}_4 \vee 1\overline{p}_2\overline{p}_4 \vee 0.8\overline{p}_2 \vee 0.8\overline{p}_4 \vee 0.5.$$

This expression means that fuzzy subgraph $\tilde{G}_1$ has 6 maximal internally stable sets $\{x_4\}$, $\{x_2, x_3\}$ and $\{x_1, x_3\}$ with the degree 1, $\{x_1, x_3, x_4\}$ and $\{x_1, x_2, x_3\}$ with the degree 0.8, and the whole set X with the degree 0.5.

It follows that the fuzzy set of internal stability of subgraph $\tilde{G}_1$ has the following form:

$$\tilde{\Psi}_1 = \{<1/1>, <1/2>, <0.8/3>, <0.5/4>\}.$$

The corresponding expression (3) for fuzzy subgraphs $\tilde{G}_2$ has the following forms:

$$\Phi_2 = (\overline{p}_1 \vee \overline{p}_2 \vee 0.5) \,\&\, (\overline{p}_2 \vee \overline{p}_3 \vee 0.6) \,\&\, (\overline{p}_2 \vee \overline{p}_4 \vee 0.1) \,\&\, (\overline{p}_3 \vee \overline{p}_4 \vee 0.8).$$

Applying the proposed algorithm, we obtain that the expression (5) for fuzzy subgraph $\tilde{G}_2$ has the form:

$$\Phi_2 = 1\overline{p}_2\overline{p}_3 \vee 1\overline{p}_2\overline{p}_4 \vee 1\overline{p}_1\overline{p}_3\overline{p}_4 \vee 0.6\overline{p}_1\overline{p}_4 \vee 0.7\overline{p}_2 \vee 0.5\overline{p}_4 \vee 0.1.$$

This expression means that fuzzy subgraph $\tilde{G}_2$ has 7 maximal internally stable sets $\{x_2\}$, $\{x_1, x_4\}$ and $\{x_1, x_3\}$ with the degree 1, $\{x_2, x_3\}$ with the degree 0.6, $\{x_1, x_3, x_4\}$ with the degree 0.7, $\{x_1, x_2, x_3\}$ with the degree 0.5, and the whole set X with the degree 0.1.

It follows that the fuzzy set of internal stability of subgraph $\tilde{G}_2$ has the following form:

$$\tilde{\Psi}_2 = \{<1/1>, <1/2>, <0.7/3>, <0.1/4>\}.$$

The corresponding expression (3) for fuzzy subgraph $\tilde{G}_3$ has the following form:

$$\Phi_3 = (\overline{p}_1 \vee \overline{p}_2 \vee 0.7) \,\&\, (\overline{p}_2 \vee \overline{p}_3 \vee 0.4) \,\&\, (\overline{p}_1 \vee \overline{p}_4 \vee 0.9) \,\&\, (\overline{p}_2 \vee \overline{p}_4) \,\&$$
$$\& \, (\overline{p}_3 \vee \overline{p}_4 \vee 0.9).$$

Applying the proposed algorithm, we obtain that the expression (5) for fuzzy subgraph $\tilde{G}_3$ has the form:

$$\Phi_3 = 1\overline{p}_2\overline{p}_4 \vee 1\overline{p}_1\overline{p}_2\overline{p}_3 \vee 1\overline{p}_1\overline{p}_3\overline{p}_4 \vee 0.7\overline{p}_3\overline{p}_4 \vee 0.9\overline{p}_2 \vee 0.4\overline{p}_4.$$

It follows that the fuzzy internally stable set of subgraph $\tilde{G}_3$ has the following form:

$$\tilde{\Psi}_3 = \{<1/1>, <1/2>, <0.9/3>, <0/4>\}.$$

From the last obtained expressions it follows that the fuzzy internally stable set of the graph $\tilde{G}$ takes the form:

$$\tilde{\Psi} = \bigcap_{i=\overline{1,3}} \tilde{\Psi}_i = \{<1/1>, <1/2>, <0,7/3>, <0/4>\}.$$

This set means, in particular, that in the considered graph at any instant of time there exists a subset of two vertices unconnected with each other; there exists a subset of three vertices with the degree of internal stability 0.7.

5 Conclusion

In this paper, we considered the concept of fuzzy internally stable set of fuzzy temporal graph, i.e., such graph, in which the degree of connectivity of the vertices is changed in discrete time. The fuzzy internally stable set is an invariant of fuzzy temporal graph, so that its characteristic is not changed under its isomorphic transformations. A fuzzy internally stable set determines the greatest degree of internal stability of temporal fuzzy graph for a given number of vertices at any time. The method and algorithm for finding internally stable set were considered. The example of finding the fuzzy internally stable set of fuzzy temporal graph was considered as well.

Acknowledgments. This work has been supported by the Ministry of Education and Science of the Russian Federation under Project "Methods and means of decision making on base of dynamic geographic information models" (Project part, State task 2.918.2017), and the Russian Foundation for Basic Research, Project № 18-01-00023a.

References

1. Ore, O.: Theory of graphs. American Mathematical Society Colloquium Publications, Providence (1962)
2. Kaufmann, A.: Introduction a la theorie des sous-ensemles flous. Masson, Paris (1977)
3. Christofides, N.: Graph Theory: An Algorithmic Approach. Academic press, London (1976)
4. Rosenfeld, A.: Fuzzy graphs. In: Zadeh, L.A., Fu, K.S., Shimura, M. (eds.) Fuzzy Sets and Their Applications, pp. 77–95. Academic Press, New York (1975)
5. Monderson, J., Nair, P.: Fuzzy Graphs and Fuzzy Hypergraphs. Heidelberg; New-York, Physica-Verl (2000)

6. Mordeson, J.N., Peng, C.S.: Operations on fuzzy graphs. Inf. Sci. **79**, 159–170 (1994)
7. Kostakos, V.: Temporal graphs. Proc. Phys. A Stat. Mech. Appl. **388**(6), 1007–1023 (2008)
8. Bershtein, L., Bozhenyuk A.: The using of temporal graphs as the models of complicity systems. Izvestiya UFY. Technicheskie nayuki, №4 (105), pp. 198–203. TTI UFY, Taganrog (2010)
9. Barzilay, R., Elhadad, N., McKeown, K.: Inferring strategies for sentence ordering in multidocument news summarization. J. Artif. Intell. Rese. **17**, 35–55 (2002)
10. Bramsen, P.J.: Doing time: inducing temporal graphs. Technical report, Massachusetts Institute of Technology (2006)
11. Baldan, P., Corradini, A., König, B.: Verifying finite-state graph grammars: an unfolding-based approach. In: Gardner, P., Yoshida, N. (eds.) CONCUR 2004. LNCS, vol. 3170, pp. 83–98. Springer, Heidelberg (2004). https://doi.org/10.1007/978-3-540-28644-8_6
12. Baldan, P., Corradini, A., Konig, B.: Verifying a behavioural logic for graph transformation systems. In: Proceedings of COMETA 2003, ENTCS, vol.104, pp. 5–24. Elsevier (2004)
13. Erten, C., Harding, P.J., Kobourov, S.G., Wampler, K., Yee, G.: Exploring the computing literature using temporal graph. https://www2.cs.arizona.edu/~kobourov/vda_final.pdf
14. Dittmann, F., Bobda, C.: Temporal graph placement on mesh-based coarse grain reconfigurable systems using the spectral method. In: Rettberg, A., Zanella, M.C., Rammig, J. (eds.) IESS 2005. IOLCS, vol. 184, pp. 301–310. Springer, Boston, MA (2005). https://doi.org/10.1007/11523277_29
15. Bershtein, L., Bozhenyuk, A., Rozenberg, I.: Determining the strong connectivity of fuzzy temporal graphs. OP&PM **18**(3), 414–415 (2011)
16. Bershtein, L., Bozhenyuk, A., Rozenberg, I.: Definition method of strong connectivity of fuzzy temporal graphs. Vestnik RGUPS, №3 (43), pp. 15–20. RGUPS, Rostov-on-Don (2011)
17. Bershtein, L., Belyakov S., Bozhenyuk, A.: The using of fuzzy temporal graphs for modeling in GIS. Izvestiya UFY. Technicheskie nayuki, №1 (126), pp. 121–127. TTI UFY, Taganrog (2012)
18. Bozhenyuk, A., Belyakov, S., Rozenberg, I.: Coloring method of fuzzy temporal graph with the greatest separation degree. Adv. Intell. Syst. Comput. **450**, 331–338 (2016)
19. Bershtein, L., Bozhenuk, A.: Maghout method for determination of fuzzy independent, dominating vertex sets and fuzzy graph kernels. Int. J. Gen Syst **30**(1), 45–52 (2001)

Prioritisation of Nielsen's Usability Heuristics for User Interface Design Using Fuzzy Cognitive Maps

Rita N. Amro[1(✉)], Saransh Dhama[1(✉)], Muhanna Muhanna[2(✉)], and László T. Kóczy[3,4(✉)]

[1] Faculty of Electrical Engineering and Informatics, Budapest University of Technology and Economics, Budapest, Hungary
rita.amro@gmail.com, saransh014@gmail.com

[2] Department of Computer Graphics, Princess Sumaya University for Technology, Amman, Jordan
m.muhanna@psut.edu.jo

[3] Department of Telecommunication and Media Informatics, Budapest University of Technology and Economics, Budapest, Hungary
koczy@tmit.bme.hu

[4] Department of Information Technology, Szechenyi Istvan University, Gyor, Hungary

Abstract. Usability Heuristics are being widely used as a means of evaluating user interfaces. However, little existing work has been done that focused on assessing the effect of these heuristics individually or collectively on said systems. In this paper, the authors propose an approach to evaluating the usability of systems that deploys a prioritised version of Nielsen's usability heuristics. Fuzzy cognitive maps were used to prioritise the original heuristics according to experts in both fields. Using either set of heuristics evaluators can identify the same number of usability issues. However, when trying to enhance the overall usability of a system, the prioritised set of heuristics can help stakeholders focus their limited resources on fixing the subset of issues that collectively has the worst effect on the usability of their systems during each iteration. To test the findings proposed by authors several websites were evaluated for various usability problems. The experimental results show that by using the proposed heuristics, evaluators were able to find a comparable number of problems to those who used Nielsen's, the prioritised heuristics resulted in an ordered list of issues based on their effect on usability. Therefore, the authors believe that heuristic evaluation in general, and their introduced heuristics in particular, are effective in dealing with issues when facing situations of limited resources.

Keywords: Fuzzy cognitive maps · Human computer interaction
Heuristic evaluation · Usability

J. Medina et al. (Eds.): IPMU 2018, CCIS 853, pp. 511–522, 2018.
https://doi.org/10.1007/978-3-319-91473-2_44

1 Introduction

Based on the usability heuristics given by Nielsen, the authors proposed a prioritised set of heuristics according to how severely they affect the usability of a user interface. The basic idea being helping decision-makers allocate their limited time and resources towards fixing the issues that would enhance their systems usability the most.

1.1 Nielsen's Usability Heuristics

Several sets of design guidelines can be found in literature, such as D. Norman's Design Principles, and Shneiderman's Eight Golden Rules of Interface Design. However, the most widely used usability heuristics were proposed by Nielsen and Mack [1]. Therefore, due to their popularity and viability in assessing the usability of systems, they were chosen for this study. Nielsen's Heuristics [1] are 10 broad rules of thumb and not specific usability guidelines. They are mainly used by experts conducting Heuristic Evaluation sessions. This type of usability evaluation is an inspection method for computer software that helps identify usability issues in the user interface design. It specifically involves evaluators/experts examining the interface and judging its compliance with recognised usability principles (the "Heuristics") [1].

The authors are convinced that there is a need in the domain of interface evaluation for a more effective and efficient method adapted to the evaluation of user interfaces. Such a method should be also inexpensive and should not require a great amount of learnability to be applied. As compared to other evaluation techniques heuristic evaluation is popularly used as it requires less time and money for implementation [2]. The heuristic evaluation can be applied to evaluating user interfaces as thoroughly yet more efficiently through the modification of Nielsen's original heuristics [3] into prioritised updated ones.

Similar to the work presented by Zhang et al. [4] as well as Mankoff et al. [5], to the domains of medical devices and ambient displays, respectively; the authors believe that modifying Nielsen's heuristics is the way to adapt heuristic evaluation. This paper explores the assumption that the ten Heuristics differ in their effect on the usability of a system/user interface as a whole. Although heuristics are presumed mutually exclusive, in the paper it is proposed that a set of underlying interconnections and relationships hold among them. These relationships can eventually render one user interface less usable than another although superficially they might both contain the same number of similarly-severity-rated yet corresponding-to-different-Heuristics usability problems.

1.2 Heuristic Evaluation

Heuristic evaluation is an easy to use usability evaluation technique used mainly to identify key usability issues in a product/software in a timely relatively inexpensive manner [1,3]. Heuristic Evaluation is usually conducted by three or more persons who independently check a product's compliance against a set of usability heuristics, identify violations of the heuristics, and assess their severity. During a heuristic evaluation, experts navigate through the interface and identify UI elements that violate usability heuristics. This inspection technique has been traditionally used to evaluate a wide range of user interfaces such as websites and desktop software applications, and it is mainly used to identify interface issues so they can be addressed in the design process.

2 Research Methodology

The proposed research methodology consists of two phases. In this investigation, the fuzzy cognitive map approach is proposed for relating the dynamic relations between Nielsen's usability heuristics with the help of experts in the field. The resulting prioritised heuristics are then put into practice in the second phase, where an experiment was designed to compare the effectiveness of Nielsen's heuristics with the proposed new set through evaluating known websites. It is worth noting that the focus of this paper is not on the usability of user interfaces, but rather on the effectiveness of applying the proposed prioritised heuristics in evaluating such systems. Furthermore, the paper presents the findings obtained in a list of issues, ordered by their effect on the overall usability of the system rather than their severity in terms of one heuristic. After discussing the results, the paper identifies some directions of future work and presents conclusions of the presented work.

2.1 Fuzzy Cognitive Maps

Fuzzy cognitive maps (FCM) offer a suitable tool to model phenomena with multiple components (so-called concepts), which mutually interact in positive and negative ways, to various degrees. (the latter being expressed by signed fuzzy membership degrees.) FCM models are suitable for simulating the behaviour of such multi-component systems starting with an initial state and converging possibly to so-called fixed-point states at each of the components. (In some FCMs, some of the concepts do not converge: they have periodical or chaotic limit behaviour. In our application field this could be avoided.) The vagueness and uncertainty occurring necessarily in such an application field, as user interface usability evaluation, it is best expressed by fuzzy membership degrees. The subjectivity of this field is very well compensated by the robustness of converging FCM models, thus the limit states will clearly express the quality of the given component or concept (in this case, the features of the modelled user interface.) Fuzzy Cognitive Maps, in general, are used to model systems consisting of multiple components [6] for system analysis and future-behaviour prediction. FCM were first introduced by Kosko [7] in 1986. However, FCM have become of significant research interest in the last decade [8] and are now widely used to analyse causal systems such as decision making, management, risk analysis, text categorisation, prediction among others [9].

In essence, a Fuzzy Cognitive Map can be developed by integrating the existing experience and knowledge of experts in a particular system by asking them to describe its structure and behaviour in different conditions [10].

The first step to modelling a given system using FCM is to define its essential components. The components are then represented by the nodes of the graph namely, concepts. These concepts are supplied with initial values determined by experts of the field in hand or using historical data. The initial values of concepts should be in the range of $[-1; +1]$, regular bipolar fuzzy graph [11], therefore, supplied values should be transformed to their representation in that interval before being fed into the FCM model.

The relationships among concepts are also specified in the range of $[-1; +1]$. Negative values are means of expressing inhibitory effects, while positive ones express amplification effects, and zero edges indicate no effect between the corresponding concepts [12].

Edge weights are usually given in the form of a quadratic matrix, called "weight matrix", where each element (wij) indicates the weight (w) concept (i) has on concept (j). No "self-loops" for a stable FCM modelling [13].

Proceeding the construction, the next states/values of the given system can be calculated iteratively until convergence [10] represented in one of the following:

1. Equilibrium at fixed numerical values.
2. Limit cycle behaviour, with the concept values falling in a loop of numerical values.
3. Chaotic behaviour, with each value reaching a variety of numerical values in a non-deterministic, random way.

2.2 Phase I: Prioritisation of Heuristics

Specifying Concepts, Relationships and Weights. Two expert heuristic evaluators were asked to specify concepts and the interconnections/connections that hold between the components. The result of their work was highlighted in a list of linguistic/verbally expressed relationships, e.g. Visibility has a high positive influence on Efficiency. Their linguistic set of connections was transformed into a weight matrix consisting of concepts, relationships and their weights. The FCM and the matrix was then constructed using FCM mapping software Mental Modeler [14]. The matrix can be seen in Fig. 1.

It can be seen that initially heuristics were assigned the same relationship and weight to usability. Relationships between the different concepts were also specified. Negative (−) values indicate a negative relationship where the increase of one Usability concept is expected to decrease the other usability concept. Whereas a positive (+) value indicates a positive relationship where the increase of one usability concept is expected to increase the other Usability concept. Zero-degree connections indicates that no relationship holds between the two.

	Visibility	Usability	Match	Control	Consistency	Prevention	Recognition	Efficiency	Aesthetic	Errors	Help
Visibility		1		0.5		1	1	1		-0.5	-0.5
Usability											
Match	1	1			0.5	1	1		0.5		
Control		1				0.5		1		-0.5	-0.5
Consistency	0.5	1	0.5			1	0.5	0.5	1	-0.5	-0.5
Prevention		1								-1	-1
Recognition	1	1	1		0.5	1				-1	-0.5
Efficiency		1		-0.2	0.5	-0.2				-0.2	
Aesthetic	1	1	0.5		0.5	0.5		0.5		-0.2	-0.5
Errors		1		-0.2		0.2	0.2	0.5			-1
Help		1								1	

Fig. 1. Concepts weight matrix

Table 1. FCM structural metrics

Structural metric	Values
Number of components	11
Number of connections	59
Connections/variables	5.36
Number of receiver components	1
Number of ordinary components	10
Density (# connections/total # of possible connections)	0.53

Construction of Corresponding FCM. Table 1 gives a overview of all the FCM entities. It can be noted that the only receiver concept is Usability; one that has incoming but no outgoing relationships/connections. After the construction of the model, and establishing its concepts, relationships and weights, the following was concluded: Level of Centrality. Centrality levels indicate how essential concepts are in a given model and can be considered a way to differentiate the importance and effect one concept holds on the model as a whole when compared to other components. When considering these levels and using Mental Modeler [14] it was found that concepts can be ordered as follows (in descending order):

1. Visibility of system status
2. Recognition rather than recall
3. Error prevention
4. Help users recognise, diagnose, and recover from errors
5. Consistency and standards
6. Match between system and the real world
7. Help and documentation
8. Aesthetic and minimalist design
9. Flexibility and efficiency of use
10. User control and freedom.

Specifying Initial Concept Values. Evaluators usually judge system's compliance to Heuristics in terms of severity levels. Therefore, they can be used to allocate the most resources to fixing the most severe problems and can also provide an indication for the need of re-evaluating design decisions. If the severity ratings indicate that several serious usability problems remain in an interface, it will probably be inadvisable to release it. However, stakeholders might consider releasing a system that has a few cosmetic problems. The severity of a usability problem is a combination of three factors:

1. How frequent is the usability problem; is it common or rare?
2. Can it be easily recovered from?
3. How persistent is the problem; is it a one-time problem that users can overcome/avoid once they know about it or will they be repeatedly bothered by it?

The outcome of heuristic evaluation sessions usually consists of a list of the complete set of usability problems that have been discovered, and each evaluator's severity rating of each problem. The following 0 to 4 rating scale can be used to rate the severity of usability problems [1]:

0 = I don't agree that this is a usability problem at all
1 = Cosmetic problem only: need not be fixed unless extra time is available on project
2 = Minor usability problem: fixing this should be given low priority
3 = Major usability problem: important to fix, so should be given high priority
4 = Usability catastrophe: imperative to fix this before product can be released.

An algorithm was developed to specify these ratings in fuzzy degrees: This algorithm[1] supposes that a heuristic/concept with no usability problems is given an initial value of 0.0 whereas, heuristics with usability problems are given a value ranging between $(0.0, -1.0]$ depending on the set usability problems reported by evaluators collectively. In scenarios without any issues reported in a given system, all initial state values are assigned a value of 0.0. As a result, the model would already be in an equilibrium with no negative effect exhibited on usability.

```
enum SeverityLevel: Double {
      case s1 = 0.0
      case s2 = -0.25
      case s3 = -0.5
      case s4 = -0.75
      case s5 = -1.0
}
func calculateInitialConceptValues(ofConcepts: [String], withProblems:
 [[SeverityLevel]], totalNumberOfProblems: Int) -> [String:Double] {
```

[1] The algorithm consists of an enumeration defining the possible severity levels in fuzzy membership degrees, a for-loop that loops through usability issues; corresponding to one heuristic at a time and calculates the total severity of all problems. The final severity level of problems corresponding to a given heuristic is calculated as a weighted average w.r.t the total number of usability issues in the system as a whole.

```
    var initialConceptValues: [String:Double] = [:]
    for (index,problemSet) in withProblems.enumerated(){
        let numberOfConceptProblems: Double = Double(problemSet.count)
        var averageSeverityOfConceptProblems = SeverityLevel.s1.rawValue
        var sumOfConceptProblemsSeverityLevels:Double = 0.0
}
for problem in problemSet {
    sumOfConceptProblemsSeverityLevels += problem.rawValue
}
averageSeverityOfConceptProblems =
sumOfConceptProblemsSeverityLevels/numberOfConceptProblems *
(numberOfConceptProblems Double(totalNumberOfProblems))
initialConceptValues[''\(ofConcepts[index])''] =
averageSeverityOfConceptProblems
      return initialConceptValues
}
```

Running Scenarios

Scenarios of Catastrophic Usability Problems. Initial tests were run to see whether the initial assumption, that each heuristic's effect on usability varies, holds. Therefore, the authors created and ran (10) different scenarios, each consisting of one catastrophic problem in only one of the concepts and measured each component's effect on usability individually. The following list enumerates which heuristics affect usability when present alone in a descending order; heuristics not included in the list showed no significant negative effect on usability when present alone:

1. Match between system and the real world
2. Visibility of system status, Recognition rather than recall, Consistency and standards
3. Aesthetic and minimalist design, Flexibility and efficiency of use.

Scenarios of Major Usability Problems. A new set consisting of (10) scenarios was run, each consisting of one major problem in only one of the heuristics and their effect on usability was measured. The following list indicates which heuristics affect usability when present alone; heuristics not included in the list showed no negative effect on usability when present alone:

– Match between system and the real world
– Visibility of system status
– Recognition rather than recall
– Consistency and standards
– Aesthetic and minimalist design.

Each of these concepts gave the same negative effect on usability although they varied in initial scenarios when specified catastrophic rather than major. It can also be noted that "Flexibility and efficiency of use" is no longer present in the above list; meaning it had negligible effect on usability with this specific severity level. The results are depicted in Fig. 2.

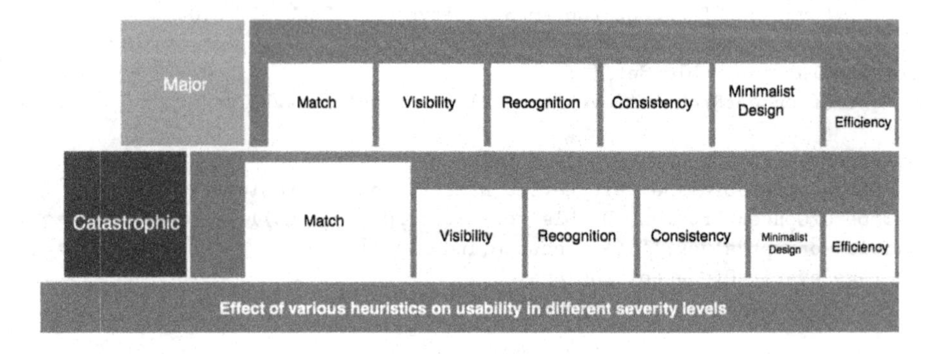

Fig. 2. Effect of heuristics on usability in different severity levels

Scenarios of Cosmetic Usability Problems. Jakob Nielsen, suggested that usability problems with cosmetic effect on a given system's usability need not be fixed unless extra time is available on hand. After running scenarios with the above hypothesis, the FCM verified the claim; cosmetic problems had negligible effect on usability.

Multiple Scenarios Consisting of Multiple Usability Problems Corresponding to Different Heuristics. From the investigation it can concluded that a usability problem in "Match between the system and the real world" has a greater negative effect on usability than one in "Recognition rather than recall". However, in the hypothesis proposed by the authors, it was discussed that there is a developed need to prioritise a set of problems for fixing when trying to increase the usability of an unusable system/user interface; since stakeholders often deal with limited resources and short product iterations. Therefore, multiple scenarios were run to see whether a certain heuristic would become of higher priority when having more usability problems than one that had greater priority in general, the values were specified using the developed algorithm.

A few scenarios were run some of which supported the proposed claims; the following is an example of how priority changes according to number issues rather than merely the severity level of given problems:

The authors ran a scenario consisting of two catastrophic problems in "Recognition rather than recall" and only one in "Match between the system and the real world". In the initial tests, "Match between the system and the real world" had a greater negative effect on usability than "Recognition rather than recall". However, when more than one problem in the latter was present, it had collectively a worse effect on usability although initially the latter was proved of higher importance to usability in general. These outcomes can prove valuable when wanting to increase the usability of a system as much as possible during an iteration when resources do not allow addressing all issues at once.

2.3 Phase 2: Initial Testing

For the scope of this paper, several websites were evaluated using the proposed model, in order to test the effectiveness of the prioritized heuristics FCM model in real-life test cases. Some of the websites evaluated indicated a minimal number of cosmetic usability issues. Therefore, their scenarios showed little-to-no effect on the overall usability of their systems when run through the model. This further supports the applicability of the proposed model since issues with low severity levels are those of low priority that do not have to be addressed immediately in cases dealing with scarce resources. The test cases was based on the following hypothesis: using the prioritised heuristics and according to the issues' severity levels, an ordered list of usability problems can be deduced according to their negative effect on usability. To conduct the required comparison, 4 evaluators were recruited, with a median of 1–2 years of evaluation experience. Each of the participants conducted an individual evaluation and provided a list of usability issues with their 4-point scale severity level rating used in [1]. The websites that will be addressed in this paper are www.irctc.co.in, www.venosc.com, and www.mohe.gov.jo, hereinafter referred to as w1, w2, and w3, respectively. These websites showed a huge number of usability violations/issues.

3 Results and Discussion

Table 2 summarises the total number of issues found in each heuristic and their severity levels in all 3 websites. Each 5-tuple enumerates the number of catastrophic issues, the number of major issues, the number of minor issues, the number of cosmetic issues, and the total number of issues, respectively, in a given heuristic.

Table 2. Heuristic evaluation summary of w1, w2 and w3

Heuristic	w1	w2	w3
Visibility of system status	2,1,0,0,3	0,1,0,0,1	1,2,1,1,5
Match between the system and the real world	1,2,0,0,3	0,0,1,0,1	1,2,0,0,3
User control and freedom	0,1,1,0,2	0,1,0,0,1	0,0,1,1,2
Consistency and standards	2,2,0,0,4	1,0,1,0,2	2,1,2,0,5
Error prevention	1,1,1,0,3	1,2,0,0,3	0,0,2,1,3
Recognition rather than recall	1,0,0,0,1	0,1,0,0,1	1,2,2,0,5
Flexibility and efficiency of use	0,2,1,0,3	0,0,0,0,0	0,3,1,0,4
Aesthetic and minimalist design	1,1,2,0,4	1,1,0,1,3	0,2,1,1,4
Help users recognize, diagnose and recover from errors	0,0,0,0,0	0,0,0,0,0	0,2,0,0,2
Help and documentation	0,0,0,0,0	0,0,1,1,2	0,2,1,0,3

Tables 3, 4 and 5 identify the heuristic whose issues would increase usability the most when addressed first in said website. They also reflect the importance of defining the set of underlying connections between the heuristics and how

Table 3. The positive effect issues would have on usability if addressed first in w1

Heuristic	Effect on usability
Match between the system and the real world	23%
Flexibility and efficiency of use	22%
Visibility of system status	21%
Aesthetic and minimalist design	20%
Consistency and standards	19%
User control and freedom	18%
Recognition rather than recall	16%
Error prevention	14%

Table 4. The positive effect issues would have on usability if addressed first in w2

Heuristic	Effect on usability
Help and documentation	25%
Aesthetic and minimalist design	24%
Consistency and standards	22%
User control and freedom	22%
Visibility of system status	22%
Match between the system and the real world	21%
Error prevention	18%
Recognition rather than recall	17%

Table 5. The positive effect issues would have on usability if addressed first in w3

Heuristic	Effect on usability
Consistency and standards	27%
Recognition rather than recall	27%
Visibility of system status	27%
Aesthetic and minimalist design	26%
Flexibility and efficiency of use	26%
Match between the system and the real world	26%
Error prevention	25%
Help and documentation	25%
Help users recognize, diagnose and recover from errors	25%
User control and freedom	25%

beneficial they can be for design decisions in a situation where stakeholders are faced with limited resources.

Percentages presented in these Tables 3, 4 and 5 do not add up to 100% due to the fact that any slight changes in initial state values affects the model's convergence results. Therefore, when issues violating a certain heuristic are fixed, its corresponding heuristic state's initial value changes pushing the entire model to iterate differently until convergence.

4 Conclusion and Future Work

4.1 Conclusion

The study was started with the assumption that Nielsen's 10 Heuristics are not mutually exclusive and some exhibit a greater effect on the Usability of systems more than others. With the use of FCM to model the behaviour of these heuristics and their relationships, and after running a variety of scenarios the following conclusions were drawn:

1. Usability Heuristics can vary in importance in the phase of Design though are equally essential when evaluating systems.
2. The prioritization of heuristics in design is subject to the issues reported in a system, the heuristics they violate and their severity ratings and cannot be generalised.
3. The prioritised version of Nielsen's Heuristics can help decision-makers make valuable design decisions when wanting to increase the usability of their systems significantly during each iteration given the available resources.
4. Using the new set of prioritised Heuristics it can be seen that they are not mutually exclusive as one might think, rather, there are implicit relationships between them. These relationships can be very useful when designing systems.

4.2 Future Work

Future work aims to further study the effects usability heuristics exhibit on one another to develop, and refine the FCM model accordingly.

Acknowledgement. This work was supported by National Research, Development and Innovation Office (NKFIH) K108405, K124055.

References

1. Nielsen, J., Mack, R.L.: Usability Inspection Methods, pp. 25–64. Wiley, New York (1994)
2. Nielsen, J.: Non command user interfaces. Commun. ACM **36**(4), 83–99 (1993)
3. Nielsen, J., Molich, R.: Heuristic evaluation of user interfaces. In: Proceedings of ACM CHI 1990 Conference, Seattle, WA, pp. 249–256 (1990)

4. Zhang, J.J., Johnson, T.R., Patel, V.L., Paige, D.L., Kubose, T.: Using usability heuristics to evaluate patient safety of medical devices. J. Biomed. Inf. **36**, 23–30 (2003)
5. Mankoff, J., Dey, A.K., Hsieh, G., Kientz, J., Lederer, S., Ames, M.: Heuristic evaluation of ambient displays. In: Proceedings of the SIGCHI Conference on Human Factors in Computing Systems, Ft. Lauderdale, FL (2003)
6. Kosko, B.: Fuzzy cognitive maps. Int. J. Man-Mach. Stud. **24**(1), 65–75 (1986)
7. Perusich, K.: System diagnosis using fuzzy cognitive maps. InTech (2010)
8. Stylos, C.D., Georgopoulos, V.C., Groumpos, P.P.: The use of fuzzy cognitive maps in modeling systems. In: Proceedings of 5th IEEE Mediterranean Conference on Control and Systems, Paphos, Cyprus (1997)
9. Papageorgiou, E.I., Salmeron, J.L.: A review of fuzzy cognitive map research at the last decade. IEEE Trans. Fuzzy Syst. (IEEE TFS) **21**(1), 66–79 (2013)
10. Papageorgiou, E.I.: Review study on fuzzy cognitive maps and their applications during the last decade. In: IEEE International Conference on Fuzzy Systems (2011)
11. Akram, M., Dudek, W.A.: Neural Comput. Appl. **21**(Suppl 1), S197–S205 (2012)
12. Hatwágner, M.F., Kóczy, L.T.: Parameterization and concept optimization of FCM models. IEEE (2015)
13. Martchenko, A.S., Ermolov, I.L., Groumpos, P.P., Poduraev, J.V., Stylios, C.D.: Investigating stability analysis issues for fuzzy cognitive maps. In: 11th Mediterranean Conference on Control and Automation - MED 2003 (2003)
14. Gray, S., Gray, S., Cox, L., Henly-Shepard, S.: Mental modeler: a fuzzy-logic cognitive mapping modeling tool for adaptive environmental management. In: Proceedings of the 46th International Conference on Complex Systems, pp. 963–973 (2003)

Discrete Bacterial Memetic Evolutionary Algorithm for the Time Dependent Traveling Salesman Problem

Boldizsár Tüű-Szabó[1(✉)], Péter Földesi[2], and László T. Kóczy[1,3]

[1] Department of Information Technology, Széchenyi István University, Győr, Hungary
{tuu.szabo.boldizsar,koczy}@sze.hu
[2] Department of Logistics, Széchenyi István University, Győr, Hungary
foldesi@sze.hu
[3] Department of Telecommunications and Media Informatics, Budapest University of Technology and Economics, Budapest, Hungary

Abstract. The Time Dependent Traveling Salesman Problem (TDTSP) that is addressed in this paper is a variant of the well-known Traveling Salesman Problem. In this problem the distances between nodes vary in time (are longer in rush hours in the city centre), Our Discrete Bacterial Evolutionary Algorithm (DBMEA) was tested on benchmark problems (on bier127 and on a self-generated problem with 250 nodes) with various jam factors. The results demonstrate the effectiveness of the algorithm.

Keywords: Traveling Salesman Problem · Time Dependent · Heuristic

1 Introduction

The Time Dependent Traveling Salesman Problem (TDTSP) is an extension of the original Traveling Salesman Problem [1] (TSP) which is one of the most famous combinatorial optimization problem. In TDTSP the edge lengths vary in time making the problem more applicable in real-life (transportation, logistics) than the TSP. The aim is to find the tour with the lowest cost which visits each node once. Like the TSP it is also an NP-hard problem.

2 The DBMEA Algorithm

The Bacterial Evolutionary Algorithm (BEA) was first introduced by Nawa and Furuhashi for discovering the optimal parameters of a fuzzy rule based system [10]. BEA is inspired by the evolution of bacteria. Since then it has been used efficiently in solving many optimization problems. In 2002 Inoue et al. used the bacterial evolutionary algorithm for an interactive nurse scheduling optimization problem [5].

In 2005 we presented a continuous Bacterial Memetic Algorithm (BMA) which combined the Bacterial Evolutionary Algorithm with the Levenberg-Marquardt method for solving continuous optimization problems [2].

J. Medina et al. (Eds.): IPMU 2018, CCIS 853, pp. 523–533, 2018.
https://doi.org/10.1007/978-3-319-91473-2_45

In 2016 we introduced the continuous Discrete Bacterial Memetic Evolutionary Algorithm (DBMEA) for solving the Traveling Salesman Problem [6]. An improved version of DBMEA was presented with accelerated local search [12].

With some modifications the DBMEA was also able to solve efficiently TSP with Time Windows instances [7].

The DBMEA is based on the Furuhashi's BEA [10], combining it with local search techniques (2-opt and 3-opt), so it belongs to the group of memetic algorithms.

Memetic algorithms [9] can often be particularly effective in solving NP-hard optimization techniques (resulting near-optimal solutions within appropriate amount of time) because it can eliminate the disadvantages of both methods. Evolutionary algorithms search in the global search causing slow convergence speed. Local search methods search in the neighborhood of the current tour, thanks to that they often get stuck in a local optimum.

The process of the DBMEA consists of the following steps:

- Creating the Initial Population

In each iteration until the stoppage criterion is met:

- Bacterial mutation for each bacterium
- Combined 2-opt and 3-opt local search
- Gene transfer performed on the population

2.1 Creating the Initial Population

In DBMEA the bacteria form a population. Each bacterium represents a possible solution for the TDTSP problem.

In DBMEA a very simple permutation encoding is used. The starting point of the tour is indexed with 0 (for each tour it is the starting node, so it does not appear in the codes), and other nodes are mapped with a unique index (*1...n*). The tours are expressed as the sequence of these indices clearly identifying the tour. Each node needs to visit once, so the length of the tours is $n - 1$. An example of encoding can be seen in Fig. 1.

The population consists of randomly created individuals and three deterministic individuals. The deterministic individuals are the following:

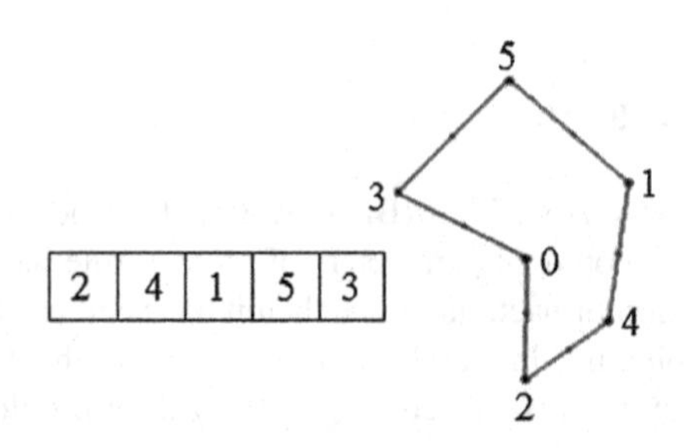

Fig. 1. The encoding of the tour

- Nearest neighbour (NN) heuristic: It creates a tour in which always the nearest unvisited city is visited.
- Secondary nearest neighbour (SNN) heuristic: in this tour always the second nearest unvisited city is visited.
- Alternating nearest neighbour (ANN) heuristic: It combines the above mentioned two methods. It represents a tour in which the nearest and second nearest unvisited cities are visited in alternating order.

2.2 Bacterial Mutation

The bacterial mutation operation – performing on every bacterium – consists of the following steps (Fig. 2):

- Creating a pre-defined number (N_{clones}) of clones from the original bacterium
- Dividing the bacterium into fixed length (not necessary coherent) segments
  The following steps are repeated until all segments are examined:
  - Selecting randomly a not yet mutated segment
  - Randomly changing the node sequence of the selected segment in the clones
  - Calculating the costs of the clones and the original bacterium
  - Selecting the best individual among the clones and the original bacterium
  - Copying the segment content of the best individual into the other clones and original bacterium
- Finally replacing the original bacterium with the best individual among the mutated original and clones in the population

The bacterial mutation results that each bacterium in the population becomes more or in worst case equally fit compared with the original bacteria.

In our algorithm two different types of mutation, the coherent segment mutation and the loose segment mutation are used.

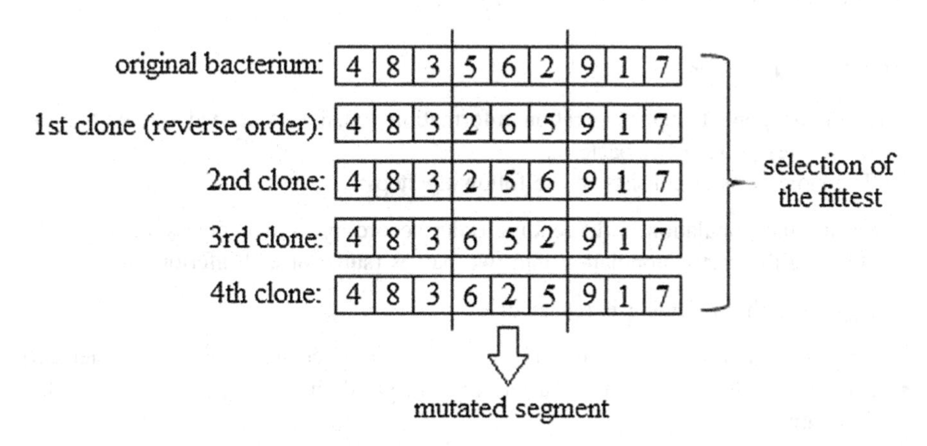

Fig. 2. Bacterial mutation

Coherent segment mutation: In this case adjacent elements form the segments. It is easy to execute: the chromosome is cut into segments with equal length (Fig. 3).

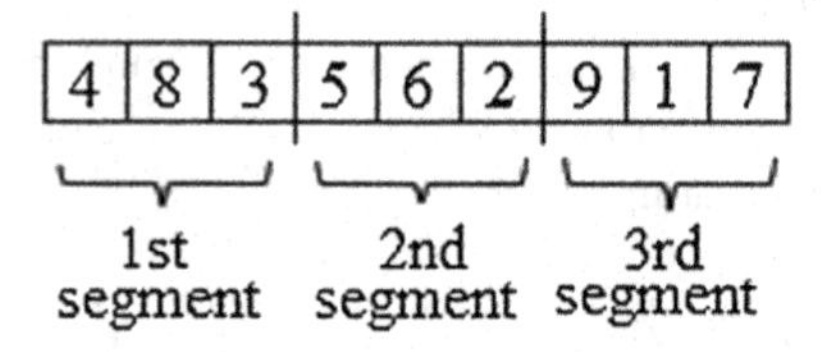

Fig. 3. Coherent segments

Loose segment mutation: As opposed to the coherent segment mutation, the segments of the bacterium do not need to consist of adjacent elements. The elements of the segments may come from different parts of the bacterium (Fig. 4).

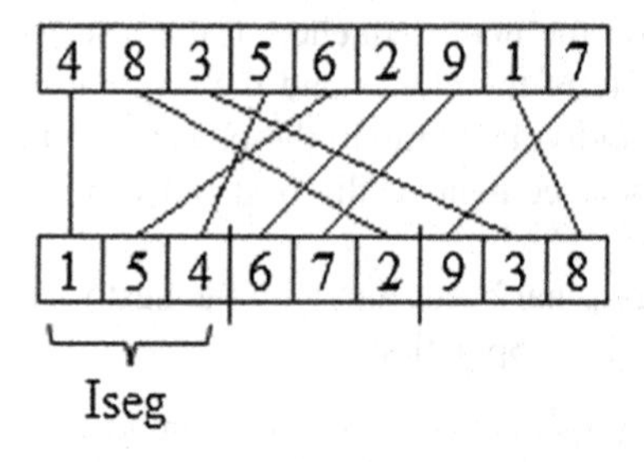

Fig. 4. Loose segments

The time complexity of the bacterial mutation is $O(N_{ind}N_{clones}n^2)$ in one generation, and the space requirement is $O(N_{ind}N_{clones}n)$ [3].

2.3 Gene Transfer

Through the gene transfer operation information transfer is carried out within the population resulting better bacteria.

The gene transfer consists of the following steps:

- Sorting the population in descending order according to their fitness values
- Dividing the sorted population into two halves (superior and inferior half)

N_{inf} times the following is repeated:

- Choosing randomly a bacterium from both parts (source and destination bacterium)
- Copying a part of the source bacterium with pre-defined ($I_{transfer}$) length into the destination bacterium
- Eliminating the double occurrence of nodes to keep the same length (Fig. 5).

The time complexity of the gene transfer consists of the following components:

- The calculation time of fitness values $O(N_{ind}n)$
- The time complexity of sorting the population in a descending order, based on the fitness values $O(N_{ind}logN_{ind})$
- The calculation time of the new fitness value of the modified bacterium, and its reinsertion into the population $O\left(N_{inf}(n + N_{ind})\right)$

The total time complexity of the gene transfer operation in each generation is $C_{GT} = O\left(N_{ind}(n + logN_{ind}) + N_{inf}(n + N_{ind})\right)$ [3].

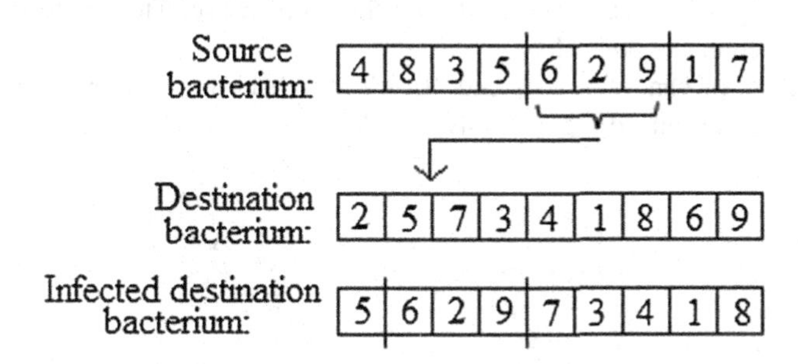

Fig. 5. Gene transfer

2.4 Local Search

Adding local search techniques to the process of an evolutionary algorithm usually greatly increases the effectiveness of the algorithm.

They search for improvements in the local environment of the candidate solutions, therefore they usually find only a local optimal solution.

In DBMEA for TDTSP a combined 2-opt and 3-opt local search is used. First, the tour is improved by 2-opt steps. And when no further improvement is possible with 2-opt, then 3-opt will be applied for the tour. The local search is stopped, when the tour cannot be improved further by 3-opt steps.

Calculation of the modified tour's cost is more time consuming, requires more operations for TDTSP than for TSP (it can be calculated from the cost of the original tour by addition of the length of the added edges and subtraction of the length of the deleted edges) because the edge exchange can also affect the length of edges that did not participate in it.

In DBMEA the following speed-up techniques were used to accelerate the local search:

- Candidate list [4]: It contains the indices of the closest vertices in ascending order and is created for all vertices. During the local search only the pre-defined number of closest vertices (candidate lists contain them) are examined for each vertex, because a shorter edge is more likely to be part of a good solution.

- "Don't look back bits" [4]: Each vertex in assigned to a "don't look back bit". If no improving was found for a given vertex v, then until an incident edge changes, do not consider v (changes its "don't look back bit" to 1).

2-opt Local Search

2-opt local search replaces two edge pairs in the original graph to improve the tour (Fig. 6).

Edge pairs (AB, CD) are iteratively replaced with AC and BD edges. If the new tour has lower cost, then edge pairs are exchanged; *AB* and *CD* edges are deleted from the graph and *AC* and *BD* edges are inserted instead (Fig. 6). One of the sub-tours between the original edges is reversed after the 2-opt move. This is stopped when no further improvement is possible.

With the use of candidate lists and "Don't look back bits" the number of possible edge changes was reduced significantly.

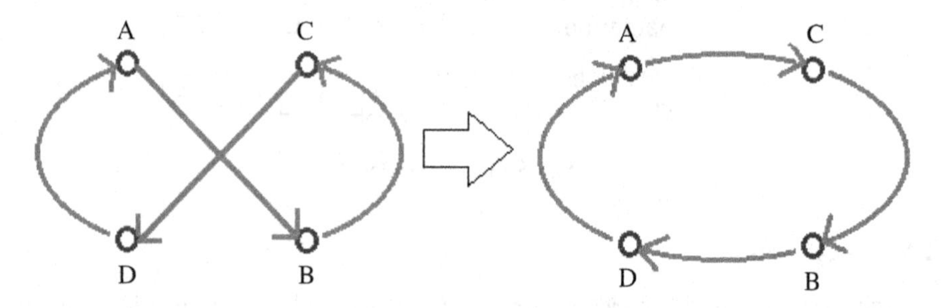

Fig. 6. 2-opt local steps

3-opt Local Search

The 3-opt local search works on edge triples. The deleting of three edges results in three sub-tours. There are four possible ways to reconnect these sub-tours (Fig. 7). The output of the 3-opt step is always the tour with the lowest cost.

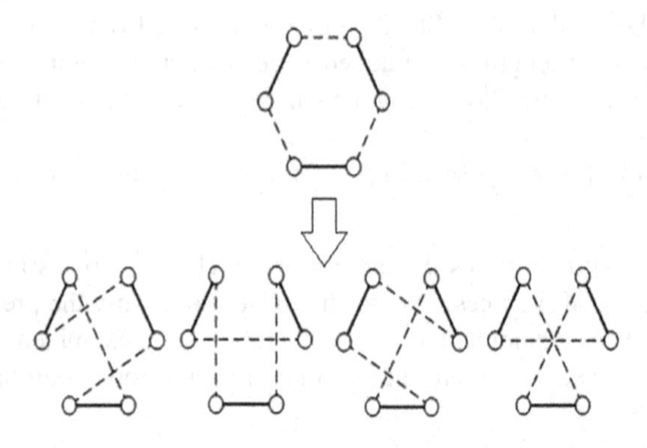

Fig. 7. 3-opt local steps

3 Computational Results

The DBMEA algorithm was tested on bier127 problem and a self-generated problem with 250 nodes.

Recently the TDTSP was examined by Schneider [11] and Li *et al.* [8]. Schneider tested his simulated annealing algorithm on bier127 problem with different jam factors. Li et al. also examined their record-to-record travel algorithm on bier127 problem. Bier127 problem consists of 127 beer gardens in the area of Augsburg.

Schneider defined a traffic jam region (coordinate of the left corner point (7080, 7200), width is 6920, and the height is 9490). The working hours of the salesman are between 9 am and 3 pm. The rush hours are between 12 pm and 3 pm. In this period the edge lengths between two nodes in the traffic region are multiplied with a jam factor. The travel speed is computed by dividing the total distance of the optimal TSP tour (118293.524) by the working hours of the salesman. The travel speed is held constant for all values of the jam factor.

DBMEA was also tested on a self-generated instance with 250 nodes (s250). A traffic jam region is defined (coordinate of the left corner point (125, 125), width is

Table 1. Comparison of results on the bier127 problem with different jam factors

Jam factor	DBMEA			Simulated annealing	RTR algorithm
	Best value	Average value	Average time [s]	Best value	Best value
1.00	118293.524	118293.524	26.878	118293.524	118293.524
1.03	118749.356	118749.356	21.993	118749.356	118796.154
1.04	118901.300	118901.300	20.237	118901.300	119971.191
1.05	119053.244	119053.244	18.802	119053.244	119503.279
1.06	119153.582	119189.167	49.288	119153.582	119857.323
1.10	119313.720	119313.720	36.190	119313.720	119957.387
1.20	119714.065	119714.065	40.395	119714.065	119714.065
1.30	120114.410	120130.594	38.964	120114.410	120637.093
1.38	120434.687	120438.749	48.904	120434.687	120434.687
1.39	120453.554	120453.554	34.028	120453.554	120453.554
1.50	120571.743	120571.743	37.150	120571.743	120617.178
1.60	120679.186	120821.535	52.399	120679.186	121108.329
1.70	120786.630	120866.307	47.350	120786.630	120898.269
1.80	120894.074	120966.040	43.558	120894.074	121195.816
1.90	121001.518	121102.499	52.044	121001.518	121148.519
2.02	121125.195	121125.195	24.668	121125.195	121298.538
3.00	121125.195	121125.195	15.839	121125.195	122222.204
10.00	121125.195	121125.195	26.750	121125.195	121167.051
100.00	121125.195	121125.195	45.636	121125.195	122280.886
2000.00	121125.195	121125.195	42.277	121125.195	121417.575

250, and the height is 250). Vertex coordinates have been generated from a uniform distribution, 70% of the nodes were placed within the traffic jam region. The working hours and the calculation of travel speed is the same as in the case of bier127.

The algorithm was tested with the following parameters:

- the number of bacteria in the population ($N_{ind} = 100$)
- the number of clones in the bacterial mutation ($N_{clones} = n_{cities}/10$)
- the number of infections in the gene transfer ($N_{inf} = 40$)
- the length of the chromosomes ($I_{seg} = n_{cities}/20$)
- the length of the transferred segment ($I_{trans} = n_{cities}/10$)
- length of the candidate lists (square root of the number of cities)

Table 2. Results on the s250 problem with different jam factors

Jam factor	DBMEA		
	Best value	Average value	Average time [s]
1.00	5593.721	5598.908	385.005
1.01	5602.693	5606.621	333.424
1.02	5607.850	5612.466	332.826
1.03	5614.172	5619.788	346.240
1.04	5618.168	5621.647	374.396
1.05	5618.168	5621.516	320.237
1.1	5618.168	5623.638	277.473
1.2	5618.168	5626.335	309.946
1.3	5618.168	5627.508	327.514
1.5	5618.168	5631.149	303.666
2.00	5618.168	5630.845	317.846

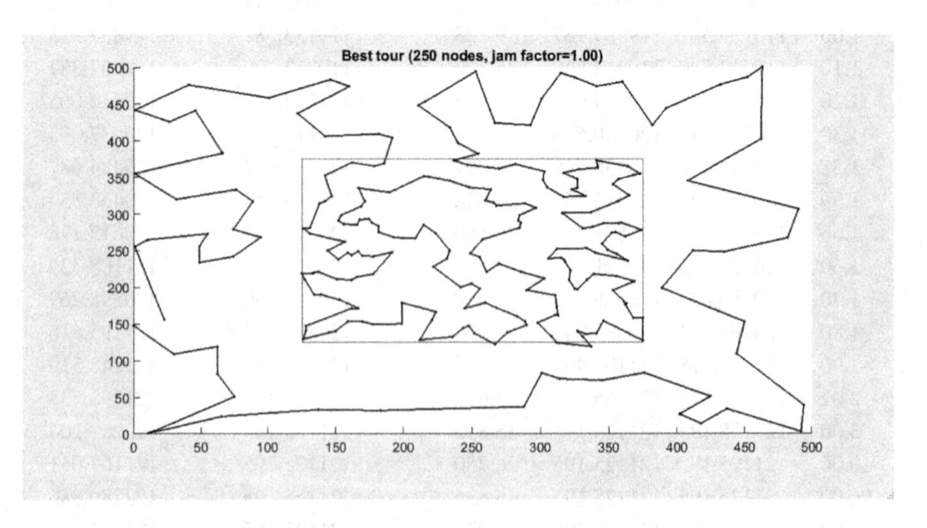

Fig. 8. Best tour (s250, jam factor = 1.00)

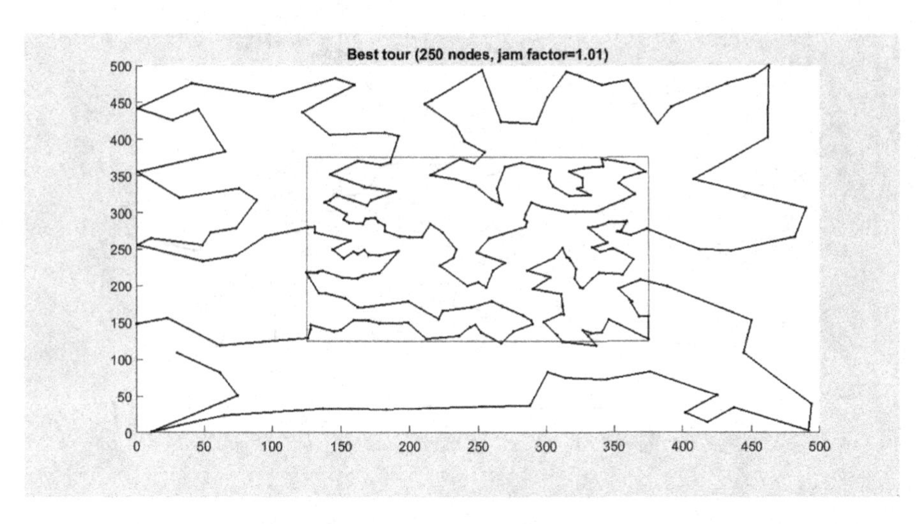

Fig. 9. Best tour (s250, jam factor = 1.01)

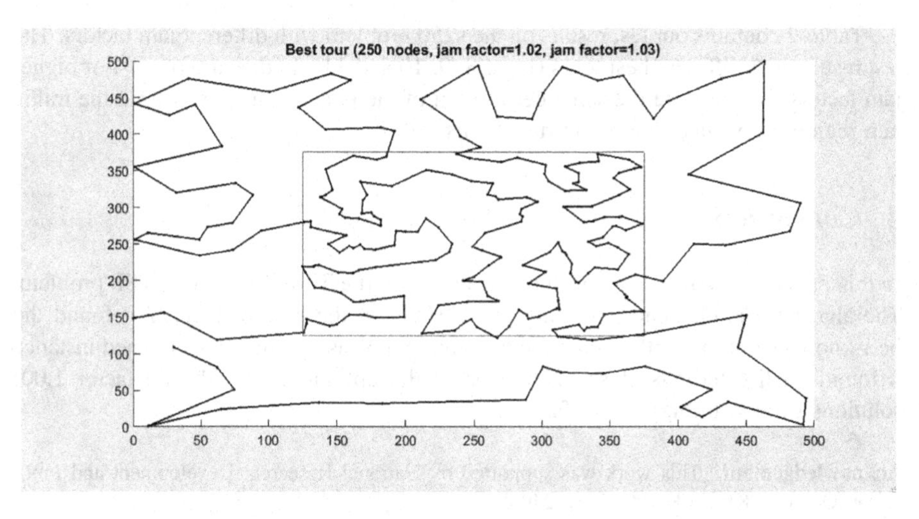

Fig. 10. Best tour (s250, jam factor = 1.02, jam factor = 1.03)

Our algorithm was tested on an Intel Core i7-7500U 2.7 GHz, 8 GB of RAM memory workstation under Linux Mint 18.2. The DBMEA was coded in C++. Our results were calculated by averaging 10 test runs.

In Table 1 comparison of our results with the Schneider's simulated annealing [11] and the record-to-record travel algorithm (RTR algorithm) [8] can be seen on the bier127 problem. In all cases the DBMEA found the same tours as the simulated annealing. In most cases the record-to-record travel algorithm failed to find the best-known tours. Unfortunately the run times are not comparable, in the case of simulated annealing the runtimes were not given, the RTR algorithm was tested on a much slower hardware than the DBMEA.

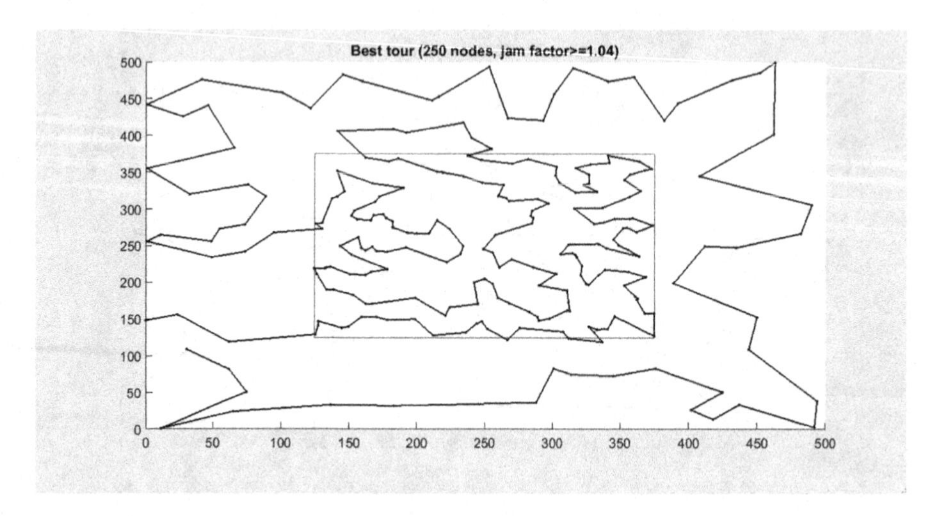

Fig. 11. Best tour (s250, jam factor $\geq$ 1.04)

Table 2 contains our test results on the s250 problem with different jam factors. The test results in 4 different best tours (Figs. 8, 9, 10 and 11). In the case of 1.04 or bigger jam factors the tours are the same because all of the nodes which are within the traffic jam region are visited before the rush hours.

4 Conclusion

In this paper our DBMEA algorithm was tested on the Time Dependent TSP problem. The algorithm is efficient in solving the TDTSP: in the case of bier127 it found the best-known solutions with all jam factors, and in the case of the self-generated instance it found quality (cost is close to the cost of the optimal tour with jam factor 1.00) solutions even with bigger jam factors.

Acknowledgement. This work was supported by National Research, Development and Innovation Office (NKFIH) K108405, K124055.

Supported by the ÚNKP-17-3 New National Excellence Program of the Ministry of Human Capacities.

References

1. Applegate, D.L., Bixby, R.E., Chvátal, V., Cook, W.J.: The Traveling Salesman Problem: A Computational Study, pp. 1–81. Princeton University Press, Princeton (2006)
2. Botzheim, J., Cabrita, C., Kóczy, L.T., Ruano, A.E.: Fuzzy rule extraction by bacterial memetic algorithms. In: Proceedings of the 11th World Congress of International Fuzzy Systems Association, IFSA 2005, Beijing, China, pp. 1563–1568 (2005)

3. Földesi, P., Botzheim, J.: Modeling of loss aversion in solving fuzzy road transport traveling salesman problem using eugenic bacterial memetic algorithm. Memetic Comput. **2**(4), 259–271 (2010)
4. Hoos, H.H., Stutzle, T.: Stochastic Local Search: Foundations and Applications. Morgan Kaufmann, San Francisco (2005)
5. Inoue, T., Furuhashi, T., Maeda, H., Takaba, M.: A study on interactive nurse scheduling support system using bacterial evolutionary algorithm engine. Trans. Inst. Elect. Eng. Jpn. **122-C**, 1803–1811 (2002)
6. Kóczy, L.T., Földesi, P., Tüű-Szabó, B.: An effective discrete bacterial memetic evolutionary algorithm for the traveling salesman problem. Int. J. Intell. Syst. **32**(8), 862–876 (2017)
7. Kóczy, L.T., Földesi, P., Tüű-Szabó, B.: Enhanced discrete bacterial memetic evolutionary algorithm-an efficacious metaheuristic for the traveling salesman optimization. Inf. Sci. (2017)
8. Li, F., Golden, B., Wasil, E.: Solving the time dependent traveling salesman problem. In: Golden, B., Raghavan, S., Wasil, E. (eds.) The Next Wave in Computing, Optimization, and Decision Technologies. Operations Research/Computer Science Interfaces Series, vol. 29. Springer, Boston (2005). https://doi.org/10.1007/0-387-23529-9_12
9. Moscato, P.: On evolution, search, optimization, genetic algorithms and martial arts -towards memetic algorithms. Technical report Caltech Concurrent Computation Program, Report. 826, California Institute of Technology, Pasadena, USA (1989)
10. Nawa, N.E., Furuhashi, T.: Fuzzy System Parameters Discovery by Bacterial Evolutionary Algorithm. IEEE Trans. Fuzzy Syst. **7**, 608–616 (1999)
11. Schneider, J.: The time-dependent traveling salesman problem. PhysicaA. **314**, 151–155 (2002)
12. Tüű-Szabó, B., Földesi, P., Kóczy, T.L.: Improved discrete bacterial memetic evolutionary algorithm for the traveling salesman problem. In: Proceedings of the Computational Intelligence in Information Systems Conference (CIIS 2016), Bandar Seri Begawan, Brunei, pp. 27–38 (2017)

Formal Concept Analysis and Uncertainty

Study of the Relevance of Objects and Attributes of *L*-fuzzy Contexts Using Overlap Indexes

Cristina Alcalde[1(✉)] and Ana Burusco[2]

[1] Department of Applied Mathematics, University of the Basque Country - UPV/EHU, Plaza de Europa 1, 20018 San Sebastián, Spain
c.alcalde@ehu.eus

[2] Departamento de Automática y Computación, Institute of Smart Cities, Universidad Pública de Navarra, Campus de Arrosadía, 31006 Pamplona, Spain
burusco@unavarra.es

Abstract. Objects and attributes play an important role in an L-fuzzy context. From the point of view of the L-fuzzy concepts, some of them can be more relevant than others. Besides, the number of objects and attributes of the L-fuzzy context is one of the most important factors that influence in the size of the L-fuzzy concept lattice. In this paper, we define different rankings for the objects and the attributes according to their relevance in the L-fuzzy concept lattice and using different overlap indexes. These rankings can be useful for the reduction of the L-fuzzy context size.

Keywords: L-fuzzy contexts · L-fuzzy concepts · Overlap indexes

1 Introduction

The L-fuzzy concept analysis [15–17] is a theory that studies the information arising from an L-fuzzy context using the L-fuzzy concepts as tools. An L-fuzzy context is a tuple (L, X, Y, R) with L a complete lattice, X and Y sets of objects and attributes respectively and $R \in L^{X \times Y}$ a fuzzy relation between the objects and the attributes. An L-fuzzy concept is a pair of L-fuzzy sets that can be interpreted as a group of elements (objects) that shares some characteristics (attributes). The set of these L-fuzzy concepts has the structure of complete lattice.

Most of the times, not all the objects neither the attributes have the same relevance from the point of view of the L-fuzzy concepts. Some of them hardly play any role in the L-fuzzy concepts.

Besides, when the cardinality of this L-fuzzy concept lattice is large, the obtained result may not be easy to handle. One of the factors that determines the size of the L-fuzzy concept lattice is the cardinality of the lattice L. The other is the size of the L-fuzzy context. The latter is analyzed in this paper.

J. Medina et al. (Eds.): IPMU 2018, CCIS 853, pp. 537–548, 2018.
https://doi.org/10.1007/978-3-319-91473-2_46

Over the past, several researchers have developed models in order to reduce the size of this lattice. In [13], Bělohlávek and Vychodil use hedges to control the size of the concept lattice. Also Wei and Qi [34], and Medina [27] have published works from the point of view of the attributes for fuzzy oriented concept lattices. Other methods to reduce the complexity of the lattice using fuzzy similarity [14] or block relations [24] have also been developed.

In [8] we have study the possibility of aggregating rows or columns of the L-fuzzy context. Sometimes, the L-fuzzy context values are independent and we can use usual aggregations as weighted means [20,21], OWA operators [22,33] and WOWA operators [31]. However, these studies are incomplete when we have values that present dependencies among them. In these situations the use of Choquet integrals [23] can be very useful as a tool for doing a proper analysis without lost of information as can be seen in [8].

This paper addresses the study of the objects and attributes of the L-fuzzy context when $L = [0, 1]$. We define different rankings for both sets according to their relevance in the L-fuzzy concept lattice and using different overlap indexes. These rankings will allow us to decide which objects and which attributes are less relevant. They will be the candidates for the elimination.

First, we are going to remember some important results about L-Fuzzy concept analysis and overlap indexes [19].

1.1 *L*-Fuzzy Concept Analysis

The Wille's Formal Concept Analysis [32] extracts information from a binary table that represents a formal context (X, Y, R) with X and Y finite sets of objects and attributes respectively and $R \subseteq X \times Y$. The hidden information consists of pairs (A, B) with $A \subseteq X$ and $B \subseteq Y$, called formal concepts, verifying $A^* = B$ and $B^* = A$, where $(\cdot)^*$ is the derivation operator that associates the attributes related to the elements of A to every object set A, and the objects related to the attributes of B to every attribute set B. These formal concepts can be interpreted as a group of objects A that shares the attributes of B.

In previous works [15,16] we have defined the L-fuzzy contexts (L, X, Y, R), with L a complete lattice, X and Y sets of objects and attributes respectively and $R \in L^{X \times Y}$ a fuzzy relation between the objects and the attributes. This is an extension of Wille's formal contexts to the fuzzy case when we want to study the relations between the objects and the attributes with values in a complete lattice L, instead of binary ones.

In our case, to work with these L-fuzzy contexts, we have defined the derivation operators 1 and 2 given by means of these expressions [16,17]:

$$\forall A \in L^X, \forall B \in L^Y, x \in X, y \in Y:$$

$$A_1(y) = \inf_{x \in X} \{I(A(x), R(x, y))\}$$

$$B_2(x) = \inf_{y \in Y} \{I(B(y), R(x, y))\}$$

with I a fuzzy implication operator defined in $(L, \leq)$.

The information stored in the context is visualized by means of the L-fuzzy concepts that are pairs $(A, A_1) \in L^X \times L^Y$ with $A \in fix(\varphi)$, set of fixed points of the operator φ, being defined from the derivation operators 1 and 2 as $\varphi(A) = (A_1)_2 = A_{12}$. These pairs, whose first and second components are said to be the fuzzy extension and intension respectively, represent a group of objects that share a group of attributes.

Using the usual order relation between fuzzy sets, that is, $\forall A, C \in L^X, A \leq C \iff A(x) \leq C(x), \forall x \in X$, we define the set $\mathcal{L} = \{(A, A_1) \mid A \in fix(\varphi)\}$ with the order relation $\preceq$ defined as: $\forall (A, A_1), (C, C_1) \in \mathcal{L}, (A, A_1) \preceq (C, C_1)$ if $A \leq C$ (or $A_1 \geq C_1$).

As φ is an order preserving operator, then the set $fix(\varphi)$ is a complete lattice and $(\mathcal{L}, \preceq)$ is also a complete lattice that is said to be the L-fuzzy concept lattice [15,16].

In addition, in the case of using a residuated implication ($I(a, b) = \sup\{x \mid T(a, x) \leq b\}$, with T a t-norm), given $A \in L^X$, (or $B \in L^Y$) we can obtain the associated L-fuzzy concept applying twice the derivation operators. In this case, the L-fuzzy concept associated to A is (A_{12}, A_1) (or (B_2, B_{21})). In the paper, residuated implication of Łukasiewicz will be used for the practical case.

Our last results are related to the use of two relations in the definition of the L-fuzzy context [2], the study of fuzzy context sequences [3,4,6,7] or the composition of L-fuzzy contexts [5]. We have also developed this theory in different areas as the treatment of incomplete information [1,10] or Mathematical Morphology [9].

Other important works that generalize the Formal Concepts Analysis using residuated implication operators are due to Bělohlávek [11,12] and Pollandt [28]. Moreover, extensions of Formal Concept Analysis to the interval-valued case are in [1,29,30] and to the fuzzy property-oriented and multi-adjoint concept lattices framework in [25,26].

1.2 Overlap Indexes

We start by recalling some basic notions about the idea of an overlap index.

Given a referential set U and $L = [0, 1]$, let L^U be the fuzzy sets of U. Bustince [19] define an overlap index as a mapping $O : L^U \times L^U \longrightarrow [0, 1]$, such that:

(i) $O(A, B) = 0$ if and only if in A and B have disjoint supports; that is, $A(i)B(i) = 0$ for every $i \in U$, and $A, B \in L^U$.
(ii) $O(A, B) = O(B, A), foreveryA, B \in L^U$.
(iii) If $B \subseteq C$, then $O(A, B) \leq O(A, C)$, for every $A, B, C \in L^U$.

An overlap index such that:

(iv) $O(A, B) = 1$ if there exists $i \in U$ such that $A(i) = B(i) = 1$ is called a normal overlap index.

Examples of overlap indexes are the following ones:

(1) Zadeh's consistency index:

$$O_Z(A, B) = \overset{n}{\underset{i=1}{max}}(min(A(i), B(i)))$$

(2) Let $M : [0,1]^2 \longrightarrow [0,1]$ be a symmetric aggregation function such that $M(x, y) = 0$ if and only if $xy = 0$. We have that:

$$O_{M,Z}(A, B) = \overset{n}{\underset{i=1}{max}}(M(A(i), B(i)))$$

is a normal overlap index that generalizes the Zadeh's index.

(3) If in the previous example, we consider a symmetric, increasing function $M : [0,1]^2 \longrightarrow [0,1]$ such that $M(1,1) < 1$ and $M(x, y) = 0$ if and only if $xy = 0$, then we obtain an overlap index which is not normal. For instance, when taking $M(x, y) = (xy)^p/2$ with $p > 0$, we arrive at the overlap index:

$$O(A, B) = \overset{n}{\underset{i=1}{max}}\left(\frac{(A(i), B(i))^p}{2}\right)$$

(4) The following is also an example of overlap index:

$$O_\pi(A, B) = \frac{1}{n}\sum_{i=1}^{n} A(i)B(i)$$

Remark 1. Formally, overlap indexes can be seen as generalized measures of fuzzy intersection of considered fuzzy sets.

Let $E \in L^U$ be a fixed non-empty fuzzy set. Given $A \subseteq U$, we define:

$$E_A(i) = \begin{cases} E(i) & \text{if } i \in A \\ 0 & \text{otherwise} \end{cases}$$

Observe that E_A is the intersection of the fuzzy set E and the crisp set A.

Now we are ready to introduce the definition of a fuzzy measure in terms of a fuzzy set and an overlap index.

Theorem 1 *[19]. If $E \in L^U$ is a fixed, non-empty fuzzy set, then the mapping $m_{O,E} : U \longrightarrow [0,1]$ given by:*

$$m_{O,E}(A) = \frac{O(E, E_A)}{O(E, E)}$$

is a fuzzy measure for every overlap index O.

2 Ranking of Objects and Attributes Using Overlap Indexes

As it has been explained in the introduction, it is interesting to establish a ranking in the object and attribute sets from the point of view of the L-fuzzy concepts. Moreover the size of an L-fuzzy context is one of the factors that determines the size of the L-fuzzy concept lattice and its manageability. The possible reduction of the size of the object and attribute sets is an interesting problem of study.

So far, some of our research lines in relation to reducing the size of the L-fuzzy context have used the method of removing rows or columns in the relation (eliminating objects or attributes). In [18] we removed the objects and/or attributes of little significance, that is, that did not appear as relevant in any L-fuzzy concept. To do this, we first obtained the L-fuzzy concept lattice, a quite laborious task.

In another different field and in order to work with missing values, in [10] infrequently appearing objects and attributes were studied. We removed them when they did not exceed a minimum support. To do this, we defined support of an L-fuzzy set $A \in L^Z$ as $supp(A) = \sum_{z \in Z} A_1(z)/|Z|$. The aim was to eliminate some rows or columns of missing values.

This definition allow us to assign a support value to every object (or attribute). To do this, for every $x_i \in X, i \in \{1, \dots, n\}$, let $\mathbf{x_i}$ be the L-fuzzy set defined by the characteristic function $\mathbf{x_i}(x_i) = 1$ and $\mathbf{x_i}(x) = 0$, for any $x \neq x_i$. Analogously for $\mathbf{y_j}, j \in \{1, \dots, m\}$.

Next, we are going to define the L-fuzzy concepts associated with the objects and the attributes of the L-fuzzy context. To do this, we will use a residuated implication operator for the definition of operators denoted by the subindexes 1 and 2.

Definition 1. *For every $x_i \in X, i \in \{1, \dots, n\}$, the pair $\mathcal{C}_{\mathbf{x_i}} = ((\mathbf{x_i})_{12}, (\mathbf{x_i})_2)$ is said to be the L-fuzzy concept derived from $\mathbf{x_i}$. Analogously $\mathcal{C}_{\mathbf{y_j}} = ((\mathbf{y_j})_2, (\mathbf{y_j})_{21}), j \in \{1, \dots, m\}$ is the L-fuzzy concept derived from $\mathbf{y_j}$.*

These concepts are the closest to the departure sets represented by $\mathbf{x_i}$ or $\mathbf{y_j}$ (study of a single object or attribute).

Then, the definitions of support of an object or an attribute are:

Definition 2. *For every $x_i \in X, i \in \{1, \dots, n\}$ and $y \in Y$:*

$$supp(x_i) = \frac{\sum_{i=1}^{n} (\mathbf{x_i})_2(y)}{|Y|},$$

and for every $y_j \in Y, j \in \{1, \dots, n\}$ and $x \in X$:

$$supp(y_j) = \frac{\sum_{j=1}^{n} (\mathbf{y_j})_1(x)}{|X|}$$

Now, given a fuzzy set E and an overlap index O, we can use Theorem 1 to associate a fuzzy measure with every object and attribute: $m_{O,E}(x_i)$ and $m_{O,E}(y_j)$ are the fuzzy measures for every $x_i \in X$ and $y_j \in Y$.

These values establish a measure of the overlap between the L-fuzzy concepts derived from the L-fuzzy sets $\mathbf{x_i}$ or $\mathbf{y_j}$ and the fuzzy set E.

We can define relations between the objects and the attributes using these fuzzy measures:

Definition 3. *For every $x_i, x_j \in X$, let C_{x_i}, C_{x_j} be the L-fuzzy concepts associated with x_i, x_j. Let E be a fuzzy set and O an overlap index.*

$$x_i \geq_{O,E} x_j \ if \ m_{O,E}(x_i) \geq m_{O,E}(x_j)$$

Analogously, $y_i \geq_{O,E} y_j$ if $m_{O,E}(y_i) \geq m_{O,E}(y_j)$, for every $y_i, y_j \in Y$.

This is a preorder relation that establishes for every fuzzy set E and overlap index O, the *Object* and the *Attribute Rankings* associated with E and O.

The election of E and O are important points. In the case of E, it represents the model we want to look like. Taking into account that the support of an object can be understood as a measure or its relevance, we take $E(x_i) = supp(x_i)$ for every $x_i \in X$. From the point of view of the attributes, we take $E(y_i) = supp(y_i)$ for every $y_i \in Y$.

3 Reducing the Size of an L-Fuzzy Context by Means of the Elimination of Rows or Columns

In previous section, we have define rankings of objects and attributes taking into account different overlap indexes. These rankings order the objects (or attributes) by means of the overlap between their derived L-fuzzy concepts and the support of the objects (or attributes). Then, our proposal is the elimination of those objects and attributes that are in the last positions of those rankings.

The advantage of this method over the one described in [18] is that it is not necessary to obtain the total L-fuzzy complete lattice. In addition, we can define a model E and eliminate the objects and attributes that do not give L-fuzzy concepts close to this model. This fact improves the idea proposed in [7].

Furthermore, if we remove a row or column in the context, we can see that the L-fuzzy concepts obtained from the non modified objects or attributes have the same membership degrees.

It is not difficult to prove the following proposition:

Proposition 1. *Let (L, X, Y, R) and $(L, X \backslash \{x_0\}, Y, \bar{R})$ be L-fuzzy contexts such that $\bar{R}(x, y) = R(x, y), \forall x \in X \backslash \{x_0\}, \forall y \in Y$. Consider $x_l \in X \backslash \{x_0\}$ and let $\mathcal{C}_{\mathbf{x_l}}$ and $\bar{\mathcal{C}}_{\mathbf{x_l}}$ be the L-fuzzy derived concepts in (L, X, Y, R) and $(L, X, Y, \bar{R})$ respectively. For any $x \in X \backslash \{x_0\}$ and for any $y \in Y$, the membership degrees in both L-fuzzy concepts are coincident.*

An analogous proposition can be proved in the case of eliminating one attribute:

Proposition 2. *Let* (L, X, Y, R) *and* $(L, X, Y \backslash \{y_0\}, \bar{R})$ *be L-fuzzy contexts such that* $\bar{R}(x, y) = R(x, y), \forall x \in X, \forall y \in Y \backslash \{y_0\}$. *Consider* $y_l \in Y \backslash \{y_0\}$ *and let* $\mathcal{C}_{\mathbf{y_l}}$ *and* $\bar{\mathcal{C}}_{\mathbf{y_l}}$ *be the L-fuzzy derived concepts in* (L, X, Y, R) *and* $(L, X, Y, \bar{R})$. *For any* $y \in Y \backslash \{y_0\}$ *and for any* $x \in X$, *the membership degrees are coincident in both L-fuzzy concepts.*

4 Practical Case

Let us see below a practical case where we will apply the results obtained in the previous sections.

Suppose that we want to do a market survey about the consumption of soft drinks in some of the major cities in Spain. To do this, we have an L-Fuzzy context (L, X, Y, R) with $L = \{0, 0.1, 0.2, 0.3, ... 1\}$, the object and attribute sets X={cola1, cola2, orangeade1, orangeade2, orangeade3, lemonade1, lemonade2, lemonade3, tonic1, tonic2} (the commercial brands are avoided) and Y={Barcelona, Bilbao, Granada, Madrid, Malaga, San Sebastian, Santander, Sevilla, Valencia, Zaragoza} and $R \in L^{X \times Y}$ the L-Fuzzy relation of Table 1 that represents the consumption of soft drinks in the different cities.

The values of the table belong to L. For instance, $R(x_1, y_4) = R(\text{cola1}, \text{Madrid}) = 0.8$ means that *cola1 is consumed in large quantities in Madrid*, but *this does not hold for tonic2 in Granada* ($R(x_{10}, y_3) = R(\text{tonic2}, \text{Granada}) = 0.1$).

In the rest of the section, the objects and the attributes will be denoted by x_i and y_j, $i, j \in \{1 ... 10\}$, respectively.

Table 1. L-fuzzy context

R	y_1	y_2	y_3	y_4	y_5	y_6	y_7	y_8	y_9	y_{10}
x_1	0.3	0.6	0.5	0.8	1.0	0.3	0.3	0.2	0.0	0.6
x_2	0.3	0.0	0.1	0.5	0.2	0.1	0.4	0.4	0.0	0.2
x_3	0.3	0.9	0.4	1.0	0.3	0.5	0.5	0.9	1.0	0.2
x_4	0.9	0.6	0.5	0.2	0.0	0.0	0.9	1.0	1.0	0.3
x_5	0.2	0.1	0.0	0.6	0.2	0.5	0.1	0.0	0.3	0.5
x_6	0.4	0.3	0.0	0.5	0.3	0.1	0.4	0.3	0.5	0.4
x_7	0.9	0.5	0.4	0.1	0.3	0.1	0.2	0.3	0.4	0.3
x_8	0.5	0.3	0.5	0.4	0.1	0.1	0.2	0.5	0.2	0.2
x_9	0.1	0.6	0.4	0.2	0.0	0.0	0.3	0.0	0.1	0.1
x_{10}	0.4	0.2	0.1	0.3	0.0	0.0	0.2	0.6	0.1	1.0

We are going to study which is the relationship among the objects and among the attributes. After this study, we will be able to reduce the size of the L-fuzzy context.

The construction of the whole L-fuzzy concept lattice has a high computational cost. So, for every $x_i, i \in \{1 \dots n\}$ and using the Łukasiewicz implication operator, we can obtain its derived L-fuzzy concept $\mathcal{C}_{\mathbf{x_i}}$. For instance, for $\mathbf{x_3}$ we have:

$$\mathcal{C}_{\mathbf{x_3}} = (\{x_1/0, x_2/0, x_3/1, x_4/0.2, x_5/0.1, x_6/0.4, x_7/0.1, x_8/0.2, x_9/0.1, x_{10}/0.1\}, \\ \{y_1/0.3, y_2/0.9, y_3/0.4, y_4/1, y_5/0.3, y_6/0.5, y_7/0.5, y_8/0.9, y_9/1, y_{10}/0.2\})$$

We can say that *Orangeade1*(x_3) *is consumed mainly in Bilbao* (y_2), *Madrid* (y_4), *Sevilla* (y_8), *and Valencia* (y_9).

For every object x_i, the fuzzy extension of the derived L-fuzzy concepts are:
$\mathcal{C}_{\mathbf{x_1}} : \{x_1/1, x_2/0.2, x_3/0.3, x_4/0, x_5/0.2, x_6/0.3, x_7/0.3, x_8/0.1, x_9/0, x_{10}/0\}$
$\mathcal{C}_{\mathbf{x_2}} : \{x_1/0.8, x_2/1, x_3/1, x_4/0.7, x_5/0.6, x_6/0.9, x_7/0.6, x_8/0.8, x_9/0.6, x_{10}/0.8\}$
$\mathcal{C}_{\mathbf{x_3}} : \{x_1/0, x_2/0, x_3/1, x_4/0.2, x_5/0.1, x_6/0.4, x_7/0.1, x_8/0.2, x_9/0.1, x_{10}/0.1\}$
$\mathcal{C}_{\mathbf{x_4}} : \{x_1/0, x_2/0, x_3/0.4, x_4/1, x_5/0, x_6/0.3, x_7/0.3, x_8/0.2, x_9/0, x_{10}/0.1\}$
$\mathcal{C}_{\mathbf{x_5}} : \{x_1/0.7, x_2/0.6, x_3/0.7, x_4/0.5, x_5/1, x_6/0.6, x_7/0.5, x_8/0.6, x_9/0.5, x_{10}/0.5\}$
$\mathcal{C}_{\mathbf{x_6}} : \{x_1/0.5, x_2/0.5, x_3/0.8, x_4/0.7, x_5/0.7, x_6/1, x_7/0.6, x_8/0.7, x_9/0.6, x_{10}/0.6\}$
$\mathcal{C}_{\mathbf{x_7}} : \{x_1/0.4, x_2/0.4, x_3/0.4, x_4/0.7, x_5/0.3, x_6/0.5, x_7/1, x_8/0.6, x_9/0.2, x_{10}/0.5\}$
$\mathcal{C}_{\mathbf{x_8}} : \{x_1/0.7, x_2/0.6, x_3/0.8, x_4/0.8, x_5/0.5, x_6/0.5, x_7/0.7, x_8/1, x_9/0.5, x_{10}/0.6\}$
$\mathcal{C}_{\mathbf{x_9}} : \{x_1/0.9, x_2/0.4, x_3/1, x_4/1, x_5/0.5, x_6/0.6, x_7/0.9, x_8/0.7, x_9/1, x_{10}/0.6\}$
$\mathcal{C}_{\mathbf{x_{10}}} : \{x_1/0.6, x_2/0.2, x_3/0.2, x_4/0.3, x_5/0.4, x_6/0.4, x_7/0.3, x_8/0.2, x_9/0.1, x_{10}/1\}$

and for the attributes:
$\mathcal{C}_{\mathbf{y_1}} : \{y_1/1, y_2/0.6, y_3/0.5, y_4/0.2, y_5/0.1, y_6/0.1, y_7/0.3, y_8/0.4, y_9/0.5, y_{10}/0.4\}$
$\mathcal{C}_{\mathbf{y_2}} : \{y_1/0.4, y_2/1, y_3/0.5, y_4/0.6, y_5/0.4, y_6/0.4, y_7/0.6, y_8/0.4, y_9/0.4, y_{10}/0.3\}$
$\mathcal{C}_{\mathbf{y_3}} : \{y_1/0.7, y_2/0.8, y_3/1, y_4/0.7, y_5/0.5, y_6/0.5, y_7/0.7, y_8/0.6, y_9/0.5, y_{10}/0.7\}$
$\mathcal{C}_{\mathbf{y_4}} : \{y_1/0.3, y_2/0.5, y_3/0.4, y_4/1, y_5/0.3, y_6/0.5, y_7/0.5, y_8/0.4, y_9/0.2, y_{10}/0.2\}$
$\mathcal{C}_{\mathbf{y_5}} : \{y_1/0.3, y_2/0.6, y_3/0.5, y_4/0.8, y_5/1, y_6/0.3, y_7/0.3, y_8/0.2, y_9/0, y_{10}/0.6\}$
$\mathcal{C}_{\mathbf{y_6}} : \{y_1/0.7, y_2/0.6, y_3/0.5, y_4/1, y_5/0.7, y_6/1, y_7/0.6, y_8/0.5, y_9/0.7, y_{10}/0.7\}$
$\mathcal{C}_{\mathbf{y_7}} : \{y_1/0.8, y_2/0.6, y_3/0.6, y_4/0.3, y_5/0.1, y_6/0.1, y_7/1, y_8/0.7, y_9/0.6, y_{10}/0.4\}$
$\mathcal{C}_{\mathbf{y_8}} : \{y_1/0.4, y_2/0.6, y_3/0.5, y_4/0.2, y_5/0, y_6/0, y_7/0.6, y_8/1, y_9/0.5, y_{10}/0.3\}$
$\mathcal{C}_{\mathbf{y_9}} : \{y_1/0.3, y_2/0.6, y_3/0.4, y_4/0.2, y_5/0, y_6/0, y_7/0.5, y_8/0.7, y_9/1, y_{10}/0.2\}$
$\mathcal{C}_{\mathbf{y_{10}}} : \{y_1/0.4, y_2/0.2, y_3/0.1, y_4/0.3, y_5/0, y_6/0, y_7/0.2, y_8/0.5, y_9/0.1, y_{10}/1\}$

We are now in condition of calculate the support values for the objects. The obtained values are shown in Table 2.

Table 2. Object support values

	x_1	x_2	x_3	x_4	x_5	x_6	x_7	x_8	x_9	x_{10}
supp	0.46	0.22	0.6	0.54	0.25	0.32	0.35	0.3	0.18	0.29

In Table 3 we can see the values obtained for the attributes.

Table 3. Attribute support values

	y_1	y_2	y_3	y_4	y_5	y_6	y_7	y_8	y_9	y_{10}
$supp$	0.43	0.41	0.2	0.46	0.24	0.17	0.35	0.42	0.36	0.38

These supports can be taken into account to study the relevance of the objects and the attributes.

Let be $U = X, Card(U) = n$ and $E(x_i) = supp(x_i)$, for every $x_i \in X$. Using overlap index O_π, we can calculate by Theorem 1 the associated fuzzy measure for each object (See Table 4).

Table 4. Fuzzy measure for objects associated with O_π and E

	x_1	x_2	x_3	x_4	x_5	x_6	x_7	x_8	x_9	x_{10}
$m_{O_\pi,E}$	0.119	0.058	0.165	0.132	0.070	0.115	0.106	0.099	0.049	0.087

These values define a relevance ranking for the objects:

$$x_3 \geq_{O,E} x_4 \geq_{O,E} x_1 \geq_{O,E} x_6 \geq_{O,E} x_7 \geq_{O,E} x_8 \geq_{O,E} x_{10} \geq_{O,E} x_5 \geq_{O,E} x_2 \geq_{O,E} x_9$$

The same classification is obtained for overlap index O_Z since the obtained fuzzy measure values are those shown in Table 5.

Table 5. Fuzzy measure for objects associated with O_Z and E

	x_1	x_2	x_3	x_4	x_5	x_6	x_7	x_8	x_9	x_{10}
$m_{O_Z,E}$	0.848	0.591	1.000	0.894	0.652	0.833	0.803	0.773	0.545	0.727

From the point of view of the attributes, with overlap index O_π, we obtain the values in Table 6.

And in Table 7 we show the values obtained with O_Z.

The same ranking is also obtained with both overlap indexes:

$$y_2 >_{O,E} y_8 >_{O,E} y_7, y_1, y_4 >_{O,E} y_3 >_{O,E} y_{10} >_{O,E} y_9 >_{O,E} y_5 >_{O,E} y_6$$

In this case, if we consider that the size of the L-fuzzy context is large and we choose as the fuzzy set E defined by the support (as the model we want to look like), we can conclude that objects x_9 and x_2 and attributes y_6 and y_5 are the candidates to be removed.

Table 6. Fuzzy measure for attributes associated with O_π and E

	y_1	y_2	y_3	y_4	y_5	y_6	y_7	y_8	y_9	y_{10}
$m_{O_\pi,E}$	0.119	0.157	0.106	0.119	0.041	0.035	0.119	0.123	0.085	0.097

Table 7. Fuzzy measure for attributes associated with O_Z and E

	y_1	y_2	y_3	y_4	y_5	y_6	y_7	y_8	y_9	y_{10}
$m_{O_Z,E}$	0.869	1.000	0.820	0.869	0.508	0.475	0.869	0.885	0.738	0.787

5 Conclusions

In this work, we have seen that overlap indexes can be useful tools to analyze the relevance of the objects and the attributes from the point of view of the L fuzzy concepts. We define different ranking associated with the different overlap indexes. These rankings can help us to remove some objects or attributes when the size of the L-fuzzy context is large.

In future works, we will study how this reduction affects to the structure of the L-fuzzy concept lattice.

Acknowledgments. This paper is partially supported by the Research Group "Intelligent Systems and Energy (SI+E)" of the University of the Basque Country (UPV/EHU), under Grant GIU 16/54, and by the Research Group "Artificial Intelligence and Approximate Reasoning" of the Public University of Navarra, under TIN2016-77356-P.

References

1. Alcalde, C., Burusco, A., Fuentes-González, R., Zubia, I.: Treatment of L-fuzzy contexts with absent values. Inf. Sci. **179**(1–2), 1–15 (2009)
2. Alcalde, C., Burusco, A.: The use of two relations in L-fuzzy contexts. Inf. Sci. **301**, 1–12 (2015)
3. Alcalde, C., Burusco, A., Bustince, H., Jurio, A., Sanz, J.A.: Evolution in time of the L-fuzzy context sequences. Inf. Sci. **326**, 202–214 (2016)
4. Alcalde, C., Burusco, A., Fuentes-González, R.: The study of fuzzy context sequences. Int. J. Comput. Intell. Syst. **6**(3), 518–529 (2013)
5. Alcalde, C., Burusco, A., Fuentes-González, R.: Some results on the composition of L-fuzzy contexts. In: Greco, S., Bouchon-Meunier, B., Coletti, G., Fedrizzi, M., Matarazzo, B., Yager, R.R. (eds.) IPMU 2012. CCIS, vol. 298, pp. 305–314. Springer, Heidelberg (2012). https://doi.org/10.1007/978-3-642-31715-6_33
6. Alcalde, C., Burusco, A.: L-fuzzy context sequences on complete lattices. In: Laurent, A., Strauss, O., Bouchon-Meunier, B., Yager, R.R. (eds.) IPMU 2014. CCIS, vol. 444, pp. 31–40. Springer, Cham (2014). https://doi.org/10.1007/978-3-319-08852-5_4

7. Alcalde, C., Burusco, A.: WOWA operators in fuzzy context sequences. In: 16th World Congress of the International-Fuzzy-Systems-Association (IFSA)/9th Conference of the European-Society-for-Fuzzy-Logic-and-Technology, EUSFLAT 2015, Gijón, Spain. Advances in Intelligent Systems Research, vol. 89, pp. 357–362 (2015)
8. Alcalde, C., Burusco, A.: On the use of Choquet integrals in the reduction of the size of L-fuzzy contexts, In: FUZZ-IEEE 2017, Naples, Italy (2017)
9. Alcalde, C., Burusco, A., Fuentes-González, R.: Application of the L-fuzzy concept analysis in the morphological image and signal processing. Ann. Math. Artif. Intell. **72**(1–2), 115–128 (2014)
10. Alcalde, C., Burusco, A., Fuentes-González, R.: Treatment of incomplete information in L-fuzzy contexts. In: Proceedings of the EUSFLAT-LFA 2005. 4th Conference of the European Society for Fuzzy Logic and Technology and 11 Reencontres Phrancophones sur la Logique Floue et ses Applications, Barcelona, pp. 518–523 (2005)
11. Bělohlávek, R.: Fuzzy Galois connections. Math. Logic Q. **45**(4), 497–504 (1999)
12. Bělohlávek, R.: Fuzzy Relational Systems. IFSR International Series on Systems Science and Engineering, vol. 20. Springer, US, New York City (2002). https://doi.org/10.1007/978-1-4615-0633-1
13. Bělohlávek, R., Vychodil, V.: Reducing the size of fuzzy concept lattices by hedges. In: Fuzz-IEEE 2005, The International Conference on Fuzzy Systems Reno, Nevada, USA, pp. 663–668 (2005)
14. Bělohlávek, R.: Similarity relations in concept lattices. J. Logic Comput. **10**(6), 823–845 (2000)
15. Burusco, A., Fuentes-González, R.: The study of the L-fuzzy concept lattice. Mathw. Soft Comput. **1**(3), 209–218 (1994)
16. Burusco, A., Fuentes-González, R.: Construction of the L-fuzzy concept lattice. Fuzzy Sets Syst. **97**(1), 109–114 (1998)
17. Burusco, A., Fuentes-González, R.: Concept lattices defined from implication operators. Fuzzy Sets Syst. **114**(1), 431–436 (2000)
18. Burusco, A., Fuentes-González, R.: Relevant information extraction in L-Fuzzy contexts. Revista Internacional de Información Tecnológica **14**(4), 65–70 (2003)
19. Paternain, D., Bustince, H., Pagola, M., Sussner, P., Kolesárová, A., Mesiar, R.: Capacities and overlap indexes with an application in fuzzy rule-based classification systems. Fuzzy Sets Syst. **305**, 70–94 (2016)
20. Calvo, T., Mesiar, R.: Weighted triangular norms-based aggregation operators. Fuzzy Sets Syst. **137**, 3–10 (2003)
21. Calvo, T., Mesiar, R.: Aggregation operators: ordering and bounds. Fuzzy Sets Syst. **139**, 685–697 (2003)
22. Fodor, J., Marichal, J.L., Roubens, M.: Characterization of the ordered weighted averaging operators. IEEE Trans. Fuzzy Syst. **3**(2), 236–240 (1995)
23. Grabisch, M.: Fuzzy integral in multicriteria decision making. Fuzzy Sets Syst. **69**, 279–298 (1995)
24. Konecny, J., Krupka, M.: Block relations in fuzzy settings. In: Proceedings of the Concept Lattice and Their Applications, pp. 115–130 (2011)
25. Medina, J., Ojeda-Aciego, M.: Multi-adjoint t-concept lattices. Inf. Sci. **180**(5), 712–725 (2010)
26. Medina, J., Ojeda-Aciego, M.: Dual multi-adjoint concept lattices. Inf. Sci. **225**, 47–54 (2013)
27. Medina, J.: Relating attribute reduction in formal object-oriented and property-oriented concept lattices. Comput. Math. Appl. **64**(6), 1992–2002 (2012)

28. Pollandt, S.: Fuzzy Begriffe: Formale Begriffsanalyse unscharfer Daten. Springer, Heidelberg (1997). https://doi.org/10.1007/978-3-642-60460-7
29. Djouadi, Y., Prade, H.: Interval-valued fuzzy galois connections: algebraic requirements and concept lattice construction. Fundamenta Informaticae **99**(2), 169–186 (2010)
30. Djouadi, Y., Prade, H.: Possibility-theoretic extension of derivation operators in formal concept analysis over fuzzy lattices. FODM **10**(4), 287–309 (2011)
31. Torra, V.: On some relationships between the WOWA operator and the Choquet integral. In: Proceedings of the Seventh International Conference on Information Processing and Management of Uncertainty in Knowledge-Based Systems (IPMU 1998), Paris France, pp. 818–824 (1998)
32. Wille, R.: Restructuring lattice theory: an approach based on hierarchies of concepts. In: Rival, I. (ed.) Ordered Sets, pp. 445–470. Reidel, Dordrecht/Boston (1982)
33. Yager, R.R.: On ordered weighted averaging aggregation operators in multi-criteria decision making. IEEE Trans. Syst. Man Cybern. **18**, 183–190 (1988)
34. Wei, L., Qi, J.J.: Relation between concept lattice reduction and rough set reduction. Knowl.-Based Syst. **23**(8), 934–938 (2010)

FCA Attribute Reduction in Information Systems

M. José Benítez-Caballero, Jesús Medina, and Eloísa Ramírez-Poussa(✉)

Department of Mathematics, University of Cádiz, Cádiz, Spain
{mariajose.benitez,jesus.medina,eloisa.ramirez}@uca.es

Abstract. One of the main targets in formal concept analysis (FCA) and in rough set theory (RST) is the reduction of redundant information. Feature selection mechanisms have been studied separately in many works. In this paper, we analyse the result of applying the reduction mechanisms given in FCA to RST, and give interpretations of such reductions.

Keywords: Formal concept analysis · Rough set · Attribute reduction

1 Introduction

Nowadays, the size of databases is usually large and the need of managing information is increasing exponentially. There exist diverse mathematical tools to treat the information given in databases. In this paper, we will establish connections between two of these theories: rough set theory (RST) and formal concept analysis (FCA). Both theories, designed to extract information from knowledge systems, work with two sets, one of them plays the role of objects and the other one plays the role of attributes. These sets are related to each other by means of a relation.

Rough sets were presented by Pawlak in [16]. The main idea was to describe sets by means of two approximations sets, the upper and the lower approximations. These sets can be considered as classical sets [12] or as fuzzy ones [10,11]. In contrast, the main target in formal concept analysis [20] is to obtain the knowledge collected in a database, using the information contained in the concepts and the existing relations among these concepts, which give rise to the algebraic structure of a complete lattice.

In addition, there are common research areas in both theories. One of them is the reduction of the set of attributes, without the elimination of important knowledge. In order to do this, the notion of reduct appears, that is, a minimal subset of attributes preserving the original knowledge. In the case of RST, the reducts keep the same capability to discern objects that the original database

Partially supported by the State Research Agency (AEI) and the European Regional Development Fund (ERDF) project TIN2016-76653-P.

J. Medina et al. (Eds.): IPMU 2018, CCIS 853, pp. 549–561, 2018.
https://doi.org/10.1007/978-3-319-91473-2_47

does. On the other hand, in FCA, the concept lattice obtained from a reduct is isomorphic to the original one. Some papers study reduction methods in both theories separately, but there are not yet many analysis on the connections of these theories from the point of view of attribute reduction [2,4,19].

In this paper, from an information system we will consider an associated context. We will reduce these contexts considering the reduction given in FCA theory. The reduction approach introduced in this paper is novelty since, for instance, tolerance relations are considered as indiscernibility relations. We will also study in this work the reduction process within a fuzzy environment. Moreover, we will present some properties and will give the interpretations of the obtained results in each case.

2 Preliminaries

In order to interpret the FCA reduction in a Rough Set environment, in this section, we will recall some definitions and results of each theories.

2.1 Rough Set Theory

Rough set theory was proposed to treat incomplete information. This information is usually presented as an information system [15,17].

Definition 1. *An* information system $(U, \mathcal{A})$ *is a tuple, where* $U = \{x_1, x_2, \ldots, x_n\}$ *and* $\mathcal{A} = \{a_1, a_2, \ldots, a_m\}$ *are finite, non-empty sets of objects and attributes, respectively. Each* $a \in \mathcal{A}$ *corresponds to a mapping* $\bar{a} : U \to V_a$, *where* V_a *is the value set of the attribute* a *over* U. *For every subset* D *of* $\mathcal{A}$, *the* D-indiscernibility relation, $Ind(D)$, *is defined by the following equivalence relation*

$$Ind(D) = \{(x_i, x_j) \in U \times U \mid \text{ for all } a \in D, \bar{a}(x_i) = \bar{a}(x_j)\}$$

where each equivalence class is written as $[x]_D = \{x_i \in U \mid (x, x_i) \in Ind(D)\}$. $Ind(D)$ *produces a partition on* U *denoted as* $U/Ind(D) = \{[x]_D \mid x \in U\}$.

The following definition presents the notion of consistent set and reduct, which are needed in order to reduce the set of attributes.

Definition 2. *Let* $(U, \mathcal{A})$ *be an information system and a subset of attributes* $D \subseteq \mathcal{A}$. *The subset* D *is a* consistent set *of* $(U, \mathcal{A})$ *if* $Ind(D) = Ind(\mathcal{A})$. *Moreover, if for each* $a \in D$ *we have that* $Ind(D \setminus \{a\}) \neq Ind(\mathcal{A})$, *then* D *is a* reduct *of* $(U, \mathcal{A})$.

The discernibility matrix is a useful tool for sorting the information between two objects [17].

Definition 3. *Given an information system* $(U, \mathcal{A})$, *its* discernibility matrix *is a matrix with order* $|U| \times |U|$, *denoted by* $M_{\mathcal{A}}$, *in which the element* $M_{\mathcal{A}}(x, y)$ *for each pair of objects* (x, y) *is defined by:*

$$M_{\mathcal{A}}(x, y) = \{a \in \mathcal{A} \mid \bar{a}(x) \neq \bar{a}(y)\}$$

Considering the discernibility function of an information system, a method to obtain reducts is presented in the next result [3,17].

Theorem 1. *Let $(U, \mathcal{A})$ be a boolean information system. An arbitrary set D, where $D \subseteq \mathcal{A}$, is a reduct of the information system if and only if the cube $\bigwedge_{a \in D} a$ is a cube in the restricted disjunctive normal form.*

2.2 Formal Concept Analysis

Formal concept analysis is another tool to study databases and extract information from them. Now, we will recall some needed notions.

Definition 4. *Let us consider a set of attributes A, a set of objects B and a relation $R \subseteq A \times B$ between them. The triple (A, B, R) is called context.*

We can also define, for a given context, two mappings $^{\uparrow} \colon 2^B \to 2^A$ and $^{\downarrow} \colon 2^A \to 2^B$, for each $X \subseteq B$ and $Y \subseteq A$ as follows:

$$X^{\uparrow} = \{a \in A \mid \text{for all } x \in X, aRx\} \tag{1}$$

$$Y^{\downarrow} = \{x \in B \mid \text{for all } a \in Y, aRx\} \tag{2}$$

These mappings are called *concept-forming operators*, which form a Galois connection. Hence, a *concept* is a pair (X, Y) satisfying $X^{\uparrow} = Y$ and $Y^{\downarrow} = X$, for all $X \subseteq B$ and $Y \subseteq A$.

The set of all the concepts of a context (A, B, R), with an inclusion order over the left argument, have the structure of a *complete lattice*, denoted by $\mathcal{B}(A, B, R)$. The following definition presents the notion of consistent set and reduct of a context, which are necessary for the reduction method.

Definition 5. *Given a context (A, B, R), if there exists a set of attributes $Y \subseteq A$ such that $\mathcal{B}(A, B, R) \cong \mathcal{B}(Y, B, R_Y)$, then Y is called a* consistent *set of (A, B, R). Moreover, if $\mathcal{B}(Y \setminus \{a\}, B, R_{Y \setminus \{a\}}) \ncong \mathcal{B}(A, B, R)$, for all $a \in Y$, then Y is called a* reduct *of (A, B, R). The* core *of (A, B, R) is the intersection of all the reducts of (A, B, R).*

According to the notion of reduct, different kinds of attributes arise.

Definition 6. *Given a formal context (A, B, R) and the set $\mathcal{Y} = \{Y \subseteq A \mid Y \text{ is a reduct}\}$ of all the reducts of (A, B, R), the set of attributes A can be split into the following three parts:*

1. *Absolutely necessary attributes (core attribute) $C_f = \bigcap_{Y \in \mathcal{Y}} Y$.*
2. *Relatively necessary attributes $K_f = (\bigcup_{Y \in \mathcal{Y}} Y) \setminus (\bigcap_{Y \in \mathcal{Y}} Y)$.*
3. *Absolutely unnecessary attributes $I_f = A \setminus (\bigcup_{Y \in \mathcal{Y}} Y)$.*

2.3 Fuzzy Formal Concept Analysis

Since the comparison will be also given in the fuzzy case, we need to recall the previous notions of FCA to a fuzzy environment. For more details see [14].

Definition 7. *A* multi-adjoint frame $\mathcal{L}$ *is a tuple*

$$(L_1, L_2, P, \preceq_1, \preceq_2, \leq, \&_1, \swarrow^1, \nwarrow_1, \dots, \&_n, \swarrow^n, \nwarrow_n)$$

where $(L_1, \preceq_1)$ *and* $(L_2, \preceq_2)$ *are complete lattices,* $(P, \leq)$ *is a poset and, for all* $i \in \{1, \dots, n\}$, $(\&_i, \swarrow^i, \nwarrow_i)$ *is an adjoint triple with respect to* L_1, L_2, P. *Multi-adjoint frames are denoted as* $(L_1, L_2, P, \&_1, \dots, \&_n)$.

The definition of adjoint triple has been studied in [5,7]. The notion of context in the multi-adjoint fuzzy case of formal concept analysis is introduced below.

Definition 8. *Let* $(L_1, L_2, P, \&_1, \dots, \&_n)$ *be a multi-adjoint frame, a* context *is a tuple* (A, B, R, σ) *such that* A *and* B *are non-empty sets (usually interpreted as attributes and objects, respectively),* R *is a* P*-fuzzy relation* $R\colon A \times B \to P$ *and* $\sigma\colon A \times B \to \{1, \dots, n\}$ *is a mapping which associates any element in* $A \times B$ *with some particular adjoint triple in the frame.*

Let us consider a multi-adjoint frame and a context for that frame. We define the concept-forming operators, denoted as ${}^{\uparrow_\sigma}\colon L_2^B \longrightarrow L_1^A$ and ${}^{\downarrow^\sigma}\colon L_1^A, \longrightarrow L_2^B$, for all $g \in L_2^B$, $f \in L_1^A$ and $a \in A$, $b \in B$, as

$$g^{\uparrow}(a) = \inf\{R(a,b) \swarrow^{\sigma(a,b)} g(b) \mid b \in B\} \tag{3}$$

$$f^{\downarrow}(b) = \inf\{R(a,b) \nwarrow_{\sigma(a,b)} f(a) \mid a \in A\} \tag{4}$$

which form a Galois connection [14]. Therefore, we can define the notion of concept as usual: a pair $\langle g, f\rangle$ is a *multi-adjoint concept* if they satisfy that $g \in L_2^B$, $f \in L_1^A$ and $g^{\uparrow} = f$ and $f^{\downarrow} = g$; with $({}^{\uparrow}, {}^{\downarrow})$ being the concept-forming operators defined above.

The following definition recalls the structure of a concept lattice into a multi-adjoint frame.

Definition 9. *The* multi-adjoint concept lattice *associated with a multi-adjoint frame* $(L_1, L_2, P, \&_1, \dots, \&_n)$ *and a context* (A, B, R, σ), *denoted by* $\mathcal{M}(A, B, R, \sigma)$, *is the set:*

$$\mathcal{M} = \{\langle g, f\rangle \mid g \in L_2^B, f \in L_1^A \text{ and } g^{\uparrow} = f, f^{\downarrow} = g\}$$

together with the ordering defined by $\langle g_1, f_1\rangle \preceq \langle g_2, f_2\rangle$ *if and only if* $g_1 \preceq_2 g_2$ *(equivalently* $f_2 \preceq_1 f_1$*).*

In this definition, we recall the notions of consistent set and reduct, needed for the reduction process.

Definition 10. *A set of attributes* $Y \subseteq A$ *is called a* consistent *set of* (A, B, R, σ) *if* $\mathcal{M}(Y, B, R_Y, \sigma_{Y\times B}) \cong_E \mathcal{M}(A, B, R, \sigma)$. *This is equivalent to say that, for all* $\langle g, f\rangle \in \mathcal{M}(A, B, R, \sigma)$, *there exists a concept* $\langle g', f'\rangle \in \mathcal{M}(Y, B, R_Y, \sigma_{Y\times B})$ *such that* $g = g'$.

Moreover, if $\mathcal{M}(Y \smallsetminus \{a\}, B, R_{Y\smallsetminus\{a\}}, \sigma_{Y\smallsetminus\{a\}\times B}) \not\cong_E \mathcal{M}(A, B, R, \sigma)$, *for all* $a \in Y$, *then* Y *is called a* reduct *of* (A, B, R, σ).

The core *of* (A, B, R, σ) *is the intersection of all the reducts of* (A, B, R, σ).

Now, we introduce the characteristic function associated with the attributes.

Definition 11. *For each* $a \in A$, *the fuzzy subsets of attributes* $\phi_{a,x} \in L_1^A$ *defined, for all* $x \in L_1$, *as*

$$\phi_{a,x}(a') = \begin{cases} x & if\ a' = a \\ 0 & if\ a' \neq a \end{cases}$$

will be called fuzzy-attributes. *The set of all fuzzy-attributes will be denoted by* $\Phi = \{\phi_{a,x} \mid a \in A, x \in L_1\}$.

The notion of meet-irreducible recalled in the next definition is essential in the attribute classification theorems [6,9] that we will consider in this work to carry out the reduction.

Definition 12. *Given a lattice* $(L, \preceq)$, *such that* $\wedge, \vee$ *are the meet and the join operators, and an element* $x \in L$ *verifying that*

1. *If* L *has a top element* $\top$, *then* $x \neq \top$.
2. *If* $x = y \wedge z$, *then* $x = y$ *or* $x = z$, *for all* $y, z \in L$.

the element x *is called* meet-irreducible ($\wedge$-irreducible) element *of* L. *A* join-irreducible ($\vee$-irreducible) element *of* L *is defined dually.*

The next theorem is the characterization of the $\wedge$-irreducible elements, in which the attribute classification theorems are based [6,8]. From this result, different attribute classification theorems were introduced.

Theorem 2. *The set of* $\wedge$*-irreducible elements of* $\mathcal{M}$, $M_F(A, B, R, \sigma)$, *is:*

$$\left\{\langle \phi_{a,x}^{\downarrow}, \phi_{a,x}^{\downarrow\uparrow}\rangle \mid \phi_{a,x}^{\downarrow} \neq \bigwedge\{\phi_{a_i,x_i}^{\downarrow} \mid \phi_{a_i,x_i} \in \Phi, \phi_{a,x}^{\downarrow} \prec_2 \phi_{a_i,x_i}^{\downarrow}\}\ and\ \phi_{a,x}^{\downarrow} \neq g_\top\right\}$$

where $\top$ *is the maximum element in* L_2 *and* $g_\top \colon B \to L_2$ *is the fuzzy subset defined as* $g_\top(b) = \top$, *for all* $b \in B$.

3 Meaning of FCA Reduction in RST

In this section, we will compute the discernibility matrices of information systems considering different kind of relations, these discernibility matrices will be interpreted as formal contexts. These contexts will be reduced using the attribute reduction methods given in FCA theory and we will give an interpretation of the obtained reductions. In the following, we will specify the procedure with more details.

Let us consider an information system $(U, \mathcal{A})$ and an indiscernibility relation, from which we can obtain a discernibility matrix. This indiscernibility relation is the usual one, presented in Definition 1, where $D = \mathcal{A}$, and that considers: two objects $x_i, x_j \in U$ are indiscernibles if $a(x_i) = a(x_j)$, for all $a \in \mathcal{A}$; and that x_i, x_j are discernible, otherwise. In this way, we can define a formal context (U, U, R), where the relation R is the indiscernibility relation.

In this environment, the concept-forming operators $({}^{\uparrow}, {}^{\downarrow})$ given in Eqs. (1) and (2) satisfy, the following equalities:

$$X^{\uparrow} = \begin{cases} [x] & \text{if } X \subseteq [x], \text{ for an } x \in X \\ \varnothing & \text{otherwise} \end{cases} \quad (5)$$

$$Y^{\downarrow} = \begin{cases} [y] & \text{if } Y \subseteq [y], \text{ for a } y \in Y \\ \varnothing & \text{otherwise} \end{cases} \quad (6)$$

for all $X, Y \subseteq U$.

Therefore in this particular case, according to the notion of concept, we have that given a subset of objects $X \subseteq U$, a concept $(X^{\uparrow}, X^{\uparrow\downarrow})$ verifies that $X^{\uparrow} = X^{\uparrow\downarrow}$, except in the case that $X^{\uparrow} = \varnothing$, in which the resulting concept is $(U, \varnothing)$. Consequently, when the attribute reduction is carried out, following the philosophy of FCA, the obtained reducts are composed of the representative elements of the equivalence classes. That is due to equivalence classes are disjunctive sets. Hence, we only need to consider one element of each class and, as a consequence, we can remove all the objects in each class of equivalence, excepting one of them. The following example will illustrate the previous idea.

Example 1. Let us consider the information system $(U, \mathcal{A})$ where the set of objects is $U = \{x_1, x_2, x_3\}$, the attributes are $\mathcal{A} = \{a_1, a_2\}$ and the following relation between them:

	a_1	a_2
x_1	1	1
x_2	0	1
x_3	1	1

In this case, the equivalence classes of the information system are $[x_1] = \{x_1, x_3\}$ and $[x_2] = \{x_2\}$. On the other hand, we can build the context (U, U, R), where R is the relation obtained from the following discernibility matrix:

$$\begin{pmatrix} \varnothing & & \\ \{a_1\} & \varnothing & \\ \varnothing & \{a_1\} & \varnothing \end{pmatrix}$$

Hence, according to the indiscernibility relation Ind(U), the relation is represented by its characteristics matrix as follows:

R	x_1	x_2	x_3
x_1	1	0	1
x_2	0	1	0
x_3	1	0	1

Considering the attribute classification given in FCA, the classification of the attributes is as follows:

$$C_f = \{x_2\}$$
$$K_f = \{x_1, x_3\}$$

Consequently, applying the attribute reduction results given in FCA, we obtain two reducts: $D_1 = \{x_1, x_2\}$ and $D_2 = \{x_3, x_2\}$, which coincide with the different representative elements of the equivalence classes $[x_1]$ and $[x_2]$. □

Therefore, if the indiscernibility relationship of the attributes is computed, we obtain an equivalence matrix from which we obtain a partition of the attributes into equivalence classes. By applying the FCA reduction, we select the representative elements of the equivalence classes.

From this reduction the computation of reducts and bireducts [1,3,18] is faster, since this computation is a NP-complete problem, and no information is missing due to once we have calculated the reducts and the bireducts, we can change any object (or attribute) by another one of its equivalence class.

We should also note that this type of reduction, removing the repeated columns and rows, is not a new mechanism since it has already been considered in several articles. What we propose in this paper is a different approach, that is, to provide a method that generalises this procedural to more general cases. For example, when we are considering an indiscernibility relationship that is not an equivalence relation or in the fuzzy case, as we will show in the rest of this section.

If we take into account a tolerance relationship instead of an equivalence relation as it has been studied in the recent years [10,11,13], we obtain a covering of the universe [4,21]. This is because a tolerance relation does not fulfill the transitivity property, therefore the different blocks that compose the covering are not disjoint and two related objects can appear in more than one block. In order to denote one block given by an object $x \in U$, we will write $[x]_t$.

Example 2. Given an information system $(U, \mathcal{A})$ which represents the weather forecast for different days, we will use a tolerance relation instead of a equivalence one in order to compute the indiscernibility relation among objects. The set of objects is $U = \{1, 2, 3, 4, 5, 6\}$, the attributes to take into account are $\mathcal{A} = \{\text{Outlook, Temperature, Humidity, Wind}\}$ and the relationship between them is defined by the following table:

In this case, we consider that two objects $x, y \in U$ are indiscernible if the values of each attribute are equal for these objects, except for at most for one

	Outlook	Temperature	Humidity	Wind
1	Sunny	Hot	High	Weak
2	Sunny	Hot	High	Strong
3	Overcast	Hot	High	Weak
4	Rain	Mild	High	Weak
5	Rain	Cool	Normal	Weak
6	Rain	Cool	Normal	Strong

attribute. If there are more than one attribute in which the objects differ, we say that they are discernible. For example, days 1 and 2 are indiscernible because they have the same values for the attributes Outlook, Temperature and Humidity, although the attribute Wind has different values. Now, if we consider days 1 and 4, we can see that they are discernible, because there exist two attributes with different values, that is, Outlook and Temperature.

Considering days 1, 2 and 3, as we said above, days 1 and 2 are indiscernible, days 1 and 3 are also indiscernibles (since they only differ in attribute Outlook), but when we compare days 2 and 3, we observe that they are discernible, because the values for the attributes Outlook and Wind are different. Therefore, this relation is reflexive and symmetric, but it is not transitive and, consequently, it is not an equivalence relation.

Thus, we obtain the associated context (U, U, R), where R is the relation between objects defined by the indiscernibility relation $\text{Ind}(U)$, obtaining the following table:

R	1	2	3	4	5	6
1	1	1	1	0	0	0
2	1	1	0	0	0	0
3	1	0	1	0	0	0
4	0	0	0	1	0	0
5	0	0	0	0	1	1
6	0	0	0	0	1	1

From the previous relation, we obtain that the covering is given by the following blocks:

$$[1]_t = \{1, 2, 3\} \qquad [4]_t = \{4\}$$
$$[2]_t = \{1, 2\} \qquad [5]_t = [6]_t = \{5, 6\}$$
$$[3]_t = \{1, 3\}$$

In addition, if we compute the concepts of the context (U, U, R), we can classify the objects as follows:

$$C_f = \{1, 2, 3, 4\} \qquad K_f = \{5, 6\}$$

From the classification we have that the obtained reducts are $D_1 = \{1, 2, 3, 4, 5\}$ and $D_2 = \{1, 2, 3, 4, 6\}$. Hence, the FCA reduction removes from the original sets those objects that appear together with all the blocks in which they belong to. For example, objects 5 and 6 appear together with the blocks $[5]_t$ and $[6]_t$, and there is no more blocks containing these objects. Therefore, we can remove one of them. Hence, this procedure remove unnecessary objects and blocks. On the other hand, we can observe that objects 2 and 3 belong to the block generated by object 1, but they appear separately in the blocks $[2]_t$ and $[3]_t$. Consequently, they must be part of the reducts. □

In the following, we will present a generalisation of the previous reduction considering a fuzzy environment, and we will give some properties and an interpretation of the obtained reduction. Specifically, we will consider a fuzzy indiscernibility relation instead of a classical one. It is important to note that the indiscernibility relation can be evaluated over any poset P, that is, $R\colon U \times U \to P$, as it was introduced in [11]. In this paper, we are interested in the meaning of the application of the fuzzy FCA reduction to RST. In order to simplify the analysis, we will define the relation over the interval $[0, 1]$.

The following definition shows how a fuzzy formal context is considered from an information system.

Definition 13. *Given an information system $(U, \mathcal{A})$, the* associated fuzzy context *is the context (U, U, R), where $R\colon U \times U \to [0, 1]$ is the relationship obtained from the indiscernibility relation of the information system.*

In order to reduce objects in fuzzy RST using fuzzy FCA, we need to recall some definitions. Considering an information system $(U, \mathcal{A})$ where U is the set of objects, $\mathcal{A}$ the set of attributes, with $|\mathcal{A}| = n$, and the mappings $a\colon U \to V_a$, for each $a \in \mathcal{A}$, where V_a is the set of values over the attribute a, we define a fuzzy indiscernibility relation $R\colon U \times U \to [0, 1]$ as:

$$R(x_i, x_j) = @(R_{a_1}(x_i, x_j), \ldots, R_{a_n}(x_i, x_j))$$

for all $x_i, x_j \in U$, where $@\colon [0, 1]^n \to [0, 1]$ is an aggregation operator and $R_{a_k}(x_i, x_j)$ is a tolerance relation between two objects with respect to the attribute a_k.

Considering a tolerance relation and fixed a value $\alpha \in [0, 1]$, we can define an α-block of an object $x \in U$ as the set:

$$[x]_\alpha = \{x_i \in U \mid \alpha \leq R(x, x_i)\} \tag{7}$$

Thenceforward, in order to present the following results, we need to fix a framework $([0, 1], [0, 1], [0, 1], \&)$ where $\&$ is the conjunctor of an adjoint triple $(\&, \swarrow, \nwarrow)$, satisfying the boundary condition $1 \,\&\, y = y$, for all $y \in [0, 1]$. Note that, this condition is satisfied by t-norms and other useful and general operators. Besides, we need to build the α-blocks, consequently, we also fix a value $\alpha \in [0, 1]$.

The following proposition shows a property about the extensions of the concepts obtained from the associated fuzzy context, over the objects belonging to an α-block generated by an object.

Proposition 1. *Let $(U, \mathcal{A})$ be an information system, (U, U, R) the associated context and $x \in U$. Then, we obtain that the inequality:*

$$\alpha \leq \phi^{\downarrow}_{x_i,\beta}(x_i)$$

holds, for all $x_i \in [x]_\alpha$ and $\beta \in [0,1]$.

The following result shows under what conditions the values of the extensions do not exceed the threshold established by α.

Proposition 2. *Given an information system $(U, \mathcal{A})$ and $x \in U$, if the conditions $\beta \,\&\, \alpha \not\leq R(x_i, x)$ and $x_i \notin [x]_\alpha$ are satisfied, then:*

$$\phi^{\downarrow}_{x,\beta}(x_i) < \alpha$$

The next proposition relates the extensions of the multi-adjoint concept lattice obtained from the associated information system to the α-blocks of an object.

Proposition 3. *Given an information system $(U, \mathcal{A})$, the associated context (U, U, R) and $\alpha \in [0,1]$, we have that, if $\beta \,\&\, \alpha \not\leq R(x_i, x_j)$ for all $x_i \notin [x_j]$, then the function $\phi_{x_j,\beta} \colon U \to [0,1]$ satisfies, for all $x_j \in U$, that $\alpha \leq \phi^{\downarrow}_{x_j,\beta}(x_i)$, if $x_i \in [x_j]_\alpha$, and $\phi^{\downarrow}_{x_j,\beta}(x_i) < \alpha$, otherwise.*

Let us clarify the previous results by means of the following example.

Example 3. In this example, we will consider the infomation system $(U, \mathcal{A})$, composed of the set of objects $U = \{x_1, x_2, x_3, x_4, x_5\}$ and the set of attributes $\mathcal{A} = \{a_1, a_2, a_3, a_4\}$, together with the relation given in the following table:

	x_1	x_2	x_3	x_4	x_5
a_1	0.34	0.21	0.52	0.84	0.83
a_2	0.13	0.09	0.36	0.16	0.15
a_3	0.31	0.71	0.93	0.69	0.69
a_4	0.75	0.5	1	1	1

We will build the discernibility matrix using the fuzzy tolerance relation between objects defined as:

$$R_{a_i}(x, y) = 1 - |R(a_i, x) - R(a_i, y)|$$

for all $a_i \in \mathcal{A}$ and $x, y \in U$ and the aggregation operator $@(l_1, l_2.l_3, l_4) = \frac{1}{6}(l_1 + l_2 + 2(l_3 + l_4)))$. That is, we will consider the following discernibility relation:

$$R_{\mathcal{A}}(x, y) = @(R_{a_1}(x, y), R_{a_2}(x, y), R_{a_3}(x, y), R_{a_4}(x, y)) \tag{8}$$

obtaining the following discernibility matrix

$$\begin{pmatrix} 1 & 0.71 & 0.64 & 0.70 & 0.71 \\ 0.71 & 1 & 0.66 & 0.71 & 0.71 \\ 0.64 & 0.66 & 1 & 0.83 & 0.83 \\ 0.70 & 0.71 & 0.83 & 1 & 1 \\ 0.71 & 0.71 & 0.83 & 1 & 1 \end{pmatrix}$$

We are going to fix a value $\alpha = 0.7$, and we will compute the 0.7-blocks for each object:

$$[x_1]_{0.7} = [x_2]_{0.7} = \{x_1, x_2, x_4, x_5\}$$
$$[x_3]_{0.7} = \{x_3, x_4, x_5\}$$
$$[x_4]_{0.7} = [x_5]_{0.7} = \{x_1, x_2, x_3, x_4, x_5\}$$

Now, we consider the associated fuzzy context (U, U, R), where R is the indiscernibility relation described in the discernibility matrix, and the frame $([0,1]_{10}, [0,1]_{10}, [0,1]_{100}, \&_G)$ where $\&_G$ is the Gödel conjunctor. In order to illustrate Proposition 3, we will choose two values for $\beta \in [0,1]$. First of all, we consider $\beta = 0.8$ and the object $x_1 \in U$, obtaining that:

$$\phi^{\downarrow}_{x_1,0.8} = \{1, 0.71, 0.64, 0.70, 0.71\}$$

We can see that the inequality $0.7 \leq \phi^{\downarrow}_{x_1,0.8}(x_i)$ holds, with $i \in \{1,2,4,5\}$, which correspond to the objects in $[x_1]_{0.7}$, as Proposition 3 asserts. Moreover, since $0.8\&_G 0.7 \not\leq R(x_1, x_3)$, Proposition 3 can also be applied to the rest of objects. The unique element which does not belong to $[x_1]_{0.7}$ is x_3 and it naturally satisfies $\phi^{\downarrow}_{x_1,0.8}(x_2) < 0.7$.

If we consider now the product t-norm $\&_P$ instead of the Gödel one, and $\beta = 0.95$, we have that

$$\phi^{\downarrow}_{x_1,0.95} = \{1, 0.75, 0.67, 0.74, 0.75\}$$

We also obtain that $0.7 \leq \phi^{\downarrow}_{x_1,0.95}(x_i)$, for all $x_i \in [x_1]_{0.7}$. In addition, since we also have $0.95\&_P 0.7 = 0.66 \not\leq 0.64 = R(x_1, x_3)$ in this case, we have that $\phi^{\downarrow}_{x_1,0.95}(x_2) < 0.7$.

Computing the concepts associated with the context (U, U, R), and applying the classification results given in [6], we can classify the attributes into the following sets:

$$C_f = \{x_1, x_2, x_3\} \qquad K_f = \{x_4, x_5\}$$

Therefore, we obtain two reducts: $D_1 = \{x_1, x_2, x_3, x_4\}$, $D_2 = \{x_1, x_2, x_3, x_5\}$.

As we can observe, the objects x_4 and x_5 appear in each 0.7-block. Therefore, we can choose one of these objects and remove the other one, similarly to the given procedural for the classical case with tolerance indiscernibility relations.

4 Conclusions and Future Work

In this work, we have shown different interpretations of the use of reduction mechanisms given in FCA to RST. Specifically, we consider information systems and indiscernibility relations, from which we obtain discernibility matrices. These matrices are seen as formal contexts in formal concept analysis. We have carried out the reductions of these contexts by means of reduction mechanisms belonging to FCA theory.

It is important to note that the reduction process proposed in this paper is a new approach, taken into account a general framework. For example, we have considered tolerance relations as indiscernibility relations, instead of equivalence relation. Furthermore, we have analysed the obtained result in the fuzzy case. In addition, we have shown different properties and we have given interpretations for such reductions.

In the future, we will continue studying the reduction mechanism proposed in this paper and we will apply it to real life problems. Besides, we will examine the influence of the FCA reductions in the computation of bireducts [3,18].

References

1. Benítez, M., Medina, J., Ślęzak, D.: Reducing information systems considering similarity relations. In: Kacprzyk, J., Koczy, L., Medina, J. (eds.) 7th European Symposium on Computational Intelligence and Mathematices (ESCIM 2015), pp. 257–263 (2015)
2. Benítez-Caballero, M.J., Medina, J., Ramírez-Poussa, E.: Attribute reduction in rough set theory and formal concept analysis. In: Polkowski, L., Yao, Y., Artiemjew, P., Ciucci, D., Liu, D., Ślęzak, D., Zielosko, B. (eds.) IJCRS 2017. LNCS (LNAI), vol. 10314, pp. 513–525. Springer, Cham (2017). https://doi.org/10.1007/978-3-319-60840-2_37
3. Benítez-Caballero, M.J., Medina, J., Ramírez-Poussa, E., Ślęzak, D.: Bireducts with tolerance relations. Inf. Sci. **435**, 26–39 (2018)
4. Chen, J., Li, J., Lin, Y., Lin, G., Ma, Z.: Relations of reduction between covering generalized rough sets and concept lattices. Inf. Sci. **304**, 16–27 (2015)
5. Cornejo, M.E., Medina, J., Ramírez-Poussa, E.: A comparative study of adjoint triples. Fuzzy Sets Syst. **211**, 1–14 (2013)
6. Cornejo, M.E., Medina, J., Ramírez-Poussa, E.: Attribute reduction in multi-adjoint concept lattices. Inf. Sci. **294**, 41–56 (2015)
7. Cornejo, M.E., Medina, J., Ramírez-Poussa, E.: Multi-adjoint algebras versus non-commutative residuated structures. Int. J. Approximate Reasoning **66**, 119–138 (2015)
8. Cornejo, M.E., Medina, J., Ramírez-Poussa, E.: On the use of irreducible elements for reducing multi-adjoint concept lattices. Knowl.-Based Syst. **89**, 192–202 (2015)
9. Cornejo, M.E., Medina, J., Ramírez-Poussa, E.: Characterizing reducts in multi-adjoint concept lattices. Inf. Sci. **422**, 364–376 (2018)
10. Cornelis, C., Jensen, R., Hurtado, G., Ślęzak, D.: Attribute selection with fuzzy decision reducts. Inf. Sci. **180**, 209–224 (2010)
11. Cornelis, C., Medina, J., Verbiest, N.: Multi-adjoint fuzzy rough sets: definition, properties and attribute selection. Int. J. Approximate Reasoning **55**, 412–426 (2014)
12. Fariñas del Cerro, L., Prade, H.: Rough sets, twofold fuzzy sets and modal logic–fuzziness in indiscernibility and partial information. In: Nola, A.D., Ventre, A. (ed.) The Mathematics of Fuzzy Systems, pp. 103–120. Verlag TUV Rheinland (1986)
13. Guan, L., Huang, D., Han, F.: Tolerance dominance relation in incomplete ordered decision systems. Int. J. Intell. Syst. **33**(1), 33–48 (2018)

14. Medina, J., Ojeda-Aciego, M., Ruiz-Calviño, J.: Formal concept analysis via multi-adjoint concept lattices. Fuzzy Sets Syst. **160**(2), 130–144 (2009)
15. Pawlak, Z.: Information systems theoretical foundations. Inf. Syst. **6**(3), 205–218 (1981)
16. Pawlak, Z.: Rough sets. Int. J. Comput. Inf. Sci. **11**, 341–356 (1982)
17. Skowron, A., Rauszer, C.: The discernibility matrices and functions in information systems. In: Słowiński, R. (ed.) Intelligent Decision Support: Handbook of Applications and Advances of the Rough Sets Theory, pp. 331–362. Kluwer Academic Publishers (1992)
18. Stawicki, S., Ślęzak, D., Janusz, A., Widz, S.: Decision bireducts and decision reducts - a comparison. In. J. Approximate Reasoning **84**, 75–109 (2017)
19. Wei, L., Qi, J.-J.: Relation between concept lattice reduction and rough set reduction. Knowl.-Based Syst. **23**(8), 934–938 (2010)
20. Wille, R.: Restructuring lattice theory: an approach based on hierarchies of concepts. In: Rival, I. (ed.) Ordered Sets, pp. 445–470. Reidel (1982)
21. Yang, B., Hu, B.Q.: On some types of fuzzy covering-based rough sets. Fuzzy Sets and Syst. **312**, 36–65 (2017). Theme: Fuzzy Rough Sets

Reliability Improvement of Odour Detection Thresholds Bibliographic Data

Pascale Montreer[1(✉)], Stefan Janaqi[2], Stéphane Cariou[1], Mathilde Chaignaud[3], Isabelle Betremieux[4], Philippe Ricoux[4], Frédéric Picard[5], Sabine Sirol[6], Budagwa Assumani[6], and Jean-Louis Fanlo[1,3]

[1] IMT Mines Alès, LGEI Laboratory, 6 av. de Clavières, 30100 Ales, France
{pascale.montreer, stephane.cariou}@mines-ales.fr

[2] IMT Mines Alès, LGI2P Laboratory, 6 av. de Clavières, 30100 Ales, France
stefan.janaqi@mines-ales.fr

[3] Olentica SAS, 17 rue Charles Peguy, 30100 Ales, France
{mathilde.chaignaud, jean-louis.fanlo}@olentica.fr

[4] Total S.A, 2, place Jean Millier, La Défense 6, 92078 Paris La Défense Cedex, France
{isabelle.betremieux, philippe.ricoux}@total.com

[5] Hutchinson S.A., Rue Gustave Nourry, 45120 Châlette-sur-Loing, France
frederic.picard@hutchinson.com

[6] Total Feluy, Zone Industrielle Feluy C, 7181 Seneffe, Belgium
{sabine.sirol, budagwa.assumani}@total.com

Abstract. Odour control is an important industrial issue as it is a criterion in purchase of a material. The minimal concentration of a pure compound allowing to perceive its odour, called Odour Detection Threshold (ODT), is a key of the odour control. Each compound has its own ODT. Literature is the main source to obtain ODT, but a lot of compounds are not reported and, when reported, marred by a high variability. This paper proposes a supervised cleaning methodology to reduce uncertainty of available ODTs and a prediction of missing ODTs on the base of physico-chemical variables.

This cleaning leads to eliminate 39% of reported compounds while conducting 84% of positive scenarios on 37 comparisons. Missing ODTs are predicted with an error of 0.83 for the train and 1.14 for the test (log10 scale). Given the uncertainty of data, the model is sufficient. This approach allows working with a lower uncertainty and satisfactory prediction of missing ODTs.

Keywords: Odour Detection Thresholds (ODT) · Data mining Reliability · Completeness · Uncertainty

1 Introduction

In the industrial environment, there is a growing need to identify compounds responsible for an unpleasant odour. This identification depends firstly on the Odour Detection Threshold (ODT) of each compound. We define ODT as the minimal concentration of a pure compound allowing to perceive its odour.

J. Medina et al. (Eds.): IPMU 2018, CCIS 853, pp. 562–573, 2018.
https://doi.org/10.1007/978-3-319-91473-2_48

The principal source of ODTs is the literature [1–6]. But, in literature, there are a lot of compounds with no reported ODTs and when ODTs data are available, they present a high variability. This situation implies a high uncertainty of ODTs. This variability can be illustrated with the butyl acetate example whose ODT values range from 0.030 mg/m^3 to 480 mg/m^3 for 14 publications [6].

The variability could potentially be explained by a set of parameters such as: difference of methods; existence and year of normalization; sample quality; environmental conditions, culture of authors; panel selection and their intrinsic diversity; panel correction, etc. Given all these potential sources of variability, there is a need to improve the reliability of these data. Unfortunately, the sparsity of these data makes most of the potential interesting statistical tools unusable. This sparsity comes from non-uniformity of the information from one author to another.

Several papers have already highlighted this issue [7, 8] and have tried to compare publications [9] but usually, researchers make a subjective sorting or simply use the mean of values [7, 10]. But, in any event, even if dataset is used without cleaning, neither the mean or the geometric mean nor the density application is justified if the ODT values distribution is not identified.

As a consequence, our paper aims to find relevant methodology allowing improving the reliability of these data. This improvement gathers an approach aiming to decrease the variability of available ODTs by a cleaning methodology. Next, we complete not reported compounds by a predictive modelling of ODTs as a function of chemical/physical variables. Explicative variables generated by the methodology are precise opposed to bibliographic ODTs.

2 Methodology

2.1 Software

The ODT values from the literature are collected in an Excel sheet and the reliability improvement is realized on the R software.

2.2 Database Construction

To analyze ODT, a database is constructed (Fig. 1). This database gathers quite a few volatile and odorous compounds. These compounds constitute the rows of the database.

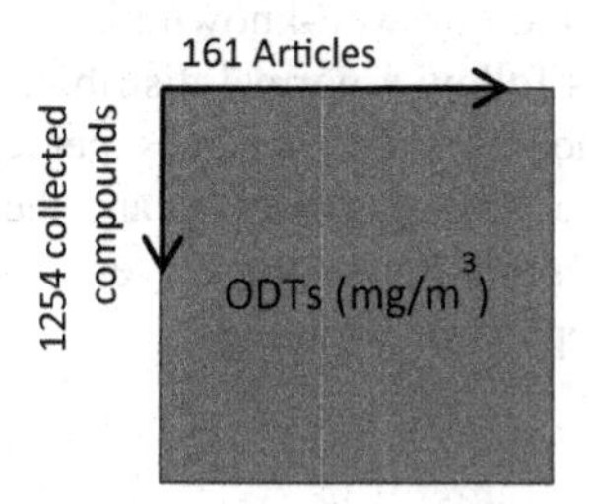

Fig. 1. ODT values (mg/m^3) database format

For each compounds, a state of art of ODTs is done to identify publications containing ODT information [6]. These publications constitute the columns.

This database will be the support of the uncertainty reduction and prediction. To date, it contains 161 publications (columns) and 1 254 compounds (rows) including only 650 compounds with at least one ODT available.

2.3 First Step of Cleaning: Provisional Publications' Isolation

The first step consists in eliminating the least reliable studies. Ideally, it would have been relevant to consider only studies containing a reliable repeatability of measurements. In this way the Cochran test based on the standard deviation of each article would have been applied [11]. Unfortunately, this information is rarely available. Considering only studies with available repeatability is therefore too restrictive. Consequently, we use the criterion of number of ODT measurements realized per study. Indeed, we have noticed that the higher the number of ODT determinations in a publication, the better the technique of analysis was described, and hence the more reliable the results. It was decided to eliminate ten percent of the total values of the database and that corresponds to publications containing less than four ODT values (Fig. 2).

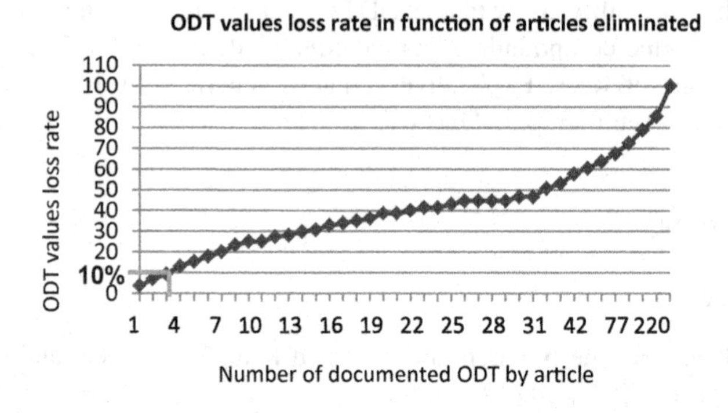

Fig. 2. ODT values loss rate in function of publications eliminated

2.4 Second Step of Cleaning Methodology: ODT Outlier Elimination

2.4.1 Outliers Definition in the Context

To define an outlier in this context, the ODT values distribution has to be defined. On the basis of an expert observation on well-known molecules, the hypothesis is that for each given compound, ODTs follow a normal distribution.

Considering the low amount of ODT values per compound (Fig. 3), the most suitable solutions to observe this distribution, are the Kernel Density Estimation (KDE) and the QQplot representation of the eleven compounds with more than ten ODT values after the first step of the cleaning.

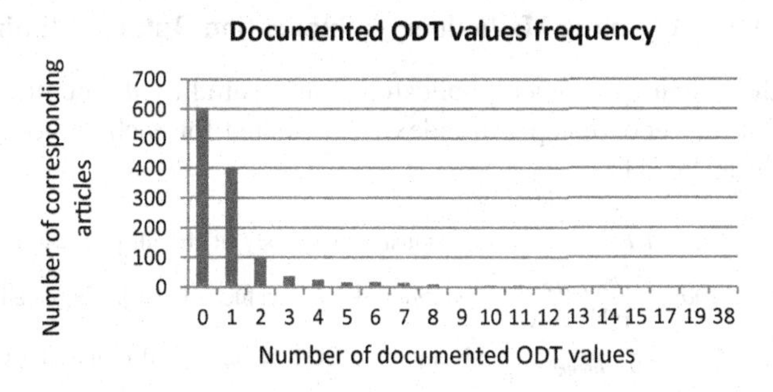

Fig. 3. Reported ODT frequency

The KDE is calculated with the "geom_density" function and the QQplot with "qqnorm" function. The correlation coefficient (CC) of QQplot of these compounds is calculated (Fig. 4).

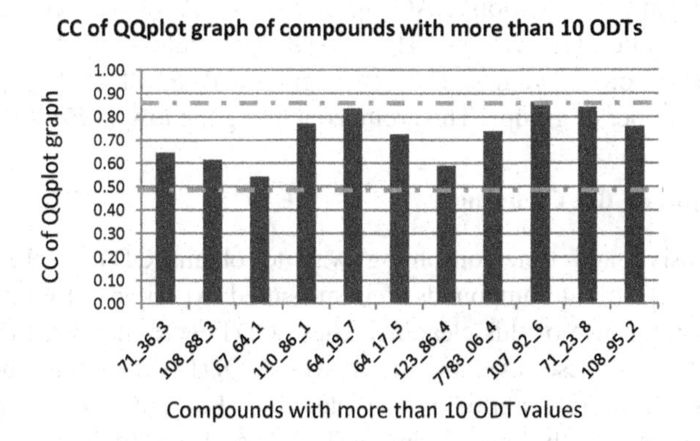

Fig. 4. Correlation coefficient of the QQplot of compounds with more than 10 ODT

On this Fig. 4 the higher the correlation coefficient is, the more likely the distribution is normal. These results encourage applying a normal test on values to detect outliers. The inter-laboratory reproducibility Grubbs test is applied [11]. The confidence level chosen is 95%.

2.4.2 Outliers Values Elimination According to Normal Distribution

The Grubbs test is not applicable with less than three values [12]. In this way, only compounds measured more than twice can be compared with the rest of the database. Thereafter, ODTs of these compounds are considered as the "tested" ODTs. That implies that some compounds' relevance cannot be analyzed at this step.

Another rule of comparison has to be set up. This is an extension of the cleaning methodology to the elimination of publications outliers. This next rule was guided by the fact that, at the previous step, the eliminated ODT values often belong to the same publications.

2.5 Third Step of Cleaning Methodology: Publication Outliers Elimination

At this step, the relevance of each publication is measured by the number of exclusion of that article at the second step. An index is calculated for each author: the exclusion frequency ratio in Eq. (1).

$$\mathbf{EF}_j = \frac{N^j_{ODT\,excluded}}{N^j_{ODT\,tested}} \times 100 \qquad (1)$$

EF_j : Exclusion frequency of the jth publication

$N^j_{ODT\,excluded}$: Number of ODT values of the jth publication excluded by Grubbs test

$N^j_{ODT\,tested}$: Number of ODT values tested in the jth publication

The ODT values of publications have not been compared in the same way. That's why a "tested rate" is calculated. For each publication, this "tested rate" is defined as the percentage of tested ODTs among all of the ODTs of this publication. Indeed, the higher the "tested rate" is; the better the $EF's$ reliability is.

First, publications with "tested rate" under 1/3 are eliminated. Then, after a Hierarchical Ascendant Classification (HAC) applied on the *EF* of selected publications, the Ward's distance index [13] is used. This classification allows to statistically separate publications based on *EF* values. The dendrogram and the SPRSQ graph are used to determine the number of groups. The group containing the lowest *EF* is finally retained.

2.6 Validation of the Cleaning

Another expensive and time-consuming way to obtain ODT is the experimental measurement. ODT of 44 compounds were measured experimentally in our laboratory to validate the relevance of this cleaning. These ODT were measured using the norm EN 13725 [14]. For these 44 compounds, whose ODT values have been measured experimentally, 40 were reported in the literature. For these 40 compounds, three barycenters are defined: the one of raw ODT values of the literature (X1), the one of remaining ODT values after the cleaning methodology application (X2) and the one of our experimental ODT values (X3). And then two differences are calculated: X1−X3 (Z1) and X2−X3 (Z2). These two differences are compared. This comparison allows to observe if the cleaning methodology leads to approach the experimental result.

To state on the relevance of the cleaning the criterion to select the "ideal" case is the mean of the log10 of the confidence interval obtained thanks the repeatability level of our experimental measurements.

2.7 Completeness of the Database by Predictive Modelling

The completeness of the database is based on a Quantitative Structure-Property Relationship (QSPR) approach. This approach consists in predicting a variable (Y) as a function of chemical and physical certain variables (X). Here the Y variable is the mean value of log10 values of cleaned ODTs. Explicative variables X are essentially calculated from the structure of compounds (66 variables). There are compositional and constitutional indexes (the number and nature of atoms, the molecular weight, the unsaturation), topological indexes which the majority are defined in Todeschini's publication [15] and electrotopological indexes as ZEP index [16].

The model approach is then divided in 2 principal steps. The first is a reduction of the number of significative input variables by a lasso technique. To predict Y as a function of X, a classical Support Vector Machine (SVM) model was constructed. Some "bad" observations was eliminated from this learning process. The partition of the dataset between the train and test sets is made in order that the train represents 75% of the dataset and reflects the variance of this one.

3 Results and Discussion

3.1 Visualization of the Proportion of Missing Values of the Database

Missing value proportion of the database (Fig. 1) is presented as a heatmap (Fig. 5).

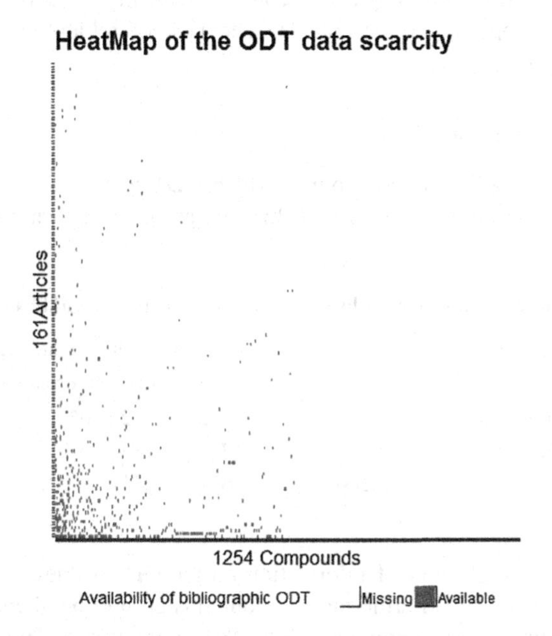

Fig. 5. Proportion of missing ODT values in the literature (a red zone corresponds to information presence; a white zone corresponds to a lack of information) (Color figure online)

The total proportion of missing values is 99%. This proportion enforces the inability to use statistical classification tools and implies the necessity to implement a more reliable methodology.

3.2 Result of the First Step

For the three "data cleaning" steps, information loss is presented from three perspectives (Table 1): the number of publications, the number of ODT and the number of

Table 1. Information loss after the first step

	Cleaning at the first step		
	Before	After	Information loss rate
Nos. of publications	161	72	55%
Nos. of ODT	1501	1367	9%
Nos. of compounds with at least 1 ODT	650	631	3%

compounds with at least one ODT. This choice is made because the most important information to monitor is the number of compounds with at least one ODT. Indeed, as it was presented, we want to predict the ODT behavior of all the compounds of the database (1254) with reported ones in literature. That's why, it's important to keep a satisfactory proportion of reported compounds.

The first step leads to isolate 55% of the publications. Even if half of the publications have been eliminated, only 3% of compounds with at least one ODT have been eliminated. This proportion is acceptable. The lack of reliability on these sources justifies to reject them.

3.3 Result of the Second Step

The Grubbs test is applied on compounds which ODT have been measured at least 3 times. Results of the information loss of this step are presented in Table 2.

Table 2. Information loss after the second step by Grubbs test

	Cleaning at the second step		
	Before	After	Information loss rate
Nos. of publications	72	72	0%
Nos. of ODT values	1367	1235	10%
Nos. of compounds with at least 1 ODT	631	631	0%

There is only a slight loss of information on the ODT values (10%) and the two other rates are still constant. Furthermore, among this proportion, there is an average of 1.65 values eliminated per compound (over 80 compounds). This low percentage supports the use of a normal distribution.

3.4 Result of the Third Step

A summary of ODT elimination at the second step on all the 72 publications, will allow applying the third step. The focus is made on the Exclusion Frequency (EF) defined by the Eq. (1) and the "tested rate" defined in the Sect. 2.5.

As it was mentioned in this section, publications with "tested rate" under 1/3 are eliminated. A first information loss summary is done after the elimination of these publications with a "tested rate" lower than 33% (Table 3).

Table 3. Information loss after the elimination of publications with tested rate lower than 33%

	Cleaning at the third step – Part 1		
	Before	After	Information loss rate
Nos. of publications	72	60	17%
Nos. of ODT values	1235	1064	14%
Nos. of compounds with at least 1 ODT	631	485	23%

The dendrogram of the HAC classification is applied on the EF of publications with a "tested rate" higher than 33% (Fig. 6).

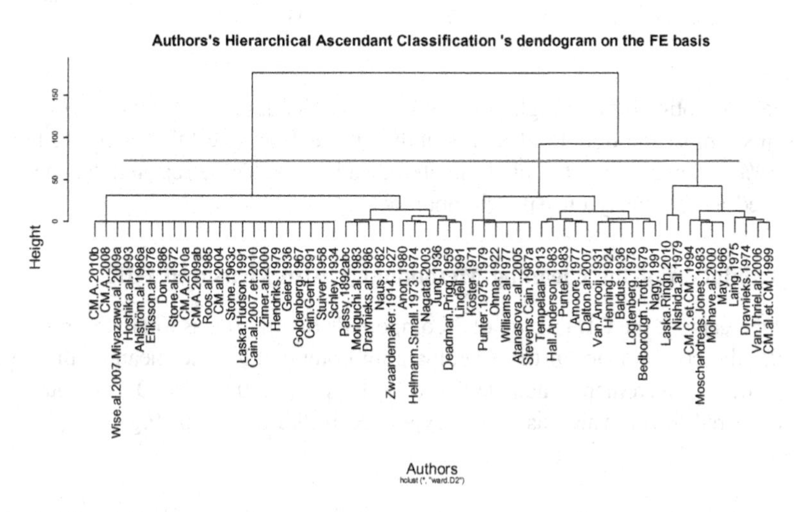

Fig. 6. HAC on EF values dendrogram (red line: the optimal classification) (Color figure online)

Thanks to the dendrogram and the SPRSQ graph, 3 groups were realized (Fig. 6). The information loss summary is done after the second part of the third step (Table 4).

Table 4. Information loss after the third step

	Cleaning at the third step - Part 2		
	Before	After	Information loss rate
Nos. of publications	60	33	45%
Nos. of ODT values	1064	701	34%
Nos. of compounds with at least 1 ODT	485	393	19%

This third step is rather drastic because it eliminates 37.7% of compounds with at least one reported ODT value in the literature.

Nevertheless, 393 reported compounds still remain, spread over 33 publications. The size of the database decreases with the increase of its reliability. Of course, we consider that it is better to work with less quantitative but more informative data.

3.5 Summary of the Cleaning Methodology

After the three cleaning methodology steps, the information loss statement is presented in Table 5.

Table 5. Information loss statement after three steps of the cleaning methodology

	Total cleaning		
	Before	After	Total information loss rate
Nos. of publications	161	33	79%
Nos. of ODT values	1501	701	53%
Nos. of compounds with at least 1 ODT	650	393	**39%**

It can be noticed that a high percentage of publications is eliminated (79%) but it represents almost the double of compounds with at least one ODT value in the literature (39%). The amount of total eliminated data is 53%. These rejected data have been considered irrelevant by our specific approach.

3.6 Validation of the Cleaning Methodology with Measured ODT

After the cleaning, the ODT value of 3 compounds out of 40 was totally eliminated and we make the comparison on the 37 remaining compounds. The mean of the repeatability of our measurement calculated in a log10 scale is 0.4. The ODTs obtained are then compared to this value as it was explained in the Sect. 2.6 (Fig. 7).

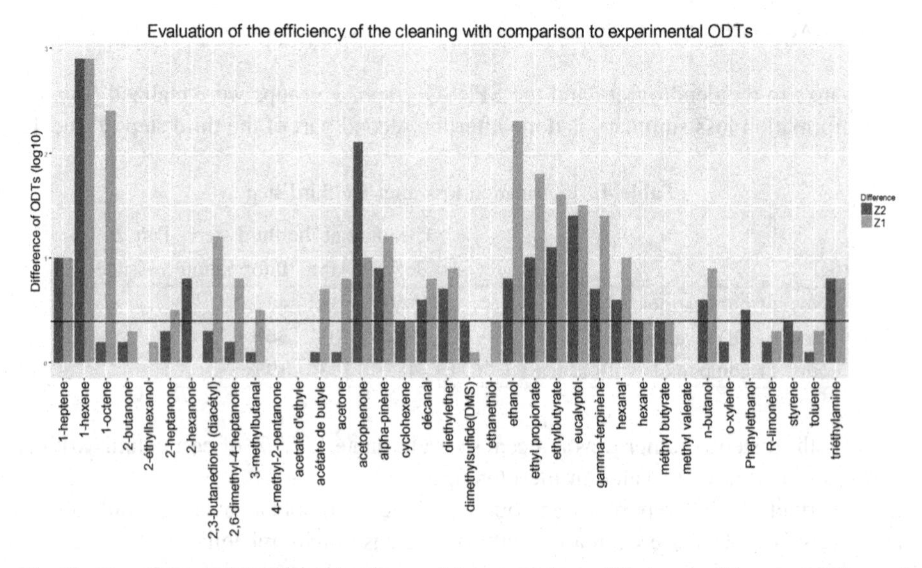

Fig. 7. Comparison of the cleaned values and the raw values of literature with our experimental values (red: cleaned ODT difference; blue: raw ODT difference) (Color figure online)

Many scenarios are recorded (Table 6) thanks to the graph of comparisons (Fig. 7).

Table 6. Summary of the evolution by the cleaning methodology

Scenario 1	Improvement of the value …	… even if it was already correct	**5 cases**
Scenario 2		… which leads to a correct value	**6 cases**
Scenario 3		… but the value still not correct	**10 cases**
Scenario 4	No improvement of the value …	… but it was already correct	**7 cases**
Scenario 5		… and still not correct	3 cases
Scenario 6	Degradation of the value …	… but it still correct	**3 cases**
Scenario 7		… but it still not correct	1 cases
Scenario 8		… becomes not correct	2 cases

Initially, 16 compounds were in the range of the measured ODT. After the cleaning, there are 21 compounds (out of 37 compared compounds). This is a first positive added value of the cleaning methodology.

Furthermore, it can be considered as positive situation after the cleaning the scenarios 1, 2, 3, 4 and 6 (Table 6). These cases represent 84% of positive cases (31 compounds out of 37). The worst scenario is the number 8. Despite of 6 compounds, this cleaning seems appropriate to decrease the uncertainty of available bibliographic ODT data.

3.7 ODTs Prediction

Applying lasso technique leads to eliminate 22 variables (out of 66) and SVM leads to a model with an error, in a log10 scale, of 0.83 on the train and 1.14 on the test with compounds eliminated (Fig. 8). As a matter of fact, among all the predictions, only less than 2.5% of observations was abnormally predicted in comparison with the others.

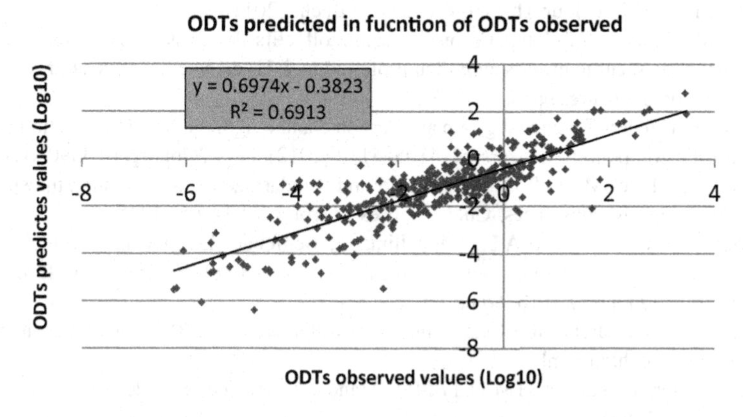

Fig. 8. ODTs predicted in function of ODTs observed

Given the uncertainty of ODT of most of compounds, the model is satisfactory as it predicts the ODT with a log10 error about 1. To our knowledge this is the first global model to predict ODT as a function of molecular characteristics on this quantity of compounds. This is an encouraging result which has to be improved thereafter.

4 Conclusion

In conclusion, the proposed cleaning methodology eliminates 39% of compounds with at least one ODT while conducting to 84% of positive scenarios on ODT values (validation on 37 compounds). The missing ODTs are predicted with an error of 0.83 for the train and 1.14 for the test (on a log10 scale). Considering the data, it's a satisfactory model. This approach allows working with a low uncertainty on available ODTs and predicts missing ODTs with a satisfactory model. This is an encouraging methodology. Thereafter, it would be interesting to strengthen the cleaning methodology with more measurements in our laboratory or a probabilistic validation.

The global predicting model of ODT seems satisfactory to odour experts. It will be improved further by complementary local approach and/or expert knowledge.

References

1. Leonardos, G., Kendall, D., Barnard, N.: Odor threshold determinations of 53 odorant chemicals. J. Air Pollut. Control Assoc. **19**, 91–95 (1969). https://doi.org/10.1080/00022470.1969.10466465
2. Ruth, J.H.: Odor thresholds and irritation levels of several chemical substances: a review. Am. Ind. Hyg. Assoc. J. **47**, 142–151 (1986). https://doi.org/10.1080/15298668691389595
3. Devos, M., Patte, F., Laffort, P., Van Gemert, L.J.: Standardized Human Olfactory Thresholds. Oirl Press, San Francisco (1990)
4. ASTM: Odor Thresholds for Chemicals with Established Occupational Health Standards (1997)
5. US EPA: Reference guide to odor thresholds for hazardous air pollutants listed in the clean air act amendements of 1990 (1992)
6. Van Gemert, L.J.: Odour Thresholds. Zeist, Utrecht (2011)
7. Falcy, M., Malard, S.: Comparaison des seuils olfactifs de substances chimiques avec des indicateurs de sécurité utilisés en milieu professionnel. Hygiène et Sécurité au travail- Cahier de notes documentaire, pp. 7–21 (2005)
8. Zarzo, M.: Effect of functional group and carbon chain length on the odor detection threshold of aliphatic compounds. Sensors **12**, 4105–4112 (2012). https://doi.org/10.3390/s120404105
9. Patte, F., Etcheto, M., Laffort, P.: Selected and standardized values of suprathreshold odor intensities for 110 substances. Chem. Senses Flavor **1**, 283–305 (1975)
10. Toropov, A.A., Toropova, A.P., Cappellini, L., Benfenati, E., Davoli, E.: Odor threshold prediction by means of the Monte Carlo method. Ecotoxicol. Environ. Saf. **133**, 390–394 (2016). https://doi.org/10.1016/j.ecoenv.2016.07.039
11. ISO17025: DémarcheISO17025. http://www.demarcheiso17025.com/fiches_techniques/test_grubbs_cochran.html
12. ASTM International: ASTM E178-00 Standard practice for dealing with outlying observation (1994)

13. Ward, J.H.J.: Hierarchical grouping to optimize an objective function. J. Am. Stat. Assoc. **58**, 236–244 (1963)
14. AFNOR: Norme NF EN 13725 (2003)
15. Todeschini, R., Consonni, V., Mannhold, R., Kubinyi, H., Folkers, G.: Molecular Descriptors for Chemoinformatics. Wiley-VCH, Germany (2009)
16. Berinde, Z., Berinde, M.: On a matrix representation of molecular. Carpath. J. Math. **20**, 205–209 (2004)

Formal Concept Analysis and Structures Underlying Quantum Logics

Ondrej Krídlo[1] and Manuel Ojeda-Aciego[2(✉)]

[1] University of Pavol Jozef Šafárik, Košice, Slovakia
[2] Departamento de Matemática Aplicada, Universidad de Málaga, Málaga, Spain
aciego@uma.es

Abstract. A Hilbert space H induces a formal context, the Hilbert formal context $\overline{H}$, whose associated concept lattice is isomorphic to the lattice of closed subspaces of H. This set of closed subspaces, denoted $\mathcal{C}(H)$, is important in the development of quantum logic and, as an algebraic structure, corresponds to a so-called "propositional system," that is, a complete, atomistic, orthomodular lattice which satisfies the covering law. In this paper, we continue with our study of the Chu construction by introducing the Chu correspondences between Hilbert contexts, and showing that the category of Propositional Systems, PropSys, is equivalent to the category of ChuCors$_{\mathcal{H}}$ of Chu correspondences between Hilbert contexts.

Keywords: Formal concept analysis · Chu correspondence Quantum logic

1 Introduction

Since its introduction, quantum mechanics raised profound conceptual problems, among which the principal was to find an adequate formalization. John von Neumann [12] investigated the mathematical mechanism underlying quantum mechanics, leading to the use of Hilbert spaces as its most natural language. Later, Birkhoff and von Neumann [5] studied the logical structure which is likely to be found in physical theories and, in particular, in quantum mechanics; this led to the discovery that the formalism based on Hilbert spaces also had a (non-classical) logical structure, which nowadays is called *quantum logic*. The main conclusion of Birkhoff and von Neumann was

O. Krídlo—Partially supported by the Slovak Research and Development Agency contract No. APVV-15-0091, University Science Park TECHNICOM for Innovation Applications Supported by Knowledge Technology, ITMS: 26220220182 and II. phase, ITMS2014+: 313011D232, supported by the ERDF.

M. Ojeda-Aciego—Partially supported by the Spanish Science Ministry project TIN2015-70266-C2-P-1, co-funded by the European Regional Development Fund (ERDF).

J. Medina et al. (Eds.): IPMU 2018, CCIS 853, pp. 574–584, 2018.
https://doi.org/10.1007/978-3-319-91473-2_49

> ... *that one can reasonably expect to find a calculus of propositions which is formally indistinguishable from the calculus of linear subspaces with respect to* set products, linear sums, *and* orthogonal complements—*and resembles the usual calculus of propositions with respect to* and, or, *and* not. (Quoted from [5])

Since then, a number of researchers have continued this line of research, and found alternative descriptions of different interesting subsets of linear subspaces. For instance, in quantum logic, ortholattices have often been used, where the (topologically) closed subspaces of a separable Hilbert space represent quantum propositions; these spaces can be represented in purely lattice-theoretical terms, by Piron's representation theorem [13], as irreducible, complete, atomistic, orthomodular lattices satisfying the covering law. Thus, it makes sense that these special types of lattices are called *propositional systems* in [14].

For the purposes of this work, we will use an additional characterization, in which closed subspaces turn out to be algebraically characterized in terms of the double orthogonal operator $(\ )^{\perp\perp}$, i.e., a linear subspace of a Hilbert space is (topologically) closed if and only if it is a fixpoint of $(\ )^{\perp\perp}$.

The lattice of closed subspaces of a Hilbert space has a solid relationship with Formal Concept Analysis (FCA), as stated in [8, p. 55], where one can read that

> *if H is a Hilbert space and $\perp$ is the orthogonality relation, then the concept lattice of the context $(H, H, \perp)$ is isomorphic to the orthomodular lattice of the closed subspaces of H.* (Quoted from [8])

In this paper, we continue our research line on the Chu construction [6] applied to different generalizations of FCA [9,10]. It is worth noting that the closely related notion of Chu space has already been applied to represent quantum physical systems and their symmetries [1,2].

Specifically, the goal of this work is to highlight the importance of the Chu construction with respect to quantum logic, by constructing a category on Hilbert formal contexts $(H, H, \perp)$ and Chu correspondences between them, and proving that it is equivalent to the category PropSys of propositional systems of Hilbert spaces.

The structure of this paper is the following: in Sect. 2 some preliminary notions related to Hilbert spaces, the lattice-theoretical approach to the subset of closed linear subspaces, and formal concept analysis are introduced; then, in Sect. 3, we give the contextual representation of propositional systems in terms of Hilbert formal contexts and Chu correspondences; finally, in Sect. 4, we draw some conclusions and present some prospects for future work.

2 Preliminary Definitions

In this section, we introduce some basic notions about Hilbert spaces, lattices, and formal concept analysis.

2.1 Hilbert Spaces

Definition 1. *A* Hilbert space H *is a real or complex vector space with an inner product which is also a complete metric space with respect to the distance function induced by the inner product.*

The induced metric topology allows to talk about the closed linear subspaces of H. The set of closed linear subspaces of H will be denoted by $\mathcal{C}(H)$.

The inner product in a Hilbert space naturally induces the orthogonality relation: two vectors $v, w \in H$ are orthogonal, written $v \perp w$, if their inner product is zero.

Definition 2. *Let H be a Hilbert space, let $S \subseteq H$ be a subspace of H and $v \in H$. We write $v \perp S$ if and only if $v \in S^{\perp}$ or, in other words, $v \perp u$ for all $u \in S$.*

It turns out that topological closure of a linear subspace can be directly rephrased in terms of the orthogonality relation, as stated below:

Lemma 1. *Let $A \subseteq H$ be a linear subspace of H, then A is closed if and only if $A = A^{\perp\perp}$.*

Definition 3. *A* ray *in a Hilbert space H is any one-dimensional linear subspace of H. Let $u \in H$ be an arbitrary vector of H, then we denote*

$$\rho(u) = \{v \in H \mid v = \lambda u \text{ for some scalar } \lambda\}.$$

The set of all rays in H is denoted by $\mathcal{P}(H)$, whereas the set of all linear subspaces of H is denoted by $\mathcal{L}(H)$.

It is worth noting that, although the main interest of Hilbert spaces is on the merging of algebraic (since it is a vector space) and topological properties (since the inner product induces the metric topology), the main tools used in this paper belong almost exclusively to linear algebra.

2.2 Lattices

A lattice L is said to be:

- *complete* if suprema and infima exist for any subset of L.
- *atomistic* if every element of L is the join of finitely many atoms.
- *orthomodular* if it has zero element 0 and unit element 1, and for any element a there is an *orthocomplement* $a^{\perp}$, i.e. an element satisfying
  1. $a \vee a^{\perp} = 1, \quad a \wedge a^{\perp} = 0, \quad (a^{\perp})^{\perp} = a$
  2. $a \leq b$ implies $a^{\perp} \geq b^{\perp}$
  3. $a \leq b$ implies $b = a \vee (b \wedge a^{\perp})$ (orthomodular law)

Definition 4. *A* propositional system *is a complete, atomistic, orthomodular lattice* $(L, \leq, (\_)^{\perp})$ *which satisfies the* covering law, *i.e., for any* $x \in L$ *and any atom* $a \in L$ *we have that* $a \wedge x = 0$ *implies* $x < a \vee x$ *and there is no element between them.*[1]

Lemma 2. *The algebraic structure of the set* $C(H)$ *of closed linear subspaces of a Hilbert space* H *is that of a propositional system.*

Definition 5. *Let* C_1 *and* C_2 *be propositional systems. A map* $h\colon C_1 \to C_2$ *is a* morphism of propositional systems *if it preserves arbitrary suprema and maps atoms of* C_1 *to either atoms or the bottom element of* C_2.

Proposition 1. *Propositional systems and their morphisms form a category* PropSys.

2.3 FCA and Intercontextual Structures

Definition 6. Formal context *is a triple* (B, A, R) *where* $R \subseteq B \times A$. *Let us define two mappings* $(-)^{\uparrow}\colon 2^B \to 2^A$ *and* $(-)^{\downarrow}\colon 2^A \to 2^B$ *as follows*

- $X^{\uparrow} = \{a \in A \mid (\forall b \in X)(b, a) \in R\}$
- $Y^{\downarrow} = \{b \in B \mid (\forall a \in Y)(b, a) \in R\}$

for any $X \subseteq B$ *and* $Y \subseteq A$. *Such mappings are called* derivation (concept-forming) operators *of the context* (B, A, R).

Proposition 2. *Pair of mappings* $((-)^{\uparrow}, (-)^{\downarrow})$ *forms a Galois connection between* $(2^B, \subseteq)$ *and* $(2^A, \subseteq)$.

Definition 7. *Let* $C = (B, A, R)$ *be a formal context. A pair of sets* $(X, Y) \in 2^B \times 2^A$ *is a* formal concept *if* $X^{\uparrow} = Y$ *and* $Y^{\downarrow} = X$. *The object part* X *is said to be the* extent, *and the attribute part* Y *is the* intent *of the formal concept. The set of all extents and intents of a formal context* C *are denoted by* $\mathrm{Ext}(C)$ *and* $\mathrm{Int}(C)$, *respectively.*

Formal concepts (X, Y) are fixpoints of Galois connection $((-)^{\uparrow}, (-)^{\downarrow})$, i.e. pairs of closed subsets made by closure operators of derivation operators.

We now recall the basic notions about the category ChuCors of formal contexts and Chu correspondences.

Definition 8. *Let* $C_1 = (B_1, A_1, R_1)$ *and* $C_2 = (B_2, A_2, R_2)$ *be two formal contexts with pairs derivation operators* $((-)^{\uparrow_1}, (-)^{\downarrow_1})$ *and* $((-)^{\uparrow_2}, (-)^{\downarrow_2})$. *Let us define the following pair of mappings* $\varphi = (\varphi_L, \varphi_R)$ *from* C_1 *to* C_2, *which we will denote as* $\varphi\colon C_1 \to C_2$

[1] This relation is called the *covering relation* and the condition would be usually denoted by $x \prec a \vee x$.

- $\varphi_L\colon B_1 \to \mathrm{Ext}(C_2)$ *and* $\varphi_R\colon A_2 \to \mathrm{Int}(C_1)$
- *and for any* $(b_1, a_2) \in B_1 \times A_2$ *holds that*

$$a_2 \in \varphi_L(b_1)^{\uparrow_2} \Longleftrightarrow b_1 \in \varphi_R(a_2)^{\downarrow_1}$$

with composition of $\varphi_1\colon C_1 \to C_2$ *and* $\varphi_2\colon C_2 \to C_3$ *defined as follows*

- $(\varphi_2 \circ \varphi_1)_L(b_1) = \left(\bigcup_{b_2 \in \varphi_{1L}(b_1)} \varphi_{2L}(b_2) \right)^{\downarrow_3\uparrow_3}$
- $(\varphi_2 \circ \varphi_1)_R(a_3) = \left(\bigcup_{a_2 \in \varphi_{2R}(a_3)} \varphi_{1R}(a_2) \right)^{\uparrow_1\downarrow_1}$

and with identity Chu correspondence $\iota\colon C \to C$ *for any* $C = (B, A, R)$ *with derivation operators* $((-)^{\uparrow}, (-)^{\downarrow})$

- $\iota_L(b) = (\{b\})^{\downarrow\uparrow}$ *for any* $b \in B$
- $\iota_R(a) = (\{a\})^{\uparrow\downarrow}$ *for any* $a \in A$

Such pair of mappings $\varphi = (\varphi_L, \varphi_R)$ *is* Chu correspondence. *The set of all Chu correspondences from* C_1 *to* C_2 *is denoted by* $\mathrm{ChuCors}(C_1, C_2)$.

Theorem 1 ([11]). *Formal contexts and Chu correspondences form a category* ChuCors*, which is equivalent to the category of complete lattices and supremum preserving maps.*

3 Contextual Representation of Propositional Systems

3.1 Hilbert Formal Contexts and Hilbert-Chu Correspondences

It is known that if H is a Hilbert space and $\perp$ is the orthogonality relation, then the concept lattice of the context $(H, H, \perp)$ is isomorphic to the orthomodular lattice of the closed subspaces of H, since $(U, U^{\perp})$ is a concept for each such subspace U, see [8, p. 55]. This justifies the following definition:

Definition 9. *Let* H *be a Hilbert space, the tuple* $\overline{H} = (H, H, \perp)$ *is said to be the* Hilbert formal context *associated to* H.

Lemma 3. *Let* H *be a Hilbert space, the concept lattice associated to the Hilbert formal context* $\overline{H}$ *is a propositional system.*

Proof. Follows from Proposition 2 and the construction of the concept lattice of $\overline{H}$. □

Lemma 4. *Let* H *be a Hilbert space. Then, the trivial subspace* $\{\mathbf{0}\}$ *and the one-dimensional subspaces of* H *are closed under* $(-)^{\perp}$.

Proof. The first statement is trivial.

Given an arbitrary vector $u \in H$, we have to prove that $\rho(u) = \rho(u)^{\perp\perp}$. It is not straightforward to check that $\rho(u) \subseteq \rho(u)^{\perp\perp}$. Now, assume that there exists $v \in H$ such that $v \in \rho(u)^{\perp\perp}$ and $v \notin \rho(u)$; from $v \notin \rho(u)$ it follows that v is not a multiple of u, and from $v \in \rho(u)^{\perp\perp}$ we obtain that $v \notin \rho(u)^{\perp}$, therefore there should exist some non-null vector $w \in H$ such that $u \perp w$ and $v \not\perp w$. On the other hand, $v \in \rho(u)^{\perp\perp}$ is equivalent to $v \perp \rho(u)^{\perp}$ that means that everything orthogonal to $\rho(u)$ is orthogonal to v that contradicts to existence of vector w. □

Definition 10. *The category* ChuCors$_{\mathcal{H}}$ *has Hilbert formal contexts as objects and, given two Hilbert spaces H_1 and H_2, the morphisms between the corresponding Hilbert formal contexts are pairs of mappings $\varphi = (\varphi_L, \varphi_R)$ where $\varphi_L\colon H_1 \to \mathcal{P}(H_2) \cup \{\mathbf{0}\}$, $\varphi_R\colon H_2 \to \mathcal{C}(H_1)$ satisfying*

$$\varphi_L(v_1) \perp v_2 \iff v_1 \perp \varphi_R(v_2)$$

In the rest of this section we will prove that ChuCors$_{\mathcal{H}}$ is, indeed, a category. To begin with, note that φ_L is well-defined since the rays and the trivial subspace are closed and, hence, extents by Lemma 4; furthermore, the image of φ_R is also closed by construction and, hence, an intent.

Lemma 5. *Any $\varphi_L \in$* ChuCors$(\overline{H_1}, \overline{H_2})$ *preserves linear dependence of vectors.*

Proof. Consider $u_i, v_i, w_i \in H_i$ for $i \in \{1,2\}$ and let us write $\rho_i(-)$ for the "ray" operator defined on H_i. Assume that $w_1 = \alpha_1 u_1 + \beta_1 v_1$ for some scalars α_1, β_1, and that $\varphi_L(u_1) = \rho_2(u_2)$, $\varphi_L(v_1) = \rho_2(v_2)$ and $\varphi_L(w_1) = \rho_2(w_2)$ for some $u_2, v_2, w_2 \in H_2$ (since the cases in which some of the images is the $\{\mathbf{0}\}$ subspace is trivial). Now we would like to prove that there exist scalars α_2, β_2 such that $w_2 = \alpha_2 u_2 + \beta_2 v_2$.

By *reductio ad absurdum*, let us assume that $w_2 \neq \alpha_2 u_2 + \beta_2 v_2$, for any $\alpha_2, \beta_2 \in K_2$. This means that there should exist $q_2 \in H_2$ such that $q_2 \perp u_2$, $q_2 \perp v_2$ and $q_2 \not\perp w_2$, that is, $q_2 \perp \varphi_L(u_1)$, $q_2 \perp \varphi_L(v_1)$ and $q_2 \not\perp \varphi_L(w_1)$. From Definition 10, we would obtain $u_1 \perp \varphi_R(q_2)$, $v_1 \perp \varphi_R(q_2)$ and $w_1 \not\perp \varphi_R(q_2)$ contradicting the fact that w_1 depends linearly from u_1 and v_1. □

Lemma 6. *Given two Hilbert spaces H_1, H_2 and $\varphi \in$* ChuCors$_{\mathcal{H}}(\overline{H_1}, \overline{H_2})$*, consider $u, v \in H_1$ satisfying $u = \lambda v$ for some $\lambda \in K_1$. Then $\varphi(u) = \varphi(v)$.*

Proof. We have $\varphi_L(u) = \varphi_L(v)$ straightforwardly from Lemma 5.

Now, if $\varphi_R(u) \neq \varphi_R(v)$ then there would exist $q \in H_1$ such that $q \perp \varphi_R(u)$ and $q \not\perp \varphi_R(v)$; this is equivalent to $\varphi_L(q) \perp u$ and $\varphi_L(q) \not\perp v$, which is not possible since $u = \lambda v$. □

Corollary 1. *By the previous Lemma, we can see any left side φ_L of any Hilbert-Chu correspondence $\varphi : \overline{H_1} \to \overline{H_2}$ as a mapping between atoms (rays) of Hilbert space H_1 and atoms (rays) or bottom (trivial subspace) of H_2.*

Lemma 7. *Let $\overline{H_1}$, $\overline{H_2}$, $\overline{H_3}$ be Hilbert formal contexts, and consider two Chu correspondences $\varphi_1 \in \text{ChuCors}_{\mathcal{H}}(\overline{H_1}, \overline{H_2})$ and $\varphi_2 \in \text{ChuCors}_{\mathcal{H}}(\overline{H_2}, \overline{H_3})$. Consider the following mappings*

$$(\varphi_2\varphi_1)_L(v_1) = \left(\bigcup_{v_2 \in \varphi_{1L}(v_1)} \varphi_{2L}(v_2)\right)^{\perp\perp}$$

$$(\varphi_2\varphi_1)_R(v_3) = \left(\bigcup_{v_2 \in \varphi_{2R}(v_3)} \varphi_{1R}(v_2)\right)^{\perp\perp}$$

then $\big((\varphi_2\varphi_1)_L, (\varphi_2\varphi_1)_R\big) \in \text{ChuCors}_{\mathcal{H}}(\overline{H_1}, \overline{H_3})$.

Proof. Firstly, note that by Lemma 5, the union $\bigcup_{v_2 \in \varphi_{1L}(v_1)} \varphi_{2L}(v_2)$ is closed because, indeed, it is a ray given by the value of any $\varphi_{2L}(v_2)$ (all of them coincide). The reason to give the definition as the double orthogonal of the union is to clarify that, actually, we are just applying the usual composition of Chu correspondences. For the right part, its definition is the closure (double orthogonal) of certain set and, hence, is closed.

For the proof of the Chu equivalence, consider the following equivalences:

$$v_1 \in \big((\varphi_2\varphi_1)_R(v_3)\big)^{\perp} \iff$$

$$v_1 \in \left(\bigcup_{v_2 \in \varphi_{2R}(v_3)} \varphi_{1R}(v_2)\right)^{\perp\perp\perp} = \left(\bigcup_{v_2 \in \varphi_{2R}(v_3)} \varphi_{1R}(v_2)\right)^{\perp}$$

$$\iff v_1 \perp u_1 \text{ for all } u_1 \in \bigcup_{v_2 \in \varphi_{2R}(v_3)} \varphi_{1R}(v_2)$$

$$\iff v_1 \perp u_1 \text{ for all } v_2 \in \varphi_{2R}(v_3) \text{ and for all } u_1 \in \varphi_{1R}(v_2)$$

$$\iff v_1 \in \varphi_{1R}(v_2)^{\perp} \text{ for all } v_2 \in \varphi_{2R}(v_3)$$

$$\iff v_2 \in \varphi_{1L}(v_1)^{\perp} \text{ for all } v_2 \in \varphi_{2R}(v_3)$$

$$\iff v_2 \perp u_2 \text{ for all } v_2 \in \varphi_{2R}(v_3) \text{ and for all } u_2 \in \varphi_{1L}(v_1)$$

$$\iff u_2 \in \varphi_{2R}(v_3)^{\perp} \text{ for all } u_2 \in \varphi_{1L}(v_1)$$

$$\iff v_3 \in \varphi_{2L}(u_2)^{\perp} \text{ for all } u_2 \in \varphi_{1L}(v_1)$$

$$\iff v_3 \perp u_3 \text{ for all } u_2 \in \varphi_{1L}(v_1) \text{ and for all } u_3 \in \varphi_{2L}(u_2)$$

$$\iff v_3 \perp u_3 \text{ for all } u_3 \in \bigcup_{u_2 \in \varphi_{1L}(v_1)} \varphi_{2L}(u_2)$$

$$\iff v_3 \in \left(\bigcup_{u_2 \in \varphi_{1L}(v_1)} \varphi_{2L}(u_2)\right)^{\perp} = \left(\bigcup_{u_2 \in \varphi_{1L}(v_1)} \varphi_{2L}(u_2)\right)^{\perp\perp\perp}$$

$$\iff v_3 \in \big((\varphi_2\varphi_1)_L(v_1)\big)^{\perp}$$

□

Lemma 8. *Composition is associative.*

Proof. Associativity of composition is proved in [11]. New definition has no impact on composition. The construction is the same. □

Lemma 9. *Let H be a Hilbert space and $S \in \mathcal{C}(H)$ arbitrary. If $u \in S$ then $\rho(u) \subseteq S$.*

Proof. $\{u\} \subseteq S$ and $\{u\}^{\perp\perp} = \rho(u) \subseteq S^{\perp\perp} = S$. □

Lemma 10. *There exist a unit morphism $\iota^{\overline{H}}$ for any object $\overline{H}$ in* ChuCors$_{\mathcal{H}}$ *which is a neutral element of arrow composition.*

Proof. Given a Hilbert formal context $\overline{H}$, the unit morphism is defined by $\iota = (\iota_L, \iota_R)$ where $\iota_L(v) = \iota_R(v) = \{v\}^{\perp\perp} = \rho(v)$ for all $v \in H$. It is easy to check that $\iota \in$ ChuCors$_{\mathcal{H}}$: it is indeed a morphism, since $\iota_L(v)\colon H \to \mathcal{P}(H)$, and $\iota_R(v)\colon H \to \mathcal{C}(H)$ and, moreover,

$$u \perp \iota_L(v) \iff u \perp \{v\}^{\perp\perp} \iff u \perp v \iff \{u\}^{\perp\perp} \perp v \iff \iota_R(u) \perp v$$

Consider $\varphi \in$ ChuCors$_{\mathcal{H}}(\overline{H_1}, \overline{H_2})$ and $\iota_i\colon H_i \to H_i$ for $i \in \{1, 2\}$. Given, $v_1 \in H_1$, by Lemma 5, we have that $\varphi_L(u) = \varphi_L(v_1)$ for all u in the ray $\rho(v_1)$ (which coincides with $\iota_{1L}(v_1)$). Then, we have

$$(\iota_1 \circ \varphi)_L(v_1) = \left(\bigcup_{u_1 \in \iota_{1L}(v_1)} \varphi_L(u_1) \right)^{\perp\perp} = \varphi_L(v_1)^{\perp\perp} = \varphi_L(v_1)$$

$$\begin{aligned}(\iota_1 \circ \varphi)_R(v_2) &= \left(\bigcup_{u_1 \in \varphi_R(v_2)} \iota_{1R}(u_1) \right)^{\perp\perp} \\ &= \left(\bigcup_{u_1 \in \varphi_R(v_2)} \rho(u_1) \right)^{\perp\perp} = \big(\varphi_R(v_2)\big)^{\perp\perp} = \varphi_R(v_2)\end{aligned}$$

Hence $(\iota_1 \circ \varphi)_L = \varphi_L$, $(\iota_1 \circ \varphi)_R = \varphi_R$ and $\iota_1 \circ \varphi = \varphi$. Now, similarly,

$$\begin{aligned}(\varphi \circ \iota_2)_L(v_1) &= \left(\bigcup_{u_2 \in \varphi_L(v_1)} \iota_{2L}(u_2) \right)^{\perp\perp} \\ &= \left(\bigcup_{u_2 \in \varphi_L(v_1)} \rho(u_2) \right)^{\perp\perp} = \big(\varphi_L(v_1)\big)^{\perp\perp} = \varphi_L(v_1)\end{aligned}$$

$$(\varphi \circ \iota_2)_R(v_2) = \left(\bigcup_{u \in \iota_{2R}(v_2)} \varphi_R(u) \right)^{\perp\perp} = \left(\bigcup_{u \in \rho_2(v_2)} \varphi_R(u) \right)^{\perp\perp}$$

due to Lemma 6

$$= \left(\bigcup_{u \in \rho(v_2)} \varphi_R(v_2) \right)^{\perp\perp} = (\varphi_R(v_2))^{\perp\perp} = \varphi_R(v_2)$$

Hence $(\varphi \circ \iota_2)_L = \varphi_L$, $(\varphi \circ \iota_2)_R = \varphi_R$, $\varphi \circ \iota_2 = \varphi$, and moreover ι_i are the neutral elements for composition of arrows. □

As a consequence of the previous results, we obtain.

Proposition 3. ChuCors$_{\mathcal{H}}$ *forms a category.*

3.2 Connection to PropSys

In this section we will describe the functorial relation between the categories ChuCors$_{\mathcal{H}}$ and PropSys. The construction is as follows:

Definition 11. *The functor* Π: ChuCors$_{\mathcal{H}} \to$ PropSys *is defined by:*

- *Given a Hilbert formal context* $\overline{H}$, *we define* $\Pi(\overline{H})$ *as the concept lattice of* $\overline{H}$.
- *Let* φ *be a morphism in* ChuCors$_{\mathcal{H}}(\overline{H_1}, \overline{H_2})$. *Then* φ *induces a mapping* $\overline{\varphi}\colon \mathcal{P}(H_1) \to \mathcal{P}(H_2) \cup \{\mathbf{0}\}$, *defined by* $\overline{\varphi}(\rho(v)) = \varphi_L(v)$ *for all* $v \in H_1$ *and ray* $\rho(v) \in \mathcal{P}(H_1)$. *Now,* $\Pi(\varphi)$ *is defined as the homomorphic extension of* $\overline{\varphi}$.[2]

The following lemma proves that Π is, indeed, a functor.

Lemma 11. Π : ChuCors$_{\mathcal{H}} \to$ PropSys *is a functor.*

Proof. Lemma 3 states that Π maps objects of ChuCors$_{\mathcal{H}}$ on objects of PropSys.

Consider an arbitrary $\varphi \in$ ChuCors$_{\mathcal{H}}(\overline{H_1}, \overline{H_2})$. By definition, $\overline{\varphi}$ maps rays of H_1 (atoms of $\mathcal{C}(\mathcal{L}(H_1))$) to either rays or zero vector singleton (respectively, atoms or bottom of $\mathcal{C}(\mathcal{L}(H_2))$) and, moreover, preserves suprema; hence Π maps Hilbert-Chu correspondences to morphisms between propositional systems.

Now, $\Pi(\iota)$ is an identity morphism in PropSys, since $\overline{\iota}(\rho(v)) = \iota_L(v) = \rho(v)$ for all $v \in H$.

Finally, consider $\varphi_i \in$ ChuCors$_{\mathcal{H}}(\overline{H_i}, \overline{H_{i+1}})$ for $i \in \{1, 2\}$, and let us prove that $\Pi(\varphi_1 \circ \varphi_2) = \Pi(\varphi_1) \circ \Pi(\varphi_2)$. Given $v_1 \in H_1$ we have that

$$(\varphi_1 \circ \varphi_2)_L(v_1) = \left(\bigcup_{v_2 \in \varphi_{1L}(v_1)} \varphi_{2L}(v_2) \right)^{\perp\perp}$$

[2] We will abuse the notation and write $\overline{\varphi}$ to refer to this homomorphic extension.

By definition, $\varphi_{1L}(v_1)$ is either $\{\mathbf{0}\}$ or a ray of H_2, and Lemma 6 ensures that for all $u_2, w_2 \in \varphi_{1L}(v_1) = \overline{\varphi_1}(\rho(v_1))$ we have $\varphi_{2L}(u_2) = \varphi_{2L}(w_2)$, in other words $\overline{\varphi_2}(\rho(u_2)) = \overline{\varphi_2}(\rho(w_2))$. Hence

$$\bigcup_{v_2 \in \overline{\varphi_1}(\rho(v_1))} \overline{\varphi_2}(\rho(v_2)) = \overline{\varphi_2}(\overline{\varphi_1}(\rho(v_1)))$$

□

We recall now some necessary notions which will be used in order to prove that the previous functor satisfies the conditions to define a categorical equivalence.

Definition 12

1. *A functor $F : \mathcal{C} \longrightarrow \mathcal{D}$ is* faithful *if for all objects A, B of a category $\mathcal{C}$, the map $F_{A,B} : \mathrm{Hom}_{\mathcal{C}}(A, B) \longrightarrow \mathrm{Hom}_{\mathcal{D}}(F(A), F(B))$ is injective.*
2. *Similarly, F is* full *if $F_{A,B}$ is always surjective.*

The proof of the categorical equivalence will be done by using the following characterization:

Theorem 2 (See [3]). *The following conditions on a functor $F\colon \mathcal{C} \longrightarrow \mathcal{D}$ are equivalent:*

- *F is an equivalence of categories.*
- *F is full and faithful and "essentially surjective" on objects: for every $D \in \mathcal{D}$ there is some $C \in \mathcal{C}$ such that $F(C) \cong D$.*

Theorem 3. *Categories* $\mathrm{ChuCors}_{\mathcal{H}}$ *and* PropSys *are equivalent.*

Proof. By the definition of $\mathrm{ChuCors}_{\mathcal{H}}$ and Π it is not difficult to see that the mapping $\Pi(H_1, H_2)\colon \mathrm{ChuCors}_{\mathcal{H}}(\overline{H_1}, \overline{H_2}) \rightarrow \mathrm{PropSys}(\Pi(\overline{H_1}), \Pi(\overline{H_2}))$ is bijective. Hence Π is full and faithful.

Let $\mathcal{L} = (L, \leq, (-)^{\perp})$ be an arbitrary propositional system. Essential surjectivity of functor Π means that there exist a Hilbert formal context whose concept lattice is isomorphic to $\mathcal{L}$.

Let L_{at} be the set of atoms of $\mathcal{L}$, and let us prove that concept lattice of the formal context $(L_a, L_a, \perp)$ where $a \perp b \Longleftrightarrow (a \leq b^{\perp}) \Longleftrightarrow (b \leq a^{\perp})$, for any two atom $a, b \in L_{at}$, is isomorphic to $\mathcal{L}$. Since $\mathcal{L}$ is atomistic, we have that every element $x \in L$ is the supremum of a finite set of atoms, say $x = \bigvee_{i \in I_x} a_i$. This means the set of atoms L_{at} is supremum-dense in L. Moreover

$$x = x^{\perp\perp} = \left(\bigvee_{i \in I_x} a_i\right)^{\perp\perp} = \left(\bigwedge_{i \in I_x} a_i^{\perp}\right)^{\perp}.$$

Furthermore, every element $y \in L$ can be represented as $y = x^{\perp} = \bigwedge_{i \in I_x} a_i^{\perp}$. Hence the set $\overline{L_{at}} = \{a^{\perp} \mid a \in L_A\}$ is infimum-dense in L. Therefore, by the second part of the basic theorem of concept lattices, the concept lattice of $(L_{at}, \overline{L_{at}}, \leq)$ or $(L_{at}, L_{at}, \perp)$ is isomorphic to $\mathcal{L}$. □

4 Conclusions

Continuing with the study of generalized Chu correspondences, we have introduced the new category $\mathsf{ChuCors}_{\mathcal{H}}$, whose objects are Hilbert formal contexts and whose morphisms are Chu correspondences between them. The notion of *Hilbert formal context* $(H, H, \perp)$ associated to a Hilbert space H was already present in Ganter and Wille's book [8], stating its close relationship with the ortholattice of closed linear subspaces and, hence, to the theory of *propositional systems* as algebraic structures underlying quantum logic. The main result in this work is the proof that the category $\mathsf{ChuCors}_{\mathcal{H}}$ is equivalent to the category PropSys of propositional systems.

This opens the way to future work oriented to the use of Chu correspondences to analyze more structures related to quantum logics, such as those in Abramsky's big toy models [1].

References

1. Abramsky, S.: Big toy models: representing physical systems as Chu spaces. Synthese **186**(3), 697–718 (2012). https://doi.org/10.1007/s11229-011-9912-x
2. Abramsky, S.: Coalgebras, Chu spaces, and representations of physical systems. Jo. Philos. Log. **42**(3), 551–574 (2013)
3. Awodey, S.: Category Theory. Oxford University Press, Oxford (2010)
4. Barr, M.: *-Autonomous Categories. Lecture Notes in Mathematics, vol. 752. Springer, Heidelberg (1979). https://doi.org/10.1007/BFb0064579
5. Birkhoff, G., von Neumann, J.: The logic of quantum mechanics. Ann. Math. **37**(4), 823–843 (1936)
6. Chu, P.-H.: Constructing *-autonomous categories. Appendix to [4], pp. 103–107
7. Engesser, K., Gabbay, D.M., Lehmann, D. (eds.): Handbook of Quantum Logic and Quantum Structures: Quantum Structures. Elsevier, New York (2007)
8. Ganter, B., Wille, R.: Formal Concept Analysis. Springer, Heidelberg (1999). https://doi.org/10.1007/978-3-642-59830-2
9. Kridlo, O., Krajči, S., Ojeda-Aciego, M.: The category of L-Chu correspondences and the structure of L-bonds. Fundam. Inf. **115**(4), 297–325 (2012). https://doi.org/10.3233/FI-2012-657
10. Kridlo, O., Ojeda-Aciego, M.: On L-fuzzy correspondences. Int. J. Comput. Math. **88**(9), 1808–1818 (2011). https://doi.org/10.1080/00207160903494147
11. Mori, H.: Chu correspondences. Hokkaido Math. J. **37**, 147–214 (2008)
12. von Neumann, J.: Mathematical Foundations of Quantum Mechanics. Princeton University Press, Princeton (1955). Reprinted from the original version published in 1932
13. Piron, C.: Foundations of Quantum Physics. W. A. Benjamin Inc., New York (1976)
14. Stubbe, I., Van Steirteghem, B.: Propositional systems, Hilbert lattices and generalized Hilbert spaces. In: [7], pp. 477–524 (2007)

Directness in Fuzzy Formal Concept Analysis

Pablo Cordero, Manuel Enciso, and Angel Mora(✉)

Andalucía Tech, Universidad de Málaga, Málaga, Spain
pcordero@uma.es, enciso@lcc.uma.es, amora@ctima.uma.es

Abstract. Implicational sets have showed to be an efficient tool for knowledge representation. An active area is the definition of some canonical sets (basis) to efficiently specify and manage the information specified with implications. Unlike in classical formal concept analysis, in the fuzzy framework it is an open issue to design methods to efficiently compute the corresponding basis from a given set of fuzzy implications and, later, manage it in an automatic way. In this work we use Simplification Logic to tackle this issue. More specifically, we cover the following stages related to this problem: the generalization of the Simplification logic to an arbitrary complete residuated lattice changing its semantic, the introduction of the syntactic closure and an algorithm to compute it, the definition of a fuzzy direct basis with minimum size, providing the so-called directness property as well, and, finally the design of an algorithm to compute this basis.

Keywords: Logic · Fuzzy Formal Concept Analysis · Basis

1 Introduction

Formal Concept Analysis (FCA) [7] provides a framework to deal with the complete life cycle of information, ranging from the data storage, knowledge discovery and knowledge representation to its efficient manipulation. Data are stored in a dataset, named formal context, that can be viewed as a relationship between a set of objects (rows) and a given set of attributes (columns). Some methods are proposed to automatically extract the knowledge allocated in the data set [10]. This information can be equivalently specified by using a concept lattice or a set of implications.

Implications are considered to be a binary relationship between two subsets of attributes, represented with the formula $A \to B$. A very similar notion appears in databases (the so-called functional dependencies) and in logic programming (by means of if-then rules). A formal context is a model of a given implication whenever for each row, if all the attributes in A holds, then all attributes in B also holds. Although concept lattices and sets of implications are two equivalent representations of the information, the second one allows a symbolic management. The usual way to do this is by using methods based

J. Medina et al. (Eds.): IPMU 2018, CCIS 853, pp. 585–595, 2018.
https://doi.org/10.1007/978-3-319-91473-2_50

on the well-known Armstrong's Axioms [1]. We have introduced an alternative logic, named Simplification Logic, which constitutes a better approach because it allows the development of automated methods strongly based on the logic (see [6,9] for more details).

Several approaches to the definition of fuzzy implications (in its functional dependency view) were proposed in the literature [11,14]. We also developed several extensions of fuzzy implications [4,5,12]. In addition, we also introduced their corresponding sound and complete axiomatic systems and some automated reasoning methods to manage fuzzy implications. The existence of different approaches are due to the different levels of fuzzification that can be considered, ranging from the attribute to the implication itself. For an a exhaustive review of the fuzzy approaches of implications, see [8].

The first step to design efficient methods for any symbolic computation is to define some kind of canonical representation in the language formulae. Regarding implications, this issue is approached with the idea of bases: sets of implications achieving minimality and optimality criteria. Such a representation allows the design of methods taking advantages of the good properties of the basis and, in this way, providing a lower cost in their management and transformation. In the classical FCA, the outstanding work of Bertet and Monjardet in [3] collected five previous definitions of basis and established their equivalence and introduced in this way the direct-optimal basis. This basis definition relied on two main pillars: the smallest number of implications (minimality) and a unique traverse of its implications to compute attribute closure (directness).

In [3,13] the authors proposed several methods to calculate the direct-optimal basis considering optimality, minimality and directness properties. To complete the basis issue, a method to get them from an arbitrary set of implications has to be developed.

Specifically, in this work, we propose the Fuzzy Simplification Logic ($\mathbb{FSL}$), extending the work presented in [5], which is a Pavelka style fuzzy logic in the framework of Fuzzy Formal Concept Analysis. We have also introduced the notion of syntactic closure, providing a solution to the so-called implication problem.

Moreover, we extend the direct-optimal basis to this new framework. Henceforth, we go over the complete way of optimal and direct basis in the fuzzy extension of FCA: we provide its definition and an automated method to built it.

Thus, in Sect. 2 some preliminaries are introduced. Section 3 shows Fuzzy Simplification Logic ($\mathbb{FSL}$), the seed which has given rise to the methods to compute and manage fuzzy implication basis. We prove the completeness of $\mathbb{FSL}$ in Sect. 4. Later, in Sect. 5, we present the closure issue and, finally, Sect. 6 approaches the obtention of the direct-optimal basis in fuzzy FCA.

2 Preliminaries

The structure of degrees will be a complete (commutative) residuated lattice, which is defined as an algebra $\mathbf{L} = \langle L, \wedge, \vee, \otimes, \rightarrow, 0, 1\rangle$ such that $\langle L, \wedge, \vee, 0, 1\rangle$ is a complete lattice with 0 and 1 being the least and greatest element, respectively; $\langle L, \otimes, 1\rangle$ is a commutative monoid (i.e. $\otimes$ is commutative, associative) and $a \otimes 1 = 1 \otimes a = a$ for each $a \in L$; $\otimes$ and $\rightarrow$ satisfy the following adjointness property:

$$a \otimes b \leq c \quad \text{iff} \quad a \leq b \rightarrow c \tag{1}$$

for any $a, b, c \in L$.

We will interpret elements in L as degrees (of truth) with the following comparative meaning: if $a = ||\varphi||$ and $b = ||\psi||$ are degrees from L assigned to formulas φ and ψ and if $a \leq b$, then φ is less true than ψ. Operations $\otimes$ and $\rightarrow$ (residuum) represent truth functions of logical connectives "fuzzy conjunction" and "fuzzy implication".

An *ideal* in $\mathbf{L}$ is a non-empty subset $I \subseteq L$ that satisfies the following two conditions: (1) $i \in I$ and $j \leq i$ imply $j \in I$; (2) $i, j \in I$ implies $i \vee j \in I$.

Considering $\mathbf{L}$, we define an $\mathbf{L}$-set A (a fuzzy set with degrees in $\mathbf{L}$) in universe M as any map $A\colon M \rightarrow L$, $A(m)$ being interpreted as "the degree to which $m \in M$ belongs to A".

The collection of all $\mathbf{L}$-sets in universe M is denoted by L^M. $A \in L^M$ is called crisp if $A(m) \in \{0, 1\}$ for all $m \in M$. We can identify a crisp $\mathbf{L}$-set in M with an ordinary subset of M.

Operations with $\mathbf{L}$-sets we use in this paper are induced componentwise by the operations of $\mathbf{L}$. For instance, the intersection of $\mathbf{L}$-sets $A, B \in L^M$ is an $\mathbf{L}$-set $A \cap B \in L^M$ such that $(A \cap B)(m) = A(m) \wedge B(m)$ for all $m \in M$. The structure of all $\mathbf{L}$-sets in M together with the induced operations is in fact a direct power $\mathbf{L}^M = \langle L^M, \cap, \cup, \otimes, \rightarrow, \emptyset, M\rangle$ of $\mathbf{L}$ which is also a complete residuated lattice where the order relation is defined as: $A \subseteq B$ iff $A(m) \leq B(m)$ for all $m \in M$.

3 Fuzzy Simplification Logic

In this work we address the adjustment of Fuzzy Simplification Logic ($\mathbb{FSL}$) [5], which is a Pavelka style fuzzy logic, to the framework of Fuzzy Formal Concept Analysis (see [2]). In [5] we consider this logic in the unit interval and here we have generalized the logic to an arbitrary complete residuated lattice changing the semantic.

Here, the semantic interpretations are given in terms of L-context in the framework of Fuzzy Formal Concept Analysis (see [2]).

To formally present $\mathbb{FSL}$, we now establish its language, the semantics and the axiomatic system:

Language: Given a non-empty finite set M whose elements will be named *attributes*, the set of (*attribute*) *implications* is defined as the set of formulas

$$\mathcal{I}m = \{A \rightarrow B \mid A, B \subseteq M\}.$$

The sets A and B will be named *premise* and *conclusion* of the implication, respectively. In addition, we will use lowercase characters for attributes

$(a, b, c, \ldots)$ and uppercase characters to denote set of attributes $(A, B, C, \ldots)$. Moreover, inside of a formula, A-B denotes the set difference and AB denotes the union $A \cup B$.

Finally, fixed a complete lattice $\mathbf{L} = \langle L, \wedge, \vee, \otimes, \rightarrow, 0, 1\rangle$, the language is defined as $\mathcal{L} = \mathcal{I}m \times L$. In order to simplify the notation, instead of writing $\langle A \rightarrow B, \vartheta\rangle$, we will write $A \xrightarrow{\vartheta} B$. Thus,

$$\mathcal{L} = \{A \xrightarrow{\vartheta} B \mid A, B \subseteq M \text{ and } \vartheta \in L\}.$$

For a formula $A \xrightarrow{\vartheta} B \in \mathcal{L}$, the value ϑ will be called the *degree* of the implication. Finally, sets of formulas $\Gamma \subseteq \mathcal{L}$ will be named *theories.*

Semantics: Concerning the semantics, the models are given by using the notion of L-context from Fuzzy FCA. In this framework, an L-context (fuzzy context) is a triplet $\mathbf{K} = \langle G, M, I\rangle$ where $I \in L^{G \times M}$ is a fuzzy relation. Elements of G are named *objects*, elements of M are named *attributes* and the relation I is named *incidence relation* where, for all $g \in G$ and $m \in M$, $I(g, m) \in L$ means the truth degree to which "object g has attribute m".

Given an L-context $\mathbf{K} = \langle G, M, I\rangle$, for each set of attributes $A \subseteq M$, an L-fuzzy set $A' \in L^G$ is defined as follows:

$$A'(g) = \bigwedge_{a \in A} I(g, a)$$

This leads to the notion of valuation, which is introduced as follows: the L-context $\mathbf{K}$ defines the following valuation mapping $\| \;\|_{\mathbf{K}} \colon \mathcal{I}m \rightarrow L$ with

$$\|A \rightarrow B\|_{\mathbf{K}} = \bigwedge_{g \in G} A'(g) \rightarrow B'(g)$$

Finally, the L-context $\mathbf{K}$ is said to be a *model* for a formula $A \xrightarrow{\vartheta} B \in \mathcal{L}$, denoted $\mathbf{K} \models A \xrightarrow{\vartheta} B$, if

$$\vartheta \leq \|A \rightarrow B\|_{\mathbf{K}}$$

This definition is extended to theories in the standard way: for a theory $\Gamma \subseteq \mathcal{L}$, the L-context $\mathbf{K}$ is a model for Γ, also denoted $\mathbf{K} \models \Gamma$ if $\mathbf{K} \models A \xrightarrow{\vartheta} B$ for all $A \xrightarrow{\vartheta} B \in \Gamma$.

In addition, we say that a formula $A \xrightarrow{\vartheta} B$ is *semantically derived* from a theory Γ, denoted $\Gamma \models A \xrightarrow{\vartheta} B$, if $\mathbf{K} \models \Gamma$ implies $\mathbf{K} \models A \xrightarrow{\vartheta} B$ for all L-context $\mathbf{K}$.

Inference System: Once we have presented the syntax and semantics of $\mathbb{FSL}$, we now introduce its axiomatic system:

Definition 1. *The axiomatic system for $\mathbb{FSL}$ has one axiom scheme and four inference rules:*

[Ax] $\vdash A \xrightarrow{1} A$

[InR] $A \xrightarrow{\vartheta_1} B \vdash A \xrightarrow{\vartheta_2} C$, *when* $\vartheta_2 \leq \vartheta_1$ *and* $C \subseteq B$.

[CoR] $A \xrightarrow{\vartheta_1} B,\ C \xrightarrow{\vartheta_2} D \vdash AC \xrightarrow{\vartheta_1 \wedge \vartheta_2} BD$.

[UnR] $A \xrightarrow{\vartheta_1} B,\ A \xrightarrow{\vartheta_2} B \vdash A \xrightarrow{\vartheta_1 \vee \vartheta_2} B$.

[SiR] $A \xrightarrow{\vartheta_1} B,\ C \xrightarrow{\vartheta_2} D \vdash C\text{-}B \xrightarrow{\vartheta_1 \otimes \vartheta_2} D\text{-}B$, *when* $A \subseteq C$ *and* $A \cap B = \varnothing$.

Now we present the well-known notions of *syntactic inference* ($\vdash$) and *equivalence* ($\equiv$):

Given $\Gamma \subseteq \mathcal{L}$ and $\varphi \in \mathcal{L}$, we say that φ is *syntactically derived* or *inferred* from Γ, denoted $\Gamma \vdash \varphi$, if there exist $\varphi_1, \ldots, \varphi_n \in \mathcal{L}$ such that $\varphi_n = \varphi$ and, for all $1 \leq i \leq n$, we have that φ_i belongs to Γ, is an axiom or is obtained by applying the inference rules to formulas in $\{\varphi_j \mid 1 \leq j < i\}$.

Two theories $\Gamma_1, \Gamma_2 \subseteq \mathcal{L}$ are said to be *equivalent*, denoted $\Gamma_1 \equiv \Gamma_2$, if $\Gamma_1 \vdash \varphi_2$, for all $\varphi_2 \in \Gamma_2$, and $\Gamma_2 \vdash \varphi_1$, for all $\varphi_1 \in \Gamma_1$.

The following theorem, whose proof is straightforward from the previous definitions, states the soundness of the axiomatic system.

Theorem 1 (Soundness). *For all $\Gamma \subseteq \mathcal{L}$ and $A \xrightarrow{\vartheta} B \in \mathcal{L}$, one has that $\Gamma \vdash A \xrightarrow{\vartheta} B$ implies $\Gamma \models A \xrightarrow{\vartheta} B$.*

4 Completeness

In order to prove the completeness of the axiomatic system, we introduce the following definitions: For any pair of sets $A, B \subseteq M$ and $\Gamma \subseteq \mathcal{L}$,

$$\vartheta^{+}_{A,B}(\Gamma) = \sup\{\vartheta \in L \mid \Gamma \vdash A \xrightarrow{\vartheta} B\} \tag{2}$$

Notice that, as a consequence of [InR] and [UnR], the set $\{\vartheta \in L \mid \Gamma \vdash A \xrightarrow{\vartheta} B\}$ is an ideal in $\mathbf{L}$. In addition, if Γ is finite, $\vartheta^{+}_{A,B}(\Gamma) = \max\{\vartheta \in L \mid \Gamma \vdash A \xrightarrow{\vartheta} B\}$.

Similarly, for any $\vartheta \in L$ and any set $A \subseteq M$, the subset of 2^M defined as follows $\{B \subseteq M \mid \Gamma \vdash A \xrightarrow{\vartheta} B\}$ is an ideal of $(2^M, \subseteq)$. Since this poset is finite, and from [CoR], one has that this ideal has a maximum element that we denote as A^{+}_{ϑ}:

$$A^{+}_{\vartheta}(\Gamma) = \max\{B \subseteq M \mid \Gamma \vdash A \xrightarrow{\vartheta} B\}.$$

The following proposition is straightforward from the previous comments.

Proposition 1. *Let $A \xrightarrow{\vartheta} B \in \mathcal{L}$ and $\Gamma \subseteq \mathcal{L}$ be a finite theory. Then,*

$$\begin{aligned} \Gamma \vdash A \xrightarrow{\vartheta} B \quad &\text{if and only if} \quad \vartheta \leq \vartheta^{+}_{A,B}(\Gamma) \\ &\text{if and only if} \quad B \subseteq A^{+}_{\vartheta}(\Gamma) \end{aligned}$$

In order to simplify the notation, when no confusion arises, we will write $\vartheta^+_{A,B}$ instead of $\vartheta^+_{A,B}(\Gamma)$ and A^+_ϑ instead of $A^+_\vartheta(\Gamma)$. To end this section, we prove that the inference system is complete for finite theories.

Theorem 2 (Completeness). *Let $A \xrightarrow{\vartheta} B \in \mathcal{L}$ and $\Gamma \subseteq \mathcal{L}$ be a finite theory. If $\Gamma \models A \xrightarrow{\vartheta} B$ then $\Gamma \vdash A \xrightarrow{\vartheta} B$.*

Proof. It is proved showing that $\Gamma \nvdash A \xrightarrow{\vartheta} B$ implies $\Gamma \not\models A \xrightarrow{\vartheta} B$. From Proposition 1, if $\Gamma \nvdash A \xrightarrow{\vartheta} B$ then $\vartheta \not\leq \vartheta^+_{A,B}$ and $B \not\subseteq A^+_\vartheta$. Consider the L-context $\mathbf{K} = \langle G, M, I \rangle$ where $G = \{g\}$, and

$$I(g,m) = \begin{cases} \vartheta^+_{A,B} & \text{if } m \notin A^+_\vartheta \\ 1 & \text{if } m \in A^+_\vartheta \end{cases}$$

Then, $\mathbf{K} \models \Gamma$ but $\mathbf{K} \not\models A \xrightarrow{\vartheta} B$ because

$$\vartheta \not\leq A'(g) \to B'(g) = \bigwedge_{a \in A} I(g,a) \to \bigwedge_{b \in B} I(g,b) = 1 \to \vartheta^+_{A,B} = \vartheta^+_{A,B}$$

Therefore, $\Gamma \not\models A \xrightarrow{\vartheta} B$. □

5 Syntactic Closure and Automated Reasoning Method

$\mathbb{FSL}$ is a generalization of classical Simplification Logic and, in that framework, the notion of syntactic closure for sets of attributes plays a central role. In this section, we center on its generalization.

Given a finite theory $\Gamma \subseteq \mathcal{L}$ and a value $\vartheta \in L$, it is easy to see that the mapping $(-)^+_\vartheta \colon (2^M, \subseteq) \longrightarrow (2^M, \subseteq)$ is a closure operator, i.e. it is isotone ($A \subseteq B$ implies $A^+_\vartheta \subseteq B^+_\vartheta$), inflationary ($A \subseteq A^+_\vartheta$) and idempotent ($(A^+_\vartheta)^+_\vartheta = A^+_\vartheta$). In addition, from Proposition 1, one has that

$$\Gamma \vdash A \xrightarrow{\vartheta} B \quad \text{if and only if} \quad B \subseteq A^+_\vartheta(\Gamma)$$

Thus, the theory Γ defines the family of closure operators $\{(-)^+_\vartheta \mid \vartheta \in L\}$, which allows to solve the *implication problem*, that is, to test if $\Gamma \vdash A \xrightarrow{\vartheta} B$ for any $A \xrightarrow{\vartheta} B \in \mathcal{L}$.

In order to compute this family of closure operators, for a set $A \subseteq M$ we define a fuzzy set of attributes $A^+ \in L^M$ as follows:

$$A^+(m) = \vartheta^+_{A,m} = \max\{\vartheta \mid \Gamma \vdash A \xrightarrow{\vartheta} m\}, \text{ for each } m \in M.$$

Notice that, for each $\vartheta \in L$, the set A^+_ϑ is the ϑ-cut of the fuzzy set A^+

$$A^+_\vartheta = \mathrm{Cut}_\vartheta(A^+) = \{m \in M \mid A^+(m) \geq \vartheta\}.$$

Given a finite theory and a crisp set A, Algorithm 1 computes this closure.

Algorithm 1. Closure Algorithm

```
Data: Γ ⊆ L be a finite theory and A ⊆ M.
Result: The fuzzy set A⁺ ∈ L^M.
X := {(x,1) | x ∈ A} /* X will be A⁺, which is a fuzzy set.   */
repeat
    X_old := X; Σ := ∅;
    foreach B --ϑ--> C ∈ Γ do
        if there exists b ∈ B with b ≠ a for all (a,κ) ∈ X then η := 0;
        else η := min{κ | (b,κ) ∈ X with b ∈ B};
        if η ⊗ ϑ ≠ 0 then X := X ∪ {(c, η ⊗ ϑ) | c ∈ C};
        if η ≠ 1 and C ⊈ Cut_ϑ(X) then
            Σ := Σ ∪ {B ∖ Cut_1(X) --ϑ--> C ∖ Cut_ϑ(X)}
    Γ := Σ;
until X = X_old;
return "A⁺ is " X
```

Example 1. In this example abc^+ is going to be computed from the set Γ

$$\Gamma = \{cd \xrightarrow{0.6} e, ac \xrightarrow{0.7} def, f \xrightarrow{0.5} dg, de \xrightarrow{0.9} ch, dh \xrightarrow{0.4} a\}$$

by considering the Lukasiewicz product. The initial set X is $\{(a,1),(b,1),(c,1)\}$ and the sketch of the trace of Algorithm 1 is depicted in Fig. 1. The output is

$$abc^+ = \{(a,1),(b,1),(c,1),(d,0.7),(e,0.7),(f,0.7),(g,0.2),(h,0.6)\}$$

Example 2. Let Γ be the theory introduced in Example 1. To decide if $\Gamma \vdash abc \xrightarrow{0.5} dh$ holds, we use the Closure Algorithm just to check whether $\{d,h\} \subseteq \mathrm{Cut}_{0.5}(abc^+) = \{a,b,c,d,e,f,h\}$, providing an affirmative answer.

$B \xrightarrow{\vartheta} C \in \Gamma$	Σ	X
$\Gamma = \{cd \xrightarrow{0.6} e, ac \xrightarrow{0.7} def, f \xrightarrow{0.5} dg, de \xrightarrow{0.9} ch, dh \xrightarrow{0.4} a\}$		
	$\varnothing$	$\{(a,1),(b,1),(c,1)\}$
$cd \xrightarrow{0.6} e$	$\{d \xrightarrow{0.6} e\}$	$\{(a,1),(b,1),(c,1)\}$
$ac \xrightarrow{0.7} def$	$\{d \xrightarrow{0.6} e\}$	$\{(a,1),(b,1),(c,1),(d,0.7),(e,0.7),(f,0.7)\}$
$f \xrightarrow{0.5} dg$	$\{d \xrightarrow{0.6} e, f \xrightarrow{0.5} g\}$	$\{(a,1),(b,1),(c,1),(d,0.7),(e,0.7),(f,0.7),(g,0.2)\}$
$de \xrightarrow{0.9} ch$	$\{d \xrightarrow{0.6} e, f \xrightarrow{0.5} g\}$	$\{(a,1),(b,1),(c,1),(d,0.7),(e,0.7),(f,0.7),(g,0.2),(h,0.6)\}$
$dh \xrightarrow{0.4} a$	$\{d \xrightarrow{0.6} e, f \xrightarrow{0.5} g\}$	$\{(a,1),(b,1),(c,1),(d,0.7),(e,0.7),(f,0.7),(g,0.2),(h,0.6)\}$
$\Gamma = \{d \xrightarrow{0.6} e, f \xrightarrow{0.5} g\}$		
	$\varnothing$	$\{(a,1),(b,1),(c,1),(d,0.7),(e,0.7),(f,0.7),(g,0.2),(h,0.6)\}$
$d \xrightarrow{0.6} e$	$\varnothing$	$\{(a,1),(b,1),(c,1),(d,0.7),(e,0.7),(f,0.7),(g,0.2),(h,0.6)\}$
$f \xrightarrow{0.5} g$	$\{f \xrightarrow{0.5} g\}$	$\{(a,1),(b,1),(c,1),(d,0.7),(e,0.7),(f,0.7),(g,0.2),(h,0.6)\}$

Fig. 1. Algorithm's schema

6 Direct Basis

As we have mentioned in the introduction, the aim of this work is to establish good properties to be demanded to the set of fuzzy implications by means of a canonical form in the set of implications (basis). This basis ensures the best behavior of methods when a huge number of closures has to be computed.

We follow the idea introduced in [3], where the authors compute set of implications having the optimality property, together with a key property for implications named directness. The following definition extends the optimality and directness definitions to our fuzzy framework:

Definition 2. *Let Γ be a set of fuzzy implications. We say that Γ is* direct *if, for each subset $X \subseteq M$,*

$$X^+ = X \cup \bigcup_{\substack{A \xrightarrow{\vartheta} B \in \Gamma \\ X \subseteq A}} \{(b, \vartheta) \mid b \in B\}$$

And Γ is said to be a direct-optimal basis if, for any direct basis Γ', we have that $\Gamma \equiv \Gamma'$ implies $\|\Gamma\| \leq \|\Gamma'\|$ where $\|\Gamma\|$ denotes the size of Γ[1].

Now, next theorem establishes the existence and uniqueness of a direct-optimal basis equivalent to an arbitrary fuzzy implication set.

Theorem 3. *Let Γ be a set of fuzzy implications, then there exists a unique direct-optimal basis Γ_d such that $\Gamma_d \equiv \Gamma$.*

The proof of the above theorem follows a similar schema to the one provided in [3] for classical implications.

In the following, we propose a method computing the direct-optimal basis for fuzzy implications in the framework of Fuzzy FCA. As far as we know, the method proposed here is the first method directly based on logic.

We will use atomic formulas on the right, in the same way that Horn Clauses in Logic Programming in order to obtain a better performance of the methods. This does not suppose a strong limitation since most of the methods in the literature also consider this requirement and the transformation of an arbitrary implication into this new form can be done in just one traversal of the original set.

Definition 3. *Let Γ be a set of fuzzy implications. We say that Γ is a* proper unit theory *if, for all $A \xrightarrow{\vartheta} B \in \Gamma$, the set B is a singleton not included in A and $\vartheta > 0$.*

It is trivial to get a proper unit theory equivalent to any set of fuzzy implications Γ:

$$\Gamma_u = \{A \xrightarrow{\vartheta} a \mid \vartheta > 0, A \xrightarrow{\vartheta} B \in \Gamma, a \in B \smallsetminus A\}$$

Algorithm 2, providing a direct-optimal basis equivalent to a given fuzzy set of implications, has the following stages:

[1] Where $\|\Gamma\| = \sum_{A \xrightarrow{\vartheta} B \in \Gamma} (|A| + |B|)$ and $|A|$ denotes the cardinality of A.

1. Transform the set of fuzzy implications in a proper unit theory.
2. Compute a direct basis by applying the following derived rule, named Strong Simplification:

$$[\texttt{sSiR}]\ A\xrightarrow{\vartheta_1}a,\ aB\xrightarrow{\vartheta_2}b \vdash AB\xrightarrow{\vartheta_1\otimes\vartheta_2}b \text{ when } a,b\notin A\cup B.$$

3. Narrow the set of fuzzy implications applying the following equivalence

$$[\texttt{NarrEq}]\ \text{If } A\subseteq C \text{ and } \vartheta_1\geq\vartheta_2,\ \{A\xrightarrow{\vartheta_1}b, C\xrightarrow{\vartheta_2}b\}\equiv\{A\xrightarrow{\vartheta_1}b\}.$$

4. Apply Composition Equivalence: $\{A\xrightarrow{\vartheta}B, A\xrightarrow{\vartheta}C\}\equiv\{A\xrightarrow{\vartheta}BC\}$.

Algorithm 2. DirectOptimal

```
    input  : A set of fuzzy implications Γ in Ω
    output: The direct-optimal basis Γ_do equivalent to Γ
    begin
L1      Γ_u := {A -ϑ-> a | ϑ > 0, A -ϑ-> B ∈ Γ, a ∈ B ∖ A}
L2      foreach A -ϑ1-> a ∈ Γ_u do
            foreach Ca -ϑ2-> b ∈ Γ_u do
                if a ≠ b and b ∉ A then add AC -ϑ1⊗ϑ2-> b to Γ_u;
L3      foreach A -ϑ1-> b ∈ Γ_u do
            foreach C -ϑ2-> b ∈ Γ_u do
                if A ⊆ C and ϑ1 ≥ ϑ2 then delete C -ϑ2-> b from Γ_u;
        Γ_do := Γ_u
L4      foreach A -ϑ1-> B ∈ Γ_do do
            foreach C -ϑ2-> D ∈ Γ_do do
                if A = C and ϑ1 = ϑ2 then replace A -ϑ1-> B and C -ϑ2-> D by
                A -ϑ1-> BD in Γ_do;
        return Γ_do
```

Example 3. We consider as input for Algorithm 2 to compute the direct-optimal basis, the set of fuzzy implications of the Example 1

$$\Gamma=\{cd\xrightarrow{0.6}e, ac\xrightarrow{0.7}def, f\xrightarrow{0.5}dg, de\xrightarrow{0.9}ch, dh\xrightarrow{0.4}a\}$$

and we consider the Lukasiewicz product.

We show the first steps of the algorithm illustrating its execution:

Step 1 (label L1)

$$\Gamma=\{\ cd\xrightarrow{0.6}e, ac\xrightarrow{0.7}d, ac\xrightarrow{0.7}e, ac\xrightarrow{0.7}f, f\xrightarrow{0.5}d, f\xrightarrow{0.5}g, de\xrightarrow{0.9}c, de\xrightarrow{0.9}h, dh\xrightarrow{0.4}a\}$$

Step 2 (label L2):

Considering the first pairs of the fuzzy implications in which the rules are applied:

(1) $cd \xrightarrow{0.6} e, de \xrightarrow{0.9} h \vdash cd \xrightarrow{0.5} h$
(2) $ac \xrightarrow{0.7} d, de \xrightarrow{0.9} h \vdash ace \xrightarrow{0.6} h$
...

We add to Γ the implications deduced to obtain the direct basis.

Step 3 (label L3):

In the direct basis Γ after the previous step, we have $ade \xrightarrow{0.5} h$ as fuzzy implication which has been derived from the initial set. Then, the algorithm in this step, remove this fuzzy implication because $de \xrightarrow{0.9} h$ is in Γ (redundant).

We remove all redundant fuzzy implications applying this step.

Step 4 (label L4):

We apply composition to the fuzzy implications in the previous step.

Finally, the direct-optimal basis obtained is

$$\Gamma = \{\, cd \xrightarrow{0.6} e, cd \xrightarrow{0.5} h, cf \xrightarrow{0.1} e, ac \xrightarrow{0.7} def, ac \xrightarrow{0.3} h, ac \xrightarrow{0.2} g,$$
$$f \xrightarrow{0.5} dg, de \xrightarrow{0.9} ch, de \xrightarrow{0.3} a, dh \xrightarrow{0.4} a, dh \xrightarrow{0.4} a, ef \xrightarrow{0.4} ch,$$
$$acd \xrightarrow{0.6} h, ade \xrightarrow{0.6} f, ade \xrightarrow{0.1} g, cdh \xrightarrow{0.1} f\}$$

7 Conclusions and Further Works

We propose the use of Simplification Logic in Fuzzy Formal Concept Analysis to manage implications considering degrees for implications. We prove that Simplification Logic is sound and complete in this framework and establish the theoretical background to compute closures and, in this way, solve the implication problem.

We also extend the definition of direct-optimal basis, a very succinct specification of the knowledge stored in the L-context (fuzzy context). A method, directly based on the new logic, is proposed to calculate the direct-optimal basis for fuzzy implications from an arbitrary fuzzy set of implications.

As future work, we plan to extend this work to a more general implication notion, developing new methods to compute other kinds of bases with other good properties.

Acknowledgment. Supported by Grants TIN2014-59471-P and TIN2017-89023-P of the Science and Innovation Ministry of Spain.

References

1. Armstrong, W.W.: Dependency structures of data base relationships. In: IFIP Congress, pp. 580–583 (1974)
2. Belohlavek, R.: Fuzzy galois connections. Math. Log. Q. **45**(4), 497–504 (1999). https://doi.org/10.1002/malq.19990450408
3. Bertet, K., Monjardet, B.: The multiple facets of the canonical direct unit implicational basis. Theor. Comput. Sci. **411**(22–24), 2155–2166 (2010). https://doi.org/10.1016/j.tcs.2009.12.021
4. Cordero, P., Enciso, M., Mora, A., de Guzmán, I.P.: A complete logic for fuzzy functional dependencies over domains with similarity relations. In: Cabestany, J., Sandoval, F., Prieto, A., Corchado, J.M. (eds.) IWANN 2009. LNCS, vol. 5517, pp. 261–269. Springer, Heidelberg (2009). https://doi.org/10.1007/978-3-642-02478-8_33
5. Cordero, P., Enciso, M., Mora, A., de Guzmán, I.P., Rodríguez-Jiménez, J.M.: An efficient algorithm for reasoning about fuzzy functional dependencies. In: Cabestany, J., Rojas, I., Joya, G. (eds.) IWANN 2011. LNCS, vol. 6692, pp. 412–420. Springer, Heidelberg (2011). https://doi.org/10.1007/978-3-642-21498-1_52
6. Cordero, P., Enciso, M., Mora, A., de Guzmán, I.P.: SL_{FD} logic: elimination of data redundancy in knowledge representation. In: Garijo, F.J., Riquelme, J.C., Toro, M. (eds.) IBERAMIA 2002. LNCS (LNAI), vol. 2527, pp. 141–150. Springer, Heidelberg (2002). https://doi.org/10.1007/3-540-36131-6_15
7. Ganter, B., Wille, R.: Formal Concept Analysis: Mathematical Foundations. Springer, Heidelberg (1999). https://doi.org/10.1007/978-3-642-59830-2
8. Ježková, L., Cordero, P., Enciso, M.: Fuzzy functional dependencies: a comparative survey. Fuzzy Sets Syst. **317**, 88–120 (2017). https://doi.org/10.1016/j.fss.2016.06.019
9. Mora, A., Cordero, P., Enciso, M., Fortes, I., Aguilera, G.: Closure via functional dependence simplification. Int. J. Comput. Math. **89**(4), 510–526 (2012). https://doi.org/10.1080/00207160.2011.644275
10. Outrata, J., Vychodil, V.: Fast algorithm for computing fixpoints of Galois connections induced by object-attribute relational data. Inf. Sci. **185**(1), 114–127 (2012). https://doi.org/10.1016/j.ins.2011.09.023
11. Raju, K.V.S.V.N., Majumdar, A.K.: Fuzzy functional dependencies and lossless join decomposition of fuzzy relational database systems. ACM Trans. Database Syst. **13**, 129–166 (1988). https://doi.org/10.1145/42338.42344
12. Rodríguez-Jiménez, J.M., Cordero, P., Enciso, M., Mora, A.: Automated inference with fuzzy functional dependencies over graded data. In: Rojas, I., Joya, G., Cabestany, J. (eds.) IWANN 2013. LNCS, vol. 7903, pp. 254–265. Springer, Heidelberg (2013). https://doi.org/10.1007/978-3-642-38682-4_29
13. Rodríguez Lorenzo, E., Bertet, K., Cordero, P., Enciso, M., Mora, A.: The direct-optimal basis via reductions. In: Proceedings of the Eleventh International Conference on Concept Lattices and Their Applications, Košice, Slovakia, 7–10 October 2014, pp. 145–156 (2014)
14. Tyagi, B.K., Sharfuddin, A., Dutta, R.N., Tayal, D.K.: A complete axiomatization of fuzzy functional dependencies using fuzzy function. Fuzzy Sets Syst. **151**, 363–379 (2005). https://doi.org/10.1016/j.fss.2004.06.005

Formal Independence Analysis

Francisco J. Valverde-Albacete[1(✉)], Carmen Peláez-Moreno[1], Inma P. Cabrera[2], Pablo Cordero[2], and Manuel Ojeda-Aciego[2]

[1] Depto. Teoría de Señal y Comunicaciones, Univ. Carlos III de Madrid, Madrid, Spain
{fva,carmen}@tsc.uc3m.es

[2] Dpt. Matemática Aplicada, Univ. de Málaga, Málaga, Spain
{ipcabrera,pcordero,aciego}@uma.es

Abstract. In this paper we propose a new lens through which to observe the information contained in a formal context. Instead of focusing on the hierarchical relation between objects or attributes induced by their incidence, we focus on the "unrelatedness" of the objects with respect to those attributes with which they are not incident. The crucial order concept for this is that of maximal anti-chain and the corresponding representation capabilities are provided by Behrendt's theorem. With these tools we introduce the fundamental theorem of Formal Independence Analysis and use it to provide an example of what its affordances are for the analysis of data tables. We also discuss its relation to Formal Concept Analysis.

Keywords: Formal Concept Analysis · Independence · Tomoi

1 Introduction

The original intent of Wille on creating Formal Concept Analysis (FCA) was, in his own words, an "attempt to unfold lattice-theoretical concepts, results, and methods in a continuous relationship with their surrounding" [9, Sect. 1]. In an interesting *Final remarks* of that seminal work Wille renounces any attempt at exhaustiveness of the lattice restructuring program and recommends: "Besides the interpretation by hierarchies of concepts, other basic interpretations of lattices should be introduced; ..." In this light, we may wonder what other views of the information carried by a context might be.

In this paper we propose an alternative conceptualization for the information contained in a formal context. Since the intuitive interpretation of the analogues of formal concepts, which we have named *formal tomoi*, describe sets of objects

CPM & FVA have been partially supported by the Spanish Government-MinECo projects TEC2014-53390-P and TEC2014-61729-EXP; IPC & PC have been partially supported by the Spanish Government-MinECo project TIN2017-89023-P; MOA has been partially supported by the Spanish Government-MinECo project TIN2015-70266-C2-P-1.

J. Medina et al. (Eds.): IPMU 2018, CCIS 853, pp. 596–608, 2018.
https://doi.org/10.1007/978-3-319-91473-2_51

and attributes that have nothing to do with each other, we call this conceptualization *Formal Independence Analysis (FIA)*.

We base it in terms of the anti-chains of a certain order related to the context, and its lattice of anti-chains (Sect. 2) using Behrendt's theorem [1] which has universal representation capabilities for finite complete lattices. In this paper, we study mathematical properties of the lattice of anti-chains, in relation to the possible extension of Behrendt's theorem to a more general framework.

The set of anti-chains of a given a poset can be ordered by using two natural approaches (${}_{\star}\preccurlyeq$ and $\preccurlyeq^{\star}$) which lead to isomorphic structures, namely, a distributive lattice. The lattice of anti-chains of a distributive lattice L turns out to be isomorphic to L; however, the lattice of *maximal* anti-chains with any of the previous orderings is also a lattice, but not necessarily distributive. Now Behrendt's theorem states that any lattice (distributive or not) is isomorphic to the lattice of maximal anti-chains of *certain* poset. The focus of this paper is to study the existence of a representation theorem by using maximal anti-chains *within* the original lattice. For this, we rephrase Behrendt's theorem in terms of FCA using *tomoi*, and analyze possible extensions.

2 Preliminaries

Adjunctions and Galois Connections. Different notions of Galois connection or adjunction can be found in the literature; these notions are strongly related, but do not coincide. The transition between the two types of adjunctions (connections) relies on using the opposite ordering in *both* preordered sets, whereas the transition between adjunctions to connections and vice versa relies on using the opposite ordering in *just one* of the preordered sets. The four different types of Galois connections and adjunctions are summarized in Table 1.

See [4,6] for a revision of the genesis and importance of Galois Connections and adjunctions, as well as a discussion of the different notation and nomenclatures for these concepts. [5] is an early tutorial with mathematical applications in mind, and [8] deals fully on how to extend FCA with the different types of connections to provide different "flavors" of FCA, as well as extending it to non-binary incidences.

The Analysis of (In)Comparability in a Poset. Let $\langle P, \leq \rangle$ be a poset. We say that $x, y \in P$ are *comparable* if $x \leq y$ or $y \leq x$, and *incomparable* otherwise, and write $x \parallel y$. The initial analysis of posets is made in terms of *comparability* [2].

Definition 1. *Let $\langle P, \leq \rangle$ be a poset, and $Q \subseteq P$. Then*

1. *Q is an* (order) ideal *if $x \in Q$ and $y \leq x$, then $y \in Q$.*
2. *Q is an* (order) filter *if $x \in Q$ and $y \geq x$, then $y \in Q$.*
3. *$\downarrow Q = \{y \in P :$ there exists $x \in Q$ with $x \leq y\}$ is called the ideal* generated by Q.
4. *$\uparrow Q = \{y \in P :$ there exists $x \in Q$ with $y \leq x\}$ is called the* filter generated by Q.

Table 1. Galois connections and adjunctions: equivalent characterizations

Galois connections	
Right-Galois Connection between $\mathbb{A}$ and $\mathbb{B}$ $(f,g)\colon \mathbb{A} \leftharpoonup \mathbb{B}$	Left-Galois Connection between $\mathbb{A}$ and $\mathbb{B}$ $(f,g)\colon \mathbb{A} \rightharpoondown \mathbb{B}$
$b \leq f(a) \Leftrightarrow a \leq g(b)$ for all $a \in A$ and $b \in B$	$f(a) \leq b \Leftrightarrow g(b) \leq a$ for all $a \in A$ and $b \in B$
f and g are antitone and $g \circ f$ and $f \circ g$ inflationary	f and g are antitone and $g \circ f$ and $f \circ g$ deflationary
Adjunctions	
Adjunction between $\mathbb{A}$ and $\mathbb{B}$ $(f,g)\colon \mathbb{A} \leftrightharpoons \mathbb{B}$	Co-adjunction between $\mathbb{A}$ and $\mathbb{B}$ $(f,g)\colon \mathbb{A} \rightleftharpoons \mathbb{B}$
$f(a) \leq b \Leftrightarrow a \leq g(b)$ for all $a \in A$ and $b \in B$	$b \leq f(a) \Leftrightarrow g(b) \leq a$ for all $a \in A$ and $b \in B$
f and g are isotone, $g \circ f$ inflationary and $f \circ g$ deflationary	f and g are isotone, $g \circ f$ deflationary and $f \circ g$ inflationary

We will write $O(P)$ and $F(P)$ to denote the sets of ideals and filters (respectively) of $\langle P, \leq \rangle$. If considered as posets, ordered by set inclusion, both are distributive lattices and are dually-isomorphic.

Definition 2. *For a poset $\langle P, \leq \rangle$ and $Q \subseteq P$ define:*

1. *$a \in Q$ is a* minimal element *of Q if $a \geq x$ and $x \in Q$ imply $a = x$.*
2. *$a \in Q$ is a* maximal element *of Q if $a \leq x$ and $x \in Q$ imply $a = x$.*

We will write $\mathrm{Minl}(Q)$ and $\mathrm{Maxl}(Q)$ to denote the set of minimal, respectively maximal, elements of Q.

Definition 3. ([2]). *For a poset $\mathbb{P} = \langle P, \leq \rangle$, a set of pairwise incomparable elements of P is called an* anti-chain. *We denote the set of anti-chains of a poset as $A(\mathbb{P})$.*

Definition 4. *Given $\langle P, \leq \rangle$ a poset, it is possible to lift the ordering structure to the powerset 2^P by defining*

$$X \;{}_\star\!\preccurlyeq Y \iff \textit{for all } x \in X \textit{ there exists } y \in Y \textit{ such that } x \leq y \tag{1}$$

$$X \preccurlyeq^\star Y \iff \textit{for all } y \in Y \textit{ there exists } x \in X \textit{ such that } x \leq y \tag{2}$$

$$X \preccurlyeq Y \iff X \;{}_\star\!\preccurlyeq Y \textit{ and } X \preccurlyeq^\star Y \tag{3}$$

In general ${}_\star\!\preccurlyeq$ and $\preccurlyeq^\star$ are both simply preordering relations in $\langle 2^P, \subseteq \rangle$. Observe that in the set of anti-chains $A(\mathbb{P})$, the relations ${}_\star\!\preccurlyeq$ and $\preccurlyeq^\star$ are also antisymmetric.

There exists a relationship with the inclusion ordering of ideals and filters since given $S, T \subseteq P$ then $S \;{}_\star\!\preccurlyeq T \iff \downarrow S \subseteq \downarrow T$ and $S \preccurlyeq^\star T \iff \uparrow T \subseteq \uparrow S$.

Because of these equivalences, we will call ${}_\star\!\preccurlyeq$ as *ideal containment relation* and $\preccurlyeq^\star$ as *filter containment relation.*

Maximal Anti-chains

Definition 5. *For a poset $\langle P, \leq \rangle$, an anti-chain $\gamma \in A(\mathbb{P})$ is said to be* maximal *if every element of P is comparable to some element of γ.*

For any subset $Q \subseteq P$, the set of elements of P which are comparable to some element of Q is called the *neighborhood* of Q and it is denoted by $\updownarrow Q = \uparrow Q \cup \downarrow Q$. An anti-chain γ is maximal if and only if $\updownarrow \gamma = P$. The set of maximal anti-chains of a set is denoted as $MA(\mathbb{P})$,

$$MA(\mathbb{P}) = \{\gamma \in A(\mathbb{P}) \mid \updownarrow \gamma = P\}.$$

It is worth noting that the orderings ${}_\star\!\preccurlyeq$ and $\preccurlyeq^\star$ coincide in $MA(\mathbb{P})$.

Proposition 1 ([1]). *If $\mathbb{P}$ is a finite poset then $\langle MA(\mathbb{P}), {}_\star\!\preccurlyeq \rangle$ is a lattice.*

In [7] Reuter asserts, "Given an anti-chain A of P, the completion of A to a maximal anti-chain is not unique but there exists a unique lowest completion." This completion can be described in terms of the operators below:

Definition 6. *For a finite partial order $\langle P, \leq \rangle$, and $A, B \in 2^P$ we define the* highest anti-chain complement of *A, denoted A^-, and the* lowest anti-chain complement of *B, denoted B_-, as*

$$\begin{aligned} \cdot^- &: 2^P \to 2^P & \cdot_- &: 2^P \to 2^P \\ A &\mapsto A^- = \max(P \setminus \updownarrow A) & B &\mapsto B_- = \min(P \setminus \updownarrow B). \end{aligned} \tag{4}$$

The Analysis of Incomparability by Means of FCA. Due to the universal complete lattice representation capabilities of FCA, we must expect the lattices of anti-chains to be describable as the concept lattice of a context. The first result in this direction is due to Wille himself [10, Proposition 1, in our notation].

Proposition 2. *Let $\langle P, \leq \rangle$ be an ordered set. The concepts of the context $(P, P, \ngeq)$ are exactly the pairs $(A, P \setminus A)$ where A is an order ideal of P; especially*

$$\underline{\mathfrak{B}}(P, P, \ngeq) \cong \langle A(\mathbb{P}), {}_\star\!\preccurlyeq \rangle \cong \langle O(P), \subseteq \rangle \quad \underline{\mathfrak{B}}(P, P, \ngeq)^d \cong \langle A(\mathbb{P}), \preccurlyeq^\star \rangle \cong \langle F(P), \subseteq \rangle$$

Moreover, when focusing on maximal anti-chains, we have the following isomorphism is credited by Reuter [7] to Behrendt [1] and Wille [10].

Proposition 3. *Let $\mathbb{P} = \langle P, \leq \rangle$ be a poset. Then $\langle MA(\mathbb{P}), \preccurlyeq \rangle \cong \underline{\mathfrak{B}}(P, P, \not>)$.*

The proposition above states that maximal anti-chains can be obtained as a concept lattice for a certain context. On the other hand, Behrendt's theorem [1] is a universal representation theorem for lattices in terms of maximal anti-chains.

Theorem 1 (Behrendt). *Let $\mathbb{L} = \langle L, \leq \rangle$ be a finite lattice. Then there exists a poset $\mathbb{P} = \langle P, \leq_P \rangle$ such that $|P| = 2|L|$, where any chain has at most 2 elements and such that $\mathbb{L} \cong MA(\mathbb{P})$, i.e., $\mathbb{L}$ is isomorphic to the lattice of maximal anti-chains of $(P, \leq_P)$.*

This is our starting point for the development of Formal Independence Analysis.

3 Formal Independence Analysis (FIA)

It is straightforward that the three following structures are equivalent: formal contexts, bipartite graphs, and posets without chains with length higher than 2. As an example of the interoperability of the three structures: given a formal context (G, M, I), a bipartite graph can be obtained as $\langle G \sqcup M, I' \rangle$ where $G \sqcup M = (G \times \{0\}) \cup (M \times \{1\})$ is the disjoint union of G and M, and I' is defined as $(g, 0)\ I'\ (m, 1)$ if and only if $g\ I\ m$. From such a bipartite graph, a poset can be obtained with simply considering the reflexive closure of I'. Finally, from a poset $(P, \leq)$ without chains of length higher than 2, a formal context can be obtained by setting G and M to be, respectively, the sets of minimals and maximals of P.

To begin with, it is worth analyzing the proof of Behrendt's Theorem, using the terminology of FCA. Given a finite lattice[1] $\mathbb{L} = \langle L, \leq \rangle$, the proof considers another poset (of twice the cardinality of L) whose set of maximal anti-chains is isomorphic to $\mathbb{L}$. Specifically, given $\mathbb{L} = \langle L, \leq \rangle$, the new poset is obtained as the disjoint union $L \sqcup L = L \times \{0, 1\}$ with the ordering relation defined as the reflexive closure of

$$(z_1, 0)\ I\ (z_2, 1) \quad \text{if and only if} \quad z_1 < z_2 \text{ or } z_1 \parallel z_2.$$

The isomorphism is the following: given $z \in L$ is associated to the maximal anti-chain $\gamma_z = \alpha_z \cup \beta_z$ where

$$\alpha_z = \{(z', 0) \in L \mid z \leq z'\} = \uparrow z \times \{0\} \quad \beta_z = \{(z', 1) \in L \mid z' \leq z\} = \downarrow z \times \{1\}$$

Instead of considering γ_z as the union $\alpha_z \cup \beta_z$, one might consider the pairs (α_z, β_z), which leads to the notion *formal tomoi*[2] in a formal context.

In order to formally introduce the definition, we will use one of the alternative interpretations above in order to apply the mappings $\cdot_-$ and $\cdot^-$ within a formal context (G, M, I).

Specifically, we will consider the 2-height poset $(G \sqcup M, \leq)$ and, given $\alpha \subseteq G$ we will define $\alpha^- = M \setminus \updownarrow \alpha = M \setminus \uparrow \alpha = M \setminus \bigcup_{g \in \alpha} I(g, \cdot)$, and similarly for β_- given $\beta \subseteq M$. It is not difficult to see that there is a bijection between maximal anti-chains and pairs $(\alpha, \beta) \in 2^G \times 2^M$ satisfying $\alpha^- = \beta$ and $\beta_- = \alpha$.

The following example shows a subtle difference in the behaviour of the operators of highest and lowest complement depending on whether they are applied within a poset or within a formal context.

[1] But it is easy to see that the proof is also applicable to arbitrary complete lattices.
[2] From greek *tomos*, division pl. *tomoi*.

Example 1. Consider the following poset

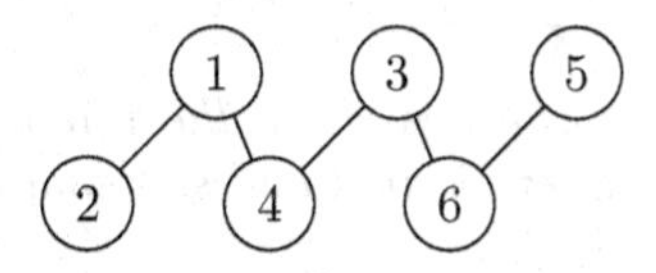

On the one hand, considering the poset structure above, given $\alpha = \{4\}$ we would obtain $\alpha^{-} = \{2, 5\}$; on the other hand, in the interpretation within a formal context, we would obtain $\alpha^{-} = \{5\}$. □

Notation. The situation above suggests to introduce a specific notation in order to avoid possible misunderstandings. Hence, we will use $\alpha^{\sim}$ and $\beta_{\sim}$ to indicate that we are assuming the construction within a formal context.

Now, the definition of a formal tomos is given as follows:

Definition 7. *Given a context (G, M, I), a* formal tomos *is a pair $(\alpha, \beta) \in 2^G \times 2^M$, such that $\alpha^{\sim} = \beta$ and $\beta_{\sim} = \alpha$. The set of formal tomoi of the context (G, M, I) will be denoted by $\mathfrak{A}(G, M, I)$.*

It is worth noting that the set of formal tomoi with the supset-subset hierarchical ordering, denoted $\underline{\mathfrak{A}}(G, M, I)$, is isomorphic to the corresponding lattice of maximal anti-chains. In fact, it turns out that $\underline{\mathfrak{A}}(G, M, I) = \underline{\mathfrak{B}}(G, M, \not I)^d$, since

$$\alpha^{\sim} = M \setminus \bigcup_{g \in \alpha} I(g, \cdot) = \{m \in M \mid g \not I m \ \text{ for all } \ g \in \alpha\} \tag{5}$$

$$\beta_{\sim} = G \setminus \bigcup_{m \in \beta} I(\cdot, m) = \{g \in G \mid g \not I m \ \text{ for all } \ m \in \beta\} \tag{6}$$

These operators adequately reflect the underlying philosophy of formal independence analysis and, in this terminology, we can obtain the following corollary of Theorem 1:

Corollary 1 (Behrendt's theorem in terms of tomoi). *Every finite lattice is isomorphic to a lattice of tomoi.*

Continuing this line of reasoning, we can state an analogue for tomoi of the basic theorem of FCA as follows:

Theorem 2 (Basic theorem of formal independence analysis).

1. *The context analysis phase: Given a formal context (G, M, I),*
   (a) *The operators $\cdot^{\sim} : 2^G \to 2^M$ and $\cdot_{\sim} : 2^M \to 2^G$ form a right-Galois connection $(\cdot^{\sim}, \cdot_{\sim}) : (2^G, \subseteq) \rightleftharpoons (2^M, \subseteq)$ whose* formal tomoi *are the pairs (α, β) such that $\alpha^{\sim} = \beta$ and $\alpha = \beta_{\sim}$.*

(b) *The set of formal tomoi* $\mathfrak{A}(G, M, I)$ *with the relation*

$$(\alpha_1, \beta_1) \leq (\alpha_2, \beta_2) \text{ iff } \alpha_1 \supseteq \alpha_2 \text{ iff } \beta_1 \subseteq \beta_2$$

is a complete lattice, which is called the tomoi lattice of (G, M, I) *and denoted* $\underline{\mathfrak{A}}(G, M, I)$*, where infima and suprema are given by:*

$$\bigwedge_{t\in T}(\alpha_t, \beta_t) = \left(\bigcup_{t\in T}\alpha_t, \left(\bigcap_{t\in T}\beta_t\right)_{\sim}^{\ \sim}\right) \quad \bigvee_{t\in T}(\alpha_t, \beta_t) = \left(\left(\bigcap_{t\in T}\alpha_t\right)^{\sim}_{\ \sim}, \bigcup_{t\in T}\beta_t\right)$$

(c) *The mappings* $\overline{\gamma} : G \to \underline{\mathfrak{A}}(G, M, I)$ *and* $\overline{\mu} : M \to \underline{\mathfrak{A}}(G, M, I)$

$$g \mapsto \overline{\gamma}(g) = (\{g\}^{\sim}{}_{\sim}, \{g\}^{\sim}) \qquad m \mapsto \overline{\mu}(m) = (\{m\}_{\sim}, \{m\}_{\sim}{}^{\sim})$$

are such that $\overline{\gamma}(G)$ *is infimum-dense in* $\underline{\mathfrak{A}}(G, M, I)$, $\overline{\mu}(M)$ *is supremum-dense in* $\underline{\mathfrak{A}}(G, M, I)$.

2. *The context synthesis phase: Given a complete lattice* $\mathbb{L} = \langle L, \leq \rangle$
   (a) $\mathbb{L}$ *is isomorphic to*[3] $\underline{\mathfrak{A}}(G, M, I)$ *if and only if there are mappings* $\overline{\gamma} : G \to L$ *and* $\overline{\mu} : M \to L$ *such that*
      - $\overline{\gamma}(G)$ *is infimum-dense in* $\mathbb{L}$, $\overline{\mu}(M)$ *is supremum-dense in* $\mathbb{L}$*, and*
      - $g\, I\, m$ *is equivalent to* $\overline{\gamma}(g) \not\geq \overline{\mu}(m)$ *for all* $g \in G$ *and all* $m \in M$*.*

   (b) *In particular,* $\mathbb{L} \cong \underline{\mathfrak{A}}(L, L, \not\geq)$ *and, if* L *is finite,* $\mathbb{L} \cong \underline{\mathfrak{A}}(M(\mathbb{L}), J(\mathbb{L}), \not\geq)$ *where* $M(\mathbb{L})$ *and* $J(\mathbb{L})$ *are the sets of meet- and join-irreducibles, respectively, of* $\mathbb{L}$*.*

Notice that the differences between the basic theorems of FIA and FCA are due to the fact that FCA focuses on the notion of "being related" which, in algebraic terms, leads to complete bipartite subgraphs and, in FCA terminology, to maximal rectangles, whereas FIA focuses on "unrelatedness", leading to completely independent subsets.

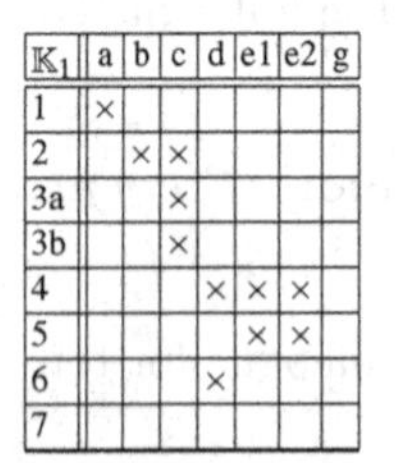

$\mathbb{K}_1$	a	b	c	d	e1	e2	g
1	×						
2		×	×				
3a			×				
3b			×				
4				×	×	×	
5					×	×	
6				×			
7							

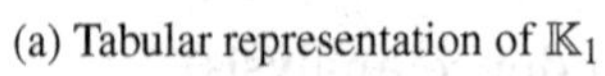

(a) Tabular representation of $\mathbb{K}_1$

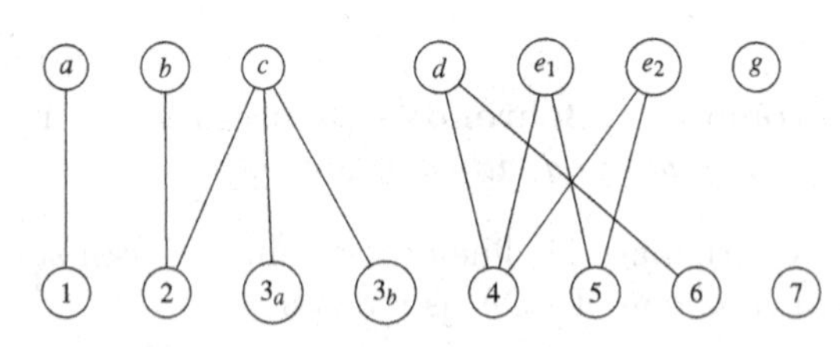

(b) Bipartite graph representation of $\mathbb{K}_1$

Fig. 1. Equivalent representations of an example context $\mathbb{K}_1 = (G, M, I)$. (a) Tabular representation. (b) Bipartite graph representation.

[3] Read *can be built as*.

Example 2. Figure 1(a) is the tabular representation of a context which admits a non-trivial block-diagonal form. Figure 1(b) is the bipartite graph representation, where this block structure is also apparent. If we represent the concept lattice, as in Fig. 2(a), these appear adjoined by top and bottom.

Consider the formal context of (G, M, I), e.g. that of Fig. 1(a): we would like to find its independence lattice $\underline{\mathfrak{A}}(G, M, I)$. Note that there is an isolated object 7 and an isolated attribute g. These are both ignored and re-introduced later. To find the meet-irreducibles, we use the object-tomos mapping $\overline{\gamma}$ over the whole of G. The result of this operation can be seen in Table 2(a). Likewise, the result of the application of the attribute-tomos mapping $\overline{\mu}$ over the whole of M can be found in Table 2(b).

Table 2. Object and attribute tomoi for $\mathbb{K}_1$ and $\underline{\mathfrak{A}}(\mathbb{K}_1)$. Not seen are object 7, appearing in each object-tomos, and attribute g, in every attribute-tomos.

$g \setminus \overline{\gamma}(g)$	γ	μ
1	$\{1\}$	$\{b,c,d,e1,e2\}$
2	$\{2,3a,3b\}$	$\{a,d,e1,e2\}$
3a	$\{3a,3b\}$	$\{a,b,d,e1,e2\}$
3b	$\{3a,3b\}$	$\{a,b,e,e1,e2\}$
4	$\{4,5,6\}$	$\{a,b,c\}$
5	$\{5\}$	$\{a,b,c,d\}$
6	$\{6\}$	$\{a,b,c,e1,e2\}$

(a) Object tomoi of $\underline{\mathfrak{A}}(G,M,I)$

$m \setminus \overline{\mu}(m)$	γ	μ
a	$\{2,3a,3b,4,5,6\}$	$\{a\}$
b	$\{1,3a,3b,4,5,6\}$	$\{b\}$
c	$\{1,4,5,6\}$	$\{b,c\}$
d	$\{1,2,3a,3b,5\}$	$\{d\}$
e1	$\{1,2,3a,3b,6\}$	$\{e1,e2\}$
e2	$\{1,2,3a,3b,6\}$	$\{e1,e2\}$

(b) Attribute tomoi of $\underline{\mathfrak{A}}(G,M,I)$

Since the object-tomoi are meet-irreducible and the attribute tomoi join-irreducible, the Dedekind-MacNeille completion using the lattice operations finds the lattice of tomoi $\underline{\mathfrak{A}}(G, M, I)$ as seen in Fig. 2(b). Note that the illustration is actually built as $\underline{\mathfrak{B}}(M, G, I^{cd})$ since the SW used to represent the lattice only understand concept lattices.

4 Generalizing the Construction of Tomoi to Posets

We have just seen that the operators $\cdot^{\sim}$ and $\cdot_{\sim}$ on a formal context, whose definition raised from a suitable modification of $\cdot^{-}$ and $\cdot_{-}$. In this section, we focus on the possible extensions of the notion of tomoi in a poset as general as possible. In the rest of the paper, we consider a poset $\mathbb{P} = \langle P, \leq \rangle$ **without infinite chains.**

Proposition 4. *For any $Q \subseteq P$, the following properties hold*

$$\mathrm{Minl}(Q),\, \mathrm{Maxl}(Q) \in A(\mathbb{P}) \tag{7}$$

$$\mathrm{Minl}(Q) \preccurlyeq^{\star} Q \;{}_{\star}\!\preccurlyeq \mathrm{Maxl}(Q) \tag{8}$$

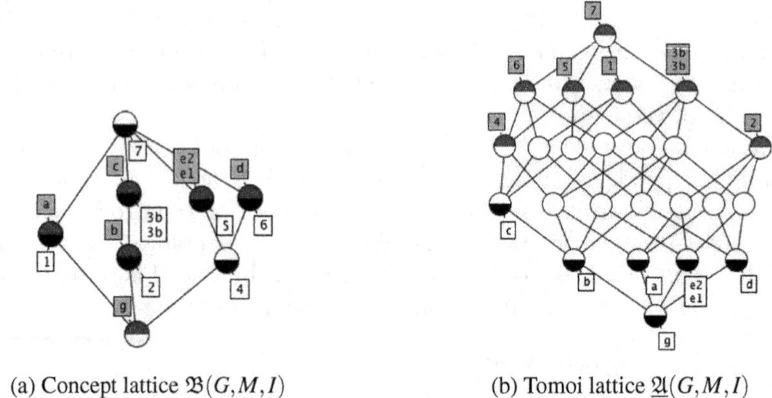

(a) Concept lattice $\mathfrak{B}(G,M,I)$ (b) Tomoi lattice $\underline{\mathfrak{A}}(G,M,I)$

Fig. 2. Two different lattices for context $\mathbb{K}_1 = (G,M,I)$; in (a) the lattice of formal concepts *showing* three adjoint sublattices; in (b) the lattice of formal tomoi *describing* the three adjoint sublattices. Notice that object-concepts are join-irreducible in (a), but object-tomoi are meet-irreducible in (b), and likewise *mutatis mutandis* for attribute-concepts and attribute-tomoi.

We will now explore the properties of the operators of highest (resp. lowest) anti-chain complement $\cdot_{-}$ and $\cdot^{-}$.

Definition 8. *A set $Q \subseteq P$ is said to be* convex *if $a, b \in Q$ and $a \leq p \leq b$ imply $p \in Q$.*

Proposition 5. *1. For any convex set $Q \subseteq P$, we have that $Q = \uparrow \mathrm{Minl}(Q) \cap \downarrow \mathrm{Maxl}(Q)$.*
2. For any $Q \subseteq P$, the set $P \smallsetminus \updownarrow Q$ is convex and, therefore,

$$P \smallsetminus \updownarrow Q = \{p \in P \mid \forall a \in Q, p \parallel a\} = \uparrow \mathrm{Minl}(P \smallsetminus \updownarrow Q) \cap \downarrow \mathrm{Maxl}(P \smallsetminus \updownarrow Q)$$

In addition, if we write $Q^{\parallel} = P \smallsetminus \updownarrow Q$, then $(\cdot^{\parallel}, \cdot^{\parallel})$ is a Galois connection in $(2^P, \subseteq)$.

Proof. Since $\mathbb{P}$ has not infinite chains, we have that $Q \subseteq \uparrow \mathrm{Minl}(Q) \cap \downarrow \mathrm{Maxl}(Q)$. By the other side, since Q is convex, any element $p \in \uparrow \mathrm{Minl}(Q) \cap \downarrow \mathrm{Maxl}(Q)$ belongs to Q.

The second item follows from the definition of $\updownarrow Q = \uparrow Q \cup \downarrow Q$ as the neighborhood of Q. Therefore $P \smallsetminus \updownarrow Q$ are those elements *not* related to any element in Q. Note that $P \smallsetminus (\uparrow Q \cup \downarrow Q) = (P \smallsetminus \uparrow Q) \cap (P \smallsetminus \downarrow Q)$ and it is convex.

Finally, for all $A, B \subseteq P$, it easy to see that $A \subseteq P \smallsetminus \updownarrow B$ if and only if $B \subseteq P \smallsetminus \updownarrow A$. □

The following result can be obtained as a consequence of the previous propositions:

Corollary 2. *For anti-chains $\alpha, \beta \in A(\mathbb{P})$, we have that:*

1. $\beta_- \preccurlyeq^\star \alpha \ {}_\star\!\preccurlyeq \beta^-$ *if and only if* $\alpha_- \preccurlyeq^\star \beta \ {}_\star\!\preccurlyeq \alpha^-$
2. $\alpha^-{}_- \preccurlyeq^\star \alpha \ {}_\star\!\preccurlyeq \alpha^{--}$
3. $\alpha_{--} \preccurlyeq^\star \alpha \ {}_\star\!\preccurlyeq \alpha_-{}^-$

Proof. 1. We have that $\beta^{\parallel} = P \setminus \updownarrow \beta = \uparrow \beta_- \cap \downarrow \beta^-$. Therefore, $\alpha \subseteq \beta^{\parallel}$ if and only if $\beta_- \preccurlyeq^\star \alpha \ {}_\star\!\preccurlyeq \beta^-$. Then, the first item is a consequence of the fact trhat $(\cdot^{\parallel}, \cdot^{\parallel})$ is a Galois connection in $(2^P, \subseteq)$.
Now, notice that (8) implies $\alpha_- \preccurlyeq^\star \alpha^-$ and $\alpha_- \ {}_\star\!\preccurlyeq \alpha^-$.
2. Since $\alpha_- \preccurlyeq^\star \alpha^- \ {}_\star\!\preccurlyeq \alpha^-$, by the first equivalence, one has $\alpha^-{}_- \preccurlyeq^\star \alpha \ {}_\star\!\preccurlyeq \alpha^{--}$.
3. Similar. □

Note that the first item of the corollary suggests the possible structure of Galois connection/adjunction of the pair operators $(\cdot_-, \cdot^-)$; moreover, from the second and third items, we have $\alpha^-{}_- \preccurlyeq^\star \alpha \ {}_\star\!\preccurlyeq \alpha_-{}^-$. This means that the only possibility for $(\cdot_-, \cdot^-)$ is to be an adjunction. Unfortunately, this is not the case, as shown in the example below.

Example 3. Given the poset $\mathbb{P}$ with the ordering depicted in the figure below

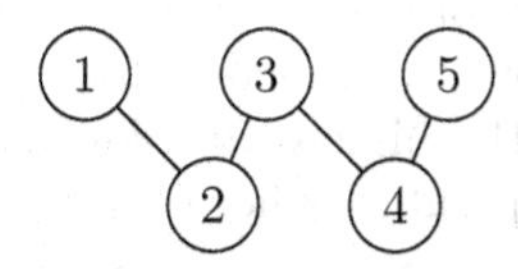

it is not the case that $(\cdot^-, \cdot_-)\colon (A(\mathbb{P}), \preccurlyeq^\star) \leftrightharpoons (A(\mathbb{P}), {}_\star\!\preccurlyeq)$, since $\{1,4\}^- \ {}_\star\!\preccurlyeq \{5\}$ holds but $\{1,4\} \preccurlyeq^\star \{5\}_-$ does not hold. □

Although, the operators do not form any kind of connection or adjunction, the notion of tomos still behaves properly, in the sense that it is strongly related to maximal anti-chains.

Proposition 6. *Let α, β be anti-chains such that $\alpha^- = \beta$ and $\beta_- = \alpha$. Then $\alpha \cup \beta$ is a maximal anti-chain.*

Proof. Since both α, β are anti-chains and $\alpha = \mathrm{Minl}\,(P \setminus \updownarrow \beta) \subseteq P \setminus \updownarrow \beta$, it is obvious that $\alpha \cup \beta$ is also an anti-chain. To show that it is maximal, assume that there exists $x \in P$ which is not related to any element of $\alpha \cup \beta$. If $x \in P \setminus \updownarrow \alpha$, there exists $b \in \beta = \mathrm{Maxl}\,(P \setminus \updownarrow \alpha)$ such that $x \leq b$, which contradicts that $x \in P \setminus \updownarrow \beta$. Analogously, assuming that $x \in P \setminus \updownarrow \beta$ also yields to a contradiction. □

The relationship, however, is not one-one, as shown in the next example.

Example 4. For poset $\mathbb{P}$ in Fig. 3, the function which merges the two components of every tomos in order to obtain a maximal anti chain need not be either one-one nor onto. The three pairs $\langle 26, 3\rangle, \langle 36, 2\rangle, \langle 6, 23\rangle$ are mapped to the same maximal anti-chain. On the other hand, no tomos leads to the maximal anti-chain 56. □

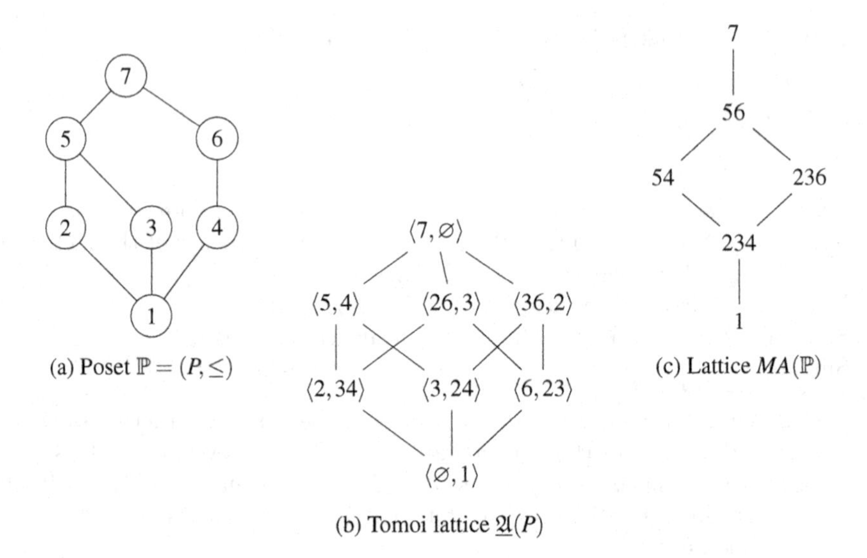

Fig. 3. Tomoi for an arbitrary poset and its maximal anti-chains do not always match.

Since operators $(\cdot^{-}, \cdot_{-})$ do not behave properly because of the previous example, it is worth considering what happens when it is complemented with the original chain, i.e., consider the highest and lowest completions of a chain. In both cases, they are closure operators whose closures are, precisely, the maximal anti-chains.

Proposition 7. *Let α be a chain in $\mathbb{P}$.*

1. *The mapping $\alpha \mapsto \alpha \cup \alpha^{-}$ is a closure operator in $(A(\mathbb{P}), {}_{\star}\!\preccurlyeq)$ whose set of closed elements is $MA(\mathbb{P})$.*
2. *The mapping $\alpha \mapsto \alpha \cup \alpha_{-}$ is a closure operator in $(A(\mathbb{P}), \preccurlyeq^{\star})$ whose set of closed elements is $MA(\mathbb{P})$.*

Proof. We only prove the first item, since the second is analogous.

The mapping is obviously inflationary and idempotent (the latter because for any anti-chain α, since $\alpha \cup \alpha^{-}$ is maximal, trivially holds that $(\alpha \cup \alpha^{-})^{-}$ is empty).

To show that the mapping is isotone, we first observe that, for any anti-chain α, the highest completion $\alpha \cup \alpha^{-}$ satisfies that

$$\downarrow(\alpha \cup \alpha^{-}) = P \smallsetminus \underline{\uparrow}\alpha \quad \text{where} \quad \underline{\uparrow}\alpha = \{x \in P : a < x \text{ for some } a \in \alpha\}.$$

As a consequence, $\alpha \cup \alpha^{-} \subseteq \operatorname{Maxl}(P \smallsetminus \underline{\uparrow}\alpha)$ which implies $\alpha \cup \alpha^{-} = \operatorname{Maxl}(P \smallsetminus \underline{\uparrow}\alpha)$ since $\alpha \cup \alpha^{-}$ is a maximal anti-chain.

Assume now that $\alpha_1 \,{}_{\star}\!\preccurlyeq\, \alpha_2$ and let us show that $\alpha_1 \cup {\alpha_1}^{-} \,{}_{\star}\!\preccurlyeq\, \alpha_2 \cup {\alpha_2}^{-}$. By Proposition 6, we know that the highest completion $\alpha \cup \alpha^{-}$ is maximal for all anti-chain α, then $\alpha_1 \cup {\alpha_1}^{-} \,{}_{\star}\!\preccurlyeq\, \alpha_2 \cup {\alpha_2}^{-}$ if and only if $\alpha_1 \cup {\alpha_1}^{-} \preccurlyeq^{\star} \alpha_2 \cup {\alpha_2}^{-}$. Hence, let us show that $\alpha_1 \cup {\alpha_1}^{-} \preccurlyeq^{\star} \alpha_2 \cup {\alpha_2}^{-}$ or equivalently $\alpha_1 \cup {\alpha_1}^{-} \subseteq \downarrow(\alpha_2 \cup {\alpha_2}^{-})$.

Since $\alpha_2 \subseteq\ \uparrow \alpha_1$, then $\underline{\uparrow}\alpha_2 \subseteq\ \uparrow \alpha_1 \subseteq \updownarrow \alpha_1$. Therefore, $P \setminus \updownarrow \alpha_1 \subseteq P \setminus \underline{\uparrow}\alpha_2$ whence

$$\alpha_1{}^- \subseteq P \setminus \updownarrow \alpha_1 \subseteq P \setminus \underline{\uparrow}\alpha_2 = \downarrow (\alpha_2 \cup \alpha_2{}^-)$$

On the other hand, if we suppose that $x \in \alpha_1 \cap \underline{\uparrow}\alpha_2$, then there exists $y \in \alpha_2$ with $y < x$, but since $\alpha_2 \subseteq \uparrow \alpha_1$ there exists $z \in \alpha_2$ such that $y < x \leq z$, which is a contradiction because α_2 is an anti-chain. Thus, $\alpha_1 \subseteq P \setminus \underline{\uparrow}\alpha_2 = \downarrow (\alpha_2 \cup \alpha_2{}^-)$.

□

5 Conclusions

In this paper we have tried to propose a new lens through which to observe the information contained in a formal context. Instead of focusing on the hierarchical relation between objects or attributes induced by their incidence we focus on the "unrelatedness" of the objects with respect to those attributes with which they are not incident.

We have named the framework that appears "Formal Independence Analysis" because it allows us to block-diagonalize formal contexts providing a means for decomposing them in terms of independent sub-contexts, in the sense that two independent sub-contexts do not share common attributes nor objects. Even if the formal context cannot be block-diagonalized the procedure still obtains joint sets of objects and attributes which are mutually unrelated, which we have named *formal tomoi* (that is "separations").

We have provided a fundamental theorem for Formal Independence Analysis and an example of use based on a formal context for which the context is effectively block-diagonalized to illustrate the possibilities of the technique.

The procedure seems to be specially interesting in data analysis where, dual to what formal concepts can glean from data describing the existence of hierarchy, formal tomoi would describe when data contexts—e.g. from genomics, contingency matrices, etc.—can be broken down into parts susceptible of independent analysis.

Further work is necessary to ascertain the relationship of lattices of formal tomoi to lattices of formal concepts, as well as to find out whether these are the only information lenses available for formal contexts.

References

1. Behrendt, G.: Maximal antichains in partially ordered sets. Ars Combinatoria **25**(C), 149–157 (1988)
2. Davey, B., Priestley, H.: Introduction to Lattices and Order, 2nd edn. Cambridge University Press, Cambridge (2002). https://doi.org/10.1017/CBO9780511809088
3. Denecke, K., Erné, M., Wismath, S. (eds.): Galois Connections and Applications. Mathematics and Its Applications, vol. 565. Kluwer Academic, Dordrecht (2004). https://doi.org/10.1007/978-1-4020-1898-5
4. Erné, M.: Adjunctions and Galois connections: origins, history and development. In: Denecke et al. [3], pp. 1–138 (2004)

5. Erné, M., Koslowski, J., Melton, A., Strecker, G.: A primer on Galois connections. Ann. N. Y. Acad. Sci. **704**, 103–125 (1993). https://doi.org/10.1111/j.1749-6632.1993.tb52513.x
6. García-Pardo, F., Cabrera, I.P., Cordero, P., Ojeda-Aciego, M.: On Galois connections and soft computing. In: Rojas, I., Joya, G., Cabestany, J. (eds.) IWANN 2013. LNCS, vol. 7903, pp. 224–235. Springer, Heidelberg (2013). https://doi.org/10.1007/978-3-642-38682-4_26
7. Reuter, K.: The jump number and the lattice of maximal antichains. Discret. Math. **88**(2–3), 289–307 (1991). https://doi.org/10.1016/0012-365X(91)90016-U
8. Valverde-Albacete, F.J., Peláez-Moreno, C.: Extending conceptualisation modes for generalised Formal Concept Analysis. Inf. Sci. **181**, 1888–1909 (2011)
9. Wille, R.: Restructuring lattice theory: an approach based on hierarchies of concepts. In: Ordered Sets (Banff, Alta., 1981), pp. 445–470. Reidel, Dordrecht-Boston (1982). https://doi.org/10.1007/978-94-009-7798-3_15
10. Wille, R.: Finite distributive lattices as concept lattices. Atti Inc. Logica Mathematica **2**, 635–648 (1985)

Fuzzy Implication Functions

Fuzzy Boundary Weak Implications

Hua-Wen Liu[1(✉)] and Michał Baczyński[2]

[1] School of Mathematics, Shandong University, Jinan 250100, Shandong, China
hw.liu@sdu.edu.cn

[2] Institute of Mathematics, University of Silesia in Katowice,
Bankowa 14, 40-007 Katowice, Poland
michal.baczynski@us.edu.pl

Abstract. An extension of fuzzy implications and coimplications, called fuzzy boundary weak implications (shortly, fuzzy bw-implications), is introduced and discussed in this paper. Firstly, by weakening the boundary conditions of fuzzy implications and coimplications, we introduce the concept of fuzzy bw-implications. And then, we investigate some of their basic properties. Next, the concept of fuzzy pseudo-negations is introduced and the natural pseudo-negations of fuzzy bw-implications are investigated. Finally, the fuzzy bw-implications generated, respectively, by aggregation operators and generator functions are discussed in details. This work is motivated by the fact that in real applications there are used some operators which are not fuzzy implications. We hope that such an extension of fuzzy (co)implications can provide a certain theoretical foundation for the real applications.

Keywords: Fuzzy connectives · Fuzzy implications
Fuzzy coimplications · Aggregation functions
Generator functions · Bw-implications

1 Introduction

It is well-known that fuzzy implications play an important role in so many fields such as fuzzy reasoning, fuzzy control, fuzzy decision making, fuzzy relational equations, fuzzy mathematical morphology, image processing, etc. According to the well-known definition of fuzzy implications (see [2]), each fuzzy implication I on the unit interval must satisfy the boundary conditions $\forall_{y\in[0,1]} I(0,y) = 1$ and $\forall_{x\in[0,1]} I(x,1) = 1$, and its N–duality J as a fuzzy coimplication (see [4]) must satisfy the boundary conditions $\forall_{y\in[0,1]} I(1,y) = 0$ and $\forall_{x\in[0,1]} I(x,0) = 0$, where N is a fuzzy negation. Such requirements for these boundary conditions are reasonable and natural from the logical point of view because the condition $\forall_{y\in[0,1]} I(0,y) = 1$ means that a falsity implies anything, while condition $\forall_{x\in[0,1]} I(x,1) = 1$ expresses that a tautology is implied by anything (see [2]). In real applications, however, we found that there are operators dissatisfying the above mentioned boundary conditions which are also used as fuzzy implications (see, e.g. [2,3]). Motivated by this fact from real applications, we weaken in the

J. Medina et al. (Eds.): IPMU 2018, CCIS 853, pp. 611–622, 2018.
https://doi.org/10.1007/978-3-319-91473-2_52

present paper the boundary conditions of fuzzy implications and fuzzy coimplications, introduce the concept of fuzzy boundary weak implications (shortly, fuzzy bw-implications), and discuss some of their basic properties. We show in this paper that such an extension of fuzzy (co)implications are also reasonable from the logical point of view. We hope that this extension can provide a certain theoretical foundation for the real applications on one hand, and on the other hand, we hope that it can enrich and perfect the theory of fuzzy (co)implications.

The rest of this paper is organized as follows. Section 2 contains the definition and analysing of basic properties of fuzzy bw-implications. In Sect. 3, we introduce the concept of fuzzy pseudo-negations and investigate the natural pseudo-negations of fuzzy bw-implications. Sections 4 and 5 discuss two classes of fuzzy bw-implications generated, respectively, by aggregation operators and unary monotonic functions. The final section contains plans for the future work.

2 Definition and Basic Properties of Fuzzy Boundary Weak Implications

Fuzzy implications have been analysed in details in [2,7], while fuzzy coimplications in [4].

Definition 2.1. *A function $I\colon [0,1]^2 \to [0,1]$ is called a fuzzy boundary weak implication (shortly, fuzzy bw-implications) if it satisfies the following conditions:*

(I1) I is decreasing in its first variable,
(I2) I is increasing in its second variable,
(I3) $I(0,1) = 1$ and $I(1,0) = 0$.

If a fuzzy bw-implication I satisfies $I(0,0) = I(1,1) = 1$, then we call it a fuzzy implication, while if a fuzzy bw-implication I satisfies $I(0,0) = I(1,1) = 0$, then it is called a fuzzy coimplication.

From Definition 2.1 we know that for a fuzzy bw-implication I, the boundary conditions $\forall_{y\in[0,1]} I(0,y) = 1$ and $\forall_{x\in[0,1]} I(x,1) = 1$ do not always hold. In other words, it allows that the weak conditions $I(0,y) < 1$ for some $y < 1$ or $I(x,1) < 1$ for some $x > 0$ occur, which differ from the fuzzy implications. From the logical point of view, we can explain them as follows: condition $I(0,y) < 1$ for some $y < 1$ means that a falsity implies a proposition with truth value y in the level of $\alpha = I(0,y)$, while condition $I(x,1) < 1$ for some $x > 0$ means that a tautology is implied in a level $\beta = I(x,1)$ by a proposition with truth value x.

It is clear that each fuzzy implication or fuzzy coimplication must be a fuzzy bw-implication, but not vice versa. If an operator is a fuzzy bw-implication but neither a fuzzy implication nor a fuzzy coimplication, we call it a proper fuzzy bw-implication. The following example shows that there exist such proper fuzzy bw-implications.

Example 2.2. Let $I_i\colon [0,1]^2 \to [0,1]$, $(i = 1, 2, \cdots, 8)$ be as follows, for all $x, y \in [0,1]$,

$$I_1(x,y) = \begin{cases} 1, & \text{if } y = 1, \\ 0, & \text{otherwise}, \end{cases} \qquad I_2(x,y) = \begin{cases} 1, & \text{if } x = 0, \\ 0 & \text{otherwise}, \end{cases}$$

$$I_3(x,y) = \begin{cases} y, & \text{if } x \le y, \\ 0, & \text{otherwise}, \end{cases} \qquad I_4(x,y) = \begin{cases} 1 & \text{if } x < y, \\ y & \text{otherwise}, \end{cases}$$

$$I_5(x,y) = y, \qquad I_6(x,y) = \max(0, y - x),$$

$$I_7(x,y) = \max(\tfrac{1}{2} - x, y), \qquad I_8(x,y) = \begin{cases} 1, & \text{if } x \le y \text{ and } x^2 + y^2 \ge \frac{1}{4}, \\ y, & \text{if } x > y \text{ and } x^2 + y^2 \ge \frac{1}{4}, \\ \frac{1}{4} - x^2, & \text{otherwise}. \end{cases}$$

Functions I_1, I_2 and I_6 are from [2,3]. It is obvious that I_i $(i = 1, 2, \cdots, 8)$ are all fuzzy bw-implications, but not fuzzy implications. In fact, each I_i with $i \neq 6$ is a proper fuzzy bw-implication, while I_6 is a fuzzy coimplication.

We denote the family of all fuzzy implications by $\mathcal{FI}$, the family of all fuzzy coimplications by $co\mathcal{FI}$ and the family of all fuzzy bw-implications by $bw\mathcal{FI}$. It is clear that $\mathcal{FI} \cap co\mathcal{FI} = \emptyset$, $\mathcal{FI} \cup co\mathcal{FI} \subseteq bw\mathcal{FI}$ and $bw\mathcal{FI} \setminus (\mathcal{FI} \cup co\mathcal{FI}) \neq \emptyset$.

Definition 2.3 (see [9]). *A function $N\colon [0,1] \to [0,1]$ is called a fuzzy negation, if it is decreasing and satisfies $N(0) = 1$ and $N(1) = 0$. Further, we say N is strong, if it is involutive, i.e., $N(N(x)) = x$ for all $x \in [0,1]$. A fuzzy negation is said to be strict, if it is continuous and strictly decreasing.*

The following result is obvious from Definition 2.1.

Proposition 2.4. *The family $bw\mathcal{FI}$ has the least member $I_\perp$ and the greatest member $I_\top$, defined as follows:*

$$I_\perp(x,y) = \begin{cases} 1, & \text{if } x = 0 \text{ and } y = 1, \\ 0, & \text{otherwise}, \end{cases} \qquad I_\top(x,y) = \begin{cases} 0, & \text{if } x = 1 \text{ and } y = 0, \\ 1, & \text{otherwise}. \end{cases}$$

Actually, $I_\perp$ is the least fuzzy coimplication, while $I_\top$ is the greatest fuzzy implication (see [2]). Another direct result of Definition 2.1 is the following proposition.

Proposition 2.5. *If $I \in bw\mathcal{FI}$ and N is a fuzzy negation, then*

(i) the N–dual I_N of I defined by $I_N(x,y) = N(I(N(x), N(y)))$, is a fuzzy bw-implication,
(ii) the N–reciprocal I^N of I defined by $I^N(x,y) = I(N(y), N(x))$, is also a fuzzy bw-implication.

For the sake of obtaining fuzzy (co)implications fulfilling different requirements, many other potential axioms have been proposed in literature. Since the fuzzy bw-implications are an extension of fuzzy (co)implications, we also consider the following important properties.

Definition 2.6. *We say that a fuzzy bw-implication I satisfies*

- NP_1 *(left neutrality property 1) if $I(1, y) = y$ for all $y \in [0, 1]$,*
- NP_2 *(left neutrality property 2) if $I(0, y) = y$ for all $y \in [0, 1]$,*
- IP_1 *(identity principle 1) if $I(x, x) = 1$ for all $x \in [0, 1]$,*
- IP_2 *(identity principle 2) if $I(x, x) = 0$ for all $x \in [0, 1]$,*
- EP *(exchange principle) if $I(x, I(y, z)) = I(y, I(x, z))$ for all $x, y, z \in [0, 1]$,*
- OP_1 *(ordering property 1) if $x \leq y \Leftrightarrow I(x, y) = 1$ for all $x, y \in [0, 1]$,*
- OP_2 *(ordering property 2) if $x \geq y \Leftrightarrow I(x, y) = 0$ for all $x, y \in [0, 1]$,*
- CB_1 *(consequent boundary 1) if $I(x, y) \geq y$ for all $x, y \in [0, 1]$,*
- CB_2 *(consequent boundary 2) if $I(x, y) \leq y$ for all $x, y \in [0, 1]$,*
- *CO (continuity) if I is a continuous mapping.*

The following results are straightforward from Definition 2.6.

Proposition 2.7. *Let $I \in bw\mathcal{FI}$, $N\colon [0, 1] \to [0, 1]$ be a strong negation and I_N be the N-dual of I defined by $I_N(x, y) = N(I(N(x), N(y)))$ for all $x, y \in [0, 1]$. Then the following statements are true.*

(i) I satisfies NP_1 *if and only if I_N satisfies* NP_2.
(ii) I satisfies IP_1 *if and only if I_N satisfies* IP_2.
(iii) I satisfies OP_1 *if and only if I_N satisfies* OP_2.
(iv) I satisfies CB_1 *if and only if I_N satisfies* CB_2.

Proposition 2.8

(i) The only fuzzy bw-implication satisfying both NP_1 *and* NP_2 *is I_5 in Example 2.2, i.e., $I_5(x, y) = y$ for all $x, y \in [0, 1]$.*
(ii) There is no proper fuzzy bw-implication satisfying IP_1 *or* IP_2.
(iii) There exist proper fuzzy bw-implications satisfying both EP and CO.
(iv) There is no proper fuzzy bw-implication satisfing OP_1 *or* OP_2.
(v) The only fuzzy bw-implication satisfying both CB_1 *and* CB_2 *is also I_5 in Example 2.2.*

Proof. (i) Suppose that I is a fuzzy bw-implication satisfying NP_1 and NP_2, i.e., $I(0, y) = I(1, y) = y$ for all $y \in [0, 1]$, then from the monotonicity of I we have $I(x, y) = y$ for any $x \in [0, 1]$.
(ii) Suppose that there exist an $I \in bw\mathcal{FI} \setminus (\mathcal{FI} \cup co\mathcal{FI})$ satisfying IP_1 or IP_2, then $I(0, 0) = I(1, 1) = 1$ or $I(0, 0) = I(1, 1) = 0$, which means that $I \in \mathcal{FI}$ or $I \in co\mathcal{FI}$, a contradiction.
(iii) It is obvious that the proper fuzzy bw-implication I_5 defined in Example 2.2 satisfies EP and CO.
(iv) Similar to the proof of (ii).
(v) It is obvious. □

3 Natural Pseudo-negations of Fuzzy Boundary Weak Implications

3.1 Fuzzy Pseudo-negations

Definition 3.1. *A function $N\colon [0,1] \to [0,1]$ is called a fuzzy pseudo-negation if it is decreasing and satisfies $N(1) = 0$. A fuzzy pseudo-negation is said to be strict if it is continuous and strictly decreasing.*

It is clear that a fuzzy negation N is a fuzzy pseudo-negation N satisfying $N(0) = 1$. We call a fuzzy pseudo-negation N with $N(0) < 1$ a proper fuzzy pseudo-negation (or proper pseudo-negation).

Remark 3.2

(i) The least continuous fuzzy pseudo-negation is $N_\perp$ defined by $N_\perp(x) = 0$ for all $x \in [0,1]$. But there does not exist the greatest continuous fuzzy pseudo-negation.
(ii) Every continuous fuzzy pseudo-negation N with $N(0) > 0$ has a unique fixed point, i.e., there exists an $e \in (0,1)$ such that $N(e) = e$.

Proposition 3.3

(i) If N is a continuous fuzzy negation, then its pseudo-inverse $N^{(-1)}$ is a strictly decreasing fuzzy pseudo-negation.
(ii) If N is a continuous fuzzy pseudo-negation, then its pseudo-inverse $N^{(-1)}$ is a fuzzy negation.
(iii) Each proper fuzzy pseudo-negation cannot be involutive.

Example 3.4 (see [2, Remark 1.4.11]*).* Consider the following continuous negation:

$$N(x) = \begin{cases} -2x+1, & \text{if } x \in [0, \frac{1}{2}], \\ 0, & \text{otherwise.} \end{cases}$$

The pseudo-inverse $N^{(-1)}(x) = \frac{1-x}{2}$ for all $x \in [0,1]$ is a fuzzy pseudo-negation.

The next result, which is a representation theorem for strict fuzzy negations, is well known in the literature. We will generalize it to the case of strictly fuzzy pseudo-negations.

Theorem 3.5 (see [5, Theorem 6.2] **or** [2, Theorem 1.4.12]**).** *A function $N\colon [0,1] \to [0,1]$ is a strict negation if and only if there exist two order automorphisms φ and ψ of $[0,1]$ such that $N(x) = \psi^{-1}(1-\varphi(x))$ for all $x \in [0,1]$.*

Theorem 3.6. *A function $N\colon [0,1] \to [0,1]$ is a strict fuzzy pseudo-negation if and only if there exist an order automorphism φ of $[0,1]$ and order isomorphism $\psi\colon [0,1] \to [0, N(0)]$ such that*

$$N(x) = \psi(1-\varphi(x)), \qquad \textit{for all } x \in [0,1]. \tag{1}$$

Proof. If φ is an order automorphism of $[0,1]$ and ψ is an order isomorphism from $[0,1]$ to $[0, N(0)]$, then the N defined by (1) is obviously a strict fuzzy pseudo-negation.

Conversely, suppose that N is a strict fuzzy pseudo-negation and write $a = N(0)$. Letting $\varphi_0 \colon [0,a] \to [0,1]$ be an order isomorphism, then $\varphi_0 \circ N \colon [0,1] \to [0,1]$ is a strict negation. It follows from Lemma 3.5 that there exist two automorphisms φ and ψ_0 of $[0,1]$ such that $\varphi_0 \circ N(x) = \psi_0(1-\varphi(x))$ for all $x \in [0,1]$. Denoting $\psi = \varphi_0^{-1} \circ \psi_0$, we get $N(x) = \varphi_0^{-1} \circ \psi_0(1-\varphi(x)) = \psi(1-\varphi(x))$ for all $x \in [0,1]$ and $\psi \colon [0,1] \to [0,a]$ is obviously an order isomorphism. □

3.2 Natural Pseudo-negations of Fuzzy Boundary Weak Implications

Proposition 3.7. *If I is a fuzzy bw-implication, then the function $N_I \colon [0,1] \to [0,1]$ defined by*

$$N_I(x) = I(x,0), \qquad \text{for all } x \in [0,1] \tag{2}$$

is a fuzzy pseudo-negation. We call it the natural fuzzy pseudo-negation (or natural pseudo-negation) of I.

Proof. N_I is decreasing because I is decreasing in its first variable. From the definition of fuzzy bw-implications we know that $N_I(1) = I(1,0) = 0$. □

It is noted that for any fuzzy coimplication I, its natural paseudo-negation N_I is the least fuzzy pseudo-negation $N_\perp$ because $N_I(0) = I(0,0) = 0$.

Example 3.8. Consider the bw-implications $I_1 - I_8$ in Example 2.2, the natural pseudo-negations of $I_i (i = 1, 2, \cdots, 8)$ are as follows:

$$N_{I_i}(x) = 0, \qquad (i = 1, \cdots, 6) \quad \text{for all } x \in [0,1],$$

$$N_{I_7}(x) = \begin{cases} \frac{1}{2} - x, & \text{if } x \le \frac{1}{2}, \\ 0, & \text{otherwise}, \end{cases} \qquad N_{I_8}(x) = \begin{cases} \frac{1}{4} - x^2, & \text{if } x \le \frac{1}{2}, \\ 0, & \text{otherwise}. \end{cases}$$

Definition 3.9. *Let I be a fuzzy bw-implication and N be a fuzzy pseudo-negation. Then I is said to satisfy*

- CP(N) *(contrapositive principle w.r.t. N) if $I(x,y) = I(N(y), N(x))$ for all $x, y \in [0,1]$,*
- L – CP(N) *(left contrapositive principle w.r.t. N) if $I(N(x),y) = I(N(y),x)$ for all $x, y \in [0,1]$,*
- R – CP(N) *(right contrapositive principle w.r.t. N) if $I(x, N(y)) = I(y, N(x))$ for all $x, y \in [0,1]$.*

Example 3.10. There exist proper fuzzy bw-implications satisfying CP(N) with some fuzzy negation N. Indeed, let us define a mapping $I \colon [0,1]^2 \to [0,1]$ as follows:

$$I(x,y) = \begin{cases} 1, & \text{if } (x,y) = (0,1), \\ 0, & \text{if } (x,y) = (1,0), \\ k, & \text{otherwise}, \end{cases}$$

where $k \in (0,1)$ is a constant. Obviously, I is a proper fuzzy bw-implication and it satisfies CP(N) for any strictly decreasing fuzzy negation N.

Proposition 3.11. *If $I\colon [0,1]^2 \to [0,1]$ is any function and N is a strict fuzzy pseudo-negation, then the following statements are true.*

(i) If I satisfies $\mathrm{L-CP(N^{(-1)})}$*, then I satisfies* $\mathrm{R-CP(N)}$.
(ii) If I satisfies $\mathrm{R-CP(N^{(-1)})}$*, then I satisfies* $\mathrm{L-CP(N)}$.

Proposition 3.12. *If $I\colon [0,1]^2 \to [0,1]$ is any function and N is a fuzzy pseudo-negation, then the following statements are true.*

(i) If I satisfies $\mathrm{NP_1}$ *and* $\mathrm{R-CP(N)}$*, then $N_I = N$.*
(ii) If I satisfies NP_1 and CP(N)*, then $N_I \circ N = id_{[0,1]}$. Furthermore, if N is a proper fuzzy pseudo-negation, then $N_I \neq N$.*
(iii) If I satisfies CP(N) *and $N_I \circ N = id_{[0,1]}$, then I satisfies* $\mathrm{NP_1}$.
(iv) If $I \in bw\mathcal{FI}$ and I satisfies $\mathrm{NP_1}$ *and* CP(N)*, then $I \in \mathcal{FI}$.*

4 Fuzzy Boundary Weak Implications from Aggregation Operators

In a similar way of constructing fuzzy implications from fuzzy logical connectives, we investigate in this section how to construct fuzzy bw-implications by means of aggregation operators and fuzzy pseudo-negations.

4.1 (A, N)–Fuzzy Boundary Weak Implications

In this paper we will use only binary aggregation operators.

Definition 4.1 (see [8]). *A function $A\colon [0,1]^2 \to [0,1]$ is called an aggregation operator, if it is increasing and satisfies $A(0,0) = 0$ and $A(1,1) = 1$. An aggregation operator A is said to be conjunctive if $A \leq \min$ and disjunctive if $A \geq \max$.*

Definition 4.2 (see [13]). *A function $U\colon [0,1]^2 \to [0,1]$ is called a uninorm, if it is commutative, associative and increasing, and has a neutral element $e \in [0,1]$, i.e., $U(e,x) = U(x,e) = x$ for all $x \in [0,1]$.*

Each uninorm U is an aggregation operator, and $U(0,1) \in \{0,1\}$. Uninorm U is said to be conjunctive if $U(0,1) = 0$, and disjunctive if $U(0,1) = 1$. From this we know that each uninorm is either conjunctive or disjunctive. But, it is noted that the structure of uninorms (see [6]) follows that any proper uninorm is neither a conjunctive aggregation operator nor a disjunctive aggregation operator.

Definition 4.3. *Let A be an aggregation operator and N be a fuzzy pseudo-negation. We call the function $I_{A,N}\colon [0,1]^2 \to [0,1]$ defined by*

$$I_{A,N}(x,y) = A(N(x),y), \qquad \text{for all } x,y \in [0,1], \tag{3}$$

an (A,N)-operator. An (A,N)-operator is called a fuzzy boundary weak (A,N)-implication (shortly (A,N)-bwimplication) if it is a fuzzy bw-implication.

The set of all (A, N)–bwimplications is denoted by $bw\mathcal{FI}_{A,N}$. It is clear that $bw\mathcal{FI}_{A,N} \subseteq bw\mathcal{FI}$.

Remark 4.4. Please note that the standard duality between aggregation functions and implications is given by $I(x, y) = n(A(x, n(y)))$. We prefer to use the formula $I(x, y) = A(n(x), y)$ for introducing link between aggregation functions and bw-implications, but it ic clear that the both equations are connected through the duality of aggregation functions.

Proposition 4.5. *Let A be an aggregation operator and N be a fuzzy pseudo-negation. Then the following statements are true.*

(i) If N is a fuzzy negation, then $I_{A,N} \in bw\mathcal{FI}_{A,N}$.
(ii) If A is disjunctive, then $I_{A,N} \in bw\mathcal{FI}_{A,N}$.
(iii) If N is a fuzzy negation and A is conjunctive, then $I_{A,N} \in bw\mathcal{FI}_{A,N} \cap co\mathcal{FI}$.
(iv) If N is a fuzzy negation and A is disjunctive, then $I_{A,N} \in bw\mathcal{FI}_{A,N} \cap \mathcal{FI}$.

For the next result the proof is same as that of (iii) and (iv) in Proposition 4.5.

Proposition 4.6. *Let $U\colon [0,1]^2 \to [0,1]$ be a uninorm and $N\colon [0,1] \to [0,1]$ a fuzzy negation. Then $I_{U,N} \in co\mathcal{FI} \cup \mathcal{FI}$.*

The following proposition shows that the bw-implications generated from uninorms and fuzzy negations cannot be proper bw-implications.

Proposition 4.7. *Let A be an aggregation operator and N be a fuzzy pseudo-negation. Then the following statements are true.*

(i) If 0 is the left neutral element of A, i.e., $A(0, y) = y$ for all $y \in [0,1]$, then $I_{A,N}$ satisfies NP$_1$.
(ii) If A is associative and commutative, then $I_{A,N}$ satisfies EP.
(iii) If 0 is the right neutral element of A, i.e., $A(x, 0) = x$ for all $x \in [0,1]$, then $N_{I_{A,N}} = N$.
(iv) If A is commutative and N is a strong negation, then $I_{A,N}$ satisfies CP(N) *w.r.t. N.*
(v) If A is commutative, then $I_{A,N}$ satisfies R − CP(N) *w.r.t. N.*

The following proposition shows that an aggregation operator A can be constructed by a bw-implication and a fuzzy negation.

Proposition 4.8. *If I is a fuzzy bw-implication and N is a fuzzy negation, then the following function*

$$A(x, y) = I(N(x), y), \qquad \text{for all } x, y \in [0,1] \tag{4}$$

is an aggregation operator.

Proof. It obviously follows from (4) that A is increasing in each variable, $A(0,0) = I(N(0), 0) = I(1, 0) = 0$ and $A(1,1) = I(N(1), 1) = I(0, 1) = 1$. □

It is clear that A defined by (4) satisfies $A(0,1) = A(1,0) = 1$ if $I \in \mathcal{FI}$ and $A(0,1) = A(1,0) = 0$ if $I \in co\mathcal{FI}$.

4.2 *RA*–Fuzzy Boundary Weak Implications

Definition 4.9. *Let A be an aggregation operator. We call the function $I_{RA}\colon [0,1]^2 \to [0,1]$ defined by*

$$I_{RA}(x,y) = \sup\{z \in [0,1] | A(x,z) \le y\}, \qquad \text{for all } x,y \in [0,1] \tag{5}$$

a residual operator generated from A (shortly, RA–operator). If I_{RA} is a fuzzy bw-implication, we call it a residual bw-implication generated from A (shortly, RA–bwimplication). We denote all of the RA–bwimplications by $bw\mathcal{FI}_{RA}$.

Now, we investigate the conditions when that RA–operators defined as above are also RA–bwimplications.

Proposition 4.10. *Let $A\colon [0,1]^2 \to [0,1]$ be an aggregation operator. Then the RA–operator I_{RA} defined by (5) is a fuzzy bw-implication if and only if $A(1,y) > 0$ for all $0 < y < 1$.*

Proof. Since A is an aggregation operator, it follows from (5) that I_{RA} is decreasing in its first variable and increasing in its second one, and $I_{RA}(0,1) = \sup\{z \in [0,1] | A(0,z) \le 1\} = 1$. Furthermore, $I_{RA}(1,0) = 0$ if and only if $A(1,y) > 0$ for all $0 < y < 1$ (see [12, Proposition 3.3]). □

In fact, the formula (5) implies such a straight result $I_{RA}(x,1) = 1$ for all $x \in [0,1]$. The following example shows that there are RA–bwimplications which are proper fuzzy bw-implications.

Example 4.11. Let us consider the following uninorm:

$$U(x,y) = \begin{cases} eT(\frac{x}{e}, \frac{y}{e}), & \text{if } (x,y) \in [0,e]^2, \\ e + (1-e)S(\frac{x-e}{1-e}, \frac{y-e}{1-e}), & \text{if } (x,y) \in [e,1]^2, \\ \max(x,y), & \text{otherwise}, \end{cases}$$

where $e \in (0,1)$, T is a t-norm and S is a t-conorm. It is obvious that $U(1,y) = 1$ for all $y \in [0,1]$. Then from the previous proposition we know that I_{RU} defined by (5) is a RU–bwimplication. But I_{RU} is neither a fuzzy implication nor a fuzzy coimplication because $I_{RU}(0,0) = \sup\{t \in [0,1] | U(0,t) \le 0\} = e \notin \{0,1\}$.

About the basic properties of I_{RA}, we have the following results from Propositions 3.5, 3.11, 3.17 and 3.21 in [12].

Proposition 4.12 ([12])**.** *Let $A\colon [0,1]^2 \to [0,1]$ be an aggregation operator. Then we have the following statements.*

(i) If 1 is a left neutral element of A, i.e., $A(1,y) = y$ for all $y \in [0,1]$, then I_{RA} satisfies NP_1.
(ii) If A is left-continuous, commutative and associative, then I_{RA} satisfies EP.
(iii) If A is continuous in its first variable, then I_{RA} is continuous if and only if the function $N_A(x) = \sup\{z \in [0,1] | A(x,z) = 0\}$ is continuous and A is strictly increasing in its second variable in the area $\{(x,y) | A(x,y) > 0\}$.
(iv) Let A be a left-continuous t-subnorm. Then I_{RA} satisfies CP(N) w.r.t. N_I under the following conditions: $N_A(x) = \sup\{z \in [0,1] | A(x,z) = 0\}$ is a continuous bijection and $A(1,x) > 0$ for all $x > 0$.

5 Fuzzy Boundary Weak Implications from Generator Functions

Similarly to the discussion of Yager's f– and g–generated implications (see [1, 14]), we define and discuss in this section the bw-implications generated by unary monotonic functions.

5.1 f–Generated Bw-implications

Definition 5.1. *Let $f\colon [0,1] \to [-\infty,+\infty]$ be a strictly decreasing and continuous function such that $f(0)\cdot f(1) \le 0$. The function $I_f\colon [0,1]^2 \to [0,1]$ defined by*

$$I_f(x,y) = f^{-1}(xf(y)), \qquad \text{for all } x,y \in [0,1], \tag{6}$$

with the convention $0\cdot(-\infty) = -\infty$, is called an f–generated operator and the function f is called an f–generator of I_f. If I_f is a fuzzy bw-implication, we call it an f–generated bw-implication. We denote by $bw\mathcal{FI}_f$ the family of all f–generated bw-implications. Obviously, we have $bw\mathcal{FI}_f \subseteq bw\mathcal{FI}$.

It is noted that the formula (6) is reasonable because $f(1) \le xf(y) \le f(0)$ for all $x, y \in [0,1]$.

Proposition 5.2. *Let f be an f–generator of I_f defined by (6). Then the following statements are true.*

(i) If $f(1) = -\infty$, then $I_f \in bw\mathcal{FI}_f \setminus co\mathcal{FI}$.
(ii) If $f(1) > -\infty$, then $I_f \in bw\mathcal{FI}_f$ if and only if $f(1) = 0$. Furthermore, if $f(1) = 0$, the I_f with the convention $0\cdot(+\infty) = 0$ is Yager's f–generated implication and hence it is a fuzzy implication in this case.

This proposition shows that $I_f \in bw\mathcal{FI}$ only if $f(1) = 0$ or $-\infty$. Thus, for any $I_f \in bw\mathcal{FI}$, it holds that $I_f(x,1) = 1$ for all $x \in [0,1]$ and hence $bw\mathcal{FI}_f \cap co\mathcal{FI} = \emptyset$.

Proposition 5.3. *Let f be an f–generator of I_f defined by (6). Then the following statements are true.*

(i) The mapping $N_{I_f}\colon [0,1] \to [0,1]$ defined by $N_{I_f}(x) = I_f(x,0)$ for all $x \in [0,1]$ is a fuzzy pseudo-negation. We call it the natural pseudo-negation of operator I_f.
(ii) The natural pseudo-negation N_{I_f} is strict if and only if $f(0) < +\infty$.

Considering Propositions 2.7 and 2.8, we now only investigate the properties NP_1, EP, CO, CB_1 and CP(N) for a bw-implication I_f.

Proposition 5.4. *Let f be an f–generator of I_f defined by (6). Then the following statements are true.*

(i) I_f satisfies NP_1 and EP.
(ii) If $f(0) < +\infty$, then I_f is continuous. Conversely, if I_f is continuous, then $f(0) < +\infty$ with the convention $0\cdot(+\infty) = 0$.
(iii) I_f satisfies CB_1.
(iv) If $N\colon [0,1] \to [0,1]$ is a strong negation and $f = N$, then I_f satisfies CP(N).

5.2 g–Generated Bw-implications

Definition 5.5. *Let* $g\colon [0,1] \to [-\infty, +\infty]$ *be a strictly increasing and continuous function with* $g(0) \le 0$ *and* $g(1) \ge 0$. *The function* $I_g\colon [0,1]^2 \to [0,1]$ *defined by*

$$I_g(x,y) = g^{(-1)}(\frac{1}{x}g(y)), \qquad \text{for all } x,y \in [0,1], \tag{7}$$

with the convention $\frac{1}{x} = +\infty$ *and* $(+\infty)\cdot 0 = +\infty$, *is called a* g*–generated operator, where*

$$g^{(-1)}(x) = \begin{cases} g^{-1}(x), & \text{if } x \in [g(0), g(1)] \\ 0, & \text{if } x \in [-\infty, g(0)] \\ 1, & \text{if } x \in [g(1), +\infty] \end{cases} = g^{-1}(\max(\min(x, g(1)), g(0))).$$

We call the function g *a* g*–generator of* I_g. *If* I_g *is a fuzzy bw-implication, we call it a* g*–generated bw-implication. We denote by* $bw\mathcal{FI}_g$ *the family of all* g*–generated bw-implications. Obviously, we have* $bw\mathcal{FI}_g \subseteq bw\mathcal{FI}$.

It is noted that the g–generated operators are actually continuous additive generators of a special class of representable semi-uninorms (see [10,11]).

Proposition 5.6. *Let* g *be a* g*–generator of* I_g *defined by (7). Then the following statements are true.*

(i) $I_g \in bw\mathcal{FI}$.
(ii) $I_g \in \mathcal{FI}$ *if and only if* $g(0) = 0$.

It follows from the proof of Proposition 5.6 (ii) that $I_g(x,1) = 1$ for all $x \in [0,1]$. Therefore, we have $bw\mathcal{FI}_g \cap co\mathcal{FI} = \emptyset$.

Proposition 5.7. *Let* g *be a* g*–generator of* I_g *defined by (7). We define the mapping* $N_{I_g}\colon [0,1] \to [0,1]$ *by* $N_{I_g}(x) = I_g(x,0)$ *for all* $x \in [0,1]$. *Then we have the following statements.*

(i) N_{I_g} *is a fuzzy pseudo-negation.*
(ii) N_{I_g} *is a fuzzy negation if and only if* $g(0) = 0$. *Further,* N_{I_g} *is of the following form in this case:*

$$N_{I_g}(x) = \begin{cases} 1, & \text{if } x = 0, \\ 0, & \text{otherwise.} \end{cases}$$

Our last result is the following.

Proposition 5.8. *Let* g *be a* g*–generator of* I_g *defined by (7). Then the following statements are true.*

(i) I_g *satisfies* NP_1.
(ii) I_g *satisfies* EP.
(iii) If $g(0) = 0$, *then* I_g *is continuous except at the point* $(0,0)$, *and if* $g(0) \neq 0$, *then* I_g *is continuous except at the point* $(0, y_0)$, *where* $y_0 \in (0,1]$ *and* $g(y_0) = 0$.
(iii) I_g *does not satisfy* CP(N) *with any fuzzy pseudo-negation* N.

6 Future Work

In this paper, we have introduced the concept of fuzzy bw-implications and we have investigated some of their basic properties and their natural pseudo-negations. In the future work, we will investigate other properties for this kind of operators such as the law of importation, the T-conditionality, the distributivity equations and intersections between different families of fuzzy bw-implications.

Acknowledgment. The work on this paper for Hua-Wen Liu was supported by the National Natural Science Foundation of China (No. 61573211). The work on this paper for Michał Baczyński was supported by the National Science Centre, Poland, under Grant No. 2015/19/B/ST6/03259.

References

1. Baczyński, M., Jayaram, B.: Yager's classes of fuzzy implications: some properties and intersections. Kybernetika **43**, 157–182 (2007)
2. Baczyński, M., Jayaram, B.: Fuzzy Implications. Studies in Fuzziness and Soft Computing, vol. 231. Springer, Heidelberg (2008). https://doi.org/10.1007/978-3-540-69082-5
3. Cordón, O., Herrera, F., Peregrin, A.: Applicability of the fuzzy operators in the design of fuzzy logic controllers. Fuzzy Sets Syst. **86**, 15–41 (1997). https://doi.org/10.1016/0165-0114(95)00367-3
4. De Baets, B.: Coimplicators, the forgotten connectives. Tatra Mt. Math. Publ. **12**, 229–240 (1997)
5. Fodor, J.C.: A new look at fuzzy connectives. Fuzzy Sets Syst. **57**, 141–148 (1993). https://doi.org/10.1016/0165-0114(93)90153-9
6. Fodor, J.C., Yager, R.R., Rybalov, A.: Structure of uninorms. Int. J. Uncertain. Fuzziness Knowl.-Based Syst. **5**, 411–427 (1997). https://doi.org/10.1142/S0218488597000312
7. Fodor, J., Roubens, M.: Fuzzy Preference Modelling and Multicriteria Decision Support. Kluwer, Dordrecht (1994). https://doi.org/10.1007/978-94-017-1648-2
8. Grabisch, M., Marichal, J.-L., Mesiar, R., Pap, E.: Aggregation Functions. Cambridge University Press, New York (2009). https://doi.org/10.1017/CBO9781139644150
9. Klement, E.P., Mesiar, R., Pap, E.: Triangular Norms. Kluwer, Dordrecht (2000). https://doi.org/10.1007/978-94-015-9540-7
10. Liu, H.W.: Semi-uninorms and implications on a complete lattice. Fuzzy Sets Syst. **191**, 72–82 (2012). https://doi.org/10.1016/j.fss.2011.08.010
11. Liu, H.W.: Distributivity and conditional distributivity of semi-uninorms over continuous t-conorms and t-norms. Fuzzy Sets Syst. **268**, 27–43 (2015). https://doi.org/10.1016/j.fss.2014.07.025
12. Ouyang, Y.: On fuzzy implications determined by aggregation operators. Inform. Sci. **193**, 153–162 (2012). https://doi.org/10.1016/j.ins.2012.01.001
13. Yager, R.R., Rybalov, A.: Uninorm aggregation operators. Fuzzy Sets and Syst. **80**, 111–120 (1996). https://doi.org/10.1016/0165-0114(95)00133-6
14. Yager, R.R.: On some new classes of implication operators and their role in approximate reasoning. Inform. Sci. **167**, 193–216 (2004). https://doi.org/10.1016/j.ins.2003.04.001

On Linear and Quadratic Constructions of Fuzzy Implication Functions

Sebastia Massanet[1,2(✉)], Juan Vicente Riera[1,2], and Joan Torrens[1,2]

[1] Soft Computing, Image Processing and Aggregation (SCOPIA) Research Group, Department of Mathematics and Computer Science, University of the Balearic Islands, 07122 Palma, Spain
{s.massanet,jvicente.riera,jts224}@uib.es

[2] Balearic Islands Health Research Institute (IdISBa), 07010 Palma, Spain

Abstract. In this paper a new construction method of fuzzy implication functions from a given one, based on ternary polynomial functions is presented. It is proved that the case of linear polynomial functions leads only to trivial solutions and thus the quadratic case is studied in depth. It is shown that the quadratic method allows many different possibilities depending on the usual properties of fuzzy implications functions that we want to preserve. Specifically, there are infinitely many quadratic functions that transform fuzzy implication functions satisfying properties like the neutrality principle, the identity principle, or the law of contraposition with respect to the classical negation, into new fuzzy implication functions satisfying them.

Keywords: Fuzzy implication function · Linear construction Quadratic construction

1 Introduction

One of the most important fuzzy logical operations in fuzzy sets and fuzzy logic are the so-called fuzzy implication functions. The importance of these operations relies into the fact that they are used to model fuzzy conditionals and also in the inference processes. Thus, fuzzy implication functions have a great quantity of applications in approximate reasoning and fuzzy control, but also in many other application fields (see [4,5,8]).

From the theoretical point of view one of the main topics concerning fuzzy implication functions is to find out new construction methods, that is, some methodologies that allow us to construct a new fuzzy implication function from one or two given ones. There are many construction methods known in the literature such as the classical ones like the conjugation, the reciprocation, the upper, lower and medium contrapositivisations, and the maximum, minimum or convex combinations of fuzzy implication functions (see [4]). More recently, other methods have appeared like some new types of contrapositivisations [1], horizontal and vertical threshold generation methods [9–11], the FNI-method

J. Medina et al. (Eds.): IPMU 2018, CCIS 853, pp. 623–635, 2018.
https://doi.org/10.1007/978-3-319-91473-2_53

[2,13], the star-composition [15], etc. (see also the book [3] and specially the chapter [12] for more details).

The interest in constructing new fuzzy implication functions comes from the necessity to have as many models as possible to manage fuzzy conditionals trying to capture their meaning in each application (see [14]). There are two main important points in these construction methods of fuzzy implication functions:

- On the one hand, they should preserve as many properties as possible, that is, if the initial fuzzy implication function I satisfies some properties the new constructed implication function should satisfy also such properties.
- On the other hand, the proper construction method should be as simple as possible with the aim that the new obtained fuzzy implication function has an expression easy to compute and to implement.

In this line, this paper presents a very easy construction method (based on quadratic functions of three variables) that will preserve many of the most usual properties of fuzzy implication functions, including the contraposition with respect to the classical negation $N_c(x) = 1 - x$, which few of the already known construction methods preserve. The idea is to make use of linear or quadratic functions $F : \mathbb{R}^3 \to \mathbb{R}$ in such a way that, starting from a fuzzy implication function I and making the change $z = I(x, y)$, we obtain a new function $I_F(x, y) = F(x, y, I(x, y))$ which turns out to be a new fuzzy implication function. The process is the same that was used by Kolesárová and Mesiar in [7] to construct new aggregation functions from old ones and, specially, new semi-copulas and quasi-copulas. Moreover, the construction methods of fuzzy implication functions presented in this paper allow us to preserve many usual properties of this kind of logical operations.

The paper is organized as follows. Section 2 is devoted to some preliminaries in order to make the paper as self-contained as possible. Section 3 deals with the mentioned construction method based on the polynomial functions. The most simple polynomials which are linear functions are studied, but in this case, it is easy to see that only trivial solutions appear. Thus, Sect. 4 is devoted to the case of quadratic functions which lead to many possibilities depending on the specific properties that we want to preserve. Finally, the paper ends with Sect. 5 devoted to some conclusions and future work.

2 Preliminaries

Let us recall some concepts and results that will be used throughout this paper. First, we give the definition of fuzzy negation.

Definition 1 ([6, Definition 1.1]). *A non-increasing function $N : [0, 1] \to [0, 1]$ is a* fuzzy negation, *if $N(0) = 1$ and $N(1) = 0$. A fuzzy negation N is*

(i) strict, *if it is continuous and strictly decreasing.*
(ii) strong, *if it is an involution, i.e., $N(N(x)) = x$ for all $x \in [0, 1]$.*

There are many different examples of fuzzy negations being the classical negation given by $N_c(x) = 1 - x$ for all $x \in [0,1]$ one of the most important ones. Next, we recall the definition of fuzzy implication functions.

Definition 2 ([6, Definition 1.15]). *A binary operator* $I : [0,1]^2 \to [0,1]$ *is called a* fuzzy implication function, *if it satisfies:*

(I1) $I(x,z) \geq I(y,z)$ *when* $x \leq y$, *for all* $z \in [0,1]$.
(I2) $I(x,y) \leq I(x,z)$ *when* $y \leq z$, *for all* $x \in [0,1]$.
(I3) $I(0,0) = I(1,1) = 1$ *and* $I(1,0) = 0$.

Let us denote by $\mathcal{I}$ the class of all fuzzy implication functions. Note that from the definition, we can deduce that for all $I \in \mathcal{I}$, $I(0,x) = 1$ and $I(x,1) = 1$ for all $x \in [0,1]$, while the symmetric values $I(x,0)$ and $I(1,x)$ are not determined from the definition. Some additional properties of fuzzy implication functions which will be used in this work are:

- The *left neutrality principle,*

$$I(1,y) = y, \quad y \in [0,1]. \qquad \textbf{(NP)}$$

- The *identity principle,*

$$I(x,x) = 1, \quad x \in [0,1]. \qquad \textbf{(IP)}$$

- The *ordering property,*

$$x \leq y \Longleftrightarrow I(x,y) = 1, \quad x,y \in [0,1]. \qquad \textbf{(OP)}$$

- The *law of contraposition* with respect to a fuzzy negation N,

$$I(N(y),N(x)) = I(x,y), \quad x,y \in [0,1], \qquad \textbf{(CP(N))}$$

  and in particular with respect to the classical negation N_c,

$$I(1-y,1-x) = I(x,y), \quad x,y \in [0,1]. \qquad \textbf{(CP(N}_\textbf{c}\textbf{))}$$

- The *exchange principle,*

$$I(x,I(y,z)) = I(y,I(x,z)), \quad x,y,z \in [0,1]. \qquad \textbf{(EP)}$$

Definition 3. *Let* I *be a fuzzy implication function. The function* N_I *defined by* $N_I(x) = I(x,0)$ *for all* $x \in [0,1]$, *is called the natural negation of* I.

As we have already mentioned in the introduction, there exist different construction methods of new fuzzy implication functions from given ones. One of them that will be useful in the paper is the following.

Proposition 1 ([4]). *Let* I_1, I_2 *be two fuzzy implication functions and* $\lambda \in [0,1]$. *The binary function* $I : [0,1]^2 \to [0,1]$ *given by*

$$I(x,y) = (1-\lambda)I_1(x,y) + \lambda I_2(x,y) \qquad \text{for all } x,y \in [0,1] \tag{1}$$

is always a fuzzy implication function.

Fuzzy implication functions constructed as in Eq. (1) are called *convex linear combinations* of I_1 and I_2. Moreover, it is well known that this construction method preserves **(NP)**, **(IP)**, **(OP)** and **(CP(**N_c**))**, but it does not preserve in general **(EP)**.

3 Construction of Fuzzy Implication Functions from Ternary Functions

Let us describe in this section the general framework for constructing new fuzzy implication functions we want to develop. Consider a ternary function $F : \mathbb{R}^3 \to \mathbb{R}$ and any fuzzy implication function $I \in \mathcal{I}$. For all $x, y \in [0,1]$ let us take $z = I(x,y)$ and let us define the function $I_F : [0,1]^2 \to \mathbb{R}$ given by

$$I_F(x,y) = F(x,y,I(x,y)) \qquad \text{for all } x,y \in [0,1]. \tag{2}$$

Although x, y and $I(x,y)$ are values in $[0,1]$, the resulting value $F(x,y,I(x,y))$ could be out of $[0,1]$. For instance, let us consider the Rescher implication

$$I_{\mathbf{RS}}(x,y) = \begin{cases} 1 & \text{if } x \le y, \\ 0 & \text{if } x > y, \end{cases}$$

and the ternary function $F(x,y,z) = x + y + z$. Thus, if we take $x = y = 1$ it follows $F(1,1,I_{\mathbf{RS}}(1,1)) = 3 \notin [0,1]$. However, we want to search for such ternary functions F that I_F not only take values in $[0,1]$, but turns to be also a fuzzy implication function. The set of functions $F : \mathbb{R}^3 \to \mathbb{R}$ satisfying this property is obviously not empty because it is clear that the projection on the third variable $F(x,y,z) = z$, which yields the original fuzzy implication function I, belongs to this set.

Moreover, since we are interested in functions F as simple as possible, we want to begin our study with ternary function $F : \mathbb{R}^3 \to \mathbb{R}$ given by polynomial expressions. Note that in this case, the trivial solution given before is not the only one as the following example shows.

Example 1. Let us give some examples of ternary polynomial functions F that transform fuzzy implication functions I into new fuzzy implication functions I_F.

1. Consider the function $F : \mathbb{R}^3 \to \mathbb{R}$ given by $F(x,y,z) = z^2$. For all $x,y,z \in [0,1]$ it holds that $z^2 \in [0,1]$ and then the function I_F is given by $I_F(x,y) = I(x,y)^2$ for all $x,y \in [0,1]$. It is easy to see that for any fuzzy implication function I the corresponding I_F is always a fuzzy implication function obtaining in this way an easy construction method. Moreover, note that in this case the construction I_F preserves **(IP)**, **(OP)** and **(CP(N))**, but it never preserves **(NP)**.
2. We can generalize the previous quadratic example to polynomial functions of degree n for all natural number n. That is, consider the family of functions $F_n(x,y,z) = z^n$ for all $n \ge 1$. This family leads to the construction methods given by
$$I_{F_n}(x,y) = I(x,y)^n \qquad \text{for all } x,y \in [0,1].$$
As above, all these constructions preserve **(IP)**, **(OP)** and **(CP(N))**, but they do not preserve **(NP)** for all $n \ge 2$. Moreover, we obtain in this way a sequence of fuzzy implication functions $(I_{F_n})_n$ starting from the given fuzzy

implication function $I_{F_1} = I$ and such that the limit is the least fuzzy implication function with the same 1-region than I. That is

$$\lim_{n\to\infty} I_{F_n}(x,y) = \begin{cases} 1 & \text{if } I(x,y) = 1, \\ 0 & \text{otherwise.} \end{cases}$$

Let us begin our study with the most simple polynomial functions: linear functions, i.e., functions $F : \mathbb{R}^3 \to \mathbb{R}$ such that

$$F(x,y,z) = ax + by + cz + d \tag{3}$$

with $a, b, c, d \in \mathbb{R}$. Unfortunately, in this simple case the trivial solution mentioned above given by the projection is the only solution as it is stated in the following proposition.

Proposition 2. *Let $F : \mathbb{R}^3 \to \mathbb{R}$ be a linear function of the form* (3)*. Then the following statements are equivalent:*

(i) For each I in $\mathcal{I}$ the function I_F is also in $\mathcal{I}$.
(ii) The function F is given by the projection $F(x,y,z) = z$, that is, $a = b = d = 0$ and $c = 1$, and therefore I_F is the proper I.

In other words, assuming F is linear, $I_F \in \mathcal{I}$ for each I in $\mathcal{I}$ if and only if F is defined as the natural projection on the third variable and then no new fuzzy implication functions appear. Thus, it is clear that for our objective we need to consider polynomial functions of higher level and we do it for quadratic functions in next section.

4 Quadratic Constructions of Fuzzy Implication Functions

We want to investigate constructions of fuzzy implication functions based on quadratic functions $F : \mathbb{R}^3 \to \mathbb{R}$ in a similar way it was done in [7] for aggregation functions. These functions F are expressed by

$$F(x,y,z) = ax^2 + by^2 + cz^2 + dxy + exz + fyz + gx + hy + iz + j, \tag{4}$$

where coefficients $a, b, c, d, e, f, g, h, i, j$ are in $\mathbb{R}$. First, we will investigate the construction of general fuzzy implication functions. In this sense, the first condition is that the function I_F for any $I \in \mathcal{I}$ must satisfy the boundary conditions inherent to fuzzy implication functions.

Proposition 3. *Let $F : \mathbb{R}^3 \to \mathbb{R}$ be a quadratic function of the form* (4) *and $I : [0,1]^2 \to [0,1]$ a fuzzy implication function. Then the following statements are equivalent.*

(i) I_F *fulfils boundary conditions* **(I3)** *and* $I_F(0,x) = I_F(x,1) = 1$ *for all* $x \in [0,1]$.
(ii) $a = b = 0, e = -(d+g), h = -f, j = -g,$ *and* $i = 1 - c + g$.

Thus, according to the previous result we see that in order to investigate quadratic functions F such that I_F is a fuzzy implication function it is necessary to consider functions F of the form

$$F(x,y,z) = cz^2 + dxy - (d+g)xz + fyz + gx - fy + (1-c+g)z - g \quad (5)$$

with $c, d, g, f \in \mathbb{R}$. It is clear that not all such functions accomplish our goal since I_F must also satisfy the monotonicities. Next example shows this fact.

Example 2. Take for instance $c = g = 1$, $f = -1$ and $d = 0$ in Eq. (5) and consider the corresponding function F that is given by

$$F(x,y,z) = z^2 - xz - yz + x + y + z - 1.$$

Consider I as the Gödel implication function which is given by

$$I_{\mathbf{GD}}(x,y) = \begin{cases} 1 & \text{if } x \le y, \\ y & \text{if } x > y. \end{cases}$$

Then the corresponding I_F satisfies

$$I_F(x,0) = F(x,0,I_{\mathbf{GD}}(x,0)) = F(x,0,0) = x - 1$$

for all $x > 0$. Thus, I_F is not a fuzzy implication function not only because I_F does not fulfil the decreasingness with respect to the first variable, but even because it takes negative values.

To ensure monotonicities, the four parameters involved in functions F of the form (5) must satisfy some complex additional properties quite difficult to manage in practice. Since we want to deal with constructions that also preserve some of the usual properties of fuzzy implication functions recalled in the preliminaries, we next investigate which parameter values are adequate in order to preserve each one of these properties.

Proposition 4. *Let* $F : \mathbb{R}^3 \to \mathbb{R}$ *be a quadratic function of the form* (5) *and* $I : [0,1]^2 \to [0,1]$ *a fuzzy implication function. Then the next items hold:*

(i) If I *satisfies* **(NP)**, *then* I_F *fulfils* **(NP)** *if and only if* $f + c = 0$.
(ii) If I *satisfies* **(IP)**, *then* I_F *fulfils* **(IP)** *if and only if* $d = 0$.
(iii) If I *satisfies* $(\mathbf{CP}(N_c))$, *then* I_F *fulfils* $(\mathbf{CP}(N_c))$ *if and only if* $d+g-f = 0$.

Note that each condition in Proposition 4 translated to Eq. (5) leads to a family of quadratic polynomial functions depending on three parameters. Thus, the discussion of the conditions that these parameters must satisfy to ensure that the corresponding I_F is a fuzzy implication function remains hard in practice.

For this reason our next steps will be to investigate the cases when two of the three conditions studied before are preserved (cases that will lead to functions F depending on only two parameters).

Each section will be devoted to each possible combination (or at least to some of the possible combinations due to space limitations). The first combination we want to deal with is the case of preserving both **(NP)** and **(IP)**. Then, we will focus our attention to the case of preserving both **(IP)** and **(CP(N_c))**.

4.1 Quadratic Constructions Preserving (NP) and (IP)

This subsection is devoted to the investigation of quadratic constructions of fuzzy implication functions preserving the identity principle **(IP)** and the left neutrality principle **(NP)**. From now on let us denote by $\mathcal{I}_{\mathbf{IP},\mathbf{NP}}$ the set of fuzzy implication functions that fulfil both properties.

First, using Proposition 4 the following result is straightforward.

Proposition 5. *Let $F : \mathbb{R}^3 \to \mathbb{R}$ be a quadratic function of the form* (5) *and I a fuzzy implication function such that $I \in \mathcal{I}_{\mathbf{IP},\mathbf{NP}}$. Then the following statements are equivalent:*

(i) I_F fulfils **(IP)** *and* **(NP)**.
(ii) $f = -c$ and $d = 0$.

Therefore, in such case, the function F depends now only on two parameters and it is given by the expression

$$F_{\alpha,\beta}(x, y, z) = \alpha z^2 + \beta xz - \alpha yz - \beta x + \alpha y + (1 - \alpha - \beta)z + \beta \tag{6}$$

where $\alpha, \beta \in \mathbb{R}$. However, this family of functions only guarantee that $I_{F_{\alpha,\beta}}$ satisfy the border conditions and **(IP)** and **(NP)**, but they can fail to be fuzzy implication functions.

Example 3. Take for instance $\alpha = 0$ and $\beta = 2$ in Eq. (6) and consider the corresponding function F that is given by $F(x, y, z) = 2xz - 2x - z + 2$. Consider I as the Fodor implication which is given by

$$I_{\mathbf{FD}}(x, y) = \begin{cases} 1 & \text{if } x \leq y, \\ \max\{1 - x, y\} & \text{if } x > y. \end{cases}$$

Then the corresponding I_F satisfies $I_F(x, 0) = -2x^2 + x + 1$ for all $x \in [0, 1]$. Thus, I_F is not a fuzzy implication function since it takes values greater than 1 and it does not fulfil the decreasingness in the first variable.

From the previous discussion, the next result provides a characterization of the quadratic constructions of fuzzy implication functions preserving **(IP)** and **(NP)**.

Theorem 1. *Let $F : \mathbb{R}^3 \to \mathbb{R}$ be a quadratic function of the form* (6). *Then the following statements are equivalent:*

(i) *For each* I *in* $\mathcal{I}_{\mathbf{IP},\mathbf{NP}}$*, the function* I_F *is also in* $\mathcal{I}_{\mathbf{IP},\mathbf{NP}}$*, that is,* I_F *is a fuzzy implication function that satisfies* **(IP)** *and* **(NP)***.*
(ii) $I_F = I_{\alpha,\beta}$*, where* $I_{\alpha,\beta}$ *is given by*

$$I_{\alpha,\beta}(x,y) = \alpha I(x,y)^2 + \beta x I(x,y) - \alpha y I(x,y) - \beta x + \alpha y + (1-\alpha-\beta) I(x,y) + \beta \tag{7}$$

with α*,* β *fulfilling the conditions* $0 \leq \alpha \leq 1$*,* $0 \leq \beta \leq 1-\alpha$ *and* $I \in \mathcal{I}$*.*

Figure 1 provides the graphical representation of the region of eligible parameter values. When $\alpha = \beta = 0$ we obtain the initial fuzzy implication function I.

Remark 1. Due to the convexity of the set of eligible parameter values P, each point $(\alpha, \beta) \in P$ is a convex combination of its vertices. The fuzzy implication function $I_{\alpha,\beta}$ corresponding to the point (α, β) can be expressed as the convex combination of fuzzy implication functions $I_{0,0}$, $I_{1,0}$ and $I_{0,1}$ corresponding to the vertices of P. As the family of all fuzzy implication functions is a convex set [4,5], $I_{\alpha,\beta}$ is a fuzzy implication function.

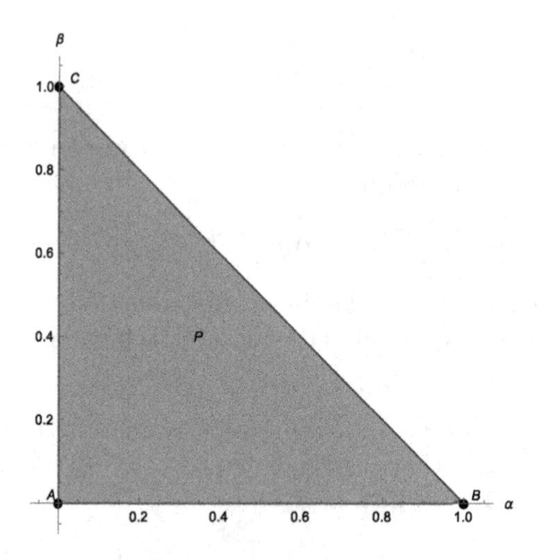

Fig. 1. The set P of all eligible pairs of coefficients (α, β) for the quadratic construction of fuzzy implication functions preserving **(IP)** and **(NP)**.

Next example shows the fuzzy implication functions obtained for each one of the three vertices of the triangle determined in Theorem 1 (and represented in Fig. 1) when we consider the Goguen implication.

Example 4. Let us consider the Goguen implication given by

$$I_{\mathbf{GG}}(x,y) = \begin{cases} 1 & \text{if } x \leq y, \\ \frac{y}{x} & \text{if } x > y, \end{cases}$$

and the quadratic construction methods given from taking the three vertices $(0,0),(1,0)$ and $(0,1)$ of the triangle P depicted in Fig. 1. According to Theorem 1, we obtain the following fuzzy implication functions:

(i) For the vertex $A=(0,0)$, we obtain the same Goguen implication.
(ii) For the vertex $B=(1,0)$, we obtain the fuzzy implication function

$$I_{1,0}(x,y)=I_{\mathbf{GG}}(x,y)^2-yI_{\mathbf{GG}}(x,y)+y=\begin{cases}1 & \text{if } x\leq y,\\ \frac{y^2-xy^2+x^2y}{x^2} & \text{if } x>y.\end{cases}$$

(iii) For the vertex $C=(0,1)$ we obtain the fuzzy implication function

$$I_{0,1}(x,y)=xI_{\mathbf{GG}}(x,y)-x+1=\min\{1,1-x+y\}$$

which is the Łukasiewicz implication $I_{\mathbf{LK}}$.

4.2 Quadratic Constructions Preserving (IP) and (CP(N_c))

To end this section, let us study which quadratic constructions of fuzzy implication functions preserve the identity principle **(IP)** and the contrapositive symmetry with respect to $N_c(x)=1-x$ **(CP(**N_c**))**. We will denote by $\mathcal{I}_{\mathbf{IP},\mathbf{CP}(\mathbf{N_c})}$ the set of fuzzy implication functions that fulfil the two considered properties.

Proposition 4 allows us to reduce the number of parameters of the quadratic function given by Eq. (5) to two parameters.

Proposition 6. *Let* $F:\mathbb{R}^3\to\mathbb{R}$ *be a quadratic function of the form* (5) *and* I *a fuzzy implication function such that* $I\in\mathcal{I}_{\mathbf{IP},\mathbf{CP}(\mathbf{N_c})}$. *Then the following statements are equivalent:*

(i) I_F *fulfils* **(IP)** *and* **(CP(**N_c**))**.
(ii) $f=-c$ *and* $g=-c-d$.

At this point, a function F satisfying the conditions provided by Proposition 6 is given by the expression

$$F_{\alpha,\beta}(x,y,z)=\alpha z^2+\beta xz-\beta yz-\beta x+\beta y+(1-\alpha-\beta)z+\beta \tag{8}$$

where $\alpha,\beta\in\mathbb{R}$. As in the study performed in Sect. 4.1, although $I_{F_{\alpha,\beta}}$ satisfies the border conditions and **(IP)** and **(CP(**N_c**))**, it is not always a fuzzy implication function in the sense of Definition 2.

Example 5. Consider $\alpha=0$ and $\beta=2$ in Eq. (8). This choice of parameter values yields the function F given by $F(x,y,z)=2xz-2yz-2x+2y-z+2$. Consider now the Łukasiewicz implication given in Example 4-(iii). Then the corresponding I_F satisfies $I_F(x,0)=1+x-2x^2$ for all $x\in[0,1]$. Thus, I_F is not a fuzzy implication function since this function is increasing for all $x\in(0,\frac{1}{4})$ taking values greater than 1.

To ensure that the quadratic construction method provides, in fact, a fuzzy implication function, the next result determines a region of eligible parameter values to achieve this fact.

Theorem 2. *Let $F : \mathbb{R}^3 \to \mathbb{R}$ be a quadratic function of the form* (8)*. For each I in $\mathcal{I}_{\mathbf{IP},\mathbf{CP}(\mathbf{N_c})}$, the function I_F is also in $\mathcal{I}_{\mathbf{IP},\mathbf{CP}(\mathbf{N_c})}$, that is, I_F is a fuzzy implication function that satisfies* **(IP)** *and* **(CP(N_c))** *if $I_F = I_{\alpha,\beta}$, where*

$$I_{\alpha,\beta}(x,y) = \alpha I(x,y)^2 + \beta x I(x,y) - \beta y I(x,y) - \beta x + \beta y + (1-\alpha-\beta)I(x,y) + \beta \quad (9)$$

with α, β such that either $-1 \le \alpha \le 0$ and $0 \le \beta \le \alpha + 1$ or $0 < \alpha \le 1$ and $0 \le \beta \le -\alpha + 1$.

Figure 2 provides the graphical representation of the region of eligible parameter values given by the previous theorem. Again, note that when $\alpha = \beta = 0$ we obtain the initial fuzzy implication function I.

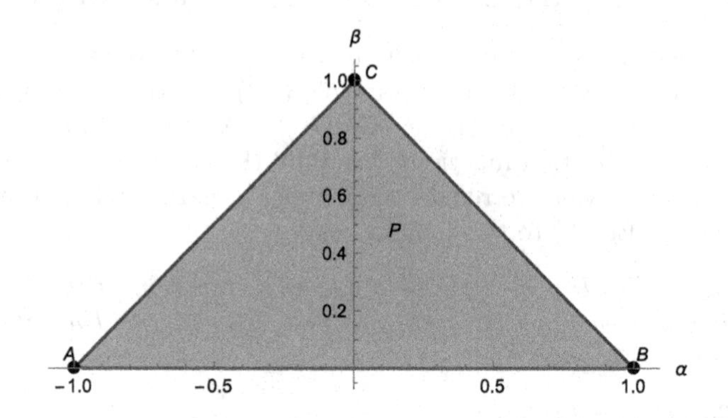

Fig. 2. A set P of eligible pairs of coefficients (α, β) for the quadratic construction of fuzzy implication functions preserving **(IP)** and **(CP(N_c))**.

Next example shows the fuzzy implication functions obtained for each one of the three vertices of the triangle determined in Theorem 2 (and represented in Fig. 2) when we consider the Łukasiewicz implication.

Example 6. Let us consider the Łukasiewicz implication given in Example 4-(iii) and the quadratic construction methods given from taking the three vertices $(-1,0),(1,0)$ and $(0,1)$ of the triangle P depicted in Fig. 2. According to Theorem 2, we obtain the following fuzzy implication functions:

(i) For the vertex $A = (-1,0)$, we obtain the fuzzy implication function

$$I_{-1,0}(x,y) = -I_{\mathbf{LK}}(x,y)^2 + 2I_{\mathbf{LK}}(x,y) = \begin{cases} 1 & \text{if } x \le y, \\ 1 - x^2 - y^2 + 2xy & \text{if } x > y. \end{cases}$$

(ii) For the vertex $B = (1, 0)$, we obtain the fuzzy implication function

$$I_{1,0}(x,y) = I_{\mathbf{LK}}(x,y)^2 = \begin{cases} 1 & \text{if } x \leq y, \\ 1 + x^2 + y^2 - 2x + 2y - 2xy & \text{if } x > y. \end{cases}$$

(iii) For the vertex $C = (0, 1)$, we obtain the fuzzy implication function $I_{0,1}(x,y) =$

$$xI_{\mathbf{LK}}(x,y) - yI_{\mathbf{LK}}(x,y) - x + y + 1 = \begin{cases} 1 & \text{if } x \leq y, \\ 1 + x^2 + y^2 - 2x + 2y - 2xy & \text{if } x > y, \end{cases}$$

the same fuzzy implication function obtained in the previous vertex.

Remark 2. Several remarks are worthy mentioning:

(i) In an analogous way to the case of the preservation of **(NP)** and **(IP)**, the fuzzy implication function $I_{\alpha,\beta}$ corresponding to the point (α, β) can be expressed as the convex combination of fuzzy implication functions $I_{-1,0}$, $I_{1,0}$ and $I_{0,1}$ yielded by the vertices of P.
(ii) Note that the point $C = (0, 1)$ yields the quadratic construction method $I_{0,1}(x,y) = I(x,y)^2$ which has been already discussed in Example 1-1. This is not unexpected since it is a construction method that preserves, among other properties, **(IP)** and **(CP(**N_c**))**.

Theorem 2 provides a whole region P of parameter values which are eligible to provide quadratic construction methods of fuzzy implication functions preserving **(IP)** and **(CP(**N_c**))**, but it is not a complete characterization since there could be more eligible parameter values. Indeed, we have the feeling that region P given in Fig. 2 could be expanded to the region P' shown in Fig. 3. To prove this fact is part of the subsequent work we want to address in the future.

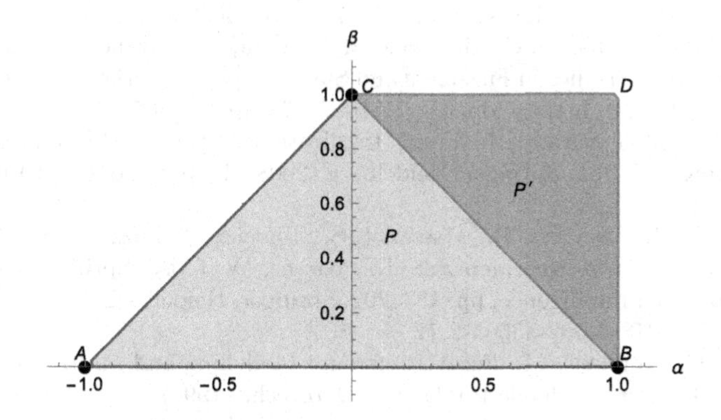

Fig. 3. Set P already described in Fig. 2 jointly with set P' of potential eligible pairs of coefficients (α, β) for the quadratic construction of fuzzy implication functions preserving **(IP)** and **(CP(**N_c**))**.

5 Conclusions and Future Work

In this paper, novel construction methods of fuzzy implication functions from a given one based on ternary polynomial functions are introduced and studied. Specifically, the linear and quadratic construction methods are analysed in depth. While the first one does not provide any new fuzzy implication functions, the quadratic method yields infinitely many different novel fuzzy implication functions. In order to cope with this situation, we have focused our efforts in the particular cases when the quadratic construction method preserves **(NP)** and **(IP)**, or **(IP)** and $(\mathbf{CP}(N_c))$. The former case is fully characterized, while some open questions remain in the latter one.

As a future work, we will focus on determining whether region P' given in Fig. 3 is also an eligible region to be chosen to generate quadratic construction methods of fuzzy implication functions preserving **(IP)** and $(\mathbf{CP}(N_c))$. Moreover, we have already started to study the case of the preservation of $(\mathbf{CP}(N_c))$ and **(NP)** with promising results. Many other combinations of two or more additional properties could be considered achieving new interesting construction methods of fuzzy implication functions.

Acknowledgments. This paper has been partially supported by the Spanish Grant TIN2016-75404-P, AEI/FEDER, UE.

References

1. Aguiló, I., Suñer, J., Torrens, J.: New types of contrapositivisation of fuzzy implications with respect to fuzzy negations. Inf. Sci. **322**, 223–236 (2015)
2. Aguiló, I., Suñer, J., Torrens, J.: A new look on fuzzy implication functions: *FNI*-implications. In: Carvalho, J.P., Lesot, M.-J., Kaymak, U., Vieira, S., Bouchon-Meunier, B., Yager, R.R. (eds.) IPMU 2016. CCIS, vol. 610, pp. 375–386. Springer, Cham (2016). https://doi.org/10.1007/978-3-319-40596-4_32
3. Baczyński, M., Beliakov, G., Bustince, H., Pradera, A.: Advances in Fuzzy Implication Functions. Studies in Fuzziness and Soft Computing Series, vol. 300. Springer, Heidelberg (2013). https://doi.org/10.1007/978-3-642-35677-3
4. Baczyński, M., Jayaram, B.: Fuzzy Implications. Studies in Fuzziness and Soft Computing, vol. 231. Springer, Heidelberg (2008). https://doi.org/10.1007/978-3-540-69082-5
5. Baczyński, M., Jayaram, B., Massanet, S., Torrens, J.: Fuzzy implications: past, present, and future. In: Kacprzyk, J., Pedrycz, W. (eds.) Springer Handbook of Computational Intelligence, pp. 183–202. Springer, Heidelberg (2015). https://doi.org/10.1007/978-3-662-43505-2_12
6. Fodor, J.C., Roubens, M.: Fuzzy Preference Modelling and Multicriteria Decision Support. Kluwer Academic Publishers, Dordrecht (1994)
7. Kolesárová, A., Mesiar, R.: On linear and quadratic constructions of aggregations functions. Fuzzy Sets Syst. **268**, 1–14 (2015)
8. Mas, M., Monserrat, M., Torrens, J., Trillas, E.: A survey on fuzzy implication functions. IEEE Trans. Fuzzy Syst. **15**(6), 1107–1121 (2007)
9. Massanet, S., Torrens, J.: On some properties of threshold generated implications. Fuzzy Sets Syst. **205**, 30–49 (2012)

10. Massanet, S., Torrens, J.: Threshold generation method of construction of a new implication from two given ones. Fuzzy Sets Syst. **205**, 50–75 (2012)
11. Massanet, S., Torrens, J.: On the vertical threshold generation method of fuzzy implication and its properties. Fuzzy Sets Syst. **226**, 32–52 (2013)
12. Massanet, S., Torrens, J.: An overview of construction methods of fuzzy implications. In: Baczyński, M., Beliakov, G., Bustince, H., Pradera, A. (eds.) Advances in Fuzzy Implication Functions. STUDFUZZ, vol. 300, pp. 1–30. Springer, Berlin Heidelberg (2013). https://doi.org/10.1007/978-3-642-35677-3_1
13. Shi, Y., Van Gasse, B., Ruan, D., Kerre, E.: On a new class of implications in fuzzy logic. In: Hüllermeier, E., Kruse, R., Hoffmann, F. (eds.) IPMU 2010. CCIS, vol. 80, pp. 525–534. Springer, Heidelberg (2010). https://doi.org/10.1007/978-3-642-14055-6_55
14. Trillas, E., Mas, M., Monserrat, M., Torrens, J.: On the representation of fuzzy rules. Int. J. Approx. Reason. **48**(2), 583–597 (2008)
15. Vemuri, N.R., Jayaram, B.: The ⊛-composition of fuzzy implications: closureswith respect to properties, powers and families. Fuzzy Sets Syst. **275**, 58–87 (2015)

On the Characterization of a Family of Generalized Yager's Implications

Raquel Fernandez-Peralta[1] and Sebastia Massanet[1,2](✉)

[1] Soft Computing, Image Processing and Aggregation (SCOPIA) Research Group, Department of Mathematics and Computer Science, University of the Balearic Islands, 07122 Palma, Spain
r.fernandezperalta@outlook.es, s.massanet@uib.es

[2] Balearic Islands Health Research Institute (IdISBa), 07010 Palma, Spain

Abstract. Over the last years, several generalizations of Yager's f and g-generated implications have been proposed in the literature expanding the number of available families of fuzzy implication functions. Among them, the so-called (f, g) and (g, f)-implications were introduced by means of generalizing the internal functions x and $\frac{1}{x}$ of the standard Yager's f and g-generated implications to more general unary functions. In particular, those generated using $\frac{x}{e}$ and $\frac{e}{x}$ with $e \in (0, 1)$ stand out due to their key role in the structure of (h, e)-implications. In this paper, the characterizations of the $(f, \frac{x}{e})$-implications are presented. These characterizations, which rely on two properties closely related to the law of importation, will be crucial in order to achieve a fully axiomatic characterization of (h, e)-implications.

Keywords: Fuzzy implication function
Generalized Yager's implications · (f, e)-generated implications
Law of importation

1 Introduction

The characterization and representation of fuzzy logical connectives have attracted the efforts of many researchers over the years. As a consequence, many results in this topic have been achieved leading to complete characterizations of whole families of these operators. Different families of aggregation functions have been characterized. For instance, some families of t-norms and t-conorms were characterized in [10] (and references therein), some families of copulas in [5,7] and [20] (and references therein) and some families of uninorms in [6,19,22]. Besides the aforementioned operators, the characterization of families of fuzzy implication functions stand out. Among the main families, (S, N)-implications with N a continuous fuzzy negation were characterized in [2], R-implications derived from left-continuous t-norms in [1] and QL and D-implications when the underlying t-conorm S and fuzzy negation N are continuous in [11]. More recently, other important families of fuzzy implication functions such as Yager's

J. Medina et al. (Eds.): IPMU 2018, CCIS 853, pp. 636–648, 2018.
https://doi.org/10.1007/978-3-319-91473-2_54

f and g-generated implications have been characterized in [16], h-implications in [17], probabilistic S-implications and survival S-implications in [12] and finally, probabilistic and survival implications in [13].

The reason for this great number of papers devoted to the characterization of fuzzy implication functions is the flexibility in the definition which allows uncountably many operators fulfilling the axioms. This great repertoire of fuzzy implication functions is very useful since it allows a researcher to pick out, depending on the context, that fuzzy implication function which satisfies the desired additional properties for a concrete application. Fuzzy implication functions play a key role in fuzzy control and approximate reasoning not only to model fuzzy conditionals, but also to perform backward and forward inferences in any fuzzy rules based system. In addition, they have been used with notable success in image processing and data mining, among many other fields. For a complete review of the applications of fuzzy implication functions, we refer the reader to [3,4].

Although many families of fuzzy implication functions have been already characterized, the characterization of some important ones remain unknown. This is the case of the so-called (h, e)-implications [14], a generalization of h-implications, which fulfil an interesting property related to a controlled increasingness with respect to the second variable of the fuzzy implication function. These operators have proved their potential in edge detection (see [9]). In [18], the structure of (h, e)-implications was presented as the threshold horizontal construction of some generalizations of Yager's f and g-generated implications. Consequently, in order to achieve an axiomatic characterization of (h, e)-implications, the first unavoidable step consists in the characterization of these generalizations of Yager's f and g-generated implications. Specifically, in this paper, we fully characterize the family of $(f, \frac{x}{e})$-implications. The results rely heavily in some modifications of the law of importation, a property which was also crucial for the characterization of Yager's f and g-generated implications [16].

The paper is organized as follows. In the next section we recall some basic definitions and properties on fuzzy implication functions. In Sect. 3, the relationship between the generalizations of Yager's implications and (h, e)-implications is recalled and described. Then, in Sect. 4, the characterization of the family of $(f, \frac{x}{e})$-implications is presented. The paper ends with some conclusions and future work.

2 Preliminaries

To make this work self-contained, we recall here some of the concepts and results which will be used throughout the paper. First of all, the definition of fuzzy negation is given.

Definition 1 ([3,8]). *A decreasing function $N : [0, 1] \to [0, 1]$ is called a fuzzy negation if $N(0) = 1$ and $N(1) = 0$. A fuzzy negation N is called*

(i) strict, if it is strictly decreasing and continuous.

(ii) strong, if it is an involution, i.e., $N(N(x)) = x$ *for all* $x \in [0,1]$.

Among fuzzy negations, the least fuzzy negation or Gödel negation N_{D_1} which is given by

$$N_{D_1}(x) = \begin{cases} 1 & \text{if } x = 0, \\ 0 & \text{if } x > 0, \end{cases}$$

will play a key role in some of the subsequent results. Next, we recall the definition of fuzzy implication function.

Definition 2 ([3,8]). *A binary operator* $I : [0,1]^2 \to [0,1]$ *is said to be a fuzzy implication function if it satisfies:*

(I1) $I(x,z) \geq I(y,z)$ *when* $x \leq y$, *for all* $z \in [0,1]$.
(I2) $I(x,y) \leq I(x,z)$ *when* $y \leq z$, *for all* $x \in [0,1]$.
(I3) $I(0,0) = I(1,1) = 1$ *and* $I(1,0) = 0$.

On the one hand, from the definition, it can be easily derived that $I(0,x) = 1$ and $I(x,1) = 1$ for all $x \in [0,1]$. On the other hand, the symmetrical values $I(x,0)$ and $I(1,x)$ are not predetermined from the definition. Fuzzy implication functions can satisfy additional properties coming from tautologies in crisp logic. We recall here those properties that will be used in this paper (see [3,8,21] for more details).

- The *exchange principle*,

$$I(x, I(y,z)) = I(y, I(x,z)), \quad x,y,z \in [0,1]. \qquad \textbf{(EP)}$$

- The *law of importation* with respect to a t-norm T,

$$I(T(x,y),z) = I(x, I(y,z)), \quad x,y,z \in [0,1]. \qquad \textbf{(LI)}$$

- The *left neutrality principle*,

$$I(1,y) = y, \quad y \in [0,1]. \qquad \textbf{(NP)}$$

- The *left neutrality principle with* $e \in]0,1[$,

$$I(e,y) = y, \quad y \in [0,1]. \qquad \textbf{(NP}_\mathbf{e}\textbf{)}$$

An important operator derived from a fuzzy implication function is its natural negation, which is always a fuzzy negation in the sense of Definition 1.

Definition 3 ([3]). *Let* I *be a fuzzy implication function. The function* N_I *defined by* $N_I(x) = I(x,0)$ *for all* $x \in [0,1]$, *is called the* natural negation *of* I.

3 From (h, e)-implications to Generalizations of Yager's Implications

As we have already stated in the introduction, there are still some families of fuzzy implication functions which have not been characterized yet. One of them is the family of (h, e)-implications [14]. This family was introduced as a generalization of the so-called h-implications [14]. Both families are generated from an additive generator of a representable uninorm. However, while h-implications satisfy **(NP)**, (h, e)-implications satisfy $(\mathbf{NP}_e)$. Let us recall the definition of (h, e)-implications.

Definition 4 ([14]). *Let $h : [0, 1] \to [-\infty, \infty]$ be a strictly increasing and continuous function with $h(0) = -\infty$, $h(e) = 0$ for an $e \in (0, 1)$ and $h(1) = +\infty$. The function $I : [0, 1]^2 \to [0, 1]$ defined by*

$$I^{h,e}(x, y) = \begin{cases} 1 & \text{if } x = 0, \\ h^{-1}\left(\frac{x}{e} \cdot h(y)\right) & \text{if } x > 0 \text{ and } y \leq e, \\ h^{-1}\left(\frac{e}{x} \cdot h(y)\right) & \text{if } x > 0 \text{ and } y > e, \end{cases}$$

is called an (h, e)-implication. The function h is called an h-generator of $I_{h,e}$.

(h, e)-implications are always fuzzy implication functions in the sense of Definition 2. Moreover, they satisfy the following interesting property.

Theorem 1 ([15]). *Let h be an h-generator with respect to a fixed $e \in (0, 1)$. Then the following properties hold:*

(i) If $x > 0$ and $y < e$, then $I^{h,e}(x, y) < e$.
(ii) If $x > 0$, then $I^{h,e}(x, e) = e$.
(iii) If $x > 0$ and $y > e$, then $I^{h,e}(x, y) > e$.

This result shows that these fuzzy implication functions allow to have a certain degree of control of the increasingness with respect to the second variable of the fuzzy implication function, which can be useful for image processing purposes [9].

Furthermore, in [18], the structure of this family of fuzzy implication functions was presented. Some definitions and results are needed before recalling this structure. First, we recall the threshold horizontal construction method of a fuzzy implication function from two given ones.

Theorem 1 ([17]). *Let I_1, I_2 be two fuzzy implication functions and $e \in (0, 1)$. Then the binary function $I_{I_1-I_2} : [0, 1]^2 \to [0, 1]$, called the e-threshold horizontal generated implication from I_1 and I_2, defined as*

$$I_{I_1-I_2}(x, y) = \begin{cases} 1 & \text{if } x = 0, \\ e \cdot I_1\left(x, \dfrac{y}{e}\right) & \text{if } x > 0 \text{ and } y \leq e, \\ e + (1 - e) \cdot I_2\left(x, \dfrac{y - e}{1 - e}\right) & \text{if } x > 0 \text{ and } y > e, \end{cases}$$

is a fuzzy implication function.

We need also to recall the definitions of the so-called (f, g) and (g, f)-generated operations.

Definition 5 ([18]). *Let $f : [0,1] \to [0,+\infty]$ be a strictly decreasing and continuous function with $f(1) = 0$ and $g : [0,1] \to [0,+\infty]$ be a continuous and strictly increasing function with $g(0) = 0$. The function $I_{f,g} : [0,1]^2 \to [0,1]$ defined by*

$$I_{f,g}(x, y) = f^{(-1)}(g(x) \cdot f(y)), \quad x, y \in [0,1]$$

*with the understanding $0 \cdot \infty = 0$, is called an (f, g)-*generated operation*, where the function $f^{(-1)}$ is the pseudo-inverse of f given by*

$$f^{(-1)}(x) = \begin{cases} f^{-1}(x) & \text{if } x \in [0, f(0)], \\ 0 & \text{if } x \in [f(0), \infty]. \end{cases}$$

Definition 6 ([18]). *Let $g : [0,1] \to [0,+\infty]$ be a strictly increasing and continuous function with $g(0) = 0$ and $f : [0,1] \to [0,+\infty]$ be a continuous and strictly decreasing function with $f(0) = +\infty$. The function $I_{g,f} : [0,1]^2 \to [0,1]$ defined by*

$$I_{g,f}(x, y) = g^{(-1)}\left(f(x) \cdot g(y)\right), \quad x, y \in [0,1]$$

*with the understanding $\frac{1}{0} = +\infty$ and $+\infty \cdot 0 = \infty$, is called a (g, f)-*generated operation*, where the function $g^{(-1)}$ is the pseudo-inverse of g given by*

$$g^{(-1)}(x) = \begin{cases} g^{-1}(x) & \text{if } x \in [0, g(1)], \\ 1 & \text{if } x \in [g(1), \infty]. \end{cases}$$

(f, g) and (g, f)-generated operations are not always fuzzy implication functions in the sense of Definition 2.

Theorem 2 ([18]). *The following statements hold:*

1. *An (f, g)-generated operation $I_{f,g}$ is a fuzzy implication function if and only if either $f(0) = +\infty$ or ($f(0) < +\infty$ and $g(1) \geq 1$).*
2. *A (g, f)-generated operation $I_{g,f}$ is a fuzzy implication function if and only if either $g(1) = +\infty$ or ($g(1) < +\infty$ and $f(1) \geq 1$).*

Whenever these operations are in fact fuzzy implication functions, (f, g) and (g, f)-generated operations are called (f, g) and (g, f)-generated implications, respectively. It is straightforward to check that (f, g)-generated implications with $g(x) = x$ are Yager's f-generated implications and (g, f)-generated implications with $f(x) = \frac{1}{x}$ are Yager's g-generated implications. Thus, these families are generalizations of Yager's implications and they play a key role in the structure of (h, e)-implications as the following result shows.

Theorem 3 ([18]). *Let $I : [0,1]^2 \to [0,1]$ be a binary function and $e \in (0,1)$. Then the following statements are equivalent:*

(i) I is an (h, e)-implication with respect to e, that is, $I = I_{h,e}$.

(ii) There exist an f-generator with $f(0) = +\infty$ and a g-generator with $g(1) = +\infty$ such that I is given by $I = I_{I_{f,\frac{x}{e}} - I_{g,\frac{e}{x}}}$.

Moreover, in this case generators h, f and g are related in the following way:

$$f(x) = -h(ex), \quad g(x) = h(e + (1-e)x), \quad h(x) = \begin{cases} -f\left(\frac{x}{e}\right) & \text{if } x \leq e, \\ g\left(\frac{x-e}{1-e}\right) & \text{if } x > e. \end{cases}$$

Thus, an (h, e)-implication is defined through the threshold horizontal construction method when we consider as initial fuzzy implication functions an (f, g)-generated implication with $f(0) = +\infty$ and $g(x) = \frac{x}{e}$ and a (g, f)-generated implication with $g(1) = +\infty$ and $f(x) = \frac{e}{x}$. For the sake of clarity and simplicity, let us define these two subfamilies $\left(f, \frac{x}{e}\right)$ and $\left(g, \frac{e}{x}\right)$ (including also the cases when the f and g-generators take finite values) denoting them for short as (f, e) and (g, e)-generated operations.

Definition 7. *Let $f : [0,1] \to [0,+\infty]$ be a strictly decreasing and continuous function with $f(1) = 0$ and $e \in (0,1)$. The function $I_{f,e} : [0,1]^2 \to [0,1]$ defined by*

$$I_{f,e}(x,y) = f^{(-1)}\left(\frac{x}{e} \cdot f(y)\right), \quad x, y \in [0,1]$$

with the understanding $0 \cdot \infty = 0$, is called an (f, e)-generated operation and f its f-generator.

Definition 8. *Let $g : [0,1] \to [0,+\infty]$ be a strictly increasing and continuous function with $g(0) = 0$ and $e \in (0,1)$. The function $I_{g,e} : [0,1]^2 \to [0,1]$ defined by*

$$I_{g,e}(x,y) = g^{(-1)}\left(\frac{e}{x} \cdot g(y)\right), \quad x, y \in [0,1]$$

with the understanding $\frac{1}{0} = +\infty$ and $+\infty \cdot 0 = \infty$, is called a (g, e)-generated operation and g its g-generator.

From Theorem 2, the following result is trivial.

Proposition 2. *The following statements hold:*

1. *An (f, e)-generated operation $I_{f,e}$ is always a fuzzy implication function.*
2. *A (g, e)-generated operation $I_{g,e}$ is a fuzzy implication function if and only if $g(1) = +\infty$.*

Whenever these conditions are fulfilled, (f, e) and (g, e)-generated operations will be called (f, e) and (g, e)-generated implications, respectively.

Example 1. *Let us show two examples of (f, e) and (g, e)-generated implications.*

(i) Let us consider the (f, e)-generated implication obtained from the f-generator $f(x) = 1 - x$ and $e = \frac{1}{2}$. This fuzzy implication function is given by:

$$I(x, y) = \begin{cases} 0 & \text{if } 2x(1-y) > 1, \\ 1 - 2x + 2xy & \text{if } 2x(1-y) \leq 1. \end{cases}$$

(ii) Let us consider the (g, e)-generated implication obtained from the g-generator $g(x) = -\ln(1 - x)$ and $e = \frac{1}{2}$. This fuzzy implication function is given by:

$$I(x, y) = \begin{cases} 1 & \text{if } x = 0, \\ 1 - (1-y)^{\frac{1}{2x}} & \text{otherwise.} \end{cases}$$

These two fuzzy implication functions are displayed in Fig. 1.

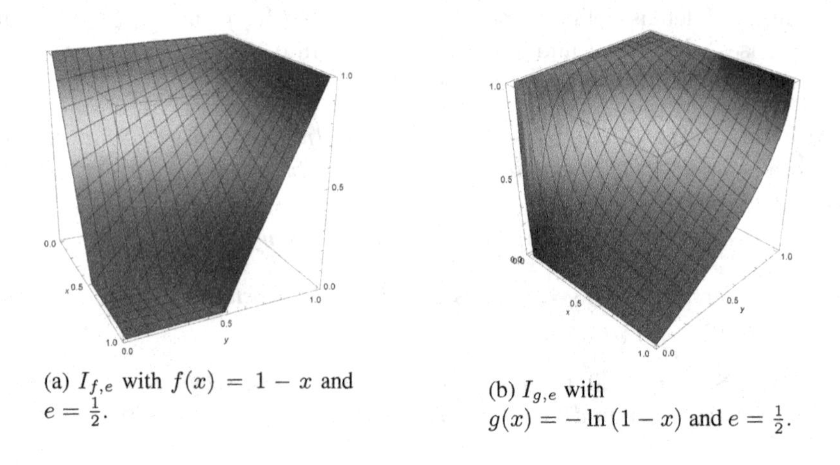

(a) $I_{f,e}$ with $f(x) = 1 - x$ and $e = \frac{1}{2}$.

(b) $I_{g,e}$ with $g(x) = -\ln(1 - x)$ and $e = \frac{1}{2}$.

Fig. 1. Plots of the fuzzy implication functions given in Example 1.

From this section it is straightforward to deduce that any possible characterization of (h, e)-implications must rely on the characterizations of these two subfamilies of fuzzy implication functions. In the next section, we will start this study by presenting some characterizations of (f, e)-generated implications.

4 Characterizations of (f, e)-generated Implications

In [18] several results related to properties of the (f, g)-generated implications were presented. Since (f, e)-generated implications are a subfamily of (f, g)-generated implications, those results can be applied to the particular case of (f, e)-generated implications. Thus, the first results of this section are corollaries of the corresponding results presented in [18] dealing with some properties which will be relevant for the characterizations of (f, e)-generated implications.

First of all, (f, e)-generated implications have a trivial 1-region as the following result shows.

Corollary 3. *Let f be an f-generator and $e \in (0,1)$. Then $I_{f,e}(x,y) = 1$ if and only if $x = 0$ or $y = 1$.*

(f,e)-generated implications do not satisfy the left neutrality principle **(NP)** but they fulfil its modified version with $e \in (0,1)$. Note that $(\mathbf{NP}_e)$ is also satisfied by (h,e)-implications, strengthening the relationship between these two families.

Corollary 4. *Let f be an (f,e) generator and $e \in (0,1)$. Then $I_{f,e}$ does not satisfy* **(NP)**, *but it satisfies* $(\mathbf{NP}_e)$.

Continuity is another important property of these fuzzy implication functions worthy to study. The following result establishes that, similarly to the case of the Yager's f-generated implications, the only possible point of discontinuity is the $(0,0)$.

Corollary 5. *Let f be an f-generator and $e \in (0,1)$. Then the following properties hold:*

1. *If $f(0) = +\infty$, then $I_{f,e}$ is continuous everywhere except at $(0,0)$.*
2. *If $f(0) < +\infty$, then $I_{f,e}$ is continuous.*

The following result studies the natural negation of (f,e)-generated implications. We will see how the form of the natural negation of these fuzzy implication functions is crucial for the characterization of these families, specially when $f(0) < +\infty$ because in such case it is strictly related to its generator.

Corollary 6. *Let f be an f-generator and $e \in (0,1)$. The following properties hold:*

1. *If $f(0) = +\infty$, then the natural negation $N_{I_{f,e}}$ is the Gödel negation N_{D_1}.*
2. *If $f(0) < +\infty$, then the natural negation $N_{I_{f,e}}$ is given by*

$$N_{I_{f,e}}(x) = \begin{cases} f^{-1}\left(\frac{x}{e} f(0)\right) & \text{if } x \leq e, \\ 0 & \text{if } x > e. \end{cases}$$

Next, let us analyse the exchange principle and the law of importation. With respect to **(EP)**, the next result shows that only those (f,e)-generated implications with $f(0) = +\infty$ satisfy this property.

Corollary 7. *Let f be an f-generator and $e \in (0,1)$. Then $I_{f,e}$ satisfies* **(EP)** *if and only if $f(0) = +\infty$.*

Since **(LI)** with respect to a t-norm T implies **(EP)**, it is clear from the previous result that only those (f,e)-generated implications with $f(0) = +\infty$ could satisfy the law of importation. However, they do not satisfy it either.

Proposition 8. *Let f be an f-generator and $e \in (0,1)$. Then $I_{f,e}$ does not satisfy* **(LI)** *with any t-norm.*

The above result shows a key difference between Yager's implications and (f, e)-generated implications. It is well-known that Yager's implications satisfy **(LI)** with respect to the product t-norm $T_{\mathbf{P}}(x, y) = xy$. Indeed, this property is used to characterize Yager's implications (see [16]). However, (f, e)-generated implications, although they do not satisfy the standard law of importation, fulfil two properties which resemble **(LI)**. Thus, let us introduce these two new properties by slightly modifying the standard law of importation.

Definition 9. *A fuzzy implication function I is said to satisfy*

1. *the (x, ey)-law of importation with a t-norm T for some $e \in (0,1)$, if*

$$I(T(x, y), z) = I(x, I(ey, z)), \text{ for all } x, y, z \in [0,1]. \qquad \mathbf{(LI)_{x,ey}}$$

2. *the (ex, y)-law of importation with a t-norm T for some $e \in (0,1)$, if*

$$I(T(x, y), z) = I(ex, I(y, z)), \text{ for all } x, y, z \in [0,1]. \qquad \mathbf{(LI)_{ex,y}}$$

Remark 1. *Note that both properties modify the standard law of importation by restricting into $[0, e]$ the domain of one of the variables of the right hand side of the equation. Obviously, we retrieve the standard law of importation when $e = 1$.*

The following proposition studies when the (f, e)-generated implications satisfy these new properties which will play an analogous role to characterize (f, e)-generated implications to the one played by the standard law of importation in the characterization of Yager's implications.

Proposition 9. *Let f be an f-generator and $e \in (0,1)$. Then the following properties hold:*

1. *$I_{f,e}$ satisfies $\mathbf{(LI)}_{x,ey}$ with respect to $T_{\mathbf{P}}$.*
2. *$I_{f,e}$ satisfies $\mathbf{(LI)}_{ex,y}$ with respect to $T_{\mathbf{P}}$ if and only if $f(0) = +\infty$.*

At first sight, this result is quite unexpected. Both properties behave in a different way depending on the value $f(0)$. This is due to the presence of the pseudo-inverse of f in the expression of these fuzzy implication functions.

At this point and using the previous results, we can obtain the characterization of (f, e)-generated implications with $f(0) < +\infty$.

Theorem 4. *Let $I : [0,1]^2 \to [0,1]$ be a binary function and $e \in (0,1)$. Then the following statements are equivalent:*

(i) I is an (f, e)-generated implication with $f(0) < +\infty$.
(ii) I satisfies $\mathbf{(LI)}_{x,ey}$ with $T_{\mathbf{P}}$ and N_I is a continuous fuzzy negation which is strictly decreasing in $(0, e)$ for some $e \in (0,1)$ and such that $N_I(e) = 0$.

Moreover, in this case the f-generator is given by

$$f(x) = N_I^{(-1)}(x) = \begin{cases} N_I^{-1}(x) & \text{if } x > 0, \\ e & \text{if } x = 0. \end{cases}$$

Once we have characterized (f, e)-generated implications with $f(0) < +\infty$, let us focus our attention to the case $f(0) = +\infty$. For such a case, some other properties are needed in order to characterize the remaining subfamily. First, let us study the monotonicity of the horizontal and vertical sections of these fuzzy implication functions.

Lemma 10. *Let f be a f-generator with $f(0) = +\infty$ and $e \in (0,1)$. Then the following properties hold:*

1. *The horizontal sections of $I_{f,e}$, $h_k(x) = I_{f,e}(x,k)$, are strictly decreasing for all $x \in [0,1]$ and $k \in (0,1)$.*
2. *The vertical sections of $I_{f,e}$, $v_k(x) = I_{f,e}(k,x)$, are strictly increasing for all $x \in [0,1]$ and $k \in (0,1]$.*

Before giving the main result of the characterization of (f, e)-generated implications with $f(0) = +\infty$ we will present some preliminary results. These results highlight that the new introduced properties $(\mathbf{LI})_{x,ey}$ and $(\mathbf{LI})_{ex,y}$ with respect to $T_{\mathbf{P}}$, jointly with the continuity except at $(0,0)$ and the property that the binary operator equals 1 if and only if $x = 0$ or $y = 1$ are powerful properties since they ensure that many other properties hold.

Lemma 11. *Let $e \in (0,1)$ and $I : [0,1]^2 \to [0,1]$ be a continuous function except at $(0,0)$ satisfying $(\mathbf{LI})_{x,ey}$ with respect to $T_{\mathbf{P}}$ and $I(x,y) = 1 \Leftrightarrow x = 0$ or $y = 1$. Then the following property holds:*

(i) I satisfies $(\mathbf{NP}_e)$.

In addition, if I satisfies also $(\mathbf{LI})_{ex,y}$ with respect to $T_{\mathbf{P}}$ then

(ii) The horizontal sections of I, $h_k : [0,1] \to [0,1]$ given by $h_k(x) = I(x,k)$, are strictly decreasing for all $k \in (0,1)$.
(iii) $N_I = N_{D_1}$.
(iv) The vertical sections of I, $v_k : [0,1] \to [0,1]$ given by $v_k(x) = I(k,x)$, are strictly increasing for all $k \in (0,1]$.

Note that the previous result makes use of both modifications of the law of importation. Recall that property $(\mathbf{LI})_{ex,y}$ with $T_{\mathbf{P}}$ does not hold for (f, e)-generated implications with $f(0) < +\infty$, but it is of paramount importance in order to characterize (f, e)-generated implications with $f(0) = +\infty$.

Finally, the following result provides the characterization of the (f, e)-generated implications with $f(0) = +\infty$.

Theorem 12. *Let $I : [0,1]^2 \to [0,1]$ be a binary function and $e \in (0,1)$. Then the following statements are equivalent:*

(i) I is an (f, e)-generated implication with $f(0) = +\infty$.
(ii) I satisfies $(\mathbf{LI})_{x,ey}$ and $(\mathbf{LI})_{ex,y}$ with respect to $T_{\mathbf{P}}$, I is continuous except at $(0,0)$ and $I(x,y) = 1 \Leftrightarrow x = 0$ or $y = 1$.

(iii) *I satisfies* $(\mathbf{LI})_{x,ey}$ *and* $(\mathbf{LI})_{ex,y}$ *with respect to* $T_{\mathbf{P}}$, $N_I = N_{D_1}$, *I is continuous except at* $(0,0)$ *and there exists* $k_0 \in (0,1)$ *such that*
- h_k *is strictly decreasing with* $h_k(0) = 1$ *and* $h_k(e) = k$ *for all* $k \in (0, k_0]$,
- v_k *are strictly increasing on the interval* $[0, k_0]$ *for all* $k \in (0,1)$.

(iv) *I satisfies* $(\mathbf{LI})_{x,ey}$ *and* $(\mathbf{LI})_{ex,y}$ *with respect to* $T_{\mathbf{P}}$, $N_I = N_{D_1}$ *and there exists* $k \in (0,1)$ *such that*
- h_k *is continuous and strictly decreasing with* $h_k(0) = 1$ *and* $h_k(e) = k$,
- $h_\bullet^{-1}(k) : (0,k) \to [0,e]$ *that assigns* $h_y^{-1}(k)$ *to some* $y \in (0,k)$ *is a well-defined, continuous and strictly increasing function satisfying* $\lim_{y \to 0^+} h_y^{-1}(k) = 0$.

Moreover, in this case the f-generator is given by

$$f(x) = \begin{cases} \dfrac{h_k^{-1}(x)}{e} & \text{if } k \leq x \leq 1, \\ \dfrac{e}{h_x^{-1}(k)} & \text{if } 0 < x < k, \\ +\infty & \text{if } x = 0. \end{cases} \tag{1}$$

Remark 2. *Recall that by Lemma 10, the horizontal sections* h_k *and the vertical sections* v_k *of an* (f,e)*-generated implication with* $f(0) = \infty$ *are continuous and strictly decreasing or increasing, respectively, for all* $k \in (0,1)$. *Thus in fact any* $k \in (0,1)$ *can be used in order to obtain the f-generator through Eq. (1).*

The previous characterization follows a similar structure to the one presented in [16] for Yager's f-generated implications with $f(0) = +\infty$ using in this case the two modified laws of importation introduced in Definition 9.

In Table 1 we give some examples of functions $I : [0,1]^2 \to [0,1]$ showing that some of the properties considered in Theorem 12-(ii) are independent from each other.

Note that the table is not complete. Up to now, we have not found either a function satisfying all the properties except $(\mathbf{LI})_{x,ey}$ with $T_{\mathbf{P}}$ for some $e \in (0,1)$ or a function satisfying all the properties other than $(\mathbf{LI})_{ex,y}$ with $T_{\mathbf{P}}$ for some $e \in (0,1)$. These two missing functions would complete the mutual independence among the properties considered in Theorem 12-(ii).

5 Conclusions and Future Work

In this paper, axiomatic characterizations of (f,e)-generated implications have been presented. These characterizations are mainly based on two properties, the so-called $(\mathbf{LI})_{x,ey}$ and $(\mathbf{LI})_{ex,y}$, obtained through a restriction of one of the variables in the equation of the standard law of importation. These characterizations will be useful in order to characterize the family of (h,e)-implications, since (f,e)-generated implications form part of the structure of these fuzzy implication functions by means of the threshold horizontal construction method.

As future work, first of all, we want to complete Table 1 to ensure the mutual independence of all the properties used in the characterization of (f,e)-generated

Table 1. The mutual independence among some properties considered in Theorem 12-(ii).

Function I	$(\mathbf{LI})_{x,ey}$ with $T_{\mathbf{P}}$ for some $e \in (0,1)$	$(\mathbf{LI})_{ex,y}$ with $T_{\mathbf{P}}$ for some $e \in (0,1)$	$I(x,y) = 1 \Leftrightarrow x = 0$ or $y = 1$	I cont $\setminus\{(0,0)\}$
$\begin{cases} \frac{xy}{e} & \text{if } xy \leq e, \\ 1 & \text{if } xy > e. \end{cases}$	✓	X	X	X
$\max\{1-x, y\}$	X	X	✓	X
$\begin{cases} \frac{xy}{x^2+y^2} & \text{if } x, y > 0, \\ 0 & \text{if } x = y = 0. \end{cases}$	X	X	X	✓
$\begin{cases} y^{x(1-y)} & \text{if } x > 0 \text{ or } y > 0, \\ 1 & \text{if } x = y = 0. \end{cases}$	X	X	✓	✓
$\begin{cases} 1 - \frac{x}{e} + \frac{xy}{e} & \text{if } x(1-y) \leq e, \\ 0 & \text{if } x(1-y) > e. \end{cases}$	✓	X	✓	X
$\begin{cases} 1 & \text{if } x = 0 \text{ or } y = 1, \\ 0 & \text{otherwise.} \end{cases}$	✓	✓	✓	X

implications with $f(0) = +\infty$. After that, we will focus on the characterization of (g, e)-implications to finally, achieve the complete characterization of the family of (h, e)-implications.

Acknowledgment. This paper has been partially supported by the Spanish Grant TIN2016-75404-P, AEI/FEDER, UE. Raquel Fernandez-Peralta also benefits from a student research collaboration grant at the Dept. of Mathematics and Computer Science of UIB conceded by the Spanish Ministry of Education, Culture and Sport.

References

1. Baczyński, M.: Residual implications revisited. Notes on the Smets-Magrez theorem. Fuzzy Sets Syst. **145**, 267–277 (2004)
2. Baczyński, M., Jayaram, B.: On the characterization of (S, N)-implications. Fuzzy Sets Syst. **158**, 1713–1727 (2007)
3. Baczyński, M., Jayaram, B.: Fuzzy Implications, vol. 231. Springer, Heidelberg (2008). https://doi.org/10.1007/978-3-540-69082-5
4. Baczyński, M., Jayaram, B., Massanet, S., Torrens, J.: Fuzzy implications: past, present, and future. In: Kacprzyk, J., Pedrycz, W. (eds.) Springer Handbook of Computational Intelligence, pp. 183–202. Springer, Heidelberg (2015). https://doi.org/10.1007/978-3-662-43505-2_12

5. de Amo, E., De Meyer, H., Díaz Carrillo, M., Fernández Sánchez, J.: Characterization of copulas with given diagonal and opposite diagonal sections. Fuzzy Sets Syst. **284**(Suppl. C), 63–77 (2016). Theme: Uncertainty and Copulas
6. Drygas, P., Ruiz-Aguilera, D., Torrens, J.: A characterization of a class of uninorms with continuous underlying operators. Fuzzy Sets Syst. **287**, 137–153 (2016)
7. Fernández-Sánchez, J., Úbeda-Flores, M.: A characterization of the orthogonal grid constructions of copulas. IEEE Trans. Fuzzy Syst. **22**(4), 1045–1047 (2014)
8. Fodor, J.C., Roubens, M.: Fuzzy Preference Modelling and Multicriteria Decision Support. Kluwer Academic Publishers, Dordrecht (1994)
9. González-Hidalgo, M., Massanet, S., Mir, A., Ruiz-Aguilera, D.: On the choice of the pair conjunction-implication into the fuzzy morphological edge detector. IEEE Trans. Fuzzy Syst. **23**(4), 872–884 (2015)
10. Klement, E., Mesiar, R., Pap, E.: Triangular Norms. Kluwer Academic Publishers, Dordrecht (2000)
11. Mas, M., Monserrat, M., Torrens, J.: QL versus D-implications. Kybernetika **42**, 351–366 (2006)
12. Massanet, S., Pradera, A., Ruiz-Aguilera, D., Torrens, J.: From three to one: equivalence and characterization of material implications derived from co-copulas, probabilistic S-implications and survival S-implications. Fuzzy Sets Syst. **323**, 103–116 (2017)
13. Massanet, S., Ruiz-Aguilera, D., Torrens, J.: On some new relations between copulas and fuzzy implication functions. In: 2017 IEEE International Conference on Fuzzy Systems, FUZZ-IEEE 2017, pp. 1–6. IEEE (2017)
14. Massanet, S., Torrens, J.: On a new class of fuzzy implications: h-implications and generalizations. Inf. Sci. **181**(11), 2111–2127 (2011)
15. Massanet, S., Torrens, J.: On some properties of (h, e)-implications. Distributivities with t-norms and t-conorms. J. Artif. Intell. Soft Comput. Res. **2**, 109–123 (2012)
16. Massanet, S., Torrens, J.: On the characterization of Yager's implications. Inf. Sci. **201**, 1–18 (2012)
17. Massanet, S., Torrens, J.: Threshold generation method of construction of a new implication from two given ones. Fuzzy Sets Syst. **205**, 50–75 (2012)
18. Massanet, S., Torrens, J.: An extension of Yager's implications. In: Montero, J., Pasi, G., Ciucci, D. (eds.) Proceedings of the 8th Conference of the European Society for Fuzzy Logic and Technology, EUSFLAT 2013, pp. 597–604. Atlantis Press (2013)
19. Mesiarová-Zemánková, A.: Characterization of uninorms with continuous underlying t-norm and t-conorm by means of the ordinal sum construction. Int. J. Approx. Reason. **83**, 176–192 (2017)
20. Nelsen, R.B.: An Introduction to Copulas (Springer Series in Statistics). Springer, New York (2006). https://doi.org/10.1007/0-387-28678-0
21. Pradera, A., Beliakov, G., Bustince, H., De Baets, B.: A review of the relationships between implication, negation and aggregation functions from the point of view of material implication. Inf. Sci. **329**, 357–380 (2016)
22. Ruiz-Aguilera, D., Torrens, J., De Baets, B., Fodor, J.: Some remarks on the characterization of idempotent uninorms. In: Hüllermeier, E., Kruse, R., Hoffmann, F. (eds.) IPMU 2010. LNCS (LNAI), vol. 6178, pp. 425–434. Springer, Heidelberg (2010). https://doi.org/10.1007/978-3-642-14049-5_44

Generalized Modus Ponens for (U, N)-implications

M. Mas[1,2], D. Ruiz-Aguilera[1,2](✉), and Joan Torrens[1,2]

[1] Department of Mathematics and Computer Science, University of the Balearic Islands, 07122 Palma, Balearic Islands, Spain
{mmg448,daniel.ruiz,jts224}@uib.es
[2] Balearic Islands Health Research Institute (IdISBa), 07010 Palma, Spain

Abstract. The Modus Ponens becomes an essential property in approximate reasoning and fuzzy control when forward inferences are managed. Thus, the conjunctor and the fuzzy implication function used in the inference process are required to satisfy this property. Usually, the conjunctor is modeled by a t-norm, but recently also by conjunctive uninorms. In this paper we study when (U, N)-implications satisfy the Modus Ponens property with respect to a conjunctive uninorm U in general, in a similar way as it was previously done for RU-implications. The functional inequality derived from the Modus Ponens involves in this case two different uninorms and a fuzzy negation leading to many possibilities. So, this communication presents only a first step in this study and many cases depending on the classes of the involved uninorms are worth to study.

Keywords: Fuzzy implication function · (U, N)-implication
Modus Ponens · t-norm · Uninorm · Natural negation

1 Introduction

Fuzzy implication functions are commonly used in fuzzy control and approximate reasoning to model fuzzy conditionals as well as to make inferences. When the Zadeh's compositional rule of inference is considered, the Modus Ponens becomes essential in the process of managing forward inferences. In these cases, the Modus Ponens translated to the framework of fuzzy logic derives into the functional inequality:

$$T(x, I(x, y)) \leq y \quad \text{for all} \quad x, y \in [0, 1], \tag{1}$$

where T is a continuous t-norm and I is a fuzzy implication function.

Due to the importance of Modus Ponens in the inference process, those t-norms T and fuzzy implication functions I that satisfy Eq. (1) have been investigated by many researchers (see for instance, [2,4,17,20,30–33]). The main studies are related to implications derived from t-norms and t-conorms. Thus, residual implications and (S, N)-implications were investigated in detail in [2,30,31],

J. Medina et al. (Eds.): IPMU 2018, CCIS 853, pp. 649–660, 2018.
https://doi.org/10.1007/978-3-319-91473-2_55

and QL and D-implications in [32]. Moreover, these results were collected and completed later in [4] (see Sect. 7.4). However, many other kinds of implication functions can be considered (see [3,4,24]). Among them, a kind of implication functions extensively studied are those derived from more general aggregation functions than t-norms and t-conorms, specially those derived from uninorms (see for instance [1,5,9,18,26–28]). Recently, the Modus Ponens has been already studied for two kinds of implications derived from uninorms: the so-called RU-implications and (U, N)-implications (see [17]).

In fact, although uninorms were firstly introduced in the framework of aggregation functions (see [11,34]), they have been also investigated as logical operators due to the fact that they must be always conjunctive or disjunctive. Taking into account their structure, uninorms have been extensively studied as generalizations of both t-norms and t-conorms. In this line, they have proved to be useful in fuzzy expert systems [10], neural networks [6] and also in fuzzy logic in general (see [25] and the references therein). In particular, conjunctive uninorms are commontly considered as fuzzy conjunctions and, in this sense, the substitution of the t-norm T by a conjunctive uninorm U in the Modus Ponens becomes natural and interesting.

Following in this line, the case of substituting the t-norm T by a conjunctive uninorm U, leading to the so-called U-Modus Ponens (or also U-conditionality), was recently proposed in [21,22]. It was proved in [21] that the implication function considered in the U-Modus Ponens must satisfy some properties that are satisfied mainly for some cases of implications derived from uninorms, that is, RU-implications and (U, N)-implications.

In this sense, the current work can be viewed as a continuation of the above mentioned papers [21,22]. In that papers the U-Modus Ponens was investigated for the class of RU-implications, whereas the idea in the current one is to deal with the same property for the case of (U, N)-implications instead of RU-implications.

The paper is organized as follows. After this introduction, Sect. 2 is devoted to some preliminaries in order to make the paper as self-contained as possible. Section 3 deals with the Modus Ponens with respect to a uninorm U, including some general results for the case of (U, N)-implications. It is proved that the disjunctive uninorm used in the construction of the (U, N)-implication can be only in some of the most usual classes of uninorms, including in particular the class of representable uninorms. The behaviour of this class of uninorms is investigated separately, whereas other classes like idempotent uninorms or uninorms continuous in the open unit square, are not included in this paper. Finally, the paper ends with Sect. 4 devoted to some conclusions and future work.

2 Preliminaries

We will suppose the reader to be familiar with the theory of t-norms, t-conorms and fuzzy negations (all necessary results and notations can be found in [14]). We also suppose that some basic facts on uninorms are known (see for instance [11]) as well as their most usual classes, that is, uninorms in $\mathcal{U}_{\min}$ and $\mathcal{U}_{\max}$, representable uninorms, uninorms continuous in the open unit square, idempotent uninorms and compensatory (or locally internal) uninorms (see the recent survey in [16] for more details on these classes of uninorms). We recall here only some facts on uninorms and implication functions in order to establish the necessary notation that we will use along the paper.

Definition 1. *A* uninorm *is a two-place function* $U : [0,1]^2 \longrightarrow [0,1]$ *which is associative, commutative, increasing in each place and such that there exists some element* $e \in [0,1]$, *called* neutral element, *such that* $U(e,x) = x$ *for all* $x \in [0,1]$.

Evidently, a uninorm with neutral element $e = 1$ is a t-norm and a uninorm with neutral element $e = 0$ is a t-conorm. For any other value $e \in]0,1[$ the operation works as a t-norm in the $[0,e]^2$ square, as a t-conorm in $[e,1]^2$ and its values are between minimum and maximum in the set of points $A(e)$ given by

$$A(e) = [0,e[\times]e,1] \cup]e,1] \times [0,e[.$$

We will usually denote a uninorm with neutral element e and underlying t-norm and t-conorm, T and S, by $U \equiv \langle T, e, S\rangle$. For any uninorm it is satisfied that $U(0,1) \in \{0,1\}$ and a uninorm U is called *conjunctive* if $U(1,0) = 0$ and *disjunctive* if $U(1,0) = 1$. On the other hand, let us recall some of the most studied classes of uninorms in the literature.

Theorem 1. ([11]) *Let* $U : [0,1]^2 \to [0,1]$ *be a uninorm with neutral element* $e \in]0,1[$.

(a) If $U(0,1) = 0$, *then the section* $x \mapsto U(x,1)$ *is continuous except in* $x = e$ *if and only if* U *is given by*

$$U(x,y) = \begin{cases} eT\left(\frac{x}{e}, \frac{y}{e}\right) & \text{if } (x,y) \in [0,e]^2, \\ e + (1-e)S\left(\frac{x-e}{1-e}, \frac{y-e}{1-e}\right) & \text{if } (x,y) \in [e,1]^2, \\ \min(x,y) & \text{if } (x,y) \in A(e), \end{cases}$$

where T *is a t-norm, and* S *is a t-conorm.*

(b) If $U(0,1) = 1$, *then the section* $x \mapsto U(x,0)$ *is continuous except in* $x = e$ *if and only if* U *is given by the same structure as above, changing minimum by maximum in* $A(e)$.

The set of uninorms as in case (a) will be denoted by $\mathcal{U}_{\min}$ *and the set of uninorms as in case (b) by* $\mathcal{U}_{\max}$. *We will denote a uninorm in* $\mathcal{U}_{\min}$ *with underlying t-norm* T, *underlying t-conorm* S *and neutral element* e *as* $U \equiv \langle T, e, S\rangle_{\min}$ *and in a similar way, a uninorm in* $\mathcal{U}_{\max}$ *as* $U \equiv \langle T, e, S\rangle_{\max}$.

Idempotent uninorms were analysed first in [7] and they were characterized in [8] for those with a lateral continuity and in [15] for the general case. An improvement of this last result was done in [29] as follows.

Theorem 2. ([29]) *U is an idempotent uninorm with neutral element $e \in [0,1]$ if and only if there exists a non increasing function $g : [0,1] \to [0,1]$, symmetric with respect to the identity function, with $g(e) = e$, such that*

$$U(x,y) = \begin{cases} \min(x,y) & \text{if } y < g(x) \text{ or } (y = g(x) \text{ and } x < g^2(x)), \\ \max(x,y) & \text{if } y > g(x) \text{ or } (y = g(x) \text{ and } x > g^2(x)), \\ x \text{ or } y & \text{if } y = g(x) \text{ and } x = g^2(x), \end{cases}$$

being commutative in the points (x,y) such that $y = g(x)$ with $x = g^2(x)$.

Any idempotent uninorm U with neutral element e and associated function g, will be denoted by $U \equiv \langle g, e \rangle_{\text{ide}}$ and the class of idempotent uninorms will be denoted by $\mathcal{U}_{\text{ide}}$. Obviously, for any of these uninorms the underlying t-norm T is the minimum and the underlying t-conorm S is the maximum.

Definition 2. *A uninorm U, with neutral element $e \in]0,1[$, is called* representable *if there exists a strictly increasing function $h : [0,1] \to [-\infty, +\infty]$ (called an* additive generator *of U, which is unique up to a multiplicative constant $k > 0$), with $h(0) = -\infty$, $h(e) = 0$ and $h(1) = +\infty$, such that U is given by*

$$U(x,y) = h^{-1}(h(x) + h(y))$$

for all $(x,y) \in [0,1]^2 \setminus \{(0,1),(1,0)\}$. We have either $U(0,1) = U(1,0) = 0$ or $U(0,1) = U(1,0) = 1$.

A representable uninorm with neutral element $e \in]0,1[$ and additive generator h will be denoted by $U \equiv \langle e, h \rangle_{\text{rep}}$ and the class of all representable uninorms by $\mathcal{U}_{\text{rep}}$.

This class is clearly contained in the class of uninorms continuous in $]0,1[^2$ which was characterized in [12] as follows (see again [16] for more details):

Theorem 3. ([12,16]) *Suppose U is a uninorm continuous in $]0,1[^2$ with neutral element $e \in]0,1[$. Then either one of the following cases is satisfied:*

(a) There exist $u \in [0,e[$, $\lambda \in [0,u]$, two continuous t-norms T_1 and T_2 and a representable uninorm R such that U can be represented as

$$U(x,y) = \begin{cases} \lambda T_1\left(\frac{x}{\lambda}, \frac{y}{\lambda}\right) & \text{if } x,y \in [0,\lambda], \\ \lambda + (u-\lambda) T_2\left(\frac{x-\lambda}{u-\lambda}, \frac{y-\lambda}{u-\lambda}\right) & \text{if } x,y \in [\lambda,u], \\ u + (1-u) R\left(\frac{x-u}{1-u}, \frac{y-u}{1-u}\right) & \text{if } x,y \in]u,1[, \\ 1 & \text{if } \min(x,y) \in]\lambda,1] \\ & \quad \text{and } \max(x,y) = 1, \\ \lambda \text{ or } 1 & \text{if } (x,y) = (\lambda,1) \\ & \quad \text{or } (x,y) = (1,\lambda), \\ \min(x,y) & \text{elsewhere.} \end{cases} \quad (2)$$

(b) *There exist* $v \in]e, 1]$, $\omega \in [v, 1]$, *two continuous t-conorms* S_1 *and* S_2 *and a representable uninorm* R *such that* U *can be represented as*

$$U(x,y) = \begin{cases} vR\left(\frac{x}{v}, \frac{y}{v}\right) & \text{if } x, y \in]0, v[, \\ v + (\omega - v)S_1\left(\frac{x-v}{\omega-v}, \frac{y-v}{\omega-v}\right) & \text{if } x, y \in [v, \omega], \\ \omega + (1-\omega)S_2\left(\frac{x-\omega}{1-\omega}, \frac{y-\omega}{1-\omega}\right) & \text{if } x, y \in [\omega, 1], \\ 0 & \text{if } \max(x,y) \in [0, \omega[\\ & \quad \text{and } \min(x,y) = 0, \\ \omega \text{ or } 0 & \text{if } (x,y) = (0, \omega) \\ & \quad \text{or } (x,y) = (\omega, 0), \\ \max(x,y) & \text{elsewhere.} \end{cases} \tag{3}$$

The class of all uninorms continuous in $]0, 1[^2$ will be denoted by $\mathcal{U}_{\text{cos}}$ [1]. A uninorm as in (2) will be denoted by $U \equiv \langle T_1, \lambda, T_2, u, (R, e)\rangle_{\text{cos,min}}$ and the class of all uninorms continuous in the open unit square of this form will be denoted by $\mathcal{U}_{\text{cos,min}}$. Analogously, a uninorm as in (3) will be denoted by $U \equiv \langle (R, e), v, S_1, \omega, S_2\rangle_{\text{cos,max}}$ and the class of all uninorms continuous in the open unit square of this form will be denoted by $\mathcal{U}_{\text{cos,max}}$.

Definition 3. *A binary operator* $I : [0,1] \times [0,1] \to [0,1]$ *is said to be a fuzzy* implication function, *or an* implication, *if it satisfies:*

(I1) $I(x, z) \geq I(y, z)$ *when* $x \leq y$, *for all* $z \in [0, 1]$.
(I2) $I(x, y) \leq I(x, z)$ *when* $y \leq z$, *for all* $x \in [0, 1]$.
(I3) $I(0, 0) = I(1, 1) = 1$ *and* $I(1, 0) = 0$.

Note that, from the definition, it follows that $I(0, x) = 1$ and $I(x, 1) = 1$ for all $x \in [0, 1]$ whereas the symmetrical values $I(x, 0)$ and $I(1, x)$ are not derived from the definition.

Definition 4. *Given a fuzzy implication function* I, *the function* $N_I(x) = I(x, 0)$ *for all* $x \in [0, 1]$ *is always a fuzzy negation, known as the* natural negation *of* I.

On the other hand, different classes of implications derived from uninorms have been studied. We recall here (U, N)-implications.

Definition 5. *Let* U *be a uninorm and* N *a fuzzy negation. The* (U, N)-operation *derived from* U *and* N *is the binary operation given by*

$$I_{U,N}(x, y) = U(N(x), y) \text{ for all } x, y \in [0, 1].$$

It is well known that $I_{U,N}$ is a fuzzy implication function if and only if U is disjunctive and then it is called a (U, N)-*implication*. Some properties of (U, N)-implications have been studied involving the main classes of uninorms: those

[1] The subindex "cos" stands here for *c*ontinuous *o*pen *s*quare.

previously recalled and also other classes of uninorms like compensatory (or locally internal) uninorms and even uninorms with continuous underlying operators (for more details see [4,5,17,19,28]). Recently, the Modus Ponens property with respect to a t-norm T has been studied in detail also for implications derived from uninorms in [17].

3 U-Modus Ponens for (U, N)-implications

In this section we want to study the Modus Ponens with respect to a conjunctive uninorm U, or U-Modus Ponens, for the class of (U, N)-implications. Let us begin by recalling the definition of U-Modus Ponens.

Definition 6. *Let I be an implication function and U a uninorm. It is said that I satisfies the* Modus Ponens property *with respect to U (U-Modus Ponens for short), or that I is an U*-conditional *if*

$$U(x, I(x, y)) \leq y \quad \text{for all } x, y \in [0, 1]. \tag{4}$$

Now, we want to study the previous inequality when the involved implication function I is an (U, N)-implication. It was proved in [21] that if U and I satisfiy Eq. (4) then U must be necessarily conjunctive. Moreover, it is well known that if $I_{U',N}$ is a fuzzy implication then the involved uninorm U' must be disjunctive. Thus, in what follows we will consider U a conjunctive uninorm and U' a disjunctive one. Let us give first some general necessary properties.

Proposition 1. *Let U and U' be a conjunctive and a disjunctive uninorm with neutral elements e and e', respectively. Let N be a fuzzy negation and $I_{U',N}$ the corresponding (U, N)-implication. If $I_{U',N}$ satisfies the U-Modus Ponens with respect to U the following items hold:*

(i) $U'(N(e), y) \leq y$ for all $y \in [0, 1]$. In particular, it must be $N(e) \leq e'$.
(ii) $U'(N(x), y) \leq e$ for all x, y such that $e \leq y < x$. In particular, it must be $U'(0, y) < e$ for all $e \leq y < 1$.
(iii) $U(x, N(x)) \leq e'$ for all $x \in [0, 1]$. In particular, if N has a fixed point e_N then it must be $U(e_N, e_N) \leq e'$.
(iv) The fuzzy negation N must be non-filling (that is, it must be $N(x) < 1$ for all $x > 0$).

The previous proposition gives some necessary conditions on the uninorms U, U' as well as on the fuzzy negation N in order $I_{U',N}$ satisfy U-Modus Ponens. From now on, we will restrict our study to the case when the disjunctive uninorm U' is locally internal on the boundary, i.e., U' satisfies $U'(0, y) \in \{0, y\}$ for all $y > e$ (see [13,16]). Note however that this is not a restrictive condition since all usual classes of uninorms are in fact, locally internal on the boundary, including all the classes recalled in the introduction, as well as compensatory uninorms and uninorms with continuous underlying operations (see [16]).

Thus, for uninorms that are locally internal on the boundary those conditions involving the uninorm U' given in the proposition above can be updated as follows.

Proposition 2. *Let U be a conjunctive uninorm with neutral element e and U' a disjunctive uninorm locally internal on the boundary with neutral element e'. Let N be a fuzzy negation and $I_{U',N}$ the corresponding (U, N)-implication. If $I_{U',N}$ satisfies the U-Modus Ponens with respect to U then:*

- *It must be $U'(0, y) = 0$ for all $y < 1$.*
- *The natural negation of $I_{U',N}$ must be drastic fuzzy negation N_D given by $N_D(x) = 0$ for all $x < 1$.*
- *U' can not be in $\mathcal{U}_{\max}$.*
- *If U' is in $\mathcal{U}_{\cos,\max}$, say $U' \equiv \langle (R, e), v, S_1, \omega, S_2 \rangle_{\cos,\max}$, then it must be $\omega = 1$.*
- *If U' is idempotent, say $U' \equiv \langle g', e' \rangle_{\text{ide}}$, then it must be $g'(0) = 1$.*

Remark 1. *From the proposition above we have that the disjunctive uninorm U' used in the construction of $I_{U',N}$ can not be anyone. However, note that many possibilities remain available between classes recalled in the preliminaries. Specifically, U' can be in any of the following classes:*

- *representable uninorms, or*
- *idempotent uninorms with $g'(0) = 1$, or*
- *uninorms in $\mathcal{U}_{\cos,\max}$ with $\omega = 1$, or*
- *uninorms in $\mathcal{U}_{\cos,\min}$ with $\lambda = 0$ ($\lambda = 0$ is necessary in order to be U' disjunctive).*

Before to deal with all these cases separately we present some more general results.

Obviously, the (U, N)-implication $I_{U',N}$ completely depends also on the fuzzy negation N used in its construction. In this line we already know that N must be non-filling in order to satisfy U-Modus Ponens. On the contrary, N needs not to be non-vanishing as the following proposition shows.

Proposition 3. *Let U' be a disjunctive uninorm in one of the classes given in Remark 1. If $N = N_D$ is the drastic fuzzy negation, then $I_{U',N}$ is given by the least fuzzy implication*

$$I_{U',N}(x, y) = I_0(x, y) = \begin{cases} 1 & \text{if } x = 0 \text{ or } y = 1 \\ 0 & \text{otherwise,} \end{cases}$$

that satisfies the U-Modus Ponens with respect to any conjunctive uninorm U.

Let us deal from now on with the case when the considered negation N is at least continuous, which is the most usual case. In this situation, the class of uninorms in $\mathcal{U}_{\cos,\max}$ can be also discarded as follows.

Proposition 4. *Let U and U' be a conjunctive and a disjunctive uninorm with neutral elements e and e', respectively. Let N be a continuous fuzzy negation and $I_{U',N}$ the corresponding (U, N)-implication. If $I_{U',N}$ satisfies the U-Modus Ponens with respect to U then U' can not be in $\mathcal{U}_{\cos,\max}$.*

Moreover, note that when the fuzzy negation N is continuous[2] it must have a fixed point that we will denote by e_N. In this case we have three specific points involved in the equation of the U-Modus Ponens. Specifically, the neutral elements of U and U' and the fixed point of N, that is, e, e' and e_N. The possible relations between them are given in the following proposition.

Proposition 5. *Let U and U' be a conjunctive and a disjunctive uninorm with neutral elements e and e', respectively. Let N be a continuous fuzzy negation with fixed point e_N and $I_{U',N}$ the corresponding (U,N)-implication. Suppose that $I_{U',N}$ satisfies the U-Modus Ponens with respect to U then*

- *If $e' < e_N$, then it must be $e_N < e$ and so $e' < e_N < e$.*
- *If $e' = e_N$, then it must be $e_N \leq e$ and so $e' = e_N \leq e$.*

The case $e' > e_N$ is not included in the previous proposition because initially, there is no restriction on the relative position of the neutral element e with respect to $e' > e_N$.

The next step in our investigation is to devote a particular study of the U-Modus Ponens for each class of uninorms that remains available for the uninorm U' according to the results stated before. That is, a case when U' is representable, when U' is in $\mathcal{U}_{\cos,\min}$ with $\lambda = 0$, and when U' is an idempotent uninorm with $g'(0) = 1$.

3.1 The Case When U' Is Representable

Let us deal in this case when U' is a disjunctive representable uninorm, say $U' \equiv \langle e', h' \rangle_{\text{rep}}$. In this case, the corresponding (U,N)-implication derived from U' and a fuzzy negation N is given by

$$I_{U',N}(x,y) = h'^{-1}(h'(N(x)) + h'(y)) \qquad \text{for all } x, y \in [0,1],$$

with the convention $\infty - \infty = +\infty$. Recall also that from a disjunctive representable uninorm $U' \equiv \langle e', h' \rangle_{\text{rep}}$, one can derive a strong negation given by $N_{h'}(x) = h'^{-1}(-h'(x))$ for all $x \in [0,1]$, which is known as the strong negation *associated to U'*.

Taking into account the considerations above, some partial results in this case are as follows.

Proposition 6. *Let U be a conjunctive uninorm with neutral element e, $U' \equiv \langle e', h' \rangle_{\text{rep}}$, N a fuzzy negation and $I_{U',N}$ the corresponding (U,N)-implication. If $U \leq U'$ and $N \leq N_{h'}$ then $I_{U',N}$ satisfies the U-Modus Ponens with respect to U.*

Moreover, the previous result can be improved in the case when the considered negation N agrees with the associated negation $N_{h'}$ as follows.

[2] Recall that continuous negations are the most usual ones. In particular, they contain the strong negations (those that are involutive) and also the strict ones (those that are strictly decreasing and continuous).

Proposition 7. *Let U be a conjunctive uninorm with neutral element e, $U' \equiv \langle e', h' \rangle_{\text{rep}}$, $N = N_{h'}$ the strong negation* associated to *U' and $I_{U',N}$ the corresponding (U, N)-implication. Then $I_{U',N}$ satisfies the U-Modus Ponens with respect to U if and only if $U \leq U'$.*

Let us give here some examples of (U, N)-implications based on disjunctive representable uninorms satisfying U-Modus Ponens.

Example 1. *From the results above we have many solutions of the U-Modus Ponens property:*

(i) Let $U' \equiv \langle e', h' \rangle_{\text{rep}}$ be a disjunctive representable uninorm with neutral element $e' \in]0, 1[$ and $N = N_{h'}$ the strong negation associated to *U'. It is well known that the underlying t-norm $T_{U'}$ and the underlying t-conorms $S_{U'}$ are then strict. Consider the uninorms in $\mathcal{U}_{\min}$ given by*

$$U_0 \equiv \langle T_{U'}, e', S_{U'} \rangle_{\min} \qquad \textit{and} \qquad U_1 \equiv \langle \min, e', S_U \rangle_{\min}.$$

Then it is clear that $U_0 \leq U'$ but $U_1 \not\leq U'$ and consequently in this case, from the theorem above, we have that $I_{U',N}$ satisfies U-Modus Ponens with respect to U_0 but it does not satisfy it with respect to U_1.

(ii) Take for instance the uninorm given by

$$U'(x, y) = \begin{cases} 1 & \textit{if } (x, y) \in \{(1, 0), (0, 1)\}, \\ \frac{xy}{(1-x)(1-y)+xy} & \textit{otherwise}, \end{cases}$$

which is well known to be a disjunctive representable uninorm with neutral element $e' = \frac{1}{2}$ and additive generator $h'(x) = \log\left(\frac{x}{1-x}\right)$. Its associated negation is the classical one $N_c(x) = 1 - x$. Thus, if we take the uninorm in $\mathcal{U}_{\min}$ given by

$$U(x, y) = \begin{cases} 2xy & \textit{if } x, y \leq 1/2 \\ 2x + 2y - 2xy - 1 & \textit{if } x, y \geq 1/2 \\ \min(x, y) & \textit{otherwise}, \end{cases}$$

we clearly have $U \leq U'$ and taking any fuzzy negation $N \leq N_c$, the corresponding (U, N)-implication $I_{U',N}$ satisfies U-Modus Ponens with respect to U.

Precisely, due to the limitations of space, in next sections we will only give examples of (U, N)-implications based on disjunctive uninorms lying in the corresponding class satisfying U-Modus Ponens, and we will leave the exhaustive study of these classes for a future work.

3.2 The Case When U' Is in $\mathcal{U}_{\cos,\min}$ with $\lambda = 0$

Let us give the following example showing many new solutions of (U, N)-implications satisfying U-Modus Ponens based on disjunctive uninorms in $\mathcal{U}_{\cos,\min}$ with $\lambda = 0$.

Example 2. *Let U' be a disjunctive uninorm in $\mathcal{U}_{\cos,\min}$ with $\lambda = 0$, say $U' \equiv \langle 0, T, u, (R, e)\rangle_{\cos,\min}$. Take U any uninorm $\mathcal{U}_{\min}$ with neutral element $e = u$ and any non-filling fuzzy negation N with fixed point $e_N = u$. Then the (U, N)-implication derived from U' and N always satisfies the U-Modus Ponens with respect to U.*

3.3 The Case When U' Is an Idempotent Uninorm with $g(0) = 1$

The following example shows again many new solutions of (U, N)-implications satisfying U-Modus Ponens, in this case based on disjunctive idempotent uninorms with $g(0) = 1$.

Example 3. *Let N be a strong fuzzy negation with fixed point $e \in]0, 1[$ and consider U and U' the conjunctive and disjunctive idempotent uninorms respectively given by*

$$U(x,y) = \begin{cases} \min(x,y) & \text{if } y \leq N(x) \\ \max(x,y) & \text{if } y > N(x), \end{cases} \qquad U'(x,y) = \begin{cases} \min(x,y) & \text{if } y < N(x) \\ \max(x,y) & \text{if } y \geq N(x), \end{cases}$$

Then the (U, N)-implication derived from U' and N always satisfies the U-Modus Ponens with respect to U.

4 Conclusions and Future Work

The Modus Ponens is the basic fuzzy rule used in approximate reasoning and fuzzy control to manage forward inferences. Thus, the conjunctor and the fuzzy implication functions that are going to be used in the inference processes are required to satisfy this property. Usually, the conjuntor is taken to be a t-norm but recently a conjunctive uninorm has been also considered leading to the so-called U-Modus Ponens.

This property was studied for residual implications derived from uninomrs in [21,22] and, following in the same line, the case of (U, N)-implications derived from disjunctive uninorms and fuzzy negations has been investigated in this paper. Our study shows that, among the most usual classes of disjunctive uninorms, only the classes of representable uninorms, uninorms in $\mathcal{U}_{\cos,\min}$ with $\lambda = 0$ and idempotent uninorms with $g(0) = 1$ are available for this purpose. Thus, the case of representable uninorms has been developed in detail, whereas the other two cases have been left for a future work. Nevertheless, many examples in the three cases have been pointed out.

As a future work, we want to extend this study to the other kinds of disjunctive uninorms that have not analyzed in depth here. Moreover, we want to deal also with other kinds of implications like h and (h, e)-implications recently introduced in [23]. Finally, a similar generalization through uninorms of the Modus Tollens property would be also worth of study.

Acknowledgments. This paper has been supported by the Spanish Grant TIN2016-75404-P AEI/FEDER, UE.

References

1. Aguiló, I., Suñer, J., Torrens, J.: A characterization of residual implications derived from left-continuous uninorms. Inf. Sci. **180**, 3992–4005 (2010)
2. Alsina, C., Trillas, E.: When (S, N)-implications are (T, T_1)-conditional functions? Fuzzy Sets Syst. **134**, 305–310 (2003)
3. Baczyński, M., Beliakov, G., Bustince Sola, H., Pradera, A. (eds.): Advances in Fuzzy Implication Functions. Studies in Fuzziness and Soft Computing, vol. 300. Springer, Heidelberg (2013). https://doi.org/10.1007/978-3-642-35677-3
4. Baczyński, M., Jayaram, B.: Fuzzy Implications. Studies in Fuzziness and Soft Computing, vol. 231. Springer, Heidelberg (2008)
5. Baczyński, M., Jayaram, B.: (U, N)-implications and their characterizations. Fuzzy Sets and Systems **160**, 2049–2062 (2009). https://doi.org/10.1007/978-3-540-69082-5
6. Benítez, J.M., Castro, J.L., Requena, I.: Are artificial neural networks black boxes? IEEE Trans. Neural Netw. **8**, 1156–1163 (1997)
7. Czogala, E., Drewniak, J.: Associative monotonic operations in fuzzy set theory. Fuzzy Sets Syst. **12**, 249–269 (1984)
8. De Baets, B.: Idempotent uninorms. Eur. J. Oper. Res. **118**, 631–642 (1999)
9. De Baets, B., Fodor, J.C.: Residual operators of uninorms. Soft Comput. **3**, 89–100 (1999)
10. De Baets, B., Fodor, J.: Van Melle's combining function in MYCIN is a representable uninorm: an alternative proof. Fuzzy Sets Syst. **104**, 133–136 (1999)
11. Fodor, J.C., Yager, R.R., Rybalov, A.: Structure of uninorms. Int. J. Uncertainty Fuzziness Knowl.-Based Syst. **5**, 411–427 (1997)
12. Hu, S., Li, Z.: The structure of continuous uni-norms. Fuzzy Sets Syst. **124**, 43–52 (2001)
13. Li, G., Liu, H.W.: On properties of uninorms locally internal on the boundary. Fuzzy Sets Syst. **332**, 116–128 (2018)
14. Klement, E.P., Mesiar, R., Pap, E.: Triangular Norms. Kluwer Academic Publishers, Dordrecht (2000)
15. Martín, J., Mayor, G., Torrens, J.: On locally internal monotonic operators. Fuzzy Sets Syst. **137**, 27–42 (2003)
16. Mas, M., Massanet, S., Ruiz-Aguilera, D., Torrens, J.: A survey on the existing classes of uninorms. J. Intell. Fuzzy Syst. **29**, 1021–1037 (2015)
17. Mas, M., Monserrat, M., Ruiz-Aguilera, D., Torrens, J.: *RU* and (U, N)-implications satisfying Modus Ponens. Int. J. Approximate Reasoning **73**, 123–137 (2016)
18. Mas, M., Monserrat, M., Torrens, J.: Two types of implications derived from uninorms. Fuzzy Sets Syst. **158**, 2612–2626 (2007)
19. Mas, M., Monserrat, M., Torrens, J.: A characterization of $(U, N), RU, QL$ and D-implications derived from uninorms satisfying the law of importation. Fuzzy Sets Syst. **161**, 1369–1387 (2010)
20. Mas, M., Monserrat, M., Torrens, J., Trillas, E.: A survey on fuzzy implication functions. IEEE Trans. Fuzzy Syst. **15**(6), 1107–1121 (2007)
21. Mas, M., Monserrat, M., Ruiz-Aguilera, D., Torrens, J.: On a generalization of the Modus Ponens: U-conditionality. In: Carvalho, J.P., Lesot, M.-J., Kaymak, U., Vieira, S., Bouchon-Meunier, B., Yager, R.R. (eds.) IPMU 2016. CCIS, vol. 610, pp. 1–12. Springer, Cham (2016). https://doi.org/10.1007/978-3-319-40596-4_33

22. Mas, M., Ruiz-Aguilera, D., Torrens, J.: On some classes of RU-implications satisfying U-Modus Ponens. In: Torra, V., Mesiar, R., De Baets, B. (eds.) AGOP 2017. AISC, vol. 581, pp. 71–82. Springer, Cham (2018). https://doi.org/10.1007/978-3-319-59306-7_8
23. Massanet, S., Torrens, J.: On a new class of fuzzy implications: h-implications and generalizations. Inf. Sci. **181**, 2111–2127 (2011)
24. Massanet, S., Torrens, J.: An overview of construction methods of fuzzy implications. In: Baczyński, M., Beliakov, G., Bustince Sola, H., Pradera, A. (eds.) Advances in Fuzzy Implication Functions. Studies in Fuzziness and Soft Computing, vol. 300, pp. 1–30. Springer, Heidelberg (2013). https://doi.org/10.1007/978-3-642-35677-3_1
25. Metcalfe, G., Montagna, F.: Substructural fuzzy logics. J. Symbolic Logic **72**, 834–864 (2007)
26. Ruiz, D., Torrens, J.: Residual implications and co-implications from idempotent uninorms. Kybernetika **40**, 21–38 (2004)
27. Ruiz-Aguilera, D., Torrens, J.: Distributivity of residual implications over conjunctive and disjunctive uninorms. Fuzzy Sets Syst. **158**, 23–37 (2007)
28. Ruiz-Aguilera, D., Torrens, J.: S- and R-implications from uninorms continuous in $]0,1[^2$ and their distributivity over uninorms. Fuzzy Sets Syst. **160**, 832–852 (2009)
29. Ruiz-Aguilera, D., Torrens, J., De Baets, B., Fodor, J.: Some remarks on the characterization of idempotent uninorms. In: Hüllermeier, E., Kruse, R., Hoffmann, F. (eds.) IPMU 2010. LNCS (LNAI), vol. 6178, pp. 425–434. Springer, Heidelberg (2010). https://doi.org/10.1007/978-3-642-14049-5_44
30. Trillas, E., Alsina, C., Pradera, A.: On MPT-implication functions for fuzzy logic. Revista de la Real Academia de Ciencias. Serie A. Matemáticas (RACSAM) 98(1), 259–271 (2004)
31. Trillas, E., Alsina, C., Renedo, E., Pradera, A.: On contra-symmetry and MPT-conditionality in fuzzy logic. Int. J. Intell. Syst. **20**, 313–326 (2005)
32. Trillas, E., Campo, C., Cubillo, S.: When QM-operators are implication functions and conditional fuzzy relations. Int. J. Intell. Syst. **15**, 647–655 (2000)
33. Trillas, E., Valverde, L.: On Modus Ponens in fuzzy logic. In: 15th International Symposium on Multiple-Valued Logic, pp. 294–301. Kingston, Canada (1985)
34. Yager, R.R., Rybalov, A.: Uninorm aggregation operators. Fuzzy Sets Syst. **80**, 111–120 (1996)

Dependencies Between Some Types of Fuzzy Equivalences

Urszula Bentkowska and Anna Król(✉)

Faculty of Mathematics and Natural Sciences,
University of Rzeszów, Rzeszów, Poland
{ududziak,annakrol}@ur.edu.pl

Abstract. The article deals with diverse types of fuzzy equivalences interpreted as fuzzy connectives. It presents some dependencies between well known fuzzy C-equivalences as well as lately examined fuzzy α–C-equivalences, fuzzy semi-C-equivalences, fuzzy weak C-equivalences, and a fuzzy equivalence defined by Fodor and Roubens.

Keywords: Fuzzy connective · Fuzzy implication
Fuzzy equivalence · Fuzzy C-equivalence · Fuzzy α–C-equivalence
Fuzzy semi-C-equivalence · Fuzzy weak C-equivalence

1 Introduction

In this contribution, recently introduced types of fuzzy equivalences, interpreted as fuzzy connectives (cf. [2,3]), are compared with the notion of a fuzzy equivalence introduced by Fodor and Roubens [11].

The concepts presented in [3], come from the idea of a fuzzy equivalence relation. Namely, definitions of a fuzzy equivalence are based on the notion of fuzzy equivalence relation which is reflexive, symmetric and transitive. Both the classical transitivity axiom of a fuzzy equivalence connective and its weaker versions are considered. These types of transitivity are defined by the use of a fuzzy conjunction or both a fuzzy conjunction and a parameter α from the unit interval and are called fuzzy α–C-equivalences. Also, other types of transitivity properties and as a result other types of fuzzy equivalences as fuzzy connectives are considered. In particular, fuzzy C-equivalences, fuzzy semi-C-equivalences as well as fuzzy weak C-equivalences are of interest.

If it comes to the notion of a fuzzy equivalence introduced by Fodor and Roubens [11], it turns out that it is compatible with the natural order in $[0,1]$. The adequate conditions are equivalent for any function on a domain $[0,1]^2$ with the values in $[0,1]$ (not necessarily a fuzzy equivalence), which will be shown in this contribution in Sect. 4. The notion of a fuzzy equivalence (in the sense of a fuzzy connective) compatible with the natural order comes from the analogous notion of a fuzzy equivalence relation on a universe X compatible with a crisp ordering on a domain X (cf. [4,5]). We recall some known dependencies between

J. Medina et al. (Eds.): IPMU 2018, CCIS 853, pp. 661–672, 2018.
https://doi.org/10.1007/978-3-319-91473-2_56

the considered classes of fuzzy equivalences as well as we present some new results and examples.

The structure of the paper is as follows. In Sect. 2, basic notions useful in the paper are presented. In Sect. 3, diverse types of fuzzy equivalences are recalled. In Sect. 4, dependencies between the fuzzy equivalences are presented.

2 Preliminaries

Here we recall basic notions and their properties which will appear in the sequel. In particular we consider fuzzy conjunctions and fuzzy implications.

2.1 Fuzzy Conjunctions

First, the definition and some properties of a fuzzy conjunction is presented. There are diverse approaches to define a fuzzy conjunction. We will apply one of the most general one.

Definition 1 ([10]). *An operation $C : [0,1]^2 \to [0,1]$ is called a fuzzy conjunction if it is increasing with respect to any variable and*

$$C(1,1) = 1, \quad C(0,0) = C(0,1) = C(1,0) = 0.$$

Corollary 1. *A fuzzy conjunction has a zero element $z = 0$. Conversely, if a binary aggregation function has a zero element $z = 0$, then it is a fuzzy conjunction.*

Corollary 2. *If a fuzzy conjunction $C : [0,1]^2 \to [0,1]$ has a neutral element 1, then it is a fuzzy conjunction fulfilling property $C \leq \min$.*

There exist many subfamilies of fuzzy conjunctions.

Example 1. Consider the following family of fuzzy conjunctions for $a \in [0,1]$

$$C^a(x,y) = \begin{cases} 1, & \text{if } x = y = 1 \\ 0, & \text{if } x = 0 \text{ or } y = 0 \\ a & \text{otherwise} \end{cases}. \tag{1}$$

Operations C^0 and C^1 are the least and the greatest fuzzy conjunction, respectively. Other examples of fuzzy conjunctions are listed below. These are well-known t-norms (i.e. increasing, commutative, associative operations with a neutral element 1): minimum, product, Łukasiewicz, drastic, which are denoted in the traditional way T_M, T_P, T_L, T_D, respectively.

$$T_M(x,y) = \min(x,y), \qquad T_P(x,y) = xy,$$

$$T_L(x,y) = \max(x+y-1,0), \; T_D(x,y) = \begin{cases} x, & \text{if } y = 1 \\ y, & \text{if } x = 1 \\ 0, & \text{otherwise} \end{cases}.$$

Example 2. The condition $C \leq \min$ is fulfilled by the following fuzzy conjunctions from Example 1: C^0, T_M, T_P, T_L, T_D.

The class of continuous Archimedean t-norms will be useful in the sequel.

Definition 2 ([12,15]). *A continuous t-norm T is called Archimedean if for any $x \in (0,1)$ it holds $T(x,x) < x$.*

Theorem 1 ([12]). *A continuous Archimedean t-norm T may be presented in the form*

$$T(x,y) = f^{-1}(\min(f(x) + f(y), f(0))),$$

where $f : [0,1] \to [0,\infty]$ is a continuous and strictly decreasing function with $f(1) = 0$ (f is called an additive generator of the t-norm T).

Among continuous Archimedean t-norms we may distinguish the class of strict t-norms (with the prominent example and generator of the class T_P) and the class of nilpotent t-norms (with the prominent example and generator of the class T_L), cf. [12].

2.2 Fuzzy Implications

Next, we focus on fuzzy implications.

Definition 3 ([1], **pp. 2,9**). *A binary operation $I\colon [0,1]^2 \to [0,1]$ is called a fuzzy implication if it is decreasing with respect to the first variable and increasing with respect to the second variable and*

$$I(0,0) = I(0,1) = I(1,1) = 1, \quad I(1,0) = 0.$$

Directly from the definition we obtain as follows.

Corollary 3. *Each fuzzy implication I is constant for $x = 0$ and for $y = 1$, i.e., it satisfies the following properties, called left and right boundary condition, respectively: $I(0,y) = 1$, for all $y \in [0,1]$ and $I(x,1) = 1$ for all $x \in [0,1]$.*

We can also consider other properties of a fuzzy implications (for more information see e.g. [1]). For this contribution one of these is of special interest.

Definition 4 ([1], **p. 9**). *We say that a fuzzy implication I fulfils the identity principle (IP) if*

$$I(x,x) = 1, \quad x \in [0,1]. \qquad \text{(IP)}$$

Example 3 ([1]). The operations I_0 and I_1 are the least and the greatest fuzzy implication, respectively, where

$$I_0(x,y) = \begin{cases} 1, & \text{if } x = 0 \text{ or } y = 1, \\ 0, & \text{otherwise,} \end{cases}$$

$$I_1(x,y) = \begin{cases} 0, & \text{if } x = 1, y = 0, \\ 1, & \text{otherwise.} \end{cases}$$

The following are well-known examples of fuzzy implications.

$$I_{\mathrm{LK}}(x,y)=\min(1-x+y,1),\ I_{\mathrm{GG}}(x,y)=\begin{cases}1, & \text{if } x\le y\\ \frac{y}{x}, & \text{if} x>y\end{cases},$$

$$I_{\mathrm{GD}}(x,y)=\begin{cases}1, & \text{if } x\le y\\ y, & \text{if } x>y\end{cases},\quad I_{\mathrm{RS}}(x,y)=\begin{cases}1, & \text{if } x\le y\\ 0, & \text{if } x>y\end{cases},$$

$$I_{\mathrm{RC}}(x,y)=1-x+xy,\qquad I_{\mathrm{YG}}(x,y)=\begin{cases}1, & \text{if } x,y=0\\ y^x, & \text{if otherwise}\end{cases},$$

$$I_{\mathrm{DN}}(x,y)=\max(1-x,y),\qquad I_{\mathrm{FD}}(x,y)=\begin{cases}1, & \text{if } x\le y\\ \max(1-x,y), & \text{if } x>y\end{cases},$$

$$I_{\mathrm{WB}}(x,y)=\begin{cases}1, & \text{if } x\le 1\\ y, & \text{if } x=1\end{cases},\quad I_{\mathrm{DP}}(x,y)=\begin{cases}y, & \text{if } x=1\\ 1-x, & \text{if } y=0\\ 1, & \text{if } x<1,\ y>0\end{cases}.$$

The implications fulfilling the property (IP) are: I_1, I_{LK}, I_{GD}, I_{GG}, I_{RS}, I_{WB}, I_{FD}, I_{DP}.

3 Fuzzy Equivalences

There exist diverse definitions of a fuzzy equivalence. In many contributions, it is used an equality $E:[0,1]^2\to[0,1]$ given by the formula

$$E(x,y)=\begin{cases}1, & \text{if } x=y\\ 0, & \text{if } x\ne y\end{cases}. \tag{2}$$

It is natural to expect that such notion of a fuzzy equivalence is a generalization of the equivalence of classical propositional calculus, that is the function $E:[0,1]^2\to[0,1]$ that fulfils conditions $E(0,1)=E(1,0)=0$, $E(0,0)=E(1,1)=1$. We will present now the approach in which definition of a fuzzy equivalence follows from the notion of a fuzzy equivalence relation (reflexive, symmetric and transitive fuzzy relation).

Definition 5 (cf. [3], [14], **p. 33).** *Let C be a fuzzy conjunction. A fuzzy C-equivalence is a function $E:[0,1]^2\to[0,1]$ fulfilling the following conditions*

$$E(0,1)=0 \qquad (\textit{boundary property}), \tag{3}$$

$$E(x,x)=1,\quad x\in[0,1] \qquad (\textit{reflexivity}), \tag{4}$$

$$E(x,y)=E(y,x),\quad x,y\in[0,1] \qquad (\textit{symmetry}), \tag{5}$$

$$C(E(x,y),E(y,z))\le E(x,z)\quad x,y,z\in[0,1] \qquad (C\textit{-transitivity}). \tag{6}$$

In the transitivity condition (6) an operation C is involved. We assume that C is a fuzzy conjunction, in particular it can be a triangular norm. We may weaken conditions given in the previous definition by replacing transitivity property with the adequate weaker transitivity conditions. Such weaker fuzzy equivalences were recently presented in [3]. Next, we will recall the notions and dependencies between them (cf. [3]).

Definition 6. *Let $\alpha \in [0,1]$, C be a fuzzy conjunction. A fuzzy weak C-equivalence is a function $E : [0,1]^2 \to [0,1]$ fulfilling conditions* (3)–(5) *and*

$$\forall_{x,y,z \in X} C(E(x,y), E(y,z)) > 0 \Rightarrow E(x,z) > 0. \tag{7}$$

A fuzzy semi-C-equivalence is a function $E : [0,1]^2 \to [0,1]$ fulfilling conditions (3)–(5) *and*

$$\forall_{x,y,z \in X} C(E(x,y), E(y,z)) = 1 \Rightarrow E(x,z) = 1. \tag{8}$$

A fuzzy α–C-equivalence is a function $E : [0,1]^2 \to [0,1]$ fulfilling conditions (3)–(5) *and*

$$\forall_{x,y,z \in X} C(E(x,y), E(y,z)) \geq 1 - \alpha \Rightarrow C(E(x,y), E(y,z)) \leq E(x,z). \tag{9}$$

The conditions (7)-(9) *will be called a weak C-transitivity, a semi-C-transitivity and an α–C-transitivity, respectively.*

Example 4. For any fuzzy conjunction C and $\alpha \in [0,1]$, the function (2) is a fuzzy α–C-equivalence and a fuzzy C-equivalence.

Dependencies between classes of fuzzy connectives from Definitions 5 and 6 (with respect to a fuzzy conjunction and with respect to a parameter α) will be recalled in the following statements.

Corollary 4. *Let C be a fuzzy conjunction. If $E : [0,1]^2 \to [0,1]$ is a fuzzy C-equivalence, then it is both a fuzzy weak C-equivalence and a fuzzy semi-C-equivalence.*

Corollary 5. *Let $C : [0,1]^2 \to [0,1]$ be a fuzzy conjunction. Function $E : [0,1]^2 \to [0,1]$ is a fuzzy 1–C-equivalence if and only if E is a fuzzy C-equivalence. Function E is a fuzzy 0–C-equivalence if and only if E is a fuzzy semi-C-equivalence.*

There is no correspondence between properties (7) and (8) what is shown in the next examples.

Example 5. Let us consider an arbitrary fuzzy conjunction C and the function

$$E(x,y) = \begin{cases} 1, & \text{if } x = y \\ 0.5, & \text{if } x \neq y, (x,y) \in (0,1)^2 \\ 0, & \text{otherwise} \end{cases}. \tag{10}$$

The function (10) is a weak C-equivalence for any fuzzy conjunction C. However, this function may not be semi-C-transitive for some fuzzy conjunction C, for example

$$C(x,y) = \begin{cases} \min(x,y), & \text{if } y \leq 1-x \\ 1, & \text{otherwise} \end{cases}. \tag{11}$$

This is why the function (10) is not a semi-C-equivalence for the given fuzzy conjunction C.

Example 6. Now, let us assume that C is an arbitrary fuzzy conjunction fulfilling strong 1–boundary condition (i.e. $C(1,1) = 1$ if and only if both inputs are equal to 1), where $a \in (0,1)$ is an idempotent element of C. Let

$$E(x,y) = \begin{cases} 1, & \text{if } x = y \\ 0, & \text{if } \{x,y\} = \{0,1\} \\ a, & \text{otherwise} \end{cases}. \tag{12}$$

The function (12) is not a weak C-equivalence but it is semi-C-equivalence.

Now we will recall dependencies between fuzzy α–C-equivalences with respect to the parameter α and the other classes of fuzzy equivalences.

Proposition 1. *Let $C : [0,1]^2 \to [0,1]$ be a fuzzy conjunction. If $E : [0,1]^2 \to [0,1]$ is a fuzzy C-equivalence, then E is a fuzzy α–C-equivalence for any $\alpha \in [0,1]$.*

Proposition 2. *Let $C : [0,1]^2 \to [0,1]$ be a fuzzy conjunction, $\alpha \in [0,1]$. If $E : [0,1]^2 \to [0,1]$ is a fuzzy α–C-equivalence, then E is a fuzzy semi-C-equivalence.*

Remark 1. If E is a fuzzy α–C-equivalence for $\alpha = 1$, then E is a fuzzy weak C-equivalence. If E is a fuzzy weak C-equivalence, then it need not be a fuzzy α–C-equivalence for any $\alpha \in [0,1]$ or it may be a fuzzy α–C-equivalence for some $\alpha \in [0,1]$. As a result there is no clear dependance between a fuzzy α–C-equivalence and a fuzzy weak C-equivalence.

Example 7. Let us consider the function E given by (10) and a fuzzy conjunction C given by (11). According to Example 5, E is a fuzzy weak C-equivalence, however it is not a fuzzy semi-C-equivalence. By Proposition 2, the latter denotes that E is not a fuzzy α–C-equivalence for any $\alpha \in [0,1]$.

The presented so far families of fuzzy equivalence are descending with respect to a fuzzy conjunction.

Proposition 3. *Let $\alpha \in [0,1]$, C_1, C_2 be fuzzy conjunctions, $C_1 \leq C_2$. If $E : [0,1]^2 \to [0,1]$ is a fuzzy α–C_2-equivalence (respectively fuzzy C_2-equivalence, fuzzy weak C_2-equivalence, fuzzy semi-C_2-equivalence), then it is also a fuzzy α–C_1-equivalence (respectively fuzzy C_1-equivalence, fuzzy weak C_1-equivalence, fuzzy semi-C_1-equivalence).*

By Proposition 3 we see that an α–C-transitivity, based on the parameter α, is a weaker property than a C-transitivity.

Corollary 6. *Let $\alpha \in [0,1]$, $C : [0,1]^2 \to [0,1]$ be a fuzzy conjunction, $C \leq \min$. If $E : [0,1]^2 \to [0,1]$ is a fuzzy* min*-equivalence, then it is a fuzzy α–C-equivalence.*

Now, let us focus on the notion of fuzzy equivalence considered by Fodor and Roubens in [11].

Definition 7 ([11], p. 33). *A fuzzy equivalence is a function $E : [0,1]^2 \to [0,1]$ which fulfils*

$$E(0,1) = 0, \tag{13}$$

$$E(x,x) = 1, \ x \in [0,1], \tag{14}$$

$$E(x,y) = E(y,x), \ x,y \in [0,1], \tag{15}$$

$$E(x,y) \leq E(u,v), \ x \leq u \leq v \leq y, \ x,y,u,v \in [0,1]. \tag{16}$$

There exists a characterization of such defined fuzzy equivalence by the use of fuzzy implications fulfilling (IP).

Theorem 2 ([11], p. 33). *A function $E : [0,1]^2 \to [0,1]$ is a fuzzy equivalence if and only if there exists such a fuzzy implication I fulfilling (IP) that*

$$E_I(x,y) = \min(I(x,y), I(y,x)), \ x,y \in [0,1]. \tag{17}$$

Corollary 7 ([11], p. 34). *A function $E : [0,1]^2 \to [0,1]$ is a fuzzy equivalence if and only if there exists such a fuzzy implication I fulfilling (IP) that*

$$E_I(x,y) = I(\max(x,y), \min(x,y)), \ x,y \in [0,1]. \tag{18}$$

Example 8. Table 1 presents examples of fuzzy equivalences generated by the use of formula (18) and these of the fuzzy implications from Example 3 that fulfil (IP). Let us see, that in the case of a fuzzy equivalence generated by the fuzzy implication I_{RS} we obtain the fuzzy equivalence E_{RS}, which is the fuzzy equality (2).

4 Dependencies Between Fuzzy Equivalences

In this section we give some dependencies between fuzzy equivalences introduced in [3] and fuzzy equivalence defined by Fodor and Roubens. First, let us notice that a fuzzy equivalence E_{GD} (cf. Table 1) may be a fuzzy C-equivalence or not, which depends on a fuzzy conjunction used in transitivity condition.

Proposition 4. *Let C be a fuzzy conjunction. A fuzzy equivalence E_{GD} is a fuzzy C-equivalence if and only if $C \leq \min$.*

Table 1. Fuzzy equivalences

I	E_I
I_{LK}	$E_{LK}(x,y) = 1 - \|x-y\|$
I_{GD}	$E_{GD}(x,y) = \begin{cases} 1, & \text{if } x = y \\ x, & \text{if } x < y \\ y, & \text{if } x > y \end{cases}$
I_{GG}	$E_{GG}(x,y) = \begin{cases} 1, & \text{if } x = y \\ \frac{x}{y}, & \text{if } x < y \\ \frac{y}{x}, & \text{if } x > y \end{cases}$
I_{RS}	$E_{RS}(x,y) = \begin{cases} 1, & \text{if } x = y \\ 0, & \text{if } x \neq y \end{cases}$
I_{WB}	$E_{WB}(x,y) = \begin{cases} 1, & \text{if } x \neq 1,\ y \neq 1 \\ x, & \text{if } y = 1 \\ y, & \text{if } x = 1 \end{cases}$
I_{FD}	$E_{FD}(x,y) = \begin{cases} 1, & \text{if } x = y \\ \max(1-y,x), & \text{if } x < y \\ \max(1-x,y), & \text{if } x > y \end{cases}$
I_{DP}	$E_{DP}(x,y) = \begin{cases} x, & \text{if } y = 1 \\ y, & \text{if } x = 1 \\ 1-x, & \text{if } y = 0 \\ 1-y, & \text{if } x = 0 \\ 1 & \text{otherwise} \end{cases}$
I_1	$E(x,y) = \begin{cases} 0, & \text{if } \{x,y\} = \{0,1\} \\ 1 & \text{otherwise} \end{cases}$

Proof. Obviously, conditions (3)–(5) are fulfilled. Let us examine the property of a C-transitivity (6). ($\Rightarrow$) Let $x, z \in [0,1]$. If $x \neq z$ then for $y = 1$ one has

$$C(x,z) = C(E(x,1), E(1,z)) \leq E(x,z) = \min(x,z).$$

On the other hand if $x = z$ then for $x = 1$ one has $C(x,x) = C(1,1) = 1 \leq 1 = x$, and for $x \neq 1$ there exists $t \geq x$ and then $C(x,x) \leq C(x,t) \leq \min(x,t) = x$. ($\Leftarrow$) Let $C \leq \min$, $x, y, z \in [0,1]$. If $x \neq y$ and $y \neq z$ then

$$C(E(x,y), E(y,z)) = C(\min(x,y), \min(y,z)) \leq \min(\min(x,y), \min(y,z)) \\ \leq \min(x,z) \leq E(x,z).$$

For $x = y$ one obtains

$$C(E(x,y), E(y,z)) = C(1, \min(y,z)) \leq \min(y,z) = \min(x,z) \leq E(x,z).$$

Similarly, for $y = z$ one has

$$C(E(x,y), E(y,z)) = C(\min(x,y), 1) \leq \min(x,y) = \min(x,z) \leq E(x,z).$$

Let us notice that by Corollary 4, Propositions 1 and 4, E_{GD} is also a fuzzy weak C-equivalence, fuzzy semi-C-equivalence and fuzzy α–C-equivalence, for any fuzzy conjunction $C \leq \min$ and $\alpha \in [0, 1]$.

Proposition 5. *Let $a \in (0, 1)$, $\alpha \in [0, 1)$. A fuzzy equivalence E_{GD} is a fuzzy α-C^a-equivalence if and only if $a < 1 - \alpha$.*

Proof. Obviously, conditions (3)–(5) are fulfilled. Let us examine the property of an α–C^a–transitivity (9).

($\Rightarrow$) For the proof by contradiction let us assume that $a \geq 1 - \alpha$ and let us consider $x \in (0, 1)$ such that $x < a$ and $y = z = 1$. We have

$$C^a(E(x, y), E(y, z)) = C^a(E(x, 1), E(1, 1)) = C^a(x, 1) = a \geq 1 - \alpha$$

and $E(x, z) = E(x, 1) = x < a$. Hence $C^a(E(x, y), E(y, z)) > E(x, z)$ and the function E is not α–C^a–transitive.

($\Leftarrow$) Let us assume that for some $x, y, z \in [0, 1]$ we have $C^a(E(x, y), E(y, z)) \geq 1-\alpha$. By the assumption that $a < 1-\alpha$ and the formula of a fuzzy conjunction C^a (1) we obtain that $C^a(E(x, y), E(y, z)) = 1$. Again by the definition of C^a and E we have $E(x, y) = E(y, z) = 1$ and $x = y = z$. Thus, we obtain $E(x, z) = 1$, what shows that inequality $C^a(E(x, y), E(y, z)) \leq E(x, z)$ holds and the function E is α–C^a–transitive.

The next example shows that a fuzzy α–C-equivalence may not be a fuzzy C-equivalence and presents examples of fuzzy α–C-equivalences.

Example 9. Let us consider a fuzzy equivalence E_{GD} and a fuzzy conjunction C^a for $a \in (0, 1)$. As C^a does not fulfil the condition $C^a \leq \min$ thus, by Proposition 4, E is not a fuzzy C^a-equivalence. On the other hand, by Proposition 5, it is a fuzzy α–C^a-equivalence for any $\alpha < 1 - a$.

Example 10. Let us consider a fuzzy equivalence E_{GD}, $a \in [0, 1]$, and a fuzzy conjunction C^a given by (1). By Example 9, E is a fuzzy α–C^a-equivalence for any $\alpha < 1 - a$ and it is easy to show that it is a fuzzy weak C^a-equivalence.

The following example shows that the notions of a fuzzy weak C-equivalence and a fuzzy semi-C-equivalence are not equivalent to the notion of a fuzzy C-equivalence for fuzzy equivalences in the sense of Fodor and Roubens.

Example 11. Let us consider the fuzzy equivalence E_{GG} (cf. Table 1) and the fuzzy conjunction $C = \min$.

Obviously E fulfils conditions (3)–(5). We shall show that it does not fulfil (6). For $x = 0.2$, $y = 0.4$, $z = 0.8$ one has

$$C(E(x, y), E(y, z)) = \min\left(\frac{\min(x, y)}{\max(x, y)}, \frac{\min(y, z)}{\max(y, z)}\right) = \min\left(\frac{0.2}{0.4}, \frac{0.4}{0.8}\right) = 0.5.$$

Moreover, $E(x, z) = \frac{\min(x,z)}{\max(x,z)} = \frac{0.2}{0.8} = 0.25$. Thus, $C(E(x, y), E(y, z)) > E(x, z)$ and the fuzzy equivalence E_{GG} is not a fuzzy min-equivalence.

Now, let us observe that from the assumption $\min(E(x,y),E(y,z)) > 0$ it follows that both $E(x,y) > 0$ and $E(y,z) > 0$. This holds if $x = y = z$ or $x, y, z > 0$. In both cases we have $E(x,y) > 0$. This is why the fuzzy equivalence E_{GG} is a fuzzy weak min-equivalence.

Let us consider a condition $\min(E(x,y),E(y,z)) = 1$. From this it follows that both $E(x,y) = 1$ and $E(y,z) = 1$ and it means that $x = y$ and $y = z$. Thus, $x = z$ and $E(x,z) = 1$. This means that the fuzzy equivalence E_{GG} is a fuzzy semi-min-equivalence.

In the next part we apply some results on a pseudo-metric from [5,8], adjusted to the domain $[0,1]$. In order to explain why we can use the results, we recall the notion of compatibility with the standard order in [0,1] and show, that the condition is equivalent to the one from definition of a fuzzy equivalence by Fodor and Roubens, namely (16).

Definition 8. *We say that a fuzzy C-equivalence E is compatible with standard order in [0,1] if for all $x,y,z \in [0,1]$ such that $x \le y \le z$ the following two inequalities hold*

$$E(x,z) \le E(y,z) \qquad \textit{and} \qquad E(x,z) \le E(x,y). \tag{19}$$

Proposition 6. *For any function $E : [0,1]^2 \to [0,1]$ the conditions (16) and (19) are equivalent.*

Proof. First, let the function E fulfil the condition (16) and let $x,y,z \in [0,1]$ such that $x \le y \le z$. Taking $t = z$ we have, by (16)

$$E(x,z) = E(x,t) \le E(y,z).$$

Similarly, taking $w = x$, by (16), we obtain

$$E(x,z) = E(w,z) \le E(x,y).$$

So, we have both inequalities in (19).

Now, let the function E fulfil the condition (19) and let $x,y,u,v \in [0,1]$ such that $x \le u \le v \le y$. By the first inequality in condition (19) we have $E(x,y) \le E(u,y)$. By the second inequality in condition (19) we obtain $E(u,y) \le E(u,v)$. Thus, $E(x,y) \le E(u,v)$, which means that E fulfils (16).

By Proposition 6 we get the following statement.

Corollary 8. *A fuzzy C-equivalence E is compatible with the standard order $\le$ in $[0,1]$ if and only if E is a fuzzy equivalence in the sense of Fodor and Roubens [11].*

One of the result from [5] states that we can transform a pseudo-metric (or a metric) into a T-equivalence if we consider a continuous Archimedean t-norm T.

Theorem 3. *Let T be a continuous Archimedean t-norm with an additive generator f. Thus for any pseudo-metric $d : [0,1]^2 \to [0,1]$, the mapping $E_d : [0,1]^2 \to [0,1]$ defined as*

$$E_d(x,y) = f^{-1}(\min(d(x,y), f(0)))$$

is a T-equivalence.

Applying additive generators in the notion of equivalence is important from practical point of view since it may simplify the computational effort. The other result from [5] gives a method how to construct an equivalence E_d from a pseudo-metric (or a metric) such that the compatibility with a classical order for $[0,1]$ is fulfilled. As it was stated before, this is equivalent to the fact that the equivalence E_d is a fuzzy equivalence in the sense of Fodor and Roubens.

Proposition 7. *Let T be a continuous Archimedean t-norm with an additive generator f and let $\leq$ be the classical order in $[0,1]$. If a pseudo-metric $d : [0,1]^2 \to [0,\infty)$ fulfils*

$$x \leq y \leq z \Rightarrow d(x,z) \geq \max(d(x,y), d(y,z)), \tag{20}$$

for all $x, y, z, \in [0,1]$, then its induced fuzzy equivalence E_d is compatible with $\leq$.

Let us consider the natural metric $d(x,y) = |x-y|$ on $[0,1]$. This metric fulfils (20). As a result of the presented dependencies, both for the diverse types of C-equivalences (and also their weaker versions) and equivalences in the sense of Fodor and Roubens, we get the following example.

Example 12 (cf. [5]). By Theorem 3 we see that for the additive generator $f(x) = 1 - x$ of T_L we get a T_L-equivalence

$$E_L(x,y) = \max(1 - \frac{1}{k} \cdot |x-y|, 0), \tag{21}$$

where $k > 0$ is a constant. Similarly, for the additive generator $f(x) = -ln(x)$ and constant $k > 0$ we get a T_P-equivalence

$$E_P(x,y) = \exp(-\frac{1}{k} \cdot |x-y|). \tag{22}$$

Both E_L and E_P are fuzzy C-equivalences (in virtue of the results from Sect. 3 they are also fuzzy α-C-equivalences, fuzzy semi-C-equivalences, fuzzy weak C-equivalences) for $C = T_L$ and $C = T_P$, respectively. They are also fuzzy equivalences in the sense of Fodor and Roubens (cf. Propositions 6, 7 and Corollary 8).

5 Conclusions

In the paper comparison between the classes of fuzzy equivalences recently introduced in [3] and the fuzzy equivalence notion introduced by Fodor and Roubens were presented. In this comparison, classes of such equivalences based on the notion of a t-norm were considered. For the future work it seems to be interesting to examine also other fuzzy conjunctions involved. For example overlap functions that may be presented by an additive generator pair [9]. Moreover, it is interesting to study connections between fuzzy equivalences presented in [3] and other fuzzy equivalences, for example the ones considered in [6,7,13].

Acknowledgment. The work on this paper was partially supported by the Centre for Innovation and Transfer of Natural Sciences and Engineering Knowledge in Rzeszów, through Project Number RPPK.01.03.00-18-001/10.

References

1. Baczyński, M., Jayaram, B.: Fuzzy Implications. Springer, Berlin (2008). https://doi.org/10.1007/978-3-540-69082-5
2. Bentkowska, U., Król, A.: Aggregation of fuzzy α-C-equivalences. In: Alonso, J.M. et al. (ed.) Proceedings of the International Joint Conference IFSA-EUSFLAT 2015, pp. 1310–1317. Atlantis Press (2015). https://doi.org/10.2991/ifsa-eusflat-15.2015.185
3. Bentkowska, U., Król, A.: Fuzzy α-C-equivalences. Fuzzy Sets Syst. https://doi.org/10.1016/j.fss.2018.01.004
4. Bodenhofer, U.: A similarity-based generalization of fuzzy orderings preserving the classical axioms. Int. J. Uncertain. Fuzziness Knowl.-Based Syst. **8**, 593–610 (2000). https://doi.org/10.1142/S0218488500000411
5. Bodenhofer, U., Küng, J.: Fuzzy orderings in flexible query answering systems. Soft Comput. **8**, 512–522 (2004). https://doi.org/10.1007/s00500-003-0308-9
6. Bustince, H., Barrenechea, E., Pagola, M.: Restricted equivalence functions. Fuzzy Sets Syst. **157**, 2333–2346 (2006). https://doi.org/10.1016/j.fss.2006.03.018
7. Bustince, H., Barrenechea, E., Pagola, M.: Image thresholding using restricted equivalence functions and maximizing the measures of similarity. Fuzzy Sets Syst. **158**, 496–516 (2007). https://doi.org/10.1016/j.fss.2006.09.012
8. De Baets, B., Mesiar, R.: Pseudo-metrics and T-equivalences. J. Fuzzy Math. **5**, 471–481 (1997). http://hdl.handle.net/1854/LU-268970
9. Dimuro, G.P., Bedregal, B., Bustince, H., Asiáin, M.J., Mesiar, R.: On additive generators of overlap functions. Fuzzy Sets Syst. **287**, 76–96 (2016). https://doi.org/10.1016/j.fss.2015.02.008
10. Drewniak, J., Król, A.: A survey of weak connectives and the preservation of their properties by aggregations. Fuzzy Sets Syst. **161**, 202–215 (2010). https://doi.org/10.1016/j.fss.2009.08.011
11. Fodor, J., Roubens, M.: Fuzzy Preference Modelling and Multicriteria Decision Support. Kluwer, Dordrecht (1994)
12. Klement, E.P., Mesiar, R., Pap, E.: Triangular Norms. Kluwer, Dordrecht (2000)
13. Król, A.: Fuzzy (C, I)-equivalences. In: Baczyñski, M. et al. (ed.) Proceedings of the 8th International Summer School on Aggregation Operators (AGOP 2015), University of Silesia, Katowice, Poland, pp. 157–161 (2015)
14. Nguyen, H.T., Walker, E.: A First Course in Fuzzy Logic. CRC Press, Boca Raton (1996)
15. Schweizer, B., Sklar, A.: Probabilistic Metric Spaces. North-Holland, Amsterdam (1983)

Selected Properties of Generalized Hypothetical Syllogism Including the Case of R-implications

Michał Baczyński[(✉)] and Katarzyna Miś

Institute of Mathematics, University of Silesia in Katowice, Bankowa 14, 40-007 Katowice, Poland
{michal.baczynski,kmis}@us.edu.pl

Abstract. In this paper we investigate the generalized hypothetical syllogism (GHS) in fuzzy logic, which can be seen as the functional equation $\sup_{z\in[0,1]} T(I(x,z), I(z,y)) = I(x,y)$, where I is a fuzzy implication and T is a t-norm. Our contribution is inspired by the article [Fuzzy Sets Syst 323:117–137 (2017)], where the author considered (GHS) when T is the minimum t-norm. We show several general results and then we focus on R-implications. We characterize all t-norms which satisfy (GHS) with arbitrarily fixed R-implication generated from a left-continuous t-norm.

Keywords: Fuzzy connectives · Fuzzy implication · T-norm
Generalized hypothetical syllogism

1 Introduction

Fuzzy relations can be seen as generalizations of characteristic functions of binary relations. The first sup-min composition of two fuzzy relations R and S was introduced by Zadeh [9] in the following way

$$(R \circ S)(x,y) := \sup_{z\in Z} \min\{R(x,z), S(z,y)\}, \qquad x \in X,\ y \in Y,$$

where R is a fuzzy relation between (standard) sets X and Z, while S is a fuzzy relation between (standard) sets Z and Y. Of course, $R \circ S$ is also a fuzzy relation between (standard) sets X and Y. Since fuzzy implications are special case of fuzzy relations, we can consider the following sup-T composition for them, replacing minimum by any t-norm T:

$$(I \overset{T}{\circ} J)(x,y) := \sup_{z\in[0,1]} T(I(x,z), J(z,y)), \qquad x,y \in [0,1]. \tag{1}$$

Note that the above sup-T composition is also noted in the literature in the algebraic notation as sup-$*$ composition as follows

$$(I \overset{*}{\circ} J)(x,y) = \sup_{z\in[0,1]} (I(x,z) * J(z,y)), \qquad x,y \in [0,1].$$

J. Medina et al. (Eds.): IPMU 2018, CCIS 853, pp. 673–684, 2018.
https://doi.org/10.1007/978-3-319-91473-2_57

It is interesting that the sup-T composition of two fuzzy implications need not be a fuzzy implication. There are known results when $I \overset{T}{\circ} J$ is such and even when particular fuzzy implications with a specified t-norm form a semigroup (see [2] or [1, Sect. 6.4]). For the sup-T composition (1) we can consider the following functional equation:

$$I \overset{T}{\circ} I = I, \tag{2}$$

where I is a fuzzy implication (or, in general, any generalization of classical implication) and T is a t-norm (or, in general, any generalization of classical conjunction). Note that (2) can be seen as I is a sup-T-idempotent fuzzy implication.

Equation (2) appeared also in 1995 in the book [7] and it was applied for the notion of generalized hypothetical syllogism.

Definition 1.1. *Let T be a t-norm and I be a fuzzy implication. We say that the pair (T, I) satisfies the generalized hypothetical syllogism if*

$$\sup_{z\in[0,1]} T(I(x,z), I(z,y)) = I(x,y), \qquad x, y \in [0,1], \tag{GHS}$$

i.e., when Eq. (2) is satisfied.

In classical logic, the hypothetical syllogism can be seen as follows

$$\frac{P \rightarrow Q,\ Q \rightarrow R}{\therefore \quad P \rightarrow R}$$

where P, Q and R are some propositions. Taking into account the compositional rule of inference (CRI), also introduced by Zadeh in [9], this schema has been translated into fuzzy logic as in (GHS). Recently, (GHS) has been analysed by Vemuri in [8] in the particular case, namely when T is the minimum t-norm. That article motivated us to deeper insight into the notion of generalized hypothetical syllogism. We would like to investigate it in a general case – with any t-norm T.

In this paper we show some properties of fuzzy implication functions preserving (GHS). Moreover, we focus on R-implications and we give conditions for them to satisfy (GHS). The paper is organized as follows. Section 2 contains basic definitions and properties regarding mainly t-norms and fuzzy implications. In Sect. 3 we show some general properties of (GHS). Finally, in Sect. 4, we present the main results for R-implications, in particular sufficient and necessary condition for R-implications generated from left-continuous t-norms.

2 Preliminaries

We assume that the reader is familiar with the classical results concerning basic fuzzy connectives (see [1,3,4,6]), but to make this work more self-contained, we recall main definitions and facts useful in the sequel.

Let us denote by Φ the family of all increasing bijections $\varphi\colon [0,1] \to [0,1]$. We say that functions $F, G\colon [0,1]^n \to [0,1]$ are Φ-conjugate, if there exists $\varphi \in \Phi$ such that $G = F_\varphi$, where $F_\varphi(x_1, \ldots, x_n) := \varphi^{-1}(F(\varphi(x_1), \ldots, \varphi(x_n)))$, for all $x_1, \ldots, x_n \in [0,1]$.

Definition 2.1. *A function $T\colon [0,1]^2 \to [0,1]$ is called a t-norm, if it satisfies the following conditions, for all $x, y, z \in [0,1]$,*

(T1) $T(x,y) = T(y,x)$,
(T2) $T(x, T(y,z)) = T(T(x,y), z)$,
(T3) $T(x,y) \leq T(x,z)$ *for* $y \leq z$, *i.e.,* $T(x,\cdot)$ *is non-decreasing,*
(T4) $T(x,1) = x$.

Another essential fuzzy connective is a fuzzy negation.

Definition 2.2. *A non-increasing function $N\colon [0,1] \to [0,1]$ is called a fuzzy negation, if $N(0) = 1$, $N(1) = 0$. Moreover, a fuzzy negation N is called strong, if it is an involution, i.e., $N(N(x)) = x$ for all $x \in [0,1]$.*

Now, we recall the definition and some important properties of fuzzy implications.

Definition 2.3. *A function $I\colon [0,1]^2 \to [0,1]$ is called a fuzzy implication, if it satisfies the following conditions:*

(I1) *I is non-increasing with respect to the first variable,*
(I2) *I is non-decreasing with respect to the second variable,*
(I3) $I(0,0) = I(1,1) = 1$ *and* $I(1,0) = 0$.

Definition 2.4 ([1, Definition 1.4.15]). *Let I be a fuzzy implication. A function $N_I\colon [0,1] \to [0,1]$ given by $N_I(x) := I(x,0)$, for all $x \in [0,1]$, is called the natural negation of I.*

Definition 2.5 (see [1]). *We say that a fuzzy implication I satisfies*

(i) the identity principle, if

$$I(x,x) = 1, \qquad x \in [0,1], \tag{IP}$$

(ii) the left neutrality property, if

$$I(1,y) = y, \qquad y \in [0,1], \tag{NP}$$

(iii) the ordering property, if

$$x \leq y \Longleftrightarrow I(x,y) = 1, \qquad x, y \in [0,1]. \tag{OP}$$

Definition 2.6. *A function $I\colon [0,1]^2 \to [0,1]$ is called an R-implication if there exists a t-norm T such that*

$$I(x,y) = \sup\{t \in [0,1] \mid T(x,t) \leq y\}, \qquad x, y \in [0,1]. \tag{3}$$

If I an R-implication generated from T, then it will be denoted by I_T.

3 Properties of Fuzzy Implications Preserving (GHS)

We know that for any fixed t-norm T and two fuzzy implications I, J, the sup-T composition $I \overset{T}{\circ} J$ is also a fuzzy implication if and only if $(I \overset{T}{\circ} J)(1,0) = 0$ (see [2, Theorem 1]. Our first result shows when the sup-T composition of two fuzzy implications satisfying (GHS) preserves this property.

Theorem 3.1. *Let T be a left-continuous t-norm and I, J be fuzzy implications. If the pairs $(T, I), (T, J)$ satisfy* (GHS)*, then also the pair $(T, I \overset{T}{\circ} J)$ satisfies* (GHS).

Proof. Let us take $x, y, t, w, z \in [0,1]$ and for simplifying the notation let us denote: $a = I(x,t), b = J(t,w), c = I(w,z), d = J(z,w), e = I(w,t), f = J(t,y)$. Now, using only associativity and commutativity of the t-norm T we obtain the following equalities

$$\begin{aligned}
&T(T(T(a,b),T(e,f)),T(d,c)) = T(T(d,c),T(T(a,b),T(e,f)))\\
&\quad = T(T(T(d,c),T(a,b)),T(e,f)) = T(T(T(a,b),T(d,c)),T(e,f))\\
&\quad = T(T(a,b),T(T(d,c),T(e,f))) = T(T(a,b),T(T(e,f),T(d,c)))\\
&\quad = T(T(a,T(T(e,f),T(d,c))),b) = T(T(a,T(T(f,e),T(d,c))),b)\\
&\quad = T(T(T(a,T(f,e)),T(d,c)),b) = T(T(T(T(a,f),e),T(d,c)),b)\\
&\quad = T(T(T(a,f),T(e,T(d,c))),b) = T(T(T(a,f),T(T(d,c),e)),b)\\
&\quad = T(T(T(T(a,f),T(d,c)),e),b) = T(T(T(a,f),T(d,c)),T(e,b)).
\end{aligned}$$

Let us rewrite the left and right sides of the above equation,

$$\begin{aligned}
&T(T(T(I(x,t),J(t,w)),T(I(w,t),J(t,y))),T(J(z,w),I(w,z)))\\
&\qquad = T(T(T(I(x,t),J(t,y)),T(J(z,w),I(w,z))),T(I(w,t),J(t,w))).
\end{aligned}$$

Taking supremum over t in $[0,1]$, from the left-continuity of T, for the left-hand side of the above equation we obtain

$$\begin{aligned}
&\sup_{t\in[0,1]} T(T(T(I(x,t),J(t,w)),T(I(w,t),J(t,y))),T(J(z,w),I(w,z)))\\
&= T(T(\sup_{t\in[0,1]} T(I(x,t),J(t,w)), \sup_{t\in[0,1]} T(I(w,t),J(t,y))),T(J(z,w),I(w,z)))\\
&= T(T((I \overset{T}{\circ} J)(x,w),(I \overset{T}{\circ} J)(w,y)),T(J(z,w),I(w,z))),
\end{aligned}$$

while for the right-hand side of the above equation we have

$$\begin{aligned}
&\sup_{t\in[0,1]} T(T(T(I(x,t),J(t,y)),T(J(z,w),I(w,z))),T(I(w,t),J(t,w)))\\
&= T(T(\sup_{t\in[0,1]} T(I(x,t),J(t,y)),T(J(z,w),I(w,z))), \sup_{t\in[0,1]} T(I(w,t),J(t,w)))\\
&= T(T((I \overset{T}{\circ} J)(x,y),T(J(z,w),I(w,z))),(I \overset{T}{\circ} J)(w,w)).
\end{aligned}$$

Again, taking supremum over w in $[0,1]$, we have for the left-hand side

$$\begin{aligned}
&\sup_{w\in[0,1]} T(T((I \overset{T}{\circ} J)(x,w),(I \overset{T}{\circ} J)(w,y)),T(J(z,w),I(w,z))) \\
&\quad = T(\sup_{w\in[0,1]} T((I \overset{T}{\circ} J)(x,w),(I \overset{T}{\circ} J)(w,y)), \sup_{w\in[0,1]} T(J(z,w),I(w,z))) \\
&\quad = T(((I \overset{T}{\circ} J) \overset{T}{\circ} (I \overset{T}{\circ} J))(x,y),(J \overset{T}{\circ} I)(z,z)),
\end{aligned}$$

while for the right-hand side we obtain

$$\begin{aligned}
&\sup_{w\in[0,1]} T(T((I \overset{T}{\circ} J)(x,y),T(J(z,w),I(w,z))),(I \overset{T}{\circ} J)(w,w)) \\
&\quad = T(T((I \overset{T}{\circ} J)(x,y), \sup_{w\in[0,1]} T(J(z,w),I(w,z))), \sup_{w\in[0,1]} (I \overset{T}{\circ} J)(w,w)) \\
&\quad = T(T((I \overset{T}{\circ} J)(x,y),(J \overset{T}{\circ} I)(z,z)),1) \\
&\quad = T((I \overset{T}{\circ} J)(x,y),(J \overset{T}{\circ} I)(z,z)).
\end{aligned}$$

Now, taking supremum over z in $[0,1]$ we have

$$\begin{aligned}
&\sup_{z\in[0,1]} T(((I \overset{T}{\circ} J) \overset{T}{\circ} (I \overset{T}{\circ} J))(x,y),(J \overset{T}{\circ} I)(z,z)) \\
&\qquad\qquad = \sup_{z\in[0,1]} T((I \overset{T}{\circ} J)(x,y),(J \overset{T}{\circ} I)(z,z)),
\end{aligned}$$

thus, using again the fact that T is left-continuous, we conclude

$$\begin{aligned}
&T(((I \overset{T}{\circ} J) \overset{T}{\circ} (I \overset{T}{\circ} J))(x,y), \sup_{z\in[0,1]} (J \overset{T}{\circ} I)(z,z)) \\
&\qquad\qquad = T((I \overset{T}{\circ} J)(x,y), \sup_{z\in[0,1]} (J \overset{T}{\circ} I)(z,z)),
\end{aligned}$$

hence

$$T(((I \overset{T}{\circ} J) \overset{T}{\circ} (I \overset{T}{\circ} J))(x,y),1) = T((I \overset{T}{\circ} J)(x,y),1),$$

therefore

$$((I \overset{T}{\circ} J) \overset{T}{\circ} (I \overset{T}{\circ} J))(x,y) = (I \overset{T}{\circ} J)(x,y),$$

which means that the pair $(T, I \overset{T}{\circ} J)$ satisfies (GHS). □

Note that the assumption of the left-continuity of a t-norm T is essential in Theorem 3.1.

Example 3.2. Let us take the drastic t-norm $T_{\mathbf{D}}$ given by

$$T_{\mathbf{D}}(x,y) = \begin{cases} 0, & x,y \in [0,1), \\ \min\{x,y\}, & \text{otherwise}, \end{cases} \qquad x,y \in [0,1], \tag{4}$$

and the following two fuzzy implications (see [8, Lemma 4.2]), for $x,y \in [0,1]$,

$$I(x,y) = \begin{cases} 1, & y = 1, \\ N(x), & y < 1, \end{cases} \qquad J(x,y) = \begin{cases} 1, & x = 0, \\ y^2, & x > 0, \end{cases}$$

where N is any fuzzy negation. It is easy to check that the pairs $(T_{\mathbf{D}}, I)$, $(T_{\mathbf{D}}, J)$ satisfy (GHS). Let us take the particular fuzzy negation N given by

$$N(x) = \begin{cases} 1, & x < 1, \\ 0, & x = 1, \end{cases} \qquad x \in [0,1].$$

By standard calculations we obtain

$$(I \overset{T_{\mathbf{D}}}{\circ} J)(x,y) = \begin{cases} 1, & x < 1, \\ y^2, & x = 1, \end{cases} \qquad x,y \in [0,1].$$

However, $((I \overset{T_{\mathbf{D}}}{\circ} J) \overset{T_{\mathbf{D}}}{\circ} (I \overset{T_{\mathbf{D}}}{\circ} J))(x,y) = 1$, for all $x,y \in [0,1]$. Hence, the pair $(T_{\mathbf{D}}, I \overset{T_{\mathbf{D}}}{\circ} J)$ does not satisfy (GHS).

We know (see [1, Theorem 6.1.1]) that the family of all fuzzy implications $(\mathcal{FI}, \leq)$ is a complete, completely distributive lattice with the lattice operations

$$(I \vee J)(x,y) := \max\{I(x,y), J(x,y)\}, \qquad x,y \in [0,1], \tag{5}$$

$$(I \wedge J)(x,y) := \min\{I(x,y), J(x,y)\}, \qquad x,y \in [0,1]. \tag{6}$$

Let us now consider two examples which show that the above lattice operations $\vee, \wedge$ of two fuzzy implications satisfying (GHS) may not preserve it.

Example 3.3. Let us take two kinds of fuzzy implications (see [8, Lemma 4.2]):

$$I(x,y) = \begin{cases} 1, & y = 1, \\ N(x), & y < 1, \end{cases} \qquad x,y \in [0,1],$$

where N is a fuzzy negation and

$$J(x,y) = \begin{cases} 1, & x = 0, \\ \varphi(y), & x > 0, \end{cases} \qquad x,y \in [0,1],$$

where $\varphi \in \Phi$. One can check that both pairs $(T_{\mathbf{LK}}, I)$ and $(T_{\mathbf{LK}}, I)$ satisfy (GHS), where $T_{\mathbf{LK}}$ is the Łukasiewicz t-norm given by $T_{\mathbf{LK}}(x,y) = \max\{x+y-1, 0\}$, for $x,y \in [0,1]$ (in fact both families satisfy (GHS) with any t-norm).

(i) By standard calculations we obtain

$$(I \wedge J)(x,y) = \begin{cases} 1, & x = 0 \text{ or } y = 1, \\ \min\{N(x), \varphi(y)\}, & \text{otherwise,} \end{cases} \qquad x, y \in [0,1].$$

Take $N(x) = 1 - x$ and $\varphi(x) = \sqrt{x}$, $x \in [0,1]$. Then we have

$$((I \wedge J) \overset{T_{\mathbf{LK}}}{\circ} (I \wedge J))(0.5, 0.25) = 0 \neq 0.5 = (I \wedge J)(0.5, 0.25).$$

Hence, the pair $(T_{\mathbf{LK}}, I \wedge J)$ does not satisfy (GHS).

(ii) Observe that

$$(I \vee J)(x,y) = \begin{cases} 1, & x = 0 \text{ or } y = 1, \\ \max\{N(x), \varphi(y)\}, & \text{otherwise,} \end{cases} \qquad x, y \in [0,1].$$

Take $N(x) = 1 - x^2$ and $\varphi(x) = x$, $x \in [0,1]$. Then we have

$$((I \vee J) \overset{T_{\mathbf{LK}}}{\circ} (I \vee J))(1, 0) = 0.25 \neq 0 = (I \vee J)(1, 0).$$

Again, the pair $(T_{\mathbf{LK}}, I \vee J)$ does not satisfy (GHS).

Below we give some conditions, when it is possible to obtain lattice operations in the family of fuzzy implications satisfying (GHS).

Theorem 3.4. *Let T be a t-norm and I, J be fuzzy implications. Further, let the pairs (T, I) and (T, J) satisfy* (GHS). *Then the following statements are equivalent:*

(i) The pair $(T, (I \wedge J))$ satisfies (GHS).

(ii) $((I \overset{T}{\circ} J) \wedge (J \overset{T}{\circ} I)) \geq (I \wedge J)$.

Proof. Let us take any $x, y, z \in [0,1]$. Using the monotonicity of T we can write

$$\begin{aligned} T&((I \wedge J)(x,z), (I \wedge J)(z,y)) \\ &= \min\{T(I(x,z), I(z,y)), T(I(x,z), J(z,y)), \\ &\qquad T(J(x,z), I(z,y)), T(J(x,z), J(z,y))\}. \end{aligned}$$

Because of the continuity of minimum, taking supremum over z in $[0,1]$ we obtain:

$$\begin{aligned} &\sup_{z \in [0,1]} T((I \wedge J)(x,z), (I \wedge J)(z,y)) \\ &\quad = \min\{\sup_{z \in [0,1]} T(I(x,z), I(z,y)), \sup_{z \in [0,1]} T(I(x,z), J(z,y)), \\ &\qquad\qquad \sup_{z \in [0,1]} T(J(x,z), I(z,y)), \sup_{z \in [0,1]} T(J(x,z), J(z,y))\}, \end{aligned}$$

thus

$$\left((I \wedge J) \overset{T}{\circ} (I \wedge J)\right)(x, y) = \min\{I, I \overset{T}{\circ} J, J \overset{T}{\circ} I, J\}.$$

Therefore, the pair $(T, I \wedge J)$ satisfies (GHS) if and only if $I \overset{T}{\circ} J \geq I \wedge J$ and $J \overset{T}{\circ} I \geq I \wedge J$, which is equivalent to the following inequality $(I \overset{T}{\circ} J) \wedge (J \overset{T}{\circ} I) \geq I \wedge J$. □

The next fact can be proven in a similar way.

Theorem 3.5. *Let T be a t-norm and I, J be fuzzy implications. Further, let the pairs (T, I) and (T, J) satisfy* (GHS). *Then the following statements are equivalent:*

(i) The pair $(T, (I \vee J))$ satisfies (GHS).
(ii) $((I \overset{T}{\circ} J) \vee (J \overset{T}{\circ} I)) \leq (I \vee J)$.

4 R-implications Satisfying (GHS)

Let us begin this section with several remarks regarding any pairs (T, I) of t-norms T and fuzzy implications I satisfying (GHS).

Lemma 4.1 (cf. [8]). *Let I be a fuzzy implication and T be a t-norm. If the pair (T, I) satisfies* (GHS), *then $T(I(1, x), I(x, 0)) = 0$, for all $x \in [0, 1]$. If, in addition, I satisfies* (NP), *then*

$$T(x, N(x)) = 0, \qquad x \in [0, 1], \tag{7}$$

for any fuzzy negation $N \leq N_I$.

Equation (7) means that the pair (T, N) satisfies the law of contradiction (see [1, Definition 2.3.14]).

Remark 4.2. Let T be a t-norm and N be a fuzzy negation. If the pair (T, N) satisfies (7), then T is not necessarily continuous. For instance, if $N = N_{D_1}$ given by

$$N_{D_1}(x) = \begin{cases} 1, & x = 0, \\ 0, & x > 0, \end{cases} \qquad x \in [0, 1],$$

then T can be any t-norm. However, if T, N are both continuous, then they should have the following forms (see [1, Proposition 2.3.15]), for any $x, y \in [0, 1]$,

$$T(x, y) = \varphi^{-1}(\max\{\varphi(x) + \varphi(y) - 1, 0\}) \text{ and } N(x) \leq \varphi^{-1}(1 - \varphi(x)),$$

where $\varphi \in \Phi$, thus T is Φ-conjugate with the Łukasiewicz t-norm.

This makes a difference between the minimum t-norm which satisfies (7) only with N_{D_1} (see [8, Lemma 2.2 (ii)]).

Lemma 4.3. *If a fuzzy implication I satisfies* (OP)*, then the pair* $(T_{\mathbf{D}}, I)$ *satisfies* (GHS)*, where* $T_{\mathbf{D}}$ *is the drastic t-norm given by* (4).

Proof. Fix arbitrarily $x, y \in [0,1]$ and let I be a fuzzy implication that satisfies (OP). We consider two cases. If $x \le y$, then $I(x,y) = 1$ and

$$1 \ge \sup_{z\in[0,1]} T_{\mathbf{D}}(I(x,z), I(z,y)) \ge T_{\mathbf{D}}(I(x,x), I(x,y)) = T_{\mathbf{D}}(1,1) = 1.$$

If $x > y$, then

$$\begin{aligned}
&\sup_{z\in[0,1]} T_{\mathbf{D}}(I(x,z), I(z,y)) \\
&\quad = \max\{\sup_{z\in[0,y]} T_{\mathbf{D}}(I(x,z), I(z,y)), \sup_{z\in(y,x)} T_{\mathbf{D}}(I(x,z), I(z,y)), \\
&\qquad\qquad \sup_{z\in[x,1]} T_{\mathbf{D}}(I(x,z), I(z,y))\} \\
&\quad = \max\{\sup_{z\in[0,y]} I(x,z), \sup_{z\in(y,x)} \min\{I(x,z), I(z,y)\}, \sup_{z\in[x,1]} I(z,y)\} \\
&\quad = \max\{I(x,y), I(x,y), I(x,y)\} = I(x,y),
\end{aligned}$$

hence the pair $(T_{\mathbf{D}}, I)$ satisfies (GHS). □

In the sequel we need the following two classical operations:

(i) the minimum t-norm $T_{\mathbf{M}}(x,y) = \min(x,y)$,

(ii) the Gödel implication $I_{\mathbf{GD}}(x,y) = \begin{cases} 1, & \text{if } x \le y, \\ y, & \text{if } x > y. \end{cases}$

In 2017 Vemuri shown the following fact.

Theorem 4.4 ([8, Theorem 3.6]). *For a left-continuous t-norm T the following statements are equivalent:*

(i) The pair $(T_{\mathbf{M}}, I_T)$ *satisfies* (GHS).
(ii) $I_T = I_{\mathbf{GD}}$.

Since the Gödel implication $I_{\mathbf{GD}}$ is an R-implication generated from the minimum t-norm $T_{\mathbf{M}}$ (see [1, Example 2.5.3]), the above theorem can be written in the following form.

Corollary 4.5. *Let T be a left-continuous t-norm. The pair* $(T_{\mathbf{M}}, I_T)$ *satisfies* (GHS) *iff* $T = T_{\mathbf{M}}$.

Now, the natural question is the following.

Question 4.6. Does, for any t-norm T, the pair (T, I_T) satisfy (GHS)?

The answer for the above question is no.

Example 4.7. Let us take the drastic t-norm $T_{\mathbf{D}}$ given by (4). The R-implication generated from it is the Weber implication (see [1, Example 2.5.3]) given by

$$I_{\mathbf{WB}}(x,y) = \begin{cases} 1, & x < 1, \\ y, & x = 1, \end{cases} \qquad x, y \in [0,1].$$

If $x = 1$ and $y < 1$, then we obtain

$$\begin{aligned}
(I_{\mathbf{WB}} \overset{T_{\mathbf{D}}}{\circ} I_{\mathbf{WB}})(x,y) &= \sup_{z\in[0,1]} T_{\mathbf{D}}(I_{\mathbf{WB}}(x,z), I_{\mathbf{WB}}(z,y)) \\
&= \max\{T_{\mathbf{D}}(I_{\mathbf{WB}}(1,1), I_{\mathbf{WB}}(1,y)), \sup_{z\in[0,1)} T_{\mathbf{D}}(I_{\mathbf{WB}}(1,z), I_{\mathbf{WB}}(z,y))\} \\
&= \max\{T_{\mathbf{D}}(1,y), \sup_{z\in[0,1)} T_{\mathbf{D}}(z,1)\} = \max\{y, \sup_{z\in[0,1)} z\} \\
&= \max\{y, 1\} = 1 \neq y = I_{\mathbf{WB}}(x,y).
\end{aligned}$$

Therefore, we analyse now the situation when T is a left-continuous t-norm.

Question 4.8. Let T be a left-continuous t-norm. Is there always a t-norm T^* such that the pair (T^*, I_T) satisfies (GHS)?

The answer for this question is yes. Since every R-implication I_T generated from a left-continuous t-norm T satisfies (OP), by Lemma 4.3 the pair $(T_{\mathbf{D}}, I_T)$ satisfies (GHS). In fact, this result holds also for the border-continuous t-norms (cf. [1, Proposition 2.5.9]. Now, the next questions occur.

Question 4.9. Let T be a left-continuous t-norm.

(i) Does the pair (T, I_T) satisfy (GHS)?
(ii) If so, is T the only one left-continuous t-norm such that the pair (T, I_T) satisfies (GHS)?

To give answers for the above questions, we recall the following result.

Theorem 4.10 ([5, Theorem 6]). *Let U be a nonempty set, F, G, H be fuzzy sets on U, $\pi, \kappa\colon [0,1]^2 \to [0,1]$. If*

(i) π is reflexive, i.e., $\pi(r,r) = 1$, for all $r \in [0,1]$,
(ii) $\kappa(1,r) = r$, for all $r \in [0,1]$,
(iii) π is κ-transitive, i.e., $\kappa(\pi(r,s), \pi(s,t)) \leq \pi(r,t)$, for all $r,s,t \in [0,1]$,
(iv) $\forall_{x\in U}\, \exists_{y\in U}\, F(x) = G(y)$,

then

$$\pi(F(x), H(y)) = \sup\{\kappa(\pi(F(x), G(z)), \pi(G(z), H(y))) : z \in U\}.$$

This general result can be applied for Question 4.9 (i), with $\kappa = T$, $\pi = I_T$, $F(x) = H(x) = G(x) = x$, $x \in U = [0,1]$. Note that for any left-continuous t-norm T, the R-implication I_T is T-transitive (see [4, Proposition 1.6]). Hence, we have showed the following fact.

Corollary 4.11. *If T is a left-continuous t-norm, then the pair (T, I_T) satisfies* (GHS).

Now, we give the answer for Question 4.9 (ii). It is main new result in our manuscript.

Theorem 4.12. *Let T^* be a t-norm and T be a left-continuous t-norm. Then the following statements are equivalent:*

(i) The pair (T^*, I_T) satisfies (GHS).
(ii) $T^* \leq T$.

Proof. ((*ii*) $\Longrightarrow$ (*i*)) It follows from the monotonicity of any t-norm and Theorem 4.10.

((*i*) $\Longrightarrow$ (*ii*)) Here we use the fact that any R-implication satisfies (NP) (see [1, Theorem 2.5.4]) and for any left-continuous t-norm T the following equality holds (see [1, Proposition 2.5.2] or [4]):

$$T(x, z) \leq y \Leftrightarrow I_T(x, y) \geq z, \qquad x, y, z \in [0, 1].$$

Since $T(x, y) \leq T(x, y)$, then

$$I_T(x, T(x, y)) \geq y, \qquad x, y, z \in [0, 1]. \tag{8}$$

Therefore, from (NP), by (GHS), by the monotonicity of T^* and from (8) we obtain, for all $x, y \in [0, 1]$,

$$\begin{aligned} T(x, y) = I_T(1, T(x, y)) &= \sup_{z \in [0,1]} T^*(I_T(1, z), I_T(z, T(x, y))) \\ &\geq T^*(I_T(1, x), I_T(x, T(x, y))) \geq T^*(x, y). \end{aligned}$$

□

As we already mentioned, for the t-norm $T_{\mathbf{M}}$ there is only one left-continuous R-implication I such that $(T_{\mathbf{M}}, I)$ satisfies (GHS) – it is the the Gödel implication $I_{\mathbf{GD}}$ (see Theorem 4.4 and Corollary 4.5). Now, we can see the reason for it and we can present the other proof of this fact. Let us take as T^* the minimum t-norm $T_{\mathbf{M}}$. From Theorem 4.12 it follows: if for a left-continuous t-norm T the pair $(T_{\mathbf{M}}, I_T)$ satisfies (GHS), then $T_{\mathbf{M}} \leq T$. But $T_{\mathbf{M}}$ is the greatest t-norm, hence $T = T_{\mathbf{M}}$. Therefore, [8, Theorem 3.6] can obtained as a corollary from our Theorem 4.12.

Finally, note that there exist pairs (T, I) satisfying (GHS), when T is not a left-continuous t-norm and I does not satisfy (IP) and (OP). For instance, $T = T_{\mathbf{D}}$ and $I = I_{\mathbf{KD}}$ the Kleene-Dienes implication given by the formula $I_{\mathbf{KD}}(x, y) = \max\{1 - x, y\}$.

5 Conclusions

In this manuscript we have investigated some general properties of fuzzy implications preserving (GHS). We dealt with R-implications generated from left-continuous t-norms and shown that for such t-norm T, the pair (T, I_T) satisfies (GHS). We also characterized all t-norms which satisfy (GHS) with arbitrarily fixed R-implication generated from a left-continuous t-norm.

Acknowledgment. This work has been supported by the National Science Centre, Poland, through Project Number 2015/19/B/ST6/03259.

References

1. Baczyński, M., Jayaram, B.: Fuzzy Implications. Studies in Fuzziness and Soft Computing, vol. 231. Springer, Berlin Heidelberg (2008). https://doi.org/10.1007/978-3-540-69082-5
2. Baczyński, M., Drewniak, J., Sobera, J.: Semigroups of fuzzy implications. Tatra Mt. Math. Publ. **21**, 61–71 (2001)
3. Fodor, J., Roubens, M.: Fuzzy Preference Modelling and Multicriteria Decision Support. Kluwer Academic Publishers, Dordrecht (1994). https://doi.org/10.1007/978-94-017-1648-2
4. Gottwald, S.: Fuzzy Sets and Fuzzy Logic: The Foundations of Application—From a Mathematical Point of View. Vieweg-Verlag, Braunschweig (1993). https://doi.org/10.1007/978-3-322-86812-1
5. Igel, C., Temme, K.-H.: The chaining syllogism in fuzzy logic. IEEE Trans. Fuzzy Syst. **12**, 849–853 (2004). https://doi.org/10.1109/TFUZZ.2004.836078
6. Klement, E.P., Mesiar, R., Pap, E.: Triangular Norms. Kluwer Academic Publishers, Dordrecht (2000). https://doi.org/10.1007/978-94-015-9540-7
7. Klir, G.J., Yuan, B.: Fuzzy Sets and Fuzzy Logic: Theory and Applications. Prentice Hall, Upper Saddle River (1995)
8. Vemuri, N.R.: Investigations of fuzzy implications satisfying generalized hypothetical syllogism. Fuzzy Sets Syst. **323**, 117–137 (2017). https://doi.org/10.1016/j.fss.2016.08.008
9. Zadeh, L.A.: Outline of a new approach to the analysis of complex systems and decision processes. IEEE Trans. Syst. Man Cyber. **3**, 28–44 (1973). https://doi.org/10.1109/TSMC.1973.5408575

Fuzzy Logic and Artificial Intelligence Problems

Interval Type-2 Intuitionistic Fuzzy Logic Systems - A Comparative Evaluation

Imo Eyoh[1(✉)], Robert John[1], and Geert De Maere[2]

[1] Automated Scheduling, Optimisation and Planning (ASAP) and Laboratory for Uncertainty in Data and Decision Making Research Groups, University of Nottingham, Nottingham, UK
{ije,rij}@cs.nott.ac.uk

[2] Automated Scheduling, Optimisation and Planning (ASAP) Research Group, University of Nottingham, Nottingham, UK
gdm@cs.nott.ac.uk

Abstract. Several fuzzy modeling techniques have been employed for handling uncertainties in data. This study presents a comparative evaluation of a new class of interval type-2 fuzzy logic system (IT2FLS) namely: interval type-2 intuitionistic fuzzy logic system (IT2IFLS) of Takagi-Sugeno-Kang (TSK)-type with classical IT2FLS and its type-1 variant (IFLS). Simulations are conducted using a real-world gas compression system (GCS) dataset. Study shows that the performance of the proposed framework with membership functions (MFs) and non-membership functions (NMFs) that are each intervals is superior to classical IT2FLS with only MFs (upper and lower) and IFLS with MFs and NMFs that are not intervals in this problem domain.

Keywords: Interval type-2 intuitionistic fuzzy logic systems
Membership functions · Non-membership functions
Decoupled extended Kalman filter

1 Introduction

A fuzzy set (FS) is a concept introduced by Zadeh [1] where elements are not restricted to binary MFs of 0 or 1 but rather is a continuum in 0 and 1. However due to the complexity and uncertainty in many applications, the ordinary FS cannot handle or minimise the uncertainty in many applications because the MF values are exactly defined. To cope with this problem, Zadeh [2] introduced type-2 FS (T2FS) where the MFs are fuzzy with the actual membership grade of an element assumed to lie within a closed interval of 0 and 1. Generally, the definition of classical FSs, both type-1 and type-2 employ only the MFs to define the two concepts: MF and NMF of an element to a set. For classical FSs, the NMF (complement) is 1 minus the membership grade. This definition may not really be the case in real life scenario as they may be some hesitation in the definition of membership degree of an element to a set such that the NMF is not complementary to the MF and vice-versa.

J. Medina et al. (Eds.): IPMU 2018, CCIS 853, pp. 687–698, 2018.
https://doi.org/10.1007/978-3-319-91473-2_58

Atanassov [3] introduced the concept of intuitionistic FS (IFS) with MF and NMF separately defined such that none is complementary to the other. In the literature, IFSs have been found to be one of the useful tools for dealing with imprecise information [4]. The reader is referred to [5] for more details on IFS. Similar to classical type-1 FSs, the MFs and NMFs of IFSs may not handle the plethora of uncertainty that fraught many applications. The T2FS introduced by Zadeh is a three dimensional structure which provides the extra degrees of freedom needed to handle higher forms of uncertainty. For the generalised T2FS, the third dimension is weighted differently which makes it complex and difficult to use [6,7]. The simpler and manageable version - interval type-2 FS (IT2FS) - have values in the third dimension equal to 1 and this makes it easier for IT2FS to be represented on a two dimensional plane. The IFSs and IT2FSs have been widely and extensively adopted by researchers in uncertainty modeling in many applications. For example, in Nguyen et al. [8], IT2 fuzzy C-mean (FCM) clustering using IFS is proposed. The authors show that the use of IFS with IT2FCM led to improved clustering quality particularly in the presence of noise. In Naim and Hagras [9,10] and Naim et al. [11], IFS and IT2FS are combined to develop a multi-criteria group decision making (MCGDM) system for the assessment of post-graduate study, selection of appropriate lighting level in intelligent environment and evaluation of different techniques for the choice of illumination in a shared environments respectively. The authors pointed out that the use of IFS and IT2FS in a MCGDM system provided decisions that are closer to the group decisions compared to some existing methods. However, in [9–11], only the IT2FS MFs are utilised and the intuitionistic fuzzy (IF) indices are evaluated on the primary MFs of the IT2FSs and no learning whatsoever is carried out on these sets.

Recently, in Eyoh et al. [12], IFS is fused with IT2FS and an interval type-2 intuitionistic FS (IT2IFS) is obtained with artificial neural network learning capability. The uncertainties in IT2IFLS that utilises IT2IFS in the IF-THEN rules are captured by the footprints of uncertainties (FOUs) of both MFs and NMFs. The developed model in [12] is applied for non-linear system prediction with encouraging results. The same authors in Eyoh et al. [13,14] applied the IT2IFLS framework for time series prediction. Results reveal that IT2IFLS exhibits superior performance to many non-fuzzy and some fuzzy approaches. The authors believe that the additional parameters provided by the NMFs give IT2IFS more design degrees of freedom thus allowing it to minimise the effects of uncertainties in many applications than the classical IT2FLS and IFLS.

Different learning methodologies have been proposed for the adaptation of the parameters of fuzzy logic systems. The study reported here utilises the DEKF to optimise the parameters of the models under investigation. The rest of the paper is structured as follows: In Sect. 1, the definitions of IFS, T2IFS and IT2IFS are given. The IT2IFLS-TSK model is formulated in Sect. 2 and parameter update rules are derived in Sect. 3. We present our experimental set-up and statistical evaluation and discussion in Sects. 4 and 5 respectively, and conclude in Sect. 6. Following are the definitions of some concepts underpinning the proposed model.

1.1 Intuitionistic Fuzzy Set (IFS)

Definition 1. *For any given finite set X, an IFS has the form: $A^* = \{(x, \mu_{A^*}(x), \nu_{A^*}(x)) : x \in X)\}$. Where $\mu_{A^*}(x) : X \to [0,1]$ is the degree of belonging of x in X and $\nu_{A^*}(x) : X \to [0,1]$ is the degree of non-belonging of $x \in X$ with $0 \leq \mu_{A^*}(x) + \nu_{A^*}(x) \leq 1$ [3].*

The IF-index, $\pi_{A^*}(x) = 1 - (\mu_{A^*}(x) + \nu_{A^*}(x))$.

1.2 Type-2 Intuitionistic Fuzzy Set (T2IFS)

Definition 2. *A T2IFS is of the form: $\tilde{A}^* = \{(x,u), \mu_{\tilde{A}^*}(x,u), \nu_{\tilde{A}^*}(x,u) \mid \forall x \in X, \ \forall u \in J_x^{\mu}, \forall u \in J_x^{\nu}\}$ where $\mu_{\tilde{A}^*}(x,u)$ is the degree of belonging and $\nu_{\tilde{A}^*}(x,u)$ is the degree of non-belonging [12].*

$$J_x^{\mu} = \left\{(x,u) : u \in \left[\underline{\mu}_{\tilde{A}^*}(x), \overline{\mu}_{\tilde{A}^*}(x)\right]\right\}. \quad (1)$$

$$J_x^{\nu} = \{(x,u) : u \in [\underline{\nu}_{\tilde{A}^*}(x), \overline{\nu}_{\tilde{A}^*}(x)]\}. \quad (2)$$

in which $0 \leq \mu_{\tilde{A}^*}(x,u) \leq 1$ and $0 \leq \nu_{\tilde{A}^*}(x,u) \leq 1$, J_x^{μ} and J_x^{ν} represent the support of the secondary MFs and NMFs on the third dimension [12].

The T2IFS $\tilde{A}^*$ can also be formulated as:

$$\tilde{A}^* = \int_{x \in X} \left[\int_{u \in J_x^{\mu}} \int_{u \in J_x^{\nu}} \{\mu_{\tilde{A}^*}(x,u), \nu_{\tilde{A}^*}(x,u)\}\right] / (x,u). \quad (3)$$

for continuous universe of discourse (UoD). For discrete UoD, $\tilde{A}^*$ becomes

$$\tilde{A}^* = \sum_{x \in X} \left[\sum_{u \in J_x^{\mu}} \sum_{u \in J_x^{\nu}} \{\mu_{\tilde{A}^*}(x,u), \nu_{\tilde{A}^*}(x,u)\}\right] / (x,u). \quad (4)$$

When $\mu_{\tilde{A}^*}(x,u) = 1$, and $\nu_{\tilde{A}^*}(x,u) = 1$, an interval type-2 intuitionistic FS (IT2IFS) is obtained. The IF-indices utilised are defined as follows [12]:

$$\pi_c(x) = max\left(0, \left(1 - \left(\mu_{\tilde{A}^*}(x) + \nu_{\tilde{A}^*}(x)\right)\right)\right). \quad (5)$$

$$\overline{\pi}_{var}(x) = max\left(0, \left(1 - \left(\overline{\mu}_{\tilde{A}^*}(x) + \underline{\nu}_{\tilde{A}^*}(x)\right)\right)\right). \quad (6)$$

$$\underline{\pi}_{var}(x) = max\left(0, \left(1 - \left(\underline{\mu}_{\tilde{A}^*}(x) + \overline{\nu}_{\tilde{A}^*}(x)\right)\right)\right). \quad (7)$$

Two footprints of uncertainties (FOUs) defined for IT2IFS are as follows [12]:

$$FOU_{\mu}\left(\tilde{A}^*\right) = \bigcup_{\forall x \in X} \left[\underline{\mu}_{\tilde{A}^*}(x), \bar{\mu}_{\tilde{A}^*}(x)\right]. \quad (8)$$

$$FOU_{\nu}\left(\tilde{A}^*\right) = \bigcup_{\forall x \in X} \left[\underline{\nu}_{\tilde{A}^*}(x), \bar{\nu}_{\tilde{A}^*}(x)\right]. \quad (9)$$

denoting MF and NMF FOUs respectively.

2 Interval Type-2 Intuitionistic Fuzzy Logic System

The structure of T2FLS rule based system is exactly as that of the classical T2FLS with components comprising of the intuitionistic fuzzifier, intuitionistic rule base, intuitionistic inference engine and intuitionistic output processing module.

2.1 Fuzzification

The fuzzification process maps the crisp input vector $x \in X$ into an IT2IFS $\tilde{A}^*$ thereby assigning to each element its MF and NMF degree in each IT2IFS partition.

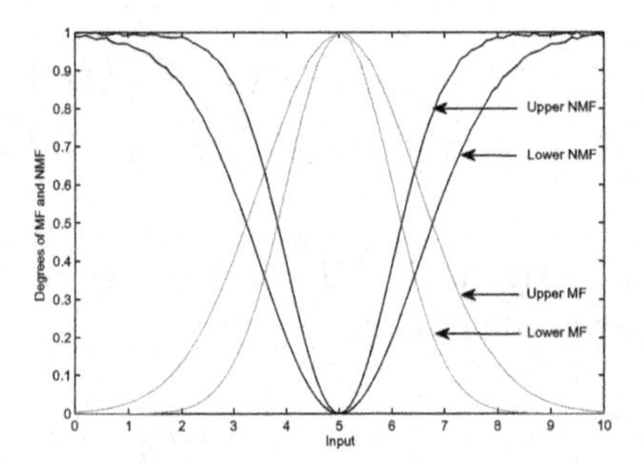

Fig. 1. A typical IT2IFS [12]

The IT2IFS is defined using a modified Gaussian function (see Fig. 1 and (10) to (13)) [12]. The scaling and shifting allows for the degree of indeterminacy to be embedded in the FOUs of IT2IFLS.

$$\overline{\mu_{ik}}(x_i) = exp\left(-\frac{(x_i - c_{ik})^2}{2\bar{\sigma}^2_{2,ik}}\right) * (1 - \pi_{c,ik}(x_i)). \tag{10}$$

$$\underline{\mu_{ik}}(x_i) = exp\left(-\frac{(x_i - c_{ik})^2}{2\underline{\sigma}^2_{1,ik}}\right) * (1 - \pi_{c,ik}(x_i)). \tag{11}$$

$$\overline{\nu_{ik}}(x_i) = (1 - \overline{\pi}_{var,ik}(x_i)) - \left[exp\left(-\frac{(x_i - c_{ik})^2}{2\bar{\sigma}^2_{1,ik}}\right) * (1 - \pi_{c,ik}(x_i))\right]. \tag{12}$$

$$\underline{\nu_{ik}}(x_i) = (1 - \underline{\pi}_{var,ik}(x_i)) - \left[exp\left(-\frac{(x_i - c_{ik})^2}{2\underline{\sigma}^2_{2,ik}}\right) * (1 - \pi_{c,ik}(x_i))\right]. \tag{13}$$

where $\pi_{c,ik}$ is the IF-index of center and $\pi_{var,ik}$ is the IF-index of variance [4].

2.2 Rules

The IF-THEN rule formation of IT2IFLS follows the general fuzzy logic rule syntax. For IT2IFLS-TSK, the antecedent are IT2IFSs while the consequent parts are linear combinations of the inputs. A typical rule structure of IT2IFLS is as shown in (14).

$$R_k : \text{IF } x_1 \text{ is } \tilde{A}^*{}_{1k} \text{ and} \cdots \text{and } x_n \text{ is } \tilde{A}^*{}_{nk} \text{ THEN } y_k = \sum_{i=1}^{n} w_{ik} x_i + b_k. \quad (14)$$

For IT2IFLS, the rules are defined for MFs in (15) and NMFs in (16) as follows:

$$R_k^\mu : \text{IF } x_1 \text{ is } \tilde{A}^*{}_{1k}^{\mu} \text{ and} \cdots \text{and } x_n \text{ is } \tilde{A}^*{}_{nk}^{\mu} \text{ THEN } y_k^\mu = \sum_{i=1}^{n} w_{ik}^\mu x_i + b_k^\mu. \quad (15)$$

$$R_k^\nu : \text{IF } x_1 \text{ is } \tilde{A}^*{}_{1k}^{\nu} \text{ and} \cdots \text{and } x_n \text{ is } \tilde{A}^*{}_{nk}^{\nu} \text{ THEN } y_k^\nu = \sum_{i=1}^{n} w_{ik}^\nu x_i + b_k^\nu. \quad (16)$$

where $\tilde{A}^*{}_{1k}, \tilde{A}^*{}_{2k}, \cdots, \tilde{A}^*{}_{ik}, \cdots, \tilde{A}^*{}_{nk}$ are IT2IFS, y_k^μ and y_k^ν are the MFs and NMFs outputs of the kth rule, w's and b's are consequent parameters.

2.3 Inference

This study adopts a TSK-inferencing system where the inputs are IT2IFS and the output of each IF-THEN rule is a linear function of the inputs, otherwise known as A2-C0 TSK-fuzzy inferencing. The inference engine of IT2IFLS is defined in (17) [12,13].

$$y = \frac{(1-\beta)\sum_{k=1}^{M}\left(\underline{f_k^\mu} + \overline{f_k^\mu}\right) y_k^\mu}{\sum_{k=1}^{M}\underline{f_k^\mu} + \sum_{k=1}^{M}\overline{f_k^\mu}} + \frac{\beta\sum_{k=1}^{M}\left(\underline{f_k^\nu} + \overline{f_k^\nu}\right) y_k^\nu}{\sum_{k=1}^{M}\underline{f_k^\nu} + \sum_{k=1}^{M}\overline{f_k^\nu}}. \quad (17)$$

and utilises the "prod" t-norm to specify the firing strength such that:

$$\underline{f_k^\mu}(x) = \prod_{i=1}^{n} \underline{\mu}_{\tilde{A}^*{}_{ik}}(x_i), \quad \overline{f_k^\mu}(x) = \prod_{i=1}^{n} \overline{\mu}_{\tilde{A}^*{}_{ik}}(x_i). \quad (18)$$

$$\underline{f_k^\nu}(x) = \prod_{i=1}^{n} \underline{\nu}_{\tilde{A}^*{}_{ik}}(x_i), \quad \overline{f_k^\nu}(x) = \prod_{i=1}^{n} \overline{\nu}_{\tilde{A}^*{}_{ik}}(x_i). \quad (19)$$

where $\underline{f}_k$ and $\overline{f}_k$ are the lower and upper firing strengths defined for both MFs and NMFs respectively, y_k^μ and y_k^ν are the outputs of the kth rule corresponding to MF and NMF respectively. The final output of IT2IFLS-TSK is a weighted average of each IF-THEN rule's output. The parameter β ($0 \leq \beta \leq 1$) weighs the contribution of the MF and NMF values in the final output.

3 Parameter Update

In this section, the parameter update rules for IT2IFLS using DEKF is derived. For comparison, the DEKF is also used to update the parameters of classical IT2FLS and IFLS.

3.1 Extended Kalman Filter Parameter Update Rule

The purpose of IT2IFLS prediction is to obtain an accurate an estimate as possible between input-output relationship of a system. Let the output of IT2IFLS be $y = f(X, \theta)$. The parameter X denotes the inputs into the system and θ is used to represent the unknown parameters of the model. The generic state equation of the non-linear system can be expressed as:

$$\theta_{t+1} = f(\theta_t) + \omega_t. \tag{20}$$

$$y_t = h(\theta_t) + \upsilon_t. \tag{21}$$

where θ is the system's state, ω is the process noise with zero mean and covariance Q and υ is the measurement noise with zero mean and covariance R. The process and measurement noise are assumed to be Gaussian and uncorrelated and:

$$\begin{aligned} E(\theta_0) = \overline{\theta}_0, \qquad & E[(\theta_0 - \overline{\theta}_0)(\theta_0 - \overline{\theta}_0)^T] = P_0, \\ E(\omega_t) = 0, \qquad & E(\omega_t \omega_l^T) = Q\delta_{tl}, \\ E(\upsilon_t) = 0, \qquad & E(\upsilon_t \upsilon_l^T) = R\delta_{tl}. \end{aligned} \tag{22}$$

where $E(.)$ is the expectation operator and δ_{tl} is the Kronecker delta. Using Taylor expansion, the state is estimated as:

$$\begin{aligned} f(\theta_t) &= f(\hat{\theta}_t) + F_t(\theta_t - \hat{\theta}_t) + H.O.T. \\ h(\theta_t) &= h(\hat{\theta}_t) + H_t(\theta_t - \hat{\theta}_t) + H.O.T. \end{aligned} \tag{23}$$

where:

$$F_t = \left.\frac{\partial f(\theta)}{\partial \theta}\right|_{\theta=\hat{\theta}_t} \qquad \text{and} \qquad H_t^T = \left.\frac{\partial h(\theta)}{\partial \theta}\right|_{\theta=\hat{\theta}_t}.$$

and H.O.T is the higher order term. The system in (23) can be approximated as in (24) when the H.O.Ts are neglected.

$$\begin{aligned} \theta_{t+1} &= F_t\theta_t + \omega_t + \phi_t. \\ y_{t+1} &= H_t^T\theta_t + \upsilon_t + \varphi_t. \end{aligned} \tag{24}$$

where:

$$\phi_t = f(\hat{\theta}_t) - F_t\hat{\theta}_t.$$

$$\varphi_t = h(\hat{\theta}_t) - H_t\hat{\theta}_t.$$

The order of the computational cost of EKF is AB^2, where A is the output dimension of the system and B is the number of parameters. An IT2IFLS with n inputs and M rules has a total of $6n+2M(n+1)$ parameters. The computational cost of EKF for IT2IFLS is $36n^2 + 4M^2(n^2 + 2n + 1) + 24nM(n + 1)$.

3.2 Decoupled Extended Kalman Filter - DEKF

In using the DEKF to learn the parameters of IT2IFLS, the antecedent and the consequent parameters for both MFs and NMFs are grouped into two vectors namely: antecedent (θ^1) and consequent (θ^2) parameter vectors. The generic parameter update rules for the parameters in the *ith* group is expressed in (25) to (27):

$$\theta_t^i = \theta_{t-1}^i + K_t^i[y_t - h(\theta_{t-1})]. \tag{25}$$

$$K_t^i = P_t^i H_t^i[(H_t^i)^T P_t^i H_t^i + R^i]^{-1}. \tag{26}$$

$$P_{t+1}^i = P_t^i - K_t^i P_t^i (H_t^i)^T + Q^i. \tag{27}$$

where K is the Kalman gain, P is the covariance matrix of the state estimation error. The unknown parameters in the antecedent are gathered into the antecedent's parameter vector and represented as (28):

$$\theta^1 = [c_{11}, c_{21}, \cdots, c_{nN}, \sigma_{11}, \sigma_{21}, \cdots, \sigma_{nN}]^T. \tag{28}$$

where n is the number of inputs and N is the number of linguistic terms. The parameters of the consequent are grouped into the consequent's parameter vector and represented as (29):

$$\theta^2 = [w_{11}, w_{21}, \cdots, w_{Mn}, b_1, b_2, \cdots, b_M]^T. \tag{29}$$

where M is the number of rules, with the MF and NMFs having separate Kalman parameters. The derivative matrix, H, is defined as:

$$H^1 = \frac{\partial y}{\partial \theta^1} \quad \text{and} \quad H^2 = \frac{\partial y}{\partial \theta^2}. \tag{30}$$

for antecedent and consequent parameters respectively. The update rule for the antecedent and consequent parameters then follow the same recursive procedures in (25) to (27). Using the DEKF algorithm reduces the computational burden of EKF in the order $36n^2 + 4M^2(n^2 + 2n + 1)$ such that the computational cost of DEKF to standard EKF is in the ratio:

$$\frac{36n^2 + 4M^2(n^2 + 2n + 1)}{36n^2 + 4M^2(n^2 + 2n + 1) + 24nM(n + 1)}$$

In the next sections, the IT2IFLS is used for the prediction of a real-world dataset - GCS dataset of a gas turbine obtained from a Nigerian-based power plant. The purpose of this simulation is to statistically analyse the performance of IT2IFLS with other existing FLSs such as the classical IT2FLS and the type-1 IFLS. The DEKF learning approach is adopted for these experimental analyses because of its theoretical strength, faster convergence and its ability at finding good solutions [15].

4 Experimental Set-Up

The GCS data is a complex dataset consisting of different operational conditions of a gas plant. The GCS data consist of 825 data points and modeled as a time series using input generating format: $[y(t-3), y(t-2), y(t-1)]$ with $y(t)$ as the output. The inputs are normalised to lie between small range of $[0, 1]$, so that larger input values do not overshadow the smaller values, thereby leading to poor prediction and learning using the embedded neural network architecture. For each run of the experiments, the data are randomly sampled and split into 70% training and 30% testing set with each data point having equal chance of being chosen for training and testing. For a clear and objective discussion and evaluation of the three models of IT2IFLS, IT2FLS and IFLS, the Kalman filter parameters R, Q and P for both MFs and NMFs are initially set as $40, 0.01I_{32}$ and $1.0I_{32}$ respectively for all experiments with 100 epochs for each run. The simulation is conducted for 30 runs. This allows for objective evaluation of the performance of the different models. Figures 2 and 3 show a single GCS input partition before and after training using IT2IFLS.

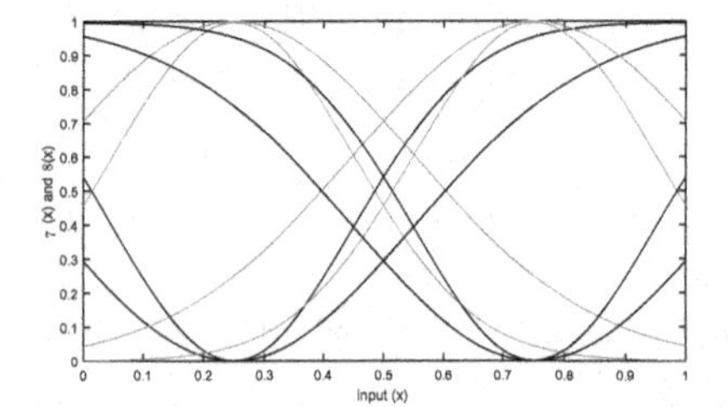

Fig. 2. GCS x_1 before training with IT2IFLS-DEKF

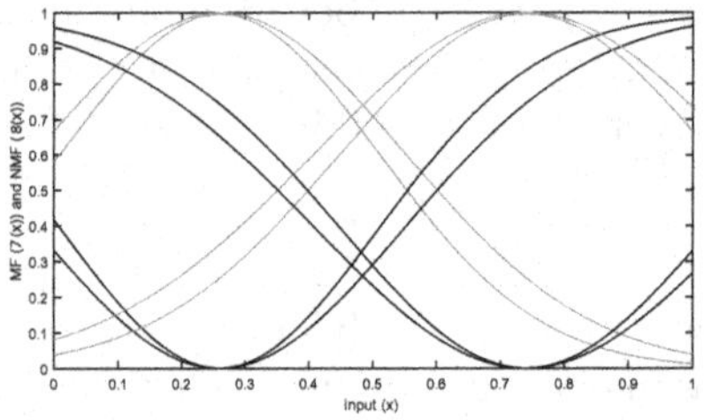

Fig. 3. GCS x_1 after training with IT2IFLS-DEKF

5 Statistical Evaluation and Discussion

In this section, statistical evaluation is conducted to test the hypothesis of this research. The main interest is to understand the effectiveness of integrating NMF and IF-indices into the classical IT2FLS (IT2IFLS). The second is to investigate the performance of the proposed framework of IT2IFLS with its type-1 counterpart. Statistical comparison is also made between IFLS and IT2FLS. To explore these analyses, three experiments are conducted. In each case, the performance metric is the root mean squared error (RMSE). Figure 4 shows the box-plot of the three models using their test RMSEs. The following hypotheses form the basis of evaluation:

- Hypothesis 1: With the integration of NMFs and IF-indices into the classical IT2FLS, the new model of IT2IFLS is able to model uncertainty in many applications than the classical IT2FLS that do not incorporate NMFs and IF-indices.

- Hypothesis 2: With MFs and NMFs that are intervals, the new model of IT2IFLS is able to model uncertainty in many applications than its type-1 variant with MFs and NMFs that are not represented as intervals values.
- Hypothesis 3: With MFs and NMFs of IFLS, the model is able to model uncertainty in many applications than the classical IT2FLS with only interval MFs (lower and upper).

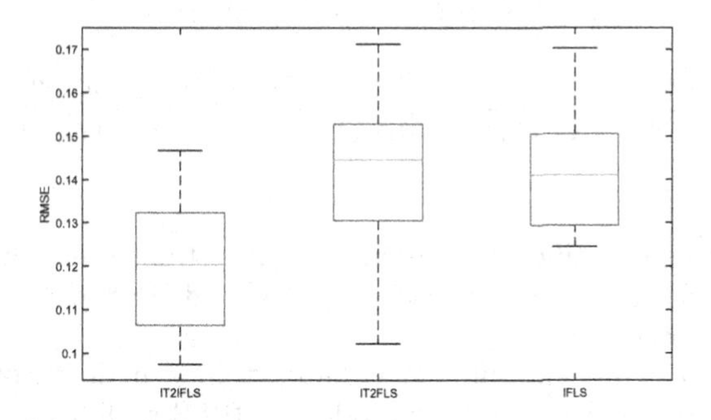

Fig. 4. Box and whisker plot showing the performance of IT2IFLS, IT2FLS and IFLS trained with DEKF.

Statistical significant differences (one-tailed) between pairs of models are carried out using Wilcoxon signed rank test (α level = 0.05), which is a non-parametric statistical hypothesis test for evaluating pairs of models. The test RMSEs averaged over 30 runs for the different fuzzy models considered here are presented in Table 1.

Hypothesis 1: The first set of experiments is focused on assessing the ability of IT2IFLS framework to provide good estimates than the classical IT2FLS. The null and the alternative hypotheses are:

- H_0: There is no significant difference in the uncertainty modeling employing IT2FLS that incorporates NMFs and IF-indices and those that do not.
- H_1: There is a significant difference in the uncertainty modeling of IT2FLS that incorporates NMFs and IF-indices and those that do not.

For the first hypothesis, the statistical analysis in Table 2 shows that there is a significant difference between the uncertainty modeling using the two FLSs (p-value = 0.0173). Based on this premise, the null hypothesis is rejected with a conclusion that there is a significant difference between the performance of IT2IFLS compared to the classical IT2FLS. As observed in the box-and-whisker plot, IT2IFLS has a smaller error value on average. This observation demonstrates the advantages of NMFs and IF-indices as an integral part of IT2FLS (IT2IFLS).

Hypothesis 2: The second set of experiments is focused on assessing the ability of IT2IFLS framework to provide good estimates than its type-1 counterpart. The null and the alternative hypotheses are:

Table 1. Gas compression system prediction

Models	Training RMSE	Test RMSE
IT2FLS-TSK	0.1504	0.1425
IFLS-TSK	0.1496	0.1423
IT2IFLS-TSK	0.1202	0.1199

Table 2. Wilcoxon's test: IT2IFLS vs IT2FLS on test data RMSE

Models	Hypothesis ($\alpha = 0.05$)	p-value
IT2IFLS vs IT2FLS	Reject H_0	0.0173

- H_0: There is no significant difference in uncertainty modeling of IT2IFLS with MFs and NMFs that are each intervals and IFLS with MFs and NMFs that are not intervals.
- H_1: There is a significant difference in uncertainty modeling of IT2IFLS with MFs and NMFs that are each intervals and IFLS with MFs and NMFs that are not intervals.

Table 3. Wilcoxon's test: IT2IFLS vs IFLS on test data RMSE

Models	Hypothesis ($\alpha = 0.05$)	p-value
IT2IFLS vs IFLS	Reject H_0	0.0091

For the second hypothesis, the statistical analysis in Table 3 suggests that there is a significant difference between the uncertainty modeling using IT2IFLS compared to IFLS (p-value = 0.0091). With this p-value, the null hypothesis is rejected. It is concluded that there is a significant difference between the IT2IFLS and IFLS. This shows that IT2IFLSs with MFs and NMFs that are intervals may be more appropriate for uncertainty modeling than those with MFs and NMFs representations that are not intervals.

Hypothesis 3: The third set of experiments is to investigate the statistical significance between IT2FLS and IFLS. The null and the alternative hypotheses are:

- H_0: There is no significant difference in the performance of IT2FLS utilising upper and lower MFs of IT2FS and IFLS utilising MFs and NMFs.
- H_1: There is a significant difference in the performance of IT2FLS utilising upper and lower MFs of IT2FS and IFLS utilising MFs and NMFs.

Table 4 shows the results of statistical comparison between classical IT2FLS and type-1 IFLS. The Wilcoxon's signed rank test at 0.05 significance level shows that there is no significant difference (p-value = 0.7336) existing between classical IT2FLS and IFLS, therefore, we fail to reject the null hypothesis.

Table 4. Wilcoxon's test: IT2FLS vs IFLS using test data RMSE

Models	Hypothesis ($\alpha = 0.05$)	p-value
IT2FLS vs IFLS	Fail to reject H_0	0.7336

6 Conclusion

In this study, the DEKF is used to optimise the parameters of IT2IFLS, classical IT2FLS and type-1 IFLS. Specifically, the following conclusions are supported:

- IT2IFLS captures more information and enables hesitation in the FS description.
- There is significant performance improvements of IT2IFLS over IT2FLS and IFLS.
- The performance of the classical IT2FLS is comparable to that of IFLS.
- The IT2IFLS with MF and NMF that are intervals can minimise the effects of uncertainties in most applications.

In the future, we intend to conduct more experiments using bench mark data sets, other fuzzy modeling functions such as triangular and trapezoidal functions and other learning algorithms.

Acknowledgement. This research work was supported by the Government of Nigeria under the Tertiary Education Trust Fund (TETFund).

References

1. Zadeh, L.A.: Fuzzy sets. Inf. Control **8**(3), 338–353 (1965)
2. Zadeh, L.A.: The concept of a linguistic variable and its application to approximate reasoning-I. Inf. Sci. **8**, 199–249 (1975)
3. Atanassov, K.T.: Intuitionistic fuzzy sets. Fuzzy Sets Syst. **20**(1), 87–96 (1986)
4. Hájek, P., Olej, V.: Intuitionistic fuzzy neural network: the case of credit scoring using text information. In: Iliadis, L., Jayne, C. (eds.) EANN 2015. CCIS, vol. 517, pp. 337–346. Springer, Cham (2015). https://doi.org/10.1007/978-3-319-23983-5_31
5. Atanassov, K.T.: On Intuitionistic Fuzzy Sets Theory. Springer, Heidelberg (2012). https://doi.org/10.1007/978-3-642-29127-2
6. Mendel, J.M., John, R.I., Liu, F.: Interval type-2 fuzzy logic systems made simple. IEEE Trans. Fuzzy Syst. **14**(6), 808–821 (2006)
7. Mendel, J.M., John, R.B.: Type-2 fuzzy sets made simple. IEEE Trans. Fuzzy Syst. **10**(2), 117–127 (2002)
8. Nguyen, D.D., Ngo, L.T., Pham, L.T.: Interval type-2 fuzzy c-means clustering using intuitionistic fuzzy sets. In: IEEE Third World Congress on Information and Communication Technologies (WICT), pp. 299–304 (2013)
9. Naim, S., Hagras, H.: A hybrid approach for multi-criteria group decision making based on interval type-2 fuzzy logic and intuitionistic fuzzy evaluation. In: 2012 IEEE International Conference on Fuzzy Systems (FUZZ-IEEE), pp. 1–8. IEEE (2012)

10. Naim, S., Hagras, H.: A type 2-hesitation fuzzy logic based multi-criteria group decision making system for intelligent shared environments. Soft. Comput. **18**(7), 1305–1319 (2014)
11. Naim, S., Hagras, H., Bilgin, A.: Employing an interval type-2 fuzzy logic and hesitation index in a multi criteria group decision making system for lighting level selection in an intelligent environment. In: 2013 IEEE Symposium on Advances in Type-2 Fuzzy Logic Systems (T2FUZZ), pp. 1–8. IEEE (2013)
12. Eyoh, I., John, R., De Maere, G.: Interval type-2 intuitionistic fuzzy logic system for non-linear system prediction. In: 2016 IEEE International Conference on Systems, Man, and Cybernetics (SMC), pp. 001063–001068. IEEE (2016)
13. Eyoh, I., John, R., De Maere, G.: Time series forecasting with interval type-2 intuitionistic fuzzy logic systems. In: 2017 IEEE International Conference on Fuzzy Systems (FUZZ-IEEE), pp. 1–6, July 2017
14. Eyoh, I., John, R., De Maere, G.: Extended Kalman filter-based learning of interval type-2 intuitionistic fuzzy logic system. In: 2017 IEEE International Conference on Systems, Man and Cybernetics, pp. 728–733 (2017)
15. Simon, D.: Training fuzzy systems with the extended Kalman filter. Fuzzy Sets Syst. **132**(2), 189–199 (2002)

Artificial Neural Networks and Fuzzy Logic for Specifying the Color of an Image Using Munsell Soil-Color Charts

María Carmen Pegalajar[1](✉), Manuel Sánchez-Marañón[2], Luis G. Baca Ruíz[1], Luis Mansilla[1], and Miguel Delgado[1]

[1] Department of Computer Science and Artificial Intelligence, University of Granada, c/Periodista Daniel Saucedo Aranda s.n., 18071 Granada, Spain
{mcarmen,bacaruiz,mdelgado}@decsai.ugr.es, luis.mansilla@gmail.com

[2] Department of Soil Science and Agricultural Chemistry, University of Granada, Campus Fuentenueva s.n., 18071 Granada, Spain
msm@ugr.es

Abstract. The Munsell soil-color charts contain 238 standard color chips arranged in seven charts with Munsell notation. They are widely used to determine soil color by visual comparison, seeking the closest match between a soil sample and one of the chips. The Munsell designation of this chip (hue, value, and chroma) is assigned to the soil under study. However, the available chips represent only a subset of all possible soil colors, in which the visual appearance for an observer is usually intermediate between several chips. Our study proposes an intelligent system which combines two Soft Computing Techniques (Artificial Neural Networks and Fuzzy Logic Systems) aimed at finding a set of chips as similar as possible to a given soil sample. This is under the precondition that the soil sample is an image taken by a digital camera or mobile phone. The system receives an image as input and returns a set of color-chip designations as output.

Keywords: Munsell soil-color charts · Artificial Neural Networks · Fuzzy logic

Abbreviations

ANN	Artificial Neural Network
MCC	Munsell Color Chart
FS	Fuzzy System
HVC	Hue, Value, and Chroma
MSE	Mean Squared Error
RGB	Red Green Blue

J. Medina et al. (Eds.): IPMU 2018, CCIS 853, pp. 699–709, 2018.
https://doi.org/10.1007/978-3-319-91473-2_59

1 Introduction

Soil science deals with soils as a natural resource of the earth's crust [1]. Among soil properties, color is one of the most evident characteristics [2]. Furthermore, soil color is key in all major soil-classification systems, providing information about soil composition, genesis, aeration, organic-matter content, fertility, and weathering [3, 4].

The standard determination of soil color in the field is made by a visual method using the Munsell soil-color charts. The first set of charts was published in the USA in 1941, and since then they have been used worldwide for measuring soil color [2]. These charts consist of artificially colored papers (chips) mounted on constant hue (shade) cards, showing value (lightness) and chroma (intensity/saturation) variations in vertical and horizontal directions, respectively. Soil colors are determined by visual comparison. An observer seeks the closest match between the soil sample and one of the standard chips. Thus, the Munsell designation of this chip (hue, value, and chroma or HVC) is assigned to the soil sample under study. The edition of the Munsell soil-color charts from the year 2000 [5] contains 238 chips in seven hue cards: 10R, 2.5YR, 5YR, 7.5YR, 10YR, 25Y, and 5Y (see Table 1). For instance, the 10YR hue chart is shown in Fig. 1.

Table 1. Number of chips per chart.

Chart	10R	10YR	5Y	5YR	2.5Y	2.5YR	7.5YR	TOTAL
Chips	35	36	31	33	31	37	35	**238**

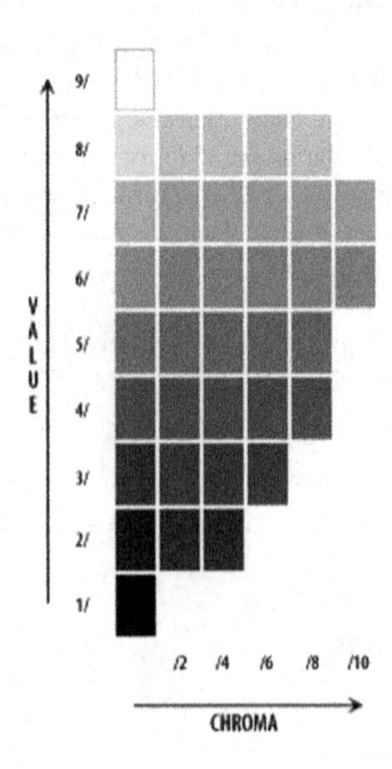

Fig. 1. Example of the 10YR Munsell color chart.

Several problems have previously been described to identify the color of soil samples using color charts, including the subjectivity of the human observer and the limited number of available chips to make a perfect match with the soil sample [6]. Other well-known difficulties include: (1) selecting the appropriate hue card, (2) determining colors that are intermediate between the hues in the charts, and (3) distinguishing between value and chroma where chromas are strong [5]. To overcome these and other limitations, the recent subdiscipline of digital soil morphometrics is promoting tools and techniques for digitally quantifying soil color from images taken with smartphones, digital cameras, and flatbed scanners [7]. The conversion from RGB values to Munsell parameters, by far the most familiar to soil scientists, uses equations that in no case reproduce the field practice of soil scientists using Munsell charts.

Little has been reported so far about models to identify soil color using machine learning. Artificial neural networks (ANN) were applied to estimate the soil moisture content from photographs taken with a digital camera, based on the fact that soil changes color according to its water content [8]. The authors used multilayer perceptron ANN varying the number of hidden neuron layers and three input variables (red, green, and blue). The networks were trained with color data of three soils in disturbed samples subjected to different water content in the laboratory. The best performing ANN had a hidden layer with 12 neurons and used the tan-sigmoid transfer function. Another specific approach for soil-color detection has also been reported using a 5-step algorithm [9]. First, a color database in RGB was created and then several segmentation methods including mean filter and k-Means were used to provide clearer images. The latter process involved matching each layer of image soil with a color in the database using Euclidean distance. Although the results were satisfactory in both study cases, none artificially reproduced the standard method of soil-color determination. The present study aims to lay the foundations for an intelligent artificial system to determine soil color using Munsell charts. In this preliminary approach, only color chips will be used to train and test the models.

2 Soil-Classification System Based on Munsell Chart Modelling

The goal of this work is to develop an intelligent system by applying Artificial Neural Networks (ANN) [10, 11] and Fuzzy Systems (FS) [12, 13] to generate an output set of Munsell color chips similar to an input image from a digital camera. The Fig. 2 diagrammatically illustrates this idea.

This study is the prologue to a more complex system in which real soil-color images will be considered in order to predict other soil features such as texture. The final objective would be to build a model that enables the study of a zone without the need of sampling or laboratory analyses [14]. Initially, however, we will try to recognize the color from an image taken by a mobile phone and a digital camera under uncontrolled lighting conditions [15].

Below, we analyze the Soft Computing techniques used, the data compiled, the design adopted, the experiments performed, and the final results found.

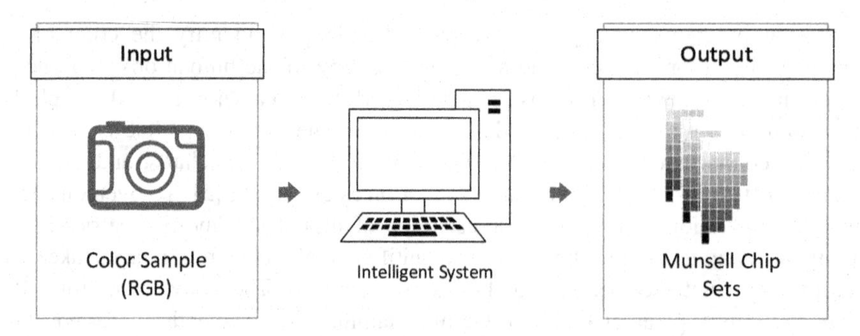

Fig. 2. General system diagram.

The proposed model should cover the following requirements and objectives:

1. The system receives the input data of color measurements expressed in terms of Red, Green, and Blue (RGB).
2. The system releases output data consisting of the most representative set of Munsell chips based on the RGB input.
3. Color measurements from a digital camera (whether independent or built into a smartphone) must be compatible with the system.

A two-phase system was used for our approach:

– *Phase 1*: Calculation of the *HVC* values related to a RGB measurement through an ANN.
– *Phase 2*: Selection of the Munsell color chips which most closely match the measurement made in *HVC* using FS.

3 Data Collection

A huge quantity of data is typically required for proper training, to ensure good system generalization. Additionally, the data should cover most of the probable classification space, so that when new data is presented to the system, it can reach the desired goal. The data training must be representative throughout the classification space.

The photographs were taken using three devices: two mobiles phones (Nokia C301 and Samsung Galaxy S2) and one digital camera (Canon EOS 1100D). An example is shown in Fig. 3. The images were taken on a grey background together with a reference white.

There are seven MCC, and seven pictures were taken for each chart and device. The charts have a specific number of chips listed in Table 1. An algorithm for image segmentation was implemented, which served as color meter according to each chip. The color meter returned the RGB coordinates. In this connection, each chip with a given HVC notation provided seven different RGB measurements.

In a controlled lighting system and assuming other constant factors such as angle or distance, the RGB values of the seven measures should be the same. Nevertheless, as

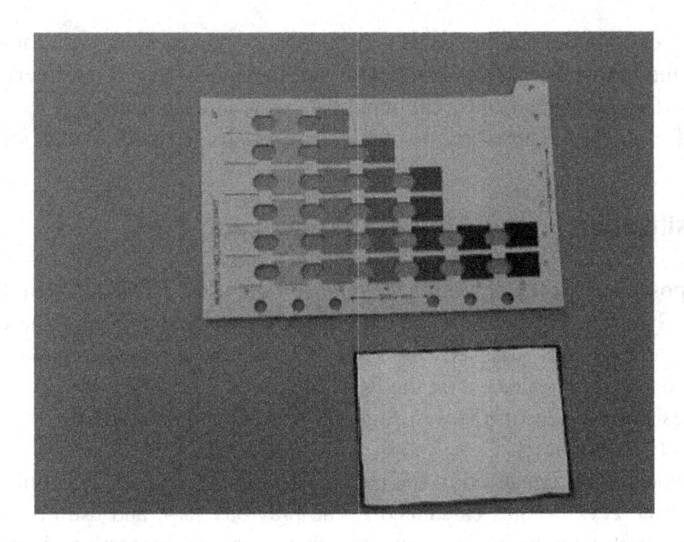

Fig. 3. Example of a photo taken from a Munsell color chart.

Table 2. Example of measurements made for the 10R 8/4 chip by the Nokia C301.

Colour	Measure						
	1	2	3	4	5	6	7
R	255	255	255	255	255	255	255
G	216	210	227	214	215	209	224
B	173	181	186	191	173	170	169

listed in Table 2, the RGB values were not constant. This happens because of the daylight conditions (lighting state), which varied between days and over one day [16]. For this reason, seven pictures were taken from each chart, for two days and different hours between 12:00 and 16:30 (Table 3).

Table 3. Days and hours of taking the images.

Day	1st	1st	1st	2nd	2nd	2nd	2nd
Hour	13:00	14:00	16:00	12:00	13:00	14:00	16:30

Hence, sufficient data were collected in order to solve the problem of the uniform color appearance. The photographs were taken without exposing the charts to direct sunlight in order to minimize specular light reflection (brightness). Sunny days were selected to ensure the appropriate brightness changes at that time, avoiding other meteorological factors. At the final point, the pictures were taken on a grey background while adding a white as a reference, in order to determine the white balance. A total of 1666 measurements were performed per device (238 chips per 7 images). The measurements were consolidated by an image-segmentation system, detailed in the following section. Note that the RGB values used to train these models came from the images of the Munsell charts. Accordingly, the human vision did not intervene at any time during the training process. The input data were labeled according to the Munsell notation of each chip. Therefore, the models are not biased with human error.

4 Classification System Based on ANN and FS

In our proposal, two Soft Computing techniques were used: Artificial Neural Networks and Fuzzy Systems. The joint use of these methods resulted in a two-phase system.

The first phase was focused on ANN [17, 18] as a regression mechanism, to compute the HVC values based on the RGB measures. The ANN directly received the RGB values from the picture (seven for each device) and returned the three numbers attached to HVC color.

The second phase consisted of the use of FS [19–21] as a classification technique. The FS got the HVC values calculated in the previous step, and estimated the set of chips most similar to the HVC input, together with a membership degree of each chip. Figure 4 shows the general functioning of the system.

The proper operation of the classification stage depended on the reliability of the HVC values calculated in the foregoing phase. With an erroneous HVC estimation, the chips selected by the FS would be incorrect. Therefore, a large part of the success of this work depended on minimizing the error in the ANN to calculate the HVC values.

For proper training and validation, the dataset has been split into two parts: 1190 color data were used to train the net (70% of 1666 measurements) and 476 for validation (30%). Each chip had 5 measurements in the training and 2 in the test set. The preparation of the datasets was the same for all devices. The cross-validation method was used for training all ANNs using the training set, and the validation involved the 476 instances which had not been seen by the model at any time.

Each measurement was composed of 6 values: 3 RGB coordinates and their 3 corresponding HVC chip values. The possible values of the hue component in the Munsell color were: 10R, 2.5YR, 5YR, 7.5YR, 10YR, 2.5Y, and 5Y. These were non-numerical values and the ANN could not work on them. Consequently, the analogous numeric values were adopted. We used the numbers recommended by the Munsell system: 10R = 10, 2.5YR = 12.5, 5YR = 15, 7.5YR = 17.5, 10YR = 20, 2.5Y = 22.5, and 5Y = 25.

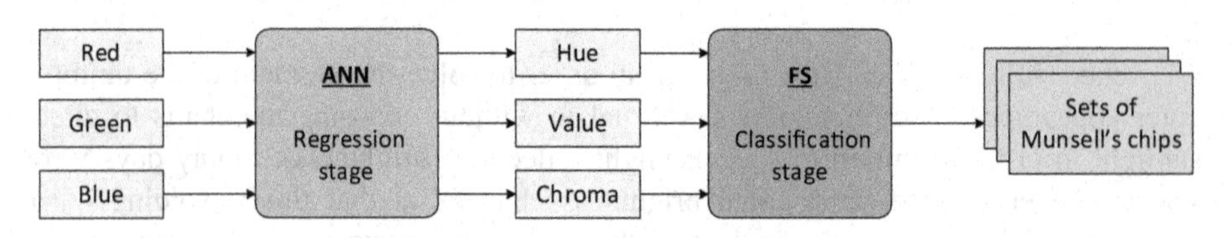

Fig. 4. Overall functioning schema of the first proposed design.

4.1 First Stage: Calculation of the HVC Values Based on ANN

This stage computed the HVC values related to RGB color measurements made for each chip. The functional requirements were to receive the RGB color samples as input and generate the equivalent HVC components as output.

Three ANN have been put into service for each device in order to compute the hue, value, and chroma associated with the RGB input. In total, 9 ANN were trained (Fig. 5).

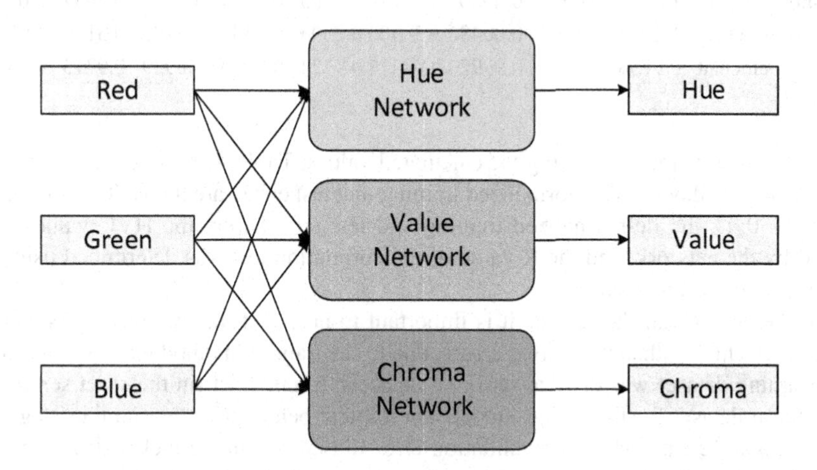

Fig. 5. Operating diagram of the suggested model in the first phase.

Neural Network Architecture

In this stage, 3 ANN per device were used, one for each Munsell coordinate (H, V, and C). All were made of three input nodes (one for each RGB component) and three output neurons (one for each H, V, and C corresponding returned value).

The operator can choose whether to establish a single or multi-hidden layer, since no universal guidelines determine the best ANN topology. For this study, one hidden layer was set, on the basis that ANNs with one hidden layer are universal approximations and that increasing the number of hidden layers lengthens the training time, which was demonstrated mathematically. The sigmoid function in the range $(-1, 1)$ has been used as the activation function in the hidden and output layers. The *Levenberg-Marquard* is the training method because, although it needs more computing resources, it converges faster than the classic *back propagation* algorithm. The mean square error (MSE) was used as the target-error goal. Finally, the input and output data were normalized within the interval $[0, 1]$.

Results of the First Stage

An optimum topology was searched for by trial and error. Special attention was placed on avoiding overtraining. The network was trained while the test error was below the training error. In fact, there were 60 hidden neurons. In addition, a correlation test was conducted for the test set, among the expected H, V, and C values, and the output of the

Table 4. ANN errors using the three devices. HN is the hue network, CN is the chroma network and VN is the value network. The first and second rows are the normalized errors for training and test. The third and fourth rows are the de-normalized errors. The last row is the correlation coefficient.

Error	Canon			Nokia			Samsung		
	HN	VN	CN	HN	VN	CN	HN	VN	CN
Norm. train	0.0030	0.0006	0.0001	0.0067	0.0044	0.0007	0.0065	0.0057	0.0006
Norm. test	0.0019	0.0011	0.0001	0.0057	0.0041	0.0006	0.0084	0.0050	0.0005
De-norm. train	1.8848	0.0363	0.0080	4.2114	0.2795	0.0439	4.0463	0.3644	0.0374
De-norm. test	1.2082	0.0694	0.0048	3.5850	0.2593	0.0392	5.2796	0.3179	0.0331
Corr. coefficient	0.9758	0.9934	0.9992	0.9247	0.9724	0.9946	0.8929	0.9713	0.9949

model, with the aim of checking the calculated values. Table 4 shows the errors of each ANN for each device. The normalized training and test errors are the HVC values in the interval $[0, 1]$, the de-normalized training and test errors were the HVC values computed by the network, and the R value of the correlation test was determined using the outputs and the test dataset.

In the analysis of the results, it is important to note that the training errors were in some cases higher than the testing errors, due to the training method used. In this study, the training dataset was divided into two datasets: training set (in the strict sense) and validation dataset for fitting the model. The typical behavior at the training stage is a continuous decrease while the validation error diminishes in parallel until it begins to increase. The training process should cease when the first minimum validation error is reached. Here, the table shows training and test errors but no validation error. If the training error is slightly higher than test, the model has reached an optimal solution.

On the one hand, the hue component registered a much greater error than value or chroma in all devices. The de-normalized test error of this component was more than 1, but the value and chroma errors were close to zero. Among the three devices, the Canon camera exhibited the lowest error, particularly in the hue value. This effect is consistent with the data analysis because the Canon measurements proved less sensitive to the light changes. In all cases, the value and chroma showed a small error, close to zero, and there was no substantial difference between the devices. On the other hand, the correlation coefficients were very high, for most of the instances, which constitutes a desired result, and it is acceptable even for the hue value, i.e. above the 0.89 (with the Samsung mobile). The results support the hypothesis concerning the hue prediction, which becomes the most important forecasting factor when classifying. Due to its higher predictive error, it is responsible for the classification made by the FS.

4.2 Second Stage: Fuzzy System

This phase determined the most representative chip set of the RGB values of the input image. The proposed design method must meet the following requirements: to use the calculated HVC values in the previous step as input and to return a set of MCC chips which are the ones most similar to HVC input.

Fuzzy Sets Assumed. As an initial step, the linguistic variables need to be identified; in this problem these are hue, value, and chroma. Afterwards, to divide the space associated with each variable, the fuzzy sets were proposed.

The hue value perceived a total of 7 possible labels: 10, 12.5, 15, 17.5, 20, 22.5 and 25. The triangle function has been used for its representation. Equation 1 defines this function, where a, b, and m are critical parameters which determine the shape and location of the functions [22]. Figure 6 illustrates an example of the fuzzy sets related to the linguistic variable value. All the membership functions of the figure follow the same Eq. (1). For instance, a is 7, b is 9 and m is 8 in the last function (blue):

$$\mu_A(x) = \begin{cases} 0, & \text{if } x \leq a \\ \frac{x-a}{m-a}, & \text{if } a < x \leq m \\ \frac{b-x}{b-m}, & \text{if } m < x < b \\ 0, & \text{if } x \geq b \end{cases} \tag{1}$$

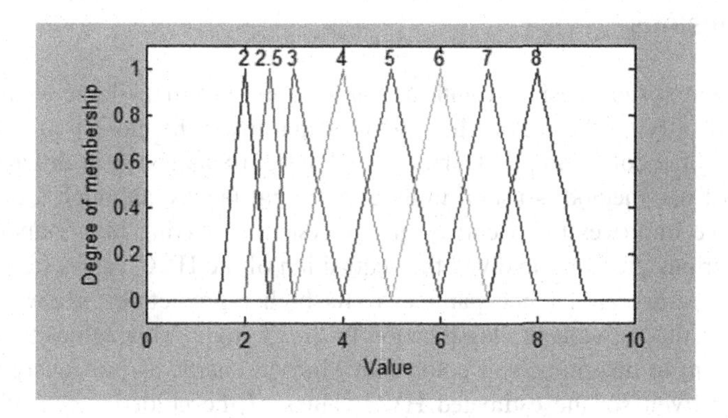

Fig. 6. Representation of the variable value using fuzzy sets.

The configuration of the value parameter has been performed in the same vein as before. In this case, the linguistic variable takes 8 different values: 2, 2.5, 3, 4, 5, 6, 7, and 8. Finally, as in previous variables, chroma is represented using the fuzzy sets with a triangular function. Its possible labels are: 1, 2, 3, 4, 6, and 8.

Furthermore, the rule-based fuzzy-logic system has a size of 238 if-then rules, one per chip. The general structure of the rules is: "*IF H is HUE and V is VALUE and C is CHROMA, then...*". For instance, the 10R 8/4 chip rule is represented as: "*IF H is 10 and V is 8 and CHROMA is 4, then 10R*".

The logic "and" indicated in the previous rule has been carried out by the product *t-norm.* In this sense, the fuzzy-logic system works with the HVC values as input taken from the previous stage by the ANN, and the values of the 238 activation rules are computed. The rule chips with a value other than 0 are positively identified. Subsequently, these values are ranked by their activation value.

Results of the Second Stage

Table 5 presents the errors made per device. The percentage of successes was estimated on the chip sets returned by the system. The chip is marked as correct if it is in the selected chips; otherwise, it is considered improperly classified. The success/failure results determine that the digital camera Canon is the best device, which is, moreover, much better compared with the other two instruments. In the performance comparisons of the two mobile phones, Nokia achieved a higher percentage of success than did Samsung, although the difference was not significant.

Table 5. Percentage of successes and failures.

Device	Percentage of successes	Percentage of failures
Canon	94.1176%	5.8824%
Nokia	76.4706%	23.5294%
Samsung	73.3193%	26.6807%

5 Conclusions

The intelligent system designed with neural-network and fuzzy-logic techniques succeeded in identifying the set of Munsell chips that made the closest color match to a RGB image of a color sample. Our proposed system was tested in different devices, showing that our method is useful in diverse circumstances, although a more sophisticated device improves the accuracy of our system by taking higher-quality photos. The most serious problem involved the acquisition of the HVC values from the neural network, in particular the hue parameter. Its higher predictive error was the main limitation for the subsequent classification by fuzzy logic. This agrees with the main visual problem to determine soil color with Munsell charts, as previously reported in soil studies. Even so, the estimated HVC values of the color image failed by only around 6% using a digital camera and around 24–27% using smartphones. Accordingly, the present study lays the foundations for the future construction of an Artificial Soil Scientist determining soil color in the field using Munsell soil-color charts.

References

1. Paz, C.G., Rodríguez, T.T., Behan-Pelletier, V.M., Hill, S.B., Vidal-Torrado, P., Cooper, M.: Encyclopedia of Soil Science. Springer, Dordrecht (2016). https://doi.org/10.1007/978-1-4020-3995-9_586
2. Soil Survey Staff: Soil survey manual. Agricultural Handbook 18 Government Printing Office, Washington, DC (1993)
3. Thwaites, R.: Color. In: Lal, R. (ed.) Encyclopedia of Soil Science, pp. 211–214. Marcel Dekkers, Inc. (2002)
4. Sánchez-Marañón, M.: Color indices, relationship with soil characteristics. In: Gliński, J., Horabik, J., Lipiec, J. (eds.) Encyclopedia of Agrophysics, pp. 141–145. Springer, Dordrecht (2011). https://doi.org/10.1007/978-90-481-3585-1_237
5. Munsell Color Company: Munsell soil color charts. Munsell color Company. Munsell Color Co., Baltimore, MD (2000)

6. Sánchez-Marañón, M., Huertas, R., Melgosa, M.: Colour variation in standard soil-colour charts. Soil Res. **43**(7), 827–837 (2005). https://doi.org/10.1071/SR04169
7. Gómez-Robledo, L., López-Ruiz, N., Melgosa, M., Palma, A.J., Capitán-Vallvey, L.F., Sánchez-Marañón, M.: Using the mobile phone as Munsell soil-colour sensor: an experiment under controlled illumination conditions. Comput. Electron. Agric. **99**, 200–208 (2013). https://doi.org/10.1016/j.compag.2013.10.002
8. Zanetti, S.S., Cecílio, R.A., Alves, E.G., Silva, V.H., Sousa, E.F.: Estimation of the moisture content of tropical soils using colour images and artificial neural networks. Catena **135**, 100–106 (2015). https://doi.org/10.1016/j.catena.2015.07.015
9. Utaminingrum, F., Robbani, I.H.: Scotect algorithm: a novel approach for soil color detection process using five steps algorithm. Int. J. Innov. Comput. Inf. Control **12**(5), 1645–1653 (2016)
10. Beucher, A., Møller, A.B., Greve, M.H.: Artificial neural networks for soil drainage class mapping in Denmark. In: 2016 7th Digital Soil Mapping (2016). http://digitalsoilmapping.org/fileadmin/digitalsoilmapping.org/Updated_book_of_abstract_for_publishing_online_260616.pdf#page=82
11. Jafarzadeh, A., Pal, M., Servati, M., FazeliFard, M., Ghorbani, M.: Comparative analysis of support vector machine and artificial neural network models for soil cation exchange capacity prediction. Int. J. Environ. Sci. Technol. **13**(1), 87–96 (2016). https://doi.org/10.1007/s13762-015-0856-4
12. Meléndez-Pastor, I., Pedreño, J.N., Lucas, I.G., Zorpas, A.A.: A model for evaluating soil vulnerability to erosion using remote sensing data and a fuzzy logic system. In: Modern Fuzzy Control Systems and its Applications. InTech (2017). https://doi.org/10.5772/67989
13. Akumu, C., Johnson, J., Etheridge, D., Uhlig, P., Woods, M., Pitt, D., McMurray, S.: GIS-fuzzy logic based approach in modeling soil texture: using parts of the Clay Belt and Hornepayne region in Ontario Canada as a case study. Geoderma **239**, 13–24 (2015). https://doi.org/10.1016/j.geoderma.2014.09.021
14. Stiglitz, R., Mikhailova, E., Post, C., Schlautman, M., Sharp, J.: Teaching soil color determination using an inexpensive color sensor. Nat. Sci. Edu, **45**(1) (2016). https://doi.org/10.4195/nse2016.03.0005
15. Stiglitz, R.Y.: Application of low-cost color sensor technology in soil data collection and soil science education (2017). http://search.proquest.com/docview/1964286298?accountid=14542
16. Sánchez-Marañón, M., García, P.A., Huertas, R., Hernández-Andrés, J., Melgosa, M.: Influence of natural daylight on soil color description: assessment using a color-appearance model. Soil Sci. Soc. Am. J. **75**(3), 984–993 (2011). https://doi.org/10.2136/sssaj2010.0336
17. Demuth, H.B., Beale, M.H., De Jess, O., Hagan, M.T.: Neural Network Design. Martin Hagan, Stillwater (2014). http://dl.acm.org/citation.cfm?id=2721661
18. Haykin, S.: Neural Networks and Learning Machines. Prentice Hall, New York (2008). http://cise.ufl.edu/class/cap6615sp12/syllabus.pdf
19. Mamdani, E.H., Østergaard, J.J., Lembessis, E.: Use of fuzzy logic for implementing rule-based control of industrial processes. In: Wang, P.P. (ed.) Advances in Fuzzy Sets, Possibility Theory, and Applications, pp. 307–323. Springer, Boston (1983). https://doi.org/10.1007/978-1-4613-3754-6_19
20. Mendel, J.M.: Fuzzy logic systems for engineering: a tutorial. Proc. IEEE **83**(3), 345–377 (1995). https://doi.org/10.1109/5.364485
21. Zadeh, L.A.: Fuzzy logic, neural networks, and soft computing. Commun. ACM **37**(3), 77–85 (1994)
22. Shi, Y., Eberhart, R.C.: Fuzzy adaptive particle swarm optimization. In: 2001 Proceedings of the 2001 Congress on Evolutionary Computation, pp. 101–106. IEEE (2001). https://doi.org/10.1109/cec.2001.934377

Fuzzy Extensions of Conceptual Structures of Comparison

Didier Dubois[1(✉)], Henri Prade[1,3], and Agnès Rico[2]

[1] IRIT, CNRS & Univ. Paul Sabatier, 118 Route de Narbonne, 31062 Toulouse Cedex 9, France
{dubois,prade}@irit.fr

[2] ERIC & Univ. Claude Bernard Lyon 1, 43 bld du 11 novembre, 69100 Villeurbanne, France
agnes.rico@univ-lyon1.fr

[3] QCIS, University of Technology, Sydney, Australia

Abstract. Comparing two items (objects, images) involves a set of relevant attributes whose values are compared. Such a comparison may be expressed in terms of different modalities such as identity, similarity, difference, opposition, analogy. Recently J.-Y. Béziau has proposed an "analogical hexagon" that organizes the relations linking these modalities. The hexagon structure extends the logical square of opposition invented in Aristotle time (in relation with the theory of syllogisms). The interest of these structures has been recently advocated in logic and in artificial intelligence. When non-Boolean attributes are involved, elementary comparisons may be a matter of degree. Moreover, attributes may not have the same importance. One might only consider *most* attributes rather than all of them, using operators such as ordered weighted min and max. The paper studies in which ways the logical hexagon structure may be preserved in such gradual extensions. As an illustration, we start with the hexagon of equality and inequality due to Blanché and extend it with fuzzy equality and fuzzy inequality.

Keywords: Square of opposition · Hexagon of opposition
Difference · Similarity · Analogy · Ordered weighted min

1 Introduction

In order to compare two objects, two images, etc. (we shall say items, more generally), we use their descriptions. In the sequel, descriptions are understood as plain lists of supposedly relevant attribute values. The two items are assumed to be described by the same set of attributes and the values of these attributes are supposed to be known.

One is then naturally led to state that two items are identical if their respective values for each relevant attribute coincide. Béziau [3] recently pointed out that identity, along with five other modalities pertaining to comparison

J. Medina et al. (Eds.): IPMU 2018, CCIS 853, pp. 710–722, 2018.
https://doi.org/10.1007/978-3-319-91473-2_60

(opposition, similarity, difference, analogy, non-analogy), form a hexagon of opposition. This notion was introduced by Robert Blanché [1,2] as a completion of Aristotle square of opposition [14] (originally introduced in connection with the study of syllogisms). By construction, a hexagon is induced by an abstract three-partition [9] and contains three squares of opposition. Blanché emphasized the point that the hexagonal picture can be found in many conceptual structures, such as arithmetical comparators, or deontic modalities [2].

The logical squares and hexagons are geometrical structures where vertices are traditionally associated to statements that are true or false, possibly involving binary modalities. The use of these structures can be extended to statements for which truth is a matter of degree [5,10]. One may then consider studying gradual comparison operators in the light of the gradual logical hexagon and the framework of fuzzy sets. The study of the compatibility between fuzzy extensions of comparison operations and the logical hexagon is the topic of this paper. In turn, this hexagon-driven approach yields an organized overview of a family of logically related operators.

This paper is organized as follows. Section 2 recalls basics of the logical square and its associated hexagon. Section 3 presents the fuzzy extension of the Blanché hexagon for inequality and equality operators. It is shown that maintaining three squares of opposition inside the hexagon induces strong constraints on aggregation operations involved. We provide two examples of quantitative hexagons for similarity indices based on cardinalities. Section 4 focuses on logical expressions agreeing with Béziau's analogy hexagon, and then on various possible fuzzy extensions. This provides a structure relating gradual indices of opposition, similarity, difference, analogy, and non-analogy. Such gradual extensions take into account approximate equality, attribute importance, and possibly fuzzy quantifiers such as "most".

2 The Square and the Hexagon of Opposition

The traditional square of opposition [14] is built with universally and existentially quantified statements in the following way. Consider a statement (**A**) of the form "all P's are Q's", which is negated by the statement (**O**) "at least one P is not a Q", together with the statement (**E**) "no P is a Q", which is clearly in even stronger opposition to the first statement (**A**). These three statements, together with the negation of the last statement, namely (**I**) "at least one P is a Q" can be displayed on a square whose vertices are traditionally denoted by the letters **A**, **I** (AffIrmative half) and **E**, **O** (nEgative half), as pictured in Fig. 1 (where $\overline{Q}$ stands for "not Q").

As can be checked, noticeable relations hold in the square:

- (i) **A** and **O** (resp. **E** and **I**) are the negation of each other;
- (ii) **A** entails **I**, and **E** entails **O** (it is assumed that there is at least one P for avoiding existential import problems);
- (iii) **A** and **E** cannot be true together, but may be false together;
- (iv) **I** and **O** cannot be false together, but may be true together.

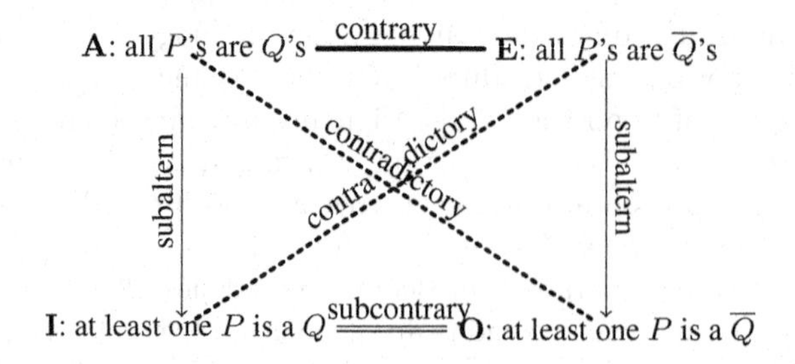

Fig. 1. Square of opposition

Blanché [1,2] noticed that adding two vertices **U** and **Y**, respectively defined as the disjunction of **A** and **E**, and the conjunction of **I** and **O**, to the square, a hexagon **AUEOYI** is obtained that contains 3 squares of opposition, **AEOI**, **YAUO**, and **YEUI**, each obeying the 4 properties above enumerated for the square. Such a hexagon exists each time a three-partition of mutually exclusive situations such as **A**, **E**, and **Y** [9] is considered. Figure 2 represents Blanché's hexagon induced by a complete preorder.

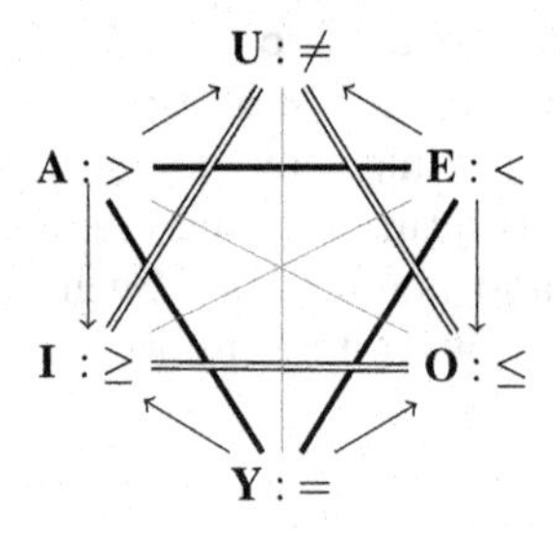

Fig. 2. Blanché's complete preorder hexagon

In the next section, a gradual extension of Blanché's hexagon with fuzzy comparison operators such as *much greater (resp. smaller) than* or *approximately equal to* is proposed.

3 The Fuzzy Blanché Hexagon

We first point out logical constraints bearing on the gradual version of the hexagon, then illustrate them on the case of fuzzy comparison operations.

3.1 Gradual Square and Hexagon

A gradual square of opposition can be defined by attaching variables $\alpha, \epsilon, o, \iota$ valued on a totally ordered set V to vertices A, E, O, I respectively, so as to respect the following constraints [12]:

- α and o (resp. ϵ and ι) are each other's negation, which requires an involutive negation n such that $o = n(\alpha)$ and $\iota = n(\epsilon)$.
- the subaltern relationship between α and ι (resp. ϵ and o) requires a multiple-valued implication operator $I : L \times L \to L$, i.e., decreasing in the first place and increasing in the second place. We must then assume $I(\alpha, \iota) = 1$ and $I(\epsilon, o) = 1$.
- there is mutual exclusion between α and ϵ, i.e., they cannot be simultaneously equal to 1, but can be both 0. We thus need a conjunction operator $C : L \times L \to L$ increasing in both places, and we must enforce $C(\alpha, \epsilon) = 0$.
- ι and o must cover all situations but they can be simultaneously 1. So we need a disjunction operator such that $D(\iota, o) = 1$

A gradual hexagon of opposition [5] is obtained by first assigning variables $\nu = D(\alpha, \epsilon)$ and $\gamma = C(\iota, o)$ to new vertices **U** and **Y**. Then we must require additional conditions to ensure that **YAUO** and **YEUI** are proper squares of opposition playing the same role as **AEOI**. Namely on top of the above conditions for the square, we must have that [5]

- **Y** and **U** are contradictory: $C(\iota, o) = n(D(\alpha, \epsilon))$;
- Subaltern relations:
  $I(\alpha, D(\alpha, \epsilon)) = I(C(\iota, o), o) = I(\epsilon, D(\alpha, \epsilon)) = I(C(\iota, o), \iota) = 1$;
- Contrariety conditions: $C(\alpha, C(\iota, o)) = C(C(\iota, o), \epsilon) = 0$
- Subcontrariety conditions: $D(D(\alpha, \epsilon), o) = D(\iota, D(\alpha, \epsilon)) = 1$
- Conditions for recovering the additional vertices to the two squares **YAUO** and **YEUI** (not mentioned in [5]):
  $\alpha = C(\iota, D(\alpha, \epsilon)), \epsilon = C(D(\alpha, \epsilon), o), \iota = D(\alpha, C(\iota, o)), o = D(C(\iota, o), \epsilon)$.

Note that the conditions in the last item ensure that the three fuzzy sets in the three-partition induced by $(\alpha, \epsilon, \gamma)$ play the same role. If we drop these conditions but preserve the mutual exclusion ones, one may still consider that we have a hexagon of opposition, we call *weak*. However, so-doing we implicitly admit that (α, ϵ) are primitive while γ is derivative, and they cannot be exchanged.

Using a conjunction C, its De-Morgan dual $D(a, b) = n(C(n(a), n(b)))$, and its semi-dual implication $I(a, b) = n(C(a, n(b)))$ then condition $C(\iota, o) = n(D(\alpha, \epsilon))$ is verified, that is γ and ν are contradictories [5]. In the sequel, we denote by (I, C, D) the triplet associated to the hexagon structure with a conjunction C, its semi-dual implication I and its De-Morgan dual D. In that case, the above additional conditions for having a hexagon reduce to $C(\alpha, C(\iota, o)) = C(C(\iota, o), \epsilon) = 0$ and $\iota = D(\alpha, C(\iota, o)), o = D(C(\iota, o), \epsilon)$.

Choosing a triangular norm for C, one may wonder if we can obtain a hexagon of opposition. This is the case if $\alpha + \epsilon + \gamma = 1$ (a fuzzy partition in the sense of Ruspini), and we choose the usual involutive negation $n(\cdot) = 1 - (\cdot)$, the Łukasiewicz t-norm $C = \max(0, \cdot + \cdot - 1)$ and the associated co-norm ($D = \min(1, \cdot + \cdot)$). In fact we prove the following, completing a proof in [5]:

Proposition 1. *If $n(a) = 1 - a$, and C is the Łukasiewicz t-norm, then the hexagon obtained from the triplet (I, C, D) is a hexagon of opposition as soon as $\alpha \leq \iota$.*

Proof. It is clear that $\alpha \leq \iota$ is equivalent to $I(\alpha, \iota) = 1$. It is also clear that $\epsilon \leq o$ follows. Moreover, as we use t-norms and co-norms for C and D, we do have that $\max(\alpha, \epsilon) \leq D(\alpha, \epsilon)$ and $C(\iota, o) \leq \min(\iota, o)$. Hence all subaltern relations hold in the hexagon. Now, consider the condition $D(\alpha, C(\iota, o)) = \iota$ at vertex **I**. It expands in $\min(\alpha + \max(\iota + 1 - \alpha - 1, 0), 1) = \iota$ indeed. Moreover $C(\alpha, C(\iota, o)) = \max(\alpha + \iota + 1 - \alpha - 2, 0) = 0$. The three other conditions at vertices **A, E, O** are obtained likewise. □

In contrast with Proposition 1, consider the Kleene-Dienes triplet (I, C, D), that is $(\max(1 - a, b), \min(a, b), \max(a, b))$, it is clear that $I(a, b) = 1$ means: if $a > 0$ then $b = 1$. The hexagon subaltern conditions for vertex **A** read: If $\alpha > 0$, then $\iota = 1$ and $\max(\alpha, \epsilon) = 1$. Assume $\alpha > 0$, then $\epsilon = 1 - \iota = 0$ and $\max(\alpha, \epsilon) = 1$ so that $\alpha = 1$. We get a Boolea hexagon. It shows that there is no weak gradual hexagon of opposition using Kleene-Dienes triplet (I, C, D).

If we relax the Kleene-Dienes triplet, by means of any implication function such that $I(a, b) = 1$ if and only if $a \leq b$, then the subaltern and mutual exclusion conditions ($\alpha \leq \iota$ and $\min(\alpha, \epsilon) = 0$) are compatible with a gradual structure, where $\alpha > 0$ enforces $\epsilon = 0, o = 1 - \alpha, \iota = 1$. However it is only a weak hexagon. Indeed, defining vertex **I** from **A** and **Y** reads $\iota = \max(\alpha, \min(\iota, o)) = \max(\alpha, 1 - \alpha) = 1$. It again enforces a Boolea hexagon. So we can state the following claim

Proposition 2. *Consider a triplet (I, C, D), where $I(a, b) = 1$ if and only if $a \leq b$, C is a triangular norm D its De Morgan dual with respect to an involutive negation n; then if mutual exclusion conditions hold as $C(\alpha, \epsilon) = C(\alpha, C(n(\alpha), n(\epsilon))) = C(C(\iota, n(\alpha)), \epsilon) = 0$, and $\alpha \leq \iota$, we get a weak gradual hexagon of opposition.*

Indeed the subaltern conditions hold in this case. The Łukasiewicz triplet and the relaxed Kleene triplet are examples where this proposition applies. However these conditions do not ensure a full-fledged symmetrical gradual hexagon. Adding condition $C(a, b) = 0$ if and only if $I(a, n(b)) = 1$ seems to be demanding, and the Łukasiewicz triplet is the only known solution then.

3.2 Fuzzy Comparison Operations and Their Hexagon of Opposition

It is possible to construct a gradual hexagon of opposition with fuzzy comparators, in such a way as to extend Blanché's hexagon of Fig. 2. To this end, we define three fuzzy relations that express notions of *approximately equal to, much greater than, much smaller than* as the ones used in the paper [8] for temporal reasoning.

A fuzzy set F on a universe U is a mapping $\mu_F : U \to [0, 1]$ where $\mu_F(x)$ represents the degree of membership of x to F. We denote by $core(F) = \{x | \mu_F(x) = 1\}$ the core of F and $supp(F) = \{x | \mu_F(u) > 0\}$ its support. The complement of F is $\overline{F}$ such that $\mu_{\overline{F}}(x) = 1 - \mu_F(x)$.

Consider for simplicity trapezoidal fuzzy intervals. Such a fuzzy set of the real line pictured on the figure below is parameterized by the 4-tuple of reals (a, b, α, β) where $core(F) = [a, b]$ and $supp(F) =]a - \alpha, b + \beta[$.

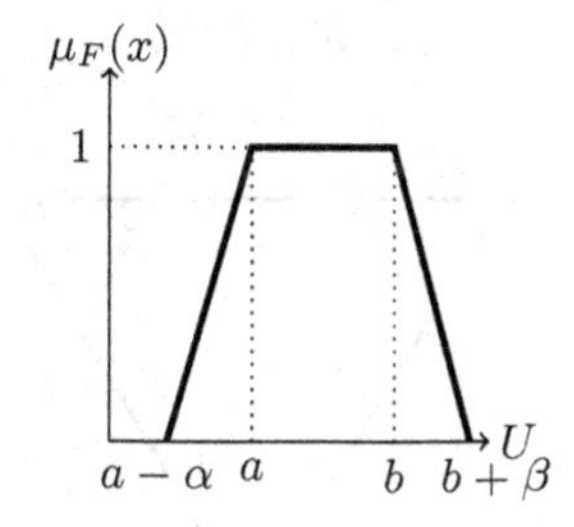

We define a translation operation consisting of adding a constant c to the 4-tuple $F = (a, b, \alpha, \beta)$:

$$F + c = (a + c, b + c, \alpha, \beta).$$

Besides, given F, its antonym is defined by F^{ant} $\mu_{F^{ant}}(x) = \mu_F(-x)$.

Let L be a symmetrical fuzzy interval with respect to the vertical coordinate axis ($L^{ant} = L$). This fuzzy set L is instrumental to define a fuzzy approximate equality relation E as $E(x, y) = L(x - y)$. Likewise, let G be a fuzzy relation representing the concept of *much greater than* of the form $G(x, y) = K(x - y)$, where μ_K is an increasing membership function whose support is in the positive real line. So, the fuzzy relation $P(x, y) = K^{ant}(x - y)$ captures the idea of *much smaller than*. Assume moreover that the three fuzzy sets K^{ant}, L, K form a fuzzy partition, namely $\mu_K(r) + \mu_{K^{ant}}(r) + \mu_L(r) = 1, \forall r \in \mathbb{R}$, as per the figure below for the trapezoidal case. To complete the hexagon, we need the fuzzy counterparts of comparators $\geq$, $\leq$, and $\neq$. To this end we consider the fuzzy set union of K and L, $K \sqcup L$, where $\sqcup$ is defined by Łukasiewicz disjunction, as the fuzzy version of $\geq$ (it is also the convex hull of $K \cup L$ where $\cup$ is modeled by max). It is easy to see that $K \sqcup L = K - 2\delta - \rho$ and $K^{ant} \sqcup L = K^{ant} + 2\delta + \rho$ (which is a fuzzy version of $\leq$). They do correspond to the concept of *approximately greater or equal* and *approximately less or equal* respectively. Finally, the fuzzy version of $\neq$ is $\overline{L} = K \sqcup K^{ant} = K \cup K^{ant}$.

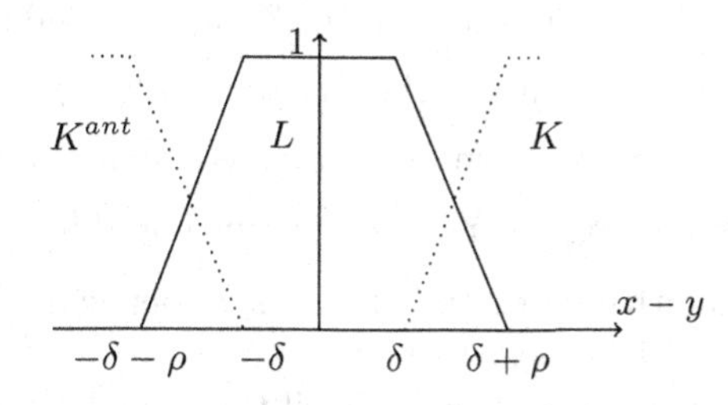

Based on results in [5], and Proposition 1, a gradual hexagon of opposition is obtained (as in Fig. 3) using negation $1 - (\cdot)$, and Łukasiewicz conjunction ($C(a, b) = \max(0, a + b - 1)$. In particular note that $\overline{K} = (K^{ant} \sqcup L)$ and $\overline{L} = K \sqcup K^{ant}$.

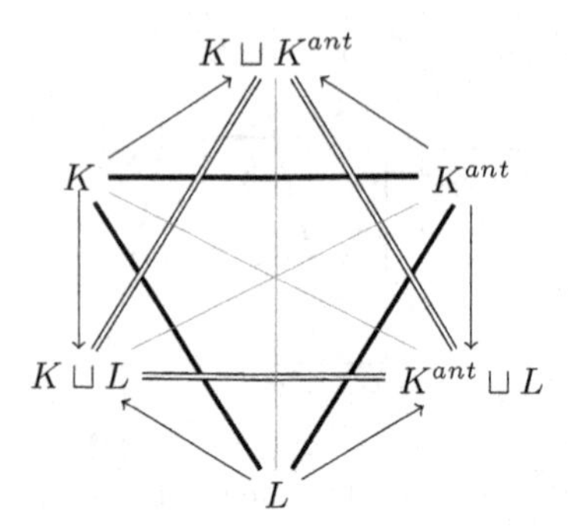

Fig. 3. Fuzzy comparator hexagon

3.3 Hexagons for Quantitative Similarities

Remember that any three-partition gives birth to a hexagon of opposition. Relying on cardinalities of subsets forming a three-partition, it is always possible to obtain a gradual hexagon of opposition. This claim will be illustrated on two cases involving quantitative similarity indices.

Let two items be described by their vectors of Boolean features $x = (x_1, ..., x_n)$ and $y = (y_1, ..., y_n)$ for a set of attributes $\mathcal{A} = \{1, \cdots, i, \cdots, n\}$.

A first partition of $\mathcal{A}$ is formed by three sets Ag^+, Ag^-, Dif:

- $Ag^+(x, y) = \{i \mid x_i = y_i = 1\}$
- $Ag^-(x, y) = \{i \mid x_i = y_i = 0\}$
- $Dif(x, y) = \{i \mid y_i \neq x_i\}$

The three-partition made of "positive identity" (Ag^+) "negative identity" (Ag^-), "opposition" (Dif) yields the hexagon of Fig. 4. It groups six indices that are all easy to interpret in terms of difference and similarity.

Another hexagon (see Fig. 5) is based on a three-partition of $X \cup Y$, where $X = \{i \mid x_i = 1\}$, $Y = \{i \mid y_i = 1\}$ (hence $X \cap Y = Ag^+(x, y)$, and $\overline{X} \cap \overline{Y} = \overline{X \cup Y} = Ag^-(x, y)$). It consists of the three sets $X \cap Y$, $\overline{X} \cap Y$ and $X \cap \overline{Y}$.

Note that $\frac{|X \cap Y|}{|X \cup Y|} = 1$ if and only if $X = Y$ if and only if $Ag(x, y) = \mathcal{A}$ where $Ag(x, y) = \{i \mid x_i = y_i\} = Ag^+(x, y) \cup Ag^-(x, y)$. Index $\frac{|X \cap Y|}{|X \cup Y|}$ is clearly Jaccard index, i.e., a well-known approximate equality measure, while $\frac{|X \triangle Y|}{|X \cup Y|}$ is a difference index (where $X \triangle Y$ is the symmetric difference). However, $\frac{|Y|}{|X \cup Y|}$ is not really a similarity index as it is not symmetrical; $\frac{|X \cap \overline{Y}|}{|X \cup Y|}$ is an opposition index "inside X", with respect to Y.

None of these two hexagons exhibit all the modalities of identity, difference, similarity, opposition and analogy that are supposed to appear in Béziau's intuitive hexagon The latter will be studied in the next section.

It is easy to verify that the two hexagons possess all properties required for being hexagons of opposition. They could be generalized, replacing relative cardinalities by weighted averages, or even Choquet integrals following a suggestion in [6].

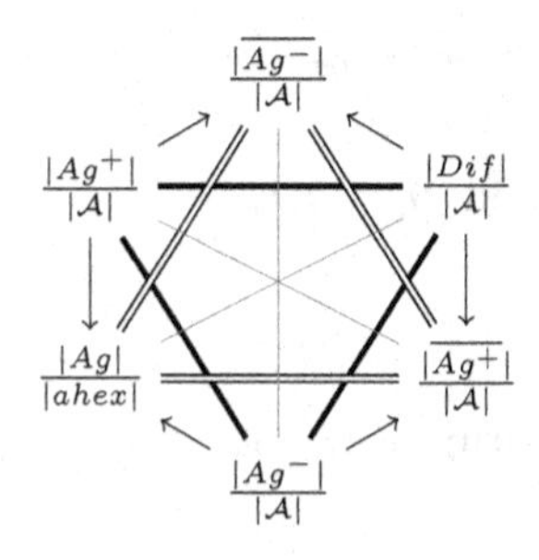

Fig. 4. Hexagon. 3-partition Ag^+, Ag^-, Dif

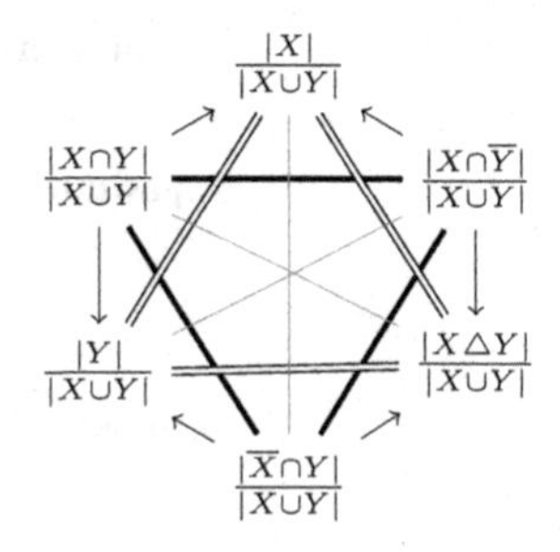

Fig. 5. Jaccard index hexagon

4 Hexagon of Opposition for the Comparison of Items

First, we recall Béziau's informal analogical hexagon [3,4] which organizes the comparison modalities between items supposedly described in terms of attribute values. Then we assume that the equality between attribute values can be approximate, that the attributes do not have the same importance, and that universal or existential quantifiers involved in the comparison modalities can become fuzzy.

4.1 Béziau's Analogical Hexagon

Consider a framework where items x and y are described by their respective attribute values x_i and y_i for attributes $i \in \{1, \cdots, n\}$. At this point attributes are assumed to be Boolean. So the attribute values are 0 or 1. Six comparison modalities between x and y can be defined in this framework:

- Identity: $\forall i,\ x_i = y_i$. Opposition: $\forall i,\ x_i \neq y_i$.
- Difference: $\exists i,\ x_i \neq y_i$. Similarity: $\exists i,\ x_i = y_i$.
- Analogy: $(\exists i, x_i \neq y_i) \wedge (\exists i, x_i = y_i)$. Non-analogy: $(\forall i, x_i = y_i) \vee (\forall i, x_i \neq y_i)$.

This provides a simple reading of the empirical analogical hexagon proposed by Béziau. Note that analogy involves at the same time ideas of similarity and difference, which agrees with the modeling proposed in [15]. It is easy to check that these six modalities are related via logical links requested to realize a hexagon of opposition, as in the analogical hexagon [3]. In particular, difference is the negation of identity and the hexagon makes it clear that analogy is the conjunction between difference and similarity (Fig. 6).

Borrowing from multiple-criteria decision evaluation methods, this hexagon can be extended to gradual modalities using weighted min and max operators on vertices, as we shall see in the sequel.

4.2 Approximate Equality and Weighted Attributes

Suppose from now on that item attributes map to a totally ordered value scale V with least and greatest elements respectively denoted by 0 and 1. The scale

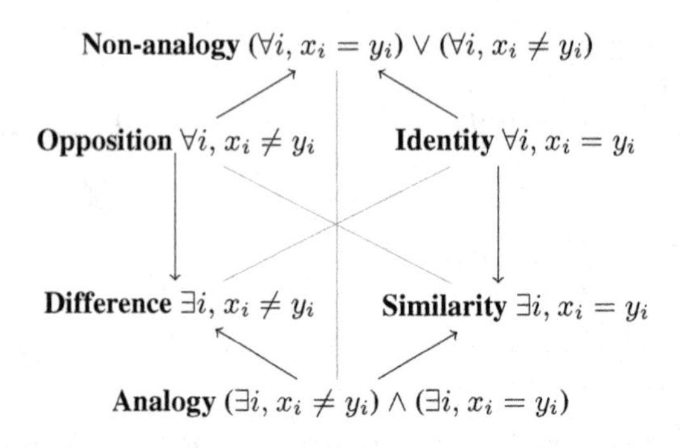

Fig. 6. Analogical hexagon

V is supposed to be equipped with an involutive negation $1 - (.)$, i.e. order-reversing on V. On each attribute, equality and difference are evaluated by means of similarity measures $\mu_{S_i} : V \times V \to V$ and dissimilarity measures $\mu_{D_i} : V \times V \to V$. It is natural to assume that $\mu_{S_i} = 1 - \mu_{D_i}$. The vector of similarities between x and y is

$$\mu_S(x, y) = (\mu_{S_1}(x_1, y_1), \cdots, \mu_{S_n}(x_n, y_n))$$

while for dissimilarity it is $\mu_D(x, y) = (\mu_{D_1}(x_1, y_1), \cdots, \mu_{D_n}(x_n, y_n))$. For any two items x and y, it is also supposed that separability holds, namely: $\mu_{S_i}(x_i, y_i) = 1$ (resp. $\mu_{D_i}(x_i, y_i) = 1$) if and only if x_i and y_i are perfectly similar (resp. dissimilar).

Proposition 2 gives conditions under which the following picture is a *weak* hexagon of opposition, in particular the conditions of mutual exclusion: $C(Opposition, Identity) = C(Opposition, Analogy) = C(Identity, Analogy) = 0$.

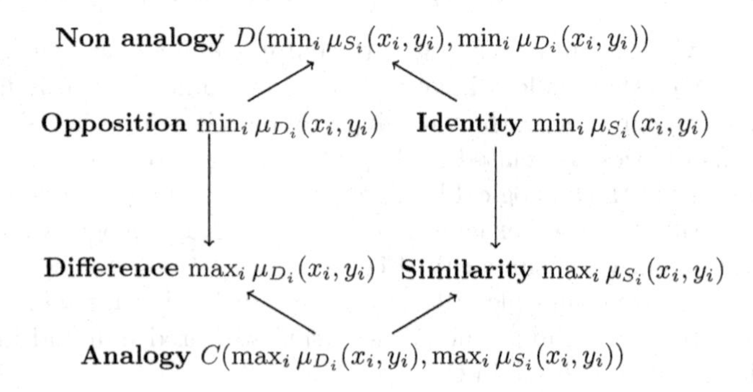

Under this representation, two items are perfectly *opposite* (resp. *identical*) if they are perfectly dissimilar (resp. similar) on each attribute. They are perfectly *different* (resp. *similar*) if there is at least one attribute for which they are perfectly dissimilar (resp. similar). For *analogy* and *non-analogy* between two items, the result depends on the conjunction C and the disjunction D involved.

Example 1. *Consider Łukasiewicz triplet* (I, C, D). *From Proposition 1, a hexagon of opposition is obtained since* $\min_i \mu_{D_i}(x_i, y_i) \leq \max_i \mu_{D_i}(x_i, y_i)$ *In this case two items are perfectly* **analogical** *if they are perfectly similar and perfectly different. They are* **non-analogical** *if* $\min_i \mu_{D_i}(x_i, y_i) + \min_i \mu_{S_i}(x_i, y_i) \geq 1$, *i.e.,* $\min_i \mu_{D_i}(x_i, y_i) \geq \max_i \mu_{D_i}(x_i, y_i)$. *In this case, similarity and dissimilarity measures are constant and equal to 1 or 0. Then, the two items are either perfectly identical or perfectly opposite.*

Operations min and max are qualitative elementary operators that can be extended by means of importance weights or priorities π_i assigned to attributes. The closer π_i to 1, the more important the attribute. Such importance weights may alter local evaluations in various ways [11], leading to operators of the form

$$MIN_{\pi}^{\rightarrow}(x) = \min_{i=1}^{n} \pi_i \rightarrow x_i, \quad MAX_{\pi}^{\otimes}(x) = \max_{i=1}^{n} \pi_i \otimes x_i,$$

where $(\rightarrow, \otimes)$ is a pair of semi-dual implication and conjunction in $\{(\rightarrow_{KD}, \otimes_{KD}), (\rightarrow_G, \otimes_G), (\rightarrow_{GC}, \otimes_{GC})\}$. $a \otimes_{KD} b = a \wedge b$ is Kleene-Dienes conjunction, and its semi-dual implication is $a \rightarrow_{KD} b = (1 - a) \vee b$. Gödel implication and conjunction are respectively defined by: $a \rightarrow_G b = 1$ if $a \leq b$ and b otherwise, and $a \otimes_G b = 0$ if $a \leq 1 - b$ and b otherwise. The contrapositive Gödel implication and conjunction are defined by: $a \rightarrow_{GC} b = 1$ if $a \leq b$ and $1 - a$ otherwise and $a \otimes_{GC} b = 0$ if $a \leq 1 - b$ and a otherwise.

We build the following hexagon where we use shorthand *Id.*, *Op.*, *Dif.*, *Sim.*, *An.* and *NonAn.* for the vertices of the analogical hexagon.

Proposition 3. *If* $n(a) = 1 - a$, *and* C *is the Łukasiewicz t-norm, then the fuzzy weighted analogical hexagon of Fig. 7, obtained from the Łukasiewicz triplet* (I, C, D), *is a hexagon of opposition as soon as there is an attribute such that* $\pi_i = 1$, *and implication* $I = \rightarrow$ *is such that* $(1 - a) \rightarrow 0 \leq 1 - (a \rightarrow 0)$.

Proof. Based on results from Proposition 17 in [12], the two conditions $\pi_i = 1$ for some i, and $(1 - a) \rightarrow 0 \leq 1 - (a \rightarrow 0)$ are sufficient to get $MIN_{\pi}^{\rightarrow}(\mu_D(x, y)) \leq MAX_{\pi}^{\otimes} \mu_D(x, y)$, that is $\alpha \leq \iota$ in the hexagon. The rest follows by Proposition 1. □

We give three examples of semi-dual pairs $(\rightarrow, \otimes)$ where this proposition applies:

1. **Kleene-Dienes.** Then: $MIN_{\pi}^{\rightarrow_{KD}}(x) = \min_{i=1}^{n} \max(1 - \pi_i, x_i) \leq MAX_{\pi}^{\otimes_{KD}}(x) = \max_{i=1}^{n} \min(\pi_i, x_i)$ is well-known. When computing the external minimum (resp. maximum), the partial evaluation of attributes of low importance is increased (resp. decreased). Such attributes will thus have limited influence

on the global rating. Then two items will be perfectly *opposite* (resp. *identical*) if for each attribute either its importance is zero, or dissimilarity (resp. similarity) between items is perfect. They will be perfectly *different* (resp. *similar*) if there exists at least one attribute with importance 1 for which there is perfect dissimilarity (resp. similarity) between items.

2. **Gödel implication:** Since $(1-a) \rightarrow_G 0 = 0$, it follows that $MIN_\pi^{\rightarrow_G}(\mu_D(x,y)) \leq MAX_\pi^{\otimes_G}\mu_D(x,y)$ if $\pi_i = 1$ for some i. The weights π_i only play the role of thresholds. Then two items will be perfectly *opposite* (resp. *identical*) if all local dissimilarities (resp. similarities) are above their thresholds. They will be perfectly *different* (resp. *similar*) if there exists at least one attribute with non-zero importance π_i and perfect dissimilarity (resp. similarity) between items.
3. **Contrapositive Gödel implication:** We can check that $(1-a) \rightarrow_{GC} 0 = 1 - a = 1 - (a \rightarrow_{GC} 0)$. It follows that $MIN_\pi^{\rightarrow_{GC}}(\mu_D(x,y)) \leq MAX_\pi^{\otimes_{GC}} \mu_D(x,y)$, i.e., $\min_{i|\pi_i > \mu_{D_i}(x_i,y_i)} 1 - \pi_i \leq \max_{i|\mu_{D_i}(x_i,y_i) > 1-\pi_i} \pi_i$. Then two items will be perfectly *opposite* (resp. *identical*) if all local dissimilarities (resp. similarities) are above their thresholds. They will be *different* (resp. *similar*) if there exists at least one attribute with $\pi_i = 1$ and non-zero dissimilarity (resp. similarity) between items.

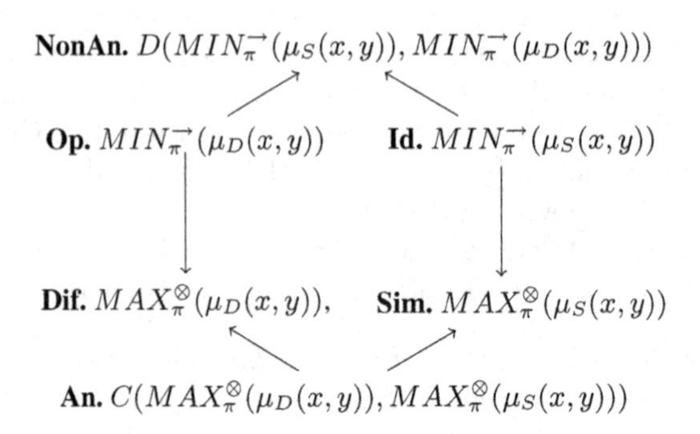

Fig. 7. Fuzzy weighted analogical hexagon

4.3 OWmin and OWmax

With operations $MIN_\pi^{\rightarrow}$ and $MAX_\pi^{\otimes}$ the result depends on all local evaluations for all attributes. One interesting issue is whether the hexagon structure can survive when the quantifiers involved are weakened using ordered weighted min and max (shorthand $OWmin$ and $OWmax$) [7]. In such operations, the quantifier *for all i* is replaced by *for the k best*, where the selection of the best ones is based on local evaluations. For a given vector $x \in V^n$, let σ be the permutation of attributes such that $x_{\sigma(1)} \geq \cdots \geq x_{\sigma(n)}$.

Define $\mu_k : \{1, \cdots, n\} \rightarrow V$ by $\mu_k(i) = 1$ if $1 \leq i \leq k$ and 0 otherwise. Then,

$$OWmin_{\mu_k}(x) = \min_{i=1}^{n} \max(1 - \mu_k(i), x_{\sigma(i)})$$

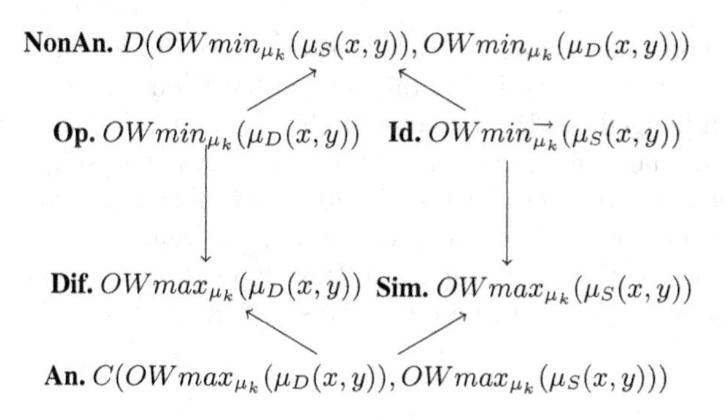

Fig. 8. OWmin-OWmax hexagon

only uses the worst among the k best attributes with respect to their partial evaluations. Likewise, from De Morgan duality,

$$OWmax_{\mu_k}(x) = 1 - OWmin_{\mu_k}(1-x) = \max_{i=1}^{n} \min(\mu_k(i), x_{\sigma(i)})$$

(where σ is now such that $x_{\sigma(1)} \leq \cdots \leq x_{\sigma(n)}$) only uses the best among the k worst attributes with respect to their partial evaluations. μ_k thus represents the quantifier *for at least the k best attributes.* Such a quantifier can vary from *for all attributes* to *for at least the best attribute.*

Unfortunately, the inequality $OWmin_{\mu_k}(\mu_D(x,y)) \leq OWmax_{\mu_k}(\mu_D(x,y))$ generally does not hold. So, the hexagon that could be constructed from these operations may not always be one of opposition. Recall that $OWmin_{\mu_k}$ and $OWmax_{\mu_k}$ are special cases of a Sugeno integral that can be written in two ways:

$$S_\gamma(x) = \max_{E \subseteq \{1,\ldots,n\}} \min(\gamma(E), \min_{i \in E} x_i) = \min_{T \subseteq \{1,\ldots,n\}} \max(1 - \gamma^c(T), \max_{i \in T} x_i).$$

where γ is a capacity and $\gamma^c(T) = 1 - \gamma(\overline{T})$ its conjugate. In the special case considered here, capacities are of the form $\gamma_k(E) = 1$ if $|E| \geq k$ and 0 otherwise. Then $OWmax_{\mu_k} = S_{\gamma_k}$. It is easy to check that the conjugate of γ_k is γ_{n-k+1}. From the above equality, it follows easily that $OWmin_{\mu_k} = OWmax_{\mu_{n-k+1}}$. Hence, the inequality $OWmin_{\mu_k} \leq OWmax_{\mu_k}$ also reads $OWmin_{\mu_k} \leq OWmin_{\mu_{n-k+1}}$ and will hold only when $k \geq (n+1)/2$. In other words, operator $OWmin_{\mu_k}$ must be sufficiently demanding to match with a necessity-like modality. Under such conditions the hexagon in Fig. 8 is a hexagon of opposition. It would be worthwhile comparing this approach with the one in [13].

5 Conclusion

We have shown that the geometrical structure called hexagon of opposition, which organizes the relationship between various comparison modalities can survive under various gradual extensions of such modalities. A study of conditions needed for the gradual hexagon of opposition (as we did for the cube of opposition in [12]) in a more general algebraic setting has been carried out, introducing weak and full-fledged graded versions of the hexagon. Further investigation is still needed to characterize proper algebraic settings that support the gradual hexagon. One may also consider the possibility of introducing an inner negation for defining from the similarity degree $\mu_S(x, y)$ a "remoteness" degree defined by $\mu_S(x, 1 - y)$ (with the requirement that $\mu_S(x, 1 - y) = \mu_S(1 - x, y)$) for $x, y \in [0, 1]$. This might lead to a cube-like structure of opposition.

References

1. Blanché, R.: Sur l'opposition des concepts. Theoria **19**, 89–130 (1953)
2. Blanché, R.: Structures Intellectuelles. Essai sur l'Organisation Systématique des Concepts, Vrin (1966)
3. Béziau, J.-Y.: The power of the hexagon. Log. Univers. **6**(1–2), 1–43 (2012)
4. Béziau, J.-Y.: An analogical hexagon. Int. J. Approx. Reason. **94**, 1–17 (2018)
5. Ciucci, D., Dubois, D., Prade, H.: Structures of opposition induced by relations - the Boolean and the gradual cases. Ann. Math. Artif. Intell. **76**(3–4), 351–373 (2016)
6. Coletti, G., Petturiti, D., Vantaggi, B.: Fuzzy weighted attribute combinations based similarity measures. In: Antonucci, A., Cholvy, L., Papini, O. (eds.) ECSQARU 2017. LNCS (LNAI), vol. 10369, pp. 364–374. Springer, Cham (2017). https://doi.org/10.1007/978-3-319-61581-3_33
7. Dubois, D., Fargier, H., Prade, H.: Beyond min aggregation in multicriteria decision: (ordered) weighted min, discri-min, leximin. In: Yager, R.R., Kacprzyk, J. (eds.) The Ordered Weighted Averaging Operators: Theory and Applications, pp. 181–191. Kluwer Academic Publishers, Dordrecht (1997)
8. Dubois, D., HadjAli, A., Prade, H.: Fuzziness and uncertainty in temporal reasoning. J. Univers. Comput. Sci. **9**, 1168–1194 (2003)
9. Dubois, D., Prade, H.: From Blanché's hexagonal organization of concepts to formal concept analysis and possibility theory. Log. Univers. **6**, 149–169 (2012)
10. Dubois, D., Prade, H.: Gradual structures of oppositions. In: Magdalena, L., Verdegay, J.L., Esteva, F. (eds.) Enric Trillas: A Passion for Fuzzy Sets. SFSC, vol. 322, pp. 79–91. Springer, Cham (2015). https://doi.org/10.1007/978-3-319-16235-5_7
11. Dubois, D., Prade, H., Rico, A.: Residuated variants of Sugeno integrals: towards new weighting schemes for qualitative aggregation methods. Inf. Sci. **329**, 765–781 (2016)
12. Dubois, D., Prade, H., Rico, A.: Graded cubes of opposition and possibility theory with fuzzy events. Int. J. Approx. Reason. **84**, 168–185 (2017)
13. Murinová, P., Novák, V.: Analysis of generalized square of opposition with intermediate quantifiers. Fuzzy Sets Syst. **242**, 89–113 (2014)
14. Parsons, T.: The traditional square of opposition. In: Zalta, E.N. (ed.) The Stanford Encyclopedia of Philosophy (2008)
15. Prade, H., Richard, G.: From analogical proportion to logical proportions. Log. Univers. **7**, 441–505 (2013)

New Negations on the Type-2 Membership Degrees

Carmen Torres-Blanc[1], Susana Cubillo[1(✉)], and Pablo Hernández-Varela[2]

[1] Departamento Matemática Aplicada a las TIC, Universidad Politécnica Madrid (UPM), Boadilla del Monte, 28660 Madrid, Spain
{ctorres,scubillo}@fi.upm.es

[2] Universidad Mayor, Núcleo Matemática, Física and Estadística, Providencia, Santiago, Chile
hernandezpab@gmail.com

Abstract. Hernández et al. [9] established the axioms that an operation must fulfill in order to be a negation on a bounded poset (partially ordered set), and they also established in [14] the conditions that an operation must satisfy to be an aggregation operator on a bounded poset. In this work, we focus on the set of the membership degrees of the type-2 fuzzy sets, and therefore, the set **M** of functions from [0, 1] to [0, 1]. In this sense, the negations on **M** respect to each of the two partial orders defined in this set are presented for the first time. In addition, the authors show new negations on **L** (set of the normal and convex functions of **M**) that are different from the negations presented in [9] applying the Zadeh's Extension Principle. In particular, negations on **M** and on **L** are obtained from aggregation operators and negations. As results to highlight, a characterization of the strong negations that leave the constant function 1 fixed is given, and a new family of strong negations on **L** is presented.

Keywords: Partially ordered sets · Functions from [0, 1] to [0, 1] Normal and convex functions · Type-2 fuzzy sets · Negations Aggregation operators

1 Introduction

Type-2 fuzzy sets (T2FSs) were introduced by Zadeh in [18] as an extension of type-1 fuzzy sets (T1FSs) [17]. While the membership degree of an element in a T1FS is a value in the interval [0, 1], the membership degree of an element in a T2FS is a fuzzy set in [0, 1]. That is, the degree in which an element belongs to the set is just a label of the linguistic variable "TRUTH". In this way, a T2FS is determined by a membership function $\boldsymbol{\mu} : X \to \mathbf{M}$, where $\mathbf{M} = [0,1]^{[0,1]}$ is the set of the functions from [0, 1] on [0, 1] (see [11,13,16]).

On the other hand, the negation and the aggregation of the information are critical elements in any inference system. For this reason their study is essential

J. Medina et al. (Eds.): IPMU 2018, CCIS 853, pp. 723–735, 2018.
https://doi.org/10.1007/978-3-319-91473-2_61

for both fuzzy sets and their extensions, and particularly for type-2 fuzzy sets. The negations in $[0,1]$ were studied and characterized by Trillas [15]. Regarding the negations in the Atanassov's intuitionistic fuzzy sets, the contributions by Bustince [1,2], Deschrijver [5] and others, have been of great value. Also, Hernández et al. in [9] introduced the definition of negation on a bounded partially ordered set (bounded poset), and they used the Zadeh's Extension Principle in order to extend the negations in $[0, 1]$ to negations in $\mathbf{L}$ (normal and convex functions of $\mathbf{M}$).

Regarding the theory about aggregation of real numbers, a complete study can be found in [12], and it has been applied to T1FSs-based logic systems. Moreover, the aggregation operators for the real numbers have been extended to other wider fields, as for example, the interval-valued fuzzy sets (see [6]). In this sense, the authors of the present work in [14] remembered the definition of aggregation operator on a bounded partially ordered set, and they applied the Zadeh's Extension Principle in order to extend the aggregation operators of type-1 to the case of T2FSs.

The main goals of this paper are, firstly, to obtain new negations on $\mathbf{L}$ from the aggregation functions as well as other strong negations on $\mathbf{L}$. Secondly, to show for the first time some negations on $\mathbf{M}$ with respect to each of the two partial orders defined in that set.

The paper is organized as follows. In Sect. 2 the definitions and properties of the T2FSs needed to understand the rest of the work are reminded. Also in this Sect. 2 the axioms of the negation in $[0,1]$ and in a bounded poset are exposed as well as some results about negations in $\mathbf{L}$ obtained by the authors in previous papers. Last part is devoted to the aggregation operators in $\mathbf{M}$. Sections 3 and 4 focus on new negations in $\mathbf{L}$. Section 5 shows the first negations obtained in $\mathbf{M}$. Finally, Sect. 6 contains some conclusions.

2 Preliminaries

Throughout the paper, X will denote a non-empty set which will represent the universe of discourse. Additionally, $\leq$ will denote the usual order relation in the lattice of real numbers.

2.1 About Type-2 Fuzzy Sets and Their Operations

Definition 1 ([13]). *A type-2 fuzzy set (T2FS),* A, *is characterized by a* membership function*:*

$$\boldsymbol{\mu}_A : X \rightarrow \boldsymbol{M} = [0,1]^{[0,1]} = Map\ ([0,1],[0,1]),$$

that is, $\boldsymbol{\mu}_A(x)$ is a fuzzy set in the interval $[0,1]$ and also the membership degree of the element $x \in X$ in the set A. Therefore,

$$\boldsymbol{\mu}_A(x) = f_x,\ where\ f_x : [0,1] \rightarrow [0,1].$$

Let $T2FS(X) = Map(X, \boldsymbol{M})$ denotes the set of all type-2 fuzzy sets on X.

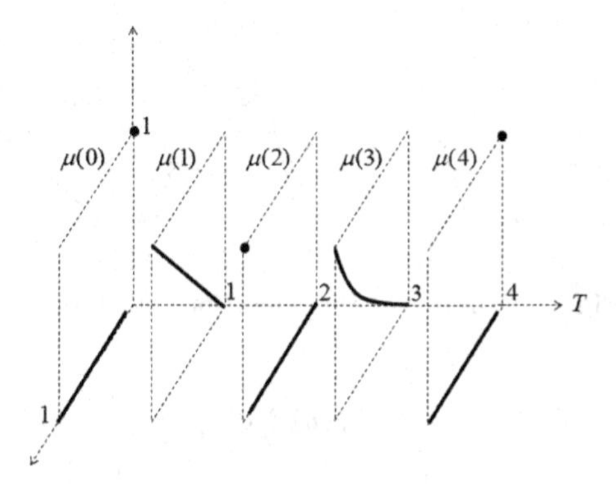

Fig. 1. Example of a T2FS.

Figure 1 shows an example of a type-2 fuzzy set on the finite universe of discourse $T = \{0, 1, 2, 3, 4\}$.

Definition 2 ([16]). *Let $a \in [0,1]$. The characteristic function of a is $\bar{a} : [0,1] \to [0,1]$, where*

$$\bar{a}(x) = \begin{cases} 1, \text{ if } x = a \\ 0, \text{ if } x \neq a \end{cases}$$

Definition 3 ([16]). *Let $[a,b] \subseteq [0,1]$. The characteristic function of $[a,b]$ is $\overline{[a,b]} : [0,1] \to [0,1]$, where*

$$\overline{[a,b]}(x) = \begin{cases} 1, \text{ if } x \in [a,b] \\ 0, \text{ if } x \notin [a,b] \end{cases}$$

Any T1FS can be equated with an interval-valued fuzzy set (IVFS). Indeed, any $\boldsymbol{\mu} : X \to [0,1]$ can be made to correspond to $\sigma : X \to I = \{[a,b] : 0 \leq a \leq b \leq 1\}$, such that, $\sigma(x) = [\boldsymbol{\mu}(x), \boldsymbol{\mu}(x)]$. Likewise, any IVFS can be equated with a T2FS. Thus, any $\sigma : X \to I$, corresponds to $\varphi : X \to \mathbf{M}$, such that $\varphi(x) = \overline{\sigma(x)}$. Therefore IVFSs extend T1FSs, and T2FSs extend IVFSs.

Walker and Walker justify in [16] that the operations on $Map(X, \mathbf{M})$ can be defined naturally from the operations on $\mathbf{M}$ and have the same properties. Thus, in the same way as in the case of the T1FSs, where the definitions and properties are given on $([0,1], \leq)$, in this paper, we will work on $\mathbf{M}$, as all the results are easily and directly extensible to $Map(X, \mathbf{M})$. Furthermore, in order to obtain adequate orders in $Map(X, \mathbf{M})$, two orders have been defined on $\mathbf{M}$. The algebraic operations (union, intersection and complementation) on $\mathbf{M}$, given in Definition 4, were determined from Zadeh's Extension Principle [13,17,18].

Definition 4 ([8,16]). *The operations of $\sqcup$ (extended maximum), $\sqcap$ (extended minimum) and $\neg$ (complementation) are defined on $\mathbf{M}$ as follows (see Fig. 2):*

$$(f \sqcup g)(x) = sup\{f(y) \wedge g(z) : y \vee z = x\},$$

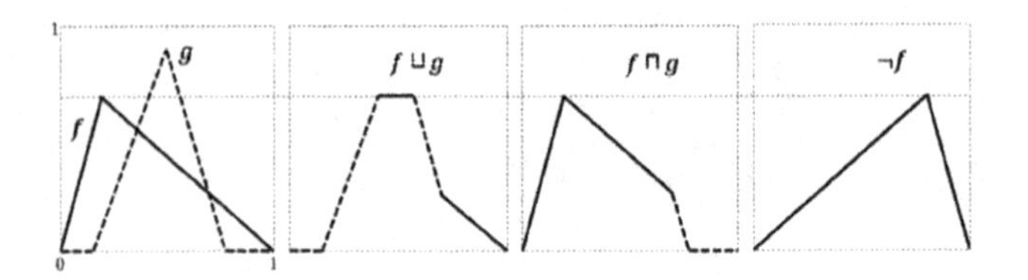

Fig. 2. Example of the operations $\sqcup$, $\sqcap$, and $\neg$.

$$(f \sqcap g)(x) = sup\{f(y) \wedge g(z) : y \wedge z = x\},$$
$$\neg f(x) = sup\{f(y) : 1 - y = x\} = f(1 - x).$$

where $\vee$ and $\wedge$ are the maximum and minimum operations, respectively, on lattice $([0,1], \leq)$.

The algebra $\mathbb{M} = (\mathbf{M}, \sqcup, \sqcap, \neg, \bar{0}, \bar{1})$ does not have a lattice structure, as it does not comply with the absorption law [8,16]. Nevertheless, the operations $\sqcup$ and $\sqcap$ satisfy the properties required for each one to define a partial order on $\mathbf{M}$.

Definition 5 ([13,16]). *The partial orders defined on* $\boldsymbol{M}$ *are as follows:*

$$f \sqsubseteq g \quad \textit{if}\ f \sqcap g = f; \quad f \preceq g \quad \textit{if}\ f \sqcup g = g.$$

Generally, these two partial orders do not coincide [13,16]. Note that $\bar{0}$ and the constant function 0 are, respectively, the minimum and maximum of $(\mathbf{M}, \preceq)$. Also, 0 and $\bar{1}$ are, respectively, the minimum and maximum of the poset $(\mathbf{M}, \sqsubseteq)$.

The following definition and theorem were given in previous papers in order to facilitate the operations in the set $\mathbf{M}$ (Fig. 3):

Definition 6 ([8,16]). *If* $f \in \boldsymbol{M}$*, we define* $f^L, f^R \in \boldsymbol{M}$ *as*

$$f^L(x) = sup\{f(y) : y \leq x\},\ f^R(x) = sup\{f(y) : y \geq x\}.$$

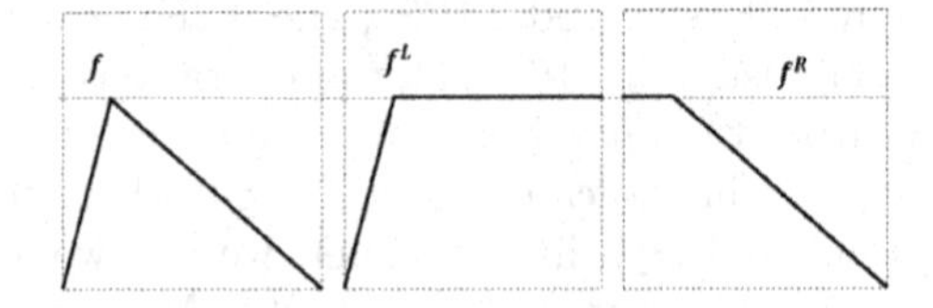

Fig. 3. Examples of f^L and f^R.

Theorem 1 ([16]). *Let* $f, g \in \boldsymbol{M}$. *Then :*

$$f \sqsubseteq g \;\Leftrightarrow\; (f^R \wedge g) \leq f \leq g^R,$$

$$f \preceq g \;\Leftrightarrow\; (g^L \wedge f) \leq g \leq f^L.$$

In the following, we will consider **L**, the subset of normal and convex functions of **M**. This set has a bounded and complete lattice structure, thanks to which negations and aggregation operators have been constructed applying the Zadeh's Extension Principle (see [9,14]).

Definition 7. *A function $f \in \boldsymbol{M}$ is normal if $sup\{f(x) : x \in [0,1]\} = 1$ and a function $f \in \boldsymbol{M}$ is convex, if for any $x \leq y \leq z$, it holds that $f(y) \geq f(x) \wedge f(z)$.*

The partial orders $\sqsubseteq$ and $\preceq$ coincide on **L**, and $(\mathbf{L}, \sqsubseteq)$ is a bounded complete lattice, $\bar{0}$ and $\bar{1}$ are the minimum and the maximum, respectively (see [7,8,13, 16]).

The following characterization will be useful for establishing new results (Fig. 4).

Theorem 2 ([7,8]). *Let $f, g \in \boldsymbol{L}$.*

$$f \sqsubseteq g \text{ if and only if } g^L \leq f^L \text{ and } f^R \leq g^R.$$

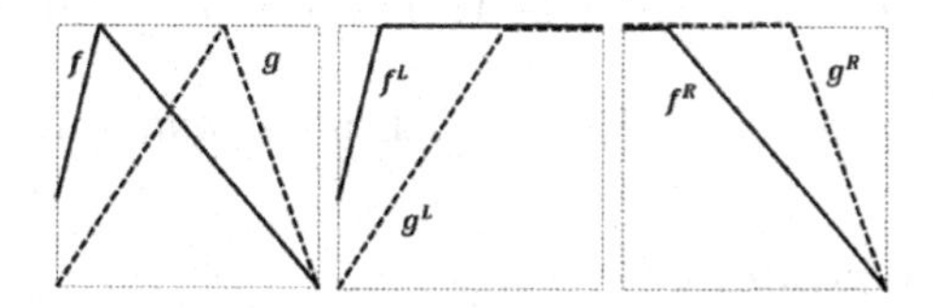

Fig. 4. Example where $f \sqsubseteq g$.

2.2 About Negations

Let us recall the definition of negation in ([0,1], $\leq$).

Definition 8. *A function $n : [0,1] \to [0,1]$ is a negation if it is decreasing respect to the order $\leq$ and satisfies $n(0) = 1$ and $n(1) = 0$. If, in addition, $n(n(x)) = x$ for all $x \in [0,1]$, then it is said to be a strong negation.*

This Definition 8 suggests us an extension to any partially ordered set (poset) with minimum and maximum elements (bounded). In this sense, Hernández et al. [9] introduced negations in partially ordered sets and gave some negations in **L**. In [10] the negations on a bounded poset were defined, but in a more restrictive way than those given in [9], since in this case the negation had to be a bijective function.

Definition 9 ([9]). *Let A be a set and $\leq_A$ be a partial order in A such that $(A, \leq_A)$ has a minimum element $0_{\leq_A}$ and a maximum element $1_{\leq_A}$. A negation in $(A, \leq_A)$ is a function $N : A \to A$ such that N is decreasing, $N(0_{\leq_A}) = 1_{\leq_A}$ and $N(1_{\leq_A}) = 0_{\leq_A}$. If, additionally, $N(N(x)) = x$ holds for all $x \in A$, it is said to be a strong negation.*

According to this definition, the function $\neg$ is a negation in $\mathbf{L}$, but not in the partially ordered sets $(\mathbf{M}, \sqsubseteq)$ and $(\mathbf{M}, \preceq)$. Moreover, in order to find negations in $\mathbf{L}$, in [9] the authors introduced and studied the following operation.

Definition 10 ([9]). *Let n be a surjective negation in $([0,1], \leq)$. The operation $N_n : \boldsymbol{M} \to \boldsymbol{M}$ is given, for any $f \in \boldsymbol{M}$, by: $(N_n(f))(x) = sup\{f(y) : n(y) = x\}$, $\forall x \in [0,1]$.*

Let us note that $N_n = \neg$, provided n is the standard negation $n = 1 - id$. Moreover, for the function N_n to be well defined it is necessary that n is surjective (and therefore, it has to be continuous).

In [9] we proved the following theorem:

Theorem 3. *Let n be a surjective negation in $([0,1], \leq)$. Then,*

1. *N_n is a negation in $(\boldsymbol{L}, \sqsubseteq)$, but it is not a negation in $\boldsymbol{M}$ with either of the two orders $\sqsubseteq$ or $\preceq$.*
2. *$(N_n(f))^L = N_n(f^R)$, $(N_n(f))^R = N_n(f^L)$, for all $f \in \boldsymbol{M}$.*
3. *$N_n(\overline{a}) = \overline{n(a)}$ and $N_n(\overline{[a,b]}) = \overline{[n(b), n(a)]}$.*
4. *$(N_n(f))(x) = f(n(x))$ for all $f \in \boldsymbol{M}$ and for all $x \in [0,1]$ if and only if n is strong.*
5. *N_n is a strong (involutive) negation in $\boldsymbol{L}$ if and only if n is strong.*

2.3 About Aggregation Operators

In [4] the aggregation operators on a bounded poset were defined. In [10] the order norms on a bounded poset, which are a particular case of aggregation operators, were introduced.

Definition 11 ([4]). *Let $(\boldsymbol{A}, \leq_A, 0_{\leq_A}, 1_{\leq_A})$ be a bounded poset. An n-ary aggregation operator on $(\boldsymbol{A}, \leq_A, 0_{\leq_A}, 1_{\leq_A})$ is a function $\chi : \boldsymbol{A}^n \to \boldsymbol{A}$ such that:*

(1) $\chi(0_{\leq_A}, ..., 0_{\leq_A}) = 0_{\leq_A}$,
(2) $\chi(1_{\leq_A}, ..., 1_{\leq_A}) = 1_{\leq_A}$,
(3) Given $f_i, g_i \in \boldsymbol{A}$, if $f_i \leq_A g_i$ for all $i = 1, ..., n$, then $\chi(f_1, ..., f_n) \leq_A \chi(g_1, ..., g_n)$ (increasing in each argument).

We define, from the Zadeh's Extension Principle, the following n-ary operators on $\mathbf{M}$.

Definition 12 ([3,14]). *Let $\phi : [0,1]^n \to [0,1]$ be a surjective n-ary operator on [0,1], and let $\star : [0,1]^n \to [0,1]$ be an n-ary operator on [0,1]. We define the n-ary operator on $\boldsymbol{M}$, $\curlywedge_{\star,\phi} : \boldsymbol{M}^n \to \boldsymbol{M}$, as*

$$\curlywedge_{\star,\phi}(f_1, ..., f_n)(x) = sup\{\star(f_1(y_1), ..., f_n(y_n)) : \phi(y_1, ..., y_n) = x\},$$

where $x, y_1, ..., y_n \in [0,1]$ and $f_1, ..., f_n \in \boldsymbol{M}$.

From now on, $\curlywedge_{\star,\phi}$ will denote the operation introduced in Definition 12, demanding ϕ be always surjective.

Theorem 4 ([14]). *Let ϕ be a continuous n-ary aggregation operator on [0,1]. And let $\star$ be an n-ary aggregation operator on [0,1], with absorbent element 0 and continuous at the point $(1,...,1) \in [0,1]^n$. Then $\curlywedge_{\star,\phi}$ is an n-ary aggregation operator on $\boldsymbol{L}$.*

Other operators different to those obtained from the Zadeh's Extension Principle were given in [3].

Definition 13. *Let $\phi : [0,1]^n \to [0,1]$, an n-ary aggregation operator on [0,1]. We define the n-ary operators on $\boldsymbol{M}$, $\chi_\phi^L : \boldsymbol{M}^n \to \boldsymbol{M}$ and $\chi_\phi^R : \boldsymbol{M}^n \to \boldsymbol{M}$, as*

$$\chi_\phi^L(f_1,...,f_n)(x) = \begin{cases} \bar{0}(x), & \text{if } f_i = \bar{0}\ \forall i, \\ \phi(f_1^L(x),...,f_n^L(x)), & \text{otherwise.} \end{cases}$$

$$\chi_\phi^R(f_1,...,f_n)(x) = \begin{cases} \bar{1}(x), & \text{if } f_i = \bar{1}\ \forall i, \\ \phi(f_1^R(x),...,f_n^R(x)), & \text{otherwise.} \end{cases}$$

where $x \in [0,1]$ and $f_1,...,f_n \in \boldsymbol{M}$.

Proposition 1. *χ_ϕ^L is an aggregation operator on $(\boldsymbol{M}, \preceq)$, and χ_ϕ^R is an aggregation operator on $(\boldsymbol{M}, \sqsubseteq)$. Moreover, χ_ϕ^L and χ_ϕ^R are aggregation operators on $\boldsymbol{L}$.*

3 New Negations in L

In this Section we present some negations in **L**, which are not going to be of type N_n given in Definition 10.

3.1 Extreme Negations in L

Proposition 2. *The operations in $\boldsymbol{L}$, $\mathbb{N}_\wedge$ and $\mathbb{N}_\vee$, given as*

$$\mathbb{N}_\wedge(f) = \begin{cases} \bar{1}, & \text{if } f = \bar{0} \\ \bar{0}, & \text{if } f \neq \bar{0} \end{cases} \qquad \mathbb{N}_\vee(f) = \begin{cases} \bar{0}, & \text{if } f = \bar{1} \\ \bar{1}, & \text{if } f \neq \bar{1} \end{cases}$$

are respectively, the minimum and the maximum negations in $\boldsymbol{L}$. Besides, they are not of type N_n for any surjective negation n in $[0,1]$.

Proof. Is trivial to prove that $\mathbb{N}_\wedge$ and $\mathbb{N}_\vee$ are negations in **L**. Furthermore, they are not of the type N_n. In fact, given the constant function 1, $N_n(1) = 1$ for any n surjective. However, $\mathbb{N}_\wedge(1) = \bar{0}$ and $\mathbb{N}_\vee(1) = \bar{1}$.

3.2 Negations in L from Aggregation Operators

The following proposition allows us to obtain new negations from others previously given.

Proposition 3. *Let $(\boldsymbol{A}, \leq_A, 0_{\leq_A}, 1_{\leq_A})$ be a bounded partially ordered set. Let λ be an m-ary aggregation operator on $\boldsymbol{A}$, and $N_1,...,N_m$ negations in $\boldsymbol{A}$. Then $\mathbb{N} : \boldsymbol{A} \to \boldsymbol{A}$,*

$$\mathbb{N}(a) = \lambda(N_1(a),...,N_m(a)) \ ,$$

is a negation in $\boldsymbol{A}$.

Corollary 1. *Let us consider the negations in* $\mathbf{L}$, $(N_{n_i}(f))(x) = sup\{f(y) : n_i(y) = x\}$, *with* n_i *surjective negations in* $[0,1]$, $i = 1, ..., m$. *And the m-ary aggregation operators in* $\mathbf{L}$, χ^R_ϕ *and* χ^L_ϕ, *given in Definition 13. Then*

$$\mathbb{N}_R(f) = \chi^R_\phi(N_{n_1}(f), ..., N_{n_m}(f)) \quad \textbf{\textit{and}} \quad \mathbb{N}_L(f) = \chi^L_\phi(N_{n_1}(f), ..., N_{n_m}(f)) \;,$$

are negations in $\mathbf{L}$, *which are not of type* N_n, *for any surjective negation n in [0,1]. That is,* $\mathbb{N}_R$ *and* $\mathbb{N}_L$ *are different from those given in the Definition 10* [9].

Proof. $\mathbb{N}_R$ and $\mathbb{N}_L$ are clearly negations on $\mathbf{L}$ according to Proposition 3.

Let us see that $\mathbb{N}_R \neq N_n$ and $\mathbb{N}_L \neq N_n$ for any n surjective negation in $[0,1]$.

Let $c \in (0,1)$ and consider the following functions in $[0,1]$:

$$f^1_c(x) = \begin{cases} c, & \text{if } x \in [0,1) \\ 1, & \text{if } x = 1 \end{cases} \quad \text{and} \quad f_{1,c}(x) = \begin{cases} 1, & \text{if } x = 0 \\ c, & \text{if } x \in (0,1] \end{cases}$$

Then, if $x \neq 1$,

$$N_n(f_{1,c})(x) = sup\{f_{1,c}(y) : n(y) = x \neq 1\} = c$$

and

$$N_n(f_{1,c})(1) = sup\{f_{1,c}(y) : n(y) = 1\} = 1,$$

and if $x \neq 0$,

$$N_n(f^1_c)(x) = sup\{f^1_c(y) : n(y) = x \neq 0\} = c$$

and

$$N_n(f^1_c)(0) = sup\{f^1_c(y) : n(y) = 0\} = 1.$$

Therefore, $N_n(f_{1,c}) = f^1_c$ and $N_n(f^1_c) = f_{1,c}$. Moreover, $f^1_c \neq 1$, $f^1_c \neq \bar{1}$, $f_{1,c} \neq 1$, $f_{1,c} \neq \bar{0}$ since $c \neq 1$ and $c \neq 0$.

On the other hand,

$$\begin{aligned}(\mathbb{N}_R(f_{1,c}))(x) &= \chi^R_\phi(N_{n_1}(f_{1,c}), \ldots, N_{n_m}(f_{1,c}))(x) \\ &= \chi^R_\phi(f^1_c, \ldots, f^1_c)(x) \\ &= \phi((f^1_c)^R(x), \ldots, (f^1_c)^R(x)) \\ &= \phi(1, \ldots, 1) = 1, \ \forall x \in [0,1],\end{aligned}$$

and

$$\begin{aligned}(\mathbb{N}_L(f^1_c))(x) &= \chi^L_\phi(N_{n_1}(f^1_c), \ldots, N_{n_m}(f^1_c))(x) \\ &= \chi^L_\phi(f_{1,c}, \ldots, f_{1,c})(x) \\ &= \phi((f_{1,c})^L(x), \ldots, (f_{1,c})^L(x)) \\ &= \phi(1, \ldots, 1) = 1, \ \forall x \in [0,1].\end{aligned}$$

So, $\mathbb{N}_R(f_{1,c}) = 1$ and $\mathbb{N}_L(f^1_c) = 1$. Then, $N_n(f_{1,c}) = f^1_c \neq 1 = \mathbb{N}_R(f_{1,c})$ and $N_n(f^1_c) = f_{1,c} \neq 1 = \mathbb{N}_L(f^1_c)$, for any n surjective negation in $[0,1]$, and so $\mathbb{N}_R \neq N_n$ and $\mathbb{N}_L \neq N_n$, for any n surjective negation in $[0,1]$.

Corollary 2. *Let* $(N_{n_i}(f))(x) = sup\{f(y) : n_i(y) = x\}$ *be the negations in* $\boldsymbol{L}$, *obtained from* n_i, $i = 1, ..., m$, *surjective negations in [0,1]. And consider the m-ary aggregation operators on* $\boldsymbol{L}$, $\curlywedge_{\star,\phi}$ *given in Definition 12. Then*

$$\mathbb{N}_{\star,\phi}(f) = \curlywedge_{\star,\phi}(N_{n_1}(f), ..., N_{n_m}(f)) \ ,$$

is a negation in $\boldsymbol{L}$. *Moreover, there exists an operator* $\star$, *for which the negation* $\mathbb{N}_{\star,\phi}$ *on* $\boldsymbol{L}$ *is different from* N_n, *for any surjective negation in [0,1], n.*

Proof. Consider, for example, the function f_c^1, with $c \in (0,1)$, as given in the proof of Corollary 1. For any surjective negation n, $N_n(f_c^1) = f_{1,c}$. Let ϕ be an m-ary continuous aggregation operator on [0,1], and let $\star$ be an m-ary aggregation operator on [0,1], with 0 as absorbent element, and continuous at the point $(1, ..., 1) \in [0,1]^m$, in such a way that if an argument of $\star$ is c, then the image of the operation $\star$ is different from c. For example, the aggregation operator $\star(a_1, ..., a_m) = (a_1 a_2 \cdots a_m)^{\frac{1}{m+1}}$. Then $\mathbb{N}_{\star,\phi}(f_c^1) \neq N_n(f_c^1)$.

4 Characterization Theorem of Strong Negations in L

The functions N_n given in Definition 10, with n strong negation in $[0,1]$, constitute a family of strong negations in $\mathbf{L}$ that transform singletons (characteristic functions of a number, $\bar{a}$) and characteristic functions of closed intervals into singletons and characteristic functions of closed intervals, respectively (see Theorem 3).

This section is devoted to investigate if any strong negation in $\mathbf{L}$, coinciding with N_n in the singletons and in the characteristic functions of closed intervals, has to be just N_n. The answer is negative. In fact, in this section, a new family of strong negations in $\mathbf{L}$ will be showed, which coincide with N_n in the singletons and in the characteristic functions of closed intervals, but they are different from the negations N_n presented in [9]. Moreover, in this section, the strong negations in $\mathbf{L}$ which leave the constant function 1 fixed, will be characterized.

Definition 14. *Let* n *be a strong negation in* $([0,1], \leq)$ *and let* α *be an order automorphism in* $([0,1], \leq)$. *The operation* $N_{n,\alpha} : \boldsymbol{L} \rightarrow \boldsymbol{L}$ *is given by,*

$$N_{n,\alpha}(f) = (\alpha \circ f^L \circ n) \wedge (\alpha^{-1} \circ f^R \circ n).$$

Theorem 5. *Let* n *be a strong negation in* $[0,1]$ *and let* α *be an order automorphism in* $[0,1]$. *Then*

1. $N_{n,\alpha}$ *is a strong negation in* $(\boldsymbol{L}, \sqsubseteq)$.
2. $N_{n,\alpha}(\overline{a}) = \overline{n(a)}$ *and* $N_{n,\alpha}(\overline{[a,b]}) = \overline{[n(b), n(a)]}$.
3. $N_{n,\alpha}(1) = 1$ *being* 1 *the constant function.*
4. $N_{n,\alpha} \neq N_{n_1}$ *for any* n *and* n_1 *strong negations in* $[0,1]$ *if* $\alpha \neq id$.

Moreover, the following result is a characterization theorem of the strong negations in $(\mathbf{L}, \sqsubseteq)$ which leave the constant function 1 fixed.

Theorem 6. $\mathscr{N} : \boldsymbol{L} \rightarrow \boldsymbol{L}$ *is a strong negation in* $(\boldsymbol{L}, \sqsubseteq)$ *such that* $\mathscr{N}(1) = 1$ *if and only if there exists a strong negation* n *and an order automorphism* α *in* $([0,1], \leq)$, *such that* $\mathscr{N} = N_{n,\alpha}$.

5 Negations in M

Let us remember that, according to Definition 9, neither $\neg$ nor N_n are negations in the partially ordered sets $(\mathbf{M}, \sqsubseteq)$ and $(\mathbf{M}, \preceq)$. Then the following results are aimed at obtaining negations in those sets.

Firstly, let us observe that the negations in $(\mathbf{M}, \sqsubseteq)$ or in $(\mathbf{M}, \preceq)$ are never negations in $\mathbf{L}$, as they are not closed in this set. In fact, for any negation in $(\mathbf{M}, \sqsubseteq)$ the function $\bar{1}$ is transformed into the constant $0 \notin \mathbf{L}$, and for any negation in $(\mathbf{M}, \preceq)$ the function $\bar{0}$ is transformed into $0 \notin \mathbf{L}$.

Now, the minimum and maximum negations in $(\mathbf{M}, \sqsubseteq)$ and in $(\mathbf{M}, \preceq)$ are showed. These negations are different from the negations $\mathbb{N}_i$, con $i = 1, ...6$, that will be given in Propositions 6–11.

Proposition 4. *The operations*

$$\mathbb{N}_{\wedge_{\sqsubseteq}}(f) = \begin{cases} \bar{1}, & \text{if } f = 0 \\ 0, & \text{if } f \neq 0 \end{cases}, \qquad \mathbb{N}_{\vee_{\sqsubseteq}}(f) = \begin{cases} 0, & \text{if } f = \bar{1} \\ \bar{1}, & \text{if } f \neq \bar{1} \end{cases}$$

are, respectively, the minimum and the maximum negations in $(\boldsymbol{M}, \sqsubseteq)$.

Proposition 5. *The operations*

$$\mathbb{N}_{\wedge_{\preceq}}(f) = \begin{cases} 0, & \text{if } f = \bar{0} \\ \bar{0}, & \text{if } f \neq \bar{0} \end{cases}, \qquad \mathbb{N}_{\vee_{\preceq}}(f) = \begin{cases} \bar{0}, & \text{if } f = 0 \\ 0, & \text{if } f \neq 0 \end{cases}$$

are, respectively, the minimum and the maximum negations in $(\boldsymbol{M}, \preceq)$.

Proposition 6. *Let n be a negation in [0,1], not necessarily surjective. Then*

$$\mathbb{N}_1(f) = \begin{cases} \bar{1}, & \text{if } f = 0 \\ n(f(1)), & \text{otherwise} \end{cases}$$

is a negation in $(\boldsymbol{M}, \sqsubseteq)$.

Proof. Firstly, note that, given two constant functions $f = a$ and $g = b$, if $a \leq b$, then $f^R \wedge g = f \leq g = g^R$, and so $f \sqsubseteq g$ (see Theorem 1). Moreover, if $f = \bar{1}$, it is $f(1) = 1$, $n(f(1)) = 0$ and then, $\mathbb{N}_1(\bar{1}) = 0$.

Regarding the decrease, if $f \sqsubseteq g$, $(f^R \wedge g) \leq f \leq g^R$, and in particular, $f(1) \leq g^R(1) = g(1)$. Therefore, $n(g(1)) \leq n(f(1))$, and finally, $\mathbb{N}_1(g) \sqsubseteq \mathbb{N}_1(f)$.

Proposition 7. *Let n be a negation (not necessarily surjective) in [0,1]. Then*

$$\mathbb{N}_2(f) = \begin{cases} \bar{0}, & \text{if } f = 0 \\ n(f(0)), & \text{otherwise} \end{cases}$$

is a negation in $(\boldsymbol{M}, \preceq)$.

Proof. Similar to the proof of Proposition 6.

Proposition 8. *Let n_1 and n_2 be two decreasing functions in $[0,1]$ such that $n_2(1) = 0$. Then*

$$(\mathrm{N}_3(f))(x) = \begin{cases} \bar{1}, & \text{if } f = 0 \\ n_1(x) \cdot n_2(f(1)), & \text{otherwise} \end{cases}$$

is a negation in $(\boldsymbol{M}, \sqsubseteq)$.

Proof. Let us observe that the function $n_1(x) \cdot n_2(f(1))$ is decreasing. If $f = \bar{1}$, it is $f(1) = 1, n_2(f(1)) = 0$, and so $n_1(x) \cdot n_2(f(1)) = 0\ \forall x$, and $\mathrm{N}_3(\bar{1}) = 0$.

Regarding the decrease, if $f \sqsubseteq g$, with f and g different from 0, $(f^R \wedge g) \leq f \leq g^R$, and in particular, $f(1) \leq g^R(1) = g(1)$. As $n_2(g(1)) \leq n_2(f(1))$, it is $n_1(x) \cdot n_2(g(1)) \leq n_1(x) \cdot n_2(f(1))$, and taking into account that $n_1(x) \cdot n_2(g(1))$ and $n_1(x) \cdot n_2(f(1))$ are decreasing, $(\mathrm{N}_3(g))^R = \mathrm{N}_3(g) \leq \mathrm{N}_3(f) = (\mathrm{N}_3(f))^R$. Finally $\mathrm{N}_3(g) \sqsubseteq \mathrm{N}_3(f)$. On the other hand, if $f = 0$ or $g = 0$, the proof of the decrease is trivial.

Proposition 9. *Let n and α be two functions in $[0,1]$, such that α is increasing, n is decreasing, and $n(1) = 0$. Then*

$$(\mathrm{N}_4(f))(x) = \begin{cases} \bar{0}, & \text{if } f = 0 \\ \alpha(x) \cdot n(f(0)), & \text{otherwise} \end{cases}$$

is a negation in $(\boldsymbol{M}, \preceq)$.

Proof. Similar to the proof of Proposition 8, taking into account that the function $\alpha(x) \cdot n(f(0))$ is increasing.

Proposition 10. *Let n_1 and n_2 be two decreasing functions in $[0,1]$, with $n_1(1) = 0$. Then*

$$(\mathrm{N}_5(f))(x) = \begin{cases} \bar{1}, & \text{if } f = 0 \\ n_1(f^R(n_2(x))), & \text{otherwise} \end{cases}$$

is a negation in $(\boldsymbol{M}, \sqsubseteq)$.

Proof. Let us observe that the function $n_1(f^R(n_2(x)))$ is decreasing. If $f = \bar{1}$, it is $f^R = 1$, and $f^R(n_2(x)) = 1$ for all x; then, $n_1(f^R(n_2(x))) = 0\ \forall x$, and $\mathrm{N}_5(\bar{1}) = 0$.

With respect to the decrease, if $f \sqsubseteq g$, with f and g different from 0, $(f^R \wedge g) \leq f \leq g^R$, where $f^R \leq g^R$. Then $f^R(n_2(x)) \leq g^R(n_2(x))$, for all x, and finally $n_1(g^R(n_2(x))) \leq n_1(f^R(n_2(x)))$. As $n_1(f^R(n_2(x)))$ and $n_1(g^R(n_2(x)))$ are decreasing, we have $(\mathrm{N}_5(g))^R = \mathrm{N}_5(g) \leq \mathrm{N}_5(f) = (\mathrm{N}_5(f))^R$. And then, $\mathrm{N}_5(g) \sqsubseteq \mathrm{N}_5(f)$. Finally, if $f = 0$ or $g = 0$, the proof of the decrease is trivial.

Proposition 11. *Let n_1 and n_2 two decreasing functions in $[0,1]$, with $n_1(1) = 0$. Then*

$$(\mathrm{N}_6(f))(x) = \begin{cases} \bar{0}, & \text{if } f = 0 \\ n_1(f^L(n_2(x))), & \text{otherwise} \end{cases}$$

is a negation in $(\boldsymbol{M}, \preceq)$.

Proof. Similar to the proof of the Proposition 10, taking into account that the function $n_1(f^L(n_2(x)))$ is increasing and that if $f = \bar{0}$, we have $f^L = 1$ and $f^L(n_2(x)) = 1$ for all x; therefore, $n_1(f^L(n_2(x))) = 0$ for all x, and $\mathbb{N}_6(\bar{0}) = 0$.

6 Conclusions

In this paper, for the first time, some negations in the partially ordered sets $(\mathbf{M}, \sqsubseteq)$ and $(\mathbf{M}, \preceq)$ have been obtained, where $\mathbf{M}$ is the set of all functions from $[0,1]$ to $[0,1]$. These functions are just the membership degrees of the type-2 fuzzy sets. In particular, the minimum and the maximum negations for each of the two orders have been showed. Moreover, a method for obtaining negations in $\mathbf{L}$ (set of the normal and convex functions of $\mathbf{M}$) from aggregation functions has been proposed. Finally, a new family of strong negations in $\mathbf{L}$ has been presented, as well as a characterization of the strong negations in $\mathbf{L}$ that leave the constant function 1 fixed.

Acknowledgements. This paper was partially supported by Universidad Politécnica de Madrid (Spain) and Universidad Mayor (Chile).

References

1. Bustince, H., Barrenechea, E., Pagola, M.: Generation of interval-valued fuzzy and Atanassov's intuitionistic fuzzy connectives from fuzzy connectives and from K_α operators. Laws for conjunctions and disjunctions, amplitude. Int. J. Intell. Syst. **23**, 680–714 (2008)
2. Bustince, H., Kacprzyk, J., Mohedano, V.: Intuitionistic fuzzy generators-application to intuitionistic fuzzy complementation. Fuzzy Sets Syst. **114**, 485–504 (2000)
3. Cubillo, S., Hernández, P., Torres-Blanc, C.: Examples of aggregation operators on membership degrees of type-2 fuzzy sets. In: Proceedings of IFSA-EUSFLAT 2015, Gijón, Spain (2015)
4. De Cooman, G., Kerre, E.: Order norms on bounded partially ordered sets. J. Fuzzy Math. **2**, 281–310 (1994)
5. Deschrijver, G., Cornelis, C., Kerre, E.: Intuitionistic fuzzy connectives revisited. In: Proceedings of the 9th International Conference on Information Processing Management Uncertainty Knowledge-Based Systems, pp. 1839–1844 (2002)
6. Deschrijver, G., Kerre, E.: Aggregation operation in interval-valued fuzzy and Atanassov's intuitionistic fuzzy set theory. In: Bustince, H., Herrera, F., Montero, J. (eds.) Fuzzy Sets and Their Extensions: Representation, Aggregation and Models, pp. 183–203. Springer, Berlin (2008). https://doi.org/10.1007/978-3-540-73723-0_10
7. Harding, J., Walker, C., Walker, E.: Convex normal functions revisited. Fuzzy Sets Syst. **161**, 1343–1349 (2010)
8. Harding, J., Walker, C., Walker, E.: Lattices of convex normal functions. Fuzzy Sets Syst. **159**, 1061–1071 (2008)
9. Hernández, P., Cubillo, S., Torres-Blanc, C.: Negations on type-2 fuzzy sets. Fuzzy Sets Syst. **252**, 111–124 (2014)

10. Komorníková, M., Mesiar, R.: Aggregation functions on bounded partially ordered sets and their classification. Fuzzy Sets Syst. **175**, 48–56 (2011)
11. Mendel, J., Jhon, R.: Type-2 fuzzy sets made Simple. IEEE Trans. Fuzzy Syst. **10**(2), 117–127 (2002)
12. Mesiar, R., Kolesárová, A., Calvo, T., Komorníková, M.: A review of aggregation functions. In: Bustince, H., et al. (eds.) Fuzzy Sets and Their Extensions: Representation, Aggregation and Models, pp. 121–144. Springer, Berlin (2008). https://doi.org/10.1007/978-3-540-73723-0_7
13. Mizumoto, M., Tanaka, K.: Some properties of fuzzy sets of type-2. Inf. Control **31**, 312–340 (1976)
14. Torres-Blanc, C., Cubillo, S., Hernández, P.: Aggregation operators on type-2 fuzzy sets. Fuzzy Sets Syst. **324**, 74–90 (2017)
15. Trillas, E.: Sobre funciones de negación en la teoría de conjuntos difusos. Stochastica **3**(1), 47–60 (1979)
16. Walker, C., Walker, E.: The algebra of fuzzy truth values. Fuzzy Sets Syst. **149**, 309–347 (2005)
17. Zadeh, L.: Fuzzy sets. Inf. Control **20**, 301–312 (1965)
18. Zadeh, L.: The concept of a linguistic variable and its application to approximate reasoning-I. Inf. Sci. **8**, 199–249 (1975)

First Steps Towards Harnessing Partial Functions in Fuzzy Type Theory

Vilém Novák(✉)

Institute for Research and Applications of Fuzzy Modeling,
Centre of Excellence IT4Innovations, University of Ostrava,
30. dubna 22, 701 03 Ostrava 1, Czech Republic
Vilem.Novak@osu.cz

Abstract. In this paper we present how the theory of partial functions can be developed in the fuzzy type theory and show how the theory elaborated by Lapierre [3] and Lepage [4] can be included in it. Namely, the latter is developed as a special theory whose models contain the partial functions in the sense introduced by both authors.

Keywords: Fuzzy type theory · Partial function
Higher-order fuzzy logic

1 Introduction

In this paper, we continue the development of the fuzzy type theory (FTT) with partial functions. We are inspired especially by the papers of Lapierre [3] and Lepage [4] in which the theory of partial functions was developed. One of their goals was to prepare grounds for introduction of the semantics of the classical type theory in which partial functions can occur. Their theory is based on a special ordering on the basis of which one can replace classical functions by partial ones that are as close to the former as much as possible.

The fuzzy type theory with partial functions (pFTT) was introduced by the author in [8,9]. The suggested solution is standard in the sense that a special value "$*$" interpreted as "undefined" is introduced. The value, however, is a special well formed formula of FTT.

It is notable that there are "undefined" values in pFTT defined for each type. Namely, we start with the definition of $*_o$ of type o being identified with the formula $\iota_{o(oo)} \cdot \lambda x_o \bot$ whose interpretation requires application of the description operator to the empty set. The result of such operation is typically undefined and so, it gives natural interpretation of $*_o$. Then we continue with definition of "undefined" for other types. First, we define $*_\epsilon$ as $\iota_{\epsilon(o\epsilon)} \cdot \lambda x_\epsilon *_o$, i.e., the description operator is here applied to a nowhere defined function. The $*_\alpha$ for other types α is then introduced analogously.

The advantage of such approach consists in the fact that $*_\alpha$ is a well formed formula and so, it can naturally be treated in the syntax of FTT. Hence,

J. Medina et al. (Eds.): IPMU 2018, CCIS 853, pp. 736–748, 2018.
https://doi.org/10.1007/978-3-319-91473-2_62

the possibility to apply the important principle of λ-*conversion* is kept. Of course, we need additional axioms assuring the intended behavior of $*_\alpha$ and its natural integration into the syntax. Since the fuzzy equality is the fundamental connective in FTT, the formula $(A_\alpha \equiv_\alpha *_\alpha)$ says that the formula A_α is equal to the element "undefined". Depending on whether this formula is true or not we naturally obtain the operator $?A_\alpha$ saying that A_α is undefined and $!A_\alpha$ saying that it is defined. Note that these operators cannot be fuzzy because we can hardly imagine any sense in saying that A_α is "partially" (un)defined. Therefore, both predicates "!, ?" are crisp.

In this paper we provide the first steps towards a sophisticated formal theory of partial functions in the "fuzzy environment". We will focus on the basic concepts and their properties introduced and proved by Lapierre and Lepage in their papers. Our goal (not completed in this paper) is to show that their theory can be naturally included in the formalism of pFTT.

2 Preliminaries

2.1 Fuzzy Type Theory

We will introduce the theory of partial functions in the fuzzy type theory with partial functions introduced in [8,9]. The fundamental algebra of truth values for FTT is a linearly ordered EQ_Δ-algebra. However, for the theory of partial functions we need a stronger algebra and so, we will work in this paper with IMTL-FTT, i.e., the fuzzy type theory on a linearly ordered IMTL_Δ algebra (a residuated lattice with involutive negation). Many properties hold also in a slightly weaker linearly ordered IEQ_Δ-algebra. We refer the reader to [5,7] for all the details of the formal system of FTT.

Let us recall that the basic syntactical objects of FTT are classical (cf. [1]), namely the concepts of *type* and *formula*. The atomic types are ϵ (elements) and o (truth values). General types are denoted by Greek letters $\alpha, \beta, \ldots$. The set of all types is denoted by *Types*. The *language* of FTT, denoted by J, consists of variables $x_\alpha, \ldots$, special constants $c_\alpha, \ldots$ ($\alpha \in \mathit{Types}$), the symbol λ, and brackets.

Semantics of FTT is based on the concept of *frame* that is a tuple

$$\mathscr{M} = \langle (M_\alpha, \doteq_\alpha)_{\alpha \in \mathit{Types}}, \mathscr{E}_\Delta \rangle \tag{1}$$

such that $\mathscr{E}_\Delta$ is a linearly ordered algebra of truth values, $M_o = E$, M_ϵ is the set of *individuals*. Otherwise, $M_{\beta\alpha} \subseteq M_\beta^{M_\alpha}$. The $\doteq_\alpha$ is a generic fuzzy equality on M_α for all $\alpha \in \mathit{Types}$.

Interpretation of a formula A_α of type α in $\mathscr{M}$ is determined with respect to an assignment $p \in \mathrm{Asg}(\mathscr{M})$ of elements from the sets M_α to variables where $\mathrm{Asg}(\mathscr{M})$ is a set of all assignments. The assignment p assures that all free variables occurring in the formula A_α are assigned specific elements so that the whole formula A_α can be evaluated in $\mathscr{M}$. The value of a formula A_α in a model $\mathscr{M}$ is denoted by $\mathscr{M}(A_\alpha)$. Clearly, $\mathscr{M}(A_\alpha) \in M_\alpha$ for all $\alpha \in \mathit{Types}$. A fuzzy set in M_α is interpretation of a formula $A_{o\alpha}$ in the model $\mathscr{M}$ being a function $A : M_\alpha \rightarrow E$. If x_α is a variable occurring free in A_α and $p(x_\alpha) = m \in M_\alpha$ then $\mathscr{M}_p(A_{o\alpha} x_\alpha) \in M_o$ is a membership degree of m in the fuzzy set $\mathscr{M}_p(A_{o\alpha})$ (recall that $M_o = E$).

A formula A_o is *crisp* if $\vdash A_o \vee \neg A_o$.

Lemma 1

(a) A formula A_o is crisp iff $\vdash A_o \equiv \boldsymbol{\Delta} A_o$.
(b) $T \cup \{A_o\} \vdash B_o$ iff $T \vdash \boldsymbol{\Delta} A_o \Rightarrow B_o$ (deduction theorem).
(c) Let A_o be crisp and T a theory in which $T \vdash (\exists x_\alpha) A_o$. Let us extend the language of T by a new constant $\mathbf{u}_\alpha$. Then the theory $T \cup \{A[x_\alpha/\mathbf{u}_\alpha]\}$ is a conservative extension of T.
(d) $\vdash (\forall x_\alpha) \boldsymbol{\Delta} A_o \equiv \boldsymbol{\Delta} (\forall x_\alpha) \boldsymbol{\Delta} A_o$
(e) $\vdash \boldsymbol{\Delta} (\forall x_\alpha) A_o \equiv (\forall x_\alpha) \boldsymbol{\Delta} A_o$

2.2 Fuzzy Type Theory with Partial Functions

This is extension of the fuzzy type theory recalled in the previous subsection. It has been published in [8,9]. Important new definitions are the following:

(i) *The values "undefined":*

$$*_o \equiv \iota_{o(oo)} \cdot \lambda x_o \bot, \tag{2}$$

$${}_{\epsilon} \equiv \iota_{\epsilon(o\epsilon)} \cdot \lambda x_\epsilon *_o, \tag{3}$$

$${}_{\beta\alpha} \equiv \lambda x_\alpha *_\beta, \qquad \alpha, \beta \in Types. \tag{4}$$

Note that the element $*_{\beta\alpha}$ represents the nowhere defined function.
(ii) *Test of undefined:* $?_{o\alpha} \equiv \lambda x_\alpha \cdot x_\alpha \equiv_\alpha *_\alpha$.
(iii) *Test of defined:* $!_{o\alpha} \equiv \lambda x_\alpha \cdot \neg ? x_\alpha$.
(iv) *Star-0 reinterpretation:* $\downarrow_{oo} \equiv \lambda x_o \cdot x_o \equiv \top$.
(v) *Star-1 reinterpretation:* $\uparrow_{oo} \equiv \lambda x_o \cdot \neg ! x_o \vee \downarrow x_o$.

These definitions give rise to the *extended algebra* of truth values. Its support is

$$E^* = E \cup \{*\}$$

where $* \notin E$ is a special "truth value" $*$ interpreted as *undefined*. The basic operations are extended as follows: Let $a, b \in E$ and $\bigcirc \in \{\wedge, \otimes\}$. Then the following tables define the operations in the extended algebra $\mathscr{E}^*_\Delta$.

The fuzzy equality $\stackrel{\circ}{=} : M^* \times M^* \rightarrow E$ for complex types is defined by

$$[h \stackrel{\circ}{=} h'] = \bigwedge_{m \in M^*_\alpha} [h(m) \stackrel{\circ}{=}_\beta h'(m)], \qquad h, h' \in M^* \tag{5}$$

where $\stackrel{\circ}{=}_\beta$ is the fuzzy equality on the set M_β. Furthermore,

$\sim$	b	$*$
a	$a \sim b$	$\mathbf{0}$
$*$	$\mathbf{0}$	$\mathbf{1}$

$\bigcirc$	b	$*$
a	$a \bigcirc b$	$*$
$*$	$*$	$*$

$\rightarrow$	b	$*$
a	$a \rightarrow b$	$\mathbf{0}$
$*$	$\mathbf{1}$	$\mathbf{1}$

$\vee$	b	$*$
a	$a \vee b$	$\mathbf{0}$
$*$	$\mathbf{0}$	$\mathbf{0}$

x	Δx
a	Δa
$*$	$*$

x	$\neg x$
a	$\neg a$
$*$	$\mathbf{0}$

Note that $\mathbf{0}$ remains the bottom element for all $a \in E$. We, at the same time, have $* \leq \mathbf{0}$ but $\mathbf{0} \not\leq *$

The logical axioms had to be modified and extended by the axioms characterizing the behavior of the value "undefined". Because of the lack of space, we do not present them in this paper. We only mention the following: unlike [8,9], it turns out necessary to add one more axiom assuring that for each type there are defined objects:

(FT-fund0) $(\exists x_\epsilon)!x_\epsilon$.

Because we also have $\vdash !\top$ and so, $(\exists x_o)!x_o$, we can extend (FT-fund0) to all types.

The definition of a frame is modified as follows.

Definition 1. *The* extended general frame *is*

$$\mathscr{M}^* = \langle (M^*_\alpha, \mathrel{\mathring{=}}_\alpha)_{\alpha \in Types}, \mathscr{E}^*_\Delta, I_o, I_\epsilon \rangle \tag{6}$$

so that the following holds:

(i) We put $M^*_o = E^*$ where the latter is the support of the extended algebra of truth values $\mathscr{E}^*_\Delta$. Furthermore, let $*_\epsilon \notin M_\epsilon$. Then we put $M^*_\epsilon = M_\epsilon \cup \{*_\epsilon\}$. For all $\alpha = \gamma\beta$, the function

$$*_{\gamma\beta} : M^*_\beta \to M^*_\gamma$$

where $*_{\gamma\beta}(m_\beta) = *_\gamma$ for all $m_\beta \in M^*_\beta$, represents "undefined". For all types $\gamma\beta$ we put $M^*_{\gamma\beta} \subseteq (M^*_\gamma)^{M^*_\beta}$, where we require that $*_{\gamma\beta} \in M^*_{\gamma\beta}$.

(ii) The $\mathscr{E}^*_\Delta$ is the extended algebra of truth degrees. We assume that the sets M^*_{oo}, $M^*_{(oo)o}$ contain all the operations discussed above.

(iii) $\mathrel{\mathring{=}}_\alpha : M^*_\alpha \times M^*_\alpha \to E$ is a fuzzy equality on M^*_α for every $\alpha \in$ Types. We define: $\mathrel{\mathring{=}}_o := \sim$ and $\mathrel{\mathring{=}}_\epsilon$ is the fuzzy equality on M^*_ϵ given explicitly. The fuzzy equality $\mathrel{\mathring{=}}_{\beta\alpha}$ for complex types is defined in (5).

(iv) $I_o : \mathscr{F}(M_o) \to M_o$, $I_\epsilon : \mathscr{F}(M_\epsilon) \to M_\epsilon$ are partial functions interpreting the basic description operators. Let $B \underset{\sim}{\subset} M_o$ and $C \underset{\sim}{\subset} M_\epsilon$. Then

$$I_o(B) = \begin{cases} a_B \in \mathrm{Ker}(B) & \text{if } B \text{ is normal,} \\ *_o & \text{otherwise,} \end{cases} \tag{7}$$

$$I_\epsilon(C) = \begin{cases} x_C \in \mathrm{Ker}(C) & \text{if } C \text{ is normal,} \\ *_\epsilon & \text{otherwise.} \end{cases} \tag{8}$$

Theorem 1 (Deduction theorem). *Let T be a theory and $A_o \in Form_o$ a closed formula such that $T \vdash !A_o$. Then*

$$T \cup \{A_o\} \vdash B_o \quad iff \quad T \vdash \mathbf{\Delta} A_o \Rightarrow B_o$$

for every formula $B_o \in Form_o$ such that $T \vdash !B_o$.

Theorem 2. *If $T \vdash !*_o$ then T is contradictory.*

Theorem 3 (Completeness of FTT with partial functions). *Let T be a theory, the special axioms of which have the form $A_o \equiv \top$. Then T is consistent iff it has a general model $\mathscr{M}$.*

In [6] we introduced two special formulas:

$$\Upsilon_{oo} \equiv \lambda z_o \cdot \neg\Delta(\neg z_o), \qquad \text{(nonzero truth value)}$$

$$\hat{\Upsilon}_{oo} \equiv \lambda z_o \cdot \neg\Delta(z_o \vee \neg z_o) \qquad \text{(general truth value)}$$

Thus $\mathscr{M}(\hat{\Upsilon}(A_o)) = 1$ iff $\mathscr{M}(A_o) \in (0,1)$ and $\mathscr{M}(\Upsilon(A_o)) = 1$ iff $\mathscr{M}(A_o) > 0$ hold in any model $\mathscr{M}$.

In FTT with partial functions the above formulas must be modified as follows:

$$\Upsilon_{oo} \equiv \lambda z_o \cdot !z_o \,\&\, \neg\Delta(\neg z_o), \tag{9}$$

$$\hat{\Upsilon}_{oo} \equiv \lambda z_o \cdot !z_o \,\&\, \neg\Delta(z_o \vee \neg z_o). \tag{10}$$

Lemma 2. *Let $\vdash ?A_o$. Then $\vdash \neg\Upsilon A_o$ as well as $\vdash \neg\hat{\Upsilon} A_o$.*

Proof. Using (9), the assumption, the provable formula $\vdash !*_o \equiv \bot$ and Rule (R) we obtain $\vdash \Upsilon A_o \equiv \bot$. The proof of the second formula is analogous.

3 Partial Functions

3.1 Additional Concepts of pFTT

In this section we will elaborate in more detail the theory of partial functions. We will work in IMTL-FTT. On many places, we will follow the results presented by Lapierre and Lepage in their papers [3,4].

Lemma 3. *For any type $\beta\alpha \in Types$ and any formula $f_{\beta\alpha}$*

$$\vdash !f_{\beta\alpha} \equiv (\exists x_\alpha)!f_{\beta\alpha}x_\alpha. \tag{11}$$

Proof. By the definition of "!" and the properties of FTT, $\vdash !f_{\beta\alpha} \equiv \neg(f_{\beta\alpha} \equiv *_{\beta\alpha})$, i.e., $\vdash !f_{\beta\alpha} \equiv \neg(\forall x_\alpha)(f_{\beta\alpha}x_\alpha \equiv *_{\beta\alpha}x_\alpha)$ which implies (11) by double negation. □

By this simple lemma, a function is defined if at least one of its functional values is defined while the other ones can be undefined. In other words, partial functions with mostly undefined functional values are, as a whole, still taken as defined. Thus, the undefined function $?f_{\beta\alpha}$ means that $f_{\beta\alpha} \equiv *_{\alpha\beta}$ and so, $f_{\beta\alpha}x_\alpha$ is never defined for arbitrary argument x_α.

The following definitions characterize the basic concepts:

(i) *Total function:* $\mathrm{TotF}_{o(\beta\alpha)} \equiv \lambda f_{\beta\alpha} \cdot (\forall x_\alpha)(!x \Rightarrow !(f_{\beta\alpha}x_\alpha))$,
(ii) *Partial function:* $\mathrm{PartF}_{o(\beta\alpha)} \equiv \lambda f_{\beta\alpha} \cdot (\exists x_\alpha)(!x_\alpha \,\&\, ?(f_{\beta\alpha}x_\alpha))$,
(iii) *Strict function:* $\mathrm{StrictF}_{o(\beta\alpha)} \equiv \lambda f_{\beta\alpha} \cdot (\forall x_\alpha)(?x \Rightarrow ?(f_{\beta\alpha}x_\alpha))$,
(iv) *Non-strict function:* $\mathrm{NStrictF}_{o(\beta\alpha)} \equiv \lambda f_{\beta\alpha} \cdot (\exists x_\alpha)(?x_\alpha \,\&\, !(f_{\beta\alpha}x_\alpha))$.

Note that all these predicates are crisp, i.e., any function is either total or partial, or either strict or non-strict.

The concept of total function above includes for higher types also functions that are defined on partial functions. However, we would like to distinguish functions that are still partial but they are defined whenever an argument is defined. This leads to the concept of *partial total function* (cf. [3]).

Definition 2 (Partial total function).

(i) $\mathrm{PTF}_{o\alpha} \equiv \lambda x_\alpha\, !x_\alpha, \qquad \alpha \in \{o, \epsilon\}$
(ii) $\mathrm{PTF}_{o(\beta\alpha)} \equiv \lambda f_{\beta\alpha} \cdot (\forall x_\alpha)(\mathrm{PTF}_{o\alpha}\, x_\alpha \Rightarrow \mathrm{PTF}_{o\beta}\, f_{\beta\alpha} x_\alpha)$.

It can be proved that the formula PTF is crisp.

The predicate PTF has the following interpretation. Let

Lemma 4. *The following holds for all $\alpha, \beta \in$ Types:*

(a) If $\vdash \mathrm{PTF}_{o(\beta\alpha)}\, f_{\beta\alpha}$ *then* $\vdash !f_{\beta\alpha}$.
(b) There is a theory T such that $T \vdash (\exists f_{\beta\alpha})(!f_{\beta\alpha} \,\&\, \neg\, \mathrm{PTF}_{o(\beta\alpha)}\, f_{\beta\alpha})$ *for all types $\alpha, \beta \in$ Types.*

Proof. (a) Let $\alpha, \beta \in \{o, \epsilon\}$. By Definition 5(ii),

$$\vdash \mathrm{PTF}_{o(\beta\alpha)}\, f_{\beta\alpha} \equiv (\forall x_\alpha)(\mathrm{PTF}_{o\alpha}\, x_\alpha \Rightarrow \mathrm{PTF}_{o\beta}\, f_{\beta\alpha} x_\alpha).$$

Using the assumption, substitution and modus ponens, we obtain $\vdash \mathrm{PTF}_{o\alpha}\, x_\alpha \Rightarrow \mathrm{PTF}_{o\beta}\, f_{\beta\alpha} x_\alpha$ which by Definition 4(i) and Rule (R) gives $\vdash !x_\alpha \Rightarrow !f_{\beta\alpha} x_\alpha$.

Now using Axiom (FT-fund0), modus ponens II and $\exists$-substitution we can write the following sequence of provable formulas

$$!!x_\alpha,\, !x_\alpha,\, !x_\alpha \Rightarrow !f_{\beta\alpha} x_\alpha,\, !f_{\beta\alpha} x_\alpha,\, !!f_{\beta\alpha} x_\alpha,\, !x_\alpha \Rightarrow (\exists x_\alpha)!f_{\beta\alpha} x_\alpha,\, (\exists x_\alpha)!f_{\beta\alpha} x_\alpha,$$

which implies $\vdash !f_{\beta\alpha}$ by (11) using Rule (R).

Induction step: Let $\vdash \mathrm{PTF}_{o\gamma}\, f_\gamma$ where $\gamma = \beta\alpha$ for some $\beta, \alpha \in Types$. Then, similarly as above, we obtain $\vdash \mathrm{PTF}_{o\alpha}\, x_\alpha \Rightarrow \mathrm{PTF}_{o\beta}\, f_{\beta\alpha} x_\alpha$ and so, $\vdash !x_\alpha \Rightarrow !f_{\beta\alpha} x_\alpha$ by the induction assumption. Then we proceed in the same way as above.

(b) Let us consider a theory T whose language $J(T)$ contains constants $\mathbf{u}_\alpha, \mathbf{v}_\alpha, \mathbf{f}_{\beta\alpha}$, $\alpha, \beta \in Types$, for which the following holds: $T \vdash !\mathbf{u}_\alpha$, $T \vdash !\mathbf{v}_\alpha$ (this is possible due to Axiom (FT-fund0)) and

$$T \vdash \boldsymbol{\Delta}(\mathbf{f}_{\beta\alpha}\mathbf{u}_\alpha \equiv \mathbf{v}_\alpha) \,\&\, (\forall x_\alpha)(\neg \boldsymbol{\Delta}(x_\alpha \equiv \mathbf{u}_\alpha) \,\&\, (!x_\alpha \Rightarrow ?\mathbf{f}_{\beta\alpha} x_\alpha)). \tag{12}$$

It is not difficult to modify the canonical model of T to fulfill (12) so that T is consistent. Moreover, $T \vdash !\mathbf{f}_{\beta\alpha}\mathbf{u}_\alpha$ by Rule (R). Then, from (12) we conclude that $T \vdash (\exists x_\alpha)(!x_\alpha \,\&\, \neg !\mathbf{f}_{\beta\alpha} x_\alpha)$ and so, $T \vdash !\mathbf{f}_{\beta\alpha} \,\&\, \neg\, \mathrm{PTF}\, \mathbf{f}_{\beta\alpha}$ which implies

$$T \vdash (\exists f_{\beta\alpha})(!f_{\beta\alpha} \,\&\, \neg\, \mathrm{PTF}\, f_{\beta\alpha}).$$

□

Example 1. Let $\mathcal{M}$ be a model due to Definition 1. Let us denote $\mathcal{M}(\mathrm{PTF}_\alpha) = \overline{\mathrm{PTF}}_\alpha \subsetneqq M_\alpha$. As PTF is crisp, $\overline{\mathrm{PTF}}_\alpha(m) \in \{\mathbf{0}, \mathbf{1}\}$ for arbitrary $m \in M_\alpha$.

Let $\alpha, \beta \in \{o, \epsilon\}$. Then:

(a) Then $\overline{\mathrm{PTF}}_\alpha(m)$ $m \in M_o \cup M_\alpha$, i.e., each truth value as well as an element from M_ϵ is taken as a "partial total function".
(b) Let $f \in (M_\beta^*)^{M_\alpha^*}$ and $\overline{\mathrm{PTF}}_{\beta\alpha}(f)$. Then $f(m) \in M_\beta$ holds for every $m \in M_\alpha$.

Note that f is still a partial function, i.e., it can assign $*_\beta$ to $*_\alpha$. However, as being partial total, it is defined on all arguments that are also partial total, which, as a special case, means that it is defined for all $m \in M_\alpha$ where $\alpha \in \{o, \epsilon\}$. □

Lemma 5. *Let T be a consistent theory and $f_{\beta\alpha}, g_{\beta\alpha}$ be functions such that $T \vdash (\exists x_\alpha)(!f_{\beta\alpha}x_\alpha \,\&\, ?g_{\beta\alpha}x_\alpha)$. Then $T \vdash \neg(f_{\beta\alpha} \equiv g_{\beta\alpha})$.*

Proof. Let us extend the language of T by a new constant $\mathbf{u}_\alpha$ and introduce its conservative extension T'. Then, by the assumption, $T' \vdash \neg(f_{\beta\alpha}\mathbf{u}_\alpha \equiv *_\beta)$ and $T' \vdash g_{\beta\alpha}\mathbf{u}_\alpha \equiv *_\beta$. From the properties of fuzzy equality, substitution and Rule (R) we obtain $T' \vdash (f_{\beta\alpha} \equiv g_{\beta\alpha}) \Rightarrow (f_{\beta\alpha}\mathbf{u}_\alpha \equiv *_\beta)$. Using this and the assumption above, we obtain $T' \vdash (f_{\beta\alpha} \equiv g_{\beta\alpha}) \Rightarrow (f_{\beta\alpha}\mathbf{u}_\alpha \equiv *_\beta) \,\&\, \neg(f_{\beta\alpha}\mathbf{u}_\alpha \equiv *_\beta)$, which is $T' \vdash (f_{\beta\alpha} \equiv g_{\beta\alpha}) \Rightarrow \bot$. As the latter is a formula of the language of T, we obtain the proposition of this lemma. □

Definition 3. *The following (crisp) formula represents a crisp ordering of objects (cf. [3,4]):*

$$\lhd_{(o\alpha)\alpha} \equiv \lambda x_\alpha\, \lambda y_\alpha \cdot ?x_\alpha \vee \boldsymbol{\Delta}(x_\alpha \equiv y_\alpha), \qquad \alpha \in \mathit{Types}. \tag{13}$$

Lemma 6. *The formula (13) has the following properties for arbitrary $\alpha \in$ Types:*

(i) $\vdash x_\alpha \lhd x_\alpha$,
(ii) $\vdash (x_\alpha \lhd y_\alpha \,\&\, y_\alpha \lhd x_\alpha) \Rightarrow \boldsymbol{\Delta}(x_\alpha \equiv y_\alpha)$,
(iii) If $\vdash (x_\alpha \lhd y_\alpha \,\&\, y_\alpha \lhd z_\alpha)$ *then* $\vdash x_\alpha \lhd z_\alpha$.

Proof

(i) follows immediately from $\vdash x_\alpha \equiv x_\alpha$ which holds both for defined as well as undefined x_α.
(ii) Let $T = \{x_\alpha \lhd y_\alpha, y_\alpha \lhd x_\alpha\}$ be a theory. As both formulas are crisp, we obtain $T \vdash \boldsymbol{\Delta}(x \equiv y)$ using (13). Then (ii) follows using the deduction theorem.
(iii) We will show that for arbitrary model $\mathcal{M}$ and an assignment p, $\mathcal{M}_p((x_\alpha \lhd y_\alpha \,\&\, y_\alpha \lhd z_\alpha)) = \mathbf{1}$ implies $\mathcal{M}_p(x_\alpha \lhd z_\alpha) = \mathbf{1}$.

If $\mathcal{M}_p(?x_\alpha) = \mathbf{1}$ then $\mathcal{M}_p(x_\alpha \lhd z_\alpha) = \mathbf{1}$. Otherwise $\mathcal{M}_p(\boldsymbol{\Delta}(x_\alpha \equiv y_\alpha) = \mathbf{1}$ which means that $\mathcal{M}_p(x_\alpha) = \mathcal{M}_p(y_\alpha)$. Furthermore, since then $\mathcal{M}_p(?y_\alpha) = \mathbf{0}$, we must have $\mathcal{M}_p(\boldsymbol{\Delta}(y_\alpha \equiv z_\alpha) = \mathbf{1}$ and we conclude that $\mathcal{M}_p(\boldsymbol{\Delta}(x_\alpha \equiv z_\alpha) = \mathbf{1}$ and so, $\mathcal{M}_p(x_\alpha \lhd z_\alpha) = \mathbf{1}$. □

It follows from this lemma that $\lhd$ is a crisp ordering.

On the basis of (13), we can define *monotonous function:*

$$\mathrm{MonF}_{o(\beta\alpha)} \equiv \lambda f_{\beta\alpha} \cdot (\forall x_\alpha)(\forall y_\alpha)((x_\alpha \lhd y_\alpha) \Rightarrow (f_{\beta\alpha} x_\alpha \lhd f_{\beta\alpha} y_\alpha)). \tag{14}$$

Lapierre and Lepage gave many arguments in favor of the idea that the functions should be monotonous in the sense of (14).

Lemma 7. *Let* $T \vdash !x_\alpha \,\&\, !y_\alpha \,\&\, (x_\alpha \lhd y_\alpha)$. *Then* $T \vdash x_\alpha \equiv y_\alpha$.

Proof. By the definition, $T \vdash ?x_\alpha \vee \boldsymbol{\Delta}(x_\alpha \equiv y_\alpha)$. The lemma then follows from $T \vdash (!x_\alpha \,\&\, ?x_\alpha) \equiv \bot$ and the properties of $\boldsymbol{\Delta}$. □

Theorem 4. *Let T be a consistent theory, $f_{\beta\alpha} \in Form_{\beta\alpha}$ be a formula such that $T \vdash TotF\, f_{\beta\alpha} \wedge NStrictF\, f_{\beta\alpha}$ and, moreover, the following is provable:*

$$T \vdash (\exists x_\alpha)(!x_\alpha \,\&\, \neg\boldsymbol{\Delta}(f_{\beta\alpha} x_\alpha \equiv f_{\beta\alpha} *_\alpha)). \tag{15}$$

Then $T \not\vdash MonF\, f_{\beta\alpha}$.

Proof. Let us assume that $T \vdash \mathrm{MonF}\, f_{\beta\alpha}$. Furthermore, let T' be a conservative extension of T by introducing a new constant $\mathbf{u}_\alpha$ that witnesses (15) (this is possible since (15) is a crisp formula). Hence, $T' \vdash !\mathbf{u}_\alpha$ and

$$T' \vdash \neg\boldsymbol{\Delta}(f_{\beta\alpha} \mathbf{u}_\alpha \equiv f_{\beta\alpha} *_\alpha). \tag{16}$$

As $T' \vdash *_\alpha \lhd \mathbf{u}_\alpha$, it follows from the monotonicity that

$$T' \vdash f_{\beta\alpha} *_\alpha \lhd f_{\beta\alpha} \mathbf{u}_\alpha. \tag{17}$$

At the same time, $T \vdash \mathrm{TotF}\, f_{\beta\alpha}$ implies $T' \vdash !\mathbf{u}_\alpha \Rightarrow !f_{\beta\alpha}\mathbf{u}_\alpha$ and consequently, $T' \vdash !f_{\beta\alpha}\mathbf{u}_\alpha$.

Furthermore, because $T \vdash \mathrm{NStrictF}\, f_{\beta\alpha}$, let T'' be a conservative extension of T' by a new constant $\mathbf{v}_\alpha$ such that $T'' \vdash ?\mathbf{v}_\alpha \,\&\, !f_{\beta\alpha}\mathbf{v}_\alpha$. This means that $T'' \vdash \neg\boldsymbol{\Delta}(f_{\beta\alpha}\mathbf{v}_\alpha \equiv *_\beta)$ and $T'' \vdash \mathbf{v}_\alpha \equiv *_\alpha$. From the latter, using Rule (R), (17) and Lemma 7 we conclude that $T'' \vdash f_{\beta\alpha}\mathbf{v}_\alpha \equiv f_{\beta\alpha}\mathbf{u}_\alpha$. But from (16) we have $T'' \vdash \neg\boldsymbol{\Delta}(f_{\beta\alpha}\mathbf{u}_\alpha \equiv f_{\beta\alpha}\mathbf{v}_\alpha)$ and we conclude that T'' is contradictory — a contradiction. □

According to this lemma, monotonicity excludes also improper functions having the properties from Theorem 4. This result is in accordance with the results of [4].

Further definition according to Lapierre and Lepage is the following.

Definition 4 (Strong difference of objects).

$$\neq^* \equiv \lambda x_\alpha\, \lambda y_\alpha\, !x_\alpha \,\&\, !y_\alpha \,\&\, \neg\boldsymbol{\Delta}(x_\alpha \equiv y_\alpha).$$

Clearly, strong difference is stronger than negation of crisp $\equiv$.

Lemma 8. *Let* $\vdash !f_{\beta\alpha} \mathbin{\&} !g_{\beta\alpha} \mathbin{\&} \neg(f_{\beta\alpha} \not\equiv^* g_{\beta\alpha})$. *Then* $\vdash (\forall x_\alpha)\boldsymbol{\Delta}(f_{\beta\alpha}x_\alpha \equiv g_{\beta\alpha}x_\alpha)$.

Proof. From the assumption, we obtain $\vdash \boldsymbol{\Delta}(f_{\beta\alpha} \equiv g_{\beta\alpha})$. The lemma then follows from the definition of fuzzy equality between functions and the provable property $\vdash \boldsymbol{\Delta}(\forall x_\alpha)A_o \Rightarrow (\forall x_\alpha)\boldsymbol{\Delta}A_o$.

By this lemma, if the functions are not strongly different then they must be fully equal (including undefined values).

Lemma 9. *Let T be a theory such that $T \vdash x_\alpha \lhd x'_\alpha$, $T \vdash y_\alpha \lhd y'_\alpha$ and $T \vdash x_\alpha \not\equiv^* y_\alpha$. Then $T \vdash x'_\alpha \not\equiv^* y'_\alpha$.*

Proof. From the definition of strong difference, we have $T \vdash !x_\alpha \mathbin{\&} !y_\alpha \mathbin{\&} \neg\boldsymbol{\Delta} (x_\alpha \equiv_\alpha y_\alpha)$. From this, the assumption, the definition of $\lhd$ and the properties of $\boldsymbol{\Delta}$ we conclude that $T \vdash x_\alpha \equiv_\alpha x'_\alpha$ as well as $T \vdash y_\alpha \equiv_\alpha y'_\alpha$. The lemma then follows from the latter using Rule (R). □

To obtain a special case of pFTT in accordance with the papers [3,4], we might extend the list of its axioms by the following one:

(FT-fund6) $(\forall f_{\beta\alpha})\,\mathrm{MonF}\, f_{\beta\alpha}, \qquad \alpha, \beta \in \mathit{Types}.$

It seems, however, that this axiom should not belong among logical axioms of pFTT and should be added only to special theories. The reason is that it would restrict pFTT too much.

The generic fuzzy equality "$\equiv$" is too strict because it was proved in [9] that $T \vdash (\exists x_\alpha)?A_{o\alpha}x_\alpha$ implies $T \vdash (\forall x_\alpha)A_{o\alpha}x_\alpha \equiv \bot$. This means that if some of x_α's is not defined then the general quantifier gives falsity. It seems reasonable, however, to weaken this strict behavior in such a way that all the undefined x_α's are ignored. Therefore, we introduce the following concept:

Definition 5 (Fuzzy equality between partial functions).

(i) $\approxeq_{(o\alpha)\alpha} \equiv \lambda x_\alpha\, \lambda y_\alpha \cdot \mathrm{PTF}_{o\alpha}\, x_\alpha \mathbin{\&} \mathrm{PTF}_{o\alpha}\, y_\alpha \mathbin{\&} (x_\alpha \equiv y_\alpha), \alpha \in \{o, \epsilon\}$
(ii) $\approxeq_{(o(\beta\alpha))(\beta\alpha)} \equiv \lambda f_{\beta\alpha}\, \lambda g_{\beta\alpha} \cdot \mathrm{PTF}_{o(\beta\alpha)}\, f_{\beta\alpha} \mathbin{\&} \mathrm{PTF}_{o(\beta\alpha)}\, g_{\beta\alpha} \mathbin{\&}$
$(\forall x_\alpha)(\mathrm{PTF}_{o\alpha}\, x_\alpha \Rightarrow (f_{\beta\alpha}x_\alpha \approxeq g_{\beta\alpha}x_\alpha))$.

Note that "$\approxeq$" is just a certain restriction of the standard "$\equiv$".

Theorem 5. *The formula "$\approxeq$" is a fuzzy equality for all $\alpha \in \mathit{Types}$, i.e., it is reflexive, symmetric and transitive w.r.t. &.*

Proof. Reflexivity and symmetry follow immediately from reflexivity and symmetry of "$\equiv$".

Transitivity: we require $\vdash (x_\alpha \approxeq y_\alpha) \mathbin{\&} (y_\alpha \approxeq z_\alpha) \Rightarrow (x_\alpha \approxeq z_\alpha)$ for all $\alpha \in \mathit{Types}$. This is immediate for $\alpha \in \{o, \epsilon\}$ due to transitivity of "$\equiv$".

Let $\alpha = \delta\gamma$. Then, since PTF is crisp and so $\vdash \mathrm{PTF}^2 \equiv \mathrm{PTF}$ is provable, we have

$$\vdash (\mathrm{PTF}_{o\gamma}\, x_\gamma \Rightarrow (f_{\delta\gamma}x_\gamma \approxeq g_{\delta\gamma}x_\gamma)) \Rightarrow ((\mathrm{PTF}_{o\gamma}\, x_\gamma \Rightarrow (g_{\delta\gamma}x_\gamma \approxeq h_{\delta\gamma}x_\gamma)) \Rightarrow (\mathrm{PTF}_{o\gamma}\, x_\gamma \Rightarrow (f_{\delta\gamma}x_\gamma \approxeq h_{\delta\gamma}x_\gamma)))$$

by the inductive assumption on the transitivity of "$\approxeq$" and the properties of FTT. Then applying generalization and the properties of quantifiers and adjunction we obtain

$$\vdash (\forall x_\gamma)(\mathrm{PTF}_{o\gamma} x_\gamma \Rightarrow (f_{\delta\gamma} x_\gamma \approxeq g_{\delta\gamma} x_\gamma)) \,\&\, (\forall x_\gamma)(\mathrm{PTF}_{o\gamma}\, x_\gamma \Rightarrow (g_{\delta\gamma} x_\gamma \approxeq h_{\delta\gamma} x_\gamma)) \Rightarrow (\forall x_\gamma)(\mathrm{PTF}_{o\gamma}\, x_\gamma \Rightarrow (f_{\delta\gamma} x_\gamma \approxeq h_{\delta\gamma} x_\gamma)). \tag{18}$$

Finally, joining (18) with

$$\mathrm{PTF}_{o(\delta\gamma)}\, f_{\delta\gamma} \,\&\, \mathrm{PTF}_{o(\delta\gamma)}\, g_{\delta\gamma} \,\&\, \mathrm{PTF}_{o(\delta\gamma)}\, g_{\delta\gamma} \,\&\, \mathrm{PTF}_{o(\delta\gamma)}\, h_{\delta\gamma} \Rightarrow \mathrm{PTF}_{o(\delta\gamma)}\, f_{\delta\gamma} \,\&\, \mathrm{PTF}_{o(\delta\gamma)}\, h_{\delta\gamma}$$

we obtain

$$\vdash \mathrm{PTF}_{o(\delta\gamma)}\, f_{\delta\gamma} \,\&\, \mathrm{PTF}_{o(\delta\gamma)}\, g_{\delta\gamma} \,\&(\forall x_\gamma)(\mathrm{PTF}_{o\gamma}\, x_\gamma \Rightarrow (f_{\delta\gamma} x_\gamma \approxeq g_{\delta\gamma} x_\gamma)) \,\&\, \mathrm{PTF}_{o(\delta\gamma)}\, g_{\delta\gamma} \,\&\, \mathrm{PTF}_{o(\delta\gamma)}\, h_{\delta\gamma} \,\&(\forall x_\gamma)(\mathrm{PTF}_{o\gamma}\, x_\gamma \Rightarrow (g_{\delta\gamma} x_\gamma \approxeq h_{\delta\gamma} x_\gamma)) \Rightarrow \mathrm{PTF}_{o(\delta\gamma)}\, f_{\delta\gamma} \,\&\, \mathrm{PTF}_{o(\delta\gamma)}\, h_{\delta\gamma} \,\&(\forall x_\gamma)(\mathrm{PTF}_{o\gamma}\, x_\gamma \Rightarrow (f_{\delta\gamma} x_\gamma \approxeq h_{\delta\gamma} x_\gamma))$$

which is transitivity of "$\approxeq$". □

The following is immediate.

Lemma 10. *If* $\vdash \mathrm{PTF}\, f_\alpha \,\&\, \mathrm{PTF}\, g_\alpha$ *then* $\vdash (f_\alpha \approxeq g_\alpha) \equiv (f_\alpha \equiv g_\alpha)$.

3.2 Examples of Partial Functions in pFTT

Square Root. We will finish this section by two examples. First, we will extend the language of pFTT by a class of constants $\Theta_{\alpha\emptyset}$, $\alpha \in Types$ fulfilling the following axiom:

$$(\forall x_o)((!x_o \,\&\, \Upsilon x_o) \equiv !\Theta_{\alpha o} x_o). \tag{19}$$

Interpretation of the function Θ in any model $\mathscr{M}$ is the following:

$$\mathscr{M}(\Theta_{\alpha o})(a) = \begin{cases} f_\alpha & \text{if } a > 0, \\ *_\alpha & \text{otherwise,} \end{cases} \qquad a \in E^*$$

where $f_\alpha \in M_\alpha^*$ is a defined element (i.e., different from $*_\alpha$). For example, if $\alpha = \gamma\beta$ then $f_{\gamma\beta}$ is a function that is defined, at least, for one argument from M_β^*.

Lemma 11. *Let T be a theory in which axiom (19) is fulfilled and A_o be a formula such that either $T \vdash ?A_o$ or $T \vdash !A_o \,\&\, \neg A_o$. Then*

$$T \vdash ?\Theta_{\alpha o} A_o.$$

Proof. It follows from (19) that $\vdash (!A_o \,\&\, \Upsilon A_o) \equiv !\Theta_{\alpha o} A_o$. If $T \vdash ?A_o$ then $T \vdash \Upsilon A_o \equiv \bot$ by Lemma 2. From it follows that $T \vdash \neg !\Theta_{\alpha o} A_o$, i.e., $T \vdash ?\Theta_{\alpha o} A_o$ by Rule (R), properties of FTT and (19).

If $T \vdash \neg A_o$ then again $T \vdash \Upsilon A_o \equiv \bot$ and we proceed as above.

Similar, slightly more general lemma is the following.

Lemma 12. *Let* $\vdash !A_\alpha \equiv !B_\beta$ *and* $\vdash ?A_\alpha$. *Then* $\vdash B_\beta \equiv *_\beta$, *i.e.*, $\vdash ?B_\beta$.

Proof. By the assumption and Rule (R) we obtain $\vdash !*_\alpha \equiv !B_\beta$. However, $\vdash \neg !*_\alpha$by the properties of FTT which implies $\vdash B_\beta \equiv *_\beta$ by Rule (R) and double negation.

Using Θ it is possible to define, e.g, the square root. Let $Sq_{\epsilon\epsilon}$ be a constant whose interpretation is the operation of "square root"[1] and assume that $\vdash \Theta_{(\epsilon\epsilon)o}\top \equiv Sq_{\epsilon\epsilon}$. Then the formula

$$\mathrm{SQRT}_{\epsilon\epsilon} \equiv \lambda z_\epsilon \cdot (\Theta_{(\epsilon\epsilon)o}\boldsymbol{\Delta}(z_\epsilon \geq 0_\epsilon))z_\epsilon$$

defines a function of square root[2] on elements of type ϵ that gives result if z_ϵ is non-negative and is undefined otherwise. For example, let us consider a model in which $M_\epsilon = \mathbb{R}$ with the standard arithmetic operations. We can then put, e.g.,

$$\mathrm{SQRT}(x - y) \equiv (\Theta\boldsymbol{\Delta}(x - y \geq 0))(x - y).$$

Interpretation of this formula is a square root of $x - y$, if it is non-negative and is undefined otherwise. More explicitly, let p be an assignment of elements to variables. Then $\mathscr{M}_p(Sq)$ is a function whose values are square roots of $x \in [0, \infty)$ (and arbitrary values for the other x) and

$$\mathscr{M}_p(\mathrm{SQRT})(\mathscr{M}_p(x)) = \begin{cases} \sqrt{\mathscr{M}_p(x)} & \text{if } \mathscr{M}_p(x) \geq 0, \\ * & \text{oterwise} \end{cases}$$

where $* = \mathscr{M}_p(*_\epsilon)$ means "undefined".

French King. Let us now consider the famous example

$$\textit{The present king of France is bald} \tag{20}$$

discussed widely in logical literature. This sentence is usually taken as false due to the fact that there is no present king of France.

We will follow here the discussion presented by Duži et al. and Tichý in [2,10]. According to their analysis, the "king of France" is an office that can be occupied by some person in the given possible world and time. Let the types of possible world and time be ω, τ, respectively and $w \in Form_\omega$, $t \in Form_\tau$ be variables of the corresponding types. Let the type ϵ represent people[3]. Then we may consider a set $I_{o\tau}$ (crisp but not necessarily) whose interpretation is a time

[1] It is required that Sq gives square root for all elements from M_ϵ, on which it is defined and may give arbitrary elements otherwise.

[2] We silently assumed that the crisp inequality $\geq$ is also defined.

[3] In fact, we can consider here some arbitrary type α provided that we will specify its properties characterizing people.

interval (this will represent the period when France had kings). Then we will define the king of France by the formula

$$\mathrm{FK}_{(\epsilon\tau)\omega} \equiv \lambda w\, \lambda t\, \Theta_{\epsilon o}(I_{o\tau}) \tag{21}$$

where $\Theta_{\epsilon o}$ is the function defined analogously as in (19). Then the formula $(\mathrm{FK}w)t$ represents the king of France in the possible world w and time t w.r.t. the period I.

Lemma 13. *Let* $\mathbf{t}^0$ *be a formula (constant) of type* τ *and assume that* $\vdash \neg\boldsymbol{\Delta}(I_{o\tau}\mathbf{t}^0)$. *Then*

$$\vdash ?(\mathrm{FK}\, w)\mathbf{t}^0$$

Proof. The lemma follows from the definition of Θ.

Due to this lemma, the king of France does not exist outside the period $I_{o\tau}$.

Let now $B_{o\epsilon}$ be a formula representing "baldness". Then the formula

$$\mathrm{Bald}_{o\epsilon} \equiv \lambda x_\epsilon\, B_{o\epsilon} x_\epsilon$$

represents a fuzzy set of bald objects. Assuming that $\mathbf{t}^0$ represents the present time, the formula

$$\mathrm{Bald}((\mathrm{FK}\, w)\mathbf{t}^0)$$

construes the sentence (20).

Further result depends on the properties of $B_{o\epsilon}$. Let us consider two axioms:

$$\neg B_{o\epsilon} *_\epsilon, \tag{22}$$

$$!B_{o\epsilon} x_\epsilon \equiv !x_\epsilon. \tag{23}$$

Clearly, (22) says that "undefined" cannot be bald, and (23) says that bald (at least in some degree) can be only defined objects.

Theorem 6

(a) Let T *be a theory in which axiom (22) holds. Then*

$$T \vdash \neg \mathit{Bald}((\mathrm{FK}\, w)\mathbf{t}^0).$$

(b) Let T *be a theory in which axiom (23) holds. Then*

$$T \vdash ?\mathit{Bald}((\mathrm{FK}\, w)\mathbf{t}^0).$$

Proof

(a) is immediate consequence of Lemma 13 and axiom (22).
(b) follows from Lemma 13 using Lemma 12.

Hence, depending on whether we accept that baldness can be either undefined or it always has some truth degree, we obtain that either sentence (20) is false, or that it has no sense.

4 Conclusion

In this paper, we demonstrated that the basic concepts of the theory of partial functions due to Lepage and Lapierre can be introduced into the formalism of the fuzzy type theory with partial functions. We also formally proved the basic properties of them. At the end we presented two examples of partial functions — the square root, and interpretation of the sentence "The present king of France is bald".

Further work will focus on extension of the theory of partial functions in pFTT. Among other concepts, we will elaborate in more detail comparison of partial functions.

Acknowledgement. This paper was supported by the grant 16-19170S of GAČR, Czech Republic.

References

1. Andrews, P.: An Introduction to Mathematical Logic and Type Theory: To Truth Through Proof. Kluwer, Dordrecht (2002)
2. Duží, M., Jespersen, B., Materna, P.: Procedural Semantics for Hyperintensional Logic. Springer, Dordrecht (2010). https://doi.org/10.1007/978-90-481-8812-3
3. Lapierre, S.: A functional partial semantics for intensional logic. Notre Dame J. Formal Log. **33**, 517–541 (1992)
4. Lepage, F.: Partial functions in type theory. Notre Dame J. Formal Log. **33**, 493–516 (1992)
5. Novák, V.: On fuzzy type theory. Fuzzy Sets Syst. **149**, 235–273 (2005)
6. Novák, V.: A comprehensive theory of trichotomous evaluative linguistic expressions. Fuzzy Sets Syst. **159**(22), 2939–2969 (2008)
7. Novák, V.: EQ-algebra-based fuzzy type theory and its extensions. Log. J. IGPL **19**, 512–542 (2011)
8. Novák, V.: Towards fuzzy type theory with partial functions. In: Kacprzyk, J., Szmidt, E., Zadrożny, S., Atanassov, K.T., Krawczak, M. (eds.) IWIFSGN/EUSFLAT -2017. AISC, vol. 643, pp. 25–37. Springer, Cham (2018). https://doi.org/10.1007/978-3-319-66827-7_3
9. Novák, V.: Fuzzy type theory with partial functions. Iran. J. Fuzzy Syst. (submitted)
10. Tichý, P.: The Foundations of Frege's Logic. De Gruyter, Berlin (1988)

On Hash Bipolar Division: An Enhanced Processing of Novel and Conventional Forms of Bipolar Division

Noussaiba Benadjimi[1(✉)], Walid Hidouci[1], and Allel Hadjali[2]

[1] Laboratoire de la Communication dans les Systmes Informatiques, Ecole nationale Supérieure d'Informatique, BP 68M, 16309 Oued-Smar, Alger, Algérie
{an_benadjimi,hidouci}@esi.dz
http://www.esi.dz

[2] LIAS - ENSMA, Poitiers, France
allel.hadjali@ensma.fr

Abstract. In this paper, two issues of bipolar division are discussed. First, we outline some new operators dealing with the bipolar division, to enrich the interpretations of bipolar queries. In this context, we propose some extended operators of bipolar division based on the connector "or else". Besides, we introduce a new bipolar division operator dealing with the "Satisfied-Dissatisfied approach". Secondly, we highlight the matter of the performance improvement of the considered operators. Thus, we present an efficient method which allows handling several bipolar divisions with a unified processing. Our idea is to design new variants of the classical Hash-Division algorithm, for dealing with the bipolar division. The issue of answers ranking is also dealt with. Computational experiments are carried out and demonstrate that the new variants outperform the conventional ones with respect to performance.

Keywords: Bipolar division · Satisfied-Dissatisfied approach
Relational division · Hash-Division algorithm

1 Introduction

Bipolar queries are a very useful type of queries, especially, when user's preferences come to play. They are very effective for many applications, such as in business intelligence applications, in recommendation systems, and in a large part of artificial intelligence applications [2]. Bipolarity refers to the distinction between two kinds of parts, called ***"poles"***, often known as positive and negative poles [2]. Bipolarity may be represented under different forms:

- **Constraint-Wish ('Required-Desired') approach**: the poles considered are: *(i) Constraint* which describes the set of mandatory values and *(ii) Wish* which defines the set of wished-for values [2].

J. Medina et al. (Eds.): IPMU 2018, CCIS 853, pp. 749–761, 2018.
https://doi.org/10.1007/978-3-319-91473-2_63

- **Satisfied-Dissatisfied approach:** the first pole specifies the values of the Y attributes set which are positively evaluated by the user (*desired*). Independently, the second pole specifies values negatively evaluated by the user (*disliked*) [3].

There are many papers on bipolarity over simple queries, particularly the selection-operator-based queries, with crisp or fuzzy preferences [5–7]. However, to the best of our knowledge, there are only a few researchers that have studied the bipolar division to express sophisticated preference queries. The bipolar division was first introduced in [4]. The authors have discussed different ways to express the bipolar division over crisp databases. They have also studied the bipolar division operator on fuzzy databases [8]. Several extended forms of bipolar division also exist in the setting of fuzzy databases, see [9,10].

On the other hand, the approach proposed in [1] for the relational division, and called **'Hash-Division'**, has proven to be an effective approach. The experimental results illustrated in [1], demonstrate that the Hash-Division, in most cases, is far better than the traditional algorithms in processing time. Some newly extended variants of this approach have been used in our previous work detailed in [11] to improve the processing time of some forms of the tolerant division. It is worthy to note that previous searches done on bipolar-division in the literature have only focused on improving the quality of query answering, by introducing new operators to model bipolarity. Although, the performance issue has not been adequately addressed by these searches. Indeed, the proposed approaches are mainly based on the nested loops-like-algorithms. In so doing, one can observe that is far from being acceptable in terms of processing time and deteriorates significantly as data size increases. In addition, queries evaluations are performed with a reduced size of data (dividend and divisor) [4,8]. This does not fit reality, especially with the advent of the Big Data. Moreover, the *Satisfied-Dissatisfied* approach of bipolarity has many uses in information systems to deal with values that can be negatively evaluated by the users. Despite this interest, no work has investigated the definition of bipolar-division in this setting, with the exception of the work of Bosc et al. detailed in [13]. In this latter, authors have interpreted the negative pole (sole) with a new operator that they called it '**anti-division**'. In this paper, we attempt to propose some solutions to the issues discussed above. So, our main contributions can be summarized as follows:

- Extend the bipolar division to deal with the Satisfied-Dissatisfied approach.
- Improve the bipolar division w.r.t the performance dimension. To this end, we suggest a novel Hash-Division algorithm.

The remainder of this paper is organized as follows. In Sect. 2, we recall some basic notions and provide related work on some existing approaches devoted to the bipolar division. In Sect. 3, we describe our new proposed operators for the bipolar division. Section 4 discusses the improvements of the proposed operators, then deep analysis and discussion of the experimental results obtained are provided in Sect. 5. Finally, Sect. 6 concludes the paper and suggests some directions for future work.

2 Background and Related Work

We provide here some basic notions needed for the reading of the paper and related work on some existing bipolar division approaches.

2.1 Relationnal Division and Bipolar Division

Relational division is used whenever an element that satisfies a whole set of requirements is sought for. In relational algebra, the division of the relation $r(X, Y)$ called ***"dividend"***, by the relation $s(Y)$ called ***"divisor"***, is a new relation $q(X)$ called ***"quotient"*** that includes some parts of the $projection(r, X)$ satisfying the following condition: x is in $q(X)$ *if and only if* x is in $Project(r, X)$ and for <u>**all**</u> y in $s(Y)$, $r(X, Y)$ contains $<x, y>$ [4]. X and Y are two compatible sets of attributes. Formally, the relational division is defined as:

$$Div(r, s, X, Y) = \{x \in projection(r, X) \mid \forall y, (y \in s) \Rightarrow (\langle x, y \rangle \in r)\} \quad (1)$$

Example 1: Consider a distribution company of some products. In its commercial activity, the company wants to select its most valued customers (buyers). The customers ranking is based on some categories of product. Let ***Customer_Order*** (*#customer*, *#product*, *#order_state*) and ***Critical_Product*** (*#product*) be two crisp relations. The figure below illustrates the relational division corresponding to the query: ***"Which customers have made an approved order for <u>all</u> critical products?"***. C_1 and C_3 are the resulting quotients because they are the customers who have made an approved order for all critical products (Fig. 1).

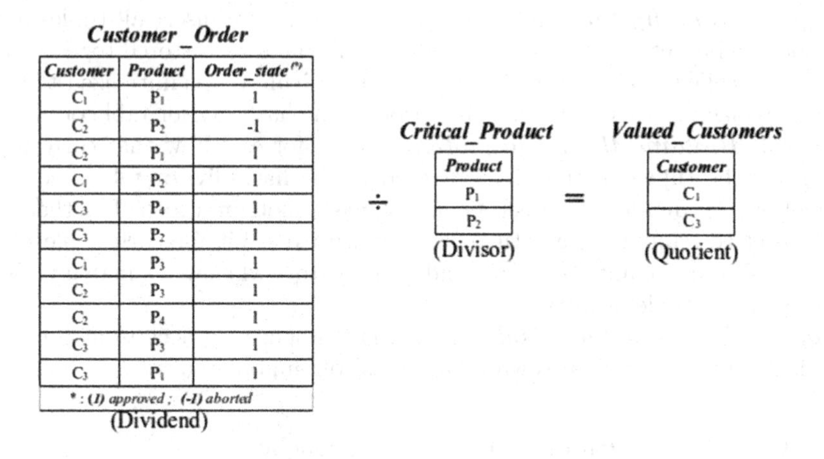

Customer_Order

Customer	Product	Order_state(*)
C_1	P_1	1
C_2	P_2	-1
C_2	P_1	1
C_1	P_2	1
C_3	P_4	1
C_3	P_2	1
C_1	P_3	1
C_2	P_3	1
C_2	P_4	1
C_3	P_3	1
C_3	P_1	1

*: (1) approved ; (-1) aborted

(Dividend)

÷

Critical_Product

Product
P_1
P_2

(Divisor)

=

Valued_Customers

Customer
C_1
C_3

(Quotient)

Fig. 1. Relational division corresponding to Example 1

The natural bipolar division operator that is straightforwardly designed for the Constraint-Wish approach consists in considering a twofold divisor

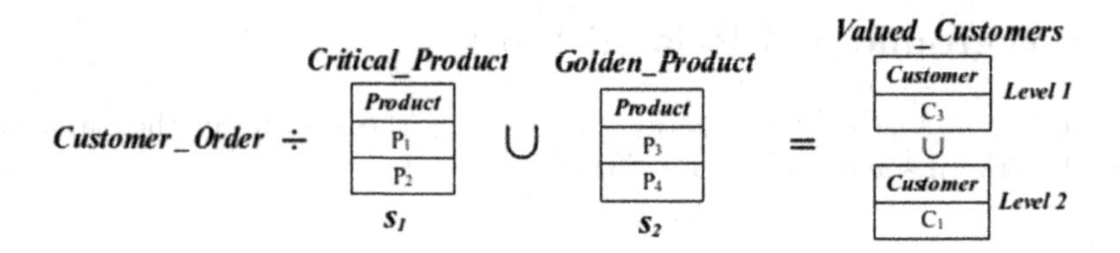

Fig. 2. Bipolar division

$s = s_1 \cup s_2$. Hence, to be satisfactory, an element x of the dividend must be associated with '<u>all</u>' of the values of s_1 (constraint); and it is even better for x to be associated with '<u>all</u>' the elements of s_2 (wish) [4].

Example 2: Let's take the relations of the previous example to which the crisp relation ***Golden _Product*** (*#product*) is added. Figure 2 illustrates the bipolar division corresponding to the query: *"Which customers have made an approved order for <u>all</u> critical products* ***and if possible*** *for <u>all</u> golden products?"*.

Here, C_1 and C_3 are no longer equally ranked. C_3 is more valued because it has made an approved order of all critical and golden products. Whereas C_1 has only made an approved order for all critical products.

2.2 Hash-Division Algorithm

In this subsection, we give a brief description of the *Hash-Division* algorithm **(HD)** (see [1] for further details). It uses two hash tables, one for the divisor and the second for the quotient. HD is proceeding in three stages:
Stage 1, Building the hash-divisor table: here, we insert all tuples of the divisor into buckets in the hash-divisor table. Each entry is stored together with an integer called **divisor number *'Num_div'***. Num_div is initialized to 0 and it is incremented whenever a new insertion in the hash-divisor table occurs.
Stage 2, Building the hash-quotient table: for each row that corresponds to one of the tuples of the divisor stored in the hash-divisor table, we insert a quotient candidate into buckets in the hash-quotient table. Together with each inserted candidate, a bitmap is kept with one bit for each tuple of the divisor. All bits are initialized to 0 and updated to 1 whenever a match with the corresponding tuple occurred.
Stage 3, Building the result: in this last stage, we select from the hash-quotient table all candidates whose bitmaps contain only ones.

2.3 Overview on Bipolar Division Approaches

In this paper, we are concerned with the bipolar division over crisp databases exclusively. In this context, there are only the works of *Bosc et al.* [4,8], and *Tamani et al.* [9,10,12] which have addressed the bipolar division problem. However, only the Constraint-Wish approach is studied in those studies.

To define the bipolar division, the work in [4] has considered a twofold divisor: a subset s_1 of values that are definitely compulsory; and a second subset of values s_2 that are merely desired. The bipolarity is expressed through the ***"and if possible"*** connector. Hence, the elements contained in the dividend relation are partitioned into three sets [4]:

- **Best**, groups elements associated with all values in both s_1 and s_2.
- **Accept**, contains elements associated with all the values of s_1 but not all those of s_2.
- **Reject**, includes elements not associated with all the values of s_1.

The slight drawback of this approach that we can point out, is that accepted elements (the second set) have all the same rank-order. Even though, a discrimination on the level of satisfaction can be used to better distinguish between accepted elements. In Subsect. 4.3 we will show how to treat this point.

A tolerant version of the bipolar division based on the "and if possible" connector is also introduced in [8, 10]. The idea is to define the constraint (resp. the wish) as the association with a proportion p_1 (resp. p_2) of elements in s_1 (resp. s_2), where p_1 and p_2 are interpreted thanks to fuzzy quantifiers such as: '***most***' and '***as many as possible***'. The general form of such a query is: "*Return elements which are associated with* ***Q1*** *elements in* s_1 *and if possible related to* ***Q2*** *of elements in* s_2", where Q1 and Q2 stand for fuzzy quantifiers.

3 New Extended Operators for the Bipolar Division

In this section, we propose some new extensions of the bipolar division. First, we propose a new compound variant based on the "or else" approach. Then, we introduce a new bipolar operator to deal with the Satisfied-Dissatisfied approach.

3.1 Bipolar Division Based on the "or Else" Connector

The connector "**or else**" was essentially proposed to handle the bipolar division over fuzzy bipolar relations[1] in [10]. Hereafter, we present how to use this connector in the bipolar division over crisp databases. Moreover, we use a fuzzy linguistic quantifier to better discriminate accepted quotients.

The semantic behind this connector is that one pole (let be the positive pole) expresses perfect associations while the other pole expresses acceptable associations. Tow main scales can be used, l_1 (***perfect***) for those associated with all values in the positive pole, and l_2 (***accepted***) for elements that are associated with all values in the negative pole. Hence, depending on the relationship between s_1 and s_2, the bipolar division based on "or else" may have two forms of interpretation:

[1] Each tuple t in a fuzzy bipolar relation r is attached with a pair of grades $(\mu_c(t), \mu_w(t))$ that expresses the degree of its satisfaction respectively to the constraint c and the wish w [10].

- $s_1 \subset s_2$: the main format of the bipolar division is: *"Return elements which are associated with* **all** *values in* s_2 ***OR ELSE*** *with* **all** *values in* s_1 *and with* ***as many as possible*** *of values in* $s_2 - s_1$*"*.
- $s_1 \cap s_2 = \emptyset$, and s_1 is more important than s_2: the main format of the bipolar division is *"Return elements which are associated with* **all** *values in* s_1 *and with* ***as many as possible*** of values in s_2 ***OR ELSE*** with **all** *values in* s_2 *and with as many as possible of values in* s_1*"*.

In these two compound forms of bipolar division, the fuzzy quantifier ***'as many as possible'***, allow to better distinguish between final quotients intra-sets (perfect and acceptable), see Sect. 4.3 for details.

Example 3: According to the example sketched in Fig. 2; when considering the critical (resp. golden) products as the positive (resp. negative) pole, the query is defined as: *"Which customers have made an approved order for all critical products and for as many as possible of golden products;* ***OR ELSE*** *customers who have made an approved order for all golden products, and for as many as possible of critical products?"*; resulting quotients are ranked as follows: $C_3 > C_1 > C_2$ (C_3 is the most valued customer).

3.2 Bipolar Division for the Satisfied-Dissatisfied Approach

To the best of our knowledge, this is the first time that the division operation is addressed in the setting of the satisfied-dissatisfied approach of bipolarity.

The *natural* way for defining the two subsets of the divisor is to consider them as two parts totally independents ($s_1 \cap s_2 = \emptyset$). The reason behind this idea is that it is little obvious to have an element which is simultaneously positively and negatively evaluated by the user. Depending on the user's preferences, several types of bipolar division on the Satisfied-Dissatisfied approach may be designed:

- **Strong symmetrical impact:** both positive and negative poles have the same impact on the result ranking. To be selected as a valid quotient, an element x must be related to all values in s_1 **and** must be not related with any value in s_2.
  Hence, the query derived from Fig. 2 can be interpreted as: *"Select customers who have made an approved order for all golden products* ***and*** *they haven't made any aborted order for the critical products?"*. Thereby C_3 is the only valid quotient. C_1 is removed from the quotient set because the golden products weren't all ordered. Similarly for C_2, it is no longer a quotient because it has made an aborted order for the critical product P_2.
- **Positive and negative poles as constraint and desire:** for this category, the negative (resp. positive) pole is just used to discriminate elements satisfying the whole set of positive (resp. negative) associations. The main format of the query is: *"Select elements which are associated with all (resp. none) values in* s_1 *(resp.* s_2*)* ***and if possible*** *associated with none (resp. all) value in* s_2 *(resp.* s_1*)"*

- **Positive and negative poles as hierarchical preferences:** here, one pole is more important than the other. Such queries are interpreted as: "*Select elements which are associated with all (resp. none) value in s_1 (resp. s_2)* ***or else*** *associated with none (resp. all) value in s_2 (resp. s_1)*".

For the two last versions, we can use a compound queries with the fuzzy quantifiers '**as many as possible**' for positive requirements and '**as little as possible**' for negative requirements, to provide better discrimination between final quotients (the principle is similar to that shown in Subsect. 3.1).

4 Improving the Bipolar Division Handling

As mentioned previously, the aspect of the performance was not dealt with in research done for the bipolar division. Indeed, the approach proposed in [4] is based on the decomposition of the bipolar division on two relational divisions which lead to a very time-consuming process. Although this decomposition is avoided in [12] which uses generic fuzzy bipolar R-Implication operator, the nested-loop based algorithm which is a very time-consuming too.
In this section, we propose to improve the effectiveness of the bipolar division relying on the Hash-Division like algorithm. In fact, we have made several changes in both data structures and processing of the three stages of the basic Hash-division, described in Subsect. 2.2, to deal with the bipolar division operators.

4.1 The First Stage

As seen in Subsect. 2.2, we store all tuples of the divisor in a hash table. Whereas in our variant, each tuple belonging to the layer s_i, $i = 1 \dots 2$, is stored in the hash-divisor table together with two integers (see Fig. 3):

- **ind_lyr:** index of the layer:
  - **'0'** for s_1 and **'1'** for s_2 if $s_1 \cap s_2 = \emptyset$.
  - **'0'** for s_1 and **'1'** for $(s_2 - s_1)$ if $s_1 \subset s_2$.

  This integer indicates the tuple offset inside the bitmap. Bits corresponding to tuples in s_1 are located in the bitmap before those belonging to s_2.
- **num_div_lyr:** the divisor number (*offset*) of the tuple in its layer (s_1 or s_2).

Divisor value	ind_lyr	num_div_lyr

Fig. 3. The data structure of a hash-divisor tuple

Hence, for each layer s_i, $i = 1 \dots 2$, we keep its own divisors counter. The two counters will be handled as in the basic version. They are initialized to 0, and whenever we insert a new tuple of the divisor of the layer s_i into the hash-divisor table, the divisors counter of s_i will be incremented.

4.2 The Second Stage

In the second stage, we have made two major differences from the basic algorithm. The first difference is how to update the bitmap. If a divisor matching with the quotient candidate occurs, we set the bit to 1 whose position in the quotient bitmap is equal to ***'offst_lyr+num_div_lyr'***, where:

- **num_div_lyr**: the divisor number stored in the matching divisor tuple.
- **offst_lyr**: is set to 0 if the label '*ind_lyr*' of the matching divisor tuple is set to 0, otherwise (*ind_lyr=1*) it is set to $|s_1|$ (the cardinality of s_1).

Therefore, depending on the relationship between s_1 and s_2, the data structure of the candidate's bitmap may be designed under two designs, as shown in Fig. 4.

The second difference is that we keep with each quotient candidate a counter of ones ($bit = 1$) in its bitmap for both layers. We called these counters ***Nb_ones***$_1$ for the layer s_1 and ***Nb_ones***$_2$ for s_2. These counters are incremented at each bit switching (0 to 1) in the corresponding layer of the quotient candidate bitmap.

It should be noted that one of the strengths of our approach is that these two first stages constitute a unified processing of all variants of the bipolar division (operators proposed in the literature and those we introduce in this paper).

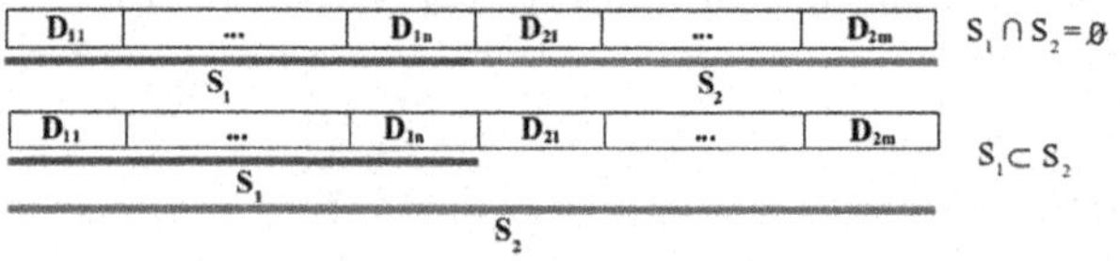

Fig. 4. The data structure for the bitmap of quotient candidates

4.3 The Third Stage

This final stage is the only not-unified phase between the different forms of the bipolar division. Indeed, for the two approaches (**"Constraint-Wish"** and **"Satisfaction-Dissatisfaction"**), we describe how to calculate satisfaction levels of quotient candidates in order to classify them thereafter. Besides, we propose novel sorting technique that involves no additional cost. It is worthwhile noting that this stage requires an additional sorting phase in traditional approaches, which can be expensive for a big number of results.

Bipolar Division Based on the Compound Connector "and if Possible". The aim is to find the *best elements* connected with all constraints (s_1) and if possible connected with *as many as possible* of wishes (s_2). Hence, in order to sort the accepted quotients depending on their satisfaction levels, we propose to use an indexed table whose size is "$|s_2| + 1$" (see Fig. 5a). During the scan of the final hash-quotient table, we proceed as follows:

- Candidates having zeros in the constraint layer in their bitmaps are immediately rejected;
- Otherwise, the candidate is selected as an accepted quotient. Then it will be stored in the indexed table, into bucket of index $class_d$ ($class_0 > class_1 > \ldots > class_{nb\_w}$) where d is the number of zeros in the wishes layer in the bitmap (representing the number of the missing wishes), and nb_w is the cardinality of wishes ($|s_2|$). Thereby, final quotients are, automatically, sorted in decreasing order according to the satisfaction levels of candidates. The cell whose index is $class_0$ points the best quotients

Bipolar Division Based on the Compound Connector "or Else"
The bipolar division based on the "or else" connector where $s_1 \cap s_2 = \emptyset$, we aim to find the *best elements* associated with all elements in either s_1 or s_2 (ideally s_1, the most important pole)[2], and it is more preferred as it is associated with as many as possible of elements in the other layer. Therefore, through the scan of the built hash-quotient table, we proceed as follows:

- We check if one of the layers s_1 and s_2 contains only ones. The candidate is thus labeled by $\{lyr, Nb\_zeros_{\overline{layer}}\}$ where:
  - **lyr:** is set to 1 if the first layer is fully satisfied, 0 if the first one if not fully satisfied but the second is, and -1 otherwise.
  - **Nb_zeros**$_{\overline{layer}}$**:** number of zeros in the other layer (in s_2 if lyr is set to 1, and in s_1 if lyr is set to 0).

In a similar way as in the previous form, we store the accepted quotients in an indexed table with two levels of index (see Fig. 5b). The first level is subdivided into two sub-levels according to the lyr label. The size of the second index is "$|s_{lyr+1}| + 1$". Quotient candidates are handled as follows:

- Candidates whose label lyr is set to -1, are rejected.
- The others are stored in the bucket whose first index is equal to ***lyr*** and the second index is equal to ***Nb*_zeros**$_{\overline{layer}}$.

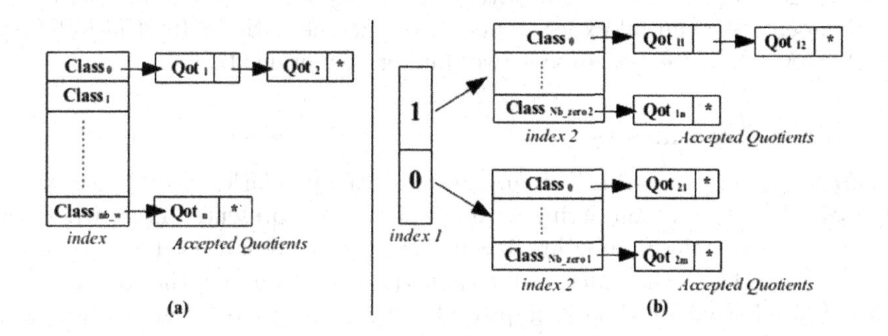

Fig. 5. Quotient candidates ranking

[2] In case $s_1 \subset s_2$ the two subsets considered are s_1 and $s_1 - s_2$.

The Satisfied-Dissatisfied Approach. As mentioned before, the bipolar division on the satisfies-Dissatisfied approach may be interpreted under three forms:

1. ***Strong symmetrical impact:*** here both positive and negative poles are mandatory. A valid quotient must have no zeros in the first pole in its bitmap, and no ones in the second pole. Thereby, all final quotients have the same satisfaction degree. No discrimination is thus needed.
2. ***Positive and negative poles as constraint and desire:*** in this case, we proceed as illustrated in Sect. 4.3. However, since there are two possible forms, the discrimination phase becomes as follows.
   - *Positive pole as a constraint*: the index of bucket in which we store candidate, takes the value of ones counter in the negative layer of the bitmap, representing the number of violated prohibitions.
   - *Negative pole as a constraint*: the index of bucket takes the value of zeros counter in the positive layer of the bitmap, representing the number of missing positive requirements.
3. ***Positive and negative poles as hierarchical preferences:*** here we proceed as shown in Sect. 4.3, where two levels of index are used. The only difference is how to compute the bucket index for the negative pole. Indeed, we check if it is fully dissatisfied (only zeros in the bitmap). Idem, the second index takes the value of the ones counter in this pole.

To select the *top* (*best*) answers in all previous forms of bipolar division, we browse the corresponding indexed table, constructed in the third stage, from the quotients with the highest satisfaction levels to the lowest ones; until the desired number of quotients are found. This sorting technique offers a better discrimination between final quotients, while no additional costs will be incurred.

5 Experimental Results

In all our implementations, the examined queries are evaluated over four dividend relations with sizes: 5.10^5, 3.10^6 and 5.10^8 tuples randomly generated[3]; and a divisor relation of different cardinality (resp. 10, 20, 50, 100 tuples) uniformly distributed over the two layers s_1 and s_2 which we consider as totally independent. We run all experiments on a machine with an Intel *i*5 CPU and 8 GB RAM. Hereafter we present two different experiments.

5.1 First Experiments

Figure 6 shows the run-times of our variant of the bipolar division based on *and if possible* connector, comparing with the classic one presented in [4] where two successive classical-divisions are involved. We can observe that our approach completes performance much faster than the classic one for the four dividend sizes. Indeed, the run-time is improved by several orders of magnitude (gain factor greater than *84* in the case of 5.10^8 dividend tuples).

[3] In the literature, to the best of our knowledge, the largest set used in the experimentations never exceed a cardinality of 30000 tuples in the dividend relation.

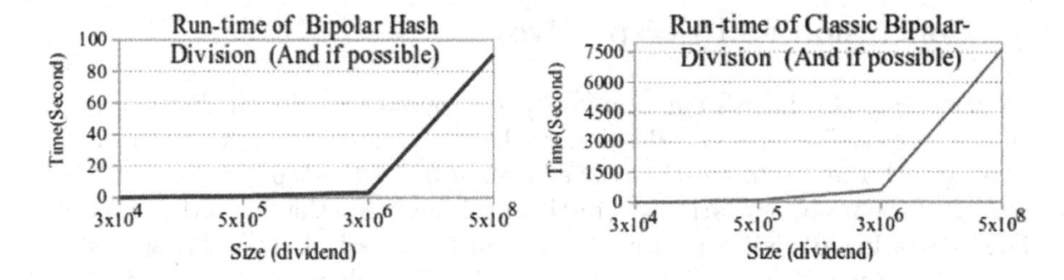

Fig. 6. Algorithm performance of the bipolar division of type "and if possible"

5.2 Second Experiments

Here, we aim to examine the performance of our new operators for the bipolar division. We describe only the empirical results of the *or else approach* where $s_1 \cap s_2 = \emptyset$, and the first two operators for *the satisfied-dissatisfied approach* described in Subsect. 3.2. Figure 7 illustrates the results obtained for these three approaches.

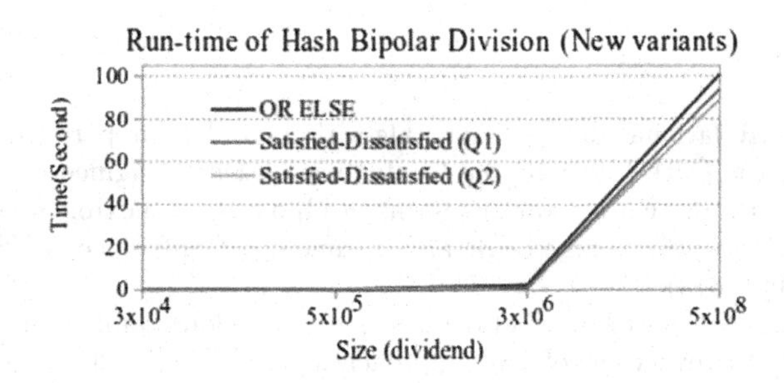

Fig. 7. Algorithm performance of the new forms of the bipolar division; Q1: Strong symmetrical impact, Q2: Positive and negative poles as constraint and desire.

An analysis of the curves above leads to the following main observations:

- Hash-division approach for the three examined queries proves its efficiency at run-time. It completes much faster compared to the classical approach dealing with bipolarity, especially for an overly large sized dividend. We observed a very small difference between the run-time of the algorithms, approximately the run-time is equal to 100 seconds for the 'or else' form and around 93 seconds for the satisfied-dissatisfied forms. Thus, the implementation of the hash bipolar division requires roughly the same run-time regardless of the form investigated.

In summary, the first results of all algorithms presented in this paper are promising. The overall approach proposed allows some benefit in effectively processing various form of bipolar division operators. Even for the complex forms, the queries are performed thanks to a single operator where no much extra cost will be needed. Moreover, the issue of answers ranking is dealt with as well.

6 Conclusion and Perspectives

We have proposed in this paper an extended operator of bipolar division based on the *"or else"* connector. Besides, we have also introduced a new operator dealing with the *"Satisfied-Dissatisfied approach"* of the bipolar division. Processing of the proposed variants is further well-improved thanks to the proposed **Hash-Bipolar division** algorithm built on the classical Hash Division algorithm. Moreover, the issue of answers ranking is dealt with as well. We have conducted some experiments, particularly for largely sized relations, and compared run-time with the original approaches (nested loop algorithms) proposed for the bipolar-division operators. As expected, the obtained performance, both for the extended versions and the new operators, is very interesting and promising. We have been able to improve the response time of the bipolar division by several orders of magnitude. However, there is still a need for multiple implementations in real SGBD and real applications eventually, to firmly validate the hash bipolar division proposed. Furthermore, we plan also to leverage the parallelism paradigm to make more efficient the presented approach.

References

1. Graefe, G.: Relational division: four algorithms and their performance. In: Proceedings of the Fifth International Conference on Data Engineering. IEEE (1989)
2. Dubois, D., Prade, H.: On various forms of bipolarity in flexible querying. In: 8th European Society for Fuzzy Logic and Technology Conference (EUSFLAT 2013), pp. 326–333 (2013)
3. Matthé, T., Nielandt, J., Zadrożny, S., et al.: Constraint-wish and satisfied-dissatisfied: an overview of two approaches for dealing with bipolar querying. In: Pivert, O., Zadrożny, S. (eds.) Flexible Approaches in Data Information and Knowledge Management, pp. 21–44. Springer International Publishing, Heidelberg (2014). https://doi.org/10.1007/978-3-319-00954-4_2
4. Bosc, P., Pivert, O.: On diverse approaches to bipolar division operators. Int. J. Intell. Syst. **26**(10), 911–929 (2011)
5. Kacprzyk, J., Zadrozny, S.: Compound bipolar queries: combining bipolar queries and queries with fuzzy linguistic quantifiers. In: 8th Conference of the European Society for Fuzzy Logic and Technology (EUSFLAT-2013). Atlantis Press (2013)
6. Kacprzyk, J., Zadrożny, S.: Hierarchical bipolar fuzzy queries: towards more human consistent flexible queries. In: 2013 IEEE International Conference on Fuzzy Systems (FUZZ), pp. 1–8. IEEE (2013)
7. Liétard, L., Tamani, N., Rocacher, D.: Fuzzy bipolar conditions of type. In: 2011 IEEE International Conference on Fuzzy Systems (FUZZ), pp. 2546–2551. IEEE (2011)
8. Bosc, P., Pivert, O.: About bipolar division operators. In: Andreasen, T., Yager, R.R., Bulskov, H., Christiansen, H., Larsen, H.L. (eds.) FQAS 2009. LNCS (LNAI), vol. 5822, pp. 572–582. Springer, Heidelberg (2009). https://doi.org/10.1007/978-3-642-04957-6_49
9. Tamani, N.: Interrogation personnalisée des systèmes d'information dédiés au transport: une approche bipolaire floue. Thèse de doctorat. Rennes 1 (2012)

10. Tamani, N., Lietard, L., Rocacher, D.: Bipolarity and the relational division. In: The Joint 7th Conference of the European Society for Fuzzy Logic and Technology (EUSFLAT 2011) and Rencontres Francophones sur la Logique Floue et ses Applications (LFA 2011), pp. 424–430 (2011)
11. Benadjmi, N., Hidouci, K.W.: New variants of Hash-Division algorithm for tolerant and stratified division. In: Christiansen, H., Jaudoin, H., Chountas, P., Andreasen, T., Legind Larsen, H. (eds.) FQAS 2017. LNCS (LNAI), vol. 10333, pp. 99–111. Springer, Cham (2017). https://doi.org/10.1007/978-3-319-59692-1_9
12. Tamani, N., Liétard, L., Rocacher, D.: A relational division based on a fuzzy bipolar R-implication operator. In: 2013 IEEE International Conference on Fuzzy Systems (FUZZ), pp. 1–7. IEEE (2013)
13. Bosc, P., Pivert, O., Soufflet, O.: Strict and tolerant antidivision queries with ordinal layered preferences. Int. J. Approx. Reason. **52**(1), 38–48 (2011)

A 2D-Approach Towards the Detection of Distress Using Fuzzy K-Nearest Neighbor

Daniel Machanje[1,2](✉), Joseph Orero[1], and Christophe Marsala[2]

[1] Faculty of Information Technology, Strathmore University, Madaraka Estate, Ole Sangale Road, P.O Box 59857, City Square, Nairobi 00200, Kenya
dmachanje@strathmore.edu

[2] Sorbonne Université, CNRS, Laboratoire d'Informatique de Paris 6, LIP6, 75005 Paris, France

Abstract. This paper focuses on a novel approach of distress detection referred to as the *2D approach*, using the fuzzy K-NN classification model. Unlike the traditional approach where single emotions were qualified to depict distress such as fear, anxiety, or anger, the 2D approach introduces two phases of classification, with the first one checking the speech excitement level, otherwise referred to as arousal in previous researches, and the second one checking the speech's polarity (negative or positive). Speech features are obtained from the Berlin Database of Emotional Studies (BDES), and feature selection done using the forward selection (FS) method. Attaining a distress detection accuracy of 86.64% using fuzzy K-NN, the proposed 2D approach shows promise in enhancing the detection of emotional states having at least two emotions that could qualify the emotion in question based on their original descriptions just as distress can be either one or many of a number of emotions. Application areas for distress detection include health and security for hostage scenario detection and faster medical response respectively.

Keywords: Speech · Emotions · Distress · 2D approach · Fuzzy K-NN

1 Introduction

In recent emotion researches, speech, other than visual signals [1] and human physiological aspects [2], has been reliably used to detect emotions [3]. The emotions in question comprise of various categories as proposed by various scholars including the popular basic emotions comprised of single emotions *anger, disgust, fear, happiness, sadness* and *surprise* by [4] to the broader categories of positive and negative emotions [5].

Categorized as a negative emotion, distress, unlike other emotions or emotional states, can be a combination of more than one emotional states. [6] describes distress as an emotional state resulting to negative effect with mood ranges that include disgust, anger, scorn, guilt, fearfulness, and depression, while [7] describes it as a mental distress, suffering or anguish resulting to unpleasant

J. Medina et al. (Eds.): IPMU 2018, CCIS 853, pp. 762–773, 2018.
https://doi.org/10.1007/978-3-319-91473-2_64

mental reactions such as fright, nervousness, anger, grief, anxiety, worry, mortification, shock, humiliation, indignity and physical pain. This paper focuses on distress as a state with the possibility of having more than one negative emotion as described by both authors, with emphasis on high excitement emotions.

Traditionally, works have been performed in single dimension where the input vector comprised of various speech features have been fed into machine learning models and an output is obtained with an accuracy indicating the possible emotion detected. This has been the case even for emotional states considered to be *compound*; possible existence of emotional states in more than one emotion such as distress. The drawbacks to the analysis of an emotion in a single dimension include simple generalization of an emotion based on few deciding characters that could give a wrong classification; use of speech character(s) that might differ in different personalities to represent various emotions (case of speaker and context dependence/independence [8]); and non-confirmation of the gradual manifestation of given emotional states within different periods in an environment. Despite these drawbacks, researchers have continually used this procedure given the ease of detecting single emotions [5], setting a poor precedence for future research.

Many classical and current machine learning techniques have been employed in the detection of emotions in speech as it is outlined hereafter in this paper, with a focus on negative emotions leaning to distress. Most of these techniques are used in their original form. This is despite [9] outlining various disadvantages of some of these techniques in detecting emotions. In the experiments focused by this paper, *fuzzy* K-NN [10] is used. Unlike the classical K-NN and other original forms of other machine learning techniques, fuzzy K-NN produces results with lower error rates, and also adds confidence measures that further enhance the description of distress as a compound emotion as originally defined. This sets a good foundation headed to the direction of classifying emotional states using multilabel approaches that have been considered near impossible, restricting most researches to either multiclass or binary classifications.

This paper introduces a new *2-dimensional* approach of detecting distress that acknowledges the multiply-shared features that various emotions exhibit and also isolate any possibility of confusing the boundaries shared by any emotions to increase the chances of accurately detecting distress. The following is the paper's plan: in Sect. 2, we present related works, outlining aspects of how distress is manifested through speech, existing speech features, machine learning approaches used in emotion experiments so far, and databases used in distress detection. In Sect. 3, we present the proposed 2D-approach, while Sect. 4 details the experiments done and results obtained using the Berlin Database of Emotional Studies. Section 5 concludes and gives a brief preview of future works.

2 Related Works

This section describes literature with regard to the direction of the paper. These include how distress manifests in speech, machine learning approaches that have been used to detect distress in the speech, speech features with regard to feature types, extraction means, preprocessing, labeling, annotation, and selection, and finally various datasets that have been used for distress detection.

2.1 Distress Manifestation Through Speech

There are several ways that distress in humans affect their voice. Muscle tension is one of the key effects imparted by distress on humans. For instance, the laryngeal musculature is adversely affected during distress situations, thus limiting adjustment of vocal cords [11], an effect that affects the voice quality in a certain way. Apart from muscular tensions, unsettledness suffered by victims in distress causes jitters [12]. Distress also generally affects the glottis, a significant body part in the area of the throat. Effect on the glottis causes voice modulation [11].

Fluctuating respiration by the subject under distress affects the subglottal pressure. This, by extension, affects the pitch, speech duration, and articulation rate [13]. Pitch can simply be defined as the highs and lows of sound judgment with association of musical melodies [14] or scientifically as sound property that can be ordered in a frequency related scale (fundamental frequency) [13]. Articulation is a sequence of signals with regard to mouth and mouth-related movements that recurs to make sounds of speech. It is also referred to as temporal pattern [13] (Fig. 1).

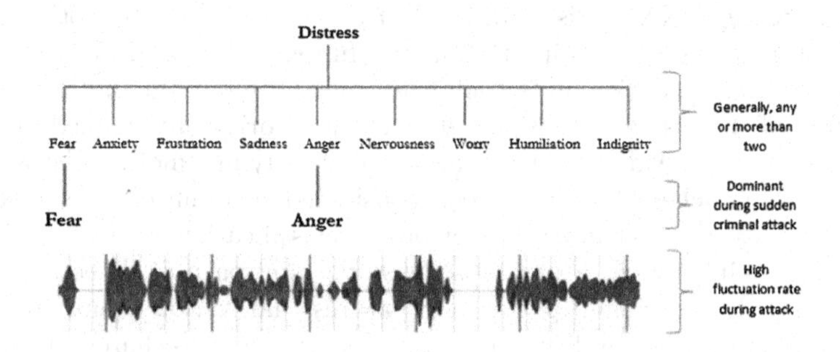

Fig. 1. Distress as a set of emotions (from [6, 7]).

2.2 Speech Features

According to [8] there are two main types of speech features that have been used for voice emotion detection. These are phonetic features, which describe the types of different sound such as vowels, consonants, and the general pronunciation, and prosodic features, which are the musical aspect of the speech. By musical aspects, the following features are implied: pitch, power/energy of the voice, speaking rates, spectral tilt, voiced trajectory duration and others.

There are various categories of speech segmentations that research in distress has focused on: whole signal versus utterances/sequences and global versus local configurations. Whole signals represent entire audio files in a database, while utterances/sequences are split segments with regard to different emotions in a given audio file [15]. Global configuration is segmentation of the audio file into utterances, while local configuration is the segmentation of speech into small fixed-size time windows. Global configurations/utterances have been known to give the best accuracy in detection [16].

The identification of emotions in an emotional speech databases is done through labeling and annotation. Usually, human labelers, based on a given criteria, are used. These could be professional or amateur labelers who could be used to eliminate biasness. Sometimes, databases are presented to the research community without labels and the researchers themselves have to label and annotate the data independently. Criteria of choosing *qualified* audio files ranges from number of agreed labelers to the percentage average of labeling accuracy.

Data extracted from speech is presented in diverse ranges that usually requires preprocessing in order to acquire optimal results from various machine learning techniques. The three main data preprocessing techniques include rescaling, normalization, and standardization.

Given the richness of features in speech audio files, sometimes too many features are extracted. Most of the features end up not being important for the final detection of the distress emotion. Therefore, to optimize for the model, various experiments utilize various feature selection tools. These include brute-force iterative feature selection algorithm [5], promising first selection (PFS) [17], forward selection [17], Fisher selection [15], ReliefF algorithm [16], and principal component analysis (PCA) [16].

2.3 Machine Learning Approaches

Various machine learning techniques have been used in general emotion detection. A detailed comparison of these approaches can be found in [9].

Hidden Markov Model (HMM) has been used in the past to detect emotion. HMM has been found to have topological restrictions that restrict analysis from only left to right and also a challenge in determining the optimal number of states within a data set. Apart from these two, it has also been found to have disparity of results when the same data is represented as discrete or continuous.

From previous research, the Gaussian Mixture Models (GMM) has been depicted as limited to modeling independent vectors. It also gives a challenge in determining the optimum number of Gaussian components.

Artificial Neural Networks (ANN) have also been used in the past in emotion detection. From these researches, ANN has been found to be effective only to nonlinear mappings. Also, ANNs have shown better performance only when the training examples are low in numbers [9]. Among the many techniques used, ANNs show relatively low classification accuracies. Finally, the diversity of ANN families have been found to be confusing, with some having been utilized only once by given researchers.

Support Vector Machine (SVM) is a family of models that in most past experiments have had an heuristic treatment of nonseparable cases. As a result, these family of models have had large accuracy ranges between speaker dependent and speaker independent classification.

Not all the machine learning techniques have been used to detect distress. Table 1 shows techniques that have been used in experiments to detect distress, using different concepts of emotions that represent distress by the given researchers (classification types and training/testing models are also shown):

Table 1. Techniques used in the detection of distress

Reference	ML technique	Classification technique	Train/test technique	Emotion (s)
[5]	Decision tree model	Binary and multilabel	75% training, 25% testing	Annoyance and frustration
[17]	Linear Discriminant Classifier (LDC) and K-nearest neighbor KNN	Binary	10LOOCV	Negative and non-negative emotions
[15]	Gaussian Mixture Model (GMM)	Binary	30LOOCV	Fear
[11]	GMM and SVM	Binary	10CV	Neutral vs. Emotional
[16]	SVM	Multiclass	75%–25% (4 iterations)	Distress (fear/anxiety)

Three classification types are common in the review of research done to detect distress: multilabel, multiclass, and binary. Binary classification is where there are two emotions to be categorized, and the result can either be one or the other. In multiclass classification, there are more than two classes, with the possibility of the result falling in any of the classes. For multilabel classification, the result falls in at least two classes.

2.4 Databases for Distress Detection

There are three types of emotional speech samples widely used in emotional databases [18]. These include *natural vocal expressions* which are recorded in the course of naturally occurring states of emotions; *induced emotional expressions* which are influenced by given circumstances, psychoactive drugs, scenarios, or words, and finally, *simulated emotional expressions* which are recorded after a given set of instructions and content regarding an emotion are given to an actor. This is usually controlled for optimal recording.

Table 2 shows the databases that have been in the literature used specifically to detect distress in variously proposed emotions.

Table 2. Databases used in distress detection

Reference	Database	DB size (utterances)	DB type	Emotions
[5]	DARPA	49,553	Simulated	**Neutral**, **Frustrated**, annoyed, tired, amused, other, and not applicable
[16]	Berlin Database of Emotional Studies	535	Simulated	Joy, sadness, boredom, neutral, anger, fear, disgust
[15]	SAFE	5,275	Simulated	**Fear**, **Neutral**, other positive and negative emotions
[11]	South Africa call center	3,000	Natural	**Neutral** and **Emotional** (anger, frustration, stress, dissatisfaction)
[17]	SpeechWorks	7,200	Natural	**Negative** (anger or frustration) vs **Non-negative** (neutral or positive emotions e.g happiness or delight.)

3 Proposed 2-Dimensional Approach

The proposed 2D approach takes a new direction in detecting distress where there are two steps of detection. The two steps are guided by the depiction of emotions in two perspectives: excitement level and polarity. This depiction of emotions is shown in Fig. 2.

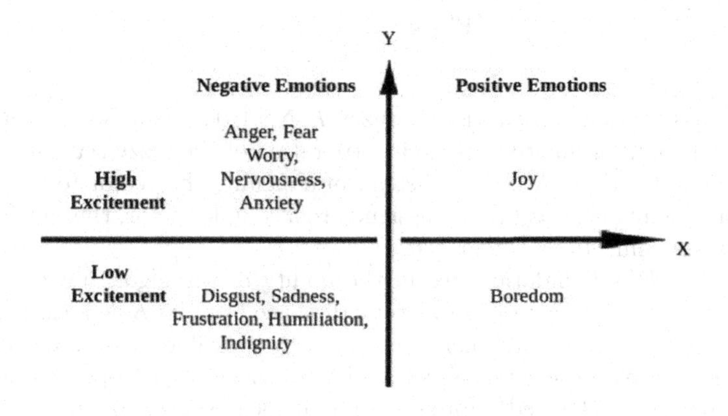

Fig. 2. Representation of the dimensions for emotions as described by [17]

In the first step, the speech is analyzed to find out the excitement level of the speech. In this category, two classifications are possible: *high* excitement level and *low* excitement level. High excitement in this situation depicts a number of emotions that include *anger, fear, worry, nervousness, anxiety* and *joy.* On the other hand, low excitement levels indicate *disgust, sadness, frustration, humiliation, indignity,* and *boredom.*

The second step is performed by a model that detects negativity and positivity of the results from the first. The negative emotions in this case include *anger, fear, worry, nervousness, anxiety, disgust, sadness, frustration, humiliation* and *indignity.* The positive emotions include *joy* and *boredom.*

Figure 3 shows the traditional emotion classification approach against what the proposed 2D approach does in Fig. 4.

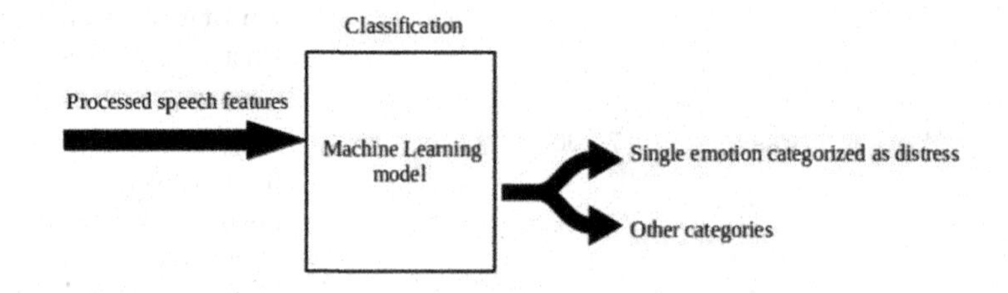

Fig. 3. Traditional emotion classification approach

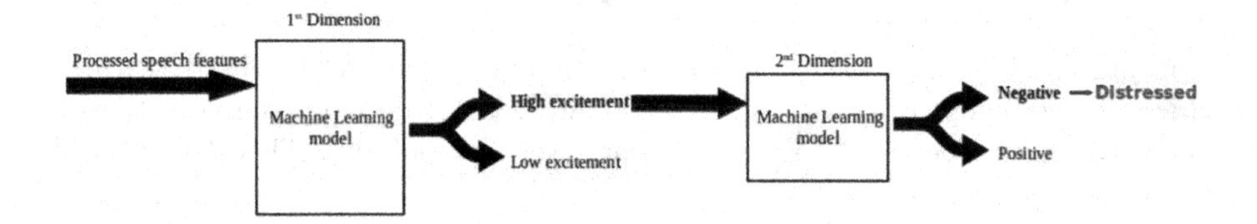

Fig. 4. 2D approach

The classification of focus used is fuzzy K-NN [10]. Fuzzy K-NN is a classification algorithm formulated by the incorporation of the fuzzy set theory [19] to the K-NN rule. In fuzzy K-NN, instead of assigning the vector to a particular class, class membership is instead assigned to a sample vector, thus avoiding any arbitrary assignments by the algorithm.

Fuzzy K-NN's foundation lies in the ability of the algorithm in assigning membership as a function of the vector's distance from its K-nearest neighbors, and assigning the neighbors' memberships in the possible classes. The extent of similarities between the *crisp* K-NN and the fuzzy K-NN stops to where both must search for the labeled sample sets for the nearest neighbors.

4 Experiments and Results on the BDES

4.1 The Berlin Database of Emotional Studies

The BDES, a simulated database, is the database of choice in the experiments in this paper. Table 3 shows the various important details of the database in focus. The 535 out of the 816 utterances were selected based on the acceptable recognition and naturalness judged by 20 subjects [20].

Table 3. The Berlin Database of Emotional Studies description [20].

Ref.	Scenario	Actors	Phrases	Total utterances	Emotions
[20]	Simulated	10 (5 male, 5 female)	10 German utterances (5 short and 5 long sentences)	816 utterances (535 selected)	Anger, fear, joy, sadness, disgust, boredom, and neutral

Table 4 shows the summary of the utterances made in the BDES. The first column represents the emotions of the databases, while the codes on the first row represent the actors who made the utterances, for instance, *a01* represents *actor number 1*. The numbers in the columns indicate the number of utterances each actor made in given emotions. The totals on the right side indicate the number of utterances made of the emotions, while the totals at the bottom indicate the number of utterances each actor was able to make.

Table 4. The Berlin Database of Emotional Studies utterance distribution [20].

	a01	a02	a04	a05	a07	b01	b02	b03	b09	b10	Totals
Anger	12	14	14	15	10	11	15	11	12	13	127
Fear	7	7	7	8	6	6	7	4	8	9	69
Joy	7	6	8	8	9	9	6	7	7	4	71
Sadness	1	7	6	10	8	3	6	10	6	5	62
Disgust	6	9	2	4	4	7	2	4	4	4	46
Boredom	6	8	7	9	11	8	6	9	9	8	81
Neutral	10	9	7	8	7	7	9	7	6	9	79
Totals	49	60	51	62	55	51	51	52	52	52	535

BDES was the choice database as it has been vastly used, thus containing good documentation. The speeches are also of high quality as they were produced under controlled and quality development environment. The emotions in the database are also adequate, and finally, the database is in distribution under the public domain.

4.2 Feature Extraction and Selection

Feature extraction was done using Praat [21]. The extraction was performed by subjecting the BDES audio files to praat scripts and storing the values of the extraction in comma separated value (csv) files.

The extraction was done at an analysis width of a window range that was 0.005 seconds. The maximum frequency was set at 5000 Hz, with the frequency step set at 20 Hz. The dynamic range was set at 70 dB. These, according to [21] provide an optimal environment to extract speech features. The spectrogram in Fig. 5 shows the properties of an angry speech utterance. The gray, cloudy appearance at the lower part represents the spectrum, the split blue lines show the speech variation, the continuous yellow line shows the speech intensity, the red, dotted lines show the speech formant, and the blue, shaded areas at the top part indicate the speech pulses.

In total, 30 features were extracted from the audio files. The features extracted included *duration*, *pitch (F0)* (medium, mean, standard deviation, maximum, minimum), *jitter* (local, local.abs, rap, ppq5, ddp), *shimmer* (local, local.dB, apq3, apq5, apq11, dda), *voicing* (fraction of unvoiced frames,number of breaks, degree of breaks), *intensity* (dB.mean, dB.minimum, dB.maximum), *pulses* (number, periods, mean period, sd of period), *harmony* (mean autocorrelation, noise to harmony ratio, harmony to noise ratio) (ppq - Period Perturbation Quotient, apq - Amplitude Perturbation Quotient, abs - absolute, dB - decibels, rap - Relative Average Perturbation). These prosodic speech features, according to [11,22], have been found to offer good results for emotion detection.

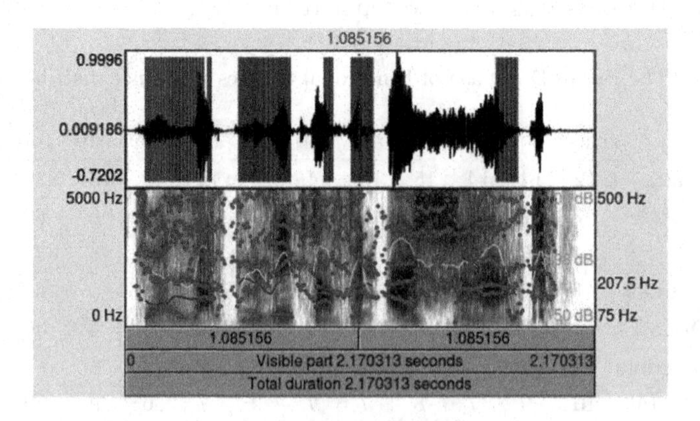

Fig. 5. Sample speech spectrogram (Color figure online)

Feature selection to choose the best performing features and reduce dimensionality was performed using the *forward selection (FS)* method as proposed by [17]. In their experiment to recognize negative emotions from speech signal, [17] find FS to have a lower error rate as compared to *promising first selection (PFS)*, another feature selection they used. Apart from a lower error rate, FS had an edge over PFS by going through all the possible combinations of features and adding them progressively whereas PFS added the new features in isolation.

As a result of the feature selection, the following 6 features were selected as the best combination: pitch mean, pitch maximum, pitch median, intensity mean, fraction of unvoiced frames, and harmony to noise ratio.

4.3 Experimental Setup and Results

Using two methods, the traditional approach in Fig. 3 and the proposed 2D approach in Fig. 4, the features selected were subjected to experiments that included three classification methods separately; K-NN, Support Vector Machine (SVM), and fuzzy K-NN, implemented with the Scikit-learn [23] toolbox.

To get the training and testing set, the data is subjected to 10-fold cross validations. The accuracy results are given in Table 5 indicating the standard deviations.

The accuracy results detected when the traditional approach in emotion detection is performed using the crisp versions of SVM and K-NN separately, there is no much disparity between the accuracies. At 70.81%, SVM produces the least accuracy, with the crisp K-NN following at 72.16%. The differences are not that much when the classification models are used on the 2D approach with 73.21% and 74.28% for SVM and K-NN respectively. It should, however, be noted that there is an improvement of the detection accuracy when the 2D approach is used.

The accuracy results take a drastic improvement on both approaches when fuzzy K-NN is used. When the classification model is applied in the traditional model, an accuracy of 78.32% results; a difference of almost 6% from the higher of the two other classification models applied on the traditional model. However, the best results are achieved when fuzzy K-NN is applied to the proposed 2D approach, with an accuracy of 86.64% being achieved.

Table 5. Experimental results

Classification	Traditional model % accuracy	2D approach % accuracy
K-NN	72.16% (4.94)	74.28% (4.35)
SVM	70.81% (5.48)	73.21% (5.17)
Fuzzy K-NN	78.32% (3.68)	86.64% (3.31)

5 Conclusion and Future Works

This paper introduces a new approach for attaining higher accuracy detection of distress as an emotional state comprised of speech with characteristic high excitement and negative polarity, using fuzzy K-NN and utilizing the 2D approach. This is unlike other researches in the past that qualified certain negative emotions as distress such as fear or anxiety by [16], and using the one-step model of detecting the emotions in question.

Using the Berlin Database of Emotional Studies (BDES) as the database of focus, features were extracted using Praat [21] and feature selection done using forward selection (FS). With an initial feature collection of 30, 6 features were selected to have the best combination in yielding the best detection accuracy, namely; pitch mean, pitch maximum, pitch median, intensity mean, fraction of unvoiced frames, and harmony to noise ratio.

To have comparative results, other than fuzzy K-NN, SVM and K-NN were also used. Fuzzy K-NN came out the superior classification model to use, attaining the highest detection accuracy in both the traditional model and 2D approach at 78.32% and 86.64% respectively. The highest accuracy attained by using fuzzy K-NN and the 2D approach qualifies the former's lower error rates and higher detection accuracy than it's counterparts, the *crisp* K-NN and SVM, and also justifies the latter's need to change of tact in detecting emotional states, especially those than can have more than one emotion such as the case in the paper, distress.

A follow up of this paper is to discover new converging features from several negative emotions that identify the distress emotion. This will be done through the study of properties of all emotions that imply distress, analyzing their boundaries and intersections, and trying to identify an intersection of properties that may be uniquely used to identify distress.

References

1. Norris, C., McCahill, M., Wood, D.: The growth of CCTV: a global perspective on the international diffusion of video surveillance in publicly accessible space. Surveill. Soc. **2**(2/3) (2002)
2. Knapp, R.B., Kim, J., André, E.: Physiological signals and their use in augmenting emotion recognition for human–machine interaction. In: Cowie, R., Pelachaud, C., Petta, P. (eds.) Emotion-Oriented Systems, pp. 133–159. Springer, Heidelberg (2011). https://doi.org/10.1007/978-3-642-15184-2_9
3. Pfister, T.: Emotion detection from speech. Gonville and Caius College (2010)
4. Ekman, P.: An argument for basic emotions. Cognit. Emot. **6**(3–4), 169–200 (1992)
5. Ang, J., Dhillon, R., Krupski, A., Shriberg, E., Stolcke, A.: Prosody-based automatic detection of annoyance and frustration in human-computer dialog. In: INTERSPEECH, Citeseer (2002)
6. Watson, D., Pennebaker, J.W.: Health complaints, stress, and distress: exploring the central role of negative affectivity. Psychol. Rev. **96**(2), 234 (1989)
7. Pearson, E.: Torts-emotional distress (1971)
8. Nicholson, J., Takahashi, K., Nakatsu, R.: Emotion recognition in speech using neural networks. Neural Comput. Appl. **9**(4), 290–296 (2000)
9. Trentin, E., Scherer, S., Schwenker, F.: Emotion recognition from speech signals via a probabilistic echo-state network. Pattern Recogn. Lett. **66**, 4–12 (2015)
10. Keller, J.M., Gray, M.R., Givens, J.A.: A fuzzy k-nearest neighbor algorithm. IEEE Trans. Syst. Man Cybern. **4**, 580–585 (1985)

11. Lefter, I., Rothkrantz, L.J., Van Leeuwen, D.A., Wiggers, P.: Automatic stress detection in emergency (telephone) calls. Int. J. Intell. Def. Support Syst. **4**(2), 148–168 (2011)
12. Johnstone, T., Scherer, K.R.: The effects of emotions on voice quality. In: Proceedings of the XIVth International Congress of Phonetic Sciences, University of California, Berkeley San Francisco, pp. 2029–2032 (1999)
13. Hansen, J.H.L., Patil, S.: Speech under stress: analysis, modeling and recognition. In: Müller, C. (ed.) Speaker Classification I. LNCS (LNAI), vol. 4343, pp. 108–137. Springer, Heidelberg (2007). https://doi.org/10.1007/978-3-540-74200-5_6
14. Kollmeier, B., Brand, T., Meyer, B.: Perception of speech and sound. In: Benesty, J., Sondhi, M.M., Huang, Y.A. (eds.) Springer Handbook of Speech Processing, pp. 61–82. Springer, Heidelberg (2008). https://doi.org/10.1007/978-3-540-49127-9_4
15. Clavel, C., Vasilescu, I., Devillers, L., Richard, G., Ehrette, T.: Fear-type emotion recognition for future audio-based surveillance systems. Speech Commun. **50**(6), 487–503 (2008)
16. Alkaher, Y., Dahan, O., Moshe, Y.: Detection of distress in speech. In: IEEE International Conference on Science of Electrical Engineering (ICSEE), pp. 1–5. IEEE (2016)
17. Lee, C.M., Narayanan, S., Pieraccini, R.: Recognition of negative emotions from the speech signal. In: IEEE Workshop on Automatic Speech Recognition and Understanding, ASRU 2001, pp. 240–243. IEEE (2001)
18. Xiao, Z., Dellandrea, E., Dou, W., Chen, L.: Features extraction and selection for emotional speech classification. In: IEEE Conference on Advanced Video and Signal Based Surveillance, pp. 411–416. IEEE (2005)
19. Zedeh, L.: Fuzzy sets. Inf. Control **8**(3), 338–353 (1965)
20. Burkhardt, F., Paeschke, A., Rolfes, M., Sendlmeier, W.F., Weiss, B.: A database of German emotional speech. In: Interspeech, vol. 5, pp. 1517–1520 (2005)
21. Boersma, P., Weenik, D.: Praat: a system for doing phonetics by computer. Report of the institute of Phonetic Sciences of the University of Amsterdam (1996)
22. Clavel, C., Devillers, L., Richard, G., Vidrascu, I., Ehrette, T.: Abnormal situations detection and analysis through fear-type acoustic manifestations. In: Proceedings of ICASSP, Honolulu (2007)
23. Pedregosa, F., Varoquaux, G., Gramfort, A., Michel, V., Thirion, B., Grisel, O., Blondel, M., Prettenhofer, P., Weiss, R., Dubourg, V., et al.: Scikit-learn: machine learning in Python. J. Mach. Learn. Res. **12**(Oct), 2825–2830 (2011)

Fuzzy Mathematical Analysis and Applications

Modified Methods of Capital Budgeting Under Uncertainties: An Approach Based on Fuzzy Numbers and Interval Arithmetic

Antonio Carlos de Souza Sampaio Filho[1(✉)], Marley M. B. R. Vellasco[2], and Ricardo Tanscheit[2]

[1] Petróleo Brasileiro S.A., Rio de Janeiro, Brazil
acsampaio@petrobras.com.br
[2] Pontifícia Universidade Católica do Rio de Janeiro, Rio de Janeiro, Brazil
{marley,ricardo}@ele.puc-rio.br

Abstract. The fuzzy modified net present value (*Fuzzy MNPV*) method for evaluation of non-conventional investment projects under uncertainty explicitly provided for the use of the opportunity costs associated with the interim cash flows of an investment project and eliminated the major problems of traditional capital budgeting methods. Based on the same assumptions that guided the development of that method, the current paper presents a unified capital budgeting solution, consisting of the modified internal rate of return (*Fuzzy MIRR*), the modified profitability index (*Fuzzy MPI*), and the modified total payback (*Fuzzy MTPB*). These methods are *MNPV*-consistent, maximize shareholder wealth and always lead to the same conditions of acceptance or rejection of investment projects.

Keywords: Modified capital budgeting methods · Triangular fuzzy numbers
Interval arithmetic

1 Introduction

Investment decisions must consider three different dimensions: the absolute return, the relative return and the liquidity of a project. There is no evaluation measurement that may capture all those dimensions at the same time.

Many deterministic and fuzzy solutions have been proposed in the last decades to deal with the problems of capital budgeting methods, but the unified approach proposed here distinguishes itself by being a generalized one regarding absolute return (*MNPV*) [1], relative return (*MIRR* and *MPI*), and liquidity (*MTPB*). All these methods lead to consistent indications of acceptance or rejection of projects. The approach takes as a basis deterministic capital budgeting methods [2], which provide an alternative structure to project analysis when the company's financing and reinvestment rates are different and the uncertain parameters can be specified as triangular fuzzy numbers (*TFNs*). The new indicators are the *Fuzzy MIRR*, the *Fuzzy MPI* and the *Fuzzy MTPB*. The application of opportunity costs and fuzzy criteria for determining the variables allow obtaining more realistic and consistent results with the market conditions. Due to the complexity of the calculations involved, new MS-Excel functions are developed by using Visual Basic for Applications (*VBA*).

J. Medina et al. (Eds.): IPMU 2018, CCIS 853, pp. 777–789, 2018.
https://doi.org/10.1007/978-3-319-91473-2_65

2 The Modified Capital Budgeting Methods

Conventional capital budgeting methods have some deficiencies that may lead to interpretation errors in companies' investment decisions. These deficiencies are mainly due to the following assumptions [3] on which those methods are based:

- intermediate cash inflows of the project are reinvested at a rate of return equal to the risk-adjusted discount rate (*RADR*) of the project, instead of the reinvestment opportunity rate of the company;
- cash outflows, after the initial investment, are discounted at the *RADR*, instead of at the financing rate (weighted average cost of capital – *WACC*) of the company.

These problems have been investigated and some authors have proposed alternative models to the conventional methods, known as modified methods. In practice, these methods are improved versions of the traditional *NPV* and *IRR* methods, eliminating problems regarding the reinvestment rate, and different versions are known by different names and acronyms [2–8]. The first one is used in this paper, since it is a unified solution and totally compatible with the methods proposed in [3, 4].

According to [2], the general equation for the modified net present value is:

$$MNPV = \frac{\sum_{t=1}^{n} CF_{it}(1+k_{rr})^{n-t}}{(1+k_{radr})^{n}} - \sum_{t=0}^{n} \frac{CF_{ot}}{(1+k_{wacc})^{t}} \tag{1}$$

where, CF_{it} is the net cash inflow at the end of period t, n is the lifetime of the project, k_{rr} is the reinvestment rate, k_{radr} is the risk-adjusted discount rate (*RADR*), CF_{ot} is the net cash outflow at the end of the period t, and k_{wacc} is the financing rate (*WACC*).

The numerator of the first term on the right-hand side of Eq. (1) is the sum of all cash inflows capitalized at the reinvestment rate until the last period (n) of the project. This sum, called terminal value (*TV*), is discounted at the risk-adjusted discount rate until period zero. The second term on the right-hand side is the sum of all cash outflows discounted at the financing rate until period zero of the project. This sum is called present value (*PV*). Thus, *TV* and *PV* can be written as:

$$TV = \sum_{t=1}^{n} CF_{it}(1+k_{rr})^{n-t} \tag{2}$$

$$PV = \sum_{t=0}^{n} \frac{CF_{ot}}{(1+k_{wacc})^{t}} \tag{3}$$

From these two unique values (*TV* and *PV*), the deterministic indicators can be established and are epresented by Eqs. (4) to (7), respectively:

$$MNPV = \frac{TV}{(1+k_{radr})^{n}} - PV \tag{4}$$

$$MIRR = \left[\frac{TV}{PV}\right]^{1/n} - 1 \quad (5)$$

$$MPI = \frac{TV(1 + k_{radr})^{-n}}{PV} \quad (6)$$

$$MTPB = \frac{PV}{TV(1 + k_{radr})^{-n}} \times n \quad (7)$$

Example 1. Consider the financing problem described in [9], regarding net cash flows associated with an oil extraction investment project. The project can be defined as a cash outflow at the value of $1,600 in year zero, for the installation of a larger pump in an operating oil well. The new pump enables the additional extraction of $10,000 of oil in year one. However, as a consequence of this installation, the well capacity partly gives out within that year. As a result, production drops by $10,000 of oil in year two. Therefore, the net cash flows for the project are −$1,600, $10,000 and −$10,000 in years 0, 1 and 2, respectively. Consider that the reinvestment rate (k_{tr}), the financing rate (k_{wacc}) and the risk-adjusted discount rate (k_{radr}) are 25%, 23% and 23% per year, respectively. As an illustration, the procedures to calculate the *MNPV, MIRR, MPI,* and *MTPB* were applied to the investment project; results are presented in Table 1 below:

Table 1. Calculation of *MNPV*, *MIRR*, *MPI* and *MTPB* for oil pump problem.

	Year		
	0--------------	**1-----------------**	**2----------------**
Net Cash Flows	–1,600.00	10,000.00	–10,000.00
Cash Outflows @ 23%	– 6,609.82	⇦⇦⇦ ⇩⇦⇦⇦⇦⇦⇦⇦	
Cash Inflows @ 25%		⇨⇨⇨	12,500.00
Present Value:	– 8,209.82	**Terminal Value:**	12,500.00

Table 1 shows that the negative cash flow of −$10,000.00 is discounted to the period zero, at a 23% financing rate, amounting to a present value (*PV*) of −$8,209.82. On the other hand, the positive cash flow of $10,000 is capitalized to the future value, in the last period of the project, based on a 25% reinvestment rate, amounting to a terminal value (*TV*) of $12,500.00. With the values for the negative and positive net cash flows concentrated in periods zero and two, respectively, the modified indices are computed by applying Eqs. 2 to 7:

$$\mathbf{MNPV} = 12,500/(1 + 0.23)^2 - 8,209.82 = \$52.46$$

$$\mathbf{MIRR} = (12,500/8,209.82)^{1/2} - 1 = 23.39\%$$

$$\mathbf{MPI} = 12,500/(1 + 0.23)^2/8,209.82 = 1.01$$

$$\mathbf{MTPB} = 8,209.82/\left[12,500/(1 + 0.23)^2\right] \times 2 = 1.99$$

The decision criteria associated with *MNPV*, *MIRR*, *MPI* and *MTPB* methods are similar to those of conventional methods and are summarized in Table 2.

Table 2. Decision criteria for the modified methods

Modified methods	Acceptance/rejection criteria		
	Acceptance	Indifference	Rejection
MNPV	$MNPV > 0$	$MNPV = 0$	$MNPV < 0$
MIRR	$MIRR > k_{radr}$	$MIRR = k_{radr}$	$MIRR < k_{radr}$
MPI	$MPI > 1$	$MPI = 1$	$MPI < 1$
MTPB	*MTPB* < Project's lifespan	*MTPB* = Project's lifespan	*MTPB* > Project's lifespan

Tables 1 and 2 show that all the modified indices lead to the same indication of project acceptance $(MNPV > 0; MIRR > k_{radr}; MPI > 1\ and\ MTPB < 2)$.

3 Modified Methods Under Uncertainties

It is assumed in Eqs. (4) to (7) that the cash flows, the reinvestment rate, the financing rate and the risk-adjusted discount rate are precisely established. However, it is not easy to determine the precise value of those parameters, and unforeseen variations may even make impractical a project initially considered as of high return.

In traditional modeling of capital budgeting problems, uncertain parameters are generally represented by probability distributions. This approach is appropriate to represent the statistical nature of uncertainties and the Monte Carlo method is often used for practical calculations. However, there are many situations where statistical data cannot be established and the parameters are determined subjectively. Moreover, the uncertainty does not always come from the parameters' random character. For example, in the sentence "this project will provide a high return in the next 10 or 15 years", different interpretations of the term "high" and the uncertainty on the useful life of the project can be a source of ambiguity and imprecision.

In such situations, modeling of capital budgeting problems can be improved by the use of an alternative approach. Fuzzy sets concepts [10] are very useful for dealing with uncertain or imprecise parameters in the form of linguistic variables. Fuzzy numbers are especially useful for the quantification of imprecise or ambiguous information. Several authors have investigated the use of fuzzy sets in decision making under uncertainty and a wide revision of the literature can be found in [11, 12].

Let's assume that cash flows, reinvestment and financing rates and the risk-adjusted discount rate of an investment project under uncertainty are represented by triangular fuzzy numbers. Such fuzzy numbers take the form shown in Fig. 1, where a hypothetical α-cut is also shown.

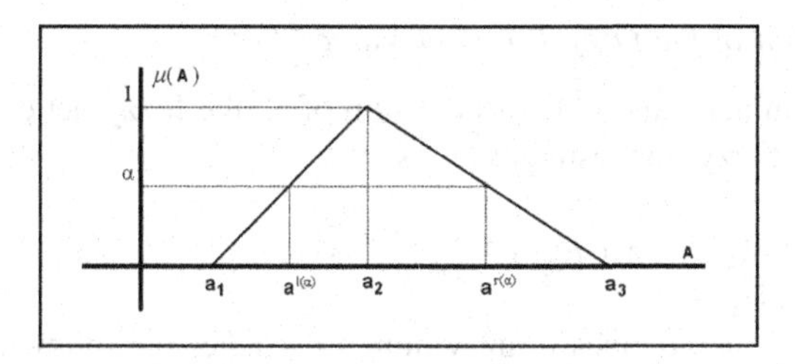

Fig. 1. Triangular fuzzy number

An L-R fuzzy number can be expressed as:

$$A^{\alpha} = \left[a^{l(\alpha)}, a^{r(\alpha)}\right] \tag{8}$$

where $l(\alpha)$ represents the line segment to the left of the fuzzy number and $r(\alpha)$ represents the line segment to the right.

The α-cut notation is also used very often:

$$\frac{a^{l(\alpha)} - a_1}{a_2 - a_1} = \alpha, \quad \frac{a_3 - a^{r(\alpha)}}{a_3 - a_2} = \alpha \tag{9}$$

Such that:

$$a^{l(\alpha)} = a_1 + (a_2 - a_1)\alpha, \quad a^{r(\alpha)} = a_3 + (a_2 - a_3)\alpha \tag{10}$$

Hence,

$$A^{\alpha} = \left[a^{l(\alpha)}, a^{r(\alpha)}\right] = \left[a_1 + (a_2 - a_1)\alpha, a_3 + (a_2 - a_3)\alpha\right] \tag{11}$$

From Eq. (8), the mathematical notations for triangular fuzzy numbers corresponding to cash flows (CF_t), reinvestment (k_{rr}), financing (k_{wacc}) and risk-adjusted discount (k_{radr}) rates can be established:

$$CF_t^{\alpha} = \left[CF_t^l + (CF_t - CF_t^l)\alpha, CF_t^r - (CF_t^r - CF_t)\alpha\right] \quad \alpha \in [0, 1] \tag{12}$$

$$k_{rr}^{\alpha} = \left[k_{rr}^l + (k_{rr} - k_{rr}^l)\alpha\,,\, k_{rr}^r - (k_{rr}^r - k_{rr})\alpha\right] \quad \alpha \in [0, 1] \tag{13}$$

$$k_{wacc}^{\alpha} = \left[k_{wacc}^l + (k_{wacc} - k_{wacc}^l)\alpha, k_{wacc}^r - (k_{wacc}^r - k_{wacc})\alpha\right] \quad \alpha \in [0, 1] \tag{14}$$

$$k_{radr}^{\alpha} = \left[k_{radr}^l + (k_{radr} - k_{radr}^l)\alpha, k_{radr}^r - (k_{radr}^r - k_{radr})\alpha\right] \quad \alpha \in [0, 1] \tag{15}$$

For the sake of simplicity, the computation of the modified methods with fuzzy cash flows and discount rates will be divided into six stages:

***Step 1**: Calculation of the fuzzy Terminal Value (TV^{α}).*

The fuzzy terminal value (TV^{α}) is the sum of all the fuzzy net cash inflows (CF_i^{α}) capitalized at the fuzzy reinvestment rate k_{rr}^{α}:

$$TV^{\alpha} = CF_{i1}^{\alpha}(1+k_{rr}^{\alpha})^{n-1} + CF_{i2}^{\alpha}(1+k_{rr}^{\alpha})^{n-2} + \ldots + CF_{in}^{\alpha}(1+k_{rr}^{\alpha})^{0} \quad CF_{it}^{\alpha} > 0 \quad (16)$$

Each interval which defines a fuzzy number satisfies the following conditions:

$$\begin{array}{ll}
[a^l, a^r] + [b^l, b^r] = [a^l + b^l, a^r + b^r] & \\
-[a^l, a^r] = [-a^r, -a^l] & \\
[a^l, a^r] - [b^l, b^r] = [a^l - b^r, a^r - b^l] & \\
[a^l, a^r]^{-1} = 1/[a^l, a^r] = [1/a^r,\ 1/a^l] & \\
[a^l, a^r]x\,[b^l, b^r] = [a^l x b^l, a^r x b^r] & a^l \geq 0\, e\, b^l \geq 0 \\
[a^l, a^r] \div [b^l, b^r] = [a^l/b^r, a^r/b^l] & a^l > 0\, e\, b^l > 0 \\
[a^l, a^r] \div [b^l, b^r] = [a^l/b^l, a^r/b^r] & a^r < 0\, e\, b^l > 0 \\
[a^l, a^r] \div [b^l, b^r] = [a^r/b^r, a^l/b^l] & a^l > 0\, e\, b^r < 0 \\
[1, 1] \div [b^l, b^r] = [1/b^r, 1/b^l] & \\
[a^l, a^r]^n = [a^{ln}, a^{rn}] & a^l \geq 0 \\
[a^l, a^r]^0 = [1, 1] &
\end{array} \quad (17)$$

Applying the conditions established in (17) to expression (16):

$$\begin{aligned} TV^{\alpha} = {} & [CF_{i1}^{l(\alpha)}x(1+k_{rr}^{l(\alpha)})^{n-1}, CF_{i1}^{r(\alpha)}x(1+k_{rr}^{r(\alpha)})^{n-1}] + \\ & [CF_{i2}^{l(\alpha)}x(1+k_{rr}^{l(\alpha)})^{n-2}, CF_{i2}^{r(\alpha)}x(1+k_{rr}^{r(\alpha)})^{n-2}] + \ldots + [CF_{in}^{l(\alpha)}, CF_{in}^{r(\alpha)}] \end{aligned} \quad (18)$$

where:

$$TV^{l(\alpha)} = CF_{i1}^{l(\alpha)}x(1+k_{rr}^{l(\alpha)})^{n-1} + CF_{i2}^{l(\alpha)}x(1+k_{rr}^{l(\alpha)})^{n-2} + \ldots + CF_{in}^{l(\alpha)} \quad (19)$$

$$TV^{r(\alpha)} = CF_{i1}^{r(\alpha)}x(1+k_{rr}^{r(\alpha)})^{n-1} + CF_{i2}^{r(\alpha)}x(1+k_{rr}^{r(\alpha)})^{n-2} + \ldots + CF_{in}^{r(\alpha)} \quad (20)$$

***Step 2**: Calculation of the fuzzy Present Value (PV^{α}).*

The fuzzy present value (PV^{α}) is the sum of all net cash outflows discounted at the fuzzy financing rate (k_{wacc}^{α}):

$$PV^{\alpha} = CF_{o0}^{\alpha} + \frac{CF_{o1}^{\alpha}}{(1+k_{wacc}^{\alpha})^1} + \frac{CF_{o2}^{\alpha}}{(1+k_{wacc}^{\alpha})^2} + \ldots + \frac{CF_{on}^{\alpha}}{(1+k_{wacc}^{\alpha})^n} \quad CF_{ot}^{\alpha} < 0 \quad (21)$$

Applying the conditions established in (17) to expression (21):

$$PV^{\alpha} = [CF_{o0}^{l(\alpha)}, CF_{o0}^{r(\alpha)}] + [\frac{CF_{o1}^{l(\alpha)}}{1+k_{wacc}^{l(\alpha)}}, \frac{CF_{o1}^{r(\alpha)}}{1+k_{wacc}^{r(\alpha)}}] + \ldots + [\frac{CF_{on}^{l(\alpha)}}{(1+k_{wacc}^{l(\alpha)})^n}, \frac{CF_{on}^{r(\alpha)}}{(1+k_{wacc}^{r(\alpha)})^n}] \quad (22)$$

where:

$$PV^{l(\alpha)} = CF_{o0}^{l(\alpha)} + \frac{CF_{o1}^{l(\alpha)}}{\left(1 + k_{wacc}^{l(\alpha)}\right)^{1}} + \frac{CF_{o2}^{l(\alpha)}}{\left(1 + k_{wacc}^{l(\alpha)}\right)^{2}} + \ldots + \frac{CF_{on}^{l(\alpha)}}{\left(1 + k_{wacc}^{l(\alpha)}\right)^{n}} \tag{23}$$

$$PV^{r(\alpha)} = CF_{o0}^{r(\alpha)} + \frac{CF_{o1}^{r(\alpha)}}{\left(1 + k_{wacc}^{r(\alpha)}\right)^{1}} + \frac{CF_{o2}^{r(\alpha)}}{\left(1 + k_{wacc}^{r(\alpha)}\right)^{2}} + \ldots + \frac{CF_{on}^{r(\alpha)}}{\left(1 + k_{wacc}^{r(\alpha)}\right)^{n}} \tag{24}$$

Step 3: *Calculation of the fuzzy Modified Net Present Value* $(MNPV^{\alpha})$.

From Eq. (4), $(MNPV^{\alpha})$ can be represented as:

$$MNPV^{\alpha} = [MNPV^{l(\alpha)}, MNPV^{r(\alpha)}] \tag{25}$$

where:

$$MNPV^{l(\alpha)} = \frac{TV^{l(\alpha)}}{(1 + k_{radr}^{r(\alpha)})^{n}} - PV^{l(\alpha)} \tag{26}$$

$$MNPV^{r(\alpha)} = \frac{TV^{r(\alpha)}}{(1 + k_{radr}^{l(\alpha)})^{n}} - PV^{r(\alpha)} \tag{27}$$

Step 4: *Calculation of the fuzzy Modified Internal Rate of Return* $(MIRR^{\alpha})$.

From Eq. (5), $(MIRR^{\alpha})$ can be represented as:

$$MIRR^{\alpha} = [MIRR^{l(\alpha)}, MIRR^{r(\alpha)}] \tag{28}$$

where:

$$MIRR^{l(\alpha)} = \left[\frac{TV^{l(\alpha)}}{PV^{l(\alpha)}}\right]^{1/n} - 1 \tag{29}$$

$$MIRR^{r(\alpha)} = \left[\frac{TV^{r(\alpha)}}{PV^{r(\alpha)}}\right]^{1/n} - 1 \tag{30}$$

Step 5: *Calculation of the fuzzy Modified Profitability Index* (MPI^{α}).

From Eq. (6), MPI^{α} can be represented as:

$$MPI^{\alpha} = [MPI^{l(\alpha)}, MPI^{r(\alpha)}] \tag{31}$$

where:

$$MPI^{l(\alpha)} = [\frac{TV^{l(\alpha)}}{(1 + k_{radr}^{r(\alpha)})^n}]/[PV^{l(\alpha)}] \quad (32)$$

$$MPI^{r(\alpha)} = [\frac{TV^{r(\alpha)}}{(1 + k_{radr}^{l(\alpha)})^n}]/[PV^{r(\alpha)}] \quad (33)$$

Step 6: *Calculation of the fuzzy Modified Total Payback Period* ($MTPB^{\alpha}$).

From Eq. (7), $MTPB^{\alpha}$ can be represented as:

$$MTPB^{\alpha} = [MTPB^{l(\alpha)}, MTPB^{r(\alpha)}] \quad (34)$$

where:

$$MTPB^{l(\alpha)} = PV^{r(\alpha)} / \frac{TV^{r(\alpha)}}{(1 + k_{radr}^{l(\alpha)})^n} \times n \quad (35)$$

$$MTPB^{r(\alpha)} = PV^{l(\alpha)} / \frac{TV^{l(\alpha)}}{(1 + k_{radr}^{r(\alpha)})^n} \times n \quad (36)$$

Example 2. Assume that in Example 1 the cash flows, the reinvestment, the financing and the risk-adjusted discount rates are established by an expert as the following triangular fuzzy numbers: $CF_0^{\alpha} = (-1600, -1600 - 1600)$, $CF_1^{\alpha} = (9000, 10000, 11000)$, $CF_2^{\alpha} = (-11000, -10000, -9000)$, $k_{rr}^{\alpha} = (23\%, 25\%, 27\%)$, $k_{wacc}^{\alpha} = (21\%, 23\%, 25\%)$ and $k_{radr}^{\alpha} = (21\%, 23\%, 25\%)$ The fuzzy *MNPV*, *MIRR*, *MPI* and *MTPB* indicators are achieved through the following steps:

(a) *Calculation of the values of fuzzy cash flows* (CF_n^{α})

From Eq. (12), the fuzzy cash flows are expressed as:

$$\begin{aligned}
CF_0^{\alpha} &= [-1600 + (-1600 - (-1600))\alpha, -1600 + (-1600 - (-1600))\alpha] \\
&= [-1600, -1600] \\
CF_1^{\alpha} &= [9000 + (10000 - 9000)\alpha, 11000 + (10000 - 11000)\alpha] \\
&= [9000 + 1000\alpha, 11000 - 1000\alpha] \\
CF_2^{\alpha} &= [-11000 + (-10000 - (-11000))\alpha, -9000 + (-10000 - (-9000))\alpha] \\
&= [-11000 + 1000\alpha, -9000 - 1000\alpha]
\end{aligned}$$

Table 3. Values of CF_0^α, CF_1^α and CF_2^α

α	CF_0^α		CF_1^α		CF_2^α	
	$CF_0^{l(\alpha)}$	$CF_0^{r(\alpha)}$	$CF_1^{l(\alpha)}$	$CF_1^{r(\alpha)}$	$CF_2^{l(\alpha)}$	$CF_2^{r(\alpha)}$
0	−1,600	−1,600	9,000	11,000	−11,000	−9,000
0.1	−1,600	−1,600	9,100	10,900	−10,900	−9,100
0.2	−1,600	−1,600	9,200	10,800	−10,800	−9,200
0.3	−1,600	−1,600	9,300	10,700	−10,700	−9,300
0.4	−1,600	−1,600	9,400	10,600	−10,600	−9,400
0.5	−1,600	−1,600	9,500	10,500	−10,500	−9,500
0.6	−1,600	−1,600	9,600	10,400	−10,400	−9,600
0.7	−1,600	−1,600	9,700	10,300	−10,300	−9,700
0.8	−1,600	−1,600	9,800	10,200	−10,200	−9,800
0.9	−1,600	−1,600	9,900	10,100	−10,100	−9,900
1.0	−1,600	−1,600	10,000	10,000	−10,000	−10,000

For $\alpha = 0$:

$$\begin{aligned} CF_0^{\alpha=0} &= [-1600, -1600] \\ CF_1^{\alpha=0} &= [9000 + 1000x0, 11000 - 1000x0] = [9000, 11000] \\ CF_2^{\alpha=0} &= [-11000 + 1000x0, -9000 - 1000x0] = [-11000, -9000] \end{aligned}$$

The values of CF_0^α, CF_1^α and CF_2^α are shown in Table 3 for different values of α.

(b) *Calculation of the values of the fuzzy rates* $k_{rr}^\alpha, k_{wacc}^\alpha$ *e* k_{radr}^α

Using the same procedures above for the calculation of the fuzzy cash flows, different values of $k_{rr}^\alpha, k_{wacc}^\alpha$ and k_{radr}^α are calculated and shown in Table 4.

Table 4. Values of k_{rr}^α, k_{wacc}^α and k_{radr}^α

α	k_{rr}^α (%)		k_{wacc}^α (%)		k_{radr}^α (%)	
	$k_{rr}^{l\alpha}$	$k_{rr}^{r\alpha}$	$k_{wacc}^{l\alpha}$	$k_{wacc}^{r\alpha}$	$k_{radr}^{l\alpha}$	$k_{radr}^{r\alpha}$
0	23.0	27.0	21.0	25.0	21.0	25.0
0.1	23.2	26.8	21.2	24.8	21.2	24.8
0.2	23.4	26.6	21.4	24.6	21.4	24.6
0.3	23.6	26.4	21.6	24.4	21.6	24.4
0.4	23.8	26.2	21.8	24.2	21.8	24.2
0.5	24.0	26.0	22.0	24.0	22.0	24.0
0.6	24.2	25.8	22.2	23.8	22.2	23.8
0.7	24.4	25.6	22.4	23.6	22.4	23.6
0.8	24.6	25.4	22.6	23.4	22.6	23.4
0.9	24.8	25.2	22.8	23.2	22.8	23.2
1.0	25.0	25.0	23.0	23.0	23.0	23.0

(c) *Calculation of the values of the fuzzy Terminal Value* (TV^{α})

From Eqs. (19) and (20), $TV^{l(\alpha)}$ and $TV^{r(\alpha)}$ for $\alpha = 0$ are expressed as:

$$TV^{l(\alpha=0)} = CF_1^{l(\alpha=0)}x(1 + k_{rr}^{l(\alpha=0)})^1 = 9,000x(1+0.23) = 11,070.00$$
$$TV^{r(\alpha=0)} = CF_1^{r(\alpha=0)}x(1 + k_{rr}^{r(\alpha=0)})^1 = 11,000x(1+0.27) = 13,970.00$$

(d) *Calculation of the values of the fuzzy Present Value* (PV^{α})

From Eqs. (23) and (24), $PV^{l(\alpha)}$ and $PV^{r(\alpha)}$ for $\alpha = 0$ are expressed as:

$$PV^{l(\alpha=0)} = CF_0^{l(\alpha=0)} + \frac{CF_2^{l(\alpha=0)}}{\left(1+k_{wacc}^{l(\alpha=0)}\right)^2} = -1,600 - \frac{11,000}{(1+0.21)^2} = -9,113.15$$

$$PV^{r(\alpha=0)} = CF_0^{r(\alpha=0)} + \frac{CF_2^{r(\alpha=0)}}{\left(1+k_{wacc}^{r(\alpha=0)}\right)^2} = -1,600 - \frac{9,000}{(1+0.25)^2} = -7,360.00$$

The values of TV^{α} *and* PV^{α} are shown in Table 5 for the different values of α.

Table 5. Values of TV^{α} and PV^{α}

	TV^{α}		PV^{α}	
α	$TV^{l(\alpha)}$	$TV^{r(\alpha)}$	$PV^{l(\alpha)}$	$PV^{r(\alpha)}$
0	11,070.00	13,970.00	(9,113.15)	(7,360.00)
0.1	11,211.20	13,821.20	(9,020.30)	(7,442.68)
0.2	11,352.80	13,672.80	(8,928.02)	(7,525.86)
0.3	11,494.80	13,524.80	(8,836.30)	(7,609.55)
0.4	11,637.20	13,377.20	(8,745.15)	(7,693.75)
0.5	11,780.00	13,230.00	(8,654.56)	(7,778.46)
0.6	11,923.20	13,083.20	(8,564.52)	(7,863.69)
0.7	12,066.80	12,936.80	(8,475.03)	(7,949.43)
0.8	12,210.80	12,790.80	(8,386.08)	(8,035.70)
0.9	12,355.20	12,645.20	(8,297.68)	(8.122.50)
1.0	12,500.00	12,500.00	(8,209.82)	(8,209.82)

(e) *Calculation of the values of the fuzzy MNPV*

From Eqs. (26) and (27), $MNPV^{l(\alpha)}$ and $MNPV^{r(\alpha)}$ for $\alpha = 0$ are expressed as:

$$MNPV^{l(\alpha=0)} = \frac{TV^{l(\alpha=0)}}{(1+k_{radr}^{r(\alpha=0)})^n} + PV^{l(\alpha=0)} = \frac{11,070.00}{(1+0.25)^2} - 9,113.15 = -2,028.35$$
$$MNPV^{r(\alpha=0)} = \frac{TV^{r(\alpha=0)}}{(1+k_{radr}^{l(\alpha=0)})^n} + PV^{r(\alpha=0)} = \frac{13,970.00}{(1+0.21)^2} - 7,360.00 = 2,181.70$$

(f) *Calculation of the values of the fuzzy MIRR*

From Eqs. (29) and (30), $MIRR^{l(\alpha)}$ and $MIRR^{r(\alpha)}$ for $\alpha = 0$ are expressed as:

$$MIRR^{l(\alpha=0)} = \left[-\frac{TV^{l(\alpha=0)}}{PV^{l(\alpha=0)}}\right]^{1/n} - 1 = \left[\frac{11{,}070.00}{9{,}113.15}\right]^{1/2} - 1 = 0.102\, ou\, 10.2\%$$
$$MIRR^{r(\alpha=0)} = \left[-\frac{TV^{r(\alpha=0)}}{PV^{r(\alpha=0)}}\right]^{1/n} - 1 = \left[\frac{13{,}970.00}{7{,}360.00}\right]^{1/2} - 1 = 0.378\, ou\, 37.8\%$$

(g) *Calculation of the values of the fuzzy MPI*

From Eqs. (32) and (33), $MPI^{l(\alpha)}$ and $MPI^{r(\alpha)}$ for $\alpha = 0$ are expressed as:

$$MPI^{l(\alpha=0)} = \left[-\frac{TV^{l(\alpha=0)}}{(1+k_{radr}^{r(\alpha=0)})^n}\right] / [PV^{l(\alpha=0)}] = \left[\frac{11,070.00}{(1+0.25)^2}\right] / [9,113.15] = 0.78$$
$$MPI^{r(\alpha=0)} = \left[-\frac{TV^{r(\alpha=0)}}{(1+k_{radr}^{l(\alpha=0)})^n}\right] / [PV^{r(\alpha=0)}] = \left[\frac{13,970.00}{(1+0.21)^2}\right] / [7,360.00] = 1.30$$

(h) *Calculation of the values of the fuzzy MTPB*

From Eqs. (35) and (36), $MTPB^{l(\alpha)}$ and $MTPB^{r(\alpha)}$ for $\alpha = 0$ are expressed as:

$$MTPB^{l(\alpha=0)} = -PV^{r(\alpha=0)} / \left[\frac{TV^{r(\alpha=0)}}{(1+k_{radr}^{l(\alpha=0)})^n}\right] \times n = 7,360.00 / \left[\frac{13{,}970.00}{(1+0.21)^2}\right] \times 2 = 1.54$$
$$MTPB^{r(\alpha=0)} = -PV^{l(\alpha=0)} / \left[\frac{TV^{l(\alpha=0)}}{(1+k_{radr}^{r(\alpha)})^n}\right] \times n = 9,113.15 / \left[\frac{11{,}070.00}{(1+0.25)^2}\right] \times 2 = 2.57$$

The values of $MTPB^{\alpha}, MTRR^{\alpha}, MPI^{\alpha}$ and $MTPB^{\alpha}$ are shown in Table 6 for the different values of α.

Table 6. Values of $MNPV^{\alpha}$, $MIRR^{\alpha}$, MPI^{α}, and $MTPB^{\alpha}$

	$MNPV^{\alpha}$		$MIRR^{\alpha}$		MPI^{α}		$MTPB^{\alpha}$	
α	$MNPV^{l(\alpha)}$	$MNPV^{r(\alpha)}$	$MIRR^{l(\alpha)}$	$MIRR^{l(\alpha)}$	$MPI^{l(\alpha)}$	$MPI^{l(\alpha)}$	$MTPB^{l(\alpha)}$	$MTPB^{l(\alpha)}$
0	**(2,028.35)**	**2,181.70**	**10.2%**	**37.8%**	**0.78**	**1.30**	**1.54**	**2.57**
0.1	(1,822.11)	1,966.25	11.5%	36.3%	0.80	1.26	1.58	2.51
0.2	(1,615.50)	1,751.40	12.8%	34.8%	0.82	1.23	1.62	2.44
0.3	(1,408.49)	1,537.13	14.1%	33.3%	0.84	1.20	1.66	2.38
0.4	(1,201.09)	1,323.43	15.4%	31.9%	0.86	1.17	1.71	2.32
0.5	(993.26)	1,110.28	16.7%	30.4%	0.89	1.14	1.75	2.26
0.6	(785.02)	897.68	18.0%	29.0%	0.91	1.11	1.80	2.20
0.7	(576.33)	685.60	19.3%	27.6%	0.93	1.09	1.84	2.15
0.8	(367.20)	474.05	20.7%	26.2%	0.96	1.06	1.89	2.09
0.9	(157.61)	263.00	22.0%	24.8%	0.98	1.03	1.94	2.04
1.0	**52.46**	**52.46**	**23.4%**	**23.4%**	**1.01**	**1.01**	**1.99**	**1.99**

The graphical representations of the fuzzy indicators are shown in Fig. 2.

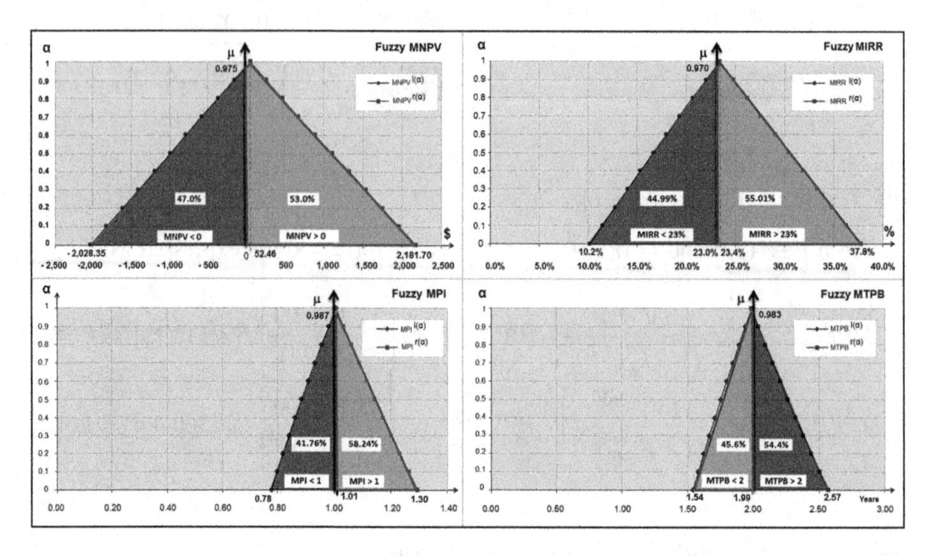

Fig. 2. Graphical representations of $MNPV^{\alpha}$, $MIRR^{\alpha}$, MPI^{α} and $MTPB^{\alpha}$ (Color figure online)

The red areas indicate that there is a possibility that the project will not to be indicated as acceptable. According to [13], this possibility can be computed by dividing the negative area by the total area, which in the *MNPV* case results in 47.0%.

As the calculation of the modified methods under uncertainty, for an investment project with cash flows and rates represented by triangular fuzzy numbers, is quite hard and complex, new financial functions were developed in *VBA* of MS-Excel. The syntaxes and codes of these functions are available in [2].

4 Conclusions

The association of fuzzy numbers to the modified methods proved to be appropriate for solving problems of investment evaluations. The proposed unified approach is simple and makes fewer assumptions than the traditional models for the treatment of uncertainty regarding the behavior of related variables, becoming very useful for decision makers. The fuzzy functions have shown good performance and are user friendly, which is an essential feature for professional use.

References

1. de Souza Sampaio Filho, A.C., Vellasco, M.M.B.R., Tanscheit, R.: Modified net present value under uncertainties: an approach based on fuzzy numbers and interval arithmetic. In: Greco, S., Bouchon-Meunier, B., Coletti, G., Fedrizzi, M., Matarazzo, B., Yager, R.R. (eds.) IPMU 2012. CCIS, vol. 300, pp. 10–19. Springer, Heidelberg (2012). https://doi.org/10.1007/978-3-642-31724-8_2
2. de Souza Sampaio Filho, A.C.: Métodos Modificados de Avaliação de Investimentos em Condições de Incerteza: Uma Abordagem Baseada em Números Fuzzy [Modified Capital Budgeting Methods under Uncertainties: An Approach Based on Fuzzy Numbers], doctoral thesis. Rio de Janeiro: Pontifícia Universidade Católica do Rio de Janeiro [in Portuguese language] (2014)
3. McClure, K.G., Girma, P.B.: Modified net present value (MNPV): a new technique for capital budgeting. Zagreb Int. Rev. Econ. Bus. **7**, 67–82 (2004)
4. Lin, S.A.Y.: The modified internal rate of return and investment criterion. Eng. Econ. **21**(4), 237–247 (1976)
5. Beaves, R.G.: Net present value and rate of return: implicit and explicit reinvestment assumptions. Eng. Econ. **33**(4), 275–302 (1988)
6. Shull, D.M.: Overall rates of return: investment bases, reinvestment rates and time horizons. Eng. Econ. **39**(2), 139–163 (1994)
7. Vélez-Pareja, I.: Ranking and optimal selection of investments with internal rate and benefit-cost ratio. A Revision. Contaduría y Administración **57**(1), 29–51 (2012)
8. Plath, D.A., Kennedy, W.F.: Teaching return-based measures of project evaluation. Finan. Pract. Educ. **4**, 77–86 (1994)
9. Solomon, E.: The arithmetic of capital-budgeting decisions. J. Bus. **29**(2), 124–129 (1956)
10. Zadeh, L.A.: Fuzzy sets. Inf. Control **8**, 338–353 (1965)
11. Kahraman, C. (ed.): Fuzzy Engineering Economics with Applications. Springer, Heidelberg (2008). https://doi.org/10.1007/978-3-540-70810-0
12. Ulukan, Z., Ucuncuoglu, C.: Economic analyses for the evaluation of is projects. J. Inf. Syst. Technol. Manag. **7**(2), 233–260 (2010)
13. Chiu, C.Y., Park, C.S.: Fuzzy cash flow analysis using present worth criterion. Eng. Econ. **39**(2), 113–137 (1994)

Solving Job-Shop Scheduling Problems with Fuzzy Processing Times and Fuzzy Due Dates

Camilo Alejandro Bustos-Tellez, Jhoan Sebastian Tenjo-García, and Juan Carlos Figueroa-García(✉)

Universidad Distrital Francisco José de Caldas, Bogotá, Colombia
{caabustost,jstenjog}@correo.udistrital.edu.co,
jcfigueroag@udistrital.edu.co

Abstract. This paper shows an iterative method for solving n-jobs, m-machines scheduling problems with fuzzy processing times and fuzzy due dates which are defined using third-party information coming from experts. We use an iterative method based on the cumulative membership function of a fuzzy set to find an overall satisfaction degree among fuzzy processing times and fuzzy due dates.

Keywords: Fuzzy scheduling · Fuzzy numbers
Cumulative membership function

1 Introduction and Motivation

Job-shop scheduling is one of the most popular tasks in production planning since it involves the sequence in which an amount of jobs must go through different machines in a predefined route/sequence (see Pinedo [13], Johnson and Montgomery [12], and Sipper and Bulfin [18]). The classical scheduling problem considers both processing times and due dates of each job as constants, so it is often solved using classical optimization techniques (unless the size of the problem forces to use metaheuristics or approximation algorithms).

In several problems processing times and due dates cannot be defined as constants, so there is a need for obtaining them from a reliable source. Those times are obtained through statistical analysis in most of cases (see Heyman and Sobel [9]), but in several situations there is no enough statistical information to obtain them. This way, the experts of the system become to a reliable information source, so their perceptions about processing times and due dates can be summarized as fuzzy sets.

This paper focuses on a job-shop, n-jobs, m-machines scheduling problem with fuzzy processing times and fuzzy due dates. Based on linear programming formulations proposed by Pinedo [13], we extend his results to a fuzzy environment using the method proposed by Figueroa-García [4] and Figueroa-García and López-Bello [7,8].

J. Medina et al. (Eds.): IPMU 2018, CCIS 853, pp. 790–799, 2018.
https://doi.org/10.1007/978-3-319-91473-2_66

This paper is divided into five principal sections. In Sect. 1, the introduction and motivation of the work are presented. In Sect. 2, the classical Linear Programming (LP) model for n-jobs and m-machines with due dates is presented; Sect. 3 presents some basics on fuzzy sets/numbers. In Sect. 4, the Fuzzy Scheduling Problem (FSP) and the proposed method for solving it are presented. Section 5 shows an application example, and finally some concluding remarks are presented in Sect. 6.

2 The Crisp Scheduling Problem

The mathematical form of an n-jobs, m-machines scheduling problem (minimizing its makespan) is as follows:

$$z = \text{Min} \sum_j +d_j^+ - d_j^- + Ms, \quad s.t. \quad x_{gj} + t_{gj} \leqslant Ms \,\forall\, j \in \mathbb{N}_n, \tag{1}$$

$$x_{rj} + t_{rj} \leqslant x_{sj} \,\forall\, j \in \mathbb{N}_n, \tag{2}$$

$$x_{ip} + t_{ip} \leqslant x_{iq} + M \cdot y_k \,\forall\, i \in \mathbb{N}_m, k \in \mathbb{N}_l, \tag{3}$$

$$x_{iq} + t_{iq} \leqslant x_{ip} + (1 - y_k) \cdot M \,\forall\, i \in \mathbb{N}_m, k \in \mathbb{N}_l, \tag{4}$$

$$x_{gj} + t_{gj} + d_j^+ - d_j^- = d_j \,\forall\, j \in \mathbb{N}_n. \tag{5}$$

Index sets:
$i \in \mathbb{N}_m$ is the set of m machines
$j \in \mathbb{N}_n$ is the set of n jobs
$k \in \mathbb{N}_l$ is the set of l conflicting jobs
Parameters:
$t_{ij} \in \mathbb{R}$ is the processing time of the j_{th} job in the i_{th} machine
$d_j \in \mathbb{R}$ is the due date of the j_{th} job
$M \in \mathbb{R}$ is a weight used to solve the conflict between two jobs $p, q \in \mathbb{N}_n$
Decision variables:
$x_{ij} \in \mathbb{R}$ is the starting time of the j_{th} job in the i_{th} machine
$Ms \in \mathbb{R}$ is the makespan of all jobs
$y_k \in \mathbb{Z}$ is a binary variable to solve the k_{th} conflict between two jobs $p, q \in \mathbb{N}_n$
$d_j^+ \in \mathbb{R}$ is the earliness of the j_{th} job
$d_j^- \in \mathbb{R}$ is the tardiness of the j_{th} job

In this model, Eq. (1) represents the makespan of the system, (2) are *sequence constraints* where the machine $r \in \mathbb{N}_m$ precedes the machine $s \in \mathbb{N}_m$ in the sequence of the j_{th} job. Equations (3) and (4) are *conflict* constraints where $p \in \mathbb{N}_n$ and $q \in \mathbb{N}_n$ are jobs that need to be processed in the i_{th} machine, so the variable y_k operates as a binary decision about starting the p_{th} job, or vice versa. Finally Eq. (5) is a *due date* constraint which is relaxation intended to guarantee feasibility of the problem (in case that we cannot meet all due dates).

3 Basics on Fuzzy Sets

A fuzzy set is denoted by emphasized capital letters $\tilde{A}$ with a membership function $\mu_{\tilde{A}}(x)$ over a universal set $x \in X$. $\mu_{\tilde{A}}(x)$ measures the membership of a value $x \in X$ regarding the concept/word/label A. $\mathcal{P}(X)$ is the class of all crisp sets, $\mathcal{F}(X)$ is the class of all fuzzy sets, $\mathcal{F}(\mathbb{R})$ is the class of all real-valued fuzzy sets, and $\mathcal{F}_1(\mathbb{R})$ is the class of all fuzzy numbers. Thus, a fuzzy set A is a set of ordered pairs of an element x and its membership degree, $\mu_A(x)$, i.e.,

$$\tilde{A} = \{(x, \mu_{\tilde{A}}(x)) \mid x \in X\}. \tag{6}$$

A fuzzy number is defined as follows:

Definition 1. *Let $\tilde{A} : \mathbb{R} \rightarrow [0, 1]$ be a fuzzy subset of the reals. Then $\tilde{A} \in \mathcal{F}_1(\mathbb{R})$ is a Fuzzy Number (FN) iff there exists a closed interval $[x_l, x_r] \neq \emptyset$ with a membership function $\mu_{\tilde{A}}(x)$ such that:*

$$\mu_{\tilde{A}}(x) = \begin{cases} c(x) & \text{for } x \in [c_l, c_r], \\ l(x) & \text{for } x \in [-\infty, x_l], \\ r(x) & \text{for } x \in [x_r, \infty], \end{cases} \tag{7}$$

where $c(x) = 1$ for $x \in [c_l, c_r]$, $l : (-\infty, x_l) \rightarrow [0, 1]$ is monotonic non-decreasing, continuous from the right, i.e. $l(x) = 0$ for $x < x_l$; $l : (x_r, \infty) \rightarrow [0, 1]$ is monotonic non-increasing, continuous from the left, i.e. $r(x) = 0$ for $x > x_r$.

The α-*cut* of a set $\tilde{A} \in \mathcal{F}_1(\mathbb{R})$ namely ${}^{\alpha}\tilde{A}$ is defined as follows:

$${}^{\alpha}\tilde{A} = \{x \mid \mu_{\tilde{A}}(x) \geqslant \alpha\} \ \forall \ x \in X, \tag{8}$$

$${}^{\alpha}\tilde{A} = \left[\inf_{x} {}^{\alpha}\mu_{\tilde{A}}(x), \sup_{x} {}^{\alpha}\mu_{\tilde{A}}(x)\right] = [\check{A}_{\alpha}, \hat{A}_{\alpha}]. \tag{9}$$

3.1 The Cumulative Membership Function (CMF)

The cumulative function $F(x)$ of a probability distribution $f(x)$ is:

$$F(x) = \int_{-\infty}^{x} f(t)\, dt, \tag{10}$$

where $x \in \mathbb{R}$. Its fuzzy version is as follows (see Figueroa-García and López-Bello [7,8], Figueroa-García [4], Pulido-López et al. [14]).

Definition 2 (Cumulative Membership Function). *Let $\tilde{A} \in \mathcal{F}(\mathbb{R})$ be a fuzzy set and $X \subseteq \mathbb{R}$, then the cumulative membership function (CMF) of $\tilde{A}$, $\psi_{\tilde{A}}(x)$ is defined as:*

$$\psi_{\tilde{A}}(x) = Ps_{\tilde{A}}(X \leqslant x), \tag{11}$$

This definition is the possibility that all elements of X are less or equal than a value x, i.e. $Ps(X \leqslant x)$. Note that in the probabilistic case $F(\infty) = 1$ while in the possibilistic case $1 < \psi(\infty) < \Lambda$, where Λ is the cardinality of $\tilde{A}$:

$$\Lambda_{\tilde{A}} = \int_{-\infty}^{\infty} \mu_{\tilde{A}}(t)\, dt. \tag{12}$$

To normalize $\psi_{\tilde{A}}(x)$ can be divide it by $\Lambda_{\tilde{A}}$:

$$\psi_{\tilde{A}}(x) = \frac{1}{\Lambda_{\tilde{A}}} \int_{-\infty}^{x} \mu_{\tilde{A}}(t)\, dt. \tag{13}$$

A graphical display of the CMF $\psi_{\tilde{A}}$ of a triangular fuzzy number is shown in Fig. 1.

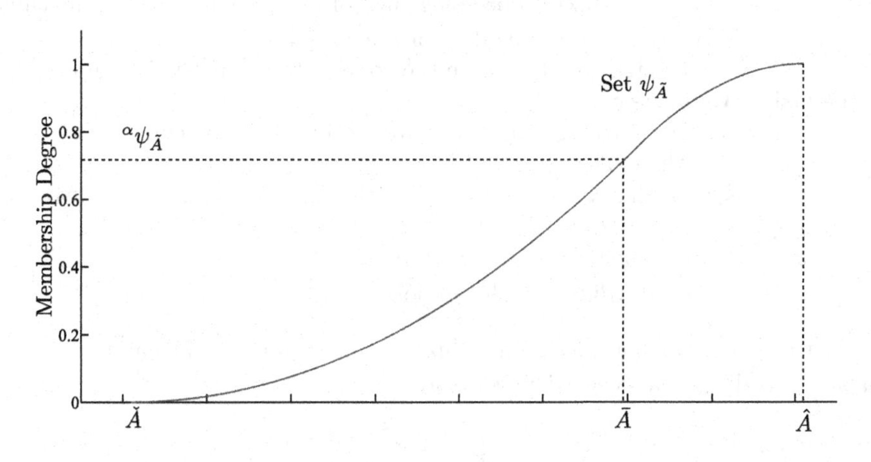

Fig. 1. Cumulative membership function $\psi_{\tilde{A}}$ of a triangular fuzzy set

4 The Proposed Fuzzy Scheduling Optimization Method

In this paper, we focus on an n-jobs, m-machines scheduling problem where its processing times and its due dates are defined by the experts of the system as fuzzy numbers just as shown in Definition 1. We take advantage of the seminal soft constraints model proposed by Zimmermann [19] since it has been proven to obtain optimal solutions for linear fuzzy constraints, and the idea is to solve an FSP with fuzzy processing times and soft constraints. To do so, we first present the classical LP model and then our proposal.

4.1 The Fuzzy Scheduling Problem (FSP)

The mathematical form of an n-jobs, m-machines scheduling problem with due dates (minimizing its makespan) is presented next:

$$
\begin{aligned}
\tilde{z} = \text{Min} \sum_{j} +d_j^+ - d_j^- + Ms,& \\
s.t.& \\
x_{gj} + \tilde{t}_{gj} \lesssim Ms \; \forall \, j \in \mathbb{N}_n,& \\
x_{rj} + \tilde{t}_{rj} \lesssim x_{sj} \; \forall \, j \in \mathbb{N}_n,& \\
x_{ip} + \tilde{t}_{ip} \lesssim x_{iq} + M \cdot y_k \; \forall \, i \in \mathbb{N}_m, k \in \mathbb{N}_l,& \\
x_{iq} + \tilde{t}_{iq} \lesssim x_{ip} + (1 - y_k) \cdot M \; \forall \, i \in \mathbb{N}_m, k \in \mathbb{N}_l,& \\
x_{gj} + \tilde{t}_{gj} + d_j^+ - d_j^- \approx \tilde{d}_j \; \forall \, j \in \mathbb{N}_n,&
\end{aligned} \tag{14}
$$

Index sets:

$i \in \mathbb{N}_m$ is the set of m machines
$j \in \mathbb{N}_n$ is the set of n jobs
$k \in \mathbb{N}_l$ is the set of l conflicting jobs

Parameters:

$\tilde{t}_{ij} \in \mathcal{F}_1(\mathbb{R})$ is the fuzzy processing time of the j_{th} job in the i_{th} machine
$\tilde{d}_j \in \mathcal{F}_1(\mathbb{R})$ is the fuzzy due date of the j_{th} job
$M \in \mathbb{R}$ is a weight used to solve the conflict between two jobs $p, q \in \mathbb{N}_n$

Decision variables:

$x_{ij} \in \mathbb{R}$ is the starting time of the j_{th} job in the i_{th} machine
$Ms \in \mathbb{R}$ is the makespan of all jobs
$y_k \in \mathbb{Z}$ is a binary variable to solve the k_{th} conflict between two jobs $p, q \in \mathbb{N}_n$
$d_j^+ \in \mathbb{R}$ is the earliness of the j_{th} job
$d_j^- \in \mathbb{R}$ is the tardiness of the j_{th} job

In this paper, we consider $\tilde{t}_{ij}$ as a fuzzy number (see Eq. (7)), and every due date $\tilde{d}_j$ as a linear fuzzy set defined as follows:

$$
\mu_{\tilde{d}_j}(x, \check{d}_j, \hat{d}_j) = \begin{cases} 1, & 0 \leqslant x \leqslant \check{d}_j \\ \dfrac{\hat{d}_j - x}{\hat{d}_j - \check{d}_j}, & \check{d}_j \leqslant x \leqslant \hat{d}_j \\ 0, & x \geqslant \hat{d}_j \end{cases} \tag{15}
$$

where $\check{d}_j \in \mathbb{R}$ is the lower bound of d_j, and $\hat{d}_j \in \mathbb{R}$ is the upper bound of d_j.

4.2 The Proposed Method

Based on fuzzy optimization concepts introduced by Rommelfanger [17], Ramík [16], Inuiguchi and Ramík [10], Fiedler et al. [1], Ramík and Řimánek [15], Inuiguchi et al. [11], we apply the proposal of Figueroa-García [4] and Figueroa-García and López-Bello [7,8], for which we present the following compact version adapted to the FSP (for minimizing Ms).

1. **Iterative method:**
   - Set $\alpha \in [0, 1]$,
   - Compute $\psi_{\tilde{t}_{ij}}$ and ${}^{\alpha}\psi_{\tilde{t}_{ij}} \; \forall \, (i, j)$,

2. **Soft constraints method:**
   - Define $\check{z} = \text{Max}\{\sum_j d_j^- + d_j^- - Ms\}$ for $t_{ij} = {}^{\alpha}\psi_{\tilde{t}_{ij}}$, $d_j = \check{d}_j$ (see Sect. 4.2),
   - Define $\hat{z} = \text{Max}\{\sum_j d_j^- + d_j^- - Ms\}$ for $t_{ij} = {}^{\alpha}\psi_{\tilde{t}_{ij}}$, $d_j = \hat{d}_j$ (see Sect. 4.2),
   - Define the set $\tilde{z}$ with the following membership function:

$$\mu_{\tilde{z}}(x^*, \check{z}, \hat{z}) = \begin{cases} 1, & c'x^* \geqslant \hat{z} \\ \dfrac{c'x^* - \check{z}}{\hat{z} - \check{z}}, & \check{z} \leqslant c'x^* \leqslant \hat{z} \\ 0, & c'x^* \leqslant \check{z} \end{cases} \tag{16}$$

   - Thus, solve the following LP model:

$$\begin{gathered} \text{Max } \{\lambda\}, \\ s.t. \\ \sum_j d_j^- + d_j^- - Ms - \lambda(\hat{z} - \check{z}) = \check{z}, \end{gathered} \tag{17}$$

$$\begin{gathered} x_{gj} + {}^{\alpha}\psi_{\tilde{t}_{gj}} + Ms \leqslant 0 \; \forall \, j \in \mathbb{N}_n, \\ x_{rj} + {}^{\alpha}\psi_{\tilde{t}_{rj}} \leqslant x_{sj} \; \forall \, j \in \mathbb{N}_n, \\ x_{ip} + {}^{\alpha}\psi_{\tilde{t}_{ip}} \leqslant x_{iq} + M \cdot y_k \; \forall \, i \in \mathbb{N}_m, k \in \mathbb{N}_l, \\ x_{iq} + {}^{\alpha}\psi_{\tilde{t}_{iq}} \leqslant x_{ip} + (1 - y_k) \cdot M \; \forall \, i \in \mathbb{N}_m, k \in \mathbb{N}_l, \\ x_{gj} + {}^{\alpha}\psi_{\tilde{t}_{gj}} + d_j^+ - d_j^- + \lambda(\hat{d}_j - \check{d}_j) = \hat{d}_j \; \forall \, j \in \mathbb{N}_n. \end{gathered} \tag{18}$$

3. **Convergence:**
   - If $\lambda^* = \alpha$ then stop and return λ^* as the overall satisfaction degree of $\tilde{z}, \tilde{t}_{ij}$, and $\tilde{d}_j$; if $\lambda^* \neq \alpha$ then set $\lambda^* = \alpha$ and go to Step 1.

5 Application Example

In this section, we solve a problem of 4 jobs in 4 machines with fuzzy due dates. We use triangular fuzzy processing times $T = (\check{t}, \bar{t}, \hat{t})$ and due dates as defined in Eq. (15). The sequences of every job through all machines are shown next.

- $J_1 : S_3 \to S_1 \to S_4 \to S_2$
- $J_2 : S_4 \to S_1 \to S_2 \to S_3$
- $J_3 : S_1 \to S_2 \to S_3 \to S_4$
- $J_4 : S_1 \to S_3 \to S_2 \to S_4$

Processing times and due dates are presented in matrix format:

$$\check{t} = \begin{bmatrix} 44 & 26 & 53 & 1 \\ 4 & 11 & 85 & 61 \\ 33 & 15 & 18 & 81 \\ 91 & 27 & 4 & 23 \end{bmatrix}; \; \bar{t} = \begin{bmatrix} 54 & 34 & 61 & 2 \\ 9 & 15 & 89 & 70 \\ 38 & 19 & 28 & 87 \\ 95 & 34 & 7 & 29 \end{bmatrix}; \; \hat{t} = \begin{bmatrix} 62 & 38 & 65 & 5 \\ 15 & 23 & 99 & 75 \\ 44 & 25 & 33 & 97 \\ 104 & 43 & 11 & 37 \end{bmatrix}$$

$$\check{d} = \begin{bmatrix} 180 \\ 184 \\ 190 \\ 189 \end{bmatrix}; \; \hat{d} = \begin{bmatrix} 268 \\ 341 \\ 341 \\ 350 \end{bmatrix}$$

Now, we apply the procedure shown in Sect. 4.2. First, we set $\alpha = 0.2$ as starting point to then compute $\psi_{\tilde{t}_{ij}}$ and ${}^{\alpha}\psi_{\tilde{t}_{ij}}\ \forall\,(i,j)$ as presented in (13) and Fig. 1. After 7 iterations of the algorithm, we found the following solution:

$$x_{21} = 139.881, x_{32} = 148.127, x_{43} = 148.127, x_{44} = 285.52, x_{11} = 81.166,$$
$$x_{12} = 70.518, x_{13} = 0, x_{14} = 142.58, x_{22} = 81.166, x_{23} = 98.917, x_{24} = 248.95,$$
$$x_{31} = 0, x_{33} = 119.76, x_{34} = 240.788, x_{41} = 136.73, x_{42} = 0$$
$$d_1^- = 31.087, d_2^- = 0, d_3^- = 0, d_4^- = 0, d_1^+ = 0, d_2^+ = 11.722, d_3^+ = 5.007, d_4^+ = 81.64$$

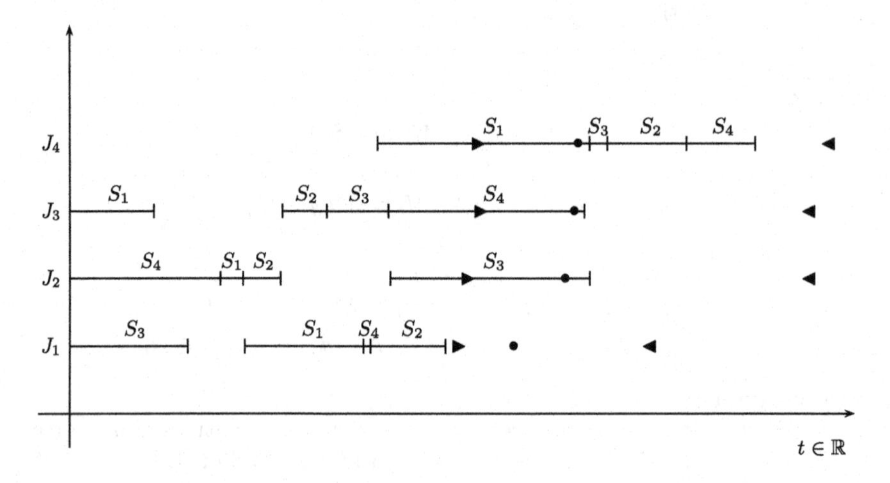

Fig. 2. Optimal schedule of all jobs through machines

Figure 2 shows the optimal sequence where the obtained optimal makespan is $Ms = 316.85$. ▶ represents $\check{d}_j$, ◀ represents $\hat{d}_j$, and • represents d_j^* after defuzzification in the last iteration i.e. $\lambda^* = 0.713$. All deffuzified processing times and due dates are shown next:

$$^{\alpha}\psi_{\tilde{t}} = \begin{bmatrix} 55.57 & 34.29 & 61.29 & 3.14 \\ 10.65 & 17.75 & 92.66 & 70.52 \\ 39.65 & 20.85 & 28.36 & 90.22 \\ 98.21 & 36.57 & 8.17 & 31.33 \end{bmatrix}; \ ^{\alpha}\tilde{d} = \begin{bmatrix} 205.26 \\ 229.06 \\ 233.34 \\ 235.21 \end{bmatrix}.$$

It is clear that 3 out of 4 jobs have tardiness while job 1 has earliness, but no any job ends after its upper due date $\hat{d}_j$. This means that all jobs will be delivered before the last date allowed by customers. Figure 3 shows the set $\tilde{z}$ of optimal makespan/earliness/tardiness for a final optimal degree $\lambda^* = 0.713$.

We point out that $\lambda^* = 0.713$ is an overall satisfaction degree of $\tilde{z}$, $\tilde{t}_{ij}$ and $\tilde{d}_j$. Also note that the algorithm finds the same overall satisfaction degree no matter the value of $\alpha \in [0, 1]$ we use to initialize it.

Figure 4 shows the values of λ^* per iteration for 2 different starting values $\alpha = \{0.2, 0.8\}$ where both points reach the same optimal value $\lambda^* = 0.713$.

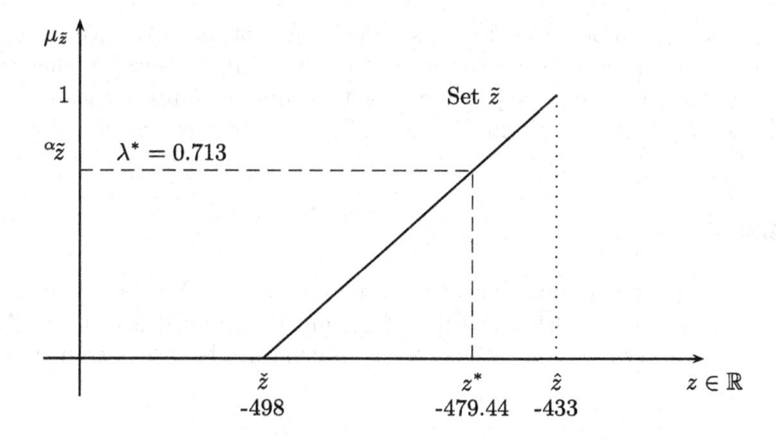

Fig. 3. Optimal Solution of the problem.

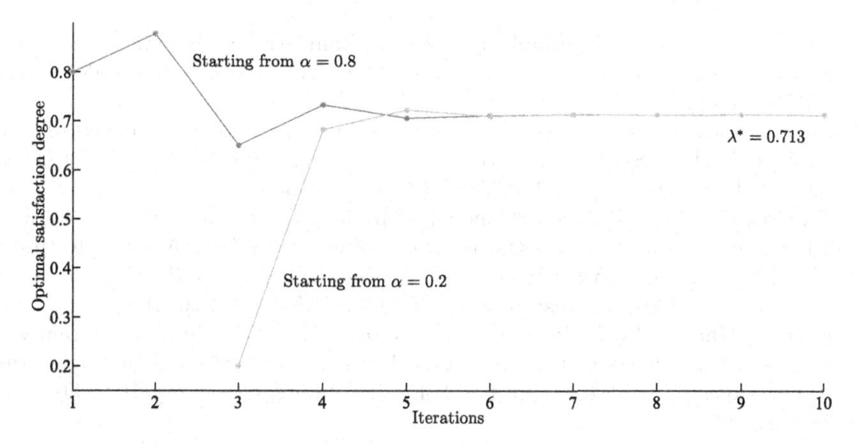

Fig. 4. Convergence of the proposed method for $\alpha = \{0.2, 0.8\}$

6 Concluding Remarks

We have presented a fuzzified version of the classical n-jobs, m-machines scheduling problem with due dates, and we applied the algorithm proposed by Figueroa-García [4] and Figueroa-García and López-Bello [7,8] with successful results. An optimal overall satisfaction degree of $\lambda^* = 0.713$ was reached, and the defuzzified solution of the problem is presented.

The algorithm can deal with any shape of convex fuzzy processing times and fuzzy linear due dates. The proposal iterates the classical Zimmermann until both fuzzy processing times and fuzzy due dates reach a single defuzzification degree. No matter what starting point α is selected, the algorithm will reach the same optimal value λ^*.

Based on the cumulative membership function $\psi_{\tilde{A}}$ of a fuzzy set, the algorithm finds an overall satisfaction degree λ^* operates as a defuzzification value for $\tilde{z}$, $\tilde{t}_{ij}$ and $\tilde{d}_j$. All defuzzified values provide the optimal sequence, and single values for t_{ij} and d_j as reference for production planning. This way, managers can handle uncertain information, provide expected shipping dates to customers, and provide expected processing times to manufacturers of every machine/station.

Future Work

Optimization problems involving Type-2 fuzzy sets (see Figueroa-García [2,3, 5,6]) are a natural extension of this model and the applied algorithm. Also a more detailed description/analysis of the algorithm will be discussed in a journal paper.

References

1. Fiedler, M., Nedoma, J., Ramík, J., Rohn, J., Zimmermann, K.: Linear Optimization Problems with Inexact Data. Springer, New York (2006). https://doi.org/10.1007/0-387-32698-7
2. Figueroa-García, J.C.: Interval type-2 fuzzy linear programming: uncertain constraints. In: IEEE Symposium Series on Computational Intelligence, pp. 1–6. IEEE (2011). https://doi.org/10.1109/T2FUZZ.2011.5949559
3. Figueroa-García, J.C.: A general model for linear programming with interval type-2 fuzzy technological coefficients. In: Proceedings of the 2012 Annual Meeting of the North American Fuzzy Information Processing Society NAFIPS 2012, vol. 1, pp. 1–6. IEEE (2012). https://doi.org/10.1109/NAFIPS.2012.6291064
4. Figueroa-García, J.C.: Mixed production planning under fuzzy uncertainty: a cumulative membership function approach. In: IEEE Workshop on Engineering Applications (WEA), vol. 1, pp. 1–6. IEEE (2012). https://doi.org/10.1109/WEA.2012.6220081
5. Figueroa-García, J.C.: On the fuzzy extension principle for LP problems with interval type-2 technological coefficients. Rev. Ing. Univ. Dist. **20**, 129–138 (2015)
6. Figueroa-García, J.C., Kalenatic, D., Lopez-Bello, C.A.: Multi-period mixed production planning with uncertain demands: fuzzy and interval fuzzy sets approach. Fuzzy Sets Syst. **206**, 21–38 (2012). https://doi.org/10.1016/j.fss.2012.03.005
7. Figueroa-García, J.C., López, C.A.: Linear programming with fuzzy joint parameters: a cumulative membership function approach. In: 2008 Annual Meeting of the IEEE North American Fuzzy Information Processing Society (NAFIPS), pp. 1–6. IEEE (2008). https://doi.org/10.1109/NAFIPS.2008.4531293
8. Figueroa-García, J.C., López, C.A.: Pseudo-optimal solutions of FLP problems by using the cumulative membership function. In: Annual Meeting of the North American Fuzzy Information Processing Society (NAFIPS), vol. 28, pp. 1–6. IEEE (2009). https://doi.org/10.1109/NAFIPS.2009.5156464
9. Heyman, D.P., Sobel, M.J.: Stochastic Models in Operations Research, Vol. II: Stochastic Optimization. Dover Publishers, Mineola (2003)
10. Inuiguchi, M., Ramík, J.: Possibilistic linear programming: a brief review of fuzzy mathematical programming and a comparison with stochastic programming in portfolio selection problem. Fuzzy Sets Syst. **111**, 3–28 (2000). https://doi.org/10.1016/S0165-0114(98)00449-7

11. Inuiguchi, M., Ramik, J., Tanino, T., Vlach, M.: Satisficing solutions and duality in interval and fuzzy linear programming. Fuzzy Sets Syst. **135**(1), 151–177 (2003). https://doi.org/10.1016/S0165-0114(02)00253-1
12. Johnson, L.A., Montgomery, D.C.: Operations Research in Production Planning, Scheduling and Inventory Control. Wiley, Hoboken (1974)
13. Pinedo, M.L.: Scheduling: Theory, Algorithms, and Systems, 5th edn. Springer, New York (2016). https://doi.org/10.1007/978-1-4614-2361-4
14. Pulido-López, D.G., García, M., Figueroa-García, J.C.: Fuzzy uncertainty in random variable generation: a cumulative membership function approach. Commun. Comput. Inf. Sci. **742**(1), 398–407 (2017). https://doi.org/10.1007/978-3-319-66963-2_36
15. Ramík, J., Řimánek, J.: Inequality relation between fuzzy numbers and its use in fuzzy optimization. Fuzzy Sets Syst. **16**, 123–138 (1985). https://doi.org/10.1016/S0165-0114(85)80013-0
16. Ramík, J.: Optimal solutions in optimization problem with objective function depending on fuzzy parameters. Fuzzy Sets Syst. **158**(17), 1873–1881 (2007). https://doi.org/10.1016/j.fss.2007.04.003
17. Rommelfanger, H.: A general concept for solving linear multicriteria programming problems with crisp, fuzzy or stochastic values. Fuzzy Sets Syst. **158**(17), 1892–1904 (2007). https://doi.org/10.1016/j.fss.2007.04.005
18. Sipper, D., Bulfin, R.: Production: Planning, Control and Integration. McGraw-Hill College, New York (1997)
19. Zimmermann, H.J.: Fuzzy programming and linear programming with several objective functions. Fuzzy Sets Syst. **1**(1), 45–55 (1978). https://doi.org/10.1016/0165-0114(78)90031-3

Author Index